EMBRYOLOGY OF FLOWERING PLANTS
TERMINOLOGY AND CONCEPTS

Embryology of Flowering Plants

Terminology and Concepts

Volume 2

Seed

EDITED BY T.B. BATYGINA

CRC Press
Taylor & Francis Group
Boca Raton London New York

CRC Press is an imprint of the
Taylor & Francis Group, an **informa** business
A SCIENCE PUBLISHERS BOOK

First published 2006 by Science Publishers

Published 2019 by CRC Press
Taylor & Francis Group
6000 Broken Sound Parkway NW, Suite 300
Boca Raton, FL 33487-2742

ISBN 13: 978-1-57808-263-6 (hbk)

Visit the Taylor & Francis Web site at
http://www.taylorandfrancis.com

and the CRC Press Web site at
http://www.crcpress.com

CIP data will be provided on request.

V.L. Komarov Botanical Institute
Russian Academy of Sciences

DEPARTMENT OF EMBRYOLOGY
AND REPRODUCTIVE BIOLOGY

RUSSIAN FOUNDATION
FOR BASIC RESEARCH

Contributors

T.B. Batygina (editor), E.V. Andronova, G.M. Anisimova, O.V. Antipova, E.A. Bragina, V.B. Brukhin, M. Cresti, M.F. Danilova, O. Erdelská, W.E. Friedman, E.N. Gerassimova-Navashina, G. Jürgens, O.P. Kamelina, O.A. Khvedynich, H. Kieft, S.N. Korobova, L.V. Kovaleva, L.G. Krasnikov, A.A.M. Van Lammeren, Y.-Q. Li, F. Ma, S.S. Medvedev, D.W. Meinke, T.N. Naumova, S.G. Nawaschin, Z.I. Nikiticheva, N.V. Obrucheva, T.A. Plyushch, V.V. Polevoy, L.M. Pozdova, M.V. Razumova, B. Rodkiewicz, S.D. Russell, T.S. Salamatova, J.H.N. Schel, I.I. Shamrov (secretary and editorial board member), S.M. Szcziparev, M.P. Solntseva, E. Szczuka, E.S. Teryokhin, G.E. Titova, V.E. Vasilyeva (editorial board member), W.L.H. Van Veenendaal, M.A. Vishnyakova, T.D. Vyshenskaya, J.L. Van Went, X.L. Ye, E.C. Yeung, S.Y. Zee, N.A. Zhinkina, G.Ya. Zhukova (editorial board member)

To S.G. Nawaschin the discoverer of double fertilization in angiosperms (100th anniversary)

Seeds are the germ of life, a beginning and an end, the fruit of yesterday's harvest and the promise of tomorrow's.

O.L. Freeman

Preface

S.G. Nawaschin's discovery stimulated numerous investigations into plant fertilization, not only in Angiospermae but also in Gymnospermae. One of his most talented pupils—E.N. Gerassimova-Navashina—developed new notions about this process, specifically the idea of fertilization types. Her investigations also became classical. Their embryological school traditions are continued nowadays in different laboratories of the world and, primarily, in the **Department of Embryology and Reproductive Biology, Komarov Botanical Institute, Russian Academy of Sciences** (head of the leading Scientific High School of the Russian Federation, T.B. Batygina).

The present staff of the Komarov Botanical Institute, Russian Academy of Sciences, considered it a pleasant duty to include in this second volume of Embryology of Flowering Plants the history of double fertilization discovery and some works of S.G. Nawaschin, published in Russian and German journals, which are bibliographical rarities. These investigations have not lost their significance over time because they contain detailed descriptions of the double fertilization process in classical objects.

The recent tendency towards consolidation of embryology with genetics, physiology and biochemistry is obvious. Progress in molecular biology has strengthened these relationships. Earlier the main genetical method was hybridological analysis, which enables recording these or those alternative features, but nowadays the attention of geneticists is focusing more and more on questions of genetical feature expression during ontogenesis, regularities of caryogamy, plasmogamy and cytokinesis, as well as histogenesis and organogenesis. For many geneticists in the world the centre of attention has become the study of mutations which affect the first developmental stages of male and female generative structures and embryogenesis (for example, the disturbance of asymmetrical initiation of the first wall in the zygote). This permits revealing the action mechanism of genes in the processes of development and organogenesis. Successful models for studying mutations have been found *(Arabidopsis thaliana, Zea mays).*

Unfortunately, specialists in the area of studying inheritance of features do not adequately consider the fundamental achievements of embryology. However, it is classical embryology, in fact, with such representatives as K. Schnarf, R. Souèges,

D.A. Johansen, P. Maheshwari, B.M. Johri, J.S. Mèstre, J.L. Guignard, A. Lebègue, S.G. Nawaschin, M.S. Yakovlev, E.N. Gerassimova-Navashina a.o., which has made significant contributions to understanding the main regularities of the origin and early stages of individual development.

This book pays great attention to the morphological and functional aspects of the fertilization process and embryogenesis, the main subjects of investigation by embryologists of St. Petersburg Academy of Science for many decades.

Besides, it is important that the very peculiar type of sexual reproduction closely connected in flowering plants with seed reproduction, has permitted such plants 'to conquer' the Earth and create the basis for the development and flourishing of our civilization. Just the seeds are the main nutritive source for mankind, which is why it is impossible to overevaluate the significance of scientific progress in their study.

Together with the traditional ideas of seed development, the original notions on the problems of origin, correlations and evolutionary transformations of seed structures are presented in this book. Data are drawn not only for angiospermous, but also for gymnospermous plants.

New data in experimental embryology, dealing with the morphogenesis of reproductive structures, specifically development of the somatic embryo, enable a discussion of the new notion of asexual reproduction—embryoidogeny.

New classifications of the developmental types of main seed structures, worked over mainly by researchers of the Botanical Institute, are presented here. Investigations on ecological embryology, the chlorophyllous embryo and its autonomy for example are generalized.

New types of embryogenesis and endospermogenesis are also described.

The task of the authors was generalization of the theoretical notions on which classical and experimental data were based, concerning the development of flower generative organs, the process of seed formation and germination, and the system of seed reproduction as a whole. This is especially urgent given the need to preserve rare and vanishing plant species, to create a seed bank and to elaborate new biotechnologies for multiplication of avaluable plant forms and varieties associated with the problems of seed reproduction.

Unification of embryological terms and notions will promote further development of ideas about the reproductive processes occurring during plant ontogenesis.

The first volume of the three-volume edition *Embryology of Flowering Plants. Terminology and Concepts* was published in English in 2002. It was devoted to the common flower structure and structure of the main generative organs—the anther and ovule. The present book is the second volume of this edition. It consists of five parts and reflects the processes which take part in seed formation. The authors took into account the long-ago matured broadening of plant embryology borders, the science which turns from descriptive into causal discipline. Besides traditional chapters—'Fertilization', 'Embryo' and 'Endosperm'—there are chapters on the nature of such storage tissue as perisperm, the seed coat structure, and also perspectives on genetic analysis of plant embryogenesis and features of post-seminal development, including seed dormancy and germination.

In preparing the book we used such well-known monographs as F. Netolitzky, *Anatomie der Angiospermen-Samen* (1926); K. Schnarf, *Embryologie der Angiospermen. Archegoniaten* (1929) and *Vergleichende Embryologie der Angiospermen* (1931); R. Souèges, *Les Lois du Développement* (1937) and *Embryogénie et Classification* (1939); P. Maheshwari,

An Introduction to the Embryology of Angiosperms (1950); D.A. Johansen, *Plant Embryology* (1950); P.A. Baranov, *The History of Plant Embryology* (1955); A.L. Takhtajan, *Foundations of Evolutionary Morphology of Angiosperms* (1964), *Outline of the Classification of Flowering Plants* (1980) and *System Magnoliophyta* (1987); V.A. Poddubnaya-Arnoldi, *General Embryology of Angiosperms* (1964b) and *Cytoembryology of the Angiosperms: Principles and Perspectives* (1976); G. Davis, *Systematic Embryology of Angiosperms* (1966); T.B. Batygina, *Wheat Embryology* (1974); T.B. Batygina, *The Grain of Cereals (Atlas)*; E.S. Teryokhin, *Parasitic Flowering Plants. The Evolution of Ontogenesis and the Mode of Life* (1977); E. Corner, *The Seeds of Dicotyledons* (1976); K. Esau, *Anatomy of Seed Plants* (1977); B.M. Johri (ed.), *Embryology of Angiosperms* (1984); T.B. Batygina and M.S. Yakovlev (eds.), *Comparative Embryology of Flowering Plants* (1981, 1983, 1985, 1987, 1990); V. Raghavan, *Experimental Embryogenesis in Vascular Plants* (1976) and *Embryogenesis in Angiosperms: A Developmental and Experimental Study* (1986); M.F. Willson, *Plant Reproductive Ecology* (1983); B.M. Johri, K. Ambegaokar, P.S. Srivastava *Comparative Embryology of Angiosperms* (1992) and others.

For more precise definition of terms meaning the following dictionaries were used: M.C. Cooke, *A Manual of Botanic Terms* [1873?]; B.D. Jackson, *A Glossary of Botanic Terms* (1916); N.N. Zabinkova and M. E. Kirpichnikov, *Latin-Russian Dictionary for Botanists* (1957); M.E. Kirpichnikov and N.N. Zabinkova, *Latin-Russian Dictionary for Botanists* (1977), *Biological Encyclopedic Dictionary* (1986, 1989); *The Oxford English Dictionary* (1989) and others.

In this encyclopedic edition the principles of constructing the dictionary with monographic description of embryonic structures and processes are combined. However, **unlike traditional dictionary structure, the terminological articles in every part of the book are disposed not in alphabetical order, but according to theme so as to create an integrated picture of each main seed structure. A subject index is given at the end of the book.**

Terminological articles comprise the main part of the text and include: the definition and semantics of the terms, their history, and major data about the origin, development, functions and classification of the structure described. In a number of cases, questions concerning evolutionary transformations and the character of distribution of embryological features among flowering plants are discussed.

The second group of articles combines conceptional, and hence the most complex articles with a discussion of questions of modern angiosperm embryology. To these are related a number of texts taken from native plant embryology classics, which have preserved their acuity to our time. These are the articles by S.G. Nawaschin and E.N. Gerassimova-Navashina. Some cases of difference between author terms and terms accepted in this edition are footnoted.

Most of the terminological and conceptional articles are complemented with illustrations such as drawings, microphotographs (LM, TEM, SEM), schemes and diagrams.

Complete bibliographical data are given in the References at the end of the volume.

T. B. Batygina
Z. I. Nikiticheva

Acknowledgements

As the editor of this volume, I express my gratitude to the team of contributing authors. I am especially thankful to colleagues from different countries of the world who accepted our proposal to participate in the preparation of this book, namely, M. Cresti, Y.-Q. Li (Italy); S.D. Russell, W.E. Friedman, D.W. Meinke (USA); A.A.M. Van Lammeren, W.Z.H. Van Veenendaal, J.H.N. Schel, H. Kieft, F. Ma (Netherlands); F.C. Yeung, S.J. Zee, X.L. Ye (Canada); G. Jürgens (Germany); B. Rodkiewicz, E. Szczuka (Poland); and O. Erdelská (Slovakia).

My special thanks also to the research workers of the Embryology and Reproductive Biology Department—V.E. Vasilyeva, as well as G.E. Titova, G.Ya. Zhukova, E.V. Andronova—for their great help in the preparation of this volume. I am also grateful to the Edition Secretary I.I. Shamrov, N.A. Zhinkina and A.A. Zaharova for beautiful illustrations, L.M. Rotenfeld for excellent photography, E.A. Bragina, O.N. Voronova, G.M. Anisimova, G.U. Vinogradova, A.A. Babro, E.E. Evdokimova, A.A. Torshilova for their valuable technical assistance.

Articles written in Russian were translated into English by I.V. Putro, O.G. Butuzova, L.A. Yakovishina, G.K. Lakhman and E.V. Dragunova.

Financial support from the Russian Foundation for Basic Research in publishing the Russian version of Volume 2, without which the book could not have materialized, is gratefully acknowledged.

T.B. BATYGINA

Contents

PART TWO—ENDOSPERM, PERISPERM

PART THREE—EMBRYO

OVERVIEW

Developmental Stages and Structures of Embryo

TYPES OF EMBRYOGENESIS

ULTRASTRUCTURAL AND HISTOCHEMICAL ASPECTS OF EMBRYOGENESIS

PHYSIOLOGICAL, BIOCHEMICAL AND GENETIC ASPECTS OF EMBRYOGENESIS

EMBRYOIDOGENY

PART FOUR—SEED

PART FIVE—SEED DORMANCY AND GERMINATION

Introduction

T.B. Batygina

The nature and mechanisms of the processes which lead to individual plant development are considered one of the most important biological problems of today. Plant embryology, which concerns the regularities of initiation and the first stages of development of a new organism, has been undergoing a flourishing period now, thanks to the general progress of natural sciences. In turn, further development of biology is unthinkable without knowledge of the first stages of ontogenesis. Embryology is an essential base for experimental and theoretical studies of reproduction. Apart from that, embryology comprises a number of fundamental processes widely applicable in practice (cytoplasm male sterility, the phenomena of apomixis, distant hybridization, etc.). That is why it has appeared in the centre of attention of geneticists, biochemists, physiologists, cytologists, biophysics and plant-breeders.

The main task set by modern embryology is to control the development of organisms. Until quite recently plant-breeders were unable to definitely predict the result of one or another cross, its usefulness in practice and the fertility of the progeny. The processes occurring in a flower, such as anther and ovule formation, fertilization, endosperm and embryo development have never yielded to control. It is quite evident that the efficiency of manipulation with separate stages of ontogenesis is directly dependent on our complete knowledge of normal morphogenetic processes. The study of embryonal development in all its aspects and at various levels, such as molecule, cell, tissue, organism and population, is an essential theoretical base for further progress of research in genetics and selection. Every work pertaining to the influence of biotic or abiotic factors on a plant demands complete knowledge of critical events in early ontogenesis, which determine further plant development.

Through many years in the Department of Embryology and Reproductive Biology of Komarov Botanical Institute, Russian Academy of Sciences the system approach has been worked out with reference to reproductive structures. From this standpoint a number of correlations between morphogenetic processes which determine the specific character of regular embryonal development have been revealed. Disturbance of such a correspondence entails disturbance of homeostasis, which results in various abnormalities of both specific organs and the whole plant, in which everything is submitted to the common laws of development. Success in the control of plant development is possible only after a complex investigation of its stages and in co-operation among scientists of different specializations.

Recently, embryology has been changing more and more from a purely morphological science into a morphophysiological and ecomorphological science. It is impossible to reveal morphogenetic and morphophysiological correlations, which determine the normal mode of embryo and seedling development, without a study of the relations between plant and environment, which change during ontogenesis.

Such discoveries of the 20th century as haploidy, parasexual hybridization and raising plants from single somatic cells are significant in terms of general biology. However, they could not be understood completely without knowledge of embryology. Probably, this was the reason for stimulating studies on morphogenesis of reproductive structures that promoted new discoveries, namely, embryoidogeny as a new type of vegetative reproduction (by somatic embryos), universality of modes of morphogenesis in natural conditions and in *in vitro* culture, different types of fertilization (pre-, post-mitotic and intermediate), Paeoniad- and Graminad-type of embryogenesis, Tubifloral-type of endosperm. Discovery of the ancestral form of double fertilization in representatives of order Gnetales made it possible to distinguish 3 types of karyogamy (G_1-, S- and G_2-) in gymnosperms and angiosperms, and later—M-type of caryogamy in angiosperms. The phenomenon of metamorphosis was discovered in ontogenesis of highly specialized parasitic flowering plants and the main regularities of adaptive evolution of reproductive structures were revealed in different taxa. A number of new hypotheses concerning the origin and evolution of monocot embryo, female gametophyte, ovule, anther and endosperm were proposed.

Progress of embryology is connected to a considerable extent with an application of both modern instrument set and new methods (E. Pacini, 1987). Use of modern techniques and tools, such as electron microscopy, autoradiography, the method of radioactive isotopes, microsurgery, modelling, 3D-reconstruction and some others, offers new perspectives for investigation at different levels of organization. Employment of fluorescence, including immunofluorescence, phase contrast, interferential contrast optics of Nomarski, electron scanning and transmission microscopy, as well as methods of modern cell biology, such as immunocytochemistry, microcinematography, enzymatic isolation of generative cell and sperms, embryo sacs and separate cells of female gametophyte, have enabled scientists to obtain a list of unique data on fine structure and functions of reproductive male and female organs. These are the peculiarities of cytoskeleton in cells of reproductive structures, special features of meiosis and sperm cell dimorphism. Quantitative studies of mRNA in pollen grains and sperms have made it possible to obtain new data on their fine structure. The mRNA in mature anthers was identified. Investigations on morphogenetic process dynamics have opened the possibility of revealing critical periods of embryonal structure development, as well as the most important embryological features, namely chlorophyll-bearing property, embryo autonomy, seed genetic heterogeneity, etc.

In the last decades of the 20th century common efforts of embryologists, geneticists and molecular biologists resulted in discovery of specific genes, which control meiosis, microsporogenesis, ovule development, as well as determination of the gene block, which control egg cell development and early stages of embryogenesis. Especial attention should be attracted to elaboration of theory of stem cells in plants and their applied aspects.

Use of reproductive structures of flowering plants (flower, anther, ovule, embryo), especially those of mutants, is a prospective approach to analysis of morphogenetic processes. In a comparative study of embryogenesis in mutants and in normal specimens, special attention should be paid to architectonics and morphofunctional organization of the egg cell at different stages of its development, namely, initial, immature and mature ones, which determine the pattem of zygote structure, as well as to that of the embryo body and the mature organism. The proteins of the egg cell cytoskeleton play a critical role in determination of polarity and symmetry axis and provide a key to understanding morphogenesis of zygote, seedling and the entire organism.

Study of integral events in plant morphogenesis at different levels and also of pattern change in the course of temporal and spatial organization of plants is one of the most important tasks facing developmental biology. From the standpoint of new ideas and facts (Gilbert *et al.*, 1996), plant embryology together with animal embryology should occupy the leading position in developmental and evolutionary biology.

During the last decades plant embryology has become a ramified area of biology. The capacity of the very notion 'embryology' has enlarged. Today plant embryologists study not only male and female generative structures, the process of fertilization, embryo and seed development in general, but also morphogenetic processes of plant reproduction at ontogenetic and population levels. Suffice it to say that organogenesis takes place through the whole ontogenesis in plants and, for example, speaking of spring wheat, the process which belongs to the scope of embryologists occupies two-thirds of the vegetation period, from the initiation of influorescence up to caryopsis formation and germination.

In general, plant embryology concerns the origin and first stages of plant body development, including seedling, and embryological structures cannot be understood without their function.

What is meant by a direction in one or another field of science is a differentiation of research according to the aims and methods and partially to objects (for example: *Zea, Arabidopsis*). In modern embryology of flowering plants several main directions can be clearly defined. At the end of the 20th century one of the rapidly developing new directions—reproductive biology—was formed, joining together embryology, anthecology, carpology, genetics, physiology and selection. Elaboration of theoretical bases of sexual and asexual reproduction, apomixis, seed and vegetative multiplication appears to be the component of reproductive biology. Reproductive biology includes the study of various interconnected stages of ontogenesis, such as flower organogenesis, blooming, pollination, fertilization, embryogenesis, seed and fruit maturation, dissemination, seed germination and seed renewal of population. All these issues are in one way or another connected with embryology in its broader sense.

The main point is to work out the problems of amphimixis and apomixis, regularities of embryogenesis, the phenomena of incompatibility and self-compatibility, regularities of potential and real seed productivity, and hence yield capacity.

Development of reproductive biology is important insofar as it concerns reintroduction and repatriation of rare, vanishing or economically valuable plants with the purpose of preserving biological diversity. This direction is acomplished by cytoembryological investigations on pollutant influence on reproductive structures within the framework of environment protection. Presently, these studies have been gaining more and more significance because reproductive structures are, most of all, subjected to the influence of unfavourable environmental factors.

Of special interest is the study of the diversity of phenomena included in apomixis. Questions pertaining to parthenogenesis, apospory, diplospory, polyembryony require more active investigation now, especially in connection with practical needs. Such forms of reproduction as vivipary and adventitious embryony deserve special attention from biologists of many specializations. All the directions of reproductive biology considered above are urgent nowadays and will undoubtedly appear in the scope of interest of biologists in the 21st century.

The study of reproductive structures has a great significance from the standpoint of the biological system of plasticity and tolerance. Thus ecological embryology is one of the most important directions among the studies of early ontogenesis. Works devoted to ecological embryology should be expanded since they deal with structural and functional bases for adaptation processes in different groups of plants. Besides, they contribute to revealing seed productivity dependence on ecological conditions in cultivated plants. This area links population embryology and reproductive biology and sets the aim to study morphogenesis and phenotype variability within a population (the study of life cycle variations and polyvariancy of reproductive systems).

The immunologic branch of embryology should play a considerable role in future in solving questions pertinenet to the problem of compatibility and incompatibility during hybridization.

The tendency to synthesize data of embryology and genetics has become more and more apparent. As long ago as 1935 N. K. Koltsov wrote '...only the consolidation of these two sciences [experimental embryology and genetics—T.B.] and also cytology and biochemistry will create a single science, which would be able to solve problems of general biology'. All problems connected with morphogenesis such as differentiation, specialization, definition of the notions 'gene and trait', 'genotype and phenotype', concern both embryology and genetics one way or another.

Experimental embryology has recently undergone vigorous development. Unlike descriptive embryology, it allows us to answer not only the question 'how does the complicated morphological process develop', but also 'why does it develop one or another way'.

The well-known embryologist P. Maheshwari foretold in 1950 that the future of embryology would undoubtedly belong to the experimental trend. The main incentive for its development has been the requirement of practical selection, for example the need to elucidate the cause of sterility and efforts to overcome it, use of adventitious embryos and so on.

The experimental trend in embryology, which was breaking through 'as separate streams' in the 19th and 20th centuries, will 'join to form a wide flood' beyond the threshold of the 21st century.

Plants regenerated in culture *in vitro* have become a model for the study of differentiation mechanisms and molecular-genetic regularities of embryonal structure morphogenesis. Among others, interesting data have been obtained on the influence of different abiotic factors on morphogenesis, possible ways of embryoid formation have been considered, etc.

Success of artificial culture depends on correct selection of the developmental stage of generative structures taken for cultivation *in vitro* and also the state of donor plants. Working out theoretical bases of *in vitro* culture of vegetative and generative structures (such as ovary, anther, ovule, embryo, endosperm, embryo sac, sperm cell and so on) has become a matter of current interest. A combination of experimental approach and study of details of embryonal structure genesis enables researchers to predict ways of regenerative morphogenesis, stage of embryo development best for embryo culture, etc. This direction is common for cell engineering and biotechnology.

Embryological data have become more and more significant for the creation of genetic and selection programmes and for elaboration of effective biotechnologies as well, since these data allow us to judge the procession of processes which yield normal seeds and the reasons for the anomalies which accompany the embryonal processes.

The system approach with reference to generative structures used in culture *in vitro* has provided some theoretical propositions (for example, embryo autonomy, critical stages of generative structure development, etc.) which are now applied to selection. Utilizing this knowledge new forms of plants have been obtained. In the laboratory of Embryology and Reproductive Biology of the Botanical Institute, Batygina is conducting work on the elaboration of the theory of critical periods, phases and stages in generative structure development. This investigative trend is considered a priority one as it permits control of definite stages of ontogenesis (Batygina, 1966, 1974, 1987, 2002; Vasilyeva, Batygina, Titova, 1987; Batygina, Vasilyeva, 2003).

The study of regularities of ontogenesis evolution, in connection with the questions of phylogeny and systematics, remains one of the main areas of modem embryology. Due to their relative conservatism, embryological features appear to be additional material for the development and closer definition of notions for ascertaining the systematic position of some orders, families, genera and even species. One of the main tasks is screening of poorly investigated taxa, which is necessary for precising the criteria of evolutionary and systematic significance of features.

New data accumulated in descriptive and experimental biology, morphology, reproductive biology and developmental biology must play their role in accelerating the work in the evolutionary embryology. What has also been serving this purpose is the number of recently published monographs, which elucidate different aspects of descriptive and comparative embryology, namely *Cytoembryology of the Angiosperms: Principles and Perspectives* (1976) and *Characteristics of Flowering Plant Families According to Cytoembryological Features* (1982) by Poddubnaya-Arnoldi, *Comparative Embryology of Flowering Plants* edited by T.B. Batygina and M.S. Yakovlev[1] (1981–1990); *Embryology of Angiosperms* edited by B.M. Johri (1984), *Comparative Embryology of Angiosperms* by B.M. Johri *et al.* (1992) and some others. More profound analysis of studies by the founders of plant embryology, namely W. Hofmeister, A.S. Faminzyn, A.T. Bolotov, R. Souèges, K. Schnarf, I.N. Gorojankin, N.I. Jeleznov, V.I. Beliaev, S.G. Nawaschin, P. Maheshwari, D.A. Johansen, E.N. Gerassimova-Navashina, V.A. Poddubnaya-Arnoldi, I.D. Romanov, M.S. Yakovlev and others should also be mentioned.

In the 5-volume book *Comparative Embryology of Flowering Plants* the embryological characteristics of families (classified according to A.L. Takhtajan's system from 1980) are given and investigations in descriptive embryology from the beginning of the 20th century to date are summarized. A broad view of the development and differentiation of embryonal structures in dicots (I–IV volumes) and monocots (V volume) in all their diversity is presented. This work can serve as a guide-book for future works on reproductive biology, as it contains generalized data on the processes of sexual and asexual reproduction, including apomixis, and the content and distribution of reserve substances in a mature seed in economically valuable, rare and vanishing plant species.

Concomitantly this publication has already proved the need for improving existing classifications of embryonal structures and processes in flowering plants

[1]see References, vol. 1, p. 222.

and their typification. For some years Russian scientists have created a number of new classifications of modes and types of reproduction in angiosperms, embryo sac developmental types, types of embryo, anther, tapetum and apomixis. The original concept of flowering plant reproduction has been worked out and an embryoidogenic type of asexual reproduction distinguished (Batygina 1989a,b,c, 1993).

Side by side with the accumulation of new data, the objectivity of classifications increases as does their conformity with real evolutionary processes. This prompts more active participation of embryologists in studies on problems of systematics and phylogeny of flowering plants. However, theoretical bases for embryological classification have yet to be sufficiently worked out.

Increased awareness of the systematic and phylogenetic significance of these classifications is closely connected with progress in our knowledge of plant evolution, its driving factors (ecological, genetic, etc.) and modes of evolutionary transformations of plant ontogenesis. This entails paying thorough attention to embryonal structures, which present the most complete reflection of the peculiarities of ontogenesis evolution; it also highlights the need for discussing both these problems and new classifications at conferences, symposia and congresses.

Evolutionary embryology has given rise to an ecomorphological approach to the study of the problems of evolutionary adaptation in plants. On the example of parasitic plants it has been stated that such general biological phenomena as metamorphosis, neoteny and reduction form the basis of evolutionary transformations of their reproductive structures. Following this direction of thinking, the hypothesis of possible mechanisms of the origin and evolution of embryo monoeotyledony has been considered within the framework of existing hypotheses of heterochrony and syncotyly.

The perspectives of applied embryology development become more obvious. The following problems can be distinguished:

—Ecological problems associated with working out approaches and methods of obtaining normal seeds of rare or economically valuable plants; preserving biological diversity—creating different banks (agricultural, genetic—seeds, embryos, sexual cells); study of soil bank of seeds by both soil scientists and mycologists; revealing the plasticity level and reproductive systems tolerance and a correlation of different ways of multiplication in different taxa under the influence of damaging factors.

—Biotechnological problems associated with creating effective technologies for mass propagation of new forms and varieties and also rare and vanishing plants by means of embryoidogenesis (somatic embryogenesis) and gemmorhizogenesis (organogenesis);

—Agricultural problems: increase of real seed productivity, creation of new forms and varieties by means of distant and parasexual hybridization, overcoming incompatibility, etc.

To realize in the 21st century all the main directions of modern embryology considered above it would be necessary to create a special complex of scientific research programmes. The aim of such programmes would be to solve an urgent, large-scale scientific or scientific-applied problem mainly on the base of previously discovered regularities with appreciation of the latest achievements of research works and the latest methodology of science and methods. The investigations of Hofmeister, Souèges, Schnarf, Dahlgren, S.G. Nawaschin and Romanov may serve as historical prototypes of special-purpose programmes.

Special-purpose scientific programmes would carry out a complex of special-purpose investigations with goal-oriented planning and involve biologists from different specializations in which programmes of embryological science should occupy a dominant position in the 21st century.

The following topics can be proposed for creating such programmes: algorithms of reproductive morphogenesis, morphological and ecological bases for seed reproduction, origin and evolution of flowering plants, and elaboration of theoretical bases for generative and vegetative structure cultivation for the purpose of preserving biological diversity (introduction, repatriation).

The appearance of monographs by Rodkiewicz (1973, 1984), Hu-Han and Yang Honjyuan, (1986), Cresti, Gori and Pacini (1988), Raghavan (1986, 1989, 1997), Bell (1992), Mukerji (1992), Ottaviano *et al.* (1992), Primack (1993) and Bhojwani and Soh, eds (2001) elucidating different aspects of embryology in the last decades of the 20th century, showed that a number of theoretical problems needed reconsideration. A need was felt to classify a number of structures which had not yet been classified and to overcome the existing contradictions in the numerous terminological questions.

It should be noted that the use of diverse, modern and classic methods has enabled us to obtain many new facts and to make a number of discoveries which ensued in a pressing demand for generalization and creation of new hypotheses and theories. In connection with this one should consider theoretical questions of embryology taking into account both the classic legacy and results of the latest investigations.

An unparalleled worldwide 3-volume encyclopaedia *Embryology of Flowering Plants. Terminology and Concepts* was published in Russia. The publication was prepared with contributions of embryologists from Russia, countries of UUS (Ukraina, Georgia, Uzbekistan, Moldova) and of scientists from the USA, the Netherlands, Sweden, China, Poland, Slovakia, India, Italy and France. This publication is a unique attempt to combine the principles of terminological vocabulary and monographic description of particular structures and processes. The text includes a number of reviews and terminological conceptual articles.

The traditional frameworks of embryology were broadened. For the first time questions concerning seed dormancy and germination, ecological aspects and systems of flowering plant reproduction were examined. Monographs by prominent botanists and embryologists from over the world were included. The latest international congresses, symposia and workshops on plant embryology (Italy, 1988; USSR, 1990; USA, 1992, 1999; Austria, 1994; Poland, 1997, 1999; Australia, 1996; Germany, 1996; Netherlands, 1998) in which aside from plant embryologists, geneticists, physiologists, cytologists and biotechnologists also participated, confirming once again the up-to-date significance of this publication.

In conclusion it should be underlined once more that the meaning of embryology will be important because it is impossible to control individual plant development without knowledge of the first ontogenetic stages. All this will require popularization of embryological science at different forums and in press, publishing new textbooks and manuals on embryology, and last but not least—training young researchers.

PART ONE—DOUBLE FERTILIZATION

HISTORY OF DISCOVERY

Double Fertilization Phenomenon

The year 1998 marks the centenary of the most notable discovery in biology, which was made by Sergius Nawaschin—the discovery of double fertilization.

The name of academician S. Nawaschin (1857-1930) is widely known and his works recognized as classics. Books containing his works are rare and the present volume is exceptional comprising his most fundamental works on the fertilization process.

An important feature of Nawaschin's work is exclusively accurate rendering of the structural pictures discovered by him under the light microscope with the help of artistically executed drawings. Nawaschin's illustrations in his works on cytology led not only to the appearance of a radically new field of botany—'fine structure of cell nucleus elements', which before his time used to be depicted as formless lumps, they are still considered masterpieces judging from the technique of realization, as they were often done in watercolour. As S. Nawaschin used to say: **'Drawings are the language of a morphologist'.** Currently, in line with the most recent technical possibilities (TEM, SEM, computer-aided reconstruction, etc.) a drawing remains a necessary and important stage in disclosing the truth.

Particularly worthy of attention is the work entitled 'Essay of Structural Representation of Sexual Nuclei' (1927). One of the most crucial issues raised by Nawaschin in this article still holds the attraction of researchers, namely, **the reasons of the ordered motion of sperms towards female sexual cells**. To resolve this issue he set forth the original **enantiomorphism hypothesis**. Nawaschin was convinced that there is mirror-like distribution of all elements in each pair of sperms and female cells, including the egg cell nucleus and the micropylar polar nucleus of the central cell. Following sperm entry into the embryo sac a complex system of four nuclei is formed therein ('differently activated bodies'). Resulting from interaction of nuclei which are oriented by pairs a certain type of gradient is formed, responsible for the direction of each sperm nucleus moving towards a certain nucleus of the female cell.

Thus, the enantiomorphism hypothesis is grounded on the assumption that in a pair of sperms, as also in female cells (the egg and the central cell), one of the nuclei is right and the other is left. Fusion therefore occurs between the corresponding nuclei (the left with the left, the right with the right). Nawaschin was the first to address one of the most significant problems in biology—cytoplasm asymmetry—having paid attention to physical and biological distinctions of D- and L-forms. He considered the problem of asymmetry in broad spectrum. He wrote: 'Similar to the asymmetric structure of a molecule or of a crystal being related with the optical activity of a given body, the asymmetry of a haploid nucleus is a condition for the specific sexually active state of the haploid cells, which is manifested in two opposite forms'. Present-day researchers pay serious attention to asymmetrical processes both in plants and animals (asymmetric cytokinesis occurring in meiosis and mitosis, e.g. during division of microspore, zygote or embryo cells). These questions are considered on cellular and subcellular levels. In this work Nawaschin is the first to pronounce dimorphism of male and female gametes, later proved also in ultrastructural studies. This 'enantiomorphism' hypothesis may currently be regarded as one of the hypotheses for elucidating the double fertilization mechanisms.

The meaning of the double fertilization phenomenon discovered by S. Nawaschin is so profound that it warrants more detailed coverage of how it evolved.

Pollenists Schleiden (1837, 1839a), Schacht (1850) and others held the opinion that the embryo is actually formed from the tip of the pollen tube. Later, however, supporters of this concept admitted they had been mistaken.

One of the world's greatest botanists, Hofmeister (1849), based on studies of 38 species of 19 genera of angiosperms, was the first to present the **complete outline of the sexual process**. Using live material (*Orchis* and *Gymnadenia*) he demonstrated pollen germination on the stigma and possible participation of the pollen tube in fertilization (see Fig. 1). He, however, maintained the view that **'the male fertilizing agent'** is admitted into the egg cell by means of **diffusion.**

Another of the most notable botanists of that time, Karsten (1860), understood that 'the fertilizing material' (French *matiere fécondante*) must enter the egg cell for the embryo to appear. Nevertheless he also carried on to explain the process of egg cell fertilization through diffusion of the 'fertilizing agent'— first through the pollen tube wall into the embryo sac, and then further into the egg cell where the fertilization act would occur. Gorozhankin (1880) in his classical works on fertilization in gymnosperms was first to refute the ruling opinion of the time regarding diffuse admittance of the sperm into the egg cell.

In 1877, Strasburger **was the first to most thoroughly describe the process of the pollen tube entry into the embryo sac through the synergid** in *Monotropa hypopitys*, having discovered that **one of the two male gametes** located in the pollen tube **enters the embryo sac** thus effecting fertilization of the egg cell. In his opinion **the other sperm was dissolved in the pollen tube**. He was the first ever to define the role of one of the sperms, i.e., he basically discovered **syngamy** ('single fertilization' in flowering plants—T.B.). What happened to the other sperm remained a mystery.

In 1886, L. Guignard while illustrating this process in representatives of the Cactaceae pointed out another way of pollen entry into the embryo sac whereby **the pollen content entered directly into the egg cell** (leaving the synergids behind).

Following the discovery of single fertilization many worldwide embryologists studied various fertilization aspects in detail including the pollen tube growth in the pistil. Treub (1891) discovered **the phenomenon of chalazogamy**. While working on the chalazogamy phenomenon in *Juglans regi*, S. Nawaschin (1895a-c) was first to establish the presence of **two sperms in the embryo sac** observable as two extended coils.

On August 24 (old style) 1898 at the 10th Convention of Russian Naturalists S. Nawaschin delivered a report entitled "New Observations of Fertilization in *Fritillaria tenella* and *Lilium martagon*". He was the first ever to demonstrate **both the male sexual cells entering the embryo sac having almost vermiform outline**, their bodies in various cases being bent 'extremely whimsically'. After that the male gametes become separated. Later on the nucleus of one of the sperms enters the egg cell and the nucleus of the other sperm 'becomes tightly appressed' to one of the not yet fused polar nuclei, i.e., according to Nawaschin, **to the sister nucleus of the egg cell** nucleus. Nawaschin noted that the male nuclei proceeded to retain their vermiform shape and that 'somewhat later the polar nucleus engaged in copulation with the other male nucleus heads towards the other polar nucleus which it encounters in the middle of the embryo sac'. It is only after a considerable period of time that complete

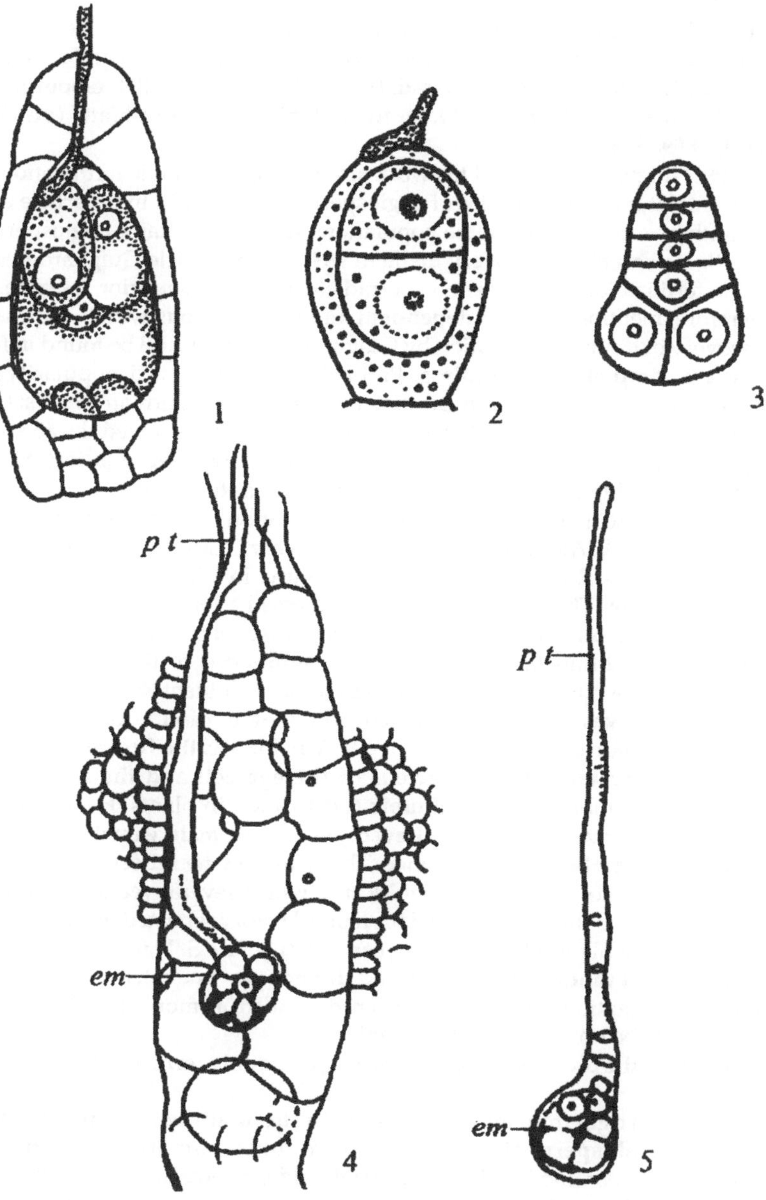

Fig. 1: Entry of pollen tube into the mature embryo sac in *Orchis militaris* (1); embryo development in *Gymnadenia odoratissima* (2, 3); embryo development from the tip of pollen tube in *Martynia lutea* (4, 5); *p t*—pollen tube, *em*—embryo. 1-3—after Hofmeister, 1849; 4, 5—after Schacht, 1850.

fusion of all nuclei occurs in the division prophase and the entire nuclear group divides as one nucleus.

This discovery (the article entitled 'Resultate einer Revision der Befruchtungsvorgänge bei *Lilium martagon* und *Fritillaria tenella*') was first published in German in Bull. de l'Acad. Imper. des Sci. de St.-Pétersbourg, T. 9, N. 4, November 1898, p. 377-382), and later in "Botanisches Centralblatt" (Bd 77, N. 2, January 4, 1899. S. 62).

In 1899, Guignard published his work also describing double fertilization in the liliaceous and later in other species of angiosperms (1900, 1922). In his article entitled 'Sur les anthérozoides et la double copulation chez les végétaux angiosperms' he mentions S. Nawaschin's discovery made in 1898. In this article Guignard describes the double fertilization phenomenon contradictory to his own prior research.

Therefore, mentionings of **independent and simultaneous** with S. Nawaschin's own discovery of double fertilization by Guignard which can still be found in foreign literature **are completely groundless**. Probably the undeserved popularity of the French botanist was rather strongly promoted by the famous German botanist Goebel, who in his journal 'Flora' (1899, Bd.86, S. 235) immediately reviewed the above work by Guignard, not giving even a cursory mention **of the three publications by S. Nawaschin, available in German and Russian long before that.**

It is necessary to mention the results of investigation on fertilization in flowering plants presented by V. Arnoldi and a French researcher O. Mottier. As quite truthfully stated by a famous Russian botanist, P. Baranov, in his 'The History of Plant Embryology' (1955): 'The history of science witnessed a lot of instances when cases which otherwise would have been significant discoveries were overlooked by researchers due to blind faith in the ruling assumptions of the time regarded as undeniable truths. A good example of this is the case of eminent Russian embryologist V. Arnoldi, which was publicly acknowledged owing to disclosure thereof by Finn (1931). Using the slide of *Scilla sibirica* ovule V. Arnoldi **was the first to see the fusion of male gamete nuclei with the nucleus of the egg cell and the nucleus of the central cell** of the embryo sac, i.e., **double fertilization** in plants. **He, however, did not quite trust his own eyes and believed the second male gamete to have been admitted into the embryo sac by chance.** That was how he missed the opportunity to make a discovery of tremendous significance which several years after was made by S. Nawaschin who thus perpetuated his name. History repeated itself with Mottier who witnessed double fertilization one year (1897) before S. Nawaschin's discovery but considered it **a random occurrence** and did not pay due attention.'

Double fertilization has currently been described in more than 300 species of flowering plants belonging to various families.

What is then **the biological significance of the double fertilization phenomenon** in angiosperms?

S. Nawaschin right from the outset adopted a strong attitude regarding **the sexual nature of the endosperm** (1900b). To confirm this assumption he adduced the experimental data obtained by De Vries (1899) and Correns (1899) on pollination by alien pollen of different races of *Zea mays*. In the opinion of Nawaschin these researchers by their outstanding experiments demonstrated that the properties of the pollen donor are transmitted not only to the embryo, but also to the endosperm. 'The origin in that case of a pair of hybrid **twins—hybrid embryo and hybrid endosperm**—the aforesaid researchers explain by the availability of real fertilization of the central cell of the embryo sac'.

Thus the essence of the 'xenia' phenomenon in flowering plants, which had intrigued many botanists since the moment of its discovery (Focke, 1881), was more fully expounded. This allowed S. Nawaschin to consider that **'there was no more justification to continue the search for the distinction in the processes occurring during the embryo and the endosperm formation'.** He arrived at his discovery through embryological study of different species. This was described in the following articles: 'On the Fertilization in Compositae and Orchidaceae' (1900a), 'Ueber die Befruchtungsvorgänge bei einigen Dicotyledonen' (1900b), 'Ueber das selbstandige Bewegungsvermogen der Spermakerne bei einigen Angiospermen' (1909), 'Näheres über die Bildung der Spermakerne bei *Lilium martagon*' (1910), 'On the Independent Mobility of Male Sexual Nuclei in Some Angiosperms' (1910), 'The Details on the Formation of Male Sexual Nuclei in *Lilium martagon*' (1911), 'On the Development of History of the Chalazogamic *Juglans nigra* and *Juglans regia*' (1912) (together with V. Finn) ('Zur Entwicklungsgechichte der Chalazogamen, *Juglans regia* und *Juglans nigra*', 1913) and 'Essay of Structural Representation of Sexual Nuclei' (1927).

In the above works the following issues are addressed for the very first time: **sexual nature of the sperm nucleus fusion with the secondary central cell nucleus, structure of sexual elements in connection with their functions, movement of sperms in the embryo sac, presence of cytoplasm in male gametes and its participation in fertilization.** Many of these aspects of double fertilization **have remained a mystery up to the present day.** Research in double fertilization in a number of monocotyledons (Liliaceae, Orchidaceae) and dicotyledons (Asteraceae, Ranunculaceae, Juglandaceae) residing in different levels of the philogenetic tree led S. Nawaschin to establish **the thesis of universality of this process for all angiosperms.**

The issue of **biological significance** of the two links of fertilization (syngamy and triple fusion), proved to be highly debatable, however.

Guignard (1899a, b) considered the second link (fertilization of the endosperm initial cell) of double fertilization to be **false fertilization** (*une sorte de pseudo-fécondation*) as opposed to fertilization proper (*une fécondation vraie*). Strasburger (1900) was of the opinion that the fusion of the sperm nucleus with the secondary nucleus of the central cell was only **a strong stimulus** to endosperm development and therefore termed it vegetative fertilization (*vegetative Befruchtung*) to be distinguished from generative fertilization or that of the egg cell and responsible, in his terms, for transmission of hereditary properties.

S. Nawaschin (1898b) himself acknowledging syngamy and triple fusion to be **a sexual act**, nevertheless, regarded polar nuclei fertilization as in a sense a polyembryony similar to that of gymnosperms. He termed the triple fusion product a 'protein zygote' and equated the forming endosperm with a modified embryo. In Nawaschin's opinion the main function of that **'twin embryo'** was **a trophic one**, i.e., the endosperm essentially functioned as a nutritive tissue.

In his appreciation of the significance of double fertilization, Takhtajan (1978) considered it a characteristic feature of flowering plants setting them apart from other plant groups. He regarded fusion of one sperm cell with the **egg cell to be fertilization proper**, and viewed the triple fusion result—**the endosperm**—as a nutritive tissue specific for flowering plants and required for nourishment of the embryo which is developed from the zygote.

Pearson (1929), Gerassimova-Navashina (1958) and a number of other researchers, on the other hand, reported that in a number of gymnosperms (e.g. in

Abies balsamea, Ephedra fragilis, Thuja, etc.) apart from the egg cell in the archegonium, the nucleus of the abdominal channelled cell is sometimes fertilized. The product of fusion of the sperm and the latter cell resembles the endosperm of angiosperms.

Poddubnaya-Arnoldi (1976) categorically denounced this hypothesis, her conviction supported by the abdominal chanelled cell nucleus not being homologous with the secondary central cell nucleus of the embryo sac. Besides, the product of fusion of the abdominal channelled cell nucleus with the sperm nucleus in gymnosperms is a diploid and not a triploid one, as in the case with most flowering plants.

Gerassimova-Navashina (1947a) held the opinion that double fertilization (as well as the embryo sac) is not a unique phenomenon found only in angiosperms as general regularities appertaining to any cell underlie it. She opposed to concepts that double fertilization arose once and at a single place. She considered this issue in line with the evolution of flowering plants and maintained the theory of their polyphyletic origin.

Smirnov (1982) and Teryokhin (1991) wrote that double fertilization is not a unique phenomenon. Smirnov held that 'only if the endosperm of flowering plants is construed as a prothallium tissue 'retarded' in its development, 'the system composition' of sexual reproduction appears to be in its essence isomeric to that of gymnosperms and other higher plants'. This opinion is similar to that of Schürhoff (1919) who maintained the view of the endosperm in flowering plants to be none other than a prothallium tissue.

Teryokhin (1991) in his work titled 'The Problems of Evolution of Seed Plant Ontogenesis' arrived at the conclusion that 'the central cell of the embryo sac in flowering plants is homologous with the initial structures of the trophic (central) prothallium part of gymnosperms, and the antipodals are homologous with its haustorial (basal) part. The central cell of the embryo sac is not therefore the second female gamete, neither is it homologous with any other sexual structure'. Based on this opinion, Teryokhin concluded that fertilization in flowering plants in its biological nature is not double. It is more likely that fertilization is single but with a trigger which is represented by the second sperm. Thus he essentially shares Strasburger's opinion who viewed the second fertilization act as a growth stimulus.

Friedman (1990a,b, 1994, 1995) showed that **double fertilization occurs not only in flowering plants but also in gymnosperms. It is most distinctly expressed in Gnetales representatives, e.g. *Ephedra, Gnetum* and *Welwitschia*.** He considers double fertilization in flowering plants and in *Ephedra* to be 'synapomorphic – evolutionary homologous'. As held by Friedman, the endosperm in flowering plants is similar to that of gymnosperms and passes through an intermediate stage called 'diploid supernumerary embryos' formed as a result of the second link of double fertilization.

One may, indeed, agree or disagree with this or that opinion on the biological significance of double fertilization and the separate stages thereof. One cannot help acknowledging, however, the fact that **the discovery itself by S. Nawaschin of the 'double fertilization' phenomenon in angiosperms as the key ontogenetic stage allowed a whole new aspect for examination of the origin both of these plants, and of gymnosperms.**

Particularly worthy of attention is the research carried out by E. Gerassimova-Navashina who not only advanced S. Nawaschin's views of double fertilization in flowering plants, but also obtained many interesting facts and propounded many remarkable ideas. Her fundamental work on the fertilization process in *Crepis* was published in 1933 and immediately gained international acclaim. **This was the first time that the mysterious process of flowering plant fertilization** was presented to researchers as a sequential series of pictorial representations demonstrating all the manifold aspects thereof. E. Gerassimova-Navashina's classical illustrations intricately presented to the most minute detail are published in many foreign monographs and textbooks in the section pertaining to Fertilization of Plants. **They once again prove the point of a drawing being the instrument of knowledge.** The culmination of the cycle of works by E. Gerassimova-Navashina on fertilization in angiosperms was her famous mitotic hypothesis. Together with two hypotheses propounded by S. Nawaschin (active sperm mobility and enantiomorphism of sexual nuclei) the mitotic double fertilization hypothesis aspiring to account for sperm behaviour in the embryo sac remains one of the most appealing up to the present day.

In the Komarov Botanical Institute, Russian Academy of Sciences, St. Petersburg, Gerassimova-Navashina established **the school of classical embryology** and founded the basis for experimental study of the fertilization process in angiosperms (influence of X-ray, temperature, distant hybridization, etc.).

It has been a hundred years since the discovery of double fertilization in flowering plants. This quite extensive period of time has once more demonstrated the complexity and the manifold character of this process. Many new interesting facts have been obtained as a result of application of different research methods (time-lapse photography, cytochemistry, TEM, SEM, etc.) and new hypotheses dwelling upon various aspects of double fertilization have been propounded.

One cannot help but marvel at the great perspicacity and intuition of S. Nawaschin who outlined the frame of issues to be addressed in relation to his discovery one hundred years ago. These issues have retained their significance to the present day: the double fertilization mechanism, the unique character of this phenomenon in angiosperms and in some gymnosperms and its biological essence.

Results of Revision of Fertilization Processes in *Lilium martagon* and *Fritillaria tenella*[1]

In complete conformity with the data obtained by the well-known researcher Guignard on fertilization in *Fritillaria meleagris* and *Lilium martagon*, as well as with the article by Overton (1891) on fertilization in the latter, I managed to show in both the species chosen for study (*Lilium martagon* and *Fritillaria tenella*) the fact that the whole process of fertilization lasts extremely long, so that we have an opportunity to observe some of its phases thoroughly.

Many of my results given below considerably differ from the data of the above-mentioned authors.

1. Every time the pollen tube was observed in contact with the embryo sac, **both of the male sexual nuclei** were observed in the embryo sac contents too. The male nuclei are cylindrical or long-conic in shape, always worm-like rolled up

[1]This article by S.G. Nawaschin was first published in German (Bull. Acad. Sci., St-Petersbourg. T. 9, No. 4, November 1898, pp. 377-382) under the title 'Resultate einer Revision der Befruchtungsvorgänge bei *Lilium martagon* und *Fritillaria tenella*'.

S.G. Nawaschin (1857-1930)

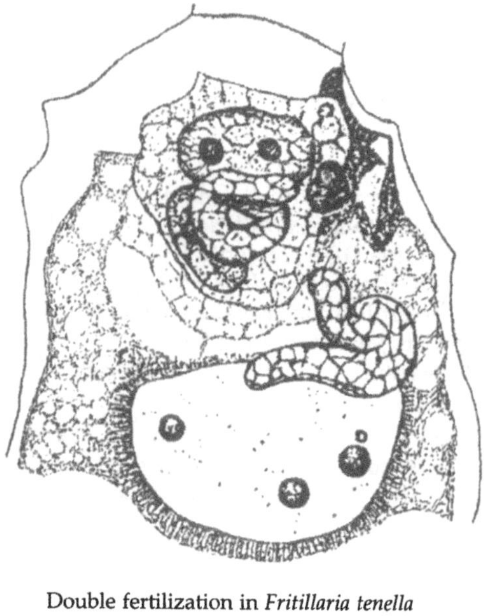

Double fertilization in *Fritillaria tenella*
(drawn by S.G. Nawaschin)

and both are located at first freely and so close to each other that in most cases they seem to compose a single whole.

2. Then the male nuclei separate from each other, and one of them reaches the egg cell and **the other comes into close contact with one of the not-yet-fused polar nuclei, in particular with the sister nucleus of the egg cell.** In addition to that, both male nuclei still keep their worm-like appearance.

   This behaviour of sexual nuclei in *Lilium martagon* was already observed by Mottier (1898). The author considers this phenomenon an abnormal one.

3. While the male nucleus, which in fact must be called generative, contact with the nucleus of the egg cell is becoming tighter and tighter, **the polar nucleus copulating with the other male nucleus directs towards the second polar nucleus, which it meets in the middle of the embryo sac.**

4. All these three nuclei, which now lie close to each other in dense protoplasm, stay isolated up to the prophase of their division and are easy to distinguish for some time; the male nucleus, which meanwhile has lost its worm-shaped appearance, is smaller than the polar nuclei, but is richer in chromatin; its chromatin skeleton is rougher than that of the polar nuclei, which differ from each other in size, and the lower 'antipodal nucleus' is considerably bigger than the upper 'egg polar nucleus'.

   I observed the composition of this group of nuclei as well as the peculiarities of some its parts in a great number of fertilized ovules, **so there is no doubt that we dealt with quite a constant phenomenon;** this fact is very easy to state in each more or less successful section.

5. Only after **the prophase of division is completed,** which occurs simultaneously in all the three nuclei, do **the nuclei fuse**, and the numerous chromosomes arrange into one common equatorial plate; thus the whole group of nuclei divides as one nucleus. The second and third divisions rapidly follow the first.

6. In the course of the first divisions of the nuclei of the endosperm the appearance of the generative nucleus copulating with the nucleus of the egg cell changes gradually and both nuclei take the appearance of a quiescent cell nucleus. **The border between the copulating nuclei can be clearly distinguished.** A full fusion of nuclei does not occur in the quiescent state. During the prophase of the first division, which in a fertilized egg cell initiates by the moment of preparation of the nuclei of the endosperm for the third division, I could not distinguish individual nuclei cavities in the nucleus of the egg.

   Consequently, judging from all the indications, fusion of a male nucleus and a female one does not occur in the quiescent state, as Mottier (1898) showed in *Lilium candidum*, but rather during the prophase of division, as was observed by Guinard in *L. martagon*.

7. It has already been mentioned that embryo and endosperm development proceeds in a regular way in the ovules of both of the plants examined. Only after the embryo in the seeds of *Fritillaria* had reached one-third its normal size, were signs of endosperm resolution and dying out of the embryo observed. On the contrary, mature seeds of *Lilium martagon* contained a normal embryo.

Some of the results mentioned here coincide closely with my previous communications on fertilization in the English walnut (Nawaschin, 1895b).

The given facts, related to plants with quite different systematic positions, such as species of *Juglans*, *Fritillaria* and *Lilium*, show that there are cases when the pollen tube does not discharge only one of its sex nuclei into the embryo sac, as was previously thought, but both its generative nuclei. Furthermore, on the basis of the shape of these nuclei, it seems possible to draw the conclusion that the male nuclei, **while still in the protoplasm of the embryo sac in a free state, are capable of self-maintained movement,** which can be compared with the movement of a wriggling worm. Finally, an amazing phenomenon is revealed: in *Fritillaria* and *Lilium,* a normally developing endosperm originates from a process which initiates by the fusion of one of the male nuclei with the sister nucleus of the egg cell nucleus, i.e., with one of the two **female nuclei**. This is why this process, just like the process of fertilization of an egg, can be called a **sexual process**. Consequently, here we deal with **a kind of polyembryony** which is a formation of a couple of twins characterized by different modes of development: one of them develops into a highly differentiated higher plant whereas the other remains in the form of a thallus and is finally absorbed by the first one.

These are all reasons to suppose that the facts described in my earlier reports concerning a number of Liliaceae, and which had been overlooked by previous researchers, will appear to be typical of other angiosperms, more complicated as a subject of investigation.

This seems to be even more likely since my interpretation of the process of fertilization in *Lilium* and *Fritillaria* as a kind of polyembryony has also a phylogenetic basis. I would like to note a remarkable analogy of the whole phenomenon with the process of fertilization in *Gnetum,* in which, according to some data (Lotsy, 1898), we deal with a similar kind of polyembryony. Here, too, fertilization consists of the following: each of the pollen tubes (in *Gnetum,* as in *Juglans,* one or more pollen tubes enter the embryo sac) discharges **both** its generative nuclei into the embryo sac and each male nucleus fuses with a female nucleus, so that each pollen tube always corresponds with a couple of products of copulation. Thus, in *Gnetum* the newly emerged 'zygotes' develop into proembryos, whereas in Liliaceae one of the zygotes turns into an 'endosperm'.

Essay of Structural Representation of Sexual Nuclei[2]

The process of fertilization in angiosperms passes in a fairly peculiar form due to which it has been termed in a special way—double fertilization. One can say that the significance of this process is so far restricted to using it only as a systematic feature. Indeed, double fertilization established in representatives of a quite considerable number of di- and monocotyledonous families represents perhaps the single clear difference between angiosperms and gymnosperms which allows drawing a natural and sharp border between them, while even such a feature which has served as a ground for designation of both these classes, namely the structure of the female generative organ, in the history of our science has repeatedly been subjected to criticism, not wholly unfounded.

Despite its undeniable significance for systematics this process, which in fact is a physiological one, has been completely ignored in physiology until now. I know only a few feeble attempts made by morphologists, including my own (Nawaschin,

---

[2]This article by S.G. Nawaschin was first published in the anniversary anthology dedicated to I.P. Borodin (1927, pp. 94-114).

1909) to expose to physiological analysis the most essential thing in this peculiar process, which is almost unique among other representatives of the vegetable kingdom and in the animal kingdom—different behaviour of male sexual nuclei, i.e., of both sperms, before fusion with female nuclei.

Meanwhile, it would seem that just in the embryo sac of an angiospermous plant fertilization passes under surprisingly simplified conditions. The female gamete is not supposed to make a choice among the multitude of gametes as most often appears to be the case with lower plants and animals; during double fertilization there is always one couple of sperms to be selected, i.e., one might think so many as is 'necessary and sufficient' (as mathematicians say) to provide the conditions for normal functioning of the whole apparatus.

Before double fertilization was discovered it was assumed that there was something imperfect lying in the conditions of the process, as one of the sperms seemed to be excessive and it was considered somehow or other delayed before entering into the embryo sac. It turned out to be quite the contrary, as development of the endosperm does not proceed without its participation. At least in hybrids the endosperm is strongly affected by the father's influence, which of course can be interpreted just as the influence of the sperm which we observe uniting with the secondary nucleus of the embryo sac. It was finally ascertained that this fusion is quite similar to the sexual union of nuclei, since it was shown (Nawaschin, 1915) that the number of chromosomes in the product of fusion is equal to the sum of chromosomes of the three nuclei involved: two female, so-called 'polar', and the male nucleus—the sperm. Therefore function of the sperm during formation of the endosperm is the same as that during formation of the embryo taken both from the genetic and morphological point of view.

The most enigmatic moment of the process in question is the divarication of the two sperms lying for some time side by side and their further movement in opposite directions: one into the egg cell, the other into the protoplasm of the embryo sac, to both polar nuclei of the latter which are already fusing. It is the manner of sperm movement that seems to me particularly mysterious and important.

It is obvious to me that the doctrine concerning the direction of self-moving cells provides no thesis which could serve as a foundation while analysing such a combination of relations as would elucidate the mechanism that distributes immovable bodies, which both sperms are, separating them one from the other. The matter seems even more complicated from the morphological point of view that both sperms represent **identical** daughter nuclei of one and the same mother nucleus, i.e., the nucleus of the generative cell of the pollen grain. The same should be assumed regarding the nuclei of the egg cell and the embryo sac, at least if the 'northern polar' nucleus, which is nearest to the egg cell, is kept in mind.

In my opinion, the reason why the sperms move in different directions lies in the fact that their essentially unlike or rather **'opposite destination'** is predetermined by the preceding division of the generative cell. If we further imagine the same concerning both nuclei being different targets of both sperm destinations and also appearing to be sister cells, then a picture of a system consisting of four bodies will emerge. As to conditions of system balance one can reason proceeding from our assumption about the opposite destination of nuclei which are of the same origin. Then we should not forget that both cytoplasms which represent the field of action though being adjacent are nevertheless to be differently directed each according to induction from both differently directed nuclei.

Let us imagine now that the sperms differ from each other in one and the same couple with their hormones, which is in connection with the assumed difference between nuclei of neighbouring cytoplasms, a sufficient condition, it seems to me, to solve our problem. It is evident that under such a condition, the cytoplasm of the egg cell in one of its parts will appear to be affected jointly by the hormone of its own nucleus and the sperm in the part which is closest to this sperm; here it will possess presumably such features which stimulate the sperm to move to this very place.

Of course, in our discussion hormones are only symbols, but whatever in fact is the interaction of male and female nuclei and their relation to corresponding cytoplasms, I believe that due to my depiction of the course of the process the following thesis has become plausible: the reason why one sperm goes to the egg cell, while the other goes to the adjacent enclosure of the embryo sac, lies in the fact that **the sperm of one and the same couple are not similar, just as female nuclei participating in the process differ from each other;** and this is so despite the fact that both couples of nuclei are sister nuclei, i.e., originate from the division of one mother nucleus, which is still assumed in morphology as equational division.

The aforesaid considerations triggered me to search for some fine, mainly concealed differences in characteristics of the nuclei originating from such division. It seemed possible to me that this difference lies not in the contents of the nucleus or the nature of its elements, considered identical, but only **in reciprocal location of the elements,** of chromosomes, in the moment of initiation of each of the daughter nuclei, i.e., in some 'structural' difference of the nuclei, in a sort of isomery, namely, stereoisomery of nuclei.

Resting in future on the analogy which, I think, can be drawn between origination, structure and features of sexual nuclei and corresponding relations of enantiomorphous stereoisomers of chemical bodies, I shall attempt to submit a depiction of behaviour and states of sexual nuclei; it is exactly **a depiction,** as there is no explanation in chemistry for similar relations and states in the field of certain physical and chemical phenomena.

The whole picture of living nature **is incompatible** with its mirror image. The original of living nature and its reflection are 'enantiomorphous', as morphologists and crystallographers, as well as chemists say regarding asymmetric systems. The prototype of such a system in its simplest form is a tetrahedron with unequal vertices or vertices of different names.

In many cases we observe location of chromosomes in diploid and haploid nuclei of cyclic character, i.e. chromosomes lie in a metaphase plate around some centre; from this the name of this stage of division in zoology becomes 'a star' (*aster*). Therefore it is reasonable to compare this system with the flat scheme used in crystallography for explanation of polyhedron symmetry, i.e. to compare with a polygon. The diploid nucleus can be represented as a polygon whose angles are pairwise the same, and the haploid corresponding nucleus—as a polygon with twice less the number of sides whose **angles are different.** Thus, the plate or star of any haploid nucleus will essentially be similar to a molecule of asymmetric structure where various atoms and radicals aggregate around one centre (carbon atom—C); in a word, the system of such a nucleus is asymmetric.

Only when a great number of chromosomes exist, especially in a nucleus of small size, is it difficult to define the individuality of chromosomes and consequently to define visually the asymmetry of the whole system. But with a moderate amount of chromosomes the problem becomes quite easy to solve and asymmetry of the nucleus plate appears as evident as asymmetry of the hand, the fingers being compared to chromosomes. The difference lies only in the fact that there is no back and palmar surface in the nucleus plate (the difference that fingers are fastened with the metacarpus is not essential), though this difference is temporal. After chromosomes have split and daughter chromosomes have dispersed to the poles there appear two plates or stars (*diaster*) with inner and outer surface, in other words, they are directed oppositely towards the centre of the mother nucleus. Chromosomes of both plates turn to be located exactly the same way as fingers of the hands being first put together by palms and then moved apart: looking at the daughter plates from the centre of the nucleus we would be certain that in one plate the chromosomes of different names—*A, B, C, D*—are arranged clockwise like fingers of the left hand (counting from the thumb) and in the other one the same chromosomes *A, B, C, D* lie in reverse order like fingers of the right hand. Therefore the plates can be named right and left plate. In other words, each of the plates, like the hands, represents mutually its own reflection: they **are enantiomorphous.**

The latter leads to the fact that a cell with the haploid set can possess the features of a system with certain polarity, i.e. with each new division it produces different elements. So, the protoplast of the pollen grain obviously divides differentially, which can easily be observed in any angiosperm where after division of the nucleus of a microspore two different cells form: a large vegetative and small generative cell. Apparently the first division of the nucleus in the embryo sac is also differential, as the micropylar nucleus when undergoing further division gives rise to nuclei of the sexual apparatus, while the opposite one initiates antipodals, i.e., the vegetative apparatus. The same peculiarity can easily be recognized during final divisions of nuclei in a ripening embryo sac: neither nuclei of its upper end undergoes equational division, as the division of one results in the formation of the future egg cell nucleus and 'northern' polar nucleus, and the other nucleus upon dividing gives rise to both synergids characterized by different properties—one matures quicker, then becomes turbid and degenerates, giving place to pollen tube contents, while the other remains much longer, but does not participate in the sexual process at all.

Hence, I think we have the right to conclude that division of the haploid nucleus in the generative cell is also not equational in essence but differential, i.e., both nuclei forming during its division, the right and the left differing substantially: one is destined or adapted to fuse with the egg cell nucleus, the other to fuse with the 'northern' polar nucleus.

So, what is the mechanism of influence of sexual nuclei upon the environment where they meet each other—the influence upon cytoplasm of the egg cell and embryo sac? It seems to me reasonable to make a supposition mentioned above that it is equally destined nuclei that should move towards one another, i.e. for example, the right nucleus (*d*) towards the right one and the left one (*l*) towards the left nucleus.

Having already pointed out the fact that haploid cells and nuclei in the embryo sac should differ in a sense, I have adduced some actual proof of such relations as well. Now comes the turn of both sperms entering the embryo sac: whether there are some visible peculiarities in this couple of sister nuclei.

From my own observations I can only mention that the sperm penetrating the egg cell stains more weakly than the sperm fusing with the polar nucleus; therefore the first one always looks more friable compared with the second one.

Data regarding the structure of sperms in Cycadaceae and *Ginkgo* in those cases where they are more authentic (Hirase, 1898; Chamberlain, 1912), i.e. where the matter concerns the real depiction of both sperms being still inside the mother cell, provide us with facts speaking in favour of my supposition that the sperm similar by birth, i.e. originated from one mother cell **are not similar** in structure.

Of course, it is difficult to search for facts concerning the idea formulated long after these facts have been established and moreover with no concrete purpose. It might be so that facts of different nature can be found in other fields of science: we can't warrant this. As for this article, my ultimate aim is to rectify the insufficiency of our knowledge about sexual cells and nuclei and to point out the morphological conditions of their activity which is directed in a definite way.

Conclusion

1. Judging from the disposition of sperms before reaching their point of destination and from their separation from 'dead globules' placed in close neighbourhood to them, it becomes obvious that it is not mechanical factors, such as for example movement of cytoplasm, that make conditions for this separation. If this were so, the 'dead globules' would have been carried away right after sperms into the egg cell and the embryo sac.
2. It is also impossible to ascribe divarication of both sperms and their direction towards two opposite destinations to any mechanical influence and should be attributed to differences between both sperms and both cytoplasms through which the sperms track their way.
3. As in this case the cytoplasm in sperms of *Phelipaea* is not observed at all, it is plausible to conclude that their differences depend first of all on the nature of their nuclei, which is consequently supposed to be different. It should be assumed as well that female nuclei towards which sperms move also different in nature and induce differently both corresponding protoplasts by their influence too.
4. It remains to suppose that the reason why sperms enter into protoplast of different cells (into the egg cell and 'cavity' of the embryo sac) lies in the fact that it is those nuclei which move towards each other that equally affect one and the same protoplast, so that their action is summed up; as a result the cytoplasm is diluted locally and the sperm moves in the direction of least resistance. The reason for the movement itself is still not clear.
5. Considering the morphological identity of nuclei of the female apparatus and both sperms in all relations except for disposition of their chromosomes, the latter circumstance should be regarded as the only condition which can initiate functional difference of nuclei of one and the same couple, i.e., difference between both sperms and difference between 'northern' polar nucleus and that of the egg cell.
6. In respect of chromosome disposition each couple of sister haploid cells can be equated to a couple of elementarily identical chemical molecules (or crystals), which differ in disposition of elements and radicals (in a crystal, of its angles) which can all be different. Such bodies are denoted as enantiomorphous. The haploid nucleus contains in its 'equatorial plate' cyclically arranged, individually different chromosomes, and after division it produces daughter nuclei whose reflection is similar in chromosome disposition (in anaphase and telophase) and which can be compared with the right and left form of the initially asymmetric system.

7. As the right form of an asymmetric body structure differs from the corresponding left form both in optical and chemical properties (rate of reaction), similarly 'the right nucleus' can differ from 'the left one' of one and the same couple in functional characteristics intrinsic to each.
8. According to the previous example (compare item 4) those sexual nuclei should meet and fuse that are of similar structure and consequently equally affect the protoplast: the right nucleus with the right one and the left one with the other left.
9. In the system of elements available during fertilization in angiosperms and participating necessarily in the process, we distinguish a couple of female nuclei with one right and one left nucleus and a couple of male nuclei, sperms, also with a right and left nucleus. This system is a necessary and sufficient condition for the egg cell and the polar nucleus 'to choose' the corresponding sperm.
10. Neither the mechanism of reductional division nor the properties of the system mentioned in item 9 should be interpreted teleologically. They are both extreme consequences of evolution of sexual relations: with the appearance of a sex reduction in number of chromosomes, characteristics of the haploid nucleus become determined; general conditions of the whole historic process—only progressive differentiation of elements and successive decrease in their number to the level of **'necessary and sufficient'**—the general law of economy in nature is evidenced.

Model Objects for Study of Pollination and Fertilization

Double fertilization is an important event in sexual reproduction, yielding a diploid embryo (arising from fertilized egg cell) and usually a triploid endosperm (arising from fertilized central cell of the embryo sac). Three phases can be distinguished in double fertilization: (i) progamic phase; (ii) fertilization proper and (iii) post-gamic phase. The second phase includes syngamy and triple fusion. Syngamy is the fusion of two haploid gametes, viz., egg and sperm cells, resulting in zygote formation. In reality, this term means sexual reproduction. Syngamy includes fusion of the cytoplasm (plasmogamy) and the nuclei (karyogamy). Triple fusion involves fusion of polar nuclei with a sperm nucleus, resulting in formation of a primary endosperm nucleus.

Gerassimova-Navashina (1947a, 1951a,b, 1960a-c, 1982) distinguished three types of syngamy, viz., premitotic, post-mitotic and intermediate. This classification is based on the stage of the cell cycle at which the male and female nuclei fuse. In the premitotic type, the gamete nuclei fuse before the onset of mitosis; the sperm nucleus completes its mitotic cycle after fusion with the egg nucleus. In the post-mitotic type, the sperm nucleus completes its mitotic cycle before gametic fusion is over; the sperm nucleus remains in contact with the egg nucleus, so that fusion occurs after mitosis has begun in the nuclei. The intermediate type of fertilization is characterized by karyogamy taking place during mitosis.

The well-known Indian embryologist P. Maheshwari (1950; p. 194) wrote: 'The time between the beginning and the end of gametic fusions is so short that one rarely succeeds in "catching" the material at the right stage'. After in-depth analysis of the technical difficulties in understanding the process of fertilization, he concluded: 'In view of these difficulties it is not surprising that our knowledge of the events concerned with fertilization has not advanced very far beyond where it stood during the early part of this century'.

The Russian school of embryologists headed by Prof. E.N. Gerassimova-Navashina made a significant contribution to the study of fertilization in different species of flowering plants. This school laid the foundation of experimental study of fertilization in flowering plants. The release of pollen tube contents in the embryo sac, the mechanism of sperm movement towards the female cells, gametic fusion (with respect to the mitotic cycle), male cytoplasm transfer and its role in fertilization, and the rates of these processes are some important aspects of syngamy and triple fusion. For their understanding it was essential to study in detail the sequence of these processes in different plant species. Evaluation of critical events in fertilization, the stimuli that guide growth of the pollen tube towards the embryo sac and the interaction of male and female gametes are also important for the study of fertilization. This required a search for model systems having features such as large embryo sac elements, long intervals between different stages of fertilization, and unique features of ovule and embryo sac, such as protrusion of embryo sac from the micropyle. *Scilla sibirica*, species of *Triticum* and *Torenia fournieri* were found to be suitable objects for the purpose Batygina and Vasilyeva, 2001).

Double Fertilization in *Scilla sibirica* (Figs. 2-4)

Before we describe the fertilization process in *Scilla sibirica*, some attention should be paid to the structure of separate elements of its embryo sac. Cells of the egg apparatus persistently show striking differences in position and character of development.

Synergids are cells with rather dense cytoplasm. They are usually located above the egg cell near the very entrance to the embryo sac and occupy its entire upper part. A highly vacuolated egg cell is located between them and is slightly lateral in position. The synergids surround the egg along their margins forming a bell-like structure, with the egg cell situated in the middle like a bell's tongue. Being always situated slightly lower than the synergids, the egg cell projects into the central cell; there appears an extracellular space between the egg and the central cell. A prominent filiform apparatus is usually present in the synergids.

Antipodal cells (usually 3) are located at the base of the embryo sac. They persist throughout fertilization, greatly increase in size and quite often 'multiply' so that their number reaches 6-10. The behaviour and number of antipodal cells vary in different species of *Scilla* (Gerassimova-Navashina and Batygina, 1958). Trabecules, which are similar to antigone (a persistent diad, which sometimes develops up to 4-nucleate stage in bisporic type of embryo sac development), occur in the chalazal part of the embryo sac in *S. sibirica* (Bhandari and Sachdeva, 1983; Bhandari and Mal, 1989). However, Gerassimova-Navashina and Batygina did not find antigone in mature embryo sacs of this plant. The lowermost antipodal cells usually resemble synergids (presence of filiform apparatus), especially at the early stage of development, when they are poorly vacuolated. The upper antipodals resemble an egg cell; they are more vacuolated and lack the filiform apparatus. These data suggest equal potential of embryo sac cells.

At the time of fertilization, the polar nuclei are generally drawn closer to each other or fuse forming a secondary nucleus; the embryo sacs with fused and unfused

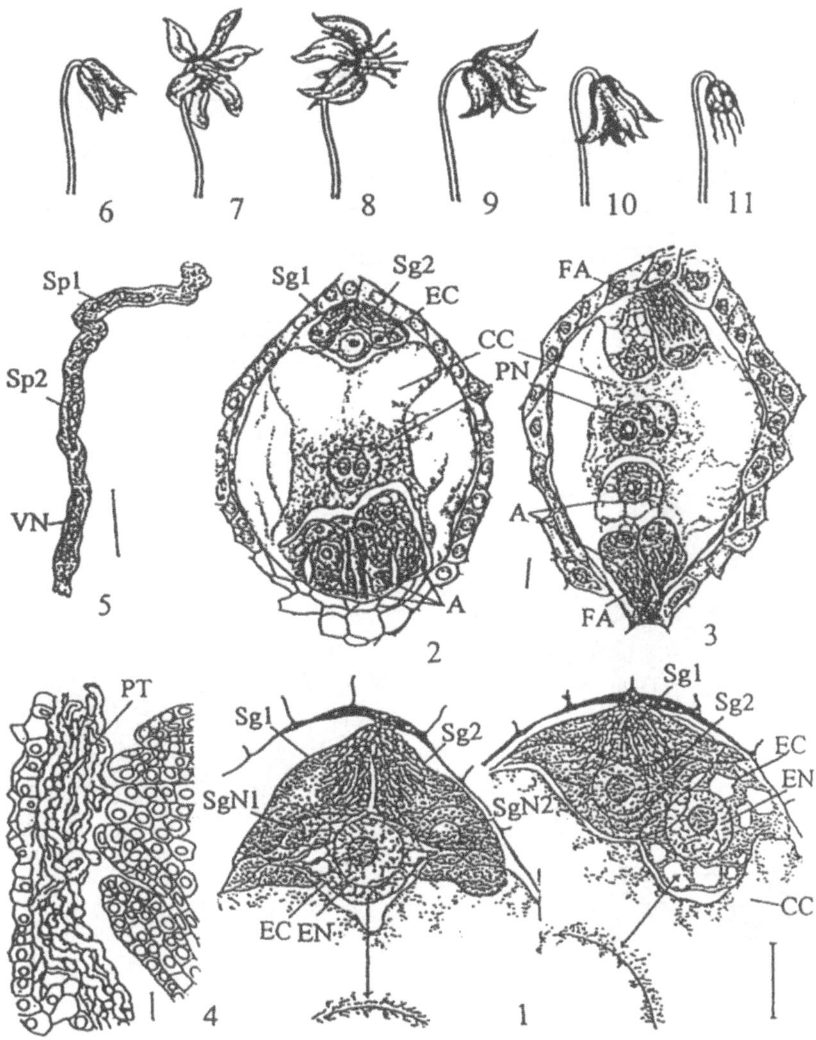

Fig. 2: Morphological changes of flower and progamic phase of double fertilization in *Scilla sibirica*: 1—upper part of an embryo sac (left, full face; right, half face); 2—structure of a mature embryo sac; 3—embryo sac soon after fertilization; 4—pollen tubes in cavity of ovary, near micropyle; 5—structure of apical part of pollen tube; 6—flower bud just before opening (castration); 7—beginning of flowering (anther dehiscence, pollination, beginning of pollen germination); 8—flowering (main stages of fertilization); 9—beginning of wilting (division of primary endosperm nucleus); 10—wilting (second and third cell division of endosperm); 11—complete wilting (first division of zygote). Bar—20 µm. A—antipodal, CC—central cell, EC—egg cell, EN—egg nucleus, FA—filiform apparatus, PN—polar nucleus, PT—pollen tube, Sg1, Sg2—synergid, SgN1, SgN2—synergid nucleus, Sp1, Sp2—sperm, VN—vegetative nucleus (after Gerassimova-Navashina and Batygina, 1958, 1964).

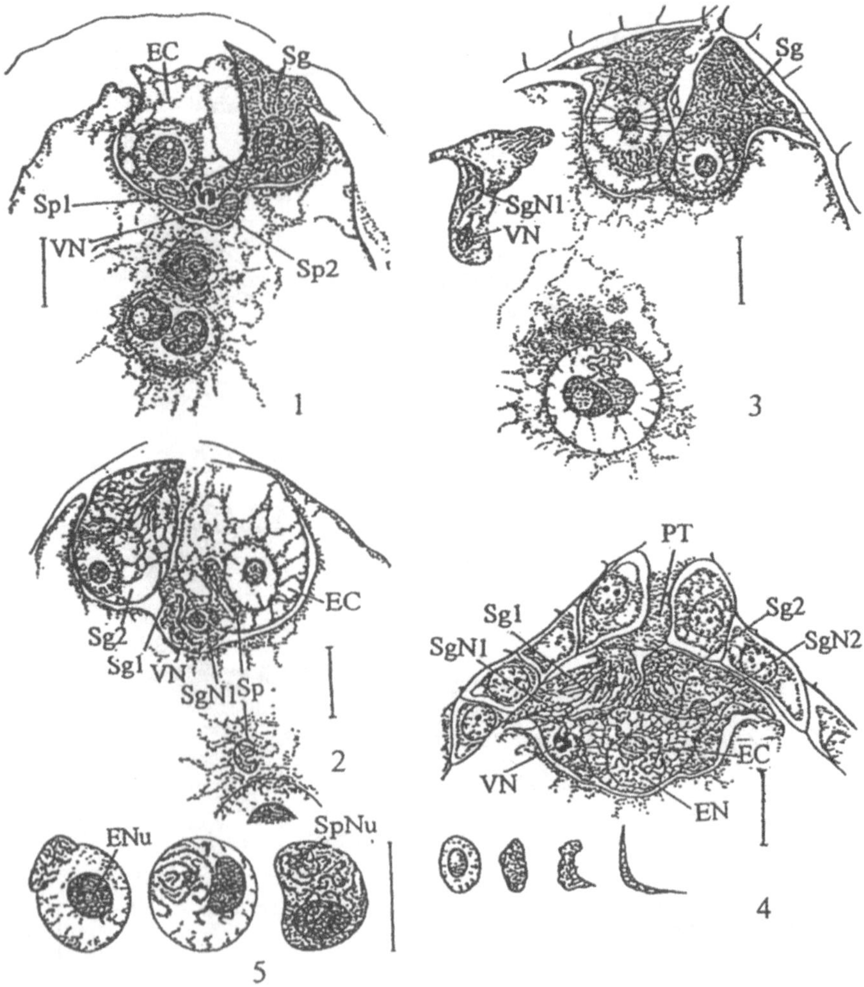

Fig. 3: Syngamy and triple fusion in *Scilla sibirica*: 1—upper part of embryo sac; pollen tube has just discharged its contents in the space between target cells, and the synergid (right of egg cell) has been damaged; sperms and vegetative nucleus can be seen; 2—sperm nuclei have reached the egg nucleus and the secondary nucleus; in the central cell the sperm is still twisted; both synergids are seen, one contains its own changed nucleus and the elongated and homogeneous vegetative cell nucleus; 3—intact synergid situated right of egg cell; destroyed synergid containing its own nucleus and vegetative nucleus can be seen above egg cell (this synergid is drawn separately); 4—typical picture of egg apparatus after entry of pollen tube; sequential stages of degeneration of synergid nucleus are drawn below; 5—sperm nucleus immerses in egg nucleus and enters a mitotic resting period (despiralization, nucleolus formation). Bar—20 μm. ENu—egg nucleolus, Sg—synergid, Sp—sperm cell, SpNu—sperm nucleolus. For other abbreviations see Fig. 2. (after Gerassimova-Navashina and Batygina, 1958).

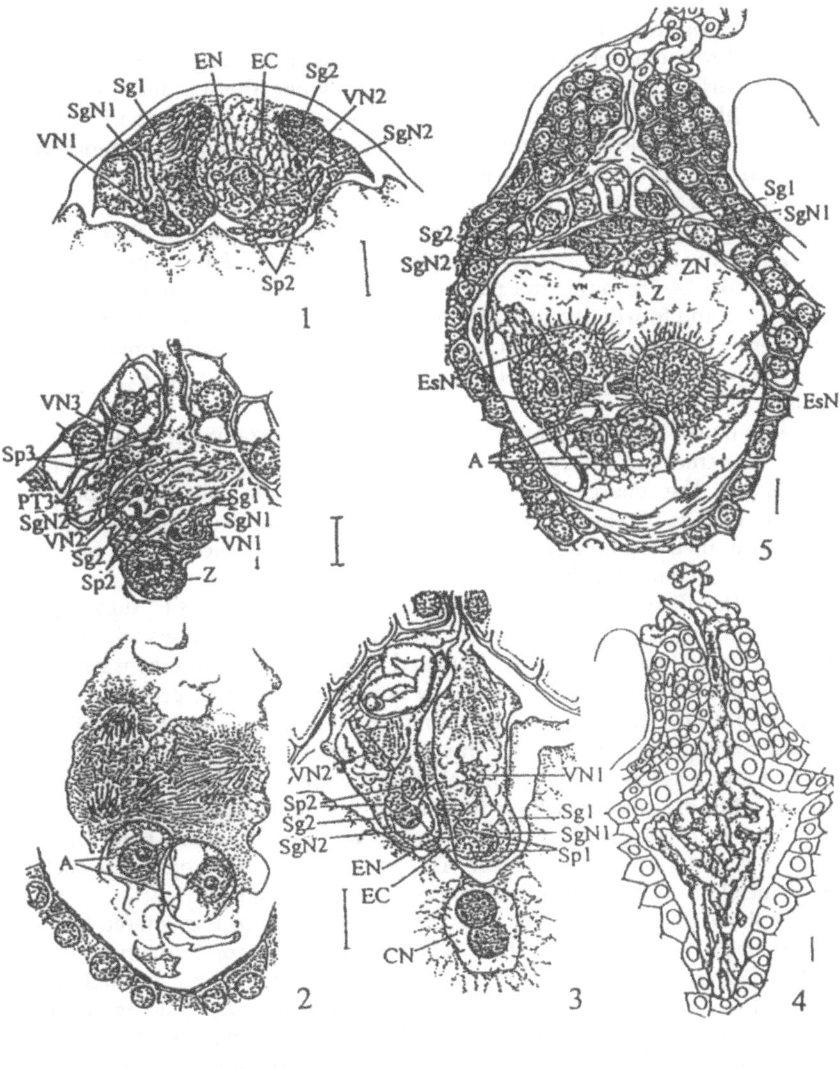

Fig. 4: Several pollen tubes entering an embryo sac of *Scilla sibirica*: 1—typical picture of an egg apparatus after entry of two pollen tubes into embryo sac, 2—upper and lower parts of embryo sac at time of second division in endosperm; three pollen tubes have entered the embryo sac, 3—two pollen tubes have entered an embryo sac; female cells remain unfertilized, 4—pollen tubes inside an abnormal embryo sac (cells were not formed in it), 5—part of an ovule with a fertilized embryo sac at time of preparing for third division in endosperm. Pollen tubes containing degenerating sperms are seen in the micropyle. Bar—20 µm. VN1, VN2, VN3—nuclei of three vegetative cells of pollen tubes, Sp2, Sp3—sperm cells of two pollen tubes, PT3—the third pollen tube, Z—zygote, ZN—zygote nucleus, CN—nucleus of central cell, EsN—endosperm nucleus. For other abbreviations see Figs. 2 and 3 (after Gerassimova-Navashina and Batygina, 1958).

polar nuclei may occur in the same ovary. Polar nuclei are located along the longitudinal axis of the embryo sac.

Fertilization in *S. sibirica* is rather prolonged and often shows great variation with respect to the onset of different stages even within the same ovary. It may be related to early-spring flowering of this plant, when the temperature fluctuates sharply. With abundant artificial pollination the different events in fertilization occur more synchronously. Within 10-20 hours after pollination the generative cell divides in the pollen tube, growing through the style. After 18-25 hours of pollination entrance of the pollen tubes into the ovary can be seen, from where they reach the ovules, enter the micropyle and embryo sac, discharge of sperm cells, or even the fusion of sexual nuclei. The first division of the primary endosperm nucleus occurs 24-28 hours after pollination, the second division after 25-48 hours, and the third and fourth divisions 90-96 hours. The zygote has a long resting period. Even after 96 hours of pollination, when the endosperm already contains 16 nuclei, a complex morphogenic process continues in the zygote; it starts to divide only 10 days later.

We shall now consider some aspects of double fertilization in more detail. When a pollen tube enters an embryo sac, it immediately discharges its contents into one of the synergids. The synergid bursts and the contents of the pollen tube appear in the space between the two target 'resting' female cells. At this stage the sperm nuclei are in telophase, their chromatin fibres are intensely stained and twisted into a dense ball, and a nucleolus is lacking.

Soon after the sperm cells are discharged in the embryo sac, they experience mutual repulsion, and this drives them away in opposite directions within the space between the two female cells to be fertilized later. In the pollen tube the sperm cells are elongated but in the embryo sac they are slightly twisted, probably due to the change in physical and chemical environment. However, after entering the female cells and coming into contact with their nuclei, sperm nuclei are no longer twisted and become more globular and vesicular. Soon the sperm nuclei 'immerse' into female nuclei. During the course of their fusion with the female nuclei, the dense ball of their chromatin becomes unwound, despiralled and loses the ability to be stained with basic stains, and a nucleolus appears. The processes of fusion of sperm nuclei with female nuclei are similar, except much faster in the central cell.

The phenomena which occur soon after the pollen tube has entered the embryo sac of *S. sibirica*, are very consistent. Invariably, the egg apparatus has an egg cell with the despiraling sperm nucleus, a healthy synergid with a normal and well-developed nucleus, and another synergid through which the contents of the pollen tube have been transported. The nucleus of the damaged synergid bursts, its shape changes, but it can always be distinguished from the vegetative cell nucleus brought by the pollen tube. The vegetative nucleus is discharged after the sperm cells and often remains in the synergid. Rarely (mentioned earlier), the discharged contents of a pollen tube bring the vegetative cell nucleus, together with sperm cells, into the space between the two target female cells.

When the fusion of egg and sperm nuclei is over and, in the central cell, the primary endosperm nucleus is about to divide or has already divided, a large number of additional pollen tubes begin to enter the ovary. These pollen tubes almost always discharge their contents into the fertilized embryo sacs, which slightly complicates the picture described above. Despite the presence of many elements brought by the tubes, they can be easily recognized. However, the egg apparatus usually shows a degenerated synergid through which the first pollen tube had grown, with its own burst nucleus and a pycnotic nucleus of the vegetative cell.

The egg nucleus is in interphase when the primary endosperm nucleus migrates to the lower part of the embryo sac and undergoes the first division, or is about to divide for the second time.

Special attention should be paid to that situation wherein pollen tubes enter both synergids but the synergids with their nuclei remain undamaged, and the egg cell and the polar nuclei are not fertilized. Mass penetration of pollen tubes into anomalous embryo sacs has been observed. This suggests that simultaneous entrance of two pollen tubes into an embryo sac reduces the pressure of their contents in both the synergids and prevents sperm discharge. In any case, the male elements delayed by a synergid lack the ability to function. All these facts suggest that success of fertilization in *S. sibirica* depends on the interaction between embryo sac elements, especially synergids, and pollen tubes. It is believed that interactions between embryo sac and pollen tube are critical for fertilization (Russell, 1996; Higashiyama *et al.*, 1998).

Pollen tubes may enter the embryo sac only until some fixed time. Immediately after fusion of the sexual nuclei rapid growth of the ovule begins, and the nucellar cells become compact, so that the pollen tubes cannot enter through the micropyle; the pollen tubes remain close to the micropyle and degenerate with all their contents.

One can assume that the discharged contents of additional pollen tubes stimulate female cell development, especially the central cell, which begins to develop into an endosperm. However, the development and division of the zygote are slightly delayed because it first undergoes a complicated process of dedifferentiation.

Fertilization in *S. sibirica* is of the premitotic type, similar to that in Asteraceae and Poaceae.The only difference is that in the former the process is much slower.

Data on fertilization in *S. sibirica,* a member of the family Liliaceae, shows it to be a classical system due to its inherent features. It presents a comprehensive picture of double fertilization and all the related processes.

Double Fertilization in Grasses

Numerous studies of fertilization in grasses *in vivo* (Batygina, 1974b, 1987a; Mogensen, 1982, 1988; Schel *et al.*, 1984; You and Jensen, 1985; Tatintseva, 1988; Russell *et al.*, 1990a; Batygina and Vasilyeva, 2001) and *in vitro* (Kranz and Lörz, 1993, 1996; Kranz *et al.*, 1995, 1998, 1999) have been published.

Progamic Phase (Fig. 5). In grasses, male gametogenesis results in the formation of two sperm cells in a pollen grain. In the majority of grasses pollen germination usually begins several minutes after pollination. Pollen tubes grow through the extracellular matrix at the receptive surface of stigmatic papillae covered with a pellicle. The pollen tubes grow through the lobes of the feathery stigma and then through the transmitting tissue, pushing through compactly arranged cells. The size and structure of sperms in *Triticum* depend, in many respects, on the ploidy of the species. The ovule in species of *Triticum* is apotropic by position in the ovary, which makes the pathway of pollen tubes from stigma to the micropyle comparatively long. From the style the pollen tubes reach the ovary and, usually, traversing between the ovary wall and the outer integument, reach the inner integument, which forms the micropyle (Fig. 6). Alternatively, the pollen tubes may reach the micropyle growing along the inner surface of the ovary, then over the integuments or along the

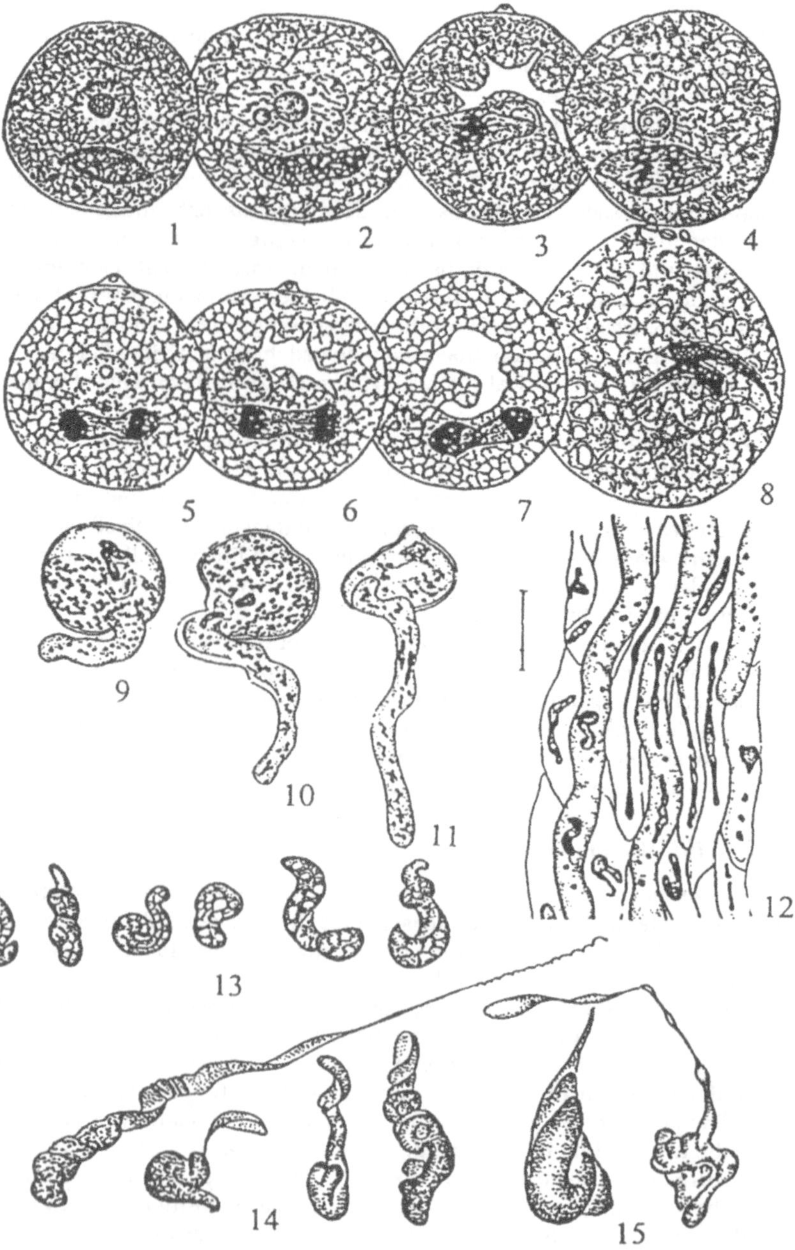

Fig. 5: Spermiogenesis in *Triticum* species: 1-8—in a pollen grain; 9-12—growing pollen tubes in the pistil; 13-15—sperms of *T. monococcum*, *T. dicoccum* and *T. aestivum* respectively. Bar—20 μm (after Batygina, 1974b).

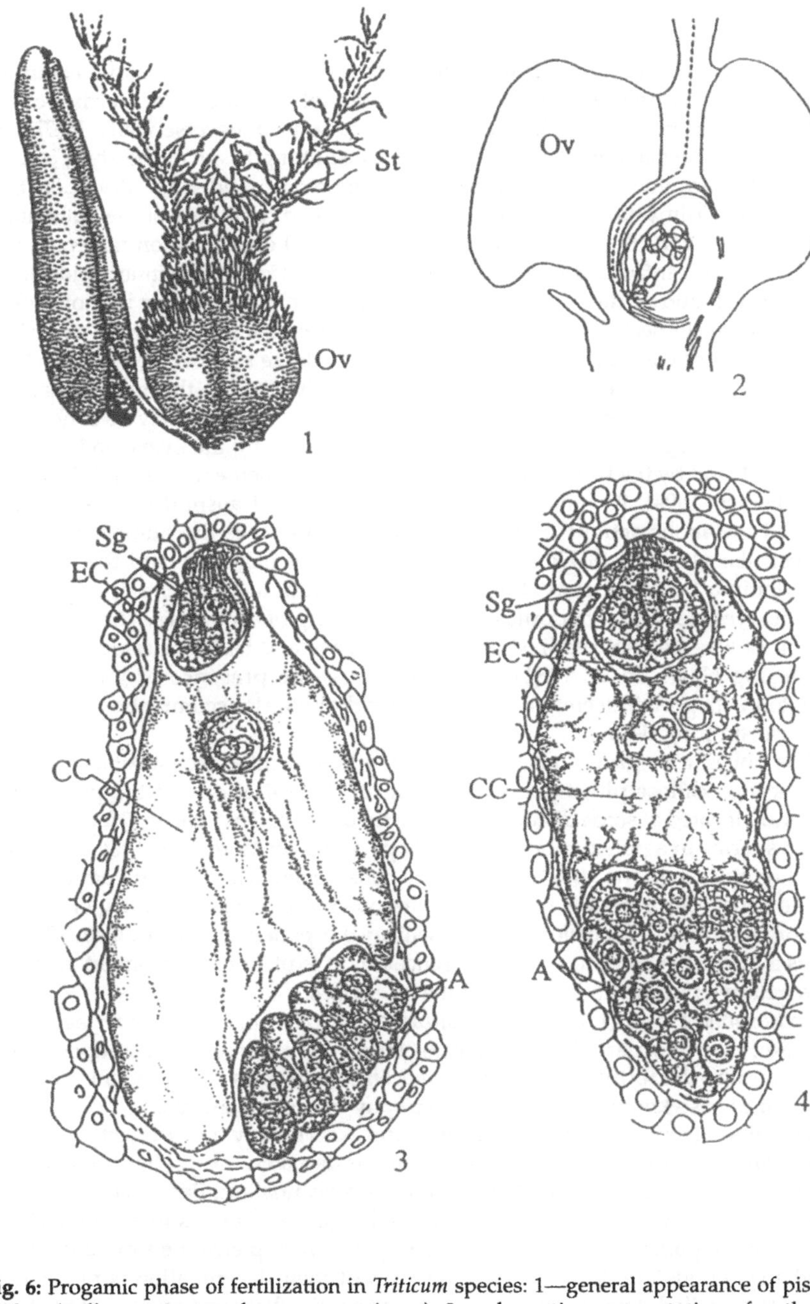

Fig. 6: Progamic phase of fertilization in *Triticum* species: 1—general appearance of pistil and anther (pollen grains can be seen on stigma); 2—schematic representation of pathway of pollen tube (indicated by dotted line) in an ovary; 3, 4—mature embryo sac before fertilization (in two projections). A—antipodal, CC—central cell, EC—egg cell, Sg—synergid, Ov—ovary, St—stigma (after Batygina, 1987a).

integumental surface only. One or more pollen tubes may enter an ovary and an embryo sac (Batygina, 1974b, 1987a).

The longitudinal and transverse cell walls of the transmitting tissue abound in plasmodesmata. As the ovary matures and the amount of extracellular matrix increases in the style, plasmodesmata disappear except those in transverse walls. In *Triticum* species plasmodesmata exist not only in the transmitting tissue of the style, but also between the terminal cells of this tissue and the ovary epidermis. Based on these observations, Vishnyakova (pers. comm.) has suggested continuity of the transmitting tissue into the ovary. She is of the opinion that a signal of pollination may pass from stigma to the placenta via a system of plasmodesmata. Cells of the inner epidermis of the ovary wall and of the transmitting tissue in the style are similar in morphology. These cells are connected to each other by plasmodesmata and also with underlying cells of the ovary wall, 2-3 layers of which contain a large number of chloroplasts.

A fibrillar material is present in the ovarian cavity between the outer integument and the epidermis of the ovary wall. Synthesis of this substance may be conditioned by the activity of dictyosomes in the epidermal cells lining the ovary and the outer integument, and by the formation of dark globules and their exocytose (Vishnyakova and Willemse, 1994). A common feature of cells in the transmitting tissue is their increased metabolic activity after pollination. In these cells the number of tubular and vesicular elements of ER and active dictyosomes increases, and the outline of plasmalemma becomes convoluted. This is accompanied by the appearance of numerous vesicles and lomasomes and the beginning of exocytosis. The presence of lomasomes was also noticed in stigmatic cells of rye and barley (Heslop-Harrison J. and Heslop-Harrison Y., 1981). Hypothetically, the presence of lomasomes is associated with protein secretion. Protein and carbohydrate accumulation is observed in the transmitting tissue.

The pollen tube grows mainly by its tip, where synthesis of substances occurs. The so–called callose plugs periodically cut off the apical part of the pollen tube. Sperms are located at some distance from the tip of the pollen tube. In *Triticum* species the two sperms in a pollen tube differ in shape and size; more often they are ribbon-shaped or spiralled. The structure of sperm cells depends both on the period of cell cycle and their chromosome number. Numerous plastids, osmiophilic globules and other organelles and inclusions are present in the pollen tube cytoplasm. Besides, there is a fibrillar substance, whose presence might explain the movement of sperms and vegetative nucleus in the pollen tube of grasses.

Fertilization Proper (Figs. 7-9). A mature embryo sac in *Triticum* species consists of an egg cell, two synergids, a binucleate central cell and numerous antipodal cells. Understanding the fertilization process and embryogenesis is dependent, to a great extent, on precise knowledge of the topography of these elements.

The first detailed study of plasmo- and karyogamy in *Triticum* species (*T. Monococcum, T. dicoccum, T. aestivum*) was undertaken using light microscopy (Batygina, 1959, 1962a,b, 1974a; Gerassimova-Navashina and Batygina, 1959). After entering the embryo sac, the pollen tube discharges its contents in a synergid. The contents of the pollen tube are often seen as 'jets' which pierce the synergid. One or both the synergids are destroyed. The process of their degeneration is rather long. Thus, at the 4-celled stage of proembryo one can see the damaged nucleus of the target synergid and intact nucleus of the persistent synergid. By the 8-celled stage of proembryo, the synergids disappear. The destroyed pycnotic nuclei of the synergids look very much like male gametes; they are often mistaken for sperm cells from additional pollen tubes.

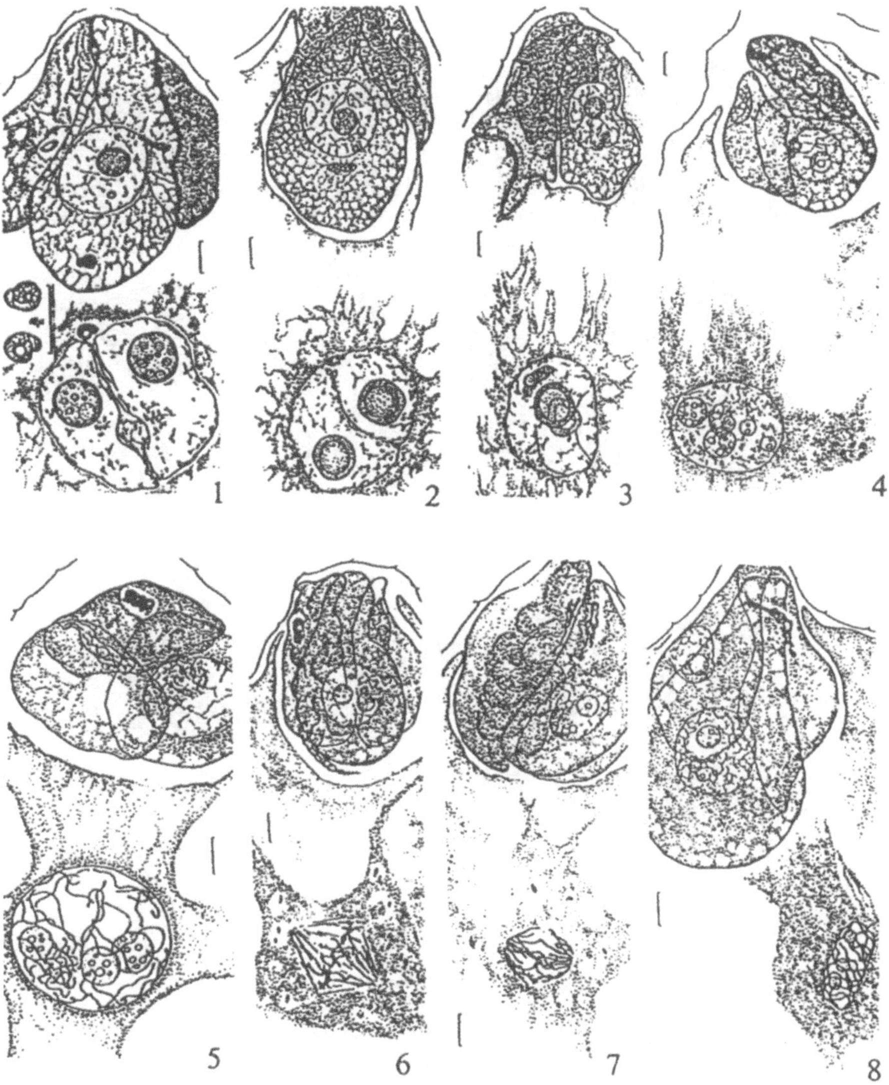

Fig. 7: Correlation between syngamy and triple fusion in wheat: 1-8—apical part of embryo sac 30 min, 1 h, 1.5 h, 2.5 h, 3 h, 5 h, 6 h and 7 h after pollination respectively. Bar—10 µm (after Batygina, 1974b).

Soon after reaching the embryo sac the sperm cells become even more spiralled than in the pollen tube. The discharged contents of the pollen tube, containing two sperms, move to the extracellular space between the egg cell and the central cell, and from this place the sperms migrate in opposite directions; one enters the central cell and the other the egg cell. In 20-40 min after pollination the sperms appear inside the cytoplasm of the two target cells.

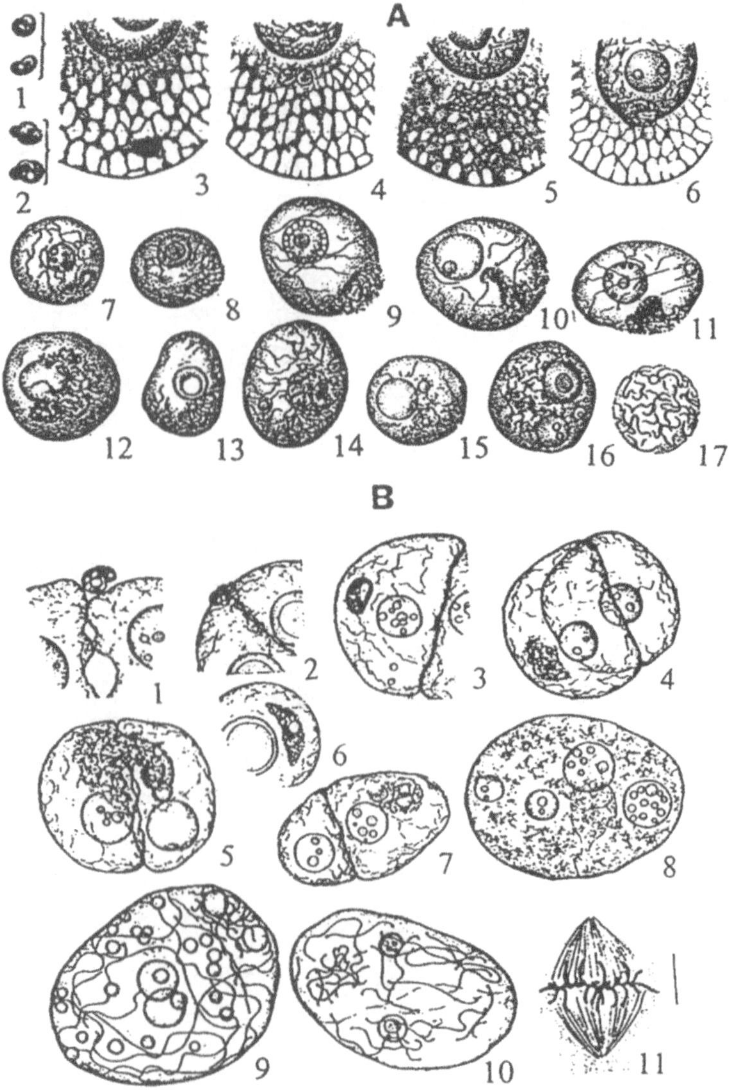

Fig. 8: Syngamy and triple fusion in *Triticum* species. A. Successive stages of transformation of a sperm shape and structure from the moment it enters embryo sac up to its complete dissolution in egg cell nucleus (premitotic type of fertilization): 1—sperms from synergids; 2—sperms from 'slot'; 3-6—sperms have entered egg cell; 7-15—fusion of sperm and egg cell nuclei; 16—zygote nucleus in interphase; 17—prophase of zygote. B. Successive stages of transformation of second sperm: 1—at time of approaching polar nuclei to their complete fusion; 2—sperm in contact with two polar nuclei; 3-6—sperm fusing with one polar nucleus; 7—sperm fuses with both polar nuclei simultaneously; 8—nucleus of primary endosperm cell in prophase (9-10) and metaphase (11). Bar—10 µm (after Batygina, 1974b).

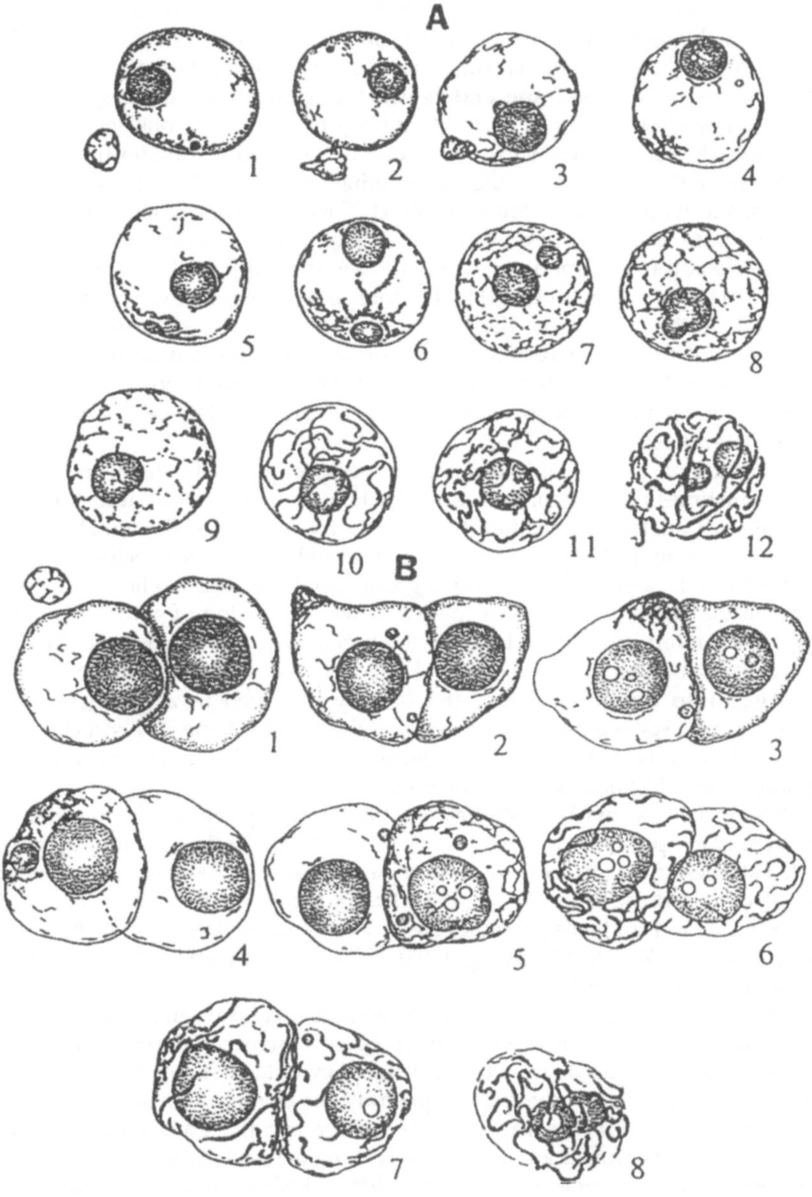

Fig. 9: Double fertilization in *Zea mays*. A. Stages of transformation of sperm shape and structure from time it enters embryo sac until its complete fusion with egg cell nucleus: 1-9—fusion of sperm and egg nuclei; 10-12—zygote nucleus in prophase. B. Stages of transformation of sperm shape and structure from time it approaches polar nuclei until their complete fusion: 1-5—fusion of sperm nucleus and one of polar nuclei; 6-8—prophase of fusing nuclei of primary endosperm cell (after Korobova, 1982).

Syngamy. After a sperm has penetrated the egg cell, it stays in the cytoplasm for an hour, which makes it possible to trace all subsequent phases of its transformation. On its way towards the egg cell nucleus, which is 'resting' at that moment (in G_0 phase), the sperm becomes loose, and its structure is clearly seen. The sperm nucleus near the egg nucleus and in contact with it has a typical shape; one end is wide and loose, while the other end is narrow and dense. Spiralled chromatin fibres are seen at both ends of the sperm. As a rule, despiralling of sperm in the egg begins from its wide end, and its contours become indistinct. There are some cases where the wide end of the sperm is already dissolved, while the narrow end is not yet in contact with the egg nucleus; thus, the sperm gradually fuses with the egg nucleus. Sperm structure in the pollen grain, pollen tube, and in the embryo sac, suggests that sperms approach female cells in the G_0 phase of the cell cycle. During its contact with female nuclei and the beginning of their fusion, the sperms do not show a nucleolus. It appears when the sperm becomes very loose and its chromatin fibre is apparently losing its continuity. Under the influence of the resting female cell, the sperm enters the G_1-phase of interphase.

Male nucleoli considerably enlarge during sperm 'dissolvement' in the female nuclei. Sometimes the nucleolus is formed when the sperm is still intact. However, the number of nucleoli may differ. In the majority of grasses, the female and male nucleoli fuse, though this process was not observed in *Triticum* species.

The entire process of sperm and egg cell fusion can last 2-6 hours, depending mainly on the habitat of the grass species. After completion of gametic fusion, the zygote enters interphase which lasts 16-18 hours in *Triticum*. At this stage the zygote shows active growth and development. The contents of the pollen tube, which were discharged into the embryo sac, are directly involved in zygote growth and development; to some extent they determine the zygote shape. The structure of the zygote nucleus and cytoplasm changes during interphase. The zygote of wheat enters prophase within 24 hours after pollination.

These processes have been studied in various grass species by electron and phase contrast microscopy and 3-dimensional reconstructions. For the first time, in the 1990s detailed studies of *in vitro* karyogamy and plasmogamy in *Zea mays* were conducted using electric impulse (Faure *et al.*, 1993; Kranz and Lörz, 1993, 1996; Kranz *et al.*, 1995). Some aspects of these processes have also been examined in *Triticum* (You and Jensen, 1985; Guo-Wei and Jia-Heng, 1992), *Triticale* (Hause and Schröder, 1987) and *Hordeum* (Mogensen, 1982).

Triple Fusion. The process of fusion of sperm and central cell is similar to that of sperm with egg cell, but the former is much faster. Whereas in the egg cytoplasm the sperm stays for about an hour, in the central cell cytoplasm it remains for only 15-30 minutes, which makes it very difficult to study the changes in a sperm during this period; sperm and polar nuclei are mostly observed in contact with each other. Before fertilization, the polar nuclei are closely placed, but not fused in the majority of grasses. When they come into contact with a sperm, they quickly unite and, as a consequence, the cytoplasm of the central cell may be trapped between them, which later degenerates. This was corroborated by other investigations, including data from electron microscopy (Mogensen, 1982, 1988). The presence of cytoplasmic 'captures' may indicate the beginning of fusion of polar nuclei. The process of their fusion with the sperm lasts for 1.5-2 hours in wheat and, as a rule, is over within 2-3 hours after pollination. During the short interphase two large nucleoli of polar nuclei and two nucleoli of the sperm can be seen. The unique features of division in the central cell are lack of male and female nucleoli fusion and asynchronous entry of sperm and polar nuclei into prophase inside the primary endosperm nucleus. Complete fusion of the chromatin of sperm and polar nuclei occurs during prophase in maize and metaphase in wheat (Batygina, 1962a,b, 1974b). Asynchronous transformation of male and female chromatin during fertilization is characteristic of all grasses (Batygina, 1987a). This phenomenon may be attributed to different durations of the cell cycle in male and female gametes and also to the fact that the two sperms from a male gametophyte are in different states (Batygina, 1961a,b, 1962a,b; Korobova, 1961).

Mechanisms of Fertilization. Studies of syngamy and triple fusion in different grasses using electron microscopy, have shown that fusion of sperms with the egg cell and the central cell starts with the establishment of contact of their plasma membranes. When a sperm enters the female cell, it loses the cytoplasm, and no sperm organelle can be seen. Possibly, during syngamy the male cytoplasm does not enter the egg cell. Mogensen (1982, 1988) reported that in *Hordeum,* the mechanisms of fertilization of the egg cell and the central cell differed. According to him, only the sperm nucleus penetrates the egg cell, while both nucleus and cytoplasm of the second sperm enter the central cell. However, he did not rule out the possibility that the process of fertilization was identical in both cases, i.e., only sperm nuclei entered the egg cell and central cell.

In vivo investigations using electron microscopy have significantly improved our understanding of the mechanisms of nuclear fusion during fertilization (Jensen, 1964b, 1973; Russell, 1983, 1985a, 1993; Mogensen and Rusche, 1985; Plyushch, 1987, 1992), as also the *in vitro* studies (Kranz and Lörz, 1993, 1996; Mattys-Rochon *et al.*, 1994; Kranz *et al.*, 1995; Kranz and Dresselhaus, 1996). Three main stages in nuclear fusion during double fertilization have been recognized: (1) the outer membranes of nuclear envelopes come into close contact or become connected by ER and subsequently fuse; (2) the inner membranes of nuclear envelopes fuse and form bridges between the nuclei; and (3) the bridges expand and capture some cytoplasm (Jensen, 1973; Faure *et al.*, 1993).

The rates of syngamy and triple fusion are not the same. Probably, they depend on the conditions existing in the egg and central cell and on the extent of chromatin condensation in the nuclei (Gerassimova-Navashina and Batygina, 1959; Batygina, 1961a,b, 1962a,b, 1974b; Khvedynich, 1969). The chromatin of polar nuclei is in a diffuse state when it comes into contact with a sperm and, therefore, the complicated reorganization, as in the egg cell nucleus does not occur, and fusion is quickly accomplished. A relation between the metabolic state of embryo sac cells and the rate of nuclear fusion has been observed (Jensen, 1965a,b; Gerassimova-Navashina and Gulyaev, 1973; You and Jensen, 1985). When an egg cell has low density of ribosomes and few organelles in its cytoplasm, the process of fusion of gametic nuclei is prolonged *(Peperomia blanda, Lilium candidum),* and if the egg cell cytoplasm is rich in ribosomes and organelles, the nuclei fuse quickly *(Crepis tectorum, Rudbeckia hirta)* (Plyushch, 1992). The rate of fusion of male and female nuclei may also depend on their ploidy: the egg nucleus is haploid and the secondary nucleus is diploid. Besides, the duration of interphase differs in these cells. This can be attributed to differences in the duration and character of dedifferentiation processes, possibly determined by different size and structure of egg and central cells.

Type of Fertilization. Syngamy in grasses is of the premitotic type. However, fusion of sperm and polar nuclei does not correspond to the classical premitotic type. The first stage, i.e., beginning of fusion of sperm and central cell nuclei precedes mitosis, which corresponds to the premitotic type. The second stage, i.e., complete fusion of male and female chromatin does not occur; the sperm nucleus enters prophase before the polar nuclei. They fuse completely by the end of prophase or during metaphase. Thus, it would be justified to classify fusion of sexual nuclei in the central cell as an intermediate type of fertilization.

Temporal Co-ordination of Syngamy and Triple Fusion. In 15-20 min after pollination, sperms are usually found in a synergid, and after 20–25 min they may be seen between the central cell and the egg cell. It is rather difficult to 'catch' this stage because in 25-30 min after pollination, in some cases even earlier, one of the sperms appears near the polar nuclei or, sometimes, in contact with them; in 45 min after pollination it is in contact with the polar nuclei. During this period the other sperm migrates towards the egg nucleus. Fusion of the sperm and polar nuclei usually begins within 1 hour after pollination. By this time the sperm has not even come into contact with the egg nucleus; fusion of sperm and egg nuclei begins only after 1.5 hours of pollination, by which time, in the polar nuclei, the sperm has almost despiralled and formed a nucleolus.

In 2–3 hours after pollination, fusion of sperm and polar nuclei is completed, and the primary endosperm nucleus enters 'dormancy'. The process of fusion of sperm and egg nuclei is fairly advanced by this time. Within 30–60 min after fusion of polar nuclei and sperm (3-4 hours after pollination), the prophase begins, first in the sperm nucleus and then in the secondary nucleus. Fusion of sperm and egg nuclei corresponds to this stage of central cell development. In 5–6 hours after pollination, when the primary endosperm nucleus is in metaphase, despiralling of the sperm in the egg cell begins and it lasts for 4-5 hours. In the polar nuclei this process takes 1.5-2 hours. The resting period also differs, lasting about 16-18 hours in the zygote and only 0.5-1 hour in the endosperm initial cell. Thus the zygote begins to divide 20-24 hours after pollination, when the endosperm is already 4-5 nucleate.

We investigated flowering, pollination and fertilization in *T. monococcum*, *T. dicoccum* and *T. aestivum* and established that each species had its own characteristics of flowering (type and rate), but premitotic type of fertilization is common to all the three species. The rates of sperm fusion with female nuclei differ. However, despite differences in the rate of the processes, the state of the central cell, egg cell, and their nuclei, as well as the zygote show a strict correlation. This correlation is slightly liable to the influence of external factors. Undoubtedly, the rate of embryological processes in general depends on the changes in environmental conditions but a correlation between syngamy and triple fusion is still maintained. If the morphogenetic correlation between syngamy and triple fusion is disturbed, this leads to anomalous development of embryo or endosperm (Batygina, 1974a,b, 1997).

Fertilization in Torenia fournieri (Figs. 10-12). It is known that in some angiosperms, such as *Phyladelphus*, *Torenia* and *Utricularia*, the embryo sac protrudes through the micropyle. *Torenia fournieri* appears to be of special interest as a model system to study double fertilization in living material; most of its embryo sac (female germ unit), including the egg apparatus and part of the central cell, is naked.

Higashiyama *et al.* (1997) studied the kinetics of the entire process of fertilization (syngamy and triple fusion), including the progamic phase, in *T. fournieri* by direct observation of living, naked embryo sacs. Ovules were excised in a liquid medium containing agarose of low gelling temperature and, after the medium had gelled, the embryos were examined by differential interference microscopy. By detailed investigations of directed growth of the pollen tube towards the embryo sac, *in vitro*,

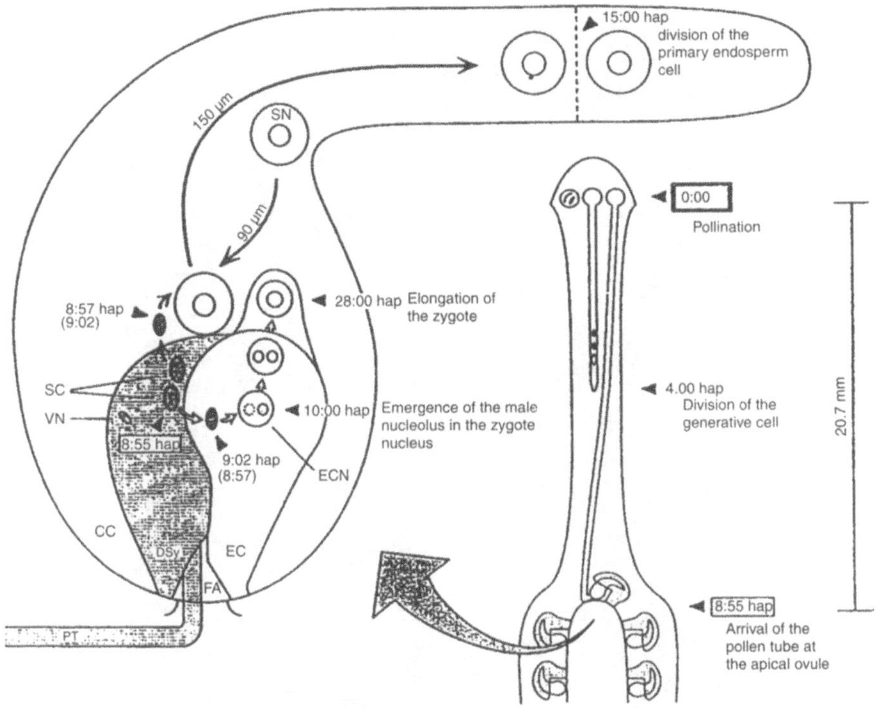

Fig. 10: Schematic representation of the process of double fertilization in *Torenia fournieri*. The time course in the pistil (right) and in the naked embryo sac of the apical ovule (left) is indicated (*hap*—hours after pollination); CC—central cell, DSy—degenerated synergid, EC—egg cell, ECN—egg cell nucleus, FA—filiform apparatus, PT—pollen tube, SC—sperm cell, SN—secondary nucleus, VN—vegetative nucleus. Courtesy Dr. T. Higashiyama, Univ. Tokyo.

they convincingly demonstrated that the region of the filiform apparatus of the living synergid is a specific attractant for a pollen tube (Higashiyama *et al.*, 1998). This supports, to some extent, the hypothesis of two types of control of the directed growth of the pollen tube towards an embryo sac (Ray *et al.*, 1997). Of great interest are the data on karyogamy and migration of the secondary nucleus of the central cell (Higashiyama *et al.*, 1997).

Modifications of Double Fertilization in Flowering Plants

Deviations from the normal course of double fertilization may be displayed in the character of syngamy or triple fusion, absence of one of them, and the extent of further development of embryo, endosperm, etc. Though double fertilization does occur in representatives of family Trapaceae, endosperm tissue is not formed; primary endosperm nucleus becomes hypertrophied and moves to the chalazal end of the embryo sac, where it gradually degenerates (Gibelli and Ferrero, 1891; Ram, 1956; Grevtsova *et al.*, 1987). In *Trapa natans*, the endosperm is formed but ephemeral (Trela-Sawicka, 1978). According to Titova (1988, Kolesova [Titova], 1996) in *T. astrachanica* the process of triple fusion is lacking due to disturbance in the mechanism

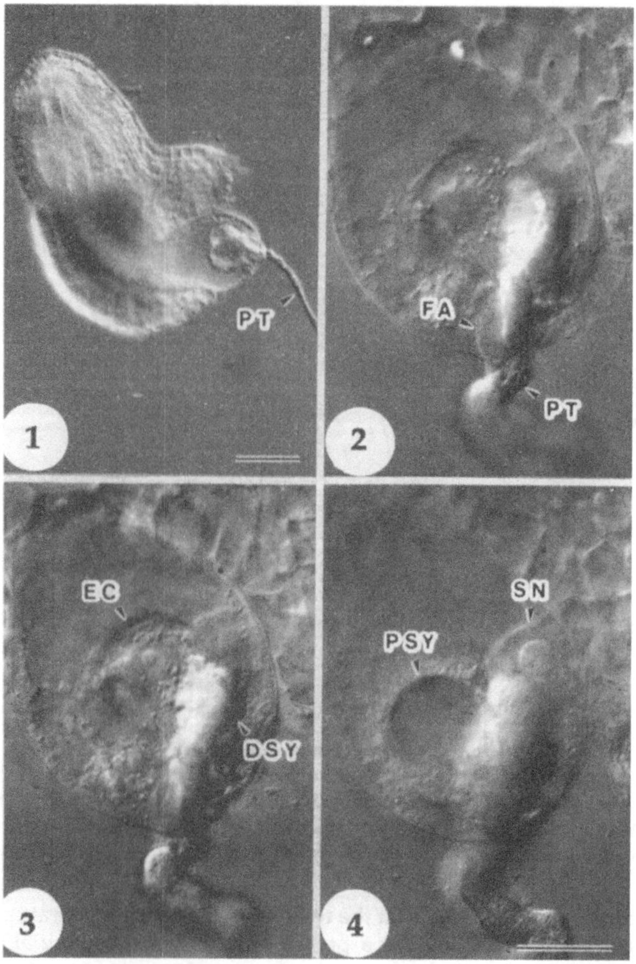

Fig. 11: Differential interference images of living ovules and naked embryo sac in *Torenia fournieri* after arrival of pollen tube, 10 h after pollination: 1—entire ovule with naked embryo sac penetrated by a pollen tube (PT); 2-4—extramicropylar region of naked embryo sac observed in different focal planes: 2—pollen tube (PT) has entered one of the synergids through the filiform apparatus (FA), 3—egg cell (EC) and degenerated synergid (DSy) with discharge of pollen tube are seen, 4—persistent synergid (PSy) and secondary nucleus (SN) with large nucleolus observed, egg cell and two synergids cannot be observed together in the same focal plane. Bar—50 µm (1) and 20 µm (2-4). Courtesy Dr. T. Higashiyama, Univ. Tokyo.

of sperm repulsion; one of the sperms is delayed in the synergid or in-between the cells of the egg apparatus.

In some members of Orchidaceae, disturbances in the fusion of sperm nucleus and central cell nucleus (nuclei) have often been observed *(Cypripedium insigne, Dendrobium nobile, Calanthe veitchii* and others) and sometimes triple fusion is lacking *(Oreorchis patens, Calanthe aristulifera, Listera nipponica)* (Savina and Poddubnaya-

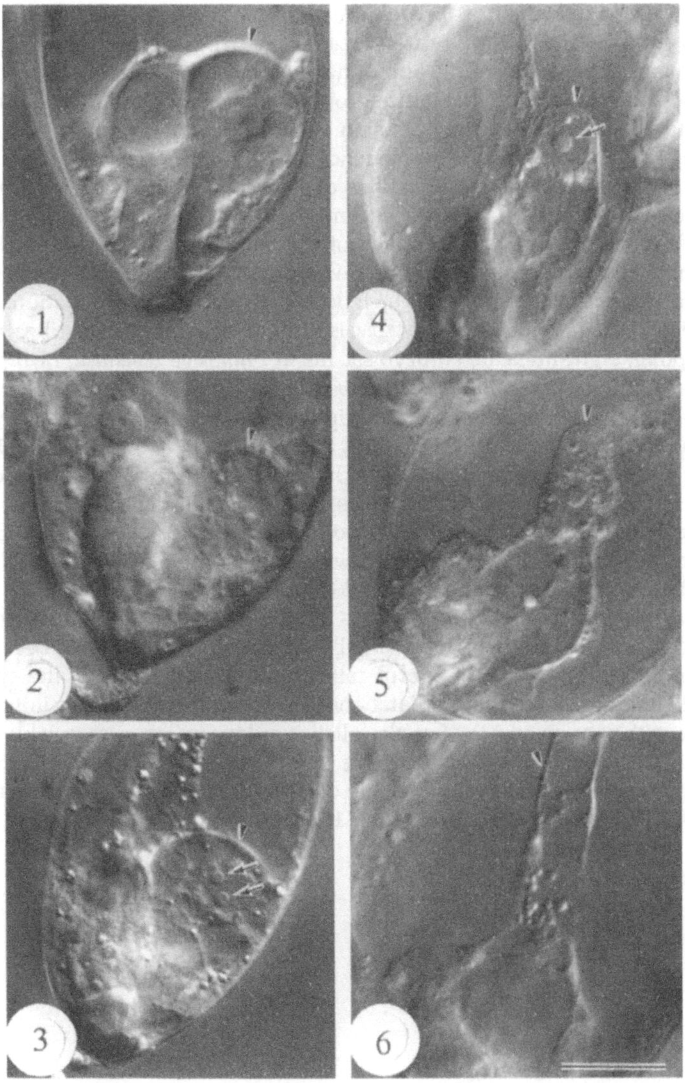

Fig. 12: Fertilization in egg cell (arrow-heads) of *Torenia fournieri* (photographs from different living embryo sacs arranged in sequence, one synergid shown left of the egg cell): 1—egg cell and synergid before pollination; 2—9 hours after pollination, just after discharge of contents of pollen tube into synergid; 3—15 h after pollination, showing two nucleoli (arrows), one from egg cell, the other from sperm cell, within the zygote nucleus; a broad cytoplasmic strand is visible above egg apparatus; in the primary endosperm cell the first division begins at that time; 4—25 h after pollination, zygote has changed shape and shows one large nucleolus (arrow) in the nucleus; nucleus and neighbouring cytoplasm have already shifted to the chalazal region of the zygote; 5—28 h after pollination, zygote has begun to elongate for embryogenesis; 6—30 h after pollination, zygote has completely elongated into the inner region before the first division. Bar—20 µm. Courtesy Dr. T. Higashiyama, Univ. Tokyo.

Arnoldi, 1990). In *Calanthe veitchii,* polar nuclei usually do not fuse and form a polar-antipodal group. This group is often located near the egg cell (Savina, 1979). The sperm nucleus approaches this group of nuclei but does not fuse with it. Thus, triple fusion does not occur. Later, the entire group of nuclei, including the sperm nucleus, degenerates. The absence of triple fusion is characteristic of family Podostemaceae (Battaglia, 1971; Nagendran, 1974; Nagendran *et al.,* 1977; Teryokhin, 1991).

One of the interesting modifications of double fertilization is the way a pollen tube enters the embryo sac in some members of family Plumbaginaceae. A characteristic feature of some members of this family, characterized by Plumbago- and Plumbagella-types of embryo sac development, is a tendency for degeneration or absence of the cells (and nuclei) which do not participate in fertilization. In extreme cases, a bicellular, monopolar embryo sac is formed by the time of fertilization instead of a tetrapolar, eight-nucleate embryo sac (as in *Plumbago zeylanica*) or bipolar, four-nucleate embryo sac (as in *Plumbagella capensis).* Such embryo sacs consist of only an egg and a central cell; synergids and antipodals are lacking (Dahlgren, 1916; Boyes and Battaglia, 1951). Cass (1972a,b), Cass and Karas (1974) and Russell (1980, 1982) stated that the egg cell in these cases has a filiform apparatus and cytoplasm resembling that of a synergid, and takes over the synergid's functions. However, unlike the synergid, a pollen tube does not penetrate such an egg cell; it only enters its filiform apparatus. Thereafter, it grows along the egg cell wall, discharging its contents between the egg and the central cell (Russell, 1980, 1982).

In *Plumbago zeylanica,* the stages of fusion have been observed at the ultrastructural level and the male organelles in the female cells have been identified and quantified (Russell, 1983). Sperm cells are deposited between the egg and the central cell and are located immediately next to each other. Fusion is initiated at one site between the sperm and the female cell. Secondary fusion sites quickly propagate and result in incorporation of the male cytoplasm when the fusion membranes vesiculate and break down in the female cytoplasm (Cass, 1972a,b; Cass and Karas, 1974; Russell and Cass, 1981a,b; Russell, 1982, 1983).

PROGAMIC PHASE OF DOUBLE FERTILIZATION

Transmitting Tract in Pistil (Plate I)

This tract is a combination of structural elements of the gynoecium responsible for its interaction with the male gametophyte in the progamic phase of double fertilization. It includes all pistil tissues (stigma, style and ovary) to which the pollen grains and the growing pollen tubes are exposed. Specialization of separate parts of the transmitting tract may be expressed to different degrees.

Structural organization of the recipient stigma surface (Plate II)

The stigma is that part of the pistil responsible for receiving pollen. Functionally, the recipient surface of the stigma must be capable of collecting pollen, providing its hydration and adhesion, and facilitating the directed growth of the pollen tube into the style. The stigma performs the recognitory function ensuring interspecific and intergeneric **incompatibility**, as well as **self-incompatibility**. Structural peculiarities, in particular the thickness of the stigma cuticle, can serve as determinants of autofertility in genera with no self-incompatibility (Lord and Heslop-Harrison, 1984).

Considerable diversity in stigma structure as well as differences in secretory activity of stigmata pointed out by plant morphologists as early as the 19th century (Knuth, 1898, cited by Y. Heslop-Harrison and Shivanna, 1977) are successfully used in systematics (Y. Heslop-Harrison, 1981). Based on the characteristics of stigmata in plants belonging to more than 1000 species of 900 genera of 250 families, several types of their structure were identified (Heslop-Harrison Y. and Shivanna, 1977). The principal functional distinction of stigmata is their separation into two large groups—**dry** and **wet**—depending on the type of secretory activity of their recipient surfaces.

Wet stigmata produce secreta or secretory mucilage (exudate) on their surface at the moment of pollen reception. The amount of secreta depends on the mechanism of its synthesis and may vary even in the same family. Thus in *Vigna* species (Fabaceae) the secreta is very abundant, whereas in *Crotalaria juncea* it is produced only as a few drops localized between the papillae of the stigma (Ghosh and Shivanna, 1982). Papillae exudate may be so plentiful that it trickles down the style as observed in *Crinum defixum* (Amaryllidaceae) (Shivanna and Sastri, 1981). The principle component of the secretory mucilage in a large group of plants (Fabaceae, Solanaceae, etc.) are lipids. They are produced in the strongly developed ER of the stigma cells and are expelled from the protoplasts in the form of drops. The lipids accumulate in the intercellular spaces and following the rupture of the stigma cuticle blend into the secreta covering the surface of the stigma. The intercellular spaces may enlarge and merge with one another into a certain type of secretory cavities as observed, for instance, in a number of pulses (Ghosh and Shivanna, 1982; Vishnyakova, 1986). Proteins also form an indispensable component of the secreta; however, their content may be rather insignificant. Predominance of the lipids and a certain amount of water define the lipophilic nature of the stigma exudate. Another group of plants (e.g. representatives of Liliaceae) is characterized by the hydrophilic secretory mucous stigma. Its main components are proteins and polysaccharides, whereas lipids are present only as sparse drops. The Golgi apparatus in such stigma cells is better developed (Knox, 1984b). Irrespective of the amount of proteins, esterase activity is always present in the secreta increasing with the ripening of the stigma (Ghosh and Shivanna, 1982). Its presence is related to the evolvement of stigma sensitivity (Mattsson *et al.*, 1974). Certain amino acids are also found in the stigma secreta (Lord and Webster, 1979). Its viscosity is caused by the presence of sugars and mucopolysaccharides.

Purportedly the functions of the secretory mucilage are protection of the stigma from dehydration (in which respect it is sometimes called 'liquid cuticle'), collection of pollen grains with their fixation on the stigma, as well as ensuring conditions for pollen tube growth (Linskens, 1978). The exudate of the stigma is utilized by the pollen tubes for biosynthesis of the pectic substances of the tube wall (Kroh and Helsper, 1974). It may also be an attractant for the pollinating insects in the flowers with no nectaries (Kenrick and Knox, 1981). Participation of the exudate components in pollen recognition has not yet been ascertained.

Dry stigmata produce no mucilage. Their recipient surface is composed of the papillate cells and is covered with a thin hydrated film—the pellicle—forming an uninterrupted layer on top of the stigma cuticle (Mattsson *et al.*, 1974; Heslop-Harrison Y., 1975). An example of the dry type stigma is that of cereals. It consists of a multitude of hairs composed of numerous unicellular papillae. Dry stigmata are found also in representatives of Asteraceae, Brassicaceae and Iridaceae.

The pellicle of dry stigmata is believed to play an important role in pollen grain hydration as well as in the activation of cutinases or cutinase precursors of the pollen tubes (Heslop-Harrison J. and Heslop-Harrison Y., 1975; Heslop-Harrison Y., 1977). Despite its insignificant thickness (e.g. 10-20 nm in Poaceae) the pellicle of dry stigmata has a complex protein composition. In *Secale cereale* (Poaceae) 24 protein components have been identified in the pellicle (Knox *et al.*, 1975). In *Pennisetum glaucum* there are more than 30 proteins, three of them with esterase activity and four others being identified as glycoproteides (Heslop-Harrison J., 1982). Representatives of *Gladiolus* (Iridaceae) have three esterase isoenzymes and three glycoproteides among 17 surface proteins in their stigma pellicle (Knox *et al.*, 1976). The presence of antigens related to expression of *S* gene alelles (self-incompatibility) was also identified therein in representatives of *Brassica* (Nasrallah and Wallace, 1967; Nasrallah *et al.*, 1972; Heslop-Harrison J., 1975). This particular feature and a number of other experimental data suggest the leading role of the pellicle in pollen recognition in self-incompatibility.

The structure of the papillate cuticle in dry stigmata is quite diverse. In cruciferous plants for instance, it is formed by radially positioned columns divided by a gap. This structure allows water to be admitted into the pellicle from the protoplast of the stigma papillae provided it is in a turgid state. As the papilla withers the cuticular columns approach one another and the water intake into the pellicle is arrested. In some representatives of Iridaceae the cuticle has a more complex chamber-like structure (Heslop-Harrison J. *et al.*, 1975). However, there are genera with uninterrupted stigma cuticle. In such cases the physiologically important stigma materials located under the cuticle are extracted to the stigma surface during contact of the latter with a pollen grain by means of cuticular transpiration typical for plants, as for instance in *Gladiolus gandavensis* (Randal, 1982)

The pellicle starts to form in the flower bud. The proteins are synthesized in the cortex layer of the papillate cytoplasm and diffuse therefrom through the pectocellulose part of the membrane and the gaps between the cuticular columns, thus forming the pellicle (Heslop-Harrison J. *et al.*, 1975). In a ripe flower the *S*-specific antigens are admitted into the pellicle in a similar manner.

Structural set-up of transmitting tract in style (Plate III)

The style is a part of the pistil above the ovary. In essence it is a continuation of the upper part of the carpel (Esau, 1977). The anatomical diversity of styles is represented by three groups: **open** (hollow), **closed** (solid) and **half-closed** (Knox, 1984b).

Open styles are formed by incomplete accretion of a conduplicate stylodium (Takhtajan, 1964) with a resultant channel in the middle of the style lined with specialized cells of adaxial carpel epidermis by origin. These cells are referred to as channelled cells and as a rule have a pronounced secretory nature. Often facing the interior of the channel, the walls of the channelled cells, exposed to which the pollen tubes grow, have numerous cell wall protuberances into the cell—so-called 'wall labyrinth'—a structure typical for transfer cells (Vassilyev, 1970; Dashek *et al.*, 1971; Vishnyakova, 1986; Kronestedt and Walles, 1986). The structure of the channelled cells may vary even in representatives of the same family. For instance, in Fabaceae apart from the wall labyrinth, the interior wall of the channelled cells by the moment of pollination may also possess other features. Specifically, in *Vigna unguiculata* the wall is thickened, whereas in *V. radiata* it is gelled and is completely absent in *Crotalaria* genus (Ghosh and Shivanna, 1982). With *Strelitzia reginae* (Strelitziaceae) taken as the example, it was demonstrated that the channelled cells are interconnected with one another and with the underlying cells by means of plasmodesmata (Kronestedt and Walles, 1986). The channelled cells may have a papillate nature and line the channel interior as a fringe. In case the channel is extended directly into the stigma as was identified in *Nelumbo nucifera* (Nelumbonaceae) (Batygina *et al.*, 1983), as well as in a number of representatives of Liliaceae, channelled cells are an extension of the papillate stigma cells. The papillate nature of the channelled cells may also be observed in half-closed styles, which have a fragment of transmitting tissue between the stigma and the channel, as was described for *Trifolium pratense* (Fabaceae) (Heslop-Harrison Y. and Heslop-Harrison J., 1982; Vishnyakova, 1991). The style channel is filled with an exudate whose content has been studied only superficially to date.

Closed styles have no channel but have one or several strands of transmitting tissue embedded in the ground tissue of the style or connected with vascular bundles. The transmitting tissue of closed styles in different taxa has its own characteristic features. It may be represented by tightly packed thick-walled cells whose layered walls are rich in pectic substances and the growing pollen tube passes through one of those layers, as for instance in *Gossypium hirsutum* (Malvaceae) (Jensen and Fisher, 1969). The most typical is the transmitting tissue whose cells are surrounded by copious amounts of an intercellular mucilagenous substance of pectic nature in which the pollen tubes grow. This substance is formed consequent to breakage of the initial contact between cells during dissolution of the middle cell lamella. Apart from pectins in this intercellular material, low-molecular carbohydrates and various enzymes or lipids have been identified (Knox *et al.*, 1986). In styles of the majority of monocotyledons the predominant components of the intercellular matrix are arabinogalactans and arabinogalactant proteins (Knox, 1984b). However, in representatives of Poaceae the transmitting tissue is devoid of this copious intercellular matrix, its cells adhering tightly to one another. The pollen tubes grow in intercellular spaces separating the cells of the transmitting tissue (Batygina, 1974b, 1987a; Vishnyakova and Willemse, 1994).

The distinctive features of the transmitting tissue cells of closed styles are their elongated shape and the presence of plasmodesmata. In immature pistils the plasmodesmata are found both in radial and in tangential walls of such cells (Knox, 1984b). As the flower develops and the contact between the cells is lost due to formation of the intercellular matrix, the plasmodesmata remain only in tangential cell walls of transmitting tissue. Their presence therein is demonstrative of the fact that the longitudinal communication between the cells is retained, which provides the opportunity for pollination signals to be transmitted. The issue as to the possibility of the pollination signals to be transmitted through the transmitting style tissue to the tissues of the placenta and the ovule remains unsolved (Knox, 1984b). The presence of plasmodesmata between the transmitting tissue cells of the style and the ovary epidermis demonstrated in *Triticum aestivum* (Vishnyakova and Willemse, 1994) indicates that such a possibility exists.

The transmitting tissue cells of the style are usually metabolically active. They contain many mitochondria and ER profiles, polysomes and amyloplasts (Jensen and Fisher, 1969; Bell and Hicks, 1976; Kenrick and Knox, 1981; Vishnyakova and Willemse, 1994), many dictyosomes and vesicles as well as lomasomes (Cresti *et al.*, 1976). The intercellular matrix is the product of the synthetic activity of these cells.

Half-closed styles combine the presence of transmitting tissue and the channel obviously representing the carpel accretion degree. For instance, in a number of representatives of Fabaceae the style is closed in its upper part and the pollen tubes grow in the intercellular matrix, whereas in the lower part of the style they grow in a developed channel. On the contrary, in *Malus domestica* (Rosaceae) the channel is found in the upper part of the style, narrowing down into a transmitting tissue strand (Vishnyakova, 1987). These facts as well as the similarity of biochemical composition of the channel liquid and the intercellular matrix in solid styles (Labarca *et al.*, 1970) point to a small functional difference of the two style types.

In styles of a number of plants, e.g. *Prunus avium* (Rosaceae), *Nicotiana alata*, *Lycopersicon peruvianum* (Solanaceae), glycoproteids were identified related to the expression of self-incompatibility *S*-gene alleles (Raff *et al.*, 1981; Mau *et al.*, 1982; Hoggart and Clarke, 1984).

As the pollen tubes grow in the style they utilize the substances of the transmitting tissue, a fact proven by electron-microscopic, cytochemical and autoradiographic researches (Loewus and Labarca, 1973; Kroh and Helsper, 1974; Vishnyakova and Willemse, 1994). Following pollination in the pistils, the amount of reserve carbohydrates and lipids reduces. The tract of a pollen tube in the transmitting tissue is frequently marked by changes in structure of the cells through which it passes (Batygina *et al.*, 1961). These cells frequently become necrotic.

Structural set-up of transmitting tract in ovary (Plate IV)

In many species, the pollen tubes subsequent to their emergence from the transmitting tissue grow in the ovarian cavity. Quite often, their passage is effected on the surface of a specialized structure—the obturator (see Obturator, vol. 1). The thin structure of the cells in this tissue points to their secretory and transfer function (Tilton and Horner, 1980). In Fabaceae the parietal obturator (arises from ovary wall epidermis) is effectively the extension of the stigmatoid style tissue and the morphology of its cells is similar to that of the channelled cells, as demonstrated for instance for *Medicago sativa* (Vishnyakova, 1986). Following fertilization the obturator cells become obliterated and die out.

In the unilocular ovary in representatives of Poaceae, pollen tubes following their emergence from the transmitting style tissue enter the ovarian cavity where they grow but are limited by the interior epidermis of the ovarian wall and the outer integument. As demonstrated by electron micrographs, this cavity is filled with a fine fibrillar osmiophilic substance (Vishnyakova and Willemse, 1994).

In multilocular ovaries, pollen tube passage to the ovules is effected through strands of specialized tissue. In the case of a bilocular cavity ovary, for instance in Brassicaceae, the tubes first grow in the septum tissue, whose central part in effect is the extension of the transmitting style tissue, afterwards passing through its cuticle and reaching the ovule micropyle through the funicular obturator (Hill and Lord, 1987).

Self-incompatibility: Structural and Functional Aspects (Plate V)

Self-incompatibility is the inability of a bisexual flower with functioning male and female gametes to form a seed after self-pollination.

Two groups of self-incompatible species are distinguished: **homomorphous** (all flowers of the self-incompatible population are identical) and **heteromorphous** (flowers in self-incompatible population differ in morphology). Most of the heteromorphous species are characterized by heterostyly (di- and tristyly) and by differences in the morphology of the stamens *(Linum, Primula, Polygonum, Fagopyrum)*. In a number of species, correlative dimorphism of pollen grains and of the stigma papillae is observed. In particular, structural dimorphism of the papilla coat of the stigma, of forms and sizes of papillae, and dimorphism of the exine structure of pollen grains are observed in the family Plumbaginaceae. In the distyled species of genus *Linum* differences can be seen in the structure and sculpture of the exine. Seed development under conditions of heteromorphous self-incompatibility is only possible in the course of additional pollination of plants of different morphological types (Vuilleumier, 1967; Dulberger, 1974, 1975; De Nettancourt, 1977).

The main factor which determines self-incompatibility is a gene (locus) *S* (East and Mangelsdorf, 1925). In some plants, in particular of family Poaceae, a feature is controlled by two genes, *S* and Z (Hayman, 1956; Lundqvist, 1956). There are examples of three- and four-gene control (Lundqvist, 1975; Osterbye, 1975). In populations of self-incompatible plants, the gene *S* is represented by a number of plural alleles. The reaction of self-incompatibility consists in pistil recognition and repulsion of pollen grains and pollen tubes containing an analogous allele of the locus *S*.

By the period of gene activity and localization of their products, one can distinguish two types of homomorphous self-incompatibility: gametophytic (East and Mangelsdorf, 1925) and sporophytic (Hughes and Babcock, 1950) (see Fig. 13). In the case of gametophytic control over self-incompatibility, the reaction of pollen depends only on its own genotype, whereas sporophytic self-incompatibility is defined by the genotype of the diploid sporophyte. Functionally, this is determined by the fact that in the first case the products of the gene of self-incompatibility synthesize after meiosis (in the protoplast of a microspore,) in the second—they go to the pollen grain from mother

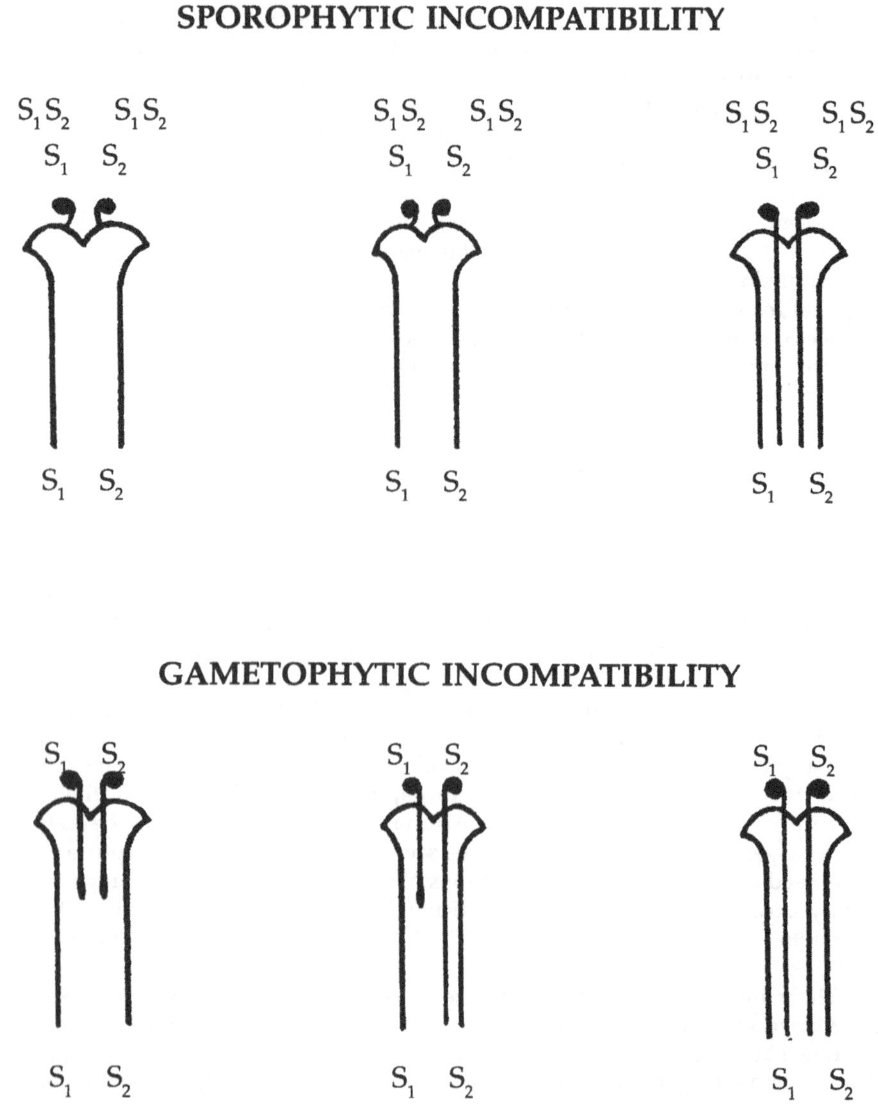

Fig. 13: Schematic representation showing morphological manifestation of two types of genetic control of self-incompatibility; S_1-S_2—genes of incompatibility (see explanation in text) (after Cornish *et al.*, 1987).

tissues of the sporophyte. There is still no common opinion on the mechanism of sporophytic control. The possibility of a premeiotic expression of the gene *S* in the cytoplasm of microsporocytes (Pandey, 1960,1970) as well as its post-meiotic expression in tapetum cells, from where substances of self-incompatibility go to the pollen grain at tapetal tissue disintegration, is debatable (Heslop-Harrison, 1968, 1978). There could be a compromise conclusion, namely, simultaneous synthesis of the products of the gene *S* in the meiocytes and the tapetum (Pandey, 1979).

Genetic control over the heteromorphous self-incompatibility is sporophytic and di-allelic (De Nettancourt, 1977). The structural principles of gene activity of heteromorphous incompatibility have not been sufficiently studied. The rest of this article concerns only homomorphous species.

In the pollen grain the products of the genes of self-incompatibility are localized in the sporoderm. In the case of gametophytic type of control they go from the protoplast of the microspore to the intine and later move into the pollen tube wall created by the intine, whereas in the case of sporophytic self-incompatibility they are kept in the exine (Heslop-Harrison *et al.*, 1973; Knox *et al.*, 1975). Molecular determinants of sporophytic self-incompatibility have also been discovered in the triphine of pollen grains in representatives of genus *Brassica* (Dickinson, 1990). Possible modes of entrance of the gene products of self-incompatibility into the sporoderm in ultrastructural studies have been the subject for special discussion (Heslop-Harrison, 1978, 1983; Vishnyakova, 1989, 1994).

The products of self-incompatibility genes are glycoproteids. On the basis of results of immunochemical tests it was determined that, in the case of sporophytic control over self-incompatibility, the products localize in the epidermal cell walls of the stigma (Kandasamy *et al.*, 1989). In the case of gametophytic self-incompatibility (for example, in *Nicotiana alata) S*-specific glycoproteids are distinguished as a component of an intercellular matrix of the conducting tissue of the style and the placentary epidermis (Cornish *et al.*, 1987), i.e., the place of their localization directly coincides with the growth direction of the pollen tubes.

It is more difficult to prove the expression of the gene *S* in the anther. Its insignificant amount and short duration were revealed in the post-meiotic period in the tapetum of *Brassica.* After tapetum disintegration this expression was no longer observed (Nasrallah *et al.*, 1991).

There are a number of steady correlations between type of genetic control, number of cells in the pollen grain, morphological peculiarities of the stigma and style and location of the self-incompatibility reaction.

Tricellular pollen grains covered with triphine, dry stigmata and closed styles are typical of species with the sporophytic self-incompatibility. In them, the reaction occurs early and takes place on the stigma (Dickinson, 1990). The recipient surface of dry stigmata is covered with a pellicle (Mattsson *et al.*, 1974). It is in the pellicle that reagents of self-incompatibility localize (Kandasamy *et al.*, 1989). Bicellular pollen grains covered with pollenkitt, wet stigmata and open styles are typical of species with gametophytic control. In this case the reaction of self-incompatibility occurs later, in the style or in the ovary.

It should be noted that, according to the terminology we are adhering to, pollenkitt is a homogenic lipid fraction and triphine—a more complicated heterogenic mixture of lipids with remnants of the cytoplasm of the tapetum, which lie on the surface of pollen grains at the final stages of anther development (Dickinson and Lewis, 1973; Dickinson, 1990).

In species with deviations from the mentioned correlations, a more complicated control over self-incompatibility is almost always observed compared to monogene or intermediate (gametophytic-sporophytic) types of control. For instance, in family Poaceae, whose representatives have a digene gametophytic control over self-incompatibility, the structural-functional characteristics of the reaction correspond to the sporophytic type. *Theobroma cacao* is a typical example of combination of reaction of the two types (Cope, 1962; Pandey, 1960). Sometimes the reaction of gametophytic self-incompatibility does not have an accurate localization. Elimination of the male gametophytes may occur throughout the progamic phase. Perhaps, this is also a reflection of the complicated genetic nature of self-incompatibility, which is supposed to occur in some representatives of family Fabaceae, in particular, *Medicago sativa* (Kazachkovskaya *et al.*, 1991).

The reaction of self-incompatibility depends on the nature of the localization by the products of the gene *S* in the pollen grain and the pistil. It is better to observe the morphological manifestation of this reaction in a luminescent microscope at the next stages of the interaction of the male gametophytes with the pistil: adhesion of the pollen to the stigma, its hydratation, germination of pollen grain, penetration of the pollen tube into the stigma tissues and growth in the conducting tract of the pistil. The variety of morphological manifestations of the self-incompatibility reaction observed in different species implies this reaction may proceed at each of these stages. Thus every *S*-genotype has a quite concrete time and place of inhibition of interaction of pollen and pollen tubes with the pistil. This allows one to speak of species- and variety-specificity of morphological manifestation of self-incompatibility.

Early manifestation of sporophytic self-incompatibility is conditioned by interaction of the triphine and/or exine substances with the pellicle of the stigma, and can manifest in no pollen adhesion to the stigma, no pollen grain germination, in inhibition of pollen tube growth after contact with the surface of the stigma or right after penetration through the pellicle and cuticle. In the case of gametophytic self-incompatibility the reaction occurs later, after interaction of the intine matter with the stigma or, more often, after exposition of the self-incompatibility reagents by the pollen tube and their reaction with the substances of the transmitting tract of the pistils along the direction of growth of the pollen tubes. It conditions a relatively long 'dialogue' between the male gametophyte and the pistil. Deceleration of growth of pollen tubes can occur in different zones of the style and even in the ovary.

Many species of flowering plants are partly self-incompatible. This manifests morphologically in that in a certain number of pistils of the sample studied some pollen tubes overpass the region of mass inhibition, penetrate the ovary and fertilize some ovules.

The reaction of self-incompatible inhibition of the pollen tubes is accompanied by various morphological deviations: twisting, spiralling, swelling and branching of tips, and intensive deposition of callose. Ultrastructural studies of such tubes showed necrosis of the cytoplasm, abnormal biosynthesis of the pollen tube wall, delineated, in particular, in aggregation of the precursors of materials of the cell wall in the pollen tube tip (De Nettancourt *et al.*, 1973; Pacini, 1981). One of the main peculiarities of the structure of self-incompatible pollen tubes is the concentric aggregation of the granular ER profile, which can be evidence of blocking of protein synthesis (De Nettancourt *et al.*, 1973; Cresti *et al.*, 1977).

In the cases of quick manifestation of the reaction of self-incompatibility—in the stigma by this time—the pollen grain and pollen tube are frequently filled with callose, as observed in sporophytic self-incompatibility in *Helianthus annuus* and in gametophytic self-incompatibility in *Hordeum bulbosum*. Often the pollen tube top bursts while reaching the blocking zone. A typical manifestation of the self-incompatible interaction of the pollen tubes with the pistil is retardation of their growth.

A morphological manifestation of the reaction of self-incompatibility in the pistil sometimes considered, is the appearance of callose deposits in the stigma papillae, contacting with the self-incompatible pollen grains under sporophytic control over self-incompatibility (Heslop-Harrison *et al.*, 1973; Dumas and Knox, 1983). However this 'biotest' on self-incompatibility must be treated critically since such callose deposits appear in the stigma papillae while applying sporogenous cells and even tissue segments of vegetative organs of the same plant (Sood *et al.*, 1982).

Molecular research on self-incompatibility in representatives of genus *Brassica* has shown that the *S*-gene is complicated and associated with other components of the genotype. *S*-allele specificity is determined by at least two genes educed from one homozygotic genotype: *SLG* (*S-locus glycoprotein*) and *SRK* (*receptor protein kinase*). It is supposed that these genes determine intercellular interactions in different tissues, which is evidently based on the activity of a common class of combined receptors of proteinkinase. Both genes *SLG* and *SRK* are similar in 94% cases at the DNA level and in 90% at the level of amino acid succession of the protein and segregate together (Nasrallah *et al.*, 1991).

However, *S*-allele specificity in Brassicaceae is not limited by expression of only these two genes. Twelve variable fields in the genome have an homology with the locus *S*. Two of them: *SLR*-1 and *SLR*-2 *(S-locus related)* are beyond its bounds. They have 65% homology with *SLG* at the level amino acid succession in the protein, similar structure and like *SLG*, are expressed in the stigma cell walls. These genes are very conservative and indifferent *S*-alleles have 99% of homology at the level of amino acids and a high level of similarity in different species. All the foregoing and also their expression in self-compatible species give reasons to suppose that they play a certain role in the process of pollination (Nasrallah *et al.*, 1991).

Gametophytic self-incompatibility is peculiar to representatives of most families, in particular: Caprifoliaceae, Chenopodiaceae, Commellinaceae, Fabaceae, Liliaceae, Poaceae, Ranunculaceae, Rosaceae, Rubiaceae, Scrophulariaceae, Solanaceae and others. Sporophytic self-incompatibility has been discovered in only six families: Asteraceae, Brassicaceae, Caryophyllaceae, Corylaceae, Sterculiaceae, and Ulmaceae (Charlesworth D. and Charlesworth B., 1979). Both types of the homomorphous self-incompatibility have not been detected within one family. One-locus gametophytic self-incompatibility is considered primitive and ancestral with respect to sporophytic (Pandey, 1962; Crowe, 1964).

Morphological manifestation of the reaction is evidence that the evolution of the structural mechanisms supplying self-incompatibility is directed to earlier mutual recognition and, correspondingly, earlier interruption of interaction of male gametophyte with the pistil. This is determined by the removal of the reacting agents into more superficial regions of the pistil and the pollen grain. Evolution of the pistils, having occurred by coalescence of the carpels (Takhtajan, 1964), can be considered a gradual moving of reagents of recognition from the female gametophyte by their transferal into the tissues of the sporophyte. The top of this process is the concentration of reagents of self-incompatibility in the stigma, viz., in its specialized structure, the pellicle, which acquired the competence of recognition. In the pollen grain the perfect mechanism of early recognition is the localization of these reagents in the triphine. In both cases self-incompatibility is determined by the sporophyte. In gametophytic control over self-incompatibility the reaction occurs as a result of the interaction of the matters of the male gametophyte with sporophytic tissues of the pistil. However, in cases of late proceeding of the reaction—in the ovary and, perhaps, the embryo sac—it cannot exclude involvement of the female gametophyte itself. It is known that recognition at the level of the gametophytes occurs in Pteridophyta and Gymnospermae, and in ancient Angiospermae occurred even at the level of the gametes (Crowe, 1964). Consequently, manifestation of reaction of the gametophytic type in the ovary and in the embryo sac can be regarded as initial, which is an argument for the primitiveness of this type.

Pollen Tube

The pollen tube is a specialized structure of the male gametophyte which is responsible for providing sperms in the embryo sac to fulfil double fertilization **(Plate VI)**. Its formation was first observed by Amici (1824) while researching germination of the pollen grain in *Portulaca deracea*. The pollen tube originated as a tubiform outgrowth of the vegetative cell of the pollen grain.

The most characteristic peculiarity of the pollen tube is the presence of sperms within its protoplast. They get into the pollen tube from the tricellular pollen grain (Apocynaceae, Cyperaceae, Menyanthaceae, Poaceae, Valerianaceae). In the plants with bicellular pollen grains it is the generative cell which get into the pollen tube and being there it divides forming two sperms (Chloranthaceae, Fagaceae, Papaveraceae, Ranunculaceae, Liliaceae). On exiting the pollen grain, sperms or the generative cell move along the pollen tube nearly to its tip.

Most plants are characterized by porogamy when the pollen tube reaches the nucellus through the micropyle. Aporogamy is a rarer phenomenon when penetration of the pollen tube into the nucellus of the ovule takes place in the chalazal region or at the side between the chalaza and micropyle (mesogamy). In all cases the pollen tube enters the embryo sac in the area of the egg apparatus.

Pollen Tube Formation and Growth

Germination of pollen grains and growth of pollen tubes pass in two stages. **The first stage** takes place during hydration mainly due to water coming in from stigma tissues and reactivation of pollen grains. In many plant species water movement is regulated by exine (Heslop-Harrison, 1971). During germination of the pollen grain intine fulfils three functions: the outer (pectin) layer regulates the opening of apertures; the middle (cellulose) layer participates in protein transport to the stigma surface during pollen hydration; the inner layer (containing glucans and pectins) is a precursor of the pollen tube cell wall (Heslop-Harrison, 1979b).

During the second stage of growth (interaction of the pollen tube and stigma) reception or rejection of pollen tubes takes place. The proteins including different enzymes play an essential role in these processes (Zinger and Petrovskaya-Baranova, 1961; Heslop-Harrison, 1979a,b, 1987; Knox, 1984a). The proteins are of different origin: sporophytic (from tapetum) in exine and gametophytic (from microspores or pollen grains) in intine. The nature of the protein secreted by the pollen tube depends on the stage of its growth (Heslop-Harrison, 1987). For example, in species with dry stigma, penetration of the pollen tube tip demands enzymatic destruction of the cuticle, fulfilled by cutinase (Linskens and Heinen, 1962). In tips of pollen tubes growing *in vitro*, various hydrolases including esterases and phosphatases as well as enzymes hydrolyzing pectin have been found cytochemically (Heslop-Harrison,

1987). Tricellular pollen grains can germinate in one minute and high intensity of respiration is observed immediately after hydration. For bicellular pollen grains typically respiration rate slows and its maximum is lower than in tricellular pollen grains (Hoekstra and Bruinsma, 1975), and germination can last for one hour. During intensive growth of the pollen tube dictyosome activity is seen in the region of the pollen tube tip, which is located near the area enriched with mitochondria. Only from this moment does the stage of intensive growth of the pollen tube begin (Dashek and Rosen, 1966).

Growth of pollen tubes at compatible stigma and in style tissues is strictly oriented and goes either through a well-distinguished channel containing exudates from surrounding cells and providing the medium for growing pollen tubes or through intercellular space between rows of tightly compacted elongated cells (where pollen tubes come into contact with products of style secretion). Pollen tubes in the ovary move through rows of secretory cells located at the inner wall of the ovary or placenta. The growth rate of pollen tubes reaches 1.5 μm per second (Shivanna *et al.*, 1978).

Both the genomes of the male gametophyte and female sporophyte participate in regulation of pollen tubes growth (Knox, 1984a; Heslop-Harrison, 1987). The mechanisms of this regulation are not absolutely known. Hypothetically there may be two types of regulation: (1) the pollen tube finds its way according to signals coming from a number of targets; (2) strict direction of pollen tube growth is provided by surrounding conditions. The first type includes the chemotropic regulation, in the second control is determined by structural peculiarities of the pathway. Possibly, pollen tubes find their way reacting simultaneously to both types of control or separately in different parts of the pistil (Heslop-Harrison, 1987).

Pollen tube growth passes under the influence of three groups of factors: (1) rate of water income and consequently pressure on the plasma membrane from inside (osmotic balance between the contents of a cell and medium influences on these factors); (2) physical influence on the cell wall; and (3) rate of synthesis of substances—precursors of the cell wall, as well as their conversions, transport and penetration into the plasma membrane of the pollen tube (Heslop-Harrison, 1987). The most essential factor regulating the growth process is water movement into the vegetative cell of the pollen tube. It is provided by different structural peculiarities of the pollen grain and the stigma surface.

The growing pollen tube keeps its characteristic zonality. It loses this feature only when growth is arrested for some reason. Four zones have been distinguished: apical, subapical, nuclear and vacuolization (Dashek and Rosen, 1966; Cresti *et al.*, 1977; Miki-Hirosige and Nakamura, 1982). Stretching of the pollen tube is associated with the apical zone. Subcellular structure of the tip is characterized by the presence of secretory vesicles inbuilt in the cell wall. The subapical zone of the pollen tube contains secretory vesicles and various organelles (mitochondria, Golgi apparatus, ER) and the nuclear zone the nucleus of the vegetative cell and sperms. At the zone of vacuolization and formation of callose plugs, separation of proximal sections of the pollen tube is observed. This event provides high metabolic potential in the first three zones. In the region of the pollen tube tip, maximal synthesis of new materials for building of the cell wall takes place. The latter consists of three layers: outer (pectin), middle (cellulose) and inner (callose) (Heslop-Harrison, 1979a). Here the main role is played by vesicles—precursors of the pollen tube cell wall forming in dictyosomes located in the zone close to the apical one. Lipid and carbohydrate resources of the pollen grain without participation of the Golgi apparatus may also serve as the source for synthesis of materials for this wall (Heslop-Harrison, 1979a, 1987; Miki-Hirosige and Nakamura, 1982). In pollen tubes of Gramineae, polysaccharide inclusions containing products of dictyosome activity have been detected (Heslop-Harrison, 1987).

Movement of dissoluble metabolites to the subapical area and subsequent transport of precursors of cell wall components to the apical zone of growth depends on the nature of movement of the cytoplasm (Iwanami, 1959). An important role in regulation of stability of pectin microfibrils of the pollen tube tip is played by calcium due to its ability to link carboxyl groups of pectins (Jaffe *et al.*, 1975). Correlations between calcium content and zonality of location of cytoplasmatic vesicles, membranes and organelles have been defined, though their exact meaning is still not clear.

Movement of the cytoplasm in the pollen tube passes in two directions—basipetal and acropetal (Litvak, 1971). It is believed that gamete movement in the pollen tube passes either due to their passive moving by means of cytoplasm current (Strasburger, 1884; Koernicke, 1903; Navashin *et al.*, 1959; Navashin, 1968) or due to their ability to move actively and independently (Kostrjukova and Chernoyarov, 1938; Kostrjukova, 1951, 1968; Steffen, 1953; Kostrjukova and Benetskaya, 1958; Polunina and Sveshnikov, 1959; Polunina, 1961; Litvak and Kolesnikov, 1965; Litvak, 1971; Heslop-Harrison J. and Heslop-Harrison Y., 1989a-c). According to Korobova (1977, 1979) movement of the generative cell or sperms in the pollen tube can be considered a particular case of interaction between the cell nucleus and cytoplasm. This opinion is based on the fact that the pollen tube with its gametes ('gamocyte'—Gerassimova-Navashina, 1951a,b) is a fully grown vegetative cell of the pollen grain, when the generative cell or sperms located inside the protoplast of the vegetative cell behave like nuclei. The nucleus of any cell has a strictly definite position at each stage of development. The place where the nucleus is located at the moment is termed by Gerassimova-Navashina (1954a,b) 'the dynamic centre' of the cell, considering it to be the place of balance of all the strengths influencing the nucleus. In the pollen tube the vegetative cell nucleus and sperms or the generative cell behaving in the same manner are also located at ' the dynamic centre'. The volume and location of the gamocyte during pollen tube growth and periodic cutting off of its tip by callose plugs are constantly changing. 'The dynamic centre' changes its location too and consequently the location of the vegetative nucleus and sperms also changes. Sperm movement in the pollen tube is associated with their location at 'the dynamic centre'of the gamocyte which, in turn, is permanently shifting in the course of pollen tube growth. According to Heslop-Harrison J. and Heslop-Harrison Y. (1989a-c), sperms in the pollen tube move independently along the elements of the actin-myosin system in the cytoplasm of the pollen tube. During growth of the pollen tube sperm location in relation to the vegetative cell nucleus is unstable (Cresti *et al.*, 1990). Each sperm is surrounded by its own plasma membrane and the plasma membrane of the vegetative cell. Sperms are linked with each other, and one of them is also linked with the nucleus of the vegetative cell (Rougier *et al.*, 1991; Yu and Russell, 1994a-c).

Movement of organelles and gametes in the pollen tube is associated with the function of the cytoskeleton (microtubules and microfilaments) (Franke *et al.*, 1972; Heslop-Harrison 1987; Heslop-Harrison *et al.*, 1988; Pierson *et al.*, 1989). The presence of myosin and kinesin has been established in the cytoplasm of the vegetative cell (Moscatelli *et al.*, 1988). The cytoskeleton of the generative cell is formed by microtubules (Tiezzi *et al.*, 1988).

Generative Cell Division and Sperm Cell Formation (Plate VII)

In higher plants the mature male gametophyte is limited to a vegetative cell enclosing a generative cell (bicellular pollen) or two sperms (tricellular pollen).

At anthesis the anthers open and mature pollen is transported by different vectors (insects, wind, water, etc.) and deposited on the stigma surface (Konar and Linskens, 1966). After pollination pollen germinates on the stigma and pollen tubes grow through the stigma and stylar transmitting tissue towards the egg cell.

Pollen tube growth is extremely fast, crossing a considerable distance from several mm (e.g. *Vitis vinifera, Olea europea*) up to 50 cm (e.g. *Zea mays*). In the category of bicellular pollen, during pollen tube growth a mitosis of the generative cell occurs and finally two sperms are formed (Goldberg *et al.*, 1993).

The pollen and pollen tubes are vectors carrying the sperm cells into the ovule and for this reason pollen tube growth is one of the key steps contributing to the fertilization process of seed plants. Recently an unbelievable amount of research on the biochemistry, morphology, physiology and molecular biology of pollen and pollen tube growth has been done (Heslop-Harrison, 1987; Pierson and Cresti, 1992; Derksen *et al.*, 1995). Pollen tubes have a specific cell shape and growth pattern (tip or polar growth). Elongation of the pollen tubes occurs at their tip by fusion of vesicles with the plasma membrane. The secretory vesicles of the Golgi apparatus are transferred to the apex by a fountain-like cytoplasmic streaming mediated by cytoskeleton motor proteins. A specific mechanism regulates constant synthesis, degradation and reorganization of cytoskeletal elements in order to maintain delivery of the vesicles to the tip and tube elongation (Derksen *et al.*, 1995).

Thanks to the introduction of new methods such as immunolabelling, freeze fixation and substitution, and devices allowing three-dimensional reconstruction of serial images, new insight has been gained into the morphology and physiology of the generative cell and sperm cells in angiosperms (for reviews see: Russell *et al.*, 1990a; Chaboud and Perez, 1992; Mogensen, 1992; Palevitz and Tiezzi, 1992; Pierson and Cresti, 1992; Russell, 1992). One of the more encouraging findings is the discovery of dimorphism of paired sperm cells (see Sperm Dimorphism, Vol. 1) in a certain number of species, including *Plumbago zeylanica* (Russell, 1984), *Spinacia oleracea* (Wilms, 1986), *Brassica campestris* and *B. oleracea* (Dumas *et al.*, 1985; McConchie *et al.*, 1985, 1987b), *Zea mays* (McConchie *et al.*, 1987a; Rusche and Mogensen, 1988), *Euphorbia dulcis* (Murgai and Wilms, 1988), *Rhododendron macgregoriae* (Taylor *et al.*, 1989) and *Gagea lutea* (Zhang *et al.*, 1995).

Dimorphism of sperm cells has been related to preferential involvement in double fertilization and the possibility of selective transmission of male cytoplasmic organelles to the next generation (see Mogensen, 1992; Russell, 1992; Zhang *et al.*, 1995).

One particular case showing a striking sperm cell dimorphism is reported in *Gagea lutea* (Zhang *et al.*, 1995) where division of the generative cell, in addition to symmetrical division, is associated with organization of the microtubular apparatus. The exceptional abundance of microtubules in the proximal part of the generative cell at the beginning of cell division is associated with translocation of the generative nucleus towards the leading end of the cell.

In *Gagea lutea* the generative cell is extremely large and its nucleus located in the distal part of the tube. The microtubular organization of the generative cell consists of fine bundles of intermingled microtubules distributed throughout the entire cytoplasm. At the late prophase the microtubules are spindle shaped and the chromosomes become longitudinally aligned relative to the tube axis. At the metaphase stage, about 3 to 4 hours after sowing, the microtubular spindle becomes shorter and discontinuous at the site of the metaphase plate. Generally the axis of the spindle is parallel to the axis of the pollen tubes. At the anaphase stage numerous bundles of microtubules interconnect the separating chromosomes. The telophase is characterized by the disappearance of the spindle and the presence of prominent bundles of microtubules between the two daughter nuclei. The end of the generative cell division occurs 8 h after sowing. Cytokinesis is associated with formation of the phragmoplast and two unequal sperm cells: a smaller leading sperm cell and a larger second sperm cell with a volume of up to 2.5 times that of the leading one.

The dimorphism of the two sperm cells of a pair has only been detected in a few, mainly tricellular species, on the basis of slight differences in cell size, number of organelles and general morphology (see Zhang *et al.*, 1995). The *Gagea lutea* dimorphism represents the first case for bicellular pollen.

Male Germ Unit (Plate VIII)

This is the functional transmission unit of sexual reproduction in flowering plants formed by the sperm cells, or their precursor generative cell, in physical association with the vegetative nucleus. The male germ unit (MGU)[3] contains all of the hereditary DNA of the male reproductive lineage, including that of the vegetative nucleus and the male germ cells (Dumas *et al.*, 1984).

MGU associations: nature and origin

Two different kinds of association are evident in the MGU: the first is the association between the vegetative nucleus and the generative cell (evident in bicellular pollen) and one of the two sperm cells (evident in tricellular pollen and late in bicellular pollen tube growth); the second is the association between the two sperm cells, initiated by generative cell division, and frequently present throughout descent in the pollen tube. The sperm cell association is based on: (1) initial sharing of the sperm common wall (formed during generative cell cytokinesis), (2) inclusion of the two sperm cells within a shared inner plasma membrane of the vegetative cell, and (3) presence of cytoplasmic junctions between the sperm cells occupied variously by microtubules (Heslop-Harrison J. and Heslop-Harrison Y., 1988), trabeculae or plasmodesmata-like connections (Russell and Cass, 1981a,b). The association between the vegetative nucleus and the generative cell usually precedes formation of the sperm cells. This second association is based on: (1) nearby position between one end of one sperm cell or the generative cell with the perinuclear cytoplasm of the vegetative nucleus, (2) tenacious maintenance of this position during movement of the MGU, and (3) presence of an elongation or projection of the associated generative or sperm cell with the vegetative nucleus. Both types of associations are evident and tenacious in many flowering plant species during MGU descent in the pollen tube. Although the association of male germ cells with the vegetative nucleus is formed first, the association of the two sperm cells is better understood and will be discussed in detail first.

[3]The term was proposed by E.N. Gerassimova-Navashina (1957); a description of sperm connection was given earlier in a number of publications (Wylie, 1904; Shattuck, 1905).

Association of the two sperm cells. The association of the two sperm cells has its origin in the division of the generative cell. Although generative cell division has often been described in the last century (Maheshwari, 1950), the descriptions reported still generate controversy (Palevitz, 1993a). Most of the variability in these descriptions seems to have as its ultimate origin the space available to contain the mitotic spindle and for the cells to undergo cytokinesis. In generative cells that divide in the pollen tube (bicellular pollen), space is usually restricted, but this is not the case in generative cells that divide in the pollen grain (tricellular pollen).

In bicellular pollen, the generative cell becomes narrow and highly elongated because of the constrained environment of the elongating pollen tube. Although the metaphase plate may be slightly inclined in the widest pollen tubes (Palevitz, 1993a), frequently the metaphase plate is highly inclined (Yu and Russell, 1993), and in the narrower ones, it may be almost vertical (Palevitz and Cresti, 1989). In bicellular pollen, generative cell cytokinesis may occur with or without a phragmoplast. In the former case, a cell plate forms between the sperm nuclei, whereas in the latter case, the cytoplasm appears to undergo cleavage by constriction. The most recent description of the former case is in *Nicotiana* where cell plate formation is essentially conventional (Palevitz, 1993a). The most noteworthy point is that gaps in the forming plate are occasionally seen under an electron microscope. Although these appear to close during later cell plate accretion (Yu and Russell, 1993), they may provide a false impression that centripetal closure is occurring. Cleavage accomplished by constriction has been most recently described in *Tradescantia* by Palevitz and Cresti (1989). In some respects, for example lack of a cell wall and evident pinching of membranes to separate the cells, generative cell cytokinesis appears similar to that in an animal; concomitantly it differs from animal cell cleavage since actin is apparently absent from the generative and sperm cells (Palevitz and Liu, 1992). This is an enigmatic model for sperm cell division which needs further analysis. That this pattern of division has been reported since the early literature (see Maheshwari, 1950), its existence seems well corroborated, but ultrastructural studies have yet to be conducted.

In tricellular pollen, cell division is not particularly spatially constrained, but the generative cell is frequently narrower in the centre of the cell forming a dumbbell-shaped profile (Charzynska, 1988). Generative cell division consistently seems to employ a phragmoplast and cell plate formation, but the pattern of cell plate expansion seems to vary among taxa. In one type, cell plate expansion follows the normal somatic pattern of centrifugal enlargement from a centrally coalesced region towards the edges of the cell, as reported in *Hordeum*, *Sambucus*, *Secale* and *Triticum* (Karas and Cass, 1976; Charzynska, 1988; Charzynska *et al.*, 1988; Charzynska and Lenart, 1989; Charzynska and Lewandowska, 1990). In the second type, cell plate expansion arises from pegs at the edge of the cell, with cell wall closure occurring centripetally, as in *Brassica* (Charzynska *et al.*, 1989), *Galium* and *Trichodiadema* (Weber, 1988a). Cell plate expansion in *Avena sativa* (oats) appears to be either centrifugal (Charzynska, 1988) or centripetal (Weber, 1988a), suggesting both a need for further study as well as the possibility that variability of this character may extend below the species level. In species in which the sperm cells separate, Weber (1988a) reported two patterns of separation: one in which accessory membranes seem to be laid down prior to separation (e.g. *Galium*) and another in which a central dense area in the pericellular space becomes a location of preferential separation (e.g. *Trichodiadema* and *Avena*). It is difficult to estimate how many sperm cells among the angiosperms may separate completely prior to descent in the pollen tube, but evidence for the reformation of MGU associations is also available later in development prior to fertilization (Mogensen and Wagner, 1987).

Association of vegetative nucleus and male germ cells. The association between the vegetative nucleus and the generative cell and its descendant sperm cells has a less obvious origin and basis. Close approaches between the vegetative nucleus and the young generative cell occur frequently after microspore division but appear to be coincidental, since many generative cells in the same anther seem to lack these associations. In *Plumbago,* the first evidence of generative cell separation from the intine seems to occur at the end of the cell nearest the vegetative nucleus. Only a small beak is present at this end of the cell rather than the long projection present later. As this beak elongates, the association seems to become more stable and more consistent. The vegetative nucleus and generative cell appear to be consistently associated from this stage onward. A greater frequency of nuclear pores seems to occur on the side of the vegetative nucleus nearest the generative cell in *Nicotiana,* as observed in freeze-fracture preparations (Wagner *et al.,* 1990), but whether these are involved in communication with the generative cell or represent anchoring points for molecules involved in the MGU association is unclear.

Localization of myosin on surface membranes of the generative cell, sperm cells and vegetative nucleus (Heslop-Harrison J. and Heslop-Harrison Y., 1989c; Tang X. *et al.,* 1989) and actin present in the vegetative cell in which they are immersed (Pierson and Cresti, 1992) may relate to the chemical basis of the MGU associations. Myosin localizations using antibodies elicited to the heads of fast skeletal muscle myosin have successfully labelled all the elements of the MGU. Antibodies against smooth muscle myosin were not as effective, but it is still quite possible that the epitopes of certain animal and plant myosins may have diverged beyond their ability to meaningfully cross-react. According to fluorescence immunocytochemistry, the specific location of the myosin label appears to be the vegetative nuclear envelope and the inner pollen plasma membrane that ensheaths the generative cell (Heslop-Harrison J. and Heslop-Harrison Y., 1989c). Based on localizations using squashed pollen contents, generative cell labelling was unexpectedly variable. Heslop-Harrison J. and Heslop-Harrison Y. (1989c) proposed that a significant number of generative cells remained coated by the pollen plasma membrane and that these were well labelled, whereas those that were presumably without the ensheathing layer did not label, and were perhaps without myosin.

According to immunogold electron microscopy, antimyosin antibodies label few sites in the generative cell: mainly the nucleus and internal organelles, not the generative cell plasma membrane (Cresti *et al.,* personal comm.). Based on these results, it seems possible that myosin is absent from the actual surface of the male germ cells, but it is also possible that it is sparse or chemically different in cross-reactivity. Preliminary results indicate that isolated sperm cells of *Plumbago* will glide in an energy-dependent manner on actin filaments of giant internode cells of the characean alga *Nitella,* although the sperm surface has not yet been successfully labelled using conventional antimyosin antibodies (Zhang, Strout, Zhu and Russell, personal communication). This suggests that myosin-like activity is present on the sperm cell surface, even if immunolabelling has not yet been successful.

Although the mechanism for association of the vegetative nucleus with the male germ cells is yet unclear, a number of observations seem relevant: (1) the vegetative nucleus and male germ cell membranes label with antibodies to myosin and (2) actin is present in a number of forms in the pollen cytoplasm. Potentially, myosin present on the vegetative nucleus and associated either directly or indirectly with the male germ cells could be cross-linked via short filaments of protein possessing actin-like properties. Potentially, this relationship could be tenacious enough to maintain a physical association between the elements of the MGU. This would constitute a non-motile myosin-actin-myosin association, but one which could promote male germ cell passage in association with active movement on actin bundles of the pollen tube (Pierson and Cresti, 1992). Actin molecules with slow turnover rates might provide linker molecules to associate the myosin-rich surfaces of these bodies. The position of the superficial myosin, in turn, may be controlled by linkers attaching to the male germ cell microtubular cytoskeleton. This model, although not proven, is not inconsistent with experimental evidence (see below).

Functional involvement of cytoskeletal elements in MGU behaviour

Evidence for a relationship between microtubules and microfilaments in the behaviour of the MGU has come from experiments that alter these cytoskeletal elements both chemically or physically. In experiments to depolymerize or disrupt microtubules using colchicine, the distance between the generative cell and the vegetative nucleus typically increased by approximately one generative cell length (Heslop-Harrison *et al.*, 1988); similar results are also reported using oryzalin (Palevitz, 1993b) and during recovery from microtubule-disrupting cold treatments (Åström *et al.*, 1991). A separation of the vegetative nucleus from the generative cell is also evident during generative cell mitosis (Palevitz, 1993a; Yu and Russell, 1993), when cortical generative cell microtubules are specifically reorganized into a spindle. At metaphase, the generative cell shape also becomes rounder, with blunter cell elongations and significant decreases in length of cell projections evident in pollen tubes grown *in vitro* (Palevitz, 1993b), *in vivo* and *semi-vivo* (Yu and Russell, 1994a-c). These observations during mitosis indicate that alterations of the MGU association are particularly sensitive to changes specifically in microtubule distribution within the generative cell. Given that the changes in generative cell microtubules affect cell shape, it is difficult to tell whether loosening of the MGU is primarily a result of the reduction of cortical microtubules, or whether it is a secondary consequence of changes in the surface or surface area of the generative cell or because of decreases in the complementarity of the surfaces between the male germ cells and the vegetative nucleus.

Cytochalasin, which disrupts actin microfilaments, does not alter associations within the MGU or change generative cell shape (Palevitz and Liu, 1992). Instead, cytochalasin appears to arrest movement of the MGU through the pollen tube, presumably through the breakdown of cortical actin bundles in the pollen tube (Heslop-Harrison J. and Heslop-Harrison Y., 1989b). In contrast, tube elongation and movement of the MGU in the tube was not radically altered by the application of colchicine, which eliminated microtubules from growing pollen tubes and generative cells (Heslop-Harrison *et al.*, 1988). Based on these results, MGU **movement** appears to be the result of a direct association between myosin on the vegetative nucleus surface and surrounding male cells, while MGU **associations** are sensitive to alterations in cortical microtubules of the male germ cells and their resultant changes in cell shape.

If actin is absent from the interior of both the generative cell and sperm cells of flowering plants, as suggested by Paleitz and Liu (1992), this finding would have considerable implications in reducing the possibility of independent cell movement and decreasing the odds of cytoplasmic movement inside male germ cells. The absence of F-actin in the phragmoplast during generative cell cytokinesis is also believed to account for the seemingly fragile partition between newly formed sperm cells.

MGU-based heterospermy

The MGU, since it involves associaion with only one of the two sperm cells, clearly establishes polarity that may be associated with both shape differences and cytoplasmic organelle differences. Shape differences are common since the sperm cell associated with the vegetative nucleus (S_{vn}) usually has an elongated projection or extension that frequently wraps around the vegetative nucleus. It is also larger in volume and surface area as a result of this characteristic and may have a flattened region that further increases the surface area of the cell. In cases where this is also related to differences in the content of heritable organelles, such as mitochondria or plastids, the sperm cells are **cytoplasmically heterospermic**. These include a significant number of species for mitochondria (Table 1). *Plumbago* (Russell, 1984) has the most extreme form of cytoplasmic heterospermy since the S_{vn} is rich in mitochondria, whereas the sperm cell unassociated with the vegetative nucleus (S_{vn}) is poorer in mitochondria and rich in plastids. This polarity in *Plumbago* is also related to preferentiality in fertilization, since the plastid-rich S_{ua} fused with the egg cell in 94% of the cases examined (Russell, 1985a,b). In plants in which there is not a clear marker for fusion, the polarity of fertilization cannot be so easily determined. Maize (*Zea mays*) sperm cells are the only flowering plant sperm to show **nuclear heterospermy**. In this plant, there is a non-disjunction of *B*-chromosomes in the generative cells, causing the two sperm cells to contain different numbers of *B*-chromosomes. The sperm cell containing more *B*-chromosomes is more likely to fuse with the egg cell (Roman, 1948). At the extremes of male cytoplasm are cotton, which contains plentiful cytoplasm (Jensen and Fisher, 1968a,b), to orchids, lacking mitochondria and plastids but containing a significant layer of cytoplasm (Cocucci and Jensen, 1969b; Yu and Russell, 1992), to umbellifers (Weber, 1988b), which appear to lack all but a thin almost film-like layer of cytoplasm.

Sperm cell motility and transport

Whether the sperm cells of flowering plants are motile has been a classical question of flowering plant embryology, with classical evidence adduced for both possibilities (Maheshwari, 1950). Modern studies have revealed that microtubules are well developed in higher plant generative cells but are not linked in any type of conventional motor protein configuration, as would be expected if they were capable of independent locomotion; they are usually even less well developed in sperm cells (Palevitz and Tiezzi, 1992). Also, actin is absent or not convincingly present in male germ cells (Palevitz and Liu, 1992). Detailed examinations of patterns of movement in living pollen tubes reflect a degree of independence of movement in the elements of the MGU (one element can even pass another in the pollen tube; Heslop-Harrison J. and Heslop-Harrison Y., 1987a,b) and even suggest amoeboid shape changes in the sperm; however, these can apparently be reconciled by actomyosin associations between the MGU and the cytoskeleton of the pollen tube (Heslop-Harrison J. and Heslop-Harrison Y., 1989a; Pierson and Cresti, 1992). Isolated sperm cells provide no evidence for independent motility (e.g., Cass, 1973; Russell and Cass, 1981a,b).

Table 1. Summary of structural and fertilization characteristics of sperm cells in angiosperms examined to date

Pollen type and species	Method	Morphology	Functional sperm	Sample (no.)	Reference
Tricellular pollen					
Monocotyledons					
Hordeum vulgaris	3-D	I	-	5	Mogensen and Rusche, 1985
Zea mays	3-D	SH	B+ (Roman, 1948)	1	McConchie *et al.*, 1987a
	3-D	SH	"	1	Rusche and Mogensen, 1988
Dicotyledons					
Brassica campestris	3-D	MCH	-	5	McConchie *et al.*, 1987b
Brassica oleracea	3-D	MCH	-	3	McConchie *et al.*, 1987b
Euphorbia dulcis	3-D	MCH	-	1	Murgia and Wilms, 1988
Gerbera jamesonii	3-D	MCH	-	1	Proovost *et al.*, 1988
Plumbago zeylanica	3-D	MPCH	P+ (Russell, 1985b)	11	Russel, 1984
Spinacia oleracea	3-D	MCH	-	7	Wilms, 1986
Bicellular pollen					
Monocotyledons					
Cymbidium goeringii	3-D	I	-	10	Yu and Russell, 1992
Gladiolus gandavensis	ImAn	SH	-	6	Shivanna *et al.*, 1988
Dicotyledons					
Nicotiana tabacum	3-D	I	-	9	Yu *et al.*, 1992
Petunia hybrida	3-D	I	-	1	Wagner and Mogensen, 1987
Rhododendron spp.	ImAn	SH	-	7	Shivanna *et al.*, 1988

Abbreviations: Methods: 3-D serial ultrathin sectioning and three-dimensional analysis; ImAn—image analysis of cell profiles and/or nuclear fluorochromatic patterns. Morphology types: I— similar complement of sperm cytoplasm; MCH— mitochondrial cytoplasmic heterospermy; MPCH —mitochondrial and plastid cytoplasmic heterospermy; SH—structural heterospermy (shape of sperms differs, condition of cytoplasmic organellar DNA unknown). Functional sperm cell: B+ —sperm cell containing supernumerary B-chromosomes; P+—sperm cell containing most plastids; Sample: number of MGUs examined.

Although the actomyosin model appears to answer fundamental questions regarding the movement of sperm cells inside the pollen tube, the final pathway of sperm cells deposited in the embryo sac appears to depend on direct actomyosin interactions with actin trails located between critical female cells (Huang *et al.*, 1993a; Huang and Russell, 1994; Huang and Sheridan, 1994). The sperm cells are hypothesized to migrate on the actin matrix until they reach the site of gamete fusion. There one sperm cell migrates on one actin trail and fuses with the egg cell to form the embryo and its brother sperm cell migrates on a different actin trail and fuses with the central cell to form the endosperm.

Sperm cell physiology

Since the work of Shivanna *et al.* (1988), there has been some doubt about whether sperm cells are completely mature before germination of the pollen grain and whether they possess any evident physiological limitations. The recent work of Zhang *et al.* (1993) and Matthys-Rochon *et al.* (1994) indicates that sperm nuclei are transcriptionally active, display electrophysiological activity and appear to be fully physiologically competent. In terms of protein and immunogenic sites, the sperm cells appear to differ from their surrounding cytoplasm (Pennell *et al.*, 1987; Geltz and Russell, 1988), suggesting that the transcripts are expressed in sperm cells and that the cells are indeed autonomous and possess a unique physiology. Further, the isolated sperm cells can be induced to fuse with isolated egg cells *in vitro* (Faure *et al.*, 1994a) and the products of fusion can be used to regenerate fertile plants (Kranz and Lörz, 1993). That the sperm cells used in these experiments are isolated from newly released pollen grains indicates that sperm cells are fully functional at anthesis in these plants. In bicellular pollen producing plants, the greater difficulty in isolating sperm cells from pollen tubes (Shivanna *et al.*, 1988) has resulted in fewer studies. Often exacting conditions are required for generative cell division; therefore, in these plants the maturity and physiological competence of sperm cells is still not known (Russell, 1991; Theunis *et al.*, 1991).

MGU as functional unit of male germ cell transmission

Evidence that the MGU is a functional unit for male reproduction includes the widespread occurrence of MGU associations that appear to facilitate transmission of the gametes (Hu, 1990; Mogensen, 1992). Among the various plants examined in detail, the majority have MGUs: 6 of 7 dicotyledons with tricellular pollen (86%), 5 of 6 dicotyledons with bicellular pollen (83%), 2 of 4 monocotyledons with tricellular pollen (50%) and 3 of 3 monocotyledons with bicellular pollen (100%) had MGUs. It seems noteworthy that among the monocotyledons, 6 of the 7 (83%) reportedly lacked an MGU at some time after reaching maturity. Among the monocotyledons with tricellular pollen, all of the examples used were grasses. In half of them, including barley (Mogensen and Wagner, 1987), the MGU was established prior to fertilization. In corn, which may lack an MGU throughout its life cycle, heterofertilization—fusion of the egg and central cells with sperm cells from different pollen tubes—is not uncommon (Robertson, 1984). The association between sperm cells and vegetative nucleus is believed to promote a rapid cotransmission of the sperm cells and to prevent heterofertilization (Russell and Cass, 1981a,b). According to P.H. Tang (pers. comm.), the vegetative nucleus may actually be actively involved in the transmission of male germ cells since laser light-induced ablation of the vegetative nucleus is reported to arrest movement of the MGU. The foregoing evidence in combination with the known necessity of both sperm cells to accomplish double fertilization (Maheshwari, 1950) constitutes strong evidence for the significance of the MGU. The MGU therefore appears to represent a functional male unit of sexual reproduction in flowering plants, as proposed earlier (Russell and Cass, 1981a,b).

POROGAMY

Porogamy (Greek *poros*—hole and *gamos*—marriage)—penetration of the pollen tube into the embryo sac of the ovule through the micropyle **(Plate IX)**. Synonym: acrogamy.

Porogamy is typical for all Gymnospermae and most Angiospermae. In spite of the multitude of pollen tubes growing in tissues of the stigma and style only one of them penetrates the micropyle and conveys two sperms to the embryo sac for fertilization of the egg cell and central cell. Penetration of several pollen tubes into the embryo sac has been noted in *Triticum* species and *Scilla sibirica,* however (Batygina, 1974b, 1987a). The pathway of the pollen tube from the stigma to the embryo sac lies through a system of various structures. Having entered the ovary it moves along the placenta, then makes a turn (up to 90° in the anatropous ovule) and reaches the micropyle. En route the pollen tube can face some additional structures (obturator, papillae of the placenta and integuments), which it has to overcome to penetrate the micropyle. Passing through the micropyle, which is formed with one or two integuments and is characterized by certain taxonomic features (length, width, presence of a cuticle and secretory cells), the pollen tube enters the nucellus.

The nucellus epidermis is usually covered with cuticle. Some sections of cuticle are destroyed under the influence of enzymes of nucellus cells (Janson, 1992). The pollen tube penetrates the nucellus through one such section. The micropyllar zone of the nucellus appears to possess the features of the secretory tissue (Bruun and Olesen, 1989; Janson, 1992). In some species (e.g. *Spinacia oleracea)* signs of change in the structure of the wall of the nucellus cells have been noted at the place of assumed growth of the pollen tube (Wilms, 1981a,b). The nucellus is the last barrier before the pollen tube enters the embryo sac.

Regardless of differences in the structure of ovule tissues through which the pollen tube is moving, several general regularities of its growth can be distinguished. Growth of the pollen tube takes place in intercellular spaces. The pollen tubes grow in the direction of the exudate and chemotropic-active agents. The composition of the exudate of the style and placenta tissues differs biochemically, due to which the pollen tubes have a different growth rate in these structures. When lacking the exudate the pollen tubes start to bend, deviating from straightforward growth or may cease to grow at all. The exudate is used as a nutrition source and concomitantly serves as a chemotropic-active factor (Tupy, 1961; Miki-Hirosige, 1964; Rosén, 1964; Linskens, 1973; Sedley, 1976; Heslop-Harrison and Reger, 1988; Janson, 1992). In cases wherein the egg apparatus is not prepared for fertilization or degeneration of its cells, the pollen tube ceases growth and does not penetrate the nucellus.

High concentrations of calcium ions have been found in the tip of the pollen tube, the micropyle, at the surface of the outer integument and in synergids. Taking into account the important role of calcium in biological processes, it was concluded that calcium is responsible for growth of the pollen tube (Mascarenhas and Machlis, 1962; Jensen *et al.*, 1983; Chaubal and Reger, 1992). In a number of plants electric and magnetic fields have been traced around the pollen tube as well as the potential difference between itself and the tissues along which it grows (Linskens and Spanjers, 1973; Weisenseel *et al.*, 1975; Sperber and Dransfeld, 1981; Nakamura *et al.*, 1991). Together with other factors directing growth of the pollen tube, it is the structure of the style (open and hollow) and trichomes which play the essential role (Wilms, 1981a,b; Kronestedt *et al.*, 1986; Lord and Kohron, 1986). There is also an opinion about the accidental growth of the pollen tube over free surfaces and its chaotic founding of the micropyle. However, experimental data mostly refute this point of view.

CHALAZOGAMY

Chalazogamy (Greek *chalaza*—knot and *gamos*—marriage)—penetration of the pollen tube into the embryo sac of the ovule through the integument (integuments) in the area of the chalaza by-passing the micropyle **(Plate IX)**. Here the pollen tube reaches the nucellus from below, then grows upwards along it. The pollen tube can ramify and change direction of growth but it enters the nucellus in the area of the micropyle, penetrating the embryo sac in the region of the egg apparatus. Synonyms: basigamy, aporogamy.

Chalazogamy was first described by Treub (1891) in *Casuarina suberosa* and later by Nawaschin (1895a-c) in *Betula, Juglans* and *Corylus.* Studies of the phenomenon continued (Nawaschin and Finn, 1912, 1913; Finn, 1936; Solntseva and Dgebuadze, 1974; Korchagina, 1981; Litvak, 1981). According to Poddubnaya-Arnoldi (1976) chalazogamy has been described in families Casuarinaceae, Betulaceae and Juglandaceae, as well as in the orders Ranales, Rosales, Rubiales, Cucurbitales, etc. Chalazogamy on the whole is found in quite a limited number of species of flowering plants and, in most cases has been noted together with porogamy (Nawaschin, 1899; Litvak, 1981). Chalazogamy has not been discovered in Gymnospermae.

Growth of the pollen tubes in chalazogamy is not strictly directed; their behaviour is more comparable to non-oriented growth in tissues, as a result of which only one pollen tube reaches the micropylar zone of the nucellus and enters it. There are intermediate variants (mesogamy) between chalazogamy and porogamy, when the pollen tube penetrates the ovule in the lateral region between the chalaza and micropyle.

Compared with porogamy, chalazogamy is less studied. However, certain peculiarities of structure and development of the ovule of chalazogamous species can be noted: strongly developed micropylar part of the nucellus with nucellar cap and parietal tissue, which moves the embryo sac deep inside the nucellus (Casuarinaceae, Betulaceae, Juglandaceae); presence of wide micropyle not closed at the moment of maturation of the embryo sac (Betulaceae) or its absence (Juglandaceae); formation of lateral outgrowths by the embryo sac which reach the base of the ovule and the apolar genesis of its elements appearing in the development of an antipodal (or antipodals) similar to the egg cell (Casuarinaceae); slowed formation of the embryo sac and penetration of the pollen tube before completion of its development (Betulaceae, Juglandaceae); delay of the fertilization for one week after pollination (Betulaceae) up to 6-7 weeks (Casuarinaceae); rather soon degradation of the pollen tube tip (Betulaceae). The whole complex of the factors mentioned can indicate the decrease in possible influence of chemotropic-active agents on growth of the pollen tube in chalazogamous plants compared with porogamic ones.

Opinions about the phylogenetic significance of chalazogamy are contradictory. Some researchers believe that this phenomenon is characteristic of evolutionarily advanced groups of Angiospermae and consider it one of the features of specialization (Takhtajan, 1966). Others consider chalazogamy a primitive feature, an initial type of pollen tube growth (Poddubnaya-Arnoldi, 1976). The analysis of several morphological peculiarities of chalazogamous plants presented above proves the opinion that chalazogamy is probably a more primitive phylogenetic feature than porogamy.

FERTILIZATION

Double Fertilization

Double fertilization **is** the process wherein one sperm fuses with the egg cell and another sperm (of the same couple) with the central cell of the embryo sac. As a result of this double fertilization a diploid embryo (from the fertilized egg cell) and typically a triploid endosperm (from the fertilized central cell) appear.

Studying the fertilization process in *Lilium martagon* and *Fritillaria tenella* S.G. Nawaschin (1898a,b, 1951) **discovered for the first time the function of both male gametes.** He showed that besides the well-known fertilization of the egg cell **one more** fertilization took place in the embryo sac in which the participants were **the second sperm** (its role was not clear before) and the central cell with **polar nuclei or its secondary nucleus. The phenomenon** was given the name 'double fertilization' because it is based on the sexual process. It has been firmly established in world science as the definition of one of the most characteristic biological peculiarities of Angiospermae, although it has also been discovered in certain representatives of Gymnospermae.

Types of Double Fertilization and Cell Cycle (Mitotic Hypothesis) (Plate X)[4]

It is now clear that the above-described diversity of the forms wherein double fertilization proceeds depends on genetically stipulated and fixed by evolution, diversity of modes of gamete development in gametophytes. The latter in angiosperms derive from sporophytes themselves and therefore depend on conditions created in them by the time the gametes have completed their formation. Sporophytes differ greatly in such a vast group of plants, which influences the development of sexual elements and subsequently future diversity of the modes of the fertilization process itself.

The main factor responsible for occurrence of **double fertilization in angiosperms is the paired arising of male and female sexual elements that are metabolically not fully developed.** This is provided with functioning of certain structures phylogenetically inherited from the preceding plant groups.

Regarding the male part, such a structure is a two-celled pollen grain which is essentially a gametophyte with shortened development that arises as a result of differential microspore development (Gerassimova-Navashina, 1951a,b). In the male gametophyte subsequently only one of its cells divides—the lesser (generative or spermiogenic) cell. Its division leads to simultaneous origin of two sperms which, however, are arrested at this or that stage of their development. This arrestation principally occurs due to the limited amount or particular quality of the cytoplasm they receive at the generative cell division in the pollen grain or later in the pollen tube, depending on how their ontogenesis proceeds.

---

[4]Editor's Note: The articles 'Double Fertilization Types' and 'Double Fertilization Process and the Mitotic Cycle of the Cell (the Mitotic Hypothesis)' have been united with a few abridgements for the purpose of reducing the size of this volume.

In case the amount of cytoplasm of the generative cell is highly limited the sperms forming therefrom contain a paltry amount of cytoplasm, barely discernible in a light-microscopic investigation. Their nuclei remain compact, highly chromatized, and often because of despiralling of the chromosomes in them, the sperms look like highly elongated bodies surrounded only by membranous covers. This was demonstrated by Cass (1973) in ultrastructural studies of barley. Such sperms usually arise in the pollen grains, which in this case are termed tricellular. Usually formed in a relatively similar medium of a pollen grain, such sperms used to be similar in appearance and are commonly found in Compositae, Gramineae and others.

Sperms formed from the generative cell with a large amount of cytoplasm acquire other features during ontogeny. Such sperms contain more cytoplasm, their nuclei become rounded or oval, and nucleoli appear in them due to commencement of decondensation of their chromatin elements.

Such sperms, however, cannot continue ontogenesis on their own due to incomplete development of the elements required. First of all, this kind of sperm does not keep the proper nucleoplasmic ratio. This, even based on visual observations, immediately affects the nuclear-nucleolar ratio necessary for normal functioning of any cell, as the nucleolus is one of the main participants in production of ribonucleic and therefore protein cell components. Other elements of the cytoplasm of such sperms also remain in a juvenile state, as proved by researches at the ultrastructural level (Cocucci, 1981). As the pollen tube progresses through the pistil tissues the latter acquire a certain gradient and hence the sperms often differ in appearance.

The second cell of the pollen grain—the larger (vegetative, or siphonogenous)—though it does not divide, provides almost all vital functions of the pollen grain at all stages of its ontogenesis. It plays the most active role in the fertilization process on the whole, and thus together with the sperms makes up **male germ unit** (Gerassimova-Navashina, 1951a,b).

As for female part sexual elements are also formed by the gametophyte with a strongly shortened development. Usually a relatively **uniform bipolar structure** appears—the embryo sac. The notable uniformity of type seems to be due to certain regularities of nuclear and cellular divisions active during development of the embryo sac (Gerassimova-Navashina, 1957, 1958).

Different development results (embryo and endospermal tissue) of the two neighbouring cells fertilized simultaneously, essentially arising from the division of the same gametophyte cell, are genetically conditioned and evolutionarily consolidated from the very beginning by the different position of each of the two cells in the whole system.

Based on research on the structure and functions of the sexual elements in Compositae and analysis of the literature, two main types of double fertilization process are recognized (Gerassimova, 1933; Gerassimova-Navashina, 1947a-c, 1951a,b): **premitotic**—when fusion of the nuclei of the sexual elements is completed **before** zygotic mitosis, as soon as the sperm nucleus comes into contact with the female nucleus, as in Compositae, Gramineae, ferns, mosses, many algae (in animals it corresponds to the so-called 'sea-urchin' type), and **post-mitotic**—when fusion of the nuclei of the sexual elements is completed **only after** zygotic mitosis, as in most Liliaceae and gymnosperms (so-called 'ascarid' type in animals) (Fig. 14)[5].

---

[5]Also see figure 16.

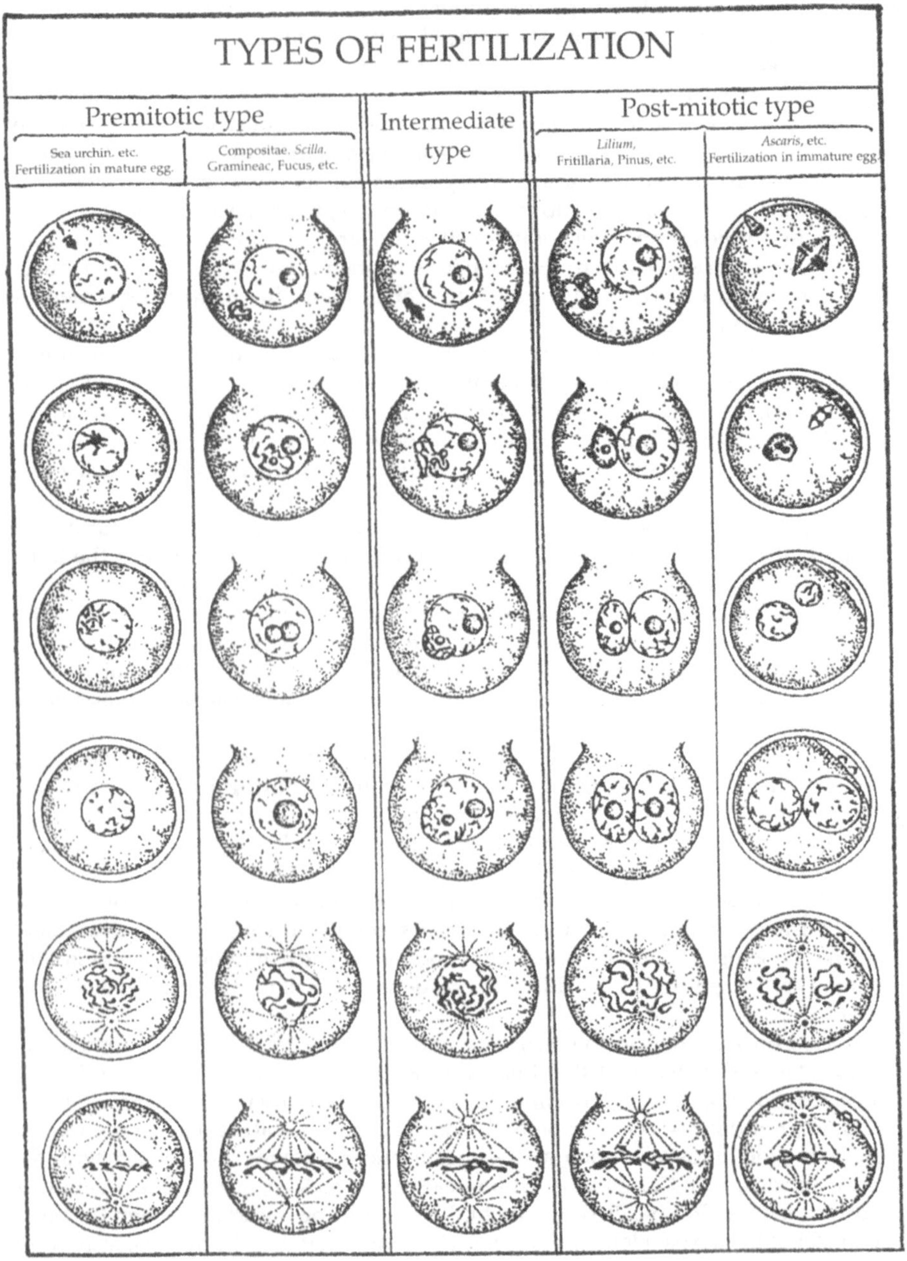

Fig. 14: Diagrammatic presentation of the fusion of sexual nuclei as based on the moment of their first contact and completion of mixing of chromatin (after Gerassimova-Navashina, 1960, 1971a).

The existence of the two main fertilization types comes from the common cytological regularity wherein nuclei in a quiescent state do not fuse because of the existence of nuclear envelopes. Nuclear fusion can only take place after the envelopes disappear. Envelope disappearance is induced either by entering the division (as in the post-mitotic type) or by presence in the sperm nucleus of a previous condition of incomplete division yet incompatible with the existence of the envelope, thus causing disappearance of the latter in the contact spots with female nuclei immediately upon contact (premitotic type).

However, further research has shown that there are several intermediate types. Intermediate fertilization forms are found in a number of plants (e.g. balsam; Steffen, 1951) and also tetraploid forms (Gerassimova-Navashina and Batygina, 1958) wherein the initial diploids are characterized by the premitotic type (e.g. *Crepis*). This is interesting as it allows a check on the above interpretation. In this fertilization form the process proceeds more slowly than in the premitotic type, but more quickly than in the post-mitotic type. The sperm nucleus having come into contact with the female nucleus partly fuses with it but externally produces its own envelope. This process in tetraploids can even undergo full transformation into the post-mitotic type, i.e., mitotic cycle arrest occurs in the sperm nucleus through the influence of the female cell.

The type of fertilization process is conditioned by the degree of readiness of the sexual elements—the structural and morphological character of their development, which essentially is the determining factor in development of variations of the process (see Fig. 14).

Recognition of fertilization types became possible after it was discovered that the sperm nuclei **undergo one mitotic cycle** from the moment of formation until they fuse with the female nuclei. Starting the process in the generative cell of this or that fertilization type, the sperm nuclei can complete the cycle and **start the new one only in female cells**, preparing new mitosis together with them. A similar process occurs in the primary endosperm cell, whose nucleus starts division together with the male nucleus to form the endosperm tissue.

Premitotic fertilization is usually characteristic of compact sperms. The nuclei of these sperms are immersed in those of the female cells soon after penetration therein, where they can restart their mitotic cycle. Here nucleoli appear in them and full decondensation of their chromatin elements occurs.

In light of the described observations a concept arose that it is **the mitotic cycle** which underlies double fertilization (Gerassimova-Navashina, 1947a).

The 'mitotic' hypothesis explains the dispersion of sperms by their mutual repulsion, similar to that of ana-telophase nuclei. Repulsion of two active nuclei appearing in one cell was first discovered by Gerassimov (1923), who observed this dynamic phenomenon in the cells of alga *Spirogyra*. The nature of the moving force remains unknown, though it is possible to assume that this force is similar to electrostatic interaction. Similar repulsion of two male nuclei apparently takes place in the embryo sac during post-mitotic fertilization, when the pollen tube discharges sperm cells: here in the area of the degenerated synergid they 'are relieved' of their cytoplasm.

The 'mitotic' hypothesis of double fertilization highlighted the connection between two biological phenomena—mitosis and the fertilization process. The mitotic life cycle of sperms appeared to be stretched out in time and space and thus evolutionarily fit for passing through such a peculiar process of fertilization in the plant kingdom.[6]

[6]Fine mechanisms of double fertilization, specifically the nature of sperm movement towards neighbouring female gametes in the embryo sac are still awaiting resolution (see 'Cytoplasmic Transmission during Fertilization' and 'Fertilization Mechanisms'—this volume).

The mitotic hypothesis was interpreted in a more detailed manner after Howard and Palk (1953), Taylor and co-authors (1957) and Wolfsberg (1964) divided the mitotic cycle phases into certain periods, which was also appropriate for the process of fertilization with sexual elements, as their development, structure and functioning had already earlier been associated with phases of the mitotic cycle (Gerassimova-Navashina, 1969b, 1971a,c).

Given the fact that in pollen grains as well as in pollen tubes, nuclei of sperms are not metabolically active, I conventionally termed their state G_0; it is only in the embryo sac (namely in female cells) that sperm nuclei can fully complete their mitotic cycle—enter the G_1-period of the zygotic interphase and together with the female cell nuclei, S- and G_2-periods—to prepare for the first zygote division, which leads to development of a new individual and division of the primary endosperm cell for development of endosperm tissue. In the latter cell, as already known, development starts sooner, probably due to a higher level of nuclear-cytoplasmic ratio. It is shifted therein towards the cytoplasm and, as demonstrated for *Crepis capillaris* in electron microscopic studies (Gerassimova-Navashina and Gulyaev, 1973) ER more extensive, which activates the cell to initiate division more promptly.

However, while entering the female nuclei the sperm nuclei of some plants quickly pass G_1-, S- and G_2-periods and sometimes enter metaphase non-simultaneously with the female partner. This often happens in the **intermediate** fertilization type. However, this phenomenon was also observed in *Lilium regale* and later in *Galtonia candicans* (Gerassimova-Navashina, 1980). Such interesting cytochemical transformations of sexual elements were studied using luminescent microscopy and other biophysical methods (Khvedynich *et al.*, 1978; Serdjuk *et al.*, 1978; Khvedynich and Bannikova, 1983).

As knowledge expunded, the mitotic hypothesis acquired slightly different interpretations in relation to changes of a physical and chemical nature, discovered during parallel analyses of the activity of both sperms and nuclei elements during mitosis in cells of the meristem (Fig. 15). These data changed the view on the nature of sperm dispersion in the **premitotic** type of fertilization (Navashin and Gerassimova-Navashina, 1958; Gerassimova-Navashina, 1978, 1980, 1982; Gerassimova-Navashina and Salnikova, 1982). It was established for the first time that during mitosis of meristematic cells in its early stages (early prophase), when chromosomes already lack RNA-components and are compounded only of DNA-proteids, these structures demonstrate mutual **repulsion** that lessens only in the late prophase, when chromosome threads are covered with the isolating RNA-component (Gerassimova, 1933; Gerassimova-Navashina, 1982).

While entering the female cells, the sperm nuclei seem to move therein according to another principle, i.e., they approach 'dynamic centres' of these cells where female nuclei are situated at the time. This is especially demonstrative in the nucleus of the central cell lying far from the egg cell, often in the narrow cytoplasm strand, sometimes near its chalazal extremity. The most characteristic example is *Galtonia candicans* Deche. (Gerassimova-Navashina, 1980) where the sperm nucleus moves along a very long strand of cytoplasm finding at the end of the like of its 'dynamic centre'. This occurs due to the structural and physiological features of the cytoplasm of the said cell which, in *Galtonia candicans*, is highly vacuolate and that is why the sperm nucleus moves here in a manner resembling that of the pollen tube medium. In the latter M. S. Navashin (1968) established, male sexual elements stop not far from the end of the pollen tube, finding there the like of a dynamically balanced state of two

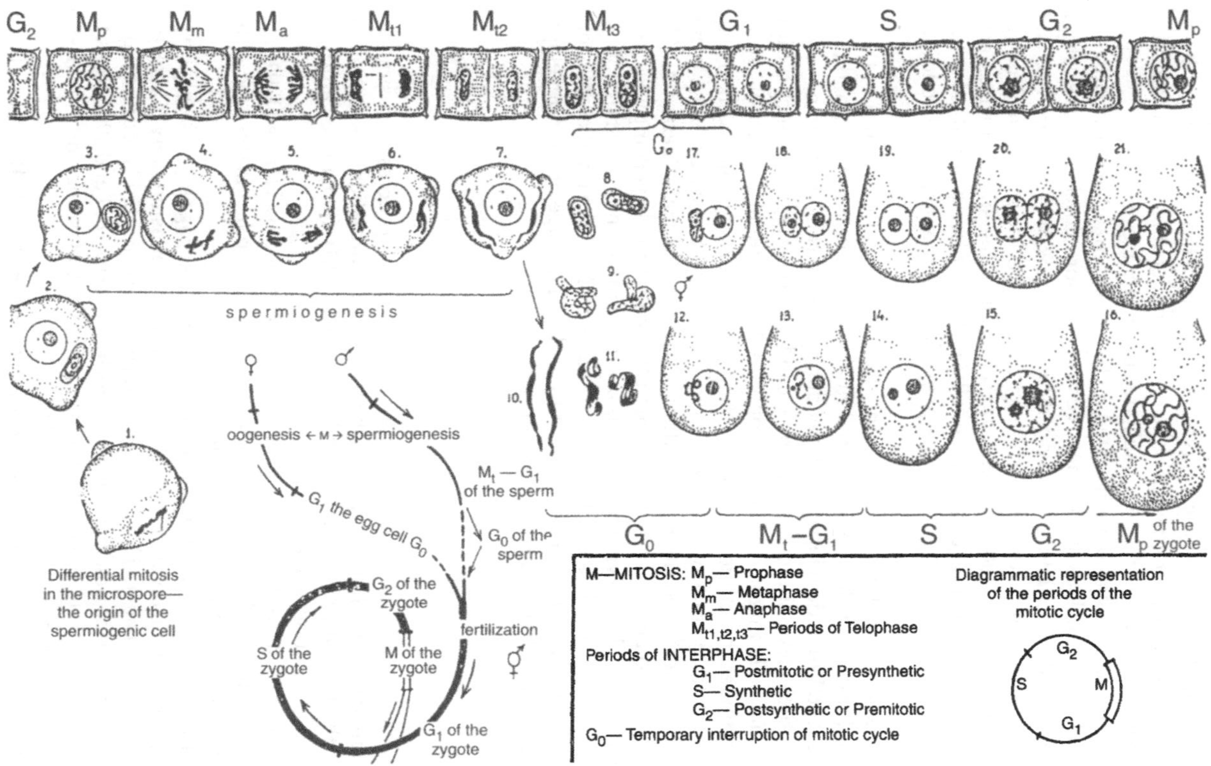

Fig. 15: Mitotic cycle of the meristematic cells (uppermost row) as compared with that of sexual cells and the behaviour of the latter in the course of fertilization (12-16—premitotic type; 17-21—postmitotic type) (after Gerassimova-Navashina, 1969b).

counterdirected cytoplasm currents. The most correct explanation of this phenomenon in the pollen tube was given by Korobova (1979). She pointed out that the sperms in the pollen tube occupy the determined and constant position, regardless of its continuous growth, just because callose locks are constantly formed in it, which preserve unchanged its protoplasmic dimensions and location of its dynamic centre, wherein sperms or the spermiogenic cell are constantly located.

Using time-lapse it was demonstrated on living material that in *Galanthus nivalis* L. the second sperm nucleus moves in the central cell more slowly than the cytoplasm currents, constantly circulating in any cell (Erdelská, 1983). Therefore it is not the cytoplasm currents that carry the sperm nucleus towards the female nucleus, as was believed by Strasburger (1900) for *Monotropa hypopitis;* the sperm nucleus actually heads towards the 'dynamic centre' of the cell, i.e., to the place where its own nucleus is located at the moment because of the physiological and structural conditions of that cell.

As the sperm nuclei while entering the female cells are in a state of an uncompleted mitotic cycle, the female cell nuclei always occur in the condition of a full mitotic rest. When developing independently, sexual cells thus come to a certain deadlock in their development, which they can leave only together. Such a different state of sexual cells might be removed only by their mutual interaction, resulting from which under the influence of the resting nucleus of the female cell, the sperm nucleus completes its mitotic cycle within it, falling into the state of rest, while the female cell being fertilized under the influence of the active **male "germ unit"** leaves the state of rest. Somehow from the moment the sperm nucleus completes its cycle within the female cell, the zygote nucleus as a whole enters a new mitotic cycle, which leads to embryogenesis. The union of sexual elements, from the point of view of such cytologic conception, occurs on the one hand as the final stage of their development, and on the other as the beginning of a new ontogenesis.

Syngamy

Syngamy (Greek *syn*—together, *gamos*—marriage) is the process of fusion of two haploid gametes (egg cell and sperm), which results in formation of a zygote. In the true sense the term means sexual reproduction **(Plate XI)**. Synonyms: amphimixis, conjugation, copulation, eugamy[7].

In its modern, narrower sense, the term was proposed by Hartog (1904) instead of the term 'fertilization'. Strasburger (1877) was the first in fact to describe syngamy (**single fertilization**) in flowering plants.

Syngamy includes the fusion of the nuclei—**karyogamy** (Greek *karyon*—nucleus+*gamos*) and fusion of the cytoplasm of gametes—**plasmogamy** (Greek *plasma, plasmat*—plasma+*gamos*). The two latter terms were probably borrowed from zoological literature (the first mentioning refers to the years 1891 and 1901; see: Oxford English Dictionary, 1989). In botanic literature the term plasmogamy was first used for gymnospermous plants (Ferguson, 1901).

The famous Indian embryologist Maheshwari (1950, p. 194) wrote: 'The time between the beginning and the end of the gamete fusions is so short that one rarely succeeds in "catching" the material at the right stage'. Having examined carefully

[7]See Battaglia, 1963.

the methodological difficulties that impede understanding the process of fertilization, he concluded: **'In view of these difficulties it is not surprising that our knowledge of the events concerned with fertilization has not advanced very far beyond where it stood during the early part of this century'.** Modilewsky (1958) was of same opinion considering that the cytological aspect of the double fertilization in angiosperms was scarcely developed.

Use of the method of temporal fixation of the material facilitated progress in study of the cytology and cytochemistry of this process (Gerassimova-Navashina, 1947a,b, 1986; Batygina, 1957, 1959, 1961a,b, 1962a,b, 1966, 1974b; Poddubnaya-Arnoldi, 1958, 1976; Korobova, 1959, 1962; Savina, 1964, 1965; Yakovlev and Yoffe, 1965; Chebotaru, 1972, 1987; Bannikova and Khvedynich, 1982). Ultrastructural investigations have also contributed their share to understanding syngamic mechanisms (Vazart, 1955; Jensen, 1964a,b, 1973; Gerassimova-Navashina and Gulyaev, 1973; Cass, 1973, 1981; Russell, 1983, 1985a,b; You and Jensen, 1985; Mogensen and Rusche, 1985; Van Lammeren, 1986a,b; Cresti *et al.*, 1988a; Zhukova, 1993; Plyushch, 1992; Faure *et al.*, 1993; Kranz and Lörz, 1993, 1994; Kranz *et al.*, 1995). However, our notion about this process remains incomplete to date. One reason lies in **the methodological complexity of obtaining** a number of ultrathin slices and computer-based reconstruction of syngamy due to the transiency of some of its stages. The other reason is lack of a **system comparative study of cytoembryological processes *in vivo* and *in vitro* at light and ultrastructural levels.** Choice of an object is of no small importance.

Before examining the process of union of the sperm and the egg cell we shall briefly recall **some peculiarities** of the structure of the egg apparatus which have **a direct relation to the mechanism of syngamy.**

Egg apparatus. Revelation of considerable differences in the structure of the egg apparatus in various species of flowering plants has initiated use of the terms **'distinctly differentiated'**, **'poorly differentiated'** and **'undifferentiated' egg apparatus (Plyushch, 1992).** Such features as the size, polarity and topography of its elements are considered.

The egg cell in most flowering plants is disposed **laterally (in profile)** and usually **slightly lower than the synergids.** The central cell wholly envelops the egg cell (except for the part where the cell is attached to the wall of the embryo sac). Fertilization does not occur until differentiation and specialization of cells of the whole egg apparatus are completed, i.e., until formation of a mature embryo sac.

Synergids are the polyfunctional cells whose main role consists of attracting pollen tubes through the micropyle into the embryo sac and in providing release of their contents and getting the sperms to the female gametes (see Synergids, Vol. I). The pollen tube can enter into one of the synergids or between synergids or passing the synergids, into the area of contact of the central cell with the egg cell. Probably, to some degree it determines the character of synergid degeneration. Within one species various ways of pollen tube entry into an embryo sac have been observed (e.g. in *Triticum aestivum;* Batygina, 1974b, 1987a, 1990). Changes (transformations) of synergids may start already from the moment when pollen reaches the stigma and even earlier, but this does not mean that the pollen tube will enter into a transforming synergid.

The pollen tube discharges its contents into the synergid through a pore (Jensen and Fisher, 1967; Chebotaru, 1972; Mogensen, 1972, 1978a,b; Wilms, 1981a,b; Russell and Cass, 1988), or through the fracturing of the pollen tube coat due to the difference in osmotic pressure inside the pollen tube and synergid (Van Went, 1970c; Poddubnaya-Arnoldi, 1976), or as a result of dilution of the pollen tube coat by hydrolases of the synergid (Godineau, 1969). Exceptionally discharge of the pollen tube occurs in the extracellular space (*Plumbago zeylanica, Plumbagella micrantha*—Russell, 1980, 1982; Russell and Cass, 1988; *Adoxa moschatellina*—Plyushch and Bannikova, 1989).

Male Germ Unit (MGU). To understand the mechanism of **simultaneous moving of sperms towards two different female cells** it is important to examine the question of the **connection between sperms inside the pollen grain, pollen tube and the extracellular space** and also the **connection of sperms with the nucleus of the vegetative cell.** At the beginning of the 20th century Shattuck (1905, p. 213) described this phenomenon while investigating formation of male gametes in *Ulmus americana*: 'It was definitely determined that the **male structures are cells,** and not **merely nuclei,** delicate limiting membranes being clearly seen. During a large part of their existence **in the pollen grain these lenticular cells** are **attached to each other by their adjoining ends** in such a manner as to make them appear in longitudinal section as if **astride** the **tube nucleus'.**

According to data by Shattuck, behaviour of these connected 'male cells' changes in the course of development: at first while in the pollen grain they possess some cytoplasm, then inside the pollen tube they lose it, and finally, in the embryo sac they acquire a small quantity of cytoplasm again. Electron-microscopic data have proved that while in the embryo sac sperms of another species *Ulmus laevis* have some cytoplasm too (Plyushch, 1989). Wylie (1904, p. 10) asserted that connection between male cells in *Elodea canadensis* was quite lasting; it remained in the pollen grain, in the pollen tube and kept safe at relatively considerable distance: 'This point of union acts like a hinge, permitting the cells to take the most varied positions with respect to one another, even swinging about so as to lie side by side.

S.G. Nawaschin and Finn (1912) observed the same picture with two male gametes connected with each other in the form of a dumbbell-like structure inside the embryo sac in *Juglans*. The authors considered this structure a 'binucleate generative cell' which corresponded to the binucleate generative cell[8] of some gymnosperms. In their opinion the fact that in *Juglans* cytoplasm remains around male nuclei inside the pollen grain, pollen tube and the embryo sac is 'an ancient feature inherited by chalazogamic plants from their gymnospermous ancestors'.

A short-term stage of the 'binucleate generative cell' ('binucleate gametic' or 'bispermatogenic cell') has been found in the process of formation of sperms in a number of species of *Triticum* genus (Batygina, 1962a,b, 1974b, 1987a). Connection between the generative cell or of sperms and the nucleus of the vegetative cell lasts during the whole period of development of the pollen grain. At present this phenomenon has been described in many plant species. The character of the connection between gametes, as well as the moment of its appearance (inside the pollen grain, pollen tube and embryo sac) can differ.

The phenomenon 'male germ unit' discovered by means of light microscopy at the beginning of the 20th century has been proven and further developed in electron-microscopic investigations (Mogensen and Rusche, 1985; Russell, 1980, 1982-1985a,b, 1991). Opinions differ about the significance of this phenomenon. One posits that 'sperm association' determines both the order of entrance of sperms into the embryo

[8]In current literature, binucleate sperm cell, see: Carmichael and Friedman, 1995.

sac and their selectivity at fertilization of the egg cell and the central cell (Heslop-Harrison J. and Heslop-Harrison Y., 1984; Russell, 1984). Other authors believe the MGU maintains co-ordination of development of all the cells of the pollen grain (Tarasenko *et al.*, 1988) and co-ordination of fertilization processes (Dumas *et al.*, 1985). To understand the biological role of the MGU it is necessary to carry out more detailed comparative embryological research on the gymnosperms and angiosperms.

Peculiarities of structure of male gametes. Discovery of the double fertilization phenomenon has made researchers think about questions concerning the **dimorphism of sperms of one couple, the presence of cytoplasm in sperms** (while they are located in pollen, pollen tube, synergid, egg cell or in the extracellular space between the egg apparatus and the central cell) and **movement of sperms**. These questions have remained open for almost one hundred years and in Nawaschin's opinion, their solution would bring us nearer to an explanation of the stages of syngamy and triple fusion.

For the first time dimorphism manifested in enantiomorphism (pair-wise difference and similarity) was observed by S. G. Nawaschin at the light-optical level in flowering plants. Nuclear heterospermy has been especially well described in gymnosperms, in particular in *Pinus* (Ferguson, 1901). S. G. Nawaschin and Finn (1912) observed dimorphism of sexual nuclei in *Juglans nigra,* explaining this difference by the fact that in flowering plants both nuclei are functional at double fertilization.

Today 'cytoplasmatic heterospermy' (different number of plastids, mitochondria, dictyosomes) and 'nuclear heterospermy' (Russell, 1991) are distinguished. However, investigations carried out by means of the technique of serial ultrathin slices and computer-based reconstruction have shown that in some species of plants sperms are isomorphous (Mogensen and Rusche, 1985; Hu, 1990; Russell, 1991; Zhu *et al.*, 1992).

In some species of flowering plants the nucleus of the smaller sperm fuses with the nucleus of the egg cell, and the nucleus of the larger sperm with the polar nucleus, in others the opposite takes place (see Maheshwari, 1950). The egg cell is usually fertilized by the sperm containing a greater amount of plastids (Russell, 1983, 1985a,b).

Examination of the movement of *Plumbago zeylanica* sperms isolated from the pollen by means of an osmotic shock has shown the bimodal distribution of speed and density of the charge of male gametes (Zhang *et al.*, 1996). The authors found correlation between this fact and the different fate of sperms of one couple during fertilization.

Experimental investigations of sperms isolated from mature pollen of *Spinacia* have revealed no differences in the structure of plasma membranes in sperms of one couple (Van Aelst *et al.*, 1990). Probably, selectivity of male gametes moving towards different female cells is not typical for all the species of flowering plants. This is proven by the presence in them of both dimorphous and isomorphous sperms.

Dimorphism of sperms may be the result of **asymmetrical cytokinesis** in the microspore, pollen grain and generative cell.

In many species of flowering plants the vegetative and generative cells contain a **different number of organelles,** in particular, plastids. The latter vary markedly in distribution (almost all the plastids move into the vegetative cell or partly into the generative cell, but are eliminated entirely during cell development, or enter the sperms but not the egg cell).

Some researchers believe that maternal plastid heredity found in most flowering plants is explained by those very cytoplasmatic changes which occur during formation of the generative cell and sperms (Hagemann, 1981, 1983; Van Went, 1984; Schröder, 1984, 1985a, b, 1986a, b, 1988). Different types of male gametes with various nuclear-cytoplasmic relations are known. For example, compact sperms with a small amount of cytoplasm are typical for plants characterized by premitotic type of fertilization.

Male gametes within an embryo sac

According to the hypothesis of enantiomorphism proposed by S.G. Nawaschin, direction of movement of male gametes towards female ones is probably determined by a special kind of polarity which occurs during division of the generative cell. He believed that the specific spiral shape of sperms serves to prove 'their independent mobility'.

For many years it has been debated whether sperms within the pollen tube and embryo sac are active or passive gametes (Strasburger, 1884; Kostrjukova and Chernoyarov, 1938; Polunina, 1961; M.S. Navashin *et al.*, 1959; Maheshwari, 1950; Steffen, 1953; Litvak, 1968; Cass, 1973; Heslop-Harrison J. and Heslop-Harrison Y., 1987a,b; Plyushch, 1992). Opinions differ regarding the mechanism of sperms entering extracellular space. According to one opinion they move together with the contents of the pollen tube and probably of the destroyed synergid (Nawaschin, 1910b; Gerassimova-Navashina and Batygina, 1958; Batygina, 1962a,b, 1966, 1974b, 1987a; Korobova, 1979, 1983). In *Scilla sibirica* and species of *Zea* and *Triticum*, the contents of the pollen tube together with sperms, often in the form of separate drops, have been observed in the extracellular space. In the pollen tube of wheat the sperms are elongated. Within the synergid and in extracellular space they become spiralled and keep that shape at the first stages of entering into the cytoplasm of the egg cell and central cell. When approaching female nuclei the sperms straighten again and become vesicular. Using the time-lapse technique the activity of movement of the sperm towards polar nuclei was proved (Erdelská, 1983). According to Jensen's concept (1972), based on detailed electron-microscopic investigations of the process of fertilization in *Gossypium*, sperms transit from the synergid into female cells not through active movement but as a result of physicochemical interaction of the cytoplasm of the pollen tube and synergid. The most interesting concept is that of the role of microtubules in sperm movement. Clustered in bunches in some species of flowering plants, they are often located parallel to the long axis of the cell (Cresti *et al.*, 1990).

Plasmogamy and karyogamy—key points of double fertilization

Discussion continues as to whether male gametes penetrate female cells as whole cells or as nuclei, or does the cytoplasm of sperms remain in the synergid (Poddubnaya-Arnoldi, 1958; Belyaeva, 1977, 1980; Gerassimova-Navashina, 1971a,b, 1980; Korobova, 1979; Smirnov, 1982; Krasnikov, 1988) or in the extracellular space between the egg cell and the central cell (Mogensen, 1982, 1988; Chebotaru, 1987; Russell *et al.*, 1990a,b). There are data to prove that sperms at the moment of uniting with female gametes contain cytoplasm (Wylie, 1904; Zamotailov, 1968; Jensen, 1973; Cass, 1981; Wilms, 1981a-d; Smirnov and Grakhantseva, 1986; Plyushch, 1992). Penetration of male cytoplasm with mitochondria and plastids of both the egg cell and the central cell has been established in *Plumbago zeylanica* (Russell, 1980, 1982, 1983). Several patterns of participation of sperm cytoplasm in syngamy and triple fusion have been considered (Van Went 1970a-c; Jensen, 1973; Wilms, 1981a-d; Mogensen, 1988; Russell *et al.*, 1990a). According to one pattern, the process of double fertilization includes plasmogamy during which plasma membranes of female and male gametes fuse. **Plasmogamy is accompanied by formation of bridges between gametes** through which nuclei of sperm cells and probably their cytoplasm pass. The authors of this hypothesis (Linskens, 1969; Van Went, 1970b; Jensen, 1972) posit that a specific membrane in the egg cell can change during syngamy so that it becomes unreceptive to the second male gamete.

In some species (for example, *Capsella bursa-pastoris*) a pore in the wall separating the synergid and the egg cell has been observed near which an open end of the pollen tube can be seen. Apparently the sperm enters the egg cell through this pore (Schulz R. and Jensen, 1968b).

Gerassimova (1933) using light microscopy gave a detailed description of karyogamy in *Crepis capillaris*. Jensen (1964) using TEM was the first to describe the mechanism of double fertilization in *Gossypium hirsutum*. Later, using other plants, including representatives of Poaceae, he showed that this complicated process consisted of a number of successive stages. Subsequently the study of karyogamy at the ultrastructural level was extended to other plant species (Van Went, 1970a-c; Schulz P. and Jensen, 1973; Kudlicka *et al.*, 1980; Wilms, 1981a-d; Mogensen, 1982; Hause and Schröder, 1987).

In 1993, Faure and colleagues investigated induction of karyogamy *in vitro* in male and female gametes of *Zea mays* by means of electric impulse. Employing electron micrography, phase-contrast microscopy and reconstruction of fusing nuclei these authors, like Jensen, described three main stages of nuclear fusion: (1) outer membranes of nuclear walls tightly adjoin each other through endoplasmatic reticulum and then fuse; (2) inner membranes of nuclear walls fuse and form bridges between the nuclei; (3) bridges enlarge and capture some amount of cytoplasm.

However, some authors hold, that various modes of fusion of membranes of nuclear walls occur during karyogamy (Szabo and O'Day, 1983; Plyushch, 1992).

The question of mechanisms of syngamy and triple fusion has long been discussed in the light of the mitotic hypothesis and the relations of these processes with the cell cycle (Gerassimova, 1933; Gerassimova-Navashina, 1947a, 1951a,b, 1957, 1960b, 1971a-c, 1980, 1982; Gerassimova-Navashina and Batygina, 1958, 1959; Korobova, 1959, 1962; Savina, 1965; Vallade, 1980; Bannikova *et al.*, 1985; Friedman, 1991) (see Fig. 15). Quantitative assessment of nuclear DNA in gametes and zygote is introduced as an important index of passing of the mitotic cycle periods (Woodard, 1956; Bennett and Smith, 1976; Bannikova and Khvedynich, 1982; Sherwood, 1995).

The process of karyogamy became especially interesting with the discovery of the ancestral form of double fertilization in some taxa of Gnetales (*Ephedra, Gnetum*) (Friedman, 1990a,b, 1991; Carmichael and Friedman, 1995), whose existence was believed in by other authors as well (Pearson, 1929; Favre-Duchartre, 1956; Gerassimova-Navashina, 1958; Teryokhin, 1991). Based on detailed investigations of nuclear DNA content in male and female gametes of *Gnetum gnemon* carried out by means of fluorescent microscopy, Carmichael and Friedman (1995, 1996) distinguished **three types of karyogamy in seed plants** (Fig. 16): **G_1-, S-phase- and G_2-karyogamy.**

These types of karyogamy reflect the character of relations between the patterns of the cell cycle and the process of fertilization. As seen from the results of investigations by the aforesaid authors, **karyogamy can pass in different periods of the cell cycle and the S-period** (i.e., period of replication of DNA) **in different moments of the process of fertilization (after fusion, before the end of fusion and before fertilization).**

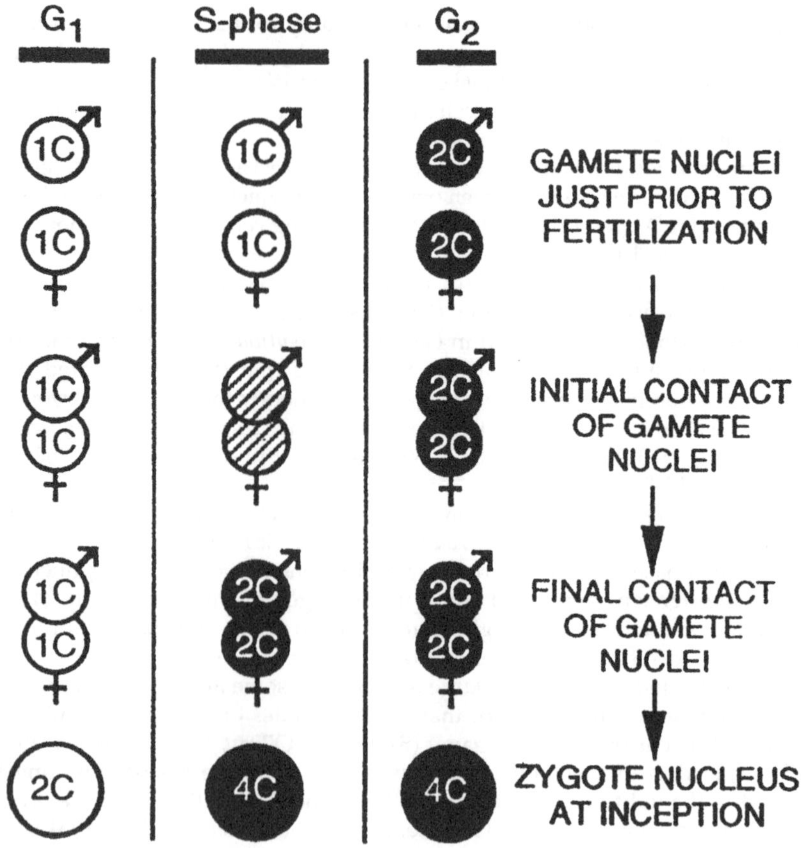

Fig. 16: Relations between cell cycle and gamete karyogamy among seed plants (after Carmichael and Friedman, 1995).

The published data on fertilization in flowering plants available at present do not allow a definite conclusion about which types of karyogamy are typical for them. Moreover, only single works provide data about the periods of the cell cycle of gametes and zygote in the object studied, duration of the interphase and so on. Comparing the concept of Gerassimova-Navashina about the types of fertilization (premitotic, post-mitotic, intermediate) with the data provided by Carmichael and Friedman about the types of karyogamy, it can be seen that **the type of fertilization, broadly speaking, indicates a temporary connection of karyogamy with the final period (M) of the cell cycle (i.e., the time of completion of nuclear fusion—before, after or during the mitosis),** while **the type of karyogamy denotes the concrete period of the cell cycle when nuclear fusion takes place.**

According to the current concept of cell cycle control, **the cell cannot enter into mitosis until replication of DNA is** fulfilled (see Jacobs, 1992, 1995). Taking into consideration these notions and cytoembryological data on the process of fertilization in different species of flowering plants, we can suppose that: (1) **premitotic type of**

fertilization can combine three types of karyogamy, 'G_1', 'S' and 'G_2'; (2) **at post-mitotic and intermediate types of fertilization only one is possible, M-karyogamy** (Fig. 17).

Data on cytology of syngamy and DNA content in male gametes and zygote serve as proof of G_1-karyogamy in flowering plants: in representatives of *Tradescantia* (Woodard, 1956), *Hordeum* (Mogensen and Holm, 1995; Mericle L. and Mericle R., 1973; Bennett and Smith, 1976), *Zea* (Moss and Heslop-Harrison, 1967; Kranz and Lörz, 1996), *Petunia* (Vallade, 1980), *Pennisetum* (Sherwood, 1995) and *Triticum* (Gerassimova-Navashina and Batygina, 1959; Batygina, 1961a,b, 1962a,b, 1987a; Gerassimova-Navashina, 1960a,c, 1982). Additional evidence of existence of G_1-karyogamy in flowering plants is found in results of investigations of sexual elements in *Crepis tectorum* obtained by X-ray (Gerassimova-Navashina *et al.*, 1963; Gerassimova-Navashina, 1971) and autoradiography (for Alimova's data, see Gerassimova-Navashina, 1980). It was determined that male and female nuclei do not enter the S-period of the cell cycle before their fusion; labelled thymidine was found only in the zygote.

An indirect indication of G_2- and S-karyogamy in some representatives of flowering plants is the fact that the DNA content of sperm nuclei doubles before fertilization (D'Amato *et al.*, 1965; Ermakov *et al.*, 1980; Friedman, 1999).

So, four types of karyogamy can be distinguished in flowering plants.

G_1-karyogamy: the egg cell and sperm pass G_0 and G_1-periods of the cell cycle, their nuclei fuse in the G_1-period and replication of DNA occurs after karyogamy, before mitosis (**in the mature zygote**).

S-karyogamy: the egg cell and sperm pass G_0 and G_1-periods, their nuclei fuse within the S-period and replication of DNA occurs during karyogamy, before mitosis (**in the immature zygote**).

G_2-karyogamy: the egg cell and sperm pass G_0, G_1 and S-periods, their nuclei fuse within the G_2-period and replication of DNA occurs before syngamy (**in gametes**).

M-karyogamy: the egg cell and sperm pass G_0, G_1, S, G_2 (and partly M)-periods, their nuclei fuse within the M-period and replication of DNA occurs immediately before completion of karyogamy and before mitosis (**in the immature zygote**).

So, **fusion of gamete nuclei can occur in different periods of the cell cycle, which determines the type of karyogamy** (Fig. 18). It can be such that passing of the cell cycle by the egg cell-zygote and the type of karyogamy correlate with the types of development: embryo sac (16 types), ovule (crassi- and tenuinucellate types) and endosperm.

Nuclei of male and female gametes in many species of plants pass the interphase at a different rate, which determines asynchronism of their entering into the periods of the cell cycle (Bannikova and Khvedynich, 1973). Overall duration of syngamy in gymnosperms is one month and more, in angiosperms it varies from several minutes to a few days. The rate of plasmogamy and karyogamy also differ in various species. So, in *Triticum* plasmogamy passes within one hour, and karyogamy within 3-4 hours (Batygina, 1961a,b, 1974b) (Fig. 19).

There is a certain correlation between the character of differentiation of gametes, metabolic state of the egg cell and rate of karyogamy (Jensen, 1965a,b; Bannikova and Khvedynich, 1973; Gerassimova-Navashina and Gulyaev, 1973; Gerassimova-Navashina, 1984; Plyushch, 1992). The higher the chromatization of female nuclei, the longer the process of karyogamy.

As shown in Figure 17, **depending on the type of fertilization and type of karyogamy, the different stages of zygotogenesis (gamete, immature or mature**

FERTILIZATION type	PREMITOTIC			INTERMEDIATE	POSTMITOTIC
Type of **karyogamy**	**G_1-**	**S-**	**G_2-**	**M-**	**M-**
Point of S-PERIOD beginning in sexual nuclei	after karyogamy	during karyogamy (the end of nuclear fusion)	before the beginning of syngamy	before the end of karyogamy, before the mitosis	
STAGE OF ZYGOTE DEVELOPMENT corresponding to the S-period of cell cycle	MATURE ZYGOTE	IMMATURE ZYGOTE	GAMETES	IMMATURE ZYGOTE	

Fig. 17: Interconnection between fertilization type, type of karyogamy, cell cycle and stage of zygote development in flowering plants. (after Batygina and Vasilyeva, 2001).

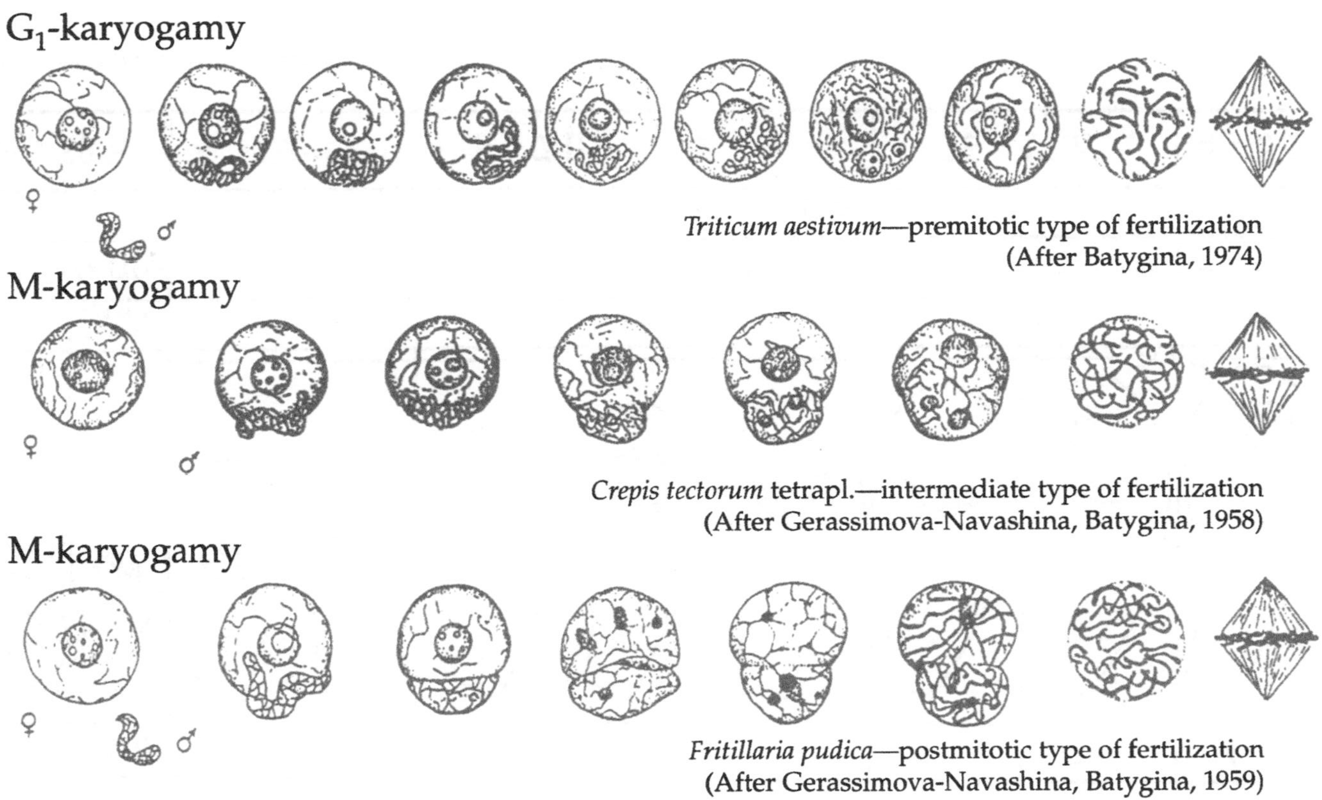

Fig. 18: Types of karyogamy in different species of flowering plants (after Batygina and Vasilyeva, 2001).

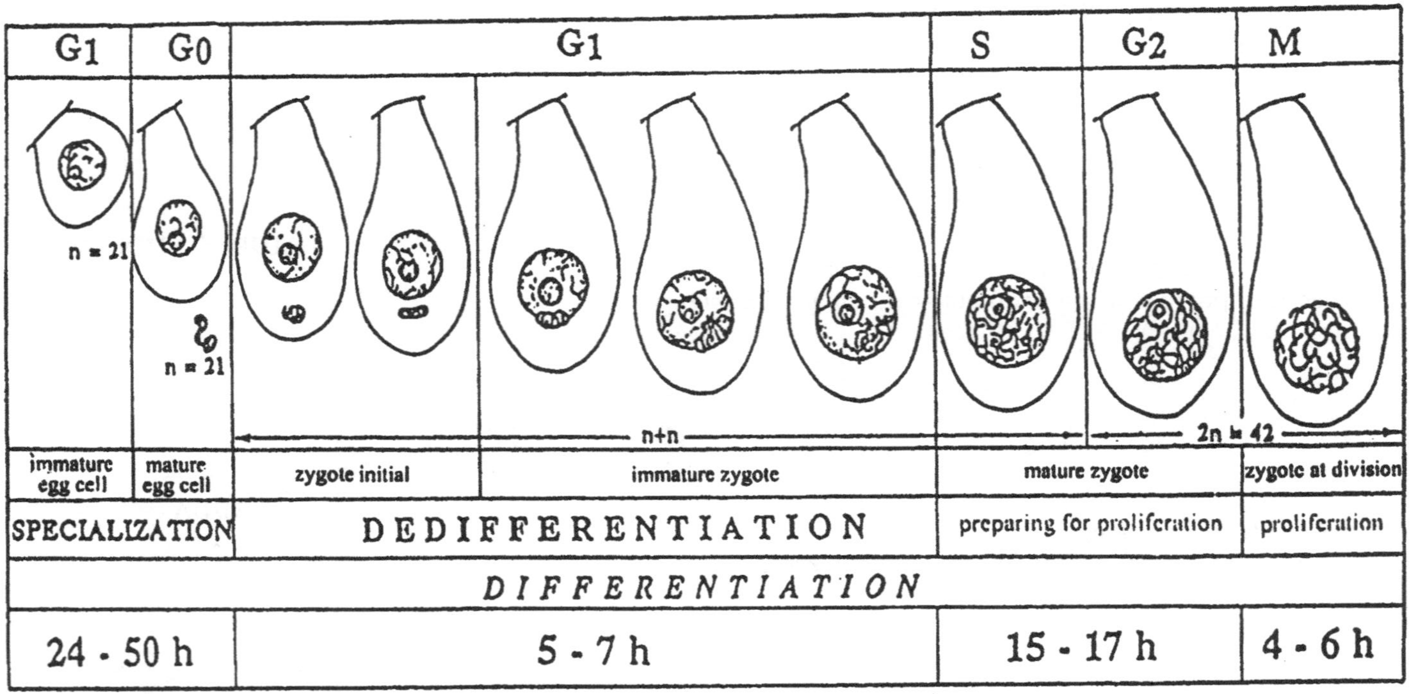

Fig. 19: Chronology of syngamy in genus *Triticum* (St. Petersburg, Pushkin) (after Batygina, 1961a,b, 1974b).

zygote) can correspond to the S-period of the cell cycle. This is another proof for **taxon-specificity of cell cycle passing in the zygote.** Probably the fact that there is no strict correlation between the stage of zygote development and the period of the cell cycle explains contradictions in data concerning the structure of the zygote and rate and character of syngamy in different species of flowering plants.

Triple Fusion

Triple fusion is the process of fusion of the polar nuclei of the central cell with the nucleus of the sperm, which results in formation of the primary nucleus of the endosperm **(Plate XII)**. It is one of the elements of fertilization of the central cell of an embryo sac. Synonyms: karyomixis[9], karyogamy.

Before discussing 'triple fusion' it is necessary to consider the structures participating in this process: polar nuclei, secondary nucleus of the central cell and primary nucleus of the endosperm[10].

Having considered all these notions Battaglia (1987) proposed a detailed terminological system. We shall dwell only on two terms suggested by him, namely, return to the term 'central nuclei' first introduced by Vesque (1897), as replacement of this term with 'polar nuclei' (Fischer, 1880) is not quite correct from either the historical or linguistic point of view. Polar nuclei, as definitively shown by Battaglia, do not occupy a polar position in the embryo sac. Further, Battaglia gives an example that is hard to refute: formation of so-called polar nuclei in tetrasporic embryo sacs of Plumbago-type of development ('cruciform polarity').

Instead of the term 'secondary nucleus' he proposed the term 'central syncarion'. Depending on the number of central nuclei taking part in syncarion formation (with different types of embryo sac development), the latter can be designated 'syndicarion', 'syntricarion', 'syntetracarion', etc.

As is known depending on the type of female gametophyte development the quantity of polar nuclei in its central cell can vary from one to several (see Polar Nuclei, Vol. 1).

Disposition of the polar nuclei in the central cell in respect of cells of the egg apparatus and antipodals before fertilization is taxon-specific. In many flowering plants **migration** of polar nuclei has been observed, as well as of the secondary nucleus of the central cell and primary nucleus of the endosperm during their formation and triple fusion. The time of fusion of polar nuclei varies: they can fuse before fertilization, during or after it. Probably, this very reason determines the fact that triple fusion can pass either **simultaneously** or **successively**. The following scheme of possible combinations of triple fusion was proposed by Poddubnaya-Arnoldi (1976).

1. The upper (P_1) and lower (P_2) polar nuclei fuse forming the **secondary** or **central nucleus of the embryo sac**, which then fuses with the nucleus of one of the sperms (*sp*), entering the embryo sac $(P_1 + P_2) + sp$. As a result **the primary endosperm nucleus** originates. This occurs in a great majority of cases.

---

[9]The term was proposed by Battaglia (1963) to denote the process of nuclear fusion during fertilization.

[10] Product of fusion of the three haploid nuclei—the nucleus of 'the germ' of the endosperm (S.G. Nawaschin, 1915) or 'trophime' (Van Tieghem, 1900).

2. Fusion of polar nuclei and nucleus of the sperm passes simultaneously ($P_1 + P_2 + sp$).
3. The nucleus of the sperm first fuses with the upper polar nucleus and then the fertilized upper polar nucleus fuses with the lower polar nucleus ($P_1 + sp) + P_2$.
4. The nucleus of the sperm first fuses with the lower polar nucleus and then the fertilized lower polar nucleus fuses with the upper polar nucleus ($P_2 + sp) + P_1$.

There are cases of apomictic development of the primary endosperm nucleus (Korobova, 1962; Chebotaru, 1972).

Sizes of the polar nuclei may differ; the smaller polar nucleus sometimes fuses with the nucleus of the sperm (*Lilium martagon, Fritillaria tenella;* S.G. Nawaschin, 1898b).

Depending on **the type of development of female gametophyte the primary endosperm nucleus can be haploid, diploid, triploid or polyploid.** With development of the embryo sac according to the Polygonum-type, the primary endosperm nucleus is usually triploid. Triploidy of endosperm was first established in *Capsella bursa-pastoris*. In the embryo sac of Oenothera- and Eriostemones-types there is only **one polar nucleus**; the **primary endosperm nucleus** originating as a result of its fertilization is **diploid.** *Tulipa* and *Melampyrum* species also have a diploid endosperm, as the chalazal polar nucleus is eliminated and the nucleus of the sperm fuses only with the micropylar polar nucleus (Yoffe, 1976).

Possible variations of the triple fusion: polar nuclei in most of the cereals studied are only **in contact** before triple fusion; their **union** occurs after **contact of the nucleus of the sperm either with one** or simultaneously **with two polar nuclei** (Guignard, 1901a-c; Sax, 1918; Batygina, 1962a,b, 1974a,b, 1987a; Chebotaru, 1972; Korobova, 1982; Mogensen, 1982). Very likely the latter takes place when the nucleus of the sperm contacts the border of both nuclei. However, in *Triticum* and *Zea* cases are known in which the polar nuclei fuse first, forming the secondary nucleus of the central cell, and then the latter unites with the nucleus of the sperm.

When the polar nuclei of *Triticum* are united, often between them can be seen the remnants of the cytoplasm captured by the polar nuclei during their closing in (Batygina, 1962a,b, 1974b). A similar phenomenon was observed in *Impatiens glandulifera* (Steffen, 1951) and *Arachis hypogaea* (Gerassimova-Navashina, 1954a,b). With complete fusion of the polar nuclei remnants of the captured cytoplasm are no longer seen. Electron-microscopic investigations carried out on cereals proved the foregoing (Mogensen, 1982).

During triple fusion the nucleus of the sperm cell changes its shape and degree of spiralling. In nuclei of sperms of *Triticum* one end is obtuse and the other tapering (Batygina, 1962a,b, 1974b, 1987a), as proved by electron-microscopic investigations. A highly spiralled nucleus of the sperm located in the cytoplasm of the central cell changes its shape on coming into contact with polar nuclei. It becomes, as it were, 'flat' and its combining with the polar nucleus (or nuclei) begins not simultaneously over the entire surface, but usually from its wide end. Analysis of electron micrographs of the process of double fertilization in *Hordeum* (Mogensen, 1982) established both the different shape and structure of the ends of the nucleus of the sperm and the non-simultaneous plunging of the sperm nucleus into the caryoplasm of the polar nucleus. Alteration of the spatial form of nuclei of the sperms bespeaks their active movement.

The process of plunging of the nucleus of the sperm into the polar nucleus (nuclei) or the secondary nucleus of the central cell is similar to that in the egg cell, but passes much quicker. Triple fusion in wheat, for example, takes 1.5-2 hours and finishes in 2-3 hours after pollination. In the formed primary endosperm nucleus, 1-2 nucleoli are seen at the place of 'dissolution' of the nucleus of the sperm. In polar nuclei of *Triticum aestivum* quite a number of additional nucleoli, up to 20, appear within some time after their fusion; however, they differ greatly in size and colouring from the nucleoli evolved by the nucleus of the sperm. Similar nucleoli are sometimes found in non-fertilized polar nuclei as well. The nature of these additional nucleoli is not yet clear but apparently they indicate increased metabolic activity of the central cell. It should be noted that insertions, very similar to these nucleoli, were frequently observed near polar nuclei in the central cell (Batygina, 1974b). During prophase of the first division a separate group of prophatic threads of the sperm is observed that slightly surpasses the beginning of the polar nuclei prophase in rates of development, which is also proved by non-contemporaneous disappearance of nucleoli. Probably, a very short period of interphase explains this phenomenon, as a result of which male chromatin fails to be evenly disposed in the female nucleus, as occurs in the zygote of wheat.

The same phenomenon, i.e., separate prophase of the male nucleus during triple fusion in wheat, was remarked by Sax (1918) who noted that male chromosomes differ from female ones in the metaphase of the first division of the endosperm as well. However, Batygina (1974) observed no difference between male and female chromatin in this phase. Due to its short duration the interphase is quite hard to spot and it is only by means of very fractional temporal fixations that the presence of this stage can be established. Evidently, this is the reason why Sax did not notice the interphase stage and arrived at the wrong conclusion that no period of dormancy occurs in the polar nuclei of the central cell of wheat and that the prophase occurs immediately, during which they commence fusion with the nucleus of the sperm which is also in the prophase stage.

Double fertilization in cereals passes according to the premitotic type, i.e, union of sexual nuclei occurs before mitosis. However, fusion of the nucleus of the sperm cell with the polar nuclei of the central cell does not correspond entirely to the premitotic type. The peculiarity of this process in the central cell lies in the fact that the nucleus of the sperm combines with nuclei of the central cell before mitosis, the nucleus of the sperm enters the prophase inside the primary endosperm nucleus separately and chromatin of the sperm cell merges completely with the polar nuclei in metaphase. As a result of this complex process of triple fusion the male and female nucleoli do not unite in the polar nuclei of the central cell. **Some researchers believe that different rates of passing of G_1, S, G_2 periods of the mitotic cycle by male and female gametes explain this phenomenon of asynchronous transformation of chromatin of male and female gametes during double fertilization in some species of plants, including cereals** (Gerassimova-Navashina, 1980, 1990; Bannikova and Khvedynich, 1982).

Hence we can suppose that it is more correct to refer fusion of nuclei in the central cell to the intermediate type of fertilization, while the whole process of double fertilization should still be characterized as the premitotic type.

The time of primary endosperm nucleus transition to division is taxon-specific: it proceeds immediately after triple fusion (e.g., Juglandaceae) or within some time. The period of interphase in the primary endosperm nucleus in wheat is shortterm and takes 0.5-1 h (Batygina, 1974b). In representatives of Butomaceae, the primary endosperm nucleus divides almost simultaneously with the zygote 24h after fertilization. The interphase of the primary endosperm nucleus of *Allium nutans* lasts for 33-34h (Stozharova and Poddubnaya-Arnoldi, 1977). In *Iris tenax*, division of the primary endosperm nucleus takes place approximately 70h after fertilization (Sokolovskaya and Shpilevoy, 1990).

Development of the endosperm can pass differently depending on conditions of vegetation of a species. So, in *Anemone nemorosa* populations in Poland it originates from the non-fertilized central cell (2n) (Trela, 1963a, b; Brouland, 1967, 1969), while in the environs of St. Petersburg in the same species it results from triple fusion (3n=45) (Yoffe, 1968, 1969, 1971b).

Representatives of Orchidaceae serve as a model object for studying the behaviour of the polar nuclei and the nucleus of the sperm during double fertilization. In some species of orchids double fertilization proceeds normally. Usually union of sexual nuclei passes according to the post-mitotic type. In a number of cases disturbances are observed in the second link of double fertilization—union of the nucleus of the sperm with the nucleus or nuclei of the central cell (e.g. in *Cypripedium insigne, Dendrobium nobile, Calanthe veitchii*). Culmination of such a disturbance is the fact that the nucleus of the sperm does not fuse with the nucleus of the central cell (*Oreorchis patens, Calanthe aristulifera, Listera nipponica*) (Savina, 1979; Poddubnaya-Arnoldi, 1982; Savina and Poddubnaya-Arnoldi, 1990). For instance, in *Calanthe veitchii* the polar nuclei usually do not merge, but together with the nucleus of an antipodal constitute a polar-antipodal group, more commonly located near the egg cell and further on, near the zygote. The nucleus of the sperm approaches this group of nuclei but does not fuse with it, i.e., triple fusion does not take place. Then, the whole group of nuclei together with the nucleus of the sperm degenerates. In this species of orchids the endosperm does not form (Poddubnaya-Arnoldi, 1964a, 1976).

Absence of triple fusion is typical for *Vencetoxicum nigrum, Mitella pentandra, Zostera marina, Taraxia ovata, Zauschneria latifolia* (Maheshwari, 1950) and for representatives of Podostemaceae (Korobova, 1985).

The question of triple fusion occurrence in the Trapaceae family is debatable. According to most authors (Gibelli and Ferrero, 1891; Ram, 1956; Grevtzova *et al.*, 1987) in representatives of this family (*Trapa natans, T. bispinosa, T. europaea* and *T. conocarpa*) **double fertilization takes place but the endosperm as a tissue does not form**—the primary endosperm nucleus becomes hypertrophied and shifts to the chalazal end of the embryo sac, where it degenerates gradually. Trela-Sawicka (1978) reported that in *T. natans* the endosperm forms and is nuclear though very ephemeral.

According to Titova (1988, Kolesova [Titova] 1996; Titova *et al.*, 1997), typically **falling out of triple fusion** is observed in the embryo sac of *T. astrachanica*, **which is a result of disturbance of the mechanism of sperm divergence:** hold-up of one of the sperms in the synergid or extracellular space. It is significant that unlike the other species of *Trapa*, in this species there are mature pollen grains of two types—two- and three-celled ones, the latter germinating more often. Usually, the nucleus of the vegetative cell does not pass into the pollen tube, as usually happens in species with three-celled pollen grains. Titova indicated that absence of triple fusion can be determined by: (1) disturbance of the connection between separate elements (vegetative nucleus and sperms) of the 'male germ unit', and (2) the fact that the secondary nucleus of the central cell in the given species is the product of fusion of the three nuclei—normal, micropylar polar nucleus and two chalazal nuclei which have not passed the third mitotic division.

Disturbances of differentiation in aposporic embryo sacs of *Poa pratensis* lead to formation of a great number of polar nuclei, which differ in behaviour during fertilization. For example, some nuclei fuse with the nucleus of the sperm forming the primary endosperm nucleus while others do not fuse and remain in the endosperm tissue (Batygina and Mametyeva, 1979). Similar peculiarities of aposporic embryo sac development were found in *Elatostema, Iris* and *Taraxacum* (Maheshwari, 1950; Poddubnaya-Arnoldi, 1976).

Jensen (1964b) was the first to give a detailed description of triple fusion in *Gossypium hirsutum,* which was later observed in other objects also (Schulz P. and Jensen, 1973; Kudlicka *et al.*, 1980; Wilms, 1981b; Hause and Schröder, 1987). Jensen believed that fusion of the outer membrane of nuclear walls with parts of the ER cisterns takes place first. Then the inner membranes of the wall fuse and bridges form between the merging nuclei which later fuse, forcing out the cytoplasm of the central cell located between them. Possibly, the cytoplasm of the central cell is not forced out by the nuclei (Bannikova and Plyushch, 1988), but undergoes the autolysis (Batygina, 1987a). Contrary to Jensen, Plyushch (1992) while studing karyogamy in *Adoxa moschatellina* showed that the nuclear wall breaks in the evagination zone, though the perinuclear space remains closed. Thus the fine mechanism of triple fusion is not quite clear because no temporal electron-microscopic investigations have been carried out.

It is still debatable whether the nuclei of sperms enter female gametes with the cytoplasm or without it. It is known that in the pollen grain and in the pollen tube sperms represent typical cells. Some authors believe that the sperm jettisons the cytoplasm when found in the synergid (Poddubnaya-Arnoldi, 1958; Krasnikov, 1988) or in the extracellular space between the egg cell and the central cell (Mogensen, 1982, 1988; Chebotaru, 1987; Russell *et al.*, 1990a,b). Other authors describe a gradual decrease in the amount of cytoplasm of sperms (e.g. in *Spinacia oleracea*—Wilms, 1981b, 1985; *Epidendrum scutella*—Cocucci, 1981) and the number of organelles (in *Hordeum*-Mogensen and Rusche, 1985).

In plants with one-parent (maternal) inheritance elimination of elements of male cytoplasmic heredity can occur during fertilization of the egg cell (Mogensen, 1988). Very likely, the same process takes place during fertilization of the central cell as well. Only as a hypothetical variant, Mogensen (1988) and Russell and colleagues (1990a,b) considered the pathways of fertilization of the egg cell and the central cell in angiosperms to differ. The difference lies in the fact that it is only the nucleus of the sperm that enters the egg cell, while in the central cell the sperm enters with cytoplasmic organelles. Probably, there are some species of plants in which the cytoplasm of the sperm passes into the egg cell and other species in which it remains beyond the bounds of the egg cell.

The question about dimorphism of sperms in flowering plants is also under discussion. S.G. Nawaschin (1927, 1951, p. 331) believed that 'sperms of one and the same couple are not similar and likewise female nuclei participating in the process differ from each other; this is so despite the fact that both couples of nuclei are sister nuclei, i.e., originate from division of one mother nucleus, while such division is still considered in morphology to be aequational'. Further investigations proved the differences of sperms in a couple (see Sperm Dimorphism, vol. 1). Pairwise differences of sperm nuclei, and of the egg nucleus and its sister, the upper polar nucleus, underlie the directed divergence, i.e., they are the differences which enable each female nucleus to choose the corresponding male one. So, in *Casuarina equisetifolia* the nucleus of the largest of sperms fuses with the nucleus of the egg cell, and the other, smaller one, with polar nuclei (or with the secondary nucleus of the central cell in *C. montana*), located near the egg apparatus during the fusion (Swamy, 1948b). Triple fusion is an obligatory stage of double fertilization. Appearance of triple fusion leads to formation of the endosperm, which due to its hybrid nature has proven to be an important source of nutrition for the embryo (Kozo-Polyansky, 1949), which increases the vitality and adaptability of angiosperms.

Cytoplasmic Transmission during Fertilization (Plate XIII)

Cytoplasmic organelles, including the mitochondria and plastids, are organelles with their own DNA, RNAs and protein synthesis machinery. Despite this feature, they still depend on the importation of numerous critical proteins, from which significant cytoplasmic compatibility problems may originate. Since some phenotypic organellar characteristics are under nuclear control (following Mendelian inheritance patterns) and others under organellar control (following non-Mendelian patterns), reciprocal breeding experiments are usually needed to distinguish whether their origin is cytoplasmic. If their pattern of inheritance differs when a plant serves as a maternal or a paternal parent, this is strong evidence of cytoplasmic inheritance. Since mitochondria and plastids are inherited independent of one another, each organelle may display different patterns of cytoplasmic inheritance.

In general, organelles may display either exclusively uniparental (maternal or paternal) inheritance or display different degrees of biparental inheritance. The most elegant evidence for biparental inheritance is the coexistence of organelles of different origins within a single cell (heteroplasmy). Frequently, however, these organelles may occupy different clonal lineages and appear as variegated pigment patterns in the plant. Organelle lineages usually segregate over time due to sorting-out so that a minority of the cells in a mature plant may have both organelle types. Variegation per se is not strong evidence of cytoplasmic control, however; in many cases a defect in chloroplast pigmentation may be under nuclear control and the variegation reflects clonal inheritance of nuclear patterns.

Plastid inheritance in the vast majority of angiosperms is uniparental or predominantly maternal with a small 'leakage' of paternal organelles (Smith, 1989). A significant minority of angiosperms display biparental plastid inheritance (Hagemann and Schröder, 1989). Only two cases of predominantly paternal inheritance of plastids have been reported (*Daucus*—Boblenz *et al.*, 1990; *Medicago*—Smith *et al.*, 1986). Mitochondrial inheritance is poorly described from classical genetics because of the paucity of useful non-lethal markers, but it is believed to be even more overwhelmingly maternal. A few instances using molecular methods have, however, indicated the presence of male mitochondrial nucleic acids in the offspring (Fairbanks *et al.*, 1988; Erickson *et al.*, 1989; Erickson and Kemble, 1990; Faure *et al.*, 1994b).

Prefertilization mechanisms reducing male cytoplasmic inheritance

Despite a common misconception about the cellular condition of the sperm, both male and female gametes are organized as complete cells containing a normal complement of organelles and surrounded by their respective plasma membranes. An interesting exception is plants with maternal plastid inheritance. These frequently have sperm cells that lack plastids according to ultrastructural (Hagemann and

Schröder, 1989) and DNA fluorescence studies of plastid nucleoids (Corriveau and Coleman, 1988). Mitochondria are also reportedly absent in two orchids studied to date, *Epidendrum* (Cocucci and Jensen, 1969b) and *Cymbidium* (Yu and Russell, 1992), although ultrastructural studies in plants other than orchids do not report as extreme a condition for mitochondria.

The presence, absence, or relative quantity of heritable cytoplasmic organelles in the male and female gametes is critical in determining the cytoplasmic complement of the zygote. One obvious mechanism for restricting the number of organelles from the male lineage at inception is the asymmetrical mitotic division of the microspore, which forms a much smaller generative cell (the immediate precursor to the two sperm cells in the pollen grain) and a larger vegetative cell (the future pollen tube). This mitotic field excludes significant numbers of organelles and in the case of plastids, their larger size may cause them to vacate the future generative cell cytoplasm. In any case, the generative cell is often less than 1% of the volume of the pollen grain. Relatively little quantitative data is available on changes in the volume of the cells of the male lineage. In *Plumbago,* the generative cell is its largest at initiation and continues to lose volume over the course of development (Russell and Strout, submitted). A relatively small number of organelles is present in the newly formed generative cell. The young generative cell of *Plumbago* contains less than five plastids and fewer than 30 mitochondria (Russell and Strout, submitted), which is fewer than one-fifth the number observed at maturity (Russell, 1984). Since this plant displays biparental transmission (Russell, 1983, 1987), the secondary increase in organelle numbers could increase the probability of inheritance if the number of organelles containing DNA is proportionate to their numbers (Bendich and Gauriloff, 1984). The multiplication of organelles from a restricted population may be evidence of a 'founder effect' that could reduce residual organellar heterogeneity, if any, within the gametes.

Diminution of male cytoplasmic organelles has been quantified in sperm cells of barley (Mogensen and Rusche, 1985) and in developmental stages of the orchid *Cymbidium* (Yu and Russell, 1992) and tobacco (Yu and Russell, 1994a,c) through the formation of enucleated cytoplasmic bodies (ECBs) which are shed from the cells. Each of these plants displays predominantly maternal cytoplasmic transmission. Interestingly, there also appears to be considerable evidence for male cytoplasmic diminution in plants that normally transmit male organelles (*Plumbago,* Russell and Strout, submitted). Possibly, such mechanisms for male cytoplasmic diminution are present in all angiosperms regardless of their eventual transmission. This tendency may reflect a need to streamline sperm cell structure during pollen maturation and pollen tube growth, or it may simply be a by-product of muscle-like actin-myosin associations that appear to be involved in the movement of the gametes within the pollen cytoplasm (Heslop-Harrison J. and Heslop-Harrison Y., 1989d; Tang X. *et al.*, 1989). Such movements may stress and stretch the cells of the male germ line to their breaking point on numerous occasions (and at numerous sites) during their descent in the pollen tube.

Modification of organelles during development has also been proposed as a factor in the inheritance of the male cytoplasm. Vaughn and colleagues (1980) proposed that organelles were directly 'altered' (i.e., incapacitated) within the male germ cells by an unspecified mechanism; however, their exclusively ultrastructural approach did not eliminate the possibility of artefacts; use of any of the readily available cytochemical methods for organellar function (e.g., Dickinson, 1986) would have removed much of the doubt about their conclusions. Molecular modification of organellar DNA has been reported in wheat microspores during tissue culture. Organellar DNA was reported to be altered in regenerated plants and this was postulated as one mechanism for male DNA exclusion (Day and Ellis, 1984). More recent work on barley has suggested that these alterations in the chloroplast genome occur during regeneration, rather than the earlier stages of reproductive committment (Mouritzen and Holm, 1994). *In vivo* organellar DNA changes in the male lineage have not been studied on a molecular level yet. It is also evident that cytoplasmic DNA visible in fluorescence microscopy disappears during development in some plants with maternal inheritance (Miyamura *et al.*, 1987), but it is not evident whether this occurs by the elimination of plastids (Clauhs and Grun, 1977), or if the DNA simply reduces to levels lower than can currently be resolved using fluorescent probes. The occurrence of nuclease C (Ca^{2+}-dependent) has been implicated in the elimination of male organellar DNA in higher plants (Nakamura *et al.*, 1992); interestingly this appears to be related to the enzyme involved in the post-zygotic elimination of organellar DNA in lower plants. During meiosis and later development, alterations in organellar structure are readily evident, suggesting that selection of organelles may indeed precede sperm cell formation (Dickinson, 1986). Although identical organelles may appear different within physiologically different cells (Russell, 1983), the possibility of meiotic and post-meiotic rearrangements of organelles is still present given the extensive associations between functional organelles and the nucleus, as observed during microsporogenesis in *Cosmos* (Dickinson, 1986).

Transmission of male organelles during fertilization

Double fertilization involves cellular fusion of one sperm with the egg cell and the other with the central cell. Both fusions use sperm cells derived from the same precursor cell, the generative cell, in order to initiate distinct and dissimilar developmental pathways. As in cytoplasmic inheritance, transmission of organelles during fertilization appears to follow three patterns: organelle transmission may be exclusively maternal or exclusively paternal, or biparental.

In plants with maternal cytoplasmic inheritance, the sperm cytoplasm is usually impoverished in organelles, especially plastids, and in this case transmission of the male cytoplasm would seem to have no effect on the egg cell. This is true of *Nicotiana* (Yu *et al.*, 1994), which shows transmission of apparently all of the male cytoplasm into both the egg and central cell, and may also be true of *Petunia* (Van Went, 1970a,b), since no ECBs indicating non-transmission were captured in this plant.

In barley, apparently the entire cytoplasmic volume is stripped from the sperm nucleus during gametic fusion (Mogensen, 1982, 1988). Since Mogensen noted that only one such ECB was found (always in contact with the egg) and that its organellar complement equalled that of intact sperm cells, he concluded that sperm cytoplasm was excluded from the egg but was transmitted into the central cell. The likelihood of male cytoplasmic exclusion from both the egg and central cells is suggested by Jensen and Fisher's observation in cotton of two large cytoplasmic regions of the approximate size of male cytoplasm within the degenerate synergid of the fertilized embryo sac (Jensen and Fisher, 1968a,b; Fisher and Jensen, 1969). Prefertilization fusion of the two sperm cells was observed in spinach by Wilms (1981a,b) who postulated transmission of the sperm nuclei in directions opposite to the egg and central cell, simultaneous with disintegration of the remaining cytoplasm.

Vacuolization of the sperm cells also appears to be a general trend prior to fertilization and may represent another mechanism that allows heritable cytoplasmic organelles to be excluded from later fertilization steps. In *Populus* (cottonwood), this trend appears to reach an extreme which may spatially separate the nucleus and the cytoplasm enough to trigger male cytoplasmic exclusion (Russell *et al.*, 1990b).

Biparental inheritance appears to be the result of sperm cells rich in cytoplasmic organelles fusing in entirety with female reproductive cells. Essentially all of the sperm cell organelles appear to be transmitted in *Plumbago* (Russell, 1983, 1984) except for a cellular projection that physically associated one sperm cell with the vegetative nucleus. ECBs presumably orginating from this projection account for 15% of the sperm organelles. In this plant, the pattern of membrane fusion and subsequent vesiculation suggest that the entire surrounding cytoplasm of the sperm may be transmitted directly into the egg and central cell cytoplasm (Russell, 1983, 1985, 1987). Other plants with predominantly maternal cytoplasmic inheritance also seem to follow this pattern (*Petunia*—Van Went, 1970a,b; *Lycopersicon*—Kadej A. and Kadej F., 1983, 1985; *Nicotiana*—Yu *et al.*, 1994).

The occurrence of strictly paternal cytoplasmic inheritance has infrequently been reported in angiosperms (*Medicago*—Smith *et al.*, 1986; *Daucus*—Boblenz *et al.*, 1990). In *Medicago* (Zhu *et al.*, 1990, 1991, 1992, 1993), careful enumeration and three-dimensional reconstruction of gametes revealed that differences in the female gamete appear to be responsible for paternal plastid inheritance, since both 'weak' and 'strong' sperm donors had numerous plastids in their generative cells and similar numbers of plastid nucleoids (Shi *et al.*, 1991). The difference appears to be in plants susceptible to paternal inheritance in which the unfertilized egg contains plastids only in the micropylar half of the cell. After fertilization, maternal plastids remain in the basal half of the zygote. Only paternal plastids are therefore present in the apical cell that becomes the embryo proper (Zhu *et al.*, 1993). The basal cell organizes the suspensor, which is a terminal embryonic lineage not expressed in the mature plant. A structural investigation of *Daucus muricatus* revealed that only a small number of plastids (8 to 12) were present in the egg and that some apparent plastids at the periphery of the egg appeared to be degenerating (Hause, 1991). These eggs further displayed a relatively high degree of vacuolization compared to the eggs of *D. carota* and *D. aureus*, which displayed greater maternal inheritance.

Post-fertilization mechanisms reducing male cytoplasmic inheritance

Control of parental cytoplasmic inheritance within the zygote most frequently appears to be the result of organellar competition in different cellular lineages, a process described as 'sorting-out' in which relaxed cellular control is exerted on the organelles. In this model, minority populations of organelles are progressively eliminated from cellular lineages through time (Birky, 1983). Given enough mitotic divisions, cells containing multiple organelle types would be reduced in frequency until one or the other organelle became extinct within that lineage. Superimposed on this pattern is the possibility of sperm dimorphism, including nuclear differences (*B*-chromosome non-disjunctive in the generative cell of corn; Roman, 1948) or cytoplasmic differences in the contents of either plastids or mitochondria or both (Russell, 1984, 1985a,b, 1993).

One instance of uniparental inheritance that has not been observed is a counterpart to the 'restriction' mechanisms reported in unicellular plants, in which a single mitochondrion and plastid of each mating type is transmitted into the zygote, but only one organelle of each type survives (Gillham, 1994). In these cases the organellar DNA of one mitochondrion and one plastid is destroyed after fertilization through proposed restriction sites in the genome. The surviving mitochondrion and plastid are determined by mating type and may be of opposite types (reviewed in Snell, 1990). Evidence for a cytoplasmic restriction of male cytoplasmic organelles is not available in angiosperms.

'Leaky' mechanisms for cytoplasmic inheritance

Frequently, cytoplasmic inheritance mechanisms allow for the occasional penetration of male characteristics into a predominantly maternal cytoplasmic pattern. Many of the models of maternal cytoplasmic inheritance in flowering plants thought to be exclusively maternal, actually express some male cytoplasmic characteristics in large-scale classical genetic studies (*Nicotiana*—Medgyesy *et al.*, 1986) and in molecular studies (*Brassica*—Erickson *et al.*, 1989; Erickson and Kemble, 1990; *Medicago*—Fairbanks *et al.*, 1988; *Musa*—Faure *et al.*, 1994b; *Populus*—Rajora *et al.*, 1992). This penetration is common enough that plants with traces of male cytoplasmic inheritance in normally maternal cytoplasm can be considered a separate class (Smith, 1989). In *Nicotiana*, there is a particularly interesting opportunity for redundant transmission of male organelles; in this plant, the sperm nucleus itself may sometimes contain identifiable mitochondria that are transmitted into the fusion nuclei of the zygote and primary endosperm (Yu and Russell, 1994b). Apparently these are incorporated in the sperm nuclei during generative cell division when organelles are not effectively excluded from the metaphase plate and cytoplasmic organelles may be incorporated in the sperm nucleus during nuclear envelope reformation. This is potentially interesting because the incorporation of organelles in the nucleus would allow organellar DNA access to the nucleus and, through recombination, could contribute to its permanent incorporation in the nuclear genome.

Numerical studies of heritable cytoplasmic organelles during fertilization

Relatively few studies enumerate the organelle content of both male and female gametes even though this may be critical if 'sorting-out' is a major method determining the nature of cytoplasmic inheritance in angiosperms. Considerable evidence to date indicates that male organelles are at a disadvantage for future expression in the offspring. Even in cases of biparental expression, male cytoplasmic organelles may be in the minority. The direct numerical impact of male and female organelles has been the subject of only three studies in flowering plants to date (Richter-Landmann, 1959; Meyer and Stubbe, 1974; Russell, 1987). These results are given in Table 2.

Table 2: Number of organelles present in male and female gametes at the time of fertilization in angiosperms

Species	Plastids		Mitochondria	
	Egg	Sperm	Egg	Sperm
Impatiens grandulifera[a]	150-480	20	1000-2500	50
Oenothera erythrosepala[b]	25-32	8-13	NA	NA
Plumbago zeylanica[c]	730 ± 110	24.3 ± 3.9	39.900 ± 3.140	39.8 ± 3.3

[a]Richter-Landmann, 1959; [b]Meyer and Stubbe, 1974; [c]Russell, 1984, 1987; NA—not available.

Impatiens, with maternal plastid inheritance, displays male:female plastid ratios of 7 to 24; *Plumbago* displays male:female plastid ratios of 30:1, while *Oenothera*, which displays frequent biparental inheritance, has male:female plastid ratios ranging from 1.92 to 4:1. Mitochondrial ratios are of the order 200:1 or higher for female mitochondria to male mitochondria (200 to 500:1 for *Impatiens* and approximately 1:1000 for *Plumbago*), which would suggest extinction of the male mitochondria in many cell lineages even during early embryogenesis, especially since somatic cells usually contain fewer than 200 mitochondria. The only additional avenue to male expression is through recombination, which is relatively frequent in mitochondria but has rarely been reported in plastids (Smith, 1989).

Although the predominant pattern of cytoplasmic inheritance is maternal, numerous mechanisms seem designed to reduce genetic variability in the cytoplasmic genome concomitant with its preservation in the nucleus. Redundant and overlapping mechanisms cause the volume of the male cytoplasm and number of organelles to be reduced. At the same time, these mechanisms are incomplete, since male organelles frequently do penetrate the cytoplasmic complement of the offspring. A remaining question is whether there is any particular pattern to organellar inheritance. The trend is towards an insular cytoplasmic genetic complement but exceptions abound. In this context, the variable nature of organelle heredity seems to provide evidence for the lack of a clear advantage in retaining cytoplasmic heterogeneity. However, current evidence also indicates that there are no mechanisms that actively promote heterogeneity either.

Types of Double Fertilization in the Light of Electron Microscopic Evidence

The difference in pace of fusion of gametes in angiospermous plants is directly linked with the existence of different types of fertilization: premitotic, post-mitotic and intermediate (Gerassimova-Navashina, 1954b, 1957, 1971a, 1980, 1982, 1990).

Subsequent research works have shown that in some plants the process of fertilization is characterized by features of two or three types of fertilization simultaneously (see Gvaladze, 1973). In the first stages it proceeds according to the post-mitotic type, but fusion of the nuclei occurs before mitosis starts, as in the premitotic type. So fertilization in species of Onagraceae (Gvaladze and Gachechiladze, 1977, 1984; Gachechiladze, 1981) and in *Galtonia candicans* (Gerassimova-Navashina, 1980) was attributed to the premitotic type on the basis of the features of its final stage. In species of genera *Rhododendron* (Yakobson, 1976) and *Peperomia* (Plyushch, 1983) it is regarded as an intermediate one, though compared with the definition given by Gerassimova-Navashina (1960a), the features of the post-mitotic type are more prominent in these plants.

We know very little about the causes of the existence of different types of fertilization in plants. Thus, on the basis of the mitotic hypothesis of double fertilization, a supposition was made that, above all, the type of fusion of gamete nuclei is determined by a functional state of sperms (Gerassimova-Navashina, 1957; Gerassimova-Navashina and Batygina, 1959; Korobova, 1959). In addition, it was especially emphasized that nuclei in 'resting state' do not fuse because of the occurrence of an envelope in them. According to these notions, in plants with the premitotic type of fertilization the nuclei of sperms enter the embryo sac with the nuclear envelope not yet formed, which allows them to fuse with nuclei of the female cells quickly. In plants with the post-mitotic type of fertilization, the nuclei of the sperms are enclosed in an envelope, which is why fusion of the nuclei of the sexual cells is delayed until the nuclei come into mitosis and their envelopes destroyed.

Evidently, if one holds this opinion, it is hard to explain the peculiarities of fertilization in *Galtonia candicans*, species of genera *Rhododendron, Peperomia*, of Onagraceae, in which the nuclei of the gametes, being in contact, do not merge for some time; nevertheless, nuclei fusion is completed before initiation of mitosis. Ultrastructural studies of the sperms of some plants, including those with premitotic type of fertilization, showed that their nuclei are fully surrounded by the envelope (Jensen and Fisher, 1968a; Cass and Karas, 1975; Cheng *et al.*, 1980; Russell and Cass, 1981a,b; Wilms and Van Aelst, 1983; McConchie *et al.*, 1985; Dumas *et al*, 1985; Mogensen and Rusche, 1985; Tarasenko and Bannikova, 1986).

Studies of the embryo sac in *Peperomia blanda* under a light microscope have shown that the nuclei of the sperm and the egg cell stay in contact for a long time. Nuclei of the central cell, as many as 7-11, come into contact with the nucleus of the sperm. Their fusion is completed before fusion of nuclei in the zygote. When formation of the primary polyploid nucleus of the endosperm is completed, the prophase of its mitosis commences. Since the early stages of karyogamy in *P. blanda* proceed in the same manner as in plants with the post-mitotic type of fertilization, and fusion of nuclei takes place before mitosis begins, the type of fertilization in this species is considered intermediate (Plyushch, 1983).

Ultrastructural study of the fusion of the gamete nuclei in *P. blanda* demonstrated that in the period described in studies under a light microscope as a period of contact of nuclei, the nuclei of the sperm and the egg cell are not isolated but joined by several bridges, i.e., they are practically a single nucleus. The bridges must form right after the contact of nuclei, since they are already observed before the visible transformations in the ultrastructure of the zygote cytoplasm. In this state (joined by the bridges) nuclei remain for quite a long time, during which in the zygote cytoplasm the number of mitochondria and plastids increases, ER (tubular and cisternoid) develops, and a polysaccharide cell wall is formed. The rounded nucleus appears in the zygote at the stage of the two-cellular endosperm (Plyushch, 1983).

Under the light microscope nuclei of the sperm and of the egg cell and tightly adjacent sperm and central cell were observed in *Gagea lutea*. No single rounded nucleus was seen either in the zygote or in the primary cell of the endosperm. The type of fertilization in *G. lutea* was defined as post-mitotic (Petrova, 1977). Electron-microscopic study of fertilization in *G. lutea* revealed that when the gamete nuclei are in contact, they are actually joined by means of the bridges of fusion. The presence of bridges between the nuclei is typical of both the zygote and the primary cell of the endosperm; the chromatin of the nuclei is mostly diffuse.

The adduced data attest to the similarity of the initial stages of fusion of the gamete nuclei in *Peperomia blanda* (intermediate type of fertilization) and *Gagea lutea* (post-mitotic type of fertilization). In both species fusion of nuclei starts from formation of bridges between them. This process proceeds in a simlar manner in other plants (Jensen, 1964, 1965b, 1973; Van Went, 1970c; Jensen and Fisher, 1967; Schulz P. and Jensen, 1973; Kudlicka *et al.*, 1981; Mogensen, 1982), including plants with the premitotic type of fertilization.

So, electron-microscopic studies have shown that in different types of fertilization fusion of nuclei occurs according to à single plan. After formation of the bridges, visible only under an electron microscope, the nuclei are no longer isolated and practically form one highly lobed nucleus. In the case of rapid joining of the gamete nuclei along the full surface of their contact area and in the case of long maintenance of the fusion bridges, differences in the functioning of the zygote nucleus and the primary endosperm nucleus are insignificant. Asynchronous transformation of the chromatin of the male and female gametes is a strong argument for this. This phenomenon was observed in tobacco and lily—plants with premitotic and post-mitotic types of fertilization respectively. Asynchronous transformation of the 'male' and the 'female' chromatin in the zygote nucleus as well as their asynchronous entry into the mitosis prophase were demonstrated in a number of animal and plant objects (see Bannikova *et al.*, 1985; Bannikova, 1986).

In spite of the fact that the general scheme of fusion of nuclei is similar in different types of fertilization, the period of absolutely full fusion of the nuclei varies in different plants. In the premitotic type of fertilization, fusion of nuclei throughout all the contact surface is completed relatively fast, and decondensation of chromatin of the sperm nucleus occurs after absolute completion of karyogamy. In the intermediate and post-mitotic types of fertilization, the process of fusion of nuclei throughout all the contact surface is prolonged. The chromatin of the sperm nucleus decondenses and a nucleolus is formed at the presence of bridges connecting this nucleus with the nucleus of the female gamete. And if in the intermediate type of fertilization, as seen in *Peperomia blanda,* fusion of the nuclei on a considerable area of contact is completed before mitosis initiation, in the post-mitotic type of fertilization there is no such process before initiation of mitosis. The description of a long contact period of nuclei in the post-mitotic type of fertilization is explained by insufficient resolution of the light microscope, which cannot distinguish the bridges of fusion between nuclei.

Perhaps, it was the difference in pace of full fusion of nuclei discovered under the light microscope that led to recognition of different types of fertilization in plants. The different duration of the processes of syngamy and of triple fusion are strong evidence for this. Thus, in wheat (premitotic type of fertilization) fusion of nuclei in the central cell lasts 1.5-2 h, and in zygote formation, 4-6 h (Batygina, 1974b). In groundnut (intermediate type of fertilization) triple fusion lasts for about 6 h and syngamy about 17 h (Zamotailov, 1958, 1960). In lily (post-mitotic type of fertilization) the period of contact of nuclei of the sperm and the central cell is 48-72 h and nuclei of the sperm and egg cell 96-120 h (Mentkovskaya, 1982). In *Cephalaria gigantea* the type of fertilization in the primary cell of the endosperm was determined as premitotic and in the zygote as intermediate (Solntseva and Kamelina, 1976). It was Gerassimova-Navashina who paid attention to the connection between rate of fusion of nuclei and type of fertilization. In tetraploids and plants grown in different experimental conditions under the influence of low temperatures and X-ray, slow-down of fertilization and, in some cases, transition from the premitotic type of fertilization to intermediate or post-mitotic was observed (Gerassimova-Navashina, 1960a,b; Gerassimova-Navashina *et al.* 1963, 1968; Gerassimova-Navashina and Batygina, 1964). Thus, the types of fertilization in angiosperms is nothing more than morphological expression (at the light microscopic level) of differences in pace of an absolutely full fusion of nuclei (throughout the full surface of their contact), i.e., there is no principal difference between the types.

Fertilization Mechanisms (Plate XIII)

Organization of male gametes. In the overwhelming majority of cases, the two sperm cells of angiosperms are identical, being derived as brother cells from the same precursor generative cell; however, in some cases the two sperm cells evidently compete to fuse with the egg and may carry differing amounts of cytoplasmic DNA-containing organelles (plastids and mitochondria) that can be expressed in the embryo. Usually the two male gametes are linked together in a pollen tube and are associated with the vegetative nucleus in a male germ unit (see Male Germ Unit). In forming the MGU, the sperm cells share a common wall or interface with each other and are usually enclosed together within the same region of surrounding pollen plasma membrane. The vegetative nucleus is found in association with this enveloping vegetative cell membrane. Since sperm cells are almost always identical, double fertilization is usually a genetically co-operative event in which brother sperm cells share in the nutrition and destiny of one offspring.

The male gametes of angiosperms are indeed true cells with a distinct, though sometimes sparse cytoplasm. The male gametes usually have fewer cellular organelles than somatic cells, and in species with maternal cytoplasmic heredity, the sperm cells are often impoverished in plastids (Hagemann and Schröder, 1989). In orchids, sperm cells may lack both plastids and mitochondria entirely (Yu and Russell, 1992), apparently deriving all of their energy anaerobically or possibly through the importation of energy-laden molecules. In Apiaceae, sperm cytoplasm may be so reduced that it is essentially non-existent (Weber, 1988a,b). Sperm cells are believed to be incapable of independent motility: instead they are passively borne through the pollen tube (PT) by means of actin-myosin associations in which the myosin-coated male gamete (Heslop-Harrison J. and Heslop-Harrison Y., 1989d; Tang X. *et al*., 1989) moves along the actin bundles present throughout much of the length of the elongated PT (Pierson and Cresti, 1992). The vegetative nucleus may also assist in PT passage because it too is coated with a myosin-like protein (Heslop-Harrison J. and Heslop-Harrison Y., 1989d) and in most species, is part of the MGU. Whether the generative or sperm cells have myosin directly on their surface or on the enveloping vegetative cell membrane awaits greater resolution of immunoelectron microscopic results. The sperm cells themselves may be capable of shape changes but are not actively motile compared with animal sperm or the sperm cells of lower plants.

The male gametes, or sperm cells, may be formed either before or after the PT germinates in different species of flowering plants. Tricellular pollen contains the two sperm cells prior to PT germination, whereas bicellular pollen contains the precursor generative cell when it is disseminated. Generative cell division to form the two sperm cells is delayed in bicellular pollen until after PT germination. This delayed mode of sperm cell formation is found in the majority (2/3 or more) of flowering plant genera and is associated with gametophytic self-incompatibility. In either type of pollen, two sperm cells are formed prior to fertilization. The male germ cell lineage is established by the meiotic division of a microsporocyte which establishes the lineage that eventually undergoes two additional mitotic divisions to form the two 'brother' sperm cells. Generative cell division results in cells with a genetically identical nuclear complement and a similar but not identical cytoplasmic genetic complement. Male gametes are formed, descend in the PT and enter the embryo sac as true cells. In rare cases wherein MGUs are loosely formed or their elements become dissociated within the tube, the possibility of fertilization from two different pollen tubes (heterofertilization) can occur (Robertson, 1984). Heterofertilization requires the arrival of two or more PTs within the embryo sac, but observation of multiple pollen tubes within an ovule are rare (Mogensen, 1978a).

Organization of female gametes. Female gametes (the egg cell, synergids and central cell) constitute a functional unit of female reproduction that has been designated the female germ unit FGU (Dumas *et al.*, 1984). This functional organization of the mature embryo sac is remarkably similar among angiosperms. Major variants include differences in the ploidy level of the central cell (based on number of polar nuclei that contribute to the central cell), and the number of synergid cells formed. Although variability in ploidy number may have consequences for the capacity of the endosperm, structural alterations of the FGU appear to be potentially more drastic in appearance. In the Peperomia type, only one synergid is formed, while synergids are absent in the Plumbago and Plumbagella embryo sac types (review: Huang and Russell, 1992). The absence of synergids in the latter two embryo sac types calls into question how a synergid, which appears to be indispensable in the vast majority of angiosperms, could be eliminated in a normal functioning FGU, but apparently the same types of cell interactions are possible as observed in the majority of angiosperms (Russell, 1983; Russell and Cass, 1988).

The egg cell is highly polarized (reviews: Huang and Russell, 1992; Russell, 1993) and appears to be quiescent, as reflected by its ultrastructural appearance (Jensen, 1965a). The synergids appear far more active and have been implicated in uptake and secretion of materials, storage of chemotropic agents for PT attraction and receipt. The micropylar cell wall is characterized in the overwhelming majority of angiosperms by prominent cell wall ingrowths, forming a so-called filiform apparatus. The central cell is far larger than the other cells and for the most part surrounds them, although the egg and synergids as well as the central cell have contacts with the external wall of the embryo sac. Based on the surface area of the central cell and the frequent occurrence of cell wall ingrowths and proliferations reflective of a transfer cell, the central cell appears to be highly involved in the uptake of nutrients into the embryo sac prior to fertilization (Huang and Russell, 1992).

The organization of FGU cell walls appears to reflect their reproductive and physiological functions. Cell walls containing prominent fibrils that presumably represent cellulose microfibrils are present in the external wall of the embryo sac, the common wall between the two synergids, and the micropylar end of the FGU. In contrast, junctions between the egg and central cell and those adjacent to the synergids have fibril-lacking or gel-like cell walls; these are the anticipated locations for gamete migration and fusion. Male gametes also lack cell walls, although such walls are frequently present during earlier stages (Russell and Cass, 1981a,b; Dumas *et al.*, 1984). The wall-less nature of both the male and female gametes undoubtedly promotes their ability to fuse and permits the plasma membranes of the gametes to contact one another directly (Dumas *et al.*, 1984), facilitating the merging of their membranes.

Pollen tube arrival and delivery of male gametes. In typical angiosperms, the pollen tube is guided to the ovary by largely mechanical forces. Once inside the ovary, however, pollen tubes make abrupt turns (at angles approaching 90°) to enter the small opening at the tip of the ovule, called the micropyle. Interestingly, regardless of the pollen load, usually only one PT enters each ovule. Pollen tube elongation continues through the micropyle until the tip of the PT contacts the embryo sac. At this location, the tip specifically enters the embryo sac through a degenerated or at least weakened synergid, by-passing its sister cell, the intact and turgid persistent synergid. Since elongation of the PT is ultimately driven by the turgor generated within the tube, it would seem logical to assume that the final path into the embryo sac would assume a course of lesser resistance and enter the 'weakened' synergid. The pollen tube tip usually weakens and forms a terminal pore through which the two sperm cells and vegetative nucleus are released. During PT discharge, the plasma membrane of the tube breaks down, exposing the plasma membranes of the two male gametes. The association of one sperm cell with the vegetative nucleus is severed by the force of the PT discharge and may generate cytoplasmic bodies.

Synergid degeneration. Degeneration of typically just one of the two synergids is believed to be a genetically preprogrammed event that may be initiated by the presence of a chemical signal produced by the growing pollen tubes (Jensen *et al.*, 1983). Synergid degeneration has been reported to occur before pollination in some species, but is usually dependent on the presence of a PT. In some plants, the actual arrival of pollen tubes within the synergid appears to be necessary to trigger synergid degeneration and stimulates the completion of synergid degeneration. In tobacco, just momentary contact with the ovule is needed to trigger synergid degeneration (Huang and Russell, 1994), whereas in *Petunia* and some other species, the synergid apparently must be directly penetrated to trigger degeneration (Russell, 1992). The receptive synergid may undergo a radical change in physiology through the discharge of the PT cytoplasm and completion of synergid degeneration may be essentially immediate.

The environment of the synergid before and after degeneration is unique. Extraordinarily high concentrations of calcium are present in the synergid which appear to attract pollen tubes through simple chemoattraction. Calcium taken up during synergid development accumulates in nearly lethal levels compared to other cells. The first evidence of this was the localization of ash in the synergid vacuole following microincineration of freeze-dried ovules of cotton (Jensen, 1965b). This was identified as calcium in wheat and pearl millet synergids using energy-dispersive X-ray spectroscopy (Chaubal and Reger, 1990, 1992). Later studies using pyroantimonate localization revealed that calcium is distributed throughout the young synergid of pearl millet initially and only later accumulates in the vacuole (Chaubal and Reger, 1995). Interestingly, the onset of degeneration also appears to be accompanied by an additional period of rapid calcium uptake. The synergid appears to secrete some calcium into the micropyle in the form of calcium pectate (Chaubal and Reger, 1995), a chemotropic attractant for PTs in pearl millet (Reger *et al.*, 1992). This sequence of events during synergid degeneration appears similar in a number of respects to apoptosis, a form of programmed cell death described in animal cells (Chaubal and Reger, 1995). The high concentration of calcium may have a chemotropic effect on pollen tubes but it is also possible that Ca^{+2} is involved in other growth phenomena relating to fertilization as well. Changes in calcium concentration serve as a trigger in some signal transduction pathways and could affect egg activation, cellular fusion and the activity of numerous enzymes within the embryo sac (Russell, 1992).

Gamete transport to female reproductive cells. An important structural feature of the synergid is that cell boundaries at the chalazal end of the synergid consist of just plasma membranes in the region adjacent to the egg and central cells. When the synergid completes degeneration, the plasma membrane of the degenerating synergid breaks down allowing materials within the synergid to leak into the intercellular region between the synergids, egg cell and central cell. It is into this region that the PT is discharged. It usually emerges from the filiform apparatus within the micropylar end of the synergid just prior to forming a terminal aperture and releases the cytoplasm of the pollen tube into the degenerating synergid. Pollen cytoplasm, including the two male gametes and the vegetative nucleus, move into the space between the egg and central cells. The problem of short distance migration of the non-motile sperm cells remains a potential problem in the region between the egg and central cells. In this respect, it seems noteworthy that a thin trail of actin is present on the surface of the egg and central cells (Huang *et al.*, 1993a; Huang and Russell, 1994; Huang and Sheridan, 1994). This extends from the middle of the synergid to the female target nucleus, providing a potential pathway for sperm passage. Separate trails exist leading to the egg cell and the central cell. If the sperm cells continue to be coated with myosin at this time, as they are earlier in development (Heslop-Harrison J. and Heslop-Harrison Y., 1989a,b; Tang X. *et al.*, 1989), they could potentially migrate on this matrix to the chalazal end of the actin trails and fuse at this site (Huang and Russell, 1994).

Gamete fusion. The major changes evident in the sperm cells prior to fertilization are that they appear more rounded in chemically and rapidly frozen preparations (Huang *et al.*, 1993b) and their cytoplasm becomes significantly more vacuolated (Russell *et al.*, 1990a). Upon their discharge from the pollen tube, the sperm cells are stripped of their enveloping PT plasma membranes, and their plasma membranes are directly exposed to cellular environment. These single membrane-bound cells become appressed to the egg and central cells with an intervening distance between the plasma membranes of the male and female gametes of 20-70 nm.

The egg and central cell are usually fertilized by brother sperm cells, in part because the two male gametes are usually linked in a male germ unit in most angiosperms, favouring cotransmission into the embryo sac (Russell, 1991; see Male Germ Unit). Even in cases in which the male gametes are originally unlinked, as in barley, the two gametes may approach each other and become connected as the PT reaches the ovule (Mogensen and Wagner, 1987). In rare cases, there are reports of the fertilization of the egg and central cell by sperm cells from different pollen tubes; some lines of corn apparently lack a traditional MGU organization throughout their passage in the PT and may therefore display an unusually high rate of heterofertilization (Robertson, 1984).

Cellular basis of gamete fusion. In normal cells, there are significant biophysical forces that resist any approach closer than 20 nm. Negative charges of the phosphate groups in membrane phospholipids and steric hindrance from other chemical groups would seem to discourage closer approach. The energy required to push unfused cells together increases significantly as the distance between them decreases (Russell, 1992). Therefore, it seems likely that the membranes of the gametes are destabilized in some fashion. Calcium, as a divalent positive cation, may link the membranes closer by neutralizing some of the negative charges. Calcium is known to be fusigenic at high concentrations (Kranz and Lörz, 1994) and appears to be required for *in vitro* fusion of isolated gametes of corn at relatively high concentrations (5 mM) (Faure *et al.*, 1994a). Fusion of gametes appears to discriminate gametes from somatic cells (Faure *et al.*, 1994a) but it is unclear whether this is due to fusigenic lipids present in gamete membranes or to recognition compounds such as those involved in viral membrane fusion (White, 1992), that may recognize specific membranes and enhance fusion. In any case, fusion of female gametes with supernumerary sperm cells does not occur and therefore barriers to polyspermy appear to be present (Faure *et al.*, 1994a).

Few ultrastructural reports have reported how the male gametes fuse with the egg and central cell (Russell, 1994; Yu *et al.*, 1994); however, two patterns appear to predominate. In some plants displaying maternal cytoplasmic inheritance, the entire cytoplasm of the sperm cell is apparently excluded from the egg cell while the sperm nucleus is permitted to enter. This pattern is **cytoplasmically restrictive**: male cytoplasmic organelles are shed or excluded in an enucleated cytoplasmic body at the edge of the egg cell. This pattern is typified by cotton, barley, spinach and cottonwood. In other plants, including some with predominantly maternal cytoplasmic inheritance, male cytoplasmic organelles are allowed to enter the egg cell. This form of gamete fusion is **cytoplasmically permissive** and male cytoplasmic inheritance is possible in this case, typified by *Petunia*, *Plumbago*, and tobacco. Still, in the latter case, significant male organelle diminution may have occurred, which may keep the chances for male cytoplasmic inheritance low in some plants (Russell, 1992; see Cytoplasmic Transmission during Fertilization).

Evidence of the exclusion of male cytoplasm in cytoplasmically restrictive gamete fusion is largely based on the detailed work of Mogensen (1982, 1988) and the presence of cytoplasmic bodies in the degenerated synergids reported in cotton (Jensen and Fisher, 1968a,b) and cottonwood (Russell *et al.*, 1990b). In barley, serial reconstructions of a large enucleated cytoplasmic body appressed to the exterior of the egg cell indicated that essentially all the male cytoplasmic organelles were excluded from the egg cell (Mogensen, 1982, 1988). Only one such body was present in each embryo sac, however, suggesting that sperm organelles are not excluded from the central cell. The cell biological basis for this pattern of gametic fusion is unclear. Cytoplasmically restrictive gamete fusion may involve formation and closure of a small channel of continuity between male and female gametes through which the male nucleus passes. The basis of this unusual membrane behaviour seems to operate against overwhelming biophysical forces; however, the occurrence of complementary contractile proteins on the membranes of male and female gametes may provide the forces necessary to establish closure. Actin is present in embryo sacs on the surface of the egg and central cell (Huang and Russell, 1994) and myosin has been reported on the surface of the generative cell (Heslop-Harrison J. and Heslop-Harrison Y., 1989d; Tang X. *et al.*, 1989); it is suspected to be present on male gametes but critical observations have yet to be made.

The mechanism of cytoplasmically permissive fusion of gametes has been documented in detail in *Plumbago zeylanica*. In this plant, many of the stages of fusion have been ultrastructurally observed and the male organelles can be identified and quantified in the female cells (Russell, 1983). Sperm cells are deposited between the egg and central cell and are located immediately next to each other. Fusion is initiated at one site between the sperm and the female cell. Secondary fusion sites quickly propagate and result in the incorporation of the male cytoplasm when the fusion membranes vesiculate and break down in the female cytoplasm. This is also the pattern in tobacco, which was recently described (Yu *et al.*, 1994). The pattern of fusion in cytoplasmically permissive plants seems to be well known but questions linger about the mechanism by which cytoplasmically restrictive plants undergo gamete fusion.

Sperm specificity. Since the plasma membrane of each male gamete is frequently in direct contact with the plasma membranes of both the egg and the central cell (Russell, 1992), it is possible and even likely that gametic recognition occurs. In the two clearest examples of sperm dimorphism, corn and *Plumbago*, one specific sperm cell has an advantage in fusing with the egg, resulting in preferential fertilization. In corn, genetic studies indicate that the sperm cell with more B-chromosomes is more likely to fuse with the egg (Roman, 1948; Carlson, 1969), whereas in *Plumbago* the smaller sperm cell containing more and sometimes all the plastids is more likely to fuse with the egg (94% of the time; Russell, 1985a,b). The factors involved in recognition are as yet not known; however, two-dimensional electrophoretic and hybridoma antibody characterization of *Plumbago* sperm cells revealed 13.5% polypeptide spots unique to the sperm (Geltz and Russell, 1988) and 18 to 28% unique immunological epitopes (Pennell *et al.*, 1987) respectively. Therefore, the possibility of unique interactions exists but the molecules involved are not known. Although few flowering plants display such distinct differences between the two sperm cells, it is possible that some signal present on the surface of male and female gametes identifies them as candidates for cellular fusion. The inhibition of spontaneous fusions between gametes of the same sex and between gametes and somatic cells seems likely based on available data, so some recognition of opposite sexed gametes seems plausible.

Nuclear fusion. Fusion of the male and female gamete nuclei swiftly ensues in some plants but is more protracted in others. The envelopes of the gamete nuclei fuse, first the outer and then the inner. This is followed by the formation of multiple nuclear bridges (Jensen, 1964a,b). In the fertilized egg cell, the sperm nucleus fuses with the egg nucleus to form the zygote nucleus, while in the central cell, the other sperm nucleus fuses with the polar nuclei to form the primary endosperm nucleus. The polar nuclei often adhere to one another prior to the arrival of the sperm nucleus in typical monosporic plants; however, in bisporic and tetrasporic plants, which may have elevated ploidy levels, the polar nuclei frequently fuse prior to the arrival of the sperm nucleus to form a secondary nucleus. The zygote ultimately forms the embryo and the derivatives of the primary endosperm nucleus organize the nutritive endosperm.

Fertilization Cytochemistry

Sperms have been studied cytochemically in greater detail (Gerassimova-Navashina, 1952, 1959, 1971a; Gerassimova-Navashina and Batygina, 1958, 1959; Vazart, 1958; Korobova, 1959; Batygina 1961a,b, 1962a,b, 1974b, 1987a). It has been shown that the sperm nuclei in the synergid, in the extracellular space between the egg cell and the central cell, in the cytoplasm of these cells, and in contact with their nuclei are represented by compact bodies which give an intensive Feulgen's reaction, which shows strong condensation of chromatin. While they are fusing with the nuclei of female cells, decrease in intensity of the Feulgen reaction occurs, and the chromatin of the sperm nuclei is hardly revealed in the fertilized nuclei of the female gametes. When the Brachet reaction was applied (Gerassimova-Navashina, 1978, 1980, 1982, 1990), the nuclei of sperms which entered the embryo sac were crimson-red (positive reaction for RNA). Afterwards, in the extracellular space between the egg cell and the central cell, they lost RNA and those in touch with the nuclei of these cells were emerald-green, consisting mostly of DNA. These transformations result in decondensation of chromatin.

In the nuclei of the egg cell and the central cell, some researchers (Vassileva, 1954; Fursov, 1964) detected no changes in nuclei chromatization, neither before nor at the moment of fertilization. Other researchers found that the Feulgen reaction intensified in the egg cell nuclei before their fusion with the sperm nuclei (Kakhidze, 1954; Solntseva, 1961; Delone and Danilina, 1963) or during consecutive phases of fusion of the nuclei of egg cells and central cells with the sperm nuclei (Batygina, 1962a,b, 1974b; Khvedynich and Bannikova, 1970a; Bannikova and Khvedynich, 1973; Bannikova *et al.*, 1985). In *Hordeum vulgare* and *Nicotiana tabacum*, after the sperm nuclei have contacted with the nuclei of female cells, especially with the nuclei of egg cells, in the latter, heterochromatin granules cease to be observed, which results in decrease of intensity of the Feulgen reaction. In the course of gamete nuclei fusion, the intensity of this reaction is reduced and chromatin does not manifest in nuclei during fertilization for some time. Then, while the fertilized nuclei are preparing for division, the Feulgen reaction becomes positive again. In polar nuclei this process occurs faster than in the egg cell nucleus, and the change of chromatin structure is less noticeable, which is associated with their weaker chromatization before fertilization.

The method of luminescent microspectral analysis of preparations stained by acridine orange (Konarev and Serdyuk, 1967; Serdjuk and Konarev, 1970), enables tracing the change of chromatin state in the female cell nuclei during their fusion with the sperm nuclei in more detail. In *N. tabacum* the greater part of the DNA of chromatin in the egg cell nucleus is combined with proteins, which shows predominance of condensed chromatin in it (Khvedynich and Serdyuk, 1977; Khvedynich *et al.*, 1978). In polar nuclei, unlike the egg cell nucleus, DNA combined with lipids is detected, which shows the predominance of decondensed chromatin. After contact with the sperm nucleus, the chromatin of the egg cell nucleus transits from a condensed state into decondensed (the quantity of DNA combined with proteins decreases, and quantity of DNA combined with lipids increases) and moves to the nuclear envelope. Only after this does the egg cell nucleus start to fuse with the sperm nucleus. In polar nuclei, decrease in connection of DNA with proteins is not observed, and the sperm nucleus starts to fuse with them right after contact.

Thus, the structural state of chromatin in the nuclei of the egg cell and of the central cell before their contact with the sperm nuclei plays a considerable role in gamete fusion. The presence of chromatin of a similar structure in nuclei of the egg cell and the central cell, conditions the same rate of fertilization in both cells. If nuclei of female cells differ in degree of chromatin condensation, the rate of the fertilization processes is not similar. Since in the majority of angiospermous plants the egg cell nucleus is mainly characterized by a compact structural state of chromatin, and the nuclei of the central cells by a diffused one, the fact of quicker passage of fertilization in the central cell rather than the egg cell becomes comprehensible.

While the sperm nucleus is fusing with the egg cell nucleus, decondensation of chromatin continues; concomitantly the amount of DNA combined with lipids increases in both nuclei, which serves as evidence of the beginning of its synthesis. In polar nuclei, DNA synthesis begins earlier than in the egg cell nucleus. Then, in the fertilized nuclei the amount of DNA combined with protein increases again and condensation of chromatin of the sperm nucleus begins because, as is known (Gerassimova-Navashina and Batygina, 1959; Batygina, 1962a,b, 1974b), its chromatin enters the prophase before the chromatin of the female cell nucleus. DNA combined with proteins is first observed in the primary nucleus of the endosperm. This is in complete conformity with the known fact of faster transition to division compared with the zygote nucleus.

Transcriptional activity of chromatin of nuclei being fertilized can be defined by the presence of free RNA synthesized *de novo*. At the beginning of fusion of the egg cell nucleus with the sperm nucleus, the amount of RNA, free and combined, in the egg cell nucleus significantly decreases compared with its amount before fertilization. This is related to its release in the cytoplasm. Then, while gamete nuclei are fusing, when the chromatin of the sperm nucleus significantly decondenses, the amount of free RNA in the nucleus increases again. Increase in amount of free RNA occurs until fertilization is completed.

The character of changes in the RNA state in polar nuclei reveals some differences compared with the egg cell nucleus. At the beginning of fertilization, when the chromatin of the sperm has already partly decondensed, the amount of free RNA does not decrease, but RNA combined with proteins and lipids starts to manifest. Later, a major part of the RNA transits to a strongly combined state and simultaneously spreads into the cytoplasm. Then, the amount of free RNA increases again but by the end of fertilization has decreased again, which is conditioned by its secondary release into cytoplasm.

Thus, after the sperm nuclei contact with the female cell nuclei, release of RNA into the cytoplasm occurs and synthesis of RNA *de novo* starts. Release of RNA into the cytoplasm from the polar nuclei occurs at a later phase of fertilization than from the egg cell nucleus. Synthesis of RNA in the egg cell nucleus, having started after its release into the cytoplasm, occurs gradually, that is why at the end of the fertilization the amount of RNA significantly increases. In polar nuclei, as a result of double release of RNA into the cytoplasm, its impulsive increase is observed. Besides, in polar nuclei, RNA strongly combined with proteins and lipids is observed in different phases of fertilization, whereas in the egg cell nucleus such RNA-lipo-proteinaceous complexes have been observed only at the beginning of fertilization.

Studies of double fertilization in *Nicotiana tabacum* and *Lilium regale* have discovered that male gametes undergo the interphase faster than female ones, irrespective of whether the male gamete nucleus penetrates the female one or they stay in contact with each other. Probably, the rate of transformation of chromatin of fusing gametes corresponds to the rate at which the predecessors of the gametes go through the periods of the mitotic cycle interphase during their formation. It was shown that the period G_1 in formation of male gametes is shorter than in formation of female ones (Bannikova *et al.*, 1981).

Earlier initiation of the prophase in the sperm nucleus compared with the female nucleus occurs as a result of asynchronous transformation of the chromatin in fusing gamete nuclei. Differences in rates of fertilization in the egg cell and in the central cell are determined by the different state of the chromatin of these cells before fertilization. As a result, the nucleus of the primary endosperm cell comes to division earlier than the zygote nucleus.

Hemigamy

Hemigamy (Greek *gemi*—half, *gamos*—marriage) is the unfinished process of syngamy, when the sperm nucleus enters the egg cell without fusing with its nucleus, but only stimulating the egg cell to further development; rarely, the process of karyogamy can take place but accompanied with considerable irregularities **(Plate XIV)**. Synonym: semigamy.

This phenomenon was discovered for the first time in 1945 by Battaglia who proposed the term 'semigamy' to denote it (Latin *semi*—half + *gamos*). However, considering a linguistic inaccuracy, the term 'hemigamy' was introduced instead (Solntseva, 1973; Battaglia, 1981) and is now used by many scholars (Nogler, 1984; Asker and Jerling, 1992). Before Battaglia explained the phenomenon, other researchers (Nawaschin, 1898b; Pace, 1913; Fagerlind, 1946) had observed hemigamy as well, but did not understand it.

Three groups of hemigamy manifestation can be distinguished: (1) the sperm nucleus enters the egg cell and stimulates its nucleus to divide, though it itself does not divide and soon becomes isolated from the other parts of the embryo by means of the cell wall; (2) the sperm nucleus, when inside the egg cell, divides simultaneously with the female nucleus, which is accompanied by formation of cell walls between all the nuclei (both the sperm and egg cell derivatives); (3) the sperm nucleus divides simultaneous with the female nucleus in the egg cell but (in contrast to the 2nd group) this division is not accompanied by formation of a cell wall.

Hemigamy of the 1st group was observed in *Cooperia drummondii*, Amaryllidaceae (Coe, 1953). The sperm nucleus, after penetrating the basal part of the egg cell, separates with the cell wall and forms the basal cell of the suspensor.

The 2nd group (some species of Amaryllidaceae) is characterized by some variants of formation of cell walls during the sperm nucleus division: (1) Sperm nucleus divides only once, and this division is accompanied by formation of cell walls between all the nuclei of the proembryo (of both paternal and maternal origin). This variant was often noted in *Cooperia pedunculata* (Coe, 1953). (2) Sperm nucleus divides one or more times, separating from other cells by a wall during the very first division, while further divisions are either accompanied by cell formation (*C. pedunculata*) or not (*Zephyranthes macrosiphon, Argyropsis candida).* In *C. pedunculata* the sperm nucleus sometimes divides twice, as a result of which four cells form in the lateral part of the proembryo, but the suspensor is formed from the cells containing the nuclei of maternal origin (Coe, 1953). In *Zephyranthes* the very first division of sperm nuclei is accompanied by laying of cell walls between all the nuclei (Solntseva and Vorsobina, 1972, 1979). Nuclei, the sperm derivatives, can be located in basal, lateral or apical parts of the proembryo. During further divisions of nuclei of paternal origin cell walls do not form and irregularities in nucleus division result in formation of restitutional nuclei.

Hemigamy of the 3rd group is characterized by the fact that the sperm nucleus, after penetrating the egg cell, does not become isolated by a cell wall from the rest of the proembryo. Its division is not accompanied by formation of cell walls (*Rudbeckia*, species Asteraceae—Battaglia, 1945a,b, 1947a,b, 1953-1954; Movsesjan, 1964, 1968; Solntseva, 1968; Kościńska-Pajak, 1985; *Arabidopsis thaliana*, Brassicaceae—Gerlach-Cruse, 1970). This case demonstrates the greatest diversity of male nucleus behaviour during hemigamy. In *R. laciniatu* the first division of male nuclei passes without deviations (Solntseva, 1968, 1971, 1973a-c, 1974). There are some irregularities during the second and subsequent divisions: overshooting of part of the chromosomes from the division spindle, which results in formation of micronuclei and aneuploid nuclei; bridges appear between nuclei under division; appearance of diploid restitutional nuclei[11]; derivatives of sperm nucleus may fuse while located inside one cell (i.e.,

[11]Battaglia (1947a,b) observed the same schemes of formation of restitutional nuclei but did not analyse them.

diploidization of nuclei of paternal origin takes place as well); typically, nuclei, derived of sperm, divide simultaneous with female nuclei situated in the same cell, and in the case of close location of figures of division, the genomes can combine, which leads to formation of 'hybrid' nuclei. Thus, as a result of hemigamy the chimeral embryo forms, whose cells contain nuclei of different origin: maternal (egg cell nucleus derivatives), paternal (sperm nucleus derivatives) and hybrid (in case of fusion of nuclei of maternal and paternal origin). Ploidity of nuclei in the embryo cells can differ. For instance, in *R. laciniata* egg cells are diploid (Ixeris-type of embryo sac development), while sperms are usually haploid (and only rarely diploid). In the mosaic embryo most nuclei are diploid (those of maternal origin), some nuclei are haploid (those of paternal origin) or diploid (in the case of restitution of nuclei of paternal origin), and some nuclei are triploid (when male and female nuclei are combined). Typically, division of triploid nuclei is accompanied by considerable irregularities, and at formation of bridges restitutional hexaploid nuclei form.

Detailed cytological researches (Kościńska, 1980; Pajak, 1982,1983; Kościńska-Pajak, 1985) have confirmed a mosaic structure in the *Rudbeckia* embryo. Haploid, diploid and triploid nuclei have been discovered in cells of *R. bicolor* embryos. As a rule, cells with nuclei, which are sperm derivatives, are located in the region of the suspensor or in the basal or lateral part of the embryo proper and do not participate in formation of the shoot apex. Occasionally they occur in the apical or lateral part of the embryo, where the apex of the future shoot initiates. Therefore, most often a new plant appears to be of the maternal type, less often the paternal type and very seldomly contains three types of tissue.

Hemigamy is quite a rare phenomenon in nature. Among flowering plants it has been discovered in dicotyledonous (14 species) and monocotyledonous (13 species) plants, and in several interspecific hybrids as well. Among dicotyledons hemigamy is typical of Asteraceae *(Rudbeckia)* and is found in Malvaceae *(Gossypium)*. Experimentally hemigamy has been induced in *Arabidopsis thaliana*. Among monocotyledons it is typical for Amaryllidaceae *(Argyropsis, Cooperia, Habranthus, Zephyranthes)* and Poaceae *(Coix* and *Zea)*.

Hemigamy is not a unique phenomenon intrinsic to Angiospermae only. It is also known in algae, for example, in species of *Acrosiphonia—A. incurva, A. sleneri, A. arcta* (Jonsson, 1964, 1971). Formation of a dikaryonic zygote in mushrooms may also be considered a variant of hemigamy.

By means of the techniques of genetic marking hemigamy has been established in *Gossypium* (Turcotte and Feaster, 1967, 1973, 1974; Barrow and Chaundhari, 1974, 1976) and *Coix* (Rao and Narayana, 1980). Hemigamy is known at both the diploid and haploid levels and can be used to produce haploids. It is hereditarily conditioned. In cotton, for example, it is passed on through both the egg cell and pollen but realized only when the nucleus of the egg cell possesses a hemigamy factor represented by one dominant gene (Chaundhari, 1975, 1978).

Hemigamy is not obligatory and does not replace or exclude amphimixis. In *Rudbeckia* it is observed in 40-42% of ovules, and in 25% of ovules, egg cell fertilization is normal while the rest degenerate. Hemigamy is always accompanied by formation of a sexual endosperm. It is detected in those plants that being adapted to cross-pollination, have to be content with self-pollination within a flower or inflorescence. On the other hand, hemigamy is evidenced not only at intraspecific, but also at interspecific crossings, for example in *Gossypium barbadense* × *G. hirsutum*. In *Rudbeckia* in distant hybridization it is not observed (Movsesjan and Kasparova, 1977a,b) but may be associated with high sterility of ovules with variants of the crossings chosen. Most likely, hemigamy is a manifestation of incompatibility at the gamete nucleus level.

Natural populations of hemigamous species are polyploid. So, in *R. laciniata* di-, tri- and tetraploids, in *R. fulgida* di- and triploids and in *R. triloba* di- and tetraploids, have been observed. All polyploids are highly polymorphous. However, polymorphism is not associated with interspecific hybridization (Perdue, 1959-1960). Evidently, polyploidy in this genus is a result of hemigamy.

Hemigamy is not seen as an accidental abnormality of fertilization; fixed genetically it plays a certain role in the evolution of some taxa (Solntseva, 1982).

Pollination and Fertilization in Gramineae (Plate XV)

Cereals differ in character (explosive, portional), duration of florescence and in type of pollination (chasmogamous, cleistogamous and others). Thus, in the environs of St. Petersburg, florescence of multiflowered wheat lasts for 7-8 days while that of uni- and diflowered ones for 2-3 days. Different species of cereals are characterized by different mechanisms of flower opening. For example, in wheat the lodicular one prevails (due to increase of lodicules), though when pollination does not occur, secondary opening of the flower takes place due to ovary growth—the ovary mechanism (Batygina, 1974a, b, 1987a; Batygina and Vasilyeva, 2001).

Numerous scientists have contributed to the investigation of fertilization of cereals (Sax, 1918; Lvova, 1950; Oksijuk and Khudyak, 1955; Symonyan, 1955; Vazart, 1955; Kluchareva, 1957; Modilewsky *et al.*, 1958; Moshkovich and Chebotaru, 1976). However, only separate stages of the process were studied and some questions, in particular, the structure of sexual nuclei, have not been sufficiently elucidated. The cell cycle of male and female gametes has not been investigated at all and development of the zygote and endosperm should be examined jointly in more detail.

While investigating fertilization in wheat, Sax (1918) focused on the cytological aspect of the phenomenon. He believed that the nucleus of one sperm, like the nucleus of the egg cell with which it 'fuses', is in the state of prophase, while the nucleus of the other sperm appears to be in the state of early metaphase by the time of its fusing with polar nuclei.

Later, Morrison (1955) insisted that during fusion with female nuclei sperms are in a state of dormancy. However, from the illustrations presented we can conclude that this is not really so.

In Vazart's work (1955), fusion of sexual nuclei in a number of cereals (*Triticum, Hordeum, Zea, Secale, Sorghum*) is examined in depth. Like Sax, Vazart was inclined to think that 'dissolution' of the sperm inside the female nucleus concurs with its prophase stage into which the egg cell nucleus enters at this time (except for *Sorghum*).

Oksijuk and Khudyak (1955) as well as with Modilewsky (1953, 1963) mentioned the structure of sexual nuclei at the moment of their fusion. They spoke of a chromatin net resembling the prophase in female nuclei and the possible stage of spireme in sperms, noted by Sax as well. However, they failed to define the state of sperms by the time they fuse with female nuclei.

Russian scientists have made a substantial contribution to the study of fertilization in cereals. Application of the temporal fixation technique (Gerassimova, 1933) promoted these investigations to a great extent. By this method successive pictures of double fertilization were obtained, which allowed a more detailed morphological description of fertilization *sensu lato* at the light level (Gerassimova-Navashina and Batygina, 1959; Batygina, 1959, 1961a,b, 1962a,b, 1974b, 1987a; Korobova, 1959, 1962, 1979; Khvedynich, 1969, 1972; Khvedynich and Bannikova, 1970a,b; Tatintseva, 1975; Bannikova and Khvedynich, 1982).

The first attempts at cytochemical investigation of fertilization were linked with use of the Feulgen reaction. Female sexual cells and sperms were studied (Vazart, 1958; Gerassimova-Navashina and Batygina, 1959; Korobova, 1959; Batygina, 1962a,b).

Application of the method of luminescent microscopic spectral analysis enabled tracing changes of the state of chromatin and nucleic acids in the nuclei of male and female gametes in cereals (Khvedynich *et al.*, 1983; Khvedynich and Serdyk, 1986). It was discovered that the nuclei of all the cells of the embryo sac and of sperms differ in structural organization of chromatin. The egg cell nucleus has a positive Feulgen reaction. The greater part of DNA is strongly associated with proteins and belongs to heterochromatin. Only a small amount of DNA forms part of diffusive chromatin. This very part is most likely to maintain the necessary level of metabolic activity of the egg cell which is relatively low before fertilization, as confirmed by the character of its organelles. Nuclei of the central cell are characterized by diffusive chromatin. Those of the egg cell and the central cell differ in amount of RNA. Thus, nuclei of the egg cell and the central cell appear to be in a different initial state before fertilization. Chromatin of the male gamete nucleus is always more condensed compared with that of the nucleus of the female sexual cell.

Investigations of fertilization in cereals at the ultrastructural level *in vivo* have recently been published (Mogensen, 1982, 1988; Schel *et al.*, 1984; You and Jensen, 1985; Tatintseva, 1988; Russell *et al.*, 1990a), and also in *in vitro* culture (Kranz and Lörz, 1993, 1996; Kranz *et al.*, 1995).

Progamic phase. Pollen tube growth usually starts at the 'sessile' stigma in most cereals within several minutes after pollination. Optimal conditions (humidity, temperature) for germination of PTs are specific for each species of cereals. After germination of the pollen at the receptive surface of the pelliculate stigma papilla, pollen tubes grow in the intercellular matrix of filaments of the stigma. Further on the tubes pass through the blades of a pinnate stigma and grow in the transmitting tissue of the style, pulling apart tightly closed cells. This results in deformation of cells and with mass germination of pollen tubes, their destruction (Batygina, 1961a,b, 1974b). Pollen tubes in the pistil often cease to grow, particularly observed when crossing various species of cereals (Batygina, 1961a,b).

The ovule of wheat is apotropic in disposition inside the ovary and hence the pathway of pollen tubes from the stigma to the micropyle is relatively long. From the style the tubes enter the ovary usually moving between its wall and outer integument, then reach the inner integument which forms the micropyle, and pass through it into the embryo sac. Pollen tubes can find their way to the micropyle otherwise also: over the inner surface of the ovary passing to the integument or only over the surface of integument. One or several pollen tubes can enter both the ovary and the embryo sac (Batygina, 1974b, 1987a).

Cells of the transmitting tissue of the style are elongated. A large oblong nucleus is located in the centre of the cell. Contents of the vacuole are electrontransparent[12].

[12]Data on ultrastructure referring to the progamic phase in wheat was kindly provided by M.A. Vishnyakova.

Dark granules are observed at the inner surface of the tonoplast. Before pollination such granules are found in the extraplasmatic space; after pollination they disappear either through inclusion in the envelope or transforming into substances of the intercellular matrix. Oblong and transverse cell walls of the transmitting tissue are pierced by numerous plasmodesmata. As the ovary ripens and the amount of intercellular matrix in the style increases, the plasmodesmata disappear and cells of the transmitting tissue remain linked by plasmodesmata only through the transverse walls (Vishnyakova and Willemse, 1994). In wheat, plasmodesmata occur not only in the transmitting tissue of the style, but also between terminal cells of this tissue and the epidermis of the ovary. This can be evidence of continuity of the transmitting tract of the pistil and of the simplast system in the carpel. It is possible that a signal of pollination passes from the stigma to the placenta through the system of plasmodesmata. The morphology and structure of cells of the inner epidermis of the ovary wall are similar to those of cells of the transmitting tissue of the style. Plasmodesmata connect these cells with each other and also with underlying cells of the ovary wall, two-three layers of which contain a great amount of chloroplasts.

In the ovary cavity between the outer integument and the epidermis of the ovary wall a fibrillar substance is observed. Probably, synthesis of this substance is associated with activity of dictyosomes in the ovary epidermis and in cells of the outer integument and also with formation in them of dark globules and their exocytosis.

According to the data of Vishnyakova, there are differences in the fine structure of cells in different zones of the transmitting tract of the pistil of wheat (stigma filaments, transmitting tissue of styloids and style, and also tissues of the ovary and ovule over which pollen tubes grow). This manifests in the form of cells and their nuclei, character of plastidoms, activity of dictyosomes, contents of vacuoles and evidently in the character of secretions as well. A general peculiarity of all the cells of the transmitting tract is their activation after pollination. This, in particular, manifests in increase of tubular and vesicular elements of AER and active dictyosomes, acquisition of a twisting contour by the plasma membrane, association of numerous vesicles with it, formation of lomasomes and also in exocytosis. Occurrence of lomasomes was also observed in the cells of stigmata of rye and barley (Heslop-Harrison J. and Heslop-Harrison Y., 1981). Presumably, occurrence of lomasomes is connected with granulocrine secretion of proteins. In cells of the transmitting tissue of the pistil of wheat, synthesis of proteins and carbohydrates is observed.

In the ovary of maize after pollination an exudate (with positive reaction on proteins, pectin, polysaccharides) is found in intercellular spaces of the transmitting tissue, and before pollination in the region of the micropyle and nucellus. It might be supposed that it participates in regulation of growth of the pollen tubes necessary for fertilization.

Growth of a PT usually proceeds from its tip where synthesis of substances takes place. From time to time the apical part of the pollen tube is cut off by so-called callose plugs and later it degenerates. Sperms are located at some distance from the tip of the tube. As the PTs grow inside the tissues of the stigma and short style, sperms undergo morphogenetic changes. In wheat, within each couple, sperms differ in size and form; most often they are ribbon-like or spiralled. Sperm structure depends not only on their phase in the cell cycle, but also on the DNA content in their nuclei.

Numerous plastids, osmiophil globules and other organelles and inclusions are found in the cytoplasm of the pollen tube. In addition a fibril substance is observed in the cytoplasm, which explains movement of sperms and the nucleus of the vegetative cell in cereals.

Double fertilization. Processes of plasmo- and karyogamy were first studied in detail at the light-optical level on the example of various cereals (Batygina, 1962a,b, 1969, 1987a; Batygina and Vasilyeva, 2001).

Penetrating the embryo sac the pollen tube meets synergids on its way and pours its contents into them. Pollen tube contents are often seen in the form of 'streams' piercing a synergid, while the PT entering one or both synergids is destroyed. Degeneration of synergids proceeds over a long time period. Thus in the embryo sac at the stage of the four-celled embryo along with the nucleus of a 'destroyed' synergid into which the pollen tube contents have been poured the whole nucleus of the second still intact synergid can be observed. Synergids are usually not observed at the stage of the eight-celled embryo. Destroyed nuclei of synergids in a picnotic state greatly resemble male gametes; they are often taken for sperms of additional pollen tubes.

At the beginning of their presence inside the embryo sac sperms become more spiralled than when in a pollen tube. Together with both sperms the PT contents being poured enter the extracellular space between the egg cell and the central cell of the embryo sac, from where sperms move in opposite directions: one goes into the egg cell **(syngamy)**, and the other into the central cell **(triple fusion)**. Within 20-40 min after pollination sperms of wheat appear in the cytoplasm of both cells. After entering the egg cell the nucleus of the sperm remains in its cytoplasm for one hour, which makes tracking all further phases of its transformation possible. On its way to the egg cell nucleus, which at this moment appears to be in a state of 'dormancy', the G_0, sperm becomes flocculent and its structure is distinctly visible. While located close to the egg cell nucleus and in contact with it, the sperm has a characteristic shape: one end is wide, the other narrow, and the wide end is looser than the narrow one (which is denser). In both ends of the sperm spiralled chromatin threads can be seen. Despiralling of the sperm inside the egg cell nucleus typically starts from the wide end, which loses its shape. There are cases in which the wide end of the sperm is already 'dissolved' while the narrow one has not yet come into contact with the egg cell nucleus. The sperm plunges into the egg cell nucleus gradually. All data on the structure of the sperm of wheat, both in the pollen grain and pollen tube, and in the embryo sac confirm that sperms reach female cells in the G_0 period. Loosening during contact and at the beginning of plunging into the female nucleus, sperms of wheat contain no nucleoli (they appear to be in the G_1 period). In the despiralling sperm, straightening occurring largely after the chromatin thread has lost its continuity, a nucleolus appears. Under the influence of a dormant female cell the sperm passes into the G_1 period of the interphase.

Male nucleoli notably increase in size as the sperm nucleus 'dissolves' in the female nucleus. Sometimes the nucleolus is already distinguishable with the sperm structure remaining. The number of nucleoli can vary. In most cereals fusion of the female nucleolus with the male one takes place, though we failed to observe the process in wheat.

In different cereals the entire process of union of the sperm with the egg cell proceeds over several hours (2-6 h), which to a great extent depends on condition of growth. After complete fusion of the sperm nucleus with that of the egg cell, the zygote enters into G_2, which in wheat lasts for 16-18 h. In the zygote it is impossible to distinguish male chromatin from female chromatin. During the dormancy period the zygote grows and develops actively. The contents of the pollen tube poured into the embryo sac plays an important role; it participates in growth and development of the zygote of wheat and is most likely to determine to some extent its shape. While the zygote is in the interphase, changes in structure of its nucleus and cytoplasm take place.

Apparently such a long dormancy results in the zygote of wheat entering the prophase of division within 24 h after pollination as a single whole (see Zygote, Part III, this volume).

These processes have been studied at ultrastructural level using phase-contrast microscopy and three-dimensional reconstruction in different species of cereals (see Model Objects for the Study of Pollination and Fertilization, Part I, this volume).

Karyogamy in cereals lasts for 1-3.5 h. A relatively similar picture of stages and time parameters were observed in *in vitro* culture during fusion of male and female gametes of *Zea mays*. In 15-18 h after fusion a zygote with two nucleoli forms which falls into a lasting dormancy (27-29 h.). Usually within 42-46 h after fusion of the egg cell nucleus with that of the sperm the first division is observed (Kranz *et al.*, 1995). Variability of this process in *Zea mays* depends to a great extent on the length of the style and environmental conditions. Sometimes fertilization starts within 22 h after pollination (Diboll, 1968; Mól *et al.*, 1993).

Triple fusion. The process of plunging of the sperm nucleus into the nucleus of the central cell is similar to that proceeding in the egg cell but passes much faster. In wheat their fusion with the sperm lasts 1.5-2 h. Already within 2-3 h after pollination it is usually completed. During the short interphase two large nucleoli of the polar nuclei and two nucleoli of the sperm can be seen. In maize complete fusion of the chromatin of the sperm and polar nuclei occurs during the prophase, in wheat during the metaphase (Batygina, 1962a,b, 1974b) (see Model Objects for the Study of Pollination and Fertilization, Part I, this volume).

Rate of syngamy and triple fusion not the same: evidently this depends on conditions in the egg cell and the central cell and also on the degree of condensation of chromatin in their nuclei. In polar nuclei at the moment of contact with the sperm nucleus chromatin is in the diffusive state, which is why the complex reconstructions seen in the egg cell nucleus do not occur and the process of fusion proceeds quickly. A connection between the metabolic state of the embryo sac cells and the rate of nuclei fusions passing in them has been reported (Jensen, 1965; Gerassimova-Navashina and Gulyaev, 1973; You and Jensen, 1985). In plants in which the egg cells have a rather low density of ribosome distribution and organelles of the cytoplasm are rare, the rate of complete fusion of gamete nuclei is protracted, for example in *Peperomia blanda* and *Lilium candidum,* while in plants in which the cytoplasm of the egg cell is rich in ribosomes and organelles (*Crepis tectorum, Rudbeckia hirta*) fusion of nuclei proceeds quickly (Plyushch, 1992). Difference in rate of fusion of sperm nuclei with female nuclei probably also depends on ploidy of cells: the egg cell nucleus has a haploid set of chromosomes, while the secondary nucleus of the central cell has a diploid one. Different sizes and structure of these two cells also determine that their period of interphase is of different duration: in the zygote it is about 12-18 h and in the primary endosperm cell 30-40 min. Different duration of the interphase in these cells can probably be explained by the different character of their differentiation and dedifferentiation (see Zygote, Part III, this volume). Rates of fusion of polar nuclei within syngamy and triple fusion should be examined from the point of view of evolution and the assumption of different functional destination of the embryo and endosperm.

Syngamy in cereals proceeds according to the premitotic type. However, fusion of the sperm with polar nuclei does not correspond with **the classical premitotic type** (see Model Objects for the Study of Pollination and Fertilization, Part I, this volume).

In this connection Plyushch's opinion (1992) should be noted. She indicated that the different types of fertilization proposed by Gerassimova-Navashina differ only in rate of absolutely complete fusion of nuclei (throughout the whole surface of their contact).

Concurrency of syngamy and triple fusion: The question of duration of space of time between pollination and fusion of gametes in wheat is under discussion. Brenchley (1909) believed that in wheat 1-2 days pass between pollination and fertilization, as he observed the prophase in the zygote within three days after pollination. Jensen (1918), Percival (1921) and Nosatovsky (1965) submitted similar data, within 24-40 h.

Temporal investigation of syngamy and triple fusion has clarified that fertilization in cereals is prolonged in time, which makes it possible to trace all its stages (for example, to observe sperms in the synergid, cytoplasm of the egg cell, etc.) and to elicit concurrency of double fertilization at different stages.

So, usually within 15-20 minutes after pollination sperms of wheat have already turned to enter the synergid, and within 20-25 minutes can be observed between the central cell and the egg cell. This stage is difficult to catch, as already within 25-30 minutes, in some cases even earlier, one sperm is located near the polar nuclei and sometimes has already come into contact with them. All this time the other sperm is located in the egg cell cytoplasm 'moving' towards its nucleus. One hour after pollination fusion of the sperm nucleus usually starts with the polar nuclei, while in the egg cell the sperm nucleus has not yet come into contact with its nucleus and only after 1.5h can commencement of fusion of the sperm nucleus with the egg cell nucleus be observed. At that time the sperm located in the polar nuclei has almost wholly despiralled and produced a nucleolus.

Within 2-3h after pollination fusion of the sperm nucleus with the polar nuclei is completed and the primary endosperm nucleus falls into a relative 'dormancy' which lasts about 0.5-1h. The process of fusion of the sperm with the egg cell nucleus is quite well expressed at this time. Between 30-60 min after fusion of the polar nuclei with the sperm, i.e., within 3-4h after pollination, as mentioned above, the prophase sets in, first in the sperm and then in the secondary nucleus. This phase of development of the central cell correlates with the phase of fusion of the sperm with the egg cell nucleus. Within 5-6h after pollination, when the metaphase or anaphase can already be observed in the endosperm, almost complete despiralling of the sperm occurs in the egg cell. Only 6-7h after pollination does the zygote nucleus enter the *S* and G_2 periods. This stage lasts for about 16-18h. Thus, the zygote begins to divide within 20-24h after pollination, when there are 4-5 nuclei in the endosperm.

Therefore, despiralling of the sperm in the egg cell nucleus continues 4-5h, while in polar nuclei of the central cell the same process takes 1.5-2h. The period of dormancy also differs, lasting about 16-18h in the zygote and only 0.5-1h in the primary endosperm cell.

However, it should be noted that we once observed fusion of sperm nuclei with female nuclei passing synchronously.

In investigating florescence and processes of pollination and fertilization in *T. monococcum, T. dicoccum, T. aestivum,* we established that each species is distinguished by a definite nature of florescence (type and rhythm), but all the species studied are characterized by one and the same type of fertilization, premitotic. Rates of fusion of sperms with female nuclei of the embryo sac are not equal. **However, despite their differences, the rate of each process separately and their correlation are determined, i.e., a definite state of the egg cell and zygote always strictly correlates with a definite state of the central cell and endosperm.** This correlation suffers little from the influence of environments and practically non-dependent on them, which has been confirmed by comparative investigations of fertilization in wheat carried out under different conditions—environs of St. Petersburg and in Kazakhstan (Gezkazgan). The rate of the general process of double fertilization has increased, but definite morphogenetic correlation of syngamy and fertilization of the central cell remains. Rate of proceeding of embryological processes on the whole depends on environments and hence the process of double fertilization can either slow down or accelerate, though their concurrency has been determined (Batygina, 1974b).

In vitro Fertilization

Fusion of isolated male and female gametes *in vitro* has been accomplished using electroporation (Kranz *et al.*, 1990, 1991a,b), induced chemical fusion (Kranz and Lörz, 1994), and in a model system using a minimal incubation medium (Faure *et al.*, 1994a). Although these experiments clearly produced 'artificial zygotes', the sperm cells in the first two systems did not fuse spontaneously with either egg cells or central cells. Faure *et al.* (1994a) used a simple medium composed of 0.5 M mannitol and 5 mM $CaCl_2$ and were able to obtain adhesion and fusion. Fewer successful fusions occurred at either 1 or 10 mM $CaCl_2$, and the effect of the $CaCl_2$ could be replaced with other calcium salts. Apparently, the event mimics natural fertilization in that fusion is spontaneous after adherence and polyspermy are precluded; addition of supernumerary sperm cells to a fertilized egg cell brought no further fusions. The exact requirements for fusion, however, are not clearly stated in this work. One question remains: Why were Kranz *et al.* (1991a,b) unable to obtain gamete fusion by aligning isolated gametes whereas Faure *et al.* (1994a) using the same plant succeeded? If successfully adhered, the gametes fused in about three minutes. Conditions inside the embryo sac may place the gametes under some pressure due to the turgor of intact embryo sac cells. *In vivo,* the pressure of male and female gamete may be a further aid to fusion.

Regeneration of plants from normally obtained (*in vivo* fertilized) zygotes has been demonstrated in corn (Campenot *et al.*, 1992; Mól *et al.*, 1993) and barley (Holm *et al.*, 1994) and from *in vitro* fertilization of corn using electroporation (Kranz and Lörz, 1993). The technical accomplishment of plant regeneration from *in vitro* and *in vivo* fusions is significant. Isolated gametes must be handled extremely carefully when removed from their native milieu, and successful regeneration provides the strongest evidence that no irreversible changes occur during this process. Modelling *in vivo* recognition and fertilization events requires fusion without the use of fusigenic factors. Isolation of male and female gametes in a diversity of species (Theunis *et al.*, 1991) may eventually provide this opportunity in a wider choice of plants.

Clearly, *in vitro* fertilization will allow wide crosses to be made between unrelated plants with greater facility than possible with somatic cells or even pollination of ovules in the 'classical' *in vitro* fertilization technique (Zenkteler, 1990). The fusigenic properties of isolated reproductive cells are possibly greater than those of somatic cells. Also, the zygote is a cell already adapted to accept new genetic combinations and may therefore be more tolerant of foreign nuclear and cytoplasmic genomes than somatic cells, potentially giving greater genetic stability. *In vitro* fertilization of isolated gametes will also provide an opportunity to develop a model of fertilization that can be modified and tested under strict experimental controls. The haploid nature of gametes and reproductive cells may also lead to precisely defined progeny. Whether knowledge of fertilization processes will allow us to better exploit this system remains to be seen, but it appears to be a promising emerging technology.

The factors that determine the final pathway of sperm cell migration within the embryo sac and the possibility of gametic competition within the embryo sac are still areas to be explored. The exact conditions within the synergid in which preprogrammed cell death is brought to completion through the arrival of the pollen tube are likely to change rapidly at the time of fertilization and have still not been completely characterized. Also, the exact mechanism by which male gametes are modified during their final descent in the tube and their method of fusion may also have an impact on the transmission and inheritance of cytoplasmically transmitted DNA, such as that of the chloroplasts and mitochondria.

PART TWO—ENDOSPERM, PERISPERM

Endosperm

The Endosperm(Greek *endon*—inside, *sperma*—seed) is a heterogenous tissue, which develops inside the embryo sac of flowering plants, having originated from **the primary endosperm cell;** the latter arises during double fertilization as a result of fusion of the central cell[1] of an embryo sac with the sperm.

Typically, endosperm cells contain triploid or more highly polyploid nuclei with genomes of paternal and maternal individuals. Ploidy of endosperm nuclei first of all depends on the number of nuclei in the central cell of an embryo sac. During fertilization of the central cell in the embryo sacs of Polygonum-, Allium- and Adoxa-type of development, where all three fusing nuclei are most often haploid, **the primary endosperm nucleus**[2], as well as its derivatives contain a triploid chromosome set. In the embryo sacs of Oenothera-type of development a 2n primary endosperm nucleus forms. The presence of a diploid endosperm has been reported in genus *Melampyrum* as well (Scrophulariaceae), which is characterized by Drusa-type of the embryo sac development. A 5n primary endosperm nucleus is typical for Fritillaria-, Plumbagella-, Penaea-, Plumbago-types, while for the Peperomia-type 9n (*Peperomia pellucida*) or 15n nucleus (*P. hispidula*) is typical. So, the primary endosperm nucleus may already be polyploid, as the nucleus of its mother cell becomes polyploid even before fertilization.

In the course of endospermogenesis nuclei may remain triploid (for example, in *Nigella damascena, N. sativa*—Kapoor, 1966; *Phytolacca acinosa, Rivina humilis*—Enzenberg, 1961) or pentaploid (*Gagea lutea*—Geitler, 1948; *Lilium* sp.—Swift, 1953). As for most plants differentiation of the endosperm is accompanied by the further rise of the degree of ploidy of its nuclei (see Yoffe, 1971a). This process usually consists in endoreduplication of chromosomes. The highest degree of ploidy (24576n), which corresponds to 13 cycles of endoreduplication, has been shown in one-nucleate chalazal haustorium of *Arum maculatum* endosperm (Erbrich, 1965). At the same time, giant (polytene) chromosomes have been observed in triploid endosperm nuclei of a number of species of plants (see D'Amato, 1984). In species of plants with an endosperm of the Cellular type of development, characterized by the presence of haustoria, degrees of nuclear ploidy differ in different parts of the endosperm. Thus, in most representatives of family Cucurbitaceae, for example, the degree of polytenization of chromosomes in the nuclei of the endosperm proper is usually higher than in haustoria nuclei (see D'Amato, 1984). On the contrary, *Cucumis sativus* (Enzenberg, 1961) and *Tozzia alpina* (Erbrich, 1965) are characterized by inverse correspondence of ploidy degrees of nuclei of endosperm and haustoria.

The endosperm of Angiospermae, like the endosperm of Gymnospermae, was initially considered as a gametophitic formation representing a reduced prothallus—which is a homologue of a prothallus of other higher plants, for instance, heterosporous ferns (Hofmeister, 1847; Sachs, 1874). Today, some authors share this idea of the nature of the endosperm (Battaglia, 1980; Smirnov, 1982; Teryokhin, 1991).

---

[1]Synonyms: endosperm mother cell, proendospermal cell.

[2]The primary endosperm nucleus is a nucleus of the primary cell of the endosperm forming as a result of fusion of polar nuclei of the embryo sac central cell (or its secondary nucleus) with the nucleus of a sperm.

After the discovery of fusion of the polar nuclei in the central cell of the embryo sac of flowering plants, which precedes formation of the primary endosperm nucleus (Strasburger, 1877), an opinion was expressed about the sexual nature of the endosperm (Le Monnier, 1887). According to such interpretation, the endosperm is an analogue of the second embryo and consequently is a sporophytic tissue. The true picture of rise of the endosperm in flowering plants was clarified after double fertilization was discovered (Nawaschin, 1898b; Guignard, 1899a, b).

S.G. Nawaschin(1898b) established that in flowering plants both male gametes participate in fertilization: one of them fuses with an egg cell, the other with the central cell containing two (or more) polar nuclei. Thus, the endosperm, like the embryo, is a new formation which arises in a majority of plants as a result of the sexual process.

Experimental researches have confirmed this concept of the endosperm nature (La Rue, 1944, 1949; Johri, 1969, 1971, 1975). While cultivating *in vitro* the endosperm of *Zea mays, Dendrophtoe falcate* and *Taxillus vestitus,* calluses, buds and roots have been derived. The number of chromosomes in the nuclei of the cells of the callus and its organs appeared to be triploid (Johri, 1971). The occurrence of endospermal haustoria is further evidence of the sexual nature of the endosperm of flowering plants, as they do not have homologues among Gymnospermae. Further, the endosperm of flowering plants is a polyploid formation as opposed to the haploid endosperm of Gymnospermae. Discovery of double fertilization has given a clue to understanding the process of rise of xenias and mosaic endosperm in the seeds of some species of flowering plants (see Poddubnaya-Arnoldi, 1976; Yakovlev, 1981; Vijayaraghavan and Prabhakar, 1984). Double fertilization has been found to occur in almost all the species studied in this respect, belonging to various families of flowering plants and differing in position in the phylogenetic system.

After completion of the process of double fertilization, the primary endosperm cell is the first cell to divide. However, exceptions have been found wherein the rapidity of division of the fertilized egg cell and central cell coincide (for example, in *Vicia faba* and *Lupinus polyphyllus*) or the zygote divides earlier than the primary endosperm cell (*Lupinus annuus* and *Trifolium alexandrinum*) (see Chubirko and Kostrikova, 1985). Other examples are known wherein a multicelled proembryo was observed in the fertilized embryo sac while the primary endosperm nucleus had not yet start to divide (*Floerkea proserpinacoides*—Maheshwari and Johri, 1956; *Alcea rosea*—Zhukova, original data—see Plate XVI, *34*).

Development of the endosperm and its histogenesis are characterized by great variability. The first researchers of the ovule of flowering plants taken after fertilization noted that there were two pathways of endosperm development initially called protein (Latin *albumen*). In the first mode a large amount of free nuclei form in the endosperm and only afterwards cell walls establish between the nuclei and the cells form (Nuclear type of endosperm development); in the second mode of formation the endosperm *ab initio* becomes cellular (Cellular type of endosperm development). Today, three main types of endospermogenesis are distinguished: Nuclear, Cellular and Helobial. The endosperm of all types of development, regardless of the mode of formation, as a result becomes cellular (however, some exceptions are known—see below: Nuclear type of endosperm development). The most widespread type of endosperm development is the Nuclear. It is typical for representatives of 159 families of dicotyledonous plants (among 240 families studied—according to the phylogenetic system of flowering plants by Takhtajan, 1980), where in 18 it was found along with the Cellular type, and in 6 families with both the Cellular and Helobial types. Further, this type of endospermogenesis is described in 38 families (among 60 studied) of monocotyledonous plants, in 11 of which it was found along with the Helobial and Cellular types of endosperm development. The Cellular type of development is characteristic of representatives of 103 families of dicotyledonous plants (in 2 along with the Helobial type and in 24, along with Nuclear or Nuclear and Helobial). In the class Monocotyledons the Cellular type of endosperm development was noted in only 3 families (in 2 along with Helobial or Nuclear and Helobial). Among the Dicotyledons the Helobial type of endosperm development has been found in only 9 families (in 2 families along with the Cellular type and in 6 families along with Cellular and Nuclear types). Mostly it is typical of monocotyledonous plants and has been described in 32 families (in one of them it was found along with the Cellular type and in 12 families along with the Cellular type, Cellular and Nuclear types or Nuclear type). All three types of endosperm formation are characterized by a considerable variability, which impedes working out their classification. Initially four main variants[3] were distinguished in the Nuclear type of endosperm development. The differences in time and space parameters of the cellularization onset in the coenocytic endosperm underlie such a subdivision (Hegelmaier, 1885). This concept was further developed by embryological study of a great number of species of Centrospermae families: Caryophyllaceae, Nyctaginaceae, Basellaceae, Portulacaceae (Rocén, 1927). Schnarf (1929) subdivided the Cellular type of endosperm development into several variants in accordance with the direction in which the first cell walls are laid down. However, in the typology of endospermogenesis a tendency to review the accepted notions gave rise to an opinion that '....the division of endosperm development into three types is quite conditional. Actually, the endosperm has a single type of development. It always passes from the nuclear phase to the cellular one' (Modilewsky, 1963, p. 325). Some researches limit the number of main types of endosperm development to two: Nuclear and Cellular (the megatypes, according to the authors terminology — Di Fulvio, 1983; Di Fulvio and Cocucci, 1986) or Cellular and Helobial types, others propose increasing the number to four types with the Cellular divided into two independent ones: Cellular proper and Tubifloral types (Kamelina, 1991, 1996). The question of the need to distinguish the Helobial type of endosperm development as an independent one also remains debatable. Some embryologists consider it a variety of the Cellular type (Swamy and Ganapathy, 1957b; Di Fulvio, 1983, 1985; Di Fulvio and Cocucci, 1986). More numerous attempts have been made to classify the modes of endosperm formation within families, for example, in the Bignoniaceae (Mauritzon, 1935; Mehra and Kulkarni, 1985), Scrophulariaceae (Glišić, 1936-1937; Krishna Iyengar, 1947; Crété 1951b, 1955b; Banerji, 1961), Boraginaceae (Di Fulvio, 1981) and Campanulaceae (see Korobova and Zhinkina, 1987). As a result of comparative embryological investigations the Vicieae-type of endosperm development has been distinguished in the Fabaceae family (Kostrikova, 1984).

In various species of flowering plants endospermogenesis passes at a different rate: in some species the period from fertilization to complete formation of the endosperm takes several days (representatives of Asteraceae), in others several weeks (Poaceae) or even months (Chenopodiaceae).

---

[3]Variants of endosperm development within the types are variously named by different authors—types, forms, variations.

The sizes and structure of the endosperm in mature seeds of different plants also vary greatly. The sizes can be much larger than those of the embryo (representatives of Magnoliaceae, Ranunculaceae, Poaceae, etc.) or be about the same volume as that of the latter (Solanaceae). In many species of plants the endosperm is consumed entirely or partly during development of the embryo (representatives of Capparales, Fabales, Geraniales, Nelumbonales, Cucurbitaceae, etc.). The endosperm often has a flat surface and resembles the shape of the seed. Further, a so-called ruminate endosperm has been found. The latter is characterized by deep folds that appear as a result of irregular growing out of the seed coat forming outgrowths in the endosperm or as a result of irregular growing out of the outer layers of the endosperm itself forming outgrowths into the nucellus. A ruminate endosperm is typical of representatives of such families as Acanthaceae, Annonaceae, Arecaceae, Scrophulariaceae, etc. (Periasamy, 1962; Vijayaraghavan and Prabhakar, 1984; Danilova and Kirpichnikov, 1985; Danilova and Savchenko, 1985).

In some representatives of flowering plants the endosperm does not form at all (as a result of exclusion of the second link of double fertilization or for other reasons) or is very weakly developed. The endosperm is not formed in the Podostemaceae family (Teryokhin, 1991) nor in some representatives of Trapaceae (Titova, 1988, Kolesova [Titova] 1996) and Orchidaceae (Teryokhin and Kamelina, 1969, 1972; Savina, 1972; Poddubnaya-Arnoldi, 1976). In some orchids (for example, *Calanthe veitchii*) the polar nuclei do not typically fuse but together with the antipodal nucleus approaching them compose a polar-antipodal group located immediately under the egg cell. Still, most representatives of family Orchidaceae are characterized by triple fusion, though quite often the product of the fusion degenerates rather quickly without dividing. In those cases in which the primary endosperm nucleus undergoes divisions, the endosperm develops in accordance with the Nuclear type. The number of endosperm nuclei vary from 2 (*Cypripedium spectabile, C. parviflorum*, etc.) to 16 (*C. septentrionalis*) (Savina and Poddubnaya-Arnoldi, 1990).

According to the classical interpretation, the endosperm is a source of food for the developing embryo and a storage of nutrients (Hofmeister, 1849; Brink and Cooper, 1947; Zinger, 1958; Wunderlich, 1959; Poddubnaya-Arnoldi, 1964b, 1976; Gerassimova-Navashina, 1971a; Yoffe, 1971a,b; Raghavan, 1976; Sokolov *et al.*, 1980; Yakovlev, 1981). According to Wardlaw (1965), it is quite difficult for an embryo to take nutrients immediately from the nucellus and other surrounding tissues. The endosperm serves as a middleman between the mother organism and the embryo. The specific intermediary role of the endosperm is conditioned by its amphimictic nature. Participation of the male gamete in the endosperm rise makes the embryo and endosperm more compatible than the embryo and sporophyte tissue of the mother plant. Boesewinkel and Bouman (1984) share a similar point of view. In the course of development and accumulation of storage reserves the endosperm participates in a complicated interaction with the surrounding somatic tissues—the places of location of storage nutrients. On the example of some representatives of Scrophulariaceae and Orobanchaceae it was shown that the structural links in 'the nutrition chain' of the embryo sac of a developing seed can vary in different species (Nikiticheva, 1979b). Some researchers emphasize that the main role of the endosperm lies in providing successful proceeding not only of the process of growth and development of the embryo, but also its germination (Teryokhin and Nikiticheva, 1981). The classical interpretation that this role mainly consists of supplying the developing embryo with nutrients should be reviewed. In *Avicennia officinale* (the Verbenaceae), for example, no storage reserves whatsoever are found in the endosperm proper. This provides grounds for supposing that it serves only for transport of nutrients from the endosperm haustorium to the suspensor (Swamy and Padmanabhan, 1961). Recent investigations have shown that at least in the early period of embryogenesis when the division processes are activated in the endosperm, it does not play an important role in providing the embryo with food. In Krishnamurthy's opinion (1988), the endosperm of flowering plants controls the changes of symmetry of the embryo in the process of its development. A study of the comparative role of the embryo and endosperm of a seed in determining the size of seedlings showed that the quantitative index of the latter depends on the endosperm mass rather than on the size of the embryo of a seed (Zhang and Maun, 1989a,b).

Phylogenetic estimation of the types of endospermogenesis still causes rather serious difficulties. Some researchers (in fact, the majority) are inclined to consider the Nuclear type of endosperm development to be phylogenetically the initial (primary) one (see review by Wunderlich, 1959; Khudyak, 1963). Others, on the contrary, consider the Cellular type of endosperm development as ancestral (see review by Wunderlich, 1959 and also Modilewsky, 1953; Swamy and Ganapathy, 1957b; Di Fulvio, 1983; Friedman, 1994, 1995). However, another opinion holds that the Nuclear and Cellular types of endosperm development should be considered two equivalent initial types. The development of one from the other could pass in different groups of flowering plants in various directions (Palm, 1915; Schnarf, 1929, 1931; Kordyum, 1978). Regarding the Helobial type of development of endosperm some researchers (Sporne, 1954; Swamy and Ganapathy, 1957b) do not consider it be an independent one. Sporne refers it to the Nuclear type, while Swamy and Ganapathy consider the Helobial endosperm to be cellular and include it in the Cellular type of endosperm development. There are other opinions represented as well (see below: Evolutionary Aspects of Endosperm Development).

TYPES OF ENDOSPERM DEVELOPMENT

Nuclear Type of Endosperm Development

Nuclear (Latin *nucleus*—nucleus) type of endosperm development[4] is characterized by the fact that at the early stages of seed formation mitotic divisions of the primary endosperm nucleus and its daughter nuclei are not accompanied by cytokinesis (free-nucleate or coenocytic phase); later on the endosperm becomes cellular (plate XVI).

This pattern of endosperm development was described as free formation of cells (German: *freie Zellbildung*) (Hofmeister, 1859; Hegelmaier, 1886; Samuelsson, 1913). The term 'nuclear endosperm' was apparently introduced by Schnarf (1929).

Division of the primary endosperm nucleus usually passes ahead of division of the zygotic nucleus. Before the division has started the primary endosperm nucleus remains at the micropylar region or moves to the central (in Onagraceae, Vitaceae species) or chalazal regions of the primary endosperm cell (Capparaceae, Tiliaceae) and, in certain cases, to the specialized outgrowth of the cell (Fagaceae). The first divisions of the nuclei in the endosperm pass simultaneously in the course of several

[4]In embryological literature a contracted variant of this term is widely used—the nuclear endosperm.

mitotic cycles. In representatives of the Euphorbiaceae family, for example, divisions remain synchronous until the stage of 16- (*Chrozophora obliqua*—Kapil, 1956) or 32-nucleate endosperm (*Acalypha indica*—Johri and Kapil, 1953; *A. lanceolata*—Thathachar, 1952), and in the Poaceae family, even up to the stage of the 64-nucleate endosperm (*Triticum monococcum*—Batygina, 1974b).

While dividing the nuclei gradually move to the periphery of the coenocytic endosperm, and are pushed aside by a large central vacuole.[5] Separate nuclei of the endosperm are observed not along the cell wall, but in cytoplasmatic stretches which cross the large central vacuole (Yakovlev and Alimova, 1976). In the middle part of the endosperm the cytoplasm is restricted to a thin layer along the cell wall. In the subchalazal, chalazal and micropylar parts, the thickness of the cytoplasm layer and density of the nuclei distribution are higher than in the middle part. Along with the large central vacuole numerous, sometimes quite large, vacuoles are observed in the layer of the cytoplasm, adjacent to the cell wall.

The typical pattern of division of the nuclei at all stages of endosperm development and in all its parts is mitosis. Most endosperm nuclei retain the same ploidy, typical round form, size and structure (Sokolov, 1980). However, Schnarf (1929) described the nuclei as differing from typical ones by their size and form. Heteromorphism of the nuclei is usually observed after completion of the free nuclei formation. The endosperm nuclei of the chalazal part start to enlarge and exceed in size the nuclei located in the micropylar part. Rarely, for example in *Fagraea fragrans* and representatives of the Bombacaceae family, an inverse picture is observed, i.e., the micropylar nuclei are larger than the chalazal (Mohrbutter, 1937; Sokolovskaya, 1983). In representatives of family Cymodoceaceae heteromorphism of the nuclei is observed throughout the whole periphery of the coenocytic endosperm; here, nuclei of different size and shape are gathered in groups (Lakshmanan, 1970). In *Echinocystis macrocarpa* (Cucurbitaceae—Scott, 1944) all endosperm nuclei are initially 7-10 mkm in diameter and contain from 1 to 3 nucleoli. Later, some greatly enlarge, reaching 150-200 mkm in diameter. The number of nucleoli in such nuclei also increases.

Observed in normal conditions of seed development the heteromorphism of nuclei not only in size and shape, but in rate of ploidy and structure, is attributed to the chalazal and to a lesser extent micropylar parts of the endosperm. So-called functional abnormalities of the mitosis underlie the heteromorphism of nuclei. Large highly polyploid nuclei of irregular shape appear in these parts of the endosperm as a result of abnormal mitoses; such nuclei form due to endomitosis, formation of restitutional nuclei, fusion of the division spindles and union of the non-sister telophatic nuclei (Petrova, 1968, 1977; Sokolov, 1968, 1975, 1978a,b, 1980; Yoffe, 1971a).

The endosperm cytoplasm at its coenocytic stage of development contains the typical set of cell organelles. The presence of numerous plastids, mitochondria and abundant polysomes has been reported. Opposite the embryo, which is chlorophyllous in approximately 40% of flowering plants (Zhukova, 1992b), the endosperm is characterized by inability to synthesize chlorophyll and, consequently, to differentiate chloroplasts. The latter have been detected only in representatives of Brassicaceae and Resedaceae (see Table 3)[6]. In the greater majority of flowering

[5]In some species the central vacuole is not observed (Oehler, 1927; Mendes, 1941; also see the illustration in the paper by Nijalingappa and Devaki, 1978).

[6]It is notable that chloroplasts in Brassicaceae and Resedaceae are observed not only in the endosperm itself, but also in its mother cell—the central cell of the embryo sac.

Table 3: List of flowering plant families in which species are characterized by the occurrence of chlorophyll in the endosperm

Families, species	References
Brassicaceae*	
Aubrieta hybrida Hort.	Zhukova, original data
Brassica juncea (L.) Czern.	Yoffe, 1952
Capsella bursa-pastoris (L.) Medik.	Schulz P. and Jensen, 1969, 1974
Diplotaxis erucoides DC.	Pacini *et al.*, 1975/1976
Hesperis matronalis L.	Yoffe, 1952
Iberis umbellata DC.	Zhukova, original data, see Plate XVI, *45, 46*
Matthiola annua Sweet.	Yoffe, 1952
Raphanus sativus L.	Yakovlev and Zhukova, 1973
	Qin De-bo and Harberd, 1988
Raphanus sativus var. *radicula* Pers.	Yoffe, 1952
Sinapis alba L.	Yoffe, 1952
Resedaceae*	
Reseda lutea L.	Chaban and Yakovlev, 1974
Reseda odorata L.	Yakovlev and Zhukova, 1973

*The first mention of the occurrence of 'amorphous' chlorophyll in the endosperm of Brassicaceae and Resedaceae without presentation of the species studied appears in a work by Hofmeister (1858).

plants, the plastid apparatus of the endosperm (not only at the coenocytic, but also at the cellular stage of development) is represented by leucoplasts. Leucoplasts are the differentiated metabolically active organelles characterized by some heteromorphism at the species level (Zhukova, 1988, 1992b). In the primary endosperm cell, plastids are concentrated around the nucleus (Schulz P. and Jensen, 1973). In *Petunia hybrida,* plastids in this period are small and devoid of starch grains (Van Went, 1970b,c), though in the central cell of the mature embryo sac they are starchy. Later, amyloplasts are observed in the endosperm cytoplasm already during the division of its primary nucleus (*Triticum monococcum*—Batygina, 1974b) or at the two-nucleate stage of development (*Gossypium hirsutum*—Schulz P. and Jensen, 1977; *Spinacia oleracea*—Wilms, 1981c; *Valeriana officinalis*[7]—Zhukova, 1988). In *Capsella bursa-pastoris,* plastids of the endosperm are represented by chloroplasts, though at the two-nucleate stage of development there is no starch in these plastids (Schulz P. and Jensen, 1977). In the successive phases of the coenocytic stage of development, the endosperm of most taxa, usually does not accumulate these reserve substances (Newcomb, 1973; Zhukova, original data—see Plate XVI, *35-37*). In wheat, on the contrary, the occurrence of reserve starch has been reported in the multinucleate coenocyte of the endosperm as well (Batygina, 1974b, 1987a). A considerable quantity of large lipid bodies is observed in the free-nucleate endosperm of *Paeonia peregrina* at the stage of the four-nucleate proembryo (Zhukova, original data—see Plate XVI, *47, 48*) and in the free-nucleate endosperm of *Alcea rosea* at the early globular stage of embryo development (Zhukova, original data—see Plate XVI, *35-37*). Accumulation of the polysaccharide grains is a characteristic of the endosperm at the cellular stage of its development (Newcomb, 1973; Vijayaraghavan and Prabhakar, 1984).

Starting from the very early stages of development of the free-nucleate endosperm a great amount of membranes of rough ER can be seen (Orchidaceae—

[7]In *Valeriana officinalis* at the two-celled stage of development also (Zhukova, 1988).

Cocucci and Jensen, 1969a,b; *Petunia hybrida*—Van Went, 1970c; *Hibiscus*—Ashley, 1975a,b; *Gossypium hirsutum*—Schulz P. and Jensen, 1977; *Paeonia peregrina*—Zhukova, original data—see Plate XVI, *47*). In the course of further development of the free-nucleate endosperm, the quantity of ER membranes increases and large stacks of rough ER are formed (*Gossypium hirsutum*—Schulz P. and Jensen, 1978; *Alcea rosea* — Zhukova, 1988). In *Alyssum maritimum* (Vijayaraghavan and Prabhakar, 1984) and *Quercus gambelii* (Singh and Mogensen, 1976) separate sections of the cytoplasm are surrounded by numerous cisternae of the rough ER. In *Valeriana officinalis* already at the two-nucleate stage of endospermogenesis long concentrically located profiles of rough ER are observed around vacuoles (Kamelina and Zhukova, 1987; Zhukova, 1988). Early stages of nuclear endosperm development are characterized by increased activity of the Golgi apparatus (*Zea mays*—Diboll, 1968; *Petunia hybrida*—Van Went, 1970b,c; *Gossypium hirsutum*—Schulz P. and Jensen, 1977; *Triticum aestivum*—Shmaraev, 1983; *Alcea rosea*—Zhukova, 1988; *Paeonia peregrina*—Zhukova, original data—see Plate XVI, *48*). The tonoplast of the large central vacuole of the endosperm is much more electron dense compared with other cytoplasmic membranes (Pacini *et al.*, 1975/1976; Zhukova and Savina, 1978; Savina and Zhukova, 1983).

According to the data of electron microscopy, the cell wall of the coenocytic endosperm has numerous outgrowths (Vijayaraghavan and Prabhakar, 1984 (review); Hu Shi-yi *et al.*, 1983; Tilton *et al.*, 1984; Oryol and Shmaraev, 1987; Bhandari and Anuradha Mal, 1989). It is only in certain species that such outgrowths of the cell wall are observed in the central cell of the mature embryo sac before fertilization and beginning of seed development (*Linum usitatissimum*—Vazart B. and Vazart J., 1966; *Helianthus annuus*—Newcomb and Steeves, 1971; *Lobelia dunnii*—Torosian, 1971).

In most taxa, after a certain number of free nuclei are formed, the cellular phase follows the initial coenocytic phase. In some taxa, the endosperm remains free-nucleate during the whole period of its existence, without passing to the cellular phase of development. This was observed in some representatives of families such as Limnanthaceae (Guignard, 1893; Maheshwari and Johri, 1956; Mathur, 1956), Juncaginaceae (Campbell, 1898; Hill, 1900), Aceraceae (Guérin, 1901), Tamaricaceae (Frisendahl, 1912; Puri, 1939), Melastomataceae (Subramanyam, 1951), Fabaceae (Rau, 1953; Kostrikova, 1955, 1956, 1984; Yoffe, 1957), Chrysobalanaceae (Tobe and Raven, 1984), Zosteraceae (Teryokhin, 1990b), and species of such genera as *Lopezia, Stenosiphon* (Onagraceae), *Cardiospermum* (Sapindaceae), *Tropaeolum* (Tropaeolaceae), *Melastoma* (Melastomaceae) (see Schnarf, 1931; Maheshwari, 1950; Poddubnaya-Arnoldi, 1982). In the Fabaceae family absence of the cellular phase in the course of endospermogenesis of the Nuclear type is typical for all representatives of the tribe Vicieae but only for one representative of tribe Hedysareae—*Stylosanthes mucronata* (Kostrikova, 1984).

The number of free-nucleate divisions before the initiation of cell wall formation varies in different taxa. In some taxa it is small and leads to the formation of 4-nucleate (*Coffea arabica*—Mendes, 1941) or 8-16-nucleate coenocyte (*Asclepias cornuti*—Frye, 1902; *Rafflesia*—Ernst and Schmid, 1913; *Leiphaimos, Cotylanthera*—Oehler, 1927; *Calotropis procera*—Sabet, 1931; *Xeranthemum*—Poddubnaya-Arnoldi, 1931; *Crepis capillaris*—Gerassimova, 1933; *Diplacrum caricinum*—Nijalingappa and Devaki, 1979). In other taxa free-nucleate divisions finish with formation of the coenocyte with several hundreds and even thousands of nuclei. So, before commencement of cytokinesis up to 1000 nuclei form in the primary endosperm cell of *Mimusops elengi* (Bhatnagar and Gupta, 1970), *Cuscuta hyalina* and *C. planiflora* (Tiagi, 1951a). In some species the process of cellularization is initiated after formation of about 1500 nuclei (*Cuscuta rostrata*—Smith, 1934), 2000-3000 (*Gossypium arboreum* var. *Bhoj*, *G. herbaceum* var. *jayadhar*—Joshi *et al.*, 1967) and more (representatives of Connaraceae—Mauritzon, 1939). The multinucleate coenocytic phase is characteristic of endospermogenesis of some genera of such families as Oliniaceae, Onagraceae, Brassicaceae, Capparaceae, Connaraceae, Primulaceae, Sapotaceae and others (see Poddubnaya-Arnoldi, 1982).

In Centrospermae depending on the pattern of transition of the free-nucleate stage of development into the cellular Silene-, Melandrium- and Heliosperma-types of cellularization were distinguished (Rocén, 1927). In species characterized by a long-term free-nucleate stage, cellularization in different regions of the endosperm does not pass simultaneously, as a result of which free-nucleate and cellular zones are observed in the endosperm of such species for some time (Alexandrov and Alexandrova, 1939; Yakovlev and Yoffe, 1965; Tiagi, 1970; Batygina, 1974b; Bannikova, 1975). A non-simultaneous transformation of the endosperm from free-nucleate organization to cellular is the consequence of the differentiated state of the multinucleate coenocyte (the latter consists of parts with dense and sparse distribution of nuclei and zones of normal and abnormal mitosis) (Sokolov, 1980). If cellularization takes place at the period when the endosperm contains only four (for example, in *Coffea arabica*—Mendes, 1941) or eight nuclei (for example, in *Leiphaimos* sp.—Oehler, 1927) and there is no vacuole in it, the whole endosperm becomes cellular at once. The direction of the cell walls establishing also varies. The process of cellularization can initiate either from the periphery of the coenocyte and move towards the centre or from one of its poles in the direction of the longitudinal micropylar-chalazal axis of the embryo sac. At the second variant of sequence of events, cytokinesis can start from the micropylar end of the endosperm becoming evident first around the embryo and extending gradually towards the chalazal end or vice versa. Less commonly, cell walls form simultaneously in all parts of the endosperm. In any of the aforesaid patterns of cellularization, the process of cytokinesis once started proceeds either up to total completion (when the endosperm on the whole becomes cellular, which is observed in the majority of plants) or does not cover all of the coenocytic zones, due to which the micropylar, chalazal and central regions of the endosperm remain in the free-nucleate stage.

In some taxa the free-nucleate region can transform into a haustorium, which is very variable in structure. Sometimes endomitotic polyploidization of nuclei takes place in it. In the Nuclear type of endosperm development haustoria appear mainly at the chalazal pole of the embryo sac (Proteaceae—Kausik, 1938a,b; 1942; Fabaceae—Rau, 1950; Aizoaceae—Ilyina and Poddubnaya-Arnoldi, 1983; Cucurbitaceae—Dzevaltovsky, 1983; Linaceae—Nikiticheva, 1985b; Polemoniaceae—Zhukova, 1987; Elaeagnaceae—Kamelina and Proskurina, 1987). In most species the haustoria are relatively short. The longest coenocytic chalazal haustoria have been found in families Euphorbiaceae and Cucurbitaceae. Thus in *Croton* species the length of haustoria is 1000 mkm, in *Cucurbita vicifolia* 12,000 mkm. In a number of species chalazal haustoria become cellular at the heart-shaped or early torpedo-like stage of embryo development (Dzevaltovsky, 1983; Kamelina and Proskurina, 1987). Micropylar endospermal haustoria have been observed in *Cyanotis tuberosa* (Chikkannaiah and Hemaraddi, 1979), in representatives of the Frankeniaceae family (Walia and Kapil, 1965; Kamelina, 1983) and in *Floerkea proserpinacoides* (Maheshwari and Johri, 1956). In *Coronaria flos-cuculi* (Caryophyllaceae) (Meyer, 1962) and in *Scleria foliosa* (Cyperaceae) (Nijalingappa and Devaki, 1979) endospermal haustoria were noted both at the micropylar and chalazal poles of the endosperm, and in *Acanthophyllum glandulosum* (Caryophyllaceae) numerous massive haustoria of various shapes were found in the lateral region of the endosperm (Turssunov and Savinova, 1979). Families in which an endospermal haustorium is a characteristic of only certain species are known. Thus, in Euphorbiaceae the 'aggressive' chalazal coenocytic haustorium is described in *Croton* species only (Kapil and Bhatnagar, 1972). Among representatives of Papaveraceae a correlation was found between the degree of development and duration of activity of antipodals and structure of the endosperm in its chalazal region: in the plants with long functioning antipodals the endosperm does not form a haustorium; on the contrary, in species with poorly developed and early degenerating antipodals the chalazal zone of the endosperm forms a haustorium at the free-nucleate stage of development (*Platystemon* and *Eschscholzia*—Ilyina, 1981). Heteromorphism of the endosperm at the cellular stage of its development does not become exhausted there. Often some of the initially one-nucleate cells of the endosperm become multinucleate due to repeated divisions of their nuclei without further cytokinesis. As a result of this process secondary coenocytes are formed. In some taxa, nuclei and cells of the endosperm entirely fill its cavity while in others more often its central part remains unfilled. One of the features of morphological differentiation of the nuclear endosperm at the cellular phase of its development is the occurrence of so-called 'vesicles'—globular evaginations of the cytoplasm which start to form as early as the coenocytic phase of endospermogenesis. They have been described in representatives of Fabaceae (Hegelmaier, 1880), Musaceae (Juliano and Alcala, 1933), Poaceae (Khudyak, 1952, 1955), Stackhousiaceae (Narang, 1953), Apiaceae (Singh and Gupta, 1956; Sehgal, 1965; Kordyum, 1967), Elaeagnaceae (Kamelina and Proskurina, 1987) and also Balsaminaceae, Brassicaceae, Cucurbitaceae, Papaveraceae and other families (see Khudyak, 1963; Grevtsova, 1987). According to Juliano and Alcala (1933), some nuclei of the free-nucleate endosperm divide more actively than others, forming separate groups of nuclei, i.e., 'nodules'. The latter gradually become isolated from the whole mass of the endosperm cytoplasm and take the shape of globes. A number of authors have described similar cytoplasmic globes without nuclei. Existing contradictions regarding the structure of the cytoplasmic globes are sometimes explained by the methodological difficulties of carrying out endosperm investigation. In serial transverse sections of the endosperm the nuclei in globes are seen only in median sections of the latter (Batygina, 1974b; Poddubnaya-Arnoldi, 1976). The role of these formations in seed development remains unclear.

As shown above, the beginning of cytokinesis is not related to any definite number of preceding free-nucleate divisions. Nor has any strict correlation been noted between transformation of the coenocytic stage to the cellular one with any certain stage of embryogenesis. Thus in *Macleaya* formation of the cell walls in the endosperm starts long before initiation of zygotic division (Ilyina, 1981). Examples are known of the beginning of cytokinesis in the endosperm in the proembryo phase of embryogenesis too: at the stage of the 3-4 celled (*Bupleurum tenue*—Gupta S. and Gupta M., 1964; *Seetzenia orientalis*—Narayana and Rao, 1963), 4-6-celled (*Phoenix pusilla*—Biradar, 1968; *Triticum monococcum*—Batygina, 1974b), 8-10-celled (*Ph. acaulis, Ph. reclinata*—Biradar, 1968; *Actinophloeus macarthurii*—Rao, 1959) and 16-celled proembryo (*Delphinium* species—Babis, 1973). The first cell walls can appear in the endosperm at later stages of embryogenesis as well—at the globular (*Peganum*—Kapil and Ahluwalia, 1963; *Phoenix sylvestris*—Mahabalé and Biradar, 1968; *Crocus thomasii*—Chichiriccó, 1989; *Glycine max*—Chamberlin and Horner, 1989), heart-shaped (*Diplotaxis erucoides*—Pacini *et al.*, 1975 / 1976; *Plumbago europaeum*—Korobova, 1983) and late heart-shaped stage (*Capsella bursa-pastoris*—Schulz P. and Jensen, 1974). Ultrastructural changes taking place in the endosperm in the period of transition from the nuclear state to the cellular one have been described using as examples *Helianthus annuus* (Newcomb, 1973), *Triticum aestivum* (Mares *et al.*, 1977), *Reseda lutea* (Chaban, 1979), *Ipomoea purpurea* (Ponzi and Pizzolongo, 1984), *Phaseolus vulgaris* (Yeung and Cavey, 1988; XuHan and Van Lammeren, 1994), *Glycine max* (Dute and Peterson, 1992), *Ranunculus sceleratus* (XuHan and Van Lammeren, 1993) and others.

The mechanisms of formation of the cell walls between the nuclei of the coenocyte at its transition to the cellular stage of endospermogenesis still remain debatable. Some authors believe cellularization passes uniformly in all parts of the endosperm —by forming cell plates (by means of the phragmoplast) (Jungers, 1931; Yakovlev and Yoffe, 1965; Petrova, 1968, 1977; Sokolov, 1968, 1980; Batygina, 1974b, 1987a; Bannikova, 1975; Fineran *et al.*, 1982; Shmaraev, 1983; Oryol and Shmaraev, 1987). Other researchers believe that in the endosperm there are patterns of cellularization not connected with mitosis, including division by constriction (Frye, 1902; Schnarf, 1926, 1929; Modilewsky, 1953; Rao, 1959; Olszewska and Gabara, 1966; Morrison and O'Brien, 1976; Mares *et al.*, 1977; Morrison *et al.*, 1978; Chaban, 1979; Chitralekha and Bhandari, 1992; Brown *et al.*, 1994). The data obtained about the ultrastructure of the *Valeriana officinalis* endosperm at the early stages of its development allow us to share the opinion of the second group of authors (Zhukova, 1986; Kamelina and Zhukova, 1987). The supposition of Rao (1959) and Mares with co-authors (1977) seems well grounded, namely, at least initiation of cytokinesis in the endosperm of the Nuclear type of development is connected with 'free', centripetal growth of cell walls between the nuclei of the by-wall layer of the cytoplasm of the endospermal coenocyte.

A positive though not strict correlation has been established between the Nuclear type of endospermogenesis with occurrence of chlorophyll in the embryo (Zhukova, 1990, 1992a,b). The latter is found in species with large embryo which is differentiated into organs (Zhukova, 1992b, 1993). Thus, endospermogenesis of the Nuclear type appears to be correlatively linked not only with chlorophyll synthesis in the embryo, but probably with formation of a large embryo in the seed.

Cellular Type of Endosperm Development

Cellular (Latin *cellula*—cell) type of endosperm development is characterized by synchronous proceeding of karyo- and cytokinesis processes starting from the first divisions in the primary endosperm cell **(plate XVII)**.

This pattern of endosperm development was described as successive cell divisions (German: *sukzessive Zellteilung*) (Hofmeister, 1859; Hegelmaier, 1886; Samuelsson, 1913).

By the time of division the nucleus is most often located at the micropylar or central part of the primary endosperm cell. The division is usually transverse, rarely longitudinal or oblique.

Cellular type of endosperm development has been observed mostly in dicotyledons but also in a number of monocotyledons (Araceae—Goldberg, 1941; Banerji, 1947; Burmanniaceae—*Thisma luetzelburgii, Th. panamensis*—Teryokhin, 1990a; Lemnaceae—Maheshwari S. and Kapil, 1963). However, some authors cast doubt on its presence in monocotyledons (Lawalrée, 1952; Maheshwari S. and Khanna, 1956; Parameswaran, 1959).

Unusual features have been found in the endospermogenesis of some parasitic flowering plants. Thus, in representatives of family Loranthaceae (Nikiticheva and Anisimova, 1987), characterized by reduced ovules, a compound endosperm forms through union of tissues of the endosperm which develops simultaneously in a number of embryo sacs after the ovary cells located between them have been destroyed. In this case several embryos start to develop, though only one, less commonly two mature. In *Viscum album*, fusion of the endosperm of two neighbouring embryo sacs has been observed (Teryokhin, 1987c).

The fully developed endosperm in many taxa is heteromorphous, which manifests in particular in the presence of haustoria often with polyploidized nuclei (see D'Amato, 1984). Besides the micropylar and chalazal haustoria, there is a distinct specialized tissue (e.g. Orobanchaceae and Scrophulariaceae families), an 'isthmus' in the endosperm—a narrow part connecting the micropylar haustorium with the central body of the endosperm (Krishna Iyengar, 1947; Tiagi, 1951b; Teryokhin and Nikiticheva, 1981). In species of the Orobanchaceae family the cells of the 'isthmus' are arranged in two tiers. They are not large, are filled with dense cytoplasm and apparently participate in supplying substances from the micropylar haustorium to the cells of the endosperm body (Tiagi and Sankla, 1963; Nikiticheva and Teryokhin, 1976; Teryokhin and Nikiticheva, 1981; Anisimova, 1983). The same topography of the endosperm is characteristic of a number of other taxa (Buddlejaceae—Dop, 1913; *Azorina vidalii* from Campanulaceae—Shamrov and Zhinkina, 1994). In the last case the endosperm 'isthmus' is represented by 2-4 tiers of cells stretched along the longitudinal axis of a seed.

Cells of the developing endosperm, especially of its peripheral layer, form secondary wall thickenings (usually the outer tangential and radial cell walls thicken). Ultrastructural and histochemical investigations of the Cellular endosperm are few and mainly devoted to the study of haustoria which play an important role in the nutrition of the endosperm and embryo. In cells of the endosperm under formation a small amount of starch, proteins, and lipids accumulates which is usually consumed in the course of development, especially if the endosperm is transitory. In case the endosperm should remain in a mature seed, reserve nutrients accumulate at the late stages of development. In some taxa the nutritive substances are represented by starch only (Callitrichaceae, Davidiaceae, Hippuridaceae), in others by starch and proteins (Chloranthaceae), by lipids and proteins (Adoxaceae, Cynomoriaceae, Gesneriaceae, Magnoliaceae, Styracaceae), by starch and lipids (Buddlejaceae, Lardizabalaceae, Viscaceae), by proteins (Gunneraceae, Saururaceae), or by lipids (Ebenaceae, Fouquieriaceae, Morinaceae, Symplocaceae, Winteraceae) (see 'Comparative Embryology of Flowering Plants', 1981, 1983, 1985, 1987).

The Cellular type of endosperm development is characterized by a variety of patterns, classified for the first time as a system by Samuelsson (1913). The system was based on orientation of the laying down of the first and second cell walls and the degree of participation of the first two cells in construction of the endosperm of the seed. Two groups are distinguished according to the disposition of the cell wall during division of the primary endosperm cell: the cell wall is longitudinal (I) or transverse (II). The second group is represented by the **A** and **B** series. **Series A:** both cells participate in formation of the endosperm of the seed: **a**—orientation of the subsequent cell walls is variable; **b**—the cells divide transversely, and **c**—in both cells divisions are longitudinal. **Series B:** only one of the cells forms the endosperm of the seed, while the other functions as a haustorium: **a**—the micropylar cell divides transversely and as a result forms a tier of three cells, of which the central one forms the endosperm of the seed and the terminal cells form haustoria; **b**—the upper cell forms the endosperm of the seed and the lower one the haustorium; and **c**—the lower cell forms the endosperm of the seed and the upper the haustorium.

The criteria which Samuelsson (1913) used for distinguishing groups and series have been used and developed further in most existent classifications. However, no single classification of the modes of development of the Cellular endosperm has been compiled to date. Various authors have defined these modes as types (Mauritzon, 1935; Rosén, 1949; Govindu, 1950; Wunderlich, 1967; Polyshchuk and Dzevaltovsky, 1971; Di Fulvio, 1983; Korobova and Zhinkina, 1987), forms (Schnarf, 1929; Glišić, 1936-1937) and variations (Kapil and Vijayaraghavan, 1962, 1965; Shamrov and Zhinkina, 1994). Analysis of the literature indicates that the specificity of development in the Cellular endosperm manifests only in the orientation of the first and several subsequent cell walls in early endospermogenesis. These constitute modifications within one type. They should be regarded as variations in the manner accepted in classifications of other generative structures (the embryo—Johansen, 1950; the microsporangium wall—Teryokhin *et al.*, 1993; the embryo sac—Teryokhin, 1996; nucellus, chalaza, obturator—Shamrov, 1997a, 1998b).

One of the first classifications, used more often than others in current literature, was proposed by Schnarf (1929) who examined all the variations (forms—according to the author) within five groups. Later, a great number of investigations on the characteristics of these variations and their presence in various flowering plants were published, so we believe it reasonable to include them in Schnarf's classification (1929).

The variations of Schnarf's classification are clustered in five groups.

The first group is characterized by **variability in orientation of laying down the first cell wall**—from the transverse through oblique to longitudinal: Diapensiaceae (*Diapensia*—Samuelsson, 1913), Valerianaceae (*Centranthus*—Asplund, 1920) and Asteraceae (*Senecio*—Afzelius, 1924).

The second group includes the following **variations characterized by longitudinal** laying down of the first cell wall. **Adoxa-variation:** the second and sometimes a number of subsequent divisions pass in the longitudinal plane (Adoxaceae—Lagerberg, 1909; Circaeasteraceae—Junell, 1931; some Chloranthaceae—Yoshida, 1957, 1959; Piperaceae—Jöhnson, 1900b; Nikiticheva, 1981a,b). **Lappula-variation:** the first cell wall is not completed and finishes at the boundary with the large central vacuole, while the second and subsequent divisions are oriented in various planes (Dipsacaceae - Doll, 1927; Kamelina, 1980, 1987; some Boraginaceae—Svensson, 1925; Swamy and Parameswaran, 1963).

In the third and fourth groups, orientation of **the first cell wall** in the endosperm is **transverse.** In the third group variations the second division in both or at least in one of the cells passes longitudinally. **The Scutellaria-variation:** the second division in both cells of the endosperm is orientated in the longitudinal plane (Lamiaceae—Schnarf, 1917; Polyshchuk and Dzevaltovsky, 1971; Gesneriaceae—Arekal, 1961; Alimova and Yakovlev, 1982; Hippuridaceae, Lentibulariaceae—Nikiticheva, 1987a,b; Vahliaceae—Mauritzon, 1933). **Prunella-variation** (=Brunella)—the micropylar cell divides longitudinally, while in the chalazal cell only karyokinesis can take place (Lamiaceae—Schnarf, 1917; Boraginaceae—Toks, 1976; Orobanchaceae—Tiagi, 1951b; Teryokhin and Nikiticheva, 1981; Anisimova, 1983; Teryokhin, 1987b; Plantaginaceae

—Souèges, 1923b; Scrophulariaceae—Meunier, 1897; Gscheidle, 1925; Nikiticheva, 1987c).

The fourth group includes **variations** which are characterized by **the second transverse division** in both or at least in one of the cells. **Ericaceae-variation:** the second division in both cells is transverse while subsequent divisions pass in various directions (Cornaceae—Alimova, 1987; Davidiaceae—Kamelina and Shevchenko, 1987; Empetraceae, Epacridaceae—Samuelsson, 1913; Ericaceae, Pyrolaceae—Samuelsson, 1913; Yacobson, 1976; Teryokhin, 1977). **Annona-variation:** the second and a number of subsequent divisions are transverse, which results in construction of a tier of 8-12 cells; further divisions are longitudinal (Annonaceae—Nicolosi-Roncati, 1903; Aristolochiaceae—Jacobsson-Stiasny, 1918; Nair and Narayanan, 1961; Marcgraviaceae—Mauritzon, 1939; Menyanthaceae—Stolt, 1921; Sarraceniaceae—Shreve, 1905). **Callitriche-variation:** the second division passes transversely only in the micropylar cell, after which in all the cells divisions are oriented in the longitudinal plane; the terminal cells are differentiated into micropylar and chalazal haustoria (*Callitriche autumnalis* – Jörgensen, 1923). However, in *C. vernalis,* according to data provided by Souèges (1952), the second division is longitudinal in both cells and then the cells divide transversely, i.e., development of the endosperm proceeds in compliance with the Scutellaria-variation. **Ceratophyllum-variation:** the second and a number of further transverse divisions pass only in the micropylar cell, while the chalazal cell does not divide further (*Ceratophyllum submersum*—Strasburger, 1902). However, according to our investigations (Shamrov and Batygina, 1984) in all the studied species of *Ceratophyllum* (*C. demersum, C. pentacanthum, C. submersum*), the chalazal cell is divided by the transverse wall and so the first stages of their endospermogenesis pass in concordance with the Ericaceae-variation.

Development of the endosperm in **the fifth group** is characterized by **laying down of the first cell wall in an oblique position.** Further divisions in both cells are oriented perpendicular to the first cell wall (Boraginaceae—Svensson, 1925). It should be noted that in representatives of several families (Cynomoriaceae—Teryokhin *et al.*, 1975; Morinaceae—Kamelina, 1980) the division is oriented across the primary endosperm cell and the forming cell walls are located obliquely with repeat to the first; the nuclei of both cells divide synchronously.

The classification by Schnarf (1929) was later supplemented with new variations. In family Lamiaceae, apart from Scutellaria- and Prunella-variations, three more variations (= types) have been distinguished: Galeopsis, Stachys (Wunderlich, 1967) and Lavandula (Polyshchuk and Dzevaltovsky, 1971). **Galeopsis-variation:** differs from Prunella-variation by weak development and early degeneration of the chalazal cell which functions as a haustorium (Galeopsis—Schnarf, 1917). **Stachys-variation:** the first division is transverse, then the micropylar cell divides obliquely and the chalazal cell becomes a haustorium without division (*Ajuga, Marrubium, Stachys*—Schnarf, 1917). **Lavandula-variation:** the micropylar cell transforms into a haustorium without division and the chalazal cell divides transversely (*Lavandula* – Polishchuk and Dzevaltovsky, 1971).

For the Scrophulariaceae family a classification of 13 variations (forms, according to Glišić, 1936-1937) was provided. **The first cell wall** in the endosperm is always **transverse.** Depending on the laying down of the second and following walls, the period of differentiation and structure of haustoria, the variations are combined in a number of groups. The author has proposed the scheme of interrelation of different variations of development of the cellular endosperm in family Scrophulariaceae. Later, with Glišić ideas taken as a basis, other systems of variations of development of the endosperm were proposed for this family (Krishna Iyengar, 1947; Crété, 1955a; Banerji, 1961; Afanasyeva, 1976).

The same principles were used by Rosén (1949) who distinguished four variations (types—according to the author) of endospermogenesis of the cellular type in representatives of Campanulaceae, Lobeliaceae and Stylidiaceae. Two variations (Codonopsis-, Sphenoclea-), similar to the Scutellaria-variation described by Schnarf (1929), are characterized by laying down of the first transverse and second longitudinal cell walls during early endospermogenesis. Here, the first four cells formed have the shape of an isobilateral tetrad. **Sphenoclea-variation:** the haustoria are four-celled and differentiated after the fourth division (Campanulaceae—*Sphenoclea*—Kausik and Subramanyam, 1945b). **Codonopsis-variation:** the haustoria are two-celled, the micropylar haustorium separates after the third and the chalazal one after the fourth division (Campanulaceae—*Codonopsis,* some *Campanula, Cyananthus;* Lobeliaceae—*Downingia, Lobelia, Pratia*—Rosén, 1932; Kausik and Subramanyam, 1945a; Stylidiaceae—Rosén, 1949). In the case of the two other variations (Phyteuma- and Isotoma-variations), the first division in the endosperm is transverse, then the micropylar cell divides longitudinally and the chalazal transversely, which results in formation of a T-shaped tetrad. **Phyteuma-variation:** the chalazal (one-celled) and micropylar (two-celled) haustoria become differentiated after the third division while the upper derivative of the chalazal cell undergoes no division (Campanulaceae—*Adenophora, Musschia, Specularia;* Lobeliaceae—*Laurentia*—Rosén, 1949). **Isotoma-variation:** in contrast to the Phyteuma-variation the upper derivative of the chalazal cell is first divided with transverse cell walls several times, and then with longitudinal ones (Lobelioideae—*Isotoma longiflora*—Kausik and Subramanyam, 1945b). In family Campanulaceae two more variations are distinguished. **Pentaphragma-variation:** the first division in the endosperm is transverse , then the micropylar cell becomes a haustorium without division, while the chalazal one undergoes several divisions forming the endosperm proper (*Pentaphragma horsfieldii*—Kapil and Vijayaraghavan, 1962, 1965). **Azorina-variation:** the first and second divisions are transverse and as a result a linear tetrad of cells forms, as in the Ericaceae-variation; the chalazal haustorium (one-celled) separates after the second division and the micropylar (two-celled) after the fourth (*Azorina vidalii*—Shamrov and Zhinkina, 1994). According to the data obtained, a modification of Rosén's scheme (1949) on the distribution of different variations (=types) of endospermogenesis of the cellular type was proposed for families Campanulaceae and Lobeliaceae (Korobova and Zhinkina, 1987).

Two special variations (= types) of development of the Cellular endosperm are distinguished in the Bignoniaceae family. **Catalpa-variation** is a characteristic of most species of this family. The first division is transverse, then in both cells longitudinal divisions take place, as a result of which 8 cells arranged in two tiers arise. Cells of the lower tier form the chalazal haustorium while cells of the upper tier give rise to the endosperm proper (Govindu, 1950). In the **Incarvillea-variation** the chalazal cell of the two-celled endosperm becomes a haustorium without division and the micropylar cell gives rise to the endosperm proper and a one-celled micropylar haustorium (*Incarvillea, Amphicome*—Mauritzon, 1935).

There are taxa in which development of the endosperm according to the Cellular type does not conform to the proposed variations. In a number of plants (representatives of Icacinaceae—Swamy and Ganapathy, 1957a; Saururaceae—

Jöhnson, 1900a; Raju, 1961; many Nymphaeaceae—Cook, 1902, 1906; Meyer, 1960; Batygina *et al.*, 1980, 1991; Winter, 1987), the first division in the endosperm is transverse and the lower cell transforms into a 'tube-like' chalazal haustorium and on the basis of the micropylar cell the endosperm proper forms.

A specific development of the endosperm is found in family Acanthaceae. As a result of the first transverse division two cells form, of which the chalazal one typically transforms into a haustorium. However, in *Thunbergia* it is the micropylar cell that becomes a haustorium, while the lower one gives rise to the endosperm proper, the secondary haustorium arising in the late stages of development (Mohan Ram and Wadhi, 1964). In many representatives of this family the micropylar cell divides further, separating the micropylar haustorium and the cell of the endosperm proper, which from the very beginning may be cellular (*Acanthus, Peristrophe*) or free nuclear (*Barleria, Chamaeranthe, Ruellia, Schaueria*) (Mohan Ram, 1962; Karlström, 1974a,b; Madhavan and Gupta, 1982).

The functions and interrelations of the Cellular endosperm with endosperms of other types of development in the process of evolutionary formation are widely discussed in the literature (Friedman, 1992a,b, 1994, 1995; see Endosperm above).

Helobial Type of Endosperm Development

Helobial type of endosperm development is characterized by formation of two cells (micropylar and chalazal) from the primary endosperm cell; in them nuclear divisions are initially not accompanied by cytokinesis. Synonym: intermediate type of endosperm development **(Plate XVIII)**.

This type was so named because it was first discovered in monocotyledonous plants of the Helobiae group. Development of the endosperm according to the Helobial type was first described in *Alisma plantago* and *Sagittaria variabilis* (Schaffner, 1896, 1897). Apparently, the term was introduced by Schnarf (1929).

The Helobial type of endosperm development is typical for many monocotyledonous plants and only for some dicotyledons, including Nymphaeaceae (Khanna, 1967), Olacaceae (Agarwal, 1963), Santalaceae (Paliwal, 1956; Bhatnagar, 1959, 1960), Saxifragaceae (Rau and Sharma, 1967; Rau, 1970). Reinvestigation of representatives of the last three families raised doubts that their endosperm is characterized by development of the Helobial type (Swamy and Krishnamurthy, 1973). Discussion has also taken place as to whether an endosperm of this type occurs in families Cabombaceae (Cook, 1906; Khanna, 1965; Davis, 1966; Padmanabhan, 1970; Batygina, 1981, 1982; Shamrov, 1981; Batygina and Shamrov, 1985; Titova, 1988; Batygina *et al.*, 1991) and Boraginaceae (Svensson, 1925; Swamy and Krishnamurthy, 1973; Vereshchagina, 1976, 1983b).

Differences exist in the genesis of the endosperm of the Helobial type of development in monocotyledons and dicotyledons. These pertain to the location of the nucleus in the primary endosperm cell and the time of initiation of cytokinesis in the micropylar and chalazal cells (Vijayaraghavan and Prabhakar, 1984).

Before division the nucleus of the primary endosperm cell in monocotyledonous plants is situated near the antipodals. Unequal division results in formation of two cells (or chambers in the terminology of some authors), of which the micropylar cell is usually larger than the chalazal one. The micropylar cell is highly vacuolate. In the first stages of its development free nuclear divisions take place, leading to formation of the multinucleate coenocyte. The coenocyte nuclei are distributed throughout the layer of the cytoplasm, adjacent to the cell wall (Maheshwari, 1950; Swamy and Parameswaran, 1963; Swamy and Krishnamurthy, 1973; Bhatnagar and Sawhney, 1981; Vijayaraghavan and Prabhakar, 1984).

The main differences in composition of both endosperm cells in the monocotyledons concern the number of nuclei. The most variable is its chalazal cell, the so-called **'basal apparatus'**. In many species it is one-nucleate and has dense cytoplasm. In *Potamogeton densus* as a result of mitosis leading to polyploidization, the nucleus enlarges 30-40 times with ploidy reaching 192n (Hasitschka-Jenschke, 1959). However, in a number of taxa (Amaryllidaceae—Dutt, 1970a,b; Bromeliaceae—Lakshmanan, 1967; Phylidraceae—Hamann, 1966; Typhaceae—Müller-Doblies, 1969; Zingiberaceae—Panchaksharappa, 1962) the chalazal cell of the endosperm is multinucleate. It contains, for example, 4 nuclei in *Asphodelus tenuifolius* (Eunus, 1952), 10-12 nuclei in *Typha latifolia* (Asplund, 1972), 16 nuclei in *Zephyranthes* and *Sternbergia* species (Vorsobina and Solntseva, 1979; Solntseva and Vorsobina, 1979) and 32 nuclei in *Hippeastrum rutilum, Pancratium longiflorum* (Dutt, 1970a,b). The variety in number of nuclei in the chalazal endosperm cell has been noted even within one family (Burmaniaceae, Juncaceae and Zingiberaceae). Thus, in family Juncaceae the chalazal cell of *Juncus bufonius* is one-nucleate (Shah, 1963), one- or two-nucleate in *J. filiformis* (Brenner, 1922), while in *J. effusus* the number of nuclei may reach 8 (Zaman, 1950), and in *Luzula pedemontana*—14-16 (Anisimova, 1990; Shamrov and Anisimova, 1993a,b). Based on differences in the structure of the chalazal cell 6 forms (more exactly variations) of development of the Helobial endosperm have been distinguished in monocotyledons (Swamy and Parameswaran, 1963), of which 5 forms are characterized by a different number of nuclei in the cell (typical form—one nucleus, **A**-form—2 nuclei, **B**-form—4, **C**-form—8, and **D**-form—16), with no cytokinesis occurring. In the 6th form (**E**) cell formation is observed and the tissue forms.

In contrast to monocotyledons, in dicotyledonous plants the nucleus of the primary endosperm cell often occupies the central position. As a result, the two formed cells appear to be equal or almost equal in size. Later, nuclei of the micropylar cell divide without cytokinesis. The chalazal cell remains one-nucleate and elongates greatly; its nucleus moves towards the antipodals (Khanna, 1967; Titova, 1988).

After reaching a certain number of nuclei in the micropylar and sometimes chalazal cell, the cellular phase of endosperm development follows the coenocytic one[8]. In those cases when both cells are coenocytes, cellularization usually starts in the micropylar cell first, although in representatives of families Bromeliaceae and Typhaceae this process first takes place in the chalazal cell (Lakshmanan, 1967; Müller-Doblies, 1969). In the micropylar cell the number of nuclei before cellularization may differ: from 16 nuclei (for example, in *Sparganium simplex*—Müller-Doblies, 1969) to 100 (in *Trillium undulatum*—Swamy, 1948d). In the chalazal cell cellularization can start already at the stage of two nuclei (*Typha latifolia*—Asplund, 1972) or on reaching 8-16 nuclei (*Juncus effusus*—Zaman, 1950; *Luzula pedemontana*—Shamrov and Anisimova, 1993b). Later, derivatives of the micropylar cell form the endosperm (according to Juel (1907)—the central endosperm). The chalazal cell or its derivatives (according to Juel (1907)—the basal endosperm) operate as a haustorium which is usually destroyed during seed development.

---

[8]There is no cellularization in representatives of families Ruppiaceae and Zannichelliaceae (Kamelina and Teryokhin, 1990a,b).

Based on ultrastructural and histochemical investigations of the endosperm of the Helobial type of development undertaken on a small number of plants species, a high functional activity of both cells was shown. In the period of free nuclear divisions these cells are characterized by features of the transfer cells and the outgrowths of cell walls are more developed in the chalazal cell (for example, in *Haemanthus katherinae*). Near these wall projections there are numerous mitochondria with well-developed cristae (Newcomb, 1978). The cytoplasm of both cells and the micropylar in particular reacts intensively on staining for proteins, DNA and RNA (in *Najas flexilis, N. marina* and *Potamogeton nodosus*—Kapoor, 1980). The dynamics of some metabolites during endosperm development has been studied on the example of *Luzula pedemontana* (Shamrov and Anisimova, 1993b). Before division already a small amount of starch and proteins has accumulated around the nucleus of the primary endosperm cell. Later, after two cells have formed there occurs a specific redistribution of substances: in the micropylar cell starch is shown in small quantities around the forming nuclei, while in the chalazal cell proteins appear (high level of content). By the stage of the late globular embryo when the cellularization of the endosperm has already passed and the chalazal cell has almost entirely degenerated, a large amount of starch accumulates in the central part of the endosperm and in its three-four peripheral layers proteins appear (relatively small content level). In the course of further development proteins are no longer present in the endosperm and the storage reserves are represented by starch.

The phylogenetic significance of the Helobial type of endosperm development is under discussion in the literature (see below: Evolutionary Aspects of Endosperm Development).

Light Microscopic Study of Endosperm Formation in *Brassica napus* L. (Plate XIX).

In a study on the *in-vivo* development of the seed of *Brassica napus* we focused on development of the endosperm applying light microscopy. The microtubular cytoskeleton in the endosperm was visualized by immunocytochemical staining of embedded and sectioned material.

Endosperm development in rape proceeds through a nuclear phase (Bhatnagar and Sawhney, 1981; Vijayaraghavan and Prabhakar, 1984). Mitotic activity of the primary endosperm nucleus and its derivatives leads to formation of a layer of free nuclei bordering the wall of the central cell. Initially the nuclei were distantly located and all were encompassed by the cytoplasm, which formed a thin layer lining the cell wall of the central cell. With increase in nuclear divisions and proliferation of the cytoplasm, the nuclei drew closer to one another but still remained free in a band of cytoplasm. A developing seed at 6 days after anthesis showed the embryo at the globular stage and the nuclear endosperm surrounding the central vacuole of the central cell. Free endosperm nuclei were scattered over the cytoplasm, sometimes at distances less than 6 μm. Immunocytochemical labelling of the butyl methyl metacrylate (BMM) sections revealed the occurrence of microtubular arrays radiating from the envelope of each interphase nucleus. Some microtubules (Mts) radiating from a nucleus interconnected with the radiating Mts of the neighbouring nuclei. Double staining with DAPI showed that a great number of DNA-containing organelles (plastids and mitochondria) occurred in the cytoplasm, especially in spherical zones surrounding the nuclei.

In the micropylar region of the embryo sac, again a thin layer of cytoplasm and nuclei surrounded the embryo proper, but around the embryo suspensor a mass of cytoplasm existed containing a cluster of nuclei not aligned in a monolayer. The distance between nuclei in the section ranged from 6 to 100 μm, indicating an uneven distribution over the cytoplasm at this early nuclear stage. In the chalazal region, the endosperm consisted of a region of dense cytoplasm with regularly spaced nuclei. Here, short arrays of Mts were distributed throughout the cytoplasm and sometimes bundles of Mts radiated from the surface of the nuclei, forming a three-dimensional network in the cytoplasm. Referring to the developmental stage of the embryo it was found that the nuclear endosperm phase persisted until the early heart-shape stage of the embryo proper, until about 12 days after anthesis. Cellularization had started in the vicinity of the embryo. The mass of endosperm cytoplasm found here in the nuclear stage, vacuolated, but the nuclei remained surrounded by cytoplasm. Cell wall formation started in the thin layer of cytoplasm surrounding the embryo proper and in the layer of cytoplasm bordering the wall of the former central cell. In the micropylar region the endosperm nuclei were completely surrounded by these cell walls. Thus real endosperm cells were formed here. In the thin layer of cytoplasm surrounding the embryo proper and in the layer of cytoplsm bordering the wall of the former central cell, however, only anticlinal cell walls developed, resulting in formation of alveolar endosperm. Alveolus formation generally started when the nuclei were closer than 30 μm. The first alveolus was about 20 μm in diameter in section. At the chalazal side the endosperm was still nuclear and contained regularly spaced nuclei.

It has recently been stressed that formation of the alveolar endosperm is strongly influenced by the microtubular cytoskeleton in a variety of plant species, such as *Triticum aestivum* (Van Lammeren, 1988), *Ranunculus sceleratus* (XuHan and Van Lammeren, 1993), *Phaseolus vulgaris* (XuHan and Van Lammeren, 1994) and *Populus nigra* (XuHan *et al.*, 1995). A model of the influence of the microtubules on alveolation and cell formation was postulated.

In the phase in which endosperm cells, alveoli and free nuclei existed simultaneously, all endosperm types exhibited specific microtubular configurations. Concomitant with vacuolation in the endosperm cytoplasm, Mts pointing to the central vacuole bent sidewards and ran immediately underneath the tonoplast of the central vacuole. Cellularization in the area of the alveolar endosperm was first seen when the embryo differentiated to the late heart- or early torpedo shape stages. Cell wall growth continued, resulting in formation of cylindrical alveoli. The nuclei migrated from the centre of the alveolus to a position near the central vacuole. The radial Mts remained prominent but pointed preferentially to the tonoplast of the central vacuole and to the ends of the growing anticlinal walls. When alveolar nuclei divided, periclinal walls appeared, and the first layer of uninucleated endosperm cells was formed. They bordered the embryo sac wall and, at the inner side, the layer of alveoli remained. In this way, more layers of endosperm cells were produced. Endosperm cells at interphase exhibited Mts in the cortical cytoplasm lining the cell walls. Some short bundles of Mts were detected around the flattened nuclei. Cellularization continued until the central vacuole of the former central cell had disappeared. A developing seed at 14 days after anthesis showed the embryo at the torpedo-shape stage. The endosperm was cellular for the greater part. In the central zone of the seed the central vacuole of the former central cell persisted. It was still lined by a layer of alveolar endosperm but the nuclei of the alveoli were less densely distributed than at 6 days after anthesis. Immunocytochemical labelling of the BMM sections revealed that the organization of the microtubular cytoskeleton in the alveoli had not changed at this stage.

Endosperm cells were not only formed by divisions in the alveoli. Existing endosperm cells divided too, especially those in the outer cell layers of the endosperm. These cells exhibited various stages of karyokinesis and cytokinesis. Technovit embedding showed the microtubular spindles at various stages of mitosis and Mts associated with the ends of new cell plates. In the alveolar zone of the endosperm many nuclei were surrounded by arrays of Mts radiating from the nuclear surface. The Mts of four neighbouring alveolar nuclei interacted at sites where the cell walls of the alveoli were being formed.

Analysing the successive stages of endosperm development in *B. napus* it appeared that there exist large differences between ovules of the same age. These differences probably depend on the position of the ovules in the fruit and the position of the fruit in the inflorescence. However, the stages of development of endosperm and embryo were related.

Cellularization of the nuclear endosperm started at the early heart-shape stage, and either occurred directly as seen in the endosperm surrounding the embryo suspensor, or was mediated by alveolation, a process first described by Hodcent (1969) in which the microtubular cytoskeleton is highly involved (XuHan, 1995). Alveolation might occur simultaneously throughout the embryo sac, as seen in the peripheral type of endosperm formation found in e.g. wheat (Fineran *et al.*, 1982, Van Lammeren, 1988), or cellularization varies depending on the site in the embryo sac. With *B. napus* cellularization occurred immediately in the area surrounding the suspensor, preceded by alveolation along the longitudinal walls of the embryo sac, and retarded at the chalazal site. Thus, cellularization of the endosperm of *B. napus* has characteristics described in the lingering type also found in e.g. bean (Yeung and Cavey, 1988; XuHan and Van Lammeren, 1994).

Asynchrony of development might be due to different distribution of cytoplasm and nuclei over the embryo sac. The micropylar area and the chalazal zone of the embryo sac are conical, likely caused by differentiation of the integument in the micropylar region and by elongation of the embryo sac at the cost of nucellar tissue in the chalazal region. Here the cytoplasm did not form a thin layer lining the wall of the embryo sac, but a mass of cytoplasm enabled the free nuclei to accumulate in more than a monolayer. Once cell wall formation is initiated, it will only result in formation of alveoli at sites where a layer of nuclei borders a large vacuole. The remaining nuclei in the cytoplasmic mass are surrounded by neighbouring nuclei and will be fully surrounded by cell walls, and thus become cells without an alveolar phase.

The first signs of cell wall formation were observed in the micropylar region. This might indicate that the signal for cellularization, if any, is located here. It is unlikely that the density of population of nuclei causes cell wall formation alone, because in the chalazal zone a high density of nuclei was also observed.

Alveolar formation was preceded by a redistribution of free endosperm nuclei, because at the late globular stage we still observed an irregular distribution of nuclei over the cytoplasm. As development proceeded, nuclei became regularly spaced. It is suggested that the regular spacing of the nuclei is caused by Mts which radiate from the nuclear surface in a manner reported for e.g. wheat (Van Lammeren, 1988).

Alveolation is the result of cell plate formation, elongation and cell wall fusion (XuHan and Van Lammeren, 1994). The formation is likely dependent on the presence of a high density of phragmoplast Mts found after karyokinesis, as well as on the presence of dense arrays of Mts between neighbouring non-daughter nuclei in the manner proposed by XuHan (1995). High densities of microtubular arrays are only obtained when nuclei are densely packed in the monolayer, that is, after repeated mitotic divisions. Consequently, the high density of nuclei, needed for the onset of cell wall formation, also results in formation of alveoli with relatively small diameters. Therefore the initial cell files at the periphery of the endosperm were slender. Increase in size of the more centripetal endosperm cells results from reduction in number of alveoli which continue to grow inwardly. Hence the distances between nuclei of neighbouring alveoli increase.

Cell formation in alveoli is caused by nuclear division and cell plate formation in a plane perpendicular to the walls of the alveoli in the manner described earlier for e.g. wheat (Van Lammeren, 1988). Once endosperm cells of *B. napus* are formed, they are still able to divide. We particularly observed cell divisions at the outer region of the endosperm. Various stages of the mitotic cycle were found as were microtubular configurations associated with various stages of karyokinesis and cytokinesis. Simultaneously cells formed in the alveoli until the anticlinal walls of alveoli at opposite sides of the central vacuole met and fused. Thus the central vacuole of the former central cell is replaced by a cellular endosperm.

In conclusion, this study enabled us to identify three stages of endosperm development in which the site within the embryo sac, density of population of nuclei and configurations of the microtubular cytoskeleton were parameters of major importance for this particular endosperm development.

Cell Differentiation in the Pericarp and Endosperm of Developing Maize Kernels (*Zea mays* L.) (Plate XX).

Structural aspects of caryopsis development have been studied for important cereals such as wheat (Buttrose, 1963; Campbell *et al.*, 1981; Cochrane, 1983; Rogers and Quatrano, 1983; Smart and O'Brien, 1983; Van Lammeren, 1988), barley (Olsen and Krekling, 1980; Felker *et al.*, 1984; Kvaale and Olsen, 1986), rye (Pomeranz, 1973) and rice (Bechtel and Pomeranz, 1977; Bechtel and Juliano, 1980; Miflin *et al.*, 1981; Oparka and Harris, 1982). A general survey related to seed development is given by Bhatnagar and Sawhney (1981; see also Friedman, 1994). Developmental studies of maize go back to the last century; many early workers analysed spikelet formation, fertilization, and the development of fruits and seeds from a structural point of view (see, among others, Guèrin, 1899; Miller, 1919; Weatherwax, 1919; Lampe, 1931; Randolph, 1936; Kiesselbach, 1949). More recently, electron microscopy stimulated subcellular analysis of fertilization (Diboll, 1964; Van Lammeren, 1986b), development of the proembryo (Van Lammeren and Schel, 1983; Schel *et al.*, 1984), the embryo proper (Van Lammeren, 1986b), the endosperm (Khoo and Wolf, 1970; Kyle and Styles, 1977) and the interaction between embryo and endosperm (Schel *et al.*, 1984).

The endosperm is considered an important nutritional factor for embryo development and germination (Lampe, 1931; Lending *et al.*, 1988; Esen and Stetler, 1992; Mikus *et al.*, 1992). Especially the outermost cell layer, the aleurone layer, which can be distinguished by the shape of its cells and the organelles within the cells, accumulates storage products during seed development and releases hydrolytic enzymes during germination. The ultrastructural development of the aleurone layer and the subaleurone layers were studied earlier (Khoo and Wolf, 1970; Kyle and Styles, 1977); however, although early interactions between the developing embryo and endosperm have been reported (Schel *et al.*, 1984), further development of cell shape, interactions between various tissues of the caryopsis and the concomitant ultrastructural changes were not investigated in detail.

The present study focuses on cellular differentiation in carpels, integument, nucellus and parts of the endosperm, starting from the moment of cellularization of the endosperm up to and including senescence of the pericarp cells. In all tissues of the caryopsis, cell differentiation resulted not only in a diverse composition of the cytoplasm, but in varying cell shapes as well.

Development of somatic tissues in the caryopsis. Development and degeneration of the carpels, integument and nucellar cells were traced within a restricted zone at the germinal face of the caryopsis during a period of three weeks after pollination. In the pericarp, three zones were distinguished at 3 DAP[9]: an endocarp zone near the ovule, consisting of about three cell layers of small, thin-walled fusiform cells which were often isodiametric in section and elongated out of the plane of sectioning; a mesocarp zone with larger, loosely arranged cells which were isodiametric to irregular in shape and an exocarp zone containing cylindrical cells which bordered each other, leaving very little intercellular space. All pericarp cells had large central vacuoles and only little cytoplasm in which amyloplasts were observed. As development proceeded, however, growth of the developing seed forced the pericarp to enclose a larger area. At 8 DAP the exocarp cells had elongated. In the mesocarp region a spongy mesophyll developed with spherical, slightly elongated or curled cells. The endocarp cells did not increase much in diameter but separated, thus enlarging the intercellular space. In cross-sections the cells were often found parallel to each other, forming fibrous sheaths covering the developing seed. The length axes of the various sheaths were not fixed. At 17 DAP the pericarp cells were still intact at an end phase of development. The diameter of the endocarp cells had not enlarged much. The exocarp cells had reached their maximal length, thereby leaving little intercellular space. Their cell walls had thickened, giving solidity to the fruit wall. The spongy mesocarp cells were compressed and so the thickness of the whole pericarp decreased. The cytoplasm of the remaining cells degenerated at about 20 DAP, thus leaving a multi-layered dead fruit wall consisting mainly of cell walls. At 22 DAP the small endocarp cells were often crushed, whereas the shape of the dead exocarp cells was still preserved.

Integuments covered the nucellus and the endosperm during the early developmental stages. Because of the retarded growth of the outer integument during ovule development, only an inner integument covered the nucellus in the sampling area. It consisted of two cell layers and a thin cuticle. The integumentary cells were cylindrical at 3 DAP. Sometimes they were only seen at this stage but occasionally the intact cytoplasm was still observed at 8 DAP. Hereafter cells were crushed between the pericarp and the expanding endosperm and their thin cell walls and cuticles disappeared completely.

The nucellus is also an ephemeral tissue. Except for the epidermis and its cuticle, the nucellus degenerated rapidly. At 3 DAP a multilayered nucellus still existed but

[9]DAP, days after pollination.

at 5 DAP only 1 or 2 layers of nucellar cells remained in the sampling area. Cell degeneration occurred near the outermost cell layer of the endosperm. First the cells exhibited cytoplasmic degeneration. Then they collapsed and eventually disappeared. The nucellar epidermis was observed as a living tissue at 8 DAP. It consisted of flattened cells covered with a thick, conspicuous cuticle which was still found in the late developmental stages. There was a large central vacuole and only a little cytoplasm. From 12 DAP onwards nucellar epidermis cells collapsed. Remnants were found up to 22 DAP because the cuticle, outer cell walls and partly thickened anticlinal cell walls persisted.

The course of cell development of the various somatic tissues described above is summarized schematically in Figure 20 with respect to cell shape development, cell wall formation and viability.

Development of the endosperm. Development of the aleurone, subaleurone and inner endosperm involves multiplication, enlargement and differentiation of cells and the accumulation of storage products. Development of the aleurone layer became prominent from 5 DAP onwards. Initially the outer endosperm cells, which

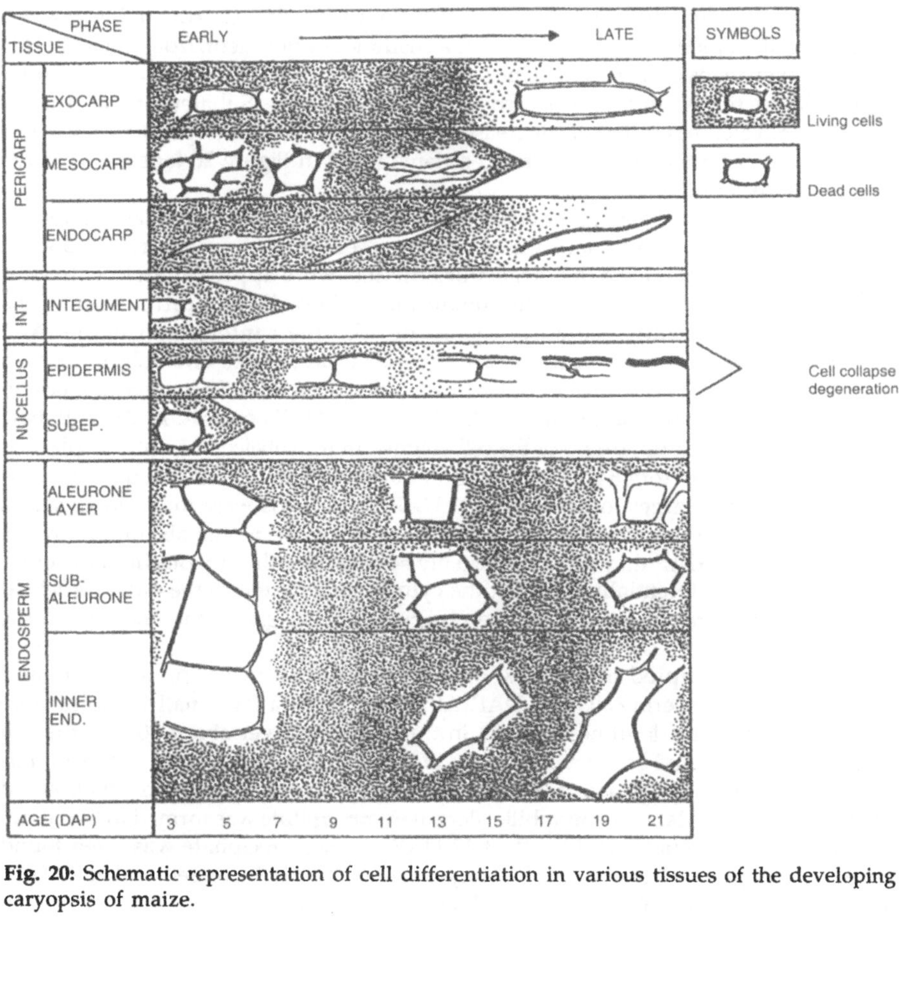

Fig. 20: Schematic representation of cell differentiation in various tissues of the developing caryopsis of maize.

developed into aleurone cells, were large (one cell of the nucellar epidermis covered 2-3 prealeurone cells). They divided in various directions but from about 7 DAP simultaneously and in only one direction. Many nuclei exhibited prophase configurations at 5-6 DAP and at 8 DAP cytokinesis had resulted in formation of an epidermal-like cell layer in which mainly anticlinal cell walls had formed. The newly formed cells were much smaller; up to 8 epidermal cells were now covered by one nucellar epidermal cell which had not increased in size. After that, the epidermal cells became rectangular, the amount of cytoplasm and organelles still increased, and accumulation of storage products started at 12 DAP. From 8 to 12 DAP the cells of the outermost cell layer of the endosperm differentiated into aleurone cells. They enlarged in the periclinal direction and up to 5 aleurone cells were covered by one nucellar cell at 12 DAP. Subaleurone cells were often found in line with the aleurone cells, indicating that periclinal cell wall formation took place repeatedly as endosperm development proceeded. These cells continued to divide, sometimes in a meristem-like fashion, forming rows of cells with new periclinal cell walls. The newly formed isodiametric cells were often smaller than the bordering aleurone cells. The inner endosperm cells, however, enlarged. During further seed development the aleurone cells enlarged in periclinal and anticlinal directions (up to 4 aleurone cells were covered by one cell of the nucellar epidermis at 17 DAP). In the ensuing development, from 17 to 22 DAP, only slight changes in sizes and shapes of the cells were seen. The anticlinal and outer cell walls increased in thickness from 17 DAP onwards. Subaleurone and inner endosperm cells, however, still increased in size. These cytomorphological data of endosperm formation are summarized schematically in Figure 20 together with the somatic tissues of the caryopsis.

Submicroscopical aspects of endosperm differentiation are summarized for the developing aleurone layer and the subaleurone endosperm in Figures 21 and 22 respectively. Nuclei, being the largest organelles, were approximately spherical to ellipsoidal and had undulating membranes. They did not differ in size and shape in aleurone and subaleurone cells, nor did their internal structure. At 3 and 5 DAP nuclei of the outermost endosperm layer were often observed near the outer cell wall. Plastids were found in all endosperm cells. They were smallest in the aleurone layer and accumulated starch from 12 DAP onwards. At 22 DAP these amyloplasts occupied up to 12% of the cell volume. In the subaleurone cells the plastids accumulated starch from 8 DAP onwards and up to 46% of the cell volume was filled with these second largest organelles at 22 DAP. Thylakoids were found in all stages but granae were never detected. Mitochondria were never abundant in the endosperm. At 3 DAP they were not equally spread over the cytoplasm in the outer cell layers but preferentially found in the cytoplasm adjacent to the outer periclinal cell walls. This phenomenon could be observed at later stages too and also in the subaleurone layer.

Vacuoles occupied about 40% of the cell volume of young aleurone and subaleurone endosperm cells at 3 DAP. At 22 DAP only a few small vacuoles were observed in the subaleurone cells but in the aleurone layer about 10% of the cell volume was occupied by vacuoles. Only in the aleurone cells did they function as aleurone vacuoles by the accumulation of storage proteins, a phenomenon observed from 5 DAP onwards. An osmiophilic, flocculent precipitate was formed in the centre of the spherical vacuoles at 8 DAP; at 12 DAP a dense precipitate was often found near the tonoplast. The amount of storage proteins clearly increased and at 22 DAP almost filled the vacuoles, then termed aleurone bodies.

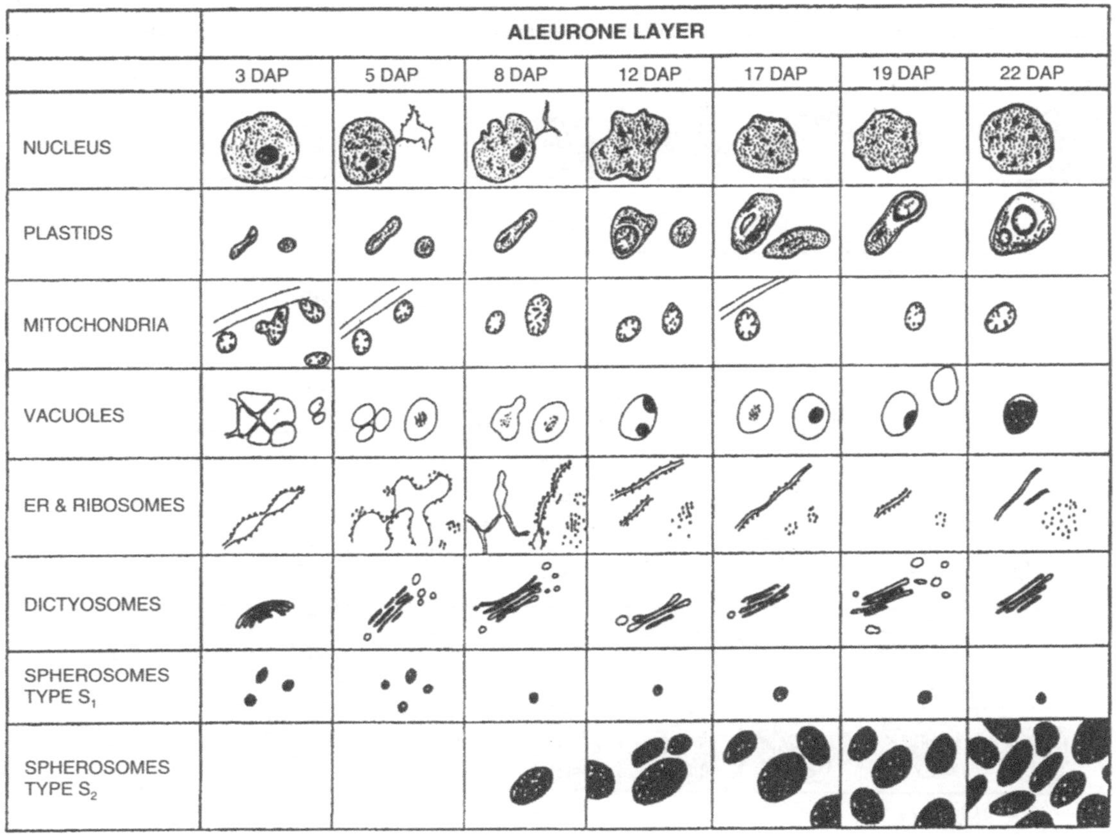

Fig. 21: Schematic representation of organelle development in differentiating aleurone cells of *Zea mays* L. from 3 to 22 days after pollination (DAP).

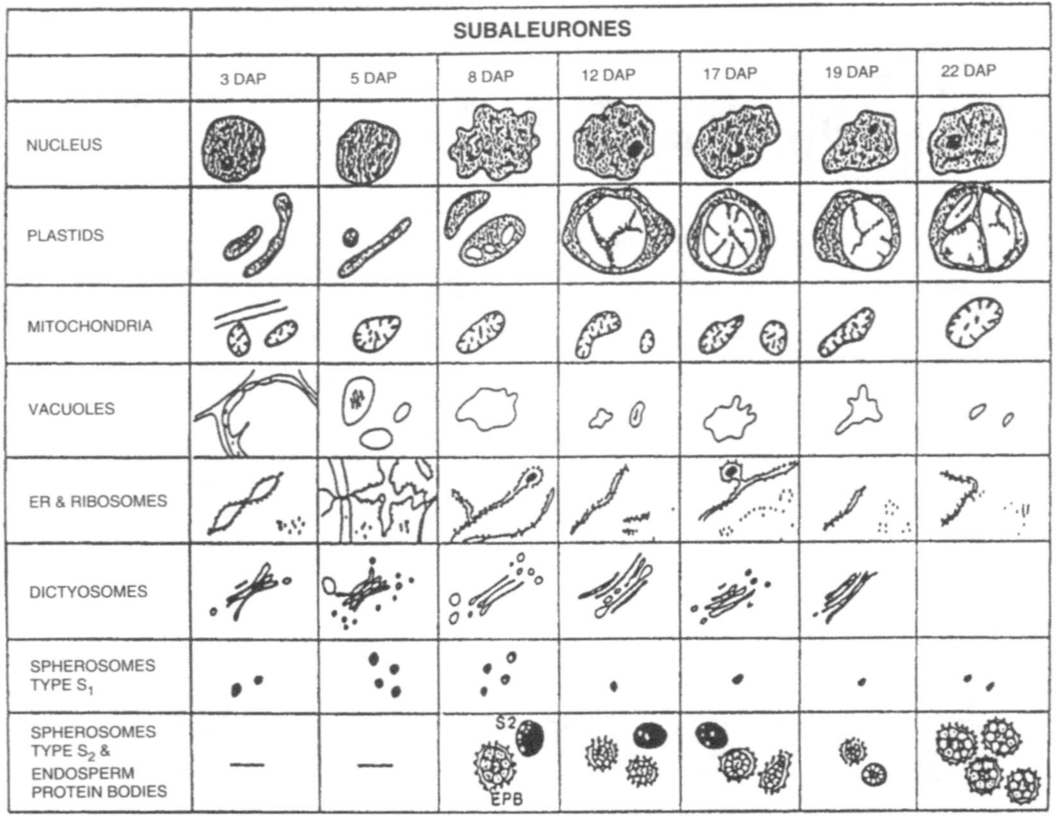

Fig. 22: Schematic representation of organelle development in differentiating subaleurone endosperm cells of *Zea mays* L. from 3 to 22 days after pollination (DAP).

The endoplasmic reticulum proved to be a conspicuous organelle system, especially in the subaleurone cells. Initially, at 3 DAP, the ER occupied but a minor part of the cell volume. At 5 DAP, however, the cisternal space increased significantly because the ER had swelled and 15% of the cell volume was occupied. In the subaleurone and inner endosperm the ER, to which many ribosomes were attached, gave rise to electron-dense globular structures, termed endosperm protein bodies (EPBs). These EPBs, surrounded by a ribosome carrying the ER membrane, were formed from 8 DAP onwards and had an average diameter of 0.14 ± 0.04 μm at 8 DAP and 0.44 ± 0.09 μm at 22 DAP. Their presence was restricted to the subaleurone and inner endosperm where they eventually occupied up to 70% of the cell volume at 22 DAP. The heavily stained ribosomes occurred singly or as short coils of polysomes. Dictyosomes were infrequently seen and in the aleurone layer their presence could not be established at 22 DAP because the cytoplasm was densely stained.

As for other storage products, two types of spherosomes were discerned in the developing endosperm. The first type (Type S_1) was mainly seen in the outer cell layers of the young endosperm up to an age of 8 DAP. Their number and total amount decreased in the aleurone layer when the first spherosomes of a second type (Type S_2) were discerned. This second type predominantly occurred in the aleurone layer. Some of these spherosomes were seen in the adjacent cell layer but none in the inner endosperm. Type S_2 spherosomes were largest (1.26 + 0.68 μm at 17 DAP) and sometimes filled over 90% of the aleurone cell cytoplasm at 22 DAP.

Cell differentiation in the caryopsis. The pericarp, known for its determined life time (see, among others, Randolph, 1936), developed various cell layers, each with cells of particular shape, wall thickness and corresponding mechanical properties. From 22 DAP onwards the exocarp consisted of rigid cell walls while the cytoplasm of the cells showed degeneration. Most dead mesocarp cells were compressed between the exocarp and the seed but the walls of the small fusiform cells of the endocarp were still recognizable. The pericarp, which initially showed a tripartite structure as is common for the caryopses of many Gramineae, protected the seed. Throughout their lifespan pericarp cells had a large central vacuole. As in other Gramineae the carpels did not contribute to food storage. The accumulation of starch was ephemeral. The intimate relation between pericarp and underlying tissues was expressed by lack of a real inner epidermis. As already mentioned for wheat and barley, the fusiform cells of the endocarp vary in direction between individual caryopses. They appeared as a network of so-called 'cellules tubulaires' (Guèrin, 1899) forming the inner face of the pericarp. Actually the nucellar cuticle is the demarcation between fruit wall and seed because the integumentary cells have already degenerated.

The lifetime of the integuments has been discussed by various authors, among them, Randolph (1936), who observed the initial stages of disintegration of the integuments at the germinal face of the caryopsis in samples examined 3 DAP. He reported that the walls of the outer cell layer disappeared first while nuclei and cytoplasm often persisted for some time. In the present study degeneration was often accomplished before 5 DAP but incidentally at 8 DAP cytoplasm with microtubules was still observed. This asynchrony in development can be attributed to differences in developmental rates in the various caryopses. The inner integument influences the pathway of the pollen tube (Heslop-Harrison J. *et al.*, 1985; Van Lammeren, 1986a) rather than the subsequent seed formation.

The nucellus is also an ephemeral tissue and it is likely that it functions in nourishing the developing endosperm when cells are digested near the endosperm epidermis. The nucellus epidermis and its cuticle existed longest and covered the endosperm. They might function as a layer providing mechanical protection because its cells lived until the cell walls of the aleurone layer started to thicken. Eventually the cells died and only remnants of the anticlinal and outer cell walls and the cuticle of the nucellus persisted. These remnants were called the nucellar membrane (Randolph, 1936; Kiesselbach, 1949). At this stage that structure is likely to function as a protective covering of the embryo and endosperm because of its semipermeable properties (Randolph, 1936).

In the caryopses of the maize inbred line A188, differentiation of the aleurone layer started at about 7 DAP with the synchronous cell divisions by which one layer of columnar cells was formed. In our experiments the phase of differentiation started 3 days earlier than reported by Kyle and Styles (1977). In contrast to e.g. barley, in which the aleurone layer consists of as many as four cell layers, one layer is usual in maize but two cell layers were found incidentally. Aleurone and subaleurone endosperm differed from each other in many aspects. Until 8 DAP cell divisions in the outer endosperm layer resulted in many small cells which developed into rectangular aleurone cells. This is in contrast with the situation found e.g. in rape-seed wherein the endosperm also exhibits small outer cells. In that case, however, their ontogeny is strikingly different because the number of these cells was already determined at the transition of the nuclear to the alveolar stage (see also Groot and Van Caeseele, 1993).

With respect to the type and location of organelles, some similarities were observed in the aleurone and subaleurone layers of this maize strain. Both layers had typically located mitochondria. The alignment along the outer periclinal cell walls points to a high and local demand for energy to permit active transmembranal export and import. This phenomenon was also observed in the synergids of maize before fertilization (Van Lammeren, 1986a) and in the modified endosperm bordering the maize embryo suspensor (Schel *et al.*, 1984). Both cell types showed remarkable proliferation of ER, though not as intense as was reported for wheat in which evaginations of the nuclear membrane took part the ER system (Kyle and Styles, 1977). Furthermore, an increasing amount in cytoplasm was noticed on all cell types and starch accumulated in all plastids, contrary to the observations of Khoo and Wolf (1970) who only found proplastids in the aleurone layer. The difference might be caused by varying developmental rates which are influenced by genotype and environmental conditions. On the other hand, organization of the cytoplasm differed between the epidermal, subepidermal and inner endosperm cells as early as 3 DAP. Differences were found in the accumulation of spherosomes (Type S_1) in the various tissues at 3 DAP; The synthesis of endosperm protein bodies from 7 DAP onwards was restricted to the subaleurone cells and inner endosperm cells, and the formation of aleurone bodies and spherosomes which started at 9 DAP occurred only in the aleurone cells.

With respect to the function of the aleurone layer, Bhatnagar and Sawhney (1981) reviewed the literature concerning hydrolytic enzymes, such as amylase that brings about the breakdown of reserve food materials in the endosperm tissue, which are utilized by the embryo during its growth and development. Protein bodies have several functions, which were summarized by Matile (1968). First they store certain reserve substances such as proteins, metals and phytin, whose synthesis might be mediated by mitochondria (Morrison *et al.*, 1975). Second, during germination the aleurone vacuoles are transformed to a 'lysosome' and their food reserves mobilized. Finally, they may form a lytic compartment in which other constituents of the cell are broken down.

The development and structure of protein bodies and spherosomes in seeds is discussed by Lott (1980), Lending *et al.* (1988) and Mikus *et al.* (1992). Khoo and Wolf (1970) reported on the development of aleurone vacuoles and protein bodies in the outer cell layer of the maize endosperm. They failed to detect protein accumulation in the vacuoles in 7-10-day kernels but found protein granules by day 12. As compared to e.g. barley and wheat (Jacobsen *et al.*, 1971), maize formed only a few aleurone protein bodies (see also Khoo and Wolf, 1970; Kyle and Styles, 1977), which occupied only a minor part of the aleurone layer, but spherosomes were abundant and took up much more space. The origin of the spherosomes was not established. The existence of spherosome membranes in maize was reported by Kyle and Styles (1977). Such a membrane was observed in the present study only when spherosomes were pressed to each other. Spherosomes show variations in staining which might be related to their heterogeneous composition (Wolf *et al.*, 1969). Aleurone protein bodies had electron dense and electron opaque inclusions. The electron dense parts, which are hard and do not penetrate well with embedded medium, often chatter during sectioning leaving holes in the section. Protein bodies of the subaleurone and inner endosperm were found to have polysomes associated with the protein body membrane. These endosperm protein bodies of maize probably accumulate the main storage protein called zein (Burr B. and Burr F., 1976). When the polysomes were isolated at a time when storage protein synthesis was active, they were able to synthesize zein. More recently the distribution of α-, β-, γ-, and δ-zein was determined by immunocytochemical methods (Lending *et al.*, 1988; Esen and Stetler, 1992).

Based on the ontogeny of aleurone protein bodies and spherosomes it cannot be excluded that the structures termed protein bodies have originated via different pathways, i.e., via spherosomes and aleurone vacuoles. This might explain the substantial variation and relative amounts of the zein classes observed among protein bodies.

Micropylar Haustorium of *Impatiens walleriana* Hook. ex Arn. (Plate XXI)

The ovule of *Impatiens walleriana* is bitegmic and tenuinucellate. Since the antipodal cells are ephemeral, the mature embryo sac contains four cells: egg cell, two synergids and large central cell. Localization and morphology of the various cells follow the classical pattern. The part of the inner integument bordering the micropyle and near the egg apparatus consists completely of thick-walled cells. The middle lamellae in this tissue are lignified. The integumentary epidermis bordering the other part of the embryo sac has differentiated into a well-developed integumentary tapetum. In the outer tangential walls of the integumentary tapetum a layer of electron-dense, acetolysis-resistant material is present.

Double fertilization is completed within two hours following pollination. After fertilization is achieved, within one day the central cell divides into a micropylar, cytoplasm-rich cell and a large chalazal, vacuolated cell. Further divisions of the micropylar cell result in a cellular endosperm, while the chalazal cell in the first instance gives rise to a nuclear endosperm. The zygote remains undivided for two days.

Directly after its formation the micropylar endosperm cell starts growing towards and into the micropyle, forming the haustorium. During this process the pollen tube and both the persistent and degenerated synergid are pushed aside and shortly after disappear. After the haustorium has protruded from the micropyle, its nucleus divides and one of the formed nuclei moves to the protrusion outside the ovule, whereas the second one remains inside the embryo sac. Subseqently a wall with many extensions is formed in the haustorium at the transmission position from embryo sac to micropyle, dividing the haustorium into external and internal parts. At this stage both haustorium parts are rich in cytoplasm, although in the outer part formation of a large vacuole has begun.

The zygote starts dividing the second or third day after fertilization. The embryo develops a short suspensor consisting of two or three cells. The suspensor stays in close contact with the internal haustorium part and both structures accumulate a large amount of starch. Meanwhile the external haustorium part invades the neighbouring funicular and placental tissues, whereas the integumentary tissue is not affected. Invasion by the haustorium is accompanied by crushing, degeneration and breakdown of cells. The external haustorium part stays unicellular and becomes very irregular in outline. Its nucleus strongly enlarges and a giant nucleolus is formed. Its cytoplasm shows a complex organization, with rough and smooth ER, abundant polysomes, and many mitochondria and dictyosomes.

Development of the embryo and endosperm is accompanied by expansion of the embryo sac volume and growth of the ovule. This involves both breakdown of integument cells and stretching of the remaining ones. The breakdown of integument cells occurs along the integumentary tapetum which itself stays intact until the sixth day of development. During this period the electron-dense, acetolysis-resistant layer in the outer tangential walls of the integumentary tapetum strongly thickens. Approximately on the sixth day after fertilization the integumentary tapetum rapidly degenerates and simultaneously the chalazal nuclear endosperm becomes almost completely cellular.

Our observations clearly establish Dahlgren's conclusion (1934) that in *Impatiens* the micropylar haustorium derives from the endosperm, and not from the embryo suspensor as stated by Jacobsson-Stiasny (1914) and Palm (1915). The haustorium is already present before the first division of the zygote has taken place. However, eventually a specific spatial and probably also functional relationship is achieved between the internal haustorium part and the embryo suspensor. The latter is indicated by the simultaneous accumulation of starch in both structures.

Formation and function of the haustorium might be related to the presence and development of the electron-dense, acetolysis-resistant layer in the outer tangential walls of the integumentary tapetum. This layer completely surrounds the embryo sac, except for the micropylar area. Structure, electron-density and acetolysis-resistance indicate the presence of sporopollenin in the layer. At the time of fertilization the amount of deposited material is rather low and the layer discontinuous. Until degeneration of the integumentary tapetum a thick and continuous layer is formed. It is likely that during this period nourishment of the developing embryo and endosperm by the surrounding integumentary tissue gradually decreases and nourishment by the micropylar haustorium changes.

Evolution of Embryogeny in Seed Plants and Developmental Origin and Early History of Endosperm.

It has long been recognized that angiosperms possess a suite of defining features associated with the process of sexual reproduction. Among the most prominent of these is the presence of an endosperm, a unique embryo-nourishing 'tissue' that is formed as the result of a second and distinctive fertilization process. For the greater part of this century, the origin and early evolution of endosperm remained an enigma. Within 2 years of the discovery of the developmental origin of the endosperm from a second fertilization event (Nawaschin, 1898b; Guignard, 1899a, b), two basic hypotheses were formulated to explain the evolutionary origin of endosperm (Thomas, 1907). One theory held that the endosperm is a delayed continuation of female gametophyte development, stimulated by the fusion of sperm nucleus with the polar nuclei of the embryo sac, and hence derived from and evolutionarily homologous with the female gametophyte of more primitive seed plants (Strasburger, 1900). An alternative hypothesis, first proposed by Sargant (1900), maintained that the endosperm represents a developmental transformation of a supernumerary embryo derived from a second fertilization event. Until recently, the evidence necessary to discriminate between these contrasting hypotheses regarding the evolution of the endosperm was lacking.

In addition to the fundamental issue of the origin of the endosperm, evolutionary biologists and plant embryologists were baffled by the question of which of the three basic patterns of endosperm development (free nuclear, *ab initio* cellular and helobial) characterized the earliest flowering plants (Maheshwari, 1950; Stebbins, 1974; Sporne, 1980; Takhtajan, 1991) (Fig. 23). Among plants with a free nuclear endosperm, a series of mitotic divisions follows formation of a typically triploid fusion nucleus, but are not initially accompanied by cytokinesis. The result is a coenocytic endosperm in which cellularization may occur later. In an *ab initio* cellular endosperm, mitosis is coupled with cell division from the outset of development (there is no initial free nuclear phase). A helobial endosperm appears to blend aspects of the nuclear and *ab initio* cellular patterns of development. Following formation of a primary endosperm nucleus, a mitotic division accompanied by cytokinesis forms a two-celled endosperm. Free nuclear divisions ensue, although the extent of free nuclear development within the two cellular compartments of the endosperm varies among and within taxa. Eventually, cellularization of one or both coenocytic cells may take place (Brink and Cooper, 1947; Maheshwari, 1950; Gifford and Foster, 1989).

It is generaly agreed that the helobial endosperm represents a derived condition among angiosperms and that character reversals among any of free types of endosperm development have been numerous over the course of flowering plant evolutionary history (Takhtajan, 1991). Indeed, many genera or families exhibit more than one type of endosperm (see Davis, 1966 for information on the systematic distribution of endosperm types). Moreover, it has been suggested that patterns of endosperm development may correlate with other features of the seed such as nucellar volume, number of integuments, or presence of endosperm haustoria (Wunderlich, 1959). However, the basic question of whether *ab initio* cellular endosperm or free nuclear endosperm characterized the earliest flowering plants has remained largely intractable.

During the past 5 years, substantial evidence (discussed below) has been presented that supports the hypothesis of Sargant (1900) that the endosperm evolved from a supernumerary embryo in the ancestors of flowering plants. As a consequence, determination of the pattern of endosperm development likely to have characterized the earliest flowering plants (*ab initio* cellular or free nuclear) can be derived from analysis of the pattern of embryogeny that characterized the immediate ancestors of flowering plants. As with the endosperm, embryos of seed plants may express either

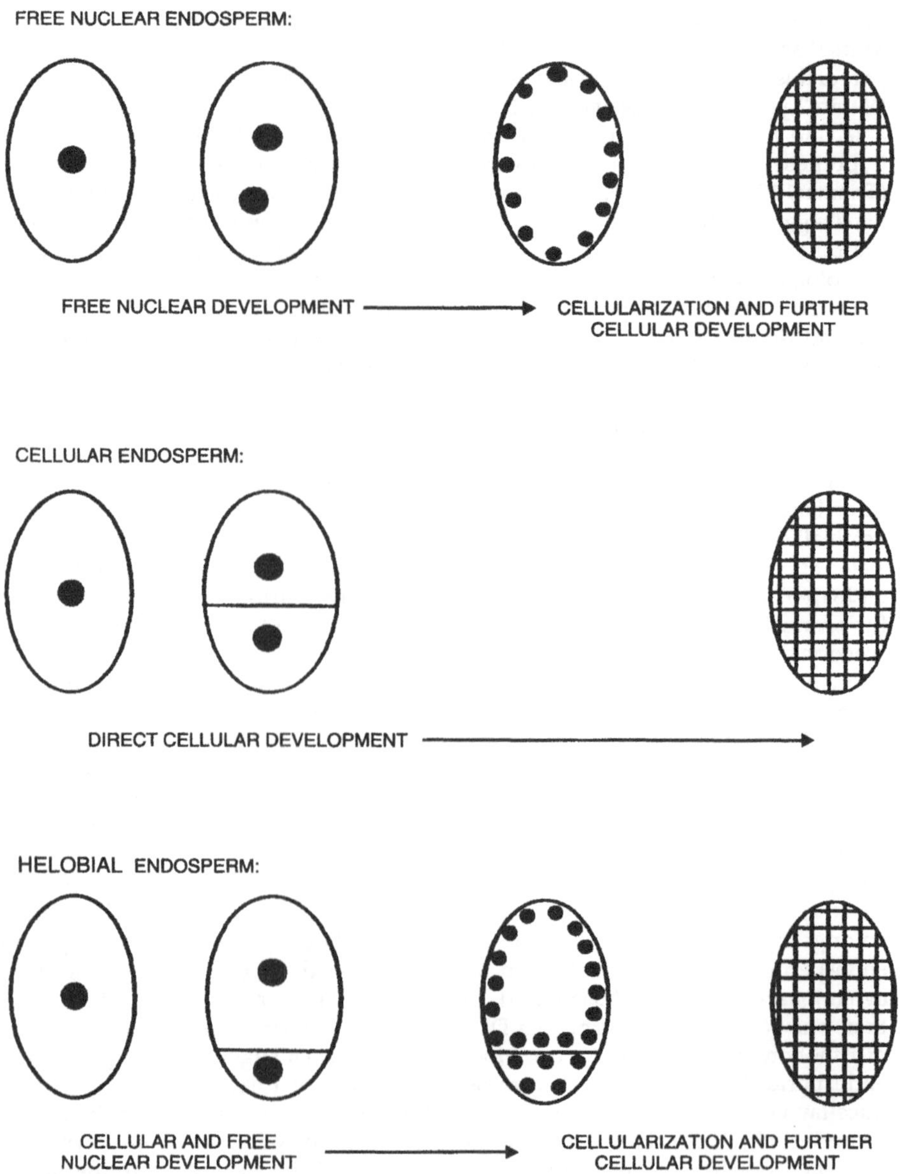

Fig. 23: Three basic patterns of endosperm development among flowering plants.

a free nuclear of *ab initio* cellular pattern of development (no seed plant embryos are 'helobial'). As will be seen, re-evaluation of the evolution of patterns of embryogeny among the ancestors of flowering plants provides information critical to determination of the developmental pattern of the endosperm that characterized the earliest flowering plants.

The goals here are threefold. First, formalisms necessary for the interpretation of comparative data within a phylogenetic context are discussed in the section 'Phylogeny-Based Character Analysis'. Current hypotheses of the phylogenetic relationships of seed plants are reviewed, and a conceptual basis for reconstruction of evolutionary history in the absence of a fossil record for the characters of interest is developed. Second, recent advances in our understanding of the evolutionary origin of double fertilization and endosperm are examined in the section 'Double Fertilization in *Ephedra* and Evolutionary Origin of the Endosperm'. Formulation of hypotheses to explain the evolution of the endosperm are shown to derive from new information on patterns of fertilization in the closest living relatives of angiosperms, the Gnetales. Third, the evolution of embryogenetic patterns among seed plants is re-evaluated in order to determine which pattern of development (free nuclear or *ab initio* cellular) was expressed by the progenitor of the endosperm, a supernumerary embryo formed as a consequence of a second fertilization event. This analysis forms the basis for the section 'Evolution of Patterns of Early Embryogenesis in Seed Plants and Its Bearing on the Developmental Nature of the Primitive Endosperm'. Through an examination of the evolution of embryogenic patterns and their relationship to the endosperm, the pattern of endosperm development that is likely to have characterized the earliest flowering plants is inferred.

Ultimately, it is hoped that this analysis will serve to stimulate further study of the nature of evolution based upon comparative data and its interpretation within a phylogenetic context. It is at this interface, between comparative and phylogenetic biology, that advances in our knowlege of the evolutionary history of land plants will be found.

Phylogeny-based character analysis. Despite extensive analysis, many aspects of the origin and early evolution of the endosperm have remained cryptic. Unfortunately, key intermediate states that might help in clarifying the origin of important reproductive features of flowering plants have not been found in the fossil record, and it is unlikely that the fossil record will ever yield significant information that can address the evolutionary history of double fertilization (Cronquist, 1988), endosperm, or patterns of embryogeny. However, this does not imply that formal means of analysis cannot be undertaken to examine and resolve questions relating to the evolution of these unique features of angiosperm reproductive biology.

From the level of genes to organismal morphology, biological innovation has almost always resulted from modification of previously existing structures or processes (Bower, 1898; Jacob, 1982). Thus, questions pertaining to the origin of identifying angiosperm features such as double fertilization and the endosperm, are ultimately questions of evolutionary equivalence or homology, and the solution of the 'abominable mystery' of the origin of flowering plants lies in the ability to recognize transformations (evolutionary homologies) between specific structures and processes expressed in angiosperms and those of ancestral or closely related taxa ('outgroups'). For this reason, clear formulation of the phylogenetic relationships of angiosperms to their outgroups and critical interpretation of comparative data are essential to the reconstruction of the evolutionary origin of flowering plants.

During the past 10 years, significant progress has been made in elucidating phylogenetic relationships among seed plants. The pioneering morphological cladistic analyses of Crane (1985a,b) and Doyle and Donoghue (1986) have, more recently, been complemented by molecular analyses based on the N-terminal amino acid sequence of rubisco (Martin and Dowd, 1991), the *rbcL* gene of the chloroplast (e.g.,

Chase *et al.*, 1993), and the ribosomal RNA gene of the nucleus (e.g., Hamby and Zimmer, 1992). Still, the precise relationships of a number of seed plant taxa (in particular extinct Palaeozoic and Mesozoic groups of seed plants, cycads, conifers, and *Ginkgo*) remain ambiguous (Crane, 1988). However, phylogenetic analyses based on morphological data (Hill and Crane, 1982; Crane, 1985a,b; Doyle and Donoghue, 1986, 1992; Loconte and Stevenson, 1990; Doyle *et al.*, 1994) and molecular data (Martin and Dowd, 1986, 1991; Zimmer *et al.*, 1989, Hamby and Zimmer, 1992; Hasebe *et al.*, 1992a,b; Chase *et al.*, 1993; Doyle *et al.*, 1994) have yielded a consensus about the relationship of angiosperms to Gnetales, a small and unusual group of seed plants that comprises three monophyletic genera: *Ephedra* (approximately 45 species), *Gnetum* (approximately 30 species) and *Welwitschia* (monotypic). In particular, these phylogenetic studies support the hypothesis that Gnetales as a whole (*Ephedra, Gnetum* and *Welwitschia*) are monophyletic and are the closest extant relatives of angiosperms. In addition, these analyses indicate that *Ephedra* is basal within Gnetales and has retained a number of primitive reproductive features (especially when compared with *Gnetum* and *Welwitschia*) that are common to cycads, conifers and *Ginkgo*.

As a consequence of the critical phylogenetic position of Gnetales, an understanding of the reproductive biology of *Ephedra, Gnetum,* and *Welwitschia* is central to interpretation of the evolution of uniquely derived (apomorphic) features of flowering plants. By comparing reproduction in angiosperms and Gnetales, it is possible to formulate explicit hypotheses of the reproductive characteristics that defined the common ancestors of these two groups of seed plants as well as to examine the further transformation of these characters in descendant clades. Based upon this approach, the historical/evolutionary events that led to establishment of the reproductive features that define flowering plants can be identified and reconstructed. As will be seen, a phylogenetically based interpretation of comparative data on the fertilization biology of members of Gnetales leads to a significant revision of implicit and explicit hypotheses governing the origin of double fertilization and endosperm, evolution of patterns of embryogenesis over the course of seed plant history and, finally, the pattern of endosperm development that is likely to have characterized the earliest flowering plants.

Double fertilization in *Ephedra* and evolutionary origin of endosperm. During the past 5 years, considerable progress has been made towards elucidation of the evolutionary origin of the endosperm (Donoghue, 1989; Friedman, 1990a,b, 1991, 1992a,b; Donoghue and Scheiner, 1992). Recent documentation of a regular process of double fertilization in two species of *Ephedra* (Friedman, 1990a, 1991, 1992b) has provided the requisite comparative and developmental information for constructing an explicit hypothesis concerning the nature of the fertilization process in the common ancestor of angiosperms and Gnetales. This, in turn, has led to formulation of a precise and detailed theory of the early history of double fertilization and the evolutionary establishment of the endosperm.

Double fertilization in Ephedra

The female gametophyte of *Ephedra* is fundamentally similar to that of most non-flowering seed plants (cycads, conifers and *Ginkgo*). The female gametophyte is monosporic, undergoes free nuclear development followed by cellularization, and eventually matures into a large (compared with the embryo sac of flowering plants) multicellular structure with two or more archegonia (Martens, 1971). At the time of fertilization, each egg cell is binucleate (Friedman, 1990a, 1992a,b) and contains an egg nucleus and a ventral canal nucleus.

In *E. nevadensis* and *E. trifurca*, two sperms from a single pollen tube regularly fertilize the egg nucleus and sister nucleus of the egg nucleus (ventral canal nucleus) within an archegonium (Friedman, 1990a,b, 1991). One sperm nucleus migrates to the egg nucleus, which invaginates as contact is established between two gamete nuclei. These two conjoined nuclei begin a co-ordinated process of migration towards the base (chalazal end) of the egg cell. Concurrent with the first fertilization event, a process of fusion is initiated between the ventral canal nucleus and the second sperm nucleus. This second fertilization event is highly similar to fertilization of the egg nucleus by the first sperm nucleus. The ventral canal nucleus invaginates as contact is made with the second sperm nucleus, and these two nuclei also undertake a chalazal migration within the former egg cell (Friedman, 1990a,b, 1991, 1992b).

At the end of the double fertilization process in *Ephedra*, two diploid nuclei reside within the cytoplasm of the former egg cell. The 'normal' zygote nucleus, derived from the fertilization of the egg nucleus by a sperm nucleus, is situated at the base of the former egg cell. A second nucleus, formed from the fusion of the ventral canal nucleus and a second sperm nucleus, resides within the micropylar half of the former egg cell (Friedman, 1990b, 1991). Microspectrophotometric data indicate that the product of the second fertilization event in *E. trifurca* is diploid and contains a 4C quantity of DNA, as does the fusion product of the egg nucleus and first sperm nucleus (Friedman, 1991). In addition, the second fertilization product is genetically identical to the normal zygote nucleus since each of the sperm nuclei that participate in double fertilization are mitotic relatives, as are the egg and ventral canal nucleus. Most significantly, the fates of the first and second fertilization products in *Ephedra* are identical: each diploid nucleus will ultimately form an embryo (Friedman, 1992b; also see section below on embryogeny in *Ephedra*). Based on the high degree of similarity with respect to fusion behaviour, cell cycle and developmental fate (i.e., production of embryos) of the two fertilization events in *E. trifurca*, the second fertilization product can be considered a 'supernumerary zygote nucleus'.

Although double fertilization is a regular feature of reproduction in *E. nevadensis* and *E. trifurca* (and possibly in *E. campylopoda* (Herzfeld, 1922), *E. foliata* (Khan, 1943) and *E. distachya* (Moussel, 1978); see Friedman (1992b) for a complete review of earlier studies in which preliminary evidence of double fertilization events in *Epherdra* is presented), it is clear that an embryo-nourishing endosperm does not result from the second fertilization event. As is the case with basal seed plants (cycads, conifers and *Ginkgo*), the large multicellular female gametophyte of *Ephedra* functions to nourish developing embryos.

Comparative and evolutionary implication of double fertilization in Ephedra

The discovery that a regular second fertilization event in *Ephedra* yields a supernumerary zygote that produces embryos, has profound implications for understanding the evolutionary origin of the endosperm. Although it may not be possible to determine whether the processes of double fertilization in *Ephedra* and angiosperms represent evolutionary homologies or are homoplasious, many aspects of double fertilization in angiosperms and *Ephedra* are congruent with the concept that a basic process of double fertilization evolved in a common ancestor of Gnetales and angiosperms and hence predates the origin of flowering plants (Friedman, 1990a).

The vast majority of flowering plants possess a monosporic ('Polygonum-type') embryo sac which is considered to be primitive (plesiomorphic) among flowering plants (Stebbins, 1974), compared with bisporic or tetrasporic embryo sacs that are found among evolutionarily derived taxa. In angiosperms with monosporic embryo sacs, one of the two polar nuclei with which the second sperm nucleus fuses is the sister nucleus of the egg nucleus (Thomas, 1907; Brink and Cooper, 1947; Huang and Russel, 1992). In *Ephedra*, the ventral canal nucleus is also the sister nucleus of the egg nucleus. In both groups of seed plants, the two sperm that engage in double fertilization are typically derived from a single pollen tube (Friedman, 1992b; Mogensen, 1992). Thus, double fertilization in basal angiosperms and *Ephedra* involves similar sets of male and female nuclei.

The key difference between *Ephedra* and angiosperms with respect to the nuclei involved in the second fertilization event, is that in flowering plants, an additional female nucleus, the chalazal polar nucleus, which is not a sister nucleus of the egg nucleus (Thomas, 1907; Brink and Cooper, 1947; Huang and Russel, 1992), participates in the second fertilization event and through its involvement results in a typically triploid fusion product. Among angiosperms with a monosporic embryo sac, both polar nuclei are genetically identical to the egg (this contrasts with bisporic and tetrasporic embryo sacs, in which polar nuclei may differ genetically from the egg). In *Ephedra*, the ventral canal nucleus is genetically identical to the egg nucleus. Although the endosperm in angiosperms with a monosporic embryo sac and the supernumerary zygote in *Ephedra* differ with respect to ploidy, the second fertilization product in each of these groups of seed plants is identical (at the level of alleles) to the 'normal' zygote resulting from the first fertilization event.

The discovery that the second fertilization product in *E. trifurca* (and probably other species of *Ephedra*, see Friedman, 1992a for review of earlier studies) regularly produces a supernumerary zygote that yields additional embryos (Friedman, 1992b), provides strong support for the hypothesis of Sargant (1900) that the endosperm represents a highly modified evolutionary derivative of a supernumerary embryo. From this perspective, it is likely that the original manifestation of double fertilization in seed plants involved the fusion of a second sperm nucleus with a fecundible ventral canal nucleus and resulted in production of a supernumerary diploid zygote with the potential to develop into an embryo. This stage of evolutionary innovation would have been established in a common ancestor of *Ephedra* and flowering plants. Within the context of this evolutionary hypothesis, *Ephedra* has maintained a generally unchanged expression of the original double fertilization process, whereas significant modifications to this process have occurred in the angiosperm lineage (i.e., triploidy and the development of a non-embryo product from the second fertilization event). It is not known whether double fertilization occurs in *Gnetum* or *Welwitschia* (Friedman, 1992b).

A further element of Sargant's (1900) original hypothesis of the origin of the endosperm involved the proposition that the developmental transformation of a supernumerary embryo (in the ancestors of angiosperms) into endosperm occurred as a consequence of the addition of a second female nucleus to the second fertilization event. While there is merit to this proposition, it is currently not possible to resolve whether the endosperm (in the physiological and developmental sense) evolved prior to or after the addition of a second female nucleus to yield a triploid second fertilization product (Queller, 1983; Friedman, 1992b). Thus, the endosperm may originally have been diploid if the addition of a third nucleus to the second fertilization event occurred after transformation of an embryo into an embryo-nourishing structure (Fig. 24).

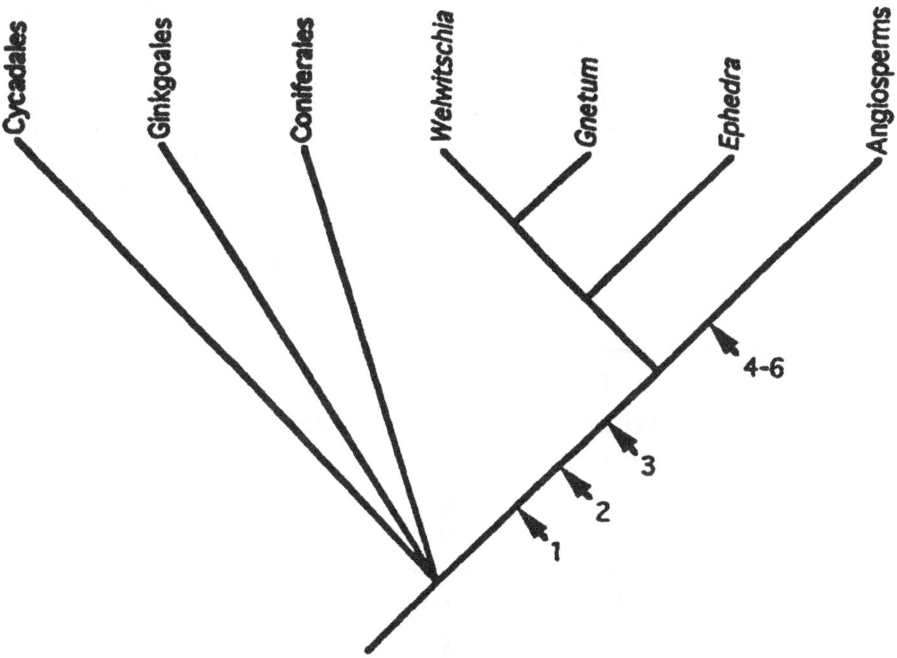

Fig. 24: Events during the evolution of double fertilization and endosperm. Character advance 1 is the evolution of an aberration of the time interval between pollination and fertilization; character advance 2, the evolution of persistence of the sister nucleus of the egg nucleus through the time of fertilization; character advance 3, the origin of double fertilization, in which the second fertilization event occurs between the sister nucleus of the egg nucleus and a sperm nucleus, and results in the production of a supernumerary zygote/embryo. Character advances 4-6 are unique features of angiosperms: addition of a second female nucleus to the second fertilization event to produce a triploid product; modification of development of the second fertilization product from an embryo to endosperm; and reduction of the female gametophyte to the embryo sac form. The order in which character advances 4-6 evolved is uncertain.

The hypothesized evolution of a regular process of fusion of a second sperm nucleus with the ventral canal nucleus to produce a supernumerary zygote in the common ancestor of angiosperms and Gnetales was likely to have been dependent on a number of specific developmental preconditions. The presence of a ventral canal nucleus at the time of fertilization in *Ephedra* is unusual among extant non-flowering seed plants (Friedman, 1992a). In most cycads, conifers and *Ginkgo*, the ventral canal nucleus degenerates between the time of its formation and entry of sperm into the egg cell (see Friedman, 1992a for some exceptions). Persistence of the ventral canal nucleus in *Ephedra* is likely to be a consequence of the extremely short period of time (less than 24 hours—Land, 1907; Moussel, 1983; Friedman, 1990b, compared with weeks or months in other non-flowering seed plants) between mitotic formation of the ventral canal nucleus and egg nucleus, and fertilization (Donoghue, 1989; Friedman, 1990a; Donoghue and Scheiner, 1992). Compared with most other seed plants, angiosperms and Gnetales display a reduction in the overall time required for reproduction and indirectly, acceleration of the reproductive cycle in the common ancestor of flowering plants and Gnetales may have led to the developmental maintenance of the ventral canal nucleus through the time of fertilization (Donoghue, 1989; Donoghue and Scheiner, 1992).

The evolution of a novel structure (endosperm in the physiological sense) capable of nourishing the young developing embryo in flowering plants or their immediate ancestors is likely initially to have created an intermediate condition in which both the female gametophyte and endosperm participated in the nourishment of the embryo (Fig. 25) (Queller, 1983, 1989).

This historical analysis leads to the prediction that the ovule of the immediate ancestors of the angiosperms possessed a female gametophyte that was still involved in the nourishment of the embryo, as well as a primitive endosperm. It will be

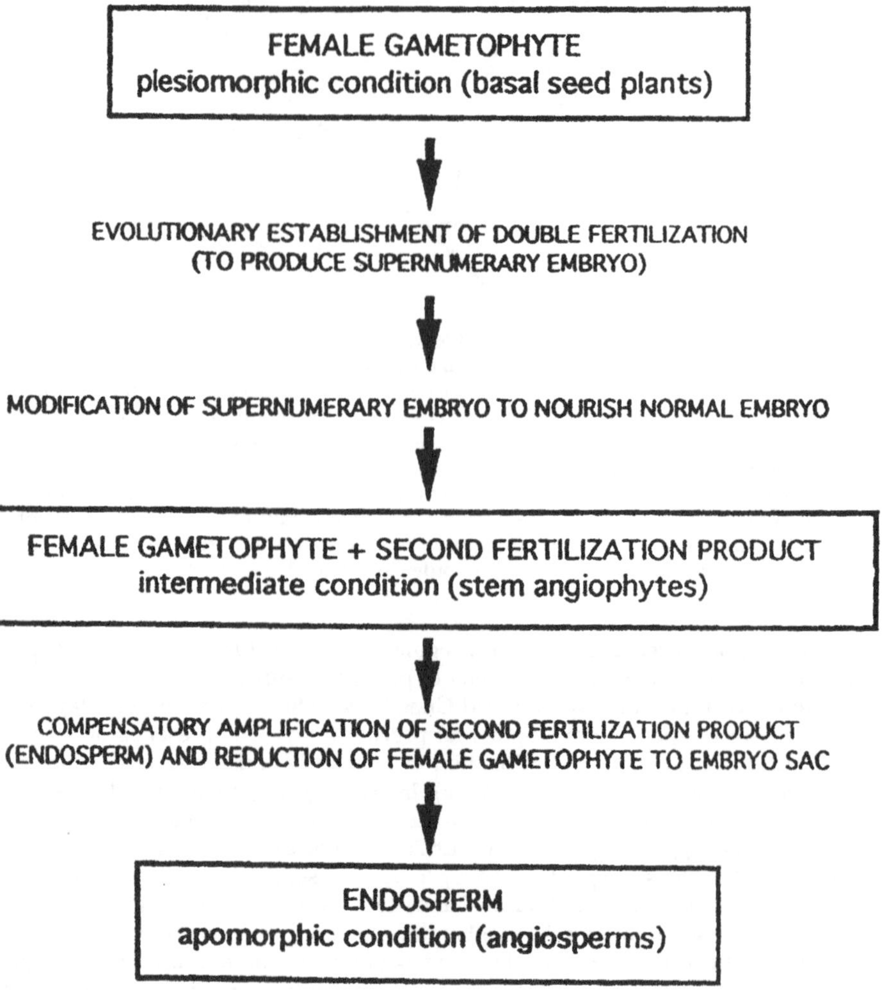

Fig. 25: Evolution of maternal resource allocation patterns to embryos in seed plants. Explanation in text.

interesting to see if future fossil discoveries of the immediate ancestors of flowering plants (members of the 'stem angiophytes' *sensu* Doyle and Donoghue, 1993) yield information with respect to the evolutionary transformation and reduction of the female gametophyte to the embryo sac form. Potential evolutionary intermediates might include fossil seeds that contain embryos associated with two separate embryo-nourishing structures.

With the progressive evolution of an increasingly important role for the endosperm in nourishment and development of the embryo, the female gametophyte of the immediate ancestors of angiosperms was ultimately freed from the developmental and physiological obligations (constraints) previously associated with nourishment of the embryo. This, in turn, made possible reduction of the female gametophyte (which in the common ancestor of Gnetales and angiosperms was large and essential to nourishment of the developing embryo) to the embryo sac form, now characteristic of all flowering plants (Friedman, 1992b).

Finally, current support for the hypothesis that the endosperm evolved from a second fertilization event that initially yielded a supernumerary embryo leads to the inevitable conclusion that the endosperm can no longer be considered a 'tissue' in the classical sense of the term. Rather, the endosperm must be seen as a highly modified evolutionary homologue of an embryo, in other words, as an *organism*, a view first espoused by Sargant (1900) and Thomas (1907). The logical extension of the discovery of the organismal status of the endosperm is that the life cycle of angiosperms can be viewed as triphasic and comprised of sporophyte, male and female gametophyte, and endosperm generations.

The evolution of patterns of early embryogenesis in seed plants and its bearing on the developmental nature of primitive endosperm. Although much of the diversity and systematic distribution of endosperm developmental patterns has been known for the greater part of this century, the question of whether free nuclear or *ab initio* cellular endosperm characterized the earliest flowering plants has generated considerable diversity of opinion (Maheshwari, 1950; Stebbins, 1974; Sporne, 1980; Takhtajan, 1991). The evolutionary derivation of the endosperm from a supernumerary embryo (resulting from a second fertilization event) in the ancestors of angiosperms provides the requisite link between patterns of embryo and endosperm development. If the endosperm evolved from a supernumerary embryo, it is reasonable to infer that the pattern of early embryo development exhibited by the common ancestor of Gnetales and angiosperms (in which double fertilization and production of a supernumerary embryo took place) is also likely to have characterized the endosperm of the first flowering plants.

Early embryogeny in extant seed plants

Extant basal seed plants, except for *Sequoia sempervirens* (Buchholz, 1939), all exhibit a pattern of free nuclear embryogeny. Following fertilization in cycads, *Ginkgo* and conifers, the zygote nucleus enters into a series of free nuclear mitotic divisions to produce free nuclei that lie within the cytoplasm of the former egg cell. In cycads, as many as 1000 proembryonal nuclei may form (Chamberlain, 1935) while among conifers, the number of free nuclei is more limited and may be as few as four. Free nuclear development of the proembryo in cycads, conifers, and *Ginkgo* is followed by a process of cellularization within the former egg cell. Cell walls are initiated between the zygotically derived nuclei and a single multicellular proembryo is organized. This critical stage of embryo development represents a highly co-ordinated aggregation of proembryo nuclei and yields a single multicellular sporophyte. Following cellularization of the proembryo, elongation of a multicellular suspensor pushes the embryo proper into the nutritive tissues of the large female gametophyte, where further embryo development occurs.

The free nuclear pattern of embryogeny displayed by primitive extant seed plants stands in marked contrast to early embryo development among all angiosperms (exept *Paeonia*, Stebbins, 1974) and *Welwitschia*, in which formation of a zygote is followed by entirely cellular development of the proembryo. The traditional evolutionary interpretation has been that the cellular embryogeny of angiosperms arose through elimination of the early free nuclear phase of embryogeny expressed in more primitive seed plants (Fig. 26). In addition, most authors have concluded that establishment of cellular embryogeny in flowering plants or their ancestors post-dated divergence of the angiosperm lineage from the common ancestor of angiosperms and Gnetales (see cladistic analyses of Crane, 1985a; Doyle and Donoghue, 1986). This conclusion was based on the proposition that early embryogenesis in *Ephedra* is free nuclear (Berridge and Sanday, 1907; Land, 1907;

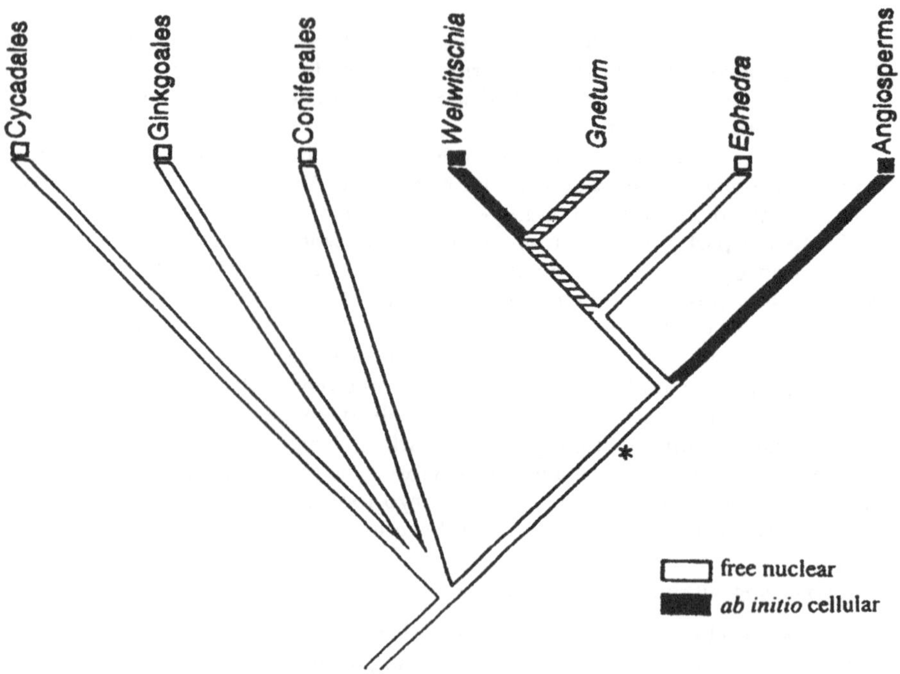

Fig. 26: Traditional hypothesis of the evolution of early embryo development among seed plants. Cycads, conifers, *Ginkgo* and *Ephedra* are considered to have had free nuclear development of the embryo. The cellular patterns of embryogeny in angiosperms and *Welwitschia* are presumed to have evolved separately. Free nuclear embryogeny is a shared derived character state (synapomorphy) of seed plants. The condition of early embryo development in *Gnetum* is uncertain. Hatched asterisk indicates that double fertilization evolved prior to establishment of a pattern of cellular embryo development.

Martens, 1971; Stebbins, 1974; Crane, 1985a, Doyle and Donoghue, 1986; Gifford and Foster, 1989), and that this free nuclear phase is evolutionarily and developmentally homologous with that of basal extant seed plants.

If early embryo development in *Ephedra* is free nuclear and homologous with that of cycads, conifers and *Ginkgo*, phylogenetic analysis indicates that the common ancestors of angiosperms and Gnetales also retained a free nuclear pattern of embryo development. This would suggest that the antecedent of the endosperm (a supernumerary embryo in the common ancestor of Gnetales and angiosperms) displayed a free nuclear phase, and that the earliest form of endosperm was also free nuclear in development. However, detailed developmental, comparative and phylogenetically based analyses of embryogenesis in *Ephedra* and, more broadly within the Gnetales, indicate that the pattern of early embryo development in these enigmatic taxa, as well as the pattern of embryogeny that characterized the common ancestor of angiosperms and Gnetales, are likely to have been misinterpreted.

Early embryogeny in *Ephedra*. Shortly after completion of sexual fusion in *Ephedra*, each of the two fertilization products enters into mitosis (Fig. 27) (Friedman, 1992b). The synchronous division of the normal zygote nucleus and the supernumerary zygote nucleus results in the production of two sets of daughter nuclei. One pair of nuclei (derived from the normal zygote nucleus) is situated at the base of the former egg cell, and a second pair of nuclei (derived from the supernumerary zygote nucleus) is positioned within the apical half of the former egg cell. A second wave of mitotic activity is initiated by each of the four nuclei within the former egg cell to yield a total of eight nuclei. These nuclei are typically arranged in two groups within the former egg cell cytoplasm: a basal set of four nuclei derived from the normal zygote nucleus and an apical set of four nuclei descended from the second fertilization product of 'supernumerary zygote nucleus' (Friedman, 1992a, b).

Subsequent to the free nuclear proliferation of each of the two fertilization products (zygotes) in *Ephedra*, elaborate arrays of microtubules (phragmoplasts) form around each of the individual nuclei and cell walls are deposited (Land, 1907; Khan,

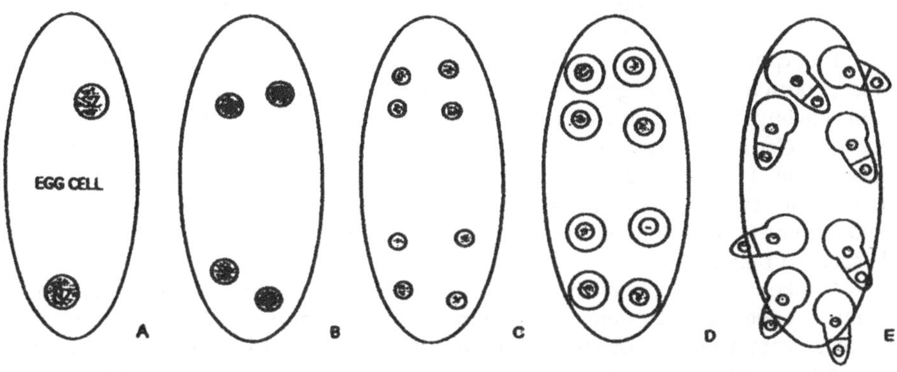

Fig. 27: Early embryogeny in *Ephedra*. Two diploid fertilization products result from gametic fusion (A): a normal zygote (from the egg nucleus and first sperm nucleus) and a supernumerary zygote (from the ventral canal nucleus and second sperm nucleus). Each fertilization product divides mitotically to produce two (B) and four diploid nuclei ('clonal' zygote nuclei) (C). Each nucleus becomes surrounded by a separate cell wall and two sets of four unicellular/uninucleate proembryos are formed (D). Each proembryo initiates a cellular pattern of development with filamentous growth (E). NZ—normal zygote nucleus, SZ—supernumerary zygote nucleus.

1943; Lehmann-Baerts, 1967; Moussel, 1977). Each spherical cell wall separates a diploid nucleus and an associated region of cytoplasm from the remainder of the former egg cell. As a consequence of this process of cellularization, two sets of four unicellular/uninucleate proembryos are formed following double fertilization in *Ephedra*. In *E. foliata*, more or fewer than eight nuclei/proembryos have been reported to form within an archegonium following fertilization (Khan, 1943), and this is likely to occur in other species of *Ephedra*. Although most, if not all, of the unicellular proembryos in *Ephedra* will undergo some further development, no more than one embryo will ultimately survive during maturation of the seed.

Following establishment of two sets of four spherical unicellular/uninucleate proembryos within an archegonium, many to all of the proembryos initiate a pattern of filamentous development (Berridge and Sanday, 1907; Land, 1907; Khan, 1943; Lehmann-Baerts, 1967; Moussel, 1977; Friedman, 1990b). Filamentous growth is initiated either prior to (*E. trifurca*; Land, 1907) or after (*E. distachya*—Berridge and Sanday, 1907 and Moussel, 1977; *E. foliata*—Khan, 1943; *E. nevadensis* and *E. trifurca*, present report) mitosis of the single nucleus of each proembryo takes place. Current observations of the first cell division in *E. trifurca* and *E. nevadensis* indicate that a cell plate may be initiated in conjunction with mitosis or cytokinesis may be delayed with respect to mitosis (Land, 1907; Lehmann-Baerts, 1967; Moussel, 1977). In either case, free nuclear development, defined as successive mitotic divisions without intervening cytokinesis (Chamberlain, 1935), does not occur during normal development of the individual proembryos of *Ephedra*.

The result of the first partitioning event in each proembryo in *Ephedra* is a two-celled proembryo. The basal cell will elongate considerably to form a unicellular primary suspensor that forces the apical cell of the proembryo into the nutritive tissue of the female gametophyte. The apical cell will proliferate and form an embryo proper, as well as additional cells that will comprise a secondary suspensor (Land, 1907; Khan, 1943; Lehmann-Baerts, 1967; Moussel, 1977).

***Evolutionary implications of embryogeny in* Ephedra**. Embryogeny in *Ephedra* appears to follow a pattern typical of most basal non-flowering seed plants: initial free nuclear development is succeeded by a process of cellularization (Fig. 28). Indeed, plant morphologists have suggested that the free nuclear phase following syngamy and production of a zygote in *Ephedra* is equivalent to the free nuclear phase of embryogeny in primitive seed plants (Martens, 1971; Gifford and Foster, 1989). More recently, cladistic analyses have codified this interpretation (Crane, 1985a; Doyle and Donoghue, 1986). However, comparative analysis suggests that formation of four free nuclei from each zygote nucleus in *Ephedra* is neither developmentally nor evolutionarily homologous to the free nuclear stage of embryogeny found among basal seed plants.

Unlike cycads, conifers and *Ginkgo*, in which cellularization of free nuclei derived from a zygote nucleus yields a single multicellular embryo, each of the free nuclei formed from a normal or supernumerary zygote nucleus in *Ephedra* establishes a separate single-celled proembryo that serves as the starting point for development of a potential sporophyte (Friedman, 1992b). Thus, the four free nuclei derived from each zygote in *Ephedra* cannot be viewed as constituent parts of a unitary organism, as is the case following free nuclear development of the proembryo among basal extant seed plants.

It has been suggested that the free nuclear phase of embryogeny in *Ephedra* represents an extreme and precocious form of cleavage polyembryony (Gifford and

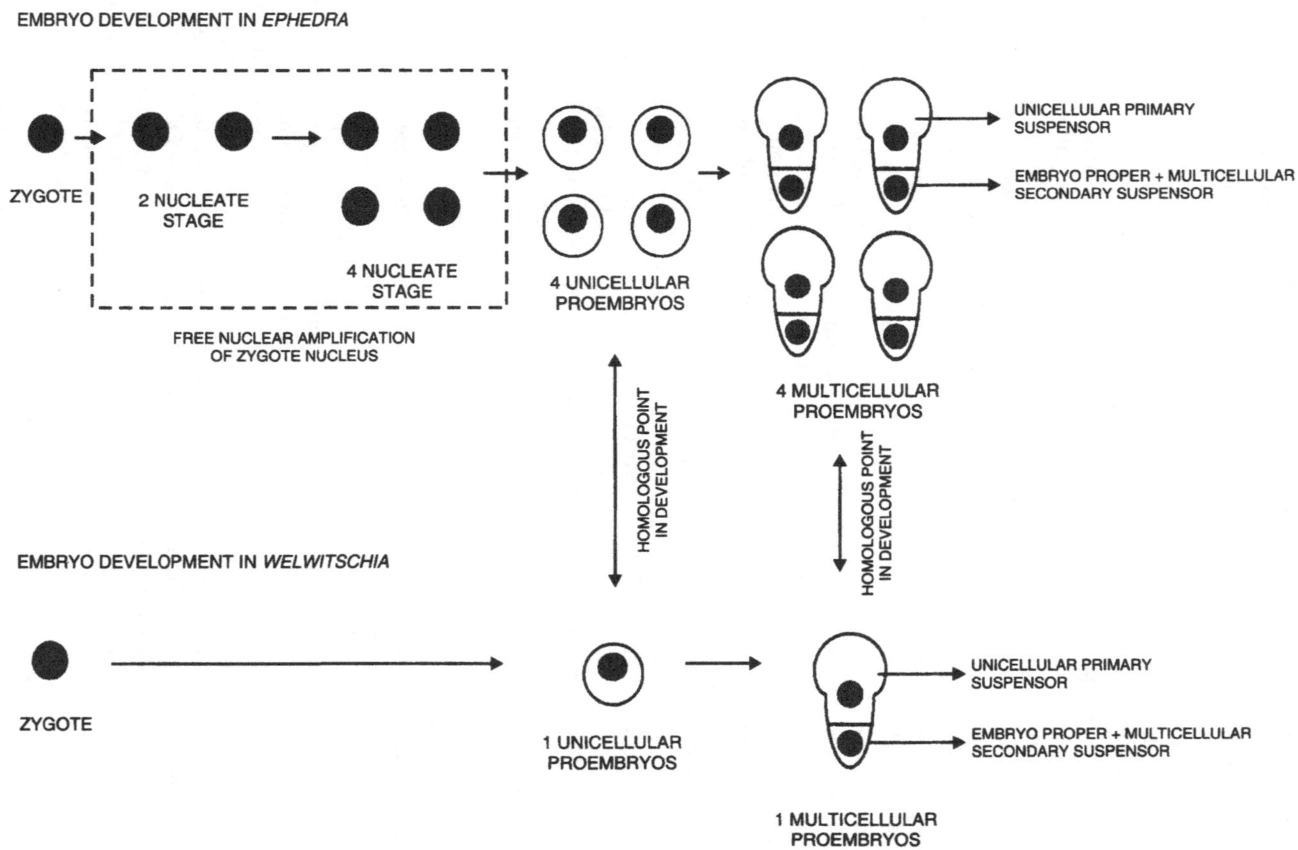

Fig. 28: Comparison of postfertilization establishment of proembryos in *Ephedra* and *Welwitschia*. Explanation in text.

Foster, 1989). Alternatively, this free nuclear phase can be viewed as a novel proliferation of a sexually formed zygote nucleus to yield four 'clonal' zygote nuclei that serve as the starting points for development of separate proembryos. In either case, the mitotic proliferation of each zygote nucleus in *Ephedra* appears to represent a uniquely derived (apomorphic) condition without parallel among extant seed plants. This, in turn, suggests that the traditional concept that *Ephedra* exhibits, and by inference, that the common ancestor of Gnetales and angiosperms exhibited, a free nuclear pattern of embryogeny must be called into question.

Additional support for the hypothesis that the common ancestors of Gnetales, and of Gnetales and angiosperms, did not display a free nuclear pattern of early embryo development can be derived from a comparative analysis of embryogenesis within the Gnetales. Embryogeny in *Welwitschia* has traditionally been characterized as cellular (Chamberlain, 1935; Martens, 1973; Crane, 1985a; Doyle and Donoghue, 1986; Gifford and Foster, 1989). It has been assumed that the loss of free nuclear development, leading to cellular embryogeny in *Welwitschia*, occurred separately and in parallel with the loss of this character in angiosperms or their immediate ancestors (Crane, 1985a; Doyle and Donoghue, 1986). However, when development of the normal sexually formed zygote and early proembryo in *Welwitschia* is carefully compared with that of each of the 'clonal' zygotes and individual unicellular proembryos of *Ephedra*, a new interpretation of the origin of cellular embryogeny in seed plants emerges—one that is congruent with a single origin of this character among seed plants.

In *Welwitschia*, following syngamy a cell wall forms around the zygote nucleus to produce a unicellular proembryo (only one proembryo forms from a zygote) (Pearson, 1909; Martens, 1973). The proembryo initiates a pattern of filamentous growth and mitosis of a single zygote/proembryo nucleus takes place shortly afterwards. Cell wall formation to form a two-celled proembryo is reported to be delayed with respect to the first mitosis (Pearson, 1909). However, free nuclear development is not expressed in the proembryo of *Welwitschia* (Chamberlain, 1935). The apical cell of the two-celled proembryo eventually produces an embryo proper and multicellular secondary suspensor, while the basal cell elongates and develops into a unicellular primary suspensor (Martens, 1971, 1973).

The striking similarities between the initial development of the single proembryo of *Welwitschia* and each of the multiple proembryos of *Ephedra* reveal important points of developmental homology. In both taxa, the starting point for development of an individual embryo is a single-celled, uninucleate proembryo. In the single proembryo of *Welwitschia* and multiple proembryos of *Ephedra*, filamentous development leads to the formation of a two-celled stage in which the basal cell will undergo further development into a unicellular primary suspensor, while the apical cell will produce a secondary suspensor and embryo proper. Lack of a free nuclear stage of development in the proembryo of *Welwitschia* and each of the multiple proembryos of *Ephedra*, again highlights the anomalous nature of mitotic proliferation of each sexually formed zygote nucleus in *Ephedra*. Although there have been a number of studies of fertilization and early embryogeny in *Gnetum*, discriptions within and among species vary considerably (Maheshwari and Vasil, 1961). At present, the pattern of early embryo development in *Gnetum* cannot be resolved with certainty (Crane, 1985a,b).

If the free nuclear proliferation of each zygote nucleus is unique (apomorphic) to *Ephedra* and not developmentally or evolutionarily homologous with the free nuclear pattern of embryogeny of basal seed plants, phylogenetic analysis indicates that the common ancestor of Gnetales and, by inference, the common ancestor of Gnetales and angiosperms, had already attained a cellular embryogeny through the loss of expression of initial free nuclear development of the proembryo (Fig. 29A). According to this interpretation, *Welwitschia* has retained a basically unchanged representation of the cellular pattern of embryo development that characterized the common ancestor of Gnetales and angiosperms. In *Ephedra*, a cellular pattern of early embryo development has also been preserved, but an additional process expressed during post-fertilization development has evolved: the free nuclear proliferation of the zygote to yield multiple 'clonal' zygote nuclei that serve as the starting points for separate unicellular proembryos. As a consequence of this new interpretation of the evolution of patterns of embryogeny, a previosly undetected synapomorphy (a shared derived character state) of Gnetales and angiosperms is revealed: in *Ephedra*, *Welwitschia* and angiosperms, individual proembryos are initiated as unicellular/ uninucleate structures and the free nuclear phase of development expressed prior to organization of a cellular proembryo in extant basal seed plants (cycads, conifers and *Ginkgo*) has been lost. Thus, the presence of cellular embryogeny in *Welwitschia* and angiosperms can no longer be viewed as the result of convergent evolution (i.e., as homoplasious) (see Fig. 29B).

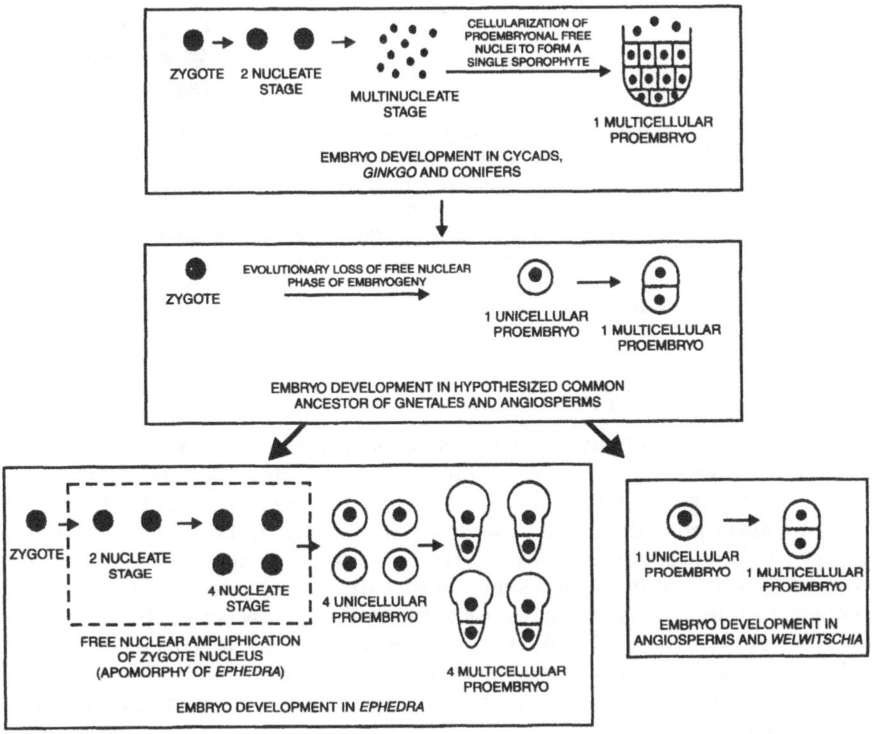

Fig. 29 A: New hypothesis of evolution of early embryo development among seed plants.

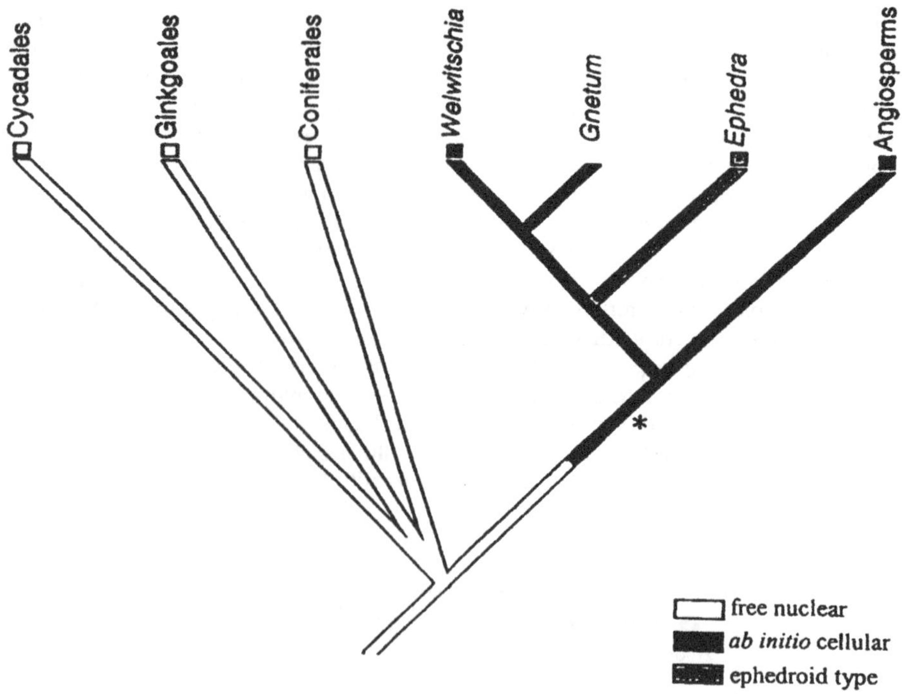

Fig. 29 B: Evolutionary trends of early embryo development among seed plants. Free nuclear development is still considered a synapomorphy of seed plants. At some point in the common ancestry of Gnetales and angiosperms, cellular embryogeny evolved through loss of an initial free nuclear phase of development. In *Ephedra*, the cellular pattern of embryogeny has been modified and a unique feature has arisen in which the zygote nucleus produces four 'clonal' free nuclei that serve as the starting points for individual cellular embryos. Hence, in *Ephedra*, embryogeny represents a modified cellular pattern. The asterisk indicates that double fertilization likely evolved *after* the establishment of a pattern of cellular embryo development. The condition of early embryo development in *Gnetum* is uncertain.

Developmental nature of primitive endosperm.

Expression of an essentially cellular development of proembryos from each of the two fertilization products in *Ephedra* indicates that the developmental loss of free nuclear embryogeny may predate the evolution of a process of double fertilization and subsequent production of a supernumerary embryo. If the common ancestor of Gnetales and angiosperms displayed a cellular pattern of embryogeny in each of its two fertilization products, this suggests that the supernumerary embryo in this common ancestor, which later gave rise to the endosperm, was also characterized by a cellular pattern of development. This, in turn, leads to the conclusion that the endosperm of the earliest flowering plants inherited and expressed an *ab initio* cellular pattern of development, just as the embryo in flowering plants is also cellular in its development.

The hypothesis that the antecedent of endosperm and earliest subsequent manifestation of endosperm were cellular in development leads to the prediction that this pattern of development should also be characteristic of extant basal flowering plants (barring any character reversals). Based upon phylogenetic analysis of the relationships of basal flowering plants (Donoghue and Doyle, 1989) and analysis of the distribution of endosperm developmental patterns among angiosperms, Donoghue and Scheiner (1992) have observed that *ab initio* cellular endosperm characterizes many of the families of flowering plants within the woody Magnoliales and other putatively basal groups of flowering plants. On the basis of this analysis, Donoghue and Scheiner (1992) have suggested that *ab initio* cellular endosperm is ancestral among angiosperms.

More recent phylogenetic analyses of basal flowering plants (Hamby and Zimmer, 1992; Taylor and Hickey, 1992; Chase *et al.*, 1993; Doyle *et al.*, 1994), derived from molecular and/or morphological data sets, have produced hypotheses of phylogenetic relationships that differ from those of Donoghue and Doyle (1989). However, if the distribution of endosperm developmental characters is analysed within the context of any of the current alternative hypotheses of angiosperm relationships, the putative basal flowering plant taxa within each of these phylogenetic hypotheses also tend to exhibit *ab initio* cellular endosperm. Thus, irrespective of which of the current major competing hypotheses for basal angiosperms may ultimately be best supported, *ab initio* cellular endosperm is likely to characterize basal taxa of flowering plants. These findings are congruent with the concept that the antecedent of the endosperm (a supernumerary embryo) was cellular in development and ultimately gave rise to an endosperm characterized by *ab initio* cellular development.

Finally, careful examination of the distribution of endosperm developmental characters among putative basal flowering plants as well as derived taxa indicates that the free nuclear endosperm has had numerous independent origins over the course of flowering plant evolutionary history. In addition, reversions to an *ab initio* cellular endosperm have been frequent (Stebbins, 1974; Cronquist, 1988; Kapil and Bhatnagar, 1991; Takhtajan, 1991).

Based upon a comparative developmental and phylogenetic analysis of fertilization and embryogenesis in angiosperms and their closest extant relatives, it is possible to construct an explicit historical hypothesis for the origin and early evolution of double fertilization and the endosperm. Formulation of this historical record is derived entirely from and dependent upon determination of the reproductive features that are likely to have characterized the common ancestors of angiosperms and their closest living relatives, the Gnetales.

Comparative analysis of reproduction in basal angiosperms and Gnetales indicates that the establishment of a process of double fertilization predates the origin of flowering plants. A regular second fertilization event is likely to have arisen as a consequence of the tendency of a second sperm nucleus to fuse with the sister nucleus of the egg nucleus (ventral canal nucleus) in the common ancestor of angiosperms and Gnetales. Initially, this second fertilization event produced a supernumerary zygote that developed into a diploid embryo genetically identical to the normal embryo derived from the fusion of a sperm and egg nucleus. Subsequent to the divergence of angiosperms and their sister lineage (which includes the Gnetales), development of the supernumerary embryo was significantly modified in the immediate ancestors of angiosperms. This led to a divergence in development and function between the two fertilization products and the metamorphosis of the supernumerary embryo into a unique embryo-nourishing structure, the endosperm (Fig. 30). The assumption of embryo-nourishing behaviour by the endosperm in turn enabled the loss of nutritive

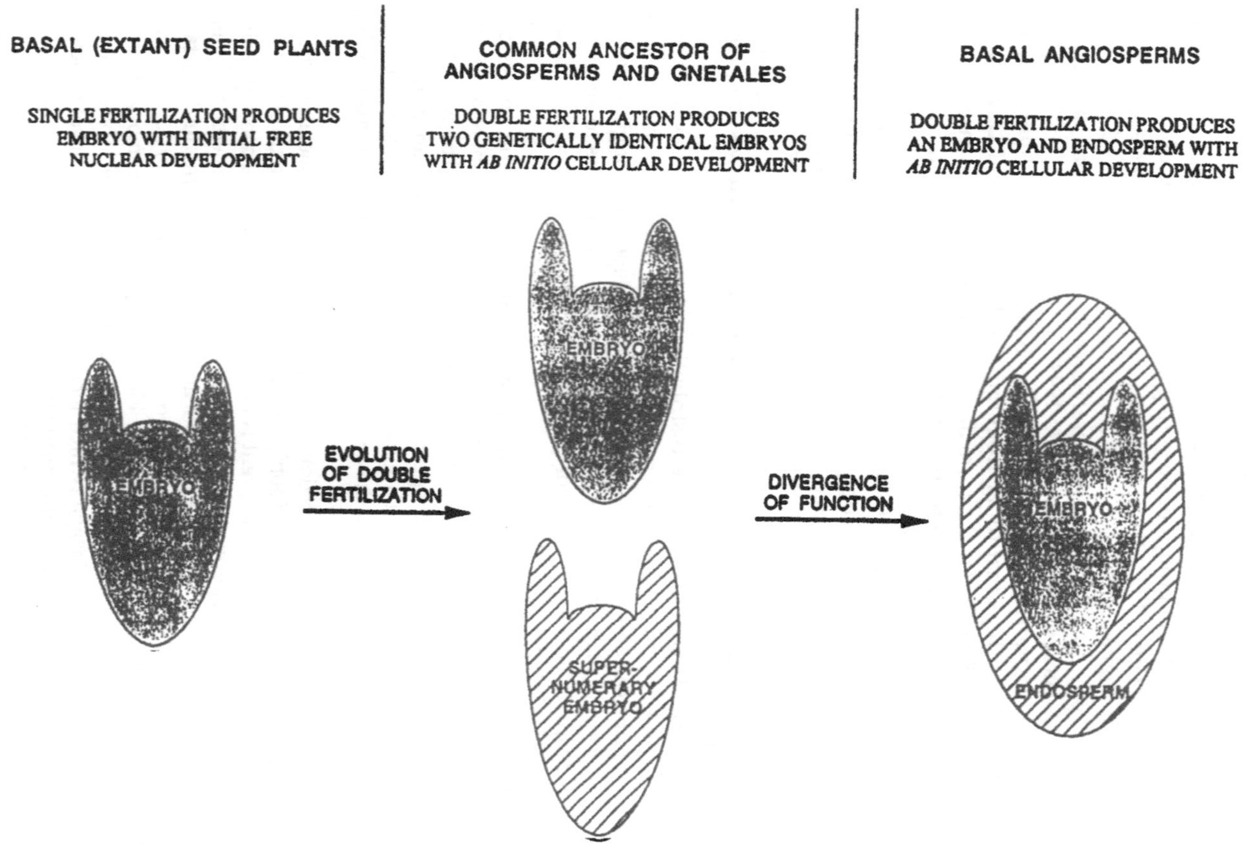

Fig. 30: Hypothesis for evolution of the endosperm from a supernumerary embryo.

function by the female gametophyte, and its subsequent reduction to an embryo sac form. It is still unclear whether the evolution of triploidy in the second fertilization event predates or post-dates development of a physiological endosperm.

From the perspective of the evolution of life cycles of land plants, the origin of the endosperm from a supernumerary embryo indicates that the endosperm can no longer be viewed as a 'tissue'. Rather, as already mentioned, the endosperm must be seen in its evolutionary context as a highly modified embryo or organism, as first espoused by Sargant (1900).

Comparative analysis of embryogeny within Gnetales leads to a significant revision of explicit and implicit assumptions about the evolution of embryogenic patterns of development among seed plants. In particular, the current evidence is most congruent with the concept that the common ancestor of angiosperms and Gnetales, in which double fertilization had been established, expressed a cellular pattern of embryogeny. If this hypothesis is correct, the antecedent of the endosperm (a supernumerary embryo) in this common ancestor of Gnetales and angiosperms was also likely to have been *ab initio* cellular in development. This, in turn, suggests that the endosperm of the earliest flowering plants inherited and expressed a pattern of cellular development.

Finally, it is hoped that as our understanding of plant phylogenetic relationships continues to expand, comparative data will be carefully re-examined and further analyses of the evolution of biological characters will be undertaken. This approach has already yielded significant new insights into the origin of land plants (Graham, 1993) and will ultimatelly allow questions concerning the origin and early evolution of major monophyletic groups of plants such as vascular plants, seed plants, and many of the key monophyletic lineages among angiosperms to be addressed. It is clear that the potential to solve some of the most intractable issues associated with the evolution of defining biological features will only be realized through an appreciation of the critical and interdependent roles of explicit developmental, comparative and phylogenetically based analyses.

Possibility of Distinguishing a Tubifloral Type of Endosperm Development (Plate XXII)

Tubifloral type of endosperm development is characterized by laying of a transversal cell wall (to the longitudinal axis of the embryo sac) during division of the primary cell of the endosperm, early differentiation of terminal haustoria, formation of the endosperm proper from derivatives of either a chalazal cell, or a micropylar cell, or from derivatives of both cells of the two-cellular endosperm.

This type of development of the endosperm (Kamelina, 1991) is analogously termed helobial or **Helobiae**-type, since this peculiar development of the endosperm occurs in families included in the group previously called Tubiflorae (Schnarf, 1917, 1929; Crété, 1951b, 1955a; Mauritzon, 1934, 1935; Krishna Iyengar, 1942, 1947).

In a formed endosperm one can distinguish the central reserve tissue, terminal haustoria, and in some cases specialized tissue in the form of a column of elongated cells (the 'isthmus' of the endosperm) located between the central part of the endosperm and micropylar haustorium (Mauritzon, 1935; Krishna Iyengar, 1940; Teryokhin and Nikiticheva, 1981). Sometimes the 'isthmus' of the endosperm is formed from the side of the chalazal haustorium (Murthi, 1946; Nikiticheva, 1979a,b).

The Tubifloral type of endosperm development is mostly typical of species with a unitegmic tenuinucellate ovule in which the integumentary tapetum only surrounds part of the embryo sac. The micropylar part of its central cell sometimes becomes haustorium-like by the moment of fertilization, which is very clearly manifested in families Escalloniaceae and Lamiaceae (Kamelina, 1984; Kamelina and Dzevaltovsky, 1987). In family Campanulaceae it was observed that both ends of the central cell of the embryo sac function as haustoria after fertilization, even before the primary nucleus of the endosperm divides (Shamrov and Zhinkina, 1994).

During endospermogenesis the terminal haustoria grow considerably and the number of nuclei and cells varies in them. The cells reveal metabolic activity, which indicates lobed nuclei and the presence of mitochondria, RER, plastids, dictyosomes and a great number of ribosomes in the cytoplasm. The presence of projections of the cell wall and destruction of adjacent cell walls of the integument indicate that the haustoria function as transfer cells (Bhandari and Sheel, 1992). Haustoria are observed in a mature seed as well.

Variation in orientation of walls in the endosperm starting from the second division, as well as differences in cell differentiation allowed to distinguish a number of 'types' of endosperm development in certain families (for example: Lamiaceae—Schnarf, 1917, Polyshchuk and Dzevaltovsky, 1971; Campanulaceae—Rosén, 1932, 1949; Crété, 1951b; Subramanyam, 1951; Kapil and Vijayaraghavan, 1962, 1965; Korobova and Zhinkina, 1987; Shamrov and Zhinkina, 1994; Acanthaceae—Mauritzon, 1934, Mohan Ram, 1960a,b; Kapil and Vijayaraghavan, 1962; Bignoniaceae—Mauritzon, 1935; Scrophulariaceae—Yamazaki, 1957; see Cellular Type of Endosperm Development above). These types can now be fairly regarded as variations of the Tubifloral type (Fig. 31).

The Tubifloral type of endosperm development occurs in 30 families of dicotyledons and one family (Araceae) of monocotyledons.

Dahlgren (1980a), while considering the distribution of types of endosperm development in connection with his system of classification of angiosperms, noted that the cellular type of endospermogenesis, typical of many families at the phylogenetic system base, is not identical to that of families of Sympetalae advanced plants of dicotyledons. Such an endosperm, cellular with terminal haustoria, is considered more advanced. Palser (1951) using the example of family Ericaceae suggested that this type of endospermogenesis should be termed 'haustorial', which is not quite appropriate since haustoria can be formed in all types of endosperm development.

The Tubifloral type of endosperm development differs considerably from the Cellular and Helobial types. Features that distinguish the Tubifloral type of development from the Cellular are: persistently transversal laying of the wall in the primary cell of the endosperm; early differentiation of the haustoria in the first and/or 2-3rd subsequent divisions; formation of endosperm proper only from derivatives of the primary micropylar or primary chalazal cells; and presence of specific tissue ('isthmus' of endosperm).

The Tubifloral type of endosperm development differs from the Helobial type by another assignment of the primary micropylar and chalazal cells: the latter may be not only an initial of the haustorium, but also, after transversal division, initials of the endosperm proper and the haustorium. Other distinctive features: the primary micropylar cell does not have a coenocytic stage and the micropylar haustorium is normally highly developed. Besides, contrary to the Tubifloral type, in the Helobial

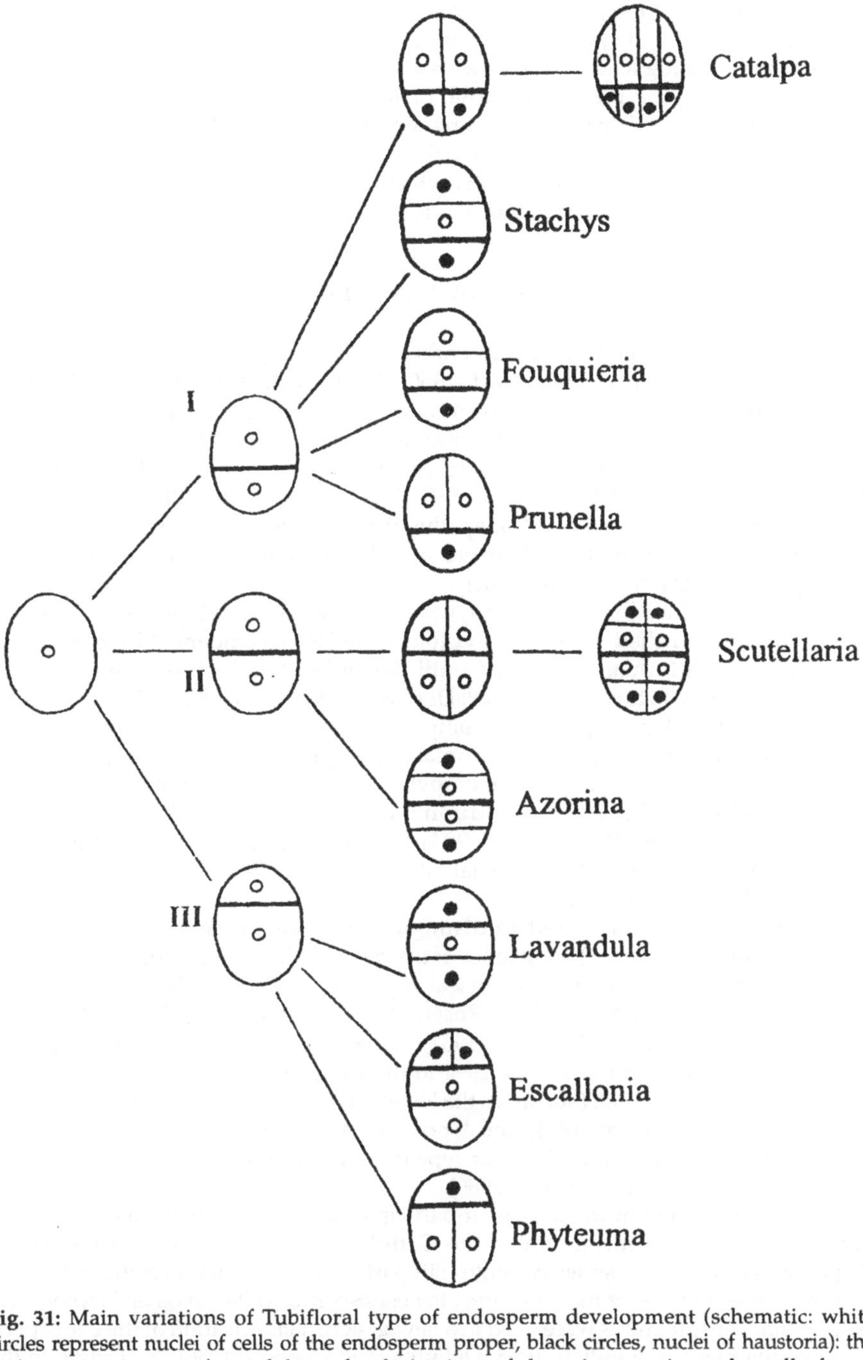

Fig. 31: Main variations of Tubifloral type of endosperm development (schematic: white circles represent nuclei of cells of the endosperm proper, black circles, nuclei of haustoria): the endosperm proper is formed from the derivatives of the primary micropylar cell of two-cellular endosperm (I), from both its cells (II) or from the primary chalazal cell (III).

type of development the primary chalazal cell does not take part in formation of the endosperm proper and specialized tissue is not formed.

All the foregoing leads to the conclussion that the Tubifloral type of endosperm development may be regarded as one of the fundamental types of endosperm development, since it is characterized by a number of features which considerably distinguish it from the other types. Further, this type of endosperm development includes a great number of variations which have similar regularities. It is quite widespread among flowering plants and is typical of the major advanced families.

Evolutionary Aspects of Endosperm Development

The evolutionary transformations in endosperm development of flowering plants are still being debated (Maheshwari, 1950; Wunderlich, 1959; Modilewsky, 1963; Poddubnaya-Arnoldi, 1976; Kordyum, 1978; Bhatnagar and Sawhney, 1981; Vijayaraghavan and Prabhakar, 1984; Kamelina, 1991; Teryokhin, 1991, 1996). The initial type is considered to be Nuclear (Schnarf, 1929; Sporne, 1954; Khudyak, 1963; Poddubnaya-Arnoldi, 1976; Battaglia, 1980; Kamelina, 1991) or Cellular (Coulter and Chamberlain, 1903; Swamy and Ganapathy, 1957a,b; Friedman, 1994, 1995). There is also an opinion (Palm, 1915; Kordyum, 1978) that both types of endosperm development appeared simultaneously.

In most classifications two main types of endosperm development in flowering plants are distinguished: Cellular (divisions of nuclei accompanied by cytokinesis) and Nuclear (cell formation delayed until the endosperm becomes multinucleate) (Samuelsson, 1913; Schnarf, 1917, 1929; Di Fulvio, 1983, 1985; Di Fulvio and Cocucci, 1986). The Helobial endosperm is usually considered a type possessing features of both the former types: division of the primary endosperm nucleus results in formation of two cells, which (typically the micropylar one) later undergo only free nuclear divisions (Khudyak, 1963; Poddubnaya-Arnoldi, 1976; Kordyum, 1978; Vijayaraghavan and Prabhakar, 1984). Some authors add a fourth type—Intermediate—along with the Helobial (Wunderlich, 1959) or the Tubifloral along with the Cellular (Kamelina, 1991).

A several criteria are used to distinguish various hierarchical categories of endosperm development: 1) occurrence of the cell wall during division of the primary endosperm nucleus (Cellular and Nuclear types); 2) position of the first and second cell walls ('types'—Mauritzon, 1935; Rosén, 1949; Wunderlich, 1967; Di Fulvio, 1983; 'forms'—Schnarf, 1929; Glišić, 1936-1937; 'variations'—Kapil and Vijayaraghavan, 1962, 1965; Shamrov and Zhinkina, 1994 in the Cellular endosperm); 3) the number of nuclei in the chalazal cell ('forms' in the Helobial type of endosperm development—Swamy and Parameswaran, 1963); and 4) position of nuclei in early endospermogenesis ('types' and 'subtypes' in the Nuclear type of endosperm development—Di Fulvio, 1983, 1985; Di Fulvio and Cocucci, 1986).

Taking into account the degree of participation of the primary micropylar and chalazal cells in the formation of the Cellular endosperm, Samuelsson (1913) distinguished two series: **series A**—both cells participate in formation of the endosperm and **series B**—only one of the cells forms the endosperm, while the other functions as a haustorium. These features were used in a new general classification of the endosperm in flowering plants (Di Fulvio, 1983; Di Fulvio and Cocucci, 1986). According to this classification the Cellular type of endosperm development as well as the Nuclear were designated as megatype, subtypes as types, and variations as subtypes. The Cellular megatype of endosperm includes 5 types and 12 subtypes. The types are subdivided into two groups which differ in position of the cell wall during the first nuclear division. The first group is characterized by longitudinal laying down of the cell wall; it combines **the isobilateral type** (both cells participate in endosperm formation) and **heteropolar type** (only one of the cells participates, while the other forms the lateral haustorium). The second group includes types wherein the first division is transverse: **isopolar** (both cells participate in constructing the endosperm), **heteropolar-micropylar** (micropylar cell participates in formation of the endosperm while the chalazal one gives rise to the haustorium) and **heteropolar-chalazal** (the chalazal cell forms the endosperm the micropylar the haustorium).

In our opinion, the criteria proposed do not entirely reflect the whole specificity of endosperm development, and in most classifications the Helobial type, as an Intermediate one, is neglected even, though it occurs in a considerable number of taxa of flowering plants. Only a few papers have noted the significance of the Helobial type in the evolution of the endosperm. In an investigation by Herr (1999) is it supposed that the Helobial type might be viewed not as an intermediate step in derivation of cellular from nuclear endosperm, but rather as a parallel evolutionary event. This perhaps indicates the need for a re-examination of typology across the whole gamut of endosperm development in angiosperms. Maheshwari (1950) had already pointed to imperfection of the classifications of the types of endosperm development, as there were transitional forms among different types. According to Teryokhin's opinion (1996), in deciding this question it is important to consider that the mode of endosperm development as an evolutionary labile structure with various haustorial formations peculiar to it, is bound to adaptive particularities of seed germination. Understanding the problem of endosperm typology is quite impossible without taking into consideration the peculiarities of evolutional appearance of various modes of its development. Earlier, a similar approach was used, for example, to study the embryogenesis of cereals (Batygina, 1987a). This forced us to review the idea of principles of endosperm typification and to reconsider the existing diversity of types and variations with respect to data on evolutional reformations. As a principle of the hierarchic classification of the modes of endosperm development we have taken that approach which allows not only typification of modes of formation of the endosperm, but also an explanation of their initiation in flowering plants. We share Teryokhin's views (1994) of the principles taken by him as a basis when working out the classification of the types of embryo sac development (morphogenetic origin, mode of formation, structural organization). The classification proposed consists of three hierarchical levels—types, subtypes and variations—each of which has its own criteria (Fig. 32). The characteristic feature of endosperm development in a considerable number of plant species is that in the first nuclear division in the endosperm is followed by formation of two primary cells—micropylar and chalazal. As a basis for distinguishing the types, the first consideration is the morphogenetic potentialities of the primary micropylar and chalazal cells of the endosperm expressed in the processes of cyto- and karyokinesis. Then, subject to which of these processes is dominant in the period of early endospermogenesis, we propose to distinguish two main **types of development**[10]**: Cellular** (karyokinesis is completed in both cells with cytokinesis) and **Helobial types** (only karyokinesis takes place in both cells and sometimes the chalazal cell remains uninucleate).

[10]The Nuclear type is considered a subtype of the Helobial type of endosperm development.

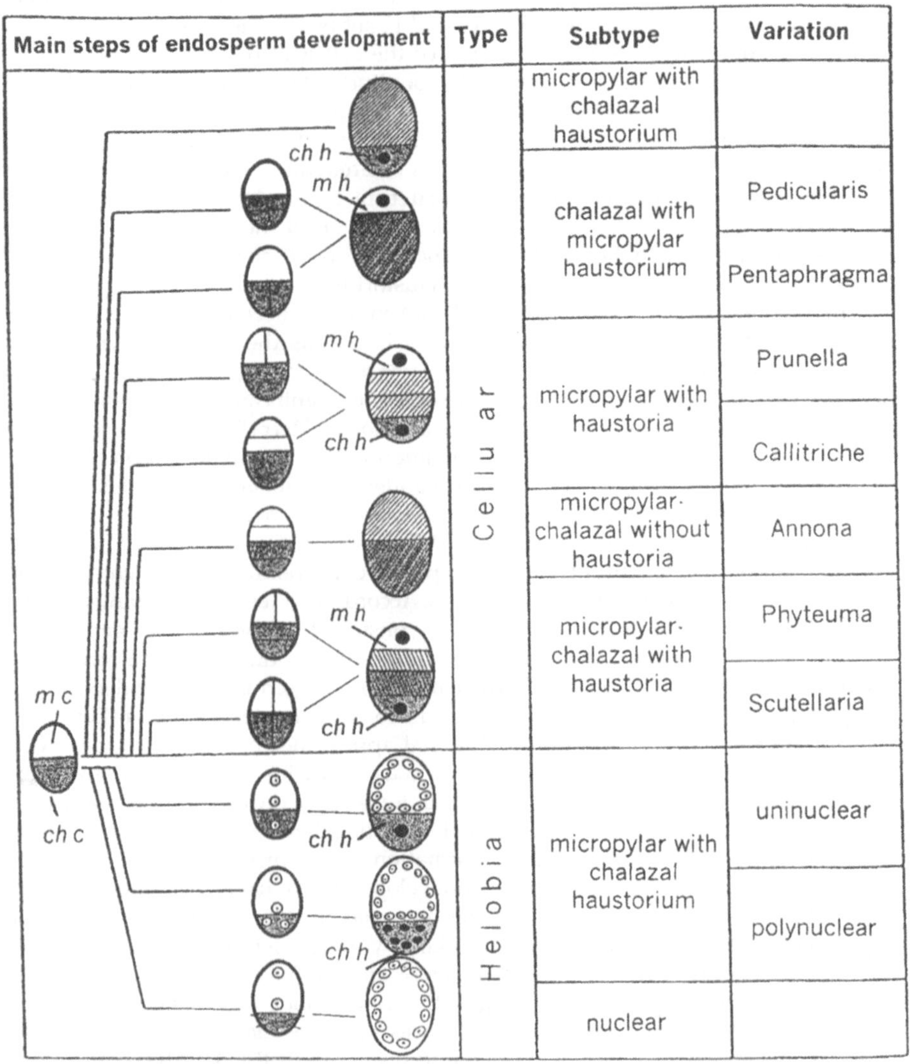

Fig. 32: Hierarchical classification of endosperm development type (subtypes and variations) in flowering plants; *ch c*—chalazal endosperm cell, *ch h*—chalazal endosperm haustorium, *m c*—micropylar endosperm cell, *m h*—micropylar endosperm haustorium.

Each type is subdivided into a number of **subtypes** according to the degree of participation of the micropylar and chalazal cells in constructing the endosperm. **In the Cellular type** the following **subtypes** are distinguished:

• **Micropylar with chalazal haustorium**—endosperm forms from the micropylar cell, and the chalazal cell functions as a haustorium (some Icacinaceae, Fouquieriaceae, Saururaceae and many of Nymphaeaceae);

• **chalazal with the micropylar haustorium**—micropylar cell becomes a haustorium, while the chalazal forms the endosperm (Loasaceae, some Campanulaceae, Lamiaceae, Scrophulariaceae);

• **micropylar with terminal haustoria**—endosperm forms out of lower derivatives of the micropylar cell, while the chalazal cell, as well as the upper derivatives of the micropylar cell, form the terminal haustoria (Buddlejaceae, Crassulaceae, Gesneriaceae, Orobanchaceae, Pedaliaceae, Plantaginaceae and many Scrophulariaceae and Verbenaceae);

• **micropylar-chalazal with terminal haustoria**—both cells of the two-celled endosperm form the endosperm and terminal haustoria (Callitrichaceae, some Campanulaceae, Lentibulariaceae, Lobeliaceae, Stylidiaceae);

• **micropylar-chalazal with no haustoria**—both cells participate in formation of the endosperm (Annonaceae, Aristolochiaceae, Ceratophyllaceae, Degeneriaceae, Monimiaceae, Nelumbonaceae, Sarraceniaceae, Winteraceae).

In the Helobial type it **two subtypes** can be singled out:

• **micropylar,** or Helobial type proper—endosperm formed out of the micropylar cell, while the chalazal transforms into a haustorium (Aponogetonaceae, Asphodelaceae, Costaceae, Hydrocharitaceae, Limnocharitaceae, Phormiaceae, Pontederiaceae, Potamogetonaceae, Ruppiaceae, Zannichelliaceae and some Cabombaceae and Nymphaeaceae);

nuclear—falling out of the first nuclear division, in consequence of which only karyokinesis is carried out during early endospermogenesis (Brassicaceae, Cactaceae, Cyperaceae, Eriocaulaceae, Fabaceae, Gentianaceae, Poaceae, Restionaceae, Rosaceae, Zosteraceae).

In each subtype of the Cellular type a number of variations[11] can be distinguished according to the position of the walls during the second division in the endosperm. Chalazal with micropylar haustorium subtype: **Pedicularis-variation**—chalazal cell divides transversely, which results in formation of a linear triad of cells (Lavandula-variation[12] included); **Pentaphragma-variation**—division in the chalazal cell is longitudinal and the triad of cells inversely T-shaped. Micropylar with terminal haustoria subtype: **Prunella-variation**—longitudinal division passes only in the micropylar cell and a T-shaped triad of cells forms (*Alectorolophus-*, *Galeopsis-*, *Limosella-*, *Linaria-* and *Veronica* VI-variations included); **Callitriche-variation**—transverse division takes place in the micropylar cell and the linear triad of cells appears (*Incarvillea-* and *Stachys*-variations included). Micropylar-chalazal with terminal haustoria subtype: **Phyteuma-variation**—micropylar cell divides longitudinally and chalazal cell undergoes transverse division, which leads to formation of the T-shaped quaternion of cells (*Isotoma*-variation included); **Scutellaria-variation**—division in both cells is longitudinal and accompanied by formation of the isobilateral tetrad of cells (*Catalpa-*, *Codonopsis-*, *Prolimosella-*, *Sphenoclea-*, *Verbascum-* and *Veronica* I-V-variations). Micropylar-chalazal without haustoria subtype: **Annona-variation**—division in both cells transverse and linear quaternion of cells forms (Azorina-, Ceratophyllum- and Ericaceae-variations included).

[11]The authors who proposed the variations are cited in the Cellular type of endosperm development.

[12]Here and further on those variations enclosed in parentheses are considered modes of formation of haustoria in the endosperm.

In the micropylar subtype of the Helobial type two variations are distinguished according to the number of nuclei in the chalazal cell of the two-celled endosperm: **uninuclear** (= only A-form according to Swamy and Parameswaran, 1963; Aponogetonceae, Hydrocharitaceae, Limnocharitaceae, Potamogetonaceae, Ruppiaceae, Zannicheliaceae and some Cabombaceae, Nymphaeaceae) and **polynuclear** (= B-E-forms according to Swamy and Parameswaran, 1963; Asphodelaceae, Costaceae, Phormiaceae, Pontederiaceae, Typhaceae).

The appearance of existing modes of formation of endosperm proceeded in relation to the character of development of the surrounding tissues of the ovule and seed (Wunderlich, 1959; Nikiticheva, 1979a,b; Erdelská, 1981a; Vijayaraghavan and Prabhakar, 1984; Teryokhin, 1991; Shamrov, 1995. The interrelations between endosperm and embryo on the one hand and antipodals on the other, must have been highly significant as also the specificity of transport of nutrients in the ovule and seed, which carried on through the system of sequentially functioning haustorial structures.

Evidently, evolution of the endosperm of flowering plants proceeded in several directions. Distribution of types of endosperm in plants is the first evidence of this fact. For monocotyledons the Helobial (34 families) and Nuclear (30 families) types of endosperm development are typical, for dicotyledons Nuclear (156 families) and Cellular (135 families) (Kamelina, 1991). However, in primitive monocotyledons the endosperm is mostly Helobial and less commonly Nuclear, while in advanced monocotyledons only the Nuclear type of endosperm is found. In primitive dicotyledons a combination of Cellular endosperm and Nuclear one was quite common, while in advanced taxa either the Nuclear type or Cellular with haustoria type are observed.

In our opinion, the initial type should be considered the endosperm development which has been characterized by features of the Cellular and Helobial types (presence of 2-celled stage after division of the primary endosperm cell, simultaneous with interconnected processes of karyo- and cytokinesis or only karyokinesis passing in the micropylar cell) (Fig. 33). As evidence for this is a peculiar endosperm development in representatives of a number of families. So, in family Aquifoliaceae (endosperm of Cellular type of development) after formation of two cells, subsequent divisions in them may not be accompanied by cytokinesis (Herr, 1961a). For representatives of family Saxifragaceae an endosperm of the Helobial type of development is typical, though it can be Cellular or Nuclear (Korobova-Semenchenko, 1977). In species of *Impatiens* (Balsaminaceae) division of the micropylar cell results in formation of a row of three cells, of which the upper one transforms into a micropylar haustorium while in the lower and middle cells free nuclear divisions pass (Dahlgren, 1934). The same is observed in the Loganiaceae family as well (Yamazaki, 1963). In family Acanthaceae the endosperm formed on the basis of the middle cell can from the very beginning be either cellular (*Acanthus, Peristrophe*) or free nuclear (*Barleria, Ruellia*) (Mohan Ram, 1962; Karlstrom, 1974a,b; Madhavan and Gupta, 1982). The endosperm in *Arabidopsis thaliana* is widely accepted as following the Nuclear type development. However, according to Herr (1999) one nucleus of the binucleate endosperm migrates into the chalazal region and further 2-8 nuclei form which become enshrouded with dense cytoplasm, creating the effect of a Helobial endosperm. The nuclei in the chalazal region of the endosperm gradually degenerate. The functional endosperm is produced primarily by nuclei of the micropylar region.

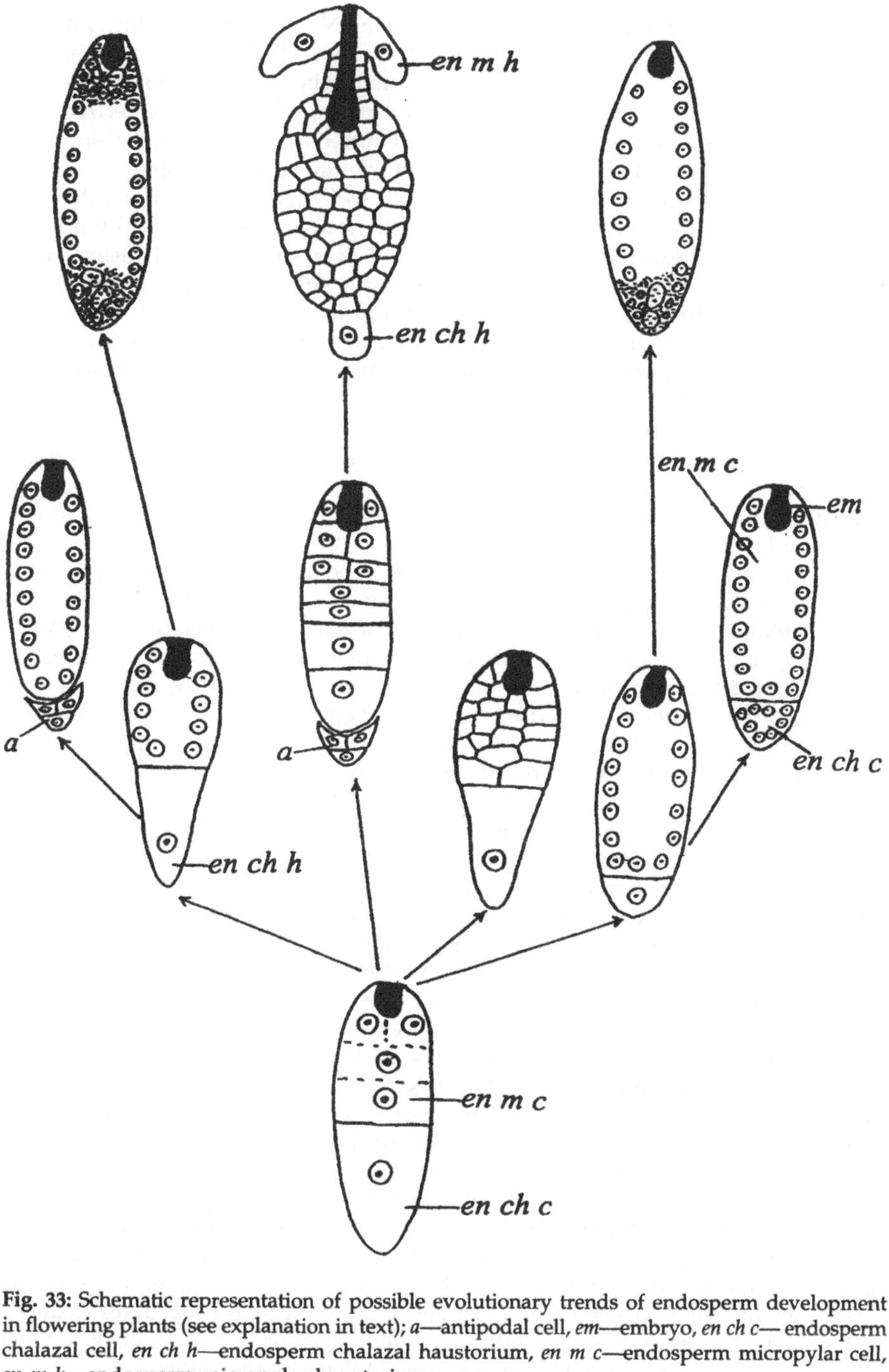

Fig. 33: Schematic representation of possible evolutionary trends of endosperm development in flowering plants (see explanation in text); *a*—antipodal cell, *em*—embryo, *en ch c*— endosperm chalazal cell, *en ch h*—endosperm chalazal haustorium, *en m c*—endosperm micropylar cell, *en m h*—endosperm micropylar haustorium.

The endosperm closest to the ancestral type of development is the cellular one (for example, Fouqieriaceae, Saururaceae and many Nymphaeaceae) which forms as a result of divisions of the micropylar cell of the two-celled endosperm, while the chalazal cell transforms into a haustorium (micropylar with chalazal haustorium subtype). Such mode of formation of the endosperm, which is characterized by a high degree of specialization of the chalazal cell from the first stages of development, has probably appeared to be a deadlock line of evolution. In our opinion, it was the endosperm formed due to divisions of both cells (primary micropylar and chalazal) that was of a greater evolutionary significance. It was characterized by a lesser degree of tissue differentiation, in that such specialized structures as haustoria were absent (micropylar-chalazal with no haustoria subtype) (Ceratophyllaceae, Monimiaceae, Nelumbonaceae, Winteraceae). In this case, absence of haustoria is compensated by formation of large, long remaining antipodals (Nelumbonaceae—Batygina *et al.*, 1982) which participate in supplying a growing endosperm with nutrients at the first stages of development. Later, on this basis the endosperm with terminal haustoria could have formed. Either upper derivatives of both cells transformed into haustoria (micropylar-chalazal with haustoria subtype—Campanulaceae, Lentibulariaceae, Lobeliaceae) or of only the micropylar cell (micropylar with haustoria subtype—Gesneriaceae, Orobanchaceae, Plantaginaceae, Scrophulariaceae).

In dicotyledons the endosperm of Nuclear type development has presumably appeared as a result of gradual reduction of the chalazal cell, whose function has been maintained by other structures. Initially formation of the endosperm passed without specific haustoria structures being shaped, which is a characteristic, for example, of a number of taxa at the bottom of the flowering plants system (Casuarinaceae, Myricaceae, Myristicaceae, Ranunculaceae, some Hamamelidaceae, Papaveraceae and other families). Further evolution was associated with heteromorphism of the endosperm and its micropylar and chalazal regions functioning as haustoria. Denser cytoplasm and aggregation of nuclei, often hypertrophied and polyploid, are observed either only at the chalazal (Aizoaceae, Gentianaceae, Polygonaceae, Rosaceae, Sapindaceae and other families) or at both poles (Brassicaceae, Caricaceae, Sterculiaceae and other families); thus there can be no cellularization passing at the chalazal pole, which later leads to formation of a multinucleate chalazal haustorium (Amaranthaceae, Elaeagnaceae, Fabaceae, Nyctaginaceae, Rhamnaceae). Representatives of the Caryophyllaceae family, for example, are characterized by a net of haustoria spread over the surface of the endosperm (Rocén, 1927; Turssunov and Savinova, 1979). A correlation can be traced between the presence of haustoria in the endosperm of the Nuclear type of development and the longevity of antipodals. In plants with antipodals that remain for a long time the endosperm forms no haustoria, while in the case of ephemeral antipodals, haustoria, first of all the chalazal one, appear.

In monocotyledons at the base of the initial type in the micropylar and sometimes in the chalazal cell of the two-celled endosperm, mainly the processes of karyokinesis pass (without cytokinesis). So, the Helobial type of endosperm had originally preferred development. Its further evolution apparently proceeded in the direction of gradual reduction of the chalazal cell of the two-celled endosperm and establishment of the nuclear mode of its development. This is proved, for example, by the presence of both types of endosperm development in a number of families (Alismataceae, Amaryllidaceae, Asparagaceae, Liliaceae, Melanthiaceae and other families). The whole chalazal pole of the nuclear endosperm has undertaken to provide the haustorial function of the chalazal cell. In the chalazal part of the endosperm of a number of taxa (Iridaceae, Juncaginaceae, Najadaceae, Zosteraceae and other families) with denser cytoplasm and aggregation of nuclei, often of irregular shape, sometimes a haustorium is formed (Arecaceae, Commelinaceae and other families). Appearance of the nuclear endosperm in monocotyledons correlates with the general reduction of the ovule (decrease of massiveness in direction from crassinucellate to tenuinucellate ovule and formation of two-layered integuments) (Shamrov, 1995, 1997a, 1998b).

According to some researchers (Smirnov, 1982; Teryokhin, 1991) in the process of appearance and further evolution of the endosperm of flowering plants, the tendency to general reduction of endosperm in a mature seed is not observed on the whole. It is only found in separate taxa, for example, in parasitic plants and can be explained by adaptation to specific conditions of existence.

Perisperm

Perisperm (Greek *peri*—around, near, *sperma*—seed)—a nutritive tissue of a seed which forms from the nucellus **(plate XXIII)**.

The term 'perisperm' was introduced by Juel (1907), though the tissue itself was first described by Brown (1826) who named it '*albumen primarium*'.

Two types of perisperm were distinguished by Juel (1907): **the true** (which remains in a mature seed) and **transitory** (which is consumed in the course of development). In current scientific literature the perisperm is usually not subdivided into two types and the notion is used only to describe those cases in which the nutritive tissue of nucellus origin remains in a mature seed. In our opinion, it is reasonable to distinguish the true and transitory types of the perisperm, which can be used for phylogenetic evaluation of taxa. So, for example, the presence of a transitory perisperm in representatives of family Ceratophyllaceae proves their connection with order Nymphaeales, characterized by the presence of a true perisperm (Batygina *et al.*, 1991; Titova *et al.*, 1993; Shamrov, 1997b).

The true perisperm is typical for taxa with crassinucellate ovules, whose nucellus **remains in a mature seed entirely or partly.** Reserve substances, most often as starch, are accumulated in the cells of the remaining nucellus. In *Hydrostemma longifolia* (Barclayaceae) starch grains are compound and consist of a multitude of small hexahedral granules of different size (Winter and Batygina, 1988). Proteins and lipids are present in the perisperm of *Yucca* (Agavaceae) species (Horner and Arnott, 1966).

A true perisperm has been discovered in some representatives of a number of orders, both of dicotyledonous (Caryophyllales, Nymphaeales, Piperales) and monocotyledonous (Zingiberales) plants, which are located at the bottom of the phylogenetic system of flowering plants.

The transitory perisperm was first described in the Saxifragaceae family (Juel, 1907). In this case (the ovule is crassinucellate too) starch is also accumulated in the nucellus, **remaining in a seed during a shorter period.** In representatives of family Ceratophyllaceae, starch deposits in the cells of the nucellus, especially in its lateral region, long before its complete destruction. This takes place at the stage of the globular embryo. Starch disappears only when passing to the organogenesis stage (Shamrov, 1995).

The genesis of the nucellus transforming into the perisperm after fertilization has not been sufficiently studied. Only the fact of a perisperm present in a seed is usually reported. In representatives of such families as Caryophyllaceae, Phytolaccaceae, Portulacaceae, the perisperm initiates from the cells of the so-called 'central body' of the amphitropous and campilotropous ovules, which is the remnant of the nucellus on the chalazal side and under which is located an embryo bent in the form of an arch or horseshoe (Cooper, 1940; Kajale, 1954; Bocquet, 1959; Kamelina and Yakovleva, 1982; Turssunov, 1986). As for representatives of the Barclayaceae and Nymphaeaceae families, it is the central part of the nucellus of the orthotropous or anatropous ovule which becomes a perisperm (Khanna, 1964, 1967; Schneider, 1978; Winter, 1987). In *Nuphar lutea* the nucellus of the ovule transforms only partly into the perisperm (Shamrov, 1998a). In this species a heterogeneous nucellus is characterized by occurrence of a number of specialized tissues, such as parietal tissue, postament and podium. In the course of seed development the postament and cells of the micropylar zone of the nucellus located around the developing endosperm as well as the parietal tissue are destroyed. The reserve starch in the form of large compound starch grains is stored in the remaining cells of the lateral region of the nucellus. Only the podium and 2-3 layers of narrow subepidermal cells moving from the podium to the embryo remain starch-free.

The size and composition of the perisperm cells in a mature seed vary. Analysing the data on perisperm presence in a seed a correlation is seen between degree of development of the perisperm on the one hand, and of the endosperm and embryo on the other. According to this fact, 4 groups of seeds with perisperm can be distinguished. **The first group**—perisperm constitutes the main part of the mature seed—there is a little endosperm and the embryo is quite small (representatives of families Barclayaceae, Cabombaceae, Cannaceae, Costaceae, Marantaceae, Nymphaeaceae, Zingiberaceae, some Saururaceae, Piperaceae). **The second group**—perisperm constitutes half of a mature seed and surrounds a large bent embryo only on the chalaza side—there is very little endosperm (representatives of families Amaranthaceae, Caryophyllaceae, Chenopodiaceae (subfamily Chenopodioideae), Basellaceae, Nyctaginaceae, Phytolaccaceae, Portulacaceae, Stegnospermataceae). **The third group**—perisperm in the form of a thin peripheral layer—the embryo is not large and the endosperm occupies the main part of a mature seed (representatives of families Heliconiaceae, Musaceae). **The fourth group**—perisperm found only around the tops of the cotyledons—the embryo is large and occupies almost the whole mature seed and reserve substances accumulate in it; the endosperm, in the form of a thin layer, is situated around the embryo root (representatives of family Chenopodiaceae, subfamily Salsoloideae).

Only a few ultrastructural investigations of perisperm cells have been done. It is reported that in representatives of genus *Yucca*, the outer cell walls of the peripheral layer of the perisperm are quite thick and contain polysaccharides. Plasmodesmata occur between the cells of the inner layers of perisperm (Horner and Arnott, 1966). In the cytoplasm of the cells of *Silene alba* (Caryophyllaceae), numerous mitochondria, free ribosomes and large plastids with a multitude of starch grains (Mohana Rao *et al.*, 1988) have been detected.

Most authors consider the presence of a perisperm in a seed a primitive feature.

PART THREE—EMBRYO

OVERVIEW

Embryo

The embryo is a rudiment of a new plant. The term appears to be borrowed from zoological literature. In the evolutionary-morphological sense, the embryo represents a rudiment of a new sporophyte generation in the life cycle of higher plants. In the embryology of seminal plants, the sporophyte at all stages of development inside the seed—from zygote up to the compound multicellular formation differentiated in tissues and organs in the mature seed, is referred to as an embryo.

In the genetic sense, angiosperm embryos can be divided into two main categories: so-called **sexual embryos**, derived from zygote or fertilized synergid, and **parthenogenetic embryos**, developed from an unfertilized egg cell (or synergid). **Parthenogenetic** embryos in agamospermous seeds may contain either haploid (in so-called reduced forms of gametophytic apomixis) or diploid (in unreduced forms) chromosome number. Embryo-like formations derived from somatic cells of the nucellus, integument, as well as the result of cleavage embryony, are termed embryoids and the process of their production referred to as embryoidogeny (Batygina, 1993a; see below: Embryoid, Embryoidogeny).

The embryo formation process taking place in the seed is termed embryogenesis. Its regular taxon-specific features permit classification of different patterns (types) of embryogenesis in flowering plants (Schnarf, 1929; Souèges, 1939c; Johansen, 1950; Batygina, 1974b). These classifications are commonly based on the diversity of succession and cell division pattern (blastomeres) in the first period of embryo development and likewise on morphogenetic potential of the blastomeres created (see below: Principles of Classification of Embryogenesis Types). Herewith investigators attach the most importance to those processes, as a result of which the proembryo (embryo so called in early stages) develops into the embryo proper (embryo 'body') and suspensor.

Attempts to attach phylogenetic significance to the classification of embryogenesis types have obviously failed to be successful (Yakovlev, 1958; Lebègue, 1966). On the other hand, within separate taxa (e.g. family level) the specific character of evolutionary modifications in embryogenesis is likely to be observed in some cases. The taxa of xenoparasitic (Scrophulariaceae, Orobanchaceae) or mycoparasitic (Pyrolaceae, Monotropaceae, Burmanniaceae, Orchidaceae) plants are of the most interest in this respect (Teryokhin, 1977).

Embryogenesis is traditionally subdivided into several stages reflecting various aspects of morphogenesis: 'proembryonal' (from zygote to differentiation of embryoderm), 'globular' (from embryoderm differentiation to initiation of cotyledons), 'heart-shaped' (initiation and first stages of cotyledon formation), 'torpedo-shaped' (the process of intensive axial growth of embryo and production of primary structures of hypocotyl and radicle) and stage of 'formed' embryo (nature and degree of development of its structures in mature seed).

The diversity of shape and structure of embryos in mature seeds is difficult to describe or classify. The peculiarities of embryo organization in ripe seeds (as Eames has already mentioned, 1961) are visibly associated with their adaptation to the mode of germination. To understand this is it suffices to compare two main categories of embryo: embryos in mature seeds of dicotyledonous and monocotyledonous plants or embryos in mature seeds of autotrophic and parasitic plants. In the latter, the embryo of a ripe seed can consist of 2-5 cells at most (*Allotropa virgata, Hypopitys monotropa*), whereas embryos of autotrophic plants represent multicellular, fairly complex organized formations.

Ordinarily the embryo of autotrophic plants comprises several main organs which ensure development of the seedling: **epicotyl** (apical part of embryo axis located between cotyledon bases), **cotyledons** proper (rudiments of first leaves, produced on lateral side of apical portion of embryonal axis), **hypocotyl** (basal part of embryonal axis situated between cotyledon base and radicle) and proper **radicle** (embryo root). In cases when the epicotyl develops enough and bears rudiments of the first true leaves, it is termed a plumule.

Embryos of monocots create only one cotyledon, at the base of which the epicotyl forms. This highly remarkable organization peculiarity is characteristic of embryos in grains. Therein the basal part of the embryo is replaced by coleorhiza (protective tissue) while one of the adventive roots fulfils the function of radicle. This and the other adventive roots break through the coleorhiza during germination. In embryos of grains a scutellum develops as well, representing the expanding hypocotyl portion and fulfilling the haustorial function in reference to the endosperm.

Evolutionary transformations of radicle (and sometimes hypocotyl) into nutritive or haustorial structures are also typical for other angiosperm groups. For example, in orchids (mycoparasitic plants) the basal part of the embryo is modified into haustorial-nutritive tissue, nourishing the embryo during germination, in whose cells digestion of mycorrhizal fungal hyphae that penetrate this tissue takes place (Burgeff, 1936). In some members of water plants also—*Utricularia* (Lentibulariaceae) and *Ruppia* (Ruppiaceae)—the basal part of the embryo transforms into reserve tissue. In the latter the adventive root produced on the stem below the plumule, carries out the role of radicle.

Transformation of the radicle into a haustorial structure is characteristic of some taxa of xenoparasitic dicots (Loranthaceae, Viscaceae) living on branches of woody plants. The group of stem parasitic plants from Lauraceae (*Cassytha*) and Cuscutaceae (*Cuscuta*) is characterized by elimination of the radicle.

A comparative analysis of embryos of modern angiosperms reveals no single evolutionary trend (Teryokhin, 1991). Evolutionary transformations of angiosperm embryos possibly reflect not only their ecological adaptive radiation, but the adaptive radiation of plants as a whole.

Two important periods of divergent evolution can be seen in the history of seminal plants. One is connected with the time of angiosperm creation, when a quick divergence of embryogenesis took place in gymnosperms (formation of primary and secondary zygotes in association with processes of concurrent development) and angiosperms (whose zygotes are homologous only to the primary zygotes of gymnosperms) (Teryokhin, 1991). The second period is associated with divergence of flowering plants of dicotyledons and monocotyledons. Relative to the question of embryo homologies it must be remembered that the original embryogenesis of *Paeonia*, according to Yakovlev (1951), represents in fact a secondary-evolutionary attempt to reproduce in angiosperms the specific model of two-staged embryogenesis of gymnosperms.

Thus, two conclusions may be drawn:

1) In the evolutionary history of seminal plants there have been two obvious periods of remarkable divergent processes.

2) The current period of angiosperm evolution reveals a great number of pathways of adaptive-ecological radiation in spite of the lack of one or a few key trends in the evolution of embryos.

Embryogeny

Embryogeny (Greek *embruon*[1]—prenatal foetus, embryo and French *génie*—feature, characteristics) is a special division of embryology describing the regularities of embryo development.

The term was introduced by Treub (1879). Souèges (1935b) defined embryogeny as a science of embryo development. Johansen (1950) regarded embryogeny as a branch of general embryology which studies kinematic (Greek kinematos—movement) aspects, i.e., the sequence of stages during embryo development.

Based on embryo development research, a number of embryogenetic laws have been established for the majority of flowering plants. Embryogenetic formulae have been suggested to represent them (Souèges, 1922a). These formulae reflect, quite precisely, the succession of blastomeres (cells of the proembryo), their number, arrangement and possible functions. A special branch of embryogeny—**embryonomy** (Greek *embruon*—embryo, *nomos*—law)—determines the extent to which embryogenetic laws are expressed in various plants. Taking into account these laws, **embryonomic types of embryo development** were suggested in Souèges's classification (1939c) (see below: Principles of Classification of Embryogenesis Types).

Embryogenetic laws.[2] The basic laws of embryo development are: law of origin, law of numbers, law of disposition, law of destination (Souèges, 1937b) (Fig. 34) and law of parsimony (Johansen, 1950).

Law of Origin. '**Embryogeny allows revelation of the origin of all the blastomeres of the embryo for each plant species very precisely and establishment of their association within each generation**'.[3]

The origin of histogens was established on the basis of results of zygote division and its derivatives during the first-second generations, the of each blastomere of eight-celled proembryo (after the third generation), of quadrants and octants (after the fourth generation). Souèges distinguished a principal branch of embryogeny which studies differentiation of histogens and origin of the embryo and its component parts and proposed that it be called embryogenesis (Greek embruon—embryo and Latin genesis—origin) in the broad sense, including gametogenesis, zygotogenesis and embryogenesis proper. Contrarily, Johansen (1950) included only embryogenesis proper in the definition of this term. In modern literature on embryology the terms embryogenesis and embryogeny are often used synonymously.

Law of Numbers. '**The number of blastomeres produced by different cell generations varies with the species and depends on the rapidity of segmentation in the cells of the same generation**'.[4] This is a law of rapidity of blastomere divisions. It

---

[1]En—in, *bruo*—growing, from Sanskrit *bhrú*—to grow, *brussa*—foetus.

[2]The laws are cited as originally interpreted by the authors.

[3]L'embryogènie permettant, dans une espèce donnée, d'établier la lignée des blastomeres, l'on peut, par rapport a l'um ou l'autre des termes de la lignée, définir très exactement, dans cette espèce d'origine des tous les blastomeres' (Souèges, 1937, p.50).

[4]Le nombre des blastoméres products aux different générations varie avec les espèces et depend de la différence de vitesse de la segmentation dans les blastoméres de même generation' (Souèges, 1937, p.58).

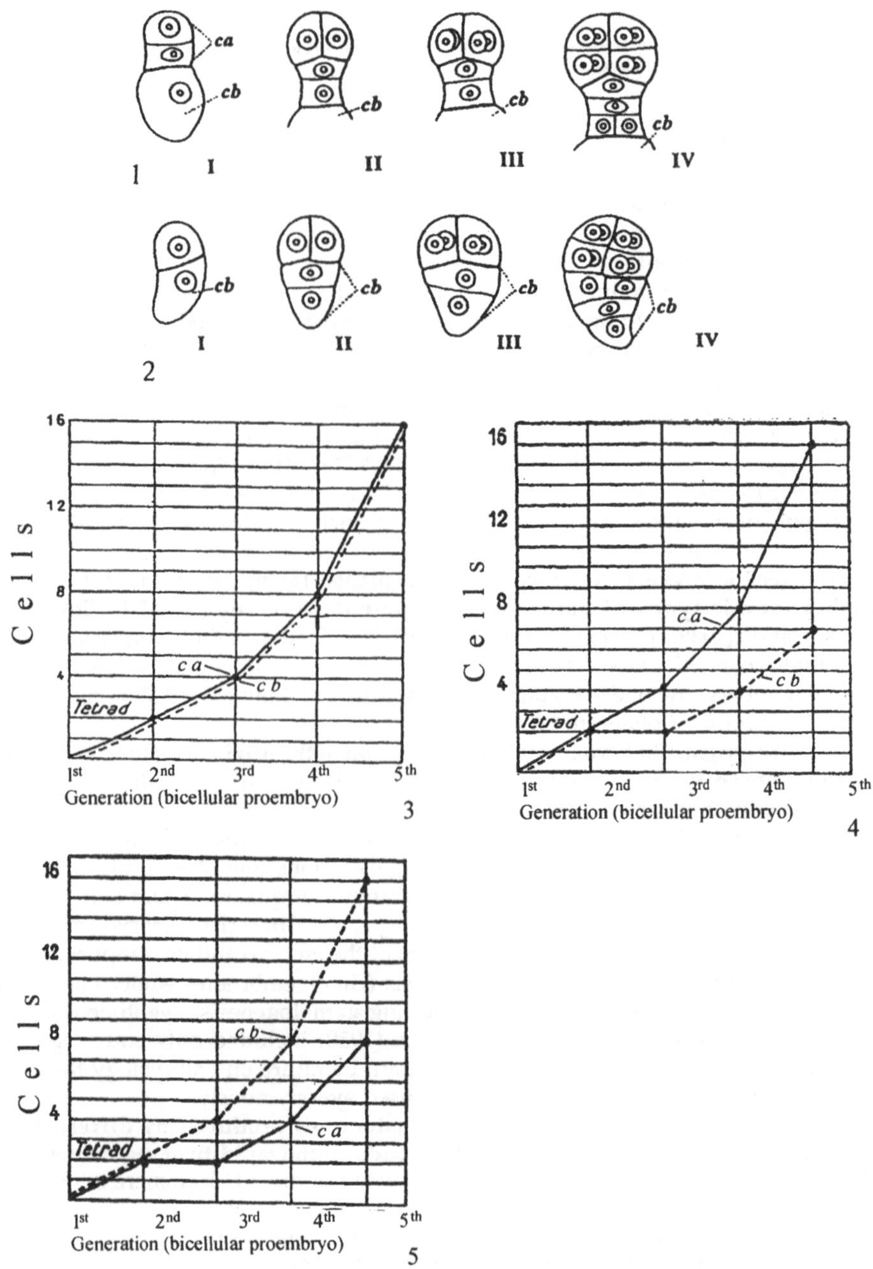

Fig. 34: Embryogenetic laws. 1, 2—the exposition of the laws of origin and disposition during the first four generations in the proembryo (I, II, III, IV) in *Saxifraga granulata* (1) and *Euphorbia exigua* (2); 3-5—the law of numbers, displaying in different rate of divisions of the apical and basal cells during the first five generations in *Polygonum persicaria* (3), *Oenothera biennis* (4), *Erodium cicutarium* (5) *ca*—apical cell, *cb*—basal cell (after Souèges, 1937b).

can be represented schematically: on the axis of abscissae—generations, on the axis of ordinates—the number of resultant blastomeres. According to Souèges, the rate of cell divisions is evidence of realization of constructive potentials of the apical and basal cells in embryo organization. Thus, if cell *ca* divides more rapidly (it has considerably more derivatives during the first generations compared to derivatives of the basal cell), this is often associated with the fact that it forms basic organs of the embryo. If the most part of the embryo is formed from cell *cb* derivatives, the state of affairs is opposite. The peculiarities of cell divisions in the embryo may be used in comparative embryological research for the purposes of taxonomy and phylogeny. However, they are not always of great significance since the rates of cell divisions may differ in different generations and, at some stages, derivatives of the apical cell may divide faster, at other stages derivatives of the basal cell.

Law of Disposition. **'In the course of normal embryonic development, the blastomeres are constituted by divisions in clearly determined directions and appear to occupy positions in accordance with the role they must play'.**[5] The field of embryogeny which studies the arrangement of blastomeres in the embryo is called **embryotectonics** (Greek *embruon*—embryo and *tektoneia*—science of peculiarities of relative position of parts).

According to Souèges (1937b), the law of origin of blastomeres allows to revelation of their interconnections **in time** and the law of disposition **in space**. Exposition of the law of disposition is clearly traced in the course of the first four generations. The number of tiers and the number of cells in each obey the following rule: 'The number of blastomeres of any tier is equal to the sum of the blastomeres composing all the lower levels'. The number of tiers and their composition are defined in different directions: vertical (according to the axis of symmetry) and horizontal (perpendicular to it). There are exceptions when, due to different rates of divisions, the number of resultant cells in the embryo varies in different taxa.

Law of Destination. **'In the course of normal development, blastomeres of one and the same species give rise to clearly determined parts, and always to the same parts of the embryonic body'.**[6] The field of embryogeny studying the constructive potentialities of blastomeres in embryo construction, Souèges termed *embryogénergie* (Greek *embruon*—embryo, *gennau*—born, *ergon*—work). Depending on which generation this law has an effect, it may differ in degree of exposition—from the first to the fourth.

Based on the expenditure of apical and basal cells in embryo construction, Souèges (1937b) proposed distinction of ascending (on the basis of cell *ca)* and descending (on the basis of cell *cb*) series of megarchetypes, proceeding from the assumption that in the course of evolution, the role of *ca* increased and the role of *cb* decreased (for conventional abbreviations of histogens, see below: Proembryo).

Ascending series:

I. $ca = pco$

II. $ca = pco + pvt$

[5]'Au cours du diveloppement embryonnaire normal, les blastoméres se constituent par divisions selon des directions bien déterminées et viennent occuper des positions en rapport avec le role qu'ils deivent jouer' (Souèges, 1937, p.68).

[6]'Les blastoméres d'une espèce donnée quand le dèest normal, donnent naissance à des parties bien déterminées, et toujours les mêmes, du corps embryonnaire' (Souèges, 1937, p.79)

III. $ca = pco + pvt + 1/2phy$
IV. $ca = pco + pvt + phy + icc$
V. $ca = pco + pvt + phy + icc + iec$
VI. $ca = pco + pvt + phy + icc + iec + co$
VII. $ca = pco + pvt + phy + icc + iec + co + s$ (partially)

Descending series:

I. $cb = pvt + phy + icc + iec + co + s$
II. $cb = phy + icc + iec + co + s$
III. $cb = 1/2phy + icc + iec + co + s$
IV. $cb = iec + co + s$
V. $cb = co + s$
VI. $cb = s$
VII. $cb = s$ (partially)

Later (Souèges, 1939c) megarchetypes V and VI were combined, since in both cases cell *cb* participates, fully or partially, only in formation of the suspensor. Examples of plants characterized by these peculiarities of embryo development are: I—many monocotyledons, II—*Malva, Muscari, Senecio, Urtica,* III—*Chenopodium, Myosotis, Polygonum, Salix,* IV—Cruciferae, *Myosurus,* Onagraceae, Plantaginaceae, Scrophulariaceae, V—Rubiaceae, Solanaceae, Umbelliferae, VI—Campanulaceae, Fabaceae, Caryophyllaqceae, Saxifragaceae.

Law of Parsimony. **'No more cells are produced by the embryo than are absolutely necessary'.**[7] This law was formulated by Johansen (1950) based on the principles of economy of material and efficiency suggested by Haberlandt (1918).

Among all the embryogenetic laws, Souèges (1937b) and later Johansen (1950) emphasized the law of origin, since embryo development is characterized by a high degree of regularity of cell divisions in most flowering plants. Souèges highlighted that it is only in the case of **normal development** that specific parts of the embryo are formed from specific blastomeres. These regularities of histogen differentiation and construction of specific organs of the embryo on their basis, are used as an instrument of accurate control over embryo development in investigations of genetic, physiological, molecular and other aspects of embryogenesis (Müller, 1963; Sheridan and Neuffer, 1982; Meinke, 1986, 1991b,c; Laux and Jürgens, 1994). Deviations from the normal order of development, which manifest in some other sequence of division of embryo cells, for example in *Hypericum,* can cause its death (Bugnicourt, 1983). Interruption of embryo development in a number of mutant plants (*Arabidopsis thaliana, Zea mays, Daucus carota)* is related to changes in expression of regulatory genes which control the course of embryogenesis and manifest as various mutations (Meinke and Sussex, 1979a,b; Sheridan and Neuffer, 1980, 1982; Breton and Sung, 1982; Meinke, 1982, 1985, 1986, 1991a; Giuliano *et al.*, 1984; Schneider *et al.*, 1989; De Jong *et al.*, 1992).

Scientists who criticize the established laws of embryogeny suggested a new, so-called dynamic approach to embryogenetic studies. An assumption was made that cells manifest certain plasticity of their potentials in the course of development; this plasticity may appear at comparatively late stages—globular stages (Foster, 1949; Swamy and Padmanabhan, 1962) or heart-shaped (Lyndon, 1990) embryo. In *Capsella bursa-pastoris* and *Arabidopsis thaliana,* no ultrastructural differences between cells are

---

[7]Johansen, 1950, p.97. The law is confirmed by formation of cell critical mass during critical periods and stages in plant ontogenesis (Batygina, 1984; Batygina and Vasilyeva, 2003).

observed up to the stage of late globular embryo which, according to some authors, indicates the fact that cells are not determined with respect to formation for specific organs of the embryo (Schulz R. and Jensen, 1968a; Mansfield and Briarty, 1991; see below: Ultrastructural Aspects of Embryogenesis).

To some extent, Teryokhin (1984, 1991) shared this point of view. He believed that at middle stages of embryogenesis the laws of blastomerization formulated by Souèges act in the form of morphogenetic zones with definite and clearly distinguishable potentials, rather than within the exact tiers of the proembryo. Vallade's study (1989) of two patterns which are characterized by different types of embryo development (leptembryate in *Petunia hybrida* and pachyembryate in *Vitis vinifera*) showed that they differ only in comparative rate of growth of the embryo when the cells obtain the minimum volume by the beginning of demarcartion of the protoderm: in *Petunia* this occurs earlier, in 150-200 mitoses, in *Vitis* in 1000-1500. Then both species enter the stage of dedifferentiation which results in formation of eumeristematic cells. Further organogenesis is, on the whole, similar at the level of morphogenetic (half-active) zones. However, the course of organogenesis is easier to trace in the leptembryate, less energy-consuming type.

Wardlaw (1955, 1963), adhering to the dynamic approach to embryo study, regarded the laws of embryogeny only in the light of data of descriptive embryology, rather than the causes of embryonic development. He believed that the gradual co-ordinated development of the embryo is conditioned by a complex of physiological factors, as well as by environmental conditions.

In some studies, the stage of formation of the histogens at the very beginning of embryo development was proved to be very important, however, the specificity of histogenesis is determined by the general strategy of development of the organism (Barlow, 1991, 1994a, b). On the example of *Arabidopsis thaliana*, it was shown that clearly pronounced tiers of cells in the proembryo do not always give rise to certain specific organs of the mature embryo. The boundaries of cells can move (Laux and Jürgens, 1994; Jürgens, 1995). A conclusion was drawn that the final structure of the embryo is established due to interactions of cells in the course of development and depends on position of structures in the embryo (Laux and Jürgens, 1994; Jürgens, 1995).

Embryo development, like ontogenesis in general, is characterized by a certain periodicity. When describing early embryo development, many authors distinguish the following stages: 1) zygote, 2) two-celled proembryo, 3) four-celled proembryo, 4) stage of quadrants (usually eight-celled proembryo), and 5) stage of octants (usually sixteen-celled proembryo). Further development of the embryo is described on the basis of its morphological characteristics, mainly its shape. Schnarf (1929) was the first to distinguish three stages: filiform (preglobular—Meinke, 1986)—the embryo consists of a row of cells; according to Padmanabhan (1996) data, the end of this stage is the first longitudinal divisions in terminal cells; globular—due to prevalence of longitudinal divisions of derivatives of the apical cell, the embryo grows and assumes the shape of a globule; heart-shaped—organogenesis starts and the embryo becomes heart-shaped, mainly due to the primordia of the cotyledons arising in the apical part. Some authors (Jürgens and Mayer, 1994) distinguish a triangular stage between the globular and the heart-shaped stages, which is characterized by flattening of the apical part of the embryo without apparent signs of organogenesis. At the heart-shaped stage, the radial symmetry changes into a bilateral one (Wardlaw, 1955, Halperin, 1966, Juguet, 1973; Tykarska, 1979; Yakovlev, 1981; Natesh and Rau, 1984;

Laux and Jürgens, 1994), which was explained by the presence of special gradients in the embryo (Wardlaw, 1955), or of certain endogenous factors (Halperin, 1966). It was discovered that transition from one type of symmetry to another is associated with polar transport of auxins which, if blocked by inhibitors, can result in formation of aberrant embryos (Liu *et al.*, 1993a; Schmidt *et al.*, 1994) (for details, see below: Sincotyly). It is important to note that the above-mentioned stages (filiform, globular, triangular, heart-shape) were initially distinguished in an embryo of dicotyledons.

For the embryo of dicotyledons, in which processes of organogenesis take place, a number of stages are distinguished after the heart-shape stage: reversed trapezium (height and breadth equal; embryo growing mainly in apical part), torpedo (cotyledons flattened against each other and constitute half the length of the embryo), walking-stick (for bent embryos—caused by the top of the ovule having turned back in cotyledons), upturned U-shape (for bent embryos: one leg formed by the hypocotyl, the other by flattened cotyledons) (Fig. 35) (Rijven, 1952; Wardlaw, 1955; Raghavan, 1976; Natesh and Rau, 1984; Raghavan and Sharma, 1995). For cereals, a number of stages are distinguished after the stage of organogenesis; they are known as stages '1-6' (they differ in different number of leaves in the plumule, from 1 to 6 respectively—Abbe and Stein, 1954; Sheridan and Neuffer, 1980; Xi and Ye, 1994). Many of the aforesaid stages are used not only in embryology, but also in physiology, genetics and molecular biology, when various aspects of embryo formation are studied.

The globular stage, during which histogenesis and preparation for organogenesis take place, is very long. Within the globular stage, a number of intermediate stages are distinguished: early (globular embryo with embryoderm), middle (embryo during histogenesis) and late (embryo before organogenesis) (Meinke and Sussex, 1979a,b; Meinke, 1985, 1986; Mansfield and Briarty, 1991; Shamrov and Nikiticheva, 1992; Shamrov and Anisimova, 1993b; Laux and Jürgens, 1994). The last stage is also termed a transitional phase (Abbe and Stein, 1954), pear-shaped (Haccius, 1972a), globular-heart-shaped stage (Meinke, 1991c) and mace-shaped stage in cereals (Xi and Ye, 1994).

The terms **'critical periods', 'critical stages'** and **'critical points'**[8] are used to evaluate the concordance of morphogenetic and morphophysiological processes in embryo development. The term 'critical periods' was introduced to define the developmental periods of plants which are more sensitive to unfavourable environmental conditions (Brounov, 1897). Then, it was used in embryology of animals. Some authors (Stockard, 1921) believed that the critical periods in embryo development are characterized by the greatest rate, others (Goldschmidt, 1938) that the critical periods are 'the focal points' of intersection of curves of the growing rates of different independent reactions. The main theses of the theory of critical periods were formulated by Svetlov (1960, 1978): 1) development of the embryo and its components consists of a number of stages, each of them starting from a relatively short period, after which visible processes occur (growing and differentiation); 2) high sensibility of the embryo cells to environmental factors activity (every environmental influence on the ontogenesis, whether internal—organism environment—or external), which is determined by a low regulative activity in these periods, is the main feature characterizing the critical periods; 3) the embryological significance of the critical periods is that during them acts of determination of

[8]New **ideas** concerning critical periods, phases and stages are expressed in the works of Batygina, 2002 and Batygina and Vasilyeva, 2003.

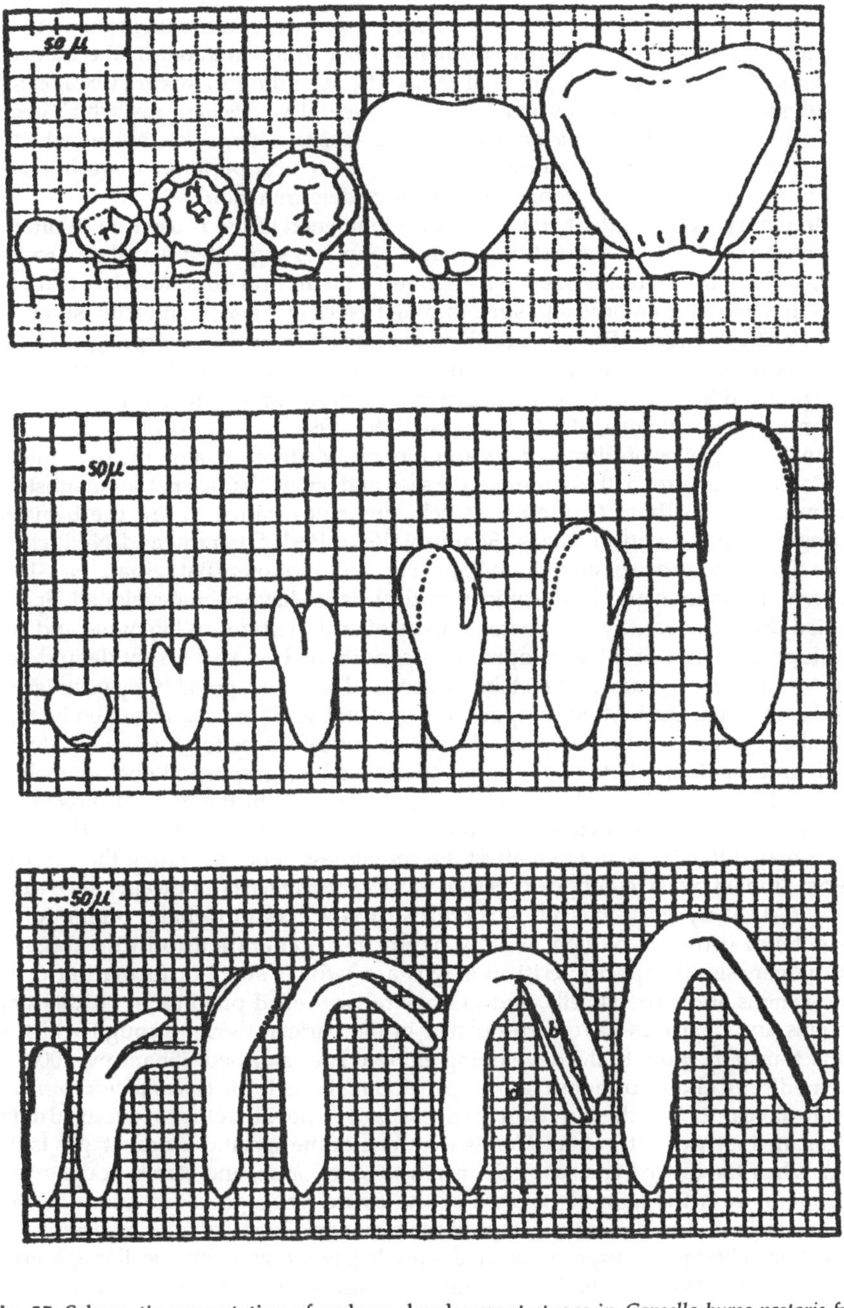

Fig. 35: Schematic presentation of embryo development stages in *Capsella bursa-pastoris* from globular to upturned U-shaped (after Rijven, 1952).

developmental processes occur. According to Wareing and Phillips (1978), successive developmental stages can be considered as a process in which a switch to alternative ways occurs in different critical points of time and space, and certain organism parts become 'determined' with respect to their further differentiation. In the theory of critical periods, some developmental moments, according to Tokin (1987), can be considered in relation to reconstruction of the integrative mechanisms, which begins in certain periods in the whole embryo and in its separate parts. It is noteworthy that in embryology of animals, along with the term 'critical periods', there is a term 'determination periods' (a stepped character of determination).

In embryogeny of plants the term 'developmental stage' is used more often in description of embryonic structures. That is why it is suggested that the parcels of time, which are characterized by change of structural-functional features in development of the embryo and surrounding tissues of the seed and fruit, should be called '**critical stages**'. The basis of their distinction comprises morphogenetic and morphophysiologic correlations in the development of reproductive structures (Batygina and Vasilyeva, 1981; Vasilyeva *et al.*, 1987; Shamrov and Nikiticheva, 1992; Shamrov and Anisimova, 1993b; Shamrov, 1995, 1996).

In some species of dicotyledons and monocotyledons, data on the dynamics of developing structures of the ovule and the seed and on the character of some substance accumulation in them (proteins, starch, dextrines, tannins) are the basis for distinguishing the critical stages (Shamrov, 1990, 1991; Shamrov and Nikiticheva, 1992; Shamrov and Anisimova, 1993b) (Fig. 36). Notions (Batygina *et al.*, 1992) concerning common and taxon-specific critical stages have been developed. Species studied differ in quantity and characteristics of critical stages in the ovule and seed development. First of all development is conditioned by various ovule and seed structure, diverse dynamics of metabolite accumulation and, possibly, by peculiarities of substance transport. More compound ovules and seeds are characterized by more critical stages in their development. Common critical stages for crassinucellate and tenui- and medionucellate ovules (archesporial cell, megasporocyte before meiosis, early and late globular embryos, mature embryo) explain the general direction of formation of the reproductive structures and surrounding tissues. Specific critical stages may differ. Some of them attest the correctness of our assigning the ovules in species studied by us to definite types. In *Luzula pedemontana* (Juncaceae) the ovule is characterized by very massive structures (it is a crassinucellate, bitegmic and mesochalazal ovule), which usually begin to degenerate at middle stages of embryogenesis. Its specific critical stages are two: 1) stage of formation of the sporogenous and parietal cells, and 2) stage of two-celled proembryo (at this stage proteins and dextrines have accumulated in the parietal tissue, through which an apical transport route to the developing embryo is established) (Shamrov, 2000). In the medionucellate, bitegmic and leptochalazal ovule of *Gymnadenia conopsea* (Orchidaceae), the destruction processes of the structures occur during megagametophyte maturation. That is why, one of the specific critical stages is that stage when the outer hypostase cells begin to deteriorate, and starch grains nearly disappear from the somatic tissues of the ovule (megagametophyte following cellularization). Another critical stage in the ovule development of *G. conopsea* is that of the two-celled proembryo, associated with degeneration of the nucellar epidermis and inner hypostase cells. In the tenuinucellate, unitegmic and mesochalazal ovule of *Gentiana cruciata* (Gentianaceae) the transformation processes of the structures take place at a still earlier developmental period. The specific critical stage in this species

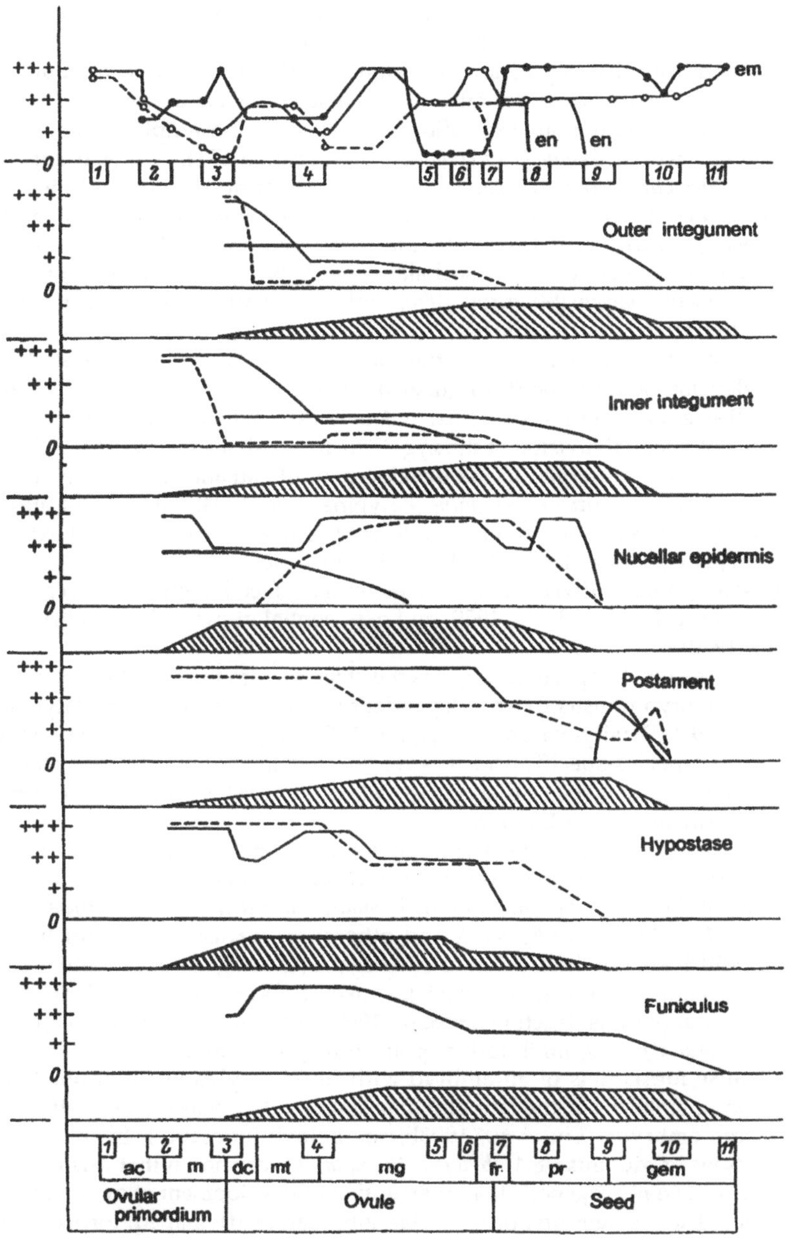

Fig. 36: Critical stages (1-11) in ovule and seed development of *Gymnadenia conopsea*. Starch is shown as a thick line (in the upper part of the figure with dark circles); proteins as a thin line and dextrines as a dashed line; shaded sectors show dynamics of developing structures (relative time of arising, development maximum, degeneration beginning and full disappearance); 0, +, ++, +++ denote levels of substance content (after Shamrov and Nikiticheva, 1992).

may be considered the stage of megaspore tetrad. The beginning of nucellus degeneration correlates with formation of procambial cell rows in the raphe and subsequent transport strengthening of metabolites through the vascular bundle and their accumulation in most cells of the ovule. On the other hand, species with various ovule types can have the same specific critical stages, for example the stage of two-celled proembryo: in *Gymnadenia conopsea* degeneration of structures occurs, while in *Luzula pedemontana* the structures show no signs of destruction and metabolism intensity grows. There are also specific critical stages which indicate the lability in their passing in different taxa with the same ovule type. In the last case the tenuinucellate, unitegmic and mesochalazal ovule of *Gentiana cruciata* and *Swertia iberica* from Gentianaceae (Shamrov, 1990, 1991) show various dynamics of substance accumulation during development.

Based on the data available in the literature, the following critical stages in embryo development can be distinguished. One is the stage of two-celled embryo whose cell sizes are determined. Thus, in *Arabidopsis gnom* the mutation is related to the change of unequal division of the zygote to an equal one (Mayer *et al.*, 1993). The same is quite often observed in the development of somatic embryos in culture *in vitro*, for example in *Triticum aestivum* (Batygina *et al.*, 1993; see below: Embryoid). The stage of differentiation of embryoderm (protoderm) may be the next critical stage (Batygina, 1968a, 1969, 1974b, 1987a; Yakovlev, 1972, 1981; Vallade, 1989; Mansfield and Briarty, 1991). It is associated with the fact that the embryo cells obtain the minimum volume (Vallade, 1989) and exponential growth of the embryo begins (Nagato, 1979).

One of the most important stages is the stage of autonomy of the embryo, from which the embryo transits to an independent mode of development (Batygina and Vasilyeva, 1979; Vasilyeva and Batygina, 1981; Raghavan, 1986). The stage of autonomy is species-specific and it is manifested in different representatives at different development stages (Batygina *et al.*, 1978b; 1983; Vasilyeva *et al.*, 1987; see below: Autonomy of the Embryo).

At present, the results of experimental research have confirmed the existence of some critical, stages sensitive to certain environmental factors in embryo development. For example, in *Eranthis hiemalis* the influence of various growing substances (cytokinins, IAA, NAA and others) on mature seeds, which include embryos which are not differentiated into organs (the stage of late globular embryo before the beginning of organogenesis), causes a wide range of variations in the cotyledonous apparatus structure (Haccius, 1959, 1960, 1972a; Haccius and Trompeter, 1960). In *Brassica juncea*, inhibitors of polar transport of auxins (HFCA, TIBA and others) cause formations of an embryo with fused cotyledons, if applied during cultivation of embryos at the late globular stage, but appears to be ineffective for heart-shape embryos (Liu *et al.*, 1993b; see below: Sincotyly). In the course of investigations of mutants of *Arabidopsis thaliana*, it was shown that it is during the critical stages of embryogenesis that interruption of development of ovules and seeds takes place. These stages are: preglobular, globular, heart-shaped, torpedo-shaped (embryo with linear cotyledons), upturned U-shaped (embryo with bent cotyledons) and mature embryo (Meinke, 1985,1986, 1991c; Health *et al.*, 1986; Franzmann *et al.*, 1989). In *Zea mays* the critical stages (phases) are characterized by disturbance of the mechanism of establishment of basal-apical polarity (preglobular stage), radial symmetry and formation of initials of different organs (mace-shaped and coleoptylar stages), as well as formation of leaf primordia in the plumule of the embryo (stages

1-6) (Sheridan and Neuffer, 1980, 1981, 1982; Clark and Sheridan, 1986, 1991; Sheridan and Clark, 1993).

The embryo is a dynamic system. In this regard, for an accurate investigation of the embryo it is necessary to consider both the peculiarities of early embryogenesis (order of divisions, topography of layers, characteristics of cells in various parts of the embryo—genealogical approach) and the changing state of the embryo, especially during the critical stages of development (dynamic approach) (Shamrov, 1996). Investigations of mutant plants serve as evidence for the foregoing; they enable evaluation of the significance of various factors in realization of the programme of embryo development during experiments.

DEVELOPMENTAL STAGES AND STRUCTURES OF EMBRYO

Zygote

Zygote (Greek *zygotos* -yoked)—a cell formed as a result of fusion of the female gamete (egg cell) with the male gamete (sperm cell). Synonyms: fertilized egg cell, syngamete and initial cell of the embryo **(Plate XXIV)**.

The term 'zygote' (like the term 'gamete') was first introduced by Strasburger (1877) to denote the result of fusion (copulation) of gametes in the algae *Spirogyra* and *Acetabularia*. Bateson (1902 cit. from Jackson, 1916) extended this term to an individual which develops from a fertilized egg cell.

Zygote formation is the initial stage of ontogenesis. In the course of a sequence of programmed complex morphogenetic transformations the zygote develops into a multicelled embryo and later into a highly specialized individual, the plant.

Maturing of the egg cell, fertilization, development of the zygote and early embryogenesis have been thoroughly studied at the molecular level in model systems in animals (e.g. fruit-fly, nematode and mouse) due to the wide range of embryonic mutants available. Discovery of cyclin, cdc2 protein and of modulators of cell division in the egg cell of the frog, sea-urchin and in the cells of yeast, promoted development of a universal model of a mitotic cycle in eukaryotes (see reviews by Murray and Kirschner, 1991; Jacobs, 1992, 1995).

Progress in molecular and cellular biology has contributed to understanding the control mechanisms of plant embryogenesis (see review by Goldberg *et al.*, 1994). Embryonic mutants of *Arabidopsis*, *Hordeum* and *Zea* were obtained and analysis of them enabled distinguishing a number of important genes of embryogenesis (Sheridan, 1988; Clark and Sheridan, 1991; Meinke, 1991a-c; Castle and Meinke, 1993). Recently some genes typical of the non-fertilized egg cell and the zygote of plants were identified (Dresselhaus *et al.*, 1996).

The problem of axial organization of the embryo body and of the plant has been studied intensively (Mayer *et al.*, 1991, 1993; Jürgens, 1995). Genetic and phenotypic analysis of the gene *gnom*, which expresses before the asymmetric zygote division in *Arabidopsis*, promoted identifying its role in the apical-basal model of the embryo structure.

Recent works on the interrelation of the cell cycle and sexual reproduction in angiosperms and gymnosperms provide a more or less detailed description of gamete structure and chronology of karyogamy (Friedman, 1991; Carmichael and Friedman, 1995; Mogensen and Holm, 1995). Taking *Zea mays* as an example, the cytological pattern of karyogamy in the culture *in vitro* was described (Faure *et al.*, 1993).

However, a whole series of questions, concerning in particular establishment of the apical-basal axis in the egg cell and zygote and three-dimensional organization of the embryo, are still debated: 1) periodization of egg cell and zygote development; 2) stage of embryo axis formation in the mature egg cell, zygote or in early embryogenesis; 3) relation between polar organization of the egg cell and zygote with expression of genes in early embryogenesis; 4) presence of hard or labile[9]

---

[9]See Huxleu and De Beer (1936); Mikhailov (1988).

phases in egg cell determination and factors (embryonic inductors[10])—gametophytic, gametic or sporophytic – which stimulate its change; 5) universality of regularities of axis establishment and embryogenetic control, described in *Arabidopsis* (typical dicotyledonous), for all the flowering plants which differ in embryological features; and 6) mechanisms on which zygote construction is based in plants and animals capable of developing both with metamorphosis and without it.

From a complex system approach (Batygina and Vasilyeva, 1981; Batygina, 1983) the strategy of studies of the zygote may be formulated as follows: 1) morphological and cytoembryological analysis of all the stages of egg cell and zygote development; 2) consideration of zygote genesis in connection with the cell cycle; 3) determination of interconnections between the character of zygote development (i.e. type of karyogamy) and the types of embryo sac development (16 types), embryo- (8 types) and endospermogenesis (usually 3 types), and the entire system of ovule-seed; 4) selecting model objects which would possess sexual elements large enough and characterized by prolonged passing through each phase.

No periodization or unified terminology covering all the phases of egg cell and zygote development are available yet. Only some stages have come to be designated by generally accepted terms (e.g. mature egg cell, young and mature zygote). A number of very important phases of syngamy escape researcher attention due to their transient nature (in most of the flowering plants). We believe that **one of these stages is establishment of fragmoplasts.** It determines cytogenesis in the embryo sac, structure of the zygote and type of karyogamy. Meanwhile, both of these cells are often used as model systems for studies of fertilization and, in particular, of karyogamy *in vitro,* upon distant hybridization, etc. The absence of precise criteria for distinguishing phases in the egg cell and zygote genesis complicates comparative analysis of data available and establishing of regularities in development of these cells. Furthermore, **a certain periodicity typical of zygote development** (as well as of development of any biological system) and **complexity of the processes** involved (formation and maturation of gametes, plasmogamy, karyogamy, etc.) **confirm the need for introducing new terms.**

We suggest a set of concepts which describe the seven main developmental stages of the egg cell and the zygote (see Fig. 37). We hope that this periodization will be useful for studies and understanding of mechanisms of fertilization and will be completed with the help of new data on ultrastructure and mitotic cycle.

I—Nucleus of a future egg cell: lower of the haploid micropylar nuclei in eight-nucleate embryo sac, sister of the upper polar nucleus.

II—Egg cell initial: one of the cells (normally, lateral) of gametophytic haploid **tissue** of the egg cell apparatus formed as a result of establishment of phragmoplasts; **totipotent, non-polar cell,** similar to meristematic cell in structure of nucleus and cytoplasm. This stage corresponds to **the beginning of formation of seven-celled, eight-nucleate embryo sac** (in Polygonum-type of embryo sac development).

III—Immature (forming) egg cell: an immature female gamete from the moment of separation from the rest of the cells of the egg apparatus up to finishing its formation; it is characterized by **decrease of totipotency, emergence of polarity** (migration of the nucleus, specific vacuolization), of the axis, of a certain type of symmetry (radial, bilateral, (dorsoventral)); it is positioned lateral to the axis of the

[10]Spemann (1938).

THE PRINCIPAL DEVELOPMENTAL STAGES OF		
EMBRYO SAC	EGG CELL AND ZYGOTE	
Formation of a coenocyte (the eight-nucleate embryo sac)	**NUCLEUS OF A FUTURE EGG CELL (a sister one of the polar nucleus in the micropylar part of embryo sac)**	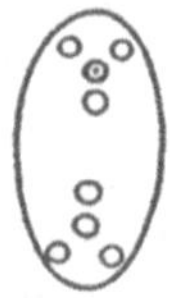
Cellularization and formation of heterogenous gametophytic tissue (the seven-celled, eight-nucleate embryo sac)	**EGG CELL INITIAL**	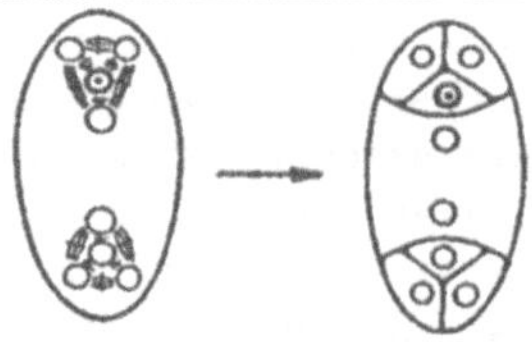
Separation of all cells in embryo sac and formation of its elements	**IMMATURE (FORMING) EGG CELL**	
Maturation of embryo sac elements (embryo sac, which is ready for fertilization)	**MATURE EGG CELL**	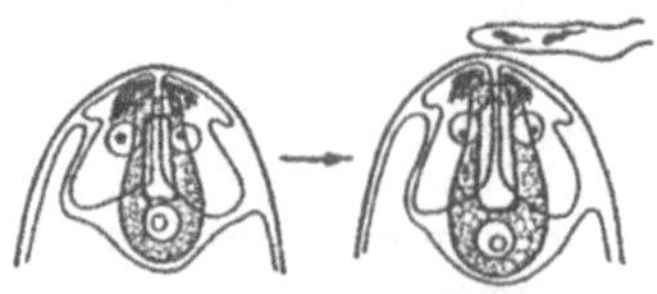
Penetration of a pollen tube into embryo sac, plasmogamy	**ZYGOTE INITIAL**	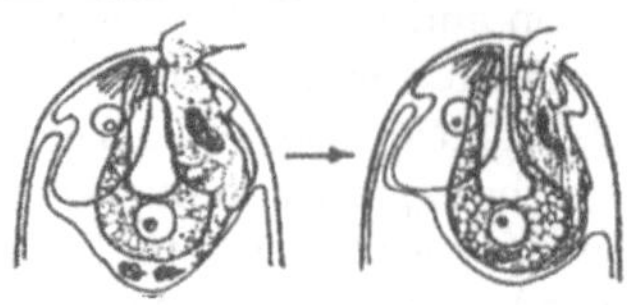
Karyogamy	**IMMATURE ZYGOTE**	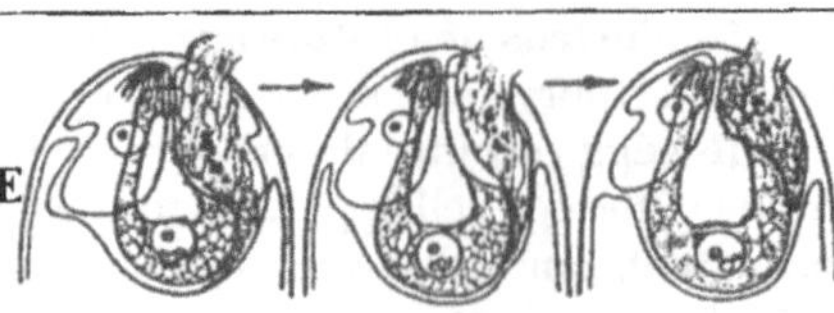
Completion of double fertilization	**MATURE ZYGOTE**	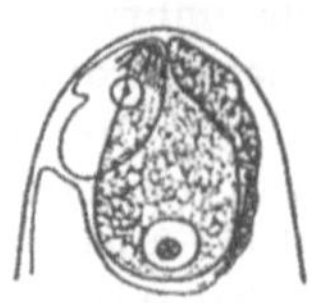

Fig. 37: Periodization of egg cell and zygote development in the embryo sac.

embryo sac. This stage normally corresponds to **the full-formed embryo sac** (all its elements are organized), **which is not yet ready for fertilization**.

IV—Mature egg cell: a mature female gamete which has passed from a formed cell to a ready-for-fertilization cell; it is a **highly differentiated and specialized, non-totipotent and polar cell** exhibiting an apical-basal organization and acquiring the final taxon-specific shape. This stage corresponds to a mature embryo sac ready for double fertilization.

V—Zygote initial: a mature egg cell whose cytoplasm has been penetrated by a sperm, from formation of the coenocytic structure[11] up to commencement of contact between sexual nuclei (contact between their outer membranes via the ER); it is a **specialized cell**, in which dedifferentiation begins under the influence of a trigger—sperm cell, **non-totipotent, polar** (apical-basal model of organization), characterized by increase (sometimes decrease) in volume, movement of nuclei, a specific type of symmetry and numerous organelles. This stage corresponds to **the embryo sac which has undergone the progamic phase and entered the next phase of fertilization.**

VI—Immature zygote: a zygote from the beginning of karyogamy (fusion of outer and inner membranes of male and female nuclei and establishment of bridges between nuclei) until complete fusion of nuclei; it is characterized by **loss of specialization, acquisition of totipotency, by polarity** (apical-basal model of organization); a special type of symmetry and a diploid nucleus; male and female nucleoli may not be fused.

VII—Mature zygote: a zygote which has passed the way from complete fusion of sexual nuclei until commencement of division; it is a **totipotent**, capable of proliferation, **polar** (apical-basal model of organization) **cell**, showing a specific type of symmetry and a diploid nucleus with amplification of nucleoli.

The duration of each developmental stage of egg cell and zygote varies in flowering plants, from a few seconds to a few days. The zygote is normally located between synergids in the micropylar part of the embryo sac and is attached to its wall laterally, which can be observed on a certain section (in profile). Sometimes the embryo sac along with the zygote comes out to the micropyle and the channel of the style. In some plant species the zygote forms haustoria.

The shape and size of the zygote are species-specific features. To a great extent these parameters are connected with the peculiarities of the suspensor structure. Zygotes are mostly pear-shaped or cylindrical. There are 'sitting' zygotes, the basal part of which is widened (e.g. *Nelumbo*—Batygina *et al.*, 1983). Zygote volume during its development varies. In some species the zygote volume considerably increases with respect to the volume of egg cell (species of genus *Crepis*—Gerassimova-Navashina, 1954a,b, 1957; *Datura stramonium*—Satina and Rietsema, 1959), in others it decreases (*Reseda lutea*—Chaban, 1975; *Gossypium*—Jensen, 1968b; Pollock and Jensen, 1964; *Hibiscus*—Ashley, 1972; *Nicotiana tabacum*—Mogensen and Suthar, 1979). In *Capsella bursa-pastoris* (Schulz R. and Jensen, 1968a) the zygote first decreases in size, then restores its initial volume.

As a new organism, the zygote is a **daughter generation**, which normally genetically differs from somatic cells of the nucellus, integuments and from the haploid cells of the gametophyte—synergids and antipodes. There are trophic and regulatory connections between the zygote and the surrounding cells and tissues.

[11]There are male and female nuclei in the cytoplasm of a fertilized egg cell; at high syngamy rate this stage is difficult to catch.

Rigid co-ordination in development of zygote, embryo and all the structures of the seed, both in time and in space is manifested in morphogenetic and morphophysiological correlations (Sinnott, 1960; Batygina, 1974b, 1987a; Wareing and Phillips, 1978; Vasilyeva *et al.*, 1984, 1987, 1988; Vasilyeva, 1988).

Like most plant somatic cells, the zygote **obeys the general laws of cell differentiation** (Gerassimova-Navashina, 1969a, 1971a,b; Jacobs, 1992). As a totipotent cell, it can produce a new organism and is not unique in this sense.

However, its origin (a product of syngamy), peculiarities of the structure and functioning of highly specialized gametes, and determination of the processes occurring in it permit claiming some unique features of this cell. The most important events are: 1) symmetric or asymmetric cytokinesis during male and female gamete formation and, consequently, equal or unequal zygote division; 2) formation of a coenocyte structure containing two nuclei of different origin (male and female) and migration of these nuclei; 3) asynchronous transformation of chromatin of the nuclei of the fusing gametes, which is conditioned by non-simultaneous gamete passing through the periods of the interphase; 4) infusion of organelles of male origin into the cytoplasm of the egg cell (biparental or uniparental inheritance).

However, whether the sperm cell infuses its organelles into the egg cell on a regular basis is still debatable (see Part I: Cytoplasmic Transmission during Fertilization). In zygotes of some representatives of flowering plants, plastids and mitochondria of male origin have been observed. For example, a non-fertilized egg cell of *Oenothera erytrosepala* has 14-32 plastids, a young zygote 33-42 (8-13 of them of male origin), a mature zygote—68-82 (14-18 of them paternal) (Schötz, 1954). This fact serves as evidence for the fact that plastids of both types have divided (Meyer and Stubbe, 1974). The number of plastids and mitochondria in the zygote increases both at the expense of their reproduction and at the expense of infusion by sperm cells (Hagemann, 1976, 1981). Some authors, in particular Possingham (1980), believe that in most cases the sperm cells have no plastids, since during division of a generative cell and formation of sperm cells they are selectively excluded, which results in only maternal inheritance of plastids. At present, the number of works is increasing which prove the presence of plastids in sperms (Tatintseva, 1988) and dimorphism of sperms. The fate of plastids of male origin in ontogenesis has not been determined yet, but there is no doubt that their infusion into the egg cell influences the behaviour of other organelles at various stages of zygote differentiation and maturation. Genetic data prove that 'male organelles' determine the character of development of an organism (biparental inheritance) (see Plyushch, 1992).

The uniqueness of the zygote is manifested in a complex of such features as its insulated state, morphological and physiological polarity, longitudinal morphological axis, cell wall structure, and pattern of the cell cycle (especially of the interphase).

Insulated state of the zygote. While the zygote is developing from an egg cell initial through syngamy to a mature zygote, the degree and character of its isolation change, which is manifested in the type of its interactions with the surrounding cells and tissues, in the structure of its wall and in the transport system. The process of isolation of the egg cell, zygote and the rest of the cells of the embryo sac encloses three critical events.

1) **Establishment of phragmoplasts and formation of a unified heterogeneous gametophytic tissue** in eight-nucleate embryo sac (stage II). The cells of this tissue are in tight contact and differ morphologically, but they are meristematic

and totipotent to some extent. However, their arrangement in the system of the embryo sac with its already established polarity and symmetry by and large determines their destiny.

2) **Beginning of isolation of cells and separation of three functionally different types of tissues**: egg apparatus, antipodal complex and central cell (initial of the endosperm) (stage III).
3) **Completion of isolation of the egg apparatus** and antipodal complex cells (stage III). The latter may persist as a tissue for a long time.

Zygote cell wall structure. Normally, an immature egg cell **is completely surrounded by a cellulose wall**. In most flowering plants, a mature egg cell, ready for fertilization, has a wall only on its basal pole, its apical pole is separated from the central cell only by a plasma membrane (Schulz R. and Jensen, 1968a; see also Zhukova, 1994). In some cases the cells of the egg apparatus have no walls at all (*Potamogeton*—Wiegand, 1900; *Peperomia blanda, Gagea lutea*—Plyushch, 1983, 1992). Construction of a **full cellulose wall** occurs again during the interphase and a mature zygote normally has a **continuous wall** though data for some species indicate its later formation (*Helianthus*—Yan Hua and Hong-Yuan Yang, 1991). The duration and character of this process are taxon-specific.

The wall of the zygote can undergo other specific changes. Thus, in *Rhododendron* and *Ledum groenlandicum*, callose accumulates in the zygote wall (Williams *et al.*, 1984). In *Wormia suffruticosa* and *Acrotrema arnottianum* (Dilleniaceae) a considerable increase in zygote size (3-4-fold) is accompanied by significant thickening of the wall, which leads to outgrowths into the endosperm (Paetow, 1931; Swamy and Periasamy, 1955; Sastri, 1958). It maintains its structure during early embryogenesis as well, forming a peculiar kind of jacket around the embryo. Electron microscopic studies of the zygote wall in *Dillenia suffruticosa, D. philippinesis, D. ovata, D. retusa* and *D. pteropoda* have confirmed that these botryoidal outgrowths are an integral part of the zygote wall. Its main component is polysaccharides, many of which are acidic polysaccharides of pectin nature. At later stages of embryo development this original jacket or mantle disappears (Yoffe and Zhukova, 1974; Yakovlev, 1983a). In *Saxifraga flagillaris*, the micropylar part of the cell wall bears numerous protuberances which penetrate it deeply (Zhgenti, 1986).

The cell wall of the zygote must be necessary for fixation of the polar axis, as was shown in *Fucus* (Kropf *et al.*, 1988), as well as for maintenance of the differentiated state of embryonic cells and for determination of the fate of cells in plant development (Berger *et al.*, 1994).

Fusion of the sperm cell with the egg cell might be the inductor which provokes a crucial change of the wall of the egg cell-zygote. At present, there are no studies which would allow determination of the exact time of commencement of formation of a new cell wall—whether it is formed right after the union of sperm and egg cell or during karyogamy, or whether its formation is completed right after fusion of male and female nuclei (in the post-meiotic and intermediate types of fertilization).

Zygote insulation provided by its wall is **relative** to some extent since, as attested by its structural organization, the zygote is connected with the cells of the gametophyte and maternal tissues (with nucellus, integuments, integumentary tapetum, etc.) through the transport system. That is why it might be necessary to consider the zygote from the point of view of 'open systems'.

Zygote polarity (Greek *polos*—pole)—space orientation of processes and structures resulting in emergence of morphophysiological differences at its opposite poles. On this basis a morphological axis with two different poles appears: **apical**, facing the central cell of the embryo sac, and **basal**. Later, zygote polarity predetermines the plane of division (transverse, oblique or longitudinal), character of cytokinesis (asymmetrical or symmetrical) and, as a result, formation of two equal or unequal daughter cells, and finally polarity of the embryo. The processes of polarization which occur in the zygote before its first division have been closelý studied in the algae *Fucus* and *Pelvetia* (Kropf, 1992, 1994).

Polarity may be **morphological** or **physiological**. **Morphological polarity** of the zygote manifests in the external (shape of the zygote) and internal structure, a result of transformations in the cell wall, as well as in plasma membrane, cytoplasm and the nucleus. Usually the nucleus and a major portion of the cytoplasm, in which most organelles are disposed, are located in its apical part, and the big vacuole in the basal part. In some plants (*Triticum, Zea, Hordeum, Avena, Myosotis*) the big vacuole is absent and small ones are mainly arranged in the peripheral zone of the zygote (Batygina, 1962a, 1974b, 1987a; Khvedynich, 1969; Chebotaru, 1972; Norstog, 1972; Poddubnaya-Arnoldi, 1976; Korobova, 1982; Mogensen, 1982; Russell, 1993). Organelles, vesicles and cytoskeletal elements undergo reorganization, too, and arrange asymmetrically in the zygote.

Physiological polarity of the zygote manifests in the gradient arrangement of ion and electric currents, of polysaccharides, proteins, phytohormones (and other matters), in different intensity of metabolism and different tropism of its parts. The apical pole of the zygote and, further, of the embryo, shows negative geotropism, whereas the basal pole shows positive geotropism (Poddubnaya-Arnoldi, 1976). Establishment of certain gradients brings a specific orientation of the zygote elements, i.e., an axial character.

Laying of the axis of polarity could be regarded as the first step of differentiation and consequently of morphogenesis (Sinnott, 1960). Polarity of the zygote is predetermined by polarity of the mature egg cell, which has an apical-basal axis (chalazal-micropylar).

A number of studies devoted to zygote development have shown that the egg cell initial possesses no gradients (Batygina, 1974b, 1987a; Kropf, 1994). But soon polarity establishes, so that the mature egg cell appears to be a polarized cell. Sperm penetration causes disorganization of its inner pattern: its vacuolization changes, microtubules are reorientated, the nuclei move, the complete cell wall is formed, etc. Thus, the zygote undergoes a process of polarization again and the mature zygote becomes even more polarized.

Polarity of the zygote (and subsequently, of the embryo) depends on the axial polarity of the embryo sac and ovule. Concomitantly one may speak of the plasticity of morphogenetic polarity, since there are instances of axis change, presence of a few axes, etc. (Batygina, 1993b). Reversal of polarity of the embryo sac, may occur, when egg cell apparatus develops on both poles, or antipodals are found at the micropylar end, and the egg cell at the chalazal one (Swamy, 1946). During fertilization and zygote development, a reversal of polarity may occur: the nucleus migrates to the basal pole of the zygote (as in *Papaver nudicaule*).

Types of symmetry. Different species of angiosperms exhibit different types of egg cell and zygote symmetry: radial and bilateral (dorsoventral). An axis and a plane of symmetry are distinguishable. Symmetry is manifested at the morphological and physiological levels. A changeover from one type of symmetry to another occurs during egg cell, zygote and embryo development. In some plant species (e.g.

Triticum, Nelumbo, Commelina—Batygina, 1968a, 1969, 1974b; Titova and Batygina, 1996), a mature zygote shows internal dorsoventral symmetry which persists in the mature embryo. Another example of dorsoventral symmetry is two different cells resulting from division of the zygote in *Fucus* (Sinnott, 1960).

Some authors (Raghavan, 1986; Jürgens, 1995) believe that the transition from radial to bilateral symmetry occurs at the time of cotyledon formation, others (Goldberg *et al.*, 1994) point out that a globular embryo is, in the formal sense, radially asymmetrical.

To the present day little, if anything, is known of the genetic components of the left-right asymmetry. Recently, a cluster of genes were identified and their asymmetric expression demonstrated, resulting in the establishment of this type of symmetry in chick embryos (see Wade Roush, 1995).

Zygote cell-cycle pattern. The mitotic cycle of a plant or animal cell consists of the following periods: G_1, S, G_2, and M (Howard and Palk, 1953). In certain cases, a period of forced rest, or mitosis arrest, a so-called G_0 period is observed (for a review, see Jacobs, 1992). The first three periods comprise the interphase.

During maturation and functioning of sexual elements, their mitotic cycle can also be subdivided into periods. In the course of their development the sperm cells fail to complete their mitotic cycle and remain temporally blocked in one of the final steps (telophase) arbitrarily designated as G_0 (Gerassimova-Navashina, 1971a, 1980). A female gamete, on completing a full mitotic cycle during its formation, enters the G_1-period and undergoes specific differentiation, then progresses to the G_0-period in which it stays until fertilization.

The transition from the G_1 to G_0 periods may be reversible: G_1 to G_0 is a pathway involving cell differentiation and arrest of mitosis, whereas G_0 to G_1 means a return to the proliferating condition. The complexity and phylogenetic diversity of mechanisms controlling the $G_0 \rightleftharpoons G_1$ transitions is due not only to reversibility but also to a variety of factors controlling the process in different plant species (Jacobs, 1992, 1995).

At the time of contact, the gamete nuclei are usually in the same period of the cell cycle. In an immature (just formed) zygote the current period of cell cycle may differ, depending on the type of karyogamy.

According to the up-to-date theory of cell cycle, we may suppose that the zygote of flowering plants is in the G_2-period before its first division. At the same time, this zygote may differ in morphology, corresponding to the type of fertilization and karyogamy: it can show completely fused nuclei (premitotic type of fertilization, G_1, G_2- and S-types of karyogamy) or partially fused (post-mitotic and intermediate types of fertillzation, M-karyogamy). All this indicates that **the pattern of cell cycle passing in the zygote is taxon-specific.**

The overall duration of the cell cycle and its individual periods in gametes and zygote vary among species of flowering plants. This appears to be due to the different structure of embryo sac, ovule and ovary, as well as species anthecology.

For eukaryotic somatic cells, duration of the S-period varies according to species but normally accounts for a relatively small portion of the mitotic cycle (Grif and Ivanov, 1975). Characteristic features of the developing female gametes are an extend duration of the S-period and a longer duration of a presynthetic period compared to the post-synthetic period (Bannikova *et al.*, 1985). The S-period in a zygote, when DNA undergoes replication, is shorter than the G_1- or G_2-period. Thus, in *Petunia hybrida* for example, the duration of G_1- and G_2-periods is 75 and 28.5 h respectively, whereas that of the S-period is 4.6 h and of the M-period 3 h (Vallade, 1980).

The chronology of double fertilization has been followed in detail in different species of *Triticum* (Batygina, 1961a, 1974,a,b, 1987a) and in *Zea mays* (Mòl *et al.*, 1994). The process of egg cell formation and maturation is relatively prolonged in *Triticum* (24-50 h). The G_1-period of zygote cell cycle (zygote formation) lasts for 5-6 h on average, *S*- and G_2-periods (zygote maturation) continue to completion for 7–12 h, and the *M*-period for 6-8 h. During the *S*-period of the cycle male chromatin is indistinguishable from female chromatin, so the zygote enters the prophase 24 h later, as a single whole. Complete fusion of nuclei in artificial fertilization *in vitro* was observed during 35 min-1 h (Faure *et al.*, 1993; Tirlapur *et al.*, 1995).

Caution must be exercised in interpretation of data in the literature on changes occurring in the egg cell prior to, during and after fertilization, since detailed studies have been conducted with a limited number of species and, often, without clearly specifying the developmental stages.

The most extensively studied zygotes are those of some species belonging to genera *Gossypium, Capsella, Epidendrum, Triticum, Zea, Hordeum,* etc. (Batygina, 1962a, 1974b, 1987a; Jensen, 1964b, 1968a-c; Schulz R. and Jensen, 1968a,b; Cocucci and Jensen, 1969a; Mogensen, 1972; Norstog, 1972; Mogensen and Suthar, 1979).

During zygote maturation, the number of ribosomes per unit of volume of cytoplasm is increased, some of them tending to become associated in polysomes (*Capsella*—Schulz R. and Jensen, 1968b; *Epidendrum*—Cocucci and Jensen, 1969a; *Petunia*—Van Went, 1970b; *Linum*—D'Alascio-Deschamps, 1973a,b). In *Gossypium,* the majority of ribosomes form helices up to 20 subunits in size (Jensen, 1968c). A new population of ribosomes is formed some distance from the zygote nucleus, the region showing numerous ER cisternae (*Lycopersicon*—Kadej F. and Kadej A., 1981). A considerable complication of the ER organization has been reported in many of the species studied.

Only after contact of the two gamete nuclei does the level of RNA (bound and unbound) in the egg nucleus decrease relative to that of the unfertilized egg cell. At the moment of fusion of nuclei, the amount of free RNA in the egg nucleus again increases.

First evidence of the nuclear DNA content of an embryo sac at the nuclear stage of development came from *Nicotiana tabacum, Lilium regale, Triticum aestivum, Galanthus nivalis* and *Triticale* (Bannikova *et al.*, 1978, 1981, 1985). The DNA content in the nuclei giving rise to generative cells was found to increase steadily throughout the *S*-period. The average DNA content per egg nucleus in these species corresponded to 1C (C = DNA content per haploid nucleus).

The process of zygote 'maturation' involves accumulation of lipids, protein, and starch. So, for example, fat droplets have been found in the basal region of the *Jasione montana* zygote, which persist until late embryogenesis stages (Erdelská, 1983).

It is widely believed that the egg cell and sperm cells are highly specialized (Batygina, 1984, 1991; Zhukova, 1994), whereas a zygote entering mitosis exhibits instead properties typical of a meristematic cell (Batygina, 1984, 1987b) and is, in a sense, totipotent. 'Totipotency is a property reproduced each time with the formation of a zygote and lost by embryo cells... during their specialization' (Tokin, 1987, p. 31). Some authors consider the zygote to be a non-specialized parenchymatous cell (Russell, 1993).

During interphase, the zygote (zygote initial, immature zygote) undergoes extremely complex morphogenetic changes which presumably result in a 'loss' of specialization by mature egg cell and acquisition by the mature zygote of proliferation capacity. This process can be regarded as egg cell-zygote 'dedifferentiation' during which cytoplasmic elements undergo 'cytoplasm reorganization'. The entry of sperm cell into an egg cell appears to be a factor triggering metabolic processes, reorganization of organelles (formation of polysomes, division of plastids and mitochondria, ER concentration, etc.), and changes in the pattern of vacuolation.

Egg cell and zygote organization, type of karyogamy and cell cycle pattern are key factors determining the sequence and pattern of embryo (and seedling) development. Studying different levels of zygote organization (subcellular, cellular, organismic) and establishing the relationships between the zygote developmental stage, cell cycle period and type of karyogamy appear to be promising lines of further enquiry into the complex, multifaceted problem of 'Zygote'. Progress in this area will largely depend on our understanding of the ontogenetic and phylogenetic relationships among different plant taxa.

Proembryo

Proembryo (Greek *pro*—before, in front of + *embruon*—embryo)—the embryo in early stages of its development before organ differentiation **(Plate XXV)**.

The term 'proembryo' (German *Vorkeim*) was introduced by Hofmeister (1849) to designate a special globular structure emerging from the upper (apical) cell following zygote division and prior to formation of the embryo differentiated into organs in angiosperms. Such a view was prompted by the fact that in the first plants to be studied (*Capsella bursa-pastoris*, several Orchidaceae) the apical cell derivatives evolve into the embryo proper (embryo body), whereas the basal cell derivatives only form the suspensor. Given this fact, for a long time many researchers referred to the basal cell as 'the initial suspensor cell'. The term 'proembryo', as construed by Hofmeister, was taken by Hanstein (1870) to characterize the embryo at the early stages of its development in some mono- and dicotyledons. Hofmeister's and Hanstein's studies on proembryo segmentation were fundamental for understanding of the main embryo formation regularities— its polarity (importance of 2-celled stage), symmetry (globular proembryo shape), etc.

Further research demonstrated that the embryo proper can originate not only from the apical, but also from the basal cell of a two-celled embryo, due to which discovery the term 'proembryo' came to be understood as the embryo with yet undifferentiated organs and tissues in the early stages of its development. However, various authors applied different criteria for distinguishing the proembryo phase.

Schnarf (1929) believed that the term 'proembryo' should have been employed to designate the embryo exclusively at the earliest stages of its development until the vertical divisions appear in its uppermost cell and the three-dimensional growth of the embryo proper is initiated. In his opinion, a T-shaped four-celled embryo can no longer be termed 'proembryo' as opposed to a linear thread-like embryo consisting of eight, nine or more cells located in one row. This view underlies his distinguishing a 'thread-like' phase in embryo development.

Souèges (1935a, 1937b, 1939c) as well as other researchers of the French embryology school held the opinion that the proembryo stage should end with the passage of the embryo to cotyledon formation, i.e., at the beginning of organogenesis. Accordingly, the term 'proembryo' is still used by some contemporary authors (Jürgens and Mayer, 1994).

According to Johansen (1950), the proembryo stage ends with formation of a sixteen-celled embryo (the fourth generation of cells) following which the suspensor cells begin to elongate and the process of histogenic differentiation of the embryo commences.

As believed by Yakovlev (1971, 1981), the important criterion for defining completion of the proembryo stage is commencement of embryoderm (epidermis) formation, which feature was taken by him to distinguish the **proembryonic** and **embryonic** periods (phases) in embryo development (see above: Embryogeny).

It should be noted, however, that a number of researchers (Batygina, 1974b; Poddubnaya-Arnoldi, 1976) justly hold application of the term 'proembryo' to be erroneous, since embryo development is a single process and it is difficult to decide when the proembryo formation stage is over and the embryo development stage begins. Nonetheless the notion of proembryo is widely used in contemporary works and only a few authors prefer other terms, for example 'preglobular embryo' (Meinke, 1986). As is evident from available data a number of important morphogenetic events (commencement of three-dimensional growth and tissue differentiation) occur during proembryo formation. That is why the proembryo stage is indeed one of the most important critical stages of embryo development (see above: Embryogeny).

The first detailed study of the succession of cellular divisions in the proembryo was conducted by Hanstein (1870), who described its development in some dicotyledons (*Capsella bursa-pastoris*) and monocotyledons (*Alisma plantago*) from the viewpoint of the **histogenic theory** (or **histogenic embryo differentiation**). According to this theory, in the early stages of its development the proembryo gives rise to the initial embryonic layers, namely the **plerome**, **periblem** and **dermatogen**, which subsequently evolve into particular tissues and organs of the embryo. In view of the fact that the histogenic theory assumes a specific morphogenetic ability in each proembryo cell, ensuing research targeted primarily the definition of regularities of cellular division in the early proembryo and their participation in formation of the principal histogens, structures and organs (Carano, 1915; Chiarugi, 1925; Schnarf, 1929; Souèges, 1910-1939).

For the cells appearing in the process of zygote division and its derivatives—histogens, which develop into separate embryo parts—a system of conventional signs was developed (Souèges, 1910, 1939c), using abbreviations of the French terms:

ca (cellule apicale)—apical cell;

cb (cellule basale)—basal cell;

m (cellule intermédiaire)—intermediate cell;

ci (cellule inférieure)—lower cell;

q—quadrant cells appearing as a result of two linear divisions;

l and *l'*—upper and lower octant cells appearing as a result of transverse divisions of the quadrant cells;

cc and *cd*—upper and lower cells appearing as a result of transverse division of the apical cell;

pco (partie cotylée)—cotyledonary area;

pvt (point végétative de la tige)—growing point of stem (stem apex);

phy (partie hypocotylée)—hypocotylar area;

icc (initiales du cilindre central de la tige et de la racine)—initials of the central cilinder of the stem and root;

iec (initiales de l'écorce de la racine)—initials of the root cortex;

co (coiffe)—root cap;

s (suspenseur)—suspensor.

To characterize proembryo development the notion of **cell generations** was introduced, the first five generations being the most important, and the major regularities of cellular division within the generations were defined (Souèges, 1939c). Consequent to zygote division (**cell generation one**), which in most angiosperms is directed transversely, two daughter cells appear—the apical cell (*ca*) and the basal cell (*cb*) located one above the other. However, there are other types of zygote division, e.g. linear or oblique. In **cell generation two** resulting from division of both dyad cells a tetrad of cells appears whose structure may likewise vary depending on the division type. So there are linear, T-shaped, isobilateral and decussate tetrads. During **cell generation three** the two upper tetrad cells form **quadrants** by means of vertical division. In **cell generation four** the dividing quadrant cells produce **octants**, which can be formed according to two major types—by means of transverse divisions of the quadrant cells accompanied by formation of **tiers** *l* and *l'*, or by diagonal partitions whereat the octants are lined up in one tier *q*. According to Souèges, the octants are the first embryo elements forming the basis for the subsequent development of its structures and organs—the cotyledons, epicotyl and in some cases the hypocotyl. With **cell generation five** underway, detachment of the protoderm (epidermis) layer occurs, resulting from the periclinal divisions of the octant cells. As a rule it is during this period that the **hypophysis** is formed in the basal proembryo area and the **epiphysis** in its apical area. According to Souèges (1934), the derivatives of these cells subsequently produce the apical meristems of the root and the stem of the embryo respectively.

In the initial cell generations the tier layout of the proembryo is most clearly expressed, however, owing to fast multidirectional division of cells this layout is quickly lost and it becomes difficult to distinguish it during later periods of proembryo development. Therefore, it is an extraordinarily complicated task to determine from which cells this or that organ of the embryo has evolved.

The above scheme of histogenic proembryo differentiation is inherent in most angiosperms; however, the time of differentiation and the modes of basic histogen formation are highly diverse. It is an established view that, on the whole, there are no principal differences in proembryo formation in the early stages of its development in mono- and dicotyledons (Souèges, 1939c; Johansen, 1950; Guignard, 1975b). However, in a number of monocotyledons (Poaceae, Commelinaceae) the embryo was determined to have a dorsoventral layout manifested in the specific oblique orientation of its first partitions and related to the type of embryo symmetry in the monocotyledons. Interestingly, to some extent a similar dorsoventral proembryo layout was found in some representatives of a number of mostly primitive dicotyledonous taxa—Juglandaceae, Nelumbonaceae, Ranunculaceae (Batygina, 1968b, 1974b, 1987a; Titova and Batygina, 1996)—and served as the basis for definition of specific embryogenic types in a number of works (Nast, 1941; Babis, 1976). There are some species in which it is difficult to single out any regularities of histogenic proembryo differentiation (so-called **irregular types of embryo development** according to Souèges (1936b), **anomalous types** according to Maheshwari, 1950 and **retarded histogenic differentiation types** according to Haccius and Bhandari, 1975). It should be noted that the classical views of the importance of cell genealogy laws in embryo development are currently being questioned. It is an accepted view that cell function is determined more by its position than by its preceding development (Christianson, 1986; Poethig, 1989; De Jong at al., 1993). However, in view of the fact that many a species displays remarkable consistency in succession of cellular divisions in proembryo development, further research is needed to resolve the issue finally (Barlow, 1994a, b). The regularities of the first four cell generations were taken as the basis for embryo type characteristics in the systems proposed by Schnarf (1929), Souèges (1939c), Johansen (1950), Periasamy (1977) and Yamazaki (1974, 1982). All these issues as well as the peculiarities of the ultrastructure of the forming proembryo, gene regulation of its development, etc. are dealt with in other articles of the present volume.

Suspensor

The suspensor (Latin *suspendo*—to suspend) is a specialized organ of the embryo, which develops from the basal cell of the proembryo *(cb)* or its derivatives **(Plate XXVI)**. The suspensor is the main pathway for nutrients and hormone transport into the embryo. It often contributes to penetration of the embryo deep into the endosperm. Morphologically the suspensor does not participate in formation of the embryo proper. When the seed matures little or nothing of the suspensor remains.

It was Meyen (1840) who proposed the term, however, he believed that the suspensor arises from the end of the pollen tube. Radlkofer (1857) was the first to propose the suspensor as a component of the embryo.

By means of the suspensor the embryo attaches to the wall of the embryo sac through which it communicates with the nucellus or integument. The suspensor possesses great morphological diversity: it may be single- or multicellular, poorly or clearly differentiated, various in shape (filiform, pillar or spherical, of complicated irregular configuration etc.) and composed either of uniform cells or differentiated into various tissues. Thus, with profuse growth of basal cells in the suspensor, between it and the embryo proper, an isthmus is normally formed, which consists of minor single-nucleate cells. The single- or multinucleate cells of the suspensor may contain di- or polyploid nuclei, including those with polytene chromosomes, and the cell wall various thickenings. Such variations among cells are even possible within the self-same suspensor.

The structure of the embryo suspensor is a characteristic feature of taxa of different rank within an entire order, or one family, genus or species. For example, a large-celled swollen suspensor is typical of representatives of orders Caryophyllales and Alismatales, a filiform suspensor typical of representatives of Scrophulariales, and a grape-cluster suspensor in Convolvulales. Some families reveal variations in the structure of the suspensor; variations are numerous in representatives of Fabaceae (Guignard, 1880, 1882; Lersten, 1983) and Orchidaceae (Treub, 1879; Swamy, 1949a; Poddubnaya-Arnoldi, 1960, 1976). In these families the structure of the suspensor is a characteristic feature of the genera.

It is difficult to compile a classification of types of suspensors due to their great morphological variability. Attempts at classification within some families, for example orchids (Swamy, 1949a), were suspended.

Usually, the shape of a formed suspensor is described, as well as some peculiarities of its cells (Guignard, 1880; Lloyd, 1902; Swamy, 1947; Maheshwari, 1950; Poddubnaya-Arnoldi, 1964b, 1976; Natesh and Rau, 1984), but this does not suffice for a complete characterization of its structure. Since there is no classification of types of suspensor in flowering plants, summarization of data on different aspects of development of this structure—histochemical, biochemical, physiological and, on the whole, functional and evolutional aspects, is hardly possible.

We suggest to take into account the complex of features of the suspensor combined into 4 groups.

The **first** group of features (A) includes morphological peculiarities of the suspensor: shape, degree of tissue differentiation, presence or absence of haustorial protrusions, number and size of cells, number and ploidy of nuclei in the cells, presence of a labyrinth of the cell wall and other features.

The **second** group of features (B) characterizes the connections of the suspensor with the surrounding structures, including direction of growth and contact with certain tissues of the developing seed. In most species the profuse growth of suspensor takes place inside the embryo sac, in the tissue of the endosperm. However, in a number of species, the suspensor completely or partially grows out from the embryo sac and penetrates somatic tissues of the seed (fruit).

Such behaviour of the suspensor must be connected with the manner of storing of reserve substances in the tissues of the developing seed and fruit, as well as with their other peculiarities. For example, in *Sedum* species, a tube-shaped haustorium is formed from the basal end of the swollen cell of suspensor. It penetrates the integument, which contains reserve carbohydrates (Nikiticheva, 1985a). In *Tropaeolum* species, the suspensor haustoria penetrate the integument, the wall of the ovary and the placenta, and reach the vascular bundle; in these tissues storage materials are abundant (Walker, 1947).

In some orchids characterized by a reduced endosperm, the suspensor grows in the micropylar direction from the very first stages, passes through the micropyle of the developing seed poor in reserve materials, then grows along the funiculus and penetrates the tissue of placenta rich in starch (e.g., in *Ophrys, Habenaria*—Swamy, 1949a; *Gymnadenia*—Shamrov and Nikiticheva, 1992).

The **third** group of features (C) concerns the origin of the suspensor from certain histogens of the early proembryo. Having taken the destiny of *ca* and *cb* cell derivatives as a basis, Souèges (1939c) distinguished 6 megarchetypes of embryogenesis, in a descendant series in which the destiny of the suspensor is reflected. The suspensor arises either from cell *cb* directly (VI megarchetype) or from its certain derivatives (I-V megarchetypes), constructive potentials of cell *cb* decreasing one after the other megarchetype. The time of suspensor differentiation in one or another taxon can be an important morphogenetic indicator of its evolutionary level. Early suspensor differentiation occurs in representatives of a number of highly organized families (Caryophyllaceae, Haloragaceae etc.).

The **fourth** group of features (D) characterizes the period of degeneration of the suspensor in relation to a stage of development of the embryo proper according to a complex of morphological and biological features.

Thus, one can propose a **clue to a morphogenetic classification of types of suspensor**, which takes into account both morphological features and peculiarities of its development and relationships with other structures.

A. Morphological features of suspensor (S) and direction of its growth.

1) S undifferentiated, several uniform unspecialized cells perform its function (*Peperomia, Cynomorium*);
2) S filiform, consists of uniform elongated vacuolated single-nucleate cells (*Lathraea, Pinguicula*, species of Loranthaceae), grows into the endosperm;
3) S filiform, consists of uniform elongated vacuolated single-nucleate cells, grows beyond the developing seed and reaches the placenta (*Gymnadenia, Ophrys*);

4) S multicellular, resembles a bunch of grapes, consists of medium-size single-nucleate cells (*Cytisus, Convolvulus*);
5) S consists of swollen multinucleate cells and a small group of cells of the 'isthmus' (*Cuscuta reflexa, C. monogyna*);
6) S complex, consists of a swollen basal cell (or two cells) with endopolyploid nucleus and a group of 'isthmus' cells (*Capsella, Scheuchzeria, Myriophyllum*);
7) S multicellular with haustorial protrusions penetrating the endosperm (*Asperula, Ipomoea obscura*);
8) S with complex structure, with haustorial protrusions running beyond bounds of the embryo sac (*Trapa, Sedum*) and seed (*Tropaeolum*).
(1, 2, 4, 5, 6, 7—suspensor located within the embryo sac)

B. **Contact of suspensor with surrounding tissues of the embryo sac, developing seed and fruit:**

1) synergids, 2) endosperm, 3) nucellus, 4) integument, 5) funiculus, 6) placenta, 7) vascular bundle of the seed, 8) vascular bundle of placenta or wall of the ovary.

C. Origin of suspensor and relative time of differentiation:

1) from undivided cell *cb* (major part) and sometimes, additionally, from cell *ca* derivatives, which form the suspensor 'isthmus' (megarchetype VI);
2) from daughter or granddaughter cell *cb* derivatives (megarchetype V, IV);
3) from distant descendants of cell *cb* (megarchetypes III, II, I).

D. Relative time of degeneration of suspensor:

1) while histogens are being formed, at the globular stage of embryo development (*Orobanche, Pinguicula*);
2) during the initial period of organogenesis, at the heart-shape stage of embryo development (*Lathraea*);
3) at the period of growth and development of embryo organs—cotyledons, apices of shoot and root (*Phaseolus, Capsella, Trapa, Polygala*);
4) at the final stages of ripening of the embryo, at partial preservation of the suspensor in the mature seed (*Sedum*).

Let us look at a number of examples, while using the numeration of the features for the brief characterization of the suspensor structure. The suspensor of the embryo in *Phaseolus coccineus* is best studied at the histochemical level; it can be characterized as multicellular, consisting of a bulbous basal part resembling a bunch of grapes and an elongated 'isthmus' ('handle', according to Avanzi *et al.*, 1970); mainly contacts with the endosperm; derivative of cell *cb* descendants; degenerates at the stage of growth and development of cotyledons.

The suspensor of *Lathraea squamaria* (Nikiticheva, 1966), like in many representatives of Scrophulariaceae, just as in *Pinguicula vulgaris* (Lentibulariaceae), is filiform, consists of uniform, elongated, one-nucleate cells, has a contact for the expanse with a well-developed cellular endosperm and, in a small section, with the integument, develops from granddaughter derivatives of the cell *cb* according to IV megarchetype of Souèges, and degenerates at the heart-shaped stage of embryo development.

The suspensor of *Sedum hispanicum* (Crassulaceae) is divided into three specialized parts: swollen basal cell (with a hypertrophied nucleus and a labyrinth of cell wall), from which long haustorial protrusions emerge. They penetrate the integument as well as a small two-layered 'isthmus'. Due to its complex structure, the suspensor contacts with different tissues; endosperm, nucellus and integuments; its origin conforms with VI megarchetype; the hypertrophied basal cell remains until seed maturation.

In the course of plant evolution, the suspensor has arisen as a specialized organ of the embryo, providing its development inside the seed. The majority of embryologists propose three basic functions of the suspensor: 1) delivery of the embryo into the most favourable environment for its development; 2) absorption and transport of nutrient materials from the surrounding somatic tissues to the developing embryo; 3) synthesis of hormones (see below: Suspensor Functions).

The function of pushing the embryo into the massive endosperm (see Maheshwari, 1950; Poddubnaya-Arnoldi, 1964b, 1976) is supported by morphological pictures of gradual elongation of the suspensor and of movement of the initial of the new sporophyte into the endosperm in many species of flowering plants (e.g., representatives of Scrophulariales). The assumption about the absorbing and nutrient functions of the suspensor is supported by a number of indirect proofs concerning its structure. Thus, Lloyd (1902) having discovered a suspensor with haustoria branching into the endosperm tissue in some Rubiaceae, regarded it as a temporary embryonic root. With its basal part, the suspensor is attached to the nucellus or integument; in taxa where the nucellus is destroyed, haustorial suspensor protrusions rich in nutritive materials and penetrating the tissues of the integument or the ovary are observed.

Biochemical, autoradiographic and electron microscopic studies have revealed important details of the fine structure of the suspensor, which proved its absorbing and transporting functions (see below: Suspensor Functions; Ultrastructural Characteristics of Suspensor).

The relative proportion of the main functions of the suspensor—mechanical (delivery of the embryo into the endosperm), transport, storing and nutrient—might depend on a complex of factors occurring in the developing seed. Certain **correlations** in the structure of the suspensor and the surrounding structures have been revealed. While analysing nutrient supply to the embryo, Mestre and Vannerau (1980) established four types of seed development, taking into consideration structural features of the suspensor and the endosperm. Type I is characterized by the absence of endosperm, its function being performed by a well-developed suspensor (Orchidaceae). Type II is characterized by a large suspensor in combination with a nuclear endosperm of small volume (*Phaseolus, Pisum, Stellaria, Tropaeolum*). Type III with abundant nuclear endosperm (*Linum*), the suspensor performs the function of an active conductor of nutrients. In Type IV a cellular endosperm is combined with an elongated suspensor. Erdelská (1981b) also suggested a type of endospermogenesis that influences the dynamics of proembryo development.

The structure of the suspensor also correlates with peculiarities of nucellus development (Nikiticheva, 1983a,b). For example, a short suspensor with a broad base is typical of taxa with small endosperm and well-developed parietal tissue of the nucellus which remains for a long time (*Scheuchzeria*). In taxa in which the endosperm destructs early or does not develop, a massive and highly differentiated suspensor is formed, which contacts with tissue of the podium (*Trapa*—Titova, 1988; Kolesova [Titova] 1996). The suspensor with haustorial protrusions running beyond the bounds of the embryo sac is normally registered in taxa with tenuinucellate ovule characterized by early reduction of the nucellus, massive integument and usually scarce endosperm (*Tropaeolum, Sedum*). A filiform suspensor with a simplified structure is typical of taxa with a tenuinucellate ovule and intensively developing cellular endosperm with haustoria. In this case, the embryo proper deepens into the centre of the massive endosperm and the suspensor can be as long as 500 µm, as in *Tozzia alpina* (Nikiticheva and Teryokhin, 1973). In this case early degeneration of the suspensor is conditioned by the fact that the embryo starts to absorb the contents of the endosperm cells by all its surface. The nutritive materials of the endosperm are replenished through endosperm haustoria, which is typical of representatives of Scrophulariaceae.

In some studies, the interaction of the embryo proper and the suspensor has also been experimentally proven (Schwartz *et al.*, 1994).

Since the suspensor is formed as a specialized organ providing the young sporophyte with nutrition in the developing seed, its evolutionary morphological transformations occurred in close interconnection with the structures of the seed and the fruit, first of all, in connection with peculiarities of localization of reserve materials in the endosperm, nucellus, integument and placenta (Nikiticheva, 1983a,b). Evolutionary transformations of the suspensor must have occurred in the direction from several weakly specialized cells of the basal part of the embryo, which attach it to the nucellus, to a specialized structure, which has a clear border with the embryo proper and which dies off after performing its function. Suspensor evolution could have been gaining of absorption and secretory properties, the signs of which are development of ER and of a labyrinth of cell wall, activity of dictyosomes, polyploidy of nuclei and so on. Secondary simplification of the structure of the suspensor in Sympetalae might have occurred in connection with an early transition of their embryo proper to active independent absorption of materials from the well-developed endosperm.

Nowadays, due to the combination of conventional cytoembryological and genetic and experimental methods, the idea of interaction of the suspensor with all the structures of the seed receives more and more powerful backing (Yeung and Meinke, 1993).

Ultrastructural Characteristics of Suspensor (Plate XXVII)

The suspensor is the derivative of either only the basal cell of the two-celled proembryo or of the basal cell with a certain degree of participation of the apical cell derivatives. Not only do the cells of the two-celled proembryo differ in size and shape, but also in the structure of their protoplasts and cell walls. The structural polarity of the egg cell which increases in the zygote before its division underlies these differences. One can consider the structural differentiation of the apical and basal poles of the two-celled proembryo proceeding from the comparative analysis of ultrastructure of its cells in *Gossypium hirsutum* (Jensen, 1964a), *Capsella bursa-pastoris* (Schulz R. and Jensen, 1968a,b; Schulz P. and Jensen, 1969), *Linum usitatissimum* (Deschamps, 1969; D'Alascio-Deschamps, 1973b), *Hordeum vulgare* (Cass and Jensen, 1970), *Quercus gambelii* (Singh and Mogensen, 1975), *Epipactis atrorubens* (Zhukova and Savina, 1978), *Petunia hybrida* (Vallade, 1980), *Alcea rosea* (Zhukova, 1981, 1988, 1992b; Kamalova *et al.*, 1983) and *Pennisetum glaucum* (Taylor and Vasil, 1995). On the basis of this analysis one can draw the following conclusion.

In most of the species studied, the apical and basal cells of the two-celled proembryo differ in size and shape of nuclei, in light- and electron-optical density of hyaloplasm, in number of ribosomes per unit of protoplast volume, in degree of its vacuolation and structure of vacuoles, in structure of plastids and their distribution in protoplast (and mitochondria, too), in number and topography of membranes ER (when the number of their profiles in both cells of the two-celled proembryo is, on the whole, limited), in structure of ER in certain species, in degree of development and activity of the Golgi apparatus, in number of starch grains and lipid bodies, and in structure of the plasma membrane and the cell walls.

Here, certain exceptions to this rule are noted. For example, in *Epipactis atrorubens* (Zhukova and Savina, 1978; Zhukova, 1992b), both nuclei of cells of the two-celled proembryo are equal in size and round shape, and the cytoplasm of both cells is of similar density under light and electron microscopy. In spite of the basal cell usually contains fewer ribosomes on the protoplast volume unit than the apical one in *Capsella bursa-pastoris* (Schulz R. and Jensen, 1968a,b) and *Epipactis atrorubens* (Zhukova and Savina, 1978), the ribosome density of both cells is the same. Polarity by ribosome density is revealed at the stage of three- and four-celled proembryo. It is plasmodesmata that are observed in the cell wall, separating the cells of the two-celled proembryo. Plasmodesmata have not been found in outer cell walls of proembryo.

In the course of further development of the proembryo, differences in the ultrastructure of the cells of the suspensor and embryo proper become still more evident. Concomitantly, in the suspensor a structural gradient (see below) clearly appears in the direction of the micropylar-chalazal axis of the embryo.

Unlike the cells of the embryo proper, the cells of the suspensor can be polynucleolus or contain polyploid nuclei, the nuclei often containing polythenic chromosomes (Cavara, 1899; Nagl, 1962a,b, 1970, 1973, 1974, 1976a-c; Schnepf and Nagl, 1970; Schulz P. and Jensen, 1969; Newcomb and Fowke, 1974; Khera Paramjit and Tilney-Bassett, 1976; Yeung and Clutter, 1979; Vereshchagina, 1983a). Generally, polyploid nuclei are formed by endoreduplication. In *Phaseolus* ploidy of the nuclei of the suspensor cells gradually increases in the direction to its micropylar pole; the nucleus of the basal cell possesses the highest level of ploidy (see Graves, 1908). In varieties of *Ph. vulgaris* the latter varies from 8x to 2048x; in *Ph. hysterinus* and *Ph. multiflorus* ploidy amounts to 4096x, and in *Ph. coccineus* giant nuclei can undergo up to 12 cycles of endoreduplication; according to the data of cytophotometry, the amount of DNA in the basal cell of the suspensor is 8192C (Brady, 1973). In this species the nuclei with polythenic chromosomes become lobed; some invaginations of the nucleus reach the chromatin (Nagl, 1976c). There are numerous pores on the evaginations of the nucleus.

In proembryonic and embryonic[12] (at the stage of the globular embryo) phases of development the suspensor cells compared to the cells of the embryo proper are more vacuolate (for example, in *Capsella bursa-pastoris*—Schulz R. and Jensen, 1968a,b; Schulz P. and Jensen, 1969; *Stellaria media*—Newcomb and Fowke, 1974; *Diplotaxis erucoides*—Simoncioli, 1974; *Tropaeolum majus*—Nagl, 1976a; *Phaseolus coccineus*—Nagl, 1977; Yeung and Clutter, 1979; *Arabidopsis thaliana*—Mansfield and Briarty, 1991; *Pennisetum glaucum*—Taylor and Vasil, 1995). In addition, the cells of the micropylar pole of the suspensor are characterized by more intensive vacuolation compared to cells of the chalazal pole (*Helianthus annuus*—Newcomb, 1973; *Pennisetum glaucum*—Taylor and Vasil, 1995; *Geranium gracile*—Zhukova, unpublished data).

The cells of the suspensor in the proembryonic phase of embryogenesis and at the stage of globular and heart-shape embryo are also notable for more developed

[12] According to Yakovlev (1972, 1981).

ER. Peculiarly abundant ER is found in the suspensor of *Capsella bursa-pastoris* (Schulz R. and Jensen, 1968a,b; Schulz P. and Jensen, 1969) and *Pelargonium* × *hortorum* Bailey (Khera Paramjit and Tilney-Bassett, 1976). Long profiles of its membranes are mainly found along the cell walls. Variability in the type of ER is observed. Some plants, such as *Phaseolus vulgaris* (Nagl, 1970; Schnepf and Nagl, 1970), *Ph. coccineus* (Nagl, 1977; Yeung and Clutter, 1979), *Ipomoea purpurea* (Ponzi and Pizzolongo, 1972) and *Stellaria media* (predominantly in the basal cell of the suspensor, Newcomb and Fowke, 1974) contain abundant smooth ER (on the background of free ribosomes), whereas in other plants (*Helianthus annuus*—Newcomb, 1973; *Stellaria media*, in all the cells of the suspensor except the basal cell—Newcomb and Fowke, 1974 and *Diplotaxis erucoides* – Simoncioli, 1974) the ER is rough.

The cells of the suspensor of most of the species studied contain more dictyosomes which are in the state of hypersecretion (Schulz P. and Jensen, 1969; Simoncioli, 1974). However, in *Helianthus annuus* (Newcomb, 1973) and *Stellaria media* (Newcomb and Fowke, 1974) the suspensor contains few dictyosomes and they seem to be relatively inactive. The protoplast of the cells of the suspensor is characterized by less electron density and fewer ribosomes than the protoplast of the cells of the embryo proper (Deschamps, 1968; Schulz R. and Jensen, 1968a,b; Schulz P. and Jensen, 1969; Simoncioli, 1974; Yeung and Clutter, 1979; Mansfield and Briarty, 1991). In addition, in the cells of the suspensor of *Diplotaxis erucoides* decrease in density of the ribosomes per unit of protoplast volume towards the micropyle was observed (Simoncioli, 1974). *Reseda lutea* is an exception: in it, the cells of the suspensor differ from the cells of the embryo proper by greater density of ribosomes (Yakovlev and Chaban, 1979).

Unlike the cells of the embryo proper, the plastids of the cells of the suspensor are represented by differentiated metabolically active organelles as early as the proembryonic phase and at the globular stage of development. In *Capsella bursa-pastoris* (Schulz R. and Jensen, 1968a,b; Schulz P. and Jensen, 1969), *Helianthus annuus* (Newcomb, 1973), *Ipomoea purpurea* (Ponzi and Pizzolongo, 1973), *Diplotaxis erucoides* (Simoncioli, 1974), *Pelargonium* × *hortorum* Bailey (Khera Paramjit and Tilney-Bassett, 1976), and *Alcea rosea* (Zhukova, 1988), the plastids of the suspensor accumulate starch. In the suspensor of *Reseda lutea* (Yakovlev and Chaban, 1979) and *Vicia faba* (Johansson and Walles, 1994), starch is not found at this period of embryogenesis. Furthermore, plastids in *C. bursa-pastoris* (Schulz P. and Jensen, 1969) and *Phaseolus coccineus* (Yeung and Clutter, 1979) increase in number. Besides, storage reserves in the form of lipid bodies are typical of the suspensor of *C. bursa-pastoris*, *H. annuus*, *P.* × *hortorum* Bailey, *Ph. coccineus*, *Tropaeolum majus* (Schulz R. and Jensen, 1968a,b; Schulz P. and Jensen, 1969; Newcomb, 1973; Khera Paramjit and Tilney-Bassett, 1976; Nagl, 1976a, 1977). As is known, the leucoplasts of the embryo in chloroembryophytes are transformed into chloroplasts during transition from globular to heart-shape stage of development (Zhukova, 1978, 1981, 1988, 1993; Whatley and Price, 1983). In addition, in some chloroembryophytes, the suspensor of the heart-shaped embryo also becomes chlorophyllous while in others, the plastids of the suspensor remain leucoplasts. Thus, differences according to the type[13] of plastids were observed in structure of the suspensor of the embryo in the flowering plants examined (Table 4).

---

[13] The basic types of mature plastids are interpreted by us as chloroplasts, leucoplasts and chromoplasts like Schimper (1885) and Wareing and Phillips (1978), the pigment set (or its absence) being the basis for their classification. The pattern of storage reserves is considered one of the criteria of the subtype, but not of the type of plastids.

Table 4. Summary of data on type of embryo proper and suspensor plastids in flowering plants[14]

Plant species	Embryo plastids type		Phase, stage of embryogenesis	Reference
	Embryo proper	Suspensor		
Alcea rosea L.	L	L	proembryonic phase	Zhukova, 1988
			embryonic phase:	
	L	L	stage of globular E	
	Ch	L	stage of heart-shaped E	
Capsella bursa-pastoris (L.) Medik.	L	L	8-celled E	Schulz R. and Jensen, 1968a,b; Schulz P. and Jensen, 1969
	L	L	globular E	
	Ch	L	heart-shaped E	
Helianthus annuus L.		L	proembryo	Newcomb, 1973
	L*[15]	L	late globular E	
	L*	L	heart-shaped E	
Ipomoea purpurea (L.) Roth	Ch*	L	in suspensor, at all stages of E development	Ponzi and Pizzolongo, 1973
Linum usitatissimum L.[16]	L	L	2-celled proembryo	Deschamps, 1969
	L	L	globular E	Deschamps, 1968; Pretová, 1985
	Ch	Ch	heart-shaped E	
Pelargonium × *hortorum* Bailey[16]	Ch	Ch	heart-shaped E	Khera Paramjit and Tilney-Bassett, 1976
	Ch	Ch	early elongated E	
	Ch	Ch	fully formed E	
	L	–	E of mature seed	
Phaseolus coccineus L.	L	L	late globular stage	Yeung and Clutter, 1979
	Ch	L	heart-shaped stage	
Pisum sativum L.	Ch	L	transition into phase of fast growth of E	Marinos, 1970
Reseda lutea L.	L	L	proembryo	Yakovlev and Chaban, 1979
	L	L	early globular E	
	L	L	globular E	
	Ch	Ch	stage of initiation of cotyledons in E	
Stellaria media (L.) Cyrill.		L	3-cellular E	Newcomb and Fowke, 1974
		L	thread-like E	
	L*	L	globular E	
	L*	L	heart-shaped E	
	L*	L	torpedo-shaped E	
Tropaeolum majus L.[17]	Ch	Ch	heart-shaped stage	Nagl, 1976a
Vicia faba L.	L	L	globular E	Johansson and Walles, 1994
	Ch not detected	L	heart-shaped E (12 days after pollination)	
	Ch	suspensor eliminated	heart-shaped E (20 days after pollination)	

[14]In denotation of stages of embyogenesis the terminology adopted by the authors was used.
[15]Abbreviations with asterisk represent the data of Yakovlev and Zhukova (1973).
[16]Chloroplasts of the embryo proper and of the suspensor in *Linum usitatissimum* (according to electron micrographs) and *Pelargonium* × *hortorum* differ by their organization.
[17]Chloroplasts of the embryo proper and the suspensor of *Tropaeolum majus* have similar ultrastructure.
Legend: Ch—chloroplasts, L—leucoplasts, E—embryo.

Among leguminous plants, for example, which are chloroembryophytes, species are known which carry chloroplasts in the suspensors of embryos as well, namely 8 species of *Crotalaria* (Anantaswamy, 1950). The suspensor of the chloroembryophyte *Lupinus albus* is also characterized by the presence of chloroplasts (Atabekova and Lin Tzyansin, 1964; Khudyak, 1964), with poorly developed grana and thylakoids (Davey, 1978). Chloroplasts have been observed in the chloroembryophyte *Pelargonium* × *hortorum* Bailey (Geraniaceae) studied from the heart-shape stage, in all the cells of the embryo, including the suspensor. However, differences were discovered in the fine structure of these organelles in the embryo and the suspensor. At all stages of development, the form of the chloroplasts is irregular but peculiar in the chloroplasts of the suspensor. Chloroplasts of the haustorium of the suspensor are notable for the presence of supergrana of large diameter (up to 40 thylakoids in a pack). Other differences have also been described (Khera Paramjit and Tilney-Bassett, 1976). In *Reseda lutea*, chloroplasts of the suspensor differ from chloroplasts of the cells of the embryo proper by more intense development of peripheral plastid reticulum (Yakovlev and Chaban, 1979). Plastids of the suspensor of chloroembryophytes *C. bursa-pastoris* (Schulz P. and Jensen, 1969), *Pisum sativum* (Marinos, 1970), *Phaseolus vulgaris* (Schnepf and Nagl, 1970), *Diplotaxis erucoides* (Simoncioli, 1974), and *Phaseolus coccineus* (Nagl, 1977; Yeung and Clutter, 1979) are represented by leucoplasts (in the embryo proper of these species, at the heart-shape stage of development, chloroplasts are observed). Plastids of the suspensor in *Linum usitatissimum* (Deschamps, 1968), *Capsella bursa-pastoris* (Schulz P. and Jensen, 1969), *Diplotaxis erucoides* (Simoncioli, 1974), *Pelargonium* × *hortorum* Bailey (Khera Paramjit and Tilney-Bassett, 1976) and *Alcea rosea* (Zhukova, 1988, 1992b) contain significant amounts of plastoglobules. Similar plastids in *Phaseolus coccineus* may be regarded as chromoplasts (Nagl, 1977; Yeung and Clutter, 1979).

Differences in the ultrastructure of the plastids of the embryo proper and of the suspensor are also observed in species with leucoplasts in all tissues of the embryo. In *Stellaria media* (Newcomb and Fowke, 1974), plastids of the suspensor are remarkable for their peculiar ultrastructure. Leucoplasts of the basal cell differ in structure from the plastids of the other two cells of the proembryo as early as at the three-celled stage of development. They are bigger than the latter. Two morphologically different zones are observed in them: an inner zone densely filled with small tubules, and the surrounding zone less densely packed with larger tubules. The authors call these plastids protein plastids. In the course of embryogenesis (this process is traced up to the stage of a torpedo-like embryo), cup-shape profiles of the plastids of the suspensor increase in number; they are particularly numerous at the globular stage of embryo development.

Unlike the embryo proper, the plastidom of the suspensor in many species examined is differentiated to a greater extent. Plastids are one of the components which determine the structural gradient in the suspensor of *Linum usitatissimum* (Deschamps, 1968), *Stellaria media* (Newcomb and Fowke, 1974), *Diplotaxis erucoides* (Simoncioli, 1974), *Tropaeolum majus* (Nagl, 1976a), *Lupinus albus* (Davey, 1978). Not only were differences observed in size of plastids, the volume of which increases towards the micropyle, but also in their ultrastructure. The degree of morphological differentiation of the suspensor plastids varies in various plants. For example, a high degree of differentiation is typical of the plastidom of the embryo suspensor in *Tropaeolum majus*. At the globular and heart-shape stages of development, in the apical 'rosette' cells of the suspensor, there are chloroplasts without starch, in cells of the middle zone chroloplasts with starch, and in the basal cells of the suspensor leucoplasts with electron-dense matrix. The ultrastructure of the chloroplasts of the chalazal cells of the suspensor possess no features which would distinguish them from the chloroplasts of the embryo proper. Their inner membrane system is only represented by several thylakoids. No other morphological characteristics of the chloroplasts are suggested. The haustoria of the suspensor contain plastids with a peculiar ultrastructure: one of them contains etioplasts and the other contains undifferentiated leucoplasts which transform into plastids with electron-dense matrix in the course of autolysis of the suspensor (Nagl and Kühner, 1976-1977).

In some species (*Capsella bursa-pastoris, Diplotaxis erucoides, Helianthus annuus*), the plastids of the suspensor are poorly developed. On the contrary, in most plants they are characterized by a high degree of inner differentiation. The plastids of the suspensor in *Pisum sativum* (Marinos, 1970), *Phaseolus vulgaris* (Schnepf and Nagl, 1970; Yeung and Clutter, 1979), *Ipomoea purpurea* (Ponzi and Pizzolongo, 1972, 1973), *Stellaria media* (Newcomb and Fowke, 1974), *Tropaeolum majus* (Nagl and Kühner, 1976-1977), *Ph. coccineus* (Yeung and Clutter, 1979), and *Vicia faba* (Johansson and Walles, 1994), both chloroplasts and leucoplasts, can be regarded as highlyspecialized plastids. In most cases, such plastids are only typical of the suspensor and are not observed in the cells of the embryo proper. In *Phaseolus vulgaris*, the plastids of the suspensor undergo a unique transformation at a later heart-shape stage. This kind of transformation has never been observed in the other species studied (Schnepf and Nagl, 1970). Some fragments of cytoplasm containing ER, polysomes and lipid bodies turn out to be enveloped by plastids. Cytochemical examinations determined high activity of acid phosphatase in such plastids (Nagl, 1977; Gärtner and Nagl, 1980). Such 'plastolisomes' transform into autophagic vacuoles at later stages of embryogenesis. They are thought to be related to senescence of the suspensor. In *Ph. coccineus* (Yeung and Clutter, 1979), such specialized plastids are especially numerous at late globular and heart-shape stages. In the course of embryo ripening, they disappear. These plastids might fulfil some specific function in embryogenesis.

On the whole, it is believed that during the proembryonic phase and the globular stage of the embryonic phase of embryo development, no manifest changes occur in the ultrastructure of the suspensor cells, as, for example, in *Capsella bursa-pastoris*. However, they can be traced in the ultrastructure of plastids. At the stage of octants (6 cells in the suspensor) the plastids are rod-shaped with dense stroma and single thylakoids, often vesiculized. They also contain ribosomes, plastoglobules and small starch grains (Schulz P. and Jensen, 1969). At the stage of globular embryo, the plastids of the suspensor become cup-shape. One more important feature of the plastidom of the suspensor becomes apparent here. The cytoplasm of the suspensor cells in *C. bursa-pastoris* appeared to remain intact up to the heart-shape stage of embryo development. Evident signs of its degeneration are only observed at the stage of a heart-shape embryo. The signs are: almost complete disappearance of ribosomes, appearance of autophagic vacuoles, loss of the ability to stain for proteins and nucleic acids and others. In addition, plastids are again an exception—they possess no signs of degeneration. There are single vesicled thylakoids, plastoglobules in plastids, and they can stain for proteins. New structural elements in the form of dense bodies packed in regular rows appear in these plastids, starch completely disappears. It is notable that plastids lose their cup-shape form (this shape is often regarded as evidence of degeneration of cytoplasm) and remain leucoplasts, whereas the plastids of the embryo proper transform into chloroplasts at the heart-shape stage.

The cells of the suspensor, especially in its basal part, have the structure of transfer cells in the majority of species (see Deschamps, 1968, 1969; Schulz P. and Jensen, 1969; Marinos, 1970; Nagl, 1970, 1973, 1976b; Newcomb and Fowke, 1974; Simoncioli, 1974; Khera Paramjit and Tilney-Bassett, 1976; Yeung and Clutter, 1979; Johansson and Walles, 1994). The cell walls between the cells of the suspensor as well as the cell wall between the suspensor and the embryo proper are pierced with numerous plasmodesmata (Schulz P. and Jensen, 1969; Newcomb, 1973; Newcomb and Fowke, 1974; Simoncioli, 1974; Johansson and Walles, 1994), but there are no protoplasmic connections between the suspensor and endosperm. In the endosperm, on the border with the lateral walls of the subbasal and middle cells of the suspensor, digitate protuberances of the cell wall are formed. At the late globular and heart-shape stages of development, the suspensor has no cuticle, unlike the embryo proper, on which the outer surface of the embryoderm is covered with cuticle at this period of embryogenesis. It is evident in *Arabidopsis thaliana*, *Capsella bursa-postoris* and *Stellaria media* (Rodkiewicz *et al.*, 1994; Yeung *et al.*, 1996).

According to the data of light and electron microscopy, the suspensor of the embryo of flowering plants has distinctive peculiarities not only at the tissue level, but also at the species level of organization. In spite of great morphological diversity of this organ of the embryo, there are a number of structural features typical of the suspensor of all or most of the species examined. These features are: its early differentiation (at the period of proembryo formation), polyploidization of nuclei, strong vacuolation of protoplast, high compared to cells of the embryo proper, development of ER and Golgi apparatus, low density of ribosomes per unit of protoplast volume, presence of highlyspecialized plastids, organization of cells according to the type transfer cells and absence of cuticle on the outer surface of the cells of the suspensor. For most of the aforesaid features, including the size of the cells of the suspensor, there exists a structural gradient in the direction of its micropylar-chalazal axis.

For a long time, the suspensor was regarded as an organ whose only function is 'pushing' of the embryo into the endosperm and its orientation in it (Maheshwari, 1950). However, as early as 1902, Lloyd noticed that the haustoria of the suspensor in some representatives of Rubiaceae function as a temporary embryo root to the suspensor. At present, functions of the suspensor of the embryo of flowering plants are considered to be equal to those of the trophoblast in mammals (see Nagl, 1973). According to the data of light and electron microscopy, the suspensor is polyfunctional. It takes part in absorption and translocation of nutrients and, possibly, of growth regulators from the surrounding seed tissues to the developing embryo. Along with the function of transfer cells (which are basically considered characteristic of the cells of the base of the suspensor), it also has storage functions. The presence of smooth ER in the cells of the suspensor allows suggesting participation of this embryo organ in synthesis of biologically active metabolites (enzymes, hormones). In this respect it is interesting to note the absence or reduction of the suspensor in embryoids whose development takes place *in vitro*. A secretory function of the suspensor is also under discussion.

Suspensor Functions (Plate XXVIII)

Several recent reviews pertaining to different aspects of suspensor biology are available (Yeung and Meinke, 1993; Goldberg *et al.*, 1994; Schwartz *et al.*, 1994, 1995;

Meinke, 1995; Raghavan and Sharma, 1995). Here, the structural development of the suspensor is highlighted.

Structural development of suspensor. The suspensor, when present, is a short-lived embryonic organ. It is important to note that derivatives of the basal cell can contribute cells to the embryo proper as in *Brassica napus* (Tykarska, 1979). On the other hand, derivatives of the terminal cell can also contribute to formation of the suspensor, as in *Stellaria media* (Pritchard, 1964). The number of cells in a suspensor is limited, ranging from a single cell in *Phaius tankervilliae*, or a few cells in *Arabidopsis* (Mansfield and Briarty, 1991) and *Alisma* (Bohdanowicz, 1987), to a few hundred in *Phaseolus vulgaris*. Morphologically, the suspensor can appear simply as a large vacuolated cell as in *Phaius tankervilliae*, a uniseriate file of cells as in *Arabidopsis* (Mansfield and Briarty, 1991), or with elaborate extensions as in *Cymbidium sinense*. During early embryogeny, development of the suspensor is rapid in terms of size compared to the embryo proper. Growth of the suspensor can be due simply to cell enlargement or both cell enlargement and increase in cell number.

Rapid growth of the suspensor may be aided in part by the cytoskeletal elements, i.e., microtubules and microfilaments. In *Arabidospsis*, during proembryo development, microtubules within the suspensor cells tend to orient transversely to direction of elongation (Webb and Gunning, 1991). In *Cymbidium sinense*, only 5-6 suspensory cells are present. The suspensor cells elongate and extend outwards, coming into close contact with the seed coat. The suspensor cells are highly vacuolated. A fine network of cortical microtubules can be found in the cytoplasm. Most of the microtubules orient transversely to the long axis of the cells except at the tip. Near the tip of the suspensor cells, microtubules tend to orient along the axis of the cell. Judging from the fluorescence intensity, more microtubules are present near the tip. In contrast, although cytoskeletal elements are present in the embryo proper, they usually take on a netlike appearance and are not as distinct as those found in the cytoplasm of the suspensor cells. The microtubular pattern as seen in the suspensor of *Cymbidium* indicates that microtubules are important in the maintenance of cell shape as well as regulating the direction of cell expansion and elongation. Furthermore, cytoskeletal elements may also be important for the translocation process. In *Cymbidium*, the rapid expansion of suspensor cells during early embryogeny suggests that the water potential within the suspensor cells must be very negative. Cytoskeletal elements are necessary for reinforcing the cell wall in order to maintain the turgor pressure within the cell. This helps in ensuring proper water balance within the cells.

Different structural specializations can be found in suspensors of flowering plants. One of the most common features of structural specialization is the suspensor cells assuming a transfer cell morphology, i.e., the presence of wall ingrowths and an abundance of organelles. Well-documented information is available for *Phaseolus* species (Yeung and Clutter, 1978, 1979; Nagl, 1974), *Tropaeolum majus* (Nagl, 1976a; Nagl and Kuhner, 1976) and *Alisma* (Bohdanowicz, 1987). Structural specializations can be observed at the proembryo stage of development. In *Alisma*, specializations begin in the basal cell as early as the two-celled stage (Bohdanowicz, 1987). Elaborate wall ingrowths can be found in suspensor cells, especially in the walls abutting the maternal tissue. The nucleus can be highly polyploid with polytene chromosomes. Smooth and rough endoplasmic reticulum, polysomes, mitochondria, and plastids are abundant within the cells. These structural specializations are not found in cells of the embryo proper. All these studies indicate that the suspensor is a unique embryonic organ and structural specializations occur at the time most likely to be the most critical period in embryo development, i.e., early growth and histodifferentiation.

Compared with the embryo proper, another feature with respect to the suspensor is the absence of cuticular material in the suspensor cell wall. In *Capsella, Arabidopsis* and *Stellaria*, a cuticular substance is present in the wall of the protoderm of the embryo proper (Rodkiewicz *et al.*, 1994). A similar observation was made in the zygotic embryo of *Cymbidium* (Yeung, Zee, and Ye, unpubl. results) and carrot (Lackie and Yeung, unpubl. results). Although the synthesis of cuticular substance may be unique to the protoderm in general, the absence of cuticular substance clearly indicates the distinct difference between the suspensor and the embryo proper cells.

Maximum development of the suspensor generally occurs at the globular-heart stage of embryo development. Soon after, the suspensor begins to degenerate and subsequently compresses as the embryo matures. Degeneration of the suspensor is a programmed process. In *Alisma*, the large basal suspensor cell begins to degenerate 7 days after pollination. Vacuoles increase in size and frequently contain dense inclusions; the organelles also begin to degenerate (Bohdanowicz, 1987). One of the features associated with the senescing suspensor is an increase in acid phosphatase activity in the plastids and cytoplasmic membrane (Nagl, 1977, Gärtner and Nagl, 1980). Nagl suggested that 'the suspensor of plant embryos undergoes programmed cell death, allowing mobilization of stored material and its utilization by the embryo proper' (Nagl, 1976b). In this respect, *Arabidopsis* suspensor mutants can be excellent tools in the study of the degenerative process. Mutants are currently available that fail to undergo programmed cell death (Schwartz *et al.*, 1995). Through the study of these mutants, it may be possible to determine the signal(s) necessary to initiate or inhibit the degenerative process.

The absence of structural specialization as seen using light and electron microscopes in some species does not necessarily mean that the suspensor is not specialized. Physiological specializations such as unique transport mechanisms may be present which simply cannot be detected by using conventional structural techniques. If the suspensor can maintain a more negative water potential, it may still be able to serve as a conduit for movement of water and dissolved material.

In monocotyledonous plants such as *Zea mays* (Schel *et al.*, 1984), the size of the suspensor is small and structural specialization is not present within it. However, when the surrounding endosperm cells are examined, it is found that they take on a transfer morphology (Schel *et al.*, 1984). Similarly, the elaborate endosperm haustorium of *Rhinanthus minor* certainly can play an important role in embryo development (Nagl, 1992). Different mechanisms must be present in seeds to ensure proper embryo development. It is not surprising to see a division of labour has taken place within seeds and some of the functions of the suspensor reside in cells surrounding the embryo.

Embryonic environment can modulate suspensor development. In somatic embryos, a suspensor may or may not be present. Even when it is present, structural specialization similar to that found in zygotic embryos is absent. Since somatic embryos, especially those derived from the liquid suspension system are bathed in a relatively uniform nutritive environment, structural specialization related to nutrient uptake is not necessary. In the *in vitro*-grown ovules from intra- and interspecific crosses in the genus *Beta*, the suspensor degenerates rapidly compared to *in situ*-grown seeds (Bruun, 1991a, b). In parthenogenetically induced embryos of Poaceae species, the endosperm fails to develop; however, larger suspensors are found (Matzk,

1991). Therefore, changes in the environment or nutrient uptake can modulate suspensor development and function.

Developmental potential of suspensor. Judging from the structural specialization and growth characteristics of suspensors in flowering plants, the suspensor is a differentiated embryonic organ. It appears from the works of Haccius (1963), Clark and Sheridan (1991) and others (see Yeung and Meinke, 1993), that further development can occur in the suspensor if the growth of the embryo proper is disrupted. Using genetic approaches, a number of suspensor mutants have been identified (Schwartz *et al.*, 1995). It was shown earlier that embryo-lethal mutants of *Arabidopsis* have abnormal suspensors (Marsden and Meinke, 1985). Recently, additional suspensor mutants were characterized. It was found that defects occurring within the embryo proper of some *Arabidopsis* mutants lead to suspensor enlargement and developing characteristics which are normally restricted to the embryo proper (Goldberg *et al.*, 1994; Schwartz *et al.*, 1994; Yadegari *et al.*, 1994). These results clearly indicate that under normal development, interaction between the embryo proper and the suspensor leads to the suppression of the embryonic programme in the suspensor, thus maintaining its identity (Schwartz *et al.*, 1995). Therefore the suspensor is not a terminally differentiated organ and is capable of further development under unusual circumstances.

Recently, another embryo mutant has been characterized which promises to provide important information concerning how suspensor development is regulated. The *Twn* mutation in *Arabidopsis* results in the formation of twin embryos. One of the twin embryos is a secondary embryo derived from the suspensor (Vernon and Meinke, 1994). Therefore, this mutant is an attractive system in the study of embryo development, especially the intercommunication between the suspensor and the embryo proper.

Suspensor functions. Morphogenetically, the presence of the suspensor is essential to development of the embryo proper. Using *in vitro* culture methods, the presence of the suspensor greatly enhances growth of the embryo proper during the early stages of development (Yeung and Sussex 1979; Monnier, 1995). The presence of the suspensor significantly increases the protein content of the embryo proper (Brady and Walthall, 1985; Walthall and Brady, 1986). In its absence, growth of the embryo proper is retarded or even arrested, leading to death of the embryo. Failure to successfully culture small proembryos may be due to the injury of the suspensor during the isolation process (Monnier, 1995).

More specifically, the suspensor functions as the main uptake site for the developing embryo. Labelling studies indicate that different metabolites and tracer compounds are preferentially taken up through the suspensor and translocated to the embryo proper (Yeung, 1980; Brady and Combs, 1988; Nagl, 1990). Orchid embryos can be excellent systems in the study of suspensor function. Unlike seeds of many flowering plants, vascular tissue is absent from the seed coat. This suggests that nutrient transfer from the maternal tissue to the embryo may not be very efficient. Thus, orchid embryos must evolve special means of obtaining nutrients from the maternal tissues in order to develop. The great diversity of suspensor morphology in Orchidaceae, as noted by Swamy (1949a), may simply be one of the mechanisms for obtaining nutrients for embryo development. In *Cymbidium sinense*, the suspensor is characterized by elaborate outgrowths of suspensor cells. The suspensor cells are tightly appressed onto the seed coat. Histochemical staining indicates the absence of cuticular substances in the suspensor wall and the cell wall remains primary in nature.

The absence of cuticular substance indicates that apoplastic transport is possible between the maternal tissue and the suspensor cells. Judging from the growth pattern of the *Cymbidium* suspensor, it is conceivable that it has developed mechanism(s) in maintaining a negative water potential within the cell. This would then account for the ability of water uptake as well as dissolved materials for growth of the embryo proper. Using orchid embryos as experimental systems, future testing of the foregoing hypothesis would provide a more definitive answer to the mechanism of uptake and transport by suspensor cells.

The suspensor can function as a site for growth substance synthesis. In *Phaseolus coccineus*, the suspensor has the ability to synthesize gibberellins, a number of which have been identified using gas chromatography-mass spectroscopy (see Alpi, 1990). Based on the qualitative and quantitative gibberellin pattern between the suspensor and the embryo proper, Piaggesi *et al.* (1989) suggested that the embryo proper may depend on the suspensor for the supply of growth substances during early stages of its development. Besides *Phaseolus coccineus*, gibberellins have also been identified in suspensors of *Tropaeolum* and *Cytisus* (Picciarelli *et al.*, 1984; Picciarelli and Alpi, 1987). Gibberellic acid has been shown to increase the protein content and protein synthesis activity of the embryo proper (Brady and Walthall, 1985; Walthall and Brady, 1986) as well as RNA synthesis in bean suspensor cells containing polytene chromosomes (Forino *et al.*, 1992). Other plant growth substances, e.g. auxin (Przybyllok and Nagl, 1977), cytokinin (Lorenzi *et al.*, 1978), and abscisic acid (Perata *et al.*, 1990) have been detected in the suspensor as well. These growth substances may play a direct role in tissue differentiation within the embryo proper. However, whether the ability to synthesize plant growth substances is common to suspensors of other flowering plants remains to be determined.

Future perspectives. In the last two decades, significant advances have been made in our understanding of suspensor biology of flowering plants. Embryo mutants have proven a useful tool in the study of suspensor biology as well as other aspects of embryo development (Meinke, 1995). Once these genes can be identified and cloned, it will be possible to proceed with molecular analysis of interactions between the embryo proper and suspensor. Some progress has already been made with respect to one of the suspensor mutants (see Meinke, 1995; Schwartz *et al.*, 1995). At present, physiological studies focus mainly on species with a relatively large suspensor such as *Phaseolus*. Whether the information obtained from the study of *Phaseolus* species is also applicable to other species awaits further study. Furthermore, little information is available on the suspensor of Gymnospermae species. Since many Gymnospermae species are important economically, greater attention should be paid to them (Yeung, 1995).

Epidermis of Embryo (Plate XXIX)

To designate the initial cells of the epidermis (Greek *epi*—on, *derma*—skin) of the embryo, Hanstein (1870) introduced the term **dermatogen** (Greek *derma* + *genos*—birth). Though the term is still sometimes used in this meaning (Nikiticheva, 1979a; Mansfield and Briarty, 1991; Guignard *et al.*, 1991), currently it is mainly applied to describe the structure of the root (primary dermatogen, see later, Embryo root). In non-Russian literature the term **protoderm** (Greek *protos*—primary + *derma*) signifies the epidermis of the embryo. However, this term is also used in another sense—to denote the outer layer of the shoot and root apexes. The special term **embryoderm** (Greek *embruon*—an embryo + *derma*) introduced by Yakovlev (1971) to designate the initial cells of the epidermis of the embryo is often found in Russian literature.

Development of the epidermis of the angiospermous embryo progresses from the initial cells to the specialized tissue. Formation of its initials is taxon-specific. In embryos with the regular type of embryogenesis, distinctive initials and establishment of the epidermal layer are normally observed after the octant stage. Periclinal divisions of octants lead to formation of outer and inner cells. The outer cells constitute the initial cells of the epidermis. The epidermis is distinguishable simultaneously in the right and left parts of the embryo, but the periclinal divisions in the upper and lower tiers never occur at the same time (Mansfield and Briarty, 1991). The process completes with formation of a clearly pronounced epidermal layer which surrounds the embryo (Hanstein, 1870; Souèges, 1934; Johansen, 1950; Yakovlev and Alimova, 1976; Nikiticheva, 1979a; Lyndon, 1990; Guignard *et al.*, 1991; Jürgens, 1995). Further growth of the epidermal layer proceeds due to anticlinal divisions of the cells. Thus, it is at the stage of octants that the position of cells of the epidermal layer is fixed and eight initials distinguishable. Their cloning by means of anticlinal divisions results in an increment of cells in the outer layer and facilitates their integrity (Jürgens, 1995). In most angiosperms the epidermis as a layer is distinctly traced in all the parts of the mature embryo including shoot and root apexes. However, in some plants a clearly defined epidermal layer is not always observed (Periasamy, 1977).

In those cases in which there is no ordering of divisions in the embryo (irregular type of embryogenesis), it is difficult to determine the moment of epidermis appearance (Poddubnaya-Arnoldi, 1976). So, in *Gossypium*, characterized by an irregular type of embryogenesis, the epidermis constitutes a distinct layer only at the stage of a 100-celled embryo (Christianson, 1986).

Typically for parasitic angiospermous plants the epidermis is not prominent at the apical pole of the embryo; this is due to disturbances in succession and orientation of cell walls in the upper tier of an octant (*Orobanche pallidiflora*—Nikiticheva, 1979a; Teryokhin and Nikiticheva, 1981). This feature is thought to correlate with peculiarities of germination of parasitic plants. In some representatives of family Orchidaceae the layer of the epidermis may be absent in the apical part (for example, *Thunia alba*—Andronova and Batygina, 1992), and also in other parts of the embryo (e.g., *Goodyera discolor*—Treub, 1879; Johansen, 1950). However, it appears in the course of development of the shoot apex during germination of the embryo and formation of the protocorm.

A possible role of periclinal divisions in establishing the epidermal layer can be observed in *Arabidopsis* and other angiosperms with ordered divisions in the first stages of embryogenesis. Recently the *KN* gene was identified, its product participating in initiation of tissues in the radial direction. In an embryo with a *kn* phenotype, epidermal initials are not distinct due to periclinal divisions and as a consequence the seedlings have no characteristic epidermal layer (Mayer *et al.*, 1991). However, the question of participation of the *KN* gene in regulation of cell wall orientation during periclinal divisions is not fully resolved.

Contrary to classical conceptions, there is an opinion that the epidermis is defined topographically as the outermost layer of the plant body (Kaplan, 1992). The first step in determining tissue orientation of the embryo—increase in number of layers in the transverse (radial) plane—is not related to regularities in the direction of cell divisions (Jürgens, 1995; Raghavan and Sharma, 1995).

Torres Ruiz and Jürgens (1994) demonstrated that in spite of the fact that cell divisions at early stages of development of the *Arabidopsis* embryo, which is an *fs* phenotype, have no definite orientation, tissues in the radial plane establish in the mutant seedling the same as in embryos of the wild phenotype. There is an assumption that cell interactions play an important role in initiation of tissue-specific gene expression and further integration of tissues. It is supposed that genetic differences between outer and inner tissues can appear due to gradual blocking of the connection between cells by means of plasmodesmata, as observed between epidermal and subepidermal cells in the seedling (Duckett *et al.*, 1994). Plasmodesmata are reportedly found in all the cells of an *Arabidopsis* embryo, starting with the octant stage and up to the late globular stage. Thus, being distinguished due to periclinal divisions the layer of the epidermis can have contact by means of plasmodesmata with inner tissues of the embryo over the whole period of the globular stage. Differences between cells of the embryo become evident at the 'heart' stage. This manifests in the fact that the cells of the epidermis at this stage of embryo development are less vacuolated compared to cells of the procambium and shoot apex. At the 'torpedo' stage tissue differences increase even more. Starch grains and numerous plastids with a well-developed system of granules appear in the epidermal cells. At this stage the embryo is able to acquire green coloration. Plastids in the epidermis are more differentiated than in other inner cells. Ribosomes are densest in the epidermal layer and cotyledons. The appearance of electron-dense bodies in all cells of the embryo has been noted, though most of them are located in the epidermis, but their role and origin are not clear (Mansfield and Briarty, 1991; see below: Ultrastructural Aspects of Embryogenesis).

Cytohistological investigation of the *Arabidopsis* embryo has revealed that though the epidermis is distinct due to periclinal divisions at the stage of octants, it forms as a layer only by the end of the globular stage (Mansfield and Briarty, 1991). The authors believe that this fact can be evidence of 'the regulation theory of organization' proposed by Wardlaw (1955) and cast doubt on Hanstein's (1870) theory, according to which the origin of embryo tissues is associated with specific 'histogens' of the proembryo.

Cell interaction can be determined not only by the origin, but also the position of cells in the embryo (Jürgens, 1995; Barlow, 1994a, b; Schmidt *et al.*, 1994). Directions of establishing of cell walls during cellular division correspond to the general direction of organism growth (Lyndon, 1990). Various types of tissues are found in the organism a result of spatial separation of cells according to their characteristics. The event takes place in the very early stages of development of the proembryo and concurs with differentiation of the epidermis, plerome and periblem. The first division of the zygote already determines the domains of the root and seedling. The phase of histogen formation at the very beginning of embryogenesis is of great importance as the new pattern of differentiation which appears in the process remains and complicates the entire space of further development of the plant.

The outline and composition of tissue are quite similar in embryos of various species though directions of establishment of cell walls can differ considerably. This means that the cells and their walls play a rather incidental role in embryo shape. At the same time, among the important morphogenetic events there are processes which do affect the pattern of growth of the embryo surface. Therefore, development of the epidermis with its cuticle which occurs in early embryogenesis can be significant for shaping at further stages (Hara, 1961, 1971, 1995; Hardham, 1982; Green, 1992; Green and Selker, 1991; Barlow, 1994a, b).

In *Daucus carota* the cuticle around the embryo was observed immediately after establishment of the epidermis (Lackie and Yeung, 1996). However, in other species the cuticle may appear in earlier stages as well. For example, in *Citrus jambhiri* it happens at the zygotic stage. Based on this fact Bruck and Walker (1985) believed that the tendency for differentiation of the epidermal layer could be traced starting from the zygotic stage. According to Kaplan (1992), cuticle formation is a characteristic of the organism but not the epidermis. The epidermis is not determined by the cuticle even though it is situated beneath it and initiated after it. It is only a positional link.

The cuticular layer participates in polarization of the embryo and formation of the axis 'shoot apex-root apex' (see below: Cuticle of Developing Embryo). At early stages of embryogenesis it isolates the embryo from the endosperm, which creates a polarized stream of substances from the suspensor to the embryo. At the late stage it probably takes part in embryo maturation.

In Yakovlev's opinion (1971), formation of the epidermis (embryoderm) evidences the transition of the proembryo period into the embryonic. The stage of embryoderm distinction is considered critical in embryo development (Nagato, 1979; Batygina and Vasilyeva, 1983b, 1987, 1988). As the epidermis becomes isolated and anticlinal divisions commence in it, divisions in all the remaining cells of the embryo become ordered (Pollock and Jensen, 1964; Christianson, 1986).

It has also been shown that the stage of epidermal formation is important during development of somatic embryos *in vitro*. Abnormalities in its formation are considered to be associated with arrestation of development of the somatic embryo. Glycoproteins, localized in the protoderm, which eliminate interruption in development and facilitate further formation of somatic embryos, have been identified (Von Engelen *et al.*, 1991; Satoh *et al.*, 1992). This highlights the interaction between the outer and inner layers carried out during development of the somatic embryo (Yeung, 1995).

The epidermis of the embryo, like other ground tissues of the plant, forms at the globular stage of the embryo. In the course of further development differentiation of the epidermis of the hypocotyl and cotyledons, the protoderm of the shoot apex and dermatogen of the root is clearly seen. Specialized cells appear (trichomes, cells of the stomatal apparatus and others). However, not all the cells are specialized; some remain totipotent and with a change in conditions are able to take another path to development. Therefore, all parts of the embryo are used as explants for the culture of tissues.

During cultivation of explants of embryonic origin (scutellum of cereals, hypocotyl, cotyledons etc.) 'direct embryoidogenesis' is typically observed. In this case development of somatic embryos usually initiates from epidermal cells (Pretová *et al.*, 1993; Yeung, 1995).

Investigations of initiation of somatic embryos from epidermal cells of the scutellum of cereals have shown that the first signal of induction of embryoidogenesis is the change of direction of cell divisions in the outer layer from anticlinal to periclinal and oblique (Jones and Rost, 1989). Such division can be found even in 24 hours of cultivation. Apparently, this fact proves poor determination of the epidermal cells of the embryo of cereals.

The epidermis of somatic embryos, for example in *Quercus*, *Daucus* and others, exhibits a tendency towards proliferation and formation of secondary somatic embryos (Shamrov *et al.*, 1992; Yeung, 1995). In Yeung's opinion (1995), the phenomenon of secondary embryoids can be explained by absence of complete differentiation of the epiderm in somatic embryos.

The embryo and protocorm of orchids tend to form secondary protocorms from the epidermis, which is determined by high potential for vegetative propagation starting from the very early stages of ontogenesis of plants (Batygina and Shevtzova, 1985).

Williams and Maheshwaran (1986) posited a hypothesis about the existence of cells differing in degree of determination and ability for 'direct embryoidogenesis' in culture *in vitro*. Meristematic cells ('pre-embryogenic determined cells', PEDC) are competent for embryogenesis and only require appropriate conditions to become determinant just like embryonic cells. As the tissue becomes more specialized, the number of cells with embryogenic determination decreases and only some epidermal cells remain capable for direct embryoidogenesis. This hypothesis explains the various modes of formation of somatic embryos from a group of cells or from one cell. It presupposes that the mode of their formation depends on the extent to which neighbouring cells of the epidermis differ in degree of determination for 'direct embryoidogenesis'. The same cells at a certain stage of their differentiation can have both embryogenic and epidermal determination.

Hypophysis

Hypophysis (Greek *hypo-, hyp*—under, below and *physo-, physi:* swelling, accretion)—a cell or a group of cells in the basal part of the proembryo from which elements of the embryo root originate (Johansen, 1950; Yakovlev, 1981) **(Plate XXX)**.

The term was introduced by Hanstein (1870) to designate the cell located at the border between the suspensor and the body of the embryo and also its derivative (a lens-shaped cell which separates at later stages). According to Hanstein, such cells serve as a connector between the embryo and the suspensor (German *Keim-Anschluss*). The primary hypophysial cell *(h)* is detached by the upper cell of the suspensor typically at the early globular stage of embryo development immediately after the dermatogen (protoderm) has been separated in it. This cell acts as a locking cell of the protoderm (German *Schlusszelle*). Afterwards it undergoes a division, which results in the formation of a lens-shaped cell *(h')*, i.e., a two-celled complex in a 'pressed' position in the proembryo, which probably suggested its name. The hypophysial cell was observed by Hanstein in embryos of a whole series of dicotyledonous plants, the most detailed examination of its development was carried out in *Capsella bursa-pastoris* (Brassicaceae).

Confirmation of the existence of the hypophysis was later provided in a number of other works on embryogenesis of *Capsella bursa-pastoris* (Coulter and Chamberlain, 1903; Schaffner, 1906; Souèges, 1919, 1934). Some authors after Hanstein give quite similar definitions of the hypophysis differing mainly on denotement of this structure and its derivatives. Coulter and Chamberlain defined the hypophysis as **an upper cell of the suspensor** denoting the latter and also its inner derivatives by the Latin letter *'p'*, and the outer by the Latin letter *'d'*, implying that they give rise to a periblem (initial cells of the cortex) and dermatogen of the root respectively.

By hypophysis Souèges meant **the upper cell of the suspensor and all its derivatives,** the inner tier of which gives rise to the cortex of the root and the outward tier to initials of the epidermis and the root cap. Souèges then proposed distinguishing **the initial cell of the hypophysis,** denoting it, like Hanstein, by the Latin letter *'h'* and **the hypophysial tissue** in which the cells forming the cortex and the root cap are denoted by the Latin letters *'iec'* (French *Initiales de l'écorce de la racine*: initials of the cortex of the root) and *'co'* (French *Coiffe:* root cap) respectively. Souèges noted that the specific lens-shaped form of the cell, which is the first derivative initial cell of the hypophysis, is determined by a 'watch crystal' form of cell wall.

Unlike the above-mentioned authors, Schaffner (1906) believed the hypophysis should be defined as **a lens-shape cell only**, forming as a result of division of the cell located **between the suspensor and the body of the embryo.** This cell is an initial of the root growing point. It remains dormant for some time, while its sister cell designated by the author by the Latin letter *'d'* functions as an initial of the root cap. The definition was accepted by Swamy and Krishnamurthy (1975).

However, most researchers (Johansen, 1950; Maheshwari, 1950; Lebègue, 1952; Yakovlev, 1958; Crété, 1963; Poddubnaya-Arnoldi, 1976; Natesh and Rau, 1984; Danilova and Kirpichnikov, 1985; Raghavan, 1990) use the term 'hypophysis' as interpreted by Souèges. At the same time, analysis of the whole variety of modes of hypophysis origination observed by Souèges and other researchers prompted some authors to propose a broader definition of this term, which is also accepted here.

Time, modes of origination and character of division of the hypophysis reveal considerable variety in different species. Souèges (1934) distinguished 6 modes of hypophysis formation.

A. Single hypophysial cell gives rise to initials of the cortex and the root cap; it can divide:

a) horizontally with formation of a cell wall in the form of a watch crystal (*Capsella bursa-pastoris, Oenothera biennis, Mentha viridis, Veronica arvensis);*
b) twice vertically with formation of four cells—'hypophysial quadrants' (*Euphorbia exiqua, Miosurus minimus, Sedum acre*).

B. Single hypophysial cell gives rise to the median part of the cap (*Nicotiana acuminata, Papaver rhoeas, Sagina procumbens*).

C. Two cells of one tier or cells of two tiers of the proembryo located one under the other give rise to initials of the cortex and the root cap respectively (*Senecio vulgaris, Urtica pilulifera*).

D. Group of cells of the proembryo differentiates into hypophysial tissue, out of which the cortex and the root cap form:

a) this group of cells originates from the single determined cell of the proembryo—the derivative of the cell *cb* (*Erodium cicutarium, Geranium molle*);
b) this group of cells separates from another group which originates from an undetermined cell of the proembryo (the embryo proper)—the derivative of the cell *ca*; the event takes place in a quite late period of embryogenesis (*Trifolium minus, Medicago lupulina, Lotus corniculatus*).

Derivatives of the hypophysis can participate not only in formation of the cortex and root cap, but also of the central cylinder, as for example in *Physalis minima* (Solanaceae), *Daucus carota* (Apiaceae) (Souèges, 1922b; Johansen, 1950). Souèges suggested distinguishing such cases from initiation of the true hypophysis which gives rise to the cortex and the root cap only.

Modes of formation of the root cap from the hypophysis differ essentially in dicotyledons and monocotyledons. In dicotyledons only the median part of the cap is formed from the hypophysis, while its lateral parts originate from the dermatogen (protoderm) of the embryo; in most monocotyledons the hypophysis gives rise to the whole cap (Souèges, 1936d; Von Guttenberg, 1960; see below: Radicle).

Of a special interest is the mode of division, the three-dimensional structure of the initial cell of the hypophysis and of the so-called lens-shaped cell separated by it. Detailed investigations of the process of hypophysis development in various representatives of families Scrophulariaceae and Orobanchaceae taken in different planes of the structure and analysis of published data (Souèges, 1919; Tiagi, 1952a,b; Crété, 1953, 1954, 1955b; Arekal, 1963, 1966) have enabled establishing that the hypophysial cell has a unique shape (Nikiticheva, 1971). It is cylindrical at the bottom and conical at the top, which determines the position and form of the cell walls that will further divide it. The initial hypophysial cell divides successively by two oblique cell walls, the first division usually taking place when the dermatogen is separating into octants, and the second division during separation of initials of the plerome and periblem at the lower tier of octants. As a result of these two divisions the *iec* cell separates. It has a classical lens-shaped form only when viewed en face; observed laterally it resembles a triangle or rhomb. After that the lower daughter cell of the hypophysis undergoes one more division. So, the three divisions of the initial hypophysial cell result in formation of the *iec* cell and three *ico* cells.

Another significant peculiarity of the hypophysis is its participation in formation of the dormant centre of the embryo root. Swamy and Krishnamurthy (1975) paid attention to an important observation made by Schaffner (1906) concerning a relatively low rate of division of the lens-shape cell during a considerable period of embryogenesis. For instance, in *Sphenoclea zeylanica* (Sphenocleaceae) this cell divides only twice and forms a plate of four cells which remain changeless even at the stage of mature embryo. Further divisions of the derivatives of this cell are observed only in the course of germination. The lens-shape cell and its derivatives are distinguished from the other cells of the proembryo by smaller size, poor staining and low protein content. Comparison of peculiarities of the structure and behaviour of this cell during embryogenesis with those of the cells of the dormant centre in roots of mature plants (Clowes, 1961) prompted Swamy and Krishnamurthy to conclude that the lens-shape cell and its derivatives possess the functions of the dormant centre of the root.

Repeated attempts were made to trace the initiation of the dormant centre of the embryo root from the hypophysis in various plants (Sterling, 1955; Alfieri and Evert, 1968; Vallade, 1972; Jones, 1977; Clowes, 1978a,b; Raghavan and O'Konkwo, 1982). However, efforts to identify at early stages of embryo development the specific cells that give rise to it were ineffective for quite a long time. According to the data of ultrastructural and histochemical investigations, cells of the hypophysis, for example in *Capsella bursa-pastoris* (Schulz R. and Jensen, 1968b), *Linum usitatissimum* (Deschamps, 1969), *Diplotaxis ericoides* (Simoncioli, 1974), are characterized by a low metabolic activity (small amount of ribosomes, low content of nucleic acids and proteins, low intensity of DNA synthesis). Evidently, cells of the dormant centre have a reduced capacity to incorporate the labelled precursors of DNA synthesis, which makes it difficult to obtain clear structural, physiological or metabolic markers to identify its cells. Only application of the method of *in-situ* hybridization to cellular polyA-RNA using [^{3}H] polyuridylic acid as a probe allowed confirmation of the origination of the dormant centre from the cells of the hypophysis, namely, from the derivatives of its inner lens-shape cell (Raghavan, 1990).

In recent years data have appeared on genetic regulation of the process of formation of the hypophysis and the apical meristem of the root of the embryo. In particular, in *gnom*-mutant embryos of *Arabidopsis thaliana* the normal hypophysis and lens-shape cell form in the period of separation of dermatogen in the proembryo, though neither further development of the derivatives of these cells nor formation of the root apex takes place. Disturbance of development of these structures and consequently absence of the root and hypocotyl in the embryo are also typical for *monopteros*-mutants of this species (Mayer *et al.*, 1993).

In conclusion, it should be noted that such features as origin, structure and time of differentiation of the hypophysis in the embryo are taxon-specific and also characterize progressive evolution within certain taxa (Souèges, 1934; Lebègue, 1952). These features underlie distinguishing the types of embryogenesis and their variations in different classifications (Johansen, 1950; Yamazaki, 1974, 1982) and are also widely used in phylogenetic constructions (see next chapter: Types of Embryogenesis: Principles of Classification of Embryogenesis Types).

Epiphysis

Epiphysis (Greek *epi*—on, above, *physo*—swelling, growth): a cell or group of cells in the apical part of the proembryo, which are the initials of bark and epidermis of a stem **(Plate XXXI)**.

The term was introduced by Souèges (1921) during investigation of embryo development in *Myosotis hispida* (family Boraginaceae); later he used it only for dicotyledons (Souèges, 1934).

Souèges (1934) summarized his ideas of the period of differentiation of the epiphysis in the following scheme.

A. One epiphysial cell is differentiated in the apical part of the proembryo:
   - a) upper cell of linear tetrad (*Papaver rhoeas, Sagina procumbens*);
   - b) upper (wedge-shape) cell of quadrants (*Geum urbanum, Erodium cicularium, Trifolium minus*);
   - c) upper (wedge-shape) cell of octants (*Myosotic hispida*).

B. Four cells, located in a circle, form epiphysial tissue; they are:
   - d) quadrants, located horizontally:
     - 1) with vertical partitions (*Geranium molle*),
     - 2) with diagonal partitions (representatives of family Asteraceae);
   - e) cells of the upper tier of octants:
     - 3) one layer of initials of the stem bark is differentiated (*Myosurus minimus, Sedum acre*),
     - 4) some layers of initials of the stem bark are differentiated (*Lepidium sativum*).

Unlike Souèges, Johansen (1950) believed the epiphysis is not differentiated in all representatives of dicotyledons. He used the presence or absence of an epiphysis in the proembryo when working out principles of classification of the types of embryogenesis in angiospermous plants. It is not distinguished in embryos whose development corresponds to the Solanad- or Piperad-type. The epiphysis is found in the proembryo developing according to Trifolium-variation of the Onagrad-type, Geum- and Erodium-variations of the Asterad-type, Polemonium- and Myosotis-variations of the Chenopodiad-type, Sagina-, Vaccaria-, Hypecoum-, Corydalis-variations of the Caryophyllad-type.

According to Swamy and Krishnamurthy (1977), the epiphysis (epicotylar locus) is a universal structure of the embryo of all angiospermous plants; however, it is differentiated in different ways in dicotyledons and monocotyledons. In dicotyledons, it is differentiated at the stage of octants and takes an axial position, while the cotyledonary locus is located in the peripheral zone of the apical part of the embryo. In monocotyledons the 'embryonic' shoot apex becomes separated into epicotyl and cotyledonary loci, starting from the first vertical division of the apical cell (*ca*), i.e., these two loci are located horizontally one after the other.

It has been determined that from the moment of initiation, the cells of the epicotyl locus (epiphysial zone) rarely divide compared with cells of other parts of the embryo. In the course of the transition of embryo from globular to heart-shape, the cell of the epiphysis remains quiescent. A quiescent centre in the axial zone of the forming shoot apex maintains histological and histochemical characteristics which are peculiar to cells of the epiphysial zone (Mèstre and Guignard, 1967; Swamy and Krishnamurthy, 1977). In *Stellaria media*, taken as an example, it was determined that apical cells of the embryo at the early stage of development (cells of epiphysis, Swamy and Krishnamurthy, 1977) differ from the rest of the cells by large nuclei with small nucleoli, low level of cytoplasmic RNA and comparatively large amount of granules, stained according to Feulgen (Pritchard, 1964).

Radicle (Plate XXXII)

Usually the embryo of seed plants develops a radicle (embryo root), which is sometimes reduced (e.g., in *Ceratophyllum*, some Orchidaceae). The presence of an embryo root, as a continuation of the embryo axis, sharply distinguishes seed plants from pteridophytes, in which the first root of the sporophyte establishes very early but is usually located slightly away from the main axis (Von Guttenberg, 1960; Voronin, 1964).

Hypocotyl-root axis

Initially, the hypocotyl and embryo root are integrated into a single whole, the so-called hypocotyl-root axis[18]. At a very early stage, due to periclinal divisions in the middle part of the embryo, a layer of protoderm or **primary dermatogen** separates (terminology by Von Guttenberg, 1960). Due to divisions in the inner layers of the embryo, which are directed parallel to its longitudinal axis, several concentric layers of cells are formed: the periblem and the plerome. At the distal end of the radicle these layers along with the dermatogen are closed to several tiers of cells developing from one cell or a group of cells of the proembryo, the hypophysis. According to Von Guttenberg (1960) these cells are called central or connective cells.

At a certain stage of embryo development (sometimes very early) differences in histogenesis of the embryo root appear in dicotyledonous and monocotyledonous plants. In dicotyledonous plants neither outer nor inner layers show any differences in the character of cell division, whereas in monocotyledons the direction of divisions

[18]Hypocotyl-root axis of the embryo is a part of the morphological axis of the embryo.

changes in the outermost layer. The inner layers of the periblem divide periclinally, due to which their number increases in breadth, whereas in the outermost layer only anticlinal divisions occur, which results in emergence of a new covering layer—secondary dermatogen forming the **rhizoderm** during germination of the seed and desquamation of the primary dermatogen.

According to Hanstein (1870), the root cap starts in the hypophysis, which consists of end-capping (connective) cells of the embryo. The hypophysis can develop from different tiers of the proembryo, mainly from *'m'* or *'n'*, more rarely from *'o'*. There are cases in which a typical hypophysis is lacking (e.g. in Fabaceae).

In *Capsella* (classical example of embryogenesis in the dicotyledons described by Hanstein), the upper derivatives of the hypophysis produce connective cells, whereas the lower one gives rise to the cap, in particular its central part, the **columella.** The flank (peripheral) parts of the cap emerge due to periclinal divisions in the dermatogen.

Von Guttenberg (1960) took *Biophytum dendroides, Brassica napus* and *Helianthus annuus* as examples of development of the embryo root in dicotyledons. The root pole in *B. dendroides* consists of several groups of cells; in the center lies a cylindrical plerome, which terminates in central cells. Under them lies the columella whose peripheral layers border the dermatogen. In a mature embryo of *Brassica napus,* two central cells are seen, which border with as yet undivided elements of the periblem, undergoing T-divisions later. The central plerome detaches initials of the pericambium (pericycle). The root cap and dermatogen are formed from one and the same initial layer of cells, the dermocalyptrogen. In *Helianthus,* the hypophysis produces only one central block of cap cells, the peripheral part of which may be formed from the dermatogen. These two parts are clearly distinguishable for some time but later the border blurs.

The hypophysial part supplies a chain of cells located more or less parallel to the axis. The cells divide two-four times and form primary columella. Then only transverse divisions take place, due to which the columella reaches 12 cells in height. The side initials of the hypophysis form wedge-shape packs of cells, which undergo anticlinal divisions and elongate tangentially.

Thus, in dicotyledons the root cap is of double origin: its central part (columella) arises from the hypophysis and its peripheral part, from the dermatogen.

The peculiarities of development of the embryo root in monocotyledons are described in a number of examples (Von Guttenberg, 1960). In *Allium giganteum* the connective cells emerge from the tier *'m'*. The initials of the primary periblem and plerome behave the same way as in dicotyledons (they undergo periclinal divisions, and the primary dermatogen forms the layers of the root cap outwards). Soon the secondary dermatogen starts separating (T-divisions stop in the outer layer of the periblem). The root cap remains small. Its central part consists of two-three rows of columella. One row of side cells, also termed side initials of the root cap, adjoins the short zone of poorly distinguishable cells of the dermatogen.

In *Galtonia canadensis,* T-divisions only localize in the outer derivatives of the primary dermatogen, due to which the meristem of the peripheral zone of the root cap is formed. After the last periclinal division, the inner layer of cells continues to divide only anticlinally, developing into the secondary dermatogen. In all cases the initial cells of the columella maintain independence. Apparently, at later stages of formation of the root meristems, they supply cell material for the peripheral zone of the root cap as well. The root cap in *Galtonia* embryo is larger. It is associated with earlier and stronger development of the hypophysis. Peripheral initials of the cap are now represented as a wedge-shape pack of cells which adjoins a region of the dermatogen. Up to complete development of the embryo, the cap is formed in the same manner as in the dicotyledons. The uppermost cells of the cap also join the cells of the primary dermatogen of the embryo.

There is another type of development of the cap in monocotyledons, which occurs without participation of the embryonic dermatogen in formation of the cap. In *Carex*, for example, the root cap is formed from the hypophysis only and surrounds the dermatogen from both sides. The latter develops into the secondary dermatogen.

Special cases of formation of primary root

Poaceae. According to Von Guttenberg (1960, 1968), in most Angiospermae, the primary root emerges exogenously. The cellular layers of the cap are formed due to periclinal divisions of the primary dermatogen. As the cap is building up, no other tissues destruct. Unlike the primary root, the branch and adventitious roots almost always emerge endogenously.

The first embryo root of the cereals is an exception, it does not develop exogenously but endogenously like the branch and adventitious roots. The cap is surrounded by another tissue, the coleorhiza, which destructs when germination occurs. The primary dermatogen does not take part in formation of the root cap (Von Guttenberg *et al.*, 1954; Philip and Haccius, 1976).

According to Pankow and Von Guttenberg (1957), purely endogenous establishment of the primary root and the presence of a coleorhiza in cereals is unique for monocotyledons. In the opinion of these authors, evolutionarily the coleorhiza is nothing but the main root, but the embryo root is an adventitious one.

Commelinaceae. In spite of Von Guttenberg's peremptory statement about the unique character of endogenous establishment of the root in cereals, it is still not clear whether other monocotyledonous plants have a coleorhiza. Solms-Laubach (1878a) and Graves (1898) gave illustrations according to which the roots emerging endogenously in Commelinaceae have a coleorhiza. In these plants the primary roots are very short. Later these data were called into question. According to Pankow and Von Guttenberg (1957), in *Tradescantia*, the cap emerges in the same way as in other plants, from the primary dermatogen, and by no means endogenously. Tillich (1977) and Lakshmanan (1978) discovered a specific structure at the base of the hypocotyl of *Commelina* seedlings, which Tillich termed 'Wurzelhals' (root neck). Development of this structure was traced by Yamashita (1983) in *Commelina communis*. The embryo of this plant does not have a suspensor. At the torpedo stage the embryo starts elongating. Then the cotyledon curves and overhangs the shoot apex. The primary dermatogen of the root pole reveals no periclinal divisions. Formation of the root cap does not start at this point, either. In the middle of the layer following the outer layer two neighbouring cells can be distinguished which develop into the central cells of the periblem (initials of root cortex, by Souèges, 1934) at later stages.

The primary root in *C. communis* establishes exogenously, as in most plants. However, certain peculiarities in cap development are observed at later stages and are associated with the general morphology of the embryo itself. The latter is enlarged at the root pole and the first layers of the root cap, which develop from the primary dermatogen, considerably extend over the periphery of the embryo. Further, the peripheral zone of the cap grows more than the central part. Soon after germination the inner tissue of the peripheral zone starts to disintegrate and, as a result, a cavity is formed inside the root cap, which divides it into inner and outer zones. The inner zone is connected with the body of the root, and the outer one destructs as the root is growing, leaving some tissue of the peripheral zone of the cap as a collar (neck) at the base of the hypocotyl. Considering the aforesaid peculiarities of development of the root cap, Yamashita (1983) suggested that this mode of root development in Commelinaceae should be regarded as intermediate between exogenous and endogenous.

Arecaceae. Observations of *Trachycarpus* and *Phoenix* showed that the basal (root) part of the embryo is differentiated very late, after germination (Pankow and Von Guttenberg, 1957). A subapical and an apical zone of the meristem are distinguished on the root pole. A dome-shape subapical zone transforms into the apical meristem of the root (zone, Gatin, 1906), which in the course of growing separates from the apical parenchyma zone ('root vagina'; Gatin, 1906). In spite of weak differentiation of the root pole, in the quiescent embryo a procambium band can be distinguished, which is directed into the root of the cotyledonary node (Vallade, 1966). Data on embryogenesis in palms are too inadequate to close the question of presence/absence of a coleorhiza, although there are some evident differences in the character of embryo root development in representatives of Arecaceae compared to Poaceae. Common features should be better searched in Commelinaceae.

Apical meristem of the root

General characteristics. Roots normally contain a number of meristematic tissues, which differ in origin, location and character of development. Before the embryo has germinated, the root is often represented only by an apical meristem consisting of initial cells and their derivatives. All the initials make a promeristem, whose cells do not differentiate and give rise to cells of another type that remain in the meristem all the time. Contrarily, the derivatives of the initials divide several times and then exit the meristem. The number of initial cells varies depending on plant species and stage of development. In most plants, the apical meristem of the main embryo root is formed in late stages of embryogenesis, though it can undergo certain changes after seed germination (Von Guttenberg *et al.*, 1955).

Structural variety of meristems has influenced some of their classifications. One such, based on histogenic analysis, was developed by Von Guttenberg (1960) (Fig. 38). He grouped all the varieties of root meristems in angiosperm plants into four types: 1) closed type of dicotyledons, 2) open type of dicotyledons, 3) closed type of monocotyledons, 4) open type of monocotyledons. The closed types have initials of various histogens separated from each other; as a result the cap and cortex, and the cortex and stele can in no way exchange cellular material. In the open types, the central (connective) cells provide material for all the histogens which have no definite initials. Apical meristems of dicotyledonous and monocotyledonous plants differ in origin of root cap and protoderm. In the former, the dermatogen and calyptrogen issue from one initial layer, the protoderm emerges as an inner derivative of this layer, and the calyptrogen as its outer derivative. The common initial layer is therefore termed the **dermocalyptrogen** and roots of dicotyledons with the closed type of meristem termed **climacorhizal**. In monocotyledons the dermatogen emerges as the outer layer of the cortex and the cap has independent initials—**liorhizal root type**. Yamashita (1983) provided a schematic drawing of the types of histogenesis in the

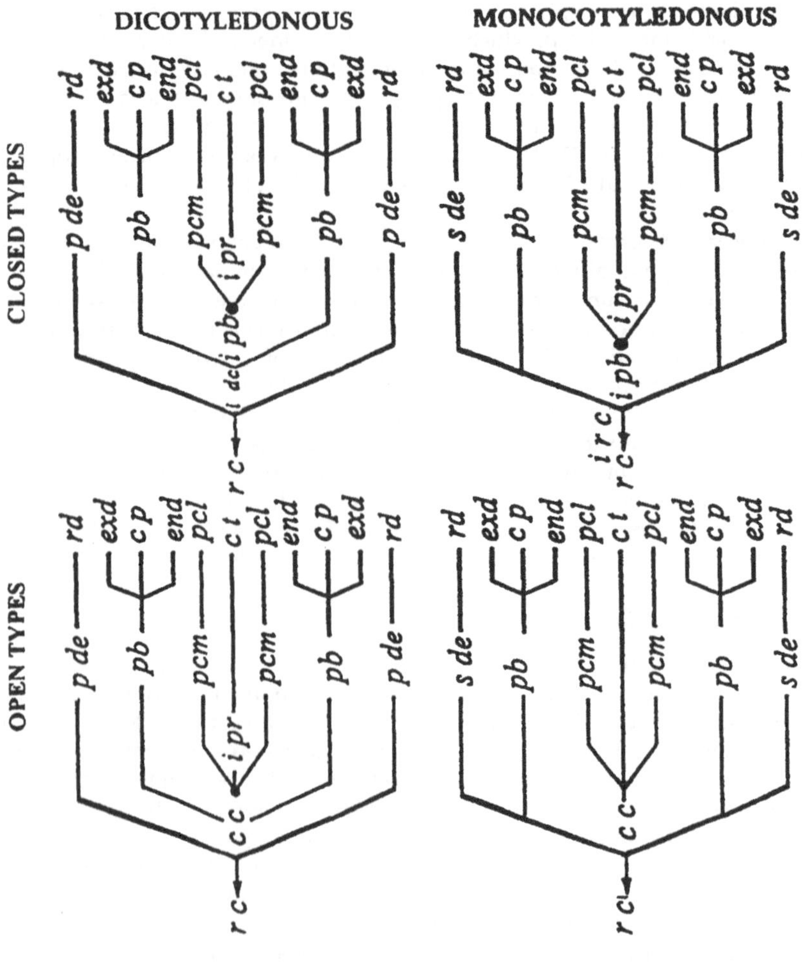

Fig. 38: Schematic representation of types of histogenesis in root of Angiospermae; *c c*—central cells, *c p*—cortical parenchyma, *c t*—conductive tissues, *end*—endoderm, *exd*—exoderm, *i dc*—initials of dermocalyptrogen, *i pb*—initials of periblem, *i pr*—initials of plerome, *i r c*—initials of root cap, *pb*—periblem, *pcl*—pericycle, *pcm*—pericambium, *p de*—primary dermatogen, *r c*—root cap, *rd*—rhizoderm, *s de*—secondary dermatogen (after Von Guttenberg, 1960, cit. from Voronin, 1964).

roots of angiosperm plants. His scheme (Fig. 39) also presupposed a possible transition from one type of meristem to another in the course of root development.

Voronin (1964) proposed his own scheme of evolutionary transformation of the apical structures of the root from Gymnospermae through the open type of dycotyledonous to the closed type. It was shown (Voronin, 1956) that the open type of meristems is closely related to more primitive orders at the base of the system. Thus, the open type is prevalent in Magnoliales, Nymphaeales and Ranales. Many of the choripetalous among dicotyledons (Rosales, Fabales, Myrtales, Sapindales,

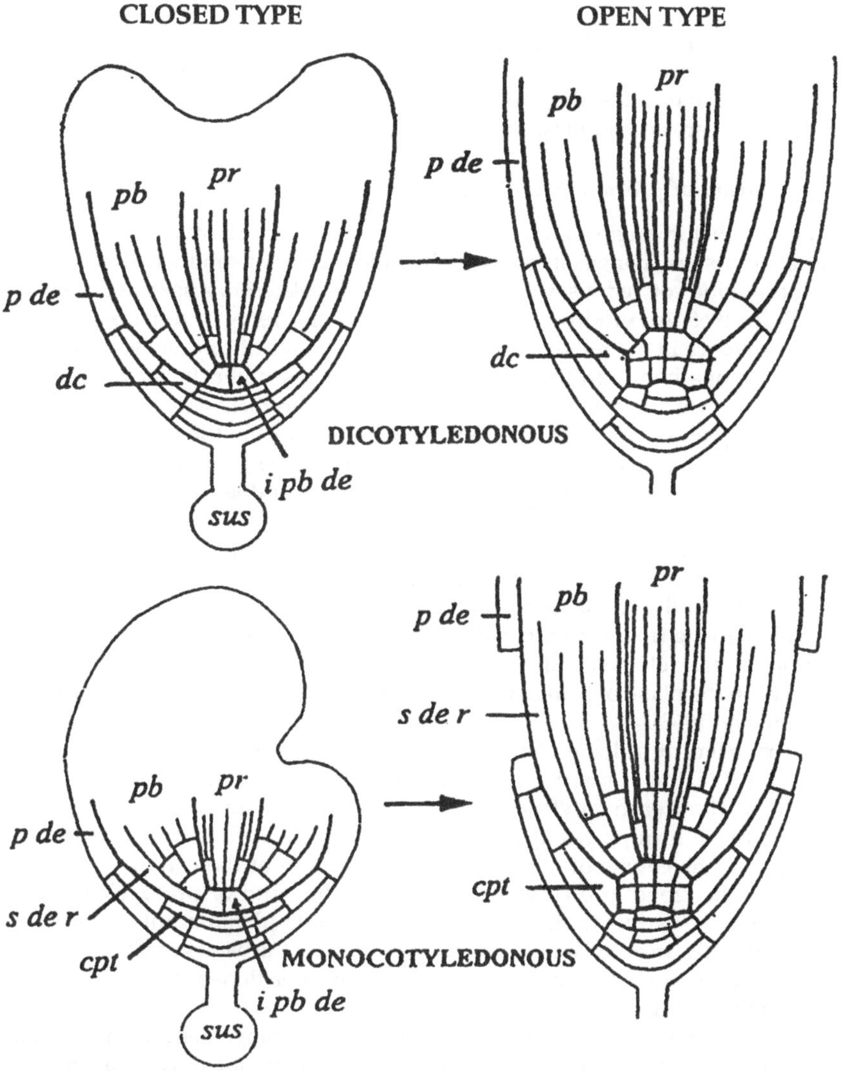

Fig. 39: Schematic representation showing primary structure of root in dicotyledonous and monocotyledonous plants; *cpt*—calyptrogen, *dc*—dermocalyptrogen, *i pb de*—initials of periblem and dermatogen, *pb*—periblem, *pr*—plerome, *p de*—primary dermatogen of the embryo, *s de r*—secondary dermatogen of root (rhizodermis), *sus*—suspensor (after Yamashita, 1983).

Malvales) fit this list and among monocotyledons—Liliales, Arales, Arecales, and Alismatales. Based on Voronin, further extension of the typology of root apices would make Von Guttenberg's scheme more detailed: the number of types would increase due to earlier distinguished and newly discovered smaller types as well as transitional forms of evolutionary importance.

Due to the complicated structure and large number of cells in meristems of higher vascular plants as well as the functional peculiarities of certain complexes of cells, three different but interacting zones have been distinguished: proximal meristem, which includes the initials of the stele, distal meristem, which includes the initials of the cap, and a quiescent centre (Feldman and Torrey, 1975).

Differentiation of procambium. Transitional zone

It is known that in angiospermous and gymnospermous plants the arrangement of conductive tissues changes in the course of movement of conductive bundles from the root into the cotyledons; this reorganization is related to a so-called transitional zone (Esau, 1965; Alexandrov, 1966). The need for this reorganization is determined by the difference in the relative position of the elements of conductive tissues in the root on the one hand, and in the cotyledons on the other. In the seedlings, the procambium, which combines the initial cells of the conductive tissues, is in the form of uninterrupted bands running from the root end to the tops of the cotyledons. Thus the procambium in the root looks like a continuous rod; in hypocotyls it divides into two or more bands extending to the cotyledons and leaves (if the embryo has leaf primordia).

The very first conductive elements of the protoxylem differentiate in the procambium cylinder of the root in an exarchial position—immediately under the pericycle, and in the cotyledons—on the inner side of the procambial band. As it is gradually moving from the outer side of the procambial bundle to the inner side, the chain of vascular elements of the protoxylem remains in a so-called cotyledonary plane, i.e., in the plane of symmetry of the cotyledons, which goes through a median rib (Danilova, 1963).

Every band of phloem elements located in the root in the intercotyledonous plane splits in the transitional zone into two bands which pass to different cotyledons. Thus, each cotyledon becomes connected with two phloem groups of the root; these groups occupy diametrically opposite positions in the latter. Therefore, in the course of reorganization of location of the conductive tissues in the transitional zone in dicotyledonous plants the band of protoxylem elements moves along the radius in the same plane (cotyledon plane), until the protoxylem elements occupy an adaxial position. Concomitantly, the bands of phloem elements in the root and cotyledons are arranged in planes positioned crosswise, and their movement is circular.

The above-described reorganization of mutual arrangement of the conductive tissues in the transitional zone is principally similar not only in all groups of dicotyledonous plants, but also in gymnospermous ones. Differences are mostly associated with the number of phloem and xylem groups in the root, which, in turn, reflects the character of the relationship between cotyledons and the first leaves and the root. Thus, for example, in *Onobrychis echidna*, which has a triarchical root structure, the third group of the xylem and the phloem is connected with the side ribs of both the cotyledons (Danilova and Zhogoleva, 1971).

In the sunflower, which has a tetrarchical root structure, the side ribs of the cotyledons also have immediate continuation in the stele of the root (Vasilevskaya, 1961). When the epicotyl develops early, the median and sometimes the side ribs are also immediately connected with the stele of the root. In all these cases the structure of the transitional zone obviously becomes complicated. However, the aforesaid type of connection between the conductive system of the root and the middle rib of the cotyledon remains unaltered.

The principal derivation from this type discovered in cereals (Danilova and Sokolovskaya, 1973) is especially striking (Fig. 40). According to Fig. 40, we may judge the character of differences in the structure of the transition zone in dicotyledons (A, B) and cereals (C). Unlike dicotyledonous plants, the cereals have a phloem band, which does not change its position in the transition zone. The specific character of the structure of the transition zone in cereals deserves special attention in a discussion of hypotheses concerning the origin of the monocotyledonous embryo.

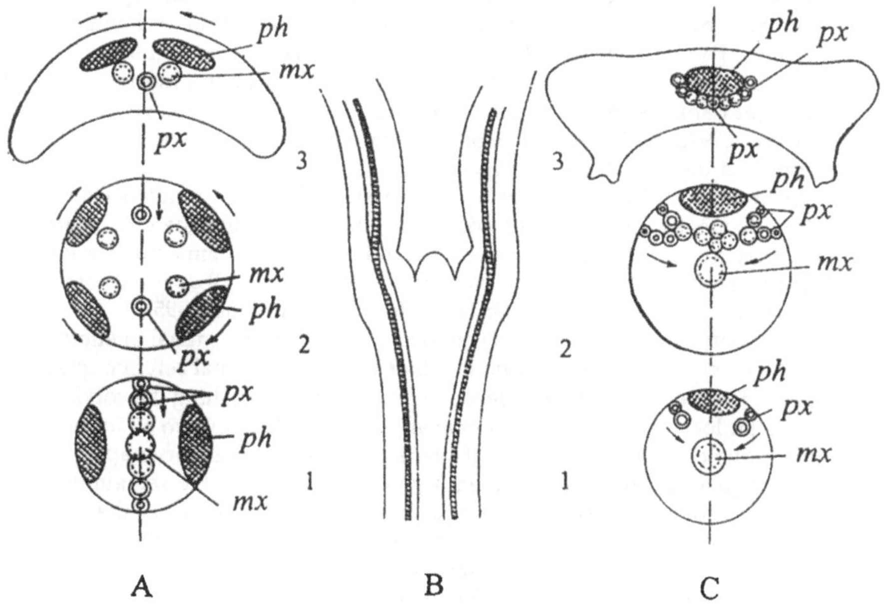

Fig. 40: Schematic representation of reorganization of mutual position of conductive tissues of the transition zone in dicotyledons (A, B) and cereals (C) (after Danilova and Sokolovskaya, 1973). 1—root, 2—transition zone, 3—cotyledon; *mx*—metaxylem, *ph*—phloem, *px*—protoxylem.

Epicotyl

The epicotyl (Greek *epi*—on, *kotuledon*—cup-shape vessel) is part of the axis of an embryo and seedling located over the cotyledonary node and which includes an apical meristem and sometimes leaf primordia. In some cases the epicotyl is interpreted as an axial part of an embryo and seedling placed between the cotyledonary node and the node of the first leaf (Eames, 1961; Yakovlev, 1981; Danilova and Kirpichnikov, 1985).

In the course of embryogenesis, two groups of stem cells form at the opposite ends of the longitudinal axis of an embryo: primary meristems of the shoot and root (Jürgens, 1995; Padmanabhan, 1996). Establishment of the axis of the future embryo occurs at the time of the first division in embryogenesis and polarization of the zygote is already being determined by polarization of the egg cell (see above: Zygote). As the first step in differentiation, isolation of the epidermal layer at the early globular stage takes place. The embryo at this period is represented by a mass of meristematic cells, whose ultrastructural differences are not always shown. During further developmental divisions, cells aggregate in the apical and basal regions of the globular embryo. These zones are considered to be direct descendants of the octants of the embryo and not the meristematic cells which are laying down *de novo* (Barlow, 1994a,b; Jürgens, 1995; Padmanabhan, 1996).

In most species studied, mainly dicotyledons, **differentiation of the shoot apex is associated with the presence of cytological differentiation of the apical domain of an embryo into peripheral and central zones** (Barlow, 1994a, b; Raghavan and Sharma, 1995; Padmanabhan, 1996). The main function of the central (or axial) zone is to provide self-renovation and to maintain integrity of the meristem. At the peripheral (or flank) zone, primordia of the lateral organs form (Steeves and Sussex, 1989; Barlow, 1994a, b). In the course of further divisions and differentiation of cells, cotyledons and the epicotyl arise (Fig. 41).

In dicotyledons, the epicotyl forms at the base of the apical (terminal) tier *l* of the proembryo (Souèges, 1921, 1934, 1936d, 1939c; Johansen, 1950). However, there is an opinion that a certain morphogenetic zonality which forms during embryo development (zones of the shoot and root apexes and hypocotyl) does not correspond to the system of tiers (Lyndon, 1990; Teryokhin, 1991; Jürgens, 1995).

In monocotyledons, the epicotyl can form at the base of terminal or subterminal tiers. In the embryo of *Halophila ovata* the division of its terminal cell into quadrants by means of two vertical cell walls, leads to division distinguishing the cotyledonary and epicotylar loci (Swamy and Lakshmanan, 1962b). In the embryo of *Sagittaria*, the epicotylar zone occupies one-fourth of the flank zone of the embryo apex, i.e. it is situated obliquely to the axis of the embryo (subterminally); the remaining three-quarters of this zone represent the base and sheath of the cotyledon, i.e., the cotyledon forms out of both terminal and subterminal tiers (Padmanabhan, 1996).

Functioning of the meristem of the shoot depends on its physiological status, which is determined by participation of key genes without which successive transformation of the apical half-sphere of the embryo into vegetative and then generative shoot is impossible. So, in seedlings of *embryonic flower* mutants of *Arabidopsis*, the vegetative shoot apex is absent and the generative shoot forms immediately during germination of the embryo (Sung *et al.*, 1992). Data on *shoot meristemless* mutants prove the existence of an embryological phase in development of the shoot apex, which permits introducing the notion 'embryonic shoot apex' (Padmanabhan, 1993a,b, 1996).

At present, it is generally accepted that the shoot apex originates in cells of the apical part of the embryo. It was first formulated in the theory of 'histogens' by Hanstein (1870) and further developed in classical works of embryology—Souèges, Johansen, Swamy. So, using the methods of classical structural analysis we can forecast the development of embryo cells starting from the stage of octants. However, in using molecular-genetic analysis, it has been shown that the fate of the cells of an embryo is not determined at the early stages of its development (Lyndon, 1990; Teryokhin, 1991; Barlow, 1994a,b; Jürgens, 1995; Padmanabhan, 1996). A supposition has been posited that the dominating mechanism is the physiological, not the cellular (genealogical) one, which is responsible for control

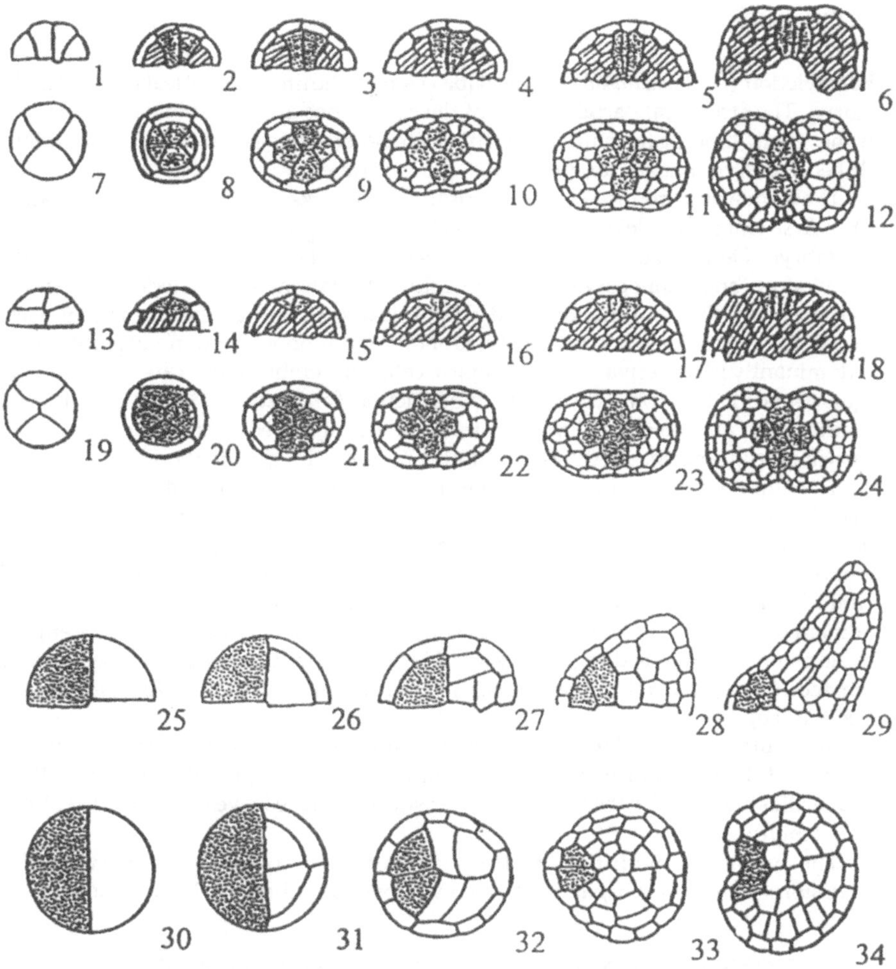

Fig. 41: Development of epicotylar (depicted by dots) and cotyledonary loci in dicotyledonous (1-24) and monocotyledonous (25-34) plants: 1-12—octants disposed at one tier, 13-24—at two tiers (1-6, 13-18—longitudinal sections; 25-29—median sections in the epicotylar-cotyledonar plane; 7-12, 19-24, 30-34—transverse sections, which correspond with certain longitudinal sections; here they are disposed one under the other) (after Swamy and Krishnamurthy, 1977).

of differentiation of tissues, organs and the whole plant organism (Barlow, 1994a, b). However most authors do not reject outright the significance of unequal divisions and succession of cell divisions in morphogenesis. In fact, it is emphasized that these processes can be more important for plant than for animal organs (Lyndon, 1990).

COTYLEDON

The cotyledon (Greek *kotuledon*—cup-shape vessel) is the first especial leaf initiated in the embryo. The structural particularities of the cotyledons are conditioned by the place (in the seed), time of formation (the first ontogenetic stages), and by the functions fulfilled (reserve, haustorial, protective, photosynthetic).

The term was introduced by Linnaeus (cit. from Jackson, 1916) for the designation of the first embryonic leaves ('seed lobes'). According to the cotyledon number in the embryo, De Jussieu (1789) suggested that plants be divided into two classes—Dicotyledons and Monocotyledons (see below: Variations in Cotyledon Apparatus Development). Shape, functions and anatomical structure of the cotyledons vary greatly. The cotyledons, both in dicotyledonous and monocotyledonous plants, arise predominantly from derivatives of the apical cell of the embryo. They are appendicular organs, modified in connection with the function fulfilled. To a great extent these modifications are related to embryo structure and the seed as a whole (presence/absence of endosperm, perisperm) and also mode of its germination.

The embryonic apical part (originating from the apical cell and its derivatives) can be likened to the shoot apical meristem. The functioning of this meristem produces one or two cotyledons (seldom many, sometimes none). Thus, the apical part of the embryo can be considered similar to the shoot apex, at which cotyledons or leaf primordia arise in various flowering plants (Batygina and Andronova, 1988, 1991; Padmanabhan, 1996). Probably, it is not worth looking for cotyledons in post-seminal development—in the seedling, if they are absent in the mature seed, because possibly their reduction occurred in the course of evolution or they did not form at all (e.g. Orchidaceae).

Molecular–genetic data about embryo development in *Arabidopsis* could be evidence of the leaf nature of cotyledons. In mutant seedlings, cotyledons lose their peculiarities and acquire the traits characteristic for foliage leaves (Meinke *et al.*, 1994; see Epicotyl).

The modes of evolutionary transformation in cotyledons remain under discussion (see below: Origin of Monocotyledonous Embryo: A New Concept).

VARIATIONS IN COTYLEDON APPARATUS DEVELOPMENT

Dicotyly

Dicotyly (Greek—*di, dis*—two, double, *kotuledon*—cup-shape vessel) is the phenomenon of development of two cotyledons in a spermatophyte embryo.

A typical dicotyledonous embryo is characterized by the establishment of bilateral symmetry during transition to organogenesis (heart-shape stage of development); formation of two cotyledon primordia of more or less equal dimensions, laying down due to the subepidermal initials, which divide according to periclinal–anticlinal type; terminal position of shoot apex, presented in mature embryo only by apical meristem; development of root apical meristem of the climacorhizal type; approximately equal ratio of length of cotyledons and hypocotyl–root axis or some exceeding the latter; epigeal type of germination, in the course of which cotyledons fulfil photosynthesizing and often a reserve function as well and are frequently divided into a plate and petiole (Haccius, 1952; Eames, 1961, Titova, orig. data).

The occurrence of two cotyledons in an angiosperm embryo is regarded as one of the features typical of the class of dicot plants (Dicotyledones = Magnoliopsida). It is correlated with a number of other structural characters of the vegetative and generative sphere: the occurrence of woody as well as herbaceous life forms, usually petioled leaves with pinnate or palmate, not closed venation; two prophylls on lateral shoots, root system of rachis-like type, shoot conductive system in the form of a siphonostele or eustele, opened vascular bundles with cambium, 5- or 4-merous flowers, simultaneous type of microspore formation, cellular or nuclear type of endosperm development etc. (see Kozo-Polyansky, 1965; Takhtajan, 1970).

Monocotyly

Monocotyly (Greek *monos*—single and *kotuledon*) is the phenomenon of development of a single cotyledon in a spermatophyte embryo.

A typical monocotyledonous embryo is characterized by the establishment of specific bilateral (dorsoventral) symmetry during transition to organogenesis; formation of a single cotyledonary primordium, laying down by both subepidermal and epidermal initials, and enveloping initially about 2/3–3/4 of the apex and later almost the whole with development of a wide sheath; shoot apex appears lateral in position in mature embryo; high capacity of embryonal plumule (up to 30-40 leaf primordia in *Cryptocoryne*, Araceae); liorhizal type of development of root apical meristem; cotyledon length considerably exceeding that of hypocotyl-root axis due to intensive intercalary basal growth of the former; hypogeal type of germination, during which the cotyledon fulfils haustorial and reserve functions and is often divided on a sheath, connective and haustorium proper.

The phenomenon of monocotyly, as well as dicotyly, is associated in evolution with the arising of a whole complex of features, according to which the class of monocot plants—Monocotyledones (= Liliopsida) has been distinguished among angiosperms. Unlike dicots, monocotyledonous plants are characterized by the presence of herbaceous life form alone (occasionally secondary woody), leaves ordinarily not subdivided on petiole and plate, parallel or arched venation of the closed type, a single prophyll on lateral shoots, root system of the racemose type, shoot conductive system in the form of an atactostele, closed vascular bundles without cambium, 3-merous flowers, successive type of production of microspore tetrads and helobial or nuclear type of endosperm development.

The question of the evolutionary interrelations between di- and monocotyly is still under discussion (see below: Origin of Monocotyledonous Embryo: A New Concept; Syncotyly; Heterocotyly).

Pseudodicotyly

Pseudodicotyly (Greek—*pseudo*—false and *kotuledon*) is the phenomenon of formation of two cotyledons in a monocotyledonous plant embryo **(Plate XXXIII)**.

The possibility of two cotyledons forming in monocotyledonous plants remains a question for discussion. Some authors consider the epiblast of the Poaceae embryo (Velenovsky, 1907; Roth, 1955; Eames, 1961; Batygina, 1974b) and peculiar scale-like organ of the embryo in the Dioscoreaceae (*Tamus, Dioscorea*) and Commelinaceae (*Commelina, Tinantia*) (Beccari, 1870; Solms-Laubach, 1878b) to be the rudiment of a second cotyledon. The embryo structure in Dioscoreaceae and some Araceae (*Arum,*

Arisaema) was interpreted as a heterocotyly expression also (Hill, 1906, 1908; Boyd, 1932; Lawton and Lawton, 1967; see below: Heterocotyly).

Reports by earlier authors about the spontaneous formation of typical dicotyledonous embryos in some Alliaceae (*Agapanthus umbellatus*—Coulter and Land, 1914), Amaryllidaceae (*Cyrtanthus sanguineus*—Farrell, 1914) and Araceae (*Colocasia antiquorum*—Suessenguth, 1921) are of special interest.

According to Coulter and Land (1914) the dicotyledonous embryos of *A. umbellatus* were able to germinate as well as monocotyledonous ones, though their structure differed. The monocotyledonous seedling consists of a cylindrical cotyledon, the upper part of which is concealed in the seed and forms a haustorium, while the lower part of this cotyledon forms an annular sheath embracing the shoot apex with the first leaf situated opposite to the cotyledon, i.e., the cotyledonary plane. The dicotyledonous seedling has an annular sheath formed by both cotyledons (in the view of cotyledonary tube) differing in thickness but not in length. It is significant that the first two leaves in dicotyledonous seedlings are situated criss-cross with reference to the cotyledons, i.e., they lie in the intercotyledonary plane. The conductive system in the seedlings of both types is similar and represented in the region of a short hypocotyl by a siphonostele, and in the region of the root has a tetrarchical organization. However, the number of bundles in the cotyledons differs: 3 in the monocotyledonous embryo and 2 in the dicotyledonous. This led authors to the conclusion that differences in the conductive system structure of cotyledons are determined by the number of their primordia initiated in the early stages of embryogenesis.

The early stages of development of monocotyledonous and dicotyledonous embryos are similar: during the transition to laying down of cotyledons in the flattened apical part of embryo the peripheral meristematic cotyledonary zone is formed, and in the upper part of this zone can be distinguished two more active points, the primordia. Later, on the lower part of the cotyledonary zone a ring (the tube) forms, and on the upper part, the cotyledons themselves, whose subsequent development can occur in two ways. Development of both primordia with equal value at the top of the ring results in formation of the dicotyledonous embryo. However, more often the cells of one of the initiated primordia gradually lose meristematic activity (cellular divisions cease) and only one cotyledon is developed. According to Coulter and Land, elimination of one of the cotyledons does not take place, but the whole peripheral cotyledonary zone is drawn into the growth of one cotyledon. Thus, the monocotyledonous embryo during development always passes the stage of dicotyledony; however, later, because of unilateral growth in one of the primordia and retardation of development of the other, only one large cotyledon is formed while the second is so small as to be hardly distinguishable.

A similar mode of embryo development was found in *Cyrtanthus sanguineus* (Farrell, 1914). However, in this species four primordia at the peripheral cotyledonary zone appear followed by transition to a dicotyledonous state and later monocotyledonous state. This was realized by the subsequent primordia 'fusion in pairs'. Like the events in *Agapanthus*, development of one of the primordia was gradually suppressed at this time.

Based on data for *Agapanthus* and *Cyrtanthus*, Coulter and Land (1914) and Coulter (1915) introduced an original hypothesis of monocotyly origin. According to this hypothesis, the monocotyledony in monocots appears not as the result of fusion of two cotyledons in the dicotyledonous embryo, or the suppression of one.

Monocotyledony is the result of undisturbed growth of the whole peripheral cotyledonary zone under the guidance of a single growing point. The authors state that monocotyly is 'simply the continuation of one growing point on the cotyledonary ring, rather than a division of growth between two growing points' (i.e., dicotyly). Simultaneously they draw a parallel with corolla formation in sympetalous plants, i.e., mechanisms of congenial coalescence of organs.

The foregoing data, considered dubious by many authors, were recently confirmed on the example of *Agapanthus praecox* Willd (= *A. umbellatus* L. Herit.) and *Cyrtanthus parviflorus*, enabling some corrections in the Coulter and Land hypothesis (Titova, 2002, 2003).

Pseudomonocotyly

Pseudomonocotyly (Greek *pseudo*—false and *kotuledon*) is the phenomenon of single cotyledon formation in the embryo of dicotyledonous plants **(Plate XXXIV)**.

The term was introduced by A. Winkler (1874), who proceeded from the notions of some earlier authors (Hofmeister, 1861; Hanstein, 1870) about the existence of principal distinctions in embryo development between monocots and dicots. They considered cotyledons of dicots always to be lateral organs, and the shoot apex the terminal one; moreover, both occur as derivatives of the apical cell (*ca*) of the two-celled proembryo; in the embryo of monocots the cotyledon occupies a terminal position as the derivative of the apical cell (*ca*) and the shoot apex a lateral one because formed by derivatives of the basal cell (*cb*). According to Winkler, compatibility of embryos in pseudomonocotyledonous plants with those in true monocots is exhibited only in the presence of a single cotyledon whereas the origin and further development of the latter (likewise the shoot apex) appear to be typical for embryos of dicots.

On the other hand, Hegelmaier (1878) demonstrated these notions to be erroneous: both cotyledons and shoot apex in the embryo of monocotyledonous, dicotyledonous and pseudomonocotyledonous plants develop from the derivatives of the *ca* cell, being by origin lateral and terminal organs respectively. Given this, he concluded that a single cotyledon in the embryo of pseudomonocotyledonous species forms as a result of suppression of the development of one of two cotyledonary primordia, which often remains as a rudiment. Hence a false impression of the occurrence in the embryo of a single cotyledon arises. Hegelmaier thus added another meaning to the term 'pseudomonocotyly'.

In the subsequent works the existence of a second cotyledonary rudiment in the embryo of many pseudomonocotyledonous plants was placed in doubt, as well as arguments in favour of origination of its single cotyledon as the result of congenial coalescence of two initial ones (Hildebrand, 1899; Sterckx, 1900; Sargant, 1903). Proceeding from this, some authors (H. Winkler, 1931; Arzt, 1933) began to interpret the term 'pseudomonocotyly' loosely, regarding the absence of the second cotyledon in pseudomonocotyledonous plants to be the result of both extreme reduction of the one of the two cotyledons and their complete fusion during embryogenesis. So-called heterocotyly, in the course of which differences in the structure and function of cotyledons, mainly during germination (e.g. in species of Peperomiaceae, Gesneriaceae families—Hill, 1906, 1938) arise, was also attributed to one of the pseudomonocotyly forms. Later, Haccius (1952) justly proposed to differentiate all these phenomena and to withdraw the term 'pseudomonocotyly' from circulation, owing to its polysemy. For cases of the second cotyledon rudiment remaining in the embryo, the author introduced the term 'anizocotyly' (Greek *anizo*—unequal and *kotuledon*); for cases, in which the embryo possesses a single cotyledon with no traces of a second one and irrespective of the mode of its arising, the term 'monocotyledonous dicotyledons'.

Representatives of 17 genera from different families of dicots, largely Ranunculaceae and Apiaceae, belong to the category 'monocotyledonous dicotyledons' (Haccius, 1952, 1953a,b, 1954; Haccius and Hartl-Baude, 1957; Haccius and Fischer, 1959; Haccius and Trompeter, 1960). The number of species, falling in the category of anizocotyledonous plants, is limited and chiefly includes members of families (e.g. Nictaginaceae, Cactaceae, Brassicaceae) of the so-called Curvembryonatae, characterized by the occurrence of a curved embryo. The authors confirmed the principle of compatibility in embryo development in monocotyledonous dicotyledons and true monocots, manifesting not only in the common origin of cotyledon and shoot apex (from derivatives of the *ca* cell), but also in the specifics of cell division during laying down and development of cotyledons, structure of conductive system, ratio of sizes of embryo organs in mature seed, their morphogenetic activity in the course of germination etc. Furthermore, the authors conducted a comparative analysis of the peculiarities of the cotyledon laying down and seedling nodal anatomy in 'monocotyledonous dicotyledons' and species with natural partial congenial fusion of cotyledons in view of the cotyledonary tube, as well as forms with teratological deviations in development of the embryo cotyledonary apparatus (poly-, anizo-, syn-, acotyly), induced by treatment with growth substances (Haccius, 1960; Haccius and Trompeter, 1960). From this analysis the authors drew the conclusion of the homology of a single cotyledon in the embryos of those plants to one of the two cotyledons in a dicotyledonous embryo and its origination as the result of gradual suppression of one of them, but not fusion. Nevertheless, the question of the origin of a single cotyledon in the embryo of pseudomonocotyledonous plants continues to be debated. Results of recent investigations on embryo development in anizocotylar *Trapa natans* as well as revision of data on embryogenesis of pseudomonocotyledonous plants enabled the conclusion that anizocotyly occurs as a consequence of the extreme degree in realizing partial unimarginal cotyledon congenial fusion, and thus to reconcile disagreements on the question of the pseudomonocotyly nature (Titova, 2000; also see Syncotyly below).

The capacity of many dicots for formation of an embryo with a single cotyledon is the reason for supporting the origination of a monocotyledonous embryo from a dicotyledonous one. However, different solutions to the question of the morphological nature of a single cotyledon in pseudomonocotyledonous plants generated several hypotheses on the origination of the cotyledon in monocots: a result of under-development of one of the cotyledons in dicots (hypothesis of underdevelopment, Hegelmaier, 1878), their fusion (hypothesis of syncotyly, Sargant, 1903) and gradual divergence of their functions in evolution (hypothesis of heterocotyly, Hill, 1906) (see below: Origin of Monocotyledonous Embryo: A New Concept; Syncotyly; Heterocotyly).

Syncotyly

Syncotyly (Greek *sun*—with, together and *kotuledon*) is the phenomenon of cotyledon 'coalescence' during embryogenesis **(Plate XXXV)**.

The term was introduced by De Vries (1895, 1903) to designate different forms and degrees of teratological coalescence of seedling cotyledons by their margins in dicotyledonous plants. The term was then used for an analogous phenomenon in a whole series of species as a norm. In later works the term 'cotyledon fusion' came into use because in the course of cotyledon initiation not a literal coalescence, but a congenial union of them takes place. However, in some forms of syncotyly, postgenial coalescence of cotyledons after their initiation is also possible (see below).

Natural and teratological syncotyly are distinguished (Weiβe, 1930). **Natural syncotyly** is most characteristic of representatives of ancient archaic taxa of dicotyledonous plants (Ranunculaceae, Berberidaceae, Papaveraceae etc.), in which it is inherent in a notable number of species. It is observed also among advanced families, such as Apiaceae, Cucurbitaceae, Tropaeolaceae. Its appearance is associated with embryo adaptation to germination in conditions of a geophytic mode of life (Sargant, 1903; Haccius, 1952, 1953b; Barykina, 1971, 1995; Tamura *et al.*, 1977; Haines and Lye, 1979). This is expressed in the development of a complex of features, which Haccius (1952) termed the 'syndrome of geophyly'. It is the formation of the shortened tubercle-like hypocotyl and so-called cotyledonary tube that arise as a result of congenial coalescence of two cotyledons at their base. The cotyledonary tube is a polyfunctional organ: its lower part, enveloping the growing point, becomes root-like, forms a rhizodermis with sucking hairs and acquires positive geotropism. Simultaneously the cells of its parenchyma serve as a reservoir of nutritive substances. Due to these transformations gradual deepening of the shoot apex into the soil occurs in the course of germination on the one hand, and creation of favourable conditions for defence of the shoot apex penetrating the soil thickness and its surviving winter cold on the other. However, formation of the long cotyledonary tube may cause difficulties during germination. For example, in *Prangos ferulacea* (=*Cachrys ferulacea*), *Chaerophyllum bulbosum* (Apiaceae) and *Polygonum bistorta* (Polygonaceae) outward emergence of most plumule leaves is realized by mechanical disruption of tube tissues on one side or only after death of the tube. In many tubular forms (*Eranthis hiemalis*, Ranunculaceae; *Smyrnium rotundifolium*, Apiaceae) the tube canal is completely reduced, and in a number of cases a special adaptation for germination has been developed—the aperture at its base through which plumule leaves emerge outwards (Bernardi, 1832; A. Winkler, 1874; Haines and Lye, 1979).

It should be noted that natural syncotyly was found not only in geophytes, but also in hydrophytes (Nymphaeaceae, Nelumbonaceae, Cabombaceae—Lyon, 1901a,b, 1905; Cook, 1902, 1906) and phanerophytes (Magnoliaceae, Fagaceae—Earle, 1938; Belostokov, 1966; Mogensen, 1970). It is characteristic also of some parasitic plants (species of Loranthaceae—Dixit, 1958; Johri *et al.*, 1957) and mangroves (*Rhizophora mangle, Ceriops candolleana*, Rhizophoraceae— Juncosa, 1982; Nikiticheva and Yakovlev, 1985).

Teratological syncotyly can arise spontaneously or be induced experimentally, by physico-chemical treatment, in representatives of various families and possesses a hereditary character (De Vries, 1903; Compton, 1913; Hanf, 1950; McIlrat *et al.*, 1950; Gavaudan *et al.*, 1953; Haccius, 1955c, 1960, 1969, 1972b; Schuster, 1956; Haccius and Trompeter, 1960; Liu *et al.*, 1993a,b). Syncotylous races are known for a whole series of species, within which a continuous row of transitional forms from typically dicotyledonous seedlings up to seedlings with completely fused cotyledons are observed.

Depending on the mode of cotyledon fusion the following **forms of syncotyly** are distinguished:

–**syncotyly in the narrow sense** (after De Vries, 1903) or **unimarginal syncotyly** (after Haccius and Trompeter, 1960): fusion of cotyledons only along one of their margins;

–**amphisyncotyly** or **amphicotyly** (after De Vries, 1903—from Greek *amphi*—around, from both sides and *syncotylia*) or **gamocotyly** (after Gavaudan *et al.*, 1953; Haccius, 1960; Danilova and Kirpichnikov, 1985—from Greek *gamos*—marriage and *kotuledon*): fusion along both margins with formation of a tubular, goblet-like structure, the so-called ascidial form of cotyledons;

–**fusion of cotyledons by their adaxial surfaces.**

Embryos of *Bunium bulbocastanum, Scaligeria cretica* (Apiaceae) are examples of unimarginal fusion of cotyledons, and *Podophyllum peltatum* (Berberidaceae), *Smyrnium olusatrum, S. perfoliatum, Chaerophyllum bulbosum* (Apiaceae) —of bimarginal fusion (Haccius, 1952, 1953b); fusion of cotyledons by their inner surfaces is described for *E. hiemalis* (Haccius, 1953b), some Loranthaceae (Johri *et al.*, 1957; Maheshwari *et al.*, 1957; Dixit, 1958) and Rhizophoraceae (Juncosa, 1982).

Fusion can be **complete** or **partial**—only at the base of cotyledons (*Anemone alpina*), or along their petioles or blades (some species of *Combretum*, Combretaceae; *S. olusatrum, S. perfoliatum*) (**hemisyncotyly**, after De Vries, 1903—from Greek *hemi*—half and *syncotylia*). Complete unimarginal cotyledonary fusion results in monocotyledonous embryo formation (*Bunium bulbocastanum*, Apiaceae; *Quercus arisonica*, Fagaceae).

In gamocotyly and **unimarginal syncotyly,** cotyledon fusion has a **congenial** character that is expressed in change of mode of their initiation. According to early authors, cotyledon initiation begins in these cases from the formation of a crescentic ridge or almost circular ring interrupted at one side by a notch or indentation. In the upper part of this ridge the primordia of two cotyledons are later formed, typical for embryos of dicotyledonous plants, but possessing different depths of sinuses (i.e., hollows, areas between their margins) that often equalize in the process of further development. Such a mode of cotyledon initiation was first described by Lyon (1901a, b; 1905) for *Nelumbo lutea* (Nelumbonaceae) and was later found in representatives of Nymphaeaceae, Ranunculaceae, Berberidaceae, Papaveraceae, Magnoliaceae, and Portulacaceae (Schmid, 1902; Cook, 1902, 1903, 1906; Lewis, 1904; Schaffner, 1904; Mottier, 1905; Earle, 1938). However, interpretation of this phenomenon differed: initiation of primordium of a single ancestral cotyledon, which bifurcates into two lobes consequent to schizocotyly (Lyon, 1901a, b; 1905), as a result of congenial fusion of cotyledons (Sargant, 1903; Suessenguth, 1921) or as an abnormal course of embryogenesis (Lewis, 1904; Mottier, 1905). In a number of later works data concerning such peculiar cotyledon initiation in the plant species mentioned above were considered dubious (see Titova and Batygina, 1996).

Haccius (1952, 1953a), having studied embryo development with cotyledonary tube, considered crescentic ridge formation during tube initiation, as well as various depth of sinuses between non-fused cotyledon parts, to be a chance impression or teratological deviation. At the same time she believed that the cotyledonary tube forms from the special basal zone, ringing the plumule region, and is not equivalent to two cotyledons of the dicotyledonous embryo.

A detailed study of *Nelumbo nucifera* embryo with light and scanning electron microscopy affirmed the fact of partial fusion of cotyledons at their base (Titova,

1988, 1990, 1992, 1998; Titova and Batygina, 1996). The process of cotyledon initiation in this species begins with flattening of the embryo apical part in apical-basal and lateral directions, characteristic of embryos of typical dicots. The shape of the apical part of the embryo changes from round to ellipsoidal and in its subepidermal layer laying down of the initials of the primordia of two cotyledons begins (transition to bilateral symmetry, three-angular stage). However, in the course of further development, the apical embryo part again becomes round, due to the change of directions of the cell divisions in the upper part of the hypocotyl, which happens now not only in the cotyledonary plane, but also in the intercotyledonary plane, along the line of future cotyledon fusion (symphysis side). Change in symmetry from bilateral again to radial, and fusion of the original cotyledonary initials into one common continuous zone in the form of a ring occur as a result. This is manifested by activation of the periclinal cell divisions in the subepidermal layer along the whole periphery of the embryo apical part. As a result of prevalence of divisions from one side of the common ring-like cotyledonary zone over the other, division of this zone by furrow at the opposite side and formation of the structure, similar to the primordium of a single cotyledon of the monocotyledonous embryo, occur; the embryo acquires a dorsoventral structure (stage of 'crescentic ridge'). Division of the originally laid down initials of the cotyledon primordia in an apical-basal direction leads to formation of their non-fused parts in the apical portion of the common cotyledonary zone; intercalary divisions of their lower derivatives results in formation of a common cotyledonary base, surrounding the shoot apex in view of the short cotyledonary tube. In 4% cases continuous intercalary growth in the common cotyledonary base was noted. The consequence of this was formation of typically monocotyledonous embryos. Thus, the ratio of cellular divisions and growth in different parts of the common cotyledonary zone determines the type of cotyledon fusion, at which the cotyledonary base can have the shape of a closed tube, cup, goblet (bimarginal fusion) or tube opened from one side by a fissure (unimarginal fusion), and most often, as in *Nelumbo*, a combination of these types. The ratio of apical and basal growth in the course of cotyledonary formation determines the degree of their fusion and, respectively, the type of embryo forming—dicotyledonous, with partially fused cotyledons or monocotyledonous (= dicotyledonous, with completely fused cotyledons).

Juguet (1990, 1992) reached a similar conclusion, according to which in dicotyledonous species with cotyledon fusion, both the crescentic ridge and cotyledonary tube appear at the basis of the initial ringlike cotyledonary zone, situated at the periphery of the embryo apical part.

Fusion of cotyledons by their inner adaxial sides is realized **post-genially** and, probably, in most cases occurs in the background of gamocotyly. The presence of the peculiar ventral, or adaxial meristem at the inner side of the cotyledon base is observed in species with a long cotyledonary tube (Haccius, 1952, 1953a). As a result of periclinal cellular divisions in this meristem in the *Podophyllum peltatum* embryo, the canal inside the tube narrows significantly. Due to the activity of this meristem in *E. hiemalis*, complete post-genial fusion of the inner, adaxial sides of cotyledons occurs. This fusion happens so early and quickly after their initiation (yet before cuticle forms at the surface) that almost no traces of the fusion in the form of a canal or seam remain, and a false impression of endogenous laying down of the plumule is created. Probably, in *S. rotundifolium* and species of Loranthaceae and Rhizophoraceae an analogous process occurs, but they have not been studied in detail.

The change in symmetry of the embryo apical part and the mode of cotyledon laying down in different forms of syncotyly influence the whole further course of embryogenesis. This change causes a shift of symmetry in the embryo and seedling as a whole, substantial changes in position and development of all their organs and also in the conductive system structure.

In seedlings of teratological syncotylous races obtained spontaneously (*Helianthus annuus*, Asteraceae—Compton, 1913; *H. annuus*, line APS-17—Titova, orig.; *Impatiens roylei*, Balsaminaceae—Holden and, Daniels, 1921), or as a result of various experimental treatments of the embryo (*E. hiemalis*—Haccius and Trompeter, 1960), at an **insignificant degree of unimarginal syncotyly,** the closer drawing together of cotyledons coalesced congenially was observed. In species with natural unimarginal syncotyly, besides cotyledonary primordia drawing together, delay in initiation of one of them and also lag in development of the forming cotyledon were noted (*N. nucifera*—Titova and Batygina, 1996; Titova, 1998; *Aconitum soongaricum*—Titova, orig.). This leads to general asymmetry of the common cotyledonary zone structure and some inequality in length of cotyledons but mainly in width. Noticeable differences in sizes of cotyledons are also apparent in embryos of some other species with partial natural syncotyly, for example, in *Actaea alba* (Mottier, 1905).

Increase in degree of unimarginal cotyledon fusion in teratological seedlings led not only to strengthening of predominance of one cotyledon over the other (asymmetry in structure of cotyledonary apparatus), but also to a corresponding change in structure of their conductive system. In teratological seedlings of *E. hiemalis* obtained as a result of 2,4-D influence on the globular embryo, gradual reduction of some cotyledonary bundles at the symphysis side was observed, namely reduction of the median bundle of the suppressed cotyledon and lateral bundles of both cotyledons. Displacement of the median bundle in the dominant cotyledon and the plane of xylem poles in the originally diarchical stele of the embryo root to the line of fusion was also observed and, correspondingly, a subsequent shift in general seedling symmetry (Haccius and Trompeter, 1960). An analogous tendency towards increase in degree of unimarginal cotyledon fusion was noted in seedlings of syncotylous races of *H. annuus* (Compton, 1913) and *I. roylei* (Holden and Daniels, 1921) as well, though in every case the peculiar, characteristic of taxon specificity was observed. **At complete unimarginal fusion** (with formation of a single cotyledon instead of two), the whole reduction of cotyledonary bundles and final shift of the plane of xylem poles in the embryo root occurred. As a result the structure of the conductive system in a single cotyledon repeated completely that of the initial non-fused cotyledons, and no traces of the transformation that occurred were present in its structure. Herewith the change from asymmetry to dorsoventral structure takes place. Among the seedlings of the teratological syncotylous race of *I. roylei* (Holden and Daniels, 1921) and line APS-17 of *H. annuus* (Titova, orig.), anizocotylous specimens with partial cotyledon fusion were found. The vascular components of such seedlings were modified principally similar to that in seedlings with completely unimarginal fused cotyledons. This allowed concluding that anizocotyly is a result of ultimate degree in expression of unimarginal syncotyly. Further, teratological syncotyly is often accompanied by schizocotyly (of one, less often of both fusing cotyledons) that brings about the condition of tri- or tetracotyly (De Vries, 1903; Holden and Daniels, 1921; Titova, orig.).

The character of the change in the conductive system in the case of **gamocotyly** is not the same. For example, the conductive system of the 'ideal' tubular structure in the *Brassica juncea* (Brassicaceae) embryo, obtained after influences on the globular embryo in culture *in vitro* with polar auxin transport inhibitors (TIBA a. o.), possessed distinct radial symmetry and was pierced by six vascular bundles of equal size, while normally every cotyledon is supplied with median and several lateral bundles (Liu *et al.*, 1993a,b). In seedlings of spontaneous teratological races, the structure of the conductive system is highly variable. This is related to the fact that in these cases the forming tube is almost never strictly radial in symmetry ('asymmetrical amphisyncotyls'; Holden and Daniels, 1921). Nevertheless, in this case common regularities are revealed, for example the tendency to formation of numerous anastomoses between conductive bundles of fusing cotyledons in the area of the common cotyledonary base (*I. roylei*; Holden and Daniels, 1921). Analogous anastomoses were noted in some species with natural cotyledon fusion, for example in *N. nucifera*.

Cotyledon fusion causes changes in the development of shoot and root apices in the embryo. At the extreme degree of teratological syncotyly (both unimarginal and gamocotyly) the majority of seedlings are dwarf-like (teratological syncotylar races of *H. annuus*; Compton, 1913; Titova, orig.). In such seedlings retardation in sloughing of the seed coat, pathological outgrowth of the hypocotyl, non-development of the plumule and embryo root, difficulties in leaf growth via the cotyledonary tube are often observed. Further shoot development is often realized at the expense of axillary meristems. In gamocotylar seedlings of *I. roylei* (Holden and Daniels, 1921), inhibition of plumule development is often accompanied by overgrowth of the cotyledonary tube canal lying higher, because of abnormal activation of the adaxial meristems at its base and emergence of leaves is realized only as a result of mechanical disruption of tissues (as observed in some naturally syncotylous species of *Podophyllum, Eranthis*). The embryos of *B. juncea* with induced tubular structure of cotyledons and of embryo-lethal mutant *emb30 of Arabidopsis thaliana*, with cotyledon fusion had a shorter and less developed hypocotyls compared with the norm. In the latter case, lack of a distinct apex in the main root and the ability for rooting was also observed even after creation of the respective conditions in culture *in vitro* (Baus *et al.*, 1986; Liu *et al.*, 1993b).

For many species with natural cotyledon fusion, contrarily, a high degree of plumule differentiation (as well as of the embryo as a whole) is characteristic. For example, in *Nelumbo* the capacity of the embryo bud is 4 leaves, in *Nymphaea* and *Rhizophora* 2, some Ranunculaceae and *Quercus* 1; herewith not uncommonly the first leaves are scale-like. Such increase in bud capacity, usually correlates with establishment of the hypogeal (or transitional to it) type of germination and, respectively, of the haustorial or preserving function of cotyledons in embryogenesis (the type of germination has been noted for some species of *Clematis, Anemone, Ranunculus, Podophyllum, Corydalis, Sanguinaria, Nelumbo, Quercus* a. o.—see Barykina, 1971, 1995; Ilyina, 1988a,b; Kolesova [Titova] and Batygina, 1988; Trifonova, 1988). In species with cotyledon fusion (Apiaceae—Denisova, 1961; Rhizophoraceae, Loranthaceae, *Nelumbo, Quercus*) the length of cotyledon significantly exceeds the length of the hypocotyl-root axis, unlike species in which cotyledon fusion is lacking and the ratio of the aforesaid organs is typical of embryos of the majority of dicots. Cases with a significant reduction in embryo root (mainly, in hydrophilous representatives of Nelumbonaceae and Nymphaeaceae), and also formation of a specific «coleorhiza» (Nelumbonaceae and Rhizophoraceae) have been recorded.

The change in shoot apex topography at cotyledon fusion is of special interest. In the embryos of *N. nucifera* (Titova and Batygina, 1996), *A. soongaricum* and also the teratological line APS-17 of *H. annuus* (Titova, orig. data) its gradual displacement from a central position to lateral, and from horizontal to the oblique relative to the morphological axis of embryo is observed (Fig. 42). In other words, the shoot apex acquires the so-called 'lateral' position, peculiar to the embryos of monocots. This displacement occurs as a result of unilateral spread of the embryo in the

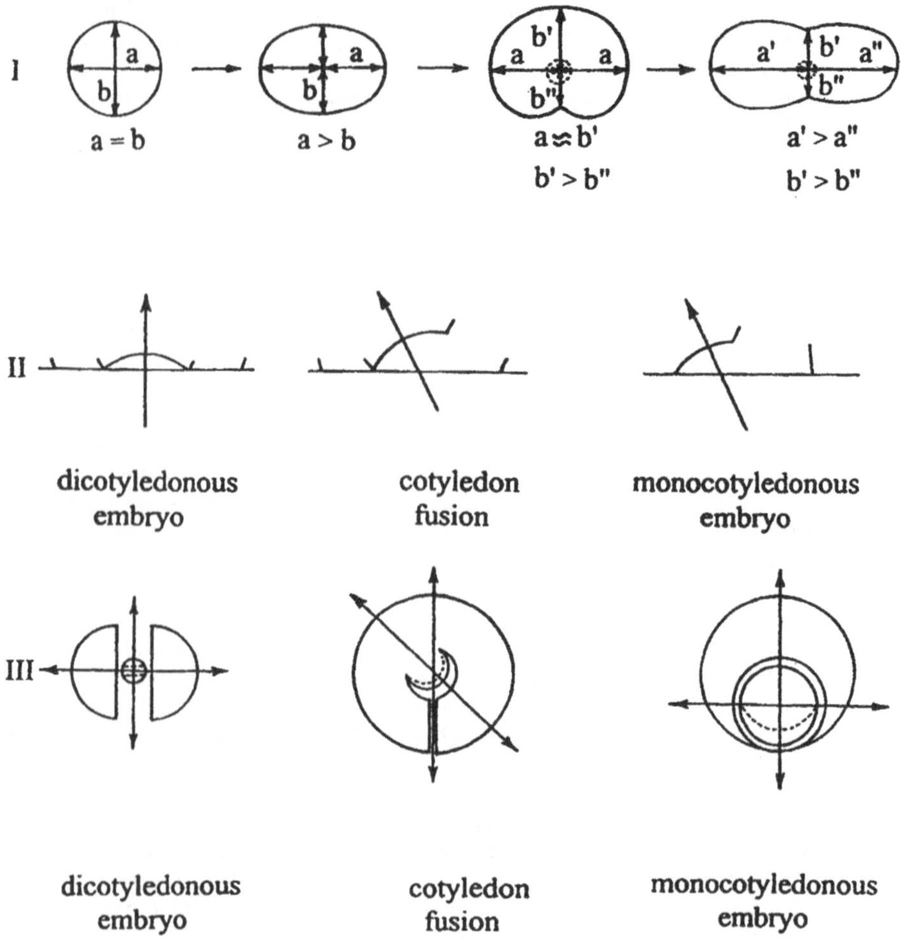

Fig. 42: Position of shoot apex and first leaf primordium in embryo with fused cotyledons compared with typical di- and monocotyledonous embryos: I—displacement of shoot apex from the central position into the lateral during cotyledon fusion; II—its displacement from horizontal position to an oblique one in respect of common cotyledonary base; III—laying down of first leaf primordium at shoot apex (at successive distichous type of phyllotaxis) opposite to symphysis side.

intercotyledonary plane in the process of cotyledon initiation (along their fusion line), and also of subsequent unilateral intercalary growth of the forming common cotyledonary base and lower adjacent hypocotyl area. It is most interesting that in typical dicotyledonous embryos of *A. soongaricum*, which form often enough in this species, such displacement of the shoot apex is lacking. Analogous processes can be traced in illustrations of embryo structure in some other species with natural (*Delphinium tricorne*—Mottier, 1905; species of Nymphaeaceae, Cabombaceae—Cook, 1906, 1909) and teratological (*H. annuus*—Krenke, 1950) fusion of cotyledons.

The first leaf position at the shoot apex in the embryo and seedling with teratological and natural unimarginal syncotyly is opposite to the result of fusion, i.e., is the single cotyledon (Sargant, 1903; Holden and Daniels, 1921; Haines and Lye, 1979). The mechanism of the change of first leaf position may differ, depending on original type of phyllotaxis in the plumule, but in all cases is probably associated with change in the divergence angle between primordia of the coalesced cotyledons. For example, in *I. roylei* with opposite leaf arrangement at the shoot apex in the seedling as a norm, at the complete unilateral teratological fusion of cotyledons, elimination of one of the two leaf rudements in the first node from the symphysis side was observed (Holden and Daniel, 1921). It is most characteristic that with increase in the degree of unilateral cotyledon fusion, drawing together and union of the axillary meristem was also observed here. In *N. nucifera* laying down of the first leaf primordium at the apex shoot of the embryo occurs not in the cotyledonary plane, as it has to be in species with alternate distichous type of phyllotaxis, but in the intercotyledonary plane, opposite to the common cotyledon base (i.e., line of cotyledon fusion), though it is shifted slightly to the side of median of the dominant cotyledon. This attests to the gradual changing of the first plumule leaf position in the direction of the symphysis side.

In turn, it is necessary to consider the fact of dorsoventrality of outer and inner embryo structure in some species with natural cotyledon fusion not only at the late stages of embryo development—from the moment of cotyledon initiation, but also at the very early, beginning with the first divisions, in the course of histogenous differentiation. For example, in *N. nucifera* this dorsoventrality is expressed in:

1) oblique position of the first cell wall in the proembryo relative to its longitudinal axis, resulting in the respective shifting of cell *ca* derivatives, and in the lateral position of cell *m* and its derivatives, which later give the initials of the main embryo root and lower hypocotyl part;
2) subsequent unilateral spread of the embryo in the area of derivatives of cells *m* and *ca*, the direction of which coincides with the line of unilateral cotyledon fusion.

The initial lateral position of the main root the oblique position of and, the morphological axis in the embryo are a consequence of these processes.

As a result of this dorsoventrality the regularities of cellular divisions in the proembryo of *N. nucifera* are traced most distinctly during the study of embryo structure in the intercotyledonary plane (coincident with the plane of unilateral cotyledon fusion) and are appropriate to the Asterad-type (Penaea-variation) of embryogenesis according to Johansen's classification (1950); the cellular divisions in the cotyledonary plane seem to be irregular.

Dorsoventrality of the embryo structure from the earliest stages, both as concerns the principal habitus and as oblique position of the first cell walls in the proembryo, was also found in some species of families Nymphaeaceae and Cabombaceae (see Titova and Batygina, 1996) and in *Delphinium* species (*D. tricolor, D. cariopetalum*—Babis, 1976), many of which are characterized by fusion of the cotyledons and passing of the 'crescentic ridge' stage at their initiation (Lubbock, 1892; Mottier, 1905). Using the early dorsoventrality of proembryo structure as a criterion, Babis distinguished two new types of embryogenesis—Valerianad and Ranunculad.

It should be mentioned that embryogenesis in some other species with natural syncotyly (*E. hiemalis, P. peltatum*) is also associated with irregular types (Haccius, 1953b). Possibly, as in *Nelumbo,* notions regarding irregularity of cell divisions believed to be wrong, i.e., associated with structural dorsoventrality of the embryo from the first stages, arose from study of its development usually through longitudinal sections only in the cotyledonary plane.

The nature of the early dorsoventrality in the embryo structure in the above-mentioned species is not yet clear, but was probably determined by gradual replacement (in evolution) of symmetry type with still earlier developmental stages.

Factors of induction and regulation of syncotyly. The first data concerning dependence of the syncotyly process on hormones were obtained during studies of plant treatment with herbicides (Hanf, 1950; McIlrath *et al.*, 1950; Gavaudan *et al.*, 1953; Schuster, 1956). For example, plants of *Gossypium, Helianthus* and others were treated with natrium salt 2,4-D, IAA, NAA at different stages of development (seedling, flower bud, flower). At germination of the seeds, which were gathered from these plants, various modes and degrees of cotyledon fusion, as well as of the first seedling leaves were observed.

In mature seeds of *E. hiemalis* containing embryos without differentiation into organs subjected to natrium and ammonium salts of 2,4-D, the developmental stage sensitive to the treatment and also the dependence of fusion mode and corresponding anatomical changes depending on concentration of the substances used were studied (Haccius and Trompeter, 1960; Haccius, 1972a). Specifically it was established that the influence of these herbicides was effective only before initiation of cotyledons (stage of globular embryo and beginning of transition to organogenesis) and had no affect after differentiation of their primordia.

More detailed studies of mechanisms of hormonal regulation of cotyledon fusion were undertaken on isolated embryos of *Brassica juncea* (Liu *et al.*, 1993a,b). While cultivating them in a medium to which inhibitors of polar auxin transport were added in graduated concentrations, blockage of the process of embryo transition from radial to bilateral symmetry and cotyledon fusion of different modes were observed. It was found that the fusion mode depended not on inhibitor type, but on its concentration. The optimal concentration at which 100% of the embryos formed a tubular cotyledonary structure was 2 μM for TIBA, 10 μM for HFCA and 80 μM for transcinnamic acid. Increasing the inhibitor concentration effected no further changes. At concentrations a bit lower than optimal (e.g., 5 μM HFCA) unilateral fusion of cotyledons was observed and the fused side usually was in contact with the medium. As in the experiments conducted by Haccius, in Liu's work too it was stated that the stage sensitive to treatment with the substances mentioned was the stage of globular embryo.

According to Liu and coauthors, embryos of *B. juncea* with fused cotyledons are phenocopies (in respect of cotyledons) of the embryos of *pin*-formed mutants *pin* 1-1 with reduced activity of auxin polar transport in the axis of inflorescence (Okada *et al.*, 1991), and also of the letal mutants *emb*30 (*gnom*) in *Arabidopsis thaliana* (Meinke,

1985; Mayer *et al.*, 1993). Liu and co-authors concluded that the absence of change in symmetry of the embryo at the transition from globular to heart-shape stage and formation of cotyledon tubular structure are associated with reduction of the auxin polar transport in the embryo, and unilateral fusion of cotyledons with its rest level. They suggested two possible models of auxin synthesis and transport during the process of cotyledon initiation and also supposed that the product of gene *emb*30 could play a direct or indirect role in auxin polar transport and, consequently, in establishing embryo symmetry during laying down of the cotyledons.

Cotyledon fusion in the embryo is also a characteristic feature of the double mutants *lec*1 *lec*2 *Arabidopsis* (Meinke *et al.*, 1994). It was shown that the locus 'fused cotyledons/lack of roots' (embryo-lethal mutant *emb*30) is closely associated with the locus 'curved trichomes' (mutant *dis*1) (Meinke *et al.*, 1989). In spite of disturbances in organ development in mutant embryos *emb*30 mentioned above, no changes in the ultrastructural characteristics of their cells, in particular the protein and lipid bodies—the main markers of normal embryo development at the cytological level, were observed (Patton and Meinke, 1990). This attests to the fact that the given mutant embryos possess the altered pattern of morphogenesis, but continue to differentiate at the cellular level, i.e., are characterized by preservation of the main 'housekeeping' function. The mutant *emb*30 warrants further study because it belongs to the group of rootless mutants, whose developmental abnormalities are often observed in somatic embryos in culture *in vitro* (Meinke, 1985).

Role of syncotyly in evolution of flowering plants. As is known, the theory of the origin of the monocot embryo as a result of fusion of the two cotyledons in a dicot embryo (**syncotyly hypothesis**) was posited by Sargant (1903, 1908). She based it on the occurrence of the first two simple or double bundles in a single cotyledon in seedlings of some monocots (Liliaceae) as well as the significant compatibility of the transition zone structure (specific Anemarrhena-type) with that of some primitive dicots with cotyledon fusion (*E. hiemalis*). In her opinion, this type of transition zone structure is primitive and original for flowering plants.

Sargant's hypothesis had many supporters among embryologists and anatomists (Holden and Daniels, 1921; Rothe, 1951; Danilova and Sokolovskaya, 1973; Dahlgren *et al.*, 1985; Juguet, 1990, 1992), who declared that cotyledon fusion explains well such peculiarities of embryo structure in monocots as the presence of a wide sheath base in the single cotyledon and its ascidiform structure, position of the first leaf of the plumule (opposite to the cotyledon), and also specific morphogenetic activity of the embryonic organs during germination (positive geotropic growth of the cotyledon base, which keeps the growing point of the shoot in the soil) as a result of adaptation to the geophilous mode of life.

Sargant's theory was also criticized, however, and the following arguments adduced against it:

1) Existence of several procambial bundles in cotyledons of monocots cannot be considered an argument because it is not known what number of them should be regarded as original; the facts of double vascular symmetry in the cotyledons of gymnosperms and some dicots without cotyledon fusion are known; it is impossible to explain all types of transition zone in angiosperms as a simple deviation of the Liliaceae model (Boyd, 1932; Eames, 1961).
2) Bundle fusion also occurs in cotyledons of monocots with double bundles (Kudryashov, 1964).

3) There are no any traces of fusion in laying down and development of the cotyledon of monocots and monocotyledonous dicotyledons (Metcalf, 1936; Eames, 1961; Haccius and Lakshmanan, 1967).
4) Syncotyly leads neither to change in leaf arrangement, nor to ontogenetic acceleration of the embryo development characteristic of embryos of monocots (Krasnikov, 1989).

Holden and Daniels (1921) partially admitted that criticism of Sargant's conclusions about seedling vascular system evolution as a whole was justified. But they noted that the similarity of vascular element behaviour in the transition zone (double vascular symmetry) in various plant groups (dicots with cotyledon fusion and without it, monocots and gymnosperms) could be due to various reasons and every concrete case have its own explanation. That is why, it appears to attest to the general direction in seedling vascular evolution. And although a double type of transition zone structure, from their point of view, actually does not have the exceptional phyletic significance attributed by Sargant, **the possibility of the origin of monocotyledonous state from a dicotyledonous one as a result of syncotyly (predominantly unimarginal) is convincingly affirmed by the data on teratological seedling anatomy.** The same data **readily explain the absence of traces of fusion in development of the single cotyledon in the majority of monocotyledonous dicotyledons** (*Ranunculus ficaria*—Metcalf, 1936; *Claytonia virginica*—Haccius, 1954) **and monocots**: these traces are present only in the intermediate stages of cotyledon fusion and in the final stage are completely lost. However, in some pseudomonocotyledonous plants (*Abronia umbellata, A. villosa*, Nictaginaceae—see Holden and Daniels, 1921) **the traces of structural changes are still preserved**. The behaviour of xylem and phloem elements in the transition zone of some cereal seedlings (*Triticum aestivum*) also accords with the cotyledon fusion mechanism (Danilova and Sokolovskaya, 1973). Arguments for monocotyly origin via syncotyly are also confirmed in studies of cotyledon initiation in a number of monocotyledonous plants (*Tamus communis*, Dioscoreaceae; *Endymion non-scriptus*, Hyacinthaceae—Juguet and Paré, 1989; Juguet, 1990, 1992). The data of some early authors about the origin of the single cotyledon in the embryo of *Agapanthus umbellatus* (Alliaceae) and *Cyrtanthus sanguineus* (Amaryllidaceae) as a result of congenial fusion of two initial cotyledons is of special interest here (Coulter and Land, 1914; Farrell, 1914; see Pseudodicotyly above). These data were confirmed in recent studies of embryogeny in these species (Titova, 2002, 2003). It is characteristic that peculiarities of laying down of cotyledons and the mechanism of transformation of two primordia into one at the peripheral cotyledonary zone in these species is rather similar to that in *Nelumbo* and other species with cotyledon fusion.

A comparative analysis of the data concerning embryo and seedling development in species with cotyledon fusion of the different modes mentioned above, demonstrated that **unimarginal syncotyly (usually in combination with partial amphisyncotyly) causes** changes in the structure of a dicotyledonous embryo, leading to the appearance of **features peculiar to the embryo of monocots:**

– symmetry type: from bilateral to dorsoventral (via asymmetry in transitive forms);

– mode of cotyledon initiation: single primordium instead of two, whose size is 2/3 the apex circumference or more;

– ratio of apical and basal growth during development of cotyledons: domination of the basal one and, accordingly, cotyledon length exceeds length of hypocotyl-root axis;

– shoot apex topography (its so-called laterality) due to unilateral intercalary growth of the common cotyledonary base and upper part of hypocotyl;

– leaf arrangement type in first node of plumule.

Further, analysis of development of the embryo and seedling in teratological syncotylous forms enables some understanding of the morphophysiological mechanism, laying in the basis of syncotyly, and supplementing Sargant's hypothesis with the morphophysiological mechanism. The arrestation ('undevelopment') and anomalies in development of the shoot apex, tendency towards activation of axillary meristems, disturbance of growth of the main embryo root and hypocotyl reduction in teratological seedlings and mutant embryos with extreme degree of syncotyly expression, are evidence of the **breakdown of the mechanism of apical domination in them.** This breakdown accords with data on the **auxin role (polar transport) in the induction of cotyledon fusion in the embryo** (Haccius and Trompeter, 1960; Liu *et al.*, 1993a,b) and explains **the specificity of the life-forms in the majority of monocotyledons**—transition to **sympodial shoot growth** at the early stages of ontogeny, unique **intercalary growth of its internodes, reduction of the main embryo root and replacement of it by adventive roots. Disturbance of the mechanism of seed coat sloughing and its long delay at cotyledon tips** (in lack of divergence of fused cotyledons to opposite sides) in teratological seedlings could result in a change in character of **morphogenetic correlations in the development of embryo and endosperm.** The set of these changes could be **the basis for establishment of the hypogeal type of germination, haustorial function of cotyledons and, as a result, increase in embryo bud capacity (so-called embryonic 'acceleration')**—the features peculiar to many species with natural syncotyly and the majority of monocots.

Suppression of one of the cotyledon rudiments (anizocotyly) at the extreme syncotyly degree, found in teratological seedlings is of indubitable importance. This corresponds with data of an analysis of embryo development in pseudomonocotyledonous species with anizocotyly (Titova, 2000). This enables **reconciliation of two basic hypotheses about the origin of the monocot embryo from that of dicots hypothesis of 'undevelopment' (Hegelmaier, 1878) and the hypothesis of 'syncotyly' (Sargant, 1903).** It further provides an explanation of the rare cases of preservation of the rudiment of the second cotyledon in some monocotyledonous plants, particularly the epiblast in cereals. As a whole, embryo development in cereals and certain species with cotyledonary fusion (*Nelumbo*) demonstrates the wonderful similarity in a number of features, for example unity of the 'mechanism' of dorsoventrality development in its structure from the very early stages (see Graminad-type embryogenesis) and even in the formation of such a specific organ as the coleorhiza.

In the light of the data presented above, Sargant's hypothesis of syncotyly appears quite realistic.

Schizocotyly

Schizocotyly (Greek *schizo*—to split, to cleave and *kotuledon*) is the phenomenon wherein the cotyledons 'divide' in the course of their development into two or more parts (see Polycotyly below).

Polycotyly

Polycotyly (Greek *polus*—much and *kotuledon*) is the phenomenon of formation of more than two cotyledons in the embryo. Synonyms: pleiocotyly (Greek *pleion*—more and *kotuledon*) **(Plate XXXVI)**.

The term, evidently, is borrowed from studies on the morphology of gymnosperms, in which the embryo has a great number with cotyledons as a rule. On analogy with gymnosperms, in angiosperms **true polycotyly** is accepted to distinguish from the so-called division of cotyledons or **schizocotyly** (Duchartre, 1848). In true polycotyly several more or less equal cotyledons, disposed symmetrically, are present in the embryo; they have own primordia **at the moment of laying down** and an independent vascular connection with the hypocotyl stele. In schizocotyly the primordia of one or both initial cotyledons 'divide' **in the course of development** more or less deeply into two, sometimes three and more cotyledons. At this time it is usually possible to trace the character of vascular connection, peculiar to the dicotyledonous embryo. However, many authors believe there is no major difference between poly- and schizocotyly and often apply the term 'polycotyly' to cover both phenomena (see De Vries, 1903; Guillaumin, 1911; Compton, 1913; Haskell, 1954b).

The term 'schizocotyly' was introduced by Worsdell (cit. by Jackson, 1916) for designation of mainly teratological division of seedling cotyledons into lobes differing in depth. Similar phenomena, observed in some plant species as a norm, were primarily referred to the category 'divided cotyledons' (Lubbock, 1892). However, the term 'schizocotyly' was later applied to them also.

The processes occurring in the case of schizocotyly should be distinguished from morphological processes, when originally the whole blade of the cotyledon, analogous to the blade of a complicated leaf, divides into lobes and even leaflets ('lobed cotyledons', Lubbock, 1892; 'complicated cotyledons', 'leaflety', Haccius, 1955a, 1972a,b), and also from the cases of formation in the apical part of cotyledons of hollows of varying depth ('emarginated cotyledons', Lubbock, 1892). As distinct from schizocotyly, these processes do not result in an increase in cotyledon number and appear to be regulated by cytokinins, but not auxins (Haccius, 1972a,b).

Poly- and schizocotyly are widely distributed among dicotyledonous plants (see reviews of Duchartre, 1848; Thiselton-Dyer, 1902; De Vries, 1903; Compton, 1913; Buchholz, 1919; Haskell, 1954b; Haccius, 1955a; Eames, 1961; Poddubnaya-Arnoldi, 1976). Polycotylous seedlings are especially characteristic of wood forms of Ranales (Eames, 1961). Among monocotyledons di- and tetracotylous embryos together with monocotyledonous ones have been found only in *Cyrtanthus sanguinensis* (Amaryllidaceae) (Farrell, 1914; see above: Pseudodicotyly). The formation of deeply bifurcated cotyledon in the embryo of *Alpinia nutans* (Zingiberaceae) mentioned by Tschirch (1891) and associated with peculiarities of seed structure—presence of postament, which intrudes into the seed cavity—is probably referred to the category of 'emarginated cotyledons'.

In most scientific works teratological phenomena (**teratological poly- and schizocotyly**) are discussed. Cases of true polycotyly normal to a species (**natural polycotyly**) are rather rare and have been recorded for primitive angiosperms mainly (e.g. *Degeneria vitiensis*, Degeneriaceae—Swamy, 1949b; *Idiospermum australiense*, Idiospermaceae—Blake, 1972). The presence of true polycotyly in a number of species remains debatable (*Acer pseudo-platanoides*, Aceraceae—Duchartre, 1848; Lubbock,

1892; Thiselton-Dyer, 1902; *Cola vera,* Sterculiaceae—Masters, 1868; Lubbock, 1892; Dalziel, 1937; some species of *Pittosporum,* Pittosporaceae—Lubbock, 1892; Thomas, 1914). Polycotyly has been observed in the embryos of many parasitic plants (some species of *Nuytsia, Loranthus,* Loranthaceae—Engler, 1889, 1926; Fletcher, 1909; Narayana, 1958; *Persoonia,* Proteaceae—Brown, 1810; Von Müller, 1882; Fletcher, 1909), and also mangrove plants (*Bruguiera eriopetala, B. caryophylloides,* Rhizophoraceae —Nikiticheva and Yakovlev, 1985). However the biological sense of polycotyly in these cases is obscure.

Natural schizocotyly has also been noted in primitive taxa of dicotyledonous plants, for example Papaveraceae (*Chelidonium majus*—Souèges, 1937c; *Eschscholzia californica*—Sachar and Mohan Ram, 1958), Ranunculaceae (Miller, 1927-1931), and also some more advanced families, for example Brassicaceae (*Schizopetalon walkeri*—Lubbock, 1892). Lubbock considered that in most cases it is a peculiar adaptation of the embryo for release from seeds with a narrow aperture, through which the thin narrow cotyledon lobes exit one after the other during germination. In some cases the presence of schizocotyly is still debated. For example, the cotyledons of the embryo in Juglandaceae are deeply divided into four lobes (Lubbock, 1892; Nast, 1941; Stone, 1970). The cotyledon lobes formed are usually similar in size but in some species (*Platycarya strobilacea*) are markedly unequal, and in a number of cases (*Pterocarya caucasica, Juglans regia, J.olanchana*) division into lobes corresponds to the character of the seed and fruit division into locules. Considering these peculiarities, the cotyledons of Juglandaceae can also be referred to the category 'emarginated cotyledons' defined by Lubbock.

Teratological polycotyly and **schizocotyly** can arise spontaneously or under physicochemical influences in representatives of most taxa of dicotyledonous plants (Duchartre, 1848; A.Winkler, 1894; Diels, 1910; Guillaumin, 1911; Tronchet, 1925a, b; Bugnon and Parrot, 1932; Datta, 1941, 1954; Barnes, 1943; Holtorp, 1944; Went, 1944; Chowdhury and Mitra, 1953; Jurkat, 1953; Reed, 1954-1955; Johri and Konar, 1955; Kapil and Bhandari, 1964; Sethumadhavan and Gopakumar, 1973). The reasons for their arising may differ. According to De Vries (1903), formation of polycotyledonous seedlings is more usual for crop plants and also for wild species that have been cultivated for a long time. An increase in per cent of polycotylous seedlings was observed as a result of crossing (*Beta vulgaris*—Munerati and Costa, 1934), and obtaining of polyploid forms (Noguti *et al.*, 1940; Janaki-Ammal—cit. by Haskell, 1954b). The appearance of polycotyly is associated with unfavourable external conditions during seed formation, i.e. lack of nutrition (Litovchenko, 1940; Straub, 1948), low temperatures (Straub, 1948; Harrison, 1964) and so on. It should be mentioned that teratological schizo- and polycotylous seedlings are not always viable. For example, in *Nicotiana tabacum* (Solanaceae) inhibition of growth of such seedlings, loss of green coloration in cotyledons and lack of plumule development were observed (Deb, 1955).

The degree of schizocotyly can vary—from a small hollow at the top of the cotyledon to deep, almost wholly separated lobes, termed the polycotylous condition. For designation of the intermediate state, between di- and polycotyly, the notions of **hemitri-** and **hemitetracotyly** (partial splitting of one or two cotyledons of the dicotyledonous embryo respectively) were introduced (De Vries, 1903). For a whole series of species, for example *Oenothera hirtella* (De Vries, 1903), *Cannabis sativa, Phacelia tanacetifolia* (Compton, 1913), *Delphinium ajacis* (Haccius, 1955b), via selection of poly- and schizocotylous races, a continuous series of transient forms (from dicotyledonous embryos to polycotyledonous via schizocotyledonous) were obtained. In such races the frequency of poly- and schizocotyly is often more than 50%.

Cases of tricotylous embryo formation (Jünger, 1869-1870; Krause, 1880; Lubbock, 1892; De Vries, 1902; Litovchenko, 1940; Holtorp, 1944; Rao and Reddy, 1979) most often pertain to teratological cases (species of *Phaseolus*—Gain, 1900; *Amomospermum*—Diels, 1910; *Morus multicaulis*—Datta, 1941; *Cochorus capsularis*—Datta, 1954; *Cinnamomum camphora*—Chowdhury and Mitra, 1953; *Daphne longilobata*—Tan, 1980). According to De Vries (1902, 1903), among tricotylous seedlings two races can be distinguished: 'half races', containing true tricotylous seedlings, the per cent of which in the progeny does not increase by the selection process, and 'intermediate races', in which the per cent of tricotylous seedlings during selection or under the influence of external conditions changes. The tricotylous state is often reached by division of one of the two original cotyledons during development. However, Holtorp (1944) established three modes of formation of tricotylous seedlings in *Brassica*: a) 'simple fission', when one of the cotyledons splits, is usually abnormal in shape and grows more slowly than the other, unsplit one; b) 'twinning' of cotyledons, probably connected with cleavage polyembryony and twin-embryo formation; and c) appearance of additional cotyledons ('supplementation'), which can appear instead of the first leaf of the epicotyl or opposite to it, and also instead of the second or some subsequent epicotylar leaves. The last mode was also found in *Indigofera linifolia* (Rao and Reddy, 1979).

In species with natural polycotyly, formation either of tricotylous (*Nuytsia floribunda*—Narayana, 1958), or tri- and tetracotylous embryos simultaneously (*Idiospermum australensis*—Blake, 1972) have also been recorded. In *Degeneria vitiensis* 87% of the polycotylous seedlings possessed three cotyledons and only 13% four (Swamy, 1949b).

In cases of natural schizocotyly synchronous, symmetrical division of both cotyledons occurs more often (*Eschscholzia californica*—Lubbock, 1892; *Schizopetalon walkeri*—Sachar and Mohan Ram, 1958; species of Juglandaceae—Lubbock, 1892; Nast, 1941; Stone, 1970).

Order of division of cotyledons in natural schizocotyly also varies and influences the number of cotyledons formed. For example, in *E. californica* **division** of cotyledons **only into two** is characteristic, which results in the formation of tetracotylous seedlings. Sometimes in this species the formation of pentacotylous seedlings is also observed; probably, this occurs due to the division of one of the original primordia **into three**. Such mode of division of both one and two cotyledonary primordia and formation of five and six cotyledons was noted in teratological seedlings of a relative species *Papaver rhoeas* (De Vries, 1903). Deeply twice bifurcated cotyledons in Juglandaceae embryos are, as mentioned above, the result of two subsequent divisions of two original cotyledon primordia into two.

Peculiarities of embryo development and initiation of cotyledons in poly- and schizocotyly have barely been studied. Embryogenesis of some species studied in this respect (*D. vitiensis, E. californica, J. regia, J. manshurica, Carya glabra*) is characterized by irregularity of cell divisions (Langdon, 1934; Nast, 1941; Souèges, 1949; Swamy, 1949b; Sachar and Mohan Ram, 1958). According to Langdon (1934) and Nast (1941) cotyledon bifurcation in the embryo of Juglandaceae (*J. regia, J. manshurica, C. glabra*) occurs as a result of cessation of activity of the apical meristem in the primary cotyledonary primordium and formation apically of a groove that divides the cotyledon in two. However, in *J. manshurica* the apical meristem remains for some time at the top of newly formed primordia.

Anatomical investigations of teratological transient forms between di- and polycotyledonous seedlings have shown that division of cotyledons leads to serious changes both in structure of the embryonic root and hypocotyl, and in sequence of leaf primordium differentiation at the shoot apex (De Vries, 1903; Compton, 1913; Holden and Bexon, 1918; Bexon, 1920; Harris *et al.*, 1921; Holden and Daniels, 1921; Tronchet, 1925b, 1928; Bexon and Wood, 1930; Munerati and Costa, 1934; Litovchenko, 1940; Banerji, 1961, 1962). So, in *Cannabis sativa* (Urticaceae) cotyledon division is followed by increase in xylem ray number in the radicle and hypocotyl steles, due to splitting of the xylem poles in the originally diarchical root (Compton, 1913). The deeper the splitt, the lower the level of division on the hypocotyl-embryo root axis. The extreme degree of cotyledon division leads to such a deep change in the vascular system structure, that polycotylous seedlings without traces of their schizocotylar origin are formed. In a number of these cases, and also in some other species (*Phacelia tanacetifolia*, Hydrophyllaceae, *Clarkia pulhella*, Onagraceae, *Amaranthus speciosus*, Amaranthaceae), another mode of change of the vascular system was observed. Partition of cotyledons does not influence the structure of the embryo root: new cotyledonous traces are not associated with splitting of the original xylem poles of the root, but appear in the cotyledons *de novo*, gradually 'extinguishing' in the hypocotyl. Cotyledons of such types are characteristic for polycotylous seedlings of some gymnosperms and belong in the category of so-called 'subsidiary cotyledons' (Hill T. and De Fraine, 1908-1910).

Deep division of the cotyledons, regardless of whether it causes the above-mentioned changes in the structure of the embryo root or not, results in an increase in leaf primordium number in the first epicotyl node. For example, in tricotylous seedlings *C. sativa* and *Ph. tanacetifolia*, formation of three leaf primordia was observed in the first epicotyl node instead of two. Similarly, in tetracotylous seedlings of *Ph. tanacetifolia* (the result of deep division of both original cotyledons), four leaf primordia were initiated in the first node of the epicotyl (Compton, 1913). In plants obtained from polycotylous seedlings, fascicular shoots were sometimes observed (Went, 1944; White, 1948). Correlations between increase in cotyledonary number and number of carpels (Tucker, 1938), and parts of the perianth (Holtorp, 1944) were established.

Factors of induction and regulation of schizo- and polycotyly. Treatment of seeds of *Eranthis hiemalis* (Ranunculaceae) in the process of their final development by some artificial analogues of growth substances (ammonium salt 2,4-D, sodium salt 2-methyl-4-chlorphenoxiacetic acid—MPCA) caused the formation of up to 80% polycotylous embryos (Haccius, 1969, 1972a,b). Half were tri- and tetracotylous, the first predominating. Another half contained five and more cotyledons; however, anatomical study showed them to be the result of partial fusion of several embryo-twins, formed at cleavage polyembryony. Polycotyly could be induced by X-ray treatment of the seeds (*Eranthis hiemalis*—Haccius and Reinholz, 1953) or whole plants (*Nicotiana tabacum*—Devreux and Scarascia Mugnozza, 1964; *Chenopodium album*— Singh and Chauchan, 1977). In experiments with *Eranthis*, the stage of late-globular embryo without features of cotyledonary initiation was sensitive to the treatments while in *Nicotiana*, polycotyly appeared after irradiation of the plants at the stage of gamete and zygote formation.

Teratological poly- and schizocotyly are hereditary features (De Vries, 1903; Brozek, 1926; Holtorp, 1944; Straub, 1948; Haskell, 1949, 1954a,b; Harrison, 1964). According to Straub (1948), the feature of tricotyly is controlled by two pairs of genes. That is why formation of the aforesaid poly- and schizocotylous races could be the result of various gene combinations, which control the quantitative expression of these features (Haskell, 1954b).

In *amp-1* mutant of *Arabidopsis thaliana*, 20% of the seedlings were tri- and tetracotylous (compared with 0.1% in the wild type). Such plants had altered phyllotaxis at the shoot apex (three leaves in the node instead of two in the wild type), lowered degree of apical domination and lowered level of fertility, stronger vegetative growth and constitutional type of photomorphogenesis (Chaudhury *et al.*, 1993). The given mutant had a high level of cytokinin content (6 times more than the wild type); this allows assumption of the important role of this class of growth substances in polycotyly induction.

Role of poly- and schizocotyly in evolution of flowering plants. According to the opinion of a number of authors (Sachs, 1882; Masters, 1868; Dorety, 1910—see Coulter and Chamberlain, 1910; Poddubnaya-Arnoldi, 1976) polycotyly is a more primitive state than dicotyly, because it is peculiar to the so-considered most ancient groups of gymnosperms (Podocarpaceae, Araucariaceae). However, in the opinion of Duchartre (1848), Dangeard (1892), Hill T. and De Fraine (1908-1910), Eames (1961), and Haccius (1955b), polycotyly represents a more advanced evolutionary state than dicotyly, and has appeared by gradual deepening of the process of bifurcation of original cotyledons of the dicotyledonous embryo (schizocotyly). This opinion was based on the fact of increased cotyledon number by division of two original cotyledons being a more or less hereditary anomaly in the great majority of dicotyledonous flowering plants and similarity of anatomy of schizocotylous seedlings (especially the presence of 'subsidiary cotyledons') to anatomy of normal polycotylous seedlings of gymnospermous plants. Supporters of the derivative character of polycotyly, some authors (e.g. Haccius, 1955b) prefer the term 'pleiocotyly' over 'polycotyly' because the latter simply means multiple cotyledons, while the former signifies excess of cotyledon number over the normal.

Buchholz (1919) formulated an original point of view about the relations among the phenomena of poly-, di- and monocotyly; according to him, evolution of the embryo occurs in the direction poly- to di- and then to monocotyly by successive fusion of pairs of numerous cotyledons in ancient plant forms. It should be mentioned, however, that the presence of at least three modes of tricotylous seedling formation in flowering plants noted above does not permit a final decision regarding the origin of polycotyly.

The similarity of schizocotyly and syncotyly in a number of features (presence of a single original primordium during cotyledon initiation, mode of vascular system transformation—Compton, 1913) has given rise to two contradictory points of view about the relations between di- and monocotyly phenomena. First the hypothesis of Lyon (1901b) about the origination of a dicotyledonous embryo from a monocotyledonous one by division of the single cotyledon into two (schizocotyly) and secondly, Sargant's hypothesis (1903, 1908) about the monocotyledonous embryo origin from dicotyledonous by fusion of the cotyledons of the latter—hypothesis of syncotyly (see above: Syncotyly).

Heterocotyly

Heterocotyly (Greek *heteros*—other, different, heterogeneous and *kotuledon*) is the distinction of cotyledons in the same embryo by size, structure and functions **(Plate XXXVII).**

One of the forms of heterocotyly and often its synonym is considered to be **anizocotyly** (Greek *anizo*—unequal and *kotyledon*), differentiation of cotyledons in the embryo mainly by size as a result of suppression of development in one of them and loss of its functions in the course of embryogenesis (although the seldom preserved rudiment can acquire another functional load, e.g. defence of the growing point during germination in *Trapa*). However, the mode of anizocotyly expression during embryogeny, as well as its consequences in onto- and phylogenesis, differ significantly from those appearing as a result of divergence in cotyledon behaviour and functions of the same embryo during germination (**heterocotyly in the narrow sense**). Proceeding from this the phenomena of anizocotyly and heterocotyly should be distinguished, as correctly pointed out by Haccius (1952). Typical anizocotylar embryos are noted in some pseudomonocotyledonous plants (see above: Pseudomonocotyly).

The differences in cotyledons of the same embryo by size and also by structure, functions and further fate in the course of germination, were first described by A. Hill (1906) in some geophilous species of *Peperomia* (Peperomiaceae)—*P. peruviana, P. parvifolia* and others (Fig. 43). As distinct from the majority of species in *Peperomia,* which have two strictly opposed cotyledons emerging from the seed during germination and functioning as assimilative organs (Jöhnson, 1900b), in these species one cotyledon remains inside the seed and fulfils the function of absorption of nutrients from the endosperm and perisperm tissues. In the mature seed of *P. peruviana,* the assimilative (epigeal) cotyledon is more massive, its blade is differentiated into epidermis with stomata and hydathodes, palisade and water-storing tissues and contains three (median and two lateral) conductive bundles. The absorbing (hypogeal) cotyledon is smaller in size, slightly peltate in shape, not differentiated into tissues typical of the assimilative cotyledon, almost lacking in stomata and hydathodes and contains only one conductive bundle. In *P. parvifolia* specialization of one cotyledon as an absorptive organ is even more strongly expressed, as manifested in absence of stomata on its surface and loss of the dorsoventral structure. The blade of the epigeal cotyledon is differentiated to a lesser degree than in *P. peruviana* and occupies a much smaller volume in the mature seed.

Hill (1906) considered that in both cases two cotyledons are present and not a single cotyledon and first leaf of the plumule. His arguments were next: 1) presence in mature seed of two cotyledons typical for dicotyledonous embryos, 2) their strictly opposed position and 3) presence of one unified node of their conductive bundles.

Differences in structure and behaviour of cotyledons during germination in representatives of family Gesneriaceae, subfamily Cyrtandroideae, are expressed otherwise (Hill, 1938). In all species of this subfamily studied, one of the two originally equal cotyledons of seedlings is assimilative in function, retarded in growth and aborts soon after germination. Concomitantly, the other cotyledon increases markedly in size and acquires an aspect similar in shape, structure, mode of pubescence and so on with foliage leaves, which appear at the shoot apex. Size increment occurs not at the expense of blade growth, but due to the activity of a special basal intercalary meristem situated in the persisting cotyledonary base, at the point of its connection with the hypocotyl. The cotyledon blade proper can barely be discerned at the end of this 'heelar' seedling leaf (so termed by Hill). It is significant that unequal growth of cotyledons is often accompanied by long delay in development of the plumule (represented only by apical meristem) and formation of the intercalary segment between them, called the mesocotyl. The further character of shoot apex

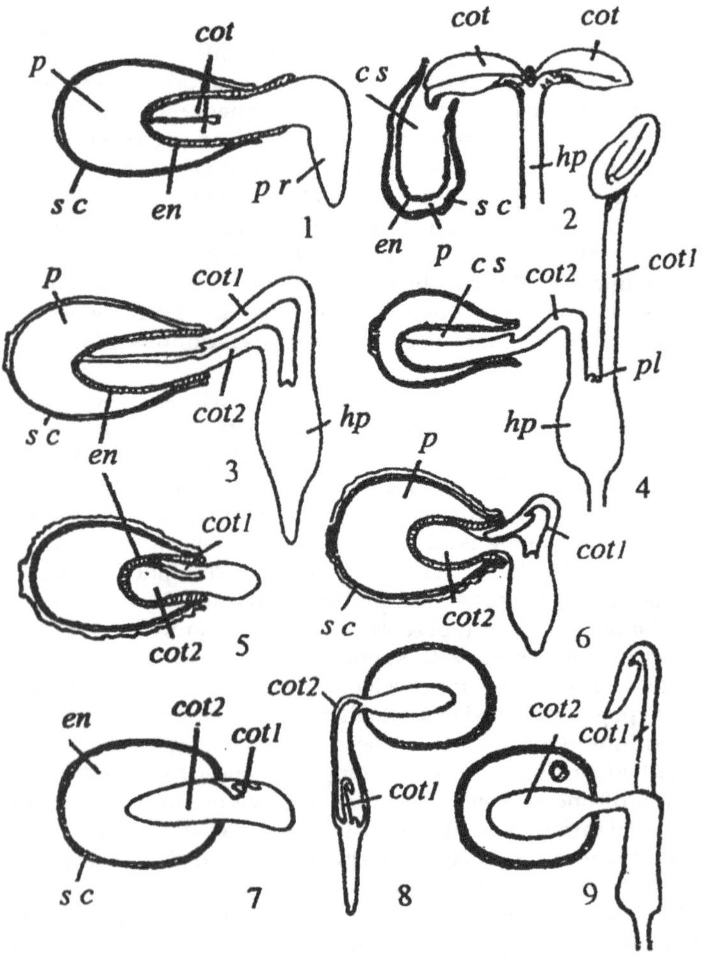

Fig. 43: Heterocotyly: (1-9—scheme illustrating monocotyledonous embryo origin by transformation of one cotyledon of dicotyledonous embryo into haustorial organ): 1, 2—seedlings of *Peperomia pellucida*, both cotyledons emerge from seed coat; 3-6—seedlings of *P. peruviana* (3, 4) and *P. parvifolia* (5, 6): one cotyledon emerges and becomes an assimilating organ (*cot1*), the other (*cot2*) remains in the seed and turns into an absorbing organ; in *P. parvifolia* absorbing cotyledon is more specialized and has mace-like form, assimilating cotyledon in rudimentary state; 7-9—seedlings of monocotyledonous plant: one cotyledon in the germinating seed—absorbing organ, the other—'the first leaf' (after Hill A., 1906). *cot, cot1, cot2*—cotyledons, *c s*—cavity in seed which remains after emergence of assimilating cotyledon, *en*—endosperm, *hp*—hypocotyl, *p*—perisperm, *pl*—plumule, *p r*—primary root; *s c*—seed coat.

development differs in different species, and accordingly three groups are distinguished: caulescent, rosette and unifoliate or 'cotyledonous'. In caulescent species (*Epithema, Klugia, Rhynchoglossum, Leptoboea, Ornithoboea, Streptocarpus*—

subgenus *Streptocarpella* and *Chirita hamosa, C. capitis*) the shoot apex forms the stem with numerous, usually, opposite leaves. In rosette species (*Saintpaulia, Ramonda, Streptocarpus*—subgenus *Streptocarpus: S. rexii, S. parviflorus*) the rosette forms from 3-4 or more leaves. Lastly, in unifoliate or 'cotyledonous' Gesneriaceae (*Streptocarpus dunnii, S. wendlandii, S. polyanthus, Chirita monophylla, Monophyllaea* species) subsequent development of the shoot apex does not occur at all throughout the life cycle, and the plant in general is represented by only one organ—overgrown cotyledon. However, in these species development of additional buds in the base of the reserving cotyledon can often be observed. According to Hill, they are of axillary and also extra-axillary origin and the appearance of extra-axillary buds is also associated with the activity of the basal meristem in the reserving cotyledon base.

Later studies (Jong, 1973, 1978; Jong and Burtt, 1975) stated that for rosette and unifoliate Gesneriaceae the presence of a unique, phyllomorphic organization is characteristic. This organization is expressed in morphological–anatomic modifications not only of the seedling structure, but also of the vegetative body of the whole plant. The intercalary segment between blades of unequal, primarily opposite cotyledons (mesocotyl, Hill), represents a typical part of the main plant axis in caulescent species. In unifoliate species its development is unusual, due to which it acquires a structure that combines the features of the stem and petiole and hence is termed a 'petiolode'. According to Jong, the petiolode results from the activity of a special intercalary meristem—'petiolode meristem'—situated near the basal meristem of the persisting cotyledon, across the base of its midrib, at the level of the shoot apical meristem. It consists of longitudinal rows of vacuolated cambiform cells, which resemble a typical rib-meristem from young internodes of ordinary stems. Dominance of intercalary growth in the petiolode from the abaxial side leads to: 1) dorsoventrality of its outer and inner structure (asymmetrical distribution of primary and secondary vascular tissues) and also of the whole seedling; 2) changes in position of the shoot apical meristem, which appears replaced at an angle of 45-60° to the longitudinal petiolode axis and situated at its adaxial side, at the bottom of a deep groove. Simultaneously the cotyledon itself changes position—from horizontal to vertical ('axial') and appears on the same axis with the petiolode hypocotyl and root, as a result the impression of its 'terminal' position is created.

In plurifoliate rosette species, the shoot apical meristem, termed 'groove meristem' by Jong because of its particular position, successively produces leaf primordia lying spirally or in two ranks. The pattern of their development completely duplicates the pattern of development of the persisting cotyledon described above. Contrarily, in unifoliate Gesneriaceae the ability of the apical meristem to form primordia is wholly suppressed. Later, in the cotyledon axils, and in rosette species also the leaf axils, the meristems are activated. Due to intercalary growth these axil meristems can move some distance away from the apical meristem of the shoot, with formation of so-called 'detached meristems', which give rise to additional vegetative buds. Their formation is accompanied by root initiation at the lower, abaxial surface of the petiolode. Thus, formation of 'extra-axillary' buds is not associated with the activity of the basal meristem at the cotyledon blade base, as thought by Hill and some other early authors.

Such an unusual mode of development led Jong to introduce the notion 'phyllomorph'—a structure consisting of enlarged cotyledon (or leaf) lamina and petiolode, which replicates the structural unit of vegetative architecture of

Gesneriaceae. The first, 'cotyledonary' phyllomorph, including the hypocotylar component after abortion of the smaller cotyledon, is the source of the subsequent, primary and additional (of axillary origin) leaf phyllomorphs. Development of each phyllomorph is controlled by a triad of meristems—basal, groove and petiolode—situated near the junction of the petiolode and lamina in the close physical association and constituting the critical centre of growth. Jong considered these interrelations of the meristems of exceptional morphogenetic importance because their co-ordinated but varied activity determines the possibility of morphogenetic innovations and reveals new perspectives in plant form evolution. One such innovation is the petiolode formation, which acquires the characters of the shoot and also the features of petiole structure due to abnormal axial development, associated with the phenomena of anizocotyly and plumule development inhibition. At the same time the mesocotyl of caulescent forms is considered an evolutionary precursor of the petiolode. The peculiar type of phyllomorphic organization and specificity of growth and development of the vegetative body in Gesneriaceae associated with this organization are the result of adaptation to seasonal climate, since the majority of events mentioned—cessation of the larger cotyledon growth and decrease in its basal meristem activity, suppression of plumule development in caulescent species—occur during unfavourable, drought conditions.

Induction of the phyllomorphic pathway of development in isocotylar species of Gesneriaceae (normally unable to enlarge cotyledons) is possible after the action of exogenous cytokinins. Suppression of basal meristem activity concomitantly in the cotyledonary base, and induction of caulescence (breaking of plumule development inhibition) can be realized with the help of exogenous GA_3 or inhibitors of polar auxin transport (Rosenblum and Basile, 1984).

Differences in size, structure and behaviour of cotyledons during germination were also marked in some woody forms of Angiospermae: Bombacaceae (*Pachira oleaginea, P. macrocarpa, Catostemma digitata, Gyranthera caribensis*—Baker, 1960; Roth, 1977; Shepherd and Alverson, 1981; Vyshenskaya, 1992), Dipterocarpaceae (*Dryobalanops aromatica, Vateria indica, Hopea ponga*—Gilg, 1925; Swaraupanahdan, 1986; Nemirowich-Danchenko, 1992) and Hyppocastanaceae (*Aesculus woerlitzensis*—List and Steward, 1965; Alimova, 1985b). It is interesting that differences in cotyledon behaviour in *A. woerlitzensis* are rather close to those observed in *Peperomia*.

In a number of studies the earlier data on seedling morphology in geophilous species of *Peperomia* and rosette and unifoliate Gesneriaceae were revised. From the point of view of Haines and Lye (1979), of the first two leaf organs of seedlings of these plants (cotyledons; Hill) only one, aborted during germination, has a cotyledonary nature and appears to be the product of fusion of two original cotyledons. The other organ, spreading during development, represents the first leaf of the plumule and its node is extremely close to cotyledonary. As an argument the authors mention Hill's data about the frequent formation of a mesocotyl between these two leaf organs in Gesneriaceae, and the peculiarities of their position relative to subsequent leaf primordia, which originate at the shoot apex of the seedling. According to Juguet (1992), there is no such phenomenon as heterocotyly in nature. It should be mentioned that the data of Jöhnson (1900b) on the existence of cotyledonary fusion in many other, non-geophilous species of *Peperomia* attest to the syncotylous nature of the embryo in geophilous species of this genus. One more argument is significant, namely, similarity in behaviour of the stem apex (analogous displacement of its position) in rosette and unifoliate Gesneriaceae species with phyllomorphic organization and species with cotyledonary fusion (see above: Syncotyly). At the same time data on Gesneriaceae embryogenesis (Crété, 1949; Arecal, 1961; Adatia *et al.*, 1971; Alimova and Yakovlev, 1982) convincingly show initiation and development of two identical cotyledons. Data on embryo development and laying down of cotyledons in geophytic species of *Peperomia* are absent.

Morphological peculiarities of seedlings of geophilous *Peperomia* were identified by Hill (1906, 1908) based on the theory of monocotyledonous embryo origin from dicotyledonous, by gradual divergence of functions between cotyledons and one of them transforming into a suctorial organ (hypothesis of heterocotyly). As a transitory form to the monocotyledonous state among monocotyledons Hill mentioned the embryo of some Araceae (*Arum, Arisaema*), in which, similar to *Peperomia*, the seedling structure and also the presence of a unified bundle node of a single cotyledon and first leaf of the plumule are observed. Establishment of the monocotyledonous state in these plants correlates with formation of a geophilous habitus (bulbous hypocotyl, shortened shoot etc.) and is the seedling adaptation to survival during lengthy drought. Divergence of functions in cotyledons during germination facilitates quick storage deposition in the tuber-like hypocotyl without plumule participation—at the expense of transport from seed perisperm by the absorbing cotyledon, as well as the expense of assimilation by photosynthesis.

One adherent of the heterocotyly hypothesis, L. Boyd (1932), interpreted the seedling structure of some other monocotyledonous plants (*Dioscorea, Thamus*—Dioscoreaceae and others) in the manner of Hill. However, considering some of the points mentioned above, a number of authors (e.g. Haines and Lye, 1979; Juguet, 1992) cast doubts on the evolutionary transformation of dicotyledonous embryos into monocotyledonous via heterocotyly.

Acotyly[19]

Acotyly (Greek *a*—against, without, and *kotuledon*) is the phenomenon of lack of cotyledon formation in the embryo. Synonym: apocotyly **(Plate XXXVIII).**

Acotyly is characteristic mainly of representatives of flowering plants with a parasitic or saprophytic mode of life: dicotyledons (Rafflesiaceae, Orobanchaceae, Scrophulariaceae, Cynomoriaceae, Balanophoraceae, Lennoaceae, Hydnoraceae, some Loranthaceae, Viscaceae) and monocotyledons (Burmaniaceae, Orchidaceae).

In the evolution of parasitic organisms, acotyly is a secondary phenomenon. In family Scrophulariaceae the transition from autotrophic nutrition (e.g. *Verbascum*) via facultative parasitism (*Odontithes verna*) to different forms of obligate parasitism (*Pedicularis, Bartsia, Lathraea squamaria, Striga orobanchoides, Harveya*) occurs and different stages of cotyledonary reduction in the embryo can be observed (Teryokhin, 1977; Teryokhin and Nikiticheva, 1981) (Figs. 44 and 45). A unipolar pattern of embryo development (with one active morphogenetic region) results from this reduction. It is known that cotyledons of dicotyledonous plants with autotrophic mode of nutrition mainly fulfil photosynthesizing and reserving functions. In parasitic plants the cotyledons became unnecessary because the nutritional function passes to a special

[19]Acotyly is often mixed with the phenomenon of retardation of organogenesis in mature seeds of species characterized by embryo post-development during the period of seed shedding from the maternal organism up to the beginning of its germination (see below: Reduced and Undifferentiated Embryos; Seed Dormancy).

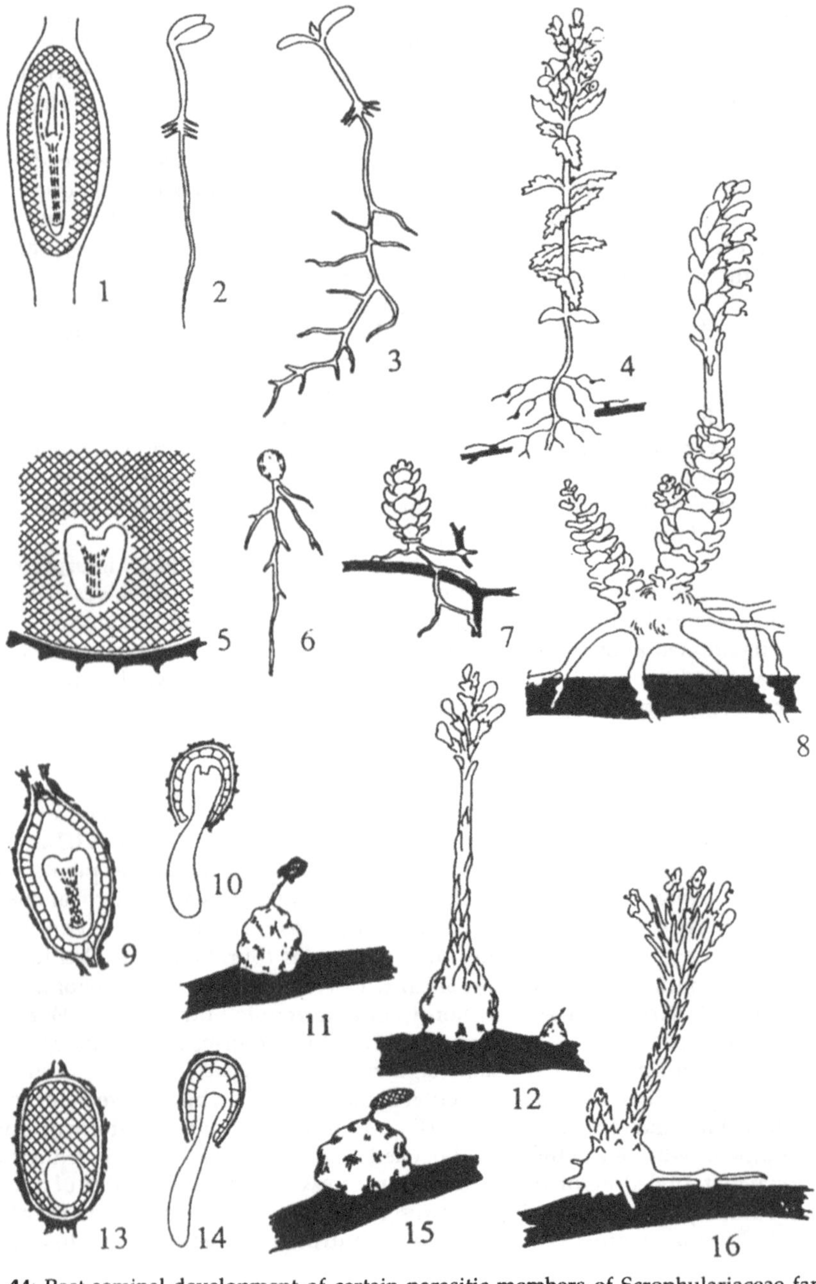

Fig. 44: Post-seminal development of certain parasitic members of Scrophulariaceae family with different degree of morphogenetic adaptation to the parasitic mode of existence: 1-4—*Euphrasia* sp., 5-8—*Lathraea* sp., 9-12—*Striga orobanchoides*, 13-16—*Hyobanche* sp. (1, 5, 9, 13—embryo in mature seed; 2, 3, 6, 7, 10, 11, 14, 15—seedling; 4, 8, 12, 16—flowering plant) (after Teryokhin, 1977).

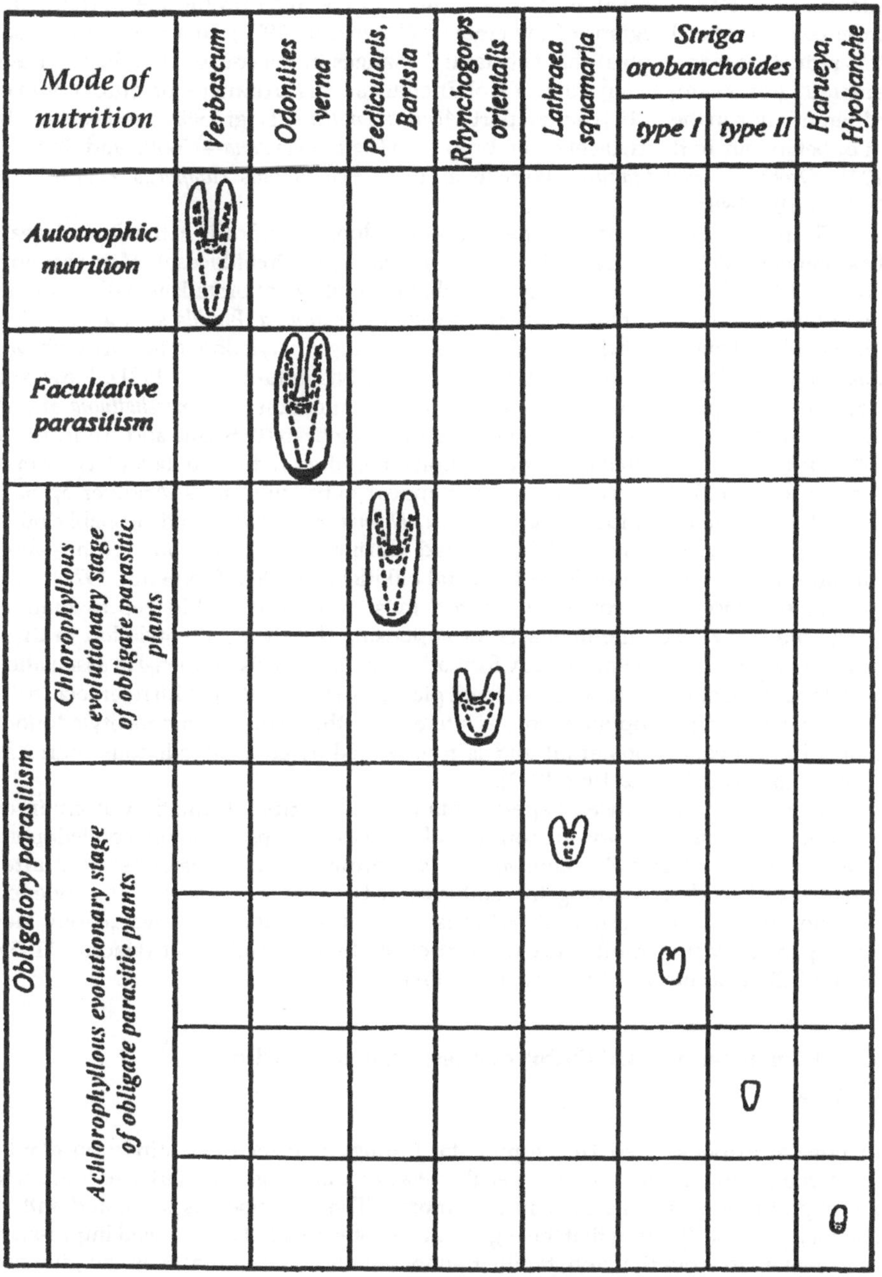

Fig. 45: Schematic representation of embryo evolution in parasitic representatives of Scrophulariaceae in connection with the transition to the parasitic mode of existence (after Teryokhin, 1977).

suckers (haustoria) formed at the roots, stems or so-called protosomes, originating from basal or apical regions of the embryo (Teryokhin, 1977). In a number of cases this reduction was so great that substantial changes in the course of embryogenesis ensued. For example, in some Orobanchaceae disturbances of embryoderm differentiation in the embryo apical part (*Aeginetia indica*—Tiagi, 1952a) and variability of *ca* behaviour in the two-celled embryo (*Epiphagus americana*—Cooke and Shively, 1941; *Orobanche pallidiflora*—Nikiticheva, 1971; *Phacellathus tubiflorus*—Teryokhin, 1973) can be observed.

The presence of acotyly in some monocotyledons, specifically in the Orchidaceae, has remained debatable (see below: Embryogenesis in Orchidaceae). Many authors (Tohda, 1968; Veyret, 1974 a. o.) consider that the embryo of orchids usually contains no cotyledons. In a number of species (*Sobralia macrantha, Bletilla striata, Arundina graminifolia, Thunia marshalliana*) a rudiment of a single cotyledon in mature embryos has been mentioned (Treub, 1879; Burgeff, 1936; Nishimura, 1981, 1991). However, detailed investigations of embryogenesis and germination in *T. marshalliana* and *B. striata* demonstrated the lack of cotyledons in the embryo (Batygina and Andronova, 1988). It was proposed that during evolution either their reduction took place or they did not form at all; the first foliar organs appear to be true leaves. Another opinion holds that in some orchids (*Bletia purpurea, Thunia marshalliana* and probably other species) the cotyledons are differentiated during seed germination (one—after Nishimura, 1991; two—after Teryokhin and Nikiticheva, 1968; Teryokhin, 1977). Data on the presence of a cotyledon in the most advanced orchid representatives (Nishimura, 1991) became the basis for supposing the so-called cotyledon in these cases to be a secondary evolutionary formation, i.e., organ arisen *de novo* in association with the germination peculiarities of these plants (Nazarov, 1995). With reference to the foregoing, the original opinion arouses interest, i.e., the embryo of monocotyledonous plants has no cotyledons at all and is represented as an acotyledonous branched protocorm (Jacques-Félix, 1958, 1982).

Acotyly can be induced experimentally. For example, formation of embryos without cotyledons in dicotyledonous *Eranthis hiemalis* and pseudomonocotyledonous *Ranunculus ficaria* and *Anemone appenina* (Ranunculaceae) was observed after treatment of seeds containing the embryos at the very early stages of cotyledon initiation with phenylboric acid. What is more, in the last two cases acotyly was accompanied by precocious development of the first plumular leaf (Haccius, 1959, 1960, 1972b; Haccius and Lakshmanan, 1967).

Axis Formation in Plant Embryogenesis: Cues and Clues
(Fig. 46)

Imagine a textbook titled **Developmental Biology** that focuses entirely on plants, mentioning animals only for their peculiar way of making germ cells by setting aside a group of precursor cells early in the embryo. The converse has been, and still is, common practice. It is true that the regenerative potential of plants, indeed impressive, sets them apart from the more familiar animal models: individual cells can give rise to embryos in culture; localized groups of stem cells, called meristems, form the adult plant in a seemingly autonomous fashion not only during normal development, but also by regeneration from lumps of undifferentiated cells in culture. These special features notwithstanding, plants do develop from a fertilized egg cell, the zygote,

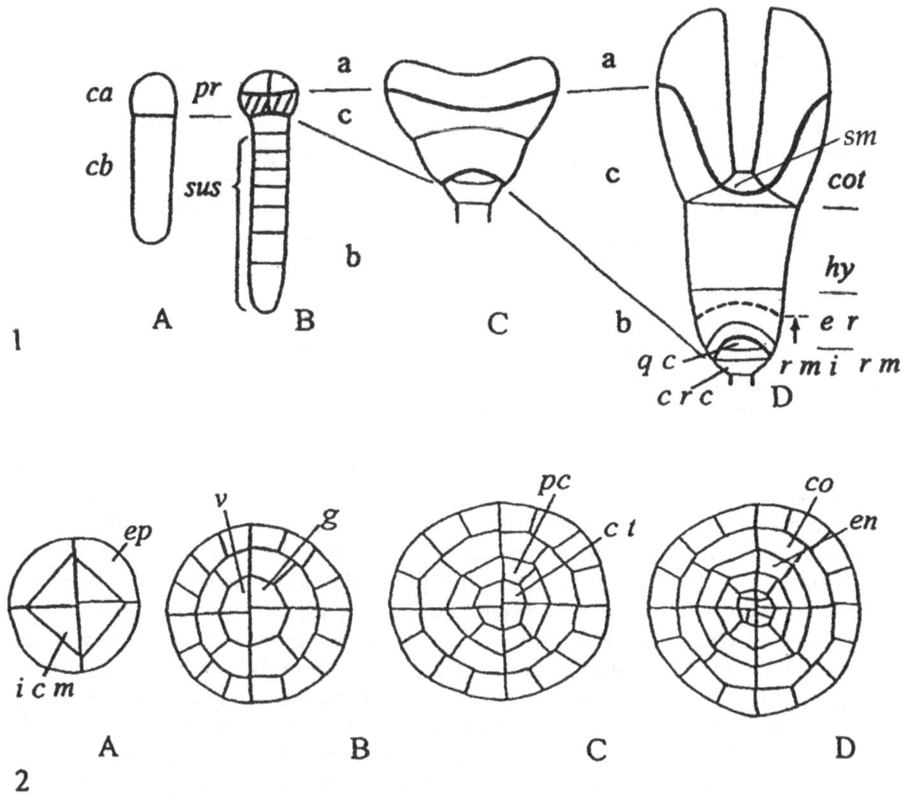

Fig. 46: Axis formation in plant embryogenesis. 1—formation of apical–basal axis in the *Arabidopsis* embryo: A—asymmetric division of zygote, B—8-celled proembryo, C—embryo at heart stage, approximate locations of cell groups that give rise to primordia of seedling structures are indicated, D—embryo at torpedo stage, clonal boundaries are marked by thick lines, broken line indicates upper end of embryonic root derived from root meristem initials, below the quiescent centre of the root meristem are initials of central root cap, primordia of seedling srtuctures: cotyledons, hypocotyl, embryonic root, root meristem, shoot meristem; 2—formation of radial pattern in the *Arabidopsis* embryo—schematic cross-sections through central region of embryo (1 A and 1 B) and through the primordium (1 C and 1 D): A—dermatogen stage, periclinal (radial) divisions within the proembryo (1B) give the outer epidermis layer and an inner cell mass, B—globular embryo, inner cell mass has split into ground tissue and the centrally located vasculated primordium, C—embryo at heart stage (1C), the pericycle layer surrounds primordium of the conductive tissue, the basic organization of the radial pattern is complete, D—embryo at torpedo stage (1D), periclinal divisions in the ground tissue generate an outer cortex and an inner endodermis layer (after Jürgens, 1995). a,b,c—apical, basal and central regions of embryo, *ca*—apical cell, *cb*—basal cell, *c r c*—central root cap, *c t*—conductive tissue, *co*—cortex, *cot*—cotyledons, *en*—endodermis layer, *ep*- epidermis layer, *e r*—embryonic root, *g*—ground tissue, *hy*— hypocotyl, *i c m*—inner cell mass, *pc*—pericycle, *pr*—proembryo, *q c*—quiescent centre, *r m*— root meristem, *r m i*—root meristem initials, *sm*—shoot meristem, *sus*—suspensor, *v*—vasculated primordium.

during the normal course of their life cycle and, like animals, have to establish the characteristic body organization of the multicellular adult form. Are the underlying mechanisms the same or different? In the past few years, genetic and molecular analysis of flower development in two plant model species, *Arabidopsis* and *Antirrhinum* (snapdragon), has established a network of MADS domain transcription factors and others that regulate this best-characterized process in plant development (Weigel and Meyerowitz, 1994). However, flower development is like putting the finishing touches to an adult plant and thus may not give clues to mechanisms that underlie earlier processes such as axis formation and generation of the overall body organization. Drawing largely on recent genetic studies in *Arabidopsis*, I shall briefly discuss post-embryonic development that eventually culminates in the formation of flowers, but mainly focus on pattern formation in the embryo that establishes the basic body organization of flowering plants.

Shoot meristem: linking the embryo with the flower. Pattern formation in animals is largely confined to embryogenesis, such that the future adult form is represented in the body organization of the mature embryo. By contrast, plant embryogenesis produces a juvenile form, the seedling, that lacks most structures of the adult plant. Embryogenesis in essence organizes two groups of stem cells at the opposite ends of the body axis, the primary meristems of the shoot and the root. These meristems then add new structures to the seedling, thus generating the species-specific adult form during post-embryonic development (Steeves and Sussex, 1989). Regardless of the appearance of the adult plant, the shoot meristem is essentially organized in the same manner in different plant species. Two functional units can be distinguished within the meristem: a central zone, required for self-renewal and integrity of the meristem, and a peripheral zone, which produces primordia of lateral organs and their associated secondary shoot meristems, such as leaves and flowers (Steeves and Sussex, 1989). Whether leaves or flowers are produced depends on the physiological state of the meristem. In **embryonic flower** (*emf*) mutant seedlings, for example, the primary shoot meristem skips altogether the vegetative phase of making leaves, producing flowers directly (Sung *et al.*, 1992). Shoot meristem identity seems to be conferred by the continuous expression of genes, such as the maize homeobox gene *Knotted* (*Kn1*) whose ectopic expression can cause formation of shoot meristems on leaves (Sinha *et al.*, 1993; Jackson *et al.*, 1994). A putative *Arabidopsis* homologue of *Kn1*, the *SHOOT MERISTEM-LESS* (*STM*) gene, is required for shoot meristem formation both in the embryo and during regeneration from tissue culture (Barton and Poethig, 1993). How the activity of such shoot meristem-specific genes is established and maintained properly is not known. However, there are candidate genes in *Arabidopsis* for performing these functions. For example, the *ZWILLE (ZLL)* gene is specifically involved in establishing the primary shoot meristem in the embryo but not in any other (non-embryogenic) context (Jürgens *et al.*, 1994). Mutations in another gene, *CLAVATA1 (CLV1)*, cause overgrowth of primary and secondary shoot meristems that also results in formation of one or more supernumerary whorls in the centre of the flower (Clark *et al.*, 1993). This phenotype suggests that the organization of the meristem is altered, due to enlargement of the central zone required for self-renewal. Although more genes need to be identified to clarify this point, it is conceivable that the primary shoot meristem may acquire a specific organization in the embryo that is subsequently maintained by cellular interactions within the meristem.

Pattern formation in *Arabidopsis* embryo. The primary meristems of the shoot and root originate at distinct positions in the embryo as part of the overall body organization of the seedling (Barton and Poethig, 1993; Dolan *et al.*, 1993). The latter may be viewed as the superimposition of an apical-basal pattern along the main body axis and a radial pattern perpendicular to this axis. The apical-basal pattern consists of a top-to-bottom array of elements: shoot meristem, cotyledons (embryonic leaves), hypocotyl (embryonic stem), radicle (embryonic root) and root meristem. The radial pattern is made up of concentric rings of tissue layers: epidermis, ground tissue (cortex and endodermis) and vascular tissue (pericycle, xylem and phloem). The origins of the apical-basal and radial pattern elements have been traced back to cell groups of the early embryo in *Arabidopsis* where the very regular patterns of cell division facilitated these analyses (Mansfield and Briarty, 1991; Jürgens and Mayer, 1994).

Formation of apical-basal axis. The apical-basal pattern formation starts with asymmetric division of the zygote that yields two daughter cells of unequal size and different fates. The small apical cell generates, by cleavage divisions, an 8-celled proembryo that will give rise to most of the embryo. The large basal cell produces a file of 7-9 cells, of which all but the uppermost one will form the extraembryonic suspensor; the uppermost cell (hypophysis) joins the proembryo later to give rise to part of the root meristem. Within one day of fertilization, three embryonic regions are thus established along the axis: apical and central, which correspond to the upper and lower tiers within the proembryo, and basal, which is represented by the adjacent hypophysis. The three embryonic regions differ in cell division patterns: apical cells divide without preferential orientation; central cells produce cell files, thus expanding the axis; and the founder cell of the basal region undergoes a stereotyped programme of division. However, the early regions (apical, central and basal) do not correspond to primordia of seedling structures. The apical region gives rise to the shoot meristem and most of the cotyledons; the central region also contributes to the cotyledons, but mainly produces hypocotyl, root and root meristem initials; and the basal region gives the remainder of the root meristem, comprising the quiescent centre and the initials of the central root cap. Data from clonal analysis support the view that cell ancestry does not play a role in generating apical-basal pattern elements: clone boundaries are variable and, moreover, can run across specific seedling structures, such as the cotyledons or the root meristem (Scheres *et al.*, 1994; Dolan *et al.*, 1994). Furthermore, regularity of cell divisions in the early embryo is deceptive since mutations in the *FASS* (*FS*) gene totally alter the pattern of cell division, without affecting pattern formation (Torres Ruiz and Jürgens, 1994). Thus, the apical-basal pattern elements appear to be established by cellular interactions in a position-dependent manner.

What is the significance of the early regions along the apical-basal axis? Evidence comes from the embryonic phenotypes of mutations in three genes, *MONOPTEROS* (*MP*), *FACKEL* (*FK*) and *GURKE* (*GK*), which delete specific seedling structures (Mayer *et al.*, 1991). In *mp* embryos, the cells in the central and basal regions divide abnormally, resulting in the absence of hypocotyl, root and root meristem (Berleth and Jürgens, 1993). In *fk* embryos, the central region is affected while the basal region is not; in the seedling, the hypocotyl is missing, such that the cotyledons are directly attached to the root (Mayer *et al.*, 1991; Jürgens *et al.*, 1994). Since the root meristem is composed of cells from both the basal and the central regions, the basal region appears to induce the adjacent cells of the central region to become root meristem initials that in turn produce the meristem-derived embryonic root. A similar argument can be made for the cotyledons, which are derived from both the apical and the central regions. The phenotype of *gk* mutants is essentially complementary to *mp*: the apical region of the embryo is altered and, later, the *gk* seedlings lack shoot meristem and cotyledons (Mayer *et al.*, 1991). This suggests that the cotyledons are initiated within the apical region, which then signals to adjacent cells of the central region to participate in cotyledon formation. Thus, the early regions may define genetically distinct groups of cells (compartments?) that generate the apical-basal pattern by cellular interactions.

How are the early regions established along the apical-basal axis? The boundary separating the central from the basal region originates with the first division of the zygote. The basal daughter cell produces a file of cells, of which the uppermost one becomes the founder cell of the basal region of the embryo. This file of cells is initially normal in *mp* embryos when the earliest defect becomes apparent in the proembryo, whereas later both the central and basal regions develop abnormally (Berleth and Jürgens, 1993). Thus, in normal development, the central region may induce the uppermost derivative of the basal cell of the zygote to become the founder cell of the basal region. By contrast, the apical and the central regions are established by the transverse cell divisions within the proembryo derived from the apical daughter cell of the zygote.

Before dividing asymmetrically, the zygote elongates about 3-fold in the future axis, and concomitantly the cortical microtubules, previously oriented at random, become aligned perpendicular to the axis (Webb and Gunning, 1991). This reorganization may reflect polarization of the zygote in response to some as yet not known signal(s). Its significance for axis formation is suggested by mutations in the *GNOM* (*GN*) gene that affect the zygote and can completely abolish regional differentiation of the seedling axis (Mayer *et al.*, 1991). The *gn* zygote expands but does not elongate, producing an enlarged apical cell at the expense of the basal cell (Mayer *et al.*, 1993). Subsequent development is highly abnormal, and the region-specific *MP* gene was shown to be ineffective in *gn* embryos. The asymmetric division of the zygote seems to be involved in fixing the apical-basal axis of the embryo, but it is not known how the *GN* gene brings about its specific effect at the molecular level. Although the predicted 160 kDa *GN* (also called *EMB30*) protein shows sequence similarity to several other proteins over a stretch of 200 amino acids referred to as the *Sec7* domain, the significance of this domain is not known, nor does the remainder of the *GN* protein readily suggest a specific function within the cell (Shevell *et al.*, 1994). Since *GN* is the only *Arabidopsis* gene known to be specifically required for apical-basal pattern formation from the zygote on, determining the primary function of the *GN* protein would be an important step in further analysis of axis formation.

Radial pattern formation. Radial pattern formation, which commences in the 8-celled proembryo, involves two choices of oriented cell division: periclinal (radial), which gives a new tissue layer, and anticlinal (circumferential), which increases the number of cells in a given layer. New tissue layers are successively formed from the periphery towards the centre. Initially, an outer layer of epidermal precursor cells is separated from an inner cell mass. The epidermis layer then expands by anticlinal cell divisions, thus maintaining its integrity in the growing embryo. Further development of the radial pattern is confined to the central region of the axis that gives rise to hypocotyl, root and root meristem initials; neither the apical nor the basal region is involved (Jürgens and Mayer, 1994; Scheres *et al.*, 1995). The inner cell mass splits into an outer layer of ground tissue and the centrally located vascular precursor cells. The latter cells again divide periclinally to give a layer of pericycle cells surrounding the precursors of the conductive tissue (phloem and xylem). This basic organization of the radial pattern is completed before the heart stage of embryogenesis. One further subdivision occurs within the ground tissue, generating an outer cortex layer and an inner endodermis layer.

By the time the root meristem has become active, clonal boundaries reflect the basic organization of the radial pattern (Dolan *et al.*, 1994; Scheres *et al.*, 1994). Are tissue-specific cell fates fixed early and then transmitted clonally? Mutant phenotypes suggest that this may be the case. For example, the epidermis cells are abnormally enlarged in early *keule* (*keu*) embryos (Mayer *et al.*, 1991) and the endodermis layer is absent in *short root* (*shr*) embryos (Benfey *et al.*, 1993; Scheres *et al.*, 1995). Furthermore, the epidermis is marked by the tissue-specific expression of the LTR gene both in the embryo and during post-embryonic development (Thoma *et al.*, 1994). The genetic distinction of different tissues might involve blocking of plasmodesmata-mediated cell communication, as was observed between the epidermis and subepidermal tissues in the seedling (Duckett *et al.*, 1994). While the available evidence seems to favour a model of genetically determined tissue types, oriented cell divisions do not appear to be required for radial pattern formation: cell divisions occur at random in early *fs* embryos, such that the characteristic radial organization of tissues is not apparent, but later the mutant seedlings display all tissues found in the wild type (Torres Ruiz and Jürgens, 1994). This flexibility suggests that cellular interactions may play an important role in initiating and (possibly) maintaining tissue-specific gene expression in a position-dependent manner.

How is the radial pattern initiated in the early embryo? To date, only one gene has been identified that may be involved in this process: in *knolle* (*kn*) embryos, the epidermis is not separated from the inner cell mass by periclinal divisions within the proembryo and the mutant seedlings appear to lack the characteristic epidermis layer (Mayer *et al.*, 1991). How the *KN* gene relates, at the molecular level, to the periclinal divisions in the early embryo remains to be determined.

What mechanisms may underlie pattern formation in the plant embryo? Pattern formation in the *Arabidopsis* embryo appears to depend largely on cell-cell communication, both in the apical-basal axis and in the radial dimension. This is formally similar to pattern formation in the familiar animal models (e.g. vulva development in *Caenorhabditis elegans* or post-blastoderm embryogenesis and imaginal disc development in *Drosophila*). However, the underlying mechanisms might differ in plants. For example, plant cells are cytoplasmically interconnected by plasmodesmata, allowing various kinds of molecules to pass freely from cell to cell. Unfortunately, it is not known whether this potential plant-specific means of intercellular communication is actually used and how it is regulated during development. Another serious problem in analysis of plant pattern formation is the shortage of markers, both morphological (which cannot be changed) and molecular. As more genes with tissue-specific or region-specific expression patterns in the embryo become available, some of the ideas presented here can be tested.

The initial events of pattern formation in the *Arabidopsis* embryo differ from the later phases since they cannot involve cell-cell communication: the asymmetric division of the zygote as well as the periclinal divisions in the 8-celled proembryo appear to segregate fates within cells. How is position translated into cell fate in these cases? One possible mechanism is suggested by the observation that in *C.elegans*, the unequal division of the zygote is associated with the different degradation of a specific uniformly distributed maternal mRNA, thus establishing territories of differential gene expression along the main body axis of the early embryo. A potentially plant-specific alternative for embryonic axis fixation is suggested by recent results from the brown alga, *Fucus*. The *Fucus* zygote also divides asymmetrically, giving rise to an upper thallus cell and a lower rhizoid cell. If the cell wall of the zygote is removed, the axis can be formed, but not fixed (Quatrano *et al.*, 1991). Moreover, cell ablation studies on early embryos suggest that the different cell fates of the thallus versus rhizoid may be imprinted in the cell wall (Berger *et al.*, 1994). Which mode, if either, of cell-fate segregation applies in the early embryo of *Arabidopsis* remains to be determined. The two proteins identified thus far do not appear to regulate gene expression directly, unlike the transcription factors that play decisive roles in flower development. Does axis formation in higher-plant embryos involve an extracellular detour on its way to the nucleus?

Morphological Classification of Embryo in Mature Seed: Critical Review of Main Systems

Morphological classifications of embryos in mature seeds are only considered within the classification of the internal morphology of the seed as a whole. The features usually considered are: position of the embryo in the seed, its shape, comparative size, the extent of differentiation and correlation of its parts and some others (see Artyushenko, 1990).

Attempts to create morphological classifications of the embryo within several large taxa of angiospermous plants have been repeatedly undertaken (e.g. Reeder, 1957; Kozhanchikov, 1967; Danilova *et al.*, 1990; Vyshenskaya, 1991; Oganezova, 1993 and others). However, it was Martin (1946) and Smirnova (1965) who managed to distinguish seed structure types to account for the morphological variety of the embryo within angiospermous plants and allow consideration of the main directions of embryo evolution.

Martin's classification. In his well-known work *'The Comparative Internal Morphology of Seeds'* (1946), Martin synthesized the results of original research on seeds of 1108 genera of flowering plants and 19 genera of gymnospermous plants, and presented them in a diagram of their longitudinal and cross-sections. The main purpose of his study was to work out a method whereby plants could be identified by their seed structure, and to create a typological classification of the internal morphology of the seed. The embryo and differences in its shape, size and position in the seed formed the basis of his classification, comprising 12 types of seed combined into 3 divisions, of which one includes three subdivisions (Fig. 47).

In describing the embryo he used the terms 'axis' (referring actually to the hypocotyl-root axis, but not distinguishing the hypocotyl nor the embryo root) and 'endosperm'(designating accumulating tissues). With respect to the embryo-endosperm ratio in the mature seed, five size designations representing volume, trically in quarter units, are given, namely: small, quarter, half, dominant and total (Fig. 48).

According to Martin, the three basic divisions of the classification – **basal, peripheral, axile**—reveal the main evolutionary trends in seed development and are of great importance in the systematics and phylogeny of flowering plants.

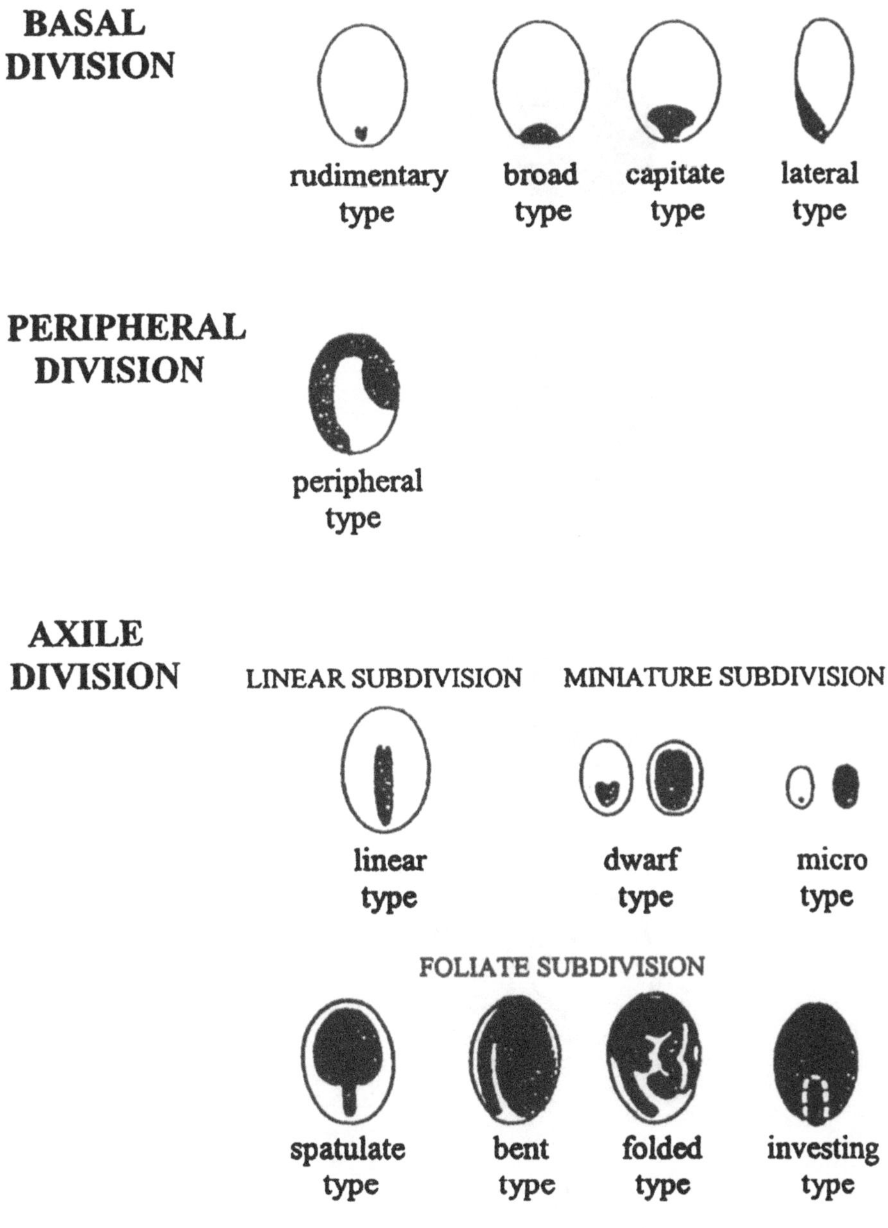

Fig. 47: Diagrammatic representation of different types of mature seed embryo (after Martin, 1946).

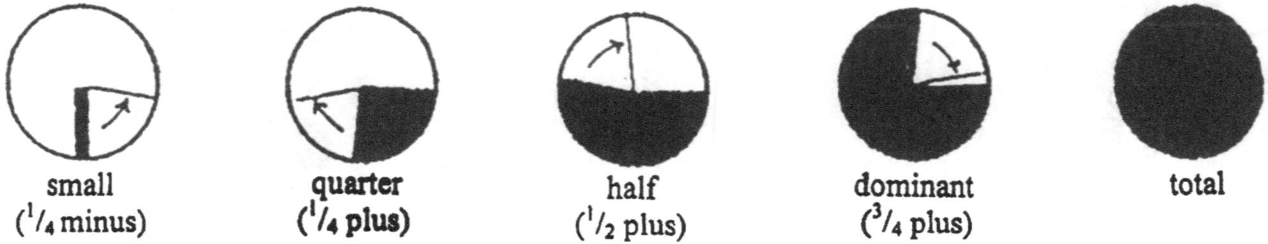

Fig. 48: Size-designations for embryos (after Martin, 1946).

Smirnova's classification. Smirnova (1965) suggested groups, levels and types as classification categories.

Depending on the nature of accumulating tissue and the degree of its development, five groups of seed structure are distinguished: endospermic, perispermal, endoperispermal, embryonic (endosperm is utilized) and proembryonic (endosperm is insignificantly developed).

On the basis of ratio of exchange in embryo and seed length, six levels are distinguished: **0** (zero) **level** 0 - ⅛; **I level** ⅛ - ¼; **II level** ¼ - ½; **III level** ½ - ¾; **IV level** more than ¾; **V level** – about 1 (embryo fills the entire seed).

Distinguishing the types of seed is based on the following complex of features: position of the embryo in the seed (embryo characterized as basal, lateral, axial, bent, peripheral, total), morphofunctional characteristics of cotyledons, predominantly functional (based on function cotyledons are characterized as haustorial, photosynthesizing, accumulating, based on shape, foliar, corymbose etc.), degree of morphological embryo differentiation and predominance of axial or cotyledonary part (estimated by ratio of lengths: cotyledon + plumule/hypocotyl + root). A total of 26 basic types are listed (with a note that there could be more), which are separately distinguished for dicotyledons and monocotyledons. The types have generic names, for example, **Alisma-, Monotropa-, Viola-, Piper- types.**

Smirnova's classification is presented in Figure 49. Six levels are presented as equal segments, set by the axis of ordinates; at the top of the scheme the groups and types of dicotyledons are distributed among these levels, on the bottom, monocotyledons. Each level and each group corresponds to a certain type of seed. Characteristics of types include characteristics of levels and groups.

Smirnova (1965) considered the endospermic group an initial group for flowering plants. This group initiates two opposite directions of adaptive evolution: a) direction typical of dicotyledons, with predominant embryo development linked with intensification of accumulating and assimilating functions of cotyledons; b) direction typical of monocotyledons, with predominant development of accumulating tissue and comparatively small embryo, linked with predomination of haustorial function of cotyledons. At the same time, she notes a commonality of groups of seeds of dicotyledons and monocotyledons by the types of seed structure, which is evidence, in her opinion, of the monophyletic character of flowering plants on the whole.

Smirnova pointed out that the seeds of representatives of various families form parallel rows of development in accordance with change in comparative size of the embryo. Hence, the larger the embryo (starting from level III), the higher the diversity of seed types (and, correspondingly, morphology of embryo) in families and orders.

It is important to note that the two classifications considered (Martin, 1946; Smirnova, 1965) complement each other in a number of aspects and undoubtedly are a substantial basis for further elaboration of the classification of the internal morphology of seeds.

Recently accumulated facts, summarized in the fundamental works *'Comparative Anatomy of Seeds'* (1985, 1988, 1991, 1996) and *'Comparative Embryology of Flowering*

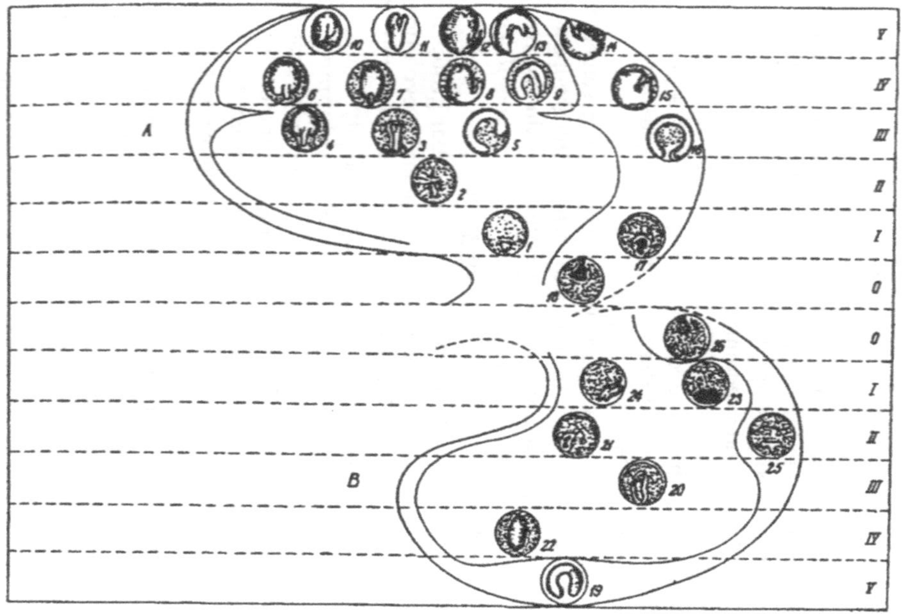

Fig. 49: Typological classification of internal morphology of seed (after Smirnova, 1965). 0- V—relative levels, A—groups and types of seeds of dicotyledons: 1-5—endospermic group: types—Apium (1), Podophyllum (2), Saxifraga (3), Viola (4), Althaea (5); 6-13—embryonic group: types—Urtica (6), Rosa (7), Lotus (8), Cannabis (9), Helianthus (10), Hippuris (11), Faba (12), Malachra (13); 14-16—perispermal group: types—Cereus (14), Echinocactus (15), Phytolacca (16); 17—endoperispermal group: type—Piper; 18—proembryonic group, seed of the type Monotropa; B—groups and types of seeds of monocotyledons; 19—embryonic group, type Alisma; 20-24—endospermic group: types—Lilium (20), Bromelia (21), Anthurium (22), Juncus (23), Poa (24); 25—endospermic group, type Musa; 26— proembryonic group, type Orchis.

Note. The figures of seed are schematic; they reflect the structure and correlation of the basic structures of the seed but not its shape.

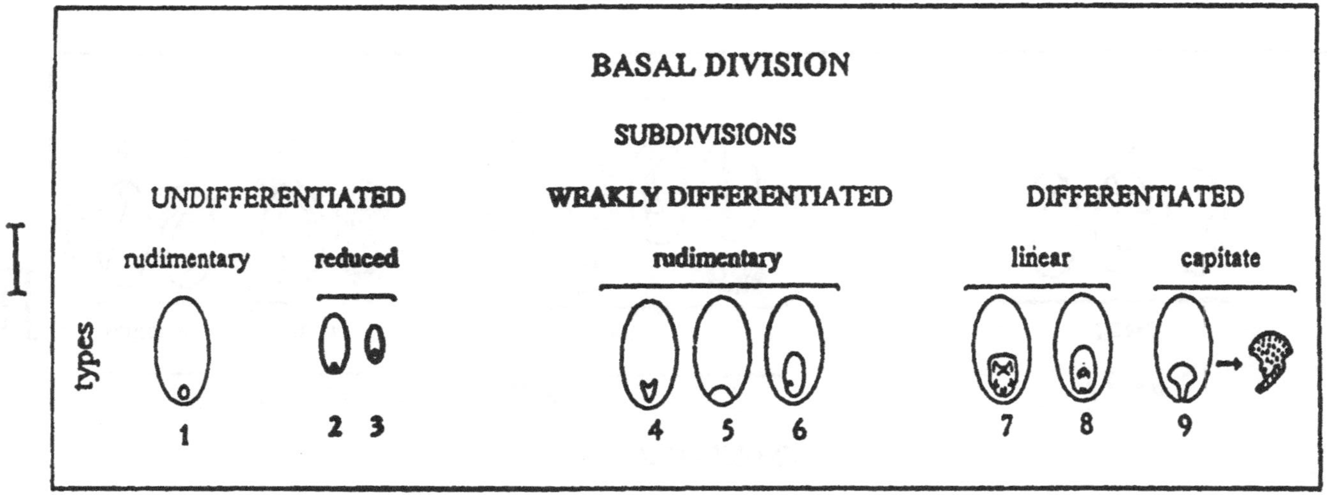

Fig. 50(I): Typological classification of internal morphology of seed. 1—*Degeneria* (Degeneriaceae, D), *Trillium* (Trilliaceae, M); 2, 3—*Piper* (Piperaceae, D), *Hydnora* (Hydnoraceae, D), *Burmannia* (Burmanniaceae, M), *Hydatella* (Hydatellaceae, M); 4—*Siparuna* (Siparunaceae, D), *Papaver* (Papaveraceae, D); 5—*Eriocaulon* (Eriocaulaceae, M), *Mayaca* (Mayacaceae, M); 6—*Juncus* (Juncaceae, M); 7—*Commelina* (Commelinaceae, M), *Musa* (Musaceae, M); 8—*Phoenix* (Arecaceae, M); 9—*Scirpus* (Cyperaceae, M); under dots there are cotyledons designated, under hatching—embryo axis; D—dicotyledons, M—monocotyledons.

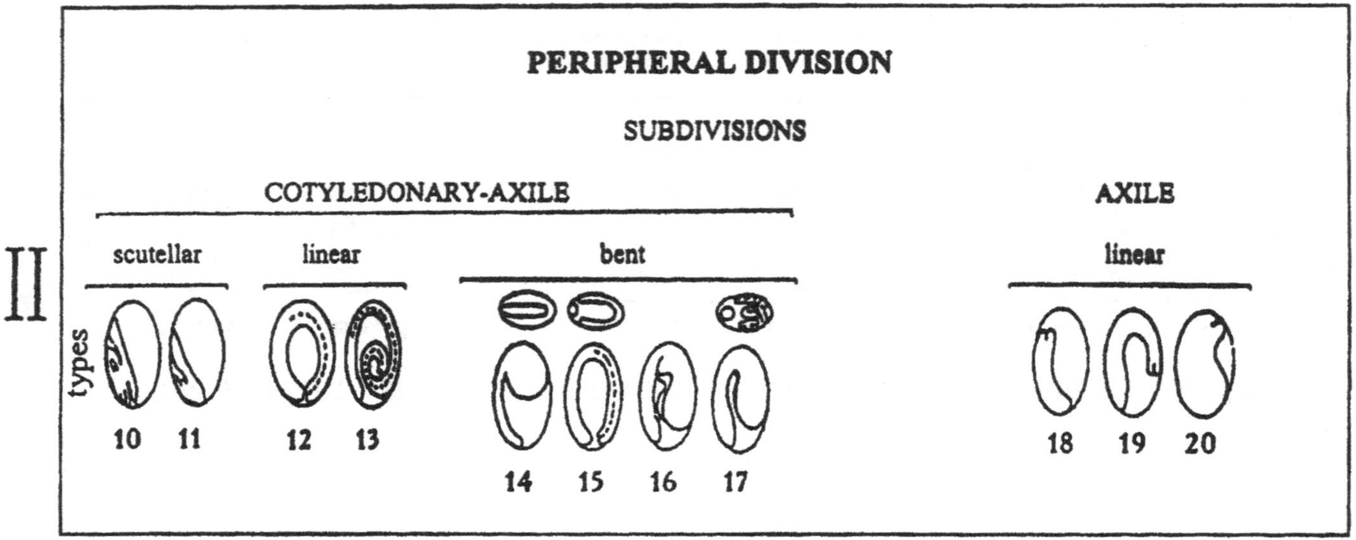

Fig. 50 (II): 10—*Zea* (Poaceae, M); 11—*Dyckia* (Bromelliaceae, M); 12—*Gypsophylla* (Caryophyllaceae, D), *Spergula* (Caryophyllaceae, D); 14—*Polygonum* (Polygonaceae, D); 15—*Mirabilis* (Nyctaginaceae, D); 16—*Pereskia* (Cactaceae, D); 17—*Althea* (Malvaceae, D); 18—*Opuntia* (Cactaceae, D); 19—*Echinocactus* (Cactaceae, D); 20—*Mamillaria* (Cactaceae, D).

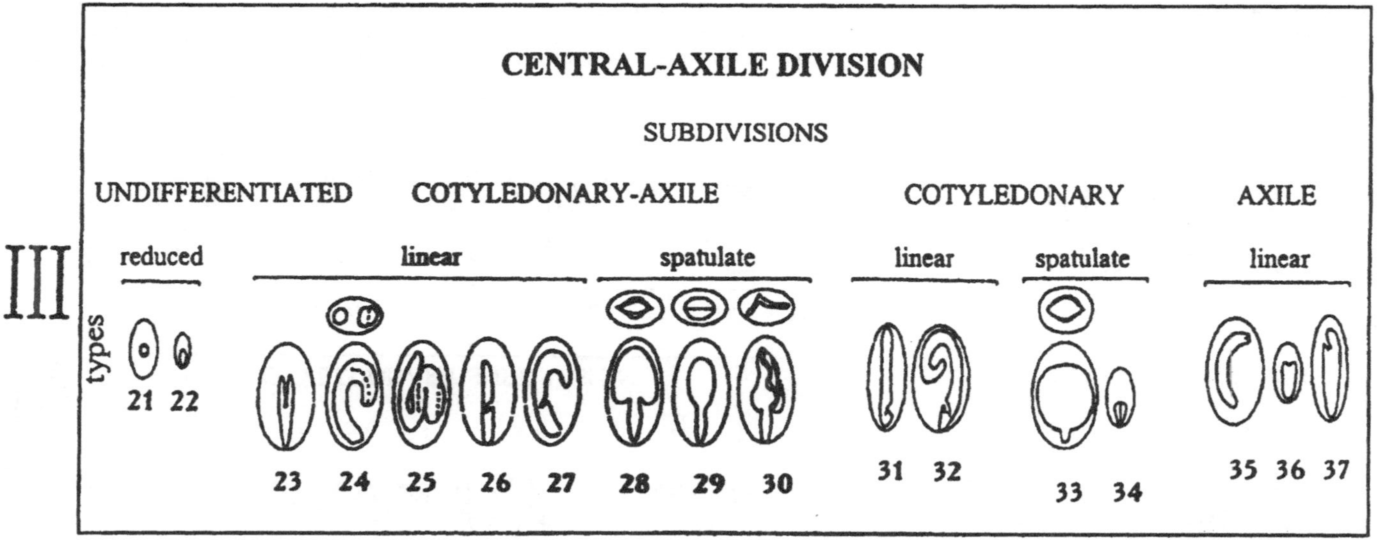

Fig. 50 (III): 21—*Balanophora* (Balanophoraceae, D); 22—*Rafflesia* (Rafflesiaceae, D), *Cytinus* (Cytinaceae, D); 23—*Actinidia* (Actinidiaceae, D), *Primula* (Primulaceae, D); 24—*Lycium* (Solanaceae, D), *Menispermum* (Menispermaceae, D); 25—*Helianthemum* (Cistaceae, D); 26—*Iris* (Iridaceae, M); 27—*Cordyline* (Dracaenaceae, M); 28—*Croton* (Euphorbiaceae, D); 29—*Oxalis* (Oxalidaceae, D); 30—*Bixa* (Bixaceae, D); 31—*Dianella* (Phormiaceae, M); 32—*Allium* (Alliaceae, M); 33—*Jatropa* (Euphorbiaceae, D); 34—*Rhododendron* (Ericaceae, D); 35—*Callitriche* (Callitrichaceae, D); 36—*Eustoma* (Gentianaceae, D); 37—*Tulipa* (Liliaceae, M).

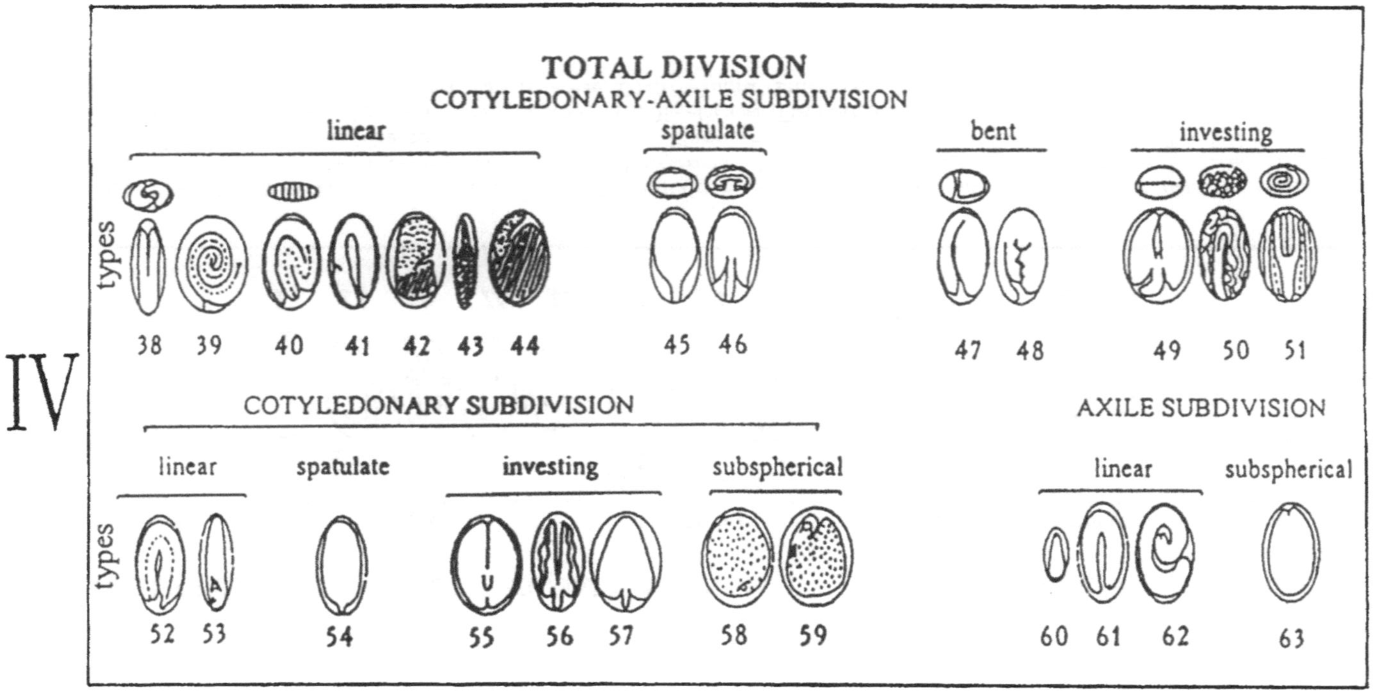

Fig. 50 (IV): 38—*Metrosideros* (Myrtaceae, D); 39—*Climacoptera* (Chenopodiaceae, D); 40—*Heliophila* (Brassicaceae, D); 41—*Hydrocleis* (Limnocharitaceae, M); 42—*Pistia* (Araceae, M); 43—*Posidonia* (Posidoniaceae, M), 44—*Ruppia* (Ruppiaceae, M); 45—*Pilea* (Urticaceae, D); 46—*Lafoensia* (Lythraceae, D); 47—*Trifolium* (Fabaceae, D); 48—*Gomidesia* (Myrtaceae, D); 49—*Cassytha* (Lauraceae, D); 50—*Pyrenaria* (Theaceae, D), *Memecylon* (Melastomaceae, D); 51—*Chimonanthus* (Calycanthaceae, D); 52—*Ouratea* (Ochnaceae, D); 53—*Aponogeton* (Aponogetonaceae, M); 54—*Rubus* (Rosaceae, D); *Clusia* (Clusiaceae, D); 55—*Camellia* (Theaceae, D); 56—*Combretum* (Combretaceae, D); *Juglans* (Juglandaceae, D); 57—*Trapa* (Trapaceae, D); 58—*Warneckea* (Melastomataceae, D); 59—*Peltandra* (Araceae, M); 60—*Hypocalymma* (Myrtaceae, D); 61—*Ternstroemia* (Theaceae, D); 62—*Campomanesia* (Myrtaceae, D); 63—*Pimenta* (Myrtaceae, D).

Plants' (1981, 1983, 1985, 1987, 1990) including the author's original data for a number of families, significantly increase our knowledge of the morphological diversity of the embryo. Taking into consideration these data and Martin's typology (1946) we propose a **new variant of classification of internal morphology of the seed**. The main purpose of this classification is to work out unified and relevant to various taxa ways of morphological description of the embryo by means of a complex of features significant for systematics and phylogeny.

The proposed classification (Fig. 50) includes three basic categories—divisions, subdivisions, types—and two additional categories—subtypes and varieties.

On the basis of **position of the embryo in the seed,** four divisions are distinguished: **basal, peripheral, central-axile and total.**

Subdivisions are distinguished for all divisions depending on **the degree and character of morphological differentiation of the embryo**. There are only six: **undifferentiated, weakly differentiated, differentiated, cotyledonary-axile, axile and cotyledonary**.

According to embryo shape, 9 types are distinguished: **rudimentary, reduced** (in these two types we have to diverge from the strict form criterion), **capitate, scutellar** (corymbose), **linear, spatulate, bent, investing, subspherical**. The types are repeated inside divisions and subdivisions. Distinguishing subtypes and varieties especially is quite arbitrary and depends on accuracy of embryo description required. Subtypes and varieties are distinguished on the basis of features **specifying the embryo shape characteristics**: linear or bending embryo and character of this bending; building and thickness of cotyledons; thickness (swelling) of the whole embryo and of its parts, and so forth.

Examples of variant occurrence are presented in the diagrams and for some genera of flowering plants the family and class indicated in the figure caption.

A morphological description of the embryo must consequently proceed on the basis of its attribution to: 1) one of four divisions, 2) one of six subdivisions, 3) one of nine types and, if necessary, 4) one of the subtypes, including the varieties. Such a description enables us to give detailed characteristics of the basic parameters of its structure.

Principles of Ecological Classification of Embryos

As studies on embryonic structures of parasitic flowering plants show (Teryokhin, 1965, 1968a,b, 1972, 1977; Teryokhin and Nikiticheva, 1968) the term 'behaviour' can be applied for description of the life mode of plant organisms. Leme (1976) pointed out that the ecology of both animals and plants includes two major aspects: (1) study of the character of the environment with which the organisms come into contact, i.e., mesological ecology or mesology and (2) study of the behaviour and reactions of organisms in this environment, i.e. ethological ecology or ethology.

The mutual effect of the environment in which plants live and their evolution can only be fairly explained in terms of ethological ecology. This allows consideration of creating an ecological classification of types of morphogenetic organization of embryos in flowering plants, analysing the adaptive and phylogenetic significance of the peculiarities of this organization in different species, and outlining the major elements of this classification.

In the embryology of plants several morphological classifications of embryos have been developed, which are more or less elaborate and based on peculiarities of embryo development or on the structure of mature embryos. Two of them (Schnarf, 1929; Souèges, 1939c) based on regularities of earlier stages of embryogenesis, are still valid nowadays. Further development of these systems (Johansen, 1950; Yakovlev, 1958; Rutishauser, 1969) has considerably completed or altered these classifications and concomitantly proved their limited potential in solving phylogenetic problems.

Souèges's classification has advantages over Schnarf's system because its subdivisions (periods, megarchetypes and series) form a well-organized system of morphologically interconnected modes of early embryogenesis. However, judging from more recent material, this orderliness reflects some regularities in paths of evolutionary transformations of embryonic structures rather than real phylogenetic connections between taxa. This becomes quite obvious when we compare Souèges's system with any currently existing phylogenetic system of angiospermous plants.

The so-called irregular types of embryogenesis undermine the phylogenetic significance of Souèges's system. As Johansen (1950) showed, the existence of 'irregular' types of embryogenesis cannot be fairly interpreted by hybridization (Souèges, 1961).

In later years attempts were made to 'liberalize' Souèges's classification (Crété, 1963; Lebègue, 1966; Méstre, 1967); however, they do not eliminate fundamental objections against the broad phylogenetic significance of general developments by Souèges.

Naturally, the question arises: Why don't the types of early embryogenesis of angiosperm plants reflect the real phylogeny of flowering plants? An answer can be found by clearing out the role and significance of early stages of embryogenesis in embryo development as well as peculiarities of their evolutionary transformations.

The list of families, characterized by the same type of early embryogenesis, evidently reveals the absence of correspondence between modern approaches to their genetic links and origin on the one hand, and uniformity of paths of early embryogenesis on the other.

It is noteworthy that two or more types of embryogenesis can be found in one and the same family (for example, in families Loranthaceae, Gentianaceae, Primulaceae, Balanophoraceae, Lauraceae, Orchidaceae). Lack of strict determination of ways of differentiation at early stages of embryogenesis has also been revealed within certain species of orchids, in particular, in *Cymbidium bicolor* (Swamy, 1942), *Epidendrum prismatocarpum* (Swamy, 1948c), *Angraecum distichum* (Veyret, 1958), *Neottia nidus-avis* (Teryokhin and Kamelina, 1969) and so on. The same effect has been found in the representatives of other families characterized by reduction of one or the other structure of a mature embryo, for example *Aeginetia indica* (Orobanchaceae), *Moneses uniflora* (Pyrolaceae), in some species of *Balanophora* (Balanophoraceae), *Cuscuta europaea* (Cuscutaceae) and some other species (Tiagi, 1952a; Teryokhin, 1962, 1977; Teryokhin and Yakovlev, 1967).

A complex analysis of embryogenesis revealed that most of the species possess a distinct structural organization of morphogenetic initials of basic organs and tissues of the mature embryo. On the contrary, in species not characterized by such a clear organization, one can register increase of variability of embryogenesis in its middle and early stages.

As Kamensky (1886) showed, reduction of the radicle and cotyledons and transformation of the epicotylar zone of the embryo goes along with corresponding reorganizations of middle and early stages of embryogenesis.

Some regularities of evolutionary transformations of embryonic structures in connection with change of ecological niche by angiospermous plants were studied in a group of families of parasitic plants as an example (Teryokhin, 1968a,b, 1972, 1977). In particular, it was revealed that transition of autotrophic flowering plants into parasitic ones triggers reconstruction of their individual development in connection with change of source and ways of obtaining nutrients. Analysis of transitional forms in families Scrophulariaceae, Orobanchaceae, Pyrolaceae, Orchidaceae showed that evolutionary transformations of embryos occurred in close connection and in accordance with the character of the functional and structural transformations of the sporophyte at the post-seminal stage of its development, during which transition to a new (parasitic) mode of nutrition occurred in ontogenesis. The peculiarities of the middle and early stages of embryogenesis changed accordingly with changes in the final stages of embryo development (Teryokhin, 1977).

Also, morphogenetic peculiarities of early stages of embryogenesis are evidently determined by the character of the structure of the ovule and the seed, which must predetermine the structure of the suspensor—a specialized haustorial apparatus of the embryo (see above: Suspensor). These interconnections manifest themselves most distinctly in representatives of families Cuscutaceae, Orchidaceae and Pyrolaceae. Interestingly, variability of the first division of the proembryo is observed only in species which lack a suspensor, for example in *Balanophora* (Teryokhin and Yakovlev, 1967).

Thus, the character of early embryogenesis is determined by two factors: peculiarities of the structure of the formed embryo and peculiarities of the ovule and the seed structure. In the evolutionary process, early embryogenesis results from a morphogenetic compromise between two complexes of factors: ways of nutrition of the sporophyte at the post-seminal stage and peculiarities of evolutionary adaptation of seeds with respect to their dispersion (since these peculiarities basically determine the structure of a mature seed).

The proposed considerations fairly explain the numerous coincidences (emphasized by Poddubnaya-Arnoldi, 1976) in modes of proembryo development in representatives of taxa which are not genetically linked, as well as their differences in representatives of closely related taxa.

There are at least three classifications of embryos in angiospermous plants based on the peculiarities of the structure of mature embryos. First of all, distinction of embryos into two groups: dicotyledonous and monocotyledonous. Considerable phylogenetic significance is attributed to the distinction in number of cotyledons, which is seen in almost every modern system of flowering plants. However, the adaptive significance of this feature is still not clear, and classification within each of the groups is not sufficiently elaborated (Grushvitsky, 1961, 1963; Smirnova, 1965). But the major factor is that cotyledons are appendicular structures and for that reason may not be able to reflect basic evolutionary transformations of axial embryo structures. Hence current data on differences in the structure of cotyledons in connection with their functions during germination suffice to serve as one element in ecological classification of embryos.

The most elaborated classification of embryos based on the study of their form, size and location in a mature seed is Martin's (1946). Since it is formalistic and does not take into account ectogenetic factors of embryo evolution, Martin's system cannot be included in ecological classification of embryos.

The basic evolutionary factors which condition changes in embryonic structures are apparently peculiarities of functioning (behaviour) of seedlings in relation to environment, their ethological ecology (Teryokhin, 1962, 1965, 1968a,b, 1972, 1975, 1977; Teryokhin and Nikiticheva, 1968, 1976; Nikiticheva, 1971). These peculiarities in turn are associated with changes in environment (mesological ecology). Any classification of embryos will reflect real evolutionary processes only if it takes into consideration the above-mentioned system of mutual dependence of interacting factors and changing structures. It should be noted that the embryo in a mature seed is functionally inactive, its interactions with the environment only become apparent when germination starts.

Various peculiarities of embryo organization are of different adaptive significance: some are of more general significance, others of more specific significance. These differences give a valuable possibility for distinguishing more or less ecologically (and, possibly, phylogenetically) significant characteristic features of embryos and enable research into the hierarchy of morphogenetic structures. The hierarchy of adaptive significance of constructive features was emphasized by Schmalhausen (1939), and his latest works on control and regulation of the evolutionary process (Schmalhausen, 1958, 1968) were one of the starting points for the approaches presented in this study.

Therefore, it is necessary to note the difficulties in creation of a phylogenetic classification of embryos mentioned by Poddubnaya-Arnoldi (1964a) and the paucity of our knowledge of processes of germination of angiospermous plants. In the creation of classifications of the processes of embryo germination, attention is mainly paid to epigeal and hypogeal germination (Klebs, 1885; Boyd, 1932; Grushvitsky, 1963). However, many unclear issues still remain. Thus, Klebs distinguished groups with epigeal and hypogeal cotyledons, whereas Grushvitsky and other authors think it necessary to distinguish basic ways of germination on the basis of functions of cotyledons: during hypogeal germination cotyledons perform one of two functions (storing or haustorial) and die off while in the seed; during epigeal germination cotyledons always perform two functions: storing (or haustorial) and assimilating. We believe it might be more reasonable to distinguish two special forms of hypogeal germination with only haustorial and only storing function rather than to divide germinations into epigeal and hypogeal. Naturally, both types of germination (meaning that cotyledons may/may not perform an assimilating function) are often registered in representatives of one and the same genus (for example in genera *Phaseolus, Acer, Caesalpinia, Prunus, Sophora* and *Peperomia*) or even in one and the same species (*Clematis recta*). This is evidence of high variability of the feature. At the same time, the storing or haustorial function of cotyledons and, accordingly, the absence or presence of an endosperm in mature seeds characterizes large taxa of angiosperms in the range of families and orders.

Klebs (1885) was the first to divide seed plants into three basic groups (with two or more cotyledons, with one cotyledon and with reduced cotyledons), attempting to link these peculiarities of embryo structure with certain peculiarities of germination.

He failed because there are several specific types with respect to structural organization of embryos and ways of germination in each of these groups. Naturally, all these differences are of adaptive character.

At the same time, an ecological and phylogenetically significant classification of embryos cannot be limited to structural or functional changes of cotyledons. It is necessary to analyse all the basic structures and functions of the sporophyte, both in the well-developed embryo and in the process of its germination.

Embryos in angiospermous plants have undergone considerable and different changes in the course of evolution. These changes concern the following aspects: (1) morphogenetic potentials of axial structures of embryos; (2) morphogenetic activity of axial structures of the embryo (rest period, temp of development); (3) degree of differentiation of axial structures of the embryo; (4) specific differentiation of axial structures of embryos; (5) structure of suspensor apparatus; (6) cotyledon structure (degree of its development, presence of reserve substances); (7) embryo size, (8) relative proportions of basic structures of the embryo; (9) shape of the embryo; (10) size ratio of endosperm and embryo as well as embryo location in the seed and (11) character of reserve materials.

Recently, differences in the paths of embryo development at early stages as well as in peculiarities of formation of the suspensor apparatus have been almost completely revealed.

The least worked out problems are: **(1) interconnections of evolutionary changes of embryos and factors of environment (factors of embryos evolution); (2) interconnections of different stages of embryogenesis in evolutionary transformations of embryo organs (morphogenetic interconnections) and (3) clarification of comparative ecological (and, probably, phylogenetic) significance of different aspects of these changes (hierarchy of structures and factors).**

As far as the first problem is concerned, it is interesting to note the lack of attempts to link the data on evolutionary changes of embryos with data on peculiarities of germination in these plants, albeit their connections are quite clear in certain cases (Salisbury, 1942; Korchagina, 1964; Teryokhin, 1968a,b, 1977; Nikiticheva, 1971).

Judging from our experience in the study of evolution of embryos in parasitic and insectivorous flowering plants, it is impossible to reveal regularities in this evolution without addressing data on germination in plants, peculiarities of their mode of life and environment (Teryokhin, 1977).

As for the second problem, its elaboration is slow since there is no summary of our knowledge in many aspects of evolutionary changes of embryos and, first of all, because of the lack of necessary data on the structure of fully formed embryos (Poddubnaya-Arnoldi, 1964a).

For some historical reasons, the third problem was worked out unidirectionally. It is widely known that at present botanists in the field of evolution and in particular, embryologists, believe the change in number of cotyledons in the embryo to be of the highest phylogenetic importance.

In our opinion many authors overlook more vital transformations of axial organs of embryos. This is probably due to the fact that such transformations are only revealed in combined study of the structure of formed embryos and their germination. One could think that, on the whole, such changes play a more important role in embryo evolution than changes of cotyledons, i.e. appendicular structures of the embryo.

On the basis of such a complex analysis we have tried to reveal some common, ecologically important changes of axial embryo organs and compare them with changes in the structure and function of cotyledons.

Embryos in flowering plants can be divided into five main types (Fig. 51), which differ in nature of morphogenetic potentials of axial structures:

type I—with bipolar morphogenetic activity of the embryo during germination at the expense of the epicotyl and radicle;

type II—with bipolar morphogenetic activity of the embryo during germination at the expense of the epicotyl and adventive roots (radicle has lost its morphogenetic potentials);

type III—with bipolar morphogenetic activity of the embryo during germination at the expense of the radicle and adventive shoots (epicotyl has lost its morphogenetic potentials);

type IV—with unipolar morphogenetic activity of the epicotylar pole of the reduced embryo during germination (the hypocotyl-radicle zone of the embryo has lost its morphogenetic potentials);

type V—with unipolar morphogenetic activity of the radicular pole of the reduced embryo during germination (the epicotylar pole of the embryo has lost its morphogenetic potentials);

Naturally, characteristics of these major types do not include comparative data on levels of organization of apical structures, on temps of morphogenetic activity during germination and a number of more specific features, which have not yet been fully studied. Structural and functional peculiarities of the appendicular organs of the embryo (cotyledons) have been studied more thoroughly; thus we compared the

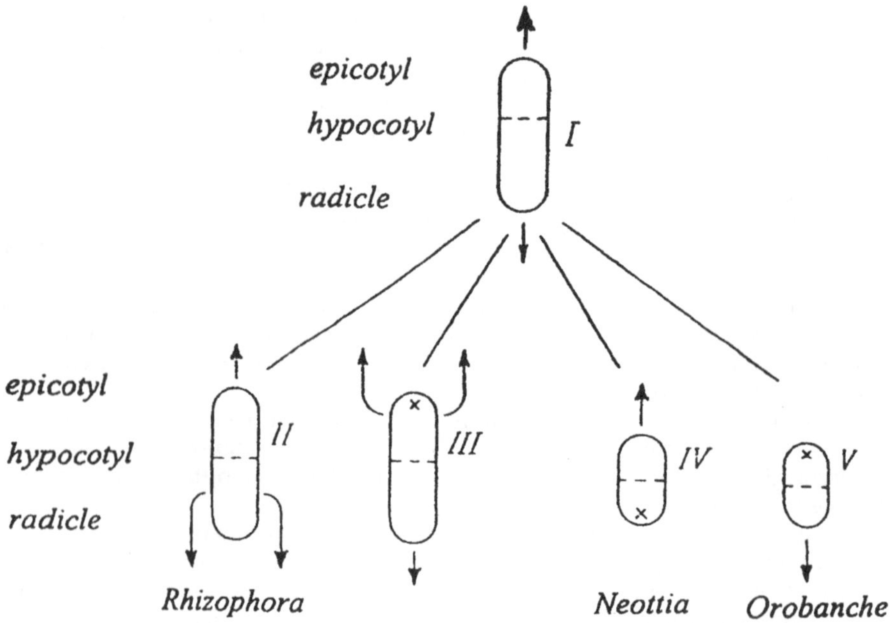

Fig. 51: Schematic representation of embryo types in flowering plants according to the morphogenetic potentials of axial structures (after Teryokhin and Nikiticheva, 1979).

major distinguished types with the data available on the structure and function of cotyledons. This comparison can be considered one of the first steps towards clarification of complex interconnections of the structures and functions of the embryo in angiospermous plants.

Each of the five listed types of morphogenetic organization of axial organs of embryos corresponds to a certain spectrum of morphogenetic and functional changes of cotyledons if formed.

The main morphogenetic changes in cotyledons include changes in their number, comparative morphogenetic potentials (isocotyly and heterocotyly) and their reduction up to complete elimination. As can be seen (Table 5), each of five basic types of morphogenetic structure of embryos includes variations of structures and functions of cotyledons and, of course, other structural and functional peculiarities.

We thought it important to show that the adaptive significance of one or the other type of morphogenetic organization of embryos can be revealed as a result of studies on its variations within one type. The first three types include different variations of autotrophic mode of life, the adaptive significance of which is perceived (through studies in cotyledons and their functions during germination) in the fullest measure in variations of type I. Some other peculiarities of seedling existence participate in the evolutionary formation of the structure of embryos of the types II and III. Types IV and V are conditioned by a parasitic mode of life.

We shall now consider some examples of correspondence between structural organization of embryos and peculiarities of their germination, which reveal the adaptive significance of the system as a whole and the leading role of ecological (ethological) factors in its formation.

Phaseolus embryos are characterized by well-developed axial organs and by developed cotyledons which have reserving and assimilating functions. Such an embryo structure corresponds to 'normal' bipolar seedling development with high morphogenetic activity of both axial apices in germination.

Rauh (1937) was the first to show that in the evolutionary process, side shoots often replace the dying main shoot in the process of individual development. Many examples of such replacements were gathered in studies of life forms of plants. Parasitic flowering plants were taken as an example and it was found that such substitutions could occur in their ontogenesis as well (Teryokhin and Nikiticheva, 1968; Nikiticheva and Teryokhin, 1976; Teryokhin, 1977). However, such crucial transformations the structure and in morphogenetic processes of axial structures of embryos are not necessarily associated with a parasitic mode of life. For example, in coastal plants (*Barringtonia* species) and mangrove plants (*Rhizophora* species) the process of substitution of the main root by adventive roots can be traced; the same is typical of cereal crops. The adaptive significance of heterocotyly is pronouncedly revealed in studies of the embryo and seedling structure in *Trapa natans* (Titova, 1992).

It is also known that embryos of certain flowering plants lost some apical structures as a result of reduction and development of such plants at the first stage of germination is strictly unipolar either at the expense of morphogenetically active structures at the epicotylar pole of the embryo (*Utricularia, Aldrovanda, Cuscuta,* Orchidaceae), or at the expense of morphogenetically active (but evolutionarily transformed) structures at the radicular pole of the embryo *(Pyrola, Orobanche).* The adaptive significance of these transformations is quite clear, just as is the adaptive significance of underdevelopment (but not of reduction) of embryos in a number of

Table 5: Classification of embryos by morphogenetic structure types and functions of cotyledons

Basic types of morphogenetic organization of embryos	Number and state of embryo cotyledons in mature seed	Cotyledon functions				
		Reserving and photosynthesizing	Reserving	Haustorial and photosynthesizing	Haustorial	Reduced
I. With bipolar morphogenetic activity	Two and more	*Phaseolus vulgari*	*Quercus*	*Euphorbia, Melampyrum*	*Tozzia*	–
	One	*Alisma*	+	*Allium*	*Canna*	–
	Reduced or absent	–	–	*Striga hermonthica*	*Lathraea*	+
II. With bipolar morphogenetic activity at the expense of shoot apex and adventive roots	Two and more	+	*Cassytha*	*Drosera, Typha*	+	–
	One	*Pinguicula*	+	*Triticum, Carex*	+	–
	Reduced or absent	–	–	–	–	*Barringtonia*
III. With bipolar morphogenetic activity at the expense of radicle and adventive shoots	Two and more	+	+	*Viscum, Bartsia*	+	–
	One	+	+	+	+	–
	Reduced or absent	–	–	–	+	+
IV. With unipolar morphogenetic activity of epicotylar pole	Two and more	+	*Aldrovanda*	+	+	–
	One	*Bletilla*	+	+	+	–
	Absent	–	–	–	–	*Utricularia, Cuscuta, Orchis*
V. With unipolar morphogenetic activity of radicular pole	Rudiments (2)	–	–	–	*Striga gesnerioides*	+
	Rudiments (1)	–	–	–	+	+
	Absent	–	–	–	–	Orobanchaceae Rafflesiaceae Cynomoriaceae

Note: The possibility for plants to exist with a certain type of morphogenetic structure of embryos with certain cotyledon function is marked by plus (+), or minus (-) impossibility for certain types of embryo structures to exist.

ephemeral plants (*Erythronium* and others). However, the correspondence between the peculiarities of germination and the structure of embryos as well as their adaptive significance has not been studied to the extent necessary in many plants.

Creation of ecological-morphological classification of embryos is not only possible but also **necessary,** since such studies can and must **contribute to understanding the evolution and phylogeny of flowering plants.**

Reduced and Undifferentiated Embryos

Investigators in the field of embryology and morphology still have no clear understanding of the differences between embryos with reduced organs and so-called 'underdeveloped' embryos which lack basic organs in the mature seed, but develop normally after seed abscission from the maternal plant (Eames, 1961; Poddubnaya-Arnoldi, 1976; Natesh and Rau, 1984). However, reduced embryos and embryos with reduced organs are the most striking examples of adaptability of evolutionary transformations.

We consider the process of reduction a process of successive (in number of generations) reduction in development of a certain structure which has lost or changed its main function. As a result of absolute loss of its functions, a structure completely disappears (eliminates) from ontogenesis; when a structure changes the main function, its rudiments (initials) remain to a certain extent. In complicated structures we distinguish two basic directions of reduction: a general reduction and a selective reduction (Teryokhin, 1968b, 1977, 1991). In case of selective reduction (with respect to embryo structures) we are speaking about embryos with reduced organs. For example, in embryos of orchid plants the radicle is reduced and initials of the hypocotyl develop into a special reserving tissue, which, apart from its main function, provides a contact with mycorrhizal fungus.

In the case of general reduction we deal with reduced embryos (Fig. 52) which lose or change functions of all the organs (epicotyl, cotyledons, hypocotyl and radicle) in many (but not all) highly specialized parasitic flowering plants. For example, loss of functions of photosynthesis leads to reduction of cotyledons and epicotyl, loss or change of functions by the radicle leads to transformation of rudiments of the radicle into haustoria-forming organs, loss or change of the basic functions by the hypocotyl leads to development of new structures from its rudiments (initials)—'tubercle' or 'tuber' (reserving organs with the function of initiation of the adventive reproductive shoots and haustoria-forming organs). The reduced embryo in the mature seed of highly specialized parasitic plants is an absolutely new structure with functions that differ fundamentally from those of the embryo of autotrophic plants.

The clue to understanding the origin of reduced embryos and embryos with reduced organs lies in the ecological peculiarities of those species of plants in which such embryos are formed. In particular, the causes of evolutionary changes of embryos leading to the reduction are change of behaviour and mode of development of a seedling.

One can successfully study these complicated evolutionary transformations only if one combines morphological and ecological investigations. That is why ecological embryology has a status of a discipline able to 'decode' the biological sense of evolutionary transformations of embryogenesis (Teryokhin, 1977, 1991).

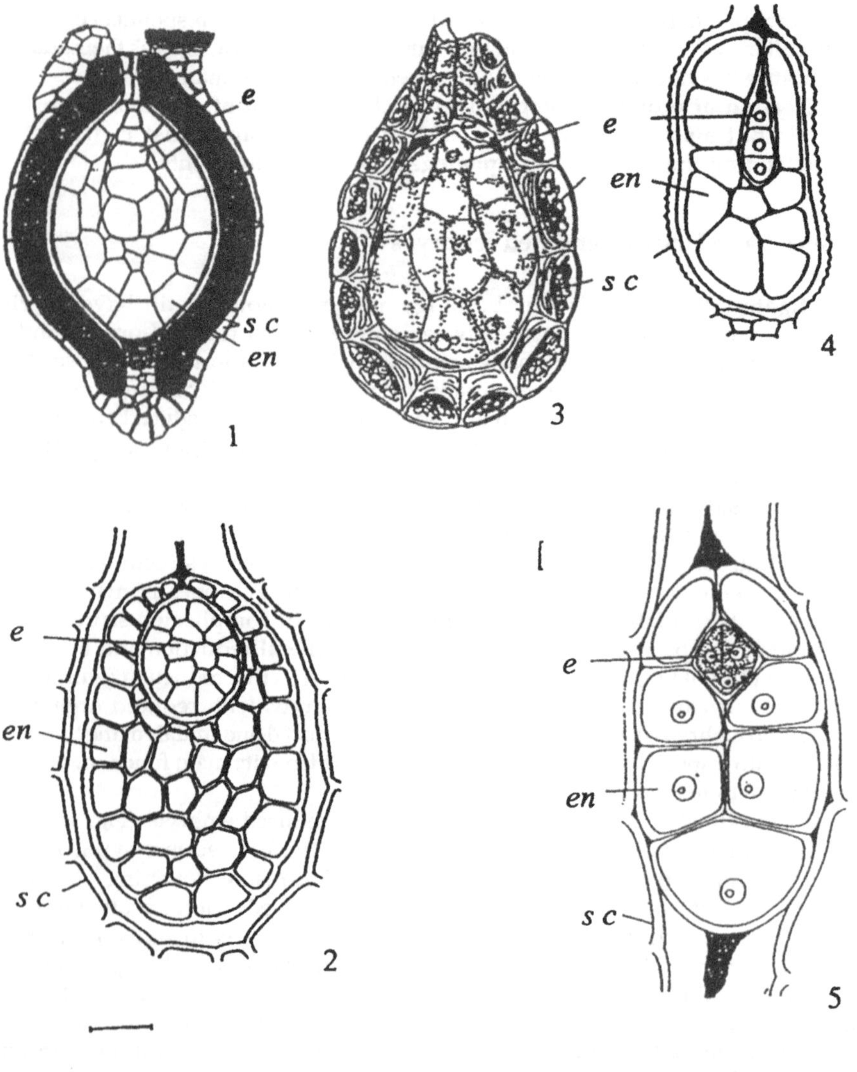

Fig. 52: Reduced embryos: 1-3—in root parasitic plants: *Apodanthes pringlei* (1), *Orobanche hederae* (2), *Mitrastemon yamamotoi* (3); 4,5—in mycoparasitic plants: *Leiphaimos auranthica* (4), *Hypopitys monotropa* (5) (1, 2, 4, 5—after Teryokhin, orig., 3—after Watanabe, 1934). *e*—embryo, *en*—endosperm, *s c*—seed coat. Bars: 2—0.05 mm, 5—10 μm.

Embryos with reduced organs (Fig. 53)

Embryos with reduced organs have been observed in parasitic plants (in xeno- and mycoparasitic as well) and in free-floating and submerged water plants. In taxa of terrestrial autotrophic plants, reduction of one of the axial structures (radicle) has only been discovered in representatives of family Poaceae, which might be determined by water or mycotrophic mode of life of their ancestor forms. There is no evidence of

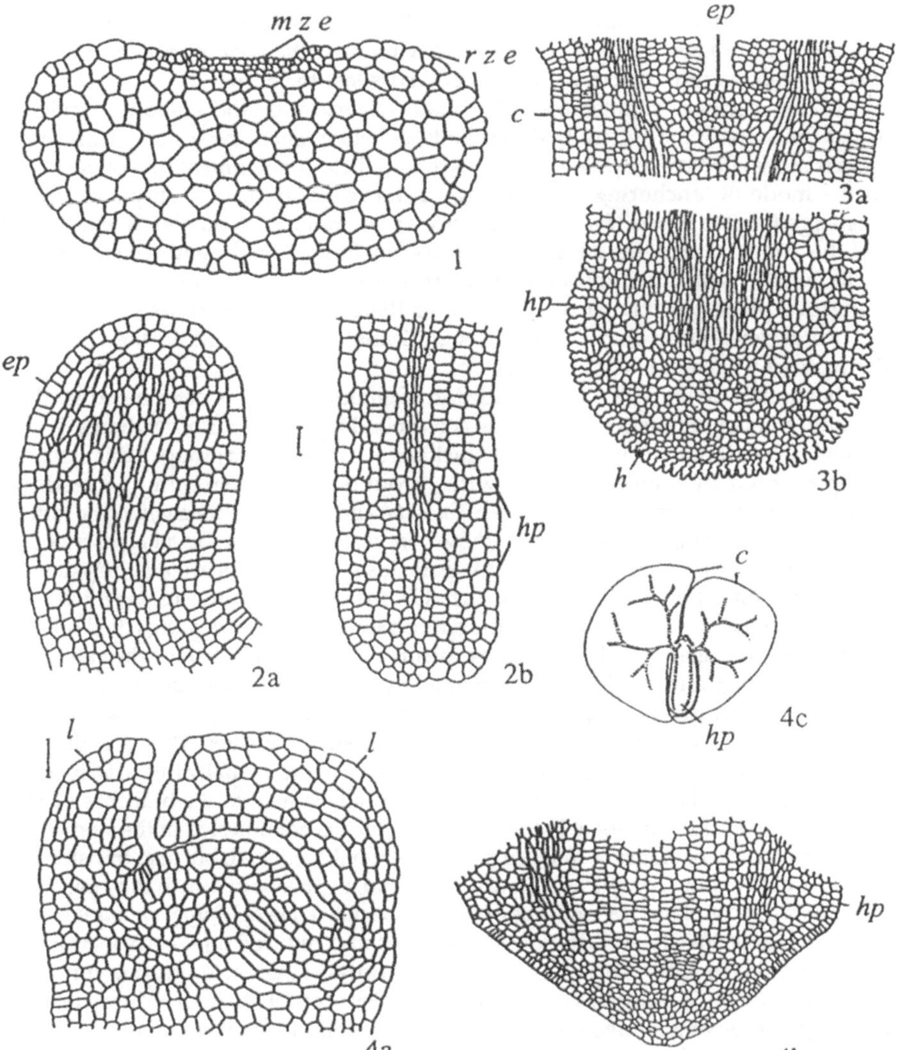

Fig. 53: Embryos with reduced organs: 1—in aquatic plant *Utricularia vulgaris*, 2-4—in stem parasitic plants: *Cuscuta europaea* (2a, b), *Viscum album* (3a, b), *Cassytha filiformis* (4a-c). *c*—cotyledon, *ep*—epicotyl, *hp*—hypocotyl, *h*—haustorium, *l*—leaf, *m z e*– micropylar zone of embryo, *r z e*—reserving zone of embryo. Bars: 1-4b—0.05 mm.

the fact that this is the insectivorous mode of life of some free-floating water plants (*Aldrovanda*, *Utricularia*) that somehow results in the reduction of embryos. And, on the contrary, the mode of life of these free-floating water plants may cause reduction and complete elimination of the radicle of their embryo, since its functions are redundant in a free-floating mode of life. On the other hand, the free-floating mode of life may favour more effective fulfillment of insectivorous nature, enlarging the territorial control of these plants over the nutrient resources. *Utricularia* advanced much further in its specialization to a new mode of life in water environment than *Aldrovanda*. In *Utricularia*, even cotyledons proved redundant and embryos lost them, too.

Many species of submerged water plants (*Ruppia* and others) show a tendency to lose functions of the radicle. In these plants, development of the system of adventive roots is more effective for living processes than radicle development (the cause might be the mode of 'anchoring' of seedlings in loose ground). In highly specialized submerged water plants (for example *Zostera*), the radicle is completely eliminated and the primordium of the adventive root is often formed on the embryo epicotyl (or on its border with the hypocotyl). Similar adaptive changes are typical of some mangrove plants. In 'viviparous' species of *Rhizophora*, in the course of intensive hypocotyl development, the radicle is significantly reduced, and in the species of *Barringtonia* it is completely missing.

In another category of plants, reduction of embryo organs is determined by their parasitic manner of nutrition and by the corresponding behaviour of seedlings. It is at the stage of seedlings that transition to parasitism occurs. The first group of this category is represented by stem parasites (with 'ground-based' beginning of seedling development) from families Cuscutaceae (*Cuscuta*) and Lauraceae (*Cassytha*). The radicle of embryos is missing in the species of these genera. In species of *Cuscuta* (described in about 100 species), more specialized for parasitism, cotyledons are also missing. In *Cassytha* the transition to parasitism occurred by means of adventive roots which were transformed into haustoria and, later, the entire basal part of the seedlings died off. The evolutionary changes of embryos of these plants are clearly adapted for the behaviour of their seedlings and, finally, are determined by transition to stem parasitism.

The second group of parasitic taxa from order Santalales (Loranthaceae, Mysodendraceae, Viscaceae) is also represented by stem parasites. They are all green plants, finishing their life cycle on branches of the host plants (usually trees). Dispersion of fruits of these plants is provided by birds, and attachment to the branches of host plants by a special sticky tissue in the pericarp, which is not digested in the stomach of birds. The radicle of embryos in these plants is not only reduced, but also transformed into a germ of a quite complicated haustorial organ, by means of which they penetrate the tissues of host plants. In embryos of higher specialized taxa (*Phrygilanthus aphyllus*, some species of *Viscum* and *Arceuthobium*) leaves also reduced and, correspondingly, cotyledons also reduced. This is caused by peculiarities of their heterotrophic nutrition.

Family Orchidaceae gives interesting examples of embryos with reduced organs. Orchid plants belong to mycoparasitic monocotyledons. In embryos of all the species examined, the radicle is reduced and the hypocotyl transformed into reserving tissue. After orchid seed germination the hyphae of mycorrhizal fungus penetrate the cells of the basal part of the seedling and here their 'digestion' begins. Mycoparasitism is a form of mutual parasitism in which through antagonistic relations a mutual exchange of different groups of nutrients occurs between the flowering plant and the fungus. According to the degree of adaptation to mycoparasitism, among orchid plants several levels and several groups of taxa corresponding to these levels were distinguished (Teryokhin, 1977). This degree is reflected not only in radicle reduction and the hypocotyl transformation, but also in level of underdevelopment of such embryo organs as cotyledon and epicotyl in a mature seed. Here we are not speaking about reduction, but rather the underdevelopment of these organs, because they either complete development or even start development already after germination, in the process of seedling development. Let us pay special attention to this since even in highly specialized species of achlorophyllous (and, consequently, leafless) orchid plants, such as the species of *Neottia, Corallorhiza, Epipogium*, morphogenetic potentials to develop a reproductive shoot in a manner typical of flowering plants are maintained in the apical area of the embryo. A list of taxa, in which certain organs are reduced is given below. The degree and form of this reduction depend on the level and direction of ecological specialization of taxa.

DICOTYLEDONS

1. Cuscutaceae: all species of *Cuscuta*.
   Ecological niche: achlorophyllous herbaceous lianas, stem parasites (with ground-based beginning of seedling development).
   Embryo: radicle and cotyledons completely lacking.
   Teryokhin, 1977.
2. Lauraceae: *Cassytha filiformis*.
   Ecological niche: stem parasite with green leaves (liana) and ground-based start of seedling development.
   Embryo: radicle apex reduced.
3. Loranthaceae: all species of *Amyema, Lysiana, Loranthus, Phrygilanthus* and others.
   Ecological niche: stem (rarely, root) parasites with green leaves; sometimes leaves reduced to scales.
   Embryo: with developed cotyledons, usually with manifested epicotyl and conductive system in an axis (sometimes, in cotyledons as well); radicle reduced and substituted by meristem tissue with papillar epidermis.
   Dixit, 1958; Prakash, 1961; Teryokhin, 1977.
4. Mysodendraceae: all species of *Mysodendron*.
   Ecological niche: stem parasites with green leaves.
   Embryo: radicle reduced and transformed into a haustorial initial.
   Kuijt, 1969.
5. Viscaceae: all species of *Viscum, Phoradendron, Korthalsella, Arceuthobium* and others.
   Ecological niche: stem parasites with green leaves; sometimes leaves reduced.
   Embryo: with a different degree of reduction of radicle and (sometimes) cotyledons.
   Kuijt, 1969; Teryokhin, 1977
6. Rhizophoraceae: species of *Rhizophora, Barringtonia*.
   Ecological niche: mangrove plants.
   Embryo: with reduced radicle, substituted by system of adventive roots.
   Teryokhin, orig.
7. Droseraceae: *Aldrovanda vesiculosa*.
   Ecological niche: free-floating water predatory plant.
   Embryo: radicle completely reduced and substituted (with a part of hypocotyl) by reserving tissue.
   Teryokhin, 1985b.
8. Lentibulariaceae: all species of *Utricularia*.
   Ecological niche: free-floating water predatory plants.
   Embryo: radicle and epicotyl completely reduced and substituted by reserving tissue, with primordia of vegetative organs in the apical area of the embryo; cotyledons eliminated.
   Nikiticheva, 1985.

Monocotyledons

1. Poaceae: all species of *Poa, Secale, Triticum, Oryza, Zea* and others.
   Ecological niche: terrestrial autotrophic herbaceous plants.
   Embryo: radicle reduced and transformed into coleorhiza and substituted by adventive roots.
   Pashkov, 1951; Von Guttenberg, 1957; Teryokhin, 1991.
2. Orchidaceae: all species of *Bletilla, Sobralia, Thunia, Habenaria, Vanda, Orchis, Ophrys, Neottia, Pogonia, Gastrodia, Corallorhyza, Epipogium* and others.
   Ecological niche: green or achlorophyllous herbaceous plant mycoparasites.
   Embryo: radicle completely eliminated and substituted by a mycohaustorial structure; with respect to degree of underdevelopment of apical structures, the embryos can be divided into 5 groups, depending on level of adaptation for mycoparasitism.
   Teryokhin and Nikiticheva, 1968; Teryokhin, 1977; and others.
3. Hydrocharitaceae: species of *Ottelia* and others.
   Ecological niche: submerged water plants.
   Embryo: radicle eliminated, basal part of swollen hypocotyl is transformed into reserving tissue.
   Teryokhin, 1985c.
4. Ruppiaceae: species of *Ruppia*.
   Ecological niche: submerged water plants.
   Embryo: radicle completely reduced and transformed (with part of hypocotyl) into reserving tissue.
   Teryokhin, 1985d.
5. Zosteraceae: all species of *Zostera* and *Phyllospadix*.
   Ecological niche: submerged water (sea) plants.
   Embryo: radicle completely reduced and substituted by original lobed reserving structure.
   Taylor, 1957a, b; Teryokhin, 1985f.

Reduced embryos

Embryos with general reduction are found only in ancient representatives of xenoparasitic and mycoparasitic plants. The reduced embryos completely lack all the organs typical of the embryos of autotrophic plants (radicle, hypocotyl, cotyledons and epicotyl). They usually lack conducting tissues as well. Thus, here, the general reduction of an embryo (with respect to its mass and structure) may reach the highest degree. For example, in the *Allotropa virgata* and *Hypopitys monotropa* (Monotropaceae) the embryo in the mature seed consists only of 2-3 undifferentiated cells.

Reduced embryos differ from embryos of other types in several principal features:

1) Such embryos are only typical of taxa with the highest level of specialization to parasitism.
2) Even at the highest degree of reduction, the basal ('radicular') pole of the reduced embryo maintains morphogenetic potential. The early stages of seedling development from reduced embryos are always characterized by unipolar development. This is associated with an urgent need to search for an appropriate

nutritive substrate (host plant or mycorrhizal fungus), since these parasitic plants are not able to form their own photosynthetic structures and obtain all their nutrition from their partners in parasitic symbiosis.

3) In connection with a high degree of adaptation for parasitism, development of seedlings from reduced embryos of the parasitic plants occurs on the basis of metamorphosis of their structure during the period of assimilation of a nutritive substrate—the partner in parasitic symbiosis (Teryokhin, 1968b, 1977, 1991).

Mycoparasitic flowering plants, for example species of *Pyrola* (Lück, 1940; Lihnell, 1942), are distinguished because the unipolar (often branched) seedling with a simplified structure lives for some time in soil, waiting for 'the attack' of mycorrhizal fungus, and its development continues only after establishment of parasitic symbiosis with this fungus.

So, all the cases of general reduction of embryos are conditioned by long-term evolutionary adaptation for the parasitic mode of life.

PLANTS—XENOPARASITES:

1. Balanophoraceae: all species of *Balanophora, Chlamydophytum, Corynaea, Dactylanthus, Ditepalanthus, Hachettea, Helosis, Langsdorffia, Mistropetalon, Rhopalocnemis, Sarcophyte, Scibalium* and others.
   Ecological niche: achlorophyllous perennial herbaceous plants, root parasites.
   Embryo: with different degrees of reduction:
   1) before stage of globular embryo (with differentiated embryoderm) composed of 40-50 cells in longitudinal section (*Helosis mexicanum, Langsdorffia hypogaea, Sacrophyte sanguinea*);
   2) before stage of spherical proembryo composed of 4-6 cells in longitudinal section, with differentiated embryoderm (*Balanophora abbreviata*).
   Fagerlind, 1945a,b; Teryokhin, 1977, 1988a.

2. Cynomoriaceae: *Cynomorium coccineum*.
   Ecological niche: achlorophyllous perennial herbaceous plants, root parasites.
   Embryo: spherical, flattened out from sides of poles, multicelled, with indistinct embryoderm and polar differentiation of cells.
   Teryokhin *et al.*, 1975; Teryokhin and Nikiticheva, 1976.

3. Hydnoraceae: all species of *Hydnora* and *Prosopanche*.
   Ecological niche: achlorophyllous, probably perennial herbaceous plants, root parasites.
   Embryo: globular, non-multicelled proembryo, without differentiation of embryoderm, but with suspensor remaining in mature seed (*Hydnora africana*).
   Teryokhin, 1988b.

4. Lennoaceae: all species of *Ammobroma, Lennoa, Pholisma*.
   Ecological niche: achlorophyllous, probably perennial herbaceous plants, root parasites.
   Embryo: spherical, with embryoderm and indistinct polar differentiation of cells (*Lennoa coerulea, Pholisma arenaria*).
   Teryokhin, 1977.

5. Orobanchaceae: all species of *Aeginetia, Boschniakia, Christisonia, Cistanche, Conopholis, Diphelypaea, Epifagus, Gleadovia, Kopsiopsis, Mannagettaea, Phacellanthus, Xylanche*.
   Ecological niche: achlorophyllous, perennial herbaceous plants, root parasites.

Embryo: with different degrees of reduction:
1) before stage of multicelled globular embryo, with or without embryoderm and initials of radicle, with polar differentiation of cells (*Cistanche phelypaea, Phacellanthus tubiflorus*);
2) before stage of oval-spherical embryo (late globular stage), with differentiated embryoderm (15-30 cells in longitudinal section), with more or less manifest polar differentiation of cells (*Cistanche tubulosa, Orobanche purpurea, Diphelypaea coccinea*).
3) before stage of oval proembryo composed of 9-10 cells in longitudinal section, no differentiation of embryoderm (*Aeginetia indica*).
Tiagi, 1952a,b; Teryokhin, 1977.

6. Rafflesiaceae: all species of *Apodanthes, Brugmansia, Cytinus, Mitrastemon, Pilostyles, Rafflesia* and others.
Ecological niche: achlorophyllous, usually perennial herbaceous plants, root or stem parasites.
Embryo: with different degrees of reduction:
1) before stage of oval proembryo with no indication of polar differentiation, composed of 8-9 cells in longitudinal median section (*Rafflesia arnoldii*);
2) before stage of octants with polar differentiation of cells (*Apodanthes pringlei, Cytinus ruber, Pilostyles thurberi, P. ingae*);
3) before stage of four-celled *proembryo (Mitrastemon yamamotoi*).
Teryokhin, 1977, 1988c.

7. Scrophulariaceae: some species of *Striga*, all species of *Harveya* and *Hyobanche*.
Ecological niche: achlorophyllous, perennial herbaceous plants, root parasites.
Embryo: globular with differentiated embryoderm and weak polar differentiation of cells (*Harveya coccinea, H. obtusifolius, Hyobanche spp., Striga gesnerioides*).
Teryokhin, 1977.

Plants—mycoparasites:

1. Pyrolaceae: all species of *Chimaphila, Moneses, Orthilia, Pyrola*.
Ecological niche: perennial herbaceous plants, often with reduced leaves, mycoparasites.
Embryo: two levels of general reduction:
1) oval multicelled embryos with differentiated embryoderm and polar differentiation of cells (*Orthilia secunda, Pyrola minor*);
2) oval non-multicelled proembryo without differentiated embryoderm, with distinct polar differentiation of cells (*Moneses uniflora*).
Teryokhin, 1977.

2. Monotropaceae: all species of *Allotropa, Hypopitys, Cheilotheca, Hemitomes, Monotropa, Monotropsis, Monotrapastium, Pleuricospora, Pterospora, Sarcodes* and others.
Ecological niche: achlorophyllous perennial herbaceous plants, mycoparasites.
Embryo: two levels of general reduction:
1) before stage of globular proembryo with differentiated embryoderm and weak polar differentiation of cells (*Pterospora andromeda, Sarcodes sanguinea*);
2) before stage of 2-5-celled proembryo (*Hypopitys monotropa, Monotropsis uniflora, Allotropa virgata*).
Copeland, 1939,1941; Teryokhin, 1977

3. Gentianaceae: *Bartonia, Cotylanthera, Leiphaimos, Obolaria, Voyria, Voyriella*.

Ecological niche: achlorophyllous perennial herbaceous plants, mycoparasites.
Embryo: two levels of general reduction:

1) before stage of few-celled oval embryo with differentiated embryoderm, 9-10-celled in longitudinal median section (*Bartonia virginica);*
2) before stage of 3-4-celled proembryo (*Voyra mexicana, Leiphaimos auranthicus*).

Oehler, 1927; Teryokhin, 1977.

4. Lobeliaceae: *Lobelia dentata.*
Ecological niche: plants with green leaves, mycoparasites.
Embryo: globular, multicelled.
Fraser, 1931.
5. Polygalaceae: *Epirhizanthes.*
Ecological niche: achlorophyllous perennial herbaceous plants, mycoparasites.
Embryo: globular, not differentiated into organs.
Van Der Pijl, 1934.
6. Burmanniaceae: all species of *Apteria, Burmannia, Dictiostega, Gymnosyphon* and others.
Ecological niche: achlorophyllous perennial herbaceous plants, mycoparasites.
Embryo: proembryo at stage of quadrants or octants (*Apteria setacea, Burmannia disticha, B. selloviana, Dictiostegia orobanchoides, Gymnosiphon squamatus*).
Teryokhin, 1977; Von Rübsamen, 1988.
7. Corsiaceae: all species of *Arachnites, Corsia.*
Ecological niche: achlorophyllous perennial herbaceous plants, mycoparasites.
Embryo: globular, with embryoderm indistinctly differentiated and persistent suspensor.
Teryokhin, 1985a.
8. Thismiaceae: all species of *Sciaphila, Thismia.*
Ecological niche: achlorophyllous perennial and annual herbaceous plants, mycoparasites.
Embryo: globular with differentiated embryoderm, composed of 10-20 cells in longitudinal median section (*Thismia americana, Th. javanica*).
Teryokhin, 1977.
9. Petrosaviaceae : all species of *Petrosavia, Protolirion.*
Ecological niche: achlorophyllous herbaceous plants, mycoparasites.
Embryo: globular.
Hutchinson, 1959.
10. Triuridaceae: all species of *Sciaphila, Triuris.*
Ecological niche: achlorophyllous perennial and annual herbaceous plants, mycoparasites.
Embryo: at octant stage (*Sciaphila maculata*).
Teryokhin, 1977.

Undifferentiated embryos

Unlike the previous groups of plants, all taxa in which the plants contain embryos undifferentiated into organs belong to autotrophic terrestrial plants. In such plants all the basic organs and tissues of the embryo (radicle, epicotyl, cotyledons and conductive system) develop during the **dormancy** period of the seeds, after their abscission from the mother plant[20]. This type of adaptation might be associated with the need for relatively quicker 'ripening' of seeds in compared to the period of embryo development. This type of adaptation must also be related to peculiarities of dissemination of the seeds and fruits of these plants. There are many ephemerals among these plants.

The gap between the beginning of dissemination and germination, associated with the dormancy period in many plants, is 'filled' in this case with the final stages of embryo development, which is necessary for normal vital activity (i.e. competitive power and survival) of seedlings. Unfortunately, peculiarities of reproductive biology (dissemination, germination and vital activity of seedlings) in these plants has not been sufficiently studied. One can expect interesting discoveries in this field. A list of taxa of plants with embryos undifferentiated into organs in mature seeds is given below.

DICOTYLEDONS

1. Chloranthaceae: species of *Chloranthus, Hedyosmum*.
   Ecological niche: autotrophic plants.
   Embryo in mature seed: globular, undifferentiated into organs (*Chloranthus japonicus, C. serratus* and others.).
   Yoshida, 1957, 1959.
2. Papaveraceae: some species of *Corydalis* and, probably, other related taxa.
   Ecological niche: autotrophic plants and ephemerals.
   Embryo in mature seed: globular (*Corydalis cava*).
   Eames, 1961.
3. Piperaceae: species of genera *Peperomia, Heckeria* and probably *Piper*.
   Ecological niche: autotrophic plants.
   Embryo in mature seed: oval, undifferentiated into organs.
   Nikiticheva, 1981b.
4. Saururaceae: species of *Houttuynia, Saururus*.
   Ecological niche: autotrophic plants.
   Embryo in mature seed: globular.
   Nikiticheva, 1981c.
5. Ranunculaceae: species of genera *Anemone, Caltha, Eranthis, Ranunculus*.
   Ecological niche: autotrophic plants.
   Embryo in mature seed: globular, undifferentiated into organs.
   Souèges, 1913; Haccius, 1957.

MONOCOTYLEDONS

1. Eriocaulaceae: species of *Eriocaulon*.
   Ecological niche: autotrophic plants.
   Embryo in mature seed: lenticular mass of cells, undifferentiated into organs (*Eriocaulon hookerianum* and others).
   Arekal and Ramaswamy, 1980.
2. Pandanaceae: species of *Pandanus*.
   Ecological niche: autotrophic plants.
   Embryo in mature seed: undifferentiated into organs.

[20] The embryo in seeds of such plants can be differentiated into organs.

Solms-Laubach, 1878a.

2. Xyridaceae: species *Xyris*.
   Ecological niche: autotrophic plants.
   Embryo in mature seed: undifferentiated into organs.
   Guignard, 1961a.

Possibility of Classifying Flowering Plants According to Occurrence of Chlorophyll in Embryo (Plate XXXIX)

The occurrence of chlorophyll in the embryo implies its capacity to synthesize leaf pigments during any period of development. The embryo is considered to be chlorophyllous irrespective of what embryo organ this pigment is present in and whether it remains during seed ripening and dehumidification. To reveal this character of the embryo one cannot be restricted by the analysis of mature seeds alone. The capacity of certain species embryo for greening in seed germination, beginning already in the fruit (e.g. some Cucurbitaceae), does not apply here (Yakovlev and Zhukova, 1973). The term **non-chlorophyllous embryo** means the lack of capacity for chlorophyll synthesis until initiation of the seed germination process.

Chlorophyllous embryos are exceedingly widespread among flowering plants. They occur in approximately 40% of the species investigated (Zhukova, 1992b). Embryological research of live material has shown that not only the embryo, but also the endosperm, closely associated with it, and the seed coat, i.e., non-leaf tissue, not specialized in photosynthesis, hidden in the thickness of other tissues and not exposed to direct sunlight, can obtain green coloration as well (see reviews of Netolitzky, 1926; Poddubnaya-Arnoldi, 1952; Zhukova, 1967, 1988, 1993, 1998, 2000; Yakovlev and Zhukova, 1973, 1975, 1980; Oganezova, 1993 etc.).

Use of paper chromatography with spectrophotometric analysis has enabled qualitative determination of the main pigment set of plastids in green embryos. It was established that the green pigment of **the embryo** of flowering plants investigated, belonging to various systematic groups, is presented by chlorophylls *a* and *b*, which do not differ feature-wise from leaf chlorophylls. All this refers to the chlorophyll of the green plumule of the embryo in the mature lotus seed, seed coat and woody pericarp whose fruit is barely transparent even at the first stages of seed formation and absolutely light-impermeable in ripening. Chlorophyll *a* and chlorophyll *b* proved to occur as well in the green endosperm of immature seeds of *Raphanus sativus* (Brassicaceae) (Yakovlev and Zhukova, 1973). Coupled with this the green pigmentation of the **seed coat** in certain Cucurbitaceae is known to present a protochlorophyll, not identical to that of etiolated seedlings.

In the majority of species investigated the base pigment set of embryo chloroplasts, including carotenoids, is similar to that of the leaf—the main photosynthesizing organ of green plants. The pigment complex of the green plumule of *Nelumbo nucifera* embryo, characterized by the absence of carotene (Zhukova, 1963, 1965, 1967, 1968; Yakovlev and Zhukova, 1973, 1980), emerges as an exception. The same peculiarity is typical for the embryo of another lotus species, *N. lutea* (Meeuse and Ott, 1962). Neoxanthin is absent in the carotenoid content of embryos of *Linum usitatissimum* at all stages of their development (Prečová, 1977, 1985). In the course of growth and organ differentiation in chlorophyllous embryos (*Haemanthus katherinae, Laburnum anagyroides, Lunaria annua, Caragana* sp., *Linum usitatissimum*), quantitative changes in chlorophylls *a* and *b* content have been observed (Ryczkowski and Szewczyk, 1972; Szewczyk and Ryczkowski, 1977; Preťová, 1985).

Ultrastructural investigations of green embryo plastids in different species of flowering plants have revealed a high degree of their differentiation, intensive development of inner membrane system, and likewise a great number of distinguishing features compared with the structure of a typical chloroplast in the leaf mesophyll (see Zhukova, 1968, 1975a,b, 1978, 1979, 1988, 1992, 1993; Khera Paramjit and Tilney-Bassett, 1976; Preťová, 1977, 1978, 1985; Yakovlev and Chaban, 1979; Yakovlev and Zhukova, 1973, 1980; Casadoro *et al.*, 1980; Zhukova and Dragunova, 1980; Whatley and Price, 1983; Casadoro and Rascio, 1987; also see electron micrographs of Zhukova in Alimova, 1985a). Despite the presence of some morphological peculiarity, these organelles have been shown to occur as chloroplasts (Zhukova and Yakovlev, 1966; Zhukova, 1968, 1972, 1978, 1979, 1993; Yakovlev and Zhukova, 1973).

The chloroplasts of angiosperm embryos, judging from their ultrastructure, have proven not to be inert organelles, losing physiological significance, as some investigators assumed *a priori*. Detected by electron microscopy and partly by biochemical investigations (see above) complex organization of these organelles, intensive development of inner membrane system, the occurrence of various modifications in ultrastructure of embryo chloroplasts in different plants, peculiarity of shape, their contact with ER elements, their own biogenesis and other features confirm the proposal that metabolic activity of chloroplasts occurs during embryogenesis, and not their functional passivity (Zhukova, 1968, 1971, 1975a, 1988, 1992; Yakovlev and Zhukova, 1973). **It was not until green embryo pigment was reliably established as a chlorophyll**, not differing in features studied in leaf chlorophyll, **and its plastids shown to be chloroplasts, that legitimacy of the term chlorophyllous embryo was proved.**

Certain regularities are distinguished in the distribution of the chlorophyllous embryo property among Angiospermae (Zhukova, 1968; Yakovlev and Zhukova, 1973): **(1) species of the same genus do not differ in character of the occurrence of chlorophyll in the embryo; (2) the capacity for chlorophyll synthesis as well as a lack of this capacity in many cases characterizes the embryo of plants united in large taxonomic groups, involving subfamilies, families and even orders.** Thus, for example, all investigated members of such families as Convolvulaceae, Brassicaceae, Dipsacaceae, Geraniaceae, Rhamnaceae, Tiliaceae, and such orders as Capparales, Fabales and Plumbaginales are typified by chlorophyllous embryos, while Boraginaceae, Caprifoliaceae, Oleaceae, Ranunculaceae, Solanaceae, likewise Araliales, Ericales, Papaverales, Saxifragales typified by non-chlorophyllous ones. The list of regularities revealed may be enlarged by the fact that **(3) among the investigated members of** such **primitive families** as Magnoliaceae, Illiciaceae, Trochodendraceae, Cercidiphyllaceae, Annonaceae, Monimiaceae, Lauraceae, Myristicaceae, Dilleniaceae, Flacourtiaceae, Proteaceae, Actinidiaceae, Paeoniaceae, Helleboraceae, Ranunculaceae, Podophyllaceae, Berberidaceae **not even one case of embryo greening was observed.** A green embryo was found only in families Ceratophyllaceae, Nelumbonaceae and Nymphaeaceae (Netolitzky, 1926; Meeuse and Ott, 1962; Zhukova, 1968; Yakovlev and Zhukova, 1973; Batygina and Shamrov, 1981; Batygina and Vasilyeva, 1981), i. e., families characterized by hydrophily, deduced as exceptions. Representatives of families mentioned to possess a number of peculiarities conditioned by hydrophily, have left traces on the structure of their vegetative organs.

It is the occurrence of chlorophyll in the embryo that applies to the series of features distinguishing these families from other primitive ones. Herewith the genus *Nelumbo*, typified by individual properties unlike the rest of Nymphaeaceae, has prompted some authors to separate it into an independent family and even order (see Takhtajan, 1959, 1966, 1970, 1987); it possesses a certain peculiarity with respect to embryo greening as well—only the first two leaves of the lotus embryo are green while the two massive half-globular cotyledons remain colourless.[21] Among monocot plants, essentially characterized by a non-chlorophyllous embryo, greening of the latter often correlates with hydrophily as well.

Of special interest are plants that differ in the feature discussed, within the same taxon of any range. A number of families, including large polymorphous groups (Araceae, Chenopodiaceae, Euphorbiaceae, Meliaceae, Myrtaceae, Polemoniaceae, Scrophulariaceae etc.) comprise species (and probably the genus as a whole) with green and colourless embryos. Thus, for example, among examined members of family Scrophulariaceae (18 genera and 31 species) 4 species were found to have a green embryo (*Limosella* sp.—Netolitzky, 1926; *Collinsia heterophylla, Penstemon humilis* —Yakovlev and Zhukova, 1973; *Euphrasia disperma*—Simpson, 1977), and the remainder a colourless one (Yakovlev and Zhukova, 1973; Zhukova, 1988). The genus *Penstemon* stands out in Scrophulariaceae by the number of its embryological properties (Afanasyeva, 1960). Based on the dispersal of mannitol and dulcitol in family Scrophulariaceae, it was concluded that the presence of these compounds is not only a generic feature, but evidently one of some subtribes and tribes. 'A strange exception in this respect is only the tribe Chelonae: among 14 genera mannitol was found only in genus *Penstemon;* the result was negative in the other genera (Monteverde, 1892, p. 448). The peculiarity of genus *Penstemon* is that chlorophyll appears in its embryo. Data on microsporogenesis and spermogenesis in some members of family Scrophulariaceae evoked the need for reviewing genetic relations among some of them within tribes and the whole family (Afanasyeva, 1960). The occurrence of members with green and colourless embryo is just one more fact attesting to the complexity of intrafamilial phylogenetic relations in family Scrophulariaceae.

In that case wherein the chlorophyllous embryo property is characteristic of the entire family, it is considered useful in the definition of related connections between families together with other features of taxonomic significance. As is known, in Engler's system family Cruciferae occurs as one of the six families in order Rhoeadales along with Papaveraceae, Capparidaceae, Resedaceae, Moringaceae and Tovariaceae. Nevertheless, many systematists (Takhtajan, 1959, 1966, 1987; Cronquist, 1965 and others.) are of the opinion that this order should be split into a number of independent orders: Papaverales for families Papaveraceae and Fumariaceae, Moringales for family Moringaceae and Capparales for the remaining families. Results of pollen analysis, as well as data on morphology and embryology (Souèges, 1939c; Yakovlev, 1958), revealed the fundamental differences between Papaverales and Capparales. Thus, Souèges (1939c) who developed a classification of embryo development types, has applied Brassicaceae coupled with Resedaceae and Capparaceae to the IV megarhetype of the first period, but Papaveraceae to the V one. Papaveraceae differs from Cruciferae,

[21] There is an opinion that the embryo of *Nelumbo nucifera* is characterized by the occurrence of not two but one cotyledon with two lobes, arisen as a result of congenital cotyledon fusion (Titova, 1988, 1992; Titova and Batygina, 1996).

Capparidaceae, and Resedaceae also by absence of chlorophyll in the embryo in the course of embryogenesis.

To discover the chlorophyll in the embryo, research is necessary not only in the mature, but also in the developing seed, because at complete seed maturation the embryo in most plant species has lost its green colour (see below). Detailed lists of families, genera and species with indication of the presence (+) or absence (–) of a chlorophyllous embryo character are given by Zhukova (1968, 1988) and Yakovlev and Zhukova (1973, 1980). The character of chlorophyllous embryo is constant (Zhukova, 1988, 2000) (see Table 6). Its occurrence in certain plants and absence in others allows subdividing all angiosperms into two large groups (Zhukova, 1968; Yakovlev and Zhukova, 1968, 1971, 1973, 1975, 1980). These groups were labelled 'chloroembryophytes' and 'leucoembryophytes' (Yakovlev, 1969). These notions were later defined (Zhukova, 1998).

A chloroembryophyte is a plant whose embryogenesis proceeds with the transformation of leucoplasts of embryo proper cells into chloroplasts (herewith a plastid-type change in the suspensor may not take place, but if it does, other regularities against plastids of the embryo proper follow). **The chloroembryophytes of the first (main) type** (Zhukova, 1988, 1992a,b, 1993), **in which embryo cotyledons first of all obtain chlorophyll** (overwhelming majority of chloroembryophytes), **and chloroembryophytes of the second type, in which only the embryo plumule possesses green coloration while cotyledons remain achlorophyllous clearly stand out among chloroembryophytes**.

The leucoembryophyte is a plant in whose embryo, unlike in chloroembryophytes, the arising of chloroplasts is observed in no phase of embryogenesis during the seed formation process. (Zhukova, 1998).

The first appearance of chloroplasts in a chloroembryophyte embryo happens in the embryonal phase of embryo development; in the majority of chloroembryophytes it is associated with transition from the globular to heart-shape stage of development (chloroembryophytes of the first type). Herewith chloroplasts are observed in all cells of the heart-shape embryo, made possible by complex comparative investigations of embryogenesis in chloro- and leucoembryophytes (Zhukova, 1978, 1981, 1988, 1992, 1993). The chlorophyll in embryos remains from the moment of its appearance and during all development 'coinciding with the period of the most intensive synthetic activity of the growing embryo' (Yakovlev and Zhukova, 1973, p. 50). It has been confirmed by research on living material of embryogenesis in 29 species from family Malvaceae (Kamalova and Abukhovskaya, 1986). At the last stages of organogenesis in some chloroembryophytes of the first type, disappearance of green coloration in some or all tissues of the embryo axial part takes place in consequence of reverse transformation of chloroplasts into leucoplasts (Brassicaceae, Fabaceae, Scrophulariaceae etc.). During maturation and in the mature airy dry seed of most chloroembryophytes of this type it is the embryo cotyledons that are non-chlorophyllous.

In chloroembryophytes of the second type, very seldom found, **transformation of leucoplasts into chloroplasts takes place in the later stages of embryogenesis and does not encompass all embryo organs**. In *Nelumbo* species (Yakovlev and Zhukova, 1973; Vasilyeva and Batygina, 1981) and members of family Ceratophyllaceae (Netolitzky, 1926; Batygina and Shamrov, 1981), only the embryo

Table 6. Basic principles of subdivision of flowering plants according to the character of the occurrence of chlorophyll in the embryo into two groups—chloroembryophytes and leucoembryophytes (additional features characterizing these groups are given in the text)

FLOWERING PLANTS

CHLOROEMBRYOPHYTES

Embryogenesis proceeds with transformation of leucoplasts of embryo proper into chloroplasts (herewith plastid-type changes in the suspensor may not take place, but if they do, then other regularities against plastids of the embryo proper follow)

→ **First (main) type**
Plastid transformation is prolonged and occurs during transition from proembryonal to embryonal phase of development, but the embryo becomes chlorophyllous visually only after globular development at the heart-shape stage (overwhelming majority of chloroembryophytes)

→ **Second type**
Transformation occurs in latest stages of embryogenesis and includes embryo plumule only (Nelumbonaceae, Ceratophyllaceae)

LEUCOEMBRYOPHYTES

Embryogenesis proceeds without change of embryo plastid-type; embryo remains non-chlorophyllous during the entire period of its development plumule possesses green coloration while cotyledons are colourless. The appearance of a slight greenish colour in one of the first two plumule leaves of *Nelumbo nucifera* embryo, as an example, was noted approximately 10-12 days after flower opening (in a greenhouse). The embryo plumule at this stage of development has already been differentiated but not yet achieved the final dimensions and shape typical for the embryo of a mature seed. The plastid apparatus is represented by organelles with prolamellar bodies and single thylakoids (Zhukova, 1968; Yakovlev and Zhukova, 1973).

The property of chlorophyllous embryo characterizes species with a massive embryo, differentiated into organs (members of Convolvulales, Capparales, Fabales, Geraniales, Malvales, Plumbaginales, Rhamnales, Tropaeolales, Aceraceae, Asclepiadaceae, Celastraceae, Cornaceae, Dipsacaceae, Linaceae, Nyctaginaceae, Simaroubaceae, Violaceae etc.). However, this correlation is not complete. Representatives of Asterales, Balsaminales, Cucurbitales, Rosales, Thymelaeales, Caryophyllaceae, Onagraceae, Solanaceae etc., having a large embryo, with organs differentiated, do not occur as chloroembryophytes. On the contrary, species with a relatively small slightly differentiated embryo, constituted only a part of the seed with an abundant endosperm (e. g. members of Dilleniales, Paeoniales, Papaverales, Ranunculales) and species with non-differentiated organs in the embryo (Orchidaceae) occurred as leucoembryophytes (reliable chloroembryophytes not observed among them) (Zhukova, 1992a,b, 1993). According to Beer (1863, see illustrations) species with seeds containing a green embryo, are mentioned in Orchidaceae. However, the author did not clearly demarcate mature dormant seeds and seeds starting germination. He appeared to observe embryo greening in the latter. The character of chlorophyllous embryo also correlates with such embryological features as Nuclear type of endosperm development (Zhukova, 1990, 1992a,b) and non-endospermous seeds (Dahlgren, 1980b). According to Browse and Slack (1985), chloroembryophytes are distinguished from leucoembryophytes by certain peculiarities of lipid metabolism as well. According to the data obtained to date, chloroembryophytes are concentrated mainly in the class Magnoliopsida (Yakovlev and Zhukova, 1973; Oganezova, 1993).

Subdivision of flowering plants into two groups (chloro- and leucoembryophytes) according to the character of occurrence of chlorophyll in the embryo, that is in common with their division by embryo cotyledon number into two classes (dicots and monocots), is verified by a complex of features. Thus, chloroembryophytes are characterized not only by the capacity for chlorophyll synthesis in the embryo in a definite period of its development and, correspondingly, by transformation of leucoplasts of the embryo proper into chloroplasts, but also by the large embryo, differentiated into organs (predominately dicotyledonous), Nuclear type of endosperm development, non-endospermous seeds and some peculiarities of lipid metabolism.

Thus, classification of flowering plants according to the character of the occurrence of chlorophyll in the embryo, likely presents classification of the same range, as subdivision of this taxon into the two classes—dicotyledonous and monocotyledonous plants. In both cases the complex of distinguishing features is responsible for the subdivision introduced.

Origin of Monocotyledonous Embryo: A New Concept

The question of the origin of mono- and dicotyledonous plants, closely associated with the origin of flowering plants on the whole, is of paramount importance. In the foreground is the problem of the origin of monocotyledony of the embryo, as one of the cardinal issues in the evolutionary embryology of plants.

Most researchers believe that the embryo of monocotyledons appeared as a result of evolutionary transformation of dicotyledons. **Pathways of transformation of the embryo during establishment of monocotyledony and homologization of its structures in various groups of plants evoke very heated discussions.** Disagreement in interpretation of the facts is most likely due to the diversity of embryo structure in present-day forms of monocotyledons, irregularity of evolution rate of various groups of properties, reappraisal of significance of separate features of development and methodological difficulties in embryogenesis studies as well. Different structures and peculiarities of embryos are often considered in isolation from one another, not taking into account the specificity of ontogenetic phases. In many respects, difficulties in solving the problem of the origin of monocotyledony have arisen from notions concerning the types of embryogenesis (Souèges, 1939c; Schnarf, 1929; Johansen, 1945, 1950). The principal question about the difference of embryogenesis in the two classes of flowering plants is often reduced to the following: which derivatives of the cells of the two-celled proembryo represent the cotyledon and shoot apex? But the problem seems to be much more complicated and its successful resolution necessitates examination of embryogenesis on the whole—mechanisms and ecology of germination included (Batygina, 1974b, 1987a). Taking into consideration all these points we have tried to provide **a critical analysis of the main hypotheses of evolutionary establishment of the monocotyledonous embryo, having worked out another approach to propose a new solution to the problem.**

In our opinion, principles and regularities of transformations in onto- and phylogenesis during embryoadaptation, types of influence of embryonic devices on definitive morphogenesis and early stages of development, shift of organ initiation in time (heterochrony) and in place (heterotropy), etc.—are universal for all multicelled organisms.

How should **a cotyledon** be interpreted? Most researchers consider cotyledons specialized leaves but not organs *sui generis*. Cotyledons, like leaves, are established as appendicular structures, whose development proceeds due to apical, basal and marginal growth, of dorsoventral symmetry, and a conductive system similar to that of the stem (hypocotyl).

Embryological data show that cotyledons are established in the apical part of the globular embryo in **the same manner** as leaf primordia at the stem apex (Arber, 1925; Mahlberg, 1960; Guignard and Méstre, 1971; Batygina, 1974b, 1987a; Batygina and Andronova, 1988; Raghavan and Sharma, 1995; Padmanabhan, 1996). **Cotyledons are probably homologous to the first couple of leaves** (which then undergo progressive embryonization in connection with the change of time of appearance in ontogenesis). They may not be united in the leaf row with further leaves, since belonging to different stages of ontogenesis they could evolve relatively independently. From the point of view of general regularities of the development the cotyledons are provisory organs (according to the terminology of Severtsov, 1939), since they develop at the embryonic stage and atrophy in the adult plant. But, as is well known, 'those serve as provisory organs which were well developed in ancestors of a species and have lost their significance in descendants' (Shmidt, 1968, p. 110).

To make the definitive form appear and develop further the embryo forms special devices that provide vitality on the one hand, and maintain qualitative transmission to the definitive state on the other. **Cotyledons** represent a typical example of such devices or **embryoadaptations.**

Thus, **cotyledons are provisory organs of the embryo. They are foliar in character but modified because of the functions they provide (protective, reserving, trophic, photosynthesizing etc). Cotyledons maintain vitality of the embryo and its transmission to the definitive state (seedling).**

In comparing embryos of the two classes of flowering plants the most attention is usually paid to **the number of cotyledons,** which in fact is shown in their name—**di- and monocotyledons.** Given such an approach to the question of their origin, researchers mainly try to solve the problem of quantitative changes in organs. However, it should be underscored that difference in number of cotyledons in embryos of these groups is neither principal, nor obligatory, as there are known embryos of dicotyledons with one or more cotyledons (*Cyclamen persicum, Ficaria verna* and others). **Besides, embryos of these groups differ in time and place of differentiation of the apical stem meristem and character of plumule development.** In dicotyledons the growing point typically remains dormant during the entire embryonic period, at best only a rudimentary bud is found. In monocotyledons the growing point, at first dormant, passes to the stem meristem in the seed, which leads to 'ontogenetic acceleration' of the development of the embryo (Batygina, 1974b; Guignard, 1984). **The second most important difference between mono- and dicotyledons is embryo behaviour during germination.** In monocotyledons it is mainly underground; thus exiting the seed envelope by the embryo is usually associated with extension of the cotyledon sheath. During germination the plumule is protected by the walls of the cotyledon sheath or by more specialized structures such as the coleoptyl in cereals.

Peculiarities of leaf establishment and order of leaf disposition (acumbent in dicotyledons and sequential in monocotyledons) as well as development of **adventive roots** with reduction of the main root in seedlings, should be treated as differences.

In our opinion, the hypothesis **of the origin of monocotyledony** should explain the appearance of **all the features mentioned** in their close interconnection with **ecological factors.**

Origin of the embryo cotyledon in monocotyledons. To clarify concrete pathways of the establishing of monocotyledony it is necessary to take into account the nature of embryo organs. Substantial factual material shows great similarity in embryogenesis of both classes of flowering plants at all stages. However, serious disputes have arisen among embryologists as to the laying down and mutual disposition of the shoot apex and cotyledon in the embryo of monocotyledons. Some researchers assert lateral (sharply different from dicotyledons) establishing of the shoot apex and terminal disposition of the cotyledon (Hofmeister, 1861; Hanstein, 1870; Famintzyn, 1879; Souèges, 1931a,b, 1940, 1954; Swamy, 1962; Tzvelev, 1975). Others believe that these structures in the mature embryo **are relatively pseudolateral and pseudoterminal in location, which appears in the result of transformations in the course of embryogenesis** (see Batygina, 1974b).

Hegelmaier (1874, 1878) had already supposed that the terminal disposition of the cotyledon in the embryo is a result of its shifting from the lateral position. Investigating the development of the embryo of *Zannichellia palustris,* Campbell (1897) was the first to show that the cotyledon and point of growth of the stem originated from the terminal segment of the proembryo. This notion was confirmed by investigation of the embryo in *Najas, Potamogeton* and *Ruppia* (Sussenguth, 1921).

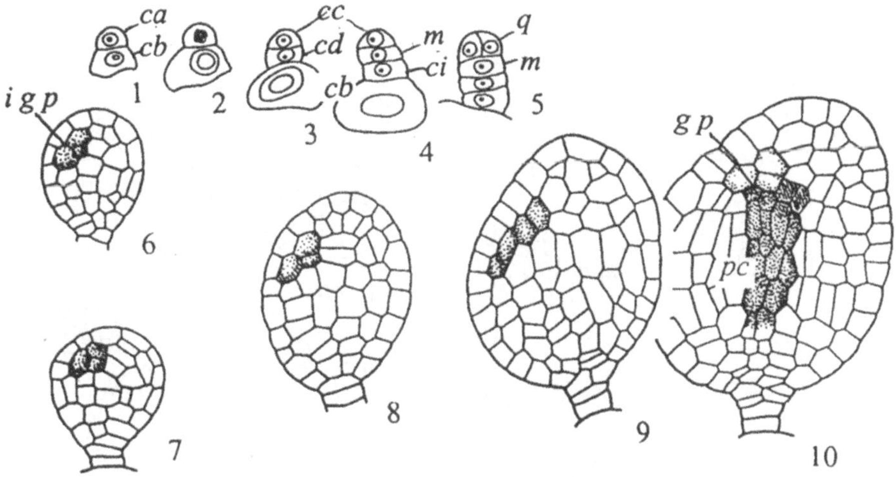

Fig. 54: Embryogenesis in *Potamogeton*; *ca*—apical cell; *cb*—basal cell; *cc, cd, ci*—derivatives of *ca* or *cb cells*; *i g p*—initials of growing point; *g p*—growing point; *m, q*—derivatives of *ca* or *cb* cells; *pc*—procambium (after Souèges, 1954).

However, these investigations were not properly developed, probably because for almost three decades one of the world's outstanding embryologists, Souèges (1924a-c, 1931a,b, 1940, 1954, 1959), maintained a completely opposite opinion (Fig. 54). After reinvestigating the embryogenesis of *Potamogeton natans* (1954), he asserted the ideas proposed earlier by him about the different character of embryogenesis in dicotyledons (shoot apex from *ca* derivatives, terminally) and monocotyledons (shoot apex from *cb* derivatives, laterally). Souèges's views were challenged by Hofmeister (1861), Hanstein (1870) and Famintzyn (1879) who expressed the idea of a major difference in embryo development in monocotyledonous (Alisma type) and dicotyledonous (Capsella type) plants. According to these authors the cotyledon of the monocotyledons is located exactly at the longitudinal axis of the embryo and progressing from the apical cell of the embryo has no symmetrical analogues. The stem apex initiates as a lateral structure. Yakovlev (1946a,b) showed one of the possible pathways of evolutionary transformation of the dicotyledonous embryo into the monocotyledonous one by means of undevelopment of one of the cotyledons. He concluded that the growth point in cereals is terminal and the cotyledon lateral. Terminal disposition of the cotyledon in the embryo is secondary due to irregular pace of cell division in different parts of the embryo.

On the example of the embryo of *Ottelia alismoides* (Haccius, 1952) (Figs. 55, 56 and 57) and *Stratiotes aloides* (Baude, 1956), it was shown that the only cotyledon and the stem apex originated from the terminal segment of the globular proembryo and that initially the cotyledon had a lateral disposition in the embryo. **Thereby the idea of ontogenetic and evolutionary 'shifting' of the cotyledon into the terminal position was grounded** as was reduction of the second cotyledon in both monocotyledons and some dicotyledons (pseudomonocotyledons) (Haccius and Hartl-Baude, 1957; Haccius and Fischer, 1959; Haccius and Lakshmanan, 1966).

Swamy (1962) put forward the hypothesis of evolutionary transformation of the dicotyledonous embryo into the monocotyledonous one. The conclusion was

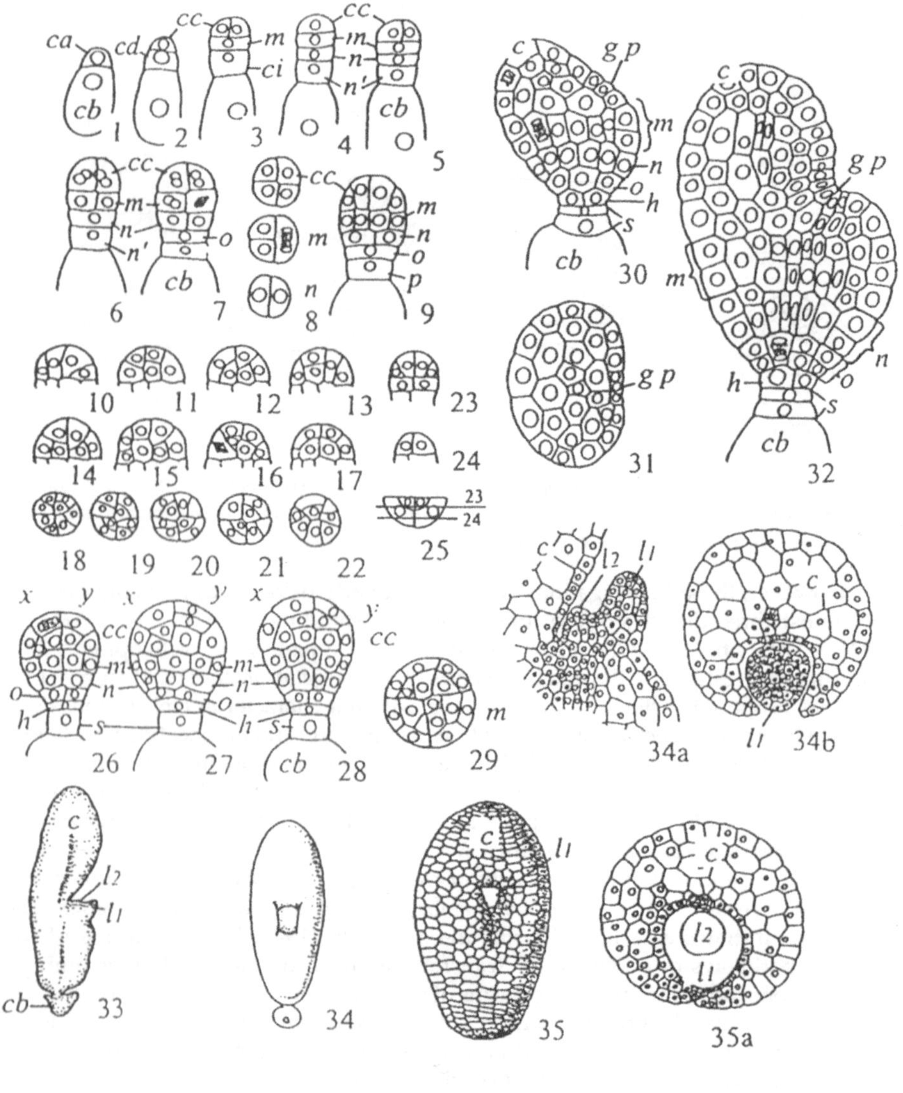

Fig. 55: Embryogenesis in *Ottelia alismoides*: 1-9—early stages of proembryo development, 10-25—different patterns of cell division in octants, 26-29—embryo before commencement of organogenesis, 30-35a—organogenesis (1-7—longitudinal sections, 8—transverse sections of the embryo, represented in Fig. 7, on its different levels, 9–17, 23–28, 30, 32, 33, 34a—longitudinal sections, 18-22, 25, 29, 31, 34b, 35a—transverse sections); *c*—cotyledon; *ca*—apical cell; *cb*—basal cell; *cc, ci*—derivatives of *ca* or *cb* cells; *g p*—growing point; *h*—hypophysial cells; *m, n, n'*, *o*—derivatives of *ca* or *cb*; *s*—suspensor; l_1, l_2—leaf (after Haccius, 1952a).

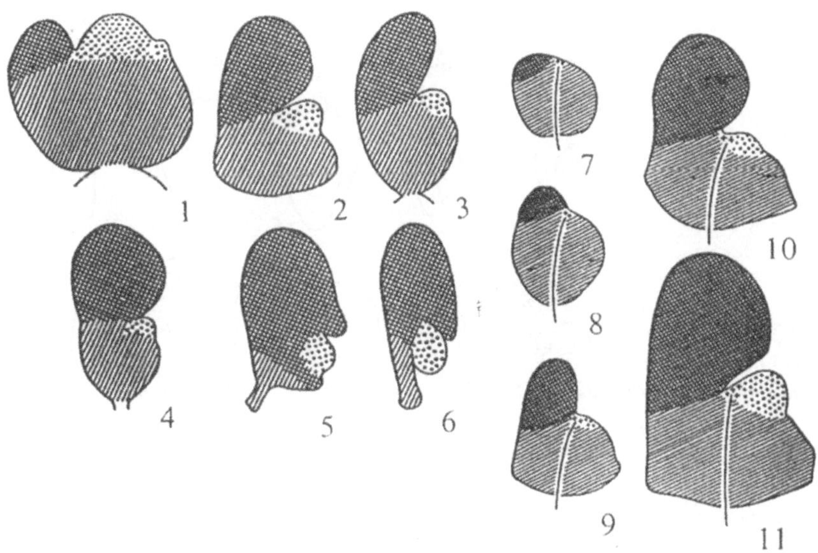

Fig. 56: Location of cotyledon and shoot apex at middle stages of embryo development: *Stratiotes aloides* (1), *Pistia stratiotes* (2, 7-11), *Ottelia aloides* (3), *Carex arenaria* (4), *Cyperus articulatus* (5), *Lemna gibba* (6) (shading—hypocotyl, double shading—cotyledon, crosses—shoot apex, dots—plumula leaf) (after Haccius and Lakshmanan, 1966).

drawn that topographically **cotyledonary** and **epicotylar loci** originate **from the common terminal locus of the proembryo** (Swamy and Lakshmanan, 1962a,b; Swamy and Parameswaran, 1962; Swamy, 1963, 1966). The apical cell of the proembryo in monocotyledons, as in dicotyledons, possesses all the potentials of the stem apex. Two types of potentials of the terminal region of the proembryo of monocotyledonous plants were discovered. In *Halophilla* (Swamy and Lakshmanan, 1962b), *Potamogeton* (Swamy and Parameswaran, 1962), *Ottelia* (Haccius, 1952; Swamy, 1963) and some other plants half of the terminal tier of the proembryo forms further on the cotyledon, while the other half functions as an epicotylar meristem. In *Sparganium* (Campbell, 1897), *Pontederia* (Souèges, 1940), *Eichhornia* (Swamy, 1966) three-quarters of the terminal tier of the embryo progresses into the cotyledon and one-quarter into the epicotyl (Figs. 58 and 59).

Data on the embryogenesis of *Agapanthus umbellatus* (Guignard and Méstre, 1969) (Fig. 60) and *Eriophorum angustifolium, Cyperus fuscus* (Juguet, 1969, 1972) illustrate the origin of the cotyledon and the growing point from *ca* derivatives and 'shifting' of the growing point into the lateral position due to the intensive extension of the cotyledon.

Kudryashov (1964) came to the conclusion that monocotyledony in the row Nymphaeaceae-Helobiae originated by means of gradual prevalence of one cotyledon over the other during ontogenetic development. The author noted that in *Alisma, Saggitaria, Butomus* and other Helobiae, the cotyledon and the growth point form from the terminal segment of the proembryo, 'shift' of the cotyledon from the lateral position to the terminal one occurs at different stages of embryogenesis, and the shoot apex is always terminal with its shifting to the lateral position an apparent event.

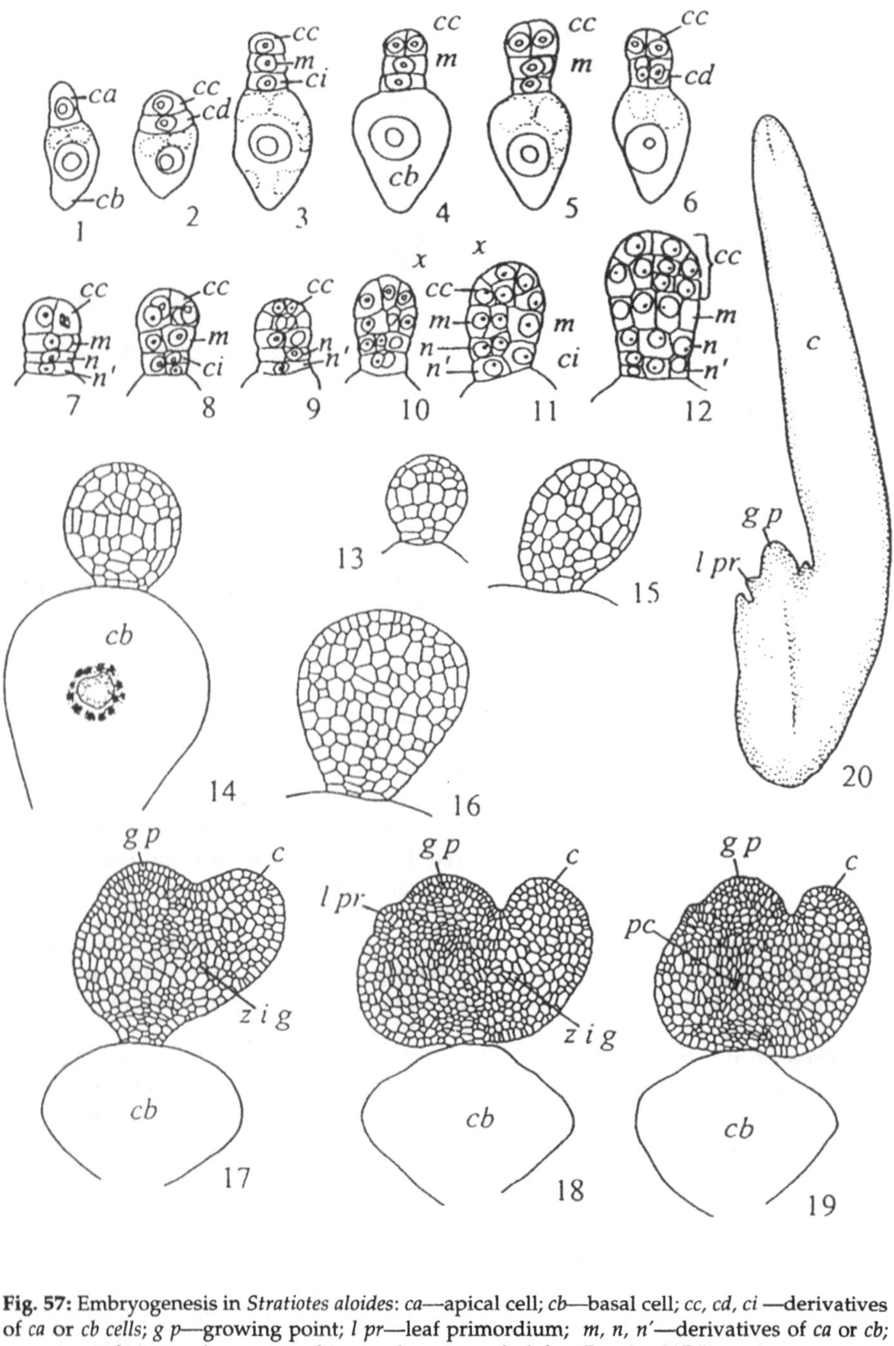

Fig. 57: Embryogenesis in *Stratiotes aloides*: *ca*—apical cell; *cb*—basal cell; *cc, cd, ci* —derivatives of *ca* or *cb cells*; *g p*—growing point; *l pr*—leaf primordium; *m, n, n'*—derivatives of *ca* or *cb*; *pc*—procambrium *z i g*—zone of intercalary growth (after Baude, 1956).

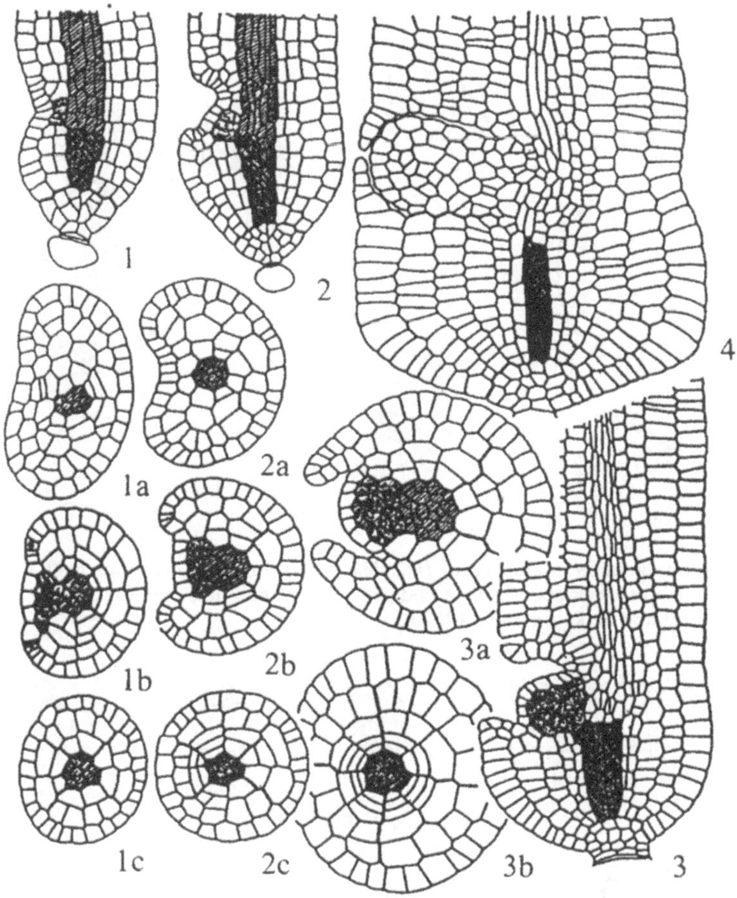

Fig. 58: Organogenesis in embryo of *Euchornia crassipes*; medial longitudial sections through cotyledon and epicotylar meristem of embryo (1-4), transverse sections in region of cotyledon (1a, 2a), epicotyl (1b, 1b, 3a), hypocotyl (1c, 2c, 3b) (shading—cotyledon procambial cells, double shading—hypocotyl procambial cells) (after Swamy, 1966).

The debatable status of the problem of monocotyledony origin in angiosperms and principally new data received by Guignard (1961a, b) concerning the embryo of cereals (cotyledon from *ca*, and not from *cb*, as was asserted earlier by Souèges) made it necessary to examine the problem of evolutionary establishment of monocotyledony with reference to cereals.

Results of our investigation on **embryogenesis** in *Triticum monococcum*, *T. dicoccum* and *T. aestivum*, and analysis of published data on sorghum, maize and barley showed that **the terminal disposition of the cotyledon and the lateral disposition of the shoot apex are apparent phenomena and are determined by changes in the course of their development. This state is secondary form both ontogenetic and evolutionary aspects** (Batygina, 1968a, b, 1969, 1974b, 1987a).

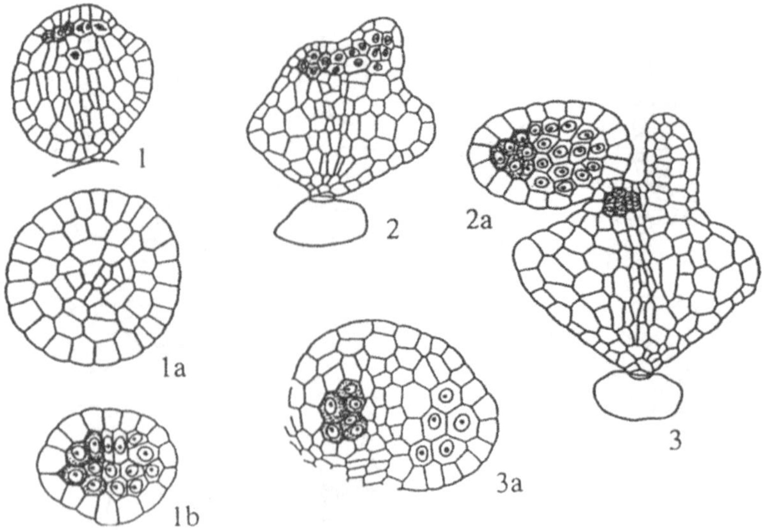

Fig. 59: Organogenesis in embryo of *Halophila ovata*, transverse sections in the ***m*** tier (1a), *q* tier (1b), area of shoot apex and cotyledon (2a, 3a) (small dots – apical zone of proembryo) (after Swamy and Lakshmanan, 1962b).

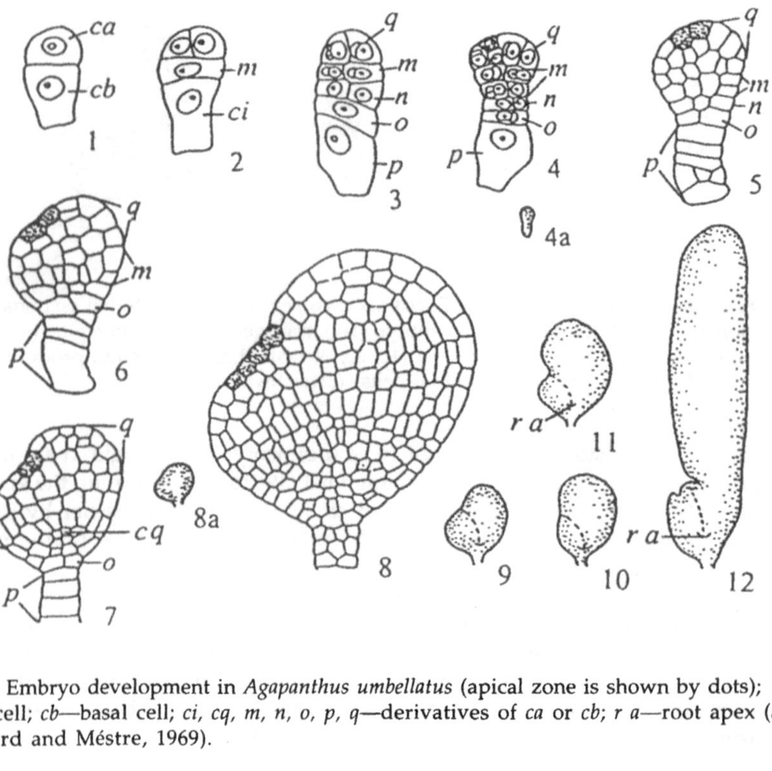

Fig. 60: Embryo development in *Agapanthus umbellatus* (apical zone is shown by dots); *ca*—apical cell; *cb*—basal cell; *ci, cq, m, n, o, p, q*—derivatives of *ca* or *cb*; *r a*—root apex (after Guignard and Méstre, 1969).

The embryo of cereals acquires the original dorsoventral structure already as a result of the first two divisions. **Invariable reproduction of the structure of the four-celled proembryo in each case and persistence of the dorsoventral type of symmetry during all further stages give grounds for thinking that the character of the first divisions of the proembryo has a very immediate relation to dorsoventrality of the embryo at further stages of its development.** As a result of primary extensions occurring as a consequence of **differences in mitotic activity, rate of growth, rhythm of cell divisions and direction of wall initiation in different zones of the embryo** of wheat, development of the scutellum takes place at its dorsal side. Thus, the peculiar disposition of organs in the mature embryo of wheat is determined by original 'redistribution' of zones of mitotic activity during embryogenesis. **The specific structure and development of embryos of cereals is a result of long evolutionary transformations of this group of plants** (see below: Graminad-type of Embryogenesis).

In connection with the study of establishment of monocotyledony, other monocots in which dorsoventrality also manifests from the first stages of embryo development are of great interest (*Commelina forskalaei*—Maheshwari and Baldev, 1958; *Cyperus fuscus*—Juguet, 1972, 1973). It is well known that the study of primitive forms (the closest to initial ones) is of great significance for establishing continuous rows of evolutionary sequences. Data on the development of the embryo of the coconut tree (Haccius and Phillips, 1979), some Helobiae (Ly Thi Ba, 1981), *Tamus communis* and *Endymion nonscriptus* (Juguet and Paré, 1989), and *Butomus umbellatus* (Krasnikov, 1989, 1992) have ascertained that in these monocotyledons the cotyledon appears as a lateral structure and the point of growth as a terminal one (Figs. 61-64 and 65).

However, considerable difference in time and character of establishment of the main organs of the embryo of different taxa of monocotyledons still evokes contrary

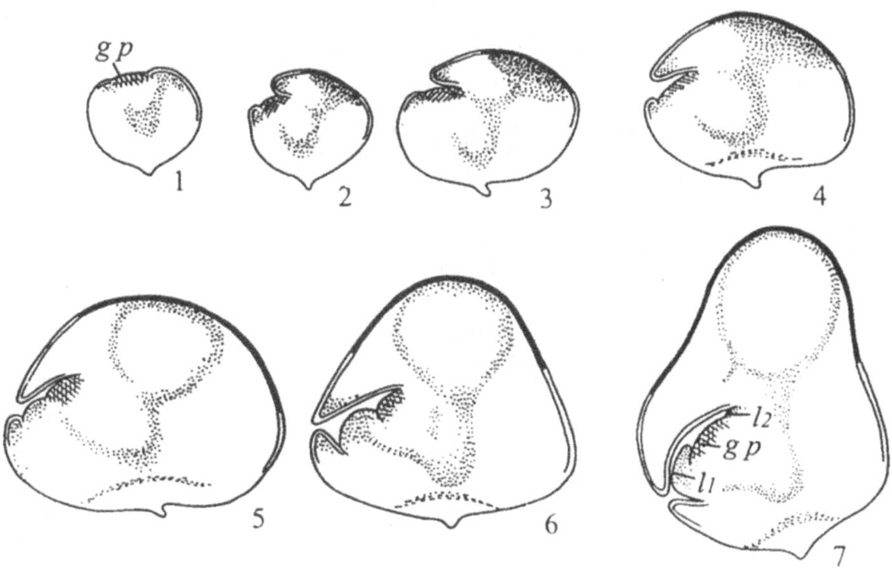

Fig. 61: Organogenesis in *Cocos nucifera* embryo (after Haccius and Philip, 1979). *g p*—growing point, l_1, l_2—leaf.

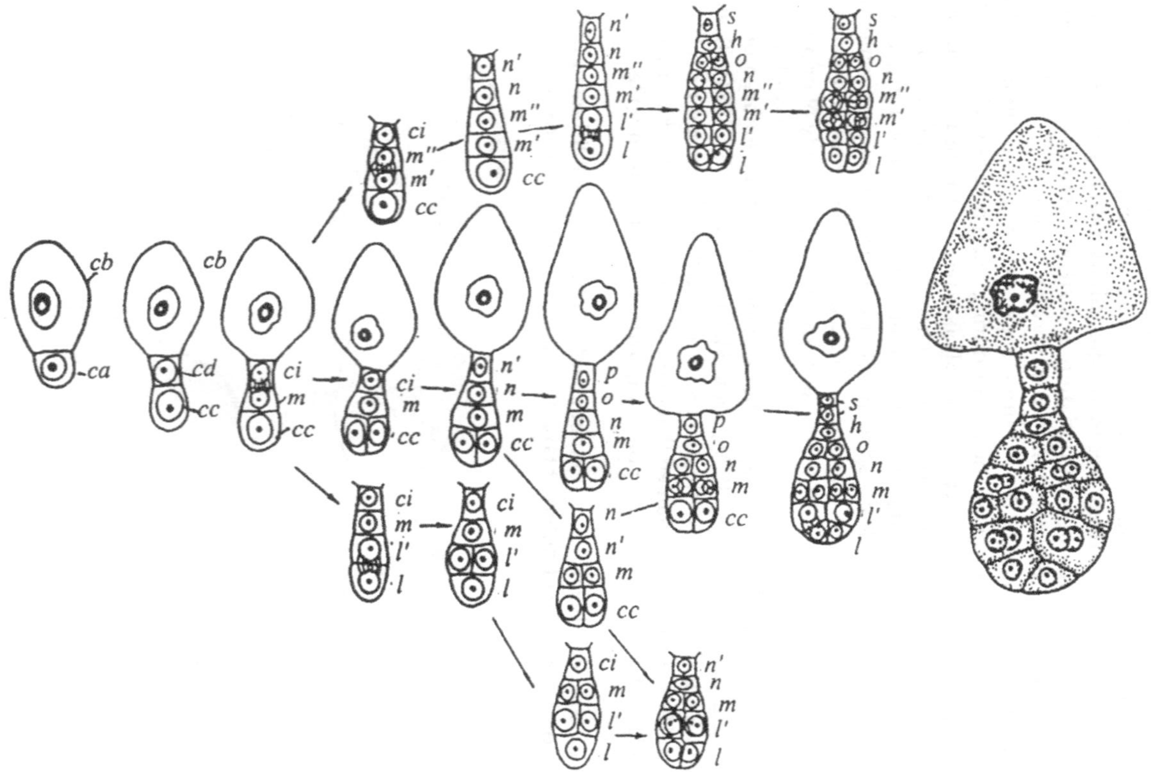

Fig. 62: Different modes of proembryo formation in *Butomus umbellatus* (after Krasnikov, 1989). *ca*—apical cell; *cb*—basal cell; rest of legend—derivatives of *ca* or *cb*.

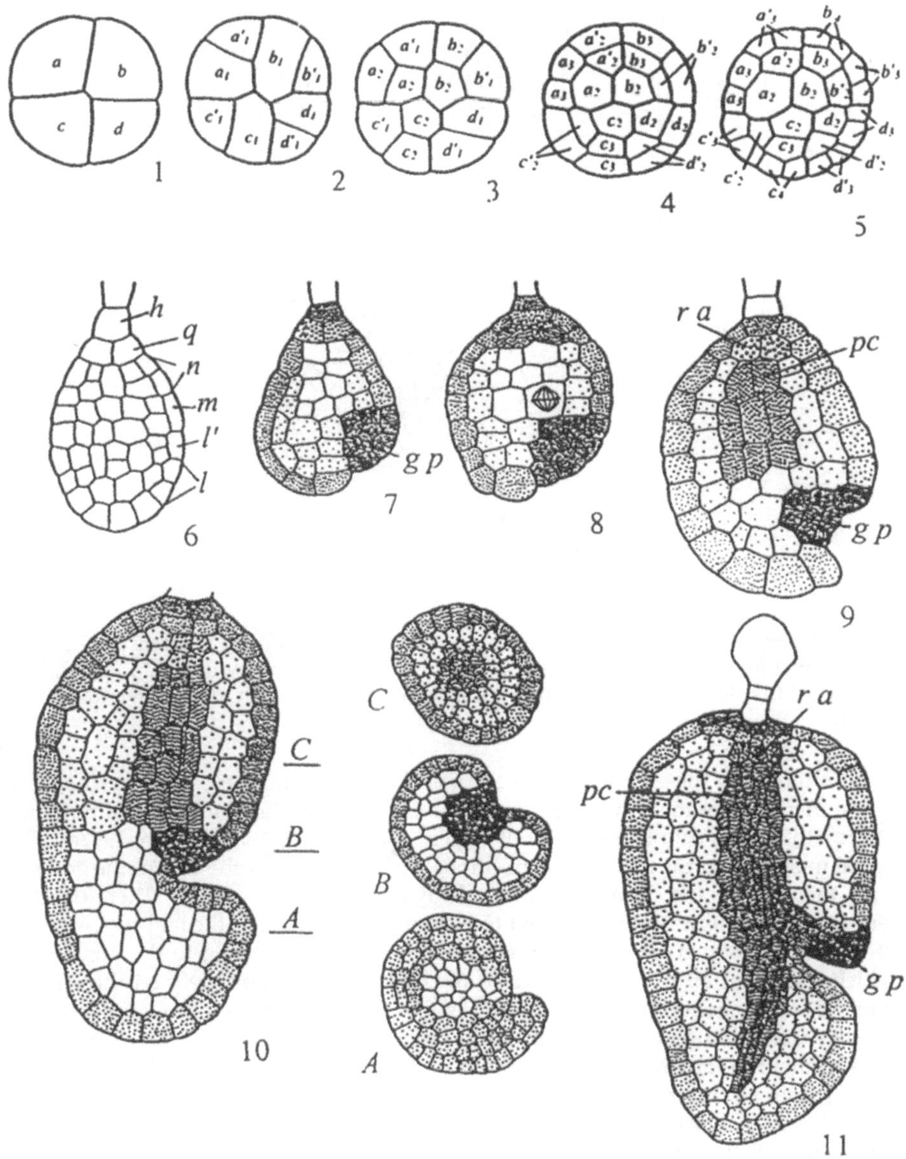

Fig. 63: Scheme of the first four divisions of quadrants in the middle embryo segments (1-5) and organogenesis (sections pass parallel to cotyledonary ridge) (6-11) in *Butomus umbellatus* (after Krasnikov, 1989). *c*—cotyledon; *g p*—growing point; *h, l, l', m, n*—derivatives of *ca* or *cb*; *pc*—procambium; *q*—derivatives of *ca*; *r a*—root apex.

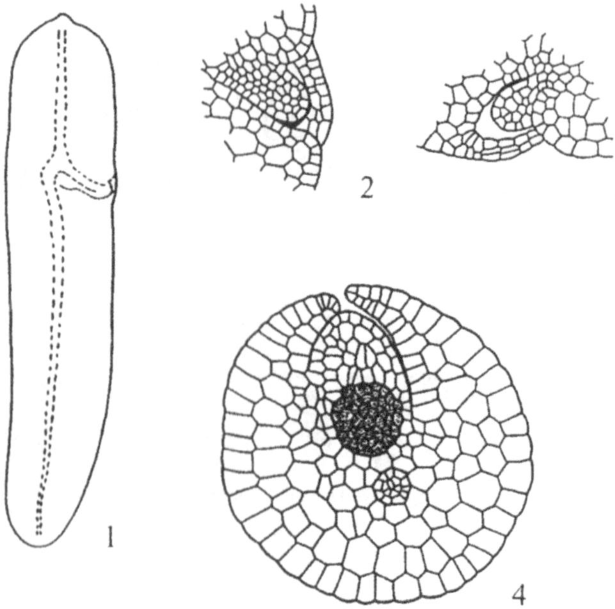

Fig. 64: Embryo in *Butomus umbellatus*: longitudinal section of embryo (1), transverse section through first leaf primordium and apical shoot meristem (shown by dots) (2), longitudinal sections of first leaf primordium and wall of cotyledonary sheath (3, 4) (after Krasnikov, 1989).

interpretations of these processes. The reason for debate about the place at which the apical meristem of the shoot appears, lies in incorrect application of the term 'point of growth' in respect of 'the dormant region' (Batygina, 1968a,b, 1974b, 1987a). The apical meristem forms in different positions with reference to the axis of the embryo from the cells of 'the dormant region' always appearing terminally. Therefore, the term 'growing point' should be used to designate the region of the embryo or group of initial cells which give rise to the true shoot apex. According to the position in which the shoot apex differentiates in the embryo, Guignard defined three main types of embryogenesis of monocotyledons: Cocos-, Agapanthus- and Cyperus-types. He also demonstrated that within these types transmission from **pachyembryate** development to **leptembryate** (according to the terminology of Vallade, 1976, 1992) takes place, i.e., there is a tendency towards decrease in number of cells of the proembryo, increase in regularity of segmentation, and early and simultaneous delineation of the protoderm.

Evolving this trend, Juguet (1989) had already defined four main types of embryogenesis in monocotyledons. The first two types are characterized by the late appearance of dorsoventral (bilateral, according to Juguet) symmetry and a massive multicelled embryo developing according to the pachyembryate type. I) **Tamus-type:** the cotyledon initiates laterally and has a leaf-like morphology (the same as in dicotyledons) with the shoot apex subaxial. II) **Cocos-** or **Elaeis-type:** cotyledon initiates laterally or laterally-apically, but in the mature embryo is terminal in position;

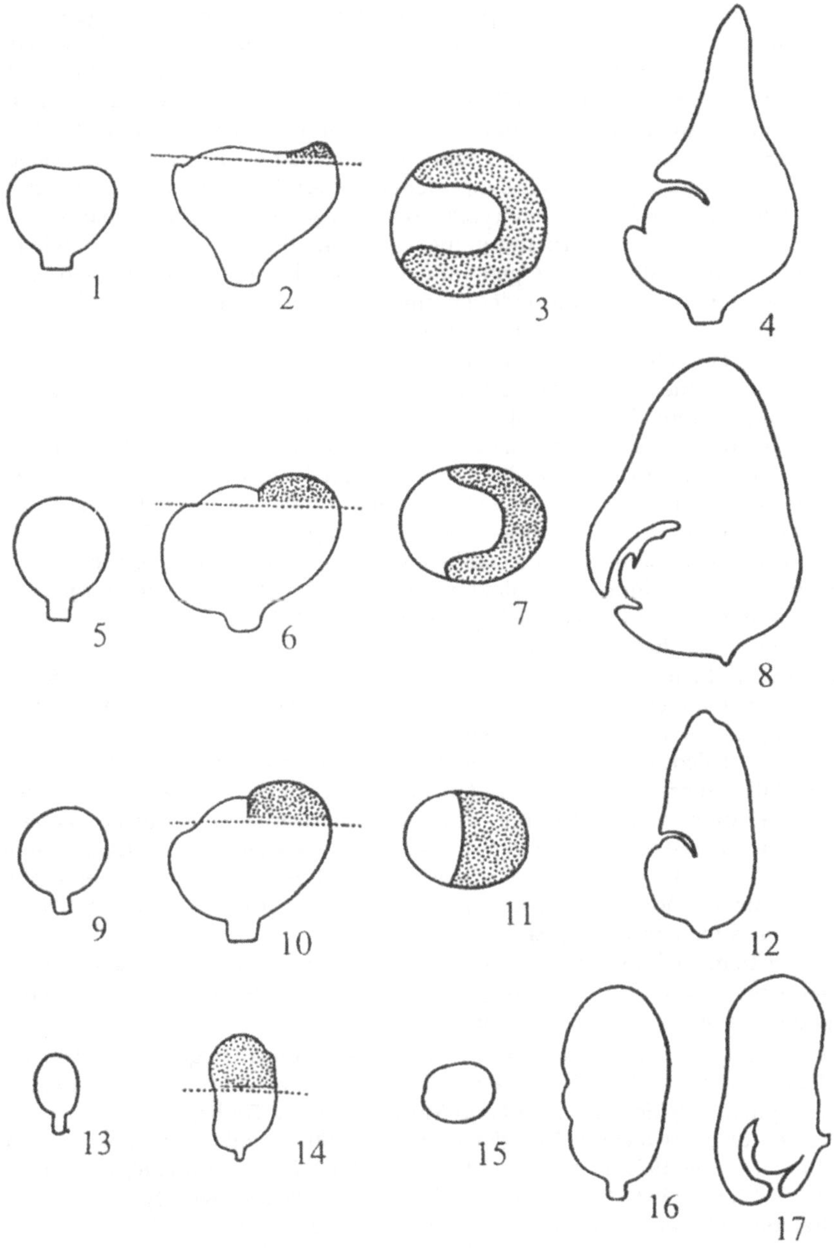

Fig. 65: Bilateral symmetry establishment in monocotyledonous embryos of *Tamus* (1-4), *Cocos* (5-8), *Endymion* (9-12), *Cyperus* (13-15, 17), *Carex* (16): 1, 5, 9, 13—embryo with radial symmetry, 2, 3, 6, 7, 10, 11, 14, 15—beginning of establishment of bilateral symmetry (transverse sections on level indicated by dotted line (3, 7, 11) and longitudinal sagittal sections (2, 6, 10, 14), cotyledon shown by dots), 4, 8, 12, 16, 17—mature embryo (after Juguet, 1989).

shoot apex initiates in the subapical position but later shifts laterally. III) **Agapanthus-** or **Endymion-type:** embryogenesis of this type proceeds faster and includes forms intermediate between pachyembryate and leptembryate development; bilaterality occurs at the stage of about 500 cells; both the cotyledon and future apical meristem establish in the lateral-apical position, but become terminal and lateral relatively. IV) **Cyperus-type:** development is rapid, leptembryate; bilaterality occurs at the stage of 40 cells; cotyledon and future apical meristem form terminally and laterally relatively. The author refers the latter to embryogenesis of Poaceae and Commelinaceae.

Thus, Batygina, Guignard and Juguet, assuming the origin of monocotyledons from dicotyledons showed the general course of 'shifting' of the cotyledon and apical meristem of the monocotyledon embryo in ontogenesis and phylogenesis. Thus Juguet thinks that the main criterion of 'the advance' of the type of embryogenesis is the time of transmission from radial symmetry to dorsoventral. Sharing entirely the general course of this reasoning we still should emphasize that, as mentioned before, dorsoventrality in the embryo of cereals appears already at the stage of the two- to four-celled proembryo, and development of the embryo proceeds faster and is characterized by a whole range of specific features. In accordance with this, one of the authors (Batygina, 1968a,b, 1974b, 1987a) has proposed distinguishing a **Graminad-type** of embryogenesis (see next chapter: Types of Embryogenesis). Probably, Juguet's classification could be completed with this type. Concerning the statement about the initially terminal establishment of the cotyledon and lateral of the shoot apex in the embryo of *Cyperus* (Juguet, 1989; Juguet and Paré, 1989), it should be noted that according to the illustrations (Juguet, 1973) a very early 'deviation' of the morphological axis is observed in the embryo of this species and evidently, it would be more correct to speak of the sublateral laying of the cotyledon and axial position of the apical meristem.

Thus, detailed investigations of embryogenesis demonstrate that **there are no principal differences in the development of embryos of the two classes of flowering plants. Variations within different groups of monocotyledons can be explained by evolutionary advance of the types of their embryogenesis.** And though highly specialized forms, for example cereals, have a few features of development in common with the dicotyledon embryo, the embryogenesis of primitive monocotyledons demonstrates their entire succession.

Main trends of investigations of the problem of evolutionary establishment of monocotyledony. The literature proposes that transmission to monocotyledony could occur as a result of the following processes:

1. **undevelopment (reduction) of one of the cotyledons** (Hegelmaier, 1874, 1878; Fleischer, 1874; Yakovlev, 1946a,b; Haccius, 1952-1979; Guignard, 1961a, b; Batygina, 1968-1987; Guignard and Méstre, 1969; Ly Thi Ba, 1981);
2. **fusion of the two cotyledons** (Sargant, 1902, 1903; Danilova and Sokolovskaya, 1973; Tzvelev, 1975; Dahlgren and Clifford, 1982; Juguet and Paré, 1989; Juguet, 1990; Titova, 1990, 1992; Titova and Batygina, 1996; see: Syncotyly);
3. **heterocotyly (anizocotyly)** (Hill, 1906, 1908; see: Heterocotyly);
4. **abrupt change of early stages of development of the embryo** (Solntseva and Yakovlev, 1964);
5. **transformation of the cotyledon into the axial organ with terminal position,** and of the shoot apex into a lateral structure (Souèges, 1931a,b; Johansen, 1945; Smirnov, 1953; Jacques-Félix, 1957; Swamy, 1962; Guignard, 1975a, b; Tzvelev, 1975; Rytova and Tsvelev, 1982);

6. **changes in type of leaf disposition and consequently of cotyledons, which are thus considered first leaves of the plant** (Arber, 1925; Batygina and Andronova, 1988; Lodkina, 1988);
7. **Absence of cotyledons in monocotyledons in general and correspondence** of their embryo to a branchy **protocorm** (Jacques-Félix, 1982);

From the point of view of the phyloembryogenesis theory by Severtzov (1939), these hypotheses can divided into two main groups. The first group includes hypotheses of undevelopment, syncotyly and heterocotyly, which imply that 'the loss' of the cotyledon has occurred according to type of deviation in compliance with the law of oligomerization of homologous organs (Dogel, 1954). All these hypotheses are based on one of the three possible mechanisms of realization of this law known for plants (reduction, fusion, substitution of organs) (Selyaninova-Korchagina, 1968).

The second group reflects the opinion that serious changes in similar archallaxis took place in monocotyledons at early stages of their development. In addition to ontogenetic changes pointing distinctly towards this type (Solntseva and Yakovlev, 1964), it is reasonable to mention all theories indicating new formations in the embryo leading to initiation of main organs of the embryo of an absolutely new type. Standing apart from the others, is the hypothesis of Arber (1925), who supposed that the change of definitive morphogenesis resulted in fundamental change of the first stages of development, as well as the hypothesis of Rytova and Tsvelev (1982), who believed that cotyledons in di- and monocotyledons are of axial nature. In our opinion, hypotheses based on ontogenetic transformations are less well founded.

It is impossible not to dwell on the hypothesis of Smirnov (1953) and another one expressed independently by Jacques-Félix (1957) concerning establishment of monocotyly in cereals. According to these authors the plumule which should be located at the main axis, is completely reduced and replaced by the lateral bud. Some other researchers rebutted this hypothesis (Tzvelev, 1975; Guignard, 1975a, b). They believed that reduction of the main axis of the embryo and formation of the adventive bud occurs in some monocotyledons, in particular, in cereals. The process leading initially to appearance of the terminal bud in the embryo, formation of another one of analogous structure (the adventive bud) as a result of its activity, and then to reduction of the first bud when the second one remains, seems to be quite speculative.

Kolomytzev (1977) expressed interesting considerations about the problem of monocotyledony of plants. On the assumption of the fact that the axis is the main organ of a higher plant, he divides all angiosperms into: **dicotyledons** (straight-stalked) and **acotyledons** (side-stalked). In the straight-stalked the embryo axis concurs with the axis of the whole plant (axis of the first order), while in the side-stalked an axis of the secondary order forms at the axis of the first order of the proembryo, and the seedling is the shoot of the secondary order. Kolomytzev attached great importance to the correlation of surface growth and volume of the embryo, thinking that the body of a complicated shape forms as a result of prevalence of surface growth over increase of its volume (the main law of morphogenesis of living organisms). With some correlation of the suface and volume cotyledons form in dicotyledons and various outgrowths in acotyledons. In his opinion, cotyledons represent lateral flattened outgrowths of limited growth at the embryo axis.

The appearance of different distortions as a result of 'flattening' of the apical part in acotyledonous embryos is a widespread phenomenon. It is observed with dichotomous branching of thallomes in algae, moss, formatting of the protallus in ferns etc. Kolomytzev considered all this a consequence not of the close relationship, but of the communion of morphogenetic laws. He assumed that **division into two classes is quite justified, because embryo size during cotyledonary formation occurs as a fundamental distinction of them; the ratio of embryo width to length in dicots never exceeds 1** (in *Capsella bursa-pastoris* it is 0.64), **but division into two classes according to the number of cotyledons he regarded as an anachronism. The closed opinion** was presented also by Jacques-Félix (1982), who considered monocots to have no cotyledon, and their embryo, in his opinion, corresponds to a **branchy protocorm**.

Probably, there are no serious reasons for holding that embryos in monocots and dicots are dissimilar by origin of cotyledon and shoot apex (Fig. 66). Artificial induction of monocotyly in dicotyledon *Eranthis hiemalis* is the most convincing experimental evidence for simple transformation of a dicotyledonous embryo into a monocotyledonous one (Haccius, 1960; Fig. 67).

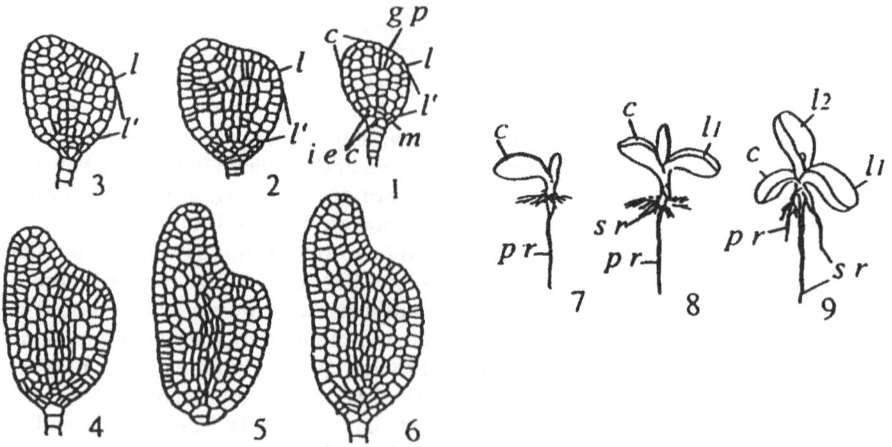

Fig. 66: Cotyledon and seedling in *Pinguicula vulgaris* at different developmental stages (after Haccius and Hartl-Baude, 1957). *c*—cotyledon, *g p*—growing point; *i e c*—initials of root cortex; l_1, l_2—leaf; *m*—intermediate cell; *p r*—primary root; *s r*—secondary root.

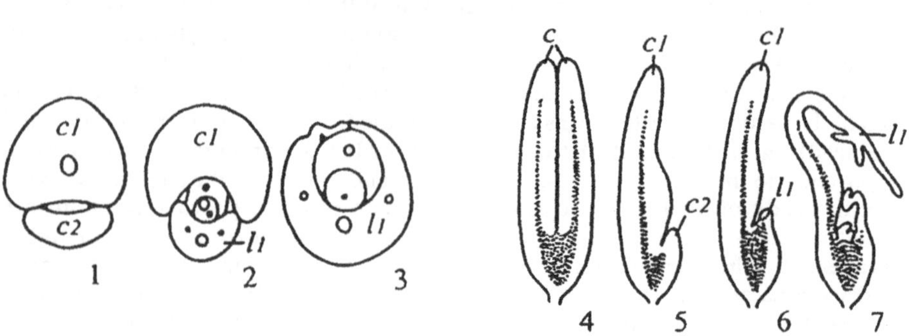

Fig. 67: Different types of embryo in *Eranthis hiemalis*: dicotyledonous (1, 4), monocotyledonous (2, 6), acotyledonous (3, 7), and anizocotyledonous (5); transverse (1-3) and longitudinal (4-7) sections (after Haccius, 1960). c_1, c_2—cotyledon, l_1—leaf.

Hypotheses consolidated on the point that quantitative changing of organs was the basal mechanism in the transition to monocotyly, appear to us to be more convincing. At present, the hypothesis of undevelopment (abort-hypothesis) has a considerable number of supporters. It supposes the gradual reduction of one cotyledon. The possibility of determining the different degree of modification of any cotyledon in dicotyledonous embryos, is considered one argument in favour of it. The hypothesis perfectly explains the change in character of shoot morphogenesis, the shape of leaf rudiment base and the order of leaf arrangement. It 'works' only in the case of acceptation that in monocotyledons no remains of the second cotyledon are preserved. Otherwise, if a cotyledon rudiment is present in any form, the first leaf of the plumule could not arise at its place (opposite to the main cotyledon) and contribute together with it to the united cotyledonary node (the latter is typical for many monocots and pseudomonocots, e.g. *Allium, Anemone*). Hence in order to explain the change in leaf arrangement, not gradual reduction, but a lump-sum falling out of the second cotyledon and its rudiment should be admitted. In this case the hypothesis of 'undevelopment' sweeps aside traditional arguments in its defense about difference in degree of cotyledon reduction in dicots. Furthermore, the hypothesis of 'undevelopment' does not explain arisal of the mechanism of 'ontogenetic acceleration' in embryo formation.

According to the hypothesis of syncotyly (Sargant, 1902, 1903), the cotyledon of monocots was derived by the fusion of cotyledon margins in a dicotyledonous ancestor, having transformed into the terminal tubular structure, adapted to penetrating soil during germination. In later years many authors supported Sargant (Danilova and Sokolovskaya, 1973; Haines and Lye, 1975, 1979; Dahlgren and Clifford, 1982; Rytova and Tsvelev, 1982; Juguet and Paré, 1989; Titova, 1990, 1992a,b; Juguet, 1992; Titova and Batygina, 1996). The structure of seedling and embryo in Nymphaeaceae, in whose cotyledonary organization traces of fusion are observed, is considered a weighty argument in favour of this hypothesis (Dahlgren and Clifford, 1982).

The following reasons against this hypothesis could be adduced. Rudimentary arisal of dicotyly in monocots that should take place to justify this hypothesis, is not observed (Haccius and Lakshmanan, 1967). There are no traces of fusion in laying down, development and structure of cotyledons in monocots. The presence of several procambial bundles in the cotyledon could not be such an argument, as it is not known what number of them should be regarded as initial (Eames, 1961). On the other hand, many cases of dicot cotyledons that while fusing are similar to the 'isthmus' of a single cotyledon in many monocots (e.g. *Eranthis*—Eames, 1961) are known to take place, but this similarity appears to be merely analogous, and in all cases syncotyly results neither in change of leaf arrangement, nor in modification of plumule structure.

According to the hypothesis of heterocotyly (anizocotyly) (Hill, 1906), the transformation of dicotyledonous embryo into monocotyledonous is realized by the functions distinguishing the cotyledons. One cotyledon transforms into a 'suctorial organ' and is retained in the seed at germination, while the second ('epigeal') becomes a leaf of the plumule. Hill presented the dicotyledonous embryo of *Peperomia parvifolia* as an example, comparing it with the monocotyledonous embryo of *Arisaema* (see: Heterocotyly).

The mentioned relation of evolutionary modifications with ecological factors, in this instance with change in germination type, is undoubtedly a strong point of the hypothesis. However, having agreed with the author, we should accept the hypogeal germination and associatively the structure of the embryo in the most advanced monocot families (Poaceae, Commelinaceae etc.) as initial. Contrastingly, in primitive monocots (*Butomus, Potamogeton, Ottelia*) the structure of the embryo (absence of suctorial cotyledon) and the character of germination (embryo exiting the seed coat right at germination) occur as derivative features. Furthermore, from the standpoint of this hypothesis, the mechanism of change in leaf arrangement and reasons for 'ontogenetic acceleration' in embryo development fail to be explained, partly conditioned by the fact that plant ontogenesis is considered a monotonous process, without taking into account differences in embryonal and definitive developmental phases.

The hypothesis of Arber (1925) implies monocotyly and dicotyly to be the result of different leaf arrangement (decussate in dicots, alternate in monocots) on the apex, arising from the very beginning of their differentiation. Not attaching special morphological importance to cotyledons and considering them the first leaves (change in connection with function), Arber regarded monocotyly as the consequence of change in leaf arrangement even in the first shoot node; thus in monocotyledons the second leaf fails to lay down because the mode of initiation has peculiarities: formation of a wide base has eliminated laying down of the second leaf on the node. In Arber's opinion, botanists, mesmerized by their terminology, coined the term 'cotyledon', considering it the organ *sui generis*. Arber questioned attempts to homologize cotyledons of dicots with the structures of a monocotyledonous embryo. In her opinion, the question of monocotyly in that form in which it is examined by most morphologists, no longer arises.

In this connection the works of Serebryakov (1952) and Danilova and Sokolovskaya (1973) should be scrutinized more carefully. Serebryakov holds that Arber's construction cannot accepted because it explains nothing, and ignores phylogenetic relations between monocots and dicots in explaining the most important embryological feature—monocotyly. According to Serebryakov, the terminality of a single cotyledon in a monocotyledonous embryo remains quite ambiguous. However, Arber was the first to attempt relating the change in position of appendicular embryo organs directly with the character of phyllotaxis on the shoot apex.

According to Danilova and Sokolovskaya (1973, p. 347), 'taking into consideration essential differences in the character of the cotyledon and leaf initiation and development, cotyledons are unlikely be regarded as the first members of the leaf row, as done by Arber, not giving independent significance to these organs. **Cotyledons arise at the first stage of proembryo differentiation from proembryo tissue and without participation of later forming apical shoot meristem.** Indicating the ascidial shape of the cotyledon as a main condition for monocotyly, Arber does not describe the shape of the cotyledon that has arisen'.

The hypothesis of Lodkina (1988) implies that monocotyledonous and dicotyledonous embryos were able to originate from a common ancestor, which had two or more alternately arranged cotyledons. Herewith the embryo of a monocotyledonous type could arise either by transformation of a second cotyledon into a coleoptyl, or as a consequence of loss of one cotyledon. In this interpretation the hypothesis appears to be a combination of the hypotheses of undevelopment, heterocotyly and Arber's hypothesis. Furthermore, a morphogenetic mechanism resulting in the appearance of two alternate cotyledonous nodes with cotyledons of the same type and internodes in the embryo is hardly imaginable. Either two cotyledons arise from the apical part of the globular embryo, and so it would be close to the dicotyledonous type (in this case the question again reduces to determination of mechanisms of monocotyly formation), or the second cotyledon has to arise at a later stage of embryo development from the other rudiment, in which case the appearance of a second organ of the same structure is hardly conceivable. The hypothesis of Lodkina also does not interpret the difference in plumule organization between these embryo classes and the morphological mechanism of transformation of a dicotyledonous embryo into monocotyledonous.

Ecological prerequisites of monocotyly arising

The opinion has already been expressed that monocots arose with the transition of their ancestors to a hydrophilic mode of life (Henslow, 1911; Takhtajan, 1964), which apparently led to simplification of their structure. This viewpoint is argued by the fact that representatives of most primitive monocot orders, as well as Nymphaeales, are water plants, and is confirmed by general embryological data.

Let us demonstrate the foregoing on the example of *Butomus umbellatus*, the most primitive member of monocots (Takhtajan, 1966). Analysis of environmental conditions of this species showed them to be very similar to those faced by the ancestors of monocots while conquering water biotopes. *B. umbellatus* grows at a shallow depth in stagnant or slowly flowing, well-warmed reservoirs. It usually occupies free areas in soft silty soil and is considered a typical hydrophyte. The same conditions characterize the life of representatives of some other primitive monocot families. What problems did these ancestors face during gradual adaptation to the littoral zone?

The stages of seed and seedling development, proceeding as a whole in new conditions, would obviously be the most vulnerable in mastering new biotopes. Probably, the monocot ancestors possessed a more primitive (typical for most dicots) epigeal type of germination. But in this environment the seed, falling on soft silty soil would inevitably be immersed in it because of permanent depositions and/or mixing of its upper layers. In other words, the seedling is confronted with hypogeal germination for survival.

– immersed in silt at the most warmed part of reservoirs (littoral zone), the seed and seedling are in conditions of critical oxygen deficit;

– after exiting the seed coat, the seedling is exposed for a long time to light deficit;

– too soft soil causes difficulties for safe attachment of the plant to the substrate.

All basic 'new acquisitions' of embryos and seedlings in *B. umbellatus* and other water monocots are probably associated with solving these problems.

Deficit of oxygen restricts embryo maturation in the seed (typical for primitive angiosperms) that promotes transformation of cotyledons into reserving organs. The mature seeds of Helobiae members and some Nymphaeales s. l. (e.g. Nelumbonales) lack an endosperm.

'Ontogenetic acceleration' in embryo development, so typical for all monocots and Nympaeaceae, appears to be necessary for rapidly reaching the surface and overcoming light and oxygen deficits. By this moment the embryo and seedling are required in prolonged consumption of nutrients, especially in the form of starch (providing the possibility of anoxic exchange) that is reflected in the intensity of reserving function in the cotyledon.

Sheltering the shoot apex inside the cotyledonary sheath, which became possible as a result of transferring to monocotyly and change in cotyledonary shape, and arising of differential growth of the cotyledon during post-seminal development, may undoubtedly be considered adaptations to germination in the thickness of silty ground. In *Butomus* the shoot apex is boosted onto the surface by lengthening of the cotyledonary sheath; what is more, the embryo is released directly from the seed coat. (Transition to hypogeal germination by means of modification of the cotyledon into a haustorium is impossible in this case due to absence of an endosperm in mature seeds.) After that the cotyledon can unfold and function for some time in common with the true leaf typical of many primitive monocots (Eames, 1961).

Interestingly, the criterion suggested by Esau (1977) for indication of hypogeal germination (cotyledon remains underground in the seed), is not applicable here. Perhaps a more reliable feature would be: surfacing of the cotyledon or not (Eames, 1961).

As in crumbly soil, very likely the adaptations here are reduction of seedling main root and arising of adventive roots. The system of numerous adventive roots would better hold a young plant on the substrate and decrease the possibility of its damage by hydrodynamic effects. Besides, the maximal quantity of humidity is concentrated therein by the soil surface and not at depth, which would make development of a long main root unprofitable.

Thus, the distinguishing peculiarities of embryos in primitive monocots could be considered as necessary adaptations, having arisen in their ancestors during conquest of reservoirs in the littoral zone.

Possible mechanisms of transformation of dicotyledonous embryo into monocotyledonous. Development of the embryo in primitive monocots up to the globular stage proceeds in the same way as in dicots. For example, in *Butomus* the cotyledon arises from the apical part of the embryo, laying down laterally, and its base occupies the lateral position in respect of the shoot apex even in the mature embryo. The procambium bundle is differentiated in the hypocotyl, cotyledon and region of the first leaf laying down. Formation of a united node of the first leaf vascular bundle with the cotyledon vascular bundle is a widespread (among monocots) phenomenon, referred to as the Allium type (Haccius and Lakshmanan, 1967). In *Butomus*, as in many of the lowest monocots (e.g. *Ottelia alismoides*—Swamy, 1962), the conductive bundle of the leaf lies down almost simultaneous with the cotyledonary bundle. The curve of the base of the vascular bundle of the cotyledon and the ventral half of the *Butomus* embryo in cross-section do not differ from the dicot embryo. The cylindrical shape of its cotyledon is secondary, arising from the ascidial one as a result of the peculiar mode of growth with spiral distribution of cell divisions. Hence, the main differences between these embryos concern only the character of morphogenesis of the dorsal embryo part. Similar to other monocots, at the globular developmental stage in *Butomus* the portion of cells of the embryo 'apical semisphere' is characterized by slowing of division intensity, forming a 'dormant zone'. Topographically this area corresponds to the locus of initial cells of the shoot apical meristem and one of the cotyledons (*pvt* + *ipco*) in the dicotyledonous embryo. The base of the initiated cotyledon surrounds this locus by a semi-annular ridge. While the cotyledon develops the ridge margins widen and lengthen (marginal growth of cotyledon), forming the cotyledonary sheath. In *Butomus* the shoot apex forms after commencement of differentiation of the first leaf primordium from cells of the 'dormant zone', which corresponds to the portion (il_1[22] + *pvt*) of the proembryo apical segment (locus *ipco* + *pvt* + il_1). The appearance of the first leaf is not associated with the activity of the apical shoot meristem.

Other monocots (e.g. *Ottelia*) exhibit similar relations between the first leaf primordium and the apical meristem (Haccius, 1952).

In *Eichhornia crassipes, Cyperus fuscus* and *Agapanthus umbellatus* (Swamy, 1966; Juguet, 1968 (1972); Guignard and Mestre, 1969), development of the axon hill in the embryo is believed to be slightly late. Its initials appear to be at the base of the semispherical projection contained in the cotyledonary sheath. Although the aforesaid authors refer to all of this structure as the apical meristem (embryo shoot apex, *pvt*), Haccius and Lakshmanan (1966) regard it as the primordium of the first leaf, which in our opinion conforms more to the facts. This rudiment completely corresponds to the rudiment of the first leaf in other monocots according to position, size and morphogenetic character. The axon hill is not visibly formed at this time and, probably, has to be represented by the same initials attributed to the portion of the 'dormant zone'. Therefore, the mature embryo of these plants possesses the differentiated primordium of the first leaf and the initials of the apical meristem, which should be marked l_1 + *pvt*.

A case of apical meristem formation opposite to that just described was recorded for *Stratioites alloides* (Baude, 1956). Development of all embryo organs in this plant begins relatively late but the rudiments of the axon hill and first leaf arise very quickly after cotyledon initiation. The axon hill is more massive and differentiates more rapidly than the first leaf primordium. But in this case both structures derive from a single locus of the proembryo apical part, as can be seen in the author's illustrations, and the leaf primordium arises before the apical meristem becomes active, although the axon hill develops faster than usual.

In all cases analyzed by us, the first leaf primordium and shoot apical meristem derive from a single locus, which has arisen at the globular stage of the embryo, whose cell division rate greatly decreases in the first period of ontogenesis compared with the rest of the proembryo parts. As already mentioned, it appears to comply with the portion of the apical semisphere of the globular embryo, involving the initials of one cotyledon and the shoot apex. With further differentiation in this locus, the character of the arising of the first leaf primordium and apical meristem is the same as in the meristem and axon hill of the dicotyledonous embryo. The conductive bundles of these structures in primitive monocots[23] and dicots combine with the united cotyledonary node. **Hence, if our reasonings are correct, the first leaf of monocots is homologous to that of dicot cotyledons and, in the early developmental stages, before the primordial first leaf arises, not the embryo shoot apex, as thought previously, but the united 'dormant locus', containing the initials of the first leaf primordium and the apical shoot meristem (il_1+*pvt*), is situated in the forming cotyledonary sheath.** Thereafter it gradually displaces into a lateral position, both in individual development and in phylogenesis (see Batygina, 1987a). The time activation of these initials relative to each other, as well as the beginning of cotyledon initiation can differ in various species due to diversity in organization of the epicotylar region

---

[22] il_1—initials of the first leaf primordium.

[23] In higher monocots the complication of the conductive system, connected with further evolution of all organs takes place.

and position of the apical meristem in the mature embryo. Our view is in consonant with the conclusions of Guignard (1984), who presumed the axon hill to be arisen in a different position relative to the embryo axis from cells of the 'dormant zone', always forming terminally. It is incorrect to apply the term 'shoot apex' to the dormant centre (Guignard, 1984), which seems to be the cause of disputes about the place of its arising (terminal or lateral). We also hold that the term 'shoot apex' should essentially be employed for embryos of dicots only, as indication of the region or group of initial cells from which the true apical shoot meristem is subsequently formed.

Insofar as the first leaf of monocots is probably homologous to one of the dicot cotyledons, it should be recognized that during monocotyly origin, displacement of the second cotyledon laying down in time (heterochrony) at later developmental stages happened. Such change per se is not something unusual and is widely familiar among embryologists of animals. Adaptive significance (usefulness) of the feature determines the time of its appearance in embryogenesis (Matveev, 1942). What adaptive importance could the delay in laying down of one cotyledon possess? We think this comparatively simple modification had a whole number of considerable **consequences**, mainly in the character of influence on morphogenesis of other organs, realized through the system of stepped dependencies (stepped correlations, Schmalhauzen, 1939). Thus each of the intermediate steps (members of the correlative chain) is thought to have no independent adaptive significance.

1. **The delay in development of one of the cotyledonary loci results in a change in cotyledon morphogenesis. The cotyledonary base widens, up to the almost annular ridge, and the shape of the growing cotyledon becomes ascidial** (cylindrical shape, e.g. in *Butomus*, is secondary). **As a result the epicotylar locus occurs and is involved in the cotyledonary sheath, which protects the shoot apex at germination.** Such a mode of cotyledon modification is accepted by all supporters of the hypothesis of underdevelopment as well.
2. Abrupt changes of embryonal adaptations are able to greatly influence the early stages of morphogenesis of definitive organs (Shmidt, 1968). In this case **the morphogenesis of the epicotylar region changes.** As a result of development delay **the loci of initials of the second cotyledon and shoot apex** differentiate with formation of the apical meristem and **primordium of leaf one, the base of which assumes the same shape as the cotyledon.** The rudiments of the following leaves accordingly arise on the opposite side of the axon hill, maximally remote from the centre of the previous one, following the law of phyllotaxis. **Thus, heterochrony in cotyledon initiation is related to change of leaf shape and leaf arrangement.**
3. **As a consequence of the delay in development, the second cotyledon exists at a later stage of ontogenesis and fails to fulfil its former functions concerning trophic relations with the endosperm. The 'replacement of functions' happens** (Dorn, 1937) (the main reserve function reduces, the minor assimilative one increases) **as a result of transformation of the provisory organ (cotyledon) into definitive—the leaf.** Such transformations represent one of the regularities of ontogenetic modifications and are often observed in invertebrates. Simultaneous with this process, activation and 'widening of functions' (Plate, 1913) of the first cotyledon take place, evoked by oligomerization and changing of its role at germination.
4. **The change at the middle stages of embryogenesis directly influences its early phases, resulting in the loss of radial symmetry in the proembryo. This**

phenomenon can be seen to a different extent in various species. In cereals, as an example, it already displays in the first cell wall orientation at the zygote division (Batygina, 1968a, b, 1974b, 1987a). The loss of symmetry at early stages naturally results in change of position of the embryo morphological axis and replacing a single cotyledon to the terminal position. In water monocots, variations in the intensity of this process have been observed. It achieves maximal expression in genera *Saggitaria* and *Alisma*, which has, perhaps, led some authors to the notion of laterality of the 'growing point' in these plants.

In the embryogenesis of invertebrates similar and more considerable influences of embryonal adaptations on the early stages of cleavage have been found. For example, in polychaetes during the transition from embryogenesis with a plankton larva to development without a larva, the animal's semispherical blastula can completely fall out (Svetlov, 1923).

5. **The shifting of laying down of one of the cotyledons in time is associated with the mechanism of 'ontogenetic acceleration'**, as illustrated in the following example. The embryos of wheat and some primitive monocots are distinguished by the time of cotyledon initiation, laying down of the first leaf and comparative temporary interruption (the degree of heterochrony) between the beginning of cotyledon initiation and differentiation in the dormant zone (il_1 + *pvt*). In this series the correlative dependence between these features and the value of 'ontogenetic acceleration' is revealed (which is greater, the quicker the second and following leaves lie down in embryogenesis).

From a comparison of embryogenesis in different systematic groups of monocots it ensued that ontogenetic acceleration is probably associated by correlative dependence with the degree of heterochrony in the cotyledon and first leaf initiation and 'shifting' of cotyledon laying down to earlier and earlier developmental stages. This dependence (if constructed graphically taking into account correlation laws) could be generalized in the Descartes system of co-ordinates.

Thus:

1. **A monocotyledonous embryo** is likely **derived from an ancestral embryo possessing two cotyledons.**
2. **All general distinguishing features of the embryo in primitive monocots occur as direct or indirect** (realizing adaptive significance through the system of morphogenetic correlations) **embryo adaptations** to environmental change.
3. **Ecological factors** (silty soil, oxygen and light deficit in conditions of the littoral zone) presumably cause the transformation of dicotyledonous embryos into monocotyledonous ones.
4. **The monocotyledonous embryo** generated **the most adaptive features** by virtue of initiation of **shifting of one of the cotyledons at a later stage of ontogenesis** (heterochrony). Lacking proper adaptive significance, this modification through the system of morphogenetic correlations instantly revealed **three characters—'ontogenetic acceleration' of development, change in leaf shape and leaf arrangement, hypogeal germination**—proving to be immediate **adaptations.**
5. **The shift in laying down of the second cotyledon** appears to be accompanied by change of its function and transformation into a definitive organ—first leaf of the seedling.
6. **The shift in laying down of the second cotyledon** results not only in a change in the last embryogenetic stages, but also reflects on its early periods, as manifested

in loss of radial symmetry in the embryo. This effect is the greater, the earlier cotyledon laying down occurs.

7. **The degree of 'ontogenetic acceleration'** correlates with the 'degree of heterochrony' and the time of cotyledon initiation in embryogenesis; what's more, relocation of initiation of the latter in earlier embryogenetic stages increases the effect of heterochrony.

The model of the transformation mechanism of a dicotyledonous embryo into monocotyledonous suggested by us, is not considered unique and may have been repeated in evolution. 'The loss' of cotyledon obviously presents an example of perspective ontogenetic differentiation, 'providing life not only in concrete conditions, but also in a much wider adaptive zone and, perhaps, providing the possibility of transition into another zone' (Timofeev-Resovsky *et al.*, 1977, p. 179). For some time it remained in different phyllums without decisive success, before finding 'serious application' in transferring to a hydrophilic life. The success was probably due to the fact that in consequence of a single, not so complicated change a number of adaptive features appeared, whose combination was ideally suited for mastering a new biotope.

TYPES OF EMBRYOGENESIS

Principles of Classification of Embryogenesis Types

19th Century Ideas. As early as in the first half of the 19th century certain stages of embryo formation in flowering plants were studied (Treviranus 1815; Brongniart, 1830; Schleiden, 1839b; Meyen, 1840; Schacht, 1850). It was Hofmeister (1849, 1859) who discovered zygote division, formation of a two-celled and, later, globular embryo from its upper (apical) cell. Hanstein (1870) was the first to trace the regularities of embryo formation, to describe the quadrant stage and to consider that the embryo structure proceeded from the three histogens inside it: dermatogen, periblem and plerome. Based on differences in embryo structure at the four-celled and quadrant stages in the differentiation period and in the structure of the aforesaid histogens as well, he described three types of embryo development in dicotyledons: **Capsella, Oenothera** and **Nicotiana**. With these three types he compared embryo development of other plants such as dicotyledons and monocotyledons. It is important to note that Nörner (1881) distinguished three types of embryo development in cereal crops on the basis of peculiarities of cell wall establishing in early embryogenesis: I—first and second divisions transversal, II—first division transversal and second oblique, III—both divisions oblique. Westermaier (1876) reinvestigated embryogenesis in *Capsella bursa-pastoris* (Brassicaceae) and made Hanstein's (1870) information on the first stages of embryo development more precise; he reconstructed the embryo structure on the basis of a series of transversal and longitudinal sections and was the first to describe the octant stage. The embryogenesis type in *C. bursa-pastoris* is considered classic for dicotyledons.

Classification by Carano (1915) and Chiarugi (1925). Carano (1915) suggested that two types of embryo development should be distinguished—**Cruciferae**-type (embryogenesis in *Capsella bursa-pastoris*) and **Asteraceae**-type (embryogenesis in representatives of the family Asteraceae—Carano, 1913, 1915)—on the basis of peculiarities of the structure of a four-celled proembryo and succession of its cell divisions during further development. Later Chiarugi (1925) justified the need for distinguishing two more types of embryo development in dicotyledons: **Solanaceae** and **Chenopodiaceae,** placing representatives of family Caryophyllaceae examined by Souèges (1922a, 1924b) in the latter type. This was based on characteristics of the structure of the tetrad of proembryo cells, as well as the contribution of the apical and basal cells to embryo construction.

Classification by Schnarf (1929). Using the criteria of Chiarugi (1925), Schnarf (1929) gave a different interpretation to the criterion concerning the contribution of the basal cell to embryo construction, whereby he added **Caryophyllaceae** as a type of embryogenesis to the classification. In the types suggested by Chiarugi the basal cell takes part in formation of the embryonic organs in contrast to the Caryophyllaceae type. The classification includes five basic types (German spelling was used to name the types, later Latin spelling was introduced.): **Cruciferen** (= Cruciferae), **Asteraceen** (=Asteraceae), **Solanaceen** (=Solanaceae), **Chenopodiaceen** (=Chenopodiaceae), **Caryophyllaceen** (=Caryophyllaceae) (Fig. 68).

Having analysed the literature, Schnarf (1929) concluded that in different species characterized by one type of embryogenesis the succession of stages (octant stage, for example) and structure of the embryo (structure of certain tiers, of the suspensor,

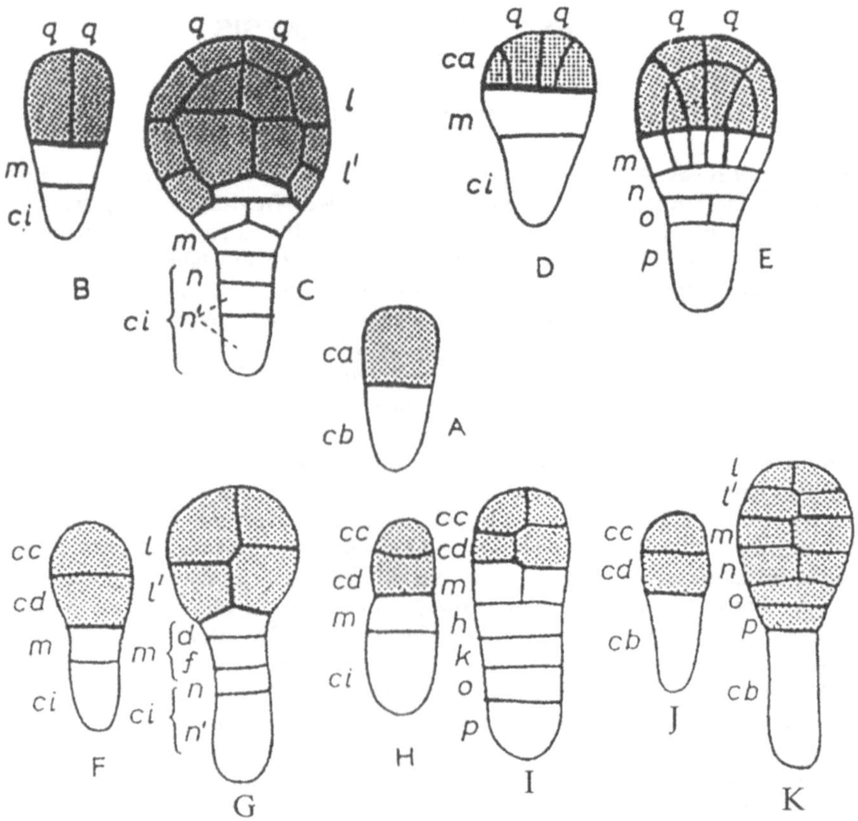

Fig. 68: Classification of embryogenesis types by Schnarf (1929): A—transverse division in zygote and formation of two-celled embryo, B-K—further development of embryo according to Cruciferae-type (B, C), Asteraceae-type (D, E), Solanaceae-type (F, G), Chenopodiaceae-type (H, I), Caryophyllaceae-type (J, K) (after Wardlaw, 1955). *ca*—apical cell; *cb*—basal cell; rest of legend—derivatives of *ca* or *cb*.

epiphysis and hypophysis) are taxon-specific (he later studied this issue in more detail, Schnarf, 1931) and can be determined as modifications. However, the modifications did not acquire an official status and hence never came into the classification. It is important to note that Schnarf emphasized the absence of fundamental differences in embryo development in dicotyledons and monocotyledons but nevertheless attributed the given types only to dicotyledons. For monocotyledons the types of embryogenesis were not determined despite the fact that detailed data on certain representatives were available. Neither were they determined for taxons with non-differentiated embryos in a mature seed (except for family Orobanchaceae).

Classification by Souèges (1939c). The basis for classification were detailed investigations carried out by the author on a large number of taxa of flowering plants. Souèges (1939c) labelled this classification **embryogenetic classification of embryonomic types** because it was based on the fundamental laws of embryogeny

(Souèges, 1937b). His investigations showed that it is difficult to identify all the blastomeres and to determine their location in the general system of the developing embryo at the late stages. This especially concerns the laws of numbers and disposition. That is why he suggested distinguishing embryogenetic groups, including all the species in which embryogenesis is subject to the same laws in the course of the first four generations.

To explain the construction of his 1939 classification, Souèges introduced such notions as **embryonomic type**. He distinguished four categories of embryonomic types: 1) **basic or archetypes**; 2) **secondary or derivatives**; 3) **overlapping**; 4) **irregular**.

In addition to the embryonomic types included in the classification, Souèges distinguished so-called geometrical types which are only determined on the basis of orientation of cell walls during proembryo segmentation, according to the laws of symmetry: 1) **orthoradial** (walls establish at a right angle to each other whereas blastomeres arrange around the longitudinal axis; radial symmetry); 2) **spiral or oblique** (walls are oblique and blastomeres displaced and do not arrange in tiers); 3) **linear or parallel** (walls lie parallel to each other and at a right angle to the axis; embryo cylindrical with axial symmetry); 4) **mixed**; 5) **irregular**.

Constructive capabilities of blastomeres were used for archetype distinction. Souèges attributed the highest importance in formation of the basic organs of the embryo to the blastomeres of the first generation (*ca* and *cb* cells). Then, in his opinion, both the number of cells increases and more discrete histogenic differentiation takes place; consequently, after the fourth generation **it is difficult to reveal features which would characterize a definite type**. In distinguishing archetypes, the laws of origin and destination, which come into power starting from the first generation, are regarded as basic. Other laws are used in the case of distinct types for comparison of number of blastomeres and tiers beginning from the second generation.

Secondary types are modifications of archetypes. The basis for their classification are features associated with convergence and manifested in different rapidity of divisions (for example, in the species of *Nicotiana* studied after the third generation, the embryo consists of six cells arranged in four tiers, and in *Hyoscyamus*, of eight cells in six tiers). The features are also manifested in different periods of differentiation of the blastomeres (for example, hypophysial cell), in different orientation of cell walls during development and in different structure of certain organs of the embryo (for example, suspensor).

For the overlapping types, typically that embryogenesis starts in accordance with certain archetypes and derivative types. Then, features of other archetypes and derivative types appear in embryo development.

The irregular types may differ among themselves. First, here belong the types of embryogenesis which lack clear regularities of development. In these types cell divisions occur along with formation of oblique walls, whereas the blastomeres of the embryo do not arrange in well-defined tiers. Second, come cases in which variability in embryogenesis is registered in a considerable number of representatives of one and the same species, and consequently embryo development can be related to different archetypes or derivative types.

Souèges (1939c) only included the archetypes and overlapping types in his embryogenetic classification. The embryogenetic system is disposed in a table, like a periodic table by Mendeleev, so that it is possible to determine precisely the place of any species in accordance with the laws of embryo development. It consists of two large divisions or periods: in the first period the laws of embryogeny are applied to the zygote, in the second only to the apical cell (the basal cell does not participate in formation of the embryo). In both periods megarchetypes are arranged vertically (major among the archetypes) and horizontally, the embryogenetic groups. In accordance with the principle of rundown, 6 megarchetypes are included in each period. Based on the difference in the structure of the tetrad of proembryo cells (series *A, B, C* in the first period and *A', B', C'* in the second), 16 embryogenetic groups are distinguished: groups 1-8 for the first and 9-16 for the second period. Thus, each period includes 48 embryogenetic types to maximum.

The first period of the system is characterized by the fact that the laws of embryogeny are only applicable to the zygote and its derivatives during the first four generations. Series *A, B, C* differ in structural types of the primary tetrads of proembryo cells emerging as a result of the second generation: *A*—T-shaped tetrad, *C*—linear, *B*—intermediate type (division oblique in the apical cell and transverse in the basal cell) (Fig. 69).

Unlike the first period, the laws of embryogeny in the second period concern the cell generations of the apical cell alone, since the basal cell does not take part in construction of the embryo proper, but only forms the suspensor. In this case the derivatives of the apical cell (the upper cell *cc* and the lower—*cd*) are likened to the two-celled proembryo: cell *cc* replaces the cell *ca* in its constructive functions and cell *cd*, cell *cb*. Transition from the first period to the second period goes by means of series *C*. It is represented by three subseries (*a, b, c*) on the ground of constructive potential of cell *cc*: *a*—cell *cc* divides vertically, *c* — transversally, *b* — obliquely. On this basis so-called secondary tetrads were distinguished. In each subseries two variants can be distinguished: (a_1, a_2, b_1, b_2, c_1, c_2) depending on the division mode of the lower derivative of cell *cc*. The variants a_2, b_2, c_2 are at the head of three new series *A', B', C'* of the second period. As a matter of fact, general regularities of tetrad formation recur on distinguishing of these series, but this process is raised, so to speak, one step up and starts anew from the apical cell of the linear tetrad of series *C*.

It is important to note that the table of the second period does not copy the first one, but makes a new next stage of the periodic system of embryonomic types. According to Souèges, there can be several periods in which the variants of the preceding period turn into the series of the following period.

Certain conceptions which Souèges assumed as a basis for his classification were criticized by several authors (Yakovlev, 1958; Crété, 1963, 1964; Lebègue, 1966; Méstre, 1967, 1981; Batygina, 1968a,b, 1969, 1974b, 1987a; Guignard, 1975a, b).

Unlike Souèges (1939c), who regarded the first megarchetype as initial (process of embryo evolution characterized by increase in the constructive potential of the apical cell which only forms the cotyledons of the embryo in this megarchetype), Yakovlev (1958) believed that primitive megarchetypes were characterized by more capabilities and, consequently, the sixth megarchetype is the most primitive (all the organs of the embryo, apart from the suspensor, arise from the apical cell).

Méstre (1964) highly appreciated Souèges's classification as perspectives for carrying out comparative research and determining the place of each newly described embryonomic type. However, he thought that an embryonomic type does not characterize separate species and genera of plants, but whole families, and is associated with general regularities of embryo evolution. Lebègue (1966) supported this point of view; he claimed that despite the variability in structure of a four-celled proembryo (Souèges distinguished different series on this basis), the general plan of development and structure of the embryo was often equal in representatives of homogeneous

A. First Major Division of First Period of Embryogenetic Classification	Series A	Series B	Series C					
	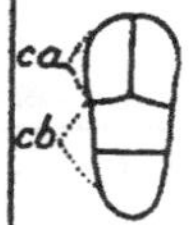	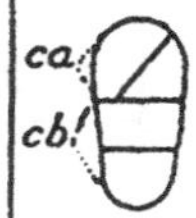	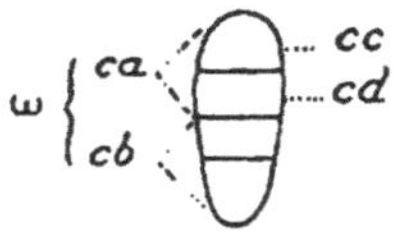					
			subseries a		subseries b		subseries c	
Megarchetypes			a1 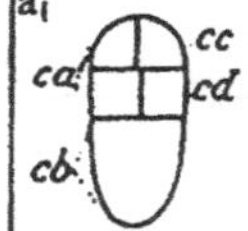	a2 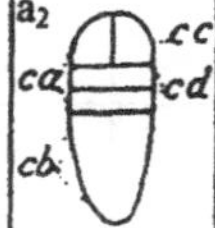	b1 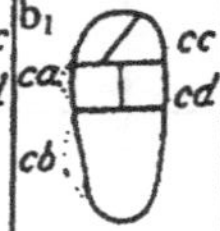	b2 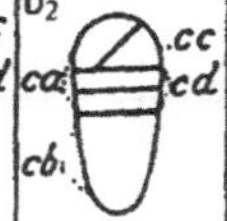	c1 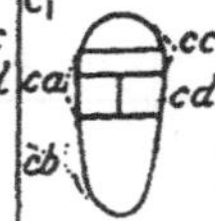	c2 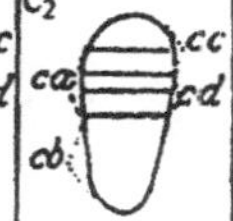
I. *cb=pvt+phy+icc+iec+co+s*	*Muscari* *Poa* (etc)							
II. *cb=phy+icc+iec+co+s*	*Senecio* *Urtica* *Lamium* (etc)	*Geum* *Ceanothus*						
III. *cb=½phy+icc+iec+co+s*	*Polygonum* *Salix*	*Viola*	*Chenopodium* *Amaranthus* (sometimes)	*Amaranthus* (sometimes)	*Myosotis*			*Amaranthus* (sometimes)
IV. *cb=iec+co+s*	*Myosurus* *Capsella* (etc)		*Helianthemum* *Radiola*					
V. *cb=co+s*			*Hyocyamus* *Papaver*	*Solanum* (some)	*Polemonium* *Echium* (sometimes)		*Echium* (sometimes)	*Solanum* (some) *Echium* (sometimes)
VI. *cb=s*		*Trifolium*						
	1st Group	2nd Group	3rd Group	4th Group	5th Group	6th Group	7th Group	8th Group

Fig. 69A: Classification of embryogenetic types by Souèges (1939c). A–First major division of first period of embryogenetic classification. See text for explanation of abbreviations.

B. Second Major Division of First Period of Embryogenetic Classification	Series A'	Series B'	Series C'					
	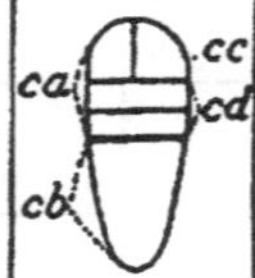	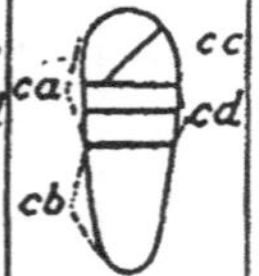	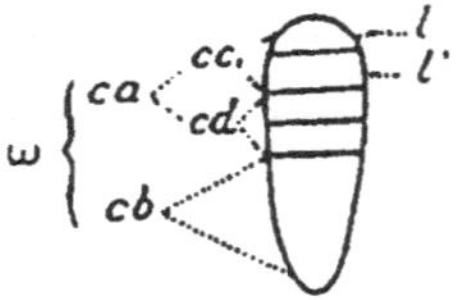					
			subseries a'		subseries b'		subseries c'	
Megarchetypes			a'₁ 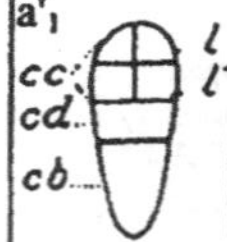	a'₂ 	b'₁ 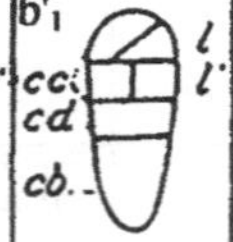	b'₂ 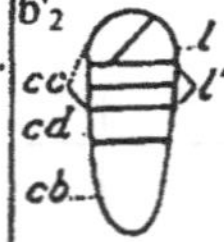	c'₁ 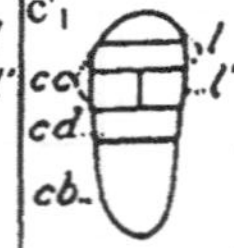	c'₂ 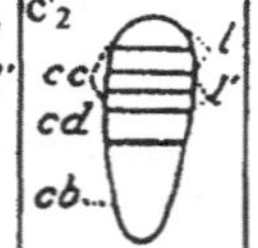
I. *cd=pvt+phy+icc+iec+co+s*	*Sagittaria*							
II. *cd=phy+icc+iec+co+s*	*Campanula Lobelia*		*Sagina Scleranthus*					
III. *cd=½phy+icc+iec+co+s*	*Samolus Portulaca*							
IV. *cd=iec+co+s*	*Hypericum Saxifraga Sedum*		*Drosera*					
V. *cd=co+s*			*Armeria Herniaria*					
VI. *cd=s*	*Medicago*							
	9th Group	10th Group	11th Group	12th Group	13th Group	14th Group	15th Group	16th Group

Fig. 69 (Contd.): B–Second major division of first period of embryogenetic classification. See text for explanation of abbreviations.

families and even orders. Further, he believed that in distinguishing megarchetypes not only the constructive capabilities of the cells of the two-celled proembryo should be taken into consideration, but also of the blastomeres of the four- and eight-celled proembryo. Not only should they be taken into consideration in the second and following periods when the lower derivative of cell *ca* hypothetically replaces in its constructive functions the basal cell, but also in characterizing of the megarchetypes of the first period.

Méstre (1967, 1981) suggested a shortened version of Souèges's system. He distinguished four periods (horizontally) and three megarchetypes (vertically). The first period was based on the regularities of zygote division, the second of apical cell division and the third and the fourth of the upper derivative of the apical cell during the third and the fourth generations. Theoretically, the number of periods is not limited, but in fact all the variations fall within the framework of the first two periods. Megarchetypes are established on the basis of constructive capabilities of the apical and basal cells in formation of the stem and root of the embryo: *A*—the stem and root emerge from the *cb* derivatives, *C*—both are formed from cell *ca, B—ca* forms the stem, *cb*—the root. In megarchetype *B* two subarchetypes are distinguished depending on whether the whole hypocotyl (B_2) or only its major part (B_1) arises from the apical cell.

Guignard (1975a, b) studied the peculiarities of initiation of the shoot and root apexes and concluded that the manner of embryo development is very similar in most of groups of flowering plants. Consequently, in his opinion embryogenetic features cannot be used to characterize major taxa, but only to specify phylogenetic interconnections. While studying embryogenesis in representatives of monocotyledons and some dicotyledons (for example, Fabaceae), he posited that megarchetypes I, V, VI must be excluded from Souèges's system since it is not cell *cb*, but cell *ca* which participates in the formation of the shoot apex in megarchetype I (monocotyledons), while in megarchetypes V and VI (Fabaceae) cell *cb* participates in both the construction of the suspensor (megarchetype VI) or in the suspensor and the root cap (V), and in the formation of the root apex.

Classification by Johansen (1945, 1950). This classification is a transformation of Schnarf 's classification (1929). Johansen changed the names of the types of embryogenesis, because he considered uniformity of names of types and families not quite right. He suggested that only the root of the Latin name of a family should be retained in the name of a type, and the suffix *ad* (= by, to) should be added to it, i.e., literally a type of embryogenesis belonging to a family. Further, he replaced the name Cruciferae-type with Onagrad-type, on the basis of Souèges conceptions (1937b) that descriptions of the Cruciferae-type are not to be based on embryogenesis in *Capsella bursa-pastoris*, the favourite embryogenetic subject, but on *Oenothera biennis* since its embryo, and especially its suspensor, is simpler in structure. According to Souèges, embryogenesis in *Capsella* is a derivative type or a variation. Classification was completed by one more criterion, character of zygote division. Using this criterion the Piperad-type was distinguished (unlike in the other types, the zygote divides longitudinally). It is interesting that despite the potential possibility of zygote division by a longitudinal cell wall as already indicated by Souèges (1937b), Johansen contrarily believed that in flowering plants such a division never occurs and that only in certain taxa, for example in *Scabiosa succisa*, can division close to longitudinal be traced.

Thus, the classification of embryogenetic types by Johansen (1950) is represented by six types: **Onagrad, Asterad, Solanad, Chenopodiad, Caryophyllad and Piperad** (Fig. 70). All the several pathways of embryo development within the types are regarded

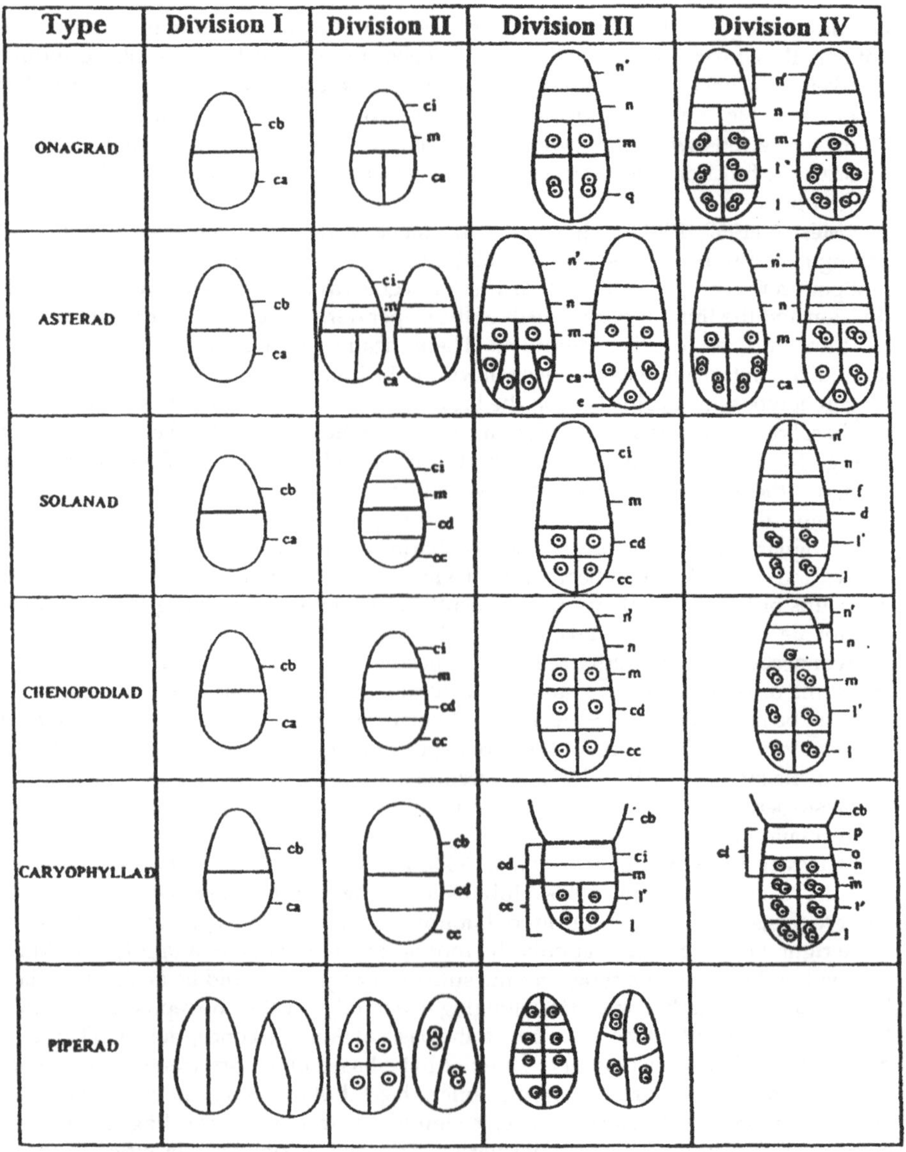

Fig. 70: Classification of embryogenetic types by Johansen (1950). *ca*—apical cell; *cb*—basal cell; rest of legend—derivatives of *ca* or *cb*.

as variations and hence named after the genera in whose representatives they were first described. It should be noted that Schnarf (1929) was the first to mention variations, more precisely modifications. On the other hand, in his classification Johansen used Souèges principles of system arrangement, since types, according to Johansen, are essentially Souèges megarchetypes and variations—derivative types.

Johansen's classification lacks a common approach to creating a base for distinguishing variations. For example, different criteria are used for different types of embryogenesis: for the Piperad-type—orientation of cell walls during the second division, structure of the suspensor and presence of cotyledons; for Onagrad-type—tiered structure of the proembryo after the third generation, the presence of quadrants, octants, epiphysis, hypophysis, and structure of the suspensor; Asterad-type—different destination of the derivatives of the basal cell, presence of epiphysis and hypophysis, structure of the suspensor; Solanad-type—number of tiers in the proembryo after the third generation, destination of tier 1, and structure of the suspensor; Chenopodiad-type—presence of epiphysis and peculiar divisions of the middle cell; Caryophyllad-type—presence of hypophysis and epiphysis, structure of the basal cell and number of cotyledons.

Batygina (1974) called into question the popular thesis about complete similarity of the initial phases of embryogenesis in monocotyledons and dicotyledons and was the first to demonstrate that differences in development of embryos do exist in some of them. She offered a new approach to determining embryogenetic types that took into consideration not only the early stages of embryo development, but also the character of embryogenesis on the whole as well as the structure of a mature embryo. On these grounds a new classification of embryogenetic types was proposed in which in addition to Johansen's classification (1950), several more unique types of embryo development were described.

One of them, **Paeoniaceae-** or **Paeoniad-type**, is characterized by a unique embryo development in the first stages (by coenocytic and later coenocytic-cellular stages of development of the sexual embryo—Yakovlev, 1951, 1958, 1983b; Yakovlev and Yoffe, 1957a-c, 1961) and by formation of a somatic embryo on this basis in the mature seed, which develops according to the Asterad-type (Batygina, 1989c, 1994; Batygina *et al.*, 1992; Brukhin and Batygina, 1994).

Another, the **Graminad-** or **Poad-type** (Batygina, 1968a, 1969, 1974b, 1987a), is distinguished based on the oblique arrangement of cell walls in the early period of embryo formation as well as organogenetic specificity in the successive stages of development.

Another classification of types of embryogenesis (Rutishauser, 1969) is a combination of the concepts posited by Schnarf (1929) and Johansen (1950), both in form and content.

Classification by Yamazaki (1974, 1982). This is mainly based on regularities of differentiation of root initials and hypocotyl and pertains only to dicotyledons. Mainly Souèges data were used and six major types of embryogenesis distinguished: **Asterad, Geraniad, Onagrad, Polygonad, Solanad, irregular** (Fig. 71). Variations were ascribed to a number of types. Though some type names were borrowed from Johansen's classification (1950), their diagnostics differs:

- Root initials are differentiated from derivatives of the basal cell **: at the expense of cell *m*** : tier *m* consists of several layers, the lower layer gives rise to root initials, the upper layer to the hypocotyl (Polygonad-type); tier *m* is multi-layered, cells of the upper layers give rise to initials of the root cortex and of the lower, initials of the root cap and part of the suspensor (Geraniad-type; in case of two-layered structure of tier *m* only the initials of the root cortex and of the root cap are differentiated—Myrtus variation); tier *m* is single-layered, the upper hypophysial cells separating as a result of transverse divisions form the root initials (Onagrad-type; cells of tier *m* may first divide longitudinally and then transversely—Ruta

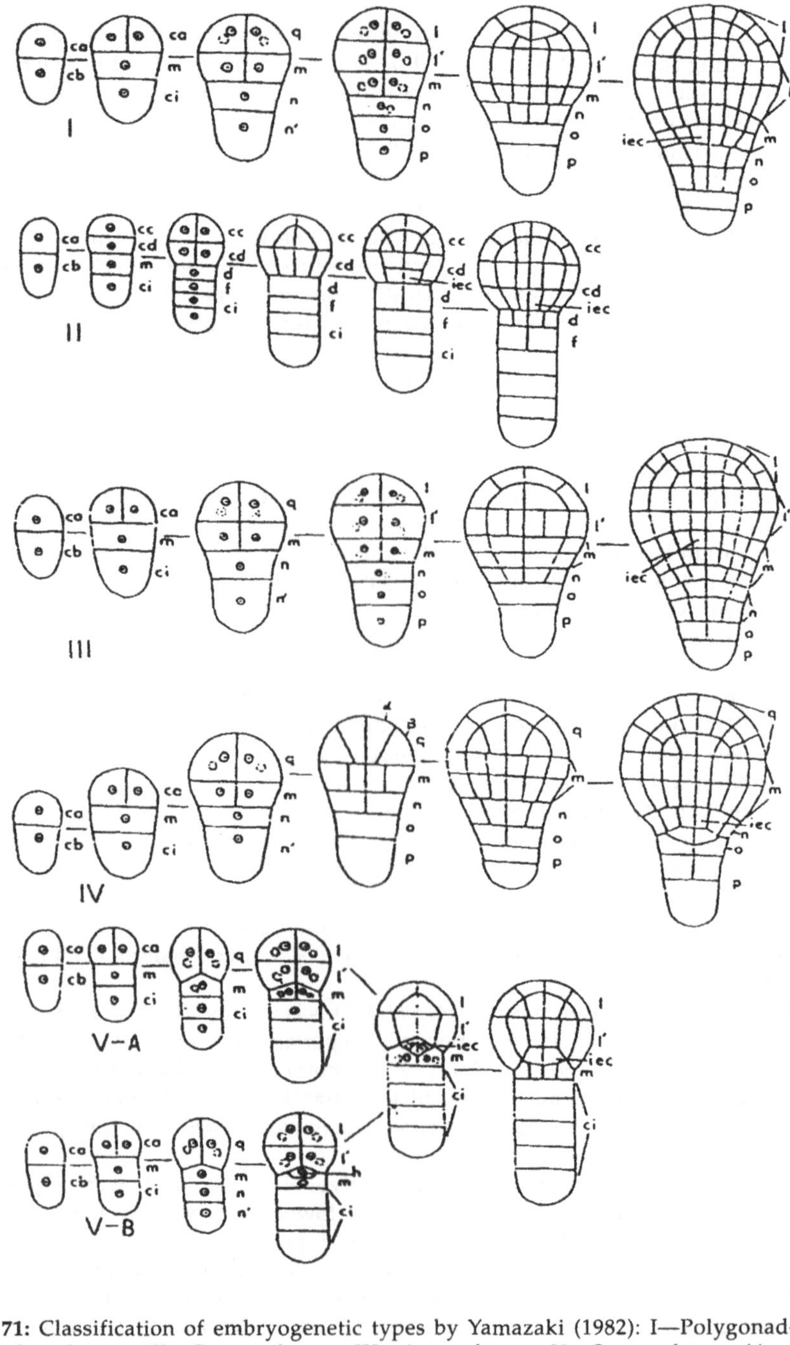

Fig. 71: Classification of embryogenetic types by Yamazaki (1982): I—Polygonad-type, II—Solanad-type, III—Geraniad-type, IV—Asterad-type, V—Onagrad-type (A—Ruta-variation, B—Onagrad-variation). *ca*—apical cell; *cb*—basal cell; rest of legend—derivatives of *ca* or *cb*; also see text.

variation); **at the expense of cell *ci***: root initials arise from cells of tier *n*, the upper *ci* derivative (Asterad-type).

- Root initials are differentiated from derivatives of the apical cell: tier *cd* consists of two layers, the upper cells become initials of the hypocotyl and the lower, initials of the root (Solanad-type).
- Irregular type characterized by the absence of regularities in differentiation of root initials.

Yamazaki considered a type of formation of root initials to be a relatively constant feature, which can characterize families and even orders.

Periasamy's classification (1977). Periasamy criticized the classifications existing at the time (Schnarf, 1929; Souèges, 1939c; Johansen, 1950), based on the regularities of proembryo formation. In his opinion, criteria at the heart of these classifications were arbitrarily chosen. He believed that in the process of embryogenesis correlation of the surface and volume of cells changes, and the main morphogenetic event, which is evidence of changes in the proembryo structure, is formation of the first inner cell. As a result of this, in the proembryo two groups of cells (outer and inner) differentiate and develop independently.

The classification of embryos includes groups, series and variations; thus, each level of hierarchy has its own criterion: the number of cell tiers in the proembryo when differentiation of the inner cells starts (group), arrangement of tiers with changing number of cells (series), arrangement of tiers in which the first inner cells are formed (variations).

So, at present there are several classifications of embryogenetic types based on different criteria. Most of them share a uniform approach to development—a **genealogical approach,** according to which embryo development is characterized by regularities which, as early as at the stage of proembryo, allow distinguishing competent cells giving rise to certain organs of the mature embryo in further development. Unlike classifications based on the genealogical approach, the classification by Periasamy (1977) is based on a **morphogenetic approach**. In his opinion, judging by initials at the 8- or 16-celled stage of development, the structure of the mature embryo might be mainly based on a topographic correspondence, rather than on actual proof of the origin of these parts of the embryo from certain initials.

In perspective, it must be dubbed a morphogenetic classification of types of embryogenesis which would include all the existing variety of modes of development and structure of the embryo. Despite certain difficulties in application of classifications for understanding the evolutionary issues of embryogeny, it is important to note that the simplest and most widely used classification is that by Johansen (1950). The suggested types and sometimes variations of embryogenesis are used in comparative embryological research in order to solve problems of systematics and phytogeny. Furthermore, a type of embryogenesis is normally characterized by strict regularities in the course of the first stages of embryo development (stages of zygote, two- and four-celled proembryo, quadrants and octants) which can also be used to control research in phylogenetic and molecular-genetic aspects of embryogenesis.

In the present work Johansen's classification (1950) was used, according to which all the distinguished types and variations of embryogenesis are described. In addition to the proposed variations, **the typical variation** has acquired an official status. According to Johansen, it is characterized by the same features as the type of embryogenesis proper.

PIPERAD-TYPE OF EMBRYOGENESIS

This type is characterized by either longitudinal or oblique arrangement of the cell wall in the zygote division, by the cruciform ('globular' according to Johansen) type of tetrad of the proembryo cells and by equal participation of the *ca* and *cb* derivatives in embryo formation; histogenic differentiation of the embryo in a mature seed comes to confinement of the protoderm, whereas a suspensor or cotyledons are not normally formed or remain rudimentary **(Plate XL)**.

For the first time this type was determined by Johansen (1945, 1950) on the basis of embryogenetic research in *Peperomia pellucida* (Jöhnson, 1900b; Fisher, 1914) and is named after family Piperaceae, to which genus *Peperomia* belongs. The special features of embryogenesis in this species are assumed as a basis for characteristics of this type (its **typical variation**).

Studies of other representatives of genus *Peperomia (P. blanda, P. eburena, P. maculosa* and *P. obtusifolia*) proved the presence of this type in family Piperaceae (Nikiticheva *et al.*, 1981) and completed the list of its characteristics. Nikiticheva (1981) suggested that this type should be called the Peperomia-type, since some authors classify genus *Peperomia* as a separate family, Peperomiaceae. The Piperad-type of embryogenesis is typical of other genera of family Piperaceae (Nikiticheva, 1981a,b). Sometimes transversal (*Peperomia*) zygote division occurs, which, however, does not change the further course of embryogenesis. In *Heckeria* and *Piper*, the suspensor is weakly manifested while *Peperomia* lacks one. In a mature seed, the embryo is globular and not differentiated into organs (*Peperomia*) or has rudiments of cotyledons (*Piper, Macropiper*) (Nikiticheva, 1981a,b). In *Peperomia peruviana* and *P. parvifolia* heterocotyly occurs, which is registered at the stage of the seedling (Hill, 1906, 1908). The Piperad-type of embryogenesis is registered in families Leitneriaceae (Johansen, 1950—according to data from Pfeiffer, 1912), Fagaceae (Brown and Mogensen, 1972), Cynomoriaceae (Teryokhin *et al.*, 1975), Ranunculaceae (Babis, 1976), Viscaceae (Teryokhin, 1987c) and Orchidaceae (Savina and Poddubnaya-Arnoldi, 1990).

Specific features of embryogenesis in representatives of other families in which zygote division is longitudinal, enabled distinguishing three more variations—**Balanophora, Scabiosa** and **Dendrophtoe**. The list of their characteristics was completed and corrected in accordance with the most recent data.

The Balanophora-variation was described on the basis of embryogenetic peculiarities in *Balanophora dioica* (Ekambaram and Panjie, 1935):

Zygote divides lengthwise, each derived cell of the two-celled proembryo also divides lengthwise, but the cell walls are curved so that both their ends attach to the primary longitudinal wall; the quadrants divide periclinally, separating the protoderm; the embryo is not differentiated into organs and the number of cells is considerably reduced.

Research in embryogenesis in *B.fungosa* subsp. *indica* var. *indica* (Teryokhin and Yakovlev, 1967; Teryokhin, 1977) proved that the type of embryo development in family Balanophoraceae is unique. However, the authors noticed that orientation of zygote division is variable and the division plane of the cells of the two-celled proembryo is always arranged at a right angle to the division plane of the zygote. Embryo development is completed at the third cell generation as well, although the cells of the protoderm may also divide anticlinally. According to one of the authors (Teryokhin, 1987a), development of the *Balanophora* embryo does not fit any existing classification of the types of embryogenesis of angiosperms.

Scabiosa-variation was described from embryogenetic data for *Scabiosa succisa* (Souèges, 1937a); this description was completed with more evidence on embryo development in other representatives of family Dipsacaceae (Souèges, 1957, 1963a,b; Kamelina, 1977, 1979, 1980):

The first zygote division occurs vertically, obliquely or almost horizontally; in the course of the two successive divisions, cell walls in each daughter cell of the two-celled proembryo are transversal and perpendicular (more rarely oblique) to the primary wall. The tetrad of cells is globular and subsequent divisions are often irregular; late histogenic differentiation of the embryo is typical; a suspensor forms, sometimes rudimentary. This variation is characterized by the presence of a mature embryo differentiated into organs. In *Cephalaria gigantea* the *ca* and *cb* derivatives do not equally participate in embryo formation, and development of the latter follows this scheme:

$ca = \frac{1}{2}\,pco + \frac{1}{2}\,pvt + \frac{1}{2}\,phy$
$cb = \frac{1}{2}\,pco + \frac{1}{2}\,pvt + \frac{1}{2}\,phy + icc + iec + co + s.$

Embryo development according to the Scabiosa-variation was described for certain representatives of families Lauraceae (Johansen, 1950—data from Cow, 1928) and Valerianaceae (Shugaeva, 1985).

The Dendrophtoe-variation is distinguished on the basis of the specific features of embryogenesis in representatives of family Loranthaceae:

Zygote divides lengthwise; both zygote and two-celled proembryo considerably lengthen; subsequent divisions are transversal, proembryo two-tiered, filiform; embryo proper forms from a pair of terminal cells, the long suspensor is two-, sometimes three-six-tiered, often spiralled. As in the Scabiosa-variation, the embryo in a seed is differentiated into organs, though its root is reduced. It is important to note that when Johansen distinguished this variation he also took into consideration such a feature as formation of multiple proembryos in the seed (due to processes of false polyembryony; Dixit, 1958), unlike other variations in which this phenomenon was not revealed.

Further investigations proved the presence of this variation in family Loranthaceae (Nikiticheva and Anisimova, 1987). Embryo development is similar in a representative of family Viscaceae, *Eubrachium ambicuum* (Bhandari and Indira, 1969), differing in that the first division of the zygote is transversal.

The Piperad-type of embryogenesis occurs both in orders at the bottom of the phylogenetic system and in highly specialized ones (Santalales and Balanophorales). Only the Scabiosa-variation is distinguished in the advanced group of sympetalous species in order Dipsacales.

ONAGRAD-TYPE OF EMBRYOGENESIS

This type is characterized by a transversal cell wall during zygote division, by the T-shaped disposition of four-celled proembryo cells and by formation of the embryo proper from derivatives of the apical cell, and the suspensor and hypophysis from derivatives of the basal cell **(Plate XLI).**

The name Onagrad-type of embryogenesis was introduced by Johansen (1950) instead of Cruciferae-type suggested earlier (Carano, 1915). Johansen distinguished this type on the basis of revision of data on embryogenesis in some representatives of family Onagraceae: *Oenothera biennis, Circaea pacifica, Godetia amoena* (Souèges, 1920c; Johansen, 1934, 1950) and others. Currently the Onagrad-type of embryogenesis has been determined for representatives of 88 families of both dicotyledonous and monocotyledonous plants. Among them are such species as *Capsella bursa-pastoris* and *Arabidopsis thaliana* (Brassicaceae), model objects for embryological and genetic research (Hanstein, 1870; Westermaier, 1876; Famintzyn, 1879; Schaffner, 1906; Souèges, 1914a,b, 1916, 1919; Yakovlev and Alimova, 1976; Meinke, 1985,1992; Laux and Jürgens, 1994; Jürgens and Mayer, 1994; Holding *et al.*, 1994) as well as a number of important agricultural plants: cereal crops (buckwheat), leguminous plants (haricot bean, soybean), vegetables (cabbage, mustard, turnip, small radish, garden radish), cucurbits (watermelon, marrow, *Luffa*), forage crops (lupine, sainfoin, lotus), and weed plants (broomrape).

General scheme of the Onagrad-type of embryogenesis:

$$ca = pco + pvt + phy + icc$$
$$cb = iec + co + s.$$

In the Onagrad-type Johansen (1950) distinguished 16 variations: **typical** (Onagrad-type itself), **Alyssum, Capsella, Catalpa, Euphorbia, Heloniopsis, Juncus, Lilium, Lotus, Lythrum, Mentha, Myosurus, Ruta, Sonerila, Trifolium** and **Veronica**. The variations are distinguished on the basis of the following criteria: presence or absence of octants, epiphysis, hypophysis; number of tiers in the embryo after the third generation; hypophysis and suspensor genesis. In our opinion, however, in descriptions of certain variations some of the criteria are not taken into consideration, which makes it difficult to carry out a comparative analysis of the characteristics of the variations and raises the question of whether some of them need to be distinguished. The clearest variations are those which differ in peculiarities of the hypophysis and suspensor formation.

Typical variation: cell *m* becomes the hypophysial cell and *ci* derivatives form a short (two-, three-celled) suspensor. This is typical of representatives of families Onagraceae and others (Johansen, 1930a, b, 193la, b, 1933, 1934; Souèges, 1935b; Lebègue, 1954, 1955).

Myosurus-variation differs from the typical variation in a uni-, double-tiered more or less long suspensor. This is typical of families Ranunculaceae, Magnoliaceae, Berberidaceae, Phytolaccaceae, Capparaceae, Moringaceae, Elaeagnaceae and others (Crété, 1951a, 1952; Phouphas, 1952; Ly Thi Ba *et al.*, 1970; Yamazaki, 1975; Dutt, 1978).

Capsella-variation: hypophysis developed at the expense of an upper derivative of cell *m*. Cell *ci* does not divide but becomes more or less swollen and vesicular, and together with the lower derivatives of cell *m* forms a long unitiered suspensor. This is typical of families Brassicaceae, Capparaceae, Datiscaceae and others (Raghavan, 1937; Crété, 1952; Rodionova, 1966, 1983a,b; Yakovlev and Alimova, 1976; Mansfield and Briarty, 1991).

Juncus-variation differs from the Capsella-variation in structure of the suspensor, which is short and two-, three-celled. This is typical of families Juncaceae, Cyperaceae, Centrolepidaceae and others (Juguet, 1969; Shamrov and Anisimova, 1993a,b; Nijalingappa and Tejavathi, 1977).

Lythrum-variation differs from Capsella- and Juncus-variations in that cell *ci* divides transversely. The suspensor is variable in number of cells and their structure. This variation is typical of families Begoniaceae and Vahliaceae (Raghavan and Srinivasan, 1942; Swamy and Parameswaran, 1960).

Among the variations distinguished on the basis of other features, the Trifolium-variation should be mentioned, which is characterized by formation of oblique cell walls in apical and basal cells and in their derivatives at the earlier stages of development. This is typical of families Lauraceae, Fagaceae, Caricaceae, Fabaceae, Apocynaceae, Arecaceae and others (Johansen, 1950; Goursat, 1969).

Later, a number of variations in the Onagrad-type were described. The **Philydrum**-variation is based on the peculiarities of cell wall formation in the course of apical cell division: the first and second divisions are not precisely median and lead to formation of unequal cells. This is typical of family Philydraceae (Kapil and Walia, 1965). The **Myriophyllum**-variation is characterized by a specific manner of formation and structure of the suspensor. It is represented by two swollen cells (with large polyploid nuclei), which form as a result of longitudinal division of cell *cb*. This is typical of *Haloragis micrantha* from family Haloragaceae (Nijalingappa, 1975).

In their study of representatives of family Orobanchaceae, Teryohin and Nikiticheva (1981) described 8 variations: **Aeginetia, Boschniakia, Christisonia, Cistanche, Diphelypaea, Epifagus, Mannagettaea** and **Orobanche**. To distinguish them they considered such features as cell wall position in the apical cell of the two-celled proembryo and in cells of the tier *l*, and the degree of differentiation of hypocotyl and embryo root initials.

ASTERAD-TYPE OF EMBRYOGENESIS

Characterized by formation of transverse cell wall in the course of zygote division, usually T-shaped disposition of the cells in the four-celled proembryo and more or less equal participation of derivatives of apical and basal cells of the two-celled proembryo in embryo construction **(Plate XLII)**.

Carano (1913, 1915) was the first to distinguish it as an Asteraceae-type on the basis data on embryogenesis for various representatives of family Asteraceae (*Bellis perennis, Calendula arvensis, Hehanthus annuus, Cichorium intybus*). Later, considering the results of investigations of Souèges (1920b, 1923a) in species of this and other families, this type was proved in systems by Chiarugi (1925) and by Schnarf (1929) (Asteraceae- and Astereen-type accordingly). Nowadays this type of embryogenesis is known for representatives of 66 families of dicotyledons and 16 families of monocotyledons. Among them are such important plants as sunflower, cotton and many others.

Scheme of the Asterad-type of embryogenesis:

$ca = pco + pvt$
$cb = phy + icc + iec + co + s.$

Johansen (1950) distinguished 10 variations in the Asterad-type: **Senecio, Geum, Erodium, Polygonum, Urtica, Lamium, Oxalis, Muscari, Poa** and **Penaea**. The main criteria used to distinguish these variations are: manner of formation of octants and presence of tiers *l* and *l'*, presence or absence of epiphysis and hypophysis, expense of *m* and *n* derivatives in construction of different tissues of hypocotyl and embryo root, and peculiarities of suspensor formation and structure.

As a typical variation of the Asterad-type, Johansen adduced the Senecio-variation described for the first time in *Senecio vulgaris* (Asteraceae) (Souèges, 1920b). In this case zygote divides transversally with formation of a two-celled proembryo. As a result of longitudinal division of the apical cell and transversal division of the basal cell a four-celled, T-shaped proembryo forms. The *ca* derivatives dividing lengthwise form quadrants (tier *q*). A characteristic peculiarity of the embryo structure in the course of transition to the next stage of development is diagonal disposition of cell walls, separating the quadrants, and as a result octants and tiers *l* and *l′* in the conventional sense do not form. The response of the derivatives of the basal cell—cells *m* and *ci*—differs in some ways: cell *m* gives the middle tier *m* whose cells afterwards divide lengthwise while cell *ci* divides transversally several times and forms cells *n*, *n′*, *o* and *p* arranged linearly one under the other. In the course of further development, the protoderm is separated due to periclinal division in tiers *q* and *m*. Later derivatives of tier *q* give cotyledons and a growing point, and those of tier *m* the upper part of the hypocotyl. Derivatives of cell *n* are initial for construction of the root cortex and derivatives of cell *o* for formation of the root cap and dermatogen; cell *p* initiates the suspensor. It should be noted that the typical variation of the Asterad-type is characterized by absence of a clearly manifested hypophysial cell. This variation is registered in some other representatives of the family Asteraceae as well as in Menyanthaceae and Apocynaceae.

Urtica-, Lamium- and Oxalis-variations as well as Senecio-variation are characterized by the presence of diagonal walls, separating quadrants into octants, and by the absence of clearly manifested hypophysis and epiphysis, but differ in the expense of cell *m* derivatives in formation of the hypocotyl and the root. The Urtica-variation is mainly typical of representatives of families Urticaceae and Bombacaceae, the Lamium-variation of Lamiaceae and Valerianaceae, and the Oxalis-variation of Oxalidaceae, Monimiaceae and some Asteraceae.

Among other variations of the Asterad-type, the Polygonum-variation is notable for the presence of classical octants, clearly manifested elements *l* and *l′* (derivatives of tier *q*) and hypophysis. This variation is registered in representatives of families Polygonaceae, Ceratophyllaceae, Fouquieriaceae, Combretaceae, Rutaceae, Asteraceae and Arecaceae.

The Geum- and Erodium-variations are basically characterized by differentiation of a particular wedge-shaped epiphysial cell which forms as a result of division of cell *ca* of the two-celled proembryo by an oblique wall. The main differences between these two variations are participation of the derivatives of cell *n* in construction of the suspensor (in the Geum-variation they participate, in the Erodium-variation they do not) and presence of the epiphysial cell (absent in Geum present in Erodium). Embryo development according to the Geum-variation occurs in representatives of families Pittosporaceae, Rosaceae, Cneoraceae, Vitaceae, Adoxaceae and others, and according to the Erodium-variation in Geraniaceae, Combretaceae and Thymelaeaceae.

According to Johansen, the Penaea-variation differs from all the other variations mainly in absence of a clearly manifested suspensor and in more or less equal participation of the *ca* and *cb* derivatives in construction of the embryo body. According to some authors (Cook, 1902; Joshi *et al.*, 1967; Souèges *et al.*, 1967), characterization of this variation can be completed with such features as oblique position of the cell wall in the course of zygote division, 'globular' (decussate or isobilateral) type of tetrad, characteristic disposition of cells *m* and *ci* in the four-celled proembryo (not one under the other, but side by side) and variability (irregularity) of cell divisions after the stage of quadrants. The Penaea-variation is known for representatives of families Penaeaceae, Nymphaeaceae, Cabombaceae, Melianthaceae, Biebersteiniaceae and some Caprifoliaceae.

Muscari- and Poa-variations were distinguished by Johansen for embryos of monocotyledons based on studies by Souèges (1924c, 1932). In both variations the *ca* derivatives only initiate the cotyledon, whereas the growing point is formed due to cells of tier *m*. Distinctions between these two variations become apparent in the third cell generation and consist in equal participation of the *m*, *n* and *o* derivatives in construction of embryo organs. In the Muscari-variation, tier *m* initiates a hypocotyl as well as a growing point, in the Poa-variation it gives the initials of the central cylinder, root cortex and partially coleoptile. However, whether to distinguish Muscari- and Poa-variations is subject to discussion. Some authors (Haccius, 1952; Guignard, 1961b; Haccius and Lakshmanan, 1966; Batygina, 1968a, 1974b, 1987a; Guignard and Méstre, 1969, 1971) showed that it is due to the *ca* derivatives that the growing point in the embryo of the cereal crops and other monocotyledons appears, and not to *cb*, as Souèges and Johansen believed (see below: Graminad-type of Embryogenesis).

For monocotyledons of the Asterad-type, based on some peculiarities of embryo development in representatives of family Commelinaceae, one more variation, **Commelina,** was distinguished (Maheshwari and Baldev, 1958; Souèges, 1958a,b; Chikkannaiah, 1962). Embryogenesis in this variation is similar to that in the Muscari-variation, but differs from the latter in absence of a suspensor, 'neighbouring' arrangement of cells *m* and *ci* in the four-celled proembryo, division of these two cells into different planes and, as a result, 'asymmetry' in the development of their derivatives, and formation of all (except the cotyledon) basic organs of the embryo from *cb* derivatives. In the opinion of Batygina (1974), such orientation of cell walls during the first divisions in the *Commelina* proembryo is manifestation of the dorsoventral character of its monocotyledonous embryo from the very first stages of development.

It is important to note that the above-mentioned features are indicative of considerable similarity of Commelina- and Penaea-variations distinguished in the Asterad-type for embryos of monocotyledonous and dicotyledonous plants respectively. Moreover, in some species of dicotyledons, in which embryogenesis belongs to the Penaea-variation (*Nelumbo nucifera*, Nelumbonaceae), from the very first stages of development the dorsoventrality typical of embryos of the monocotyledonous plants linked in particular with oblique disposition of the first cell walls in the embryo is distinct (Titova and Batygina, 1996). On the other hand, a considerable similarity in the same features is shared by the Penaea-variation of the Asterad-type, Trifolium-variation of the Onagrad-type and Scabiosa-variation of the Piperad-type, which is why it is difficult to distinguish between them (for example, Lauraceae, Fagaceae, Dipsacaceae). In the opinion of Natesh and Rau (1984), embryogenesis of the species characterized by the Penaea-variation should be studied anew. Apparently, the same could be said for all variations open to said question.

SOLANAD-TYPE OF EMBRYOGENESIS

Characterized by transverse disposition of the cell wall in the zygote, linear tetrad of the cell of the proembryo and embryo construction mainly from derivatives of the apical cell of the two-celled proembryo (only a suspensor and sometimes a hypophysis form from the basal cell) **(Plate XLIII)**.

Chiarugi (1925) was the first to distinguish the Solanaceae-type based on studies by Souèges (1920a, 1921) on embryogenesis in various species of Solanaceae. Later it appeared as the Solanaceen-type in Schnarf 's system (1929). At present, this type of embryogenesis is registered in 47 families of dicotyledons (Campanulaceae, Linaceae,

Papaveraceae, Teaceae, Hydnoraceae, some Rubiaceae and others) and monocotyledons (Asparagaceae, Trilliaceae and others).

General scheme of the Solanad-type of embryogenesis:

$ca = pco + pvt + phy + icc + iec$
$cb = co + s.$

In the classification of the types of embryo development, Johansen (1950) distinguished the following variations: **Hyoscyamus, Nicotiana, Physalis I, II, III, Datura, Papaver, Sherardia, Linum, Hydnora**. The following features were used as criteria for differentiating: number of proembryo cells at the stage after the third cell generation, presence of hypophysial cell, peculiarities of its formation and character of division, presence of the epiphysial cell, and peculiarities of construction and functioning of the suspensor. Six of the eight mentioned variations are registered within family Solanaceae.

One typical Nicotiana-variation of the Solanad-type is the embryo development in *Nicotiana tabacum* (Solanaceae) (Souèges, 1922b). In the course of zygote division the cell wall establishes transversally to the longitudinal axis of the embryo. Further transversal divisions in each cell of the two-celled proembryo lead to formation of a linear four-celled proembryo. Then octants form through division of the two upper tiers (*l* and *l'*) of the proembryo by longitudinal walls directed at a right angle to the primary wall, consequent to the protoderm is demarcated. The cotyledons, growing point, hypocotyl and embryo root form at the expense of the derivatives of the octants. The hypophysial cell is absent, and the plerome and periblem of the embryo root are formed from the lowest derivatives of tier *l'*. As a result of the transversal and sometimes longitudinal divisions of the cells of the two lower tiers of the proembryo (*m* and *ci*) a suspensor and root cap are formed.

Hyoscyamus- and Physalis I-variations are very similar to the above-mentioned variation, in fact they only differ in number of cell rows in the suspensor. The variation Datura is also similar and characterized by variability of number of cells in the proembryo after the third cell generation.

Variations Papaver, Physalis II and Physalis III are distinguished based on the amount of tier *l* derivatives expended in construction of the embryo proper. For instance, in the Papaver-variation, tier *l* derivatives only form the stem apex with preliminary separation of the epiphysial cell (at present registered in *Ostrowskia magnifica* of family Campanulaceae—Kamelina and Zhinkina, 1998). In Physalis II-variation, the cotyledons and hypocotyl are foı .ned at the expense of tier *l* derivatives, and in Physalis III-variation, the plerome and periblem of the embryo root are also formed.

The Linum-variation differs from the other nine variations in differentiation of the hypophysial cell.

The variation Hydnora is registered in representatives of genus *Hydnora* (Hydnoraceae). The two-celled proembryo continues transverse segmentation until a linear row of as many as fifteen cells is produced, but Johansen did not indicate the proportion derived from the basal and terminal cells respectively. Next, longitudinal walls are erected beginning in about the fifth cell from the terminal end and extending through the intervening cells as far as the ninth. These cells alone engender the embryo proper. The cells at the basal end gradually disintegrate, but those above the embryonic body persisted in the oldest embryos observed. The latter are undifferentiated. It is important to note that it is often difficult to distinguish variations in the Solanad-type since it is impossible to determine the character of the division of cells, which is due to variability of the location of the embryo inside a seed.

CHENOPODIAD-TYPE OF EMBRYOGENESIS

Characterized by transversal arrangement of the cell wall in the zygote, a linear tetrad of cells of the proembryo and formation of the embryo from derivatives of both the apical cell (cotyledon and stem apex) and the basal (hypocotyl, embryo root, suspensor) of the two-celled proembryo **(Plate XLIV)**.

The Chenopodiaceae-type was first determined by Chiarugi (1925) from data by Souèges (1920a) on embryogenesis in certain representatives of the family Chenopodiaceae. Later it was confirmed as the Chenopodiaceen-type in Schnarf's system (Schnarf, 1929). At present this type of embryogenesis is known in 20 families of dicotyledonous plants (Chenopodiaceae, Amaranthaceae, Polemoniaceae, Convolvulaceae, Boraginaceae and others.)

General scheme of Chenopodiad-type of embryogenesis:

$$ca = pco + pvt + \tfrac{1}{2}\,phy$$
$$cb + \tfrac{1}{2}\,phy + ice + icc + iec + co + s.$$

According to Johansen's classification (1950), three variations are distinguished in the Chenopodiad-type: **typical** (Chenopodiad-type proper), **Polemonium** and **Myosotis**. The following features were used as criteria in carrying out this differentiation: the number of proembryo cells after the stage of the third cell generation; presence of a hypophysial cell and peculiarities of its formation and character of division; presence of epiphysial cell; peculiarities of construction and functioning of the suspensor.

Embryo development in *Chenopodium bonus-henricus* (Chenopodiaceae – Souèges, 1920a) can serve as an example of a typical variation of the Chenopodiad-type. During zygote division the cell wall lies transversally, which results in formation of the apical (*ca*) and basal (*cb*) cells of a two-celled embryo. The subsequent transversal divisions in each of these two cells result in formation of a linear four-celled proembryo. In the course of further development in both derivatives of the apical cell of the proembryo, vertical walls lie at a right angle to the initial wall, and in this way tiers *l* and *l'* are formed. Cell *m* also divides twice by vertical walls and four cells result. As a result of two successive transversal divisions cell *ci* forms four cells, *h*, *k*, *o* and *p* which, except for cell *h*, give rise to a suspensor. Typically, there is no clearly differentiated hypophysis in this variation (Chenopodiaceae, Amaranthaceae, Lentibulariaceae).

Variations Polemonium and Myosotis differ from the typical variation mainly in the presence of an epiphysis. However, in the Polemonium-variation cell *m* of the tetrad divides transversely so that cells *d* and *f* are formed. Derivatives of cell *d* form the root cap and of cell *f*, part of the suspensor. From Johansen's point of view, this variation is transitional between the Chenopodiad-type and Solanad-type. The variation Polemonium was identified in *Polemonium caeruleum* (Polemoniaceae—Souèges, 1939b). *In* the Myosotis-variation, division of cell *m* is vertical, and its derivatives form a part of the hypocotyl and the central cylinder of the root. Formation of a clearly manifest hypophysis is typical. This variation can be found in representatives of families Boraginaceae, Polemoniaceae (*Polemonium pauciflorum*) and Convolvulaceae.

CARYOPHYLLAD-TYPE OF EMBRYOGENESIS

Characterized by transversal arrangement of the cell wall during zygote division, linear tetrad of proembryo cells and formation of the basic embryo organs from the apical cell of the two-celled proembryo **(Plate XLV)**.

It must have been distinguished by Schnarf (1929) as the Caryophyllaceen-type on the basis of research by Souèges (1924b) in embryogenesis of species of family Caryophyllaceae. At present this type of embryogenesis has been determined for 19 families of dicotyledonous plants (Cercidiphyllaceae, Crassulaceae, Droseraceae, Fumariaceae, Haloragaceae, Pyrolaceae and others) and 17 monocotyledonous (Alismataceae, Araceae, Ruppiaceae, Zannichelliaceae and others).

In the classification of embryo development types, 13 variations of the Caryophyllad-type were distinguished (Johansen, 1950): **Sagina, Vaccaria, Hypecoum, Corydalis, Sedum, Medicago, Myriophyllum, Drosera, Fumaria, Ruppia, Sagittaria, Zannichellia, Aglaonema**. To distinguish them, the following features are used as criteria: presence or absence of hypophysial cell, presence or absence and division mode of epiphysial cell, peculiarities of construction and functioning of the suspensor.

The Sagina-variation can be considered an example of a typical variation of the Caryophyllad-type. In the course of zygote division of *Sagina procumbens* (Caryophyllaceae) the cell wall establishes transversely. Afterwards, there is only cell *ca* which divides twice transversely and as a result a four-celled linear proembryo is formed. Cell *cb* becomes large, swollen, does not divide and takes part in construction of the embryo body. From *ca* derivatives, tiers *ci*, *m*, *l* and *l'* are formed. The cells of three tiers (*m*, *l*, *l'*) divide lengthwise, and those of tier *ci* divide transversely giving cells *n* and *n'*, as a result of which, after the third cell generation, an eight-celled proembryo of five tiers is formed. Further divisions in tiers *l*, *l'* and *m* also occur lengthwise, while cell *n'* divides transversely forming tier *o* and *p*. In the fourth cell generation a 16-celled proembryo is formed, whose cells are disposed into 6 tiers. From the cells of tier *l* a growing point is formed, from tier *l'*—cotyledons, from tier *m*—hypocotyl and from tier *n*—root cap. The suspensor forms at the expense of *cb* and *ca* derivatives (tiers *o* and *n*). It is noteworthy that the absence of a hypophysis and presence of an epiphysis are typical of the Sagina-variation.

Scheme of embryo development according to the typical variation of Caryophyllad-type:

$ca = pvt + pco + phy + icc + iec + co + \frac{1}{2}\, s$
$cb = \frac{1}{2}\, s$

The Vaccaria-variation differs from the Sagina only by presence of a long suspensor, described in some representatives of family Caryophyllaceae.

Variations Hypecoum and Corydalis differ from the Sagina in presence of a hypophysis. Further, in the Hypecoum-variation the suspensor is absent, while in the Corydalis-variation it is represented by two swollen cells with two or three nuclei. The variation Hypecoum is registered in representatives of family Hypecoaceae and the variation Corydalis in family Fumariaceae.

The Sedum-variation is characterized by the presence of a hypophysis and absence of an epiphysis. The basal cell *cb* divides actively and forms filiform ingrowths. The variation Sedum is known in representatives of family Crassulaceae.

In variations Medicago, Myriophyllum, Drosera and Fumaria, there is no hypophysis. However, unlike all the other variations of this type, in the last three variations the suspensor is normally formed only at the expense of the basal cell of the two-celled proembryo. The Medicago-variation of embryo development is described in representatives of family Fabaceae, Myriophyllum-variation in family Haloragaceae, Drosera-variation in family Droseraceae, and Fumaria-variation in family Fumariaceae.

Interestingly, the above nine variations were distinguished by Johansen for dicotyledonous embryos. For monocotyledonous embryos he distinguished variations

Ruppia, *Sagittaria* and *Zannichellia*. These variations are described for representatives of families Ruppiaceae, Alismataceae and Zannichelliaceae respectively. Lastly, variation Aglaonema Johansen described for certain acotyledonous species, in particular, representatives of family Meliaceae, noting however, that the latter is very similar to the Sagittaria-variation. Later, in addition to the 13 variations mentioned, another was distinguished in the Caryophyllad-type, namely **Phytolacca**, for one species of family Phytolaccaceae, *Phytolacca dioica* (Kajale, 1954).

New Types of Embryogenesis

GRAMINAD-TYPE OF EMBRYOGENESIS

This type of embryogenesis is characterized by a series of oblique walls (to the longitudinal axis of an embryo), which condition the dorsoventral character of the internal structure of the early proembryo and the mature embryo. It is also characterized by specific organogenesis and by a unique structure of main organs which are only typical of embryos of the cereal crops (scutellum, coleoptile, epiblast, coleorhiza, mesocotyl, plumule) **(Plates XLVI** and **XLVII)**.

This type was first distinguished while investigating of embryogenesis in *Triticum* (Batygina 1968a,b, 1969). Revision of data (Randolph, 1936; Maheshwari, 1950; Narayanaswamy, 1956; Chebotaru, 1960, 1970, 1972; Paulson, 1969; Khazova, 1970) and especially of illustrations on embryo development in other cereals (*Zea*, *Hordeum*, *Panicum* and others), as well as original data on embryogenesis in *Poa* (Batygina and Kirillova, 1971; Batygina and Mametjeva, 1979), supported our suggestion (Batygina, 1969) that the Graminad-type of embryogenesis is typical of all cereals (Batygina, 1974b, 1987a). At present, many works considering the embryo structure in cereals, justify distinction of the Graminad-type (Tzvelev, 1975; Natesh and Rau, 1984; Petrova *et al.*, 1985).

The characteristics of embryogenesis in cereals have long been discussed. Nörner (1881) made the first attempt to classify the embryo of cereals and distinguished three types of development: I) first and the second divisions transversal, II) first division transversal, second oblique, and III) both divisions oblique. He did not consider the dorsoventral character of the proembryo nor the specificity of embryo development, but took note of the two oblique walls in the proembryo (type III). Later, embryogenesis in cereals was attributed to Onagrad- (Yakovlev, 1946a,b) or Asterad-type (Poddubnaya-Arnoldi, 1976), to megarchetype I (Souèges, 1939c) or to megarchetype II (Guignard, 1961a,b). Guignard discredited Souèges idea about the origin of the growing point in cereals. He showed that the growing point is not formed from the basal cell, as Souèges (1924c) claimed, but from the apical cell of the two-celled proembryo. Guignard was the first to carry out detailed research in embryogenesis in cereals and proposed a fundamentally new scheme of embryo development in them:

ca = *pco* + *pvt*
cb = *phy* + *icc* + *iec* + *co* + *s*
versus Souèges (1924c):
ca = *pco*
cb = *pvt* + *phy* + *icc* + *iec* + *co* + *s*.

Maheshwari (1950) believed that **embryogenesis in cereals keeps within none of the existing classifications**; he argued **that there is no specific order in the direction**

of cell divisions. On that ground, embryogenesis in cereals **was determined as an abnormal type of development** ('unclassified or abnormal embryos'). To back up his conclusion Maheshwari adduced illustrations of the corn embryo (Randolph, 1936). Unfortunately, many researchers today, while studying embryogenesis in corn and other cereals continue to cite this work by Randolph. Later Norstog (1972), who studied the ultrastructure of the proembryo of *Hordeum*, supported Maheshwari's idea about irregular wall laying in the embryo. It was noted earlier that the cell walls in the proembryo of barley establish irregularly and at different angles to the embryo axis (Merry, 1941). However, Korobova (1962) persuasively showed that there is a certain regularity in laying of the first walls in the proembryo in *Zea mays*.

The zygote of cereals is a system characterized by a certain polarity of arrangement of cellular organelles and reserve materials. Its apical pole becomes a zone of protein synthesis, whereas the basal pole contains osmotically active substances.

As a result of diverse vacuolization of the zygote, the pattern of the first division in all cereals is asymmetric and displaced towards its apical end. Daughter cells developing as a result of **differentiating mitosis** prove to be **unequal**. The basal cell (*cb*) is considerably vacuolated and larger than the apical cell (*ca*).

In the zygote of wheat, symmetric mitosis with a truncated spindle of division is sometimes observed, which must be conditioned by a relatively close position of the nucleus to the cell wall of the zygote. In this case the degree of asymmetry of the spindle may determine different behaviour of chromosomes in the later anaphase and telophase (Geitler, 1935; Gerassimova-Navashina, 1951a,b). Such an asymmetric mitosis is a base for unequal cell division in the pollen of cereals (Batygina, 1974b). However, asymmetric mitosis is not an indispensable condition for unequal division since unequal cells may be formed in a symmetric pattern of division as well (Geitler, 1935; Maheshwari, 1950; Alimova, 1994; Golubovskaya, 1994).

The first cell wall, as seen in dorsoventral section, is arranged obliquely by the time of the second division. The oblique position of the first wall in the zygote in many plants (for example, in *Sorghum*) mainly depends on the oblique position of the spindle of division; in other plants (*Triticum*, *Poa*, and others) it also depends on one-sided growing of the zygote.

Thus, it is **already by a representative oblique position of the wall in the zygote that the specific dorsoventral structure of the embryo in cereals may be identified.**

The subsequent division always occurs in cell *cb*; this division is also differential and effects formation of cells *m* and *ci*. The second cell wall also establishes oblique to the longitudinal axis of the embryo, normally in the plane perpendicular to the first wall, rather than parallel to it, as was earlier believed (Souèges, 1924c; Roth, 1955; Guignard, 1961a,b), its upper end connecting with the wall between blastomeres *ca* and *cb*. The third division takes place in the apical cell of the three-celled embryo. The third cell wall is laid oblique to the first; however, its orientation to the axis of the proembryo may vary. **The specific arrangement of proembryo cells shows that the tetrad of blastomeres in cereals is of a special type.** The subsequent unequal division occurs in *ci*. The wall also establishes oblique to that dividing cells *m* and *n'*. Cell *m* divides vertically, forming two new blastomeres. The oblique division of cell *n'* effects formation of cells *o* and *p*.

A representative early dorsoventral structure of the proembryo in cereals, its developmental pattern and specific participation of blastomeres in embryo construction enable drawing up the following scheme of its genesis (Batygina, 1968a,b, 1974b):

ca = *pvt* + *pco* (partially scutellum and partially epiblast + coleoptile)
cb = *pco* (partially scutellum + partially epiblast) + *phy* + *icc* + *iec* + *co* + *s*.

According to the criteria assumed as a basis for distinction of classical types of embryogenesis (Schnarf, 1929; Souèges, 1939c; Johanses, 1950), **distinction of a new type of embryogenesis, namely Graminad,** is completely justified (Batygina, 1968a,b, 1969, 1974b).

Thus, the **Graminad-type is mainly characterized by a specific position of a series of oblique ('inclined') cell walls in the embryo at early stages of development (dorsoventral internal structure of the embryo) and by specific organogenesis,** rather than by position of the cell wall in the zygote.

The lateral position of the shoot apex and terminal position of the cotyledon in the mature embryo have been considered **secondary**, both in the ontogenetic and phylogenetic sense (Batygina, 1968a,b, 1969, 1974b, 1987a; Guignard and Mèstre, 1971; Natesh and Rau, 1984). It is important to note that the dorsoventral structure of the embryo is mainly conditioned by the dorsoventral character of those structures in which it develops (embryo sac, ovule, ovary).

Early stages of embryogenesis **in different species of cereals are similar**, whereas **the mature embryo construction differs slightly.**

Now we shall consider causes for disputes over the type of embryogenesis in cereals. We believe thát one of the major impediments to revealing a regular character of laying of the first walls in the proembryo of cereals is **methodological inaccuracy**. Most researchers have **not mentioned which plane of the embryo, embryo sac, ovule or ovary they were dealing with.** This has resulted in contradictory interpretations of embryological data on one and the same object and made it impossible to reveal specific regularities of the dorsoventral structure of the embryo in cereals. It is necessary to study the embryo structure in all the developmental stages **in three planes** (dorsoventral, bilateral and transversal) in order to obtain an accurate idea of embryogenesis in cereals.

Special attention should be paid to orientation of the embryo during electron microscopic research. When the section plane is not indicated, it is difficult to determine the accuracy of the author's conclusions regarding the character of laying of walls in the proembryo (this is also true for light microscopy). This is associated both with the specific internal structure of the embryo and the difficulties in creating a three-dimensional image due to the numerous ultrafine sections. The embryo of cereals has been a pet subject for geneticists, physiologists and plant-breeders, who pay special attention to the position of the first walls in the proembryo (Clark and Sheridan, 1986, 1988, 1991). That is why investigative techniques must be especially considered.

The other reason why the type of embryogenesis in cereals has become controversial is, undoubtedly, the **strong influence of distinguished authors** (Schnarf, 1929; Souèges, 1939c; Johansen, 1950), in particular, **their idea of complete similarity of proembryo development in all monocotyledons and dicotyledons.**

Analysis of data on early embryogenesis in cereals and in other representatives of flowering plants enabled positing a thesis about the **difference in embryo development in monocotyledons (in particular, cereals) and in most dicotyledons from the very first stages** (Batygina, 1974b).

Even if it were hypothetically possible to recognize fundamental similarity of embryogenetic laws for monocotyledons and dicotyledons, there would be strong doubts about the correctness of such an assumption when embryos of monocotyledons and dicotyledons are assigned to one embryogenetic type in existing classifications and, concomitantly, embryos of dicotyledons (even in one taxon) are assigned to different types, whereas **not a single type of embryogenesis is distinguished for embryos of monocotyledons.**

Furthermore, it should be emphasized that **in previous investigations on early embryogenesis in flowering plants the questions of how early the typical internal dorsoventral structure of the proembryo is formed, and in what way it relates to the dorsoventral character of the mature embryo were ignored**. Neither was the question of whether the proembryo in representatives of other taxa of dicotyledons and monocotyledons has a dorsoventral structure resolved (Batygina, 1973, 1974b). To some extent, we have considered all these issues in our study of formation of the monocotyledonous embryo in flowering plants (Batygina, 1974b, 1978a,b; see also Origin of Monocotyledonous Embryo: A New Concept).

Embryogenesis in *Commelina forskalai* is of utmost interest. The mature embryo of this plant has a dorsoventral structure. Specialization of the embryo in *Commelina* is so clearly manifest that it shows up at early stages of proembryo development. Analysis of illustrations and, in part, description of processes has shown that the orientation of cell walls during the first six divisions in the proembryo is equal to that in the embryo of *Triticum, Zea, Setaria, Hordeum, Poa, Avena* and others. Thus, in some other representatives of monocotyledons, a dorsoventral structure of the embryo is observed from the very first stages of development.

The same situation is observed in embryo development in *Nelumbo*—a plant placed among dicotyledons by many authors. The dorsoventral internal structure of the early proembryo in the lotus, which develops in conformity with the Penaea-variation of the Asterad-type, conditions the dorsoventral structure of the embryo at later developmental stages up to the mature state (Batygina *et al.*, 1983, Titova and Batygina, 1996). We observed a similar effect in the development of the somatic embryo in the seed of *Paeonia* (see below: Paeoniad-type of Embryogenesis), as well as the sexual embryo of *Scabiosa succisa* studied by Souèges (1937a) (Scabiosa-variation of Piperad-type). Thus, the presence of oblique walls during zygote division in the Penaea-variation of the Asterad-type and the Scabiosa-variation of the Piperad-type might be evidence of dorsoventral internal structure of the early proembryo in a number of dicotyledons. How the dorsoventral internal structure of the embryo in dicotyledons manifests in later developmental stages is still open to question.

While characterizing an embryogenetic type, earlier **researchers never took into consideration the structure of the mature embryo and peculiarities of embryogenesis as a whole**. Meanwhile, many times the idea was proposed (Poddubnaya-Arnoldi, 1964b, 1976; Batygina, 1968a,b, 1987a) **that in order to break through the formalism of existing classifications of embryogenetic types one should consider embryogenesis as a whole,** i.e., not only its early, though very important, developmental stages, but also the middle and late stages, describing the structure, shape, size and position of the embryo in the mature seed and the **ecology of germination**.

That is why, in order give more complete characteristics of the Graminad-type, we thought it necessary to consider the character of embryogenesis as a whole, including the structure of the mature embryo of cereals along with the 'classical' characteristic features used as a basis for determination of the embryogenetic type.

Cereals are distinguished among monocotyledons by shape and construction of the mature embryo. Researchers have paid special attention to different aspects of embryo development in cereals, such as morphology and comparative anatomy of embryos of *Triticum, Zea* and *Avena* (Avery, 1930; Arber, 1934), differentiation of certain organs of embryos in cereals (Roth, 1955, 1957, 1961; Pankow and Von Guttenberg, 1957, 1959; Von Guttenberg, 1960; Pankow, 1962; Modilevsky, 1963), and type of *Triticum* embryogenesis (Poddubnaya-Arnoldi, 1944, 1964b). A number of works are dedicated to the issue of evolutionary formation of monocotyledonous structure (Yakovlev, 1946a,b, 1951, 1971; Batygina, 1968a,b, 1974b, 1987a; Guignard and Méstre, 1969; Guignard, 1975a,b). The problem of homology of certain structures of embryo in cereals and of organs of other angiospermous plants is also discussed (Sokolovskaya, 1965, 1968; Danilova and Sokolovskaya, 1971, 1973).

There are many interpretations of the morphological nature of embryo organs in cereals. For example, the scutellum, coleoptile and mesocotyl are most often regarded as linked structures. At present, the scutellum is viewed as the only true cotyledon and the coleoptile as its outgrowth or as the first leaf of the plumule. As for the origin of the epiblast, there are two points of view. Some researchers (Velenovsky, 1907; McCall, 1934; Krasovskaya, 1952; Roth, 1955, 1957, 1961; Brown, 1960, 1965; Eames, 1961) treat the epiblast as a cotyledon. Others (Yakovlev, 1950; Brown, 1959; Sokolovskaya, 1965, 1968; Guignard and Méstre, 1971) consider the epiblast an outgrowth from the scutellum or the coleorhiza. We consider the scutellum, coleorhiza and epiblast to be derivatives of one cotyledon which acquired different functions in the evolutionary process.

To understand the nature of embryo organs in cereals it is necessary to know the anatomy of the conductive system of the embryo. Cereals were found to possess a special character of connection of the cotyledon vein with the conducting system of the hypocotyl (Danilova and Sokolovskaya, 1971). This enabled the aforesaid authors to endorse the idea put forward by Merry (1941) and Brown (1959, 1960, 1965) that the scutellum and the coleoptile are parts of a single cotyledon. They regarded the mesocotyl as the base of the cotyledon, which is fused with the hypocotyl.

The nature of the basal zone of the embryo in cereals is also subject to discussion. According to the classical interpretation, the coleorhiza is the bottom part of the hypocotyl, in which the main embryo root establishes endogenously (Velenovsky, 1907; Goebel, 1932). Some other researchers of cereals support this idea (Roth, 1957, 1961; Brown, 1965; Sokolovskaya, 1967, 1968).

Another point of view assumes that the coleorhiza in cereals is a metamorphosed main embryo root and the first root which establishes in the coleorhiza endogenously is an adventive one (Pashkov, 1951; Pankow and Von Guttenberg, 1957; Von Guttenberg, 1960; Guignard, 1975a,b; Guignard and Méstre, 1971; Tillich, 1977).

Sokolovskaya (1967, 1968) considered the coleorhiza to be an external part of the root cap, since these structures develop as a single unit during embryogenesis. She stated that there are no reasons for depriving the main embryo root of its proper morphological significance. It is of great interest that in cultivated embryos of *Hordeum* green leaf hairs develop on the scutellum, whereas on the coleorhiza root hairs are formed (Norstog, 1969).

The shoot apex of the embryo in cereals, as in many monocotyledons, develops on the basis of derivatives of the apical cell of the embryo. However, whether the shoot apex establishes laterally (Johri, 1984) or terminally (Batygina, 1969, 1974b, 1987a; Juguèt, 1969) is still subject to discussion. It is considered in detail below.

Phases of embryo development. Two phases of embryo development can be distinguished in cereals: (1) blastomerization (primary differentiation)—formation of critical mass of embryo cells, with differentiation of the epidermis as the final stage, and (2) organogenesis, including histogenesis—initiation of organs and their tissue differentiation.

The dorsoventral structure of the embryo manifests throughout its formation and becomes more apparent morphologically in later stages. However, during zygote and proembryo development it establishes due to different degrees of stretching of the cell walls on their dorsoventral and ventral sides, whereas during organogenesis the dorsoventral structure of the embryo manifests in a specific orientation and frequency of cell divisions and in the degree of stretching of cell walls. Due to its dorsoventral structure, the embryo of cereals is difficult to study. For example, the **lateral position of the zygote and the embryo in the embryo sac could only be observed when examined in profile; en face they are located in the centre.** The real disposition of cell walls in the embryo may likewise be revealed only through comparison of patterns of proembryo divisions in transversal, dorsoventral and bilateral sections.

Blastomerization, or phase of primary differentiation. The zygote in cereals starts dividing in 24 hours after fertilization (for example, in *Triticum*) and its shape varies in different cereals. In wheat it is normally pear-shaped but may change in the course of development. **Two unequal cells** form as a result of **asymmetric mitosis**, which occurs due to polarity of the zygote. The apical cell (*ca*) is smaller than the basal cell (*cb*), which is more vacuolate. Such an unequal division of the zygote has been observed in many species of cereals (Randolph, 1936; Yakovlev, 1946a,b; Chebotaru, 1960, 1970; Batygina, 1962a; Korobova, 1962; Khazova, 1970). The oblique position of the first cell wall in the zygote and of subsequent walls in the proembryo depends on the oblique attitude of the spindle in some cases, and on the uneven 'growing' of the zygote and proembryo as well in other cases. This 'growing' might be stimulated by the content of the pollen tube, which changes the character of growing of the zygote and embryo.

Embryo epidermis in wheat is characterized by basipetal development, i.e., it is first differentiated in the apical region and later in the basal region of the embryo.

Ultrastructural study of the first stages of embryo development in *Hordeum* revealed that all its cells are vacuolate; the vacuoles in the cells of the suspensor are bigger. The largest vacuoles are located along the periphery of the cells (Norstog, 1972). During the first stages of embryogenesis the number of organelles (ribosomes, mitochondria, plastids and others) increases and the ratio of organelles to zygote changes. At this stage of embryo development remnants of the content of the pollen tube may be observed. Concentration of ribosomes in the cytoplasm of embryo cells is lower than in the endosperm and dictyosomes have considerably smaller vesicles.

Organogenesis. As a result of pulsing divisions in certain cell clusters the embryo becomes multicelled and its organs and structures are formed. While the embryo is growing and developing, its shape changes considerably. During organogenesis the leading role belongs to successive changes in rhythm of mitotic activity and to orientation of cell divisions in different regions of the embryo, which conditions the specific position and structure of embryo organs in cereals. Wherever cell divisions take place, cells are smaller and richer in cytoplasm; wherever mitoses are rare, cells are bigger, more vacuolate and contain less cytoplasm. The extent of coloration of cells enables detections of centres of meristematic activity in the developing embryo. It is important to note that certain 'waves' of mitotic divisions are traceable. In the course of embryo development, when it have differentiated to a great extent, and when all the organs has been formed, mitoses are observed both apical and basal parts of the scutellum, as well as in some other regions of the embryo (root and others).

Several days after fertilization (in wheat, in 120-144 hours) the apical-lateral region of the proembryo starts growing towards the placento-chalaza. In this region cells actively divide; they are slightly bigger than the cells on the opposite side of the embryo, where a shoot apex will soon form. Further growth of the apical-lateral region of the embryo effects formation of the scutellum. In this period cells considerably increase in size in the upper central region of the embryo as well.

In the course of further development, complex transformation occurs and, as a result, the scutellum shifts from a lateral position to a terminal one, which is typical of a mature seed. Thus, in the embryo of wheat, **organogenesis starts with formation of the scutellum** rather than with formation of the coleoptile as Roth believed (1955, 1957, 1961). The shape and size of the scutellum vary in different species of cereals. While the caryopsis is developing and germinating, considerable changes occur in the cells of the scutellum.

The scutellum consists of heterogeneous cells. Its epidermis, adjacent to the endosperm, has a specific form and structure. In some varieties of wheat and corn, the epidermis is ruminated, in corn glandules may form in it. On the scutellum of some species a **ligule** (abdominal scale) is formed, which is an outgrowth of the scutellum. A bicarinate ligule appears in the embryo of rice; its tips are termed 'auriculae' ('bear's ears') (Saha, 1956; Tateoka, 1964). Its cells can be considered as transfer and secretory. A conducting system is differentiated in the scutellum. There are many aleurone grains and other inclusions in it. The main function of the scutellum is establishment of a connection between the endosperm and all the structures of the embryo. Transformations of substances coming from the endosperm into the embryo and further into the seedling occur in a specialized epithelial layer containing abundant hydrolytic enzymes. During germination, the scutellum functions as a source of gibberellins (Stoddart *et al.*, 1973).

Later on, after the scutellum has started growing, intensity of mitoses increases in the zone of formation of the **shoot apex** (growing point), which establishes terminally. Further development of this region leads to the appearance of a furrow separating the cotyledon and the growing point and then to separation of the **coleoptile**, differentiation of leaves and growing point of the plumule.

The coleoptile resembles a hollow cone with an aperture in the upper part through which the germ emerges during germination. The aperture is in the form of a fissure in which the inner and outer epidermis close (O'Brien and Thimann, 1967). Usually, there are many stomata on both surfaces of the coleoptile, some of which function as hydathode pores (Esau, 1977). The main function of the coleoptile is to protect the growing point while a germinating seed is 'pushing through' the ground rind. In some species of cereals the coleoptile can fulfil a photosynthetic function.

Establishment of the coleoptile and differentiation of a procambial band in the scutellum occur simultaneous with commencement of formation of the first embryo root; that is why in this period a region of mitotic activity (intensive cell coloration) can be seen in the central part of the embryo and in the base of the scutellum, where differentiation of the procambium starts. Elements of the xylem and phloem in the embryo root are grouped alternately along the circumference of the stele (Jung, 1937). Thus, root formation starts endogenously in the basal part of the embryo near the base of the scutellum. The presence of independent histogens of the plerome, periblem and root cap in the embryo allows us to attribute root development in cereals, in particular in *Triticum*, to the closed type (Von Guttenberg, 1960). The dermatogen of the root develops from the outer layer of the periblem.

During this period significant changes in the suspensor are observed: its cells significantly increase in size and become considerably vacuolated. The bottom cells of the suspensor become 'empty' and slough by the time of coleoptile differentiation (sometimes slightly earlier).

In some species of cereals the number of adventive roots established in the embryo may differ, for example in wheat 2-4, in rye 3-4 and in corn 5. Winter and spring varieties often differ in number of roots and sometimes their origin.

The root cap and **coleorhiza** (root vagina) form and develop as a single unit; only at the end of embryogenesis, when the embryo has matured, does the coleorhiza separate from the cap. According to some authors (Sokolovskaya, 1967), this separation results from the different rate of growth of the cap and the coleorhiza. In the cells of the coleorhiza of wheat, aleurone grains have been discovered. The main function of the coleorhiza is to protect the embryo root. Secondarily, it is a source of water and nutrients necessary for growth of the root (or roots) during germination. A fissure between the coleorhiza and the base of the scutellum is typical of panicled cereals (Reeder, 1957, 1961).

The epiblast is differentiated slightly later than the root and settles on the opposite side with respect to the scutellum, forming a scale-like outgrowth containing no vessels (Narayanaswamy, 1955a,b). It is similar to the coleorhiza in cell structure. In wheat (Diamant variety) aleurone grains have been discovered in some of its cells, which have been similar to those in the scutellum and coleorhiza. Judging from changes in shape and size of its cells during germination of the caryopsis, like the coleorhiza, it conducts water to the developing embryo. Some researchers consider the epiblast to be the second, reduced cotyledon. Size and shape of the epiblast differ in various species of plants; it is wanting in *Panicum* and *Zea* (Narayanaswamy, 1955a,b). Given the differences in functions of the scutellum (realizing haustorial role) and the epiblast (water absorption at the moment of seed germination) the later establishment and development of the latter is understandable. Unlike the scutellum, the epiblast does not come into contact with the endosperm, which fact might determine its function. The presence or absence of the epiblast and its size depend on growing conditions, very important at caryopsis germination.

Analysis of the different structures and organs of the embryo of the cereals has shown that there are two vascular bundles in the coleoptile. There is no vascular tissue in the epiblast, coleorhiza or ligule. The procambial band in the scutellum is branched to a great extent (Yakovlev, 1950). The conducting bundles of the scutellum and of the coleoptile enter the stele of the embryo axis in one and the same node, representing a single system (Danilova and Sokolovskaya, 1973).

Thus, the **epiblast, coleoptile** and **scutellum** constitute a rather **unified structure**. In connection with our study of the nature of the epiblast we shall propose some ideas concerning the structure of the embryonic shoot apex. Data on embryo development in wheat (Batygina, 1974b, 1987a) allow us to posit an opinion that **the apical part of the proembryo** may be treated as **an apical meristem of the stem and may be considered as the first shoot apex of a plant**. It is also possible that the shoot apex is formed after laying of the cotyledons in monocotyledons, from the *cb* derivatives. **One cannot oppose the apical pole of the embryo to the apical meristem of the shoot, since they are derivatives of the apical cell *(ca)*.** Cotyledons are formed as a result of activity of the apical meristem of the embryo; that is why the scutellum and epiblast may be respectively considered a cotyledon and the first, modified leaf. We cannot agree with the notion that the epiblast and the leaf (cotyledon) should not be considered homologous, because the epiblast has no procambial bands. In reduced leaves (squamous) procambial bands are not formed (Eames, 1961) since, due to the reduction, they have not reached the 'critical mass' at which differentiation of a conductive system starts. It is known that the conductive bundle is formed at a certain degree of differentiation of a leaf primordium. At one stage of embryo development, a plumule differentiates further. Inside it leaves lying in a ridge differentiate, the number of which differs in different species of cereals (2-4). A rudimentary shoot is an axis consisting of nodes and internodes. The internodes are proximal and look like a 'parcel' of flat discs. In each internode, a leaf is formed through intercalary growth. The size of the internodes depends on the species of plant and conditions of germination, that practically determine the plant's height. The growing point is formed at the expense of the initial cells of the apical meristem located on the very top of the shoot.

Evolutionary formation of the embryo of cereals. Evolution of cereals has brought **their specialization to a high degree**, in particular to a specific **dorsoventral** structure of embryos right **from the first stages of embryogenesis**; starting with the laying of the first wall, this specific structure is **manifest.** It is impossible to interpret normal embryogenesis in cereals without taking into account the phylogenetic position of an examined group of plants and the peculiarities of evolutionary formation of an embryo in monocotyledons on the whole. The origin of the monocotyledonous embryo has not been sufficiently studied, as evidenced from present-day debates over various aspects of this issue.

There are several fundamentally different hypotheses concerning the evolutionary formation of the cotyledon in the embryo of monocotyledons, in particular in cereals. A major issue of this problem is the question of the position of the shoot apex in the embryo of monocotyledons (lateral or terminal). Cereals are of great interest here since the dorsoventral structure of monocotyledons manifests from the very first stages of embryo development (Batygina, 1969, 1974b, 1976, 1987a). Analysis of existing data and results of research in wheat showed that actualization of displacement of the shoot apex to the lateral position during embryogenesis varies from group to group of monocotyledons (Guignard and Méstre, 1969, 1971). Guignard (1984) believed that acquisition of the monocotyledonous structure should be considered a progressive tendency. In most monocotyledons, considered to be more primitive, especially in Fluviales and in typical monocotyledons (for example Liliiflores) 'displacement' of the initial locus of the shoot apex to the lateral position occurs comparatively slowly and late during embryogenesis. Contrarily, in groups of advanced specialization, the dorsoventral structure of the embryo occurs so quickly that the growing point seems to be formed laterally at the moment of emergence. Cereals are a striking example. Detailed analysis of development of the embryo in *Triticum* enabled detection of a gradual 'displacement' of the zone of establishment of the shoot apex during ontogenesis from the terminal to the lateral position as a result of specific combination of allometric processes of growth, which are specific for embryogenesis of cereals. The lateral position of the growing point and the terminal position of the cotyledon in the mature embryo of wheat is a secondary phenomenon, both in the ontogenetic and probably the evolutionary sense (Batygina, 1968a,b, 1969; Paulson, 1969).

In the same manner noted by Guignard and Méstre, the initial shoot locus 'displaces' to the lateral position very early in embryogenesis in Pontederiaceae (Swamy, 1966), Juncaceae, Typhaceae, Cyperaceae (Guignard, 1961a, b; Haccius and Lakshmanan, 1966; Juguèt, 1969, 1972, 1973, 1989, 1992). However, in a recent work by Padmanabhan (1996) the idea was proposed that shoot apex development in some monocotyledons occurs laterally from the very beginning. This allowed the author to suppose more than one model of shoot apex formation in monocotyledons. Analysis of the literature and our own data on embryo development in monocotyledons allowed us to suppose that the apical part of the embryo, both in monocotyledons and in dicotyledons, could be treated as the apical meristem of the stem, and to consider cotyledons as the first pair of leaves forming on it just like primordia of the leaves appear on the stem apex (Guignard and Méstre, 1969, 1971; Batygina, 1974b; Batygina and Andronova, 1988; Padmanabhan, 1996). As belived by Padmanabhan, the presence of 'leaf' cotyledons with green plates, leaf stalk and axillary buds in the embryo of dicotyledons, is strong proof of the fact that the apical domain is the first meristem of the embryo shoot, and the cotyledon is a side organ relative to the polar axis. Jacques-Felix (1988) considered the cotyledon of a monocotyledonous embryo a prophyll of the lateral axis, regarding the rest of the embryo as a protocorm without the cotyledons. Padmanabhan (1993a,b) considered the cotyledon in *Commelina* and palms a prophyll, the top of which transformed into the haustorial organ.

However, the mechanism of formation of one cotyledon in monocotyledons, including cereals is still debatable (see Origin of Monocotyledonous Embryo: A New Concept).

While considering the problem of monocotyledonous embryo origin, it is important to pay special attention to the study of polarity and symmetry of the embryo (including internal symmetry), since both these features are of great importance in discussing the terminal and lateral position of the epicotyl meristem in the embryo of monocotyledons.

It is known that polarity is already established in the zygote, which, as Wardlaw (1955) believed, determines the polarity of the whole embryo. Padmanabhan (1996) upheld this concept because it is based on observations of embryos in which cell divisions run in a regular and predictable way. However, in some species of monocotyledons, derivatives of the zygote do not arrange in regular patterns, and it is hard to trace the terminal pole. In these cases the polarity of the zygote may not wholly determine the polarity of the embryo. Furthermore, new facts have recently appeared showing that the position of the axis of polarity of the egg cell and zygote can change during their development (Kropf, 1994). Undoubtedly, genetic works with mutants make an important contribution to the studies on polarity. An apical-basal model of the polarity axis formation has been suggested (the embryo of *Arabidopsis* taken as example) and one of several genes (*GNOM*) defining it (Mayer *et al.*, 1993).

Caryopsis structure. The indehiscent, dry, single-seed fruit of cereals is termed a caryopsis. It usually has a thin, closely adherent pericarp that creates the impression of their merger (Richard, 1808). The size of the caryopsis varies and the shape may differ also, from oval to ovoid to narrowly cylindrical (Tzvelev, 1982). Ventral, dorsal and lateral sides as well as a tip and hilum are distinguishable in the caryopsis. On the ventral side of the fruit in the place of merging of the margins of the carpel, a furrow is often formed, which is sometimes well manifested the entire length of the caryopsis (*Triticum aestivum, Festuca rubra*) or located only in the basal part of the fruit (tribes Paniceae, Cynodonteae); sometimes the furrow is lacking (tribe Oryzeae) (Petrova *et al.*, 1985).

The embryo is located at the base of the caryopsis near its wall and is turned towards its ventral side. Occupying but a small part of the whole volume of the caryopsis the embryo only adjoins the endosperm with the scutellum, rather than with its entire surface as in most monocotyledons.

In spite of the considerable similarity, embryos of cereals differ in genesis and structure. **Structural features of the mature embryo may be used for characterizing of structural types of cereals** (Bruns, 1892; Van Tieghem, 1987; Martin, 1946; Reeder, 1957, Guignard, 1961a,b; Kinges, 1961; Swift and O'Brien, 1972a,b; Petrova *et al.*, 1985). **Six types** have been distinguished: **true festucoids, true bambusoids, ozyzoid-olyroids, arundinoid-danthonoids, chloridoid-eragrostoids and panicoids** (Reeder, 1957). Embryos of wheat, rye, barley and oats belong to the festucoid type of development, embryos of corn to the panicoid type.

In festucoid cereals, the embryo occupies less than half of the caryopsis (*Dactylis glomerata, Festuca rubra*), in panicoid cereals (*Coix, Zea*) the embryo is large and occupies more than half of the caryopsis, in eragrostoid (chloridoid) cereals (tribe Cynodonteae: *Sporobolus, Eragrostis*) the embryo is medium in size (Petrova *et al.*, 1985). The embryo of Bambusa species is remarkable for its strongly grown cotyledon, which occupies the whole central part of the fruit, as well as for location of the plumule and the root curved in the same direction (Petrova and Yakovlev, 1968; Yakovlev and Kamelina, 1973). **Size ratio of endosperm and embryo in the fruit of cereals is of high taxonomic significance.** The endosperm of the mature caryopsis may be hard, soft, thick, loose and so on. Its consistence varies because of different proportions of protein and starch.

Cereals are characterized by hypogeal germination. The embryonic shoot, closed by the coleoptile, reaches the ground surface as a result of elongation of the first internode (Onderdonk and Ketcheson, 1972; Petrova *et al.*, 1985).

PAEONIAD-TYPE OF EMBRYOGENESIS

This type of embryogenesis is characterized in flowering plants by **unique** passing of the first developmental stages of the **sexual embryo** by formation of a coenocyte-cellular structure and incomplete embryo development (**lack of histogenic differentiation, cotyledons, shoot and root apices**), and by formation of a **somatic embryo (embryoid)** at the base of the coenocyte-cellular structure (Batygina, 1994a, 1999) **(Plate XLVIII)**.

It was formerly thought that the first zygote division and subsequent divisions in the proembryo and embryo are accompanied by formation of cell walls in all flowering plants. Yakovlev (1951) was the first to note that the first stages of peony **sexual embryo** development occur in an exceptional manner, not known for Angiospermae. In the **first division of the zygote nucleus and subsequent divisions the nuclei formed** (more than 1000) **in the 'proembryo structure' are not accompanied by the laying of cell walls and as a result the coenocyte structure forms first** (Yakovlev and Yoffe, 1957a-c). Later, at the periphery of the coenocyte-cellular proembryo 'meristematic centres' are formed (up to 25). Only one of the centres undergoes preferential development and finally develops into a **well-differentiated sexual embryo** (Yakovlev, 1983a).

Development of this new concept about the reproductive system of flowering plants (Batygina, 1978c, 1989a-c, 1990, 1992a, 1993a,b) as well as also experimental data concerning embryo development in different species of peony in natural conditions and in culture *in vitro* (Batygina *et al.*, 1978; Batygina and Butenko, 1981; Batygina *et al.*, 1992; Brukhin and Batygina, 1994) enabled major corrections in our understanding of peony embryogenesis and a new insight into the mature embryo which forms in the course of peony seed development (Batygina, 1999).

From our point of view, the Paeoniad-type of embryogenesis is characterized by **incompleteness.** The peony sexual embryo **lacks the developmental stages**

characteristic of the majority of angiosperms: two-celled proembryo, tetrad, quadrants, octants, histogenic differentiation of the embryo, and formation of cotyledons, shoot and root apices. Embryo formation ends in the early stages of development—at the stage of coenocyte-cellular structure, in which the epidermis is formed; viewed classically, the mature seed of peony lacks a sexual embryo.

One more peculiarity of the peony sexual embryo must be considered—**the tendency towards asexual (vegetative) reproduction expressed in the ability to clone,** the end result of which is this: at the base of the coenocyte-cellular structure in the developing seed a sexual embryo does not form (as reported by Yakovlev, 1951, 1983b), but rather **a set of somatic embryos (embryoids),** only one of which remains in the mature seed (Batygina, 1989a-c, 1994a; Batygina *et al.*, 1992; Brukhin and Batygina, 1994).

Thus, in the first stages of early ontogenesis the transition from heterophasic reproduction to homophasic occurs (Batygina, 1989 a-c, 1990, 1992, 1993a,b) (Fig. 72). Such switching-off programmes take place during formation of the coenocyte-cellular structure. In connection with this we think that at the **early** stages of ontogenesis (during **seed** or **post-seminal** development) in some species of flowering plants (peonies, orchids, some parasitic plants) a **peculiar 'metamorphosis'** occurs. It is expressed in the changing of morphogenetic pathways: embryogenesis (sexual) embryoidogenesis or somatic embryogenesis (asexual) (Batygina, 1993c). The asexual pathway of embryo formation in peony seed, as also in some other plant species (orchids etc.) for which **cleavage (monozygotic-embryonal embryoidogeny)**, as well as **nucellar and integumental (ovular)** are characteristic, represents **reproductive embryogeny** (see: Embryoid, Embryoidogeny).

From our analysis of the peony seed **two types of embryos form successively, sexual and somatic (embryoid), which differ in origin.** The sexual embryo forms as a result of the **sexual process** (syngamy)—**heterophasic reproduction. The end** of its development is **formation of the coenocyte-cellular structure** with epidermis; **successive differentiation of the sexual embryo does not occur.** Later, **formation of somatic embryos begins** (homophasic reproduction—2n = 2n), arising from certain cells of the coenocyte-cellular sexual proembryo. In the course of subsequent development usually one of the somatic embryos differentiates into organs while the rest perish. Thus, the end result of these complicated morphogenetic transformations in the mature peony seed is usually a **somatic embryo** (and generally only one), and **not a sexual embryo,** as reported by Yakovlev (1951, 1958, 1983b).

Formation of numerous somatic embryos at the coenocyte-cellular structure in peony is, in our opinion, no more than the process of **sexual embryo cloning** in an early stage of its development (cleavage, monozygotic embryoidogeny) (see Embryoidogeny).

Let us take a closer look at those features of embryogenesis commonly found in peony.

Development of sexual embryo. Like most flowering plants, peony has the process of double fertilization. No deviations during syngamy and triple fusion were noted. At the zygote division karyokinesis is not accompanied by cytokinesis, and hence a two-nucleate coenocyte is formed. In the second division, in which nuclei divide synchronously, a four-nuclear coenocyte forms. During nuclear divisions sometimes in the middle of the fragmoplast a 'granular little road' is initiated—the rudiment of the future cellular plate; however, the plate per se does not form. In the course of further development the coenocytic structure with 16 nuclei and more appears. At the

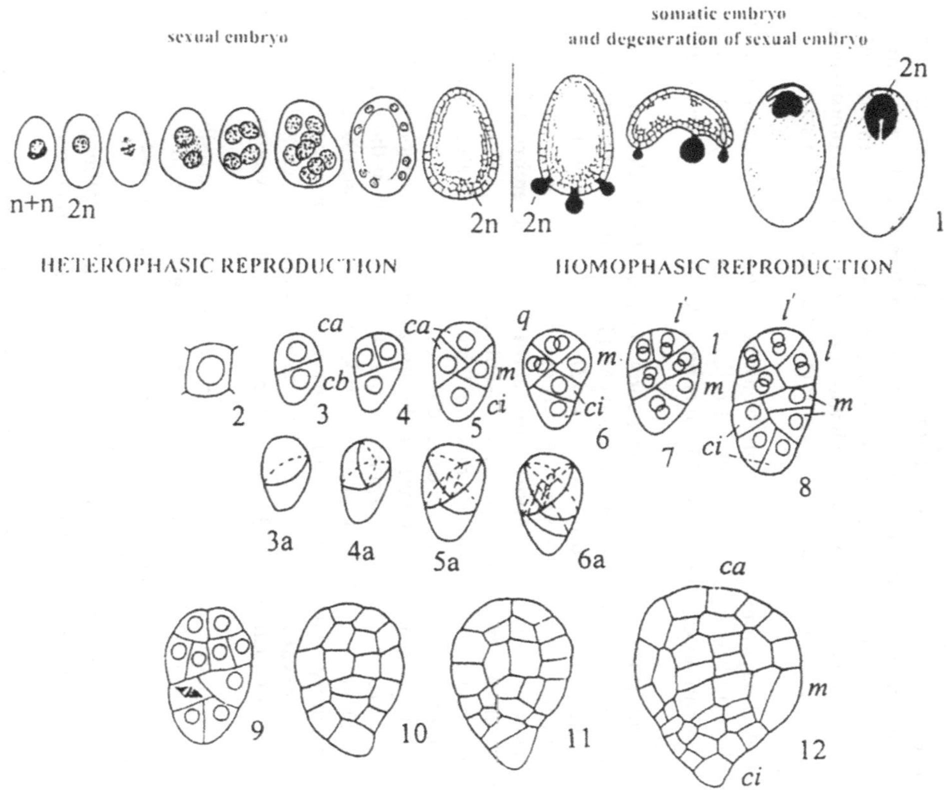

Fig. 72: Embryo development in seed of *Paeonia* species: 1—switch-over of developmental programme from heterophasic reproduction to homophasic reproduction; 2- 12— subsequent stages of embryoid development (3a- 6a— stereoscopic image of embryo). Bar— 50 µm. 1— after Batygina, 1997 a, with modifications; 2- 12— after Brukhin, Batygina, 1994. *ca*— apical cell; *cb*— basal cell; rest of legend— derivatives of *ca* or *cb*.

beginning of development coenocytic nuclei are regularly distributed throughout the embryo cell. However, commencing with the eight-nucleate stage, the large central vacuole is formed in the embryo (consequent to small vacuole fusion), and nuclei are distributed at the coenocyte periphery. The synchronous mode of division is disturbed after approximately 7-8 successive nuclear divisions.

Coenocytic-cellular embryogenesis stage. On reaching a certain critical mass (stage of approximately 400-500 nuclei) cell wall formation begins. In *Paeonia anomala* this occurs 15-18 days and in *P. lactiflora* 18-19 days after pollination. Cellularization in the coenocyte passes basipetally, from the peripheral part of the embryo to the centre, where nuclear divisions continue without cell wall formation. This is accompanied by gradual diminishing of the central vacuole volume and appearance of a set of rather large vacuoles.

Between 19-21 days after pollination, in the formed coenocyte-cellular structure, due to periclinal divisions of the peripheral cells, the epidermic layer begins to differentiate asynchronously; the cells of this layer have denser cytoplasm and slightly increased nuclei.

Development of the sexual embryo in peony is completed with formation of the epidermal layer, and this is the only feature of its **histogenic differentiation.**

Development of somatic embryo (embryoid). After initiation of the cell walls and isolation of the epidermal layer in the coenocyte-cellular embryo, cellular differentiation begins, and these cells are **the initials of the somatic embryos**. They are distinguishable among the other cells by meristematic character, cytoplasm more densely coloured, high nuclear-cytoplasm ratio and thicker cell wall. Initial cell cytoplasm contains more protein than the surrounding cells.

Embryoid initiation occurs preferably in the apical part of the coenocyte-cellular sexual embryo, exogenously, at the expense of periclinal divisions in the epidermal layer cells (Batygina *et al.*, 1992; Brukhin, 1993; Brukhin and Batygina, 1994).

The time of embryoid initiation as well as their number and developmental rhythm in various species are variable. For example, in *P. anomala* initiation occurs '22-24 days after pollination, in *P. lactiflora* 23-30 days, and the number of embryoids is 7-10 and 9-17 respectively.

In spite of the initiation of such a great embryoid number, usually only one develops to the state of differentiation into organs, the rest perish in the early developmental stages. Only a few cases of the presence in the mature seed of *P. anomala* of two completely formed embryoids have been recorded.

From the very moment of initiation up to almost complete differentiation the embryoid conserves the connection with the coenocyte-cellular structure of the sexual embryo from which it originated.

However, it is necessary to note that from the initiation stage of embryoids a gradual loss of turgor is seen in the coenocyte-cellular structure, lysis of interior cells and diminishing of its volume; at the stage of mature seed it usually degenerates completely. The coenocyte-cellular structure degenerates at different stages of seed development in different species.

Development of peony embryoids in the seed is subject to certain regularities. The first transverse division in the embryoid initials on the coenocyte-cellular structure leads to formation of apical (*ca*) and basal (*cb*) cells in a two-celled embryoid. Later, the apical cell is divided by an almost oblique partition, while the basal cell gives rise to two cells—middle *m* and lower *ci*. Every derivative of the apical cell is divided by a partition oriented perpendicularly to the first oblique wall, and forms quadrants (tier *q*). At the next division these quadrants are the origin of octants and tiers *l* and *l'* are distinguished. The upper tier later forms embryo cotyledons. Cell *m* similarly divides into four quadrants, which later give rise to the hypocotyl and root plerome initials. Following divisions in cell *cb*, derivatives effect formation of the root periblem and root cap initials. Since basal and apical cells participate almost equally in the embryoid body formation, and there is no suspensor, it may be taken that development of the somatic embryo in peony passes according to the Asterad-type, Penaea-variation by Johansen classification (1950).

Usually formation of the epidermal layer in the embryoid happens at the globular developmental stage. Then in the subepidermal layer of the embryoid apical part, periclinal divisions can be observed and cotyledons are formed (heart-shape stage, 35-38 days after pollination). As cotyledons continue to grow (torpedo-shape stage, 38-40 days after pollination), procambial bundles differentiate in them and the shoot apex is weakly expressed. Some asymmetry in the developing embryoid (cotyledons differ in size) draws attention. The embryoid occupies approximately one-tenth of the mature seed.

The somatic embryo, which has received preferable development among other embryoids **formed on the coenocyte-cellular structure**, passes the stages **peculiar to the sexual embryo of dicotyledonous plants,** i.e., **globular, heart-shape** and **torpedo-shape** stages.

Experimental data affirm **the identity of peony somatic embryo** development *in situ* (in the seed) and **in culture** *in vitro* (Batygina *et al.*, 1992; Brukhin, 1993; Brukhin and Batygina, 1994; Batygina, 1999).

EMBRYOGENESIS IN ORCHIDACEAE (Plate XLIX)

Orchidaceae is one of the families characterized by high diversity in morphology of vegetative and generative structures. It numbers about 25,000 species (Dressler, 1981). In spite of great interest shown in these remarkable plants, barely more than 60 species have been studied embryologically.

Embryogenesis. The most considerable achievements in the study of embryogenesis and post-seminal development of orchids were realized in investigations by Veyret (1965, 1974). Based on analysis of published and original data from research on the embryo and seedling of 34 species of 21 subtribes[24] a conclusion was drawn that early stages of development of the embryo in orchids progress in the same manner as in other representatives of flowering plants. The zygotic division which is usually transverse leads to formation of a two-celled proembryo. As a result of division of the apical (ca) and basal (cb) cells a four-celled proembryo forms (stage of tetrad, or primary tetrad[25]). Further differentiation of the proembryo can pass on the basis of the primary (series of the first period) or secondary tetrad which forms due to divisions of the derivatives of the apical cell of the four-celled proembryo, i.e., in this case the *cb* cell does not participate in construction of the embryo proper (series of the second period). Four types of primary and secondary tetrads are found in orchids, A_1, A_2, C_1,

[24] Veyret (1965) used Schlechter's system of classification of orchid species. At present the taxonomy of orchids has been substantially reviewed (Dressler, 1981).

[25] Here and further on according to the classification by Souèges (1939); see also Principles of Classification of the Types of Embryogenesis.

C_2[26] (or isobilateral, T-shape, inversely T-shape and linear types respectively). Formation of *B* type tetrads (i.e. with the oblique cell wall during division of the *ca* cell) is not reported in orchids. In a number of cases different types of tetrads were observed in one and the same species. Thus, linear and T-shaped tetrads are found in *Polystachya microbambusa* (Veyret, 1965, 1974), *Dactylorhiza baltica* and *D. incarnata* (Andronova, 1988), *Orchis purpurea, O. provincialis* (Lagutova and Chebotaru, 1992) and *Calypso bulbosa* (Yeung and Law, 1992); isobilateral and T-shape types in *Epidendrum vitellinum* and *Polystachya geraensis* (Veyret, 1965, 1974).

Further development of the proembryo is characterized by stages of quadrants and octants. Depending on the species, the derivatives of the *ca* cell as well as some of those of the *cb* cell pass these stages. Tiers *l* and *l'* are represented by derivatives of the apical cell. The embryo can also have the *m, ci* and other tiers subject to which derivatives of the *cb* cell participate in formation of its body. No features of histogenic differentiation except for distinguishing of the epiderm initials due to periclinal divisions of cells of all tiers were revealed in the investigations by Veyret. Heterogeneity of cells in size was detected in the epidermis layer (Johansen, 1950).

It was shown in *Gymnadenia conopsea* that longitudinal divisions in cells of the tiers in octants (*l* and *l'* tiers) correspond to the divisions as a result of which the initials of the periblem and plerome get separated (Shamrov and Nikiticheva, 1992). The *m* cell and its derivatives undergo transverse divisions forming a row of cells. The upper cell of this row becomes the hypophysary (*h*), while the other give rise to a filiform suspensor. However, the authors have emphasized that they distinguished these histogens according to their position in the embryo proper. No structural differentiation of cells of various histogens was observed. In Johansen's opinion (1950) the presence of a hypophysary cell in embryos of *Orchis, Gymnadenia* and other species of orchids similar to them is debatable. He argued that though a cell exists which by analogy with the embryo of other flowering plants has a position typical for the hypophysary, its role in the embryo of orchids is reduced to participation in building the mass of suspensor cells.

There is an opinion concerning zonality which is intrinsic to the embryo of orchids and distinguished on the basis of differences in cell size (Veyret, 1965, 1974; Batygina and Vasilyeva, 1983c; Batygina and Shevtsova, 1985; Shamrov and Nikiticheva, 1992) or ultrastructural organization (Zhukova and Savina, 1978). Zonality is formed at the early stage of development given the fact that the tendency to divide remains only in cells of the epidermis and underlying cells of the apical part of the embryo, while the other cells soon lose their ability to divide and grow considerably in size, turning into parenchymal cells. Correlation between size of zones (small-celled and large-celled ones) varies depending on the species of plant and the stage of embryo development. In the period of intensive growth of the embryo the small-celled zone consitutes a part that amounts from ½ to ⅓ of the embryo volume, its cells dividing rapidly. As the seed ripens divisions in the apical part of the embryo cease. In some species zonality still remains at the stage of the mature embryo. The small-celled zone in *Goodyera repens, Polystachya microbambusa* (Veyret, 1965), *Thunia alba* (Andronova, 1988) makes up ⅓ of the embryo volume or less, and in *Bletilla striata* (Andronova, 1988) about ½ of it. In other species—*Epipactis palustris, E.atrorubens, Limodorum abortivum, Cephalanthera ensifolia*—such a zonality in the embryo of the mature seed is feebly marked or totally absent (Veyret, 1965).

---

[26] Designation of the tetrad types of embryo according to the classification by Souèges (1936,1939) modified by Veyret (1965, 1974).

In most species the embryo in the mature seed has histological differentiation manifested in distinguishing just three tissues: epidermis, parenchyma and meristem in the apical part. There is no differentiation of the elements of the vascular system.

Localization and dynamics of nutrients during embryo development. In *Calypso bulbosa* some starch grains first appear in the basal cell of the three-celled proembryo. By the time the seed has ripened it is the main reserve product for all the cells of the embryo. Along with starch some lipid drops also occur in the cells. All the cells of the mature embryo except for the basal have a dense cytoplasm and give a positive reaction to proteins during staining by amido black 10*B*. However, no single protein bodies were observed in the cytoplasm of cells of the embryo (Yeung and Law, 1992).

In *Gymnadenia conopsea,* accumulation of starch and proteins is associated with different tiers of the embryo: the small-celled apical zone is characterized by a high starch content, while proteins accumulate in the cells of the basal region (Shamrov and Nikiticheva, 1992).

In *Dactylorhiza baltica*, starch appears in the cells of the embryo at the 4-6-celled stage and gradually increases during the course of its further development. However, when the embryo passes to ripening, the starch has been completely utilized and synthesis of reserve proteins commences. In the mature seed large protein globules are evenly distributed over all the cells of the embryo. As starch is used, lipid drops together with proteins start to appear in the cells of the embryo. Starch grains do not always accumulate in embryos of other species of genus *Dactylorhiza*. However, the dynamics of protein and lipid accumulation is similar to that of *D. baltica* (Andronova, 1988). In mature embryos of *Cypripedium acaule* an aggregation of proteins and lipids has been observed but no starch (Leroux *et al.*, 1995).

Thus, the present data argue that by the time the seed has completely ripened all the cells of the embryo, regardless of size and location, are filled with reserve nutrients (proteins and lipids or starch and lipids). This supports the current opinion that the embryo of a mature seed is devoid of a meristematic tissue (Nishimura, 1991). However, despite the fact that in the apical part of the embryo the meristem stops functioning in the period of seed ripening, meristematic activity recommences in this region during its germination.

Peculiarities of embryo differentiation in orchids. Weak differentiation of the embryo is associated with absence or poor development of the endosperm (Poddubnaya-Arnoldi, 1964a). Endosperm reduction is typical for the whole family of orchids. Here two situations can progress: (1) double fertilization occurs normally, and the primary endosperm nucleus divides, forming several nuclei; (2) double fertilization is disturbed and the following variants are possible: a) nucleus of the sperm fuses with nuclei of the central cell, although the primary endosperm nucleus degenerates without dividing; b) the nucleus of the sperm comes into contact with nuclei of the central cell but does not fuse with them and later the whole group destructs; c) the sperm does not enter the central cell. According to data by Savina (1972), in the ovary of *Ophrys insectifera* sometimes the primary endosperm nucleus did not divide and concomitantly sometimes the primary endosperm nucleus did not form. The author believed that both structural and functional reduction of endosperm may occur because even if several nuclei do form, their participation in embryo nutrition is not at all essential. However, other researchers (Teryokhin and Kamelina, 1969) have posted that only structural reduction of the endosperm is typical for orchids. It actively carries on its trophic function in a limited space of time during development of a young sporophyte, though represented solely by a cell with one nucleus.

One opinion holds (Teryokhin and Kamelina, 1969, 1972; Teryokhin, 1977) that reduction of the endosperm in orchids is directly related to degree of reduction of the embryo and inversely related to degree of development of the suspensor. For example, the multinuclear endosperm is mainly found in those species of orchids whose embryos have a poorly developed suspensor or totally lack one. Further, the number of endosperm nuclei decreases with lessening of mass, amount of cells and degree of differentiation of the embryo of the mature seed. But this rule is not always observed. For instance, in spite of the fact that embryos of *Thunia alba* and *Bletilla striata* consist of a number of cells, differ in degree of differentiation (in *Thunia*, without a primordium of a leaf-like organ, in *Bletilla,*—with such) and possess a well-developed suspensory apparatus, these two species are characterized by the occurrence of a multinuclear endosperm containing from 4 to 10 nuclei (Andronova, 1988). Probably, the degree of endosperm reduction is associated with duration of zygote ripening. In *Thunia* and *Bletilla* the time from fertilization up to division of the zygote is one month or less. Afterwards an embryo develops with a long filiform (*Thunia*) or strongly branched (*Bletilla*) suspensor forming rapidly. In these species division of the primary endosperm nucleus takes place before the zygotic division and later from 4 to 10 nuclei form which function at early stages of embryo development. After a swift development of the suspensor by means of which the embryo receives nutrition from the mother tissues, the endosperm nuclei degenerate. In other species (e.g. *Dactylorhiza* sp.) the zygotic division occurs merely 1-2 days after fertilization. In the course of development of the embryo in which a suspensor similar to that in *Thunia* and *Bletilla* forms rapidly, the primary endosperm nucleus gradually degenerates without dividing.

Absence of differentiation of organs in the embryo of orchids can be associated with the ecology and physiology of seed germination and mycoparasitic mode of life (Teryokhin, 1977, 1991).

In a small number of the species of orchids studied, formation of a **'cotyledonary' primordium** in the apical part of the embryo was recorded (*Polystachya microbambusa*—Veyret, 1965, 1974; *Bletilla striata* (= *B. hyacintina*)—Bernard, 1909; Tohda, 1968; Nishimura, 1991; *Epidendrum vitellinum*—Veyret, 1965, 1974; *Sobralia macrantha*—Treub, 1879; Nishimura, 1991; *Arundina graminifolia*—Nishimura, 1991; *Thunia alba* (= *T. marshalliana*)—Teryokhin and Nikiticheva, 1968; Teryokhin, 1977, 1991). Data on the genesis of this 'cotyledon' are scant, however.

Considering the question of the presence of the cotyledon in orchids two aspects may be noted. First of all the background of the question shows that the main criterion for attributing the embryo to the group 'cotyledonous' is its form. As a rule, embryos of mature seeds of most orchid species taken in a longitudinal section are oval in form, and termed acotyledonous (Veyret, 1965, 1974; Poddubnaya-Arnoldi, 1976; Batygina and Vasilyeva, 1983a). Embryos of some representatives of the family are spindle-shape. In its apical part several epidermal cells are prominently larger in size and form a protuberance, thereby differing substantially from other epidermal cells (*Manniella gustavi, Polystachya microbambusa, Spiranthes autumnalis*—Veyret, 1965; *Thunia alba*—Andronova, 1988). In the apical part of embryos of other species a groove occurs (*Rangaeris muscicola, Vanilla fragrans*—Veyret, 1965). Some representatives of orchids in which the shape of the embryo differs from oval are ascribed to 'cotyledonous', others to acotyledonous species. Data on *Thunia alba* are conflicting. Some authors (Teryokhin, 1977, 1991; Nishimira, 1991) attribute this species to the 'cotyledonous'. According to other researches (Andronova and Batygina, 1992) the embryo of *Thunia* has no cotyledon (more details given later).

The second aspect of the problem of presence of the cotyledon in orchids relates to the discussion concerning the term 'cotyledon' itself (see Cotyledon; Acotyly; Origin of Monocotyledonous Embryo: A New Concept).

To establish the morphological and functional character of embryo structures in orchids the process of germination was accurately traced in culture *in vitro* of species contrasting in degree of differentiation of the embryo.

Germination of embryos which are not differentiated into organs (example of *Dactylorhiza baltica* and *Phalaenopsis hybridum*). *Dactylorhiza* and *Phalaenopsis* and closely related species are characterized by absence of the cotyledon (Veyret, 1965, 1974; Nishimira, 1981, 1991; Batygina and Vasilyeva, 1983; Batygina and Shevtsova, 1985; Shevtzova *et al.*, 1986; Batygina and Andronova, 1988; Batygina and Andronova, 1991).

During the first stage of embryo germination (swelling) almost all its cells expand considerably, which leads to rupture of the seed coat. A seedling is formed, whose early stage is termed the 'protocorm'. Some epidermal cells initiate hairs. In the case of *D. baltica* hairs are one-celled and uniseriate. At this stage the protocorms represent a whole zone, as both are characterized by the presence of epidermal hairs (zone I). At a further stage a new zone (not found in the embryo of the mature seed) is formed due to divisions of the cells of the apical region of the protocorm (zone II). As the cells of zone II grow and differentiate, some epidermal cells become capable of initiating epidermal hairs of the same type as those in zone I. The non-differentiated protocorm is radial-symmetrical in shape. In the apical part of zone II a green shoot apex forms (zone III). Zone III is devoid of epidermal filaments. The first leaf organ establishes as a ridge. The time of its formation concurs with differentiation of the vascular bundle of the protocorm. The vascular bundle never enters zone I and is laid down at the border between zone II and zone III. At the first stages of its development the shoot of *Dactylorhiza* has closed internodes. The base of the shoot (zone I + zone II) enlarges. It is milk-white and thickly covered with epidermal hairs. As the shoot grows (zone III) adventive roots establish endogenously in nodes of the first and further leaves. The type of hairs forming at the root complies with the type of hairs in zone I and the enlarged base of the shoot (zone II) (Fig. 73).

The first stages of protocorm formation are associated with consumption of reserve proteins and lipids by its cells. By the time the protocorm leaves the seed coat protein globules have almost completely disappeared while in basal cells of the seedling starch grains appear and increase in quantity. In cells of the shoot apex starch is not found either before initiation of primordia or afterwards.

Germination of the embryo of *Phalaenopsis* is characterized by a number of peculiarities. At the early stage the protocorm is radial-symmetrical in shape and green. In the basal part of the protocorm multiseriate hairs form. A groove appears in the apical part of the protocorm. It is surrounded by clavate glands with heads directed towards the centre of the groove, which indicates the beginning of differentiation of the shoot apex. With this stage the shape of the protocorm starts to change. It acquires a dorsoventral structure. At its upper side a 'ridge' covered with hair-glands forms. Hairs at the lower side and also at the adventive roots are arranged in groups (Fig. 74).

The seedling of *Dactylorhiza* and *Phalaenopsis* can be divided into two parts: the lower one (zone I + zone II) imitates the root appearance and maintains its functions (reserving of substances, contact with soil and fungi) and the upper one represents the shoot apex (zone III).

Germination of embryos with 'cotyledonary' primordium (example of *Bletilla striata* and *Thunia alba*). Investigation of the structure of protocorms in *Bletilla*

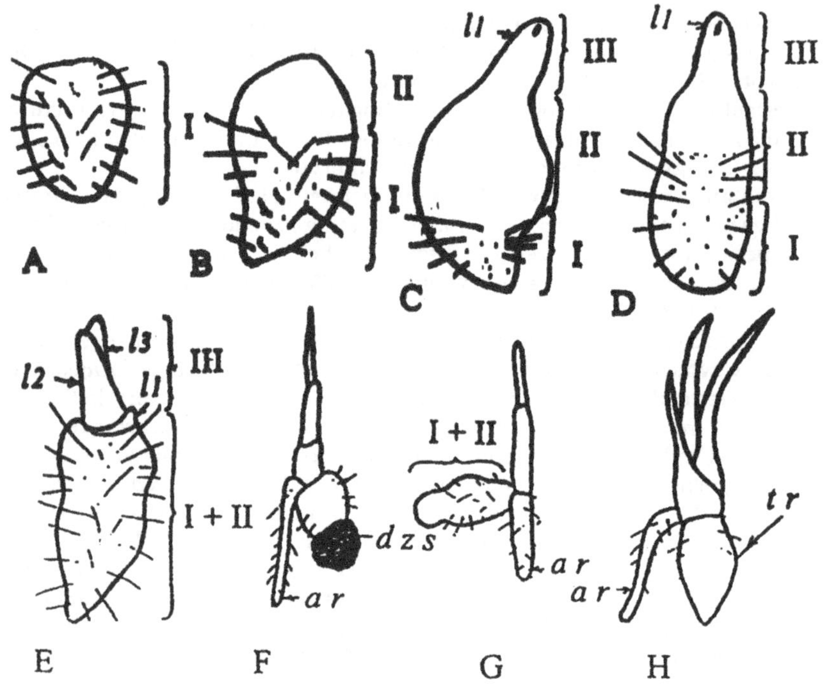

Fig. 73: Development of *Dactylorhiza baltica* seedling in culture *in vitro*. *a r*—adventive root, *d z s*—degenerated zone of seedling, l_1—l_3—1-st-3rd leaves, *t r*—thickened root.

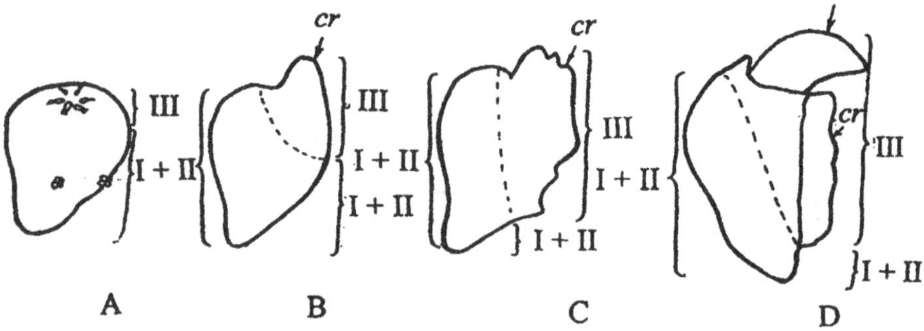

Fig. 74: Development of *Phalaenopsis hybridum* seedling in culture *in vitro*. *cr*—crest-like organ, *l*—leaf.

striata carried out by scanning electron microscope showed that the first leaf-like organ (cotyledon) establishes as a ridge or crest (Batygina and Andronova, 1988; Batygina and Andronova, 1991). In longitudinal sections the ridge can take the form of one protuberance or of two protuberances. Bernard (1909) designated one of the two 'protuberances' as the 'cotyledon' and the other as 'the first leaf'. As a matter of fact, the two 'protuberances' do not represent two separate organs, but constitute the opposite sides of one and the same organ which has the shape of a ridge (Batygina *et al.*, 1994). The idea of the leaf-like organ (cotyledon) being a 'protuberance' or 'prominence' has led to misunderstanding of the structure of the embryo and seedling not only in *Bletilla*, but also in some other species of orchids particular, *Thunia alba*. In the longitudinal section the embryo of *Thunia* has a 'protuberance' in the apical part which, as in *Bletilla striata*, is considered a 'cotyledon'. However, genesis and structure of this part of the embryo and seedling in *Thunia* differ greatly from *Bletilla*. In the embryo of *Thunia* a ridge is not formed, though large pyramidal cells are seen in its apical part in the protoderm. These cells seem to interrupt the layer of epidermis consisting of regular cells similar in structure and size. At the first stage of germination a spindle-like protocorm develops with the apical protuberance remaining. Division of protocorm cells starts in the zone beyond this protuberance, but later its cells also become involved. Consequently the protuberance gradually disappears and a clear-cut epidermal layer arises in the apical part, the apex flattens and then the ridge of the first leaf organ forms (Andronova and Batygina, 1992).

The nature and function of the protuberance in *Thunia* have yet to be fully clarified. Proceeding with the data available on development of the protocorm of these two species regarded as 'cotyledonous', it is evident that 'the protuberance' in the apical part of the embryo in *Thunia* and the ridge of the embryo in *Bletilla* are not homologous structures. On the other hand, the first leaf-like organ establishing in the post-seminal period in *Thunia* is homologous to the same one that forms in the course of embryo development during formation of the seed in *Bletilla*.

Thus, 'the cotyledon' in different species of orchids is considered either the apical end of the axis of the embryo or protocorm, or the first leaf organ forming in their apical part. The protracted debate as to whether a cotyledon occurs or not in orchids is still not finished (Veyret, 1965, 1974; Teryokhin, 1977, 1991; Nishimura, 1981, 1991; Jacques-Félix, 1982; Batygina and Vasilyeva, 1983; Guignard, 1984; Batygina and Andronova, 1988; Batygina and Andronova, 1991; Andronova and Batygina, 1992).

Classification of embryos of orchids. Probably, the first classification of embryos of orchids was proposed by Swamy (1949a). He believed that only the first two or three cell divisions during embryo formation were ordered in these plants, while further divisions obeyed no particular laws whatsoever. In his opinion, embryos of orchids can be divided into two groups: group A—embryos without suspensor, in which derivatives of the basal and apical cells participate more or less equally in construction of the embryo proper; group B—embryos with suspensor, in which the basal cell plays quite a minimal role in construction of the embryo body. Swamy's A and B groups correspond to the Asterad- and Onagrad-types of embryogenesis in Johansen's classification (1945, 1950). That case in which divisions of the apical and basal cells are irregular was designated the Cymbidium-form by Swamy.

Swamy (1949a) distinguished 5 types of suspensor in orchids: type 1—initial cell of the suspensor neither divides nor grows much, elongating only in some species; such a suspensor does not exceed the bounds of the seed coat; type 2—initial cell of the suspensor undergoes a division giving rise to a filiform suspensor which consists of 5-10 cells; the suspensor elongates, exits the micropyle, penetrates tissues of the placenta and transforms into a haustorium; type 3— initial cell of the suspensor and its derivatives divide in transverse and longitudinal directions, so that the suspensor formed resembles 'a bunch of grapes'; type 4—initial cell of the suspensor undergoes three divisions and an 8-celled suspensory apparatus forms; its cells usually grow towards the chalaza encircling the body of the embryo; type 5—divisions of the zygote and derivatives of *ca* and *cb* cells occur irregularly, with 6-10 being formed; most of these cells, those located closest to the micropyle, elongate and form the suspensory apparatus, while a filiform embryo originates from a cell that does not lengthen.

Considering the laws of blastomerization and degree of participation of the basal cell in construction of the embryo body, Veyret (1965, 1974) proposed three types of embryo development in orchids:

1. main type, which implies that development passes according to certain rules;
2. irregular type, which implies that development passes according to various patterns, but that concrete laws of development exist for each of them;
3. secondary type, which implies that the early stage of development differs from the main type in: (a) rate of segmentation of homologous blastomeres, (b) direction of establishment of certain cell walls, which results in early differentiation of the embryo and (c) differences in differentiation of most of the cells of the embryo body which are not critically important.

According to the classification by Johansen (1950), development of embryos in orchids falls under the Onagrad-, Asterad- and Piperad-types (Savina and Poddubnaya-Arnoldi, 1990). However, application of Johansen's classification to orchids has been associated with two circumstances:

1. Development of embryos in one and the same species can proceed in compliance with different patterns starting from the first divisions (presence of several types of tetrads). For example, in *Calypso bulbosa* two types of embryo development are observed—Onagrad and Chenopodiad (Yeung and Law, 1992).
2. The formula of embryogenesis in Johansen's classification (1950) shows that organs of the embryo and seedling originate from derivatives of *ca* and *cb* cells. For example, in the Onagrad-type the cotyledons, shoot apex and hypocotyl form from derivatives of the *ca* cell, while derivatives of the *cb* cell give rise to the embryo root. As mentioned above, the embryo in most orchids is not differentiated into organs. It is quite difficult to distinguish histogens since no structural differences have been found between cells of various zones of the embryo (Shamrov and Nikiticheva, 1992).

It is generally accepted that only the morphogenetic character of the terminal (*l*) tier of the embryo which generates the shoot apex can be reliably defined (Veyret, 1965, 1974; Poddubnaya-Arnoldi, 1976; Batygina and Vasilyeva, 1983a; Batygina and Shevtsova, 1985; Shevtsova *et al.*, 1986). The embryo root in orchids is not expressed morphogenetically. Basal cells of the embryo soon lose their ability to divide and it is not renewed during germination. The meristem of the root does not form.

As shown above, the embryo in most of the species studied represents a whole zone with respect to structure and functional meaning (except for several axial apical cells). It increases during post-seminal development as the protocorm grows. Considering structure and function, the enlarged base of the seedling is an accumulating organ. There are two points of view concerning its morphogenetic character. One opinion holds that the basal part of the embryo and protocorm is a hypocotyl (Nishimura, 1991; Shamrov and Nikiticheva, 1992). Bernard (1909) and Veyret (1965, 1974) believed the protocorm in orchids imitative of the root. It should be noted that in accordance with the Onagrad-type of embryogenesis, derivatives of the *ca* cell participate in formation of the hypocotyl, while those of the *cb* cell take part in formation of the root. However, as mentioned above, the basal part of the embryo and protocorm in orchids is represented by both derivatives of *ca* and *cb* cells in different proportions.

This contradicts the principles underlying the characteristics of embryogenesis of the Onagrad-type compiled by Johansen. Thus, the aforesaid peculiarities of embryogenesis of orchids do not allow use of existing classifications of embryos in describing it, as their main criterion is direction of establishment of cell walls during cell division (i.e., cell genealogy).

On comparing the structure of the embryo of orchids and other monocotyledonous and dicotyledonous plants it becomes clear that in spite of its weak morphological differentiation, the embryo of orchids does have a special differentiation, namely physiological. While in most plants that accumulate organs in the form of rhizomes and tubers these appear in the post-seminal period during ontogensis, in orchids these organs form in the first stages of ontogenesis, already in embryogenesis, as a result of transformation of the basal zone of the embryo into an accumulating organ (Veyret, 1965, 1974; Batygina and Vasilyeva, 1983). Seemingly the function of cells of the embryo and protocorm in orchids is defined not by their origin from *ca* and *cb* cells derivatives, but by their position in the embryo proper. Thus, formation of zonality is not a genealogical but a physiological character. Therefore, in the case of orchids, it is reasonable to apply to the classification considering presence of morphogenetic zones responsible for formation of seedling organs: shoot apex, hypocotyl and root, as assumed in recent years in a number of works on morphogenetic and genetic aspects of embryogenesis (Teryokhin, 1991; Laux and Jürgens, 1994; Jürgens, 1995; see Embryogeny). In the embryo of orchids two morphogenetic zones can be distinguished giving rise to the shoot apex and 'accumulating organ'. The latter imitates the structure and function of the root, though apparently it is a hypocotyl since it develops from the middle zone of the embryo, which is characteristic of other flowering plants.

ULTRASTRUCTURAL AND HISTOCHEMICAL ASPECTS OF EMBRYOGENESIS

Ultrastructural Aspects of Embryogenesis (Plate L)

The number of species in which embryogenesis has been studied by electron microscope is many times less than those examined under the light microscope. Given the keen interest shown nowadays in the problem of embryogenesis, new experimental approaches are used to solve it. Study of the genetic regulation of embryogenesis on model systems (*Arabidopsis, Zea*), in particular: processes of asymmetric division of a zygote, synthesis of reserve substances, etc. attract special attention among researchers (Patton and Meinke, 1990; Meinke, 1991a-c; Lindsey and Topping, 1993; De Jong *et al.*, 1993; West and Harada, 1993). Unfortunately, results of ultrastructural investigations of embryogenesis are hardly involved in discussing these issues (Thomas, 1993). The problem of embryogenesis periodization has been considered in numerous works. But the criteria for distinguishing key states of a developing embryo and the terminology used by authors differ markedly. Data of ultrastructural investigations are almost not employed in working out classifications of types of embryogenesis. However, due to specificity of the method they can provide quite important data characterizing the structural and functional state of cells and tissues of the embryo. The ultrastructural characteristics of successive development of the embryo of flowering plants including zygotic, proembryonic and embryonic periods (after Yakovlev, 1981) with subdivision into stages (globular, heart-shape, torpedo-like, formed embryo) or smaller gradations (if introduced by authors) are presented below.

Zygotic period. The zygote of most flowering plants is a polarized and functionally active cell. It is characterized in many species by the occurrence of a large vacuole located at the basal part of the cell and shifting of the nucleus with the greater part of cell organelles into the apical zone. Thus, formation of the vacuole is accompanied by a considerable increase in size of the zygote compared with the egg cell. Zygotes of the species studied in the following genera can serve as examples: *Zea, Linum, Capsella, Epidendrum, Quercus, Arabidopsis, Pennisetum* (Vazart B. and Vazart J., 1966; Diboll, 1968; Schulz R. and Jensen, 1968b; Cocucci and Jensen, 1969a; Chebotaru, 1972; Singh and Mogensen, 1975; Mansfield and Briarty, 1991; Taylor and Vasil, 1995).

In species studied of some other genera of flowering plants, such as *Hordeum, Triticum, Helianthus, Reseda, Pennisetum*, and also in the sexual allopolyploid line of *Triticum* no large vacuole is found in the zygote; its vacuum is represented by a multitude of vacuoles of various size, with cell organelles more or less evenly distributed throughout the cytoplasm of the cell (Cass and Jensen, 1970; Norstog, 1972; Yakovlev and Chaban, 1979; You and Jensen, 1985; Yan Hua and Hong-Yuan Yang, 1991; Taylor and Vasil, 1995; Naumova and Matzk, 1998). The nucleus either occupies the central position or is slightly removed towards the basal part of the cell. Thus, in most cases no intensive growth of the zygote is observed and on maturation it remains similar to the egg cell in size.

In a limited number of species, including the species studied in *Alcea, Cephalaria, Gossypium, Hibiscus, Nicotiana* (Belyaeva, 1980; Pollock and Jensen, 1964; Ashley, 1972; Mogensen and Suthar, 1975; Kamelina, 1977, 1980; Zhukova, 1988), the zygote decreases in size compared with the egg cell due to disappearance of the large vacuole. In the case of deviations from the characteristic vacuolization of the zygote arising as a result of hybridization or other influences, disturbances in the further course of embryogenesis may occur (hybrid *Hibiscus*) (see Zygote).

The zygotic division in most flowering plants is transverse. This division can be symmetrical or asymmetrical. With symmetric division both sister cells of the embryo are of the same size; with asymmetric division they are dissimilar and the apical cell appears to be smaller than the basal one. Normally division of the zygote is asymmetric in those cases when it has a large vacuole and even before division represents a structure with distinctly expressed polarity. In some cases when the zygote lacks the large vacuole, asymmetric division may also prevail (species of *Triticum, Reseda, Helianthus, Pennisetum*—Batygina, 1974b, 1996; Yakovlev and Chaban, 1979; You and Jensen, 1985; Yan Hua and Hong-Yuan Yang, 1991; Taylor and Vasil, 1995). An exceptional significance of mechanism of the asymmetric division in plants has been shown on the example of embryogenesis of the *gnom*-mutant of *Arabidopsis*. It is supposed that a genetic control exists, co-ordinating sizes of the basal and apical cells of the embryo of plants (Kirk *et al.*, 1991).

Proembryonic period. A two-celled proembryo the same as the zygote exhibits distinct polarity which remains at all further stages of its development. Size and ultrastructural organization of sister cells can differ in different species of flowering plants. This depends on whether the zygotic division is symmetric or asymmetric (see Zygote). As the basal cell mainly participates in formation of the suspensor, the ultrastructure of this cell and its derivatives has been observed in the corresponding part (see: Ultrastructural Characteristics of Suspensor). Here only the description of the apical cell and its derivatives giving rise to the embryo proper is provided.

The apical cell formed as a result of **asymmetric division of the zygote** is smaller in size than the basal one and has an oblong nucleus of irregular shape. A two-celled proembryo typically is not larger than the zygote. In *Gossypium* it does not exceed the size of the zygote up to the stage of 75 cells (Pollock and Jensen, 1964). The apical cell contains considerably more cell organelles than the basal one. Uneven distribution of cell organelles between the sister cells of the proembryo both in quantity and content, probably determines the difference in their genetic abilities (Nagl, 1973; Newcomb and Fowke, 1974; Vallade, 1980; Meinke, 1991b,c).

The apical cell has a dense cytoplasm and reacts positively on staining for proteins and RNA (Schulz R. and Jensen, 1968a,b). It is devoid of the large vacuole, which in Jensen's opinion (1976) is important for development of the embryo proper. Ribosomes increase in number compared with the zygote (Vallade, 1980; Mansfield and Briarty, 1991; Zhukova and Savina, 1978). Mitochondria morphologically resemble those of the zygote (Singh and Mogensen, 1975; Taylor and Vasil, 1995). Structure of the plastids compared with plastids of the zygote does not change greatly, although their sizes increase (Singh and Mogensen, 1975). Synthesis of starch and lipids is more intensive in the apical cell than in the zygote (Schulz R. and Jensen, 1968a,b; Taylor and Vasil, 1995). A few starch grains are found in the apical cell of *Quercus* (Singh and Mogensen, 1975). The endoplasmic reticulum (ER) with respect to degree of development and activity is similar to the ER of the zygote (Schulz R. and Jensen, 1968a,b), though it is better developed and active in *Quercus* (Singh and Mogensen, 1975). As for the number and activity of dictyosomes considerable interspecific differences have been noted. They are scant and weakly active in *Capsella* and *Petunia* (Schulz R. and Jensen, 1968a,b; Vallade, 1980), and abundant and active in *Quercus* (Singh and Mogensen, 1975) and *Arabidopsis* (Mansfield and Briarty, 1991). Vacuoles are small and located evenly over the entire apical cell. The outward wall of cells of a two-celled proembryo may be irregularly thickened (Zhukova and Savina, 1978; Mansfield and Briarty, 1991) or more homogeneous in thickness (Taylor and Vasil, 1995). It lacks plasmodesmata in all the species studied while the cell wall between the apical and basal cells is thinner than the outward one and has numerous plasmodesmata.

The apical cell formed as a result of **symmetric division of the zygote** (e.g. *Hordeum* —Norstog, 1972) is the same size as the basal one and characterized by predominance of single ribosomes, the system of cristae in mitochondria appears to be better developed, and the ER is rather more active compared with the zygote and is represented by elongated cisterns. Dictyosomes are not numerous and do not show high activity. The cell wall surrounding the embryo is thickened and lacks plasmodesmata; in certain regions between the plasma membrane and the cell wall deposition of an electron-dense substance has been observed. In the allopolyploid line of *Triticum* (Naumova and Matzk, 1998) sister cells of a two-celled proembryo are similar in size and ultrastructural organization.

Ultrastructural data on development of the embryo at the proembryo stage in different species are far from consistent (Schulz R. and Jensen, 1968a,b; Norstog, 1972; Zhukova and Savina, 1978; Mansfield and Briarty, 1991; Yan Hua and Hong-Yuan Yang, 1991). The main tendencies in transformation of cells of the embryo proper at the proembryo stage in the course in its development include: increase in density of ribosomes and predominance of single ribosomes compared with few polysomes. Such data may be evidence of weakening of protein synthesis intensity at the proembryo stage compared with the zygotic stage. The presence of abundant mitochondria with poorly developed inner membrane system and tendency towards growth of abundant mitochondria in some species due to their divisions, probably point to a relatively low energetic potential of cells of the proembryo and an opportunity for its augmentation at further stages. A great number of plastids capable of starch synthesis are typical for many species, though occurrence of starch in the proembryo is not observed in all of them. It is most likely that synthesized starch is actively employed by cells of the proembryo when undergoing divisions. Evidently, metabolism of carbohydrates plays an important role in early embryogenesis (Schulz R. and Jensen, 1968a,b). At the proembryo stage it is carbohydrates which are basically used as an energy source for cell divisions. Dictyosomes and endoplasmic reticulum do not express high activity, though the outward wall of cells continues to thicken and new walls appear between cells inside the proembryo; participation of ER in this process is quite possible.

The size of the proembryo in most flowering plants despite a considerable number of cells does not greatly exceed the size of the zygote. Growth of numerous cells with an almost changeless volume naturally leads to decrease in cell size during each further division. So, in *Capsella* the embryo at the stage of octants is half its size at the stage of quadrants (Schulz R. and Jensen, 1968a,b). However, in *Reseda* the size of the embryo in the octant phase does not change compared with previous phases (Yakovlev and Chaban, 1979). The twenty-celled embryo of *Hordeum* is not larger than at earlier stages (Norstog, 1972). It was supposed that only after reaching its minimal size can the cell differentiate (Jensen, 1976). Therefore, at the proembryo stage despecialization of its cells takes place; they gradually become meristematic, losing accordingly, features of the zygote which, as is well known, represents a highly specialized cell. However, another opinion posits that the zygote beginning to divide represents a totipotent meristematic cell (Batygina, 1978a,b, 1987a, 1996, see Zygote).

Investigations have also shown that cells of the embryo proper of different taxa of flowering plants, both monocotyledons and dicotyledons, manifest no distinct ultrastructural differences at the proembryo stage (Schulz R. and Jensen, 1968a,b; Norstog, 1972; Singh and Mogensen, 1975; Yakovlev and Chaban, 1979; Yan Hua and Hong-Yuan Yang, 1991; XuHan, 1995). Consequently the ultrastructural characteristic of the proembryo is not likely to be used as a taxon-specific feature. According to Natesh and Rau (1984) existing classifications based on succession of cell divisions do not reflect inner factors of morphogenesis of embryos at the proembryo stage which are intrinsic to flowering plants on the whole. Recent investigations on zygotic embryogenesis more and more confirm that in plants it is the position of a cell in the embryo rather than its preceding history which determines its functions (Christianson, 1986; Poethig, 1989; Irish and Sussex, 1992; De Jong *et al.*, 1993).

Embryonic period. In its present description the globular stage has been subdivided into early globular, middle globular and late globular stages (Singh and Mogensen, 1975).

The early globular stage is characterized by completion of embryoderm formation. In *Capsella* it occurs at the stage of a 16-celled embryo. At this stage no differences between cells of the embryo proper are found either in the ultrastructure or in distribution of proteins and nucleic acids in them. Plastids, mitochondria, ribosomes, dictyosomes and ER do not differ morphologically from similar organelles of the embryo at the proembryo stage. However, in *Capsella* differences appear at the early globular stage in the form of cells and vacuolization in connection with their position in the embryo (Schulz R. and Jensen, 1968a,b). In *Quercus* some cells have large and small vacuoles simultaneously while other cells contain only small vacuoles (Singh and Mogensen, 1975). In *Helianthus* poor vacuolization persists in all cells (Newcomb, 1973). In *Petunia* gradual and progressive decrease in size of vacuoles from 4-5 μm in a two-celled proembryo to less than 0.5 μm in a globular one has been recorded (Vallade, 1980). The nucleus holds a nucleolus with small nucleolar vacuoles (*Pennisetum*—Taylor and Vasil, 1995). The outer membrane of the nuclear envelope has a multitude of ribosomes (*Quercus*—Singh and Mogensen, 1975). In the cytoplasm ribosomes are plentiful, most of them present as polysomes (*Quercus*—Singh and Mogensen, 1975). Numerous ribosomes, partly combined with polysomes, are typical for *Helianthus* (Newcomb, 1973). Mitochondria are also abundant which morphologically have not changed much compared with proembryo stage mitochondria (*Quercus*—Singh and Mogensen, 1975); they are small in *Helianthus* (Newcomb, 1973) while in *Pennisetum* (Taylor and Vasil, 1995) the cristae look slightly better developed than in the earlier stage. Plastids are numerous. They have changed in several aspects: increase in size, more intensive starch synthesis and stroma becomes transparent. In *Helianthus* and *Arabidopsis* plastids still remain poorly differentiated with electron-dense stroma, their form varies from cyathiform to oval (Newcomb, 1973; Mansfield and Briarty, 1991). In *Pennisetum* plastids the inner lamellar system starts to differentiate (Taylor and Vasil, 1995). Rough ER prevails in *Quercus* (Singh and Mogensen, 1975). Only a few dictyosomes are found in *Helianthus* (Newcomb, 1973). Reserve proteins and lipids are not found. Structure of the outer wall of cells of the embryo does not differ from that at the proembryo stage.

Middle globular stage. At this stage of development the embryo of *Capsella* consists of 32 cells. No ultrastructural differences between cells are observed in it.

Nuclei, plastids, mitochondria, dictyosomes, ribosomes, ER and vacuoles are similar morphologically in all the cells of the embryo. In *Reseda* also (Yakovlev and Chaban, 1979) no marked ultrastructural differences between cells of the embryo are found at the globular stage, though the number of ER cisterns and dictyosomes increases slightly, the outer wall of the cells thickens and the embryo increases notably in size. There are no histochemical differences between cells of the embryo (Schulz R. and Jensen, 1968a,b).

Late globular stage in *Quercus* (Singh and Mogensen, 1975) is characterized on the whole by increase in cell size due to increase in number and size of vacuoles. Most vacuoles have various inclusions. Hydratation of the cell grows and accordingly membranes of all organelles become transparent. Concentration of ribosomes, polysomes and also ER reaches its peak. Dictyosomes are numerous and active. Plastids and mitochondria are also plentiful; mitochondria are relatively changeless while plastids enlarge and most of them contain starch. No other reserve substances are found in the cells. Cell walls between sister cells inside the embryo thicken, but the middle plate is hardly traced. Similar alterations are typical for *Arabidopsis* as well (Mansfield and Briarty, 1991).

So, the main structural-functional transformations of the embryo at globular stage of development relate to increase in size of cells of the embryo and as a consequence of size of the embryo itself due to growth in volume of the vacuome, which is evidence of intensification of metabolic processes. Reaching of the highest density of ribosomes and predominance of polysomes is a characteristic of this stage, which can argue for a sharp growth in protein synthesis.

Part of the proteins accumulate in vacuoles as inclusions. Along with synthesis of insoluble proteins produced by polysomes intensified synthesis of glycoproteins is generated by the ER and dictyosomes, their number and activity at the globular stage of development abruptly increasing in many species. The necessary supply of carbohydrates (starch) is maintained by plastid activity. Other storage reserves are still not found. As in the proembryo stage, no ultrastructural differences between cells of the embryo proper are found at the globular stage, except for the character of vacuolization and form of cells (*Capsella*). Isolation of the embryo continues to intensify. It manifests in thickening of the outer wall of cells and deposition of callose and cutin in some species (Williams *et al.*, 1984).

Heart-shape stage. The number of cells in the embryo during transition to initiation of cotyledons defining commencement of the heart-shape stage can vary in different taxa of flowering plants. In *Capsella* some ultrastructural and histochemical differences between cells of different parts of the embryo intensify at the heart-shape stage. These differences consist of stronger vacuolization of cells of the procambium (plerome) and the ground meristem (periblem) compared with cells of the embryoderm. An electrodense cuticle is distinctly visible on the surface of the embryo. Those parts of the embryo which grow actively, especially cotyledons, are characterized by a high activity of enzymes and a higher RNA content (Schulz R. and Jensen, 1968a,b). Similar differentiation takes place in *Diplotaxus erucoides* (Simoncioli, 1974), *Reseda lutea* (Yakovlev and Chaban, 1979) and *Arabidopsis thaliana* (Mansfield and Briarty, 1991). In *R. lutea* considerable changes occur in the ultrastructure of cells of the embryo: quantity of rough ER increases, plastids and mitochondria become larger, the lamellar system is formed in plastids and most of them synthesize starch.

At the heart-shape stage of embryogenesis transformations occur in the plastidome leading to greening of the embryo in those species of flowering plants having to chloroembryophytes.

Greening of the embryo is associated with differentiation of the lamellar system in plastids and the possibility of photosynthesis. Plastids of cells of the ground meristem and embryoderm form the lamellar system which is more intensive compared to that generated by plastids of cells of the procambium (*Linum*—Deschamps, 1968; *Capsella*—Schulz R. and Jensen, 1968a,b; Jensen, 1976). Issues concerning the plastid apparatus of the embryo of flowering plants are examined in detail in a number of works (Yakovlev and Zhukova, 1973; Zhukova, 1979, 1993; Yakovlev and Zhukova, 1980; Rodkiewicz *et al.*, 1994; see also: Possibility of Classifying Flowering Plants According to Occurrence of Chlorophyll in Embryos).

The main distinguishing features of the heart-shape stage of embryogenesis are differentiation of tissues and appearance of ultrastructural and histochemical differences between them. Cells of the procambium (plerome) and the ground meristem (periblem) become more vacuolized than cells of the embryoderm. Intensity of protein and nucleic acid synthesis continues to grow (Vasilyeva *et al.*, 1987). Metabolism of cells of the embryo on the whole changes towards synthesis of storage reserves. Apart from starch, protein globules and lipid bodies are found in cells.

Investigations carried out on lethal mutants of *Arabidopsis* have shown that at the heart-shape stage of development, reduction in level of reserve protein synthesis in the embryo occurs and further, the time of initiation of their synthesis shifts towards a later stage of embryogenesis. Appearance of protein bodies in cells of the embryo was used by authors as a marked feature allowing delineation of time parameters of protein synthesis (Patton and Meinke, 1990). Probably, since in other flowering plants the intensity and time of initiation of reserve substance synthesis is genetically regulated the same as in *Arabidopsis*, they can also be used as significant features.

The **torpedo-like stage** is characterized by intensive growth of cotyledons (Schulz R. and Jensen, 1968a,b; Yakovlev and Zhukova, 1973; Tang P.-H. *et al.*, 1989; Mansfield and Briarty, 1991). In *Capsella* and *Reseda* further increase in ultrastructural differences between tissues of the embryo has been observed. Cells of the plerome have the highest degree of vacuolization. It is also well documented that direction of metabolism of the embryo on the whole has changed towards accumulation of storage reserves. Starch accumulation takes place in all parts of the embryo simultaneously, though most intensively in cotyledons. In *Capsella* along with starch a great amount of lipids synthesize. In *Arabidopsis* numerous spherical electron-dense bodies with homogeneous matrix of unknown nature are reported to appear in all the cells of the embryo (Mansfield and Briarty, 1991). The heart-shape and torpedo-like stages are regarded by some authors as a single whole and are described together as a stage of the main physiological functions when storage reserves synthesize and accumulate (*Nelumbo*—Tang P.-H. *et al.*,1989).

Stage of embryo maturation. As the seed ripens significant structural-functional transformations associated with the seed passing into a state of dormancy take place in the embryo. The process of embryo ripening is accompanied by dehydration and compression of its cells (Tang P.-H. *et al.*,1989). Then the ultrastructural transformations mentioned have a physiological but not destructive character. They are supposed to serve as an adaptation to dehydration of the seed. Such simplifications of the ultrastructure of cell organelles are considered typical for tissues with a low metabolic activity (Hallam, 1972).

Histochemical and Immunohistochemical Aspects of Embryogenesis (Plate LI)

Histochemical and immunohistochemical methods enable revelation of qualitative changes and dynamics of metabolites in the cells of an embryo and surrounding structures during seed development.

Histochemical methods of research have been used in botany since the 1920s-1930s. Their main advantage is that they demand less material than is necessary to carry out standard biochemical procedures.

While investigating the early embryogenesis of *Linum catharicum* it was shown that the apical cell of the two-celled proembryo stains more intensively on RNA and proteins than the basal one (D'Alascio-Deschamps, 1973b). However, in some plant species it is evident that the basal cell contains more proteins and RNA than the apical one (Pritchard, 1964; Singh and Mogensen, 1975).

Pritchard (1964), who studied nuclear DNA at early stages of development of *Stellaria media* embryo, noted weakening of staining of the nuclei at the stage of the 30-60-celled proembryo accompanied by the appearance of Feulgen-positive granules in the cytoplasm. In *Gossypium hirsutum* during the whole period of embryogenesis the amount of overall DNA in a nucleus remained more or less constant, while no DNA was found in the cytoplasm (Fisher and Jensen, 1972).

Differential distribution of starch and proteins is found in tissues of the globular embryo of *Gymnadenia conopsea* (Shamrov and Nikiticheva, 1992). In the apical part of the embryo represented by small cells starch accumulates, and in the basal part formed with large cells, proteins, which are also found in the suspensor. In the embryo of *Stellaria media* no considerable changes in distribution of RNA and proteins take place during several cell generations, while at the same time in the globular embryo all the cells are stained with equal intensity on RNA and proteins (Pritchard, 1964). At the stage of the heart-shape embryo accumulation of nucleic acids and proteins in cotyledon primordia is greater than that observed in the hypophysis and suspensor. In *Luzula pedemontana* in the cells of the embryo at the globular stage of development starch disappears, while dextrines and proteins are present. During initiation of different organs in the embryo (one of the critical stages) the amount of proteins increases and dextrines disappear (Shamrov and Anisimova, 1993b).

Along with the traditional histochemical methods, such as staining by Feulgen (to reveal DNA), with procion dyes (for simultaneous detection of proteins and insoluble carbohydrates), PAS-reaction (to define the content of polysaccharides of cell walls and reserve starch), methods of quantitative definition of enzymes have been widely used for investigation of embryogenesis (see Raghavan, 1976; Natesh and Rau, 1984). The latter are not concerned with defining the enzyme content in tissues but rather their activity: degree of use of a substrate by an enzyme and accumulation of terminal product of enzyme reaction. A study of localization of enzymes showed that for example, alkaline phosphatase hydrolyzing phosphoric ethers is found both in the embryo proper and the suspensor of *Stellaria media* (Pritchard and Bergster, 1969). In the cells of the embryo proper this enzyme is equally distributed, while in the cells of the suspensor it is localized in plastids. Contrarily, activity of acid phosphatase is minimal in the suspensor and maximal in the embryo proper, especially in the embryoderm. The hypophysis shows no phosphatase activity whatsoever. With respect to the activity of cytochromoxidase—the key respiratory enzyme—in *Stellaria media* it is higher in the meristematic regions. On the example of *Tropaeolum majus* it was also shown that the highest activity of respiratory enzymes is observed in the globular proembryo while in the course of ripening of the embryo it decreases, which is probably associated with its entering physiological dormancy. The reductive-oxidative enzyme peroxidase which regulates the level of auxin in tissues, has its peak activity in the early stages of embryo development and gradually decreases in late stages (Bhalla *et al.*, 1979).

In the process of development of the seed changes occur in the content of plastic substances in its structures. So, for example, in *Gymnadenia conopsea* after fertilization starch disappears from the cells of integuments and appears in great amount in the apical part of the zygote, in the intact synergid and in the primary endosperm cell (Shamrov and Nikiticheva, 1992). On the whole the reaction on dextrines and proteins decreases in tissues of the ovule but remains at an average level in the epidermis of the nucellus and postament right up to the stage of octants. Dextrines are evident in the destroyed synergid and the only antipodal cell, while proteins are present in the zygote and around the nucleus of the primary endosperm cell. In *Luzula pedemontana* during fertilization and first divisions in the endosperm, starch accumulates in the outer integument and lateral part of the nucellus, while in the inner integument tannins are laid down (Shamrov and Anisimova, 1993b). In the course of cellularization in the endosperm, callose thickenings form at the tangential walls of the cells of the parietal tissue. During initiation of organs in the embryo the reaction on proteins in the epidermal layer of the nucellus becomes positive, and in the cells of the outer epidermis of the seed, secretion of mucilage of polysaccharide nature begins. During ripening of the embryo tannin-containing layers of cells form in the inner integument, hypostase and chalaza. Thus, gradual redistribution of nutrients takes place in favour of the ripening embryo.

Immunohistochemical methods are refined methods of investigation of embryogenesis. They are based on the ability of an animal organism to produce specific antibodies against an alien material (antigen). Antigen and antibody interact with each other with high specificity. To define localization of the substance under study a conjugate of antibodies with fluorescent stains is received without causing weakening of the affinity of antibodies to an antigen. When placing such a conjugate on a tissue section an antigen-antibody reaction takes place which is clearly visible in ultraviolet light with the help of a fluorescent microscope. There are methods of direct localization of an antigen when labelled antibodies are put straight onto a section and a specific link between them and the antigen occurs and indirect localization when serologically active layers are used, i.e, the 'sandwich method' (Fig. 75). When using the latter method, sensitivity of the reaction increases and fluorescence becomes more intense because each antibody is capable of connecting several molecules of the fluorescing anti-γ-globulin.

Immunohistochemical methods play an important role in the study of cytoskeleton and mitotic division of cells. In turn, disposition of preprophase stretches of tubuline associated with actin and of the mitotic spindle in a cell determine direction and plane of cell division, which finally leads to certain morphogenetic consequences.

Spectrine, one of the most important proteins of the cortical net of animal cells and one of the components of the so-called membrane skeleton has been also found in the plasma membrane of plant cells. Taking a somatic embryo received in the suspension culture of carrot as a model it was found by means of immunocytochemical methods with application of polyclonal antibodies to spectrine that the initial cell of the somatic

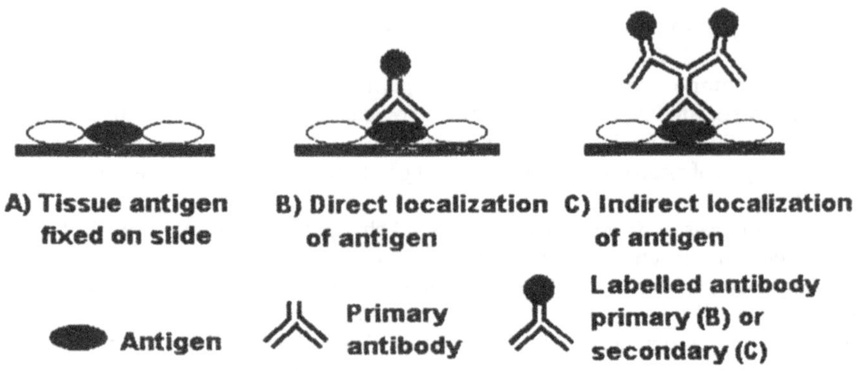

Fig. 75: Scheme showing direct and indirect localization of antibody in cells.

embryo shows strong association with the fluorescing antibody mostly in the region of the plasma membrane (Ruijter and Emons, 1993). Degenerating and dead cells examined by means of DAPI (4,6-diamine-2-phenylindol) staining under a differential interference contrast microscope did not display fluorescence.

In the cells of the somatic embryo of carrot unequal distribution of spectrine has been observed. In the cells of the proembryonic cell complex the label was found in the plasma membrane, just as it was in the single cell of the suspension culture of carrot. In somatic embryos at the globular stage of development, especially in their embryoderma, the label was connected with the plasma membrane and nuclei. Usually, when nuclei were labelled the nucleoli remained negative. In epidermal cells of embryos at the torpedo stage the label was seen not only in the plasma membrane, but also in the cytoplasm and nuclei.

Another example of tissue-specific connection with the spectrine label was found in zygotic embryos of maize (Ruijter and Emons, 1993). In all the cells of the scutellum many plastids were labelled, though in adjusting tissues of the coleoptile the label was scarcely observed despite the fact that the presence of plastids in the cells of the coleoptile was distinguished by means of phase-contrast microscopy. In amyloplasts the label was not observed. In smaller plastids, which shifted to the borders of cells, the label was found in their membranes.

It has been postulated that spectrine attaches other proteins to the membrane and thus stabilizes the organelles. Probably the occurrence of spectrine in protoplasts is explained by the need for the spectrine net for stabilization of the cell membrane. Single cells and small clusters of cells in the cell suspension have similar features as they are almost completely devoid of cell contacts and have a non-rigid cell wall liable to deformation (Ruijter and Emons, 1993).

Seed tissues surrounding the embryo were also studied by means of immunofluorescent microscopy. Thus, injection of the cells of the endosperm of *Haemanthus katherinae* while in the state of mitosis with fluorescent phalloidin enabled tracing *in vivo* the dynamics of *F*-actin cytoskeleton during migration of the chromosomes (Schmit and Lambert, 1990). Short actin microfilaments located parallel to the spindle axis of nuclear division have been found in the equatorial region during transition from anaphase to telophase, which were connected with the cortical *F*-actin net. Further, it was determined that actin was organized *de novo* during development of the cell plate, and that it was not a passive redistribution of fragments of actin polymers due to the activity of microtubules.

Using double labelling with first an antibody mixture of anti α- and anti β-tubuline and a second antibody GaM-FITC, the influence of taxol on microtubules (Mts) during development of the endosperm in *Brassica napus* was studied (Brukhin and Van Lammeren, unpubl). Taxol is an alkaloid of diterpenoid nature, which prevents depolymerization of tubuline. It was found that taxol treatment substantially increases the number of MTs in endosperm cells. Tubuline arrays became longer and thicker compared to wild type control. Taxol produced abnormalities in the arrangements of MTs and prevented their depolymerization, especially at the early stage of endosperm development.

Cuticle of Developing Embryo (Plate LII)

Several authors, in particular Yakovlev and Chaban (1979) and Chamberlin *et al.* (1993) investigating embryogenesis respectively in *Reseda lutea* and *Glycine max*, have described an electron-dense layer, presumably a cuticle, covering an embryo proper at the late globular and heart stages. The cuticle was not present on the suspensor.

We confirmed these observations following embryogenesis in *Stellaria*. The outer cell wall of the early globular embryo is initially homogeneously dense for electrons. There is no distinguishable cuticle at this stage. The cuticle is visible on the more developed globular embryo proper and on the heart-shape embryo but not on the suspensor cell walls.

In order to follow in detail the course of cuticular development in embryogenesis we applied fluorescence microscopy (Rodkiewicz *et al.*, 1994). Cuticular cutin was visualized by auramine O in embryos of various species of dicotyledonous and one species of monocotyledonous plants (Table 7).

Table 7: List of species investigated

Alismataceae:	*Alisma plantago-aquatica*
Campanulaceae:	*Campanula rotundifolia*
Caryophyllaceae:	*Melandrium album, Stellaria media*
Cruciferae:	*Arabidopsis thaliana, Berteroa incana, Brassica campestris, Brassica napus, Capsella bursa-pastoris, Iberis umbellata, Lepidium ruderale, Sinapis alba*
Fabaceae:	*Lathyrus vernus, Pisum sativum*
Geraniaceae:	*Geranium pratense*
Linaceae:	*Linum usitatissimum*
Malvaceae:	*Malva silvestris*
Plantaginaceae:	*Plantago media*
Tiliaceae:	*Tilia cordata*
Violaceae:	*Viola arvensis, Viola tricolor*

Formation and extension of fluorescing and non-fluorescing areas of the embryo surface change during embryogenesis according to a pattern similar in all the species investigated. We may assume therefore that this pattern is common to a great number of plants (Szczuka *et al.*, 1995).

The first fluorescing cuticle appears on the apical part of a proembryo, just before delimitation of the protoderm. This cuticle gradually extends over the whole embryo proper while the suspensor remains non-fluorescent. The cuticle persists during the heart stage. Again there was no fluorescence on the suspensor and the walls between the embryo proper and the suspensor.

The fluorescing cuticle disappears during transition to the torpedo stage or in the early torpedo embryo but reappears on the entire embryo by the end of the torpedo stage, being particularly strong at the radicular tip. Subsequently, in the U-shaped embryo fluorescence disappeared from the radicular part, but a fluorescing cuticle was distinctly observed on the shoot apex, cotyledons and upper part of the hypocotyl.

A cuticle formed over the surface of an embryo may be impervious to solute exchange, as suggested by Chamberlin *et al.* (1993). We assumed that a very thin cuticle may only decrease exchange between the embryo and endosperm. A layer of cutin, however, isolates to a certain extent the globular and heart-shape embryos. Therefore, some substances must be transported only by the suspensor whose cell walls lack cutin. The stream of substances entering a globular embryo from one direction may enhance polarization of the embryo and formation of the root-shoot axis.

The cuticle's presence over the embryo proper and its absence on the suspensor supports the prevalent view that in a number of angiosperms the suspensor plays an active role as an important route for the transport of substances into the early embryo (Yeung and Meinke, 1993).

Obliteration of the suspensor in the late heart or early torpedo stages apparently coincides with or is followed by the disappearance of fluorescing cutin. Such being the case, the torpedo embryo should absorb materials directly from the endosperm. It has been shown by several authors that after transition from suspensorial to surface nutrition, the cotyledons appear to be the major uptake region of the embryo (Erdelská, 1984; Johansson and Walles, 1993).

During the late torpedo stage or during transition from the torpedo stage to the next stage of embryogenesis, the embryo forms a new fluorescing cutin layer over its entire surface. This layer should to some extent isolate the embryo from influence of the endosperm and limit exchange of solutes. Cuticular isolation at this stage may act as a factor for induction of early maturation processes in the embryo. Once these processes have begun, the cutin layer on the radicular part of the embryo is degraded and unimpeded communication between embryo and endosperm is partially restored. This apparently promotes further maturation of the embryo.

We assume that cuticular changes during embryogenesis are linked with differentiation and maturation of embryos.

PHYSIOLOGICAL, BIOCHEMICAL AND GENETIC ASPECTS OF EMBRYOGENESIS

Autonomy of the Embryo

Autonomy of the embryo (Greek *autos*—self and *nomos*—law) is a feature of the embryo which characterizes its capability for self-regulation and development independent from the maternal plant **(Plate LIII)**. Synonyms: self-sufficiency, independence.

The terms 'autonomy', 'autonomous' and 'autonomous development' were broadly used to denote development of a plant organism independent of exogenous factors (Cooke, [1873?]; Roux, 1912; Kirpichnikov and Zabinkova, 1977; Chailakhyan *et al.*, 1982). Progressive autonomization of development has been traced in vertebrates through comparative-embryological material complemented by experimental data (Schmalhausen, 1968). Even individual parts of an organism reveal some autonomy.

Schmalhausen (1968, p. 352) considered autonomy one of the most pronounced characteristics of all living organisms: 'All biological systems are characterized by a relative capability for self-regulation, i.e., homeostasis. By means of auto-regulation the existence proper of each given system is maintained, as well as its composition and structure with its typical inner relations and regular transformations of the whole system in space and time'. In Schmalhausen's opinion, autonomization of ontogenesis is one of the directed processes of evolution of the organic world.

With respect to an embryo of plants, the term 'autonomy' was first proposed by Batygina (1975, pers. comm.) and later experimentally proved and further specified (Batygina *et al.*, 1978b; Batygina and Vasilyeva, 1979, 1987; Lukyanyuk and Ignatova, 1980; Batygina and Butenko, 1981; Vasilyeva and Batygina, 1981). Different, mostly one-sided, interpretations of this feature of the embryo were proposed. Thus, the dependence of the embryo on a maternal plant, was considered either in terms of physiological self-sufficiency or maturity of embryo (seed), its capability for autotrophic existence and germination (Zubkus, 1950; Zinger, 1958; Harrington, 1972; Raghavan, 1976), or only in terms of its morphogenetic interactions with the endosperm (basically correlation of size of the embryo and endosperm in a mature seed) (Takhtajan, 1966; Poddubnaya-Arnoldi, 1976).

Autonomy of the embryo broadly speaking means its independence from all environmental factors, including the maternal organism. Here we shall consider this feature only in terms of interactions of the embryo with the maternal organism. The embryo becomes autonomous at a certain developmental stage specific for every taxon (Batygina and Vasilyeva, 1979, 1981, 1983a,b; Batygina *et al.*, 1983). Among the critical periods of ontogenesis, this is the most important one, because from this moment the transition of the embryo (new sporophyte) to a relatively independent mode of development begins, i.e., **autonomization of ontogenesis** occurs. This feature is peculiar to the embryo developing in a seed under natural conditions.

Establishment of autonomy of the embryo is a complicated multi-stage process. While the embryo is developing in the maternal plant, the **'degree of autonomy'** increases. Two types of autonomy are distinguished: **complete and relative** (Batygina, 1987a; Batygina and Vasilyeva, 1988).

Complete autonomy of the embryo is established after germination is completed and a seedling formed, when all connections (structural and functional) with the maternal organism (fruit, seed) disappear. In various species this connection is interrupted at different stages of ontogenesis, which is conditioned by specificity of the structure of the mature seed and the fruit (presence of endosperm, perisperm, structure of seed coat, pericarp etc.) and, finally, by the entire complex of adaptive characteristics of a certain species of plant.

The embryo acquires relative autonomy[27] after it becomes independent of a number of factors of the maternal organism (hormones, carbohydrates, vitamins etc.). It can be revealed experimentally, in particular, by means of embryoculture—**the ability of the embryo to terminate normal embryogenesis and to give rise to a normal seedling on a simple non-hormonal medium,** containing mineral salts, sucrose and vitamins. Relative autonomy might also appear at different times in different parts of the embryo (for example, in *Brassica napus*—Fernandez and Bisgrove, 1992).

For a long time, the most widespread point of view was that the stage from which the embryo could commence formation of a normal plant is similar for all dicotyledons; it is characterized by differentiation of cotyledons and the shoot apex (Modilewsky, 1958; Yoffe and Zhukova, 1965; Raghavan, 1976; Raghavan and Srivastava, 1982). However, vast embryological data and, in particular, from embryo culture allow the conclusion that autonomy of the embryo in various representatives of flowering plants manifests at different stages of development: globular (Orchidaceae), heart-shape (*Capsella bursa-pastoris, Arabidopsis thaliana*), and torpedo-shape with different degrees of plumule development (beginning of formation of the shoot apex—*Vicia faba*, well-developed plumule with two forming leaves – *Nelumbo nucifera*); in cereals, autonomy takes place at the stage of the scutellum and shoot apex formation (*Triticum aestivum, Hordeum vulgare, Oryza sativa*) (see Table 8).

The **species-specificity** of the autonomy stage of the embryo must be caused not only by the genotype of a new sporophyte (a type of embryogenesis, the pattern of seedling development) and the genotype of the maternal organism (conditions inside the developing seed), but also morphogenetic and morphophysiological correlations in the embryo and surrounding tissue development, which have been formed in the course of evolution of every plant taxon.

This stage is characterized by a complex of morphogenetic, physiological and biochemical features. Among them, the main ones are: size (mainly critical mass of cells) and differentiation degree of the embryo and surrounding tissues; rate of growth, differentiation and metabolic processes; activity of certain enzymes; need for different nutrients and growth regulators; and sensitivity of embryo to environmental factors. Such features of the autonomous embryo as independence of exogenous hormones and its capability for germination are common for all flowering plants.

In *Nelumbo nucifera* taken as a model object, the gradual 'happening' of embryo autonomy in a seed *in vivo* was traced (Vasilyeva *et al.*, 1987; Titova, 1988; Vasilyeva, 1993a; Morozova, 1993). Changes of embryo differentiation were **strictly correlated in time with the degree of development (or degeneration) of the surrounding tissues and with certain maxima (or minima) of relative rate of growth and changes of level of dissoluble carbohydrates, total protein and nucleic acids.**

At the autonomy stage (10-11 DAP[28]; entire process of seed development and maturation, 30-40 DAP) the embryo attained 3/5 of its final size, had a **formed cotyledon** with two lobes and **conductive system** in them and a **plumule** consisting of the lamina

---

[27] Further we shall use the term "autonomy" for short.
[28] Here and further DAP means days after pollination.

Table 8. Stage of manifestation of capability of embryo for germination in culture *in vitro* on non-hormonal medium[29]

Family	Species	Stage	References
		Z Y G O T I C E M B R Y O	
Asteraceae	*Helianthus annuus*	Shoot apex differentiation	Kalinin, 1956
Brassicaceae	*Arabidopsis thaliana*	Heart-shape	Meinke, 1994
	Capsella bursa-pastoris	Early heart-shape	Monnier, 1976
	Raphanus sativus	Shoot apex differentiation	Kalinin, 1956
Convolvulaceae	*Cuscuta lupuliformis*	(embryo diameter, 1 mm)	Guzowska and Zenkteler, 1969
Fabaceae	*Glycine max*	Beginning of cotyledon differentiation	Eisenberg and Mascarenhas, 1985
	Phaseolus coccineus	Heart-shape	Yeung and Sussex, 1979
	Vicia faba	Torpedo-shape	Batygina and Vasilyeva, orig.
Magnoliaceae	*Magnolia soulangeana*	Torpedo-shape	Le Page Degivry, 1970
	M. grandiflora	– – – // – – –	– – – // – – –
Malvaceae	*Abelmoscus esculentum*	(11 DAP)	Patil, 1966
	A. manihot	(10 DAP)	– – – // – – –
Nelumbonaceae	*Nelumbo nucifera*	Torpedo-shape (embryo with well-developed plumule)	Batygina and Vasilyeva, 1979 Vasilyeva and Batygina, 1981
Orchidaceae	*Bletilla striata*	Globular (embryo with well developed first leaf organ)	Batygina *et al.*, 1988
	Cypripedium calceolus	Globular	– – – // – – –
	Dactylorhiza baltica	– – – // – – –	– – – // – – –
	D. maculata	– – – // – – –	Batygina and Vasilyeva, 1979, 1980
	Thunia alba	– – – // – – –	Batygina *et al.*, 1988
Poaceae	*Hordeum vulgare*	Differentiation of scutellum and shoot apex	Lukyanyuk and Ignatova, 1980; Lukyanyuk, 1983;
	Oryza sativa	– – – // – – –	Davoyan and Smetanin, 1979;
	Triticum aestivum	– – – // – – –	Kalinin, 1956; Sharma and Gill, 1982; Symons *et al.*, 1983

Contd.

Table 8. *contd.*

Solanaceae	*Datura innoxia*	Differentiation of shoot apex	Kalinin, 1956
	Petunia hybrida	– – – // – – –	
Ulmaceae	*Ulmus americana*	Beginning of cotyledon differentiation	Anagnostakis, 1977
Valerianaceae	*Valeriana officinalis*	Late heart-shape	Batygina and Kravtsova, orig.
		S O M A T I C E M B R Y O	
Paeoniaceae	*Paeonia anomala* *P. lactiflora* *P. rubra* *P. suffruticosa*	Early torpedo-shape (beginning of cotyledon differentiation, shoot apex morphologically unexpressed)	Batygina *et al.*, 1978b; Vasilyeva *et al.*, 1978; Batygina and Butenko, 1981

[29]i.e. embryo autonomy stage (V.V., T.B.).

of the first leaf and the primordium of the second. Here the following features were observed: (1) thickening of polysaccharide cell walls of the epidermis in tops of the cotyledon lobes and differentiation of xylem elements in the root apex; (2) maximal relative rate of growth of dry mass and embryo length (correspondingly, the second and third peaks on the growth rate curve); (3) second (in the course of embryogenesis) increase in content of soluble saccharides; (4) appearance of starch grains in tops of cotyledon lobes and in the plumule; (5) plumule begins to turn green; (6) synthesis of proteins specific for this stage (molecular weight – 96.6; 94.0; 56.2; 47.9 kD); (7) appearance of ABA in the embryo; and (8) capability for the germination on a simple non-hormonal medium.

The readiness of the lotus embryo for independent development correlates with significant reduction of the endosperm and the inner integument in size and with the intensification of differentiation of conductive system elements in the seed. Further, by the autonomy stage transformation of the endosperm cells is completed. Its cells become multinuclear, and the nuclei multinucleolar; the cell walls of polysaccharide nature thicken.

Critical events in *Capsella bursa-pastoris* and *Arabidopsis thaliana* embryogenesis described by Meinke (1994), can probably be considered as a characterization of embryo autonomy (although the author himself does not use the term 'autonomy'). He mentioned that in the period from globular to heart-shape stage the following took place: changes in symmetry of the embryo related to initiation of the cotyledons; appearance of a provascular tissue; embryo begins to turn green; establishment of autotrophy, evinced in the ability of the embryo to grow in the culture on a simple defined medium; and cellularization in the endosperm (Meinke, 1991a, 1994).

At the early stages of plant ontogenesis, already in the zygote a taxon-specific development programme can be found, characterized by polyvariant determinancy of the growth and development processes (Batygin and Demyanchuk, 1995). One example of determinancy is the potential **polyvariant capability of the embryo for germination, which is manifested in the autonomy, 'precocious germination' and normal germination of the mature seed**. The germination programme is overlapped by a programme of embryonic growth, which must be regulated by a vascular system connecting the embryo with the maternal plant (Dure, 1975).

According to the scheme (Fig. 76), given by Ihle and Dure (1972), in *Gossypium* embryogenesis there are at least two periods in which the embryo is capable of the precocious germination. In the course of isolation of the embryo in the first period (20-30 DAP) a slow precocious germination is observed. In such an immature embryo, mRNA of germination is absent, but after elimination of the seed tissues the necessary genes activate there, and transcription and then translation of information for germination occur. In the second period (30-50 DAP), when the vascular connection between the seed and the maternal plant vanishes, synthesis of DNA ceases (as is the case of embryo isolation at the earlier stage), and transcription of mRNA of germination starts, but without translation. The latter would lead to vivipary—the lethal outcome of which is hindered under natural conditions by accumulation of ABA in the seed tissues. The embryo, isolated in this period, also reveals precocious germination.

Autonomy is not only specific for zygotic, but also for somatic embryos. For example, in *Paeonia anomala* and *P. suffruticosa* the somatic embryo, formed in the seed, is able to germinate in non-hormonal medium and hence becomes autonomous at the early torpedo-shape stage (Batygina *et al.*, 1978a; Vasilyeva *et al.*, 1978; Batygina and Butenko, 1981; Brukhin and Batygina, 1994).

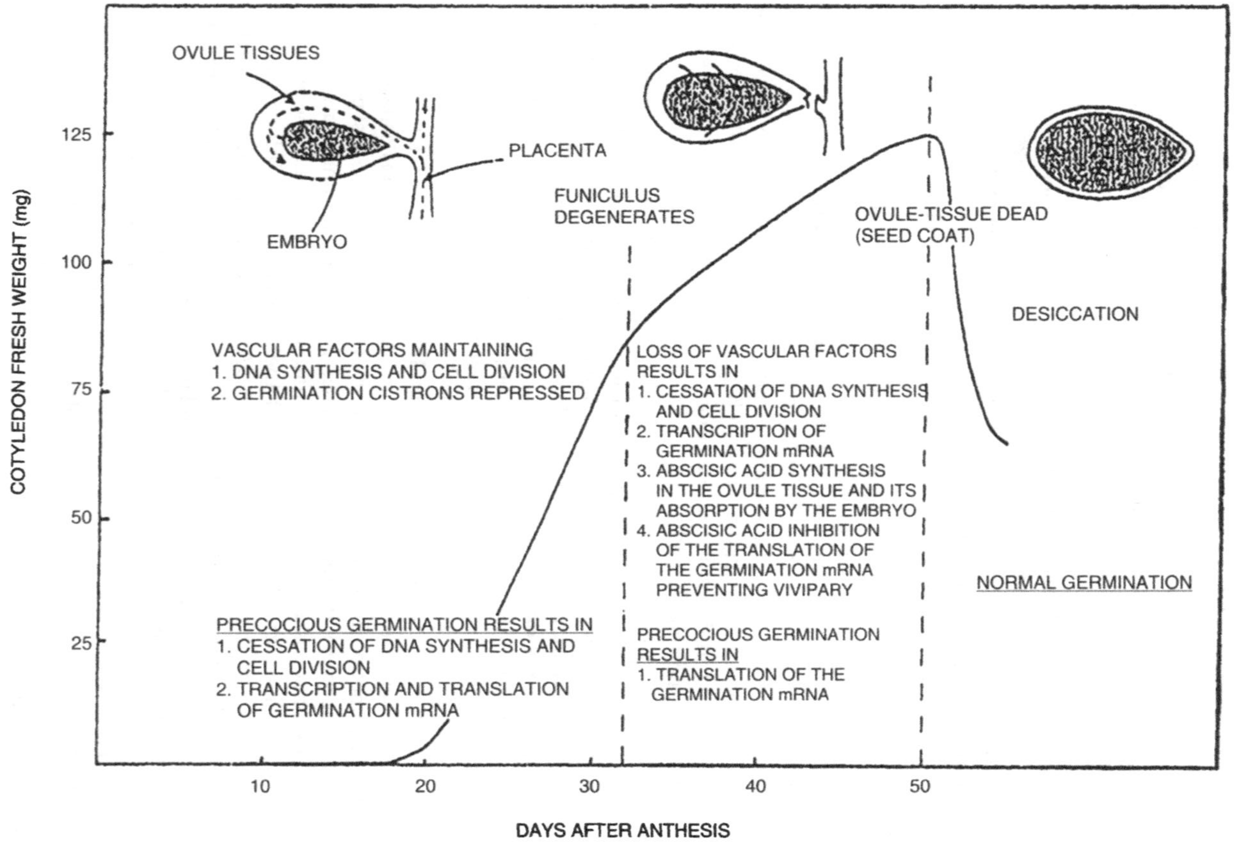

Fig. 76: Postulated scheme of developmental events in cotton cotyledon embryogenesis (after Ihle and Dure, 1972).

In viviparous species of *Bryophyllum* the autonomization process is protracted, and revealed in formation of a detached layer between the somatic embryo[30] and the maternal tissues of the leaf (Bragina *et al.*, 1995; Batygina *et al.*, 1996). Probably, an embryoid in *B. daigremontianum* becomes autonomous at the torpedo-shape stage (cotyledons formed and plumule with two leaf primordia, establishment of adventive roots in stele of hypocotyl). An embryo isolated at this stage develops into a normal plant in a standard medium. Autonomy of the somatic embryo in *B. calycinum* also establishes at the torpedo-shape stage, but as a result of exit of the roots from the maternal leaf tissues.

It is known that in the course of embryo development a gradual transition from heterotrophic mode of living to autotrophic occurs. With this the requirement of the embryo for organic nitrogen, carbohydrates, hormones, vitamins and nutrients specific for every plant species (for example, phenol compounds in *Papaver somniferum*—Pontowich *et al.*, 1976) gradually changes. In the early stages of embryogenesis these substances usually come from the surrounding maternal tissues. In the course of further formation of the seed, the embryo acquires the ability to synthesize them. Starting with the stage of autonomy, the strict accordance of rhythmics of different processes in the embryo and surrounding tissues is violated, and consequently the embryo acquires the capability for self-regulation (Vasilyeva, 1993b).

Mechanisms of embryo autonomization have not yet been sufficiently studied. One such mechanism could be consecutive raising of the blockage of metabolic pathways for autotrophic existence (Raghavan, 1976). This notion has been worked out in the detail in embryology of animals. The activation or inhibition of various enzymes at a certain stage of development, based on regulation of activity of genes, serve as a reflection of autonomous regulation in plants on the intracellular level (Wareing and Phillips, 1978). The unblocking of enzyme systems needed for autonomous development of the embryo in plants, can be traced, for example, by such key enzymes as nitrate reductase and ribulose-bisphosphate carboxylase (Raghavan and Sharma, 1995).

Embryo autonomy can be revealed not only by means of a culture *in vitro*. It is known that immature embryos of the main cereals (*Triticum, Hordeum, Avena*) are able to germinate after sowing in natural conditions (Modilewsky, 1958). This capability of immature embryos is very important, as it reduces seed dormancy and hastens germination. In adverse weather conditions (increased air and soil moisture) or in the course of experimental influence, germination of immature embryos as early as inside the seed on the maternal plant was observed (Lemann and Aikhele, 1936; Pope and Brown, 1943). Mangelsdorf (1923) distinguished the race of *Zea mays* under the name 'germinating seed', in which germination usually proceeds in the ears. Germination of mutant maize embryos on the heterozygotic ears coincides with loss of a regulating factor, which inhibits germination during seed development (Robertson, 1955).

Use of viviparous mutants of another model object, *Arabidopsis thaliana*, enabled identification of a very important block of genes—*LEC*1, *LEC*2 and *FUS*3, which in combination with *ABI*3 play a critical role in regulation of late embryogenesis, inhibition of vivipary (precocious germination) (Meinke *et al.*, 1994). The immature embryos of these viviparous mutants start to germinate precociously, from the torpedo-shape stage. In the course of cultivation in a simple medium, they yield viable plants of normal appearance, except for the presence of foliate cotyledons with trichomes on

[30] This might be a transitional form between the embryoid and the bud.

the adaxial side. It should be recalled that, as mentioned before, in the wild type *Arabidopsis*, the capability for embryo germination (autonomy, in our interpretation) is manifested earlier—at the heart-shape stage.

These investigations initiate a number of issues to be heeded by embryologists: how the multivariate capability of the embryo for germination fits into the evolutionary process; whether the embryos of viviparous mutants have an autonomy stage; and which genes are expressed at the stage of autonomy.

A study of all the stages of embryogenesis and, in particular, the embryo autonomy stage is a necessary theoretical prerequisite for a number of applied investigations: overcoming incompatibility in distant hybridization, achieving haploproduction, reduction of the period of seed dormancy, selection of optimal time of yield and others. This is one of the main methods for increasing the effectiveness of selection processes and for conservation of rare, vanishing and economically valuable plant species.

Embryo autonomy as a **genetic feature** which determines the **universal capability of plants for vivipary**, deserves special attention and further study.

Observations on Living Embryo (Plate LIV)

The study of living embryos through direct observation and monitoring of their life processes is the irreplaceable part of efforts to elucidate the basic embryological problems for scientific and educational purposes. Besides this, the study of living embryos has also been carried out for various practical purposes. Development of living plant embryos may be studied *in vivo* and *in vitro*. In addition, methods have recently been elaborated for gamete isolation, *in vitro* fusion (fertilization) and embryo development from zygotes originated from fused gamete protoplasts.

Embryos *in vivo*. The short-term study of embryos *in vivo* is limited to plant species with thin and relatively transparent seed coats (Orchidaceae, some Campanulaceae etc.) and also to early embryogenesis because of seed transparency decreasing with development of the endosperm. Observation of living embryos in seeds may be realized in some nutrient fluids or media (sucrose, agar-sucrose media), nearly inert fluids (silicone oils, mineral oils) or in clearing fluids (methyl salicylate, NaOH, chloralhydrate, solution by Herr, 1961b, and so forth). Clearing fluids increase seed transparency but shorten survival of seeds and embryos and also observation to just a few hours or even minutes. Observation with simple nutrient or inert media is possible for some hours and even some days. Complete nutrient solutions composed from macro- and microelements, sucrose, vitamins and growth substances are usually not very suitable for direct microscopical observation of embryos inside seeds. They promote the further cell development of seed-coat layers and thereby decrease seed transparency.

The main significance of an *in-vivo* study is the possibility of direct observation and measurement of events taking place in early embryogenesis in semicontrolled conditions. For instance, by time-lapse technique the time course of zygote and linear embryo development of *Jasione montana* L. (Campanulaceae) was measured. The zygote formed a tubular outgrowth, 60-80 mm long, growing to the centre of the embryo sac shortly after fertilization. Growth of the zygote lasted 12-16 h with growth speed 5-6 mm h^{-1} (Erdelská and Klasová, 1978). The speed of zygote growth was correlated and strictly adjusted to the speed of the first cell wall development in the cellular endosperm of this species.

Further time-lapse monitoring yielded data on duration of the mitotic cycle in individual cells of the linear embryo. The mitotic cycle of the basal cell of a two-celled proembryo was shorter than that of the apical cell. This typical time gradient in duration of the mitotic cycle, reflecting polarity of the embryo, persisted throughout the whole linear phase of embryo development. It was also associated with increasing length and vacuolization of cells in direction from embryo apex to its base. The shorter the mitotic cycle, the more differentiated the embryonal cell. Cells of the future suspensor with advanced cell differentiation had a shorter mitotic cycle than that of the future main embryo.

Zygote growth and time course of individual phases of the first mitosis of the zygote in correlation with development of the free nuclear endosperm of *Galanthus nivalis* has also been followed by the time-lapse technique (Erdelská and IWF, 1982).

The more advanced phases of embryo development have been studied *in vivo* in some orchids (Poddubnaya-Arnoldi, 1976) in connection with parallel histochemical studies.

The seeds of chloroembryophytes (Brassicaceae, Linaceae etc.) are often used for observation of embryos *in vivo* for educational purposes or to distinguish the developmental stage of the embryo for further study. After clearing, it is possible to characterize the size, shape and degree of differentiation of globular, heart- and torpedo-shape dicotyledonous embryos and also in some leucoembryophytes. This method aids in ascertaining the appropriate stage for further experimental procedure, for instance for excision and further cultivation *in vitro*, for fixation following methods of optical or electron microscopy etc.

An *in-vivo* study, compared to an *in vitro* one, enables taking into account developmental correlations of all tissues and cells in the embryo sac concomitantly.

Embryos *in vitro*. The study of the embryo *in vitro* requires its excision from tissues of the seed and also from correlative influences which act in the seed. Cultivation *in vitro* supposes the metabolic and developmental adaptation of embryos to the new life conditions. Competency for adaptation and further development *in vitro* is dependent on genotype and developmental phase of the embryo at the moment of the excision. The transition of embryos from heterotrophy to autotrophy, associated with competence of development outside the seed, is reached at different developmental phases in different plant species (Batygina and Vasilyeva, 1981). But in general, the more advanced the development of the embryo at the moment of excision, the more successful its further development in artificial conditions.

It is important to adjust the medium composition to the developmental phase of the embryo at the time of excision. Older (mature or nearly mature) embryos can develop and germinate on relatively simple media consisting of essential inorganic salts (macro- and microelements) and carbon source, usually sucrose, solidified with agar-agar. Cultivation of younger embryos requires in addition a suitable, well-balanced solution of macro-, micronutrients and sucrose as well as organic nitrogen, vitamins and growth substances. The younger the embryo at the time of excision, the higher its requirements for micro-environmental conditions of specific quality during cultivation (osmotic value, consistence and pH value of the medium, position on the medium, aeration, temperature and illumination).

The suitability of life conditions for embryos *in vitro* consists in their ability to:

a) maintain the meristematic character of cells and to promote cell divisions in early embryogenesis (during cell determination and early differentiation);

b) promote growth processes in the central phase of embryo development; and
c) promote maturation in the last phase of embryo development.

Due consideration of the various physiological requirements of an embryo in individual phases of its development *in vitro* is of great importance for the success of an embryo culture. It is useful to moderate the concentration of mineral salts, sucrose and amino acids during cultivation and during very dramatic changes in the morphology of embryos (Monnier, 1976). The idea of altering the medium during cultivation was recently implemented in experiments with proembryos of *Brassica juncea* also (Liu *et al.*, 1993a,b). The culture system was composed of two agar layers, with the top layer containing higher osmolarity than the bottom one. Proembryos were placed on the top layer, whose osmotic pressure was gradually decreased during culture because of the diffusion of osmotically active compounds into the bottom layer.

Developmental analysis of embryos *in vitro* revealed certain characteristics. The growth curve of the embryo *in vitro* and the morphological characteristics of development of embryos *in vitro* differ in some features compared with similar features *in vivo*. Embryos *in vitro* display a tendency to overcome the phase of maturation by direct transition of the embryonal to the post-embryonal phase of development, bypassing reserve substances accumulation and entrance into dormancy.

Differences in growth rhythm. The typical growth curve of embryos *in vivo* exhibits three phases: early slower phase (globular and early heart-shape embryo), second phase of relatively fast longitudinal growth (from heart-shape to late torpedo stage of the embryo) and a third phase with very slow culmination of growth. An embryo *in vitro* mimics the same growth tendency but at a notably lower speed and for a longer time. For instance, the daily values of longitudinal growth of flax embryos excised as heart-shaped (8 days after fertilization) were lower than those of embryos *in* vivo, whereas duration of the growth phase *in vitro* was longer. Increase in length of the embryo represents during 7 days of cultivation (from 8-15 day *in vivo*) only 10-15% that of an embryo *in vivo*. However, embryos *in vitro* very often do not achieve the full length of seed embryos at all, because of very quick decrease in longitudinal growth in favour of width growth of the embryo (cotyledon).

Detailed microcinematographical analysis of the growth course of excised flax embryos *in vitro* revealed oscillations partly regular but not strictly linked to the diurnal rhythm. All growth differences between embryos *in vivo* and *in vitro* are evidently caused by different life conditions. In microcinematographical time-lapse investigations have to be considered as well as regular light impulses during expositions.

Morphological differences. Development of embryos *in vivo* is strictly regular, particularly with respect to number and order in cell divisions. This is manifested in regular longitudinal and radial patterning and typical shape of the embryo and its parts (tissues and organs) correlated with the developmental stage of tissues in the seed and fruit. The character of epigenetic control of embryo development changes after excision. This is structurally manifested by decrease in regularity of development and by changes in proportionality of individual tissues of the embryo. The degree of abnormalities increases in direction from mature to very young excised embryos and depends on suitability of the cultivation medium for different developmental stages of embryos. A well-balanced nutrient medium of high osmotic value, enriched with amino acids, vitamins and some growth regulators (kinetin) was succesfully used for the cultivation of globular proembryos of several angiosperm species excised after protoderm establishment. Embryos of most dicotyledons can be cultivated *in vitro* after establishment of cotyledons, closely following establishment of both patternings (radial = establishment of dermatogen, periblem and plerome and longitudinal = establishment of the embryo axis). This constitutes a crucial point in embryogenesis. If the basic patternings do not become established, the embryo continues development as a globular or callus-like structure, or dies.

In the cultivation of excised embryos, the embryo length:width ratio ('coefficient of embryo differentiation,' Monnier, 1968) decreases compared with *in-vivo* development. The value of this ratio is lower, the younger the embryo at the time of excision. *In-vitro* cotyledons are shorter, wider, usually deviate from each other at an obtuse angle and often with irregulary ordered cell rows. Their cells often lose the meristematic character very soon, enlarge and vacuolize. The same is true of the scutella of grasses. Cotyledons develop as more or less direct structures even in species with U-shaped embryos in the seeds (e.g. *Capsella bursa-pastoris*).

Whereas for cultivation of proembryos, establishment of cotyledons following differentiation of the plumula-radicular axis is decisive, for cultivation of torpedo-shape embryos, maintenance of functionality of established radicular and plumular tissues is important. The gradual loss of the meristematic character of the embryo axis cells, mainly after excision of older heart-shape or torpedo-shape embryos, is related to cessation of cell divisions and tendency to precocious germination (Erdelská *et al.*, 1973). Sometimes original meristems (e.g.radicular meristem) degenerate and are replaced by laterally originated new ones.

Plants originating from excised embryos *in vitro* are usually smaller than those cultivated from seeds. This is associated with imperfect completion of the process of cell determination and differentiation *in vitro*. However, the more suitable the cultivation conditions, the more the structure of embryos *in vitro* resembles that *in vivo*.

Embryo maturation *in vitro*. Embryo maturation inside the seed is characterized by decrease in growth, synthesis and accumulation of reserve substances, decrease in water content and entrance into dormancy. Embryos *in vitro* tends to avoid these processes and to germinate without accumulation of reserve substances and without acquisition of the physiological characteristics important for entrance into dormancy (e.g. drought and frost resistance). Embryos *in vitro* sometimes mimic one or two processes of all the characteristics important for maturation. For instance, in a medium of relatively low osmotic value (medium with macro-, micronutrients and 5 g l^{-1} of sucrose), flax embryos excised 14 days after fertilization (torpedo stage) showed some signs of entering dormancy. Temporary loss of chlorophyll and reduction in growth associated with aquisition of the ultrastructural features typified dormant embryos for about 14 days. The embryos then gradually became green and began to germinate even on the unchanged cultivation medium (Pretová *et al.*, 1981). A high osmotic concentration of the cultivation medium and/or presence of abscisic acid (ABA) in the medium for older embryos will prevent premature germination and allow embryogenesis to be completed in most plant species. Sometimes the presence of suitable precursors in the cultivation medium is necessary for reserve substances synthesis in the phase of maturation. In this way, for instance, excised flax embryos synthesize storage lipids if they obtain acetate and biotin from the culture medium (Kurtz and Miramon, 1957). ABA content and high osmotic value of the medium enabled reserve substance accumulation in *Brassica* embryos (Finkelstein and Crouch, 1984). Frost resistance was achieved by maturation of *Capsella* embryos on a medium with high osmotic value (Monnier and Leddet, 1978).

Embryo cultures are used as a tool for elucidation of some fundamental embryological problems such as embryo nutrition, regulation of embryo development, role of different embryo organs, tissues and cells in different phases of plant development, or to solve various tasks of plant improvement and breeding (breaking of seed dormancy, 'embryo rescue' in some hybrids, utilization of immature embryos as initial explants for initiation of callus formation for organogenesis or somatic embryogenesis, utilization of immature embryos in experiments on cellular genetic manipulations etc.). The number of papers and reviews published on this topic since the beginning of the 20th century, especially in its second half is considerable (e.g. Raghavan, 1980; Monnier, 1995) and attest to living plant embryos as objects of paramount importance in biological research activities.

Embryos from zygotes cultivated *in vitro*. Attempts to cultivate a zygote and linear or early globular embryo *in vitro* were unsuccessful for a very long time. Only recently was a methodological breakthrough achieved in this area of research. First, partially isolated embryo sacs of *Zea mays* were cultured *in vitro* shortly after fertilization. The zygotes underwent normal embryo development and embryos grew to maturity and regenerated into plants (Campenot *et al.*, 1992; Mòl *et al.*, 1993). Also, success was achieved in the isolation of male and female gamete protoplasts, their cultivation and *in vitro* fusion (fertilization) to zygote. Individual phases of embryo and plant development from zygote were also observed in detail and described (Kranz and Lörz, 1993). This process, in most structural features, seems to resemble that of the development of embryos from isolated protoplasts of somatic cells.

These successes predicated considerable improvement in understanding higher plant reproduction, including embryo formation, over the next few years.

Prospects for Genetic Analysis of Plant Embryogenesis

Embryogenesis plays a central role in the life cycle of flowering plants. Ever since the process of double fertilization was documented at the turn of the century, plant embryologists have attempted to characterize the cellular and biochemical changes that occur within developing seeds (Maheshwari, 1950; Raghavan, 1976). For many years, emphasis centred on comparative morphology and analysis of cell division patterns during early stages of embryo development (Johansen, 1950; Wardlaw, 1955). Attention then shifted to experimental studies of somatic embryogenesis, embryo culture and haploid embryos derived from microspores (Raghavan, 1986). Manipulation of zygotic embryos was limited by the presence of surrounding maternal tissues. Electron microscopy provided additional details on cellular changes associated with embryogenesis but generally failed to identify the mechanisms responsible. Plant embryologists believed that genetic factors played an important role in morphogenesis, but mutants with altered patterns of embryo development were largely ignored (Meinke, 1986).

Recent advances in molecular biology have led to renewed interest in plant embryology and the underlying patterns of gene expression that occur throughout seed development (Goldberg *et al.*, 1989). Many genes transcribed during embryonic maturation have now been examined at the molecular level. Genetics provides a complementary approach to the study of plant embryo development by enabling identification of genes with essential functions during this critical stage of the life cycle (Meinke, 1986). This review explores the potential benefits and limitations of mutant analysis in relation to plant embryo development. Additional information is available in recent reviews on plant development (Steeves and Sussex, 1989; Lyndon, 1990), experimental plant embryogenesis (Johri, 1984; Raghavan, 1986; Williams and Maheswaran, 1986) and plant developmental genetics (Sheridan, 1988; Meinke, 1991b,c).

Overview of plant embryogenesis. Plant embryologists initially attempted to explain characteristic patterns of cell division observed early in development by establishing fundamental laws of embryogeny (Johansen, 1950). These laws of parsimony, numbers, origin, disposition and destination were thought to reflect the fundamental organization of embryos from various species. This view was gradually replaced by the realization that cell division patterns are determined more by genetic and biophysical factors than by laws of embryogeny. Examination of developmental pathways in different angiosperms nevertheless revealed a number of common features. The most critical events appear to be formation of apical meristems, establishment of basic patterns of symmetry and cellular organization, and transition from a heterotrophic zygote dependent upon nutrient reserves of surrounding maternal tissues to an autotrophic embryo capable of surviving desiccation and producing a viable seedling after germination. Early stages of plant embryo development are characterized by cell division and morphogenesis. This is followed by a period of cell specialization and embryonic maturation in preparation for dormancy and germination. Several features distinguish embryogenesis in angiosperms from related pathways in animal systems: (1) the process of double fertilization and subsequent interactions between embryo and endosperm tissues; (2) the apparently minor role played by maternal mRNAs stored in unfertilized eggs; (3) absence of cellular migration during embryo development; (4) absence of a germ line established early in development; (5) activation of large numbers of zygotic genes during very early stages of embryogenesis; (6) presence of a male gametophyte active in transcription that leads to elimination of many deleterious alleles before fertilization; (7) establishment of relatively few specialized cell types; (8) formation of apical meristems that ultimately produce the adult plant; and (9) the small size of the zygote and its location deep within maternal tissues. The zygote in flowering plants is also not unique; a wide range of somatic cells can be induced to produce embryos either as part of the normal life cycle or through experimental manipulation. Genetic studies of embryogenesis in *Drosophila* and *C. elegans* might not therefore be directly applicable to higher plants.

Capsella (Cruciferae) served for many years as a model for embryo development in dicots (Maheshwari, 1950; Schulz R. and Jensen, 1968a,b). A similar pattern of development was later found in *Arabidopsis* (Mansfield and Briarty, 1991; Webb and Gunning, 1991). Molecular signals responsible for double fertilization have yet to be identified but recent studies have shown that in many plants the two sperm cells involved are dimorphic (Russell, 1991; see also S.G.Nawaschin, 1927; Sperm Dimorphism, vol. 1).

Several critical events occur during the transition from globular to heart stages of embryogenesis in *Capsella*: (1) the embryo proper establishes bilateral symmetry with initiation of cotyledons, (2) cellular differentiation becomes apparent with formation of incipient vascular tissue, (3) the embryo proper begins to turn green and becomes visible through the translucent seed coat, and (4) the triploid endosperm tissue present as a syncytium of free nuclei during early stages of embryogenesis starts to become cellular. Both the suspensor and cellular endosperm begin to degenerate during later stages of development and have disappeared by maturity. Further expansion of the cotyledons and hypocotyl occurs as the embryo passes through the torpedo, early curled, and mature cotyledon stages of development. The shoot apex forms a cluster of small cells between the expanding cotyledons but remains quiescent until the initial stages of germination.

Although the pattern of development observed in *Capsella* can serve as a model for plant embryogenesis, it cannot reflect the diversity of embryonic pathways observed throughout angiosperms (Wardlaw, 1955). Development of the megagametophyte itself is highly variable (Maheshwari, 1950). Some plants lack synergids, whereas others differ in number and placement of antipodals and polar nuclei. The endosperm tissue is not always triploid; it can range from 2n to 15n depending on the number of polar nuclei produced during megagametogenesis. Two or more megaspores may remain functional and contribute to the mature megagametophyte. This can result in genotypic differences among the haploid nuclei present at the time of fertilization.

The initial division of the zygote is often transverse but subsequent division patterns clearly differ between taxonomic groups. Particular emphasis has been placed on division of the terminal cell at the two-cell stage and the relative contributions of the basal cell and its descendants to the embryo proper (Johansen, 1950; Maheshwari, 1950). Many angiosperms can be classified according to predictable patterns of cell division during early stages of embryogenesis. Other species exhibit irregular patterns even at the earliest stages of development. Gymnosperm embryos differ in a more dramatic fashion because they often progress through a free nuclear phase before the globular stage of development. The size and appearance of the suspensor also differ greatly among angiosperms. Suspensors range from a single cell to a massive column of several hundred cells. Some contain polytene chromosomes, others are polyploid or multinucleate, and a few produce large haustoria that invade surrounding maternal tissues and apparently facilitate the transfer of nutrients to the developing embryo proper. The extent of differentiation at the time of seed maturation also varies. In a few cases, the 'mature' seed contains an undifferentiated globular embryo that completes the remaining stages of development after being shed from the plant. The shoot apex is quiescent and barely visible in some embryos at maturity, whereas in others it produces several leaves before germination. Most embryos prepare for extended periods of desiccation and dormancy following completion of embryogenesis, but in viviparous mangroves, precocious germination occurs within the fruit. Endosperm development among angiosperms also differs in timing and extent of cellularization, presence of specialized haustoria and amount of tissue present at maturity. Models of gene function in relation to plant embryo development must therefore account for both the common features and interesting variations observed among angiosperm embryos.

Genetic analysis of plant embryogenesis. Mutants have been used in several ways to study plant development (Meinke, 1991c). One approach has been to analyse cell division patterns by genetically marking single cells or cell layers and noting the distribution of resultant clonal sectors (Tilney-Bassett, 1986; Poethig, 1989). Three major questions have been addressed through clonal analysis: (1) how many cells within the meristem contribute to the formation of a particular plant structure; (2) when do meristematic cells become restricted to predictable developmental fates; and (3) how variable are the lineages and division patterns of meristematic cells at different stages of development? Clonal analysis has provided extensive information on the origin and developmental fates of cells within the embryonic shoot apex of maize (Poethig *et al.*, 1986; McDaniel and Poethig, 1988). The conclusion from these studies has been that meristematic cells become restricted in their developmental fates but do not always divide in the same manner or contribute to precisely the same portion of the adult plant. Similar studies with other plants have supported the conclusion that cell lineages play a minor role in plant development (Jegla and Sussex, 1989). Clonal analysis has not been widely used to study cell lineages during early stages of embryogenesis (Christianson, 1986). One problem has been the identification of cell-autonomous genetic markers that can be readily scored throughout embryonic development. Another limitation has been marking specific cells at precisely the same stage of development. It should nevertheless be possible to apply the principles of clonal analysis to embryogenesis in *Arabidopsis* and to characterize in detail the nature of cell lineages during early stages of embryo development.

Mutants can also be used to identify genes with important developmental functions. The field of plant developmental genetics has expanded dramatically in recent years with the identification of mutants defective in almost every aspect of development (Meinke, 1991c). With floral mutants of *Arabidopsis* (Bowman *et al.*, 1991) and *Antirrhinum* (Schwarz-Sommer *et al.*, 1990), considerable progress has been made towards identifying genes with direct roles in the regulation of morphogenesis (Coen, 1991). Emphasis in this case was placed on a small number of mutants with homoeotic changes in floral structure. Embryogenesis presents a greater challenge with respect to mutant analysis because a large number of genes are first expressed at this stage of the life cycle. Jürgens *et al.* (1991) followed the *Drosophila* approach and focused on a small number of pattern mutants with defects in both embryo development and establishment of apical-basal pattern elements in adult plants. My laboratory has taken the approach that a wide range of mutants with different phenotypes must be examined to understand the relationship between gene function and embryo development.

Many embryonic mutants are likely to be defective in essential housekeeping functions (Haughn and Somerville, 1988). Several strategies have been proposed to facilitate the identification of genes involved in the regulation of morphogenesis (Meinke, 1991c). These include focusing on mutants with phenotypes that appear to result from changes in cellular identity, mutants defective in genes with embryo-specific functions, mutants with defective embryos that are green rather than white, and mutants that consistently block at a critical stage of morphogenesis. Mutations in housekeeping genes expressed during both gametogenesis and embryogenesis may be eliminated before fertilization (Meinke, 1982; Ottaviano and Mulcahy, 1989). The long style of maize provides a greater selective barrier to mutant pollen tubes than the short style of *Arabidopsis*. Housekeeping defects that result in embryonic lethality may still provide valuable information on the genetic control of cellular functions. For example, a biotin auxotroph of *Arabidopsis* that becomes lethal during embryogenesis has provided the most definitive information to date on biotin synthesis in plants (Schneider *et al.*, 1989; Shellhammer and Meinke, 1990). Other mutants should facilitate the identification of genes that function in cell division and metabolism. Many laboratories are utilizing molecular approaches to identify housekeeping genes with important functions. Analysis of embryonic mutants will enable a complementary genetic approach.

Some embryonic lethals may be defective in genes with important developmental functions. Early stages of embryo development are characterized by patterns of morphogenesis that differ from subsequent iterative events associated with the shoot apex (Poethig, 1990). Defects in genes regulating early embryogenesis may result in lethality, whereas defects in genes regulating the shoot apex may result in homoeosis or heterochrony. Some transcription factors required for completion of embryogenesis may also serve global functions throughout growth and development. Loss of a single factor through mutation may consequently disrupt a wide range of essential functions and result in lethality during embryogenesis. Although pattern mutants are clearly interesting and should provide valuable insights into the genetic control of plant embryogenesis, there are several reasons why research with other embryonic mutants should also be continued. Perhaps the most compelling reason is that polarity and pattern formation are not the only questions that merit analysis in relation to plant embryogenesis. Many genes are likely to play a critical role in the complex interactions that occur between different parts of a developing seed. Understanding the nature of interactions between the embryo proper, suspensor, endosperm, seed coat, nucellus, and maternal plant is as important from a developmental perspective as understanding how meristems become positioned in the proper location. Other genes may play important roles in determining cell division patterns in the early embryo, cellularization of endosperm tissues, nuclear changes in the suspensor, establishment of the dermatogen, initiation of cotyledons, and differentiation of provascular tissues. Identifying these genes may be difficult, but defining their functions during embryogenesis will be required to answer fundamental questions in plant development. Embryonic lethals, defectives, and pattern mutants should therefore play complementary roles in the genetic analysis of plant embryo development.

Diversity of embryonic mutants. Embryonic mutations have been identified in a wide range of flowering plants including *Arabidopsis* (Müller, 1963; Franzmann *et al.*, 1989; Jürgens *et al.*, 1991; Meinke, 1991a), *Zea* (Sheridan and Neuffer, 1982; Dolfini and Sparvoli, 1988; Clark and Sheridan, 1991), *Hordeum* (Bosnes *et al.*, 1987; Felker *et al.*, 1987), *Daucus* (LoSchiavo *et al.*, 1988; Schnall *et al.*, 1988), *Oryza* (Nagato *et al.*, 1989), and *Pisum* (Bhattacharyya *et al.*, 1990). Many of these mutations were induced with chemical mutagens (Neuffer and Sheridan, 1980; Meinke, 1985; Jürgens *et al.*, 1991), X-irradiation (Müller, 1963), transposable elements (Clark and Sheridan, 1991), or T-DNA from *Agrobacterium* (Errampalli *et al.*, 1991). Important features of maize and *Arabidopsis* as model systems for genetic analysis of plant embryogenesis are summarized in Table 9.

Various strategies have been employed to identify mutants defective in embryo development: (1) screening ears of maize produced by M_1 plants after pollen mutagenesis (Neuffer and Sheridan, 1980); (2) screening immature siliques of *Arabidopsis* produced by chimeric M_1 plants after seed mutagenesis (Müller, 1963; Meinke and Sussex, 1979a,b) and (3) screening M_2 seedlings of *Arabidopsis* germinated on agar plates (Jürgens *et al.*, 1991). Variant lines of carrot defective in somatic embryogenesis have also been identified in culture (LoSchiavo *et al.*, 1988; Schnall *et al.*, 1988). Recessive embryonic lethals are generally maintained as heterozygotes that produce 25% aborted seeds after self-pollination. Heterozygous and wild-type plants are distinguished at maturity by screening for the presence of defective seeds.

The high frequency of mutants defective in embryogenesis is consistent with a large number of target genes performing essential functions at this stage of the life cycle. Molecular studies of RNAs present in developing seeds indicate that 20,000 genes may be expressed during embryogenesis (Goldberg *et al.*, 1989). Many of these genes may perform non-essential functions, be duplicated in the genome, or be required during gametogenesis. The precise number of genes identified through loss-of-function mutations is therefore difficult to predict. Jürgens *et al.* (1991) estimated that *Arabidopsis*

Table 9: Maize and *Arabidopsis*—model systems for genetic analysis of plant embryogenesis

General feature	*Arabidopsis*	Maize
Chromosome number	n = 5	n = 10
Haploid genome size	7×10^4 kb	3×10^6 kb
Generation time	40 days	120 days
Fruits per plant	Many	Few
Seeds per fruit	30-60	400-600
Seed length	0.5 mm	8 mm
Seed weight	30 μg	200 mg
Shoot apex (seed)	Small	Large
Leaf primordia (seed)	Absent	Present
Endosperm at maturity	Absent	Present
Target of mutagenesis	Seeds	Pollen
Insertional mutagenesis	Available	Available
Embryo mutants	Numerous	Numerous
Endosperm mutants	Not certain	Numerous
Chromosome walking	Practical	Impractical
Plant transformation	Established	Limited
B-A translocations	Absent	Available
Somatic embryogenesis	Difficult	Difficult
Clonal analysis	Possible	Established
Relevance of model	Dicots	Monocots

contains approximately 4000 genes essential for embryogenesis. This number is based on the relative frequencies of embryonic lethals and pigment mutants recovered following seed mutagenesis. The actual number of essential genes may be significantly lower because spontaneous embryo abortion in large populations of M_2 plants may lead to overestimates of the frequency of embryo-lethal mutations. We estimate that only 500 genes may be identified through embryonic lethals and defectives in *Arabidopsis*. Existing collections of embryonic mutants have therefore not approached saturation.

Most embryonic mutants are defective in either (1) early stages of cell division and morphogenesis, (2) accumulation of pigments and storage materials during embryonic maturation, or (3) preparation for dormancy and germination. Included in the first class are many defective-kernel (*dek*) and germless (*emb*) mutants of maize (Sheridan, 1988; Clark and Sheridan, 1991) and embryonic (*emb*) mutants of *Arabidopsis* (Meinke, 1991a). The second class includes a wide range of mutants defective in endosperm development in maize (Coe *et al.*, 1988) and barley (Bosnes *et al.*, 1987). Tester lines of maize carrying B-A chromosomal translocations have been particularly useful in demonstrating which mutants are defective only in endosperm development (Neuffer and Sheridan, 1980). A wrinkled seed mutant of pea defective in starch-branching enzyme (Bhattacharyya *et al.*, 1990) and mutants of soybean with altered levels of abundant seed proteins (Goldberg *et al.*, 1989) have also been described. Albino and pale green mutants identified following seed mutagenesis in *Arabidopsis* (Müller, 1963; Feldmann, 1991; Jürgens *et al.*, 1991) can be viewed as embryonic mutants because wild-type embryos remain green before maturation. *Fusca* mutants that accumulate anthocyanin during embryogenesis and become lethal at germination have also been described in *Arabidopsis* (Weiland and Müller, 1972). The final class of mutants defective in seed dormancy includes viviparous mutants of maize (McCarty *et al.*, 1989) and hormone mutants of *Arabidopsis* (Finkelstein *et al.*, 1988).

A wide range of interesting phenotypes has been identified among *Arabidopsis* mutants defective in embryo development. Jürgens *et al.* (1991) divided their pattern mutants into four classes: (1) pattern deletions that lack specific regions in the mature embryo, (2) deletions associated with duplications of embryonic structures, (3) multiplication of cotyledons and (4) transformation of cotyledons into shoots. The first class is further divided into phenotypic groups that differ with respect to deletion pattern. Included in this class is the *gnom* mutant, which results in deletion of the root apical meristem. The second class includes a particularly interesting mutant known as *doppelwurzel*, in which the apical portion of the embryo, including the shoot meristem, is deleted and replaced with a mirror image of the basal portion. The resultant seedling is well differentiated but produces roots from both apices. Genetic studies indicate that after screening 44,000 M_2 families on plates and identifying 250 pattern mutants representing perhaps 40 to 50 different genes, Jürgens *et al.* (1991) may have saturated for putative pattern mutants of *Arabidopsis*. Further analysis of these pattern mutants should provide dramatic insights into genetic control of plant development.

Interesting phenotypes have also been identified among embryonic lethals and defectives of *Arabidopsis* (Meinke, 1991a). These include the presence of fused cotyledones, twin embryos, abnormally large suspensors, green blimps, distorted epidermal layers, single cotyledons, enlarged shoot apices, split hypocotyls, embryos with altered patterns of symmetry, bloated embryos with giant vacuolated cells, reduced hypocotyls that fail to produce roots, and embryos that protrude through the seed coat late in maturation. A female gametophytic factor that results in 50% aborted seeds regardless of pollen genotype has also been identified. These mutants may be used to study the establishment and maintenance of symmetry during embryo development, relationship between cell size and embryogenesis, genetic control of polyembryony, fusion of plant tissues during development, regulation of cell division patterns in the dermatogen, control of apical development during embryogenesis, relationship between embryo development and seed maturation, and nature of interactions between the developing embryo proper and suspensor.

Two approaches to mutant isolation are likely to become more widespread in future: transposon tagging in maize (Chandlee, 1991) and T-DNA tagging in *Arabidopsis* (Feldmann, 1991). Over 50 *emb* mutants of maize have recently been isolated from lines containing Robertson's *Mutator* (Clark and Sheridan, 1991). Other collections of putatively tagged *dek* mutants are being analysed by Donald Robertson (Iowa State), Mary Tierney (Ohio State), and Carol Rivin (Oregon State). Screening of 6000 transgenic *Arabidopsis* families produced at Du Pont after *Agrobacterium*-mediated seed transformation yielded at least 100 additional mutants defective in embryo development. Analysis of these mutants is being performed in my laboratory at Oklahoma State University and in the laboratories of Robert Goldberg, Robert Fisher and John Harada, University of California. Populations of transgenic *Arabidopsis* plants generated through other methods (Koncz *et al.*, 1989, 1990; Van Lijsebettens *et al.*, 1991) may also yield tagged embryonic mutants. Mutagenesis with diepoxybutane and molecular cloning of deletion mutants through genomic subtraction represents an approach to gene isolation that may become more widespread in future (Straus and Ausubel, 1990).

Two questions will need to be addressed as embryonic mutants become more abundant: (1) which mutants should be examined in detail? and (2) which mutants are defective in the same gene? The first question may be addressed from opposite directions. Some tagged mutants will be chosen simply because their pattern of insertion facilitates gene cloning. Others will be selected on the basis of interesting phenotypes, sequence similarities with known genes, or existence of multiple alleles. The second question is not trivial in cases where a large number of target genes is involved. Over 10.000 pairwise crosses would be required to complete allelism tests among the 150 embryonic mutants currently maintained in my laboratory. This number could be reduced if crosses were limited to mutants with similar phenotypes, but alleles with different patterns of abnormal development would be missed. The alternative approach is to map the chromosomal locations of mutant genes and limit complementation tests to genes that map to similar regions. Sixteen genes essential for embryo development in *Arabidopsis* have already been mapped relative to visible markers (Patton *et al.*, 1991). Another 75 lines including T-DNA insertional mutants are currently being mapped in my laboratory. Many *dek* and *emb* loci of maize have also been localized to chromosome arms (Neuffer and Sheridan, 1980; Clark and Sheridan, 1991; Scanlon *et al.*, 1991).

Molecular analysis of embryonic mutants. Detailed analysis of embryonic mutants will require molecular isolation of the defective genes. The most promising approach with maize involves transposon tagging (Sheridan, 1988). This approach has already been used to isolate genes with important functions (Chandlee, 1991) and should be applicable to analysis of embryonic mutants (Clark and Sheridan, 1991). Several methods of gene isolation are currently employed with *Arabidopsis*: (1) chromosome walking from adjacent restriction fragment length polymorphism markers, (2) insertional mutagenesis with T-DNA or maize transposons followed by recovery of flanking plant sequences and (3) deletion mutagenesis and subsequent gene isolation through genomic subtraction. T-DNA insertional mutagenesis offers several advantages for the near future. Numerous mutants have already been isolated following *Agrobacterium*-mediated seed transformation (Errampalli *et al.*, 1991; Feldmann, 1991; R.B. Goldberg, pers. comm.) and additional families are being generated that should contain new mutations. Flanking plant sequences can be isolated through either plasmid rescue (Yanofsky *et al.*, 1990) inverse polymerase chain reaction (Earp *et al.*, 1990), or screening of genomic libraries. The primary disadvantages of seed transformation are that approximately half the embryonic mutants identified to date are not tagged and most have duplications and rearrangements of T-DNA that complicate molecular analysis. Transposon tagging could become a valuable alternative to seed transformation should maize elements transpose at high frequency to linked chromosomal sites in *Arabidopsis*. Chromosome walking is likely to remain tedious for several years but should become the method of choice once a physical map of the genome is completed.

Molecular complementation of embryonic lethals presents a special challenge because homozygous mutant tissues are not readily available. This problem might be circumvented were multiple alleles isolated and shown to differ in nucleotide sequence. An alternative approach might be to transform heterozygotes and screen for a distortion of segregation ratios in the next generation. The disadvantage in this case might be the limited number of seeds produced by primary transformants. The best approach might be to introduce the putative wild-type allele into wild-type plants through transformation, cross F_1 plants carrying this inserted gene with plants heterozygous for embryonic mutation, and test for a reduction in the per cent mutant seeds produced after self-pollination. The frequency of defective seeds in rescued transformants should be reduced from 25% to 6% if the introduced wild-type gene is unlinked to the mutant locus and restores normal function to homozygous mutant embryos. Another advantage to this strategy is that any wild-type ecotype can be chosen for transformation.

Research with mutants defective in plant embryogenesis has reached a critical transition point. Previous studies with maize and *Arabidopsis* have clearly demonstrated that a wide range of mutants can be isolated. The challenge now is to demonstrate the value of these mutants to plant biology. The most significant long-term contribution will be the identification and molecular isolation of a large number of essential genes in higher plants. These genes are likely to play critical roles in cell division, metabolism and differentiation.

The immediate concern is how specific mutants can provide information relevant to plant embryo development. The presence of abnormal suspensors in embryonic mutants of *Arabidopsis* illustrates how lethality can be informative. The suspensor is known to play an active role in regulating development of the emrbyo proper during early stages of plant embryogenesis (Yeung, 1980). Much less attention has been placed on the role of the embryo proper in regulating development of the suspensor. Several studies have shown that destruction of the embryo proper at early stages of development is often followed by abnormal growth of the suspensor (Raghavan, 1976). One model proposed to explain this phenomenon states that continued growth of the suspensor is normally inhibited by the embryo proper, and that the developmental potential of cells in the suspensor is revealed when this inhibitory effect is removed through a mutation lethal only to the embryo proper (Marsden and Meinke, 1985). This model is consistent with the relatively high frequency of mutants found with abnormal suspensors. It also suggests that variations in suspensor morphology found in angiosperms may be caused in part by different levels of inhibition by the embryo proper. This model can be tested at a molecular level now that several T-DNA insertional mutants of *Arabidopsis* have been shown to produce abnormal suspensors (Errampalli *et al.*, 1991).

Embryonic mutants with a non-random distribution of aborted seeds in heterozygous siliques have been equally valuable in providing information on the extent of overlap between gametophytic and sporophytic gene expression in higher plants (Meinke, 1982). Similar patterns of gene expression have been described in maize (Clark and Sheridan, 1988; Ottaviano *et al.*, 1988). The hypothesis that *dek* mutants represent null alleles of quantitative trait loci in maize (Lee *et al.*, 1991) is being tested by comparing map locations of *Mutator*-induced *dek* mutants and known quantitative trait loci (Scanlon *et al.*, 1991). Molecular isolation of embryonic genes from maize and *Arabidopsis* will provide access to a valuable collection of promoters active at different stages of embryo development. Suicide genes fused to these promoters could be used to examine the developmental consequences of selective cell death at different stages of embryogenesis. This represents a powerful new approach to experimental plant embryology. These promoters could also be used to change the timing of accumulation of major storage materials, introduce novel gene products into specialized cell types, or eliminate unwanted seeds during fruit production. Expression of essential genes during somatic and zygotic embryogenesis could be compared by inducing mutant cells to undergo embryogenesis in culture and determining whether defects occur at the same stage of development. Embryonic mutants might also be used to identify genes responsible for embryo abortion in interspecific hybrids (Brink and Cooper, 1947) and spontaneous embryo abortion in nature (Wiens *et al.*, 1987, 1989).

Continued advances in related disciplines will be required to complement genetic approaches to plant embryogenesis. Recent studies on the microtubular cytoskeleton in *Arabidopsis* (Webb and Gunning, 1991) and double fertilization in a non-flowering seed plant (Friedman, 1990a,b) serve as reminders that microscopy will continue to play an important role in plant embryology. The potential application of confocal light microscopy to this field remains to be explored. Experimental manipulation of somatic embryos (Schiavone and Racusen, 1990), *in vitro* fertilization of ovules (Zenkteler, 1990) and growth of immature seeds in culture (Raghavan, 1986) are methods that should complement genetic analysis of embryonic mutants. *In-situ* localization of mRNAs (Fernandez *et al.*, 1991) and β-glucuronidase fusions with *cis*-acting elements (Benfey *et al.*, 1990) may provide additional information on patterns of cellular differentiation. With this combination of genetic, experimental and molecular approaches, it should be possible over the next decade to make dramatic advances in our understanding of plant embryology.

Polarity and Embryogenesis

Polarity, an axial organization of a plant organism is one of the essential elements providing its integrity. Axial symmetry is the basic principle which ensures: formation of the specific three-dimensional structure of the organism, plant spacing, co-ordination of functions and adequate response to various stimuli (Medvedev, 1996). The processes of polarization of cells and tissues in morphogenesis play a defining role, since the gradients of morphogenetic factors emerging at that are the basis for differential activity of the genome at various stages of plant development.

In plants, mutual interaction of at least three factors is essentially important during formation of the physiological polarization axis. These factors are: process of active basipetal transport of auxin (IAA), gradients of bioelectric potential (BEP), as well as gradients of calcium ions which are formed due to polar streams of this cation (Medvedev *et al.*, 1989a,b, 1991; Medvedev, 1994, 1996). Studies of the principles of formation and mutual effect of these gradients in plants allow understanding not only the mechanisms of polar growth and tropisms, but also the mechanisms of attracting effect of phytohormones, apical dominating, donor–acceptor interaction, regeneration, embryogenesis and morphogenesis as a whole.

In differentiation induction, the first step is polarization of an object, which can be initiated due to physical and/or chemical effects (environmental factors and signals and processes occurring inside the organism). The distinctive feature of polarizing effects is their vectorial, directed character. Thus both the type of polarizing effect and its quantitative parameters are of major significance.

Gravitation is one of the most manifested external polarizing effects, since the direction and value of gravitation stimulus do not change in time in each particular point of space and it accompanies the plant during its entire ontogenesis (Merkis, 1976,1990). Thus, plants obviously formed intricate mechanisms which allow control and correction of the body position relative to the vector of gravity creating new axes of polarity.

Light is another important external factor determining vital functions of a plant organism. Photosynthesis, photomorphogenesis and phototropic reaction are linked with light absorption of various parts of electromagnetic spectrum by some special structures (receptors, pigments, proteins) (Volotovsky, 1992). Vector effect of light is of importance in induction of the zygote differentiation in some algae. Apparently, at early stages of development formation of axial light gradients along the overground part and the root system occurs in plants since plant cells and tissues can function as light conductors. Thus the root system is often under the effect of red and infrared rays due to the intensive absorption of the high energy part of the spectrum by tissues of the overground part.

In natural conditions plant organisms are usually under the effect of several external polarizing factors at one time. Normally, when the direction of light and gravitation stimulus are the same, the curvature increases in a lateral direction. When light and gravitation acted in opposite directions, the antagonistic character of their mutual effect was observed (Mirza, 1987). When the phototropic reactions of green and etiolated plants were compared, it turned out that the gravitropic reaction dominated during etiolation. When a seedling is greening, the phototropic reaction increases to a higher extent, so that photostimulation represses and even reverses the gravitropic curvature, despite the increase in gravitation ability (Hart and MacDonald, 1981).

In many plants, polarity of cell distribution in the embryo sac along the micropylar-chalazal axis is observed. The egg cell and synergids locate near the micropyle, and the antipodals are located at the chalazal pole (Maheshwari 1950; Poddubnaya-Arnoldi, 1976; Polevoy and Salamatova, 1991; West and Harada, 1993). Thus, in the embryo sac the egg cell can be polarized along the axis of the micropyle-chalaza. Polarization is the basis for all the following stages of embryogenesis. In the zygote of many species of plants, polarity of plastid, mitochondrion and endoplasmic reticulum distribution in the axial direction was observed (West and Harada, 1993). That is why when the polarized zygote divides transversely, the formed cells of the embryo have cytoplasm with various organelle sets. In the embryo of cotton, the apical cell contains more plastids and mitochondria whereas the basal cell is more vacuolated. It has been shown, for instance, that cereals are characterized by hard polarization of the forming embryo since the ovule and embryo sac are initially asymmetric, bipolar (Batygina, 1974b, 1987a, 1993b; Esau, 1977). In the seed coat, endosperm and, partially, in the embryo proper of *Haemanthus katherinae* gradients of osmotic value, non-reduced sugars and amine nitrogen occur in the direction of the chalaza-micropyle (Ryczkowski, 1980).

In the course of differentiation of carrot somatic embryos, as transition to the heart-shape and torpedo-shape stages is under way, specific proteins are observed: some are localized in the basal part of the embryo, others in the apical part (Schiavone and Racusen, 1989). In embryogenesis of a number of plants (lotus, orchids, parasitic plants and others) in the evolutionary process, the 'primary' bipolarity can be lost and monopolarity establishes (Batygina, 1993b). In further development secondary bipolarity emerges, as a result of formation of adventive organs. Thus, a complete description of morphogenesis can only be provided through analysis of the whole development cycle, which includes both early polar differentiation and the following morphological changes along the main axis of the organism.

Germination of the zygote in algae *Fucus* and *Pelvetia* can be a model subject for study in early stages of growth and differentiation (Sinnot, 1960; Bentrup, 1984; Goodner and Quatrano, 1993). Non-fertilized egg cells are radially symmetrical. After fertilization, the zygote moves down to the bottom and in 12-24 hours a rhizoidal protrusion is formed on its bottom side, if conditions are normal. The first division is perpendicular to the axis of the protrusion. Two daughter cells emerging (rhizoidal and thalloid) that differ considerably in structure and character of further development. During the first zygote division a constant axis of polarity of the future organism is formed. Polarization of the zygote can be induced in experiments: by one-side light, additional gravitation effects, gradients of phytohormones and some ions (K^+, H^+, Ca^{2+}), electric field and a number of other influences. The ability to polarize is species-specific to some extent. Thus, for the zygote of *Fucus*, which germinates at comparatively little depth, light is an important factor of polarization, whereas in the egg cell of *Cystosira barbata*, which is related to *Fucus* but develops in darkness, the rhizoid emerges on the spot of spermatozoid penetration (Sinnott, 1960; Robinson and Jaffe, 1975; Robinson and McCaig, 1980; Bentrup, 1984).

Gradients of bioelectric potentials. As the zygote and embryo are differentiating, as well as in the course of rhizogenesis, 'embryonic' electric streams arise, and a specific arrangement of field lines is formed, which indicates points with the maximum rate of growth and zones of laying of future organs of the plant.

Discovery of transcellular ion streams (Jaffe, 1980) has become important for understanding cell polarization processes. In plant cells characterized by a polar type of growth (zygote, germinating pollen, root hairs, rhizoids and so on), electrical currents (movement of positive charges) are registered, which enter the growing parts of a cell and come out at the parts that have already grown (Jaffe *et al.*, 1975; Weisenseel *et al.*, 1975, 1979; Jaffe and Nuccitelli, 1977; Robinson, 1977; Quatrano, 1978; Brawley *et al.*, 1984). These electrical streams precede morphological changes and are connected with such charge carriers as the ions Ca^{2+}, K^+ and H^+.

Thus, in the brown alga *Pelvetia* (in 30 seconds after fertilization) 'electrical' polarization of the zygote occurs and transcytoplasmic current arises, which enters the area of the future rhizoid and comes out at the opposite, thallus side. Here, the highest current density has been registered in the field of the most intensively growing rhizoid pole (Nuccitelli and Jaffe, 1975, 1976). Similar current loops have been registered in the course of embryo development (Brawley *et al.*, 1984; Goldsworthy and Rathore, 1985).

In wild carrot the electrical field is registered around differentiating embryoids at the globular stage; this field is characterized by a specific lines arrangement, which is maintained throughout development (Brawley *et al.*, 1984). The highest input current density has been registered at the apical pole in the zones of forming cotyledons. The maximum output current was attributed to the zone of laying of the apical meristem of the root. The current density in the aforesaid zones increased during embryo development. In similar studies of embryoids generated in the callus culture of tobacco and in the suspension culture of carrot cells, apart from an electrical gradient between forming cotyledons and the root, a gradient pH was observed along the axis of the embryoid (Goldsworthy and Rathore, 1985; Goldsworthy, 1986; Rathore *et al.*, 1988; Rathore and Robinson, 1989). In wheat, oat, haricot, pea and potato electrical currents were registered during formation of the root system; the currents entered the apical meristem of the main root and the tension zone, coming out at the zone of mature grown cells (Ezaki *et al.*, 1988; Miller and Gow, 1989a,b; Toko *et al.*, 1989).

Polar transport of IAA. First of all, endogenous gradients of electric potentials must effect distribution of phytohormones in the plant tissues (Medvedev *et al.*, 1980; Medvedev and Markova, 1990). Adequate provision of phytohormones in tissues is of determinative importance in embryogenesis. For example, when callus proceeds to form an embryoid, the quantity of endogenous IAA and ABA increases. In tissues of non-embryogenic callus twice as many cytokinins accumulate as in the embryogenic callus (Rajasekaran *et al.*, 1987). The polar auxin transport is likely connected also with the electric polarization discovered in embryos of haricot and maple (Fry and Wangermann, 1976).

The process of active polar transport of IAA must be necessary for embryogenesis starting from as early as the globular stage, since inhibitors of transport of auxin suppressed transition to the heart-shape stage of embryo development (West and Harada, 1993). It was found that the external electric field can influence the active basipetal transport of both exogenous ^{14}C-IAA and endogenous auxin (Medvedev and Markova, 1990). In both cases, transmission of electric current through the plant tissue stimulated transport of IAA in the direction of a negatively charged electrode, which could be explained within the framework of the theory of tropisms by Kholodny-Vent. We supposed that both native and externally applied electric gradients could influence the work of membrane proteins, carriers of auxin, changing their activity or localization in the plasma membrane, which, finally, is the real cause of changes of parameters of active basipetal transport of IAA (Medvedev *et al.*, 1980; Medvedev and Markova, 1990).

Polar currents and gradients of calcium ions. In plant tissues, along with the polar basipetal transport of IAA, acropetal currents of Ca^{2+} ions have been found. Active polar currents of calcium ions must be necessary for formation of gradients of this cation, in the first place, in cells and tissues with a polarized type of growth, as well as in tissues with an undeveloped vascular system, in the absence of transpiration, in tropisms. Polar movement of calcium is registered in segments of the hypocotyl of the sunflower (De Guzman and De La Fuente, 1984, 1986), in young fruits of tomato and in leaf-stalks of salad (Banuelos *et al.*, 1987, 1988), in coleoptiles of corn and in internodes of pea (Medvedev *et al.*, 1989a,b). Overvoorde and Grimes (1994) studied the role of Ca^{2+} and calmodulin in somatic embryogenesis in carrot. They found that the concentration of calcium in a medium must be no less than 200 µM to induce formation of embryoids. Using fluorescent probes it was found that at the heart-shape and torpedo-shape stages the level of Ca-calmodulin complex increases in zones of forming cotyledons and root. Application of blockers of Ca-channels verapamil and nifedipine, as well as of Ca-ionophore A23187 suppressed development of embryoids. This means that a definite gradient of Ca^{2+} ions on the plasma membrane of developing cells is necessary to enable normal embryogenesis.

Ca^{2+} ions are an essential factor in polarization. The ratio of calcium level in apoplast and cytoplasm, functioning of membrane systems of transport of calcium, cell and tissue gradients of this cation, participation of Ca^{2+} ions in work of the system of secondary agents are of great importance. During experiments carried out by Tirlapur *et al.* (1993), the character of distribution of membrane-combined calcium and calmodulin in embryo sacs of petunia and tobacco was analysed with the help of fluorescent probes of chlorotetracycline and fluorphenazine. Calcium and calmodulin appeared to concentrate in synergids and the central cell during formation of the embryo sac, whereas in the egg-cell and in antipodal cells the content of Ca^{2+} ions was lower. An axial gradient of calcium distribution was found in synergids, which increased in the direction of the micropylar end of the embryo sac. Later, when the *Petunia* embryo sac was maturating, a high level of calcium was only registered in one of the synergids and in the egg cell.

Apparently, formation of polar currents of Ca^{2+} ions in plant tissues occurs at the expense of concordant functioning of the Ca-channels and Ca–ATPs (Medvedev and Shtonda, 1989; Medvedev *et al.*, 1989b; Medvedev, 1994; Markova *et al.*, 1995). Initiation of morphogenesis in plants must start with formation of polar currents of Ca^{2+} ions, which create an electric axis of polarity and influence the distribution of phytohormones in tissues. Thus, it might be calcium which carries primary information about the appearing vector of polarization. The appearance of the gradient of calcium ions is followed by formation of gradients of the activity of the calcium-calmodulin system, other messenger systems and, finally, the activity of numerous enzymes. This is the manner in which establishment of secondary gradients (secretory, membrane, metabolic and others) must occur. And lastly, elements of cytoskeleton sensitive to change in Ca^{2+} ion activity complete the structural polarization of the cell.

Cytoskeleton. Cytoarchitectonics of the egg cell, especially those parts of the cytoskeleton are associated with the plasma membrane, are the carriers of information about spatial organization and future development of the egg cell (Isaeva, 1994). Application of cytochalazin, while depressing polymerization and functioning of actin filaments, disturbs the normal orientation of the cortical cytoskeleton, for example, in eggs of the sea-urchin and in egg cells of the alga *Pelvetia wrightii*, resulting in an amoeboid character of the protoplast and unnatural shapes of these organisms. Thus polarization of both the egg cell and the embryo is disturbed (Isaeva, 1990, 1994). Analysis of the influence of cytochalazin on differentiation of the zygotes of *Fucus* and *Pelvetia* showed that fibrillar actin plays the key role in cell polarization and in maintenance of axial anisotropy of its morphogenesis (Isaeva, 1994). The complex of actin and actin-binding proteins must fulfil mechanical-chemical fixation of the contents of the cytoplasm, in particular, of the carriers of morphogenetic information, which can exert a defining influence on the spatial determination of the differential activity of the genome not only in the course of embryogenesis, but also in further development of the plant (Davidson, 1986, 1994).

The most important factors influencing the organization of cortical microtubules are Ca^{2+} ions, hormones and gradients of BEP (Hush and Overall, 1991; Warren Wilson *et al.*, 1991; Shibaoka, 1991, 1993). Concentration of Ca^{2+} ions in the cytoplasm and pH are of great importance in the process of assembling (and disassembling) of actin and intermediate filaments, as well as tubuline. Warren Wilson J. and Warren Wilson P. (1993) emphasized the fact that the elements of phloem, xylem and tracheids are normally arranged parallel to the current of auxin. They proposed that the polar transport of auxin can induce transverse orientation of microtubules with respect to the main axis of the cell; this in turn results in regular arrangement of cellulose microfibrils and, further, to axial orientation of cells. Weak electric fields (1 mv per cell) cause reorientation of cortical microtubules, the orientation of which becomes perpendicular to the applied field (Hush and Overall, 1991). In polarized cells, the asymmetry of the complex of plasma membrane and cytoskeleton associated with asymmetric distribution of ion channels and ion pumps of the membrane manifests itself respectively in vectorization of ion transport (Hopkins, 1991) and in generation of electric gradients (Almers and Stirling, 1984). This might be the reason why the cytochalazins depress the electrophysiological polarization of the zygote completely but reversibly (Brawley and Robinson, 1985).

Oscillations. Initiation of morphogenesis can be evoked by many, mostly stress influences. The reaction normally does not develop in linear mode but an oscillatory one. In most cases, stress influences (internal and external) initiate oscillations of the electrophysiological characteristics of plant tissues. That is why oscillatory effects play an important role in the processes of growth, morphogenesis and regeneration in plants. For example, in the process of regeneration of the umbel in the alga *Acetabularia*, electric waves are observed, which propagate upwards at a speed of 400 μm s^{-1} (Kof and Kefeli, 1994). At the basis of many oscillatory effects are chemical processes, the intermediate ingredients of which may change auto-oscillatory. The oscillations of ion concentration (K, Ca, Na, Mg and others) in the developing endosperm (Ryczkowski and Reczyński, 1988) and of ionized calcium in the cytosol are examples of this effect. Thus, after the spermatozoid has penetrated the egg cell, a calcium wave arises, which propagates in the cytoplasm (Tsien R.W. and Tsien R.Y., 1990; Gillot and Whitaker, 1993; Miyazaki *et al.*, 1993).

Our experiments have shown that the process of polar movement of the Ca^{2+} ions in the segments of the coleoptiles of corn runs in an impulse, oscillatory mode in an axial direction (Medvedev, 1996). It is necessary to note that the rhythmic character of the polar transport of calcium coincided with the characteristics of the impulse polar movement of IAA in cambium in wood plants, discovered by Wodzicki and Zajaczkowski (1989). Since plant tissues are provided with auxin in a rhythmical mode, these authors believed that the regulation of IAA of the growing processes in plants is fulfilled on an oscillatory, impulse basis. This may be supported by data compiled by Kristie and Jolliffe (1986), who registered 10-minute oscillations in the speed of growth of mustard seedlings. As far as these examples show, it is the character of oscillations of the morphogene that carries morphogenetic information, rather than the concentration of morphogene per se.

EMBRYOIDOGENY

Embryoid

An embryoid (Greek *embryon*—embryo, *oides*—like) is a germ of an individual which is formed asexually *in situ, in vivo* and *in vitro* **(Plate LV)**. At all stages of its formation, typically the embryoid is bipolar in organization, just like the sexual embryo. Synonyms: somatic embryo, embryo-like structure, adventive embryo.

Vasil and Hildebrandt (1966) were the first to suggest the term 'embryoid' for embryo-like structures, which are formed in culture *in vitro*, and for nucellar and foliar embryos too. Haccius and Lakshmanan (1969) and Haccius (1971) came to the conclusion that this term may be used for all formations, including 'adventive' (asexual) embryos, in which the radicular pole is typically closed and located in front of the stem one, as well as for embryo-like structures in culture *in vitro*.

An embryoid can be formed **exogenously** or **endogenously**, usually from one somatic cell, rarely from an embryonic cellular complex, in **homophasic reproduction** (i.e., in the absence of meiosis and fertilization).

It is typical of an embryoid **to form a new axis**, which joins the shoot and root apices. As a rule, it **has no common vascular system with the maternal organism** (closed radicular pole). The genesis of the embryoid, its shape and size are **taxon-specific**. The basic features which have been established for sexual embryos (polarity, cellular and histogenic differentiation, autonomy and others) are typical of the embryoid as well. In embryoid germination, a new individual 'is born' (Batygina, 1987a, 1989a,b, 1993a) and a short-term connection with the maternal organism has been registered. **The duration of the contact and the stage of embryoid development, at which this contact occurs, differs in several species**. A bud, with which the somatic embryo is often tangled, is a structure with monopolar development and only during further regeneration does it transform into a bipolar structure giving a new individual.

In the light of new concepts concerning the modes and types of reproduction and propagation of flowering plants (Batygina, 1978, 1987, 1992, 1993b, 1994, 1996), the embryoid is regarded as an elementary structural unit of embryoidogenic type of vegetative propagation (see Embryoidogeny below). **At all stages of ontogenesis, starting from the zygote, a tendency towards asexual mode of reproduction exists in somatic embryos.**

Embryoids may arise in natural conditions on various structures and organs: **on the basis of zygote and sexual embryo** (e.g., in *Paeonia, Triticum, Zea* and representatives of family Orchidaceae), from **nucellar** (*Citrus*, some species of *Poa* and in Orchidaceae), **integumentary** (*Allium*) cells, on the leaf (*Bryophyllum, Hammarbia palludosa*), **stem** (*Ranunculus*) or **root**.

In some taxa (e.g. *Paeonia*), the process of formation of a somatic embryo instead of a sexual one in the seed was fixed during evolution (see: Paeoniad-type of Embryogenesis). In others (e.g. *Eranthis*—Haccius, 1965) fulfillment of the embryoidogenic mode of morphogenesis only occurs **in stress situations**. The formation of embryoids may run simultaneous with formation of the sexual embryo (e.g. *Bryophyllum daigremontianum, B. calycinum*). The main idea of the concept is **the universality of morphogenetic processes** for embryoids, **which take place in embryoid formation both in natural conditions and in culture *in vitro***.

A striking example of this is formation of the embryoid in the peony seed. The genesis of the embryoid runs in accordance with the Penaea-variation, **Asterad-type** of embryogenesis (Brukhin and Batygina, 1994). Apical and basal cells share almost equal parts in embryo formation. The somatic embryo of peony develops in a similar way in culture *in vitro* (see: Paeoniad-type of Embryogenesis).

In *Ranunculus sceleratus,* development of the embryoid in culture *in vitro* and of the sexual embryo in the seed also runs in accordance with the **Asterad-type** (Konar and Nataraja, 1965; Konar *et al.*, 1972).

Disturbances of regularities in **embryoid** as well as in sexual embryo development may occur **at any stage of formation**, affecting both first divisions in the embryoid and organogenesis. They often lead to anomalies in development of the embryoid as well as of the seedling and the adult plant, both *in situ* and *in vitro*.

Great similarity in development of embryoids in tissue culture and of sexual embryos *in vivo* has been shown for *Aconitum heterophyllum* (Ranunculaceae) (Ahuja, Titova and Batygina, orig. data). Embryoids are formed from single initial cells of the callus and undergo all the main developmental stages peculiar to sexual embryos—quadrants, octants, globular, heart- and torpedo-shape. On the whole, embryoidogenesis in *A. heterophyllum* runs in accordance with the Onagrad-type of embryogenesis, peculiar to the family Ranunculaceae (Sokolovskaya, 1981). One should note the presence of a well-differentiated suspensor in the embryoid. In conditions of an experiment, anomalies in development of the embryoids were distinguished, the most frequent precocious differentiation of xylem elements at the stage of heart-shape embryo. This anomaly significantly decreased the output of plant-regenerants, and may be associated with deviation of balance of carbohydrates and hormones in the medium and in the callus tissue.

The same situation was observed in the culture of callus taken from immature embryos of *Triticum aestivum* (Batygina *et al.*, 1993). The embryoids were formed endogenously, from single cells of a special meristematic zone formed inside the globular callus. In spite of the fact that, on the whole, the development of embryoids was similar to the development of sexual embryos in *Triticum in vivo* (Batygina, 1968b, 1969, 1974b, 1987a), it was characterized by several deviations from regularities from the very first stages, namely: different orientation of the initial cells of the embryoids and respectively of their apical and basal poles towards the callus surface; delay in divisions in the apical cell (*ca*) derivatives compared with cell *cb* derivatives and, later, deviation in sequence of their divisions, which led to underdevelopment and 'linear' structure of the apical pole of the embryoid. Later, such embryoids degenerated and did not produce plant-regenerants.

Comparative morphological studies of sexual and somatic embryos formed *in situ* and *in vitro*, distinguished the existence of intermediate forms between the classical type of embryo and bud. In their development, these intermediate forms also underwent globular, heart-shape and torpedo-shape stages; they were also characterized by the presence of cotyledons, plumule and hypocotyl but nevertheless exhibited some specific features (Batygina, 1992a). They usually arose on vegetative organs and had no main root. Adventive roots formed and at some stage established a connection with the conductive system of the plant. Such an intermediate form appears to be the structure that forms on the leaves of *Bryophyllum diagremontianum* and *B. calycinum* (the so-called 'propagule'). On the whole, morphogenesis of these structures is similar: they are formed **from derivatives of the 'dormant** meristem', proliferating during leaf development. The 'globular' stage of the propagule in

Bryophyllum differs in morphology from the classical globular stage of the sexual embryo. Formation of the propagule cotyledons in both species occurs in the same manner as in the sexual embryos of dicotyledons: usually, one of the cotyledons in the early stages of development is smaller than the other, after which the two become almost equal in size. The cells of the cotyledons contain starch. The formed and mature propagule is bipolar, has well-differentiated cotyledons, plumule and some adventive roots. Detachment of seedlings from the maternal leaf occurs due to formation of separating layers of cells, some of which are filled with tannins. The time of detachment of the propagule from the maternal organism depends on conditions in which the mature plant grows. The propagule appears to become autonomous when it loses the connection with the conductive system of the leaf and when the adventive roots are formed. In *B. diagremontianum* it occurs at the late torpedo-shape stage (Batygina *et al.*, 1996; see also: Autonomy of Embryo).

In spite of certain clarity of interpretation of the term 'embryoid', the question of expediency of its usage is still debated. Thus, the famous Indian embryologists Swamy and Krishnamurthy (1981) suggested rejection of the term, based on the fact that the embryoid has no initials of epiphysis and hypophysis and some other features which are typical of the sexual embryo (see: Polymorphism of Sexual and Somatic Embryos as Manifestation of their Developmental Parallelism under Natural Conditions and in Tissue Culture).

It should be noted that, historically, the term 'embryo' is used to designate the germ of a new organism, resulting in fertilization. In this connection the word-combination 'somatic embryo', mostly used in research works on tissue culture, appears contradictory because it expresses opposite concepts: sexual ('embryo') and asexual ('somatic') processes.

From the position of the modern state of embryological science, this contradiction is smoothened because 'embryoid' is a germ of a new individual, which differs from the sexual embryo **mostly by its origin** (from the somatic cell or cells) (Batygina, 1996). At present, a comparative analysis of the genesis of sexual and somatic embryos allows claiming parallelism in their development (Batygina, 1996).

As far as the term 'adventive embryo' is concerned, it is hardly worth using for embryoids formed in callus and suspension cultures, because we do not know which is the main one and which the additional. This is also applicable to embryoids formed on vegetative organs.

Embryoidogeny

Embryoidogeny (Greek *embryon* + *oides* + *genus*—origin) is a type of asexual (broadly speaking, vegetative) reproduction of flowering plants *in situ, in vivo, in vitro,* the elementary structural unit of which is the embryoid **(Plate LVI)**. Propagation in this case is fulfilled by seeds in which embryoids are formed, or by miniature seedlings, which develop on various plant organs from embryoids.

We were the first to propose the term 'embryoidogeny' as a new category of vegetative reproduction, to consider the role of this phenomenon in the reproductive system of flowering plants (Figs. 77 and 78) (Batygina, 1987b, 1989a,b, 1990, 1991, 1992b, 1993a,b, 1994, 1996).

It is known that formation of a new individual in flowering plants occurs either in a **sexual manner**, i.e., the **sexual process** precedes embryogenesis (**heterophasic**

<table>
<tr><td colspan="13">MODES AND TYPES OF REPRODUCTION AND MULTIPLICATION IN ANGIOSPERMS</td></tr>
<tr><td colspan="2">with alternation of generations</td><td colspan="11">without alternation of generations</td></tr>
<tr><td colspan="2">sexual</td><td colspan="11">asexual</td></tr>
<tr><td colspan="4">SEED</td><td colspan="9">VEGETATIVE</td></tr>
<tr><td>gamo-spermy</td><td colspan="3">agamospermy</td><td colspan="9">aspermy</td></tr>
<tr><td>amphy-mixis</td><td>apomixis (gameto-phytic)</td><td colspan="11">amixis</td></tr>
<tr><td colspan="2" rowspan="3">EMBRYOGENY</td><td colspan="5">EMBRYOIDOGENY</td><td colspan="6">GEMMORHIZOGENY</td></tr>
<tr><td colspan="2">floral</td><td colspan="3">vegetative</td><td>floral</td><td colspan="3">vegetative</td><td rowspan="4">sar-men-tation</td><td rowspan="4">parti-cula-tion</td></tr>
<tr><td>embryo-nic (cleavage)</td><td>ovular (nucel-lar. integu-mentary)</td><td>foliar</td><td>cauli-genic</td><td>rhizo-genic</td><td></td><td>foliar</td><td>cauli-genic</td><td>rhizo-genic</td></tr>
<tr><td>Mangrove, Hordeum, Zea, Arabi-dopsis</td><td>Antennaria, Hieracium, Taraxacum etc.</td><td>Or-chis, Paco-nia, Eritro-nium etc.</td><td>Citrus, Allium etc.</td><td colspan="3">Crassula, Bryophyllum, Ranunculus etc.</td><td>Allium, Festuca, Poa, etc.</td><td colspan="3">Lilium, Hammarbia etc.</td></tr>
<tr><td>viviparity</td><td colspan="3"></td><td colspan="7">VIVIPARITY</td></tr>
</table>

Fig. 77: Interconnection between different types of reproduction and multiplication in angiosperms.

reproduction), or asexually, which means that formation of somatic embryos or buds and roots[31] proceeds **from somatic cells without gametes and sexual process (homophasic reproduction)**.

According to our conception, there are **two forms of embryoidogeny: floral (reproductive) and vegetative. Floral embryoidogeny** is a formation of embryo-like structures in the seed; it includes **embryonic**[32] **(monozygotic), ovular (integumentary**

[31] Organogenesis: gemmogenesis, rhizogenesis and gemmorhizogenesis.

[32] Since the term 'cleavage' does not reflect the origin of embryoids, it is **replaced by the term 'embryonic embryoidogeny'**. We introduced the term **'ovular** (Latin *ovulum*) **embryoidogeny'**, which combines the phenomena of **integumental and nucellar embryony (embryoidogeny)**. Introduction of these terms allows determination not only of the origin of these somatic embryos (from the somatic cells of the embryo or the ovule), but also the stages of ontogenesis at which they are formed.

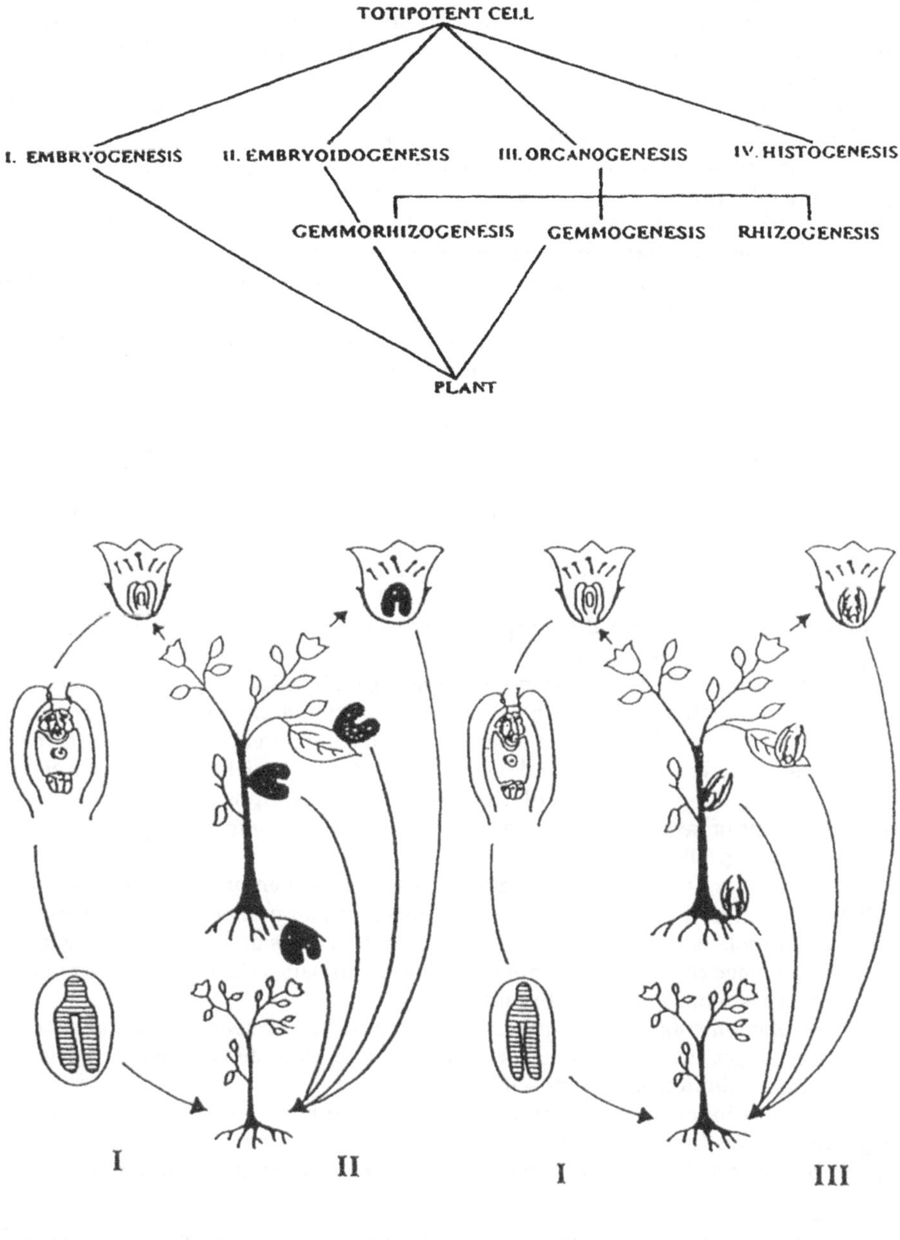

Fig. 78: Morphogenetic pathways and types of sexual (embryogeny, I) and asexual (embryoidogeny, II; gemmorhizogeny, III) reproduction and multiplication in flowering plants.

and nucellar) and some forms of homophasic vivipary[33], the position and role of which in the reproductive system have not yet been determined or properly considered. **Vegetative embryoidogeny is a formation of embryo-like structures on the vegetative organs (leaf, stem, root)**; it is represented **by foliar, cauligenous and rhizogenous forms** respectively.

Usually, embryoidogeny is characterized by establishment of many embryoids (polyembryoidogeny = polyembryony) in generative or vegetative organs. That is why the **phenomenon of polyembryoidogeny** in the reproductive system of flowering plants **must be considered broader**, including all the aforesaid forms (Poddubnaya-Arnoldi, 1976; Batygina, 1978c, 1993a; Grant, 1981) and **must not be shortened to the adventive one (nucellar and integumentary)**, as some researchers do (Naumova, 1989). The term '**adventive embryo**' should only be used for **those somatic embryos** which are formed alongside the **sexual embryo** in the seed.

For seedlings arising from embryoids both in seed and vegetative organs **we** use **the term 'clone' (or 'microclone')**. This term is not unusual, but quite correct; it reflects the essence of embryoidogenic type of vegetative reproduction. It is important to note that in some representatives of flowering plants (Citroideae, Orchidaceae and others) **two different clones** can be formed in one seed—**two generations**: **clone of the maternal organism** (at the expense of nucellar and integumentary embryoidogeny) and **clone of the daughter individual** (at the expense of embryonic embryoidogeny) (Fig. 79).

Theoretically, one can believe that at the expense of monozygotic embryony (embryoidogeny) and vivipary (embryoidogenic) during one season a **genetically homogeneous population differing in age** can be observed. Concomitantly, at the expense of nucellar and integumentary embryoids, which are **derivatives of the same zygote** as the embryoids formed in **monozygotic embryoidogeny and vivipary**, the parental (maternal) genotype is maintained in the population for several seasons.

It should be noted again, that when embryoids of different origins can be formed in one seed—**maternal (ovular)** and **new daughter (embryonic) origin**—plants of two genotypes develop in the population. This is a presupposition for creation of heterogeneity of seeds, which ultimately determines genetic heterogeneity of the population (Fig. 80).

Some forms of vegetative vivipary are also forms of embryoidogeny (Batygina, 1989a,b, 1990). In this case, since both in embryonic and ovular embryoidogeny are involved, bipolar structures—embryoids, which are clones of the maternal organism at the later stage of ontogenesis, are formed from somatic cells of the leaf, stem and root (vis-à-vis monozygotic embryoids).

The phenomenon of vivipary is detected in different species of plants and in various ecological zones. Formation of the propagule is attributed to a certain vegetative organ; this attribution is taxon-specific.

Embryoidogeny, represented by different forms, has a bigger **adaptive significance** for flowering plants than another type of vegetative reproduction—gemmorhizogenesis. Its advantage is that in the first, **the propagation unit is a germ**

[33] The term 'vivipary' (Greek *viviparia*—viviparity) we use for both the viviparous structures formed on the vegetative organs from the somatic cells (homophasic reproduction represented by either the embryoids, or the buds, or intermediate forms), and for the viviparous proper, which are formed in mangroves (heterophasic reproduction) (Batygina, 1996). Here only cases of homophasic reproduction are considered.

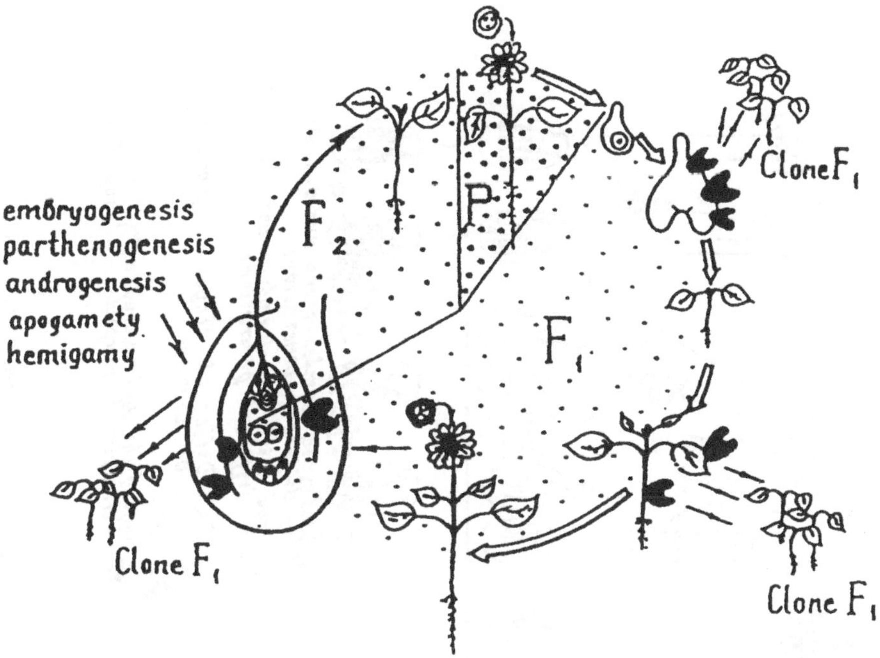

Fig. 79: Embryoidogenic type of reproduction in angiosperms.

of the whole organism (embryoid or small seedling) formed in the seed or on the vegetative organs, **while in gemmorhizogeny it is a part of an organism, i.e., separate organs** (bud or root). In the latter, special conditions and time are required for regeneration up to the whole individual. In vivipary, a valuable bipolar organism (miniature seedling) is formed; in some species of plants it is able not only to root at once, but even to spread with wind or water short and long distances, like seeds.

Many flowering plants are characterized by **vegetative reproduction represented by different forms of embryoidogeny and gemmorhizogeny**, side by side with seed (sexual) reproduction. Correlation between these forms of reproduction is taxon-specific. Thus, in representatives of family Orchidaceae (*Nigritella nigra, Eulophea epidraea, Spiranthes australis*) all the forms of embryoidogeny, gemmorhizogeny and sexual reproduction are found (Afzelius, 1928; Swamy, 1943, 1948c; Batygina, 1989c). In different species of family Crassulaceae, seminal reproduction and vivipary occur. In representatives of family Poaceae, equally with formation of sexual embryos, ovular (nucellar) and embryonic (monozygotic) embryoidogeny and gemmorhizogeny are distinguished. In representatives of subfamily Citroideae, ovular (nucellar) embryoidogeny occurs equally with the sexual process (Batygina and Mamentyeva, 1979). In some onions (e.g. in *Allium odorum*), both integumentary embryoidogeny (Modilewsky, 1931) and gametophytic apomixis (Wojciechowicz and Strzelecka, 1997) have been revealed.

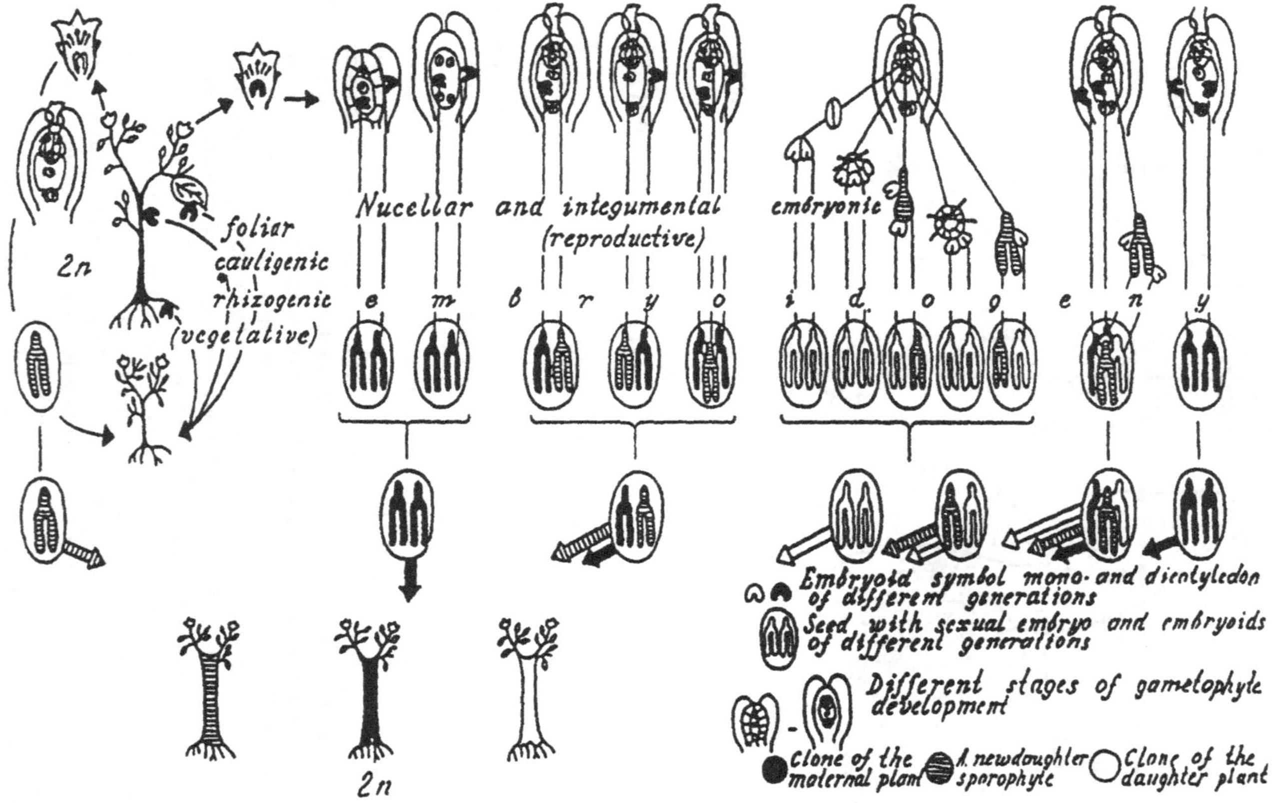

Fig. 80: Embryoidogenic type of asexual reproduction and multiplication of flowering plants. Heterogeneity of seeds.

Thus, **embryoidogeny** has definite biological significance in the reproductive system of flowering plants, employing the pathway of asexual (vegetative) **reproduction and propagation**. Such fixed forms of reproduction as **ovular (nucellar, integumentary), embryonic (monozygotic, cleavage) and some forms of vivipary are only elements of one total type of reproduction—the embryoidogenic one**. Forms of embryoidogeny are taxon-specific in a certain sense. **The basis of the phenomenon described above is the exclusive feature of the plant cell-totipotency**.

In this connection, it is hard to agree with **the statement of some authors, who attribute a special role to certain forms of embryoidogeny in the system of reproduction of flowering plants**. This is not quite correct because in order **to determine the role of each form of embryoidogeny, it must be considered vis-à-vis and in connection with other forms, as well as with embryogeny, gemmorhizogeny and gametophytic apomixis**.

There is no doubt that forms of embryoidogeny being diverse, we are faced with an **original type of vegetative reproduction**. It correlates with Grant's opinion (1981) that both adventive embryony and vivipary (homophasic one—T.B.) are forms of vegetative reproduction (i.e., the individual is formed without the sexual process). The phenomenon of embryoidogeny is of great significance in the evolution of the system of reproduction of flowering plants; however, some aspects have not been determined because of the low level of investigation of the different forms of reproduction it embraces.

Polymorphism of Sexual and Somatic Embryos as Manifestation of Their Developmental Parallelism under Natural Conditions and in Tissue Culture (Plates LVII to LIX)

The new approach and strategy of investigation have permitted revelation of the existence of the somatic embryo **not only *in vitro*, but also in natural conditions.**

The plant organism is able to form somatic embryos at all stages of its development and on different organs (vegetative and generative), side by side with sexual reproduction. This ability enlarges the plasticity and tolerance of the reproductive system. The system approach has also permitted introduction of the new notion—'embryoidogeny', as a special category of asexual reproductive (vegetative one) *in situ, in vivo* and *in vitro*, and consideration of its role in the reproductive system of flowering plants (Batygina, 1989a-c, 2000). Embryoidogeny includes the following forms: **ovular** (**nucellar** and **integumentary**), **embryonic** (**monozygotic-cleavage**) and **homophasic viviparity** (**foliar, cauligenic** and **rhizogenic**). The somatic embryo occurs as the **elementary structural unit of all these forms of reproduction**.

Development of sexual and somatic embryos proceeds in parallel, manifesting in their great polymorphism. The only **difference between them is origin:** the zygotic embryo results from **heterophasic reproduction, the embryoid from homophasic reproduction. Various transitional forms** from embryo, embryoid to bud can be revealed.

Thus **sexual embryo, embryoid and bud appear as three elementary** structural units of reproduction and propagation.

The correlation of different modes of reproduction in plants, i.e., the reproductive strategy of species (Batygina, 1992a,b), is predominantly determined by the process of adaptive evolution. This prolonged process has led to the emergence of a great diversity of reproductive structures. These are zygotic embryos, adventive buds, propagules, somatic embryos etc. To encompass the diversity of embryo-like structures appearing in plant tissue culture *in vitro* and on plants *in situ* and *in vivo*, the general (blanket) term 'embryoid' was proposed (Vasil and Hildebrandt, 1966). Later, the following definition was created to expand the term: embryoid (somatic embryo)—the originative structure of an individual arising asexually *in situ*, *in vivo* and *in vitro*. The embryoid is bipolar at all stages of development as is the sexual embryo. Unlike these two structures, a bud develops as a pole itself and can give rise to an individual only after regeneration of roots (Batygina, 1992a,b, 1996).

Similarity of sexual and somatic embryo development is considered one of the main points under discussion of terms of morphogenetic pathways investigation such as embryogenesis, embryoidogenesis and gemmorhizogenesis. The point of discussion is: to what extent does the morphogenesis of the somatic embryo correspond to that of the sexual embryo, i.e., to what extent does it obey the laws of cell division posited by Errera, Sachs and Hertwig and the laws of embryogeny (Souèges, 1937b; Johansen, 1950). The attention paid to the role of embryoidogeny in the general system of plant reproduction is insufficient. The terminology pertaining to embryo-like structures and their classification also remains debatable.

Only separate aspects of somatic embryo morphogenesis *in vitro* have been studied in terms of comparative embryology. A comparison of somatic embryos under natural conditions (cleavage, nucellar, integumentary, cauligenic, rhizogenic, foliar) with sexual embryos and with somatic embryos from *in vitro* culture is practically absent.

The well-known Indian embryologists Swamy and Krishnamurthy (Swamy and Krishnamurthy, 1981) came to the conclusion that development of the somatic embryo *in vitro* would never stand comparison with that of the zygotic. They stated that the embryoid is devoid of the main features of the sexual embryo and from the morphological viewpoint more closely related to a shoot bud. Their arguments are these: the embryoid is devoid of internal differentiation because its initial cell lacks polarization; the embryoid lacks an orderly pattern of cell divisions; the laws of cell divisions of Errera and Sachs are not obeyed; a suspensor does not form; the protoderm formation is belated and also incomplete; the embryoid lacks typical centres of polar organization—the hypophysis and epiphysis; the embryoid often lacks the initials of hypocotyls as well; organization of the main root is suppressed, as neither plerome and periblem develop, and so on.

The latest data on development of the somatic embryo *in situ* and *in vitro* provide evidence for regarding the-above-mentioned signs proposed by Swamy and Krishnamurthy to distinguish sexual embryos from somatic, as optional. They cannot be used in all cases of definition of the morphological status of the sexual embryo as well as the embryoid.

The system approach (Batygina and Vasilyeva, 1983a,b) allowed for the first time a comparison of the modes of sexual and somatic embryo formation on model objects *in situ*, *in vivo* and *in vitro*. It revealed that development of a somatic embryo *in situ* and *in vitro* recapitulates that of a sexual embryo, the only difference being origin (sexual embryo results from heterophasic reproduction, somatic embryo from homophasic reproduction). A new type of asexual reproduction in flowering plants—embryoidogeny —was established. The somatic embryo occurs is an elementary structural unit of embryoidogeny (Batygina, 1987b, 1993a,b, 1994; Vasil and Hildebrandt, 1966).

In connection with this it seemed necessary to carry out a comparative analysis of sexual and somatic embryo development (*in situ*, *in vivo* and *in vitro*). The following structures were chosen for comparison: **(1) initial cells of sexual and somatic embryos** (Batygina and Vasilyeva, 1997); **(2) sexual embryos of plants with different types of embryogenesis,** and of seedling development with respect to their reproductive biology and ecology; **(3) somatic embryos developing on different parts of the plant** (seed, leaf, stem, root); **(4) somatic embryos, obtained from different explants** (plants from diverse taxa), with various modes of initiation and development **in *in vitro* culture.**

This paper examines sexual and somatic embryos in terms of certain characteristics, for example polarity, symmetry, formation of epiphysis and hypophysis, development of main organs, shoot and root apices, and others.

***Initial cells of sexual and somatic embryos* (Fig. 81)**

Morphogenesis of initial cells of both sexual and somatic embryos determines to a great extent proceeding of embryo- and embryoidogenesis. Characterization of initial cells still remains a major problem of morphogenesis of the embryo.

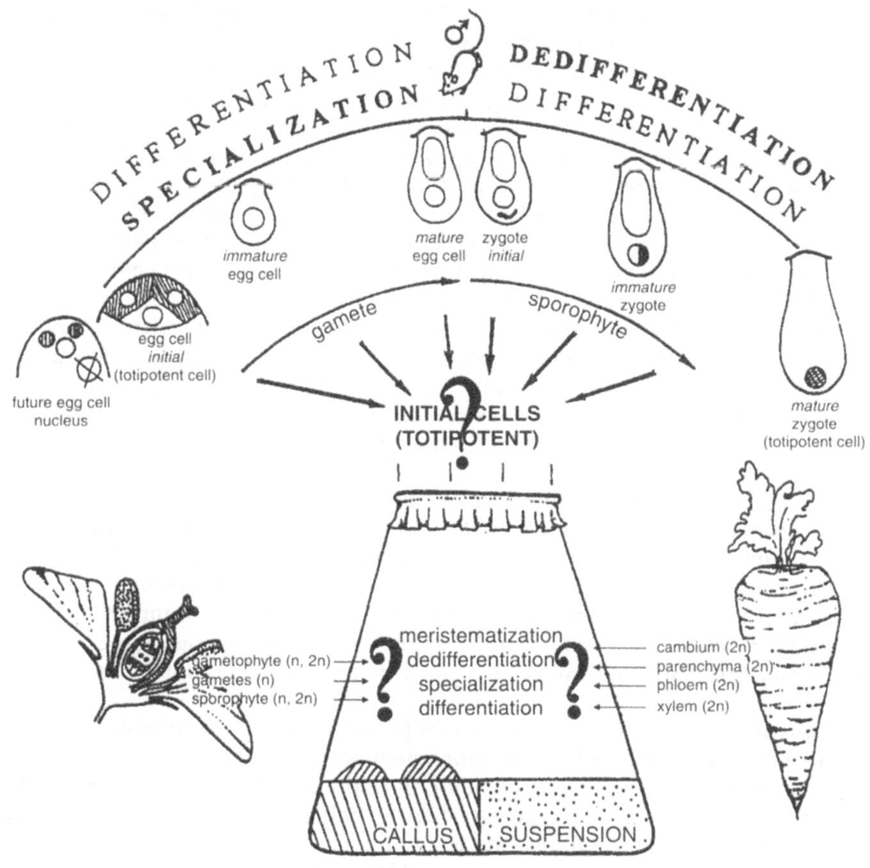

Fig. 81: Formation of sexual and somatic embryo initial cells *in vivo* and in culture *in vitro* (schematic).

The system approach suggests comparing **initial cells of the somatic embryo** *in vitro* not only with **the egg-cell and zygote,** but also with **initial cells** which **under natural conditions** generate **somatic embryos in the seed** (cleavage, nucellar, integumentary) or **on vegetative organs** (foliar, cauligenic and rhizogenic). In this respect the following questions have first priority:

1. What cell in the embryo sac can be considered an analogue of the initial cell of the somatic embryo (*in situ* and *in vitro*): the egg-cell (gamete) or zygote (sporophyte)?
2. To what stage of the mitotic cycle do these cells belong and what is the rate of passing of all these periods (for example, of the interphase)?

Comparison of the somatic initial cell both with the egg-cell and zygote is reasonable. It is well known that the egg-cell of some species of plants in natural conditions gives rise to parthenogenetic embryos which develop in accordance with certain laws. However, the question of whether the egg-cell taking the pathway of parthenogenesis can really be considered the egg-cell concerning its inner state or in this sense is it closer to the zygote remains enigmatic. Besides, comparison of initials of somatic embryos forming *in vivo* with the egg-cell or zygote is more correct than comparison with initials of the somatic embryo *in vitro,* since initials of the somatic and sexual embryos are determined genetically and develop in the system of the organism as a whole. It is easier to establish their genesis and hence easier to identify them.

A thorough comparative investigation of the genesis of the egg-cell and zygote *in vitro* in various species of flowering plants has provided grounds for specifying division of their development into periods. This allows characterizing in more detail the complex structural transformations that occur in the egg-cell and zygote (see: Zygote).

The first stages of formation of the egg-cell in plant species whose embryo sac develops according to the Polygonum-type are relatively uniform. Initials of egg-cells are morphogenetically similar and totipotent. They have the centrally disposed nucleus typical of meristematic cell vacuolization and envelope.

As 'the egg-cell initial' progresses into a 'formed' and then 'mature' egg-cell, its size increases, shape alters, and polarity and symmetry establish, consequent to the property of totipotency is partially lost. Typically the mature egg cell exhibits: specific vacuolization of the cytoplasm, shifting and redisposition of the nucleus, and transformation of the envelope (in the apical part). These features are taxon-specific.

The specialized mature female gamete 'is not able' to develop further by itself. It is only after a sperm (trigger) enters the embryo sac that a complicated process of dedifferentiation starts in the egg-cell. During fusion of female and male nuclei the zygote initial is formed and dedifferentiation continues right through the zygote interphase, in which the cytoplasm, envelope and nucleus undergo a number of changes that lead to formation of a taxon-specific contour of a mature zygote on the one hand, and regaining of the meristematic state (totipotency) and ability for proliferation on the other.

Thus, in the 'egg-cell-zygote' system the pulsating character of ongoing processes can be observed: meristematization, differentiation, specialization, dedifferentiation and again meristematization. Apparently, only at the stages of meristematization and proliferation, can the 'egg cell-zygote' system be compared with initials of the somatic embryo both in natural conditions and in culture *in vitro.*

Development of the initial cells is strongly influenced by the external environment. The egg cell initial and also the immature and mature egg-cell, like the zygote, constitute part of the complex system of the egg apparatus. In the course of genesis the egg-cell and zygote become isolated and relatively autonomous. Besides, the egg apparatus itself constitutes a part of the complex system of the embryo sac, which differs greatly in developmental pattern in various taxa (16 types of embryo sacs). The type of embryo sac largely determines the polarity and symmetry of the egg-cell, zygote and character of the first division (symmetrical/asymmetrical).

To date it has not been ascertained whether such processes take place in cells which give rise to somatic embryos *in situ*.

Initial cells of somatic embryos of various species of flowering plants taken under natural conditions are relatively uniform regarding morphology and structure. It was established that in all systems embryogen cells, which originate from embryoids, possess a number of features which are typical of fast-dividing meristematic cells, namely, small size, dense cytoplasm, large nucleus with a notably increased nucleous and small vacuoles. Their histochemical and ultrastructural peculiarities are evidence of intensive synthesis of RNA and metabolic activity (Batygina, 1984; Williams and Maheswaran, 1986). Meristematic epidermal cells of the sexual embryo (in various species of Paeoniaceae) and of the stem (in *Ranunculus sceleratus*) from which the somatic embryo forms, acquire polarity before the first division (the vacuole forms and the nucleus is often disposed eccentrically).

Somatic initial cells of the embryoid developing in nature and initial cells of the sexual embryo have their own pathways of development (see: Zygote). However, they obey general laws of cell division and embryogenesis as they are included in the system of the whole individual. This, as shall be seen later, maintains normal development of the somatic embryo, seedling and plant *in vivo*.

Investigation of initial cells of somatic embryos in culture *in vitro* is considered one of the most complicated problems of morphogenesis, since in the majority of cases an explant is heterogenic in structure, i.e., initial cells differ in origin. There are two model systems in tissue culture—callus and suspension cultures which differ in character of interaction of cells. The callus culture is characterized by more tough links between cells. This manifests in the fact that peripheral cells more often than underlying ones become competent and give rise to somatic embryos. Polarity and division in these cells is usually determined by their disposition with respect to surface of the callus (e.g. in cereals).

In the literature there are suggestions that isolation of separate cells of the callus (e.g., in *Citrus*—Button *et al.*, 1974; *Daucus carota*—Petru, 1970) promotes their transition to embryogenic development, which would probably correlate with the relatively isolated state of a mature zygote in the embryo sac. However, other observations did not prove this phenomenon to be a common character (Konar *et al.*, 1972; Maheswaran and Williams, 1985).

Another hypothesis (Sharp *et al.*, 1980, 1982; Evans *et al.*, 1981) holds that those cells which produce direct embryoidogenesis *in vitro* were already determined before their explantation; thus there should be a trigger that stimulates their morphogenesis. Indirect embryoidogenesis *in vitro* initiates from the cells, demanding proliferation and differentiation. In both systems differentiation and specialization of initial cells and of the embryo itself are determined to a great extent by the critical mass of these cells, callus and structures forming in culture *in vitro* (Halperin, 1967, 1970; Batygina *et al.*, 1978b; Batygina, 1984).

The data on initial cells in a suspension culture of *Daucus carota* (Komamine *et al.*, 1992) are interesting indeed. These cells are polar before division, have a large vacuole and a nucleus shoved back to the cell wall. Whether this process corresponds to formation of the egg-cell or zygote or initials of the somatic embryo under natural conditions has yet to be clarified.

As of now it is difficult to compare development of initials of somatic embryos *in vitro* and the egg-cell-zygote, as facts concerning morphogenetic processes in cells of the explant are quite insufficient. However, even the scant data available do allow asserting the similarity in morphology of initial cells of the sexual and somatic embryos *in vivo* and *in vitro*. Obtaining a great number of somatic embryos of various plant taxa in culture *in vitro* depends on thoroughness of our knowledge of the genesis of initial cells and their main characteristics (phase of mitotic cycle, structure of envelope, state of cytoplasm, quantity and disposition of organelles etc.). This would allow a more correct estimation and comparison of initial cells of the somatic embryo *in vitro* with initial cells of sexual and somatic embryos under natural conditions.

Sexual embryos under natural conditions

A high totipotency peculiar for the mature zygote is preserved in both *ca* and *cb* cells. Souèges took the degree of participation of these cells in embryo formation as a basis for classification of embryogenic types and established six basic types (megarchetypes) (Souèges, 1939c). Two other embryogenic types were established later (**Paeonad-** and **Graminad-types**), bringing the total to eight (Batygina, 1974b). Totipotency gradually reduces during subsequent embryo development in the course of histogenic differentiation and specialization. Only separate cell loci remain totipotent in the embryo, seedling and plant. High totipotency in the first stages of development may determine the great polymorphism of embryos.

Souèges (1934) revealed the initials and loci of hypophysis and epiphysis in the embryos of different angiosperm species. Their derivatives give rise to the shoot and root apices. Formation of hypophysis and epiphysis initials and loci takes place at the early stages of embryogenesis, but the point of their differentiation is taxon-specific.

The ontophylogenetic approach allows to reveal at least 5 groups of zygotic embryos varying by this feature:

1) the embryo exhibits **typical** initials and **loci of hypophysis and epiphysis** (for example, *Geum urbanum*—**Asterad-type** of embryogenesis (Souèges, 1948));
2) the embryo exhibits **only a typical initial of hypophysis and its locus** (for example, *Arabidopsis thaliana* and *Capsella bursa-pastoris*—**Onagrad-type** of embryogenesis (Yakovlev and Alimova, 1976; Maheshwari, 1950));
3) the embryo exhibits **only a typical epiphysis initial and its locus** (for example, *Polemoniutn caeruleum*—**Chenopodiad-type** of embryogenesis (Souèges, 1934; Kapil *et al.*, 1968));
4) the embryo **lacks typical initials and loci of hypophysis and epiphysis (for** example, *Triticum aestivum*—**Graminad-type** of embryogenesis (Batygina, 1969, 1987a, 1996));
5) **the initials of hypophysis and (or) epiphysis arise** in the embryo, but **do not develop the typical loci of hypophysis and (or) epiphysis** (for example, *Platantera bifolia* and *Gymnadenia conopsea*—**Onagrad-type** of embryogenesis (Veyret, 1965; Shamrov and Nikiticheva, 1992).

Polarity and morphological axis are normally established in the embryos of the first four groups, and shoot and root apices develop in an orderly manner (in the first three groups, the apex of the main root, and in the fourth the apices of adventive roots). The mature embryo of the fifth group exhibits morphological polarity, but shoot and root apices are not distinguished exomorphically. The embryos of all groups can be classified as normal types, not irregular, though they lack traits of the classical embryo according to Swamy and Krishnamurthy.

Epicotyl development varies greatly from taxon to taxon (for example, from epicotylar locus to a well-developed plumule). The same situation applies to the root (a developed main root, or just its initials, or the adventive roots, or total absence of roots can be observed in the embryo).

A comparative analysis of structure and genesis of the mature embryo in different angiosperm taxa provided evidence that almost all the signs proposed by Swamy and Krishnamurthy (see above) and many other scientists to characterize the sexual embryo vary greatly.

Somatic embryos under natural conditions

Asexual (homophasic) reproduction in flowering plants is accomplished by two elementary structural units of different morphological essence: a somatic embryo (individual as a whole) and an adventive bud (only a part of an organism). The majority of botanists when speaking about vegetative reproduction bear in mind only gemmorhizogenesis. Both structures seem to develop in parallel during evolution, and now they are realized side by side with the sexual embryo (heterophasic reproduction) to varying degrees in different taxa. Embryoids under natural conditions could be provisionally divided into two groups: embryoids developing in the flower and those developing on vegetative organs. Morphogenesis of the somatic embryo and subsequently of a seedling *in situ* occurs in the whole system of the parental organism and under its influence. The situation is similar for the sexual embryo. This peculiarity of development probably accounts for the orderly embryogenesis and a high per cent of seedlings that arise. On the contrary, development of somatic embryos *in vitro* never gives a high quantity of normal plant-regenerants.

The unique form of embryoidogeny (monozygotic-cleavage) can be observed in *Paeonia* (Batygina, 1998). The zygotic embryo phase in the seed of all *Paeonia* species finishes at the stage of protoderm formation in the coenocyte-cellular structure (this being heterophasic reproduction). Then the sexual embryo is cloned when its epidermal cells give rise to somatic embryos. The somatic embryo does not differ from the sexual one of a typical dicot (**Asterad-type** of embryogenesis) in morphogenesis. Establishment of the polar axis takes place from the stage of initial cell. Orderly development of root and shoot apices occurs subsequently. The genesis of *Paeonia* somatic embryo and its structure seems to be determined by conditions within the embryo sac and ovule, where it takes place. Probably this also guides the development of nucellar and integumentary embryoids. However, unlike the *Paeonia* embryo, these develop during cloning of a parental sporophyte. The first divisions in them are irregular, the initials and loci of the hypophysis and epiphysis are not observed, but development of the protoderm, shoot and root apices and the differentiation of the main root with all its elements correspond to those of sexual embryos (Naumova, 1987).

The external shape and internal structure of the *Paeonia* somatic embryo and of nucellar embryos (ovular embryoidogeny) are similar to those of the zygotic embryo.

Development of the somatic embryo on the stem of *Ranunculus sceleratus in situ* takes place in a similar manner (Konar *et al.*, 1972). It occurs according to the **Onagrad-type.** The authors mentioned the absence of suspensor differentiation, unlike sexual embryos of this species. Formation of somatic embryos from epidermal leaf cell derivatives was observed in *Crassula multicava* (McVeign, 1938).

The inner structure in these embryos, particularly of shoot and root apices, may be compared with that of the majority of zygotic embryos, in spite of the absence of typical initials and loci of hypophysis and epiphysis. Development of sexual and somatic embryos of the same species under natural conditions occurs according to the same type of embryogenesis.

Thus, the analysis of undertaken somatic embryos in different taxa, developing in natural conditions, showed notable variety among them. The structure and genesis of somatic embryos *in situ* are taxon-specific and to a considerable extent determined by the place of their formation as well as the environment. For example, somatic embryos that arise in the seed more often develop the main root. Those formed on vegetative organs usually develop adventive roots.

Somatic embryos in in vitro *culture*

As is known *in-vitro* culture gives researchers two different model systems (callus and suspension) to obtain somatic embryos. These systems are distinguished by many parameters that result in different structure and behaviour of initiating cells and determine the whole genesis of somatic embryos. Development *in vitro* of somatic embryos may be traced on the example of several species, contrasted by a number of features (Thorpe and Stasolla, 2001).

In the callus culture of *Triticum aestivum*, **the endogenous initiation of meristematic zone** was revealed. This zone occurs as a special tissue, **which consists of cell rows strictly oriented to the callus surface** (Batygina *et al.*, 1978b; Titova *et al.*, 2002). **Later the cells of this zone become the initials of somatic embryos.** It is noteworthy that in wheat microspore culture, embryoids arise **exogenously** (from callus epidermal cells). At the early stages of development the sequence and plane of divisions in the sexual embryo and embryoids are relatively similar **(Graminad-type)** (Batygina, 1969, 1996), although the embryoids are seen to have some variability, preserved at subsequent stages of their development. The initials are differently oriented inside the cluster and the polar axes of young embryoids are situated with different planes with reference to the callus surface. Delay in divisions of *ca* derivatives has also been observed compared with *cb* derivatives. It subsequently causes disturbance of their sequential divisions and vacuolization character; as a result, embryoids appear with 'linear' structure of the apical pole and abnormal histogenesis. In zygotic embryogenesis the apical pole gives rise to most of the scutellum (cotyledon) and the plumule, while in abnormal embryoids apical cells are destroyed. Lastly, the majority of abnormal embryoids degenerate. In some cases normal plumule formation was observed, though development of adventive roots was disturbed (these embryoids might possibly be considered transitional forms). Usually in the cultivars of *Triticum* studied, regeneration of plants by means of embryoidogenesis did not occur.

A comparative analysis of the development of the somatic embryo in callus culture and the sexual embryo *in situ* in *Aconitum heterophyllum* (Ranunculaceae) showed their relative similarity (Titova, Ahuja and Batygina, orig. data). Single initial cells of the callus gave rise to embryoids which subsequently passed all the main stages of normal embryogenesis—tetrad (T-shaped), quadrants, octants, globule, heart- and torpedo-shaped. Embryoidogenesis in *A. heterophyllum* occurs on the whole according to the **Onagrad-type** of embryogenesis, which is peculiar for

Ranunculaceae (Sokolovskaya, 1981). It was noted that embryoids exhibit well-differentiated suspensors. In the course of the experiment different abnormalities were observed in embryoid development, the most frequent being precocious differentiation of the xylem in the plerome area at the heart-shape stage. This was evidently associated with the imbalance of carbohydrates and hormones in the culture medium and callus tissue. As a result, the number of regenerants was markedly low.

A comparative analysis of data on structure and development of embryoids in *Daucus carota* (produced in callus and suspension culture) revealed its relative similarity to that of the zygotic embryo (Batygina *et al.*, 1996; Komamine *et al.*, 1992).

Thus a considerable resemblance of sexual and somatic embryo development has been revealed in the main morphogenetic regularities: polarity (bipolarity), symmetry (radial, bilateral, dorsoventral), pattern of cell divisions and histogen differentiation (formation of morphogenetic fields), morphogenetic and morphophysiological correlations and allometry. The same critical stages are exhibited in the development of sexual and somatic embryos: laying down of the first cell wall, protoderm formation, differentiation of organs, autonomy etc.

Embryoidogenesis does not recapitulate zygotic embryogenesis only in the case of abnormal development (first divisions are irregular in the embryo and laws of cell divisions (Errera's, Sachs' and Hertwig's) and the laws of embryogeny are disobeyed; protoderm formation is abnormal, formation of root and shoot apices is distorted and so on).

Transitional forms from embryo to bud

Analysis of the literature and original data on the development of sexual and somatic embryos in natural conditions and *in vitro* led us to the conclusion that there are structures which differ from typical sexual and somatic embryos and from the bud in their morphology. In connection with this we now introduce a term the 'transitional form' (in the terms of evolution), by which we mean a structure exhibiting traits of an embryo (for example, globular, heart-shape and torpedo-shape stages of development) and of a bud (formation of adventitious roots during regeneration) (Batygina, 1993a,b, 1996). Embryos of Nelumbonaceae (*Nelumbo nucifera*), Ceratophyllaceae (*Ceratophyllum demersum*), Poaceae, Orchidaceae and Orobanchaceae illustrate the possible modes of such transition. A similar phenomenon can be observed among somatic embryos.

Embryos of *Nelumbo* (hydrophyte) and *Ceratophyllum* (hydatophyte) lack an epiphysis and hypophysis. In the embryo of *Nelumbo* the main root is substituted by adventive roots during germination. The latter originate at the base of the plumule leaves in later stages of embryogenesis. However, the embryo of *Nelumbo* is bipolar from the very beginning of its development (Snigirevskaya, 1964; Titova and Batygina, 1996).

The embryo of *Ceratophyllum* seems to be bipolar in the first stages of its genesis, as it exhibits a group of cells which can be taken for initials of a main root. However, in the mature embryo only a well-developed plumule can be found and no main root. As for the adventive roots, they arise neither in the seed, nor in the seedling. The seedling of *Ceratophyllum*, thus lacks typical bipolarity (Shamrov and Batygina, 1984, 1988).

In the majority of Poaceae (xerophytes) the mature embryo exhibits a well-formed plumule and developed adventive roots (their number depends on species, and where and when the parental plant grows). The main root was transformed into a coleorhiza in the course of evolution (Pashkov, 1951; Pankow and Von Guttenberg, 1957;

Von Guttenberg, 1960; Guignard, 1961b). However, a number of investigators are of the opinion that the main root exists in the embryo of grasses. The embryo is bipolar and dorsoventral from the first stages of development, although the question remains as to whether the embryo of grasses preserves primary polarity, or is it replaced by a secondary polarity in the course of embryogenesis.

In Orchidaceae species the shoot and root apices are not exhibited morphologically in the mature embryo. Later, during protocorm formation the bud and adventive root arise and secondary polarity establishes (Teryokhin and Nikiticheva, 1968; Teryokhin, 1977; Batygina and Vasilyeva, 1983c).

The embryo of parasitic plants serves as the best model for investigation of the reduction of typical initials of the epiphysis and hypophysis of shoot and root apices. A notable characteristic of its genesis is the great variability of the first developmental stages. This is displayed in the diverse contribution of *ca* and *cb* derivatives to the formation of the embryo body. For example, in *Aeginetia indica* even the first divisions are irregular and hence the type of embryogenesis cannot be elucidated.

However, the structure of a mature embryo gives evidence that allies embryogenesis in *A. indica* as well as in other Orobanchaceae to the **Onagrad-type**. Differentiation of the hypocotyl and radicle initials is distorted to a different extent in Orobanchaceae. Besides that, in holoparasitic plants protoderm formation is irregular in some parts of the embryo (Teryokhin and Nikiticheva, 1981; Teryokhin, 1977).

So, analysis of sexual embryos in parasitic plants with diverse degree of reduction, proved that bipolarity can be observed from the very first stages of embryogenesis. But it vanishes in the course of development and secondary polarity is established only during sprouting (in contact with the host plant).

A comparison of morphology and embryology of the above-mentioned sexual embryos provided evidence that they can be considered transitional forms (because they exhibit morphological features of the sexual embryo and the bud). The whole complex of enumerated characteristics can be interpreted as evidence of evolutionary tendencies to a transition from a normal sexual embryo to a bud (or vice versa).

Such reproductive structures as propagules arising on vegetative organs of plants (homophasic reproduction) bear similarity to the aforementioned transitional forms (heterophasic reproduction). In *Bryophyllum* **bipolar propagules** arise on the leaf. They **lack the initials and loci of hypophysis and epiphysis. Globular, heart- and torpedo-shape stages are observed** in their development (and this **brings them closer to somatic embryos). However, absence of a main root and development of adventive roots show that propagules are similar to the adventive bud.** Morphogenesis of these structures may be compared with that of the sexual embryo in Poaceae (Batygina *et al.*, 1993), which may also be considered a transitional form.

The investigation of somatic embryogenesis of *Triticum* in culture *in vitro* revealed that some of the embryoids obtained are transitional forms. They have a well-developed plumule, but the morphogenesis of their adventive roots is distorted, compared to sexual embryos (Batygina *et al.*, 1978b).

Polymorphism of sexual and somatic embryos (*in situ, in vitro*) shows that morphogenetic reorganizations may occur at the level of cell, organ and organism and can affect different stages of embryo and seedling development. At the earliest stages of embryoidogenesis these reorganizations may be displayed as irregular divisions and varying contribution of *ca* and *cb* derivatives in embryo formation; in the middle stages the hypophysis and epiphysis may vanish; and in the later stages the embryo and even the seedling may lack the main or adventive roots, and embryo differentiation on the whole may be reduced from the beginning of its development. In the sexual embryo the point of primary polarity establishment may have shifted as a result of adaptive evolution, or be substituted by secondary polarity during seedling formation (Batygina, 1993).

Development of sexual and somatic embryos (*in situ* and *in vitro*) obeys the laws of cell division (Errera's, Sachs' and Hertwig's) and the laws of embryogeny (Souèges' and Johansen's). However, variability of the first zygote division and of subsequent stages of embryogenesis (eight types of embryogenesis and more than 50 variations (Johansen, 1950)), and the existence of transitional forms provide evidence that these laws should not be taken as absolute.

In every initial cell, zygote, embryo, regardless of embryogenetic type and plant species, all the genetic information responsible for cell division pattern in the proembryo and for subsequent morphogenetic events is available. In the course of evolution certain types of embryogenesis were determined for each taxon, but the initial morphogenetic potential of both sexual and somatic embryos is usually more completely realized under stress conditions (hybridization, mutations, tissue culture etc.).

PART FOUR—SEED

Seed

A seed is a unit of dissemination peculiar for seed-plants; it develops on the maternal plant from the ovule and contains an embryo (rudimentary daughter organism[1]) and specialized storage tissue (endosperm, perisperm and so on) enveloped in a protective cover, the seed coat **(Plate LX)**.

Seeds are distinguished by shape, size, surface texture, colour and interior structure. Their size fluctuates greatly. Usually they are not large, rarely reaching 1 cm in length. Shallow, powder-like seeds are characteristic, especially of species of family Orchidaceae and many parasitic flowering plants. Coarse seeds are observed in most Fabaceae (especially in lianes of genus *Entada*), in species of genus *Aesculus*, in some Cucurbitaceae (e.g. *Sechium edule*) and others. Representatives of family Arecaceae are known to have the largest seeds; in *Lodoicea maldivica* they reach 50 cm in diameter and weigh up to 15-20 kg. The shape, size and mass of the seed may fluctuate visibly even in the same species. Seeds of some species exhibit exceptional constancy in characters, e.g. *Ceratonia siliqua*; jewellers in the past used the seeds of this plant as a measure of weight (carat).

Seeds may be lustrous, glossy as though varnished, or devoid of any sheen whatsoever, i. e. mat. They may to be non-pubescent, bare or with silky, velvety, woolly hair, covered with separate, sometimes leathery hairs etc. The surface of seeds varies greatly; it may be absolutely smooth or rough, chuted, channelled, ribbed, wrinkled, cancellate, cellular, pitted, bossed, pointed etc. Seeds differ in colour as well: black, greyish-brown, grey, yellow, green, white or any other colour and shade.

Usually the seed is fastened to the fruit wall by the funiculus, and seldomly it is attached directly to the placenta, in which case it is termed sessile. Occasionally mature seeds hang down on the filamentous funicula. According to their arrangement in the fruit, vertical and horizontal seeds are distinguished.

Seeds frequently bear different kinds of appendages (arils, arilodes) and protrusions, for instance 'wings'; occasionally the seed bears a fascicle of hairs.

In the micropyle zone integumentary or even nucellar tissues often transform during seed development, resulting in formation of an embryotega, micropylar collar and other structures, being distinguished in various species by their origin and organization. In certain taxa the chalazal region of the seed also possesses specific features of structure.

The origin and structure of seed covers, occurrence, degree of development, nature and structural peculiarities of storage tissues (perisperm, endosperm), degree of embryo differentiation, its shape, location in seed, proportion of embryo and endosperm tissues, their relative contribution in substance storage and other features are taken into account in comparative anatomical investigations of the seed. Structural details of such seminal parts as hilum and raphe, micropyle and chalaza are of special significance in characterizing definite taxonomic groups of plants.

---

[1]In seeds of flowering plants the embryo of maternal origin can be formed alongside the daughter one (see: Embryoidogeny).

Seed Coat

The seed coat represents the outer tissues (testa and tegmen) of a seed, in whose formation integuments as well as occasionally some other portions of the seed (nucellus, chalazal tissue etc.) contribute **(Plate LXI)**.

The term 'testa' (Latin *testa* = shell) was introduced by Gaertner (1788) for determination of the outer hard membrane of the seed coat. The term 'tegmen' (French *tégument* = membrane, cover) was suggested by Mirbel (1829) to designate the inner sheath of the seed coat (see Shamrov, 1995). Following Corner's proposal (1976) most contemporary authors designate by the term 'testa' only that part of the seed coat derived from the outer integument, unlike the 'tegmen', which is a derivative of the inner one.

Depending on predominant participation of either integument in seed coat development, testal and tegminal seeds are discerned. In testal seeds the inner integument disintegrates completely during the seed formation process and is retained in the seed coat simply as a destructed film, in which the cell wall contours of the compressed cells, devoid of content, are hardly conspicuous. Complete disturbance of the outer integument and specialization of the inner one in the course of seed maturation typify the group of tegminal seeds. Participation of both integuments in seed coat formation (testal-tegminal) is also widespread. Occasionally aril tissues are naturally involved in the seed cover.

Rare is the case when all the integumentary tissues transform into the seed coat. A portion of the integumentary cellular layers is obliterated and only one or a few layers produce a proper protective seed coat.

Seeds could be classified according to which layer of testa or tegmen (otherwise both these components of the seed coat) occurs as a protective (mechanical) one. On the whole, more than 20 various combinations according to the arrangement of protective layers in the seed coat have been defined. Let us mention a few: (1) exotestal seed (mechanical layer is formed on the basis of outer epidermis of the outer integument, producing exotesta); (2) exo–mesotestal seed (mechanical layers derive from the outer integument and integumentary parenchyma); (3) exo–endotestal seed (mechanical layers are comprised from the outer and inner epidermis of the outer integument); (4) endotestal seed (mechanical layer develops on the basis of inner epidermis of the outer integument); (5) exotegminal seed (mechanical layer differentiates in outer layer of the inner integument); (6) endotegminal seed (endotegmen predominantly contributes to formation of the mechanical layer of the seed coat); (7) exotestal–endotegminal seed (mechanical layers in outer epidermis of the outer integument and in inner epidermis of the inner integument); (8) the seed, whose coat devoid of mechanical layers exits as a destructed film.

During transformation of certain integumentary layers into mechanical tissues the cell walls are encrusted with substances of different nature. In most dicotyledons the main encrusted compound is lignin, in monocotyledons phytomelan and phlobaphens are mentioned more often. The black pigment phytomelan preferentially accumulates in cells of the outer testal epidermis, where the cellular structure often becomes indistinguishable and a solid melanin cortex arises. In some monocots this cortex obtains a considerable thickness, when integumentary parenchyma layers spread under the epidermis are included in its structure (*Astella, Cordyline, Geitenoplesium, Dianella, Hypoxis, Hereria* etc.). Phlobaphens hardly ever accumulate in outer epidermis cells; they are encountered in testal parenchyma or in the endotesta (Hypoxidaceae, Philydraceae, Stemonaceae, Dioscoreaceae, Flagellariaceae, Musaceae), but never in the exotesta. They occur most frequently in tegminal tissue.

The correlation between development of multilayered testa and accumulation of melanin concomitant with arising of the melanin cortex was examined in order Liliales as an example. In most families it is the testal seeds in which the inner integument collapses during the course of seed development. Seeds in families in which the melanin cortex is absent as a matter of fact, ordinarily obtain one way or another a developed tegmen with phlobaphens as encrusted substances (Danilova *et al.*, 1990). Being derivatives of tannin substances, phlobaphens and melanins stain the seeds reddish-brown or black. Most dicotyledons as well as some monocotyledons are characterized by a colourless seed coat consisting of cells deficient of contents, whose shape is maintained due to lignification of the cell walls. Hüber (1969) estimated this feature to be the most primitive in order Liliales. It has been mentioned in families Melanthiaceae, Iridaceae, Velloziaceae, Alstroemeriaceae and some others from this order. However lignin has been discovered in the seed coat of other monocot orders also, in particular Burmanniales, Bromeliales, Commelinales, Pandanales and Arales. It can be seen from this list that lignin appears in seeds of highly organized taxa as well, where in several cases it completely displaces other encrusted elements. Given this, the opinion about the primitive nature of the lignified seed coat appears dubious.

Silica is also mentioned among the other materials encrusted in the seed coat and pericarp. Crystals and raphides of calcium oxalate are most common as inclusions in seed coat cells. Specific crystal-bearing layers are usually taken as a good diagnostic feature of some monocotyledonous families (for example Dioscoreaceae and Agavaceae). Among dicotyledons a crystal-bearing layer is observed in seeds of *Magnolia, Myristica, Aristolochia, Ribes, Liquidambar, Carica, Drosophyllum, Oenothera,* as well as members of Geraniaceae, Oxalidaceae, Meliaceae, Polygalaceae, Aceraceae, Myrtaceae, Vitaceae, Tiliaceae, Proteaceae, Lythraceae, Primulaceae, Violaceae and others.

Seeds with a multilayered and juicy seed coat named 'sarcotesta' have been distinguished (Zazhurilo, 1940; Takhtajan, 1948; Van Der Pijl, 1955; Corner, 1976). Sarcotestal seeds sometimes occur among monocots but are more prominent among dicots. Certain botanists consider this feature characteristic of the initial type of seed organization.

Among monocotyledons seeds with sarcotesta are observed in *Ornithogalum, Pushkinia, Eucomis, Bomaria* and others. Species of family Araceae have a mucous sarcotesta. In general, seeds with mucous covers—predominantly the outer epidermis of testa—are represented in both classes of Magnoliophyta. This phenomenon was termed 'mixospermy' (Grübert, 1974). Two functions attribute to mixospermy: (1) dispersal of seeds for long distances and their attachment to animal skin and 2) fixing of diaspores (seeds and fruits) in soil close to the maternal plant, which is of great importance for plants in arid regions for example (Fahn and Cutler, 1992).

Mucification may be typical for all epidermal cells of the seed coat or only specialized glandular cells or glandular filaments alone. Electron microscopic investigations have shown these cells to display the structure customary for secretory cells, synthesizing polysaccharide mucus (Hyde, 1970). During seed development epidermal cells usually increase in size and secrete the mucus in periplasmatic space. By the time of seed maturation the mucus usually almost fills the volume of the mucous cell, but the protoplast remains by one of the walls in the form of a diminutive lump (Fahn, 1990). While drying off the mucous volume diminishes, but on soaking the seeds the mucus swells and often bursts through the cell walls to envelop the entire seed surface. The perennial desert plant *Blepharis ciliaris*, in which only apical cells of multicellular hairs have a glandular function, exemplifies a specific mixospermy adaptation to the conditions of seed germination. Drying these cells facilitates close adherence of hairs to the seed surface; with moistening the hairs smoothen and move apart (Gutterman and Witztum, 1977).

Seed formation is also accomplished by the contribution of structures belonging by their nature to the funiculus or integuments but differentiated into different kinds of appendages: arils, arilodes, wings, samaras, pouches and so on, being commonly of great biological importance and therewith comprising notable taxonomic features.

In some groups of plants the seeds obtain unique characteristics because of morphogenetic activity of their micropylar and chalazal regions. Such a specialized structure as an embryotega is known in certain monocot and dicot plants. Thus the embryotega in members of orders Zingiberales and Commelinales emerges mainly on the outer integument, and its development is accompanied by formation of the so-called micropylar collar arising due to the local activity of cell divisions in the micropylar zone of the outer integument. Otherwise development of the embryotega is not accompanied by micropylar collar formation (Philydraceae, Araceae, Lemnaceae). The embryotega in representatives of Hydatellaceae and Mayacaceae develops on the score of the inner integument. Among dicotyledons the embryotega is observed in families Haloragaceae, Nymphaeaceae, Penaeaceae and others.

In some taxa of dicotyledons seed coats are produced with participation more or less of the chalaza. A perichalazal seed is one in which a conducting bundle passes from the funiculus up to the micropyle, while integuments fuse with each other in a median plane throughout almost all the seed except the micropylar pole. This type of seed organization is assumed to develop due to intercalary growth of the chalaza in a median plane. A pachychalazal seed is one whose cover is generally formed during growth of the chalaza, expanded in all directions while integuments, either one or both, remain only on the micropylar end. Herewith the seed coat may be penetrated by a large number of conducting bundles and the endosperm is often ruminated.

In the 1980s features of surface sculpture became of special interest. Depending on origin, primary, secondary and tertiary surface sculptures were distinguished (Barthlott, 1981, 1984). The primary sculpture pertains to the shape and arrangement of the seed coat epidermal cells; different kinds of cuticular productions as well as various protrusions of the outer walls of the epidermal cells constitute the secondary sculpture; lastly, micromorphology of epicuticular wax depositions characterizes the tertiary sculpture of the seed surface.

PART FIVE—SEED DORMANCY AND GERMINATION

Seed Dormancy (Plate LXII)

Plant growth begins with seed germination. Mature seeds located in adequate conditions (humidity, temperature, aeration) sprout rather quickly. Nevertheless seeds of an overwhelming majority of wild, as well as many cultivated plants undergo a state of dormancy, of which two types have been distinguished: **conditional** and **organic dormancy** (Grebinsky, 1961; Nikolaeva, 1967).

Conditional dormancy is induced by external causes (lack of moisture, temperature and light conditions), unfavourable for germination. Seeds under organic dormancy fail to germinate even in the presence of all conditions suitable for this process. In some species seed dormancy is so deep that it is a special preparation before sowing needed for their germination. In natural environments the germination of such seeds commences for several months or not infrequently for 1 or 2 years after sowing, and seedling emergence may be strained for several years. The ability to gain the state of conditional or organic dormancy prevents premature seed germination, encouraging survival of species and creation of a seed bank in the soil.

The following point, usually overlooked in the literature devoted to this problem, should be emphasized. Having arisen in the evolutionary process as an adaptive character for survival in unfavourable environmental conditions, passing through dormancy (latent period) became a necessary stage of ontogenesis. Normal plants cannot appear from seeds bearing dormancy until appropriately treated. Otherwise the seedlings occasionally formed from dormant seeds are characterized by dwarfism and disturbance of normal internode development. Ordinarily they give rise to plants of the rosette type, as shown on the example of seeds in *Acer tataricum* (Nikolaeva *et al.*, 1974).

The depth of organic seed dormancy, the reasons for its inducement and the conditions for its interruption vary greatly in different species. Because of this, systematizing the diversity of seed dormancy is a Herculean task approached by the stalwart. Crocker (1916) offered the first classification of types of seed dormancy distinguishing 7 types according to the causes determining restricted seed germination. Shumilina (1949) worked out a similar classification of dormancy types for woody plant seeds. All subsequent attempts to indicate dormancy types (Vegis, 1964) sought to understand certain aspects of this complex phenomenon, but adduced no distinct notions as to particular types.

At present the classification elaborated by Nikolaeva (1967, 1977, 1982) appears to be the most complete (see Table 10). Interactions between the causes of germination inhibition and germination conditions were taken as the principles of this classification. Accordingly the entire diversity of organic dormancy existing in nature was subdivided into 3 groups: **exogenous, endogenous** and **combined dormancy**.

The first group comprises three types of exogenous dormancy (**physical**, **faint** and **drastic**) defined solely by the properties of external covers, primarily the pericarp. It should be emphasized that the authors of this classification examined the seeds in the form they exhibit while separating from the maternal plant to disperse in nature. There may be seeds as they are, fruits or their portions. The pericarp appears to be an additional or in certain cases (fruit trees, nuciferous plants) the main inhibitor of seed germination. Hard seed coat is observed as the most characteristic example of exogenous

Table 10: Classification of types of seed organic dormancy (Nikolaeva *et al.*, 1985, modified)

Formula and type of dormancy	Causes of dormancy	Conditions of dormancy breakdown	Examples
Exogenous (A):			
A_1—faint	inhibiting effect of pericarp, faint	removing or disturbance of pericarp, fruit washing, long-term warm or cold stratification, some other temperature influences	*Fraxinus japonica*
A_2—drastic	inhibiting effect of pericarp, drastic	the same except fruit washing	*Pistacia vera*
A_{ph}—physical	water impermeability of covers	scarification, temperature influence, H_2SO_4 treatment etc.	*Gleditsia triacanthos*, *Robinia pseudoacacia*
Endogenous (B, C):			
B—morphological	underdeveloped embryo	warm stratification	*Annona crassiflora*
C—physiological:	physiological inhibiting mechanism	disturbance of covers, short cold stratification,	*Triticum aestivum*, *Pinus*
C_1—shallow	(PhIM) of germination, faint	dry storage, sprouting in light, GA_3 treatment etc.	*sylvestris*, *Impatiens balsamina*
C_2—intermediate	PhIM of germination, moderate	long-term cold stratification and sometimes other treatments	*Acer pseudoplatanus* (without pericarp), *Fraxinus viridis* (without pericarp)
C_3—deep	PhIM of germination, drastic	only long-term cold stratification (in certain cases GA_3 and cytokinin treatment reduces stratification, extends its temperature range or even replaces it)	*Malus domestica*, *Acer tataricum*, *Impatiens parviflora*
Morpho-physiological simple **(B-C):**	combination of underdeveloped embryo and PhIM of germination		
B-C_1—shallow	PhIM of germination, faint	two-stage stratification: warm then cold or GA_3 treatment	*Actinidia kolomikta*
B-C_2—intermediate	PhIM of germination, moderate	same, sometimes GA_3 treatment	*Aralia mandshurica*
B-C_3—deep	PhIM of germination, drastic	same, but GA_3 treatment stimulates embryo post-development alone	*Panax ginseng*, *Euonymus europaea*

contd.

Table 10. contd.

Formula and type of dormancy	Causes of dormancy	Conditions of dormancy breakdown	Examples
$B\text{-}C_3^e$—morpho-physiological deep epicotylar	combination of underdeveloped embryo and drastic PhIM of epicotyl development	two-stage stratification:warm for seed germination and root formation, then cold for shoot growth	*Viburnum opulus* (without stone)
$B\text{-}C_3^d$—morpho-physiological deep double	combination of underdeveloped embryo and drastic PhIM of germination and epicotyl development	four-stage stratification: warm for embryo post-development, cold for germination, again warm for root growth, again cold for epicotyl development	*Trillium erectum*
BC-C—morpho-physiological complicated:	combination of underdeveloped embryo and PhIM of embryo post-development and germination		
$BC\text{-}C_1$—shallow	PhIM, faint	cold stratification or GA_3 treatment	*Aralia continentalis*
$BC\text{-}C_2$— intermediate	PhIM, moderate	cold stratification (sometimes GA_3 treatment)	–
$BC\text{-}C_3$—deep	PhIM, drastic	cold stratification alone	*Magnolia kobus*
Combined dormancy (A-B-C)	combinations of reasons for exogenous and endogenous dormancy	compound preparation before sowing or long-term stratification	*Tilia cordata*

dormancy. This phenomenon is widespread among flowering plants (Mimosaceae, Fabaceae, Malvaceae, Tiliaceae, Caesalpiniaceae, Cannaceae etc.). To overcome exogenous dormancy the seeds were exposed to various temperature influences as well as physical and chemical treatments (Poptsov, 1976.)

Delay of germination, associated mainly with features of the embryo and the covers enveloping it, is attributed to the second group (types of endogenous dormancy). This group encloses morphological dormancy, physiological and morphophysiological types, differed by dormancy depth. The third group presents the types of combined dormancy conditioned by the combination of reasons for exogenous and endogenous dormancy.

Physiological type of dormancy is the most indicative for plants inhabiting temperate climate regions. This dormancy is conditioned by low growing activity of the embryo, which in combination with low gas permeability of covers realizes a double inhibiting mechanism (physiological inhibiting mechanism, PhIM). The influence of lower temperatures on swollen seeds (cold stratification) is the universal factor breaking down PhIM. Depending on PhIM efficacy, physiological dormancy has divided into three types with rather relative borders: **shallow, intermediate** and **deep**.

Shallow physiological dormancy is exhibited in temporary delay of germination or narrowing of temperature range, providing the possibility of germination. This type of dormancy is widely prevalent in seeds of agricultural (cereals, species of *Lactuca, Helianthus annuus*), weedy (species of *Xanthium, Hyoscyamus* and others) as well as some woody plants (species of genera *Pinus, Betula, Corylus* etc.). Shallow physiological dormancy is associated with low growing activity of the embryo, that fails to overcome the inhibitory effect of the surrounding covers. With removal or just partial breakage of the latter, seeds with shallow dormancy germinate well. It is commonly reimposed during dry storage and under the influence of such factors as short cold stratification, germinating at fluctuating temperatures, exposition of swollen seeds to light, their treatment with nitrates, nitrites, thiourea, gibberellic acid (GA_3) and likewise with high atmospheric oxygen content. In certain species the type of physiological dormancy is not so distinctly expressed: on the one hand rather prolonged cold stratification is required for its breakdown, and on the other embryos isolated from such seeds can germinate without preliminary preparation. This dormancy type is observed in seeds of *Acer negundo, A. pseudoplatanus, Fraxinus lanceolata, Prunus armeniaca*, species of genus *Amaranthus* and others. Dry storage of such seeds usually reduces their dormancy, while GA_3 treatment stimulates seed germination on condition that the pericarp or a stone is removed.

Deep physiological dormancy is peculiar for seeds of numerous woody (*Acer, Malus, Pyrus, Sorbus* etc.) and herbaceous (certain species of *Impatiens, Ambrosia* etc.) plants. Embryos isolated from dormant seeds either fail to grow absolutely or start with slow growth and in most cases, as mentioned above, are characterized by **physiological dwarfism**. Here one or both cotyledons reveal growth, sometimes a hypocotyl develops with swollen protuberances, but the root is mostly stunted. Only exposure of intact seeds to lower temperatures offers the possibility of normal development. Promotion of germination of such seeds can also be evoked through hormonal treatments.

Certain generalizations (Razumova and Nikolaeva, 1981) have been drawn regarding the influence of hormones on germination of seeds remaining in physiological dormancy. The response of gibberellin treatment of seeds, whose dormancy is associated with PhIM of germination, depends on the depth of dormancy, i.e. efficacy of PhIM. It is well known that in seeds with shallow dormancy gibberellins accelerate germination, while in seeds with deep dormancy only growth of isolated embryos can be stimulated by gibberellins. GA_3 treatment does not replace the cold stratification requirement in intact seeds staying in deep dormancy, but in some cases (species of *Impatiens*) can expand the temperature range within which germination takes place. Information on effects of cytokinins on the seeds with shallow dormancy is confirmed by data on the promoting influence of kinetin on germination of seeds sensitive to light (species of genus *Lactuca* etc.). Kinetin treatment of seeds has proven highly perspective for breaking deep dormancy in some *Acer* species, but not until the pericarp is removed. By means of kinetin treatment combined with thiourea and dimethylsulphoxide the state of dormancy in the pericarp can be broken as well (*Acer, Sorbus*—Pozdova, 1986; Razumova, 1987). Such physiologically active substances as fuzicoccin and jasmine acid are also rather effective in the removal of physiological dormancy (Nikolaeva *et al.*, 1987a).

The next and the most complicated group of seeds with restrict germination is characterized by underdevelopment of the embryo at the moment of dissemination.[1] Before we examine the dormancy phenomenon in these seeds some questions concerning embryo post-development beyond the maternal plant should be touched upon. Seeds with underdeveloped embryo are considered the least investigated among all types of seeds. According to Grushvitsky (1961) 'The process of embryo development in seeds shed from the maternal plant proves to be, as it were, in the neutral zone between positions of embryologists, who observe embryogenetic stages up to seed maturation on the plant, and morphologists in the narrow sense, studying ontogenetic stages usually only from the seed germination point'. However, it must be noted that seeds with an underdeveloped embryo have not been out of the embryologist's mind, but rather these seeds were examined without reference to their type of dormancy.

The degree of embryo development in mature seeds may differ in various species. In some plants it is a globular embryo with no signs of differentiation of organs (*Hepatica nobilis, Paraquilegia microphyla, Acanthopanax sessiliflorum*), in others it occurs at the heart-shape stage (*Adonis vernalis, Helleborus purpurascens*, some species of genus *Aconitum*). At the moment of dissemination embryo formation can be completed at later stages, with well-pronounced cotyledons, constituting sometimes more than half the embryo length (*Corydalis cava, Fumaria capreolata, Trollius altaicus, Heracleum sosnowskyi*). Lastly, the embryos may possess not only a considerable degree of differentiation but large dimensions as well (some species of *Fraxinus* and *Euonymus*—Nikolaeva, 1979).

In phylogenetic aspects seeds with underdeveloped embryos are typical for most ancient families. Martin (1946) in his work based on investigations of proportions between embryo and seed dimensions in 1287 species, mainly North American, found 11 such families. This type prevailed in 5 of them (Aquifoliaceae, Araliaceae, Magnoliaceae, Papaveraceae, Ranunculaceae) but only in particular members of Apiaceae, Berberidaceae, Oleaceae, Liliaceae, Arecaceae and Smilacaceae. Grushvitsky (1961) enlarged this list (up to 38 families), including therein most families from the orders Magnoliales, Laurales, Ranales besides others.

It has been established that embryo size and degree of differentiation depend on what species the seeds belong to and the geographical spread of this species

[1]The peculiarities of dormancy in seeds with reduced embryo typified for epiphyte and parasitic plants are not reviewed here (see: Reduced and Undifferentiated Embryos).

(Poptsov *et al.*, 1981). Seeds with underdeveloped embryo are well distinguished by their germination biology. Their germination capability is known to be associated first of all with embryo post-development. Some authors (Valishina and Zinger, 1952; Poptsov *et al.*, 1981) have indicated a direct correlation between embryo size and the terms of seed germination. Others refute such a correlation. Furthermore the same period of post-development is essential for several plant embryos of different size (*Panax, Euonymus*).

The next more complicated problem is the longevity of the entire period of preparation before sowing (involving embryo post-development and breakdown of dormancy) that results ultimately in seed germination. The classification of seed dormancy types stated above (Nikolaeva, 1967) elucidates this question. Therein the underdeveloping embryo is contemplated as a factor of seed dormancy. This dormancy is referred to as **morphological** and is more often observed in members of tropical flora and only a few plants of the temperate zone (some species of genera *Aquilegia, Anemone, Bupleurum, Thalictrum* and *Lilium*). Embryo post-development in these seeds commonly takes place at 20-30ºC during a short (rarely prolonged) period, after which the seeds sprout immediately.

In plants of temperate and subtropical latitudes, underdevelopment of the embryo often combines with PhIM of germination. Such dormancy is termed morphophysiological. Depending on the depth of physiological dormancy, i.e. on 'efficacy' of PhIM, Nikolaeva *et al.* (1985) indicated several types of **morphophysiological** seed dormancy. Overcoming **simple morphophysiological dormancy** (shallow, intermediate, deep) usually happens in the course of two-stage stratification: warm (for embryo post-development) and cold ones at 0-10ºC (for PhIM removing). While touching upon the conditions of germination of seeds with **complicated types of morphophysiological dormancy**, it should be emphasized that in natural environments often after seed shedding the process of embryo post-development fails to be completed within autumn; hence breakdown of seed dormancy cannot occur in the winter period. Embryo after-ripening in these seeds finishes as late as the following summer and hence the PhIM of germination overcomes only in the second winter. Removal of the PhIM that took place only during cold stratification, as in the case of physiological dormancy, should always be preceded by embryo post-development.

Among the methods for accelerating germination of seeds with underdeveloped embryo, phytohormone treatment has received the most attention. GA_3 treatment of seeds accelerates embryo afterripening and encourages extension of the temperature range that triggers this process (Araliaceae, Celastraceae, Ranunculaceae, Schisandraceae). However, GA_3 effects are unequal in different species, both on the embryo post-development process and breakdown of dormancy. In cases in which embryo underdevelopment combines with shallow physiological dormancy ('faint PhIM') (*Aralia cordata* and *A. continentalis*), GA_3 treatment proves effective at both stages, i.e., GA_3 not only substitutes warm stratification but even eliminates the need for a cold one. In seeds of *Aralia mandshurica*, in which dormancy is alternatively fairly deep, soaking them in a GA_3 solution also induced rapid germination of a goodly portion in warmth (Bogdanova, 1971).

In cases of deep morphophysiological dormancy, when embryo post-development takes place in warmth and PhIM removing in cold (e.g. some species of genera *Bupleurum, Eleutherococcus, Euonymus, Panax*), GA_3 treatment of seeds permits to exclude the warm stratification completely but not influence the timing of subsequent cold one.

Seeds with morphophysiological dormancy include those in which embryo post-development through removing of PhIM occurs just at low temperature, about 0-3ºC (complicated deep morphophysiological dormancy). In such seeds (species of genera *Tulipa, Heracleum*—Razumova and Nikolaeva, 1981; *Fritillaria, Ornithogalum*—Pozdova and Razumova, 1994; *Osmorhisa, Erythronium* and *Delphinium*—Baskin C. and Baskin Y., 1994, Baskin C. *et al.*, 1995) embryo afterripening and successive germination are not accelerated by GA_3 treatment. Furthermore, in seeds of *Fritillaria eduardii* and *F. raddeana* low temperatures are needed not only for embryo post-development and germination, but for removal of epicotyl dormancy as well (Pozdova and Razumova, 1994). Seeds of some *Trollius* species, characterized by deep morphophysiological dormancy, germinate under cold effects (0-3ºC) during not less than 3-8 months. But as soon as they are exposed to GA_3 the seeds successfully germinate at 18-20ºC for 7-30 days (Nikolaeva *et al.*, 1987b).

In several species with morphophysiological dormancy (likewise some species with physiological dormancy), GA_3 expands the temperature range of seed germination. Thus, it has been shown on the example of *Aconitum* that GA_3 evokes seed germination at 9-10ºC, just as in water control at this temperature embryo post-development occurs alone (Pozdova and Balmasova, 1994).

Accelerating germination of seeds with morphophysiological dormancy can be induced by means of kinetin after completion of embryo post-development process either during warm stratification (e.g. in *Panax*—Bogdanova, 1971) or under the influence of GA_3 treatment of seeds (e. g. in *Euonymus*—Razumova and Nikolaeva, 1981). At the same time kinetin treatment does not stimulate germination of seeds bearing complex deep morphophysiological dormancy, as was demonstrated on the example of *Tulipa tarda* (Nikolaeva and Razumova, 1973), as well as species of genera *Heracleum* and *Magnolia* (Razumova, 1978).

Information on morphoanatomical modifications during the embryo post-development process is scant; however, Lodkina *et al.* (1971) and Lyashuk (1971, 1972) did provide data earlier that fundamental changes occur in the structure of the embryo conductive system, differentiation of shoot apex, cotyledonary tissues etc. during this period.

It was demonstrated on the example of *Aconitum soongaricum* (Pozdova *et al.*, 1994), that by the time of seed shedding from the maternal organism the embryo has entered the early torpedo-shape stage of development, with cotyledons, shoot apex, embryonal root and hypocotyl differentiated. In the process of embryo post-development, considerable expansion of size, differentiation of the palisade layer of mesophyllic cotyledons, and development of primordium of the first leaf in plumule were observed. The most substantial modifications were seen in the conductive system of the embryo. Only the apical meristem of the embryo root showed no visible changes. The process of embryo post-development in the species studied began in a wide temperature range (20-22, 9-10 and 0-3ºC). However, normal proceeding of differentiation stages in the course of post-development was only observed at 9-10ºC.

A special role in the embryo after-ripening process is thought to be played by the complex of so-called tabular cells formed in its different tissues and organs. Kovalev (1977) reported the occurrence of these cellular complexes in the axial parts of embryos in the mature seeds of *Quercus* and *Castanea.* In the embryo of *Acer tataricum* (physiological dormancy) complexes of tabular cells arise from within the hypocotyl and embryonal root even during the maturation process (Lodkina *et al.*, 1971). In species whose seeds are characterized by morphophysiological dormancy, complex formation takes place in the same region during embryo post-development under the specific (depending on species) conditions: in *Euonymus europaea* and *Panax ginseng* mainly at the end of warm stratification (18-20ºC) and in *Fraxinus excelsior* after passing through warm stratification and transferral of seeds to cold (0-3ºC) (Lyashuk, 1971). In *Aconitum soongaricum* tabular cells also arise during embryo post-development but within cotyledonary bases. The assumption confirmed (Pozdova *et al.*, 1998) is that tabular cell differentiation is necessary for realizing seed germination: rapid and simultaneous elongation of cells within complexes enables the embryo to rupture easily the restraining seed covers and thereby release the seedling on surface.

With perfection of methods for studying the problem of seed dormancy and accumulation of experimental data, notions regarding implication of processes occurring in seeds during dormancy breakdown were modified (Khan, 1977). It should be emphasized that investigation of processes taking place in dormant seeds under the effect of low temperature, considered as mentioned before, a main factor leading to endogenous dormancy removal, is of special interest. Since the 1930s-40s the great importance of phytohormones in seed dormancy removal has been widely recognized. Concrete expression was given later to the hypothesis of the chief role of hormone balance (Kefeli, 1974; Khan, 1977; Nikolaeva, 1977). It is associated with temporary succession of hormone changes (abscisic acid, cytokinins, gibberellins) happening in seeds while coming out of deep dormancy. Thus, it was shown in some *Acer* species as an example, that at the beginning of cold stratification abscisic acid content decreases greatly, cytokinin activity increases, after which the content of gibberellins rises (Webb *et al.*, 1973a, b). It should be mentioned that until present times the role of abscisic acid in this balance remained highly controversial (Le Page Degivry *et al.*, 1990; Le Page Degivry and Garello, 1992) and was rejected outright by some authors (Nikolaeva, 1991).

It is a fact of special attention that in studies on phytohormone regulation of seed dormancy the role of such stimulators as indoleacetic acid (IAA) is almost ignored. Nikolaeva *et al.* (1974) did reveal the high content of IAA and its irregular distribution in embryos of *Acer tataricum* to be the cause of the physiological dwarfism phenomenon. Accumulation of IAA in superoptimal concentrations the authors showed is an appreciable factor in dormancy induction. Its content decreases in the seed with deep physiological dormancy as soon as germination under low temperature influence becomes possible.

According to the present standpoint, phytohormone contents and activity change under the influence of enzymes. Data on definition of enzyme activity in seeds with shallow as well as deep physiological dormancy verify this (Lewak *et al.*, 1975; Biswas *et al.*, 1978; Nikolaeva, 1982; Bekker and Nikolaeva, 1984; Polyakova *et al.*, 1986). What is more, the example of acidic phosphatase shows that a new isoenzymes appears under the effect of cold stratification in *A. tataricum* seeds (Polyakova and Nikolaeva, 1990).

Elaboration of the problem examined thus far is hardly complete. Experimental data on which the hypothesis of the leading role of phytohormones in induction and breakdown of seed dormancy is based, are still contrary and mainly comprise information on the dynamics and possible significance of particular hormones in seeds of certain species. The mechanism of enzyme apparatus activation exposed to lower temperatures also remains vague. And lastly, as mentioned above, there are many 'blank spaces' in the research on germination biology of seeds with underdeveloped embryo. The question of how and under what external effects post-embryonal development in the group of seeds given is realized, has yet to be thoroughly investigated. Revelation of the main embryogenetic stages and factors and determination of their passage are of great importance both for comprehending the nature of dormancy types, typified for seeds of a definite genus or species, and for working out methods of seed preparation before sowing.

Morphology and Physiology of Seed Germination

Germination can be defined as that sequence of events which forces an air-dried seed in conditional dormancy, to answer to water inflow by a rise in total metabolic activity and development of a seedling (Mayer and Poljakoff-Mayber, 1989).

Germination includes:

1) **Period of 'early germination'**—from the beginning of seed swelling up to root appearance (breakage of seed coat by root tip); during this period activation of metabolism and preparation of axial organs of an embryo for initiation of cell growth by stretching.
2) **Period of seedling formation**—from root appearance to formation of autotrophic seedling. It includes transition to active growth of axial organs, resulting in emergence of a plumule above the soil surface, and in some cases cotyledons, and root deepening into soil. Leaf greening and growth and also formation of the root system are completed in this period.

Mobilization of reserve substances provides the beginning of growth in the first period and intensive growth during the second.

Types of germination (Table 11). In **hypogeal** germination (Greek *hyp, hypo*—under and *geo, geus*—soil), cotyledons remain inside the soil, do not grow and play the role of a container of reserve deposits. The hypocotyl starts to grow with the first pushing out of an embryo root from a seed. The hypocotyl, initially short, increases in length several times. The epicotyl immediately grows intensely, the last spurt yielding a plumule with leaf primordia. Hypogeal germination is typical, for example, of seeds of pea, beans, oak and horse chestnut.

In **epigeal** germination (Greek *epi*—on + *geo, geus*), cotyledons arise above ground and turn into cotyledonous leaves. The embryo root starts to extend first, breaking

Table 11. Types of germination

	A root appearing				
Hypogeal germination	Beginning of hypocotyl cell growth by stretching	Beginning of root cell growth by stretching	Beginning of root cell divisions	Beginning of epicotyl cell growth by stretching	Root appearance results from growth of hypocotyl apical cells by stretching
Epigeal germination	Beginning of root cell growth by stretching	Beginning of root cell divisions	Beginning of hypocotyl cell growth by stretching	Beginning of hypocotyl cell divisions	Root appearance results from growth of root basal cells by stretching

through the seed coat. The hypocotyl quickly starts to grow after acceleration of embryo root growth, bearing the cotyledons above the soil and the plumule between them. Seeds typified by epigeal germination are: string bean, cucumbers, tomatoes, cotton, garden radish and sunflower.

In monocotyledonous plants, formation of a seedling also occurs on hypogeal (Poaceae) or epigeal (Alliaceae) type.

Initiation of growth of both types of seedling occurs similarly, due to stretching of cells, without participation of cell division (Table 11).

In hypogeal germinating seeds, cell divisions in the root meristem always begin a little after the beginning of hypocotyl growth (for example, in horse chestnut through 40 h, fodder bean 20 h, pea 4 h). In epigeal germinating seeds, cell divisions in the root occur after stretching in it has begun and also in different periods after root appearance (for example, in cotton through 10 h, string bean almost simultaneously).

Growth of cells by stretching is obligatory for initiation of growth of all embryo organs (Obrucheva, 1982) and commences when the critical water saturation level is reached. This results in accumulation of osmotically active substances and softening of cell walls. Cytokinins and auxins, and also dihydrofuronan-oxymethylpyrrol-carboxyaldehyde are more often considered starting factors of cell divisions. Synthesis and doubling of DNA, and formation of proteins necessary for passage of the mitotic cycle, occur during transition of cells to divisions.

The root, as the source of water, is the first of the embryonic organs to commence intensive growth after germination. Other organs begin to grow later and provide access to light, the second obligate condition of seedling formation.

Basal cells are stretched in the embryo root at the first, forming a zone of stretching. **Formation of this zone structure** provides the co-ordination between formation of cells and their stretching along an axis of a root that is necessary for ensuring quick lengthening (Obrucheva, 1982).

Another structural feature of a root is the presence in it of a **quiescent centre.** This arises in early embryogenesis when the root cap initials are forming, and manifests during germination at the beginning of mitotic activity in the root; in the process of acceleration of division its volume decreases due to transition of peripheral initial cells to division (Rondet, 1965; Byrne and Heimsch, 1970; Vallade, 1972; Obrucheva, 1982; Ivanov, 1987; Barlow, 1995).

Acceleration of divisions is probably connected with change of ratio between cytokinins, auxins and abscisic acid. Acceleration of stretching can be interpreted as change of ratio between auxins and gibberellins synthesized in a zone of stretching (Jones and Phillips, 1966; Sitton *et al.*, 1967; Robbins and Harvey, 1969; Barlow, 1976; Zavala and Branden, 1983; Barlow *et al.*, 1991; Tanimoto, 1992; Müller *et al.*, 1994).

Epicotyl growth at hypogeal germination begins with slow stretching of basal cells, which gradually amplifies in the direction of the apex, involving again formed cells. Cell divisions in pea epicotyl occur below a first leaf primordium during 2–5 days. Thus fast epicotyl lengthening is achieved and promoted through soil. Under light, epicotyl growth is inhibited both due to reduction in duration of period of divisions and stretching of cells to a lesser length (Nougarede and Rembour, 1980; Miyamoto and Kamisaka, 1988).

Hypocotyl growth at epigeal germination is also long and intensive, occurring in a manner similar to epicotyl: slow stretching of basal cells, followed by intensive cell divisions in the apical part. Further lengthening occurs only due to stretching of cells and it extends to the apex. Light inhibits both division and stretching of cells (Mohr and

Peters, 1960; Hacker *et al.*, 1964; Sakurai *et al.*, 1974; Stuart *et al.*, 1977; Taylor and Cosgrove, 1989).

In cereals, realization out of aboveground organs through a layer of soil occurs due to growth of the **mesocotyl**, which pushes out the apex of a stalk with leaves protected by the coleoptile. With light, the cereal seedling grows due to elongation of the leaves and stalk. Fast (frequently prolonged) hypocotyl, epicotyl or mesocotyl growth in darkness, occurring by intensive stretching of available and formed cells, provides emergence of a seedling to light through a dense and thick layer of soil, that guarantees its survival.

Light detains hypocotyl, epicotyl and mesocotyl growth but stimulates growth of leaves. In dicotyledonous plants the leaf grows as a result of divisions of stretching cells, thus achieving 60–70% of the final size; further accretion of a leaf plate occurs due to stretching of cells only. In cereals, the zone of cell divisions lies in the basal part of a leaf; cells initially divide, then stretch. Under exposure to light, the number of leaf palisade parenchymal cells and stomas increases. In leaf greening the etioplasts turn into chloroplasts that then divide. The chlorophyll and other pigments synthesize concomitantly (Maksymowych, 1973; Dale, 1982; Ivanov, 1987). Under light, photomorphogenesis processes begin in seedlings. Thus autotrophic seedling formation from a viable seed is completed.

Let us now consider retrospectively the entire series of processes occurring in an air-dried seed from the moment of its contact with water to its ultimate fulfillment with appearance of a root, i.e., initiation of growth.

Water inflow in a seed. Germination begins from the moment of water inflow in a seed. Humidity of air-dried seeds before initiation of swelling consists on average 10% of fresh weight. For most cultivated plants, the lowest critical level of humidity at which seed viability remains is considered to be 5%.

Initially the seed coat swells, followed by gradual saturation of the cotyledons and axial organs with water. Speed of swelling depends on availability of water, seed size, a contact surface of the seed with water, degree of permeability of the seed coat, and also temperature.

Water content in seeds differs at the moment of root appearance and is defined by seed composition. The basic water-absorbing substances are proteins and next polysaccharides of the cellular wall, starch and mucilage. Starchy and oleiferous seeds absorb less water than proteinaceous ones. Axial organs absorb water faster and reach a higher moisture level than storage organs and tissues; the water saturation range of axial organs at root appearance amounts to 63–80% (fresh weight) (Obrucheva, 1982; Mayer and Poljakoff-Mayber, 1989).

The typical curve of water inflow response over time into seeds has a three-phase character: 1) sharp rise, fast swelling; 2) lag period, slow swelling up to seed root appearance; 3) gradual rise associated with beginning of cell growth by stretching in axial organs (Bewley and Black, 1978, 1985).

Driving force of water absorption by seeds is the difference between water potentials of a soil solution and a seed. The water potential of air-dried seeds is very low and varies from -400 up to -50 MPa. On the first phase of swelling it is quickly increased, i.e., becomes less negative and therefore ability of seeds to pump water attenuates. During the lag-period, the water potential rises slowly, coming nearer to zero. Seeds start to germinate successfully at water potentials in the range –1.0 up to –0.1 MPa. The water potential of growing organs declines a little after the root appears (Bewley and Black, 1978; Ross and Hegarty, 1979; Askochenskaya, 1982).

During the period 'early germination' the ratio between water potential components, i.e., matrix, osmotic and turgor potentials, changes also (Bewley and Black, 1978; Askochenskaya, 1982; Obrucheva *et al.*, 1993).

The period of 'early germination' is subdivided into two stages: (1) activation of metabolism and (2) preparation of axial organ cells for the beginning of stretching. The first stage of early germination occurs both in cotyledons and in axial organs of the embryo. It consists of sharp intensification of all basic processes of metabolism. At swelling hydration and activation of all enzymes occur and change in conformation of their molecules is apparent (Bewley and Black, 1985; Mayer and Poljakoff-Mayber, 1989).

Intensification of respiration indicates activation of the basic metabolic systems. Air-dried seeds respire very slightly. Respiration intensity of axial organs is increased 10-fold 15 minutes after unimpeded access of water. The dominant respiratory mode from the outset of swelling is oxidation of the substrate through glycolysis and the Krebs cycle. Phosphorylation amplifies together with rise of respiration in swelling seeds, providing energy to the basic metabolic reactions and initiation of growth processes (Pradet, 1982; Asahi and Maeshima, 1983; Obrucheva and Kovadlo, 1985).

In axial organs activation speed of initial metabolic processes is higher than in cotyledons because their water saturation increases faster; but the character of activation and sequence of process initiation are identical. Fast water inflow in embryos causes both initial activation of respiration and transformation of amino acids and Krebs cycle acids. Synthesis of sucrose and lipids begins later (Collins and Wilson, 1972). Synthesis of protein in swelling seeds is also apparent very early (in seeds without coat, within 20 min.). Reserved mRNA, synthesized still in forming seeds, codes synthesis of two types of proteins: (1) those providing further fulfillment of basic metabolism and (2) those typical for its germination (Brucker *et al.*, 1982; Dyur, 1982; Aspart *et al.*, 1984; Bewley and Black, 1985; Grzelczak *et al.*, 1985; Sato and Furuya, 1985; Cosgrove and Zhen-Chang Li, 1983). Synthesis mRNA *de novo* at swelling begins early and provides intensification of protein synthesis. Synthesis DNA begins usually after RNA and protein synthesis; it precedes the beginning of cell division (Osborn, 1982).

Thus, activation of the basic metabolic systems is engendered and completed during the first phase of swelling. In the presence of water these forces provide automatic receipt in seeds of 55–60% moisture thus full activation of metabolism. In cotyledons which have remained underground during germination, this level of moisture and achieved activation of metabolism is maintained during further germination. Their specificity consists only in induced synthesis of the hydrolytic enzymes providing intensive mobilization of basic reserve substances. The achieved degree of water saturation and metabolic activation does not suffice for the beginning of growth processes in axial organs.

The processes preparing the stretching of cells (the second stage of 'early germination'), run only in axial organs during a lag phase of swelling. They start to function against a background of intensive basic metabolism, energy provision and substance inflow from self-reserves. The physiological mechanisms preparing the start of cell stretching, should provide both water inflow into cells and their vacuolization, and increase in cell wall plasticity.

The first physiological mechanism is accumulation of **osmotic-active substances** that allow a cell to absorb new portions of water and raise its own water saturation. Their accumulation occurs due to hydrolysis of starch, reserve protein, and also phytin and after root appearing due to inflow of sucrose, amides and potassium from the cotyledons (Pernollet, 1982; Obrucheva *et al.*, 1993). However, intensification of osmotic water inflow does not suffice for actual commencement of cell stretching.

The second physiological mechanism preparatory to cell stretching, consists of **activation H⁺-ATPase in the plasma membrane** of axial organ cells that provides transport of protons from the cytoplasm into cell wall and its softening (Obrucheva and Antipova, 1994; Kutschera, 1994).

Mobilization of storage substances at germination. The content of reserve substances in seed of cultivated plants considerably exceeds the seedling requirements necessitated by the selection of plants specifically to increase the nutritional value of their seeds for man. Proportions of storage proteins, fats and carbohydrates are species-specific, though seeds are conditionally subdivided on starchy, oily and proteinaceous. In starchy seeds the proportion of proteins, fats and starch makes up respectively 10-15 : 2-4 : 60-75 (cereals, buckwheat), in highly oily 17-31 : 48-68 : 0.1-12 (sunflower, peanut, castor-oil plant), and in proteinaceous 23-31 : 1-6 : 47-54 (leguminous).

The trigger of reserve protein and starch mobilization at germination is water saturation increase up to threshold levels. For example, in fodder bean splitting of starch begins at a 45% moisture level (fresh weight), and splitting of protein at 50%. The course of proteolysis varies in monocotyledonous and dicotyledonous plant seeds due first of all to species-specific composition of storage proteins and their packing (Khavkin, 1982; Preston and Kruger, 1986; Wilson, 1986; Obrucheva *et al.*, 1988; Vaintraub *et al.*, 1990).

As water saturation at swelling rises faster in axial organs of embryo than in cotyledons, mobilization of reserves in them also begins earlier and provides preparation and commencement of growth before inflow of products mobilized from cotyledons. In axial organs of the pea embryo half of its stocks of proteins (albumins) is exposed to hydrolysis right up to root appearance, and the other half (globulins) hydrolyzed more slowly, during actual root appearance. In axial organs of the corn embryo, globulins are initially exposed by proteolysis and prolamins and glutelins of the endosperm used as long-term reserves. Active mobilization of reserve substances in cotyledons begins round about the time that mobilization of reserves in the axial organs has been exhausted. Sucrose, amides and potassium ions forming at hydrolysis of phytin translocate from storage organs to axial organs (Simola, 1974; Khavkin, 1982; Laidman, 1982; Gumilevskaya *et al.*, 1993).

A generalized model depicting the physiological processes in early germination and in initial stage of seedling formation is given in Figure 82 (Obrucheva, 1991, 1993).

The starting role of water saturation is manifested only during the period of early germination. At the activation stage of basic metabolism (seed swelling up to 20% moisture) hydration of enzyme proteins occurs, modification of their molecule conformation and transition of some enzymes into an active state. Achievement of 21–25% moisture content results in spasmodic increase of respiration (in soybean seeds), to occurrence of activity of cytochrome-reductase (in salad seeds) and glutamate-decarboxylase (in wheat seeds). Swelling up to 40–45% starts a new mRNA transcription (in rape-seeds), results in mitochondrion biogenesis completion and sharp intensification of breath (in pea seeds). Increase of water saturation up to 45–50% initiates synthesis of protein on reserved mRNA (in wheat seeds), and also hydrolysis of starch and spare proteins (in pea seeds). Hence, a particular starting level of hydration is necessary for each metabolic system. Actuating of basic metabolism by water coming from physical factors is possible only because enzymes, substrates and mRNA have synthesized beforehand, at formation and ripening of a seed.

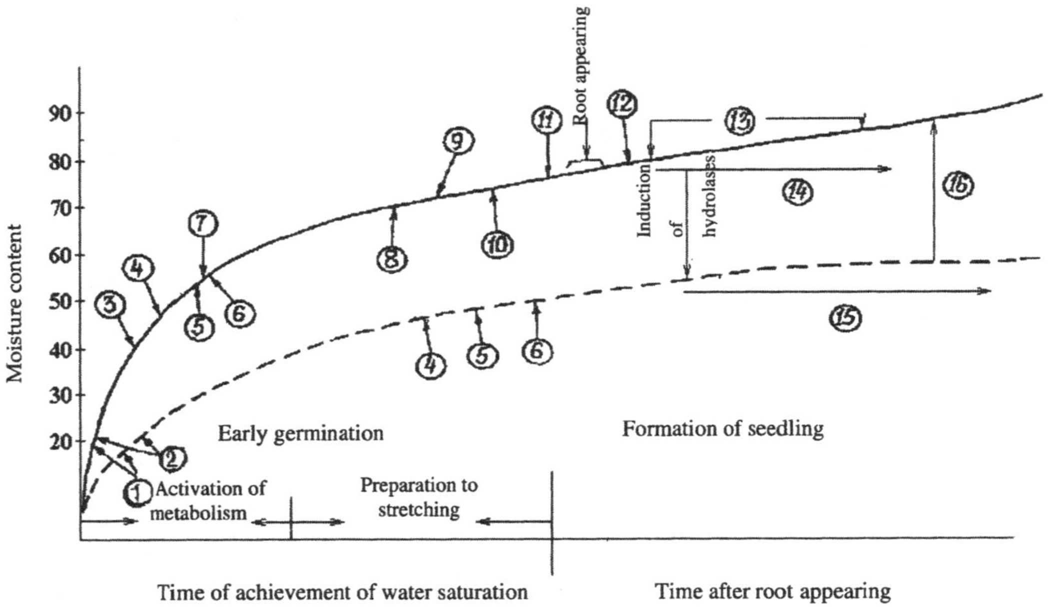

Fig. 82: General scheme of physiological processes at 'early germination' of seeds which have been in conditional dormancy. Abscissa: the normal time scale (in hours); Ordinate: moisture of axial organs and cotyledons (%, on a fresh weight).

Continuous line—axial organs; dashed line—cotyledons; arrows show inclusion of various metabolic systems: 1—activation of respiration; 2—beginning of amino acid transformation; 3—beginning of transcription of mRNA synthesized *de novo*; 4—finishing of mitochondrion genesis and intensification of respiration; 5—beginning of starch mobilization; 6—beginning of storage protein mobilization; 7—beginning of synthesis of proteins on reserved mRNA; 8—accumulation of osmotic substances; 9—increase of vacuolar system volume; 10—beginning of cell wall acidification; 11—beginning of cell stretching in hypocotyl (hypogeal germination) or in root (epigeal germination); 12—beginning of cell stretching in root (hypogeal germination); 13—initiation of cell divisions in various plant root; 14—synthesis of gibberellins and abscisic acid; 15—mobilization of basic stores in cotyledons; 16—beginning of substance inflow in axial organs from cotyledons.

The stage of preparation for growth by stretching (for example, in axial organs of fodder beans) is characterized by accumulation of osmotic active substances (moisture 65–68%), increase in vacuole system volume and activation of plasma membrane H^+-ATPase at further increase of the water saturation up to 68–70%. Thus the starting role of water saturation is realized in preparation for stretching, which occurs only at water inflow up to 72–73%. Role of water saturation as a trigger is confirmed by the fact that with water loss by a swollen seed at any level of water saturation, processes already begun are arrested, but with water restoration restart from the level attained before dehydration. Should any process not have begun before dehydration, in the course of rehydration it will proceed at the renewed water saturation level from the level it would have attained from the first swelling.

In Figure 82 the time of achievement of the given water saturation level is accepted as a time scale prior to root appearance, whereas in all other studies only time of seed swelling is generally taken into account. Our approach has allowed co-ordination of a course of events at the start of germination in various kinds of seeds based on their water content, and has shown the general pattern of physiological processes starting through achievement of water saturation threshold levels.

There are no convincing proofs that early germination is benefited by phytohormones at the time of emerging from exogenous dormancy, though seeds contain various phytohormones in both a free and conjugate state. The known facts of breaking of seed organic dormancy by gibberellins and cytokinins have no relation to seeds in exogenous dormancy. Their treatment by different phytohormones usually exerts no influence on speed of root appearance. Content ABA in seeds in exogenous dormancy is much lower than the level which inhibits growth processes; therefore it is impossible to explain that for initiation of growth during swelling, washing away of ABA from the seed is necessary (Obrucheva, 1982; Wang *et al.*, 1995). Phytohormones simply cannot be considered endogenous triggers of early germination.

In ripening seeds these phytohormones are capable of providing metabolic readiness of seeds for germination, which remains during seed drying and exogenous dormancy and is realized during early germination due to water inflow and consecutive start of metabolic systems. During early germination the synthesis of new phytohormones occurs only when intensive growth processes in axial organs begin.

In growing axial organs an intensive synthesis of numerous gibberellins, abscisic acid and other phytohormones occurs; gibberellins and cytokinins move from growing axial organs to cotyledons (or endosperm), causing induction of some hydrolases there. Synthesis of hydrolytic enzymes on new matrixes occurs due to amino acids formed at initial mobilization of reserve protein. These enzymes realize mobilization of the basic stocks in cotyledons, in addition to initial mobilization of starch and protein, begun due to increase in seed water saturation up to a threshold level. As a result a fund of sugars and amides is formed and their inflow in axial organs begins. Thus intensive growth of seedling axial organs is supported (Sponsel and MacMillan, 1980; Laidman, 1982; Obrucheva *et al.*, 1985, 1988; Bewley and Black, 1985).

Scutellum and its Role in Germination

The **scutellum** is one of the important seed structures in cereals, directly bordering the endosperm. Its morphological nature is still under discussion (see: Graminad-type of Embryogenesis). Usually three types of cells are distinguished in the scutellum.

Parenchymatous cells of isodiametric form constitute the essential mass of the scutellum. Numerous intercellular spaces are located between them. In dry kernels the cytoplasm is dense, vacuolated, and contains abundant lipid drops, protein bodies with phytin, ribosomes and mitochondria with slightly developed cristae (Nieuwdorp, 1963; Swift and O'Brien, 1972a,b).

Epithelial cells (cells of scutellum epidermis) are arranged in a single layer on the side of the scutellum adjoining the endosperm. They are elongate cylindrical in shape, with dense cytoplasm containing numerous spherosomes along the cell wall and protein bodies located, usually, in the apical part of the cells. Fragments of ER and mitochondria have been observed in the epithelial cells. The epithelial cell walls are of primary origin; in wheat, barley and oats they contain no pectin substances and lignin (Nieuwdorp, 1963; Swift and O'Brien, 1972).

Cells of conductive system. The conductive system of the scutellum is represented by a large bundle derived from the stele of the embryo hypocotyl, and highly branched in the region of the epithelial cells. Its particular elements are laid down during embryo differentiation, while complete formation occurs during germination of kernels. Mature elements of the sieve tubes arose in the scutellum 3-6 h later and the first tracheid elements after 18 h from initiation of grain swelling. Another 12 h later the entire complex of the conductive system in the scutellum was considered wholly formed (Swift and O'Brien, 1970, 1971).

In germination of kernels the scutellum fulfils **reserve, secretory, absorptive** and **conductive** functions. **The reserve function** of the scutellum is exhibited in accumulation of proteins, starch, lipids and phytin. Therein, soluble saccharides, organic acids, vitamins (ascorbic acid and thiamine), RNA, DNA and mineral elements have been found (Panchaksharappa and Rudramunijappa, 1975; Ovcharov, 1976).

In the course of germination reserved polymers within cells disintegrate under hydrolasic influence into a simple soluble compound (intracellular digestion). The question of how acidic pH indications are maintained in the organelles of cells in which the autolysis of food reserves takes place, remains unresolved. At the early stages of germination (first 2 days) the embryo is in fact solely ensured nutrients at the expense of scutellum reserves, a considerable portion of which goes firstly to respiration, providing secretory and transport processes with energy and metabolites, as well as to synthesis *de novo* of hydrolases in epithelial cells of the scutellum (Dure, 1960).

The secretory function of the scutellum is that of the digestion of endosperm stored polysaccharides and proteins during germination. The mechanism of this process is compatible to that proceeding during digestion of animal food in insect-eating plants or that in the animal stomach, wherein proteins are splitting under the influence of acids and acidic hydrolases (acidic extracellular digestion). The epithelial cells of the scutellum are the ones responsible for the secretion. They undergo divisions in swelling corn seeds, considerably extend and penetrate deeply into the endosperm, increasing thereby its absorptive surface. Herewith the contacts between neighbouring epithelial cells begin to weaken and intercellular space appears. Cells of the epithelium become papillate and stretch towards the endosperm. Invaginations of epithelial cells inside the scutellum have also been detected but their meaning is not still clear (Nieuwdorp and Buys, 1964; Swift and O'Brien, 1972a,b; Zamski, 1975).

With commencement of germination measurable changes take place in the ultrastructure of epithelial cells as well. Reserve substances of protein bodies and spherosomes contribute to the formation of ER components. A vacuole appears, the Golgi apparatus develops, the number of mitochondria increases and their functionality is enhanced. In such a manner epithelial cells turn to intensive synthesis of hydrolases and to secretion.

Secretion of H^+ ions by the scutellum into the endosperm is considered one of the earliest events in a seed that has commenced swelling. The ability of the scutellum to acidify external environment was discovered comparatively recently in maize, barley and wheat (Lado *et al.*, 1975; Szcziparev *et al.*, 1976; Muchtar and Laidman, 1982; Drozdowicz and Jones, 1995). Release of protons by the maize scutellum is an active process, requiring energy. Respiration poisons—malonate, sodium fluoride and dinitrophenol—suppressed the secretion of H^+ ions, while K^+, Ca^{2+} and Na^+ considerably increased its entrance. The latter probably indicates the electrogenous nature of the H^+-pump on the plasma membrane of scutellar epithelial cells. Isolated scutella secreted into incubating medium phosphoric, gluconic, citric, acetic, propionic and butyric acids (Szcziparev and Chuprova, 1980). Such a set of acids is hardly accidental. Some proved to be promoters of seed germination; citric acid displayed the capacity even for restoring the vitality of seeds that had lost germination ability (Radua *et al.*, 1967). What role these acids play in the maize endosperm after secretion by the scutellum is still not known. However, it is doubtless that acidification of the endosperm in swelling grains of maize occurs due to the scutellum. Experiments on the pH measurement in endosperm homogenates extracted from dry grains and from intact swelling ones support this fact. The pH decreased greatly (from 5.6 after 2 h of swelling to 4.8 after 14 h) in intact endosperms only, whereas in isolated ones diminution of pH was insignificant and smooth (Szcziparev *et al.*, 1976). Endosperm acidification during germination was realized unevenly in different parts. In cross-sections of maize, wheat and barley grains stained with pH indicators, the endosperm even one day after swelling began showed an acidic reaction (pH 5.2-5.8); moreover, the wave of acidification derived from the scutellum, close to which the medium reaction is more acidic (pH 4.5) than in other endosperm zones (Philimonov and Szcziparev, 1995). Acidification of the endosperm in wheat and barley is likely also affected by the aleuronic layer, but not earlier than 2-3 days of germination (Micola and Virtanen, 1981).

And what are the mechanisms of acid secretion by the scutellum? ATPase has been detected in epithelial cells of barley and maize but its contribution to the acidifying activity of the scutellum has not been elucidated (Wheeler and Humphreys, 1979; Chaffey and Harris, 1985). Later works using inhibitors of transport plasma membrane ATPases (ortovanadate, diethylstilbestrol) have demonstrated that H^+-ATPase participates in secretion of H^+ by maize scutella. Inhibitors decreased the acidifying capacity of scutella up to 20-40% only, even though they were applied in a concentration higher than ordinary; moreover, they hardly affected scutella extracted a day after swelling had begun (Szcziparev *et al.*, 1993). Along with ATPase there may be some other mechanisms of H^+ secretion by the scutellum, for example the redox-chain in the plasma membrane of epithelial cells. Thus, using the histochemical method the capacity of extracted maize scutella for restoring exogenous ferricyanide—an indicator of redox-chain, was revealed (Szcziparev and Myachina, 1990). Herewith the scutellar surface adjoining the endosperm exhibited the most ferricyanide reductase activity. Reduction of ferricyanide in maize scutella was accompanied by enhancement of acidification in the external medium, while the addition of ethanol increased both processes. All these results evidence occurrence in the plasma membrane of the epithelial cells of redox-chains that are H^+-pump, and concurrently participate in the acidifying activity of maize scutella. Unfortunately, reports on redox-chain function in other cereals are still not available in the literature. The question of the possibility of acid secretion by the scutellum in vesicles of the Golgi apparatus together with hydrolases deduced in the endosperm, also remains unsolved.

CO_2, released from the scutellum during respiration and dissolving in a liquid phase in the endosperm, appears to play a definite role in endosperm acidification. Actually, the intensity of respiration in the maize scutellum is one order higher than in the endosperm; the respiration coefficient is more than 1 (1.5-1.8). Decarboxylized enzymes—malic, pyruvatdecarboxylase—have been observed in the scutellum, and their activity is much higher in the scutellar part adjoining the endosperm than in the rest of the embryo (Belozerova and Plotnikova, 1984; Belozerova *et al.*, 1986). Here the question arises as to whether the bicarbonate ion, appearing during CO_2 solution in the endosperm, occurs as an activator of H^+-ATPase of epithelial cell plasma membrane in the maize scutellum, as happens in the mucosa of the human stomach.

Releasing of acids in the endosperm from the scutellum and acidification of its tissues is considered to be the first and most important stage in the digestion of endosperm nutrient reserves. Owing to it, firstly, the optimal medium pH for the activity of acidic hydrolases in the endosperm is created (in dry grains the endosperm has a neutral reaction). Secondly, the functional activity of protein inhibitors of proteases and amylase is suppressed (Polanovski, 1967; Jin-a-Jin *et al.*, 1993). Thirdly, the capacity of transport systems in the plasma membrane of the epithelial cells is elevated (their optimal activity is observed in the acidic medium) (Steward, 1971).

As soon as acidification occurs, intensive secretion of hydrolases by the scutellum into the endosperm commences. Cellulase, α- and β-amylases, phosphatase, proteases, nuclease etc., have been discovered among hydrolases, under the effect of which endosperm reserves dissolve (Fincher, 1989). The secretory process is retarded by respiration inhibitors, colchicine, and promoted by Ca^{2+} ions and GA_3 (Polevoy *et al.*, 1981; Szcziparev and Ilchukov, 1983). It is calcium that is required for the fusion of secretory vesicle membranes with the plasma membrane during exocytosis, as well as for synthesis of amylase. It is probably significant in combining stimulus and secretion (Salamatova and Zauralov, 1991). Hydrolases are synthesized *de novo* by the granular ER and in vesicles of the Golgi apparatus and secreted through the plasma membrane into the cell wall. The mechanism of their passing through the latter has not yet been studied.

Absorptive activity of the scutellum increases rapidly 2-3 days after germination. Soluble products of hydrolysis of endosperm reserves are absorbed by the scutellum epithelium. Transport of nutrients from the endosperm to the embryo occurs through the conductive system of the scutellum. Glutamine and sucrose appear to be the main transport forms of nitrogen and carbon in the scutellum, synthesizing them in its parenchymatous cells. The mechanism of passing through the scutellum phloem is presumably compatible to that of assimilates in the stem. The transport of sugars through the plasma membrane of epithelial cells has been thoroughly investigated in maize. Its scutellum best absorbs sucrose and maltose at an acidic (not more than 5) pH while the pH value exerts no influence on the absorption of glucose and fructose (Humphreys and Garrard, 1964; Szcziparev *et al.*, 1990). The mechanism of transport of neutral compounds (e. g. sucrose) in the scutellum cells comprises the system of H^+ ion and sucrose symport, negatively charged free carrier and neutral complex carrier-sucrose-H^+. Transport is supported by the gradient of sucrose and electrochemical gradient of H^+, created by the redox-chain and H^+-ATPase of the epithelial cell plasma membrane (Humphreys, 1978). A similar system of connected transport appears to exist for aminoacids and peptides, whose absorption is optimal in an acidic medium (pH 5.0). In the scutellum of various cereal species carriers of both non-specific, common for major amino acids, and specific for some amino acids have been detected (Salmenkallio and Sopanen, 1989). Whether the transport of disaccharides and peptides through the plasma membrane is accompanied by their hydrolysis, inducing additional energizing of the membrane, is not yet clear.

The scutellum ceases functioning 8-10 days after inception of germination. In this time degeneration of its tissues and cell autolysis are observed. The scutellum has accomplished its function at the heterotrophic stage of seedling development.

Hormonal Regulation of Seedling Growth and Development

Phytohormones are endogenous regulators required for interaction of cells, tissues and organs in the course of ontogenesis and for adequate response of the organism to the signals received from the external environment. The hormonal system of a new sporophyte starts with its formation during embryogenesis. In some dicotyledonous plants the embryo, at the torpedo-shape stage, already possesses hormonal independence (autonomy), i.e., synthesizes all the phytohormones necessary and is not short of introducing them from outside itself (Pontowich, 1978; Batygina and Vasilyeva, 1983b).

Role of phytohormones in swelling and root emergence in seeds in etiolation phenomena. Phytohormones in seeds appear to be associated with soluble saccharides, amino acids, proteins and polysaccharides. Auxin associated (IAA–indole-3-acetic acid) and cytokinins in cereal kernels have been detected in the endosperm and scutellum and gibberellins and ABA (abscisic acid) in the scutellum and other parts of the embryo.

During seed imbibition the content of free IAA, gibberellins and cytokinins rapidly increases consequent to hydrolysis of associated forms. The further course of germination depends on the ratio of free phytohormone concentrations. A high concentration of cytokinins and gibberellins in tissues activates germination. On the contrary, ABA inhibits germination as is usually observed when treating germinating seeds with exogenous ABA.

The growth of embryonal axial cells by stretching facilitates the breakdown of seed coats and emergence of the root tip (see: Morphology and Physiology of Seed Germination). In dicotyledonous plants growth by straining of the hypocotyl cells activates, i.e., induced by great enhancing of ethylene synthesis in embryo tissues and by elevation of CO_2 concentration as well (Abeles *et al.*, 1992). Carbon dioxide encourages acidification of the cytoplasm that intensifies H^+-pump (H^+-ATPase) in the plasma membrane of hypocotylar cells. Active release of H^+ ions from the cytoplasm into the cell walls effects 'acidic growth' as a result of increase in plastic stretching of cell walls during their acidification.

Seedling growth in soil is realized in the absence of light. In these conditions an intensive elongation of growing internodes is observed. Seedling organs are white or yellow colour since chlorophyll synthesis in cells does not occur (in angiosperms). Such seedlings are referred to as etiolated. In many species internode straining is determined by the high content of gibberellins and auxin (Salamatova, 1994; Okamoto *et al.*, 1995).

Role of phytohormones in regulation of heterotrophic mode of seedling nutrition. Given the absence of light growing seedling cells receive nutrients from accumulative organs. The mechanism of nutrition regulation has been best studied in cereal seedlings. Phytohormones (IAA, cytokinins, gibberellins), stored in the endosperm, activate the digestive function of scutellar epithelial cells, while changing from the associated form to the free one (see: Scutellum and Its Role in Germination). This results in the 'digestion' of both reserved substances and the endosperm cells themselves adjacent to the scutellum.

The 'digestion' of peripheral endosperm cells is realized 'by means' of cells in the aleuronic layer. Two to three days after germination gibberellins diffuse from the scutellum and embryonal axis into the cells of the aleuronic layer. Under gibberellin influence dissimilation of storage proteins, synthesis of RNA, inducing of mRNA required for intensive hydrolase synthesis, are activated in aleuronic cells. These hydrolases (amylase, cellulase, RNAase, proteinases etc.) are secreted from aleuronic cells to the endosperm (Jacobsen and Chandler, 1988). Along with that activity of the proton pump (H^+-ATPase) in the plasma membrane of aleuronic cells increases, resulting in acidification of endosperm tissues and extracellular acidic digestion, i.e., phenomena compatible to that of the digestive function of the scutellum are observed. The presence of cytokinin is essential for inducing the effect of gibberellins on the functional activity of aleuronic cells. ABA depresses the action of these phytohormones. Products of digestion, accumulated in the endosperm (sugars, amino acids etc.), are absorbed by scutellum epithelial cells, fill the phloem ends and through the vascular bundles are delivered to the growing parts of the seedling. All these processes are also exposed to hormone control (Micola and Virtanen, 1981; Polevoy *et al.*, 1990).

Phytohormones and motility of the seedling. For orientation in space the root and shoot of the seedling developing in darkness (in soil), are thought to use the gravitational field. Here the root grows towards the centre of the Earth (positive gravitropism) and the shoot upwards from the centre of the Earth (negative gravitropism). The gravitational signal is perceived by the cells of the root cap (columella), coleoptile (in grasses) and upper growing internode (in dicots). The perception of gravity direction occurs in cells (statocytes) with the participation of intracellular organelles possessing high density (statolytes). In consequence of the pressure of statolytes on the ER and perhaps vacuolar membranes, the concentration of Ca^{2+} ions, which accomplish the function of the secondary intracellular messenger (mediator), enhances locally in the cytoplasm. Ca^{2+} ions promote lateral shifting of IAA transport (mechanism not known). Auxin concentration elevates on the lower side of the horizontally disposed axial organs. In the shoot axial organs it leads to acceleration of cell growth from lower side of the organ believed to be a basis for negative geotropism. Alternatively, growth of lower cells in the root straining zone will be restricted because of the high sensitivity of these cells to auxin and the inhibitory effect in its higher concentration on growth. As a result the root will curve towards the Earth's centre. The part in which the growing (tropistic) bend took place, draws itself up some time later (phenomenon of autotropism). Due to this mechanism the vertically growing shoot and root permanently produce circular movements (nutations), which may be regarded as a correction of growth direction and as searching motions in space (Merkis, 1990; Konings, 1995).

Elongation of shoot and root, owing to stretching of cells in appropriate growth zones, occurs as the mode of movement (irreversible motional activity) of these organs, which could be examined as a primary mechanism of transference of axial organs in multicellular plants. With the deficit in mineral nutrition, especially nitrogen, synthesis of nitrogen-containing IAA and cytokinins reduces in a great number of plants; as a consequence shoot growth decreases while root growth accelerates. The expediency of this behaviour is quite apparent because it permits more rapid movement of the root in new soil horizons in search of mineral substances (Kuiper *et al.*, 1988; Kudoyarova and Usmanov, 1991).

Strategy of shoot and root movement in soil is varied. The root tip does not exert much pressure on soil particles, searching rather for spaces between them. Correction of these searching movements is realized by the root cap with participation of IAA and, presumably, ABA. If the root apex is brought up short against a hindrance, synthesis of the stress hormone—ethylene—in the straining growth zone accelerates. It is ethylene that is heuristic for disorganizing microtubules in the cortex cell layer, as a result of which cells begin to stretch not in length, but isodiametrically, and thickenings are formed in the zone of straining growth. These thickenings move soil particles apart, allowing the root tip to advance forward again (Darwin, 1941; Abeles *et al.*, 1992).

The growing shoot, on the contrary, exerts vigorous pressure on soil particles. It is a coleoptile in cereals and falcate apical part of the hypocotyl or epicotyl in dicotyledons, —an effective shape for pushing upwards in soil. Its curved shape is determined by active synthesis of ethylene in the subapical portion of the stem (Abeles *et al.*, 1992).

Phytohormones and transition to autotrophic mode of nutrition. Etiolated seedlings are highly sensitive to light. Apprehension of light signals is realized in plants by molecular photoreceptors located in membranes. The phytochrome is a receptor of red and far red light, while the cryptochrome and probably phytochrome as well receptors of blue. The phytochrome concentration is especially high in the apical part of the shoot and root tip. Therefore, these predominant centres are the first to respond to the appearance of light. At local lighting, rays can spread into all parts of a plant because tissues display the property of light transfer (Vogelmann, 1993). Removal from light results in the following alterations in seedlings: internode lengthening is rapidly hindered, the subapical hook in dicots straightens, temporary acceleration of coleoptile growth occurs in grasses, chlorophyll synthesis and chloroplast formation are induced, and growth and unfolding of the first leaves proper activated. A light signal switches off the genetic programme from scotomorphogenesis[2] to photomorphogenesis, which ensures plant switchover to the phototrophic mode of nutrition (De Greef and Frèdérico, 1983; Mohr, 1983).

In all these processes the hormonal system plays an important role. Light hindrance of lengthening of a growing internode was demonstrated to be associated with inhibition of polar IAA transport and decrease in gibberellin reception (Scott and Wilkins, 1969; Toyomasu *et al.*, 1992). Straightening of the subapical hook is conditioned by quick diminution of ethylene synthesis in cells of this zone of the stem (Abeles *et al.*, 1992). Growth and unfolding of leaves in cereals are attributable to increase in active gibberellin content in them (Appleford and Lenton, 1991; Kopcewicz and Madela, 1992). Growth of leaf plates in dicotyledonous plants requires augmentation of auxin, gibberellin and cytokinin concentrations (Wareing and Phillips, 1978; Goodwin and Erwee, 1983). Functional activity of a green leaf dramatically depends on permanent cytokinin flow from roots (Kulaeva, 1982).

---

[2] Greek *scotos*—darkness.

General characteristics of plant hormonal system. The shoot apex is the place of occurrence of auxin IAA synthesis. Synthesis of cytokinin takes place in the root tip. IAA from the shoot apex is transmitted through the living cells of vascular bundles up to the root tip at the rate of 0.5-1.5 sm h^{-1}. Cytokinins (zeatin and its conjugates—zeatin riboside, zeatin ribotide etc.) and gibberellins are transported from the root tip, where they synthesize, passively along vascular bundles as a part of xylem sap. Auxin is needed for regulation of cell division (together with cytokinin), for cell growth by straining and for differentiation of vascular bundles. It activates the division of cambium cells and the extension of axial organs in thickness, inhibits growth of axile buds, likewise regulates the rate of division and streching of cells in the distal part of the root and the laying down of lateral roots, and lastly is the inductor of the genetic programme of root formation. Cytokinin, in turn, is required for cell division (coupled with auxin), reducing cell polarization and promoting isodiametrical growth. It hinders the laying down of lateral roots, induces development of axile buds, activates growth of mesophyll cells by means of straining (in dicots), prevents leaf senescence. Cytokinin is necessary for laying down and development of shoot apex, as **the main inductor of the genetic programme of shoot formation.**

Thus, IAA and cytokinin constitute **the chief pattern of the plant hormonal system, functioning according to the principle of positive feedback that provides progressive seedling development.** In the course of shoot formation gibberellins begin to synthesize in leaves, abscisic acid (ABA) mostly in leaves, but also in the root cap. Synthesis of ethylene may be realized in all cells, but especially those in which concentration of IAA and cytokinin is high. GA is required for shoot development, plant transition to flowering and for regulation of many other physiological and morphogenetic programmes. ABA usually fulfils the function of hormone inhibitor. The concentration of ABA and ethylene in tissues increases greatly in stress conditions and in the course of organ senescence (Polevoy, 1982; Derfling, 1985; Polevoy and Salamatova, 1991; Bewley and Black, 1994).

BIBLIOGRAPHY

Abbe, E. C. and O. L. Stein. 1954. The growth in the shoot apex in maize: embryogeny. Amer. J. Bot. 41: 285-293.

Abeles, F. B., P. W. Morgan and M. E. Saltveit, Jr. 1992. Ethylene in Plant Biology. Academic Press, San Diego, CA.

Adatia R. D., Y. B. Sharma and M. R. Vijayaraghavan. 1971. Studies in the Gesneriaceae. I. Morphology and embryology of *Platystemma violoides* Wall. Bot. Notiser, 124(1): 25-38.

Afanasyeva, N. G. 1960. Knowledge of phylogeny in Scrophulariaceae family. Ph.D. thesis, Kazan, Russia.

Afanasyeva, N. G. 1976. Embryology of the genus *Veronica* L. Kazan University Press, Kazan, Russia.

Afzelius, K. 1924. Embryologische und zytologische Studien in *Senecio* und verwandten Gattungen. Acta Horti Bergiani, 8(7): 123-219.

Afzelius, K. 1928. Die Embryobildung bei *Nigritella nigra*. Svensk Botanisk Tidskrift, 22: 82-91.

Agarwal, S. 1963. Morphological and embryological studies in the family Olacaceae—I. *Olax* L. Phytomorph., 13(2): 185-196.

Alexandrov, V. G. 1966. Anatomy of Plants. Vysschaya schkola, Moscow, Russia.

***Alexandrov, V. G. and O. G. Alexandrova. 1939.** Endosperm formation in durum and soft wheats. Reports USSR Acad. Sci., 25 (4): 320-323.

Alfieri, J. R. and R. F. Evert. 1968. Analysis of meristematic activity in the root tip of *Melilotus alba* Desr. New Phytologist, 67: 641-647.

Alimova, G. K. 1985a. Family Aceraceae, pp. 183-185. *In* M. S. Yakovlev (ed.). Comparative Embryology of Flowering Plants. Brunelliaceae—Tremandraceae. Nauka, Leningrad, Russia.

Alimova, G. K. 1985b. Family Hyppocastanaceae, pp. 185-189. *In* M. S. Yakovlev [ed.]. Comparative Embryology of Flowering Plants. Brunelliaceae—Tremandraceae. Nauka, Leningrad, Russia.

Alimova, G. K. 1987. Family Cornaceae, pp. 18-21. *In* T. B. Batygina and M. S. Yakovlev (eds). Comparative Embryology of Flowering Plants. Davidiaceae—Asteraceae. Nauka, Leningrad, Russia.

Alimova, G. K. 1994. Microspore, pp. 82-83. *In* T. B. Batygina (ed.). Embryology of Flowering Plants. Terminology and Concepts. 1. Generative Organs of Flower. Mir i Semya, St. Petersburg, Russia.

***Alimova, G. K. and M. S. Yakovlev. 1982.** On the embryology of *Streptocarpus rexii* (Gesneriaceae). Bot. Journal (St. Petersburg), 67(4): 470-479.

Almers, W. and C. Stirling. 1984. Distribution of transport proteins over animal cell membranes. J. Membrane Bio., 77(3): 169-186.

Alpi, A. 1990. Gibberellins in embryo development, pp. 468-475. *In* R. P. Pharis and S. B. Rood (eds). Plant Growth Substances. Springer-Verlag, Berlin.

Amici, G. B. 1824. Observations microscopiques sur diverses espèces de plantes. Ann. Sci. Natur. Bot., 2: 41–70, 211–248.

Anagnostakis, S. L. 1977. In vitro culture of immature embryos of American elm. Hort Science, 12(1): 44.

---

*Astrisked works are in Russian—General Editor.

Anantaswamy, R. 1950. The suspensor haustoria of some species of *Crotalaria*. Linnean, N. S., 14(56): 557.

Andronova, E. V. 1988. Embryogenesis and postseminal development of orchids (on the example of *Dactylorhiza baltica, D. incarnata, Thunia marshalliana, Bletilla striata*). Ph.D. thesis, Leningrad, Russia.

Andronova, E. V. and T. B. Batygina. 1992. Uniformity of initiation of the first leaf during germination of the orchids (on example of the *Thunia marshalliana* Rchb.f.), pp. 37-38. *In* Proc. XI Int. Symp. "Embryology and Seed Reproduction", July 3-7, 1990, Leningrad. Nauka, St. Petersburg, USSR.

***Anisimova, G. M. 1983.** Development and structure of seed in *Orobanche cumana* and *O. alsatica* (Orobanchaceae). Bot. Journal (St. Petersburg), 68(4): 473-482.

Anisimova, G. M. 1990. Family Juncaceae, pp. 183-187. *In* M. S. Yakovlev (ed.). Comparative Embryology of Flowering Plants. Butomaceae—Lemnaceae. Nauka, Leningrad, Russia.

Appleford, N. E. J. and J. R. Lenton. 1991. Gibberellins and leaf expansion in near-isogenic wheat lines containing Rht 1 and Rht 3 dwarfing alleles. Planta, 183(2): 229-236.

Arber, A. 1925. Monocotyledons: A Morphological Study. Cambridge Univ. Press, Cambridge, UK.

Arber, A. 1934. The Gramineae. A Study of Cereal, Bamboo and Grass. Cambridge Univ. Press, Cambridge, UK.

Arekal, G. D. 1961. Embryology of *Klugia notoniana*. Bot. Gazette (Chicago), 123(2): 144-150.

Arekal, G. D. 1963. Embryological studies in Canadian representatives of the tribe Rhinantheae, Scrophulariaceae. Can. J. Bot., 41(2): 267-302.

Arekal, G. D. 1966. Embryology of *Veronica serpyllifolia* L. Proc. Indian Acad. Sci., 64(5): 241-258.

Arekal, G. D. and S. N. Ramaswamy. 1980. Embryology of *Eriocaulon hookerianum* Staff., and the systematic position of Eriocaulaceae. Bot. Notiser, 133: 295-309.

Artjushenko, Z. T. 1990. Atlas on Descriptive Morphology of Higher Plants. Seed. Nauka, Leningrad, Russia.

Arzt, T. 1933. Über die Embryobildung von Pseudomonocotylen (*Podophyllum emodii* und *Eranthis hiemalis*). Beihefte Bot. Centralblatt, 50(1): 671-696.

Asahi, T. and M. Maeshima. 1983. Biogenesis of mitochondria in higher plant cells, pp. 119-132. *In* M. Nijhoff and W. Yunk (eds). The New Frontiers in Plant Biochemistry. Japan Sci. Soc. Press, Tokyo, Japan.

Ashley, T. 1972. Zygote shrinkage and subsequent development in some *Hibiscus* hybrids. Planta, 108(4): 303-317.

Ashley, T. 1975a. Alterations in the fine structure of developing endosperm of *Hibiscus* hybrids. Caryologia, 28(1): 73-80.

Ashley, T. 1975b. Fine structure of early endosperm development in *Hibiscus*. Caryologia, 28(1): 63-71.

Asker, S. and L. Jerling. 1992. Apomixis in Higher Plants. CRC Press, Boca Ration, FL, USA.

Askochenskaya, N. A. 1982. Water regime of seeds, pp. 184-222. *In* A. A. Prokofyev (ed.). Seed Physiology. Nauka, Moscow, Russia.

Aspart, L., Y. Mayer, M. Laroche and P. Penon. 1984. Developmental regulation in the synthesis of protein encoded by stored mRNA in radish embryos. Plant Physiol., 76(3): 664-673.

Asplund, E. 1920. Studien über die Entwicklungsgeschichte der Blüten einiger Valerianaceen. Kongliga Svenska Vetenskapsakademiens Handlingar, N. S., 61(3): 3-66.

Asplund, J. 1972. Embryological studies in the genus *Typha*. Svensk Botanisk Tidskrift, 66(1): 1-17.

Åström, H., L. Vitranen and M. Raudaskoski. 1991. Cold stability in the pollen-tube cytoskeleton. Protoplasma, 160: 99-107.

***Atabekova, A. I. and Lin Tzyansin. 1964.** Embryogenesis in lupine. Reports Tymiryazev Agric. Academy, 4: 38-42.

Avanzi, S., P. G. Clionini and F. D'Amato. 1970. Cytochemical and autoradiographic analyses on the embryo suspensor cells of *Phaseolus coccineus*. Caryologia, 23(4): 605-638.

Avery, G. S. 1930. Comparative anatomy and morphology of embryos and seedlings of maize, oats and wheat. Bot. Gazette (Chicago), 89(1): 1-39.

Babis, H. 1973. General description of a mature ovule and embryo sac, early stages of development of the endosperm and of embryo development in some species of *Delphinium* L. Genetica Polonica, 14(1): 29-36.

Babis, H. 1976. Development of the embryo in some species of the genus *Delphinium* L. Acta Soc. Bot. Poloniae, 45, (1-2): 135-142.

Baker H. G. 1960. Apomixis and polyembryony in *Pachira oleaginea* (Bombacaceae). Amer. J. Bot., 47(4):296-301.

Banerji, I. 1961. The endosperm in Scrophulariaceae. J. Indian Bot. Soc., 40(1): 1-11.

Banerji, J. 1947. Life history of *Typhonium trilobatum*. Proc. Natl. Inst. Sci. India, 13(3): 207-230.

Banerji, M. 1962. On anatomy of teratological seedlings. II. *Calendula officinalis* L. Bull. Bot. Soc. Bengal, 16(1): 24-29.

***Bannikova, V. P. 1975.** Cytoembryology of Interspecific Incompatibility in Plants. Naukova Dumka, Kiev, Ukraine.

***Bannikova, V. P. 1986.** Interspecific Incompatibility in Plants. Naukova Dumka, Kiev, Ukraine.

***Bannikova, V. P. and O. A. Khvedynich. 1973.** Fertilization process in *Nicotiana rustica*. Ukrainian Bot. Journal, 30(4): 668-670.

***Bannikova, V. P. and O. A. Khvedynich. 1982.** Principles of Plant Embryology. Naukova Dumka, Kiev, Ukraine.

***Bannikova, V. P. and T. A. Plyushch. 1988.** Ultrastructure of embryo sac and embryo of *Calanthe vestite* (Orchidaceae). Bot. Journal (St. Petersburg), 73(10): 1401-1408.

***Bannikova, V. P., O. A. Khvedynich and O. V. Ovsyannikova. 1978.** Research of mitotic cycles of gametogenesis in some angiosperms. Reports Ukrainian SSR Acad. Sci., Series Biology, 240(4): 963-966.

***Bannikova, V. P., O. A. Khvedynich, S. P. Shpilevaya, E. K. Ostapenko and O. V. Ovsyannikova. 1981.** Research of DNA synthesis dynamics during gametogenesis in plants. Zh. obshchei biol., 42(2): 226-233.

***Bannikova, V. P., O. A. Khvedynich, S. P. Shpilevaya, E. K. Ostapenko, N. V. Sidorova and T. A. Plyushch. 1985.** Sexual Cells and Fertilization in Angiosperms and Algae. Naukova Dumka, Kiev, Ukraine.

Banuelos G. S., F. Bangerth and H. Marschner. 1987. Relationship between polar basipetal auxin transport and acropetal Ca^{2+} transport into tomato fruits. Physiol. Plantarum (Copenhagen), 71(3): 321-327.

Banuelos G. S., F. Bangerth and H. Marschner. 1988. Basipetal auxin transport in lettuce and its possible involvement in acropetal calcium transport and incidence of tipburn. J. Plant Nutr., 11(5): 525-533.

Baranov P. A. 1937. Structure and Development of Cotton. Atlas. Ogiz-Izogiz, Moscow-Leningrad, Russia.

Baranov, P. A. 1955. The History of Plant Embryology. USSR Acad. Sci. Press. Moscow-Leningrad, Russia.

Barlow, P. W. 1976. Towards an understanding of the behaviour of root meristems. J. Theor. Biol., 57: 433-451.

Barlow, P. W. 1991. From cell wall networks to algorithms. The simulation and cytology of cell division patterns in plants. Protoplasma, 162: 69-85.

***Barlow, P. W. 1994a.** Cell division in meristems and the significance of this process for organogenesis and plant morphogenesis. Ontogenez, 25(5): 5-28.

Barlow, P. W. 1994b. Rhythm, periodicity and polarity as bases for morphogenesis in plants. Biol. Rev., 69: 475-525.

Barlow, P. W. 1995. Structure and function at the root apex—phylogenetic and ontogenetic perspectives on apical cells and quiescent centres, pp. 3-18. *In* P. Baluska et al. (eds). Structure and Function of Roots. Kluwer Acad. Publ., Dordrecht, Netherlands.

Barlow, P. W., P. Brain and J. B. Parker. 1991. Cellular growth in roots of a gibberellin-deficient mutant of tomato (*Lycopersicon esculentum* Mill.) and its wild-type. J. Exper. Bot., 42: 339-351.

Barnes, B. 1943. Seedling abnormalities in *Anagallis arvensis* L. Proc. Linnean Soc. London, 154: 88-91.

Barrow, J. and H. Chaundhari. 1974. Cotton haploids produced by semigamy from specific breeding lines, p. 274. *In* Haploids in Higher Plants: Advances and Potential. University of Guelph, Guelph, Canada.

Barrow, J. R. and H. K. Chaundhari. 1976. A homozygous interspecific F2 hybrid of *Gossypium barbadense* × *Gossypium hirsutum* via the semigametic haploid method. Crop Science (Madison), 16(3): 441-442.

Barton, M. K. and R. S. Poethig. 1993. Formation of the shoot apical meristem in *Arabidopsis thaliana*: an analysis of development in the wild type and in the shoot meristemless mutant. Development, 119: 823-831.

***Barykina, R. P. 1971.** Peculiarities of the first stages of ontogenesis in *Podophyllum emodi* Wall. and *P. peltatum* L. Bot. Journal (St. Petersburg), 56(7): 921-931.

Barykina, R. P. 1995. Morphologo-ecological regularities of the somatic evolution in Ranunculaceae family. Ph.D thesis, Moscow, Russia.

Baskin, C. C. and J. M. Baskin. 1994. Deep complex morphophysiological dormancy in seeds of the mesic woodland herb *Delphinium tricorne* (Ranunculaceae). Int. J. Plant Sci., 155(6): 738-743.

Baskin, C. C., S. E. Meyer and J. M. Baskin. 1995. Two types of morphophysiological dormancy in seeds of two genera (*Osmorhiza* and *Erythronium*) within Acto-Tertiary distribution pattern. Amer. J. Bot., 82(3): 293-298.

Battaglia, E. 1945a. Fenomeni citologici nuovi hella embriogenesi ("Semigamia") e hella microsporogenesi. ("Doppio nucleo di restituzione") di *Rudbeckia laciniata* L. (nota, preventiva). Nuovo Giornale Bot. Ital., N. S. 52(1): 34-38.

Battaglia, E. 1945b. Semigamie chez *Rudbeckia sullivantii* Boynton et Beadle (Compositae). Proc. 8th Int. Bot. Cong. (Paris). Sec. 7-8, pp. 245-247.

Battaglia, E. 1947a. Ricerche cariologiche ed embriologiche sul genere *Rudbeckia* (Asteraceae). XI. Semigamia in *Rudbeckia speciosa* Wender. Nuovo Giornale Bot. Ital., N. S., 54(3-4): 531-559.

Battaglia, E. 1947b. Le "semigamia", singolare comportamento del nucleo spermatico helle nova diploidi del be specie apomictiche del genere *Rudbeckia* (Asteraceae) e consequente embriogenesi di tipo ĉhimarico. C.R. l'Acad. de Sci. Lincei, Roma, Ser. 8, 2: 63-67.

Battaglia, E. 1953-1954. Ricerche citotassonomiche in *Dipcadifulium* Webb. Caryologia, 6 (I): 63-76.

Battaglia, E. 1963. Apomixis, pp. 221-264. *In* P. Maheshwari (ed.). 1963. Recent Advances in the Embryology of Angiosperms. Univ. Delhi, Delhi, India.

Battaglia E. 1971. The embryo sac of Podostemonaceae—An interpretation. Caryologia, 24: 403-420.

Battaglia, E. 1980. Embryological questions: 2. Is the endosperm of angiosperms sporophytic or gametophytic? Ann. Bot. (London), 39(1): 9-30.

Battaglia, E. 1981. Embryological questions: 3. Semigamy, hemigamy and gynandroembryony. Ann. Bot. (London), 39(2): 173-175.

Battaglia, E. 1987. Embryological questions: 10. Have the expressions polar nuclei "and secondary nucleus" been rightly established? Appendix: Hofmeister W. (1847), "Untersuchungen bei Oenothereen". Atti della Soc. Toscana Sci. Natur. Pisa, Ser. B, 94: 127-150.

Batygin, N. F. and A. M. Demyanchuk. 1995. Calculation of wheat ontogenesis (methodic recommendations). All-Russian Inst. Plant Industry, St. Petersburg, Russia.

Batygina, T. B. 1957. On the condition of sexual elements in fertilization process in wheats, pp. 45-46. *In* Abstracts of VI Congress All-Union Bot. Soc., 8. Moscow-Leningrad, Russia.

Batygina, T. B. 1959. Cytology of fertilization process at distant hybridization in wheats, pp. 19-20. *In* Abstracts Coord. Conf. Problem "Key Questions on Cytology". Leningrad, Russia.

Batygina, T. B. 1961a. Some data on cytology of fertilization process in wheat, pp. 334-340. *In* V. G. Alexandrov (ed.). Morphogenesis of Plants. Moscow State University Press, Moscow, Russia.

***Batygina, T. B. 1961b.** Transformation of male gametes during fertilization in wheats. Reports USSR Acad. Sci., 137(1): 220-223.

***Batygina, T. B. 1962a.** Fertilization process in wheats. Trans. Komarov Bot. Inst. USSR Acad. Sci., Series VII, 5: 260-293.

***Batygina, T. B. 1962b.** Microsporogenesis and development of pollen grain in wheats. Reports USSR Acad. Sci., 142(5): 1205-1208.

***Batygina, T. B. 1966.** Fertilization process at distant hybridization in the genus *Triticum* L. Bot. Journal (St. Petersburg), 51(10): 1461-1479.

***Batygina, T. B. 1968a.** Embryogenesis in the genus *Triticum* L. (in connection with questions on monocotyly and distant hybridization in cereals). Bot. Journal (St. Petersburg), 53(4): 480-490.

***Batygina, T. B. 1968b**. On the possibility of distinguishing of new type of embryogenesis in Angiospermae. Reports USSR Acad. Sci., 186(6): 1499-1502.

Batygina, T. B. 1969. On the possibility of separation of a new type of embryogenesis in Angiospermae. Rev. Cytol. Biol. Végétales, 32(3-4): 335-341.

Batygina, T. B. 1973. Embryology of wheat. Ph.D. thesis, Leningrad, Russia.

Batygina, T. B. 1974a. Fertilization process in cereals, pp. 205-220. *In* H. F. Linskens (ed.). Fertilization in Higher Plants. North-Holland Publ. Co., Amsterdam, Netherlands.

Batygina, T. B. 1974b. Wheat Embryology. Kolos, Leningrad, Russia.

Batygina, T. B. 1976. Embryological processes related to distant hybridization in Gramineae, pp. 18-20. *In* Abstracts of Indo-Soviet Symposium "Embryology of Crop Plants", March 2-6, Delhi, India.

Batygina, T. B. 1978a. Embryological processes at distant hybridization in Gramineae family. Proc. Natl. Acad. Sci. India, Section B., Biol Sci. 4:30-42.

Batygina, T. B. 1978b. Embryology of wheat. Proc. Natl. Acad. Sci. India, Section B., Biol. Sci., 44: 13-29.

Batygina, T. B. 1978c. On the certain regularities of morphogenesis at plant regeneration *in vitro*, pp. 13-14. *In* Abstracts of Regional Conference "Theoretical Problems of Plant Regeneration". Makhachkala, Russia.

Batygina, T. B. 1981. Family Nelumbonaceae, pp. 116-120. *In* M. S. Yakovlev (ed.). Comparative Embryology of Flowering Plants. Winteraceae—Juglandaceae. Nauka, Leningrad, Russia.

Batygina, T. B. 1982. On the position of Nymphaeales and Nelumbonales orders in the system of angiosperms (according to the embryological data), pp. 18-21. *In* Materials VI Moscow Conf. Plant Phylogeny. Moscow State Univ., Moscow, Russia.

***Batygina, T. B. 1983.** The system approach to the problem of embryo differentiation in angiosperms, pp. 25-26. *In* Abstracts VII All-Union Conf. Plant Embryology "Problems of Gametogenesis, Fertilization and Embryogenesis". "Phan", Tashkent, Uzbek Republic.

Batygina, T. B. 1984. Problems of morphogenesis in situ, in vivo and in vitro, pp. 43-56. *In* F. J. Novák, L. Havel and J. Doležel (eds). Proc. Int. Symp. "Plant Tissue and Cell Culture Application to Crop Improvement". Olomouc, 24-29 September, 1984. Czechoslovak Acad. Sci., Prague, Czechoslovakia.

Batygina, T. B. 1987a. The Grain of Cereals (Atlas). Nauka, Leningrad, Russia.

Batygina, T. B. 1987b. New concept of asexual reproduction in flowering plants, p. 5. *In* Abstracts XIV Int. Bot. Cong., Berlin.

Batygina, T. B. 1989a. New approach to the system of reproduction in flowering plants. Apomixis Newsletter, 1: 52-55.

Batygina, T. B. 1989b. A new approach to the system of reproduction in flowering plants. Phytomorphology, 39(4): 311-325.

Batygina, T. B. 1989c. New concept of asexual reproduction in flowering plants, pp. 28-44. *In* J. Paré and M. Bugnicourt (eds). Some Aspects and Actual Orientation in Plant Embryology. Universite Picardie, Amiens, France.

Batygina, T. B. 1990. Embryoidogenic type of reproduction in flowering plants. Apomixis Newsletter, 2: 58-66.

Batygina, T. B. 1991. New hypothesis about the initials and genesis of embryoids (somatic embryos) and a position of embryoidogeny in the reproduction system. Apomixis Newsletter, 3: 19-24.

Batygina, T. B. 1992a. Embryological bases of plasticity and the adaptive abilities of reproductive systems in flowering plants, pp. 201-212. *In* B. A. Ursev [ed.]. Biological Diversity: Approaches to the Study and Conservation. St. Petersburg, Russia.

Batygina, T. B. 1992b. Relation between embryoidogeny (somatic embryogeny) and apomixis, pp. 333-334. *In* Book of Poster Abstracts, XIII Eucarpia Congress

"Reproductive Biology and Plant Breeding", July 6-11, 1992, Angers, France. Imprimiere MALGOGNE, Angers, France.

***Batygina, T. B. 1993a.** Embryoidogeny—a new category of reproduction in flowering plants, pp. 15-25. *In* E. S. Teryokhin (ed.). Trans. Komarov Bot. Inst. Russian Acad. Sci. Problems of Reproductive Biology of the Seminal Plants, 8: 15-25., Komarov Bot. Inst., St. Petersburg, Russia.

Batygina, T. B. 1993b. Some aspects of morphogenetic polarity in plant ontogenesis, p. 258. *In* Abstracts III Cong. All-Union Bot. Soc., Publ. St. Petersburg Univ., St. Petersburg, Russia.

Batygina, T. B. 1993c. Totipotency of ovule and seed cells in reproduction system of angiosperms, p. 362. *In* Abstracts XV Int. Bot. Cong., August 28-September 3, 1993. Yokohama, Japan.

Batygina, T. B. 1994a. Apomixis, agamospermy and viviparity and their role in flowering plant reproductive system, pp. 16-18. *In* Trans. Int. Symp. "Apomixis in Plants: State of the Problem and Perspectives of Researches". Publ. Saratov State Univ., Saratov, Russia.

Batygina, T. B. 1994b. The ecology role of vivipary and pseudovivipary in the adaptive evolution of reproductive systems in angiosperms, p. 311. *In* Proc. VI Int. Cong. Ecol., 21-26 August 1994, Manchester, UK.

Batygina, T. B. 1996. Parallel development of somatic and sexual embryos, p. 4. *In* Abstracts 14th Int. Cong. Sexual Plant Reproduction "Plant Reproduction '96", February 18-23, 1996, Lorne, Victoria, Australia. Cumberland Resort, Lorne, Victoria, Australia.

Batygina, T. B. 1998. Morphogenesis of somatic embryos developing in natural conditions. Biologija, 3: 61-64.

Batygina, T. B. 1999. Genetic heterogeneity of seeds. Acta Biol. Cracovensia, Series Bot., 41: 39-50.

Batygina, T. B. 2000. Embryoidogeny is a new type of vegetative propagation, pp. 334-350. *In* T. B. Batygina (ed.). Embryology of Flowering Plants. Terminology and Concepts, vol. 3. Reproductive Systems. World and Family Publ., St. Petersburg, Russia.

Batygina, T.B. 2002. Critical periods used for embryonal structures, 33. *In* J. Bednara (ed.). Abstracts XVII Int. Cong. Sexual Plant Reproduction "Sexual Plant Reproduction in Nature and the Laboratory". Maria Curie Sklodowska Univ. Press, Lublin, Poland.

***Batygina, T. B. and I. M. Kirillova. 1971.** Embryogenesis of bluegrass, pp. 15-16. *In* M. S. Yakovlev (ed.). Materials V All-Union Conf. on Plant Embryology. Shtiintsa, Kishinev, Moldova.

Batygina, T. B. and T. B. Mametyeva. 1979. On the embryology of the genus *Poa* L., pp. 89-95. *In* M. S. Yakovlev (ed.). Actual Problems of Embryology of Angiosperms. Nauka, Leningrad, Russia.

***Batygina, T. B. and V. E. Vasilyeva. 1979.** Behaviour of some angiosperm embryos in culture *in vitro*, pp. 125-130. *In* Abstracts III All-Union Conf. "Culture of Plant Cells". Publ. Armenian SSR Acad. Sci., Abovjan, Armenia.

***Batygina, T. B. and V. E. Vasilyeva. 1980.** System of reproduction in orchids, pp. 107-110. *In* Abstracts All-Union Symposium "Preservation and Cultivation of Orchids". Tallinn, Estonia.

*Batygina, T. B. and R. G. Butenko. 1981. Morphogenetic potencies of angiosperm embryo (on the example of genus *Paeonia* representatives, family Paeoniaceae). Bot. Journal (St. Petersburg), 66(11): 1531-1547.

Batygina, T. B. and I. I. Shamrov. 1981. Family Ceratophyllaceae, pp. 110-115. *In* M. S. Yakovlev (ed.). Comparative Embryology of Flowering Plants. Winteraceae—Juglandaceae. Nauka, Leningrad, Russia.

Batygina, T. B. and V. E. Vasilyeva. 1981. Experimental study of embryo differentiation in angiosperms. Acta Soc. Bot. Poloniae, 50(1-2): 257-264.

*Batygina, T. B. and V. E. Vasilyeva. 1983a. Development of embryo and seedling in some orchids, pp. 73-75. *In* Materials II All-Union Conf. "Preservation and Cultivation of Orchids", Kiev, Ukraine.

*Batygina, T. B. and V. E. Vasilyeva. 1983b. On the experience of system approach to the problem of embryo differentiation in angiosperms. Ontogenez, 14(3): 304-311.

Batygina, T. B. and V. E. Vasilyeva. 1983c. System of reproduction of Orchidaceae (on example of *Dactylorhiza maculata* (L.) Soó.), pp. 27-33. *In* O. Erdelská (ed.). Proc. VII Int. Cytoembryol. Symp."Fertilization and Embryogenesis in Ovulated Plants". High Tatra (Račkova dolina), 14–17 June 1982. Veda, Bratislava, Czechoslovakia.

*Batygina, T. B. and I. I. Shamrov. 1985. Comparative embryology of Nymphaeales and Nelumbonales orders and the problems of their systematics and phylogeny. Bot. Journal (St. Petersburg), 70(3): 368-373.

*Batygina, T. B. and G. G. Shevtsova. 1985. Metamorphosis in the ontogenesis of orchids (on the example of *Cymbidium hybridum,* Orchidaceae). Bot. Journal (St. Petersburg), 70(12): 1614-1621.

*Batygina, T. B. and V. E. Vasilyeva. 1987. Applied aspects of embryology. Autonomy of embryo and embryoculture of flowering plants. Bot. Journal (St. Petersburg), 72(2): 155-161.

Batygina, T. B. and E. V. Andronova. 1988. Is there a cotyledon in the Orchidaceae? Reports USSR Acad. Sci., 30(4): 1017-1019.

Batygina, T. B. and V. E. Vasilyeva. 1988. Some aspects of autonomy of embryo in flowering plants. Phytomorph., 38(4): 293-297.

Batygina, T. B. and E. V. Andronova. 1991. Is there a cotyledon in the Orchidaceae? An opinion, pp. 369-374. *In* C. P. Malik and J. P. Abrol (eds). Recent Advances in Plant Biology. Narendra Publ. House, Delhi, India.

Batygina, T. B. and V. E. Vasilyeva. 1997. Sexual reproduction of flowering plants: periodization of egg cell and zygote development and possible types of caryogamy, pp. 170-198. *In* B. Bhatia (ed.). Plant Form and Function. Angkor Publishers, New Delhi, India.

Batygina, T. B and V. E. Vasilyeva. 2001. In vivo fertilization, pp. 101-142. *In* S. S. Bhojwani and W. Y. Soh (eds). Current Trends in the Embryology of Angiosperms. Kluwer Acad Publ., Dordrecht, Netherlands.

Batygina, T. B and V. E Vasilyeva. 2003. Periodization in the development of flowering plant reproductive structure: critical periods. Acta Biologica Cracoviensia. Ser. Bot. 45(1): 27-36.

*Batygina, T. B., O. A. Dolgova and S. N. Korobova. 1961. Behaviour of pollen tubes at intra- and interspecific hybridization. Reports USSR Acad. Sci., 136(6): 1482-1485.

Batygina, T. B., T. E. Freiberg and V. E. Vasilyeva. 1978a. Development of excised embryos in *Paeonia* species, p. 17. *In* Abstracts Indo-Soviet Symp. "Embryology of Crop Plants," August 23-26, 1977, Leningrad. Komarov Bot. Inst. Leningrad, USSR.

***Batygina, T. B., V. E. Vasilyeva and T. B. Mametyeva. 1978b.** The problems of morphogenesis *in vivo* and *in vitro*. Embryogenesis in Angiospermae. Bot. Journal (St. Petersburg), 63(1): 87-111.

***Batygina, T. B., T. I. Kravtsova and I. I. Shamrov. 1980.** On the comparative embryology of the members of Nymphaeales and Nelumbonales orders. Bot. Journal (St. Petersburg), 65(8): 1071-1088.

***Batygina, T. B., I. I. Shamrov and G. E. Kolesova. 1982.** Embryology of Nymphaeales and Nelumbonales. II. The development of female embryonic structures. Bot. Journal (St. Petersburg), 67(9): 1179-1195.

***Batygina, T. B., G. E. Kolesova and V. E. Vasilyeva. 1983.** Embryology of Nymphaeales and Nelumbonales. III. Embryogenesis of *Nelumbo nucifera* Gaertn. Bot. Journal (St. Petersburg), 68(3): 311-326.

Batygina, T. B., G. E. Titova and I. I. Shamrov. 1991. Comparative embryology of Nymphaeaceae (Nymphaeales s. l.) in connection with the questions on their phylogeny and systematics, pp. 11-14. *In* Materials V Moscow Conf. Plant Phylogeny. Nauka, Moscow, Russia.

Batygina, T. B., G. E. Titova and V. B. Brukhin. 1992. Is the embryo in the mature seed of *Paeonia* sexual? pp. 4-5. *In* Abstracts XII Int. Cong. Sexual Plant Reproduction "Plant Reproductive Biology. Pollen, Ovules and Seeds". July 19-23, 1992. Columbus, OH, Ohio State Univ., Columbus, OH, USA.

Batygina, T. B., I. I. Shamrov and G. E. Titova. 1993. Somatic embryogenesis in cereals (comparative embryological approach), p. 564. *In* Abstracts XV Int. Bot. Cong., August 28-September 3, 1993. Yokohama, Japan.

Batygina, T. B., E. V. Andronova and L. A. Kartseva. 1994. Morphogenesis of the embryo and seedling of flowering plants (an example of orchid), pp. 1039-1040. *In* Proc. 13th Int. Cong. Electron Microscopy, July 17-22, 1994, Paris, France.Vol. 3B; Applications in Biological Science, Paris, France.

Batygina, T. B., E. A. Bragina and G. E. Titova. 1996. Morphogenesis of propagules in viviparous species *Bryophyllum daigremontianum* and *B. calycinum*. Acta Soc. Bot. Poloniae, 65(1-2): 127-133.

Batygina, T. B., E. V. Andronova, P. V. Kulikov and E. G. Philippov. 1988. Embryoculture as a way of repatriation and reintroduction of rare plant species (on the example of orchids), p. 231. *In* Abstracts Int. Conf. "Biology of Cultivated Cells and Biotechnology". Nauka, Novosibirsk, Russia.

Baude, E. 1956. Die Embryoentwicklung von *Stratioides aloides*. Planta, 46(6): 649-671.

Baus, A. D., L. Franzmann and D. W. Meinke. 1986. Growth in vitro of arrested embryos from lethal mutants of *Arabidopsis thaliana*. Theor. Appl. Genet., 72: 577-586.

Beccari, O. 1870. Nota sull'embrione delle Dioscoreaceae. Nuovo Giornale Bot. Ital., 2: 150.

Bechtel, D. B. and Y. Pomeranz. 1977. Ultrastructure of the mature ungerminated rice (*Oryza sativa*) caryopsis. The caryopsis coat and the aleurone cells. Amer. J. Bot., 64: 966-973.

Bechtel, D. B. and B. O. Juliano. 1980. Formation of protein bodies in the starchy endosperm of rice (*Oryza sativa* L.): a re-investigation. Ann. Bot. (London), 45: 503-509.

Beer, J. G. 1863. Beiträge zur Morphologie und Biologie der Familie Orchideen. Carl Gerold's Sohn, Wien.

*Bekker, A. M. and M. G. Nikolaeva. 1984. Change of lipolitic activity of apple tree seeds during stratification. Physiologia i biokhimya kulturnych rastenyi, 16: 523-528.

Bell, J. and G. Hicks. 1976. Transmitting tissue in the pistil of tobacco: light and electron microscopic observations. Planta, 131: 187-200.

Bell, P. R. 1992. Green plants. Their origin and diversity. Dioscorides Press, Portland, OR, USA.

***Belostokov, G. P. 1966.** Structure of seedling conductive system in foliar woody plants with hypogeal germination. Bot. Journal (St. Petersburg), 51(5): 705-716.

***Belozerova, L. S. and T. P. Plotnikova. 1984.** Some enzymes of anaerobic metabolism in germinating maize kernel. Vestnik Leningrad. Gosudarst. Univ. Biol., 9(2): 98-101.

***Belozerova, L. S., G. M. Dreval and E. V. Zhelezkova. 1986.** Malic enzyme in scutellum of germinating kernels in maize. Vestnik Leningrad. Gosudarst. Univ. Biologia, 11(1): 62-66.

***Belyaeva, N. S. 1977.** Embryo sac development and fertilization in angiosperms (on the example of Turkmen flora representatives). Ph.D. thesis, Ashkhabad, Turkmenistan.

***Belyaeva, N. S. 1980.** Interrelation of male gametes with female gametophyte cells in angiosperms. Hypothesis of double fertilization. Proc. Turkmen SSR Acad. Sci., Biol. Series, 5: 43-51.

Bendich, A. J. and L. P. Gauriloff. 1984. Morphometric analysis of cucurbit mitochondria: the relationship between chondriome volume and DNA content. Protoplasma, 119: 1-7.

Benfey, P. M., L. Ren and N.-H. Chua. 1990. Tissue-specific expression from CaMV 35S enhancer subdomains in early stages of plant development. EMBO Journal, 9: 1677-1684.

Benfey, P. M., P. J. Linstead, K. Roberts, J. W. Schiefelbein, M.-Th. Hauser and R. A. Aeschbacher. 1993. Root development in *Arabidopsis*: four mutants with dramatically altered root morphogenesis. Development, 119: 57-70.

Bennett, M. D. and J. B. Smith. 1976. The nuclear DNA content of the egg, the zygote and young proembryo cells in *Hordeum*. Caryologia, 29: 435-446.

Bentrup F.-W. 1984. Cellular polarity, pp. 473-490. *In* H. F. Linskens and J. Heslop-Harrison (eds). Cellular interactions. Berlin.

Berg, R. Y. 1958. Seed dispersal, morphology of *Trillium*. Skr. Narske vid.-akad, Oslo, 1: 1-36.

Berger, F., A. Taylor and C. Brownlee. 1994. Fate-determining positional information provided by the plant cell wall during embryogenesis. Science, 263(5152): 1421-1423.

Berleth, T. and G. Jürgens. 1993. The role of the monopteros gene in organizing the basal body region of the *Arabidopsis* embryo. Development, 118: 575-587.

Bernard, N. N. 1909. L'Evolution dans la symbiose les orchidées et leurs champignons commensaux. Ann. Sci. Natur., Botanique, 9: 1-196.

Bernhardi, J. J. 1832. Über die merkwurdegsten Verschiedenheiten des entwickelten Pflanzenembryo und ihren Wert für die Systematik. Linnaea, 7: 561-613.

Berridge, E. M. and E. Sanday. 1907. Oogenesis and embryogeny in *Ephedra distachya*. New Phytologist, 6: 128-134.

Bewley, J. D. and M. Black. 1978. Physiology and Biochemistry of Seeds in Relation to Germination, vol. 1. Development, Germination and Growth. Springer, Berlin.

Bewley, J. D. and M. Black. 1985. Seeds: Physiology of Development and Germination. Plenum Press, New York, NY, USA.

Bewley, J. D. and M. Black. 1994. Seeds: Physiology of Development and Germination. Plenum Press, New York, NY, USA.

Bexon, D. 1920. Observations on the anatomy of teratological seedlings. II. On the anatomy of some polycotylous seedlings of *Centranthus ruber*. Ann. Bot. (London), 34: 81-94.

Bexon, D. and A. E. Wood. 1930. Observations on the anatomy of the teratological seedlings. VII. The anatomy of some polycotylous seedlings of *Impatiens roylei* Walp. Ann. Bot. (London), 44(174): 297-309.

Bhalla, R. L., M. B. Singh and C. P. Malik. 1979. Physiology of sexual reproduction. VI. Embryogenesis in *Tropaeolum majus* L. Enzyme changes. Acta Bot. Indica, 7(1): 72-86.

Bhandari, N. N. and Anuradha Mal. 1989. Wall ingrowth or trabecula in the free nuclear embryo sac of *Scilla sibirica* Sato. Phytomorph., 39(1): 21-28.

Bhandari, N. N. and Sh. Asnani. 1968. Studies in the family Ranunculaceae. XI. Morphology and embryology of *Ceratocephalus falcatus* Pers. Beitrage zur Biologie der Pflanzen, 45(2): 271-290.

Bhandari, N. N. and K. Indira. 1969. Studies in the Viscaceae. IV. Embryology of Eubraehion (Hook. et Arn.) Engl. Bot. Notiser, 122(2): 183-203.

Bhandari, N. N. and A. Sachdeva. 1983. Some aspects of organization and histochemistry of the embryo sac of *Scilla sibirica* Sato. Protoplasma, 117: 170-178.

Bhandari, N. N. and Km. Sheel. 1992. Ultrastructural studies on endosperm haustoria in *Andrographis serpyllifolia*, pp. 76-77. *In* T. B. Batygina (ed.). Proc. of the XI Int. Symp. "Embryology and Seed Reproduction", 3-7 July, 1990, Leningrad, USSR. Nauka, St. Petersburg, Russia.

Bhatnagar, S. P. 1959. Some observations on the post-fertilization development of the embryo sac of *Santalum*. Phytomorph., 9(1): 87-91.

Bhatnagar, S. P. 1960. Morphological and embryological studies in the family Santalaceae. IV. *Mida salicifolia* A.Cunn. Phytomorph., 10(2): 198-207.

Bhatnagar, S. P. and M. Gupta. 1970. Morphology and embryology of *Mimusops elengi* L. Bot. Notiser, 123(4): 455-473.

Bhatnagar, S. P. and V. Sawhney. 1981. Endosperm—its morphology, ultrastructure, and histochemistry. Int. Rev. Cytology, 73: 55-102.

Bhattacharyya, M. K., A. M. Smith, T. H. N. Ellis, C. Hedley and C. Martin. 1990. The wrinkled-seed character of pea described by Mendel is caused by a transposon-like insertion in a gene encoding starch-branching enzyme. Cell (Cambridge), 60: 115-122.

Biradar, N. V. 1968. Studies on palms: embryology of *Phoenix pusilla* Gaertn., *P. acaulis* Buch. and *P. reclinata* Jacq. Proc. Indian Acad. Sci., 67(4): 165-173.

Birky, C. W. 1983. Relaxed cellular controls and organelle heredity. Science, 222: 468-475.

Biswas, P. K., A. Devi, P. K. Roy and K. B. Paul. 1978. Enzyme activity in dormant and nondormant large crabgrass (*Digitaria sanguinalis*) seeds following hydration. Weed Science, 26(1): 90-93.

Blake, S. T. 1972. *Idiospermum* (Idiospermaceae), a new genus and family for *Calycanthus australiensis*. Contribution from Queensland Herbarium, 12: 1-37.

Boblenz, K., T. Nothnagel, and M. Metzlaff. 1990. Paternal inheritance of plastids in the genus *Daucus*. Molec. Gen. Genet., 220: 489-491.

Bocquet, G. 1959. The campylotropous ovule. Phytomorph., 9(3): 222—227.

Boesewinkel, F. D. and F. Bouman. 1984. The seed: structure, pp. 567-610. *In* B. M. Johri (ed.). Embryology of Angiosperms. Springer-Verlag, Berlin.

Bogdanova, V. M. 1971. Conditions of seed germination of Far East Araliaceae and pathways of its acceleration. Ph.D. thesis, Leningrad, Russia.

Bohdanowicz, J. 1987. *Alisma* embryogenesis: the development and ultrastructure of the suspensor. Protoplasma, 137(23): 71-83.

Bosnes, M., E. Harris, L. Aigeltinger, and O. A. Olsen. 1987. Morphology and ultrastructure of 11 barley shrunken endosperm mutants. Theor. Appl. Genet., 74: 177-187.

Bower, F. O. 1898. The British Association. Bristol meeting. Section K (Botany). Opening address by Professor F. O. Bower, Sc.D., F.R.S., President of section II. Nature, 59: 88-91.

Bowman, J. L., D. R. Smyth and E. M. Meyerowitz. 1991. Genetic interactions among floral homeotic genes of *Arabidopsis*. Development, 112: 1-20.

Boyd, L. 1932. Monocotylous seedlings. Morphological studies in the post-seminal development of the embryo. Trans. Proc. Bot. Soc. Edinburgh, 31: 1-224.

Boyes, J. W. and E. Battaglia. 1951. The tetrasporic embryo sacs of *Plumbago coccinea, P. scandens* and *Ceratostigma willmottianum*. Bot. Gazette, 112(4): 485-489.

Brady, J. 1973. Feulgen cytophotometric determination of the DNA content of the embryo proper and suspensor cells of *Phaseolus coccineus*. Cell Differentiation, 2: 65-75.

Brady, T. and E. D. Walthall. 1985. The effect of the suspensor and gibberellic acid on *Phaseolus vulgaris* embryo protein content. Develop. Biol., 107: 531-536.

Brady, T. and S. H. Combs. 1988. The suspensor is a major route of nutrients into proembryo, globular and heart stage *Phaseolus vulgaris* embryos, pp. 419-424. *In* M. Cresti, P. Gori and E. Pacini (eds). Sexual Reproduction in Higher Plants. Springer-Verlag, Berlin.

Bragina, E. A., G. E. Titova and T. B. Batygina. 1995. The models of the formation of new sporophyte in genus *Bryophyllum* (Crassulaceae), 8. Abstracts VII Conf. Plant Embryologists of Czech Republic, Slovakia and Poland, 14-16 September. Lublin, Poland.

Brawley, S. H. and K. R. Robinson. 1985. Cytochalasin treatment distract the endogenous currents associated with cell polarization in fucoid zygotes: studies of the role of F-actin in embryogenesis. J. Cell Biol., 100(4): 1173-1184.

Brawley, S. H., D. F. Wetherell and K. R. Robinson. 1984. Electrical polarity in embryos of wild carrot precedes cotyledon differentiation. Proc. Natl. Acad. Sci., U.S.A, 81(19): 6064-6067.

Brenchley, W. E. 1909. On the strength and development of wheat, *Triticum vulgare*. Ann. Bot. (London), 23: 117-139.

Brenner, W. 1922. Zur Kenntnis der Blutenentwicklung einiger Juncaceen. Acta Soc. Scient. Fennica, 50(4): 1-37.

Breton, A. M. and Z. R. Sung. 1982. Temperature-sensitive carrot variants impaired in somatic embryogenesis. Develop. Biol., 90: 58-66.

Brink, R. A. and D. C. Cooper. 1947. The endosperm in seed development. Bot. Rev., (Lancaster), 13(8-9); 423-541.

Brongniart, A. 1830. Die Zeugung und Entwicklung des Embryo in den phanerogamischen Pflanzen, Bot., 4, s. 167-326. *In* R. Brown (ed.). Vermischte botanische Schriften. Nurnberg.

Brouland, M. 1967. A propos de la formation de l'albumen à partir du noyau secondaire non fécondé chez l'*Anemone nemorosa* L. Ann. Scient. l'Univ. de l'A.R.E.R.S, 5(4): 165-169.

Brouland, M. 1969. Recherches cytologiquc sur la fécondation chez quelques Ranunculacées: *Caltha palustris* L., *Anemone nemorosa* L., *Ranunculus acris* L., *Helleborus foetidus* L. Rev. Cytol. Biol. Végétales, 32(3-4): 309-317.

Brounov, P. I. 1897. The Significance of Agricultural-Meteorological Observations and a Brief Handbook for Their Performance. St. Petersburg, Russia.

Brown, R. 1810. On the natural order of plants called Proteaceae. Trans. Linnean Soc. London, 10: 15-226.

Brown, R. 1826. Sur la structure de l'ovule antérieurement a l'impregnation dans les plantes phanérogames et sur la fleur femelle des Cycadées et des Coniferées. Ann. Sci. Natur., Botanique, 7: 211-244.

Brown, R. C. and H. L. Mogensen. 1972. Late ovule and early embryo development in *Quercus gambelii*. Amer. J. Bot., 59(4): 311-316.

Brown, R. C., B. E. Lemmon and O.-A. Olsen. 1994. Endosperm development in barley: microtubule involvement in the morphogenetic pathway. Plant Cell, 6(9): 1241-1252.

Brown, W. 1959. The epiblast and coleoptile of the grass embryo. Bull. Torrey Bot. Club, 86(1); 13-16.

Brown, W. 1960. The morphology of grass embryo. Phytomorph., 10(3): 215-233.

Brown, W. 1965. The grass embryo—a rebuttal. Phytomorph., 15(3): 274-284.

Browse, J. and C. R. Slack. 1985. Fatty-acid synthesis in plastids from maturing sunflower and linseed cotyledons. Planta, 166(1): 74-80.

Brozek, A. 1926. Inheritance in the Monkey Flower. J. Heredity, 17: 113-129.

Bruck, D. K. and D. B. Walker. 1985. Cell determination during embryogenesis in *Citrus jambhiri*. I. Ontogeny of the epidermis. Bot. Gazette (Chicago), 146(2): 188-197.

Brucker, J. D., S. P. Cheng, and A. Markus. 1982. Protein synthesis and seed germination, pp. 387-396. *In* M. G. Nikolaeva and N. V. Obrucheva (eds). Physiology and Biochemistry of Seed Dormancy and Germination. Kolos, Moscow, Russia.

Brukhin, V. B. 1993. The development of peony embryo *in vivo* and *in vitro*. Ph.D. thesis, St. Petersburg, Russia.

Brukhin, V. B. and T. B. Batygina. 1994. Embryo culture and somatic embryogenesis in culture of *Paeonia anomala*. Phytomorph., 44(3-4): 151-157.

Bruns, E. 1892. Der Grasembryo. Flora, 76: 1-33.

Bruun, L. 1991a. Histological and semi-quantitative approaches to in vitro cellular responses of ovule, embryo and endosperm in sugar beet, *Beta vulgaris* L. Sexual Plant Reprod., 4: 64-72.

Bruun, L. 1991b. A statistical analysis of some genetical, physiological and anatomical parameters of the development of in situ and in vitro-grown ovules from intra- and interspecific crosses in the genus *Beta*. Sexual Plant Reprod., 4: 118-125.

Bruun, L. and P. Olesen. 1989. A structural investigation of the ovule in sugar beet, *Beta vulgaris*: the micropylar nucellus. Nordic J. Bot., 9: 81–97.

Buchholz, J. T. 1919. Studies concerning the evolutionary status of polycotyledony. Amer. J. Bot., 6: 106-119.

Buchholz, J. T. 1939. The morphology and embryogeny of *Sequoia gigantea*. Amer. J. Bot., 26: 248-257.

Bugnicourt, M. 1983. Embryogénie des *Hypericum* de France. Bull. Soc. Bot. France, 130: 195-205.

Bugnon, P. and A. Parrot. 1932. Sur la valeur morphologigue du cotyledon chez les Ombelliferes monocotyledones. C.R Hebdomadaires Séances l'Acad. Sci., 95: 332.

Burgeff, H. 1936. Samenkeimung der Orchideen und Entwicklung ihrer Keimpflanzen. G. Fischer, Jena.

Burr, B. and F. A. Burr. 1976. Zein synthesis in maize endosperm by polyribosomes attached to protein bodies. Proc. Natl. Acad. Sci., USA, 73: 515-519.

Butnik, A. A. 1991. Family Chenopodiaceae, pp. 77-82. *In* A. L. Takhtajan (ed.). Comparative Anatomy of Seeds, vol. 3. Dicotyledons. Caryophyllidae—Dilleniidae. Nauka, Leningrad, Russia.

Button, J., J. Kochba, and C. H. Borman. 1974. Fine structure of an embryoid development from embryogenic ovular callus of "Shamonti" orange (*Citrus sinensis* Osb.). J. Exper. Bot., 25: 446-457.

Buttrose, M. S. 1963. Ultrastructure of developing aleurone cells of the wheat grain. Australian J. Biol. Sci., 16: 768-774.

Byrne, J. M. and C. Heimsch. 1970. The root apex of *Malva sylvestris*: 2. The quiescent center. Amer. J. Bot., 57:1179-1184.

Campbell, D. H. 1897. A morphological study of *Najas* and *Zannichellia*. Proc. California Acad. Sci., 3(1): 1-67.

Campbell, D. H. 1898. The development of the flower and embryo in *Lilaea subulata* H.B.K. Ann. Bot. (London), 12:1-28.

Campbell, W. P., J. W. Lee, T. P. O'Brien and M. G. Smart. 1981. Endosperm morphology and protein body formation in developing wheat grain. Australian J. Plant Physiol., 8: 5-19.

Campenot, M. K., G. Zhang, A. J. Cutler and D. D. Cass. 1992. *Zea mays* embryo sacs in culture. I. Plant regeneration from 1 day after pollination embryos. Amer. J. Bot., 79: 1368-1373.

Carano, E. 1913. Alcune osservazioni sull'embriogenesi delle Asteraceae. Ann. Bot. (Rome), 11(2): 313-315.

Carano, E. 1915. Ricerche sull'embriogenesi delle Asteraceae. Ann. Bot. (Rome), 13(2): 251-301.

Carlson, W. R. 1969. Factors affecting preferential fertilization in maize. Genetics, 62: 543-554.

Carmichael, J. S. and W. E. Friedman. 1995. Double fertilization in *Gnetum gnemon*: the relationships between the cell cycle and sexual reproduction. Plant Cell, 7(12): 1975-1988.

Carmichael, J. S. and W. E. Friedman. 1996. Double fertilization in *Gnetum gnemon* (Gnetaceae): its bearing on the evolution of sexual reproduction within the Gnetales and the anthophyte clade. Amer. J. Bot., 83(6): 767-780.

Casadoro, G. and N. Rascio. 1987. Cotyledonal chloroplasts in the hypogeal seeds of clementine. Planta, 170: 300-307.

Casadoro, G., P. M. Colombo, and N. Rascio. 1980. Plastids in the quiescent embryo and young seedlings of the chloroembryophyte *Citrus nobilis* × *Citrus aurantium amara pumila*. Ann. Bot. (London), 45(4): 415-418.

Cass, D. D. 1972a. Occurrence and development of a filiform apparatus in the eggs of *Plumbago capensis*. Amer. J. Bot., 59: 279-283.

Cass, D. D. 1972b. Ultrastructure of the egg of *Plumbago zeylanica*. Amer. J. Bot., 59: 648.

Cass, D. D. 1973. An ultrastructural and Nomarski-interference study of the sperms of barley. Canadian J. Bot., 51(3): 601-605.

Cass, D. D. 1981. Structural relationships among central cell and egg apparatus cells of barley as related to transmission of male gametes. Acta Soc. Bot. Poloniae, 50(1/2): 177-180.

Cass, D. D. and W. A. Jensen. 1970. Fertilization in barley. Amer. J. Bot., 57(1): 62-70.

Cass, D.D. and J. Karas. 1974. Ultrastructural organization of the egg of *Plumbago zeylanica*. Protoplasma, 81: 49-62.

Cass, D. D. and J. Karas. 1975. Development of sperm cells in barley. Canadian J. Bot., 53(10): 1051-1162.

Castle, L. A. and D. W. Meinke. 1993. Embryo-defective mutants as tools to study essential functions and regulatory processes in plant embryo development. Seminars Developm. Biol., 4: 31-39.

Cavara, F. 1899. Studi sul de richerche intorno allo svilluppo del frutto della "Thea chinensis" sims., cultivata nel orto botanico di Pavia. Atti Inst. Bot. Univ. Pavia, Ser. 2, 5: 265-326.

Chaban, I. A. 1975. Embryogenesis and endospermogenesis of *Reseda lutea* L. and *Chelidonium majus* L. according to light and electron microscopy. Ph.D. thesis, Leningrad, Russia.

Chaban, I. A. 1979. Electron microscopy investigation of endosperm in *Reseda lutea* L., pp. 58-69. *In* M. S. Yakovlev (ed.). Actual Questions on Embryology of Angiosperms. Nauka, Leningrad, Russia.

***Chaban, I. A. and M. S. Yakovlev. 1974.** Embryology of *Reseda lutea* L. I. Megasporogenesis and embryo sac development. Bot. Journal (St. Petersburg), 59(1): 24-37.

Chaboud, A. and R. Perez. 1992. Generative cells and male gametes: isolation, physiology and biochemistry. Int. Rev. Cytol., 140: 205-232.

Chaffey, N. J. and N. Harris. 1985. Localization of ATPase activity of scutellar epithelial cells of germinating barley (*Hordeum vulgare* L.). J. Exper. Bot., 36(171): 1612-1619.

Chailakhyan, M. Kh., R. G. Butenko, O. N. Kulaeva, V. I. Kefeli and N. I. Axenova. 1982. Terminology of Growth and Development of Higher Plants. Nauka, Moscow, Russia.

Chamberlain, C. J. 1912. Morphology of *Ceratozamia*. Bot. Gazette (Chicago), 53(1): 1-19.

Chamberlain, C. J. 1935. Gymnosperms, Structure and Evolution. Univ. Chicago Press, Chicago, IL, USA.

Chamberlin, M. A. and H. T. Horner. 1989. A study of endosperm development in *Glycine max*. Amer. J. Bot., 76(6), suppl.: 27.

Chamberlin, M. A., H. T. Horner and G. R. Palmer. 1993. Nutrition of ovule, embryo sac and young embryo in soybean: an anatomical and autoradiographic study. Canadian J. Bot., 71: 1153-1168.

Chandlee, J. M. 1991. Analysis of developmentally interesting genes cloned from higher plants by insertional mutagenesis. Develop. Genet., 12: 261-271.

Charlesworth, D. and B. Charlesworth. 1979. The evolution and breakdown of S-allele systems. Heredity, 43: 41-55.

Charzynska, M. 1988. Some aspects of development and division of generative cell in tricellular type pollen grains. Ann. Scient. l'Univ. Reims et l'ARERS, 23: 81-84.

Charzynska, M. and N. Lenart. 1989. Pollen grain of barley (*Hordeum vulgare* L.)—pattern of development. Acta Soc. Bot. Poloniae, 58: 313-320.

Charzynska, M. and F. Lewandowska. 1990. Generative cell division and sperm cell association in the pollen grain of *Sambucus nigra*. Ann. Bot. (London), 65: 685-689.

Charzynska, M., F. Ciampolini, and M. Cresti. 1988. Generative cell division and sperm cell formation in barley. Sexual Plant Reprod., 1: 240-247.

Charzynska, M., M. Murgia, C. Milanesi and M. Cresti. 1989. Origin of sperm cell association in the "male germ unit" of *Brassica* pollen. Protoplasma, 149: 1-4.

Chase, M. W., D. E. Soltis, R. G. Olmstead, D. Morgan, D. H. Les, B. D. Mishler, M. R. Duvall, R. A. Price, H. G. Hills, Y. L. Qui, K. A. Kron, H. H. Rettig, E. Conti, J. and D. Palmer, J. R. Manhart, K. J. Sytsma, H. J. Michaels, W. J. Kress, K. G. Karol, W. D. Clark, M. Hedren, B. S. Gaut, R. K. Jansen, K. J. Kim, C. F. Wimpee, J. F. Smith, G. R. Furnier, S. H. Strauss, Q. Y. Xiang, G. M. Plunkett, P. S. Soltis, S. Swensen, S. E. Williams, P. A. Gadek, C. J. Quinn, L. E. Eguiarte, E. Colenberg, G. H. Learn, S. W. Graham, S. C. H. Barret, S. Dayanandan and V. A. Albert. 1993. Phylogenetics of seed plants: an analysis of nucleotide sequences from the plastid gene rbcL. Ann. Missouri Bot. Garden, 80: 528-580.

Chaubal, R. and B. J. Reger. 1990. Relatively higher calcium is localized in synergid cells of wheat. Sexual Plant Reprod., 3: 98-102.

Chaubal, R. and B. J. Reger. 1992. Calcium in the synergid cells and other regions of pearl millet ovaries. Sexual Plant Reprod., 5: 34–46.

Chaubal, R. and B. J. Reger. 1995. Localization of calcium in young synergid cells of pearl millet by antimonate: indications of calcium influx, programmed cell death, and calcium release. Sexual Plant Reprod., 8.

Chaudhury, A. M., S. Letharn, S. Craig and E. S. Dennis. 1993. Amp1—a mutant with high cytokinin levels and altered embryonic pattern, faster vegetative growth, constitutive photomorphogenesis and precocious flowering. Plant Journal, 4(6): 907-916.

Chaundhari, H. K. 1975. The production of haploids by semigamy and their potential utilization in breeding cotton. Ph.D. Abstr. Int.B., 1534-B.

Chaundhari, H. K. 1978. Use of semigamy in the production of cotton haploids. Bull. Torrey Bot. Club, 105(2): 98-103.

***Chebotaru, A. A. 1960.** The influence of some environmental factors on the fertilization and embryogenesis processes in maize, pp. 49-74. *In* Trans. 1st Scientific Conf. Moldavian Young Scientists, Kishinev, Moldova.

***Chebotaru, A. A. 1961.** The influence of some environmental factors on pollination, fertilization and early embryogenesis processes in maize. PhD thesis, Kiev, Ukraine.

***Chebotaru, A. A. 1970.** Cytoembryological and electron microscopy investigation of maize (*Zea mays* L.). Ph.D thesis. Central Committee Moldavian Communist Party Press, Kishinev, Moldova.

***Chebotaru, A. A. 1972.** Embryology of Maize. Shtiintsa, Kishinev, Moldova.

***Chebotaru, A. A. 1987.** Cytological aspects of conception on zygote neoplasm in algae, mosses, ferns and seed plants, pp. 3-14. *In* Embryology and Anatomy of Reproductive and Vegetative Organs of Certain Lower and Higher Plants. Shtiintsa, Kishinev, Moldova.

Cheng, Z., H. Shiyi, X. Liyun, L. Xiuru and S. Jiaheng. 1980. Ultrastructure of sperm cell in mature pollen grain of wheat. Scientia Sinica, 23(3): 371-379.

Chiarugi, A. 1925. Embryologia della Cistaceae. Nuovo Giornale Bot. Ital., N. S. 32: 223-314.

Chichiriccó, G. 1989. Embryology of *Crocus thomasii* (Iridaceae). Plant System. Evol., 168(1-2): 39-47.

Chikkannaiah, P. S. 1962. Morphological and embryological studies in the Commelinaceae, pp. 23-36. *In* "Plant Embryology". A Symp. Council Scientific and Industrial Research, November 11-14, 1960. Dept. Bot., Univ. Delhi, New Delhi, India.

Chikkannaiah, P. S. and B. Hemaraddi. 1979. A note on the megasporogenesis and endosperm of some Commelinaceae. Current Sci., 48(8): 363.

Chitralekha, P. and N. N. Bhandari. 1992. Cellularization of endosperm in *Asphodelus tenuifolius* Cav. Phytomorph., 42(3-4): 185-193.

Chowdhury, J. K. and J. N. Mitra. 1953. Abnormal tricotyledonous embryo and the morphological structure of normal fruit and seed of *Cinnamomum camphora*. Sci. Cult., 19: 159-160.

Christianson, M. L. 1986. Fate map of the organizing shoot apex in *Gossypium*. Amer. J. Bot., 73: 947-958.

Chubirko, M. M. and L. N. Kostrikova. 1985. Family Fabaceae, pp. 67-77. *In* M. S. Yakovlev (ed.). Comparative Embryology of Flowering Plants. Brunelliaceae—Tremandraceae. Nauka, Leningrad, Russia.

Clark, J. K. and W. F. Sheridan. 1986. Developmental profiles of the maize embryo-lethal mutants dek22 and dek23. J. Heredity, 77: 83-92.

Clark, J. K. and W. F. Sheridan. 1988. Characterization of the two maize embryo-lethal defective kernel mutants rgh*-1210 and fl*-1253B: effects on embryo and gametophyte development. Genetics, 120: 279-290.

Clark, J. K. and W. F. Sheridan. 1991. Isolation and characterization of 51 embryo-specific mutants of maize. Plant Cell, 3: 935-951.

Clark, S. E., M. P. Running, and E. M. Meyerowitz. 1993. CLAVATA I, a regulator of meristem and flower development in *Arabidopsis*. Development, 119: 397-418.

Clauhs, R. P. and P. Grun. 1977. Changes in plastid and mitochondrial content during maturation of generative cells of *Solanum* (Solanaceae). Amer. J. Bot., 64: 377-383.

Clowes, F. A. L. 1961. Apical Meristems. Blackwell, Oxford, UK.

Clowes, F. A. L. 1978a. Origin of quiescence at the root pole of pea embryos. Ann. Bot., (London), 42: 1237-1239.

Clowes, F. A. L. 1978b. Origin of the quiescent centre in *Zea mays*. New Phytologist, 80: 409-419.

Cochrane, M. P. 1983. Morphology of the crease region in relation to assimilate uptake and water loss during caryopsis development in barley and wheat. Australian J. Plant Physiol., 10: 473-491.

Cocucci, A. E. 1981. Aspectos ultrastructurales de la fertilization en angiosperms. Kurtziana, 14: 41-62.

Cocucci, A. E. and W. A. Jensen. 1969a. Orchid embryology: megagametophyte of *Epidendrum scutella* following fertilization. Amer. J. Bot., 56(6): 629-640.

Cocucci, A. E., and W. A. Jensen. 1969b. Orchid embryology: pollen tetrads in *Epidendrum scutella* in the anther and on the stigma. Planta, 84: 215-229.

Coe, C. E. 1953. Cytology of reproduction in *Cooperia pedunculata*. Amer. J. Bot., 40(5): 335-343.

Coe, E. H. Jr., M. G. Neuffer and D. A. Hoisington. 1988. The genetics of corn, pp. 81-258. *In* G. F. Sprague and J. W. Dudley (eds). Corn and Corn Improvement. Amer. Soc. Agronomy, Madison, WI, USA.

Coen, E. S. 1991. The role of homeotic genes in flower development and evolution. Ann. Rev. Plant Physiol., Plant Molec. Biol., 42: 241-279.

Collins, D. M. and A. T. Wilson. 1972. Metabolism of the axis and cotyledons of *Phaseolus vulgaris* seeds during early germination. Phytochem., 11(6): 1931-1935.

Compton, R. H. 1913. An anatomical study of syncotyly and schizocotyly. Ann. Bot. (London), 27(108): 793-821.

Cook, M. T. 1902. Development of the embryo sac and embryo of *Castalia odorata* and *Nymphaea advena*. Bull.Torrey Bot. Club, 29: 211-220.

Cook, M. T. 1903. The development of the embryo sac and embryo of *Claytonia virginica*. Ohio Naturalist, 3(3): 349-353.

Cook, M. T. 1906. The embryology of some Cuban Nymphaeaceae. Bot. Gazette, 42(5): 376-392.

Cook, M. T. 1909. Notes on the embryology of the Nymphaeaceae. Bot. Gazette, 48: 56-59.

Cooke, E. A. and A. F. Shively. 1941. Observations on the structure and development of *Epiphagus virginiana*. Trans. Proc. Bot. Soc. Pennsylvannia, 2: 352-398.

Cooke, M. C. [1873?]. A Manual of Botanic Terms. W. H. Allen a. Co, London, UK.

Cooper, D. C. 1940. Macrosporogenesis and embryology of *Portulaca oleracea*. Amer. J. Bot., 27(5): 326-330.

Cope, F. W. 1962. The mechanism of pollen incompatibility in *Theobroma cacao*. Heredity, 17: 157-182.

Copeland, H. F. 1939. The structure of *Monotropis* and the classification of the Monotropoideae. Madroño, 5(4): 105-119.

Copeland, H. F. 1941. Further studies on Monotropoideae. Madroño, 6(4); 97-119.

Corner, E. J. H. 1976. The Seeds of Dicotyledons, vols. 1, 2. Cambridge Univ. Press, Cambridge, UK.

Cornish, E. C., J. M Pettitt, I. Bonig and A. E. Clarke. 1987. Developmentally-controlled tissue-specific expression of a gene associated with self-incompatibility in *Nicotiana alata*. Nature, 326: 99-102.

Correns, C. 1899. Untersuchungen über die Xenien bei *Zea mays*. Berichte Deutschen Bot. Gesellschaft. S. 410.

Corriveau, J. L. and A. W. Coleman. 1988. Rapid screening method to detect potential biparental inheritance of plastid DNA and results for over 200 angiosperm species. Amer. J. Bot., 75: 1443-1458.

Cosgrove, D. J. and Li Zhen-Chang. 1993. Role of expansins in cell enlargement of oat coleoptiles. Analysis of developmental gradient and photocontrol. Plant Physiol., 103(4): 1321-1328.

Coulter, J. M. 1915. The origin of monocotyledony, II. Monocotyledony in grasses. Ann. Missouri Bot. Garden, 2(1-2): 175-183.

Coulter, J. M. and C. J. Chamberlain. 1903. Morphology of Angiosperms. Appleton & Co., New York, NY, USA.

Coulter, J. M. and C. J. Chamberlain. 1910. Morphology of Gymnosperms. Univ. Chicago Press, Chicago, IL, USA.

Coulter, J. M. and W. J. G. Land. 1914. The origin of monocotyledony. Bot. Gazette, 57: 509-519.

Cow, G. V. 1928. Morphology of sassafras in relation to phylogeny of angiosperms. Bot. Gazette, 86: 149-171.

Crane, P. R. 1985a. Phylogenetic analysis of seed plants and the origin of angiosperms. Ann. Missouri Bot. Garden, 72: 716-793.

Crane, P. R. 1985b. Phylogenetic relationships in seed plants. Cladistics, 1: 329-348.

Crane, P. R. 1988. Major clades and relationships in the "higher" gymnosperms, pp. 218-272. *In* C. B. Beck (ed.). Origin and Evolution of Gymnosperms. Columbia Univ. Press, New York, NY, USA.

Cresti, M., P. Gori and E. Pacini. 1988. Sexual Reproduction in Higher Plants. Springer-Verlag, Berlin.

Cresti, M., M. Murgia and C. H. Theunis. 1990. Microtubule organization in sperm cells in the pollen tubes of *Brassica oleracea* L. Protoplasma, 154(2/3): 151–156.

Cresti, M., E. Pacini, F. Ciampolini and G. Sarfatti. 1977. Germination and early tube development in vitro of *Lycopersicum peruvianum* pollen: ultrastructural features. Planta, 136: 239–247.

Cresti, M., J. L. Van Went, E. Pacini, and M. T. M. Willemse. 1976. Ultrastructure of transmitting tissue of *Lycopersicon peruvianum* style: development and histochemistry. Planta, 132: 305-312.

Crété P. 1949. Recherches embryologiques sur les Gesnériacées. Développement de l'albumen et de l'embryon chez le *Chirita lavandulacea* Stapf. Bull. Soc. Bot. France, 96(7-9): 234-235.

Crété, P. 1951a. Embryogénie des Capparidacées. Développement de l'embryon chez le *Cleome graveolens* Rafin. C.R. Hebdomadaires Séances l'Acad. Sci., 233(9): 562-564.

Crété, P. 1951b. Répartition et intérêt phylogénétigue des albumens. Formations haustoriales chez les Angiospermes et plus particulièrement chez les gamopétales. Ann. Sci. Naturelles. Bot. Biol. Végétale, 12: 131-191.

Crété, P. 1952. Embryogénie des Datiscacées. Développement de l'embryon chez le *Datisca cannabina* L. C.R. Hebdomadaires Séances l'Acad. Sci., 234(10): 1082-1084.

Crété, P. 1953. Embryologie des Scrophulariacées. Développement de l'albumen et de l'embryon chez le *Digitalis purpurea* L. Phytomorph., 3(3): 168-172.

Crété, P. 1954. Embryologie de l'*Erinus alpinus* L. (Scrophulariacées). Les relations entre les genres *Erinus* L. et *Digitalis* L. Phytomorph., 4(3-4): 325.

Crété, P. 1955a. L'application de certains donées embryologiques a la systématique des Orobanchacées et de quelques familles voisines. Phytomorph., 5(4): 422-435.

Crété, P. 1955b. L'origine du sac embryonnaire et le développement de l'albumen chez l'*Allopletus songuineus* Mart. (Gesnériacées). Bull. Soc. Bot. France, 102(5-6): 205-207.

Crété, P. 1963. Embryo, pp. 171-220. *In* P. Maheshwari (ed.). Recent Advances in the Embryology of Angiosperms. Univ. Delhi, Delhi, India.

Crété, P. 1964. L'embryogénie et son rôle dans les essais de classification phylogénétique. Phytomorph., 14: 70-78.

Crocker, W. 1916. Mechanics of dormancy in seeds. Amer. J. Bot., 3: 99-120.

Cronquist, A. 1965. The status of the general system of classification of flowering plants. Ann. Missouri Bot. Garden, 52(3): 281-303.

Cronquist, A. 1988. The Evolution and Classification of Flowering Plants. New York Botanical Garden, Bronx, NY.

Crowe, L. K. 1964. The evolution of outbreeding in plants, I. The angiosperms. Heredity, 19(3): 435-457.

Dahlgren, K.V.O. 1916. Zytologische and embryologische Studien über die Reichen Primulales und Plumbaginales. Kongliga Svenska Vetenskapsakademiens Handlingar. 56:1-80.

Dahlgren, K. V. O. 1934. Die Embryologie von *Impatiens raylli*. Svensk Bot. Tidskrift, 28(1): 103-125.

Dahlgren, R. 1980a. A revised system of classification of the angiosperms. Bot. J. Linnean Soc., 80(2): 91-124.

Dahlgren, R. 1980b. The taxonomic significance of chlorophyllous embryos in angiosperm seeds. Bot. Notiser, 133: 337-341.

Dahlgren, R. M. T. and H. T. Clifford. 1982. The Monocotyledons: A Comparative Study. Academic Press, London, UK.

Dahlgren, R. M. T., H. T. Clifford and P. F. Yeo. 1985. The families of the Monocotyledons. Structure, Evolution and Taxonomy. Springer-Verlag, Berlin.

D'Alascio-Deschamps, R. 1973a. Modification ultrastructurales liées à la fécondation chez le lin. Bull. Soc. Bot. France, 120(5-6): 137-146.

D'Alascio-Deschamps, R. 1973b. Organisation du sac embryonnaire du *Linum catharticum* L., espèce récoltèe en station naturelle; ètude ultrastructurale. Bull. Soc. Bot. France, 120(5-6): 189-195.

Dale, J. E. 1982. The Growth of Leaves. Arnold, London.

Dalziel, J. M. 1937. The Useful Plants of West Tropical Africa. Crown agents for the colonies, London, UK.

D'Amato, F. 1984. Role of polyploidy in reproductive organs and tissues, pp. 519-566. *In* B. M. Johri (ed.). Embryology of Angiosperms. Springer-Verlag, Berlin.

D'Amato, F., M. Devreux and G. T. Scarascia Mugnozza. 1965. The DNA content of the nuclei of the pollen grains in tobacco and barley. Caryologia, 18: 377-381.

Dangeard, P.-A. 1892. Les maladies du pommier et du porier. Libraire Blanchier, Poitiers — Paris, France, pp. 1-64.

***Danilova, M. F. 1963.** On the principles of conductive apparatus structure in dicotyledon seedlings. Bot. Journal (St. Petersburg), 48(9): 1249-1270.

***Danilova, M. F. and T. B. Sokolovskaya. 1971.** Anatomy of cotyledonary node in cereal embryo, pp. 47-48. *In* Materials V All-Union Conference on Plant Embryology. Shtiintsa, Kishinev, Moldova.

***Danilova, M. F. and E. P. Zhogoleva. 1971.** Anatomy of seedling in *Onobrychis echidna* Lipsky. Bot. Journal (St. Petersburg), 56(9): 1307-1317.

***Danilova, M. F. and T. B. Sokolovskaya. 1973.** Seedling anatomy of some cereal species and the question on monocotyledon origin. Bot. Journal (St. Petersburg), 58(3): 337-349.

Danilova, M. F. and M. E. Kirpichnikov. 1985. The glossary of terms, pp. 8-31. *In* A. L. Takhtajan [ed.]. Comparative Anatomy of Seeds, vol. 1. Monocotyledons. Nauka, Leningrad, Russia.

Danilova, M. F. and M. I. Savchenko. 1985. Family Arecaceae, pp. 236-253. *In* A. L. Takhtajan (ed.). Comparative Anatomy of Seeds, vol. 1. Monocotyledons. Nauka, Leningrad, Russia.

***Danilova, M. F., E. N. Nemirowich-Danchenko, G. A. Komar and M. M. Lodkina. 1990.** Some trends of structural evolution of monocot seeds. Bot. Journal (St. Petersburg), 75(6): 755-773.

Darwin, Ch. 1941. The ability of plants to move, pp. 153-317. *In* Transactions, 8. USSR Acad. Sci., Moscow-Leningrad, Russia.

Dashek, W. V. and W. G. Rosen. 1966. Electron microscopical localisation of chemical components in the grown zone of lily pollen tubes. Protoplasma, 61: 192–203.

Dashek, W. V., H. R. Thomas and W. G. Rosen. 1971. Secretory cells of lily pistils. II. Electron microscope cytochemistry of canal cells. Amer. J. Bot., 58(10): 909-920.

Datta, R. M. 1941. Tricotyledony in *Morus multicaulis* Perr. Current Sci., 10: 440.

Datta, R. M. 1954. Tricotyledony in *Corchorus capsularis* Linn. Sci. and Cult., 20: 240-241.

Davey, J. E. 1978. The white lupin suspensor. Proc. Electron Microscopy Soc. South Africa, 8: 113-114.

Davidson, E. N. 1986. Gene Activity in Early Development. Academic Press, New York, NY, USA.

Davidson, E. N. 1994. Molecular biology of embryonic development: how far have we come in the last ten years? Bioessays, 16(9): 603-615.

Davis, G. L. 1966. Systematic Embryology of Angiosperms. John Wiley & Sons, Inc., New York, NY, USA.

***Davoyan, E. I. and A. P. Smetanin. 1979.** Obtaining of callus and regeneration of rice plants. Physiologia rastenyi, 26(2): 323-328.

Day, A. and T. H. N. Ellis. 1984. Chloroplast DNA deletions associated with wheat plants regenerated from pollen: possible basis for maternal inheritance of chloroplasts. Cell (Cambridge), 39: 359-368.

De Greef, J. A. and H. Frederico. 1983. Photomorphogenesis and hormones, pp. 401–427. *In* A. Pirson and M. H. Zimmermann (eds). Springer, Berlin. Encyclopedia of Plant Physiology, New Ser., 16A: Photomorphogenesis.

De Guzman, C. C. and R. K. De La Fuente. 1984. Polar calcium flux in sunflower hypocotyl segments. I. The effect of auxin. Plant Physiol., 76(2): 347-352.

De Guzman, C. C. and R. K. De La Fuente. 1986. Polar calcium flux in sunflower hypocotyl segments, 2. The effect of segment orientation, growth and respiration. Plant Physiol., 81(2): 408-412.

De Jong, A. J., E. D. L. Schmidt and S. C. De Vries. 1993. Early events in higher plant embryogenesis. Plant Molec. Biol., 22: 367-377.

De Jong, A. J., J. Cordewener, F. LoSchiavo, M. Terzi, J. Vandekerckhove, A. Van Kammen and S. C. De Vries. 1992. A carrot somatic embryo mutant is rescued by chitinase. Plant Cell, 4: 425-433.

De Jussieu, A. L. 1789. Genera Plantarum secundum ordines naturales disposita. Herrisant, Paris, France.

De Nettancourt, D. 1977. Incompatibility in Angiosperms. Springer-Verlag, Berlin.

De Nettancourt, D., M. Devreux, A. Bozzini, M. Cresti, E. Pacini and G. Sarfatti. 1973. Ultrastructural aspects of self-incompatibility mechanism in *Lycopersicon peruvianum*. J. Cell Sci.,12: 403-419.

De Vries, H. 1895. Over de Erfelijkheid van Synfisen. Bot. Jaarb., 7: 129-197.

De Vries, H. 1899. Sur la culture des monstruosites. Gauthier-Villars, Paris, 4 pp.

De Vries, H. 1902. Über tricotyle Rassen. Berichte Deutschen Bot. Gesellschaft, 20(2): 45.

De Vries, H. 1903. Die Mutationstheorie, vol. 2. Verlag von Veit & Comp., Leipzig.

Deb, D. B. 1955. The phenomenon of tri- and schizocotyly in *Nicotiana tabacum* L. Sci. Cult., 21(6): 329-330.

***Delone, N. A. and A. N. Danilina. 1963.** Cytochemical characteristics of interphasic nuclei of cells with different physiological functions. Reports USSR Acad. Sci., 151(5): 1195-1198.

*Denisova, G. A. 1961.** On the peculiarities of the development of embryos in the seeds of some Apiaceae. Reports USSR Acad. Sci., 139(4): 999-1000.

Derfling, K. 1985. Hormones of Plants. System Approach. Mir, Moscow, Russia.

Derksen, J., T. Rutten, T. Van Amstel, A. De Win, F. Doris and M. Steer. 1995. Regulation of pollen tube growth. Acta Bot. Neerlandica, 44: 93-119.

Deschamps, R. 1968. L'embryon du lin, *Linum usitatissimum* L., étude au microspore electronique. Rev. Cytol. Biol. Végétales, 31: 357-360.

Deschamps, R. 1969. Premier stades du développément de l'embryon et de l'albumen du lin: étude au microscope électronique. Rev. Cytol. Biol. Végétales, 32(3/4): 379-390.

Devreux, M. and G. T. Scarascia Mignozza. 1964. Effects of gamma radiation on the gametes, zygote and proembryo in *Nicotiana tabacum* L. Radiation Bot., 4: 373-386.

Diboll, A. G. 1964. Electron microscopy of female gametophyte and early embryo development in *Zea mays*. Ph.D. thesis, University Texas, Austin, TX, USA.

Diboll, A. G. 1968. Fine structural development of the megagametophyte of *Zea mays* following fertilization. Amer. J. Bot., 55(7): 797-806.

Dickinson, H. G. 1986. Organelle selection during flowering plant gametogenesis, pp. 37-60. *In* S. H. Mantell, G. P. Chapman and P. F. S. Street (eds). The Chondriome—Chloroplast and Mitochondrial Genomes. Longman, London, UK.

Dickinson, H. G. 1990. Self-incompatibility in flowering plants. Bioessays, 12(4): 155-161.

Dickinson, H. G. and D. Lewis. 1973. Cytochemical and ultrastructural differences between intraspecific compatible and incompatible pollinations in *Raphanus*. Proc. Royal Soc. London. Series B. Biol. Sci., 183: 21-38.

Diels, L. 1910. Menispermaceae. *In* A. Engler (ed.). Das Pflanzenreich, vol. 4. Engelmann, Leipzig.

Di Fulvio, T. E. 1981. La embriologia en la sistematica de angiospermas. Kurtziana, 14: 21-39.

Di Fulvio, T. E. 1983. Los "tipos" de endosperma y de haustorios endospermicos. Su clasificacion. Kurtziana, 16(1-4): 7-31.

Di Fulvio, T. E. 1985. El sistema EODP en el ordenamiento de Tubiflorae y en la endospermogenesis nuclear. An. Acad. Nac. Cienc. Exactas, Físicas Naturales Buenos Aires, 37(1-4): 111-119.

Di Fulvio, T. E. and A. E. Cocucci. 1986. La endospermogenesis nuclear y el sistema EODP. Kurtziana, 18: 13-21.

Dixit S. N. 1958. Morphological and embryological studies in the family Loranthaceae. IV. *Amyema* Van Tiegh. Phytomorph., 8(3-4): 346-364.

Dnyansagar, V. R. and D. C. Cooper. 1960. Development of the seed of *Solanum phureja*. Amer. J. Bot., 47: 176-186.

Dogel, V. A. 1954. Oligomerization of Homologous Organs. Leningrad State Univ. Press, Leningrad, Russia.

Dolan, L., K. Janmaat, V. Willemsen, P. Linstead, S. Poethig, K. Roberts and B. Scheres. 1993. Cellular organization of the *Arabidopsis thaliana* root. Development, 119: 71-84.

Dolan, L., C. M. Duckett, C. Grierson, P. Linstead, K. Schneider, E. Lawson, C. Dean, S. Poethig and K. Roberts. 1994. Clonal origin and patterning in the root epidermals of *Arabidopsis*. Development, 120: 2465-2474.

Dolfini, S. F. and F. Sparvoli. 1988. Cytological characterization of the embryo-lethal mutant dek-1 of maize. Protoplasma, 144: 142-148.

Doll, W. 1927. Beiträge zur Dipsaceen und Dipsaceen-änlicher Pflanzen. Bot. Archiv., 17: 107-146.

Donoghue, M. J. 1989. Phylogenies and the analysis of evolutionary sequences, with examples from seed plants. Evolution, 43: 1137-1156.

Donoghue, M. J. and J. A. Doyle. 1989. Phylogenetic analysis of angiosperms and the relationships of Hamamelideae, pp. 17-45. *In* P. R. Crane and S. Blackmore (eds). Evolution, Systematics and Fossil History of the Hamamelideae. Clarendon, Oxford, NY, USA.

Donoghue, M. J. and S. M. Scheiner. 1992. The evolution of endosperm: a phylogenetic account, pp. 356-389. *In* R. E. Wyatt (ed.). Ecology and Evolution of Plant Reproduction. Chapman and Hall, New York, NY, USA.

Dop, P. 1913. Recherches sur le développement et la nutrition du sac embryonnaire et de l'endosperme des *Buddleja*. Bull. Soc. Bot. France, 60: 9-16, 45-50, 92-98.

Dorn, A. 1937. Origin of Vertebrates and the Principle of Function Change. Biomedgiz, Moscow, Russia.

Doyle, J. A. and M. J. Donoghue. 1986. Seed plant phylogeny and the origin of angiosperms: an experimental cladistic approach. Bot. Rev. (Lancaster), 52: 321-431.

Doyle, J. A. and M. J. Donoghue. 1992. Fossils and seed plant phylogenies reanalyzed. Brittonia, 44: 89-106.

Doyle, J. A. and M. J. Donoghue. 1993. Phytogenies and angiosperm diversification. Paleobiology, 19: 141-167.

Doyle, J. A., M. J. Donoghue and E. A. Zimmer. 1994. Integration of morphological and ribosomal RNA data on the origin of angiosperms. Ann. Missouri Bot. Garden, 81: 419-450.

Dresselhaus, T., S. Fuchs, S. Heuer, E. Kranz and H. Lörz. 1996. Gene isolation from egg cells and in vitro produced zygotes of maize, p. 12. *In* Abstracts 14th Int. Cong. Sexual Plant Reproduction "Plant Reproduction '96", February 18-23, 1996, Lorne, Victoria, Australia. Cumberland Resort, Lorne, Victoria, Australia.

Dressler, R. L. 1981. The Orchids: Natural History and Classification. Harvard Univ. Press, Cambridge, MA, USA.

Drozdowicz J. M. and R. L. Jones. 1995. Hormonal regulation of organic and phosphoric acid release by barley aleurone layers and scutella. Plant Physiol., 108: 769-776.

Duchartre, P. 1848. Sur les embryons dits polycotyles. Ann. Sci. Naturelles. Botanique, ser. 3, X: 207.

Duckett, C. M., K. J. Oparka, D. A. M. Prior, L. Dolan and K. Roberts. 1994. Dye-coupling in the root epidermis of *Arabidopsis* is progressively reduced during development. Development, 120: 3247-3255.

Dulberger, R. 1974. Structural dimorphism of stigmatic papillae in distylous *Linum* species. Amer. J. Bot., 61: 238-243.

Dulberger, R. 1975. Intermorph structural differences between stigmatic papillae and pollen grains in relation to incompatibility in Plumbaginaceae. Proc. Royal Soc. London. Series B. Biol. Sci., 188: 257-274.

Dumas, C. and R. B. Knox. 1983. Callose and determination of pistil viability and incompatibility. Theor. Appl. Genetics, 67(1): 1-10.

Dumas, C., R. B. Knox and T. Gaude. 1985. The spatial association of the sperm cells and vegetative nucleus in the pollen grain of *Brassica*. Protoplasma, 124(3): 168–174.

Dumas, C., R. B. Knox, C. A. McConchie and S. D. Russell. 1984. Emerging physiological concepts in fertilization. What's New Plant Physiol., 15: 17-20.

Dure, L. III. 1960. Cross nutritional contributions of maize endosperm and scutellum to germination and growth of axis. Plant Physiol., 35(6): 919-925.

Dure, L. III. 1975. Seed formation. Ann. Rev. Plant Physiol., 26: 259-278.

Dute, R. R. and C. M. Peterson. 1992. Early endosperm development in ovules of soybean, *Glycine max* (L.) Merr. (Fabaceae). Ann. Bot. (London), 69(3): 263-271.

Dutt, B. S. M. 1970a. Amaryllidaceae. Bull. Indian Natl. Sci. Acad., 41: 365-367.

Dutt, B. S. M. 1970b. Embryology of *Pancratium longiflorum*. Phytomorph., 20(1): 1-5.

Dutt, B. S. M. 1978. Embryo development in *Moringa concanensis* Nimno. Current Sci., 47(18): 693.

Dyur, L. S. 1982. Reserved messenger ribonucleic acid and seed germination, pp. 373-386. *In* M. G. Nikolaeva and N. V. Obrucheva (eds). Physiology and Biochemistry of Seed Dormancy and Germination. Kolos, Moscow, Russia.

Dzevaltovsky, A. K. 1983. Family Cucurbitaceae, pp. 127-138. *In* M. S. Yakovlev (ed.). Comparative Embryology of Flowering Plants. Phytolaccaceae—Thymelaeaceae. Nauka, Leningrad, Russia.

Eames, A. J. 1961. Morphology of the Angiosperms. McGrawhill, New York, NY.

Earle, T. T. 1938. Embryology of certain Ranales. Bot. Gazette, 100(2): 257-275.

Earp, D. J., B. Lowe and B. Baker. 1990. Amplification of genomic sequences flanking transposable elements in host and heterologous plants: a tool for transposon tagging and genome characterization. Nucleic Acids Res., 18: 3271-3279.

East, E. M. and A. Mangelsdorf. 1925. A new interpretation of the hereditary behaviour of self-sterile plants. Proc. Natl. Acad. Sci., USA, 11: 1174-1175.

Eisenberg, A. J. and J. P. Mascarenhas. 1985. Abscisic acid and the regulation of synthesis of specific seed proteins and their messenger RNAs during culture of soybean embryos. Planta, 166(4); 505-514.

Ekambaram, T. and R. R. Panije. 1935. Contribution to our knowledge of *Balanophora*, II. Life history of *B. dioica*. Proc. Indian Acad. Sci. Section B, Biol. Sci., 1: 522-543.

Engler, A. 1889. Proteaceae, pp. 1-289. *In* A. Engler and K. Prantl (eds). Die natürlichen Pflanzenfamilien. Verlag von W. Engelmann, Leipzig.

Engler, A. 1926. Interabteilung Angiospermae, pp. 1-167. *In* A. Engler and K. Prantl (eds). Die natürlichen Pflanzenfamilien. Verlag von W. Engelmann, Leipzig.

Enzenberg, U. 1961. Beiträge zur Karyologie des Endosperms. Österreichische Bot. Zeitschrift, 108(3): 245-285.

Erbrich, P. 1965. Über endopolyploidie und kernstrukturen in endospermhaustorien. Österreichische Bot. Zeitschrift, 112(3): 197-262.

Erdelská, O. 1981a. Dynamics of early embryo development in reference to endosperm and embryo types in angiosperms. Acta Soc. Bot. Poloniae, 50(1-2): 269–272.

Erdelská, O. 1981b. Embryológia krytosemenných rastlin. Veda, Bratislava, Czéchoslovákia.

Erdelská, O. 1983. Microcinematographical investigation of the female gametophyte, fertilization and endosperm development, pp. 49-54. *In* O. Erdelská (ed.). Proc. VII Int. Cytoembryological Symp. "Fertilization and Embryogenesis in Ovulated Plants". High Tatra (Račkova dolina), 14-17 June, 1982. Veda, Bratislava, Czechoslovakia.

Erdelská, O. 1984. Some structural features of the embryo/endosperm interaction in angiosperms, pp. 168-170. *In* M. T. M. Willemse and J. L. Van Went (eds). Proc. 8th Int. Symp. "Sexual Reproduction in Seed Plants, Ferns and Mosses", August 20-24, 1984. PUDOC, Wageningen, Netherlands.

Erdelská, O. 1991. Embryo development in vitro, pp. 336-344. *In* J. Sebanek, Z. Sladky and S. Prochazka (eds). Experimental Morphogenesis and Integration of Plants. Academia, Praha.

Erdelská, O. and Inst. Wiss. Film (Göttingen). 1971. *Jasione montana* (Campanulaceae) — Entwicklung des Endosperm, pp. 1-10. Film E 1751 des IWF. Publ. von O. Erdelská, Gottingen.

Erdelská, O. and A. Klasova. 1978. La région micropylaire du sac embryonnaire de *Jasione montana* L. avant et après la fécondation. Bull. Soc. Bot. France. Actualities Bot., 1-2: 249-252.

Erdelská, O. and Inst. Wiss. Film (Göttingen). 1982. Befruchtung und frühe Entwicklung von Embryo and Endosperm beim Schneeglöckchen, p. 15. Film C 1463 des IWF. Gottingen: Publ. von O. Erdelská, Publ. Wiss. Film, 1985. Sekt. Biol. Ser. 17,20/C 1463.

Erdelská, O., D. Kobetieova and A. Pretová. 1973. The development of excised flax embryos in vitro. Biologia (Bratislava), 28: 235-239 (in Slovak).

Erickson, L. and R. Kemble. 1990. Paternal inheritance of mitochondria in rape-seed (*Brassica napus*). Molec. Gen. Genetics, 222: 135-139.

Erickson, L., R. Kemble, and E. Swanson. 1989. The *Brassica* mitochondrial plasmid can be sexually transmitted. Pollen transfer of a cytoplasmic genetic element. Molec. and Gen. Genet., 218: 419-422.

***Ermakov, I. P., E. M. Morozova and L. V. Karpova. 1980.** On the contents of DNA in nuclei of male gametophyte in certain flowering plants. Reports USSR Acad. Sci., 251(4): 1023-1024.

Ernst, A. and E. Schmid. 1913. Über Blüte und Frucht von *Rafflesia*. Ann. Jardin Bot. Buitenzorg, 2(12): 1-58.

Errampalli, D., D. Patton, L. Castle,L. Mickelson, K. Hansen, J. Schnall, K. Feldmann and D. Meinke. 1991. Embryonic lethals and T-DNA insertional mutagenesis in *Arabidopsis*. Plant Cell, 3: 149-157.

Esau, K. 1965. Plant Anatomy. J. Wiley & Sons, New York, NY, USA.

Esau, K. 1977. Anatomy of Seed Plants. J. Wiley & Sons, New York, NY.

Esen, A. and D. A. Stetler. 1992. Immunocytochemical localization of d-zein in the protein bodies of maize endosperm cells. Amer. J. Bot., 79(3): 243-248.

Eunus, A. M. 1952. Contributions to the embryology of Liliaceae. 3. Embryogeny and development of the seed of *Asphodelus tenuifolius* Cav. Lloydia, 15(3): 149-155.

Evans, D. A., W. R. Sharp and C. E. Flick. 1981. Growth and behaviour of cell cultures. Embryogenesis and organogenesis, pp. 45-113. *In* T. A. Thorpe (ed.). Proc. UNESCO Symp., San Paulo, Brazil, 1978. "Plant Tissue Culture: Methods and Applications in Agriculture". Academic Press, New York, NY.

Ezaki, B., K. Toko, K. Yamafuji and F. Irie. 1988. Electric potential patterns around a root of the higher plant. Trans. Inst. Electronics and Communication Engineers in Japan, Sect. E., 71(10): 965-967.

Fagerlind, F. 1945a. Bau der floralen Organe bei der Gattung *Langsdorffia*. Svensk Bot. Tidskrift, 39(2): 197-210.

Fagerlind, F. 1945b. Blüte und Blutenstand der Gattung *Balanophora*. Bot. Notiser, 39: 330-350.

Fagerlind, F. 1946. Sporogenesis, Embryosackentwicklung und pseudogame Samenbildung bei *Rudbeckia laciniata*. Acta Horti Bergiani, 14(3): 39-90.

Fahn, A. 1990. Plant Anatomy. Pergamon Press, Oxford, UK.

Fahn, A. and D. F. Cutler. 1992. Xerophytes. Stuttgart, Berlin.

Fairbanks, D. J., S. E. Smith, and J. K. Brown. 1988. Inheritance of large mitochondrial RNAs in alfalfa. Theor. Appl. Genetics, 76: 619-622.

***Famintsyn, A. S. 1879.** Embryological researches. Reports of Russian Acad. Sci., 7(26): 1-19.

Farrell, M. E. 1914. Ovary and embryo of *Cyrtanthus sanguineus*. Bot. Gazette, 57(5): 428-436.

Faure, J.-E., C. Digonnet and C. Dumas. 1994a. An in vitro system for adhesion and fusion of maize gametes. Science, 263: 1598-1600.

Faure, J.-E., H. L. Mogensen, Ch. Dumas, H. Lörz and E. Kranz. 1993. Caryogamy after electrofusion of single egg and sperm cell protoplasts from maize: cytological evidence and time course. Plant Cell, 5: 747-755.

Faure, S., J. L. Noyer, F. Carreel, J. P. Horry and C. Lanaud. 1994b. Maternal inheritance of chloroplast genome and paternal inheritance of mitochondrial genome in banana (*Musa acuminata*). Current Genetics, 25: 265-269.

Favre-Duchartre, M. 1956. Contribution á l'étude de la reproduction chez le *Ginkgo biloba*. Rev. Cytol. Biol. Végétales, 17(1-2): 1-218.

Favre-Duchartre, M. 1989. General view over plant proembryos, pp. 167-180. *In* J. Pare and M. Bugnicourt (eds). Some Aspects and Actual Orientations in Plant Embryology. Univ. Picardie, Amiens, France.

Fedotova T. A. 1985. Family Stemonaceae, pp. 128-129. *In* A. L. Takhtajan (ed.). Comparative Anatomy of Seeds, vol. 1. Monocotyledons. Nauka, Leningrad, Russia.

Feldman, L. J. and J. G. Torrey. 1975. The quiescent centre and primary vascular tissue pattern formation in cultured roots of *Zea*. Canadian J. Bot., 23: 2796-2803.

Feldmann, K. A. 1991. T-DNA insertion mutagenesis in *Arabidopsis*: 1. Mutational spectrum Plant Journal, 1: 71-82.

Felker, F. C., D. M. Peterson and O. E. Nelson. 1984. Development of tannin vacuoles in chalaza and seed coat of barley in relation to early chalazal necrosis in the segl mutant. Planta, 161: 540-549.

Felker, F. C., D. M. Peterson and O. E. Nelson. 1987. Early grain development of the seg2 maternal-effect shrunken endosperm mutant of barley. Canadian J. Bot., 65: 943-948.

Ferguson, M. C. 1901. The development of the egg and fertilization in *Pinus strobus*. Ann. Bot. (London), 9: 435-479.

Fernandez, D. E. and S. P. Bisgrove. 1992. Accumulation of mRNAs during embryo maturation and precocious germination in *Brassica napus*, 19. *In* Abstracts XII Int. Cong. Sexual Plant Reproduction "Plant Reproductive Biology. Pollen, Ovules and Seeds". July 19-23, Columbus, OH, Ohio State Univ., Columbus, OH, USA.

Fernandez, D. E., F. R. Turner and M. L. Crouch. 1991. In situ localization of storage protein mRNAs in developing meristems of *Brassica napus* embryos. Development, 111: 299-313.

Fincher, G. B. 1989. Molecular and cellular biology associated with endosperm mobilization in germinating cereal grains. Ann. Rev. Plant Physiol., Molec. Biol., 40: 305-346.

Fineran, B. A., D. J. C. Wild and M. Ingerfeld. 1982. Initial wall formation in the endosperm of wheat, *Triticum aestivum*: a reevaluation. Canadian J. Bot., 60(9): 1776-1795.

Finkelstein, R. R. and M. L. Crouch. 1984. Precociously germinating rape-seed embryos retain characteristics of embryogeny. Planta, 162: 125.

Finkelstein, R., M. Estelle, J. Martinez-Zapater and C. Somerville. 1988. *Arabidopsis* as a tool for the identification of genes involved in plant development, pp. 1-25. *In* D. P. S. Verma and R. B. Goldberg (eds). Temporal and Spatial Regulation of Plant Genes. Springer-Verlag, New York, NY.

Finn, A. A. 1936. On the history of the development of chalasogamic, *Ostrya carpinifolia* Scop. J. Inst. Bot. UkrSSR Acad. Sci., 8(16): 15-28 (in Ukrainian).

Fischer, A. 1880. Zur Kenntnis der Embryosackentwicklung einiger Angiospermen. Jenaische Zeitschrift für Naturwissenschaft herausgegeben von der Medicinisch-naturwissenschaftlichen Gesellschaft zu Jena, 14: 9-132.

Fisher, D. B. and W. A. Jensen. 1969. Cotton embryogenesis: the identification as nuclei of the x-bodies in the degenerated synergid. Planta, 84: 122-133.

Fisher, D. B. and W. A. Jensen. 1972. Nuclear and cytoplasmic DNA synthesis in cotton embryos: a correlated light and electron microscope autoradiographic study. Histochemie, 32: 1-22.

Fisher, G. C. 1914. Seed development in the genus *Peperomia*. Bull. Torrey Bot. Club, 41(3): 137—156, (4): 221-241.

Fleischer, E. 1874. Beiträge zur Embryologie der Monocotyledon. Flora, 57: 369-448.

Fletcher, J. J. 1909. Illustrations of polycotyledony in the genus *Persoonia* with some reference to *Nuytsia*. Proc. Linnean Soc. New South Wales, 33: 867.

Forino, L. M. C., A. M. Tagliasacchi, A. Cavallini, G. Cionini, E. Giraldi and P. G. Cionini. 1992. RNA synthesis in the embryo suspensor of *Phaseolus coccineus* at two stages of embryogenesis, and the effect of supplied gibberellic acid. Protoplasma, 167: 152-158.

Foster, A. S. 1949. Practical Plant Anatomy. Van Nostrand, New York, NY.

Franke, W. W., W. Herth, W. J. Van Der Wonde and D. J. Morr. 1972. Tubular and filamentous structures in pollen tubes: possible involvement as guide elements in protoplasmic streaming and vectorial migration of secretory vesicles. Planta, 105(4): 317–341.

Franzmann, L., D. A. Patton and D. W. Meinke. 1989. In vitro morphogenesis of arrested embryos from lethal mutants of *Arabidopsis thaliana*. Theor. Appl. Genetics, 77: 609-616.

Fraser, L. 1931. An investigation of *Lobelia gibbosa* and *L. dentata*. 1. Mycorrhiza, latex system and general byology. Proc. Linnean Soc. London, 56: 497-525.

Friedman, W. E. 1990a. Double fertilization in *Ephedra*, a nonflowering seed plant: its bearing on the origin of angiosperms. Science, 247: 951-954.

Friedman, W. E. 1990b. Sexual reproduction in *Ephedra nevadens* (Ephedraceae): further evidence of double fertilization in a nonflowering seed plant. Amer. J. Bot., 77: 1582-1598.

Friedman, W. E. 1991. Double fertilization in *Ephedra trifurca*, a non-flowering seed plant: the relationship between fertilization events and the cell cycle. Protoplasma, 65: 106-120.

Friedman, W. E. 1992a. Double fertilization in nonflowering seed plants and its relevance to the origin of flowering plants. Int. Rev. Cytology, 140: 319-355.

Friedman, W. E. 1992b. Evidence of a pre-angiosperm origin of endosperm: implications for the evolution of flowering plants. Science, 255: 336-339.

Friedman, W. E. 1994. The evolution of embryogeny in seed plants and the developmental origin and early history of endosperm. Amer. J. Bot., 81(11): 1468-1486.

Friedman, W. E. 1995. Organismal duplication, inclusive fitness theory, and altruism: understanding the evolution of endosperm and the angiosperm reproductive syndrome. Proc. Natl. Acad. Sci., USA, 92: 3913-3917.

Friedman, W. E. 1999. Expression of the cell cycle in sperm of *Arabidopsis:* Implication for understanding patterns of gametogenesis and fertilization in plants and other eukaryotes. Development 126: 1065-1075.

Frisendahl, A. 1912. Cytologische und entwicklunggeschichtliche Studien über *Myricaria germanica* Desv. Kongliga Svenska Vetenskapsakademiens Handlingar, N. S., 48(7): 1-62.

Fry, S. C. and E. Wangermann. 1976. Polar transport of auxin through embryos. New Phytologist, 77(2): 313-317.

Frye, T. C. 1902. A morphological study of certain Asclepiadaceae. Bot. Gazette, 34: 389-413.

Fursov, V. I. 1964. Dynamics of nucleic acids in ontogenesis of wheat flower and kernel, pp. 100-102. *In* V. G. Konarev et al. (eds). Biology of Nucleic Metabolism in Plants. Nauka, Moscow, Russia.

***Gachechyladze, M. I. 1981.** Embryology of some species of *Onagraceae* family. Ph.D. thesis, Tbilisi, Georgia.

Gaertner, J. 1788. De fructibus et seminibus plantarum, vol. 1. Accedunt seminum centuriae quinque piores. Stuttgardiae typ. Akad. Carolinae, sumt. Auctories, Stuttgart, Leipzig.

Gain, E. 1900. Sur la tricotylie et l'anatomie des plantules de *Phaseolus tricotyles*. Rev. Gén. Bot., 12: 1-369.

Gatin, C. L. 1906. Recherches anatomiques et chimiques sur la germination des palmiers. Ann. Sci. Natur., Bot., 3:191-314.

Gavaudan, P., G. Debraux and A. Bonnenfaut. 1953. La gamocotylie experementale. C. R. Torr. Lab. Biol. Ved. Fac. Sci., Poitiers, 4.

Geitler, L. 1935. Beobachtungen die erste Teilung in Pollenkorn der Angiospermen. Planta, 24(3): 361-386.

Geitler, L. 1948. Notizen zur endomitotischen Polyploidisierung in Trichocyten und Elaiosomen sowie über Kernstrukturen bei *Gagea lutea*. Chromosoma, 3: 271-281.

Geltz, N. R. and S. D. Russell. 1988. Two-dimensional electrophoretic studies of the proteins and polypeptides in mature pollen grains and the male germ unit of *Plumbago zeylanica*. Plant Physiol., 88:764-769.

Gerassimov, I. I. 1923. Some notes on cell nucleus function, pp. 123-129. *In* V. M. Arnoldi. (ed.) Russian Classics of Plant Morphology. Gosizdat, Moscow, Russia.

Gerassimova, H. N. 1933. Fertilization in *Crepis capillaris*. Cellule, 42(1): 103-148.

***Gerassimova-Navashina, H. N. 1947a.** Mitotic hypothesis of double fertilization. Reports USSR Acad. Sci., 57(4): 395-398.

***Gerassimova-Navashina, H. N. 1947b.** Morphological data on cytoplasm of male gametophyte in *Crepis*. Reports USSR Acad. Sci., 56(4): 415-418.

***Gerassimova-Navashina, H. N. 1947c.** On the behaviour of sperms in pollen tube of *Crepis*. Reports USSR Acad. Sci., 57(3): 285-288.

*Gerassimova-Navashina, H. N. 1947d. On the development and structure of sperms in *Crepis*. Reports USSR Acad. Sci., 56(6): 643-646.

*Gerassimova-Navashina, H. N. 1951a. Pollen grain, gametes and sexual process in angiospermous plants. Trans. Komarov Bot. Inst. USSR Acad. Sci., Series VII, 2: 294-355.

Gerassimova-Navashina, H. N. 1951b. Pollen grain, gametes and sexual process in angiospermous plants, pp. 28-33. *In* Abstracts Delegate Conference All-Union Botanical Society. Moscow-Leningrad, Russia, 4.

*Gerassimova-Navashina, H. N. 1952. On the cytological and embryological understanding of pollination process. Trans. Komarov Bot. Inst. USSR Acad. Sci., Series VIII, 3: 165-211.

Gerassimova-Navashina, H. N. 1954a. Double fertilization of angiosperms, its nature and origin. Ph.D thesis, Leningrad, Russia.

*Gerassimova-Navashina, H. N. 1954b. Embryo sac development, double fertilization and the question about origin of angiospermous plants. Bot. Journal (St. Petersburg), 39(5): 655-681.

*Gerassimova-Navashina, H. N. 1957. Fertilization as an ontogenetic proccess. Bot. Journal (St. Petersburg), 42(11): 1654-1673.

*Gerassimova-Navashina, H. N. 1958. On the gametophyte and the main salient features of development and functioning of reproducing elements in angiospermous plants. Probl. Bot., 3: 125-167.

*Gerassimova-Navashina, H. N. 1959. Embryological observations on *Arachis hypogaea* L. Bot. J. (St. Petersburg), 44(10): 1453-1466.

Gerassimova-Navashina, H. 1960a. A contribution to the cytology of fertilization in flowering plants. Nucleus, 3(1): 111-120.

*Gerassimova-Navashina, H. N. 1960b. Influence of temperature conditions on the course of embryological processes in plants. Reports Ukrainian SSR Acad. Sci., 131(3): 688-691.

Gerassimova-Navashina, H. N. 1960c. Fertilization process in tetraploid forms, pp. 39-40. *In* Abstracts III All-Union Conference Embryologists, Izd. Moskovskogo Univ., Moscow, Russia.

Gerassimova-Navashina, H. 1969a. Influence of low temperature on the process of fertilization in *Crepis capillaris*. Phytomorph., 19(1).

Gerassimova-Navashina, H. 1969b. Some cytological aspects of double fertilization. Rev. Cytol. Biol. Végétales, 32(3-4): 301-308.

*Gerassimova-Navashina, H. N. 1971a. Double fertilization in angiosperms and some of its theoretical aspects, pp. 113-152. *In* V. P. Zosimowich (ed.). Problems of Embryology, Naukova Dumka, Kiev, Ukraine.

*Gerassimova-Navashina, H. N. 1971b. On the periods of mitotic cycle of angiosperm gametes. Vestnik Moskov. Univ., Series Biol., 3: 93-94.

*Gerassimova-Navashina, H. N. 1971c. On the regularity of fertilization process in angiosperms, pp. 34-36. *In* Abstracts V All-Union Conference Plant Embryology. Shtiintsa, Kishinev, Moldova.

*Gerassimova-Navashina, H. N. 1978. Some additional data on the course of mitosis of plant cell. Reports USSR Acad. Sci., 243(1): 265-267.

*Gerassimova-Navashina, H. N. 1980. Double fertilization and mitosis. Cytol. Genet., 14(6): 69-82.

Gerassimova-Navashina, H. N. 1982. Process of double fertilization in angiosperms and mitotic cycle of the cell. Phytomorph., 32(1-3): 222-233.

*Gerassimova-Navashina, H. N. 1984. Main features of development and structure of female gametophyte—gametes in angiosperms and some principles of the diversity of their functioning, 14. *In* G.E. Gvaladze (ed.). Materials All-Union Symp. "Morphofunctional Aspects of Development of Female Generative Structures in Seed Plants", Telavi, Georgia. Georgran Acad. Sci., Telavi, Georgia.

*Gerassimova-Navashina, H. N. 1986. Fertilization in angiosperms. Trans. Appl. Bot., Genet., Select., 99: 97-100.

*Gerassimova-Navashina, H. N. 1990. Fertilization in angiospermous plants. Bot. Journal (St. Petersburg), 75(8): 1061-1071.

*Gerassimova-Navashina, H. N. and T. B. Batygina. 1958. The process of double fertilization in *Scilla sibirica* Andr. Bot. Journal, 43(7): 959-988.

*Gerassimova-Navashina, H. N. and T. B. Batygina. 1959. On the course of sexual nucleus fusion during fertilization in cereals. Reports USSR Acad. Sci., 124(1): 223-226.

Gerassimova-Navashina, H. N. and T. B. Batygina. 1964. Some additional data on sexual apparatus and fertilization process in angiospermous plants (on the example of *Scilla sibirica* Andr.), pp. 86-99. *In* L.V. Kudryashov (ed.). Problems of Modern Embryology. Izd. Moskov. Univ., Moscow, Russia.

*Gerassimova-Navashina, H. N. and V. A. Gulyaev. 1973. Some data on ultrastructure of embryo sac cells after pollination in *Crepis capillaris* (L.) Wallr. Bull. Main Bot. Garden USSR Acad. Sci., 89: 14-20.

*Gerassimova-Navashina, H. N. and T. V. Salnikova. 1982. Influence of ethylenimine and dimethylsulfate on plant cell nuclei. Cytol. Genet., 4: 7-11.

Gerassimova-Navashina, H. N., T. B. Batygina, S. N. Korobova and O. A. Dolgova. 1963. Fertilization and embryo development at X-ray treatment of sexual elements, 12. *In* Abstracts IV Conference Embryologists. Nauka, Moscow-Leningrad, Russia.

*Gerassimova-Navashina, H. N., R. N. Kapil, S. N. Korobova and G. I. Savina. 1968. Double fertilization process at the lower temperatures. Bot. Journal (St. Petersburg), 53(5): 614-627.

Gerlach-Cruse, D. 1970. Experimentelle Auslosung von Semigamie bei *Arabidopsis thaliana* (L.) Heynh. Biol. Zentralblatt, 89: 435-456.

Ghosh, S. and K. R. Shivanna. 1982. Anatomical and cytochemical studies of the stigma and style in some legumes. Bot. Gazette, 143(3): 311-318.

Gibbs, L. S. 1907. Notes on the development and structure of the seed in the Alsinoideae. Ann. Bot. (London), 21(31): 25-55.

Gibelli, G. and F. Ferrero. 1891. Ricerche di anatomia e morphologia intorno allo sviluppo e des seme delle *Trapa natans* L. Malpighia, 5: 1-64.

Gifford E. M. and A. S. Foster. 1989. Comparative Morphology of Vascular Plants. W. H. Freeman & Co., San Francisco, CA, USA.

Gilbert, S. F., J. M. Opitz and R. A. Raff. 1996. Resynthesizing evolutionary and developmental biology. Develop. Biol., 173: 357-372.

Gilg, E. 1925. Dipterocarpaceae, pp. 237-269. *In* A. Engler (ed.). Die natürlichen Pflanzenfamilien, vol. 21. Verlag von Wilhelm Engelmann, Leipzig, Germany.

Gillham, N. M. 1994. Organelle Genes and Genomes. Oxford Univ. Press, New York, NY.

Gillot, I. and M. Whitaker. 1993. Imaging calcium waves in eggs and embryos. J. Exper. Biol., 184: 213-219.

Giuliano, G., F. LoSchiavo and M. Terzi. 1984. Isolation and developmental characterization of temperature-sensitive carrot cell variants. Theor. Appl. Genetics, 67: 179-183.

Glišić, L. M. 1936-1937. Ein Versuch der Verwertung der Endospermmerkmale fur typologische und phylogenetische Zwecke innerhalb der Scrophulariaceen. Bull. Inst. Jardin Bot. l'Univ. Belgrade, 4: 42-73.

Godineau, J. C. 1969. Ultrastructure du sac embryonnaire de *Crepis tectorum* L.: état apres la cellularisation et la fusion des noyaux polaires. Ann. Scient. l'Univ. l'A.R.E.R.S, 9(1): 78-88.

Goebel, K. 1899. Flora, 86, S. 235.

Goebel, K. 1932. Organographie der Pflanzen. III, I. Samenpflanzen, pp. 1379-1820. G. Fischer, Jena.

Goldberg, B. 1941. Life history of *Peltandra virginica*. Bot. Gazette, 102: 641-662.

Goldberg, R. B., S. J. Barker and L. Peres-Grau. 1989. Regulation of gene expression during plant embryogenesis. Cell (Cambridge), 56: 149-160.

Goldberg, R. B., T. P. Beals and P. M. Sanders. 1993. Anther development; basic principles and practical applications. Plant Cell, 5: 1217-1229.

Goldberg, R. B., G. De Paiva and R. Yadegari. 1994. Plant embryogenesis: zygote to seed. Science, 266(5185): 605-614.

Goldschmidt, R. 1938. Physiological Genetics. McGraw-Hill Book Co., New York, NY.

Goldsworthy, A. 1986. The electric compass of plants. New Scientist, 109(1489): 22-23.

Goldsworthy, A. and K. S. Rathore. 1985. The electrical control of growth in plant tissue cultures: the polar transport of auxin. J. Exper. Bot., 36(168): 1134-1141.

Golubovskaya, I. N. 1994. Meiosis, pp. 70-76. *In* T. B. Batygina (ed.). Embryology of Flowering Plants. Terminology and Concepts. 1. Generative Organs of Flower. Mir i Semya, St. Petersburg, Russia.

Goodner, B. and R. S. Quatrano. 1993. *Fucus* embryogenesis: a model to study the establishment of polarity. Plant Cell, 5(10): 1471-1481.

Goodwin, P. B. and M. G. Erwee. 1983. Hormonal influences on leaf growth, pp. 207-232. *In* J. E. Dale and E. L. Milthorpe (eds). The Growth and Functioning of Leaves. Cambridge Univ. Press, Cambridge, UK.

***Gorozhankin, I. N. 1880.** On corpuscules and sexual process in gymnosperms. Uchenye Zap. Moskov. Univ., Natural-historical Dept., 1: 1-177.

Goursat, M. 1969. Recherches sur l'embryogénie des Papilionacées. Ph.D. thesis, Univ. Paris, Paris, France.

Govindu, H. C. 1950. Studies in the embryology of some members of the Bignoniaceae. Proc. Indian Acad. Sci., 32(4): 164-178.

Graham, L. E. 1993. Origin of Land Plants. John Wiley & Sons, New York, NY.

Grant, V. 1981. Plant Speciation. Columbia Univ. Press, Columbia, NY.

Graves, A. 1898. Recherches anatomiques et physiologiques sur le *Tradescantia virginica* L. Bruxelles.

Graves, A. H. 1908. The morphology of *Ruppia maritima*. Trans. Connecticut Acad. Arts Sci., 14(2): 59-170.

***Grebinsky, S. O. 1961.** Growth of Plants. Lvov Univ. Press, Lvov, Ukraine.

Green, P. B. 1992. Pattern formation in shoots: a likely role for minimal energy configurations of the tunica. Int. J. Plant Sci., 153(3): 59-75.

Green, P. B. and J .M. L. Selker. 1991. Mutual alignments of cell walls, cellulose and cytoskeleton: their role in meristems, pp. 277-288. *In* C. W. Lloyd (ed.).The Cytoskeletal Basis of Plant Growth and Form. Academic Press, London, UK.

Grevtsova, N. A. 1987. Family *Apiaceae*, pp. 35-44. *In* T. B. Batygina and M. S. Yakovlev (eds). Comparative Embryology of Flowering Plants. Davidiaceae—Asteraceae. Nauka, Leningrad, Russia.

***Grevtsova, N. A., V. Yu. Nazarova and V. N. Tikhomirov. 1987.** On the embryology of the genus *Trapa* L. Comparative embryological investigation of female generative sphere in *Trapa europaea* Fler. and *T. conocarpa* (Aresch.) Fler. Biologicheskie Nauki, 7: 89-93.

***Grif, V. G. and V. B. Ivanov. 1975.** Time parameters of mitotic cycle in flowering plants. Cytologia, 17(6): 694-717.

Groot, E. P. and L. A. Van. Caeseele. 1993. The development of the aleurone layer in canola (*Brassica napus*). Canadian J. Bot., 71: 1193-1201.

Grootjen, C. J. 1983. Development of ovule and seed in Marantaceae. Acta Bot. Neerlandica, 32(1-2): 69-86.

Grootjen, C. J. and F. Bouman. 1981. Development of ovule and seed in *Costus cuspidatus* (Zingiberaceae) with special reference to the operculum. Bot. J. Linnean Soc., 83(1): 27-39.

Grübert, M. 1974. Studies on the distribution of myxospermy among seeds and fruits of Angiospermae and its ecological importance. Acta Biol. Venezuelica, 8: 315-534.

Grushvitsky, I. V. 1961. Role of embryo underdevelopment in evolution of flowering plants. Komarov Readings, XIV. USSR Acad. Sci., Moscow-Leningrad, Russia.

***Grushvitsky, I. V. 1963.** Hypogeal germination and functions of cotyledons. Bot. Journal (St. Petersburg), 48(6): 906-915.

Grzelczak, Z. F., S. Rahman, T. D. Kennedy and B. G. Lane. 1985. Compartmentation of the protein, its translatable mRNA, and its biosynthesis among roots, stems and leaves of wheat seedlings. Canadian J. Biochem. Cell Biol., 63(9): 1003-1013.

Gscheidle, A. 1925. Über Haustorienbildung in der Gattung *Veronica*. Flora, 117(1): 144-172.

Guérin P. 1899. Recherches sur le développement du tégument seminal et du péricarpe des Graminées. Ann. Sci. Naturelles. Bot., 8(9): 1-59.

Guérin, P. 1901. Développement de la graine et en particulier du tégument séminal de quelques Sapindacées. J. Bot., appl. Agric., Pharm., Méd., Arts, 15: 336-362.

Guignard, J. L. 1961a. Embryogénie des Palmiers. Développement de l'embryon chez le *Chamaerops humilis* L. C.R. Hebdomadaires Séances l'Acad. Sci., 253: 1834-1936.

Guignard, J. L. 1961b. Recherches sur l'embryogénie des Graminées; rapports des Graminées avec les autres Monocotylédones. Ann. Sci. Naturelles. Bot., 12(2,3): 491-610.

Guignard, J. L. 1975a. Du Cotylédon des Monocotylédones. Phytomorph., 25(2): 193-200.

Guignard, J. L. 1975b. Embryogénie et classification embryogénique. Bull. Soc. Bot. France, 122: 281-294.

Guignard, J. L. 1984. Le cotylédon des Monocotylédones. Bull. Muséum Natl. d'Histoire Naturelle, (Paris), Ser.4., 1: 19-27.

Guignard, J. L. and J. C. Mèstre. 1969. L'origine du cotylédon et du cone végétatif chez les Monocotylédones. Bull. Soc. Bot. France, 116: 207-214.

Guignard, J. L. and J. C. Méstre. 1971. L'embryon des Graminées. Phytomorph., 20(2): 190-197.

Guignard, J. L., C. Agier and M. Bury. 1991. Développement de l'embryon chez zygotique chez l'*Arabidopsis thaliana* Schur. Bull. Soc. Bot. France, Letters Bot. 138(2): 149-154.

Guignard, L. 1880. Notc sur la structure et les functions du suspensor chez quelques Legumineuses. Bull. Soc. Bot. France, 27: 253–257.

Guignard, L. 1882. Recherches sur le sac embryonnaire des phanerogames angiosperms. Ann. Sci. Naturelles. Bot., France, 13(6): 136–199.

Guignard, L. 1886. Observation sur les ovules et la fécondation des Cactées. Bull. Soc. Bot. France, 33: 276-280.

Guignard, L. 1893. Recherches sur le développement de la graine et en particulier du tégument séminal. J. Bot., appl. l'Agric., Pharmacie, à la Médecine et aux Arts, 7: 1-311.

Guignard, L. 1899a. Sur les anthérozoides et la double copulation sexuelle chez les végétaux angiosperm. C.R. Hebdomadaires Séances l'Acad. Sci., 128: 864-871.

Guignard, L. 1899b. Sur les anthérozoides et la double copulation sexuelle chez les végétaux angiosperm. Rev. Gén. Bot., 11: 129-135.

Guignard, L. 1900. L'appareil sexuel et la double fécondation dans les tulipes. Ann. Sci. Naturelles. Bot., 11: 365-384.

Guignard, L. 1901a. La double fécondation dans le mais. J. Bot., appl. l'Agric., Pharm., Méd., Arts, 15(2): 37-50.

Guignard, L. 1901b. La double fécondation dans le *Najas major*. J. Bot., appl. Agric. Pharm. Méd., Arts, 15(7): 205-213.

Guignard, L. 1901c. La double fécondation chez les Ranunculacées. J. Bot., appl. l'Agric. Pharm., Méd., Arts, 15(12): 394-408.

Guignard, L. 1922. La fécondation et la polyembryonie chez les *Vincetoxicum*. Mémoires l'Acad. Sci. l'Inst. France, 57: 1-25.

Guillaumin, M. A. 1911. Germinations anormales. Bull. Soc. Bot. France, 58: 481-488.

Guo-Wei, T. and S. Jia-Heng. 1992. Ultrastructural observations of fertilization in wheat (*Triticum aestivum*), I. The fusion of the sperm and egg nuclei. Chinese J. Bot., 4: 87-91.

Gupta, S. C. and M. Gupta. 1964. Some observations on the embryology of *Bupleurum tenue* Buch. Ham ex D.Don. Current Sci., 33(18): 563-564.

Gutterman, Y. and A. Witztum. 1977. The movement of integumentary hairs in *Blepharis ciliaris* (L.) Burtt. Bot. Gazette, 138(1): 29-34.

Guzowska, J. and M. Zenkteler. 1969. In vitro development of immature embryos of *Cuscuta lupuliformis* Krock. Bull. Soc. Sci. Lettres de Lødz; CL. 3, 9: 37-38.

***Gvaladze, G. E. 1973.** Comparative embryological investigation of embryo sacs of Polygonum-, Oenothera-, Allium-and Fritillaria-types in some plants. Ph.D. thesis, Tbilisi, Georgia.

***Gvaladze, G. E. and M. I. Gachechyladze. 1977.** On the embryology of *Onagraceae*. Reports GeorgSSR Acad. Sci., Serie Biol., 3(6): 514-520.

***Gvaladze, G. E. and M. I. Gachechyladze. 1984.** On the mitotic hypothesis of double fertilization in angiosperms. Reports GeorgSSR Acad Sci., Serie Biol., 115(3): 613-616.

Haberlandt, G. 1918. Physiologische Pflanzenanatomie. Verlag von Wilhelm Engelmann, Leipzig.

Haccius B. 1952a. Die Embryoentwicklung bei *Ottelia alismoides* und das Problem des terminalen Monokotylen. Planta, 40: 443-460.

Haccius, B. 1952b. Verbreitung und Ausbildung der Einkeimblättrigkeit bei den Umbelliferen. Österreichische Bot. Zeitschrift, 99(4): 483-505.

Haccius, B. 1953a. Embryologische und histogenetische Studien an "monocotylen Dikotylen". Bau und Funktion des Keimblatt-Stiels. Berichte der Deutschen Bot. Gesellschaft, 66: 17-19.

Haccius, B. 1953b. Histogenetische Untersuchungen an Wurzelhaube und Kotyledonarscheide geophiter Keimpflanzen *Podophyllum* und *Eranthis*. Planta, 41: 439-458.

Haccius, B. 1954. Embryologishe und histogenetische Studien an "monocotylen Dikotylen", 1. *Claytonia virginica* L. Österreichische Bot. Zeitschrift, 101(3): 285-303.

Haccius, B. 1955a. Experimentally induced twinning in plants. Nature, 176: 355-357.

Haccius, B. 1955b. Schizokotylie und Pleiokotylie bei *Delphinium ajacis*. Zeitschrift inductive Abstammungs- und Vererbungslehre, 86(4): 498-520.

Haccius, B. 1957. Embryologische und histogenetische Studien an "monocotylen Dicotylen", 2. *Pinguicula vulgaris* L. und *P. alpina* L. Österreichische Bot. Zeitschrift, 103(5): 567-587.

Haccius, B. 1959. Morphoregulatorische Beeinflussung pflanzlicher Embryonen durch Phenylborsaure. Naturwissenschaftliche Wochenschrift, 46: 153.

Haccius, B. 1960. Experimentell induzierte Einkeimblättrigkeit bei *Eranthis hiemalis*. II. Monocotylie durch Phenylborsaure. Planta, 54(5): 482-497.

Haccius, B. 1963. Restitution in acidity-damaged plant embryos: regeneration or regulation? Phytomorph., 13: 107-115.

Haccius, B. 1965. Untersuchungen über Somatogenese aus den Suspensorenzellen von *Eranthis hiemalis* Embryonen. Planta, 64: 219-224.

Haccius, B. 1969. Anomalien der pflanzenlichen Embryogenese nach Einwirkung von 2,4-D, TIBA und Morphaktinen, ein Vergleich. Vorträge aus dem Gesamtgebiet der Botanik. Hrsg, von der Deutschen botanischen Gesellschaft. Neue Folge, 3: 89-101.

Haccius, B. 1971. Zur derzeitigen Situation der Angiospermen Embryologie. Bot. Jahrb. für Systematik, Pflanzengeschichte und Pflanzengeographie, 91(2/3): 309-329.

Haccius, B. 1972a. Experimentelle Embryologie der Pflanzen. Forschungsberichte der Johannes Gutenberg—Universität Mainz, 1: 68-73.

Haccius, B. 1972b. Primärblatt—ähnliche Fiederkotyledonen nach Cytokinin—Behandlung junger Embryonen von *Eranthis hiemalis*. Beitrage zur Biol. Pflanzen, 48: 301-311.

Haccius, B. and E. Reinholz. 1953. Somatisch induzierte Veranderung der Keimblattzahl bei *Eranthis hiemalis* durch Röntgenstrahlen. Naturwissenschaften, 40(20): 533.

Haccius, B. and E. Hartl-Baude. 1957. Embryologische und histogenetische Studien an "monokotylen Dikotylen", 2. *Pinguicula vulgaris* L. und *Pinguicula alpina*. Österreichische Bot. Zeitschrift, 103(5): 567-587.

Haccius, B. and E. Fischer. 1959. Embryologische und histogenetische Studien an "monocotylen Dicotylen", 3. *Anemone apennina* L. Österreichische Bot. Zeitschrift, 106(5): 373-389.

Haccius, B. and G. Trompeter. 1960. Experimentell induzierte Einkeimblattrigkeit bei *Eranthis hiemalis.* 1. Synkotylie durch 2,4-Dichlorphenoxyessigsaure. Planta, 54(5): 466-481.

Haccius, B. and K. K. Lakshmanan. 1966. Vergleichende Untersuchungen der Entwicklung von Kotyledon und Sprosscheitel bei *Pistia stratiotes* und *Lemna gibba*, ein Beitrag zum Problem der sogenanten terminalen Blattorgane. Beitrage zur Biol. Pflanzen, 42(3): 425-443.

Haccius, B. and K. K. Lakshmanan. 1967. Experimental studies on monocotyledonous dicotyledons: phenylboric acid-induced "dicotyledonous" embryos in *Cyclamen persicum*. Phytomorph., 17(1-4): 488-494.

Haccius, B. and N. N. Bhandari. 1975. Delayed histogen differentiation as a common primitive character in all types of nonzygotic embryos. Phytomorph., 25(1): 91-94.

Haccius, B. and V. J. Philip. 1979. Embryo development of *Cocos nucirera* L. Plant System. Evol., 132(1-2): 91-106.

Hacker, M., K. M. Hartmann and H. Mohr. 1964. Zeilteilung und Zeilwachstum im Hypokotyl von *Lactuca sativa* L. unter dem Einfluss des Lichtes. Planta, 63(3): 253-268.

Hagemann, R. 1976. Plastid distribution and plastid competition in higher plants and the induction of plastom mutations by nitroso-urea-compounds, pp. 331-338. *In* Th. Bucher, W. Neuper, W. Siebald and S. Werner (eds). Genetics and Biogenesis of Chloroplasts and Mitochondria. North Holland, Amsterdam, Netherlands.

Hagemann, R. 1981. Unequal plastid distribution during the development of male gametophyte of angiosperms. Acta Soc. Bot. Poloniae, 50: 321-327.

Hagemann, R. 1983. The formation of generative and sperm cells without plastids in angiosperms and the underlying mechanisms, pp. 97-99. *In* O. Erdelská (ed.). Proc. VII Int. Cytoembryological Symp. "Fertilization and Embryogenesis in Ovulated Plants". High Tatra (Račkova dolina), 14-17 June, 1982. Veda, Bratislava, Czechoslovakia.

Hagemann, R. and M-B. Schröder. 1989. The cytological basis of the plastid inheritance in angiosperms. Protoplasma, 152: 57-64.

Hagemann, W. 1959. Vergleichende morphologische, anatomische und entwicklungsgeschichtliche Studien an *Cyclamen persicum* Mill. sowie einigen weiteren Cyclamen-Arten, pp. 1-88. *In* W. Troll and H. Von Guttenberg (eds). Botanishe Studien vol. 9, Fischer Jena.

Haines, R. W. and K. A. Lye. 1975. Seedlings of Nymphaeaceae. J. Linnean Soc., Bot., 70: 255-265.

Haines, R. W. and K. A. Lye. 1979. Monocotylar seedlings: a review of evidence supporting an origin by fusion. J. Linnean Soc. Bot., 78(2): 123-140.

Hallam, N. D. 1972. Embryogenesis and germination in rye (*Secale cereale* L.), 1. Fine structure of developing embryo. Planta, 104(2): 157-166.

Halperin, W. 1966. Alternative morphogenetic events in cell suspension. Amer. J. Bot., 53: 443-453.

Halperin, W. 1967. Population density effects on embryogenesis in carrot cell cultures. Exper. Cell Res., 48(1): 170-173.

Halperin, W. 1970. Embryos from somatic plant cells. *In* H.A. Padykula (ed.). Control Mechanism in the Expression of Cellular Phenotypes. (Symp. Int. Soc. Cell. Biol.). Academic Press, New York, NY. 9(4): 169-191.

Hamann, U. 1966. Embryologische, morphologische, anatomische und systematische Untersuchungen an Philydraceen. Willdenowia, 4(1): 1-178.

Hamby, R. K. and E. A. Zimmer. 1992. Ribosomal RNA as a phylogenetic tool in plant systematics, pp. 50-91. *In* P. S. Soltis D. E. Soltis, and J. J. Koyle (eds). Molecular Systematics of Plants. Chapman and Hall, New York, NY.

Hanf, M. 1950. Keimung von Unkrautern und Kulturpflanzen nach Behandlung des Bodens mit 2,4-D-Mitteln. Nbl. Dtsch. Pflanzen-Schutzdienst, 2(6).

Hanstein, J. 1870. Die Entwicklung des Keimes der Monokotylen und Dikotylen. Botanische Abhandlungen an Gebiet der Morphologie und Physiologie. 1, Hf.1. A. Marcus, Bonn.

Hara, N. 1961. Apical views of vegetative shoot apex. Bot. Magazine (Tokyo), 74: 267-268.

Hara, N. 1971. Structure of the vegetative shoot apex of *Clethra barbinervis*. I. Superficial and transectional views. Bot. Magazine (Tokyo), 84(991): 8-17.

Hara, N. 1995. Developmental anatomy of the three-dimensional structure of the vegetative shoot apex. J. Plant Res., 108: 115-125.

Hardham, A. R. 1982. Regulation of polarity in tissue and organs, pp. 377-403. *In* C. W. Lloyd (ed.). Cytoskeleton in Plant Growth and Development. Academic Press, London, UK.

Harrington, J. F. 1972. Seed storage and longevity, pp. 145-245. *In* T. T. Kozlowski (ed.). Seed Biology, Vol. III. Academic Press, New York, NY.

Harris, J. A., E. W. Sinnott, J. Y. Pennypacker and G. B. Durham. 1921. The vascular anatomy of dimerous and trimerous seedlings of *Phaseolus vulgaris*. Amer. J. Bot., 8: 63.

Harrison, B. J. 1964. Factors affecting the frequency of tricotyly in *Antirrhinum majus*. Nature, 201(4317): 424.

Hart, J. W. and I. R. Macdonald. 1981. Phototropism and geotropism in hypocotyls of cress (*Lepidium sativum* L.). Plant, Cell Environ., 4(3): 197-201.

Hartog, M. 1904. Some problems of reproduction. Quar. J. Microscop. Sci., N.S., 47(188): 583-608.

Hasebe, M., M. Ito, R. Kofuji, K. Iwatsuki and K. Ueda. 1992a. Phylogenetic relationships in Gnetophyta deduced from rbcL gene sequences. Bot. Magazine (Tokyo), 105: 385-391.

Hasebe, M., M. Kato, R. Kofuji, M. Ito, K. Iwatsuki and K. Ueda. 1992b. Phylogeny of gymnosperms inferred from rbcL gene sequences. Bot. Magazine (Tokyo), 105: 673-679.

Hasitschka-Jenschke, G. 1959. Bemerkenswerte Kernstrukturen in Endosperm und im Suspensor zweier Helobiae. Österreichische Bot. Zeitschrift, 106: 301-314.

Haskell, G. 1949. Changes accompanying selection for pleiocotyly. Heredity, 3: 382.

Haskell, G. 1954a. Influence of position of truss on pleiocotyly in tomato. Nature, 173: 311.

Haskell, G. 1954b. Pleiocotyly and differentiation within angiosperms. Phytomorph., 4: 140-152.

Haughn, G. W. and C. R. Somerville. 1988. Genetic control of morphogenesis in *Arabidopsis*. Develop. Genet., 9: 73-89.

Hause, G. 1991. Ultrastructural investigations of mature embryo sacs of *Daucus carota*, *D. aureus* and *D. muricatus*—possible cytological explanations of paternal plastid inheritance. Sexual Plant Reprod., 4: 288-292.

Hause, G. and M. B. Schröder. 1987. Reproduction in *Triticale*, 2. Karyogamy. Protoplasma, 1939(2/3): 100-104.

Hayman, D. L. 1956. The genetical control of incompatibility in *Phalaris caerulescens*. Australian J. Biol. Sci., 9: 321-331.

Health, J. D., R. Weldon, C. Monnot and D. W. Meinke. 1986. Analysis of storage proteins in normal and aborted seeds from embryo-lethal mutants of *Arabidopsis thaliana*. Planta, 169: 304-312.

Hegelmaier, F. 1874. Zur Entwicklungsgeschichte monokotyledoner Keime nebst Bemerkungen über die Bildung der Samendeckel. Bot. Zeitung (Berlin), 32(39-44): 631-639, 648-656, 657-672, 673-688, 689-699, 705-711.

Hegelmaier, F. 1878. Vergleichende Untersuchungen über Entwicklung dicotyledoner Keime mit Berucksichtigung der Pseudomonokotyledonen. E. Schweizerbartische Verlagshandlung (E. Koch), Stuttgart.

Hegelmaier, F. 1880. Zur Embryogenie und Endospermentwicklung von *Lupinus*. Bot. Zeitung (Berlin), 38(5-9): 65-73, 81-91, 97-104, 121-151.

Hegelmaier, F. 1885. Untersuchungen über die Morphologie des Dicotyledonen-Endosperms. Nova Acta Leopoldina, 49(1): 1-104.

Hegelmaier, F. 1886. Zur Entwicklungsgeschichte endospermatischer Gewebekorper. Bot. Zeitung (Berlin), 44: 529-596.

Henslow, H. 1911. The origin of monocotyledons from dicotyledons through self-adaptation to a moist or aquatic habit. Ann. Bot. (London), 25: 717-744.

Herr, J. M. 1961a. Endosperm development and associated ovule modifications in the genus *Ilex*. J. Elisha Mitchell Sci. Soc., 77: 26-32.

Herr, J. M. 1961b. A new clearing squash technique for the study of ovule development in angiosperms. Amer. J. Bot., 58: 785-790.

Herr, J. M. 1999. Endosperm development in *Arabidopsis thaliana* (L.) Heynh. Acta Biol. Cracovensia, Series Bot., 41: 103-109.

Herzfeld, S. 1922. *Ephedra campylopoda* Mey. Morphologie der weiblichen Blüte und Befruchtungsvorgang. Denkschriften der Kaiserlichen Akademie der Wissenschaften, Wien. Mathematisch-Naturwissenschaftliche Klasse, 98: 243-268.

Heslop-Harrison, J. 1968. Ribosome site and S-gene action. Nature, 218: 90-91.

Heslop-Harrison, J. 1971. Sporopollenin in the biological context, pp. 1–30. *In* J. Brooks, P. R. Grant, M. Muir, P. Van Gijzel and G. Shaw (eds). Sporopollenin. London, UK.

Heslop-Harrison, J. 1975. Incompatibility and the pollen stigma interaction. Ann. Rev. Plant Physiol., 26: 403-425.

Heslop-Harrison, J. 1978. Genetics and physiology of angiosperms incompatibility systems. J. Cell Sci., 202: 73-92.

Heslop-Harrison, J. 1979a. Aspects of the structure, cytochemistry and germination of the pollen of rye (*Secale cereale* L.). Ann. Bot., Suppl. 1, 44: 1–47.

Heslop-Harrison, J. 1979b. Pollen walls as adaptive systems. Ann. Missouri Bot. Garden, 66(4): 813–829.

Heslop-Harrison, J. 1982. Pollen-pistil interaction and cross-incompatibility in grasses. Science, 215(4538): 1358-1364.

Heslop-Harrison, J. 1983. Self-incompatibility: phenomenology and physiology. Proc. Royal Soc. London. Series B. Biol. Sci., 218: 371-395.

Heslop-Harrison, J. 1987. Pollen germination and pollen tube growth, pp. 1–78. *In* K. L. Giles and J. Prakash (eds.). Int. Review Cytology. Pollen: Cytology and Development. London, UK.

Heslop-Harrison, J. and Y. Heslop-Harrison. 1975. Enzymic removal of the proteinaceous pellicle of the stigma papilla prevents pollen tube entry in the Caryophyllaceae. Ann. Bot., (London), 39: 163-165.

Heslop-Harrison, J. and Y. Heslop-Harrison. 1981. The pollen-stigma interaction in the grasses. 2. Pollen tube penetration and the stigma response in *Secale*. Acta Bot. Neerlandica, 30(4): 289-307.

Heslop-Harrison, J. and Y. Heslop-Harrison. 1984. The disposition of gamete and vegetative cell nuclei in the extending pollen tubes of grass species *Alopecurus pratensis* L. Acta Bot. Neerlandica, 33(1): 131-134.

Heslop-Harrison, J. and Y. Heslop-Harrison. 1987a. An analysis of gamete and organelle movement in the pollen tube of *Secale cereale* L. Plant Science, 51: 203-213.

Heslop-Harrison, J. and Y. Heslop-Harrison. 1987b. The disposition of gamete and vegetative cell nuclei in the extending pollen tubes of a grass species *Alopecurus pratensis* L. Ann. Bot. (London), 60(5): 595-601.

Heslop-Harrison, J. and Y. Heslop-Harrison. 1988. Tubulin and male gamete interconnections in the pollen tube of grass *Alopecurus pratensis*. Ann. Bot. (London), 61: 249-254.

Heslop-Harrison, J. and Y. Heslop-Harrison. 1989a. Actomyosin and movement in the angiosperm pollen tube: an interaction of some recent results. Sexual Plant Reprod., 2: 199–207.

Heslop-Harrison, J. and Y. Heslop-Harrison. 1989b. Actomyosin and movement of the vegetative nucleus of the angiosperm pollen tube: association with the actin cytoskeleton. J. Cell Sci., 93: 229–308.

Heslop-Harrison, J. and Y. Heslop-Harrison. 1989c. Cytochalasin effects on structure and movement in the pollen tube of *Iris*. Sexual Plant Reprod., 2: 27–37.

Heslop-Harrison, J. and Y. Heslop-Harrison. 1989d. Myosin associated with the surfaces of organelles, vegetative nuclei and generative cells in angiosperm pollen grains and tubes. J. Cell Sci., 94: 319–325.

Heslop-Harrison, J., Y. Heslop-Harrison and R. B. Knox. 1973. The callose rejection reaction: a new bioessay for incompatibility in Cruciferae and Compositae. Incompatibility Newsletter, 3: 75-76.

Heslop-Harrison, J., Y. Heslop-Harrison and J. Barber. 1975. The stigma surface in incompatibility response. Proc. Royal Soc. London. Series A, 188: 287-297.

Heslop-Harrison, J., Y. Heslop-Harrison, M. Cresti, A. Tiezzi and A. Moscatelli. 1988. Cytoskeleton elements, cell shaping and movement in the angiosperm pollen tube. J. Cell Sci., 91: 49–60.

Heslop-Harrison, Y. 1977. The pollen-stigma interaction: pollen tube penetration in *Crocus*. Ann. Bot. (London), 41: 913-922.

Heslop-Harrison, Y. 1981. Stigma characteristics and angiosperms taxonomy. Nordic J. Bot., 1(3): 401-420.

Heslop-Harrison Y. and K. R. Shivanna. 1977. The receptive surface of angiosperm stigma. Ann. Bot. (London), 41: 1233-1258.

Heslop-Harrison, Y. and J. Heslop-Harrison. 1982. Pollen-stigma interaction in the Leguminosae: the secretory system of the style in *Trifolium pratense*. Ann. Bot. (London), 50: 635-645.

Heslop-Harrison, Y. and J. Reger. 1988. Tissue organisation, pollen receptivity and pollen tube guidance in normal and mutant stigmas of the grass *Pennisetum typhoides* (Biam.) Stapf et Hubb. Sexual Plant Reprod., 1: 182–193.

Heslop-Harrison, Y., J. Heslop-Harrison and B. J. Reger. 1985. The pollen-stigma interaction in the grasses. 7. Pollen-tube guidance and the regulation of tube number in *Zea mays* L. Acta Bot. Neerlandica, 34(2): 193-211.

Higashiyama, T., H. Kurowia S. Kawano, and T. Kurowia. 1997. Kinetics of double fertilization in *Torenia fournieri* based on direct observations of the naked embryo sac. Planta, 203: 101-110.

Higashiyama, T., H Kuroiwa, S. Kawano and T. Kuroiwa. 1998. Guidance in vitro of the pollen tube to the naked embryo sac of *Torenia fournieri*. Plant Cell, 10: 2019-2031.

Hildebrand, F. 1899. Die Keimung der Samen *Anemone apennina*. Berichte der Deutschen Bot. Gesellschaft, 17: 161-166.

Hill, A.W. 1906. The morphology and seedling structure of the geophilous species of *Peperomia*, together with some views on the origin of monocotyledons. Ann. Bot., 20(80): 395-427.

Hill, A.W. 1908. The origin of Monocotyledons. Ann. Bot., 22(88): 713-714.

Hill, A. W. 1938. The monocotyledon seedling of certain Dicotyledons with special reference to the Gesneriaceae. Ann. Bot. (London), N. S. 2: 127-143.

Hill, C. R. and P. R. Crane. 1982. Evolutionary cladistics and the origin of angiosperms, pp. 269-361. *In* K. A. Joysey and A. E. Friday (eds). Problems of Phylogenetic Reconstruction. Academic Press, New York, NY.

Hill, H. W. 1920. Studies in seed germination. Experiments with *Cyclamen*. Ann. Bot. (London), 34: 417-423.

Hill, J. P. and E. M. Lord. 1987. Dynamics of pollen tube growth in the wild radish *Raphanus raphanistrum* (Brassicaceae). II. Morphology, cytochemistry and ultrastructure of transmitting tissues and path of pollen tube growth. Amer. J. Bot., 74(7): 988-997.

Hill, T. G. 1900. The structure and development of *Triglochin maritimum*. Ann. Bot. (London), 14(53): 83-107.

Hill, T. G. and E. De Fraine. 1908-1910. On the seedling structure of Gymnosperms. Ann. Bot. (London), 1908. 22(88): 689-712; 1909. 23(90): 189-227; 1910. 23(91): 433-457.

Hirase, S. 1898. Études sur la fécondation et l'embryogénie du *Gingko biloba*. J. Coll. Sci. (Tokyo), 12, part.2:

Hodcent, E. 1969. Recherche sur les facteur entrant en jeu lors de l'alveolization des prothalles femelles de l'*Araucaria araucana*. Rev. Cytol. Biol. Végétales, 32: 155-164.

Hoekstra, F. A. and J. Bruinsma. 1975. Respiration and vitality of binucleate and trinucleate pollen. Physiol. Plantarum, 34: 221–225.

Hofmeister, W. 1847. Untersuchungen des Vorgangs bei der Befruchtung der Oenotheren. Bot. Zeitung (Berlin), 5: 785-792.

Hofmeister, W. 1849. Die Entstehung des Embryo der Phanerogamen. Eine Reihe microscopischer Untersuchungen von Wilhelm Hofmeister. Verlag von Fr. Hofmeister, Leipzig.

Hofmeister, W. 1858. Neuere Beobachtungen über Embryobildung der Phanerogamen. Jahrb. Wissenschaftliche Bot., 1: 82-186.

Hofmeister, W. 1859. Neuere Beitrage zur Kenntnis der Embryobildung der Phanerogamen. I. Dikotyledonen mit ursprunglich einzelligem, nur durch Zellteilung wachsendem Endosperm. Abhandlungen Königlich Sächsechen Gesellschaft der Wissenschaften, 6: 535-672.

Hofmeister, W. 1861. Neuere Beitrage zur Kenntniss der Embryobildung der Phanerogamen. II. Monocotyledonen. Abhandlungen Königlich Sächsechen Gesellschaft der Wissenschaften, 7: 631-760.

Hoggart, R. M. and A. E. Clarke. 1984. Arabinogalactans are common components in angiosperm styles. Phytochemistry, 23: 1571-1573.

Holden, H. S. and D. Bexon. 1918. Observation on the anatomy of teratological seedlings. I. On the anatomy of some polycotylous seedlings of *Chieranthus cheiri*. Ann. Bot. (London), 32: 513-530.

Holden, H. S. and M. E. Daniels. 1921. Observation on the anatomy of teratological seedlings. IV. Further studies on the anatomy of atypical seedlings of *Impatiens roylii* Walp. Ann. Bot. (London), 35(140): 461-492.

Holding, D.R., R.J. MacKenzie and S.A. Coomber. 1994. Genetic and structural analysis of five *Arabidopsis* mutants with abnormal root morphology generated by the seed transformation method. Ann. Bot., 74(3): 193-204.

Holm, P. B., S. Khudsen, P. Mouritzen, D. Negri, F. L. Olsen and C. Roue. 1994. Regeneration of fertile barley plants from mechanically isolated protoplasts of the fertilized egg cell. Plant Cell, 6: 531-543.

Holtorp, H. E. 1944. Tricotyledony. Nature, 153(3870): 13-14.

Hopkins, C. R. 1991. Polarity signals. Cell (Cambridge), 66: 827-829.

Horner, H. T. and H. J. Arnott. 1966. A histochemical and ultrastructural study of pre- and postgerminated *Jucca* seeds. Bot. Gazette, 127(1): 48-64.

Howard, A. and S. R. Palk. 1953. Synthesis of RNA in normal and irradiated cells and its relation to chromosome breakage. Heredity, 6: 261-273.

Hu, S. Y. 1990. Male germ unit and sperm heteromorphism: the current status. Acta Bot. Sinica, 32(2): 230-240.

Hu, Shi-yi, Zhu Cheng and S. Y. Zee. 1983. Transfer cells in suspensor and endosperm during early embryogeny of *Vigna sinensis*. Acta Bot. Sinica, 25(1): 1-7.

Huang, B. Q. and S. D. Russell. 1992. Female germ unit: organization, reconstruction and isolation. Int. Rev. Cytol., 140: 233-293.

Huang, B. Q. and S. D. Russell. 1994. Fertilization in *Nicotiana tabacum*. Cytoskeletal modifications in the embryo sac during synergid degeneration. A hypothesis for short-distance transport of sperm cells prior to gamete fusion. Planta, 194: 200-214.

Huang, B. Q. and W. F. Sheridan. 1994. Female gametophyte development in maize. Microtubular organization and embryo sac polarity. Plant Cell, 6: 845-861.

Huang, B. Q., E. S. Pierson, S. D. Russell, A. Tiezzi and M. Cresti. 1993a. Cytoskeletal organization and modification during pollen tube arrival, gamete delivery and fertilization in *Plumbago zeylanica*. Zygote, 1: 143-154.

Huang, B. Q., G. W. Strout and S. D. Russell. 1993b. Fertilization in *Nicotiana tabacum*: ultrastructural organization of propane jet-frozen embryo sacs in vivo. Planta, 191: 256-264.

Hüber, H. 1969. Die Samenmerkmale und Verwandtschafsverhaltnisse der Liliifloren. Mitteilungen Bot. Staatssammlung München, 8(2): 219-538.

Hughes, M. B. and E. B. Babcock. 1950. Self-incompatibility in *Crepis foetida* subsp. *rhaeadifolia*. Genetics, 35: 570-588.

Hu-Han and Yang Honjyuan. 1986. Haploids of Higher Plants in vitro. Acad. Publ. Beijing, China. Springer-Verlag, Berlin.

Humphrey, J. E. 1896. The development of the seed in the Scitamineae. Ann. Bot. (London), 10(37): 1-40.

Humphreys, T. E. 1978. A model for sucrose transport in the maize scutellum. Phytochemistry, 17(4): 679-684.

Humphreys, T. E. and L. A. Garrard. 1964. Glucose uptake by the corn scutellum. Phytochemistry, 3(4): 647-656.

Hush, J. M. and R. L. Overall. 1991. Electrical and mechanical fields orient cortical microtubules in higher plant tissues. Cell Biol. Int. Reports, 15(7): 551-559.

Hutchinson, J. 1959. The Families of Flowering Plants. Clarendon Press, Oxford, UK, 2 vols.

Huxleu, J. S. and G. R. De Beer. 1936. The principles of experimental embryology. Biomedgiz, Moscow-Leningrad, Russia.

Hyde, B. B. 1970. Mucilage-producing cells in the seed coat of *Plantago ovata*: developmental fine structure. Amer. J. Bot., 57(210): 1197-1206.

Ihle, J. N. and L. Dure III. 1972. The developmental biochemistry of cotton seed embryogenesis and germination. III. Regulation of the biosynthesis of enzymes utilized in germination. J. Biol. Chem., 247(16): 5048-5055.

Ilyina, G. M. 1981. Family Papaveraceae, pp. 142-150. *In* M. S. Yakovlev (ed.). Comparative Embryology of Flowering Plants. Winteraceae—Juglandaceae. Nauka, Leningrad, Russia.

Ilyina, G. M. 1988a. Family Fumariaceae, pp. 227-231. *In* A. L. Takhtajan (ed.). Comparative Anatomy of Seeds. Dicotyledons. Magnoliidae, Ranunculidae. Nauka, Leningrad, Russia, vol. 2.

Ilyina, G. M. 1988b. Family Papaveraceae, pp. 208-221. *In* A. L. Takhtajan (ed.). Comparative Anatomy of Seeds. Dicotyledons. Magnoliidae, Ranunculidae. Nauka, Leningrad, Russia, vol. 2.

Ilyina, G. M. and V. A. Poddubnaya-Arnoldi. 1983. Family Aizoaceae, 16-19. *In* M. S. Yakovlev (ed.). Comparative Embryology of Flowering Plants. Phytolaccaceae—Thymelaeaceae. Nauka, Leningrad, Russia.

Irish, V. L. and J. M. Sussex. 1992. A late map of the *Arabidopsis* embryonic shoot apical meristem. Development, 115: 745-753.

***Isaeva, V. V. 1990.** Breaking of zygote polarization in brown alga *Pelvetia wrightii*; effect of cytochalazin. Ontogenez, 21(4): 422-428

Isaeva, V. V. 1994. Cells in Morphogenesis. Nauka, Moscow, Russia.

Ivanov, V. B. 1987. Proliferation of cells in plants, pp. 1-219. *In* Results of Science and Technology, series Cytology, 5, Moscow, Russia.

***Ivanov, V. B. and T. K. Litinskaya. 1967.** Simultaneous staining of proteins and carbohydrates with procion dye-stuffs. Cytology, 9: 112-116.

Iwanami, Y. 1959. Physiological studies of pollen. J. Yokohama Municipal Univ. Series C, Natural Sciences, 116: 1–137.

Jackson, B. D. 1916. A Glossary of Botanic Terms. Duckworth & Co.; J.B. Lippincott Co., London, UK.

Jackson, D., B. Veit and S. Hake. 1994. Expression of maize KNOTTED-*1* related homeobox genes in the shoot apical meristem predicts patterns of morphogenesis in the vegetative shoot. Development, 120: 405-413.

Jacob, F. 1982. The Actual and the Possible. Univ. Washington Press, Seattle, WA, USA.

Jacobs, T. W. 1992. Control of the cell cycle. Developm. Biol., 153: 1-15.

Jacobs, T. W. 1995. Cell cycle control. Ann. Rev. Plant Physiol., Molec. Biol., 46: 317-339.

Jacobsen, J. V. and P. M. Chander. 1988. Gibberellin and abscisic acid in germinating cereals, pp. 164-193. *In* P. J. Davies (ed.). Plant Hormones and Their Role in Plant Growth and Development. Kluwer Acad. Publ., Dordrecht, Netherlands.

Jacobsen, J. V., R. B. Knox and N. A. Pyliotis. 1971. The structure and composition of aleurone grains in the barley aleurone layer. Planta, 101: 189-209.

Jacobsson-Stiasny, E. 1914. Versuch einer phylogenetischen verwertung der Endosperm- und Haustorialbildung bei den Angiospermen. S. B. Akad. Wissenschaften, Wien, 123: 467-603.

Jacobsson-Stiasny, E. 1918. Zur Embryologie der Aristolochiaceae. Denkschriften der Kaiserlichen Akad. Wissenschaften, Wien. Mathematisch-Naturwissenschaftliche Klasse, 95(1): 65-77.

Jacques-Félix, H. 1957. Sur une interprétation nouvelle de l'embryon des Graminées. La nature axillaire de la gemmule. C.R. Hebdomadaires Séances l'Acad. Sci., 245: 1260-1263.

Jacques-Félix, H. 1958. Évolution des idees sur l'embryon des Gramineae. 83e Congres des Sociétes Savantes. Poitiers, pp. 323-337.

Jacques-Félix, H. 1982. Les monocotylédones n'out pas de cotylédon. Bull. Mus. Nat. d'Histoire Naturelle (Paris), Ser. 4., 1-2: 3-40.

Jacques-Félix, H. 1988. Les Liliopsida (ex Monocotylédones) n'out pas de cotylédon. II. La préfeuille de la planluse: ses rapports avec cells des axes feluilés. Bull. Mus. Natl. d'Histoire Naturelle (Paris), 4(10): 275-333.

Jaffe, L. A. and R. Nuccitelli. 1977. Electrical control of development. Ann. Rev. Biophys. Bioeng., 6: 445-476.

Jaffe, L. A., M. N. Weisenseel and L. F. Jaffe. 1975. Calcium accumulations within the growing tips of pollen tubes. J. Cell Biol., 67: 488–492.

Jaffe, L. F. 1980. Calcium explosion as triggers of development. Ann. New York Acad. Sci., 339: 86-101.

Janson, J. 1992. Pollen tube-pistil interaction and fertilization in *Lilium longiflorum*. Landbouwuniversiteit, Wageningen.

Jegla, D. E. and I. M. Sussex. 1989. Cell lineage patterns in the shoot meristem of the sunflower embryo in the dry seed. Developm. Biol., 131: 215-225.

Jensen, G. H. 1918. Studies on the morphology of wheat. Wash. State Agric. Exp. Sta. Bull., pp. 1-150.

Jensen, W. A. 1964a. Cell development during plant embryogenesis. Meristems and Differentiation. Brookhaven Symp. Biol., 16: 179-202.

Jensen, W. A. 1964b. Observation on the fusion of nuclei in plants. J. Cell Biol., 23(3): 669-672.

Jensen, W. A. 1965a. The ultrastructure and composition of egg and central cell of cotton. Amer. J. Bot., 52(8): 781-797.

Jensen, W. A. 1965b. The ultrastructure and histochemistry of the synergids of cotton. Amer. J. Bot., 52: 238-256.

Jensen, W. A. 1968a. Cotton embryogenesis. Polysome formation in the zygote. J. Cell Biol., 36: 403-406.

Jensen, W. A. 1968b. Cotton embryogenesis: the tube-containing endoplasmic reticulum. J. Ultrastructure Res., 22: 296-302.

Jensen, W. A. 1968c. Cotton embryogenesis: the zygote. Planta, 79(4): 346-366.

Jensen, W. A. 1972. The embryo sac and fertilization in Angiosperms. Univ. Hawaii, Harold L. Lyon Arboretum lecture, 3: 2-31.

Jensen, W. A. 1973. Fertilization of flowering plants. BioScience, 23(1): 21-27.

Jensen, W. A. 1976. The role of cell division in angiosperm embryology, pp. 391-405. *In* Cell Division of Higher Plants, London, UK.

Jensen, W. A. and D. B. Fisher. 1967 (1968). Cotton embryogenesis: double fertilization. Phytomorph., 17(1-4): 261-269.

Jensen, W. A. and D. B. Fisher. 1968a. Cotton embryogenesis: the entrance and discharge of the pollen tube in the embryo sac. Planta, 78(2): 158-183.

Jensen, W. A. and D. B. Fisher. 1968b. Cotton embryogenesis: the sperm. Protoplasma, 65(3): 277-286.

Jensen, W. A. and D. B. Fisher. 1969. Cotton embryogenesis: the tissue of the stigma and style and their relation to the pollen tube. Planta, 84: 97-121.

Jensen, W. A., M. E. Ashton and C. A. Beasley. 1983. Pollen-tube-embryo sac interaction in cotton, pp. 67–72. *In* Mulcahy, D. L. and E. Ottaviano (eds). Pollen: Biology and Implications for Plant Breeding. Elsevier Sci. Publ. Co., Elsevier Biomed. Press, New York, NY.

Jin-a-Jin, C. J. Boihon, and J. H. Wong. 1993. Tioredoxin-linked changes in regulatory properties or barley p-amilase/subtilisin inhibitor protein. Plant Physiol. Biochem., 31(6): 1112-1121.

Johansen, D. A. 1930a. Studies on the morphology of the Onagraceae. II. Embryonal manifestations of fasciation in *Clarkia elegans*. Bot. Gazette (Chicago), 90(1): 75-91.

Johansen, D. A. 1930b. Studies on the morphology of the Onagraceae. IV. *Stenosiphon linifolium*. Bull. Torrey Bot. Club, 57(5): 315-326.

Johansen, D. A. 1931a. Studies on the morphology of the Onagraceae. III. *Taraxia ovata*. Ann. Bot. (Chicago), 45(177): 111-124.

Johansen, D. A. 1931b. Studies on the morphology of the Onagraceae. V. *Zauschneria latifolia*, typical of a genus characterized by irregular embryology. Ann. New York Acad. Sci., 33: 1-26.

Johansen, D. A. 1933. Studies on the morphology of the Onagraceae. VII. *Gayophytum ramosissimum*. Bull. Torrey Bot. Club, 60(1): 1-8.

Johansen, D. A. 1934. Studies on the morphology of the Onagraceae. VIII. *Circaea pacifica*. American J. Bot., 21: 508-510.

Johansen, D. A. 1945. A critical review of the present status of plant embryology. Bot. Review, 11: 87-107.

Johansen, D. A. 1950. Plant Embryology. Chronica Botanica, Waltham, MA, USA.

Johansson, M. and B. Walles. 1993. Functional anatomy of the ovule in broad bean (*Vicia faba* L.) I. Histogenesis prior to and after pollination. Int. J. Plant Sci., 154: 80-88.

Johansson, M. and B. Walles. 1994. Functional anatomy of the ovule in broad bean (*Vicia faba* L.): ultrastructural seed development and nutrient pathways. Ann. Bot. (London), 74(3): 233-244.

Jöhnson, D. S. 1900a. On the development of *Saururus cernuus*. Bull. Torrey Bot. Club, 27(7): 365-372.

Jöhnson, D. S. 1900b. On the endosperm and embryo of *Peperomia pellucida*. Bot. Gazette, 30(1): 1-11.

Johri, B. M. 1935. Studies in the family Alismataceae. 3. *Sagittaria quanensis* H.B.K. and *S. batifolia* Willd. J. Indian Bot. Soc., 22(1): 33-48.

Johri, B. M. 1969. Morphogenetic response of mature endosperm, p. 103. *In* Abstracts XI Int. Bot. Cong., Seattle, WA, USA.

Johri, B. M. 1971. Differentiation in plant tissue cultures, pp. 159-186. *In* Proc. 58th Indian Sci. Cong. Assoc. Bangalore, Pt. II.

Johri, B. M. 1975. Organogenesis in the tissue cultures of endosperm and embryo, p. 295. *In* Abstracts XII Int. Bot. Cong., Leningrad, July 3-10, 1975. Nauka, Leningrad, USSR.

Johri, B. M. [ed.] 1984. Embryology of Angiosperms. Springer-Verlag, Berlin.

Johri, B. M. and R. N. Kapil. 1953. Contribution to the morphology and life history of *Acalypha indica* L. Phytomorph., 3(3): 137-151.

Johri, B. M. and R. N. Konar. 1955. A contribution to the morphology and embryology of *Ficus religiosa* Linn. Current Sci., 24(11): 382-385.

Johri, B. M., J. S. Agrawal, and G. Sudha. 1957. Morphological and embryological studies in the family Loranthaceae—I. *Helicanthes elastica* (Desr.) Dans. Phytomorph., 7(3-4): 336-354.

Johri, B. M., K. Ambegaokar and P. S. Srivastava. 1992. Comparative Embryology of Angiosperms, vol. 1, 2. Springer-Verlag, Berlin.

Jones, P. A. 1977. Development of the quiescent center in maturing embryonic radicules of pea (*Pisum sativum* L. ev. Alaska). Planta, 135: 233-240.

Jones, R. L. and I. D. J. Phillips. 1966. Organs of gibberellin synthesis in light-grown sunflower plants. Plant Physiol., 41(8): 1381-1386.

Jones, T. J. and T. L. Rost. 1989. Histochemistry and ultrastructure of rice (*Oryza sativa* L.) zygotic embryogenesis. Amer. J. Bot., 76: 504-513.

Jong, K. 1973. *Streptocarpus* (Gesneriaceae) and the phyllomorph concept. Acta Bot. Neerlandica, 22: 217.

Jong, K. 1978. Phyllomorphic organisation in rosulate *Streptocarpus.* Notes Royal Bot. Garden, 36(2): 369-396.

Jong, K. and B. L. Burtt. 1975. The evolution of morphological novelty exemplified in the growth patterns of some Gesneriaceae. New Phytologist, 75: 297-311.

Jonsson, S. 1964. Nouveau type de parthenogènese haploide chez les Algues: la partenogénese plasmogamic. C.R. Hebdomadaires Séances l'Acad. Sci., 258: 2145-2148.

Jonsson, S. 1971. Des lignées apocaryogamiques et segregation dans la population de l'*Acrosiphonia arcta,* Chlorophyceé marine d'Heligoland. Helgoländer Wissenschaftliche Meeresuntersuchungen, 22: 281-294.

Jörgensen, C. A. 1923. Studies on Callitrichaceae. Bot. Tidsskrift, 38(1): 81-126.

Joshi, P. C., A. M. Wadhwani and B. M. Johri. 1967. Morphological and embryological studies of *Gossypium* L. Proc. Natl. Inst. Sci. India, 33(1-2): 37-93.

Juel, H. O. 1907. Studien über die Entwicklungsgeschichte von *Saxifraga granulata.* Nova Acta Regiae Soc. Scient. Upsaliensis, 4(1): 1-41.

Juguet, M. 1969. Embryogénie des Juncacées. Développement de l'embryon chez le *Luzula pilosa* (L.) Willd. C.R. Hebdomadaires Séances l'Acad. Sci., 268(25): 3036-3039.

Juguet, M. 1972. Observations embryogenesis chez les Cyperacées et les Juncacées. Abstr. 93 Congr. des Societes Savantes, Tours, 1968, Séc. Sci. 3 Paris, pp. 547-572.

Juguet, M. 1973. Expression précoce de la monocotylédonie et mise en place du cotylédon et du point végétatif de la tige chez quelques Monocotylédones avec quelques remarques sur les types de symétrie des embryons de Monocotylédones et de Dicotylédones. Bull. Soc. Bot. France, Mém. publiés Soc. Bot. France, Colloque Morphologie, PP. 337-353.

Juguet, M. 1989. Not all the Monocotyledons are as monocotyledonous as it is believed, pp. 206-222. *In* J. Paré and M. Bugnicourt (eds). Some Aspects and Actual Orientations in Plant Embryology. Univ. Picardie, Cedex, Amiens, France.

Juguet, M. 1990. Origin and stages of monocotyly in the monocotyledons, p. 68. *In* Abstracts XI Int. Symp. "Embryology and Seed Reproduction". July 3-7, 1990. Leningrad. Nauka, St. Petersburg.

Juguet, M. 1992. Origin and stages of monocotyly in the monocotyledons. I. A preliminary report, pp. 236-237. *In* Proc. XI Int. Symp. "Embryology and Seed Reproduction". July 3-7, 1990. Leningrad. Nauka, St. Petersburg.

Juguet, M. and J. Paré. 1989. Étude au microscope électronique à balayage de l'edification du cotylédon et du futur méristeme, caulinaire chez deux Monocotyledones, avec quelques remarques sur les divers états et l'origine de la monocotylie. Bull. Soc. Bot. France, 136(3): 225-233.

Juliano, J. B. and P. E. Alcala. 1933. Floral morphology of *Musa errans* (Blanco) Teodoro var. Botoan Teodoro. Philippine Agriculturist, 22: 91-126.

Juncosa, A. M. 1982. Developmental morphology of the embryo and seedling of *Rhizophora mangle* L. (Rhizophoraceae). Amer. J. Bot., 69(10): 1599-1611.

Junell, S. 1931. Die Entwicklungsgeschichte von *Circaeaster agrestis*. Svensk Bot. Tidskrift, 25: 238-270.

Jung, C. T. 1937. Developmental anatomy of the seedling of the rice plant. Bot. Gazette, 99: 786-802.

Jünger, E. 1869-1870. Über tricotiledone Embryonen. Uebersicht der Arbeiten und Veränderungen der Schlesischen Gesellschaft für Vaterlandishe Kultur. Breslau, 46: 137-142; 48: 95.

Jungers, V. 1931. Figures caryocinétiques et cloisonnement du protoplasme dans l'endosperme d'*Iris*. Cellule, 40(3): 293-354.

Jürgens, G. 1995. Axis formation in plant embryogenesis: cues and clues. Cell (Cambridge), 81: 467-470.

Jürgens, G. and U. Mayer. 1994. *Arabidopsis*, pp. 7-21. *In* J. B. L. Bard (ed.). Embryos. Colour Atlas of Development. Wolf Publ., London, UK.

Jürgens, G., U. Mayer, R. A. Torres Ruiz, T. Berleth and S. Misera. 1991. Genetic analysis of pattern formation in the *Arabidopsis* embryo, pp. 27-38. *In* K. Roberts (ed.). Molecular and Cellular Basis of Pattern Formation. Company of Biologists, Ltd., Cambridge, UK.

Jürgens, G., R. A. Torres Ruiz,T. Laux, U. Mayer and T. Berleth. 1994. Early events in apical-basal pattern formation in *Arabidopsis*, pp. 95-103. *In* G. Coruzzi and P. Puigdomenech (eds.). Plant Molecular Biology: Molecular-genetic Analysis of Plant Development and Metabolism. Springer, Berlin.

Jurkat, G. 1953. Anomalien bei *Pulsatilla*-Keimlingen. Diss. Tubinggen.

Kadej, A. and F. Kadej. 1983. The ultrastructure of the egg cell in *Lycopersicon esculentum* Miller at the moment of fertilization, pp. 249-252. *In* O. Erdelská (ed.). Proc. VII Int. Cytoembryological Symp. "Fertilization and Embryogenesis in Ovulated Plants". High Tatra (Račkova dolina), 14-17 June, 1982. Veda, Bratislava, Czechoslovakia.

Kadej, A. and F. Kadej. 1985. Sperm cells in the fertilization process in *Lycopersicon esculentum*, p. 149. *In* M. T. M. Willemse and J. L. Van Went (eds). Proc. 8th Int. Symp. "Sexual Reproduction in Seed Plants, Ferns and Mosses". August 20-24, 1984. PUDOC, Wageningen, Netherlands.

Kadej, F. and A. Kadej. 1981. Ultrastructural organization of the tomato embryo sac at the time of fertilization. Acta Soc. Bot. Poloniae, 50: 139-142.

Kajale, L. 1954. A contribution to the embryology of the Phytolaccaceae. 2. Fertilization and the development of embryo, seed and fruit in *Rivina humilis* and *Phytolacca dioica*. J. Indian Bot. Soc., 33(3): 206-225.

*Kakhidze, N. T. 1954. Change of embryo sac elements during fertilization in tomatoes. News USSR Acad. Sci., Series Biology, 1: 71-78.

Kalinin, F.L. 1956. Physiological-biochemical peculiarities of embryonal development of plants. Ph.D. thesis, Moscow, Russia.

*Kamalova, G. V. and A. P. Abukhovskaya. 1986. Embryology of the species of Malvaceae family. FAN, Tashkent, Uzbek SSR.

Kamalova, G. V., G. Ya. Zhukova and V. A. Rumi. 1983. Family Malvaceae, pp. 260-266. *In* M. S. Yakovlev (ed.). Comparative Embryology of Flowering Plants. Phytolaccaceae—Thymelaeaceae. Nauka, Leningrad, Russia.

Kamelina, O. P. 1977. Comparative embryological investigation of representatives of Dipsacaceae and Morinaceae families. Ph.D. thesis, Leningrad, Russia.

Kamelina, O. P. 1979. Embryo development in representatives of Dipsacaceae and Morinaceae families, pp. 82-89. *In* M. S. Yakovlev (ed.). Actual Questions of Angiosperm Embryology. Nauka, Leningrad, Russia.

Kamelina, O. P. 1980. Comparative Embryology of Families Dipsacaceae and Morinaceae. Nauka, Leningrad, Russia.

Kamelina, O. P. 1983. Family Frankeniaceae, pp. 175-179. *In* M. S. Yakovlev (ed.). Comparative Embryology of Flowering Plants. Phytolaccaceae—Thymelaeaceae. Nauka, Leningrad, Russia.

*Kamelina, O. P. 1984. On the embryology of genus *Escallonia* L. (Escalloniaceae). Bot. Journal (St. Petersburg), 69(10): 1304-1316.

Kamelina, O. P. 1987. Family Dipsacaceae, pp. 184-192. *In* T. B. Batygina and M. S. Yakovlev (eds). Comparative Embryology of Flowering Plants. Davidiaceae—Asteraceae. Nauka, Leningrad, Russia.

*Kamelina, O. P. 1991. Comparative embryological analysis as a method of phylogenetic systematics of flowering plants. Ph.D. thesis, Tashkent, Uzbek Republic.

*Kamelina, O. P. 1994. Embryology and systematics of the genus *Tetradiclis* (Tetradiclidaceae). Bot. Journal (St. Petersburg), 79(5): 11-27.

Kamelina, O. P. 1996. Endosperm of seed. On the typology and phylogenetic estimation, pp. 5-7. *In* Materials Conf. "Problems of Reproductive Biology of Plants". Perm State Univ., Perm, Russia.

Kamelina, O. P. and A. K. Dzevaltovsky. 1987. Family Lamiaceae, pp. 225-236. *In* T. B. Batygina and M. S. Yakovlev (eds). Comparative Embryology of Flowering Plants. Davidiaceae—Asteraceae. Nauka, Leningrad, Russia.

Kamelina, O. P. and O. B. Proskurina. 1987. Family Elaeagnaceae, pp. 105-110. *In* T. B. Batygina and M. S. Yakovlev (eds). Comparative Embryology of Flowering Plants. Davidiaceae—Asteraceae. Nauka, Leningrad, Russia.

Kamelina, O. P. and S. V. Shevchenko. 1987. Family Davidiaceae, pp. 7-12. *In* T. B. Batygina and M. S. Yakovlev (eds). Comparative Embryology of Flowering Plants. Davidiaceae—Asteraceae. Nauka, Leningrad, Russia.

Kamelina, O. P. and E. S. Teryokhina 1990a. Family Ruppiaceae, pp. 39-44. *In* T. B. Batygina and M. S. Yakovlev (eds). Comparative Embryology of Flowering Plants. Butomaceae—Lemnaceae. Nauka, Leningrad, Russia.

Kamelina, O. P. and E. S. Teryokhina 1990b. Family Zannichelliaceae, pp. 44-50. *In* T. B. Batygina and M. S. Yakovlev (eds). Comparative Embryology of Flowering Plants. Butomaceae—Lemnaceae. Nauka, Leningrad, Russia.

*Kamelina, O. P. and G. V. Yakovleva. 1982. Megasporogenesis and embryo sac development in *Dianthus caryophyllus* L. (Caryophyllaceae). Bot. Journal (St. Petersburg), 67(1): 80-87.

Kamelina, O. P. and G. Ya. Zhukova. 1987. Family Valerianaceae, pp. 171-177. *In* T. B. Batygina and M. S. Yakovlev (eds). Comparative Embryology of Flowering Plants. Davidiaceae—Asteraceae. Nauka, Leningrad, Russia.

***Kamelina, O. P. and N. A. Zhinkina. 1998.** On the embryology of *Ostrowskia magnifica* (Campanulaceae): the ovule and seed. Bot. Journal (St. Petersburg), 83(3): 9-20.

***Kamensky, F. 1886.** Comparative investigations on the development and structure of *Urticularia*. Trans. St. Petersburg Soc. Nature Experience, 17: 1-71.

Kandasamy, M. K., D. J. Paolillo,C. D. Faraday, J. B. Nasrallah and M. E. Nasrallah. 1989. The S-locus specific glycoproteins in *Brassica* accumulate in the cell wall of developing stigma papillae. Developm. Biol., 134: 462-472.

Kapil, R. N. 1956. A further contribution to the morphology and life history of *Chrozophora* Neck. Phytomorph., 6(3-4): 278-288.

Kapil, R. N. and M. R. Vijayaraghavan. 1962. Embryology and the systematic position of *Pentaphragma horsfieldii* (Mig.) Airy. Current Sci., 31(7): 270-272.

Kapil, R. N. and K. Ahluwalia. 1963. Embryology of *Peganum harmala* Linn. Phytomorph., 13(2): 127-140.

Kapil, R. N. and N. N. Bhandari. 1964. Morphology and embryology of *Magnolia*. Proc. Natl. Inst. Sci. India, 30(5-6).

Kapil, R. N. and M. R. Vijayaraghavan. 1965. Embryology of *Pentaphragma horsfieldii* (Mig.) Airy Shaw with a discussion on the systematic position of the genus. Phytomorph., 15(1): 93-102.

Kapil, R. N. and K. Walia. 1965. The embryology of *Philydrum lanuginosum* and the systematic position of the Philydraceae. Beiträge Biol. Pflanzen, 41(3): 381-404.

Kapil, R. N. and A. K. Bhatnagar. 1972. Endosperm in Euphorbiaceae—a critical appraisal. Advances of Plant Morph., 376-393.

Kapil, R. N. and A. K. Bhatnagar. 1991. Embryological evidence in angiosperm classification and phylogeny. Botanische Jahrb. Systematik, Pflanzengeschichte, Pflanzengeographie, 113: 309-338.

Kapil, R. N., P. N. Rustagi and V. Rukmani. 1968. A contribution to the embryology of Polemoniaceae. Phytomorph., 18(3): 403-412.

Kaplan, D. R. 1992. The relationship of cells to organisms in plants: problem and implications of an organismal perspective. Int. J. Plant Sci., 153(3): 28-37.

Kapoor, B. M. 1966. Contributions to the cytology of endosperm in some angiosperms. X. *Nigella damascena* and *N. saliva*. Caryologia, 19: 73-83.

Kapoor, T. 1980. Embryological and histochemical studies in some aquatic monocotyledons— Potamogetonaceae and Najadaceae. Ph.D. thesis, Univ. Delhi, Delhi, India.

Karas, I. and D. D. Cass. 1976. Ultrastructural aspects of sperm cell formation in rye: evidence for cell plate involvement in generative cell division. Phytomorph., 26: 36-45.

Karlström, P. O. 1974a. Embryological studies in Acanthaceae. III. The genera *Barleria* L. and *Grobbea* Harv. Svensk Bot. Tidskrift, 68(2): 121-135.

Karlström, P. O. 1974b. Embryological studies in Acanthaceae. IV. The genera *Asystasia* Bl. and *Chamaeranthemum* Nees. Svensk Bot. Tidskrift, 68(3): 325-328.

Karsten, H. 1860. Das Geschlechtsleben der Pflanzen und die Partenogenesis. Verlag von R. Decker, Berlin.

Kausik, S. B. 1938a. Studies in the Proteaceae. I. Cytology and floral morphology of *Grevillea robusta* Cunn. Ann. Bot. (London), 2(8): 899-910.

Kausik, S. B. 1938b. Studies in the Proteaceae. II. Floral anatomy and morphology of *Macadamia ternifolia* F. Muell. Proc. Indian Acad. Sci., 8(1): 45-62.

Kausik, S. B. 1942. Studies in the Proteaceae. VII. The endosperm of *Grevillea robusta* Cunn. with special reference to the structure and development of the vermiform appendage. Proc. Indian Acad. Sci., 16(5): 121-140.

Kausik, S. B. and K. Subramanyam. 1945a. Development of endosperm in *Lobelia nicotianaefolia* Heyne. Current Sci., 21: 269-278.

Kausik, S. B. and K. Subramanyam. 1945b. An embryological study of *Isotoma longiflora* Presl. Proc. Indian Acad. Sci., 21: 269-278.

***Kazachkovskaya, E. B., M. A. Vishnyakova and L. I. Oryol. 1991.** Self-incompatibility and ovule abortion before fertilization as a factor of reducing of seed productivity in *Medicago sativa* and *M. varia* (Fabaceae). Bot. Journal (St. Petersburg), 76(2): 41-53.

Kefeli, V. I. 1974. Natural Inhibitors of Growth and Phytohormones. Nauka, Moscow, Russia.

Kenrick, J. and R. B. Knox. 1981. Structure and cytochemistry of the stigma and style of some Australian species of *Acacia*. Australian J. Bot., 29: 733-745.

Khan, A. A. 1977. Seed dormancy: changing concepts and theories, pp. 29-45. *In* A. A. Khan (ed.). The Physiology and Biochemistry of Seed Dormancy and Germination. North-Holland Publ. Co., Amsterdam, Netherlands.

Khan, R. 1943. Contributions to the morphology of *Ephedra foliata* Boiss. II. Fertilization and embryogeny. Proc. Natl. Acad. Sci. India, 13: 357-375.

Khanna, P. 1964. Morphological and embryological studies in Nymphaeaceae. I. *Euryale ferox* Salisb. Proc. Indian Acad. Sci., 59(4): 237-247.

Khanna, P. 1965. Morphological and embryological studies in Nymphaeaceae. II. *Brasenia schreberi* Gmel. and *Nelumbo nucifera* Gaertn. Australian J. Bot., 13(3): 379-387.

Khanna, P. 1967. Morphological and embryological studies in Nymphaeaceae. III. *Victoria cruziana* D'Orb and *Nymphaea stellata* Willg. Bot. Magazine. (Tokyo), 80(950): 309-312.

Khavkin, E. E. 1982. Metabolism in germinating seeds, pp. 275-310. *In* A. A. Prokofiev (ed.). Seed Physiology. Nauka, Moscow, Russia.

***Khazova, I. I. 1970.** On the embryology of cultured species of sorghum. Bot. Journal (St. Petersburg), 55(1): 93-102.

Khera Paramjit K. and R A. E. Tilney-Bassett. 1976. Fine structural observation of embryo development in *Pelargonium × hortorum* Bailey with normal and mutant plastids. Protoplasma, 88(1): 7-23.

Khoo, U. and M. J. Wolf. 1970. Origin and development of protein granules in maize endosperm. Amer. J. Bot., 57: 1042-1050.

***Khudyak, M. I. 1952.** New data on cell formation in the endosperm of spring wheats. Reports USSR Acad. Sci., 84 (1): 143-145.

Khudyak, M. I. 1955. Peculiarities of endosperm development in wheat Kakhetinskaya. Bot. J. Ukrainian Acad. Sci., 12(3): 52-67 (in Ukrainian).

***Khudyak, M. I. 1963.** Endosperm of angiosperms (peculiarities of the development and the role in fruitification). Publ. Ukrainian SSR Acad. Sci., Kiev, Ukraine.

***Khudyak, M. I. 1964.** On the question about the role of unassimilating chloroplasts in the process of fructification. Zh. obshchei biol., 25(5): 364-370.

Khvedynich, O. A. 1969. Double fertilization process in barley (*Hordeum vulgare* L.). Ukrainian Bot. J., 26(5): 123-126 (in Ukrainian).

***Khvedynich, O. A. 1972.** Fertilization process and early stages of embryo and endosperm development at distant hybridization. Ph.D. thesis, Kiev, Ukraine.

***Khvedynich, O. A. and V. P. Bannikova. 1970a.** Change of chromatization of female and male sexual nuclei during fertilization in barley. Cytol. Genet., 4(5): 387-391.

***Khvedynich, O. A. and V. P. Bannikova. 1970b.** Disturbance in fertilization process at hybridization of barley with rye. Bot. Journal (St. Petersburg), 55(8): 1111-1120.

Khvedynich, O. A. and L. S. Serdyuk. 1977. Physical and chemical state of chromatin in fusing gametes of tobacco, pp. 83-84. *In* Abstracts III Cong. All-Union Soc. Geneticists and Plant-breeders Devoted to N. I. Vavilov. Vavilov Inst. Plant Breeding, Leningrad, Russia.

Khvedynich, O. A. and V. P. Bannikova. 1983. Chromatin transformations in fertilized nuclei of embryo sac, pp. 257-259. *In* O. Erdelská (ed.).Proc. VII Int. Cytoembryol. Symp."Fertilization and Embryogenesis in Ovulated Plants". High Tatra (Račkova dolina), 14-17 June, 1982, vol. 6. Veda, Bratislava, Czechoslovakia.

***Khvedynich, O. A. and L. S. Serdyuk. 1986.** The state of chromatin in nuclei of embryo sac cells in *Triticale*. Bot. Journal (St. Petersburg), 71(5): 617-622.

Khvedynich, O. A., L. S. Serdyuk and V. P. Bannikova. 1978. On the rate of fertilization process, pp. 85-96. *In* N.E. Bulygin et al. (eds). Abstracts VI Cong. of All-Union Bot. Society, Kishinev, Moldova, Nauka, Leningrad, Russia.

Khvedynich, O. A., V. P. Bannikova, and L. S. Serdyuk. 1983. Cytochemical investigation of male gametophyte in *Lilium regale* Wils. Ukrainian Bot. J., 40(2): 57-59 (in Ukrainian).

Kiesselbach, T. A. 1949. The structure and reproduction of corn. Nebraska Agric. Exp. St. Bull., 161: 1-96.

Kihara, H. and J. Nishiyama. 1932. Different compatibility in reciprocal crosses of *Avena* with special reference to tetraploid hybrids between hexaploid and diploid species. Japanese J. Bot., 6(2): 245-305.

Kinges, H. 1961. Merkmale des Gramineenembryos. Bot. Jahrb. Systematik, Pflanzengeschichte, Pflanzengeographie, 81(1): 50-93.

Kirk, D. L., M. R. Kaufman, R. M. Keeling and K. A. Stamer. 1991. Genetic and cytological control of the asymmetric divisions that pattern the *Volvox* embryo. Development (suppl.), 1: 67-82.

Klebs, G. 1885. Beiträge zur Morphologie und Biologie der Keimung, pp. 536-635. *In* W. Pfeffer (ed.). Untersuchungen aus dem Botanischen Institut zu Tübingen, Bot. 1 part IV. Verlag von Wilhelm Engelmann, Leipzig.

***Kluchareva, M. V. 1957.** Fertilization in wheat. Zh. obsczcei biol., 18(4): 288-297.

Knox, R. B. 1984a. The pollen grain, pp. 197-271. *In* B. M. Johri (ed.). Embryology of Angiosperms, vol. 1. Springer-Verlag, Berlin.

Knox, R. B. 1984b. Pollen–pistil interactions, pp. 508-608. *In* A. Pirson and M.H. Zimmermann (eds.). Encyclopedia of Plant Physiology, New Ser., 17: Cellular Interactions. H.F. Linskens and J. Heslop-Harrison (eds). Springer, Berlin.

Knox, R. B., J. Heslop-Harrison and Y. Heslop-Harrison. 1975. Pollen wall proteins: localisation and characterisation of gametophytic and sporophytic fractions. Biol. J. Linnean Soc., 7(1): 177-187.

Knox, R. B., E. G. Williams and C. Dumas. 1986. Pollen, pistil and reproductive function in crop plants. Plant Breeding Rev., 4: 7-79.

Knox, R. B., A. Clarke, S. Harrison, P. Smith and J. J. Marchalonis. 1976. Cell recognition in plant determinants of the stigma surface and their pollen interactions. Proc. Natl. Acad. Sci., USA 73(8): 2788-2792.

Koernicke, M. 1903. Der heutige Stand der pflanzlichen Zellforschung. Berichte der Deutschen Bot. Gesellschaft, 21(1): 66–134.

Kof, E. M. and V. I. Kefeli. 1994. The Growth and Morphogenesis of *Acetabularia.* Nauka, Moscow, Russia.

Kolesova [Titova], G. E. 1996. Family Trapaceae, pp. 242-259. *In* A. L. Takhtajan (ed.). Comparative Anatomy of Seeds. Dicotyledons, 5. Mir i semya-95, St. Petersburg. Nauka, Leningrad, Russia.

Kolesova, G. E. and T. B. Batygina. 1988. Family Nelumbonaceae, pp. 157-162. *In* A. L. Takhtajan (ed.). Comparative Anatomy of Seeds, vol. 2. Dicotyledons. Magnoliidae—Ranunculidae. Nauka, Leningrad, Russia.

***Kolomytsev, G. G. 1977.** On the development of monocotyledonous plants and on the nature of their cotyledons. Trans. Leningrad Soc. Naturalists, 73(3): 78-99.

***Koltsov, N. K. 1935.** The role of gene in development physiology. Biol. Zh., 4: 753-774.

Komamine, A., R. Kawahara, M. Matsumoto, S. Sunabori, T. Toya, A. Fujiwara, M. Tsukahara, J. Smith, M. Ito, M. Fukuda, K. Nomura and T. Fujimura. 1992. Mechanisms of somatic embryogenesis in cell cultures: physiology, biochemistry and molecular biology. In Vitro Cell Development Biology, 28(1): 11-14.

Komar, G. A. 1985a. Family Alliaceae, pp. 78-82. *In* A. L. Takhtajan (ed.). Comparative Anatomy of Seeds. Monocotyledones. Nauka, Leningrad, Russia, vol. 1.

Komar, G. A. 1985b. Family Smilacaceae, pp. 132-133. *In* A. L. Takhtajan (ed.). Comparative Anatomy of Seeds. Monocotyledons. Nauka, Leningrad, Russia, vol. 1.

Konar, R. N. and K. Nataraja. 1965. Experimental studies in *Ranunculus sceleratus* L. Development of embryos from the stem epidermis. Phytomorph., 15(2): 132-137.

Konar, R. N. and H. F. Linskens. 1966. Physiology and biochemistry of the stigmatic fluid of *Petunia hybrida.* Planta, 71: 372-387.

Konar, R. N., E. Thomas, and H. E. Street. 1972. Origin and structure of embryoids arising from epidermal cells of the stem of *Ranunculus sceleratus* L. J. Cell Sci., 11: 77-93.

Konarev, V. G. and L. S. Serduyk. 1967. Determination of DNA state in cell, pp. 21-26. *In* V. G. Konarev (ed.). Methods of Investigations of Nucleic Acids in Plants. Nauka, Leningrad, Russia.

Koncz, C., N. Martini, R. Mayerhofer, Z. Koncz-Kalman,H. Korber, G. P. Redeui and J. Schell. 1989. High-frequency T-DNA mediated gene tagging in plants. Proc. Natl. Acad. Sci. USA 86: 8467-8471.

Koncz, C., R. Mayerhofer, Z. Koncz-Kalman, C. Nawrath, B. Reiss, G. P. Redeui and J. Schell. 1990. Isolation of a gene encoding a novel chloroplast protein by to T-DNA tagging in *Arabidopsis thaliana.* EMBO Journal, 9: 1337-1346.

Konings, H. 1995. Gravitropism of roots: an evaluation of progress during the last three decades. Acta Bot. Neerlandica, 44(3): 195-223.

Kopcewisz, J. and K. Madela. 1992. Phytochrome-controlled level of growth substances in etiolated oat seedlings. Acta Soc. Bot. Poloniae, 61(3-4): 381-388.

***Korchagina, I. A. 1964.** Comparative investigation on the germination mode and internal seed morphology of some birches. Bot. Journal (St. Petersburg), 49(10): 1487-1496.

Korchagina, I. A. 1981. Family Betulaceae, pp. 201-210. *In* M. S. Yakovlev (ed.). Comparative Embryology of Flowering Plants. Winteraceae—Juglandaceae. Nauka, Leningrad, Russia.

***Kordyum, E. L. 1967.** Cytoembryology of the Umbellate Family. Naukova dumka, Kiev, Ukraine.

***Kordyum, E. L. 1978.** Evolutionary Cytoembryology of Flowering Plants. Naukova dumka, Kiev, Ukraine.

***Korobova, S. N. 1959.** On the course of fertilization in maize (*Zea mays* L.). Reports USSR Acad. Sci., 127(4): 921-924.

Korobova, S.N. 1961. The process of fertilization in maize (*Zea mays* L.), pp. 514-519. *In* V. G. Alexandrov (ed.). Morphogenesis of Plants, vol. 1. Moscow State Univ., Moscow.

***Korobova, S. N. 1962.** On the embryology of maize. Trans. Komarov Bot. Inst. USSR Acad. Sci., Series VII, 5: 294-314.

***Korobova, S. N. 1977.** The movement of sperms in the pollen tube of angiosperms. Bot. Journal (St. Petersburg), 62(11): 1585-1589.

Korobova, S. N. 1979. The movement of sperms in the pollen tube and embryo sac of angiosperms, pp. 5-19. *In* M. S. Yakovlev (ed.). Actual Questions of Angiosperm Embryology, Nauka, Leningrad, Russia.

Korobova, S. N. 1982. Formation of female gametophyte, fertilization, development of embryo and endosperm in maize, pp. 151-176. *In* G. E. Shmaraev et al. (eds). Cultured Flora of USSR. Kolos, Moscow, Russia.

Korobova, S. N. 1983. Family Plumbaginaceae, pp. 59-64. *In* M. S. Yakovlev (ed.). Comparative Embryology of Flowering Plants. Phytolaccaceae—Thymelaeaceae. Nauka, Leningrad, Russia.

Korobova, S. N. 1985. Family Podostemaceae, pp. 80-83. *In* M. S. Yakovlev (ed.). Comparative Embryology of Flowering Plants. Brunelliaceae—Tremandraceae. Nauka, Leningrad, Russia.

Korobova, S. N. and N. A. Zhinkina. 1987. Family Campanulaceae, pp. 300-308. *In* T. B. Batygina and M. S. Yakovlev (eds). Comparative Embryology of Flowering Plants. Davidiaceae—Asteraceae. Nauka, Leningrad, Russia.

***Korobova-Semenchenko, L. V. 1977.** On the embryology of the family Saxifragaceae. VII. Departure during endosperm development in *Bergenia crassifolia* (L.) Fritsch. Vestnik Moskov. Gosudarst. Univ., Series 16, Biology, 2: 81-84.

Kościńska, M. 1980. Embryo and endosperm in *Rudbekia bicolor* Nutt. Acta Biol. Cracoviensia, Series Bot., 22(1): 71-79.

Kościńska-Pajak, M. 1985. Caryological instability in semigamic *Rudbeckia bicolor* Nutt. Acta Biol. Cracoviensia, Series Bot., 27: 39-55.

***Kostrikova, L. N. 1955.** Embryological investigation of *Vicia faba* L. Bull. Moscow Soc. Naturalists, Biological Dept., 60(6): 101-106.

***Kostrikova, L. N. 1956.** Embryological investigation of *Vicia faba* L. Vestnik Moskov. Gosudarst. Univ., Series Geol., Geogr., Biol., Soil Science, 2: 59-80.

***Kostrikova, L. N. 1984.** On the endosperm of angiosperms. Vestnik Moskov. Gosudarstv. Univ., Series Geol., Geogr., Biol., Soil Science, 4: 3-9.

***Kostrjukova, K. Ju. 1951.** On the question of heterogeneity of sperms-cells, derivated from the single pollen tube. Bot. Journal (St. Petersburg), 8(3): 17-21.

***Kostrjukova, K. Ju. 1968.** Sperm heterogeneity in double fertilization, pp. 107-111. *In* V. P. Bannikova (ed.). Materials All-Union Symp. Plant Embryology. Naukova dumka, Kiev, Ukraine.

Kostrjukova, K. Ju. and M. V. Chernoyarov. 1938. The observation of the pollen tube germination in *Clivia miniata* Hort. *in vitro*. *In* The Anthology in Memory of Academician A. F. Fomin. Kiev, Ukraine. (in Ukrainian).

***Kostrjukova, K. Ju. and G. K. Benetskaya. 1958.** Do the further development of embryology prove the truth of S. G. Nawaschin's theory about the independent mobility of male gametes in angiosperms. Reports Armenian SSR Acad. Sci., 11(9): 7-24.

Kovalev, A. G. 1977. Growth of roots in *Quercus robur* L. and *Aesculus hippocastanum* L. seedlings (cellular analysis). Ph.D. thesis, Moscow State Univ., Moscow, Russia.

***Kozhanchikov, V. I. 1967.** Morphological characters of seeds in family Caryophyllaceae and possible trends of their evolution. Bot. Journal (St. Petersburg), 52(9): 1277-1286.

***Kozo-Polyansky, B. M. 1949.** The modes of representation of plant evolution system. Bot. Journal (St. Petersburg), 34(3): 245-252.

Kozo-Polyansky, B. M. 1965. Handbook of Higher Plant Systematics. Publ. Voronezh Univ., Voronezh, Russia.

Kranz, E. and H. Lörz. 1993. In vitro fertilization with isolated single gametes results in zygotic embryogenesis and fertile maize plants. Plant Cell, 5: 739-746.

Kranz, E. and H. Lörz. 1994. In vitro fertilization of maize by single egg and sperm cell protoplast fusion mediated by high calcium and high pH. Zygote, 2: 125-128.

Kranz, E. and T. Dresselhaus. 1996. In vitro fertilization with isolated higher plant gametes. Trends Plant Sci., 1: 82-89.

Kranz, E. and H. Lörz. 1996. In vitro fertilization with angiosperm gametes, p. 27. *In* Abstracts 14th Int. Cong. Sexual Plant Reprod. "Plant Reproduction '96", February 18-23, 1996, Lorne, Victoria, Australia. Cumberland Resort, Lorne, Victoria, Australia.

Kranz, E., J. Bautor and H. Lörz. 1990. In vitro fertilization of single, isolated gametes, transmission of cytoplasmic organelles and cell reconstitution of maize (*Zea mays* L.), pp. 252-257. *In* A. H. J. Nijkamp, L. H. W. Van Der Plas and J. Van Aartrijk (eds). Progress in Plant Cellular and Molecular Biology. Kluwer Acad. Publ., Dordrecht, Netherlands.

Kranz, E., J. Bautor and H. Lörz. 1991a. Electrofusion-mediated transmission of cytoplasmic organelles through the in vitro fertilization process, fusion of sperm cells with synergids and central cells, and cell reconstitution in maize. Sexual Plant Reprod., 4: 17-21.

Kranz, E., J. Bautor and H. Lörz. 1991b. In vitro fertilization of single, isolated gametes of maize mediated by electrofusion. Sexual Plant Reprod., 4: 12-16.

Kranz, E., P. Von Wiegen and H. Lörz. 1995. Early cytological events after induction of cell division in egg cells and zygote development following in vitro fertilization with angiosperm gametes. Plant Journal, 8: 9-23.

Kranz, E., G. Kumlehn and T. Dresselhaus. 1999. Fertilization and zygotic embryo development in vitro, pp. 337-349. *In* M. Cresti *et al.* (eds). Fertilization in Higher Plants. Molecular and Cytological Aspects. Springer-Verlag, Berlin.

Kranz, E., P. Von Wiegen, H. Quader and H. Lörz. 1998. Endosperm development after fusion of isolated, single maize sperm and central cells in vitro. Plant Cell, 10: 511-524.

Krasnikov, L. G. 1988. Fertilization in *Butomus umbellatus* L., 22-23. *In* Trans. Youth Conf. Leningrad Botanists, Leningrad, Russia. VINITI 14.07.88 N 5684-B 88, 3.

Krasnikov, L. G. 1989. Embryology of the family Butomaceae. Ph.D thesis, Leningrad, Russia.

Krasnikov, L. G. 1992. New conception of monocotyledon embryo origin, pp. 284-285. *In* Proc. XI Int. Symp. "Embryology and Seed Reproduction". Leningrad, 3-7 July, 1990. Nauka, St. Petersburg, Russia.

***Krasovskaya, I. V. 1952.** Anatomical-morphological regularities during the development of root system in cereals. Uchenye zapiski Saratovskogo. Univ., 35: 15-70.

Krause, K. E. H. 1880. Drei Kotyledonen. Archiv Ver. Freunde Naturg. Mecklenburg, 236 P.

Krenke, N. P. 1950. Regeneration in Plants. Publ. USSR Acad. Sci., Moscow-Leningrad, Russia.

Krishna Iyengar, C. V. 1940. Development of the embryo sac and endosperm-haustoria in some members of Scrophulariaceae. V. *Ilysanthes hyssopioides* Benth and *Bonnaya tenuifolia* Sprengh. J. Indian Bot. Soc., 19: 5-17.

Krishna Iyengar, C. V. 1942. Development of the seed and its nutritional mechanism in Scrophulariaceae. Part. 1.*Rhamphicarpa longifolia* Benth., *Centranthera hispida* Br., and *Pedicularis zeylanica* Benth. Proc. Natl. Inst. Sci. India, 8: 249-261.

Krishna Iyengar, C. V. 1947. Development of the seed and its nutritional mechanism in Scrophulariaceae. Pt. II. J. Mysore Univ., N. S., B., 7B(2): 82-98.

Krishnamurthy, K. V. 1988. Endosperm controls symmetry changes in the developing embryos of angiosperms. Proc. Indian Acad. Sci., 98(4): 257-259.

Kristie, D. N. and P. A. Jolliffe. 1986. High resolution studies of growth oscillations during stem elongation. Canadian J. Bot., 64(11): 2399-2405.

Kroh, M. and J. P. F. G. Helsper. 1974. Transmitting tissue and pollen tube growth, pp. 167-178. *In* H. F. Linskens (ed.). Fertilization in Higher Plants. North-Holland Publ. Co., Amsterdam, Netherlands.

Kronestedt, E. and B. Walles. 1986. Anatomy of *Strelitzia reginae* flower (Strelitziaceae). Nordic J. Bot., 6: 307-320.

Kronestedt, E., B. Walles and J. Alkemar. 1986. Structural studies of pollen tube growth in the pistil of *Strelitzia reginae*. Protoplasma, 131: 224–232.

Kropf, D. L. 1992. Establishment and expression of cellular polarity in fucoid zygotes. Microbiol. Rev., 56: 316-339.

Kropf, D. L. 1994. Review: Cytoskeletal control of cell polarity in a plant zygote. Developm. Biol., 165(2): 361-371.

Kropf, D. L., B. Kloareg and R. S. Quatrano. 1988. Cell wall is required for fixation of the embryonic axis in *Fucus* zygotes. Science, 239: 187-190.

Kudlicka, K., M. Kuras and B. Rodkiewicz. 1980. The ultrastructure of the nucellus and the gametophyte in developing ovules of *Stellaria media*. Acta Medica Polona, 21(4): 363-364.

Kudlicka, K., M. Kuras and B. Rodkiewicz. 1981. Embryo sac development in *Stellaria media*. Acta Soc. Bot. Poloniae, 50(1-2): 143-146.

***Kudoyarova, G. R. and I. Yu. Usmanov. 1991.** Hormones and mineral nutrition. Physiol. Biochem. Cultured Plants, 23(3): 232-244.

***Kudryashov, I. V. 1964.** The origin of monocotyledony in monocots in Helobiae as an example. Bot. Journal (St. Petersburg), 49(4): 473-486.

Kuijt, J. 1969. The Biology of Parasitic Flowering Plants. Univ of California, Berkeley, CA, USA.

Kuiper, D., J. Schuit and P. J. C. Kuiper. 1988. Effects of internal and external cytokinin concentration on root growth and shoot to ratio of *Plantago major* ssp. pleiosperma at different nutrient conditions. Plant Soil, 3(1): 231-236.

Kulaeva, O. N. 1982. Hormone regulation of physiological processes in plants on the level of RNA and protein synthesis, pp. 1-84. *In* 41st Timiryazev Readings. Nauka, Moscow, Russia.

Kurtz, E. B. and A. Miramon. 1957. A system for the study of fat synthesis in flax embryos. Plant Physiology, 37: 32.

Kutschera, U. 1994. The current status of the acid-growth hypothesis. New Phytologist, 126(10): 549-569.

Kvaale, A. and O. A. Olsen. 1986. Rates of cell division in developing barley endosperm. Ann. Bot. (London), 57: 829-833.

Kyle, D. J. and E. D. Styles. 1977. Development of aleurone and sub-aleurone layers in maize. Planta, 137: 185-193.

La Rue, C. D. 1944. Regeneration of endosperm of gymnosperms and angiosperms. Amer. J. Bot., 31(8): 45.

La Rue, C. D. 1949. Cultures of the endosperm of maize. Amer. J. Bot., 36, suppl. 10: 798.

Labarca, C., M. Kroh and F. Loewus. 1970. The composition of stigmatic exudate from *Lilium longiflorum*. Labelling studies with myo-inositol, d-glucose and L-proline. Plant Physiol., 46: 150-156.

Lackie, S. and E. C. Yeung. 1996. Zygotic embryo development in *Daucus carota*. Canadian J. Bot., 74: 990-998.

Lado, P., F. Rasi-Caldogno and R. Colombo. 1975. Acidification of the medium associated with normal and fusicoccin induced seed germination. Physiol. Plantarum (Copenhagen), 34(4): 359-364.

Lagerberg, T. 1909. Studien über die Entwicklungsgeschichte und systematische Stellung von *Adoxa moschatelina* L. Kongliga Svenska Vetenskapsakademiens Handlingar, N. S., 44(4): 1-86.

***Lagutova, O. I. and A. A. Chebotaru. 1992.** Study of embryogenesis in *Orchis purpurea* and *Orchis provincialis* (Orchidaceae). Bot. Journal (St. Petersburg), 77(12): 53-57.

Laidman, D. L. 1982. Control mechanisms in the mobilization of stored nutrients in germinating cereals, pp. 371-392. *In* The Physiology and Biochemistry of Seed Development, Dormancy and Germination. Elsevier Biomed, Amsterdam, Netherlands.

Lakshmanan, K. K. 1967. Embryological studies in the Bromeliaceae. I. *Lindmannia penduflora* (C.H. Wright) Stapf. Proc. Indian Acad. Sci., 65(2): 49-55.

Lakshmanan, K. K. 1970. Potamogetonaceae. Bull. Indian Natl. Sci. Acad., 41: 348-351.

Lakshmanan, K. K. 1978. Studies on the embryo development of *Commelina benghalensis*. III. Cotyledons. Phytomorph., 28: 253-261.

Lampe, L. 1931. A microchemical and morphological study of the developing endosperm of maize. Bot. Gazette, 91: 337-376.

Land, W. J. G. 1907. Fertilization and embryogeny in *Ephedra trifurca*. Bot. Gazette (Chicago), 44: 273-292.

Langdon, L. M. 1934. Embryogeny of *Carya* and *Juglans*, a comparative study. Bot. Gazette (Chicago), 96(1): 93-117.

Laux, T. and G. Jürgens. 1994. Establishing the body plan of the *Arabidopsis* embryo. Acta Bot. Neerlandica, 43(3): 247-260.

Lawalrée, A. 1952. L'embryology des Lemnaceae. Observation sur *Lemna minor* L. Cellule, 54(3): 305-326.

Lawton, J. R. S. and J. R. Lawton. 1967. The morphology of the dormant embryo and young seedling anatomy of five species of *Dioscorea* from Nigeria. Proc. Linnean Soc. London, 178(2): 153-159.

Le Monnier, G. 1887. Sur la valeur morphologique de l'albumen chez les Angiosperms. J. Bot., appl. l'Agric., Pharm., Méd., Arts, 1: 140-142.

Le Page Degivry, M. T. 1970. Dormance de graine associée à une immaturité de l'embryon: étude en culture in vitro chez *Magnolia soulangeana* Soul. Bod. et *Magnolia grandiflora* L. Planta, 90(3): 267-271.

Le Page Degivry, M. T. and G. Garello. 1992. In situ abscisic acid synthesis. A requirement for induction of embryo dormancy in *Helianthus annuus*. Plant Physiol., 98(4): 1386-1390.

Le Page Degivry, M. T., P. Barthe and G. Garello. 1990. Involvement of endogenous abscisic acid in onset and release of *Helianthus annuus* embryo dormancy. Plant Physiol., 92(4): 1164-1168.

Lebègue, A. 1952. Recherches embryogéniques sur quelques Dicotylédones Dialypetales. Ann. Sci. Naturelles. Bot. Biol. Végétale, Ser. XI, 13: 1-160.

Lebègue, A. 1954. Embryogénie des Oenotheracées. Développement de l'embryon chez le *Gaura biennis* L. C.R. Hebdomadaires Séances l'Acad. Sci., 239(9): 616-618.

Lebègue, A. 1955. Le *Lopezia coronata* Audr.; Son embryogénese et sa position systématique chez les Oenotheracées. Bull. Soc. Bot. France, 102(9): 503-505.

Lebègue, A. 1966. Nouveau aspects d'une classification embryologique. Bull. Soc. Bot. France, 113(3-4): 129-133.

Lebègue, A. and R. Poix. 1956. Le devéloppement embryonnaire chez l'*Urtica urens* L. Bull. Soc. Bot. France, 103(9-10): 587-589.

Lee, M., L. Veldboom, M. Scanlon, and A. Myers, M. James and D. Robertson. 1991. Mutator-tagged defective kernel mutants as putative QTL for kernel size. Maize Genet. Coop. Newsletter, 65: 10-11.

Lehmann-Baerts, M. 1967. Etude sur les Gnetales — XII. Ovule, gametophyte kernelle et embryogénese chez *Ephedra distachya* L. Cellule, 67: 53-87.

Lemann, E. and F. Aikhele. 1936. Physiology of Grain Seed Germination. Selkhozgiz, Moscow-Leningrad, Russia.

Leme, J. 1976. Principles of Biogeography. Progress, Moscow, Russia.

Lending, C. R., A. L. Kriz, B. A. Larkins and C. E. Bracker. 1988. Structure of maize protein bodies and immunocytochemical localization of zeins. Protoplasma, 143: 51-62.

Leroux, G., D. Barabé and J. Vieth. 1995. Morphogenèse comparée de protocormes du *Cypripedium acaule* (Orchidaceae) cultivés in vitro avec ou sans sucre. Canadian J. Bot., 73(9): 1391-1406.

Lersten, N. R. 1983. Suspensors in Leguminosae. Bot. Review, 49(3): 233–257.

Lewak, S., A. Rychter and B. Zarska-Maciejewska. 1975. Metabolic aspects of embryonal dormancy in apple seeds. Physiol. Végétale, 13(1): 13-22.

Lewis, C. E. 1904. Studies on some anomalous dicotyledonous plants. Bot. Gazette (Chicago), 37: 127-138.

Lihnell, D. 1942. Keimungsversuche mit *Pyrola* Samen. Symbolae Bot. Upsalienses, 6(3): 1-37.

Lindsey, K. and J. F. Topping. 1993. Embryogenesis: a question of pattern. J. Exper. Bot., 44(259): 359-374.

Linskens, H. F. 1969. Fertilization mechanisms in higher plants, pp. 189-253. *In* C. B. Metz and A. Monray (eds). Fertilization—Comparative Morphology, Biochemistry and Immunology, vol. 4. Academic Press, New York, NY.

Linskens, H. F. 1973. Activation of the ovary. Caryologia, 25(suppl.): 27–41.

***Linskens, H. F. 1978.** Breaking reaction at incompatible pollination and its overcoming. Physiol. Rastenyi, 19(1): 192-204.

Linskens, H. F. and W. Heinen. 1962. Cutinase—Nachweise in Pollen. Zeitschrift Bot., 50(4): 338–347.

Linskens, H. F. and A. W. Spanjers. 1973. Changes of the electrical potential in the transmitting tissue of *Petunia* styles after cross- and self-pollination. Plant Cell Incompatibility Newsletter, 3: 81–85.

List, A. Jr. and F.C. Steward. 1965. The nucellus, embryo sac, endosperm and embryo of *Aesculus* and their interdependence during growth. Ann. Bot. (N. S.), 29 (113): 1-15.

***Litovchenko, A. G. 1940.** The role of threecotyly in socialist industry of plants. Reports USSR Acad. Sci., 27: 816-820.

Litvak, A. I. 1968. Karelian birch (*Betula verrucosa* var. *carelica* Sae) in the Ukrainian forests. Ukrainian Bot. J., 25(1): 103-106 (in Ukrainian).

***Litvak, A. I. 1971.** Luminescent microscopy in the investigation of pollination and pollen tube growth in pistil. *In* Genetics and Selection in Moldova, Kishinev, Moldova.

Litvak, A. I. 1981. Family Juglandaceae, pp. 216-223. *In* M. S. Yakovlev (ed.). Comparative Embryology of Flowering Plants. Winteraceae–Juglandaceae. Nauka, Leningrad, Russia.

***Litvak, A. I. and S. M. Kolesnikov. 1965.** The using of microphotography in the research of male gametophyte development processes in angiosperms, pp. 186-206. *In* A. E. Kovarsky (ed.). Biology of Fertilization and Heterosis of Cultivated Plants. Kartya Moldavenyask, Kishinev, Moldova.

Liu, C.-M., Z.-H. Xu and N.-H. Chua. 1993a. Auxin polar transport is essential for the establishment of bilateral symmetry during early plant embryogenesis. Plant Cell, 5: 621-630.

Liu, C.-M., Z.-H. Xu, and N.-H. Chua. 1993b. In vitro development of early globular stage zygotic embryos from *Brassica juncea*. Plant Journal, 3: 291-300.

Lloyd, F.E. 1902. The comparative morphology Rubiaceae. Memoirs of the Torrey Bot. Club, 8: 1–112.

Lo Schiavo, F., G. Guiliano and Z. R. Sung. 1988. Characterization of a temperature sensitive carrot cell mutant impaired in somatic embryogenesis. Plant Sci., 54: 157-164.

Loconte, H. and D. W. Stevenson. 1990. Cladistics of the Spermatophyta. Brittonia, 42: 197-211.

***Lodkina, M. M. 1988.** Evolutionary relationship between monocots and dicots according to the data on embryo and seedling research. Bot. Journal (St. Petersburg), 73(5): 617-629.

***Lodkina, M. M., A. I. Lyashuk and M. G. Nikolaeva. 1971.** Anatomical changes in embryos of *Acer tataricum* L. during seed maturation and stratification. Bot. Journal (St. Petersburg), 56(3): 321-334.

Loesener, Th. 1930. Zingiberaceae, pp. 541-650. *In* A. Engler (ed.). Die natürlichen Pflanzenfamilien. 2 Auflösung, Leipzig, vol. 15.

Loewus, F. and C. Labarca. 1973. The nutritional role of pistil exudate in pollen tube wall formation in *Lilium longiflorum*. II. Production and utilization of exudate from stigma and stylar canal. Plant Physiol., 52: 87-92.

Lord, E. M. and B. D. Webster. 1979. The stigmatic exudate of *Phaseolus vulgaris* L. Bot. Gazette, 140: 266-271.

Lord, E. M. and Y. Heslop-Harrison. 1984. Pollen-stigma interaction in Leguminosae: stigma organisation and the breeding systems in *Vicia faba* L. Ann. Bot. (London), 54(6): 827-836.

Lord, E. M. and L. N. Kohorn. 1986. Gynoecial development, pollination, and the path of pollen tube growth in the tepary bean, *Phaseolus acutifolius*. Amer. J. Bot., 73: 70–78.

Lorenzi, R., A. Bennici, P. G. Cionini, A. Alpi and F. D'Amato. 1978. Embryo-suspensor relations in *Phaseolus coccineus*: cytokinins during seed development. Planta, 143: 59-62.

Lotsy, P. 1898. Resultate einer Untersuchung über die Embryologie von *Gnetum gnemon* L. Bot. Centralblatt, 75(9): 257-261.

Lott, J. N. A. 1980. Protein bodies. *In* P. K. Stumpf, E. E. Conn and N. E. Tolbert (eds). The Biochemistry of Plants, vol. 1. The Plant Cell. Academic Press, New York, NY.

Lubbock, J. 1892. Contribution to our Knowledge of Seedlings. Kogan Paul, Trench Trubner and Co. Ltd., London, UK, vols. 1, 2.

Lück, R. 1940. Zur Biologic der heimischen *Pirola*-Arten. Schriften Physikalisch-Ökonomischen Gesellschaft zu Königsberg, 71(2): 300-334.

Lukyanyuk, S. F. 1983. Elaboration of methods *in vitro* for the obtaining of barley and triticale haploids. Ph.D. thesis, Moscow, Russia.

***Lukyanyuk, S. F. and S. A. Ignatova. 1980.** Methods of Tissue and Organ Culture and Plant Selection. Methodical Recommendations. VSGI, Odessa, Ukraine.

Lundqvist, A. 1956. Self-incompatibility in rye. I. Genetic control in the diploid. Hereditas, 42(3-4): 293-348.

Lundqvist, A. 1975. Complex self-incompatibility systems in angiosperms. Proc. Royal Soc. London. Series B. Biol. Sci., 188: 235-245.

***Lvova, I. N. 1950.** Some questions on cytophysiology of fertilization in cereals. Selektsya i semenovodstvo, 9: 11-23.

Ly Thi Ba. 1981. Comparative embryogeny and phylogeny of Ranales and Helobiales. IV. The conclusions. Rev. Gén. Bot., 88 (1049-1051): 347-373.

Ly Thi Ba, J. L. Guignard and J. C. Méstre. 1970. Embryogénie des Magnoliacées. Développement de l'embryon chez le *Magnolia grandiflora* L. C.R. Hebdomadaires Séances l'Acad. Sci., 271(15): 1272-1275.

***Lyashuk, A. I. 1971.** Anatomical changes in embryos of *Fraxinus excelsior* L. during stratification. Bot. J. (St. Petersburg), 56(11): 1689-1693.

Lyashuk, A. I. 1972. Anatomy of embryos in *Fraxinus excelsior* L. and *Acer tataricum* L. during seed stratification process. Ph.D. thesis, Leningrad, Russia.

Lyndon, R. F. 1990. Plant Development. Unwin Hyman, London, UK.

Lyon, H. L. 1901a. Observation on the embryogeny of *Nelumbo*. Minnesota Bot. Stud., 2(5): 643-655.

Lyon, H. L. 1901b. The phylogeny of cotyledon. Postelsia. Minnesota seaside station, pp. 57-86.

Lyon, H. L. 1905. The embryo of the angiosperms. Amer. Naturalist, 39: 13-34.

Madhavan, R. and S. C. Gupta. 1982. Histochemical studies on postfertilized ovules of *Peristrophe bicalyculata*. 1. Total carbohydrates of insoluble polysaccharides. Beitrage Biol. Pflanzen, 57(2): 309-321.

Mahabalé, T. S. and N. V. Biradar. 1968. Studies on palms: embryology of *Phoenix sylvestris* Roxb. Proc. Indian Acad. Sci., 67(2); 77-96.

Maheshwari, P. 1950. An Introduction to the Embryology of Angiosperms. McGraw-Hill Book Co., New York, NY.

Maheshwari, P. and B. M. Johri. 1956. The morphology and embryology of *Floerkea proserpinacoides* Willd. with a discussion on the systematic position of the family Limnanthaceae. Bot. Magazine. (Tokyo), 69(820-821): 410-423.

Maheshwari, S. C. and P. P. Khanna. 1956. The embryology of *Arisaema wallichianum*, and the systematic position of the Araceae. Phytomorph., 6(3-4): 379-388.

Maheshwari, S. C. and B. Baldev. 1958. A contribution to the morphology and embryology of *Commelina forskalai* Vahb. Phytomorph., 8(3-4): 277-298.

Maheshwari, P. and V. Vasil. 1961. Gnetum. Council of Scientific and Industrial Research, New Delhi, India.

Maheshwari, S. C. and R. N. Kapil. 1963. Morphological and embryological studies on the Lemnaceae. 2. The endosperm and embryo of *Lemna paucicostata*. Amer. J. Bot., 50: 907-914.

Maheshwari, P., B. M. Johri and S. N. Dixit. 1957. The floral morphology and embryology of the Loranthoideae (Loranthaceae). J. Madras Univ., 27: 121-136.

Maheswaran, G. and E. G. Williams. 1985. Origin and development of somatic embryos formed directly on immature embryos of *Trifolium repens* in vitro. Ann. Bot. (London), 57: 109-117.

Mahlberg, P. G. 1960. Embryogeny and histogenesis in *Nerium oleander* L. I. Organization of primary meristematic tissues. Phytomorph., 10(2): 118-131.

Maksymowych, R. 1973. Analysis of Leaf Growth. Cambridge University Press, Cambridge, UK.

Mangelsdorf, P. C. 1923. The inheritance of defective seeds in maize. J. Heredity, 14: 119-125.

Mansfield, S. D. and L. G. Briarty. 1991. Early embryogenesis in *Arabidopsis thaliana*. II. The developing embryo. Canadian J. Bot., 69: 461-476.

Mares, D. J., B. A. Stone, C. Jeffery and K. Norstog. 1977. Early stages in development of wheat endosperm. II. Ultrastructural observations on cell wall formation. Australian J. Bot., 25(6): 599-613.

Marinos, N. G. 1970. Embryogenesis of the pea (*Pisum sativum*) II. An unusual type of plastids in the suspensor cells. Protoplasma, 71(3): 227-233.

***Markova, I. V., A. Yu. Batov, A. V. Moshkov, G. B. Maximov and S. S. Medvedev. 1995.** Calcium transport systems of maize coleoptile plasmalemma. Physiol. Rastenyi, 42(2): 262-267.

Marsden, M. P. F. and D. W. Meinke. 1985. Abnormal development of the suspensor in an embryo-lethal mutant of *Arabidopsis thaliana*. Amer. J. Bot., 72: 1801-1812.

Martens, P. 1971. Les Gnetophytes. Encyclopedia of Plant Anatomy, 12. Gebrüder Borntraeger, Berlin.

Martens, P. 1973. Recherches sur *Welwitchia mirabilis*, V. Evolution ovulaire et embryogénese. Cellule, 70: 165-258.

Martin, A. C. 1946. The comparative internal morphology of seeds. Amer. Midland Naturalist, 36(3): 513-660.

Martin, P. G. and J. M. Dowd. 1986. A phylogenetic tree for some monocotyledons and gymnosperms derived from protein sequences. Taxon, 35: 469-475.

Martin, P. G. and J. M. Dowd. 1991. Studies of angiosperm phylogeny using protein sequences. Ann. Missouri Bot. Garden, 78: 296-337.

Mascarenhas, J. P. and Machlis L. 1962. The pollen tube chemotropic factor from *Antirrhinum majus*: bioassay, extraction and partial purification. Amer. J. Bot., 49: 482–489.

Masters, M. T. 1868. Outlines of elementary botany. *In* Oliver D. (ed.). Flora of Tropical Africa, vol. 1. L. Reeve and Co., London, UK, 479 pp.

Mathur, N. 1956. The embryology of *Limnanthes*. Phytomorph., 6(1): 41-51.

Matile, P. 1968. Aleurone vacuoles as lysosomes. Zeitschrift Pflanzenphysiol., 58: 365-368.

Matthys-Rochon, E., R. Mól, P. Heizmann and C. Dumas. 1994. Isolation and microinjection of active sperm nuclei into cells and central cells of isolated maize embryo sacs. Zygote, 2: 152-157.

Mattsson, O., R. B. Knox, J. Heslop-Harrison and Y. Heslop-Harrison. 1974. Protein pellicle of stigmatic papillae as a probable recognition site in incompatibility reaction. Nature, 247: 298-300.

***Matveev, B. S. 1942.** On adaptation to lactation in mammals. Zool. Journal (St. Petersburg), 21(3): 47-68.

Matzk, F. 1991. A novel approach to differentiated embryos in the absence of endosperm. Sexual Plant Reprod., 4: 88-94.

Mau, S.-L., J. Raff and A. E. Clarke. 1982. Isolation and partial characterisation of components of *Prunus avium* L. styles, including an antigenic glycoprotein associated self-incompatibility genotype. Planta, 156: 505-516.

Mauritzon, J. 1933. Studien über die Embryologie der Familien Crassulaceae und Saxifragaceae. Ph.D. thesis, Botanischen Inst. Lund., H. Ohlssons Buchdruckerei, Gleerupska Univ., Bokhandeln, Lund, Sweden.

Mauritzon, J. 1934. Die Endosperm und Embryoentwicklung einiger Acanthaceen. Acta Univ. Lundensis, 30: 1-41.

Mauritzon, J. 1935. Etwas über die Embryologie der Bignoniaceen. Bot. Notiser, pp. 60-77.

Mauritzon, J. 1936. Samenbau und Embryologie einiger Scitamineen. Acta Univ. Lundensis, 31(9): 1-31.

Mauritzon, J. 1939. Über die Embryologie von *Marcgravia*. Bot. Notiser, 1: 249-255.

Mayer, A. M., A. Poljakoff-Mayber. 1989. The Germination of Seeds. Pergamon Press, Oxford, UK.

Mayer, U., G. Buttner and G. Jurgens. 1993. Apical-basal pattern formation in the *Arabidopsis* embryo: studies on the role of the *gnom* gene. Development, 117: 149-162.

Mayer, U., R. A. Torres Ruiz, and T. Berleth, S. Misera and G. Jürgens. 1991. Mutations affecting body organization in the *Arabidopsis* embryo. Nature, 353: 402-407.

McCall, M. A. 1934. Development, anatomy and homologies in wheat. J. Agric. Res., 48(4): 283-321.

McCarty, D. R., C. B. Carson, P. S. Stinard and D. S. Robertson. 1989. Molecular analysis of viviparous-1: an abscisic acid-insensitive mutant of maize. Plant Cell, 1: 523-532.

McConchie, C. A., S. Jobson and R. B. Knox 1985. The structure of sperm cells in *Brassica*, pp. 147-148. *In* M. T. M. Willemse and J. L. Van Went (eds). Sexual Reproduction in Seed Plants, Ferns and Mosses. Proc. 8th Int. Symp. Sexual Reproduction. August 20-24, 1984. PUDOC, Wageningen, Netherlands.

McConchie, C. A., T. Hough, and R. B. Knox. 1987a. Ultrastructural analysis of the sperm cells of maize, *Zea mays*. Protoplasma, 139: 9-19.

McConchie, C. A., and S. Russell, C. Dumas and R. B. Knox. 1987b. Quantitative cytology of the mature sperm cells of *Brassica campestris* and *B. oleracea*. Planta, 170: 446-452.

McDaniel, C. N. and R. S. Poethig. 1988. Cell-lineage patterns in the shoot apical meristem of the germinating maize embryo. Planta, 175: 13-22.

McIlrath, W. J., D. R. Erogle and A. A. Dunlap. 1950. Persistance of 2,4 D-stimulus in cotton plants with reference to its transmission to the seed. Bot. Gazette, 112(4): 511-518.

McVeign, J. 1938. Regeneration in *Crassula multicava*. Amer. J. Bot., 25: 7-11.

Medgyesy, P., A. Pay and L. Marton. 1986. Transmission of paternal chloroplasts in *Nicotiana*. Molec. Gen. Genet., 204: 195-199.

Medvedev, S. S. 1994. The nature of polarity in plants. Biologija (Vilnius), 3: 99-106.

Medvedev, S. S. 1996. Physiological bases of plant polarity, pp. 1-159. *In* Trans. St. Petersburg Soc. Naturalists. Kol'na, St. Petersburg, Russia.

***Medvedev, S. S. and I. A. Shtonda. 1989.** On the role of calcium ions in gravitropistic reaction. Biol. nauki, 6: 94-97.

Medvedev, S. S. and I. V. Markova. 1990. How can the electrical polarity of axial organs regulate plant growth and IAA transport. Physiol. Plantarum (Copenhagen), 78(1): 38-42.

***Medvedev S. S., G. B. Maximov and V. V. Polevoy. 1980.** Significance of potential difference between mesocotyle and root system for growth processes and accumulation of mineral elements by maize seedlings. Vestnik Leningrad. Univ., Biol., 4: 85-90.

***Medvedev, S. S., G. B. Maximov and I. V. Markova. 1991.** Role of calcium ions in gravitropism regulation. Exper. Biol., 4: 71-92.

***Medvedev, S. S, I. V. Markova, A. Yu. Batov and G. B. Maximov. 1989a.** Polar flows of calcium ions and plant tissue growth. Physiol. rastenyi, 36(5): 990-997.

***Medvedev, S. S, I. V. Markova, M. F. Shishova and T. M. Sopova. 1989b.** On polar transport of calcium in plant tissues. Vestnik Leningrad. Univ., Series 3, 1(3): 79-83.

Meeuse, A. D. J. and E. C. J. Ott. 1962. The occurrence of chlorophyll in *Nelumbo* seeds. Acta Bot. Neerlandica, 11(2): 227-228.

Mehra, K. R. and A. R. Kulkarni. 1985. Embryological studies in Bignoniaceae. Phytomorph., 35(3-4): 239-251.

Meinke, D. W. 1982. Embryo-lethal mutants of *Arabidopsis thaliana*: evidence for gametophytic expression of the mutant genes. Theor. Appl. Genet., 63: 381-386.

Meinke, D. W. 1985. Embryo-lethal mutants of *Arabidopsis thaliana*: analysis of mutants with a wide range of lethal phases. Theor. Appl. Genet., 69: 543-552.

Meinke, D. W. 1986. Embryo-lethal mutants and the study of plant embryo development. Oxford Surveys Plant Molec. Cell Biology, 3: 122-165.

Meinke, D. W. 1991a. Embryonic mutants of *Arabidopsis thaliana*. Developm. Genet., 12: 382-392.

Meinke, D. W. 1991b. Genetic analysis of plant development, pp. 437-490. *In* F. C. Steward and R. G. S. Bidwell (eds). Plant Physiology: A Treatise, vol. 10. Growth and Development. Academic Press, New York, NY.

Meinke, D. W. 1991c. Perspectives on genetic analysis of plant embryogenesis. Plant Cell, 3(9): 857-866.

Meinke, D. W. 1992. A homeotic mutant of *Arabidopsis thaliana* with leafy cotyledons. Science, 258: 1647-1650.

Meinke, D. W. 1994. Seed development in *Arabidopsis thaliana*, pp. 253-295. *In* E. M. Meyerowitz and C. R. Somerville (eds). *Arabidopsis*. Cold Spring Harbor Laboratory Press, Cold Spring Harbor, New York, NY.

Meinke, D. W. 1995. Molecular genetics of plant embryogenesis. Ann. Rev. Plant Physiol., Molec. Biol., 46: 369-394.

Meinke, D. W. and I. M. Sussex. 1979a. Embryo-lethal mutants of *Arabidopsis thaliana*: a model system for genetic analysis of plant embryo development. Developm. Biol., 72: 50-61.

Meinke, D. W. and I. M. Sussex. 1979b. Isolation and characterization of six embryo-lethal mutants of *Arabidopsis thaliana*. Developm. Biol., 72: 62-72.

Meinke, D. W., L. M. Franzmann, T. C. Nickle and E. C. Yeung. 1994. Leafy cotyledon mutants of *Arabidopsis*. Plant Cell, 6(6): 1049-1064.

Meinke, D. W., D. Patton, J. Shellhammer, A. Reynolds-Duffer, L. Franzmann, T. Scheider and K. Robinson. 1989. Developmental and molecular genetics of embryogenesis in *Arabidopsis thaliana*, pp. 121-132. *In* R. Goldberg (ed.). The Molecular Basis of Plant Development. Alan Liss, New York, NY.

Mendes, A. J. T. 1941. Cytological observations in *Coffea*. VI. Embryo and endosperm development in *Coffea arabica* L. Amer. J. Bot., 28(9): 784-789.

***Mentkovskaya, E. A. 1982.** Peculiarities of fertilization process in lily, pp. 243-245. *In* V. P. Bannikova (ed.). Cytological–Embryological and Genetic–Biochemical Bases of Pollination and Fertilization in Plants. Naukova dumka, Kiev, Ukraine.

Mericle, L. W. and R. P. Mericle. 1973. Confounding the quandary of zygotic DNA. Barley Genetics Newsletter, 3: 39-42.

***Merkis A. I. 1976.** Significance of gravity in processes of plant growth, pp. 116-173. *In* Problems of Cosmic Biology, 33. Nauka, Moscow, Russia.

***Merkis A. I. 1990.** Gravity in processes of plant growth, pp. 1-185. *In* G. B. Maximov (ed.). Problems of Cosmic Biology, 68. Nauka, Moscow, Russia.

Merry, J. 1941. Studies on the embryo of *Hordeum sativum*. I. The development of the embryo. Bull. Torrey Bot. Club, 68(8): 585-598.

Méstre, J. C. 1964. Recherches d'embryogénie comparée: les rapports phylogénétiques des Composées. Thése, Paris, France.

Méstre, J. C. 1967. La signification phylogénetique de l'embryogénie. Rev. Gén. Bot., 74(877): 273-324.

Méstre, J. C. 1981. Embryogénie et phylogénie des végétaux superieurs. Ann. Biol., 20(1): 61-96.

Méstre, J. C. and J.-L. Guignard. 1967. La mise en place des méristèmes caulinaire et radiculaire au cours de l'embryogénése du *Cerastium pumilum* Curt. Bull. Soc. Bot. France, 114(9): 387-396.

Méstre, J. C. and A. Vannereau. 1980. La nutrition de l'embryon au cours de l'embryogénése des Angiospermes. Bull. Soc. Bot. France. Actualities Bot., 3/4: 39–50.

Metcalf, C. R. 1936. An interpretation of the morphology of a single cotyledon of *Ranunculus ficaria* based on embryology and seedling anatomy. Ann. Bot. (London), 50: 103-120.

Meunier, A. 1897. Le développement seminal dans le genere *Veronica*. Cellule, 12: 297-333.

Meyen, F. J. F. 1840. Noch einige Worte über den Befruchtungsakt und die Polyembryonie bei den höheoren Pflanzen. Haude und spenersche Buchhaus, Berlin.

Meyer, B. and W. F. Stubbe. 1974. Das Zahlenverhalthis von mutterlichen und vaterlichen Plastiden in den Zygoten von *Oenotera erythrosepala* Borbas (syn. *O. lamarckiana*). Berichte Deutschen Bot. Gesellschaft, 87: 29-38.

*Meyer, K. I. 1960. On the embryology of *Nuphar luteum* Sm. Bull. Moscow Soc. Naturalists, Biological Dept., 65(6): 48-59.
*Meyer, K. I. 1962. On the embryology of *Coronaria floscuculi* (L.) A. Br. Bot. Journal (St. Petersburg), 47(6): 847-852.
Micola, I. and M. Virtanen. 1981. Regulation of pH in the starchy endosperm of barley grain during reserve protein mobilization. Abhandlungen Acad. Wissenschaften DDR, Abteilung Mathematik, Naturwissenschaften, Technik, 5: 257-258.
Miflin, B. J., S. R. Burgess and P. R. Shewry. 1981. The development of protein bodies in the storage tissues of seeds: subcellular separations of homogenates of barley, maize, and wheat endosperms and of pea cotyledons. J. Exper. Bot., 32: 199-219.
Mikhailov, A. T. 1988. Embryonal Inductors. Nauka, Moscow, Russia.
Miki-Hirosige, H. 1964. Tropism of pollen tubes to the pistils, pp. 152–158. *In* H. F. Linskens (ed.). Pollen Physiology and Fertilization. North-Holland Publ. Co., Amsterdam, Netherlands.
Miki-Hirosige, H. and S. Nakamura. 1982. Process of metabolism during pollen-tube wall formation. J. Electron Microscopy, 31(1): 51–62.
Mikus, M. M., M. Bobak and A. Lux. 1992. Structure of protein bodies and elemental composition of phytin from dry germ of maize (*Zea mays* L.). Bot. Acta, 105: 26-33.
Miller, A. L. and N. A. R. Gow. 1989a. Correlation between profile of ion-current circulation and root development. Physiol. Plantarum (Copenhagen), 75(1): 102-108.
Miller, A. L. and N. A. R. Gow. 1989b. Correlation between root-generated ionic currents, pH, fusicoccin, indoleacetic acid, and growth of the primary root of *Zea mays*. Plant Physiol., 89(4): 1198-1206.
Miller, E. C. 1919. Development of the pistillate spikelet and fertilization in *Zea mays* L. J. Agric. Res., 18(5): 255-293.
Miller, E. W. 1927-1931. On the occurrence of schizocotyly in certain ranunculaceous seedlings. Trans. Proc. Bot. Soc. Edinburgh, 30: 21.
Mirbel, C. F. B. 1829. Nouvelles recherches sur la structure et les développements de l'ovule végétale. Ann. Sci. Naturelles. Bot., 17: 302-318.
Mirza, J. I. 1987. The effect of light and gravity on the horizontal curvature of roots of gravitropic and agravitropic *Arabidopsis thaliana*. Plant Physiol., 83(1): 118-120.
Miyamoto, K. and S. Kamisaka. 1988. Stimulation of *Pisum sativum* epicotyl elongation by gibberellin and auxin. Different effects of two hormones on osmoregulation and cell walls. Physiol. Plantarum (Copenhagen), 74: 457-466.
Miyamura, S., T. Kuroiwa and T. Nagata. 1987. Disappearance of plastid and mitochondrial nucleoids during the formation of generative cells of higher plants revealed by fluorescence microscopy. Protoplasma, 141: 149-159.
Miyazaki, S., H. Shirakawa, K. Nakada and Y. Honda. 1993. Essential role of inositol 1,4,5,-triphosphate receptor, Ca^{2+} release channel in Ca^{2+} waves and Ca^{2+} oscillations at fertilization of mammalian eggs. Developm. Biol., 158: 62-78.
Modilewsky, Ya. S. 1931. On the embryo development in *Allium odorum* L. Vestnik Kiev. Bot. Sada, 12-13: 27-48 (in Ukrainian).
*Modilewsky, Ya. S. 1953. Embryology of Angiosperms. Publ. Ukrainian SSR Acad. Sci., Kiev, Ukraine.

Modilewsky, Ya. S. 1958. Ontogenesis of higher plants and theory of stage development. Ukrainian Bot. J., 15(1): 3-21 (in Ukrainian).

***Modilewsky, Ya. S. 1963.** Cytoembryology of Higher Plants. Publ. Ukrainian SSR Acad. Sci., Kiev, Ukraine.

***Modilewsky, Ya. S., P. F. Oxijuk, M. I. Khudyak, L. K. Dzjubenko and R. A. Bejlis-Vyrovaja. 1958.** Cytoembryology of Main Cereals. Publ. The Ukrainian SSR Acad. Sci., Kiev, Ukraine.

Mogensen, H. L. 1970. Syncotyly in *Quercus arizonica*. Amer. J. Bot., 57(101): 1207-1210.

Mogensen, H. L. 1972. Fine structure and composition of the egg apparatus before and after fertilization in *Quercus gambelii*: the functional ovule. Amer. J. Bot., 59(9): 931-941.

Mogensen, H. L. 1978a. Pollen tube-synergid interactions in *Proboscidea lousianica* (Martyniaceae). Amer. J. Bot., 65(9): 953-964.

Mogensen, H. L. 1978b. Synergids of *Proboscidea louisianica* (Martyniaceae) before fertilization. Phytomorph., 28(1): 114-122.

Mogensen, H. L. 1982. Double fertilization in barley and the cytological explanation for haploid embryo formation, embryoless caryopses and ovule abortion. Carlsberg Res. Comm., 47: 313-354.

Mogensen, H. L. 1988. Exclusion of male mitochondria and plastids during syngamy as a basis for maternal inheritance. Proc. Natl. Acad. Sci., USA, 85: 2594-2597.

Mogensen, H. L. 1992. The male germ unit: concept, composition, and significance. Int. Rev. Cytol., 140: 129-147.

Mogensen, H. L. and H. K. Suthar. 1975. Ultrastructure of the egg apparatus of *Nicotiana tabacum* (Solanaceae) before and after fertilization. Bot. Gazette (Chicago), 140(2): 168-179.

Mogensen, H. L. and M. L. Rusche. 1985. Quantitative ultrastructural analysis of barley sperm. I. Occurrence and mechanism of cytoplasm and organelle reduction and the question of sperm dimorphism. Protoplasma, 128(1): 1-13.

Mogensen, H. L. and V. T. Wagner. 1987. Association among components of the male germ unit following in vivo pollination in barley. Protoplasma, 138: 161-172.

Mogensen, H. L. and P. B. Holm. 1995. Dynamics of nuclear DNA quantities during zygote development in barley. Plant Cell, 7: 487-494.

Mohan Ram, H. Y. 1960a. The development of the seed of *Andrographis serpyllifolia*. Amer. J. Bot., 47(3): 215-219.

Mohan Ram, H. Y. 1960b. Post-fertilization studies in the ovule of *Ruellia tuberosa*. Lloydia, 23(1): 21-27.

Mohan Ram, H. Y. 1962. Post-fertilization development of the ovule in *Barleria cristata*. J. Indian Bot. Soc., 41(2): 288-296.

Mohan Ram, H. Y. and M. Wadhi. 1964. Embryology and the delimination of the Acanthaceae. Phytomorph., 14(3): 388-413.

Mohana Rao, P. R., J.-L. Guignard and S. Duret. 1988. An ultrastructural study of perisperm and endosperm in *Silene alba* Miller E. H. L. Krause. Bull. Soc. Bot. France, 135(2): 123-130.

Mohr, H. 1983. Pattern specification and realization in photomorphogenesis, pp. 336-357. *In* A. Pirson and M. H. Zimmermann (eds). Encyclopedia of Plant Physiology, New Ser., 16A: Photomorphogenesis. W. Shropshire jr. and H. Mohr (eds). Springer, Berlin.

Mohr, H. and E. Peters. 1960. Der Phototropismus und das lichtabhangige Langenwachstum des Hypokotyls von *Sinapis alba* L. Planta, 55(6): 637-646.

Mohrbutter, C. 1937. Embryologische Studien an Loganiaceen. Planta, 26(1): 64-80.

Mól, R., E. Matthys-Rochon and C. Dumas. 1993. In vitro culture of fertilized embryo sacs of maize. Zygotes and two-celled proembryos can develop into plants. Planta, 189: 213-217.

Mól R., E. Matthys-Rochon and C. Dumas. 1994. The kinetics of cytological events during double fertilization in *Zea mays* L. Plant Journal, 5(2): 197-206.

Monnier, M. 1968. Comparaison du développement des embryons immatures de *Capsella bursa-pastoris,* in vitro et in situ. Bull. Soc. Bot. France, 115: 15-29.

Monnier, M. 1976. Culture in vitro de l'embryon immature de *Capsella bursa-pastoris* Moench.(l). Rev. Cytolog. Biol. Végétales, 39(1-2): 1-120.

Monnier, M. 1995. Culture of zygotic embryos, pp. 117-153. *In* T. A. Thorpe (ed.). In vitro Embryogenesis in Plants. Kluwer Acad. Publ., Dordrecht, Netherlands.

Monnier, M. and C. Leddet. 1978. Sur l'aquisition de la rêsistance au froid des embryons immaturès de Capsella bursa-pastoris. C.R. Hebdomadaires Séances l'Acad. Sci., 287: 615.

***Monteverde, N. A. 1892.** On the dispersal of mannitol and dulcitol in plant kingdom. Botanical reports, Botanical Garden St. Petersburg Univ., III: 431-451.

Morozova, N. M. 1993. Investigation of morphophysiological correlations in developing embryo (on model systems *Nelumbo nucifera* Gaertn. and *Daucus carota* L.). Ph.D. thesis, St. Petersburg, Russia.

Morrison, I. N. and T. P. O'Brien. 1976. Cytokinesis in the developing wheat grain; division with and without a phragmoplast. Planta, 130(1): 57-67.

Morrison, I. N., J. Kuo and T. P. O'Brien. 1975. Histochemistry and fine structure of developing wheat aleurone cells. Planta, 123: 105-116.

Morrison, I. N., T. P. O'Brien and J. Kuo. 1978. Initial cellularization and differentiation of the aleurone cells in the ventral region of the developing wheat grain. Planta, 140(1): 19-30.

Morrison, J. W. 1955. Fertilization and postfertilization development in wheat. Canadian J. Bot., 33(2): 168-176.

Moscatelli, A., A. Tiezzi, R. Vigniani, G. Cai, A. Bartalesi and M. Cresti. 1988. Presence of kinetin in tobacco pollen tube, pp. 205–209. *In* M. Cresti, O. Gori and E. Pacini (eds). Sexual Reproduction in Higher Plants. Springer, Heidelberg.

***Moshkovich, A. A. and A. A. Chebotaru. 1976.** Rye: Caryology, Embryology, Cytology. Shtiintsa, Kishinev, Moldova.

Moss, G. I. and J. Heslop-Harrison. 1967. A cytochemical study of DNA, RNA and protein in the developing maize anther. Ann. Bot., 31(4): 555-572.

Mottier, D. M. 1898. Ueber das Verhalten der Kerne bei der Entwicklung des Embryosacks und die Vorgänge bei Befruchtung. Jahrb. wissenschaftliche Bot., 31: 125-158.

Mottier, D. M. 1905. The embryology of some anormalous dicotyledons. Ann. Bot. (London), 19(76): 447-463.

Mouritzen, P. and P. B. Holm. 1994. Chloroplast genome breakdown in microspore cultures of barley (*Hordeum vulgare* L.) occurs primarily during regeneration. J. Plant Physiol., 144: 586-593.

Moussel, B. 1977. Structures protoplasmiques au sein de la cavite archegoniale et modalites de la proembryogenese chez l'*Ephedra distachya* L. Rev. Cytol. Biol. Végétales, 40: 73-123.

Moussel, B. 1978. Double fertilization in the genus *Ephedra*. Phytomorph., 28: 336-345.

Moussel, B. 1983. Données temporalles concernant les diverses phases du cycle de reproduction sexuee chez *l'Ephedra distachya* L. Rev. Cytol. Biol. Végétales, 6: 103-109.

***Movsesjan, S. N. 1964.** Fertilization process in *Rudbeckia maxima*. Reports Armenian SSR Acad. Sci., 17(2): 77-83.

***Movsesjan, S. N. 1968.** Comparative embryological investigation of some *Rudbeckia* species, pp. 138-140. *In* Proc. All-Union Symp. Plant Embryology. Naukova dumka, Kiev, Ukraine.

***Movsesjan, S. N. and I. P. Kasparova. 1977a.** Hemigamy in *Rudbeckia laciniata* L. under the different modes of pollination. Bot. Journal (St. Petersburg), 62(3): 422-429.

***Movsesjan, S. N. and I. P. Kasparova. 1977b.** On some peculiarities of hemigamy, pp. 151-159. *In* D. F. Petrov (ed.). Cytogenetic Principles of Plant Selection. Nauka, Novosibirsk, Russia.

Muchtar, N.O. and P. L. Laidman. 1982. Mineral ion transport in the embryos of germinating wheat (*Triticum aestivum*). J. Exper. Bot., 33(135): 643-655.

Mukerji, K. G. 1992. Seed Biology. Naya Prokash, Bidhan Sarani, Calcutta, India.

Müller, A. J. 1963. Embryonentest zum Nachweis rezessiver Letalfaktoren bei *Arabidopsis thaliana*. Biol. Zentralblatt, 82: 133-163.

Müller, M. L., P. W. Barlow and P.-E. Pilet. 1994. Effect of abscisic acid on the cell cycle in the growing maize root. Planta, 195: 10-16.

Müller-Doblies, U. 1969. Über die Blütenstande und Blüten sowie zur Embryologie von *Sparganium*. Bot. Jahrb. Systematik, Pflanzengeschichte, Pflanzengeographie, 89(3): 359-450.

Munerati, O. and T. Costa. 1934. Di alcune forme teratologiche della *Beta vulgaris* L. e loro eredita. Zeitschrift für Inductive Abstammungs, Vererbungslehre, 66(3/4): 463-489.

Murgai, M. and H. L. Wilms. 1988. Three-dimensional image and mitochondrial distribution in sperm cells of *Euphorbiadulcis*, pp. 75-79. *In* H. J. Wilms and C. J. Keijzer (eds). Plant Sperm Cells as Tools for Biotechnology. PUDOC, Wageningen, Netherlands.

Murray, A. W. and M. W. Kirschner. 1991. What controls the cell cycle. Scient. Amer., 264(3): 34-41.

Murthi, S. N. 1946. Studies in the Labiatae. II. Contribution to the morphology of *Ocimum adscendens* Wind. J. Univ. Bombay, 14(5): 37-46.

Nagato, J. 1979. Incorporation of H^3-uridine and H^3-leucine during early embryogenesis of rice and barley in caryopsis culture. Plant Cell Physiol., 20(4); 765-773.

Nagato, Y., H. Kitano, O. Kamijima, S. Kikuchi and H. Satoh. 1989. Developmental mutants showing abnormal organ differentiation in rice embryos. Theor. Appl. Genet., 78: 11-15.

Nagendran, C.R. 1974. Is the embryo sac of Podosemonaceae bisporic? Current Sci., 4: 259-260.

Nagendran, C.R., G.D. Arecal and K. Subramanyam. 1977. Embryo sac studies in three Indian species of *Polypleurum* (Podostemaceae). Plant System. Evol., 128: 215-226.

Nagl, W. 1962a. 4096-Ploidie und "Riesenchromosomen" im Suspensor von *Phaseolus coccineus*. Naturwissenschaftliche Wochenschrift, 49: 261-262.

Nagl, W. 1962b. Über Endopolyploidie, Restitutionskernbildung und Kernstrukturen im Suspensor von Angiospermen und einer Gymnosperme. Österreichische Bot. Zeitschrift, 109: 431-494.

Nagl, W. 1970. Inhibition of polytene chromosome formation in *Phaseolus* by polyploid mitoses. Cytologia, 35: 252-258.

Nagl, W. 1973. The angiosperm suspensor and the mammalian trophoblast: organs with similar cell structure and function. Bull. Soc. Bot. France, Mém. publ. Soc. Bot. France, Colloque Morphol., pp. 289-301.

Nagl, W. 1974. The *Phaseolus* suspensor and its polytene chromosomes. Zeitschrift Pflanzenphysiol., 73(1): 1-44.

Nagl, W. 1976a. Early embryogenesis in *Tropaeolum majus* L.: ultrastructure of the embryo-suspensor. Biochem. Physiol. Pflanzen, 170(3): 253-260.

Nagl, W. 1976b. Ultrastructural and developmental aspects of autolysis in embryo-suspensors. Berichte Deutschen Bot. Gesellschaft, 89(23): 301-311.

Nagl, W. 1976c. Ultrastructure of the polytene nuclei in the embryo-suspensor of the bean, *Phaseolus coccineus*: toothing of cytoplasm and nucleoplasm. J. Biol. Sci. 6: 11-17.

Nagl, W. 1977. "Plastolysomes" — plastids involved in the autolysis of the embryo-suspensor in *Phaseolus*. Zeitschrift Pflanzenphysiol., 85(1): 45-51,

Nagl, W. 1990. Translocation of putrescine in the ovule, suspensor and embryo of *Phaseolus coccineus*. J. Plant Physiol., 136: 587.591.

Nagl, W. 1992. The polytene endosperm haustorium of *Rhinanthus minor* (Scrophulariaceae): functional ultrastructure. Canadian J. Bot., 70: 1997-2004.

Nagl, W. and S. Kühner. 1976–1977. Early embryogenesis in *Tropaeolum majus* L.: diversification of plastids. Planta, 133(1): 15-19.

Nair, N. C. and K. R. Narayanan. 1961. Studies on the Aristolochiaceae. 2. Contribution to the embryology of *Bragantia wallichii*. Lloydia, 24(2): 199-203.

Nakamura, S., A. Fukushima, H. Iwayama and H. Suzuki. 1991. Electrotropism of pollen tubes of *Camellia* and other plants. Sexual Plant Reprod., 4: 138–143.

Nakamura, S., T. Ikenara, H. Uchida, T. Suzuki and Sodmergen. 1992. Fluorescence microscopy of plastid nucleoids and a survey of nuclease C in higher plants with respect to the mode of plastid inheritance. Protoplasma, 169: 68-74.

Narang, N. 1953. The life-history of *Stackhousia linariaefolia* A. Cunn. with a discussion on its systematic position. Phytomorph., 3(4): 485-493.

Narayana, H. S. and C. G. P. Rao. 1963. Floral morphology and embryology of *Seetzenia orientalis* Decne. Phytomorph., 13(2): 197-205.

Narayana, R. 1958. Morphological and embryological studies in the family Loranthaceae. 3. *Nuytsia floribunda* (Labill.) R. Br. Phytomorph., 8(3-4): 306-323.

Narayanaswamy, S. 1955a. The structure and development of the caryopsis in some Indian millets. 4. *Panicum miliare* Lamk. and *P. miliaceum* L. Phytomorphology, 12(2): 61-73.

Narayanaswamy, S. 1955b. The structure and development of the caryopsis in some Indian millets. 5. *Echinochloa frumentacea* Link. Phytomorph., 18(5): 161-181.

Narayanaswamy, S. 1956. Structure and development of the caryopsis in some Indian millets. IV. *Setaria italica*. Bot. Gazette, 118: 112-121.

Nasrallah, M. E. and D. H. Wallace. 1967. Immunogenetics of self-incompatibility in *Brassica oleracea*. Heredity, 22: 519-527.

Nasrallah, M. E., D. H. Wallace and R. M. Savo. 1972. Genotype, protein phenotype relationships in self-incompatibility in *Brassica*. Genet. Res. (Cambridge), 20: 151-160.

Nasrallah, J. B., T. Nishio and M. E. Nasrallah. 1991. The self-incompatibility genes of *Brassica*: expression and use in genetic ablation of floral tissues. Ann. Rev. Plant Physiol., 42: 393-422.

Nast, C. G. 1941. The embryogeny and seedling morphology of *Juglans regia*. Lilloa, 6(1): 163-206.

Natesh, S. and M. A. Rau. 1984. The embryo, pp. 377–443. *In* B. M. Johri (ed.). Embryology of Angiosperms. Springer-Verlag, Berlin.

Naumova, T. N. 1987. The family Celastraceae, pp. 49-54. *In*: T. B. Batygina and M. S. Yakovlev (eds). Comparative Embryology of Flowering Plants. Davidiaceae—Asteraceae. Nauka, Leningrad, Russia.

Naumova, T. N. 1989. Nucellar and integumentary embryony in the system of seed reproduction of flowering plants. Ph.D. thesis, Leningrad, Russia.

Naumova, T. N. and F. Matzk. 1998. Differences in the initiation of the zygotic and parthenogenetic pathway in the Salmon lines of wheat: ultrastructural study. Sexual Plant Reprod., 11: 121-130.

***Navashin, M. S. 1968.** The pollen tube of angiosperms as a polar-differentiated system, pp. 141-146. *In* Reports All-Union Symp. Plant Embryology, Naukova umka, Kiev, Ukraine.

***Navashin, M. S. and H. N. Gerassimova-Navashina. 1958.** On the research of cellular processes on fixed material. Bot. Journal (St. Petersburg), 43(2): 167-177.

Navashin, M. S., L. M. Makushenko and Z. V. Bolkhovskykh. 1959. Morphological study of pollen tube, p. 17. *In* Materials Coordination Conf. Problem "The Main Questions of Cytology". Nauka, Leningrad, Russia.

Nawaschin, S. 1895a. Ein neues Beispiel der Chalazogamie. Bot. Centralblatt, 63(12): 353-357.

***Nawaschin, S. G. 1895b.** On the fertilization in elm and nut-grove. Trans. St. Petersburg Soc. Naturalists, Dept. Bot., 25: 23-25.

***Nawaschin, S. G. 1895c.** On the ordinary birch and on the morphological significance of chalazogamy. Trans. St. Petersburg Soc. Naturalists, Dept. Bot., 25: 26-61.

***Nawaschin, S. G. 1898a.** New observations of fertilization in *Fritillaria tenella* and *Lilium martagon*, pp. 16-21. *In* Diary X Congr. Russian Naturalists and Physicians, vol. 6, Kiev, Ukraine.

Nawaschin, S. G. 1898b. Resultate einer Revision der Befruchtungesvorgänge bei *Lilium martagon* und *Fritillaria tenella*. Bull. l'Acad. Impér. Sci. St. Pétersburg, 9: 377-382.

Nawaschin, S. 1899. Neue Beobachtungen über Befruchtung bei *Fritillaria tenella* and *Lilium martagon*. Bot. Centralblatt, 77(2): 62.

***Nawaschin, S. G. 1900a.** On the fertilization in Compositae and Orchidaceae (Preliminary report). Reports Imper.Acad. Sci., 13(3): 335-340.

Nawaschin, S. G. 1900b. Über die Befruchtungsvorgänge bei einigen Dicotyledonen. Berichte Deutschen Bot. Gesellschaft, 18(5): 224-230.

Nawaschin, S. G. 1909. Über das selbstständige Bewegungsvermögen der Spermakerne bei einigen Angiospermen. Österreichische Bot. Zeitschrift, 59(12): 457-467.

Nawaschin, S. G. 1910a. Nöheres über die Bildung der Spermakerne bei *Lilium martagon*. Ann. Jardin Bot. Buitenzorg, 3, Suppl. 2: 871-904.

***Nawaschin, S. G. 1910b.** On the independent mobility of male sexual nuclei. Reports Kiev Soc. Naturalists, 20(4): 321-336.

***Nawaschin, S. G. 1911.** The details on the formation of male sexual nuclei in *Lilium martagon*. Reports Kiev Soc. Naturalists, 21(4): 119-151.

***Nawaschin, S. G. 1915.** Haploid, diploid and triploid nuclei in *Crepis virens*. Reports Kiev Soc. Naturalists, 25(1): 139-152.

Nawaschin, S. G. 1927. Essay of Structural Representation of Sexual Nuclei, pp. 94-114. *In* A. A. Yachevsky (ed.). Dedicatory volume to I. P. Borodin. Leningrad, Russia.

Nawashin, S. G. 1951. Selected transactions, 1. Publ. USSR Acad. Sci., Moscow-Leningrad, Russia.

***Nawashin, S. G. and V. V. Finn. 1912.** On the developmental history of the chalazogamic *Juglans nigra* and *Juglans regia*. Reports Kiev Soc. Naturalists, 22(3-4): 1-85.

Nawaschin, S. G. and V. V. Finn. 1913. Zur Entwicklungsgeschichte der Chalazogamen, *Juglans regia* und *Juglans nigra*. Mém. l'Acad. Impér. Sci. St. Pétersburg, 31(9):1–59.

Nazarov, V. V. 1995. Reproductive biology of the Crimean orchids. Ph.D. thesis, St. Petersburg, Russia.

Nemirowich-Danchenko, E. N. 1992. Family Dipterocarpaceae, pp. 263-278. *In* A. L. Takhtajan (ed.). Comparative Anatomy of Seeds. Nauka, Leningrad, Russia, vol. 4.

Netolitzky, F. 1926. Anatomie der Angiospermen-Samen. Handbuch der Pflanzenanatomie, 10: 1-364.

Neuffer, M. G. and W. F. Sheridan. 1980. Defective kernel mutants of maize. I. Genetic and lethality studies. Genetics, 95: 929-944.

Newcomb, W. 1973. The development of the embryo sac of sunflower *Helianthus annuus* after fertilization. Canadian J. Bot., 51(5): 879-890.

Newcomb, W. 1978. The development of cells in the coenocytic endosperm of the African blood lily *Haemanthus katherinae*. Canadian J. Bot., 56(5): 483-501.

Newcomb W. and T. A. Steeves. 1971. *Helianthus annuus* embryogenesis: embryo sac wall projections before and after fertilization. Bot. Gazette, 132(4): 367-371.

Newcomb, W. and K. C. Fowke. 1974. *Stellaria media* embryogenesis: the development and ultrastructure of the suspensor. Canadian J. Bot., 52(3): 607-614.

Nicolosi-Roncati, F. 1903. La formazione dell'endosperma nell *Anona cherimolia*. Boll. Soc. Bot. Ital., 4: 115-117.

Nieuwdorp, P. J. 1963. Electron microscopic structure of the epithelial cells of the scutellum of barley. I. The structure of the epithelial cells before germination. Acta Bot. Neerlandica, 12(3): 295-301.

Nieuwdorp, P. J. and M. C. Buys. 1964. Electron microscopic structure of the epithelial cells of the scutellum of barley. II. Cytology of the cells during germination. Acta Bot. Neerlandica, 13(4): 59-65.

Nijalingappa, B. H. M. 1975. Embryological studies in *Haloragis micrantha*. Phytomorph., 25(2): 146-155.

Nijalingappa, B. H. M. and D. H. Tejavathi. 1977. Embryo development in *Scirpus mucronatus* L. Current Sci., 46(3): 84-85.

Nijalingappa, B. H. M. and N. Devaki. 1978. Embryological studies in *Diplacrum caricinum* R. Br. Beitrage Biol. Pflanzen, 54(2): 215-225.

Nijalingappa, B. H. M. and N. Devaki. 1979. Endosperm of *Scleria foliosa* Hochstetter ex A. Richard. Current Sci., 48(10): 451-452.

*__Nikiticheva, Z. I. 1966.__ On embryology of *Lathraea squamaria* L. (Scrophulariaceae). Bot. Journal (St. Petersburg), 51(11): 1607-1617.

*__Nikiticheva, Z. I. 1971.__ Embryogenesis of some parasitic species of families Scrophulariaceae and Orobanchaceae. Bot. Journal (St. Petersburg), 56(1): 93-105.

Nikiticheva, Z. I. 1979a. The development of embryo, endosperm and seed in parasitic members of Scrophulariaceae, pp. 130-157. *In* M. S. Yakovlev (ed.). Actual Questions on the Embryology of Angiosperms, Nauka, Leningrad, Russia.

*__Nikiticheva, Z. I. 1979b.__ On the interconnections in the development of endosperm and ovule tissues in Scrophulariaceae and Orobanchaceae species. Bot. Journal (St. Petersburg), 64(11): 1567-1582.

Nikiticheva, Z. I. 1981a. Embryological features of some Piperales. Acta Soc. Bot. Poloniae, 50(1-2): 329—332.

Nikiticheva, Z. I. 1981b. Family Piperaceae, pp. 84-90. *In* M. S. Yakovlev (ed.). Comparative Embryology of Flowering Plants. Winteraceae—Juglandaceae. Nauka, Leningrad, Russia.

Nikiticheva, Z. I. 1981c. Family Saururaceae, pp. 81-84. *In* M. S. Yakovlev (ed.) Comparative Embryology of Flowering Plants. Winteraceae—Juglandaceae. Nauka, Leningrad, Russia.

*__Nikiticheva, Z. I. 1983a.__ Morphogenetic transformations of embryo suspensor in different groups of flowering plants, pp. 13-14. *In* Materials 8th All-Union Conf. Plant Embryology. FAN, Tashkent, Uzbeck Republic.

Nikiticheva, Z. I. 1983b. The suspensor structure in some angiosperms, pp. 279-282. *In* O. Erdelská (ed.). Proc. VII Int. Cytoembryol. Symp. "Fertilization and Embryogenesis in Ovulated Plants". High Tatra (Račkova dolina), 14-17 June, 1982. Veda, Bratislava, Czechoslovakia.

Nikiticheva, Z. I. 1985a. Family Crassulaceae, pp. 29-34. *In* M. S. Yakovlev (ed.). Comparative Embryology of Flowering Plants. Brunelliaceae—Tremandraceae. Nauka, Leningrad, Russia.

Nikiticheva, Z. I. 1985b. Family Linaceae, pp. 196-201. *In* M. S. Yakovlev (ed.). Comparative Embryology of Flowering Plants. Brunelliaceae—Tremandraceae. Nauka, Leningrad, Russia.

Nikiticheva, Z. I. 1987a. Family Hippuridaceae, pp. 295-299. *In* T. B. Batygina and M. S. Yakovlev (eds). Comparative Embryology of Flowering Plants. Davidiaceae—Asteraceae. Nauka, Leningrad, Russia.

Nikiticheva, Z. I. 1987b. Family Lentibulariaceae, pp. 283-288. *In* T. B. Batygina and M. S. Yakovlev (eds). Comparative Embryology of Flowering Plants. Davidiaceae—Asteraceae. Nauka, Leningrad, Russia.

Nikiticheva, Z. I. 1987c. Family Scrophulariaceae, pp. 251-259. *In* T. B. Batygina and M. S. Yakovlev (eds). Comparative Embryology of Flowering Plants. Davidiaceae—Asteraceae. Nauka, Leningrad, Russia.

Nikiticheva, Z. I. 1990a. Family Alismataceae, pp. 16-20. *In* T. B. Batygina and M. S. Yakovlev (eds). Comparative Embryology of Flowering Plants. Butomaceae—Lemnaceae. Nauka, Leningrad, Russia.

*__Nikiticheva, Z. I. and E. S. Teryokhin. 1973.__ On embryology of *Tozzia alpina* L. (Scrophulariaceae). Bot. Journal (St. Petersburg), 58(5): 680-691.

*__Nikiticheva, Z. I. and E. S. Teryokhin. 1976.__ Development of seed and seedling of *Orobanche pallidiflora* Wimm. et Grab. (Orobanchaceae). Bot. Journal (St. Petersburg), 61(5): 690-700.

Nikiticheva, Z. I. and M. S. Yakovlev. 1985. Family Rhizophoraceae, pp. 120-125. *In* T. B. Batygina and M. S. Yakovlev (eds) Comparative Embryology of Flowering Plants. Brunelliaceae—Tremandraceae. Nauka, Leningrad, Russia.

Nikiticheva, Z. I. and G. M. Anisimova. 1987. Family Loranthaceae, pp. 74-82. *In* T. B. Batygina and M. S. Yakovlev (eds). Comparative Embryology of Flowering Plants. Davidiaceae—Asteraceae. Nauka, Leningrad, Russia.

Nikiticheva, Z. I. and E. S. Teryokhin. 1987. Family Cuscutaceae, pp. 201-206. *In* T. B. Batygina and M. S. Yakovlev (eds). Comparative Embryology of Flowering Plants. Davidiaceae—Asteraceae. Nauka, Leningrad, Russia.

Nikiticheva, Z. I. and O. B. Proskurina. 1989. Embryology of species of the genus *Convolvulus* (Convolvulaceae). Ovule and seed development. Bot. Journal (St. Petersburg), 74(8): 1108-1128.

***Nikiticheva, Z. I. and O. B. Proskurina. 1992.** Embryology of *Scheuchzeria palustris* (Scheuchzeriaceae). Bot. Journal (St. Petersburg), 77(1): 3-18.

***Nikiticheva, Z. I., M. S. Yakovlev and T. A. Plyushch. 1981.** The development of ovule, embryo sac, embryo and endosperm in some species of the genus *Peperomia* (Piperaceae). Bot. Journal (St. Petersburg), 66(10): 1388-1397.

Nikolaeva, M. G. 1967. Physiology of Deep Seed Dormancy. Nauka, Leningrad, Russia.

***Nikolaeva, M. G. 1977.** Some results of seed dormancy investigation. Bot. Journal (St. Petersburg), 62(9): 1350-1368.

Nikolaeva, M. G. 1979. Accelerated Germination of Dormant Seeds. Nauka, Leningrad, Russia.

Nikolaeva, M. G. 1982. Seed dormancy, pp. 125-183. *In* A. A. Prokofiev (ed.). Physiology of Seeds. Nauka, Moscow, Russia.

***Nikolaeva, M. G. 1991.** Does the abscisic acid participate in regulation of seed formation and germination? Bot. Journal (St. Petersburg), 76(8): 1144-1150.

***Nikolaeva, M. G. and M. V. Razumova. 1973.** On the influence of temperature and growth substances on germination of tulip seeds. Bull. Main Bot. Garden USSR Acad. Sciences, 89: 73-75.

***Nikolaeva, M. G., M. M. Lodkina and A. I. Lyashuk. 1974.** On the question of physiological dwarfism nature of dormant seed embryos. Bot. Journal (St. Petersburg), 59(1): 14-23.

Nikolaeva, M. G., M. V. Razumova, and V. N. Gladkova. 1985. Reference Book on Dormant Seed Germination. Nauka, Leningrad, Russia.

***Nikolaeva, M. G., T. V. Daletskaya, L. M. Pozdova, M. V. Razumova and V. L. Tikhonova. 1987a.** Conditions of germination of dormant seeds in *Trollius* species (Ranunculaceae) and some other rare and preservation needed plants. Bot. Journal (St. Petersburg), 72(2): 238-244.

***Nikolaeva, M. G., L. M. Pozdova, E. N. Polyakova, T. V. Daletskaya, M. V. Razumova and G. A. Firsov. 1987b.** Investigation of possibility of accelerated germination of woody plant seeds with deep dormancy. Bot. Journal (St. Petersburg), 72(4): 480-489.

Nishimura, G. 1981. Comparative morphology of *Cattleya* and *Phalaenopsis* (Orchidaceae) seedlings. Bot. Gazette, 142(3): 360-365.

Nishimura, G. 1991. Comparative morphology of cotyledonous orchid seedlings. Lindleyana, 6(3): 140-146.

Nogler, G. A. 1984. Gametophytic apomixis, pp. 476-518. *In* B. M. Johri (ed.). Embryology of Angiosperms. Springer-Verlag, Berlin.

Noguti, Y., U. Oka and T. Otuka. 1940. Studies of polyploids in *Nicotiana* induced by the treatment with colchicine. II. Growth rate and chemical analysis of diploid and its autotetraploid in *Nicotiana rustica* and *N. tabacum*. Japanese J. Bot., 10: 343-364.

Nörner, C. 1881. Beitrag zur Embryoentwicklung der Gramineen. Ph.D. thesis, Regensburg, Germany.

Norstog, K. 1969. Morphology of coleoptile and scutellum in relation to tissue culture responses. Phytomorph., 19(3): 235-241.

Norstog, K. 1972. Early development of the barley embryo: fine structure. Amer. J. Bot., 59(2): 123-132.

Nosatovsky, A. I. 1965. Wheat [Biology]. Kolos, Moscow, Russia.

Nougarede, A. and J. Remebour. 1980. Extension cellulaire, division cellulaire et niveaux de ploidie des cellules corticales de l'epicotyle du pois au cours de l'elongation. Canadian J. Bot., 58(4): 486-501.

Nuccitelli, R. and L. F. Jaffe. 1975. The pulse current pattern generated by developing fucoid eggs. J. Cell Biol., 64(4): 636-643.

Nuccitelli, R. and L. F. Jaffe. 1976. Current pulses involving chloride and potassium efflux relieve excess pressure in *Pelvetia* embryos. Planta, 131: 315-320.

O'Brien, T. P. and K. V. Thimann. 1967. Observation on the fine structure of oat coleoptile. 3. Correlated light and electron microscopy of the vascular tissues. Protoplasma, 63(4): 443-478.

Obrucheva, N. V. 1982. Seed germination, pp. 223-274. *In* A. A. Prokofiev (ed.). Seed Physiology. Nauka, Moscow, Russia.

Obrucheva, N. V. 1991. Physiology of first stages of seed germination in dicotyledons. Ph.D. thesis, Moscow, Russia.

Obrucheva, N. V. 1993. Early germination: growth initiation and role of water, pp. 275-280. *In* Fourth Int. Workshop Seeds, vol. 1. Univ. P. et M. Curie, Paris, France.

***Obrucheva, N. V. and L. S. Kovadlo. 1985.** Two stages of respiration intensification of germinating pea seeds while increasing water saturation. Physiol. rastenyi, 32(4): 753-761.

***Obrucheva, N. V. and O. V. Antipova. 1994.** Starting of axial organ growth and its preparation during germination of seeds being in forced dormancy. 2. Initiation of "acidic growth" in axial organs of bean seeds. Physiol. rastenyi, 41(3): 443-447.

***Obrucheva, N. V., G. Zembdner and V. Date. 1985.** Endogenic abscisic acid in germinating pea seeds. Reports USSR Acad. Sci., 280(1): 254-256.

***Obrucheva, N. V., L. S. Kovadlo and A. A. Prokofiev. 1988.** Level of water saturation as a trigger of starch and protein mobilization during pea seed germination. Physiol. rastenyi, 35(2): 322-328.

***Obrucheva, N. V., O. V. Antipova and I. M. Ivanova. 1993.** Starting of axial organ growth and its preparation during germination of seeds being in forced dormancy. 1. Accumulation of osmotic active substances in axial organs of bean seeds. Physiol. rastenyi, 40(5): 742-748.

Oehler, E. 1927. Entwicklungsgeschichtlich-zytologische Untersuchungen an einigen saprophytischen Gentianaceen. Planta, 3(4): 641-733.

***Oganezova, G. G. 1993.** On embryo structure in relationship of Liliaceae and possibilities of its use in systematics. Bot. Journal (St. Petersburg), 78(12): 9-23.

Okada, K., J. Keda, M. K. Komaki, C. J. Bell and Y. Shimura. 1991. Requirement of the auxin polar transport system in early stages of *Arabidopsis* floral bud formation. Plant Cell, 3: 677-684.

Okamoto, A., M. Katsumi, H. Nojiri, N. Murofushi and H. Okamoto. 1995. The relationship between levels of endogenous gibberellins and the response of *Vigna* hypocotyls to exogenous indole-3-acetic acid. Plant Cell Physiol., 36(1): 165-171.

Oksijuk, P. F. and M. I. Khudyak. 1955. New data on fertilization in wheat. Reports USSR Acad. Sci., 105(4): 835-837.

Olsen, O. A. and T. Krekling. 1980. Grain development in normal and high lysine barley. Hereditas, 93: 147-160.

Olszewska, M. J. and B. Gabara. 1966. Recherches sur les cytocinèses dans l'endosperme d'*Iris pseudoacorus* et d'*Iris sibirica*. I. Les cytocinèses au cours du développement de l'endosperme. Acta Soc. Bot. Poloniae, 35(4): 557-573.

Onderdonk, I. J. and J. W. Ketcheson. 1972. A standartization of terminology for the morphological description of corn seedlings. Canadian J. Plant Sci., 52(6): 1003-1006.

Ono, T. 1929. Embryologie der Liliaceae, mit besonderer Rücksicht auf die Endospermbildung. I. Melanthioideae und Aletroideae. Science Reports Tôhoku Imper. Univ. Ser. 4 (Biology), 4(2): 381-393.

Oparka, K. J. and N. Harris. 1982. Rice protein body formation: all types are initiated by dilation of the endoplasmic reticulum. Planta, 154: 184-188.

***Oryol, L. I. and I. G. Shmaraev. 1987.** Ultrastructure of embryo sac wall in *Triticum aestivum* (Poaceae) after fertilization. Bot. Journal (St. Petersburg), 72(6): 753-757.

Osborn, D. J. 1982. Nucleic acids and seed germination, pp. 357-372. *In* M. G. Nikolaeva and N. V. Obrucheva (eds). Physiology and Biochemistry of Seed Dormancy and Germination. Kolos, Moscow, Russia.

Osterbye, U. 1975. Self-incompatibility in *Ranunculus acris* L. Hereditas, 80: 91-112.

Ottaviano, E. and D. L. Mulcahy. 1989. Genetics of angiosperm pollen. Advances Genet., 26: 1-64.

Ottaviano, E., D. Petroni and M. E. Pe. 1988. Gametophytic expression of genes controlling endosperm development in maize. Theor. Appl. Genet., 75: 252-258.

Ottaviano, E., D. L. Mulcahy, M. Sari-Gorla and G. B. Mulcahy. 1992. Angiosperm Pollen and Ovules. Springer-Verlag, Berlin.

Ovcharov, K. E. 1976. Physiology of Seed Formation and Germination. Kolos, Moscow, Russia.

Overton, E. 1891. Beitrag zur Kenntniss der Entwicklung und Vereinigung der Geschlechtsprodukte bei *Lilium martagon*. Zürich.

Overvoorde, P. J. and H. D. Grimes. 1994. The role of calcium and calmodulin in carrot somatic embryogenesis. Plant Cell Physiol., 35(2): 135-144.

Pace, L. 1913. Apogamy in *Atamosco*. Bot. Gazette (Chicago), 56: 371-394.

Pacini, E. 1981. Pollen-stigma interactions in plants with gametophytically controlled self-incompatibility. Phytomorph., 31: 175-180.

Pacini, E. 1987. New techniques and recent findings in embryological researches. *In* F. Garbari and E. Pacini (eds). Plant Embryology. The Tuscan Contribution. Soc. Bot. Italiana, Pisa, Italy.

Pacini, E., C. Simoncioli and M. Cresti. 1975 (1976). Ultrastructure of nucellus and endosperm of *Diplotaxis erucoides* during embryogenesis. Caryologia, 28(4): 525-538.

Padmanabhan, D. 1970. Nymphaeaceae. Bull. Indian Natl. Sci. Acad., 41: 60-62.

Padmanabhan, D. 1993a. Dicot and monocot embryos. A comparative study. J. Swamy Bot. Club, 10: 173-182.

Padmanabhan, D. 1993b. Embryogenesis in *Cocos nucifera* L. A new interpretation. J. Swamy Bot. Club, 10: 169-171.

Padmanabhan, D. 1996. Tracing the shoot apex in angiosperm embryos, pp. 39-51. *In* K. G. Mukerji (ed). Advances in Botany. APH Publ. Corp., New Delhi, India.

Paetow, W. 1931. Embryologische Untersuchungen on Taccaceen, Meliaceen und Dilleniaceen. Planta, 14: 441-470.

Pajak, M. 1982. Development of embryo and chromosome mosaicism in *Rudbeckia bicolor* Nutt., p. 52. Abstracts 7th Int. Symp. 'Fertilization and Embryogenesis in Ovulated Plants". High Tatra (Račkova dolina), June 14-17, 1982. Veda, Bratislava, Czechoslovakia.

Pajak, M. 1983. Development of embryo and chromosome mosaicism in *Rudbeckia bicolor* Nutt., pp. 295-298. *In* O. Erdelská (ed.). Proc. 7th Int. Symp. "Fertilization and Embryogenesis in Ovulated Plants" High Tatra (Račkova dolina), June 14-17, 1982, Veda, Bratislava, Czechoslovakia.

Palevitz, B. A. 1993a. Organization of the mitotic apparatus during generative cell division in *Nicotiana tabacum*. Protoplasma, 174: 25-35.

Palevitz, B. A. 1993b. Relationship between the generative cell and vegetative nucleus in pollen tubes of *Nicotiana tabacum*. Sexual Plant Reprod., 6: 1-10.

Palevitz, B. A. and M. Cresti. 1989. Cytoskeletal changes during generative cell division and sperm cell formation in *Tradescantia virginiana*. Protoplasma, 150: 54-71.

Palevitz, B. A. and B. Liu. 1992. Microfilaments (F-actin) in generative cells and sperm: an evolution. Sexual Plant Reprod., 5: 89-100.

Palevitz, B. A. and A. Tiezzi. 1992. Organization, composition, and function of generative cell and sperm cytosceleton. Int. Rev. Cytol., 140: 149-185.

Paliwal, R. L. 1956. Morphological and embryological studies in some Santalaceae. Agra Univ. J. Res. (Science), 5(1): 193-284.

Palm, B. T. 1915. Studien über Konstruktionstypen und Entwicklungswege des Embryosackes der Angiospermen. Ph.D. thesis, Stockholm.

Palser, B. F. 1951. Early endosperm development and embryogeny in *Cassiope hypnoides*. Trans. Illinois State Acad. Sci., 44: 51-57.

Panchaksharappa, M. G. 1962. Embryological studies in some members of the Zingiberaceae, pp. 224-238. *In* "Plant Embryology". Symposium Council Scientific and Industrial Research. November 11-14, 1960, Dept. Bot., Univ. Delhi, New Delhi, India.

Panchaksharappa, M. G. and C. K. Rudramuniyappa. 1975. Scutellum and endosperm in the seedllings of Gramineae—a histochemical study. Proc. Indian Acad. Sci., 82(6): 211-217.

Pandey, K. K. 1960. Evolution of gametophytic and sporophytic systems of self-incompatibility in Angiosperms. Evolution, 14: 98:110.

Pandey, K. K. 1962. A theory of S-gene structure. Nature, 196: 236-238.

Pandey, K. K. 1970. Time and site of S-gene action, breeding systems and relationships in incompatibility. Euphytica, 19: 364-372.

Pandey, K. K. 1979. Overcoming incompatibility and promoting genetic recombination in flowering plants. New Zealand J. Bot., 17(4): 645-663.

Pankow, H. 1962. Histogenetische Studien an den Blüten einiges Phanerogamen. Bot. Studien (Jena), 13: 1-106.

Pankow, H. and H. Von Guttenberg. 1957. Vergleichende Studien über die Entwicklung monokotyler Embryonen und Keimpflanzen. Bot. Studien (Jena), 7: 1-39.

Pankow, H. and H. Von Guttenberg. 1959. Studien über die Anlage der Achselknospen und Blattprimordien bei Gramineen. Planta, 52(6); 629-643.

Parameswaran, N. 1959. A contribution to the embryology of *Theriophonum minutum*. Proc. Indian Acad. Sci., 50(1): 15-25.

***Pashkov, G. D. 1951.** On the morphological nature of root sheath in cereals. Bot. Journal (St. Petersburg), 36(6): 597-606.

Patil, J. A. 1966. The culture in vitro of immature embryos of okra (*Abelmoscus esculentus*). Indian J. Plant Physiol., 9(1): 59-65.

Patton, D. A. and D. W. Meinke. 1990. Ultrastructure of arrested embryos from mutants of *Arabidopsis thaliana*. Amer. J. Bot., 77(5): 653-661.

Patton, D. A., L. H. Franzmann and D. W. Meinke. 1991. Mapping genes essential for embryo development in *Arabidopsis thaliana*. Molec. Gen. Genet., 227: 337-347.

Paulson, I. W. 1969. Embryogeny and caryopsis development of *Sorghum bicolor* Moench. Crop Sci., 9(1): 97-102.

Pearson, H. H. W. 1909. Further observations on *Welwitschia*. Phil. Trans. Royal Soc. London, (B), 200: 331-402.

Pearson, H. H. W. 1929. Gnetales. Cambridge Univ. Press, Cambridge, New York, NY.

Pennell, R. I., N. R. Geltz, E. Korner and S. D. Russell. 1987. Production and partial characterization of hybridoma antibodies elicited to the sperm of *Plumbago zeylanica*. Bot. Gazette, 148: 401-406.

Perata, P., P. Picciarelli and A. Alpi. 1990. Pattern of variations in abscisic acid content in suspensor, embryos, and integuments of developing *Phaseolus coccineus* seeds. Plant Physiol., 94: 1776-1780.

Percival, J. 1921. The Wheat Plant. London, UK.

Perdue, R. E. 1959 (1960). The somatic chromosomes of *Rudbeckia* and related genera of the Compositae. Contrib. Gray Herbarium Harvard Univ., 185, 186: 129-162.

Periasamy, K. 1962. The ruminate endosperm: development and types of rumination, pp. 61-74. *In* "Plant Embryology". Symposium Council Scientific and Industrial Research. November 11-14, 1960, Dept. Bot., Univ. Delhi, New Delhi, India.

Periasamy, K. 1977. A new approach to the classification of angiosperm embryos. Proc. Indian Acad. Sci., 86(1): 1-12.

Pernollet, Y. C. 1982. Les corpusculus proteiques des graines, stade transitoire de vacuoles specialesées. Physiol. Végétale, 20(2): 259-276.

***Petrova, L. R. and M. S. Yakovlev. 1968.** Morphology and anatomy of fruits and seeds of bamboos *Melocanna bambusoides* Trin. and *Phyllostachys bambusoides* Sieb. et Zucc. Bot. Journal (St. Petersburg), 53(12): 1688-1703.

Petrova, L. R. and L. K. Safina. 1985. Family Dioscoreaceae, pp. 133-135. *In* A. L. Takhtajan (ed.). Comparative Anatomy of Seeds. Monocotyledons. Nauka, Leningrad, Russia, vol. 1.

Petrova, L. R., T. A. Fedotova and T. S. Nikolaevskaya. 1985. Family Poaceae, pp. 192-205. *In* A. L. Takhtajan (ed.). Comparative Anatomy of Seeds. Monocotyledons. Nauka, Leningrad, Russia, vol. 1.

Petrova, T. F. 1968. Cytoembryological investigations of Lilioideae. Ph.D. thesis, Leningrad, Russia.

Petrova, T. F. 1977. Cytoembryology of Liliaceae. Subfamily Lilioideae. Nauka, Moscow, Russia.

Petru, E. 1970. Development of embryoids in carrot root callus culture (*Daucus carota* L.). Biol. Plantarum, 12: 1-5.

Pfeiffer, W. M. 1912. The morphology of *Leitneria floridana*. Bot. Gazette, 53(3): 189-203.

***Philimonov, S. V. and S. M. Szcziparev. 1995.** On the acidification of endosperm in cereal caryopsises. Vestnik Sankt-Peterburg. Univ., 10: 90-95.

Philip, V. J. and B. Haccius. 1976. Embryogenesis in *Bambusa arundinacea* and structure of the mature embryo. Beitrage zur Biol. Pflanzen, 52: 83-100.

Phouphas, C. 1952. Développement du proembryon chez le *Gynandropsis pentaphylla* D.C. (Capparaceae). Bull. Soc. Bot. France, 99(1-3): 18-20.

Piaggesi, A., P. Picciarelli, R. Lorenzi and A. Alpi. 1989. Gibberellins in embryo-suspensor of *Phaseolus coccineus* seeds at the heart stage of embryo development. Plant Physiol., 91: 362-366.

Picciarelli, P. and A. Alpi. 1987. Embryo-suspensors of *Tropaeolum majus*: identification of gibberellin A63. Phytochemistry, 26: 329-330.

Picciarelli, P., A. Alpi, L. Pistelli and M. Scalet. 1984. Gibberellin-like activity in suspensors of *Tropaeolum majus* L. and *Cytisus laburnum* L. Planta, 162: 566-568.

Pierson, E. S. and M. Cresti. 1992. Cytoskeleton and cytoplasmic organization of pollen and pollen tubes. Int. Rev. Cytol., 140: 73-125.

Pierson, E. S., H. M. P. Kengen and J. Derksen. 1989. Microtubules and actin filaments co-localize in pollen tubes of *Nicotiana tabacum* L. and *Lilium longiflorum* Thunb. Protoplasma, 150: 75—77.

Plate, L. 1913. Selektionsprinzip und Probleme der Artbildung. Ein Handbuch des Darwinismus. W. Engelmann, Leipzig, Berlin, vol. 15.

Plyushch, T. A. 1983. Submicroscopic characteristics of reproductive structures of *Peperomia blanda* (*Piperaceae*) ovule. Ph.D. thesis, Leningrad, Russia.

Plyushch, T. A. 1987. Ultrastructure of cells of embryo sac in *Lilium regale* and *Lilium candidum (Liliaceae)* before fertilization. Bot. Journal (St. Petersburg), 72: 651-657.

***Plyushch, T. A. 1989.** Ultrastructure of embryo sac in *Ulmus laevis* (Ulmaceae). Bot. Journal (St. Petersburg), 74(10): 1450-1455.

***Plyushch, T. A. 1992.** The Ultrastructure of Embryo Sac in Angiosperms. Naukova dumka, Kiev, Ukraine.

***Plyushch, T. A. and V. P. Bannikova. 1989.** Ultrastructure of embryo sac and fertilization in *Adoxa maschatellina* (Adoxaceae). Bot. J. (St. Petersburg), 74(1): 59-65.

Poddubnaja-Arnoldi, V. A. 1931. Ein Versuch der Anwendung der embryologischen Methode bei der Losung einiger systematischer Fragen. Beihefte zum Botanischen Centralblatt, 48b: 141-237.

Poddubnaya-Arnoldi, V. A. 1944. Embryology of angiosperms and its significance for systematics, plant breeding and genetics. Ph.D. thesis, Moscow, Russia.

***Poddubnaya-Arnoldi, V. A. 1952.** Investigation of embryos in angiosperms on living material. Bull. Main Bot. Garden USSR Acad. Sci., 14: 3-12.

***Poddubnaya-Arnoldi, V. A. 1958.** The investigation of fertilization process in some angiosperms on living material. Bot. J. (St. Petersburg), 43(2): 178-193.

***Poddubnaya-Arnoldi, V. A. 1960.** On suspensor haustorium of embryos in orchid seeds. Bull. Main Bot. Garden USSR Acad. Sci., 37: 47-52.

***Poddubnaya-Arnoldi, V. A. 1964a.** Comparative embryological study of representatives of Orchidaceae family. Bull. Main Bot. Garden USSR Acad. Sci., 54: 51-62.

Poddubnaya-Arnoldi, V. A. 1964b. General Embryology of Angiosperms. Nauka, Moscow, Russia.

Poddubnaya-Arnoldi, V. A. 1976. Cytoembryology of the Angiosperms: Principles and Perspectives. Nauka, Moscow, Russia.

Poddubnaya-Arnoldi, V. A. 1982. The Characteristics of Flowering Plant Families According to the Cytoembryological Features. Nauka, Moscow, Russia.

Poethig, R. S. 1989. Genetic mosaic and cell lineage analysis in plants. Trends Genet., 5: 273-277.

Poethig, R. S. 1990. Phase change and the regulation of shoot morphogenesis in plants. Science, 250: 923-930.

Poethig, R. S., E. H. Coe, Jr. and M. M. Johri. 1986. Cell lineage patterns in maize embryogenesis: a clonal analysis. Developm. Biol., 117: 392-404.

Polanovski, A. 1967. Tripsin inhibitor from rye seeds. Acta Biochim. Polonica, 14(4): 389-395.

Polevoy, V. V. 1982. Phytohormones. Publ. Leningrad State Univ., Leningrad, Russia.

Polevoy, V. V. and T. S. Salamatova. 1991. Physiology of Plant Growth and Development. Publ. Leningrad State Univ., Leningrad, Russia.

***Polevoy, V. V., S. M. Shchyparev and V. V. Ilchukov. 1981.** The effect of indoleacetic acid and calcium on protein secretion by scutellums of maize. Reports USSR Acad. Sci., 256(3): 763-765.

***Polevoy, V. V., S. M. Shchyparev and O. V. Moskaleva. 1990.** Hormone regulation of seed germination, pp. 119-130. *In* Seed Physiology. Dushanbe, Tajikistan.

Pollock, E. G. and W. A. Jensen. 1964. Cell development during early embryogenesis in *Capsella* and *Gossypium*. Amer. J. Bot., 51(9): 915-921.

Polunina, N. N. 1961. Microphotography in the research of pollen in some plants, pp. 459-462. *In* V. G. Alexandrov (ed.). Morphogenesis of Plants, 2. Publ. Moscow State Univ., Moscow, Russia.

***Polunina, N. N. and A. I. Sveshnikov. 1959.** Microphotography in the research of pollen and pollen tubes in some Amarillidaceae species. Reports USSR Acad. Sci., 127(1): 217-219.

***Polyakova, E. N. and M. G. Nikolaeva. 1990.** Change of isoenzyme structure of acidic phosphatase during the breaking of seed dormancy in Tatarian maple. Physiol. rastenyi, 37(4): 707-711.

***Polyakova, E. N., M. G. Nikolaeva and T. V. Daletskaya. 1986.** Activity of acidic phosphatase during the breaking and induction of dormancy in Tatarian maple seeds. Physiol. rastenyi, 33(6): 1130-1137.

Polyshchuk, V. S. and A. K. Dzevaltovsky. 1971. On the embryology of lavender (*Lavandula spica* L.). Ukrainian Bot. J., 28(6): 778-781 (in Ukrainian).

Pomeranz, Y. 1973. Structure and mineral composition of cereal aleurone cells as shown by scanning electron microscopy. Cereal Chem., 50: 504-511.

Pontowich, V. E. 1978. Early embryogenesis of angiosperms and its hormone regulation, pp. 205-234. *In* V. I. Kefeli (ed.). Plant Growth. Initial Mechanisms. Nauka, Moscow, Russia.

***Pontowich, V. E., O. B. Ivanova and T. F. Marchenko. 1976.** Influence of some phenol compounds of oily poppy placenta on embryogenesis in isolated ovules. Reports USSR Acad. Sci., 288: 1489.

Ponzi, R. and P. Pizzolongo. 1972. The ultrastructure of suspensor cells of *Ipomoea purpurea* Roth. J. Submicroscopic Cytol., 4: 199-204.

Ponzi, R. and P. Pizzolongo. 1973. Ultrastructure of plastids in the suspensor cells of *Ipomoea purpurea* Roth. J. Submicroscopic Cytol., 5(3): 257-263.

Ponzi, R. and P. Pizzolongo. 1984. Ultrastructural study of change from free-nuclear to cellular endosperm in *Ipomoea purpurea* Roth and *Cytinus hypocistis* L. Giornale Bot. Italiano, 118(3-4): 147-154.

Pope, M. N. and E. Brown. 1943. Induced vivipary in three varieties of barley possessing extreme dormancy. J. Amer. Soc. Agron., 35: 161-163.

Poptsov, A. V. 1976. Biology of Hard Coat Seeds. Nauka, Moscow, Russia.

Poptsov, A. V., V. I. Nekrasov and I. A. Ivanova. 1981. Articles on Seed Studies. Nauka, Moscow, Russia.

Possingham, J. V. 1980. Plastid replication and development in the life cycle of higher plants. Ann. Rev. Plant Physiol., 31: 113-129.

***Pozdova, L. M. 1986.** Breaking of deep dormancy in seeds of Tatarian maple by the kinetin treatment in the presence of dimethylsulfoxid. Physiol. rastenyi, 33(2): 338-343.

***Pozdova, L. M. and M. A. Balmasova. 1994.** Peculiarities of latent period in some species of genus *Aconitum*, Ranunculaceae, p. 194. *In* I. Sykura (ed.). Abstracts "Studies of Plant Ontogenesis of Natural and Cultured Floras in Botanical Institutes of Eurasia". Natl. Ukrainian Acad. Sci., Kiev, Ukraine.

***Pozdova, L. M. and M. V. Razumova. 1994.** Dormancy and germination of some rare plants in family Liliaceae. Bot. Journal (St. Petersburg), 79(6): 69-73.

Pozdova, L. M., O. G. Butuzova and G. E. Titova. 1994. Peculiarities of embryo postdevelopment in *Aconitum soongaricum* Stapf (Ranunculaceae) at different temperature conditions, p. 29. *In* Abstracts of X Conf. Introductory Seed Breeding "Peculiarities of Development and Germination of Introductory Seeds". Moscow, Russia.

***Pozdova, L. M., O. G. Butuzova and G. E. Titova. 1998.** The embryo development in seeds with morpho-physiological type of dormancy in *Aconitum soongaricum* (Ranunculaceae) as an example. Bot. Journal (St. Petersburg), 83(11): 63-74.

Pradet, A. 1982. Oxidative phosphorylation in seeds during the initial phases of germination, pp. 347-369. *In* The Physiology and Biochemistry of Seed Development, Dormancy and Germination. Elsevier Biomed. Press, Amsterdam.

Prakash, S. 1961. Morphological and embryological studies in the family Loranthaceae. VII. *Atkinsonia ligustrina* (Cunningh). F.V. Muell. Phytomorph., 11: 4-325.

Preston K. R. and J. E. Kruger. 1986. Mobilization of monocot protein reserves during germination. In Plant Proteolytic Enzymes. CRC Press, Boca Raton, 2: 2-18.

Pretová, A. 1977. Pigments in young embryos of *Linum usitatissimum* L. Photosynthetica, 11(2); 217-219.

Pretová, A. 1978. Embryogenesis of the flax: changes in the contents of pigments and fats in vitro and in situ. Biologia (Bratislava), 33(1): 29-34.

Pretová A. 1985. Asimilačné pigmenty, tuky a ultraštruktúra plastidov počas embryogenézy l'anu in situ a in vitro. Acta Bot. Slovaca, B. 4, 1-73.

Pretová, A., M. Eiamporova and O. Erdelská. 1981. Chlorophyll content and culture of flax embryos in vitro. Acta Soc. Bot. Poloniae, 50: 291-293.

Pretová, A., M. G. Ostrolucka, I. I. Shamrov and T. B. Batygina. 1993. Morpho-histological examination of calli, somatic embryos and other structures derived in vitro in *Quercus* sp. Biologia (Bratislava), 48(4): 451-456.

Primack, R. B. 1993. Essentials of Conservation Biology. Sinaver Assoc. Inc. Publ., Sunderland, MA, USA.

Pritchard, H. N. 1964. A cytochemical study of embryo development in *Stellaria media*. Amer. J. Bot., 51(3): 472-479.

Pritchard, H. N. and K. A. Bergster. 1969. The cytochemistry of some enzymatic activities in *Stellaria media* embryos. Experientia (Basel), 25(1): 1116-1117.

Proovost, E., D. Southworth and R. B. Knox. 1988. Three-dimensional reconstruction of sperm cells and vegetative nucleus in pollen of *Gerbera jamesonii*, pp. 69-73. *In* H. J. Wilms, C. J. Keijzer (eds). Plant Sperm Cells as Tools for Biotechnology. PUDOC, Wageningen, Netherlands.

Przybyllok, T. and W. Nagl. 1977. Auxin concentration in the embryo and suspensors of *Tropaeolum majus*, as determined by mass fragmentation (single ion detection). Zeitschrift Pflanzenphysiol., 84: 463-465.

Puri, V. 1939. Studies in the order Parietales. l. A contribution to the morphology of *Tamarix chinensis* Lour. Beihefte Bot. Centralblatt, 59A: 335-349.

Qin, De-bo and D. J. Harberd. 1988. Ultrastructural study of early embryo development, *Raphanus sativus* L. Acta Bot. Sinica, 30(5): 473-479.

Quatrano, R. S. 1978. Development of cell polarity. Ann. Rev. Plant Physiol., 29: 487-510.

Quatrano, R. S., L. Brian, J. Aldridge and T. Schultz. 1991. Development, 113 (Suppl. l): 11-16.

Queller, D.C. 1983. Kin selection and confict in seed maturation. J. Theor. Biol., 100: 153-172.

Queller, D.C. 1989. Inclusive fitness in a nutshell, pp. 73-109. *In* P. H. Harvey and L. Partridge (eds.). Oxford Surveys in Evolutionary Biology, vol. 6: Oxford Univ. Press, Oxford, UK.

Radlkofer, L. 1857. Der Befruchtungsprocess im Pflanzenreiche und sein Verhaltniss zu dem Thierreiche. Verlag von Wilhelm Engelmann, Leipzig.

Raff, J. W., R. B. Knox and A. E. Clarke. 1981. Style and S-allele-associated antigenes of *Prunus avium*. Planta, 153(2): 124-129.

Raghavan, T. S. 1937. Studies in the Capparidaceae. I. The life history of *Cleome chelidonii* Linn. fil. J. Linnean Soc. Bot., 51: 43-72.

Raghavan, V. 1976. Experimental Embryogenesis in Vascular Plants. Academic Press, London, UK.

Raghavan, V. 1980. Embryo culture. Int. Rev. Cytol., suppl. 11 B: 209-240.

Raghavan, V. 1986. Embryogenesis in Angiosperms: a Developmental and Experimental Study. Cambridge Univ. Press, Cambridge, UK.

Raghavan, V. 1989. Developmental Biology of Fern Gametophytes. Cambridge Univ. Press, Cambridge.

Raghavan, V. 1990. Origin of the quiescent center in the root of *Capsella bursa-pastoris* (L.) Medik. Planta, 181: 62-70.

Raghavan, V. 1997. Molecular Embryology of Flowering Plants. Cambridge Univ. Press, Cambridge, UK.

Raghavan, V. and V. Srinivasan. 1942. A contribution to the life history of *Vahlia viscosa* and *V. oldenlandioides*. Proc. Indian Acad. Sci., 15(2): 83-105.

Raghavan, V. and S. N. C. O'Konkwo. 1982. Studies on the germination of seeds of the root parasites, *Alectra vogelii* and *Striga gesnerioides*. II. DNA synthesis and development of the quiescent center in the radicle. Amer. J. Bot., 69(10): 1646-1656.

Raghavan, V. and P. S. Srivastava. 1982. Embryo culture, pp. 195-230. *In* B. M. Johri (ed.). Experimental Embryology of Vascular Plants. Springer, Berlin.

Raghavan, V. and K. K. Sharma. 1995. Zygotic embryogenesis in gymnosperms and angiosperms, pp. 73-115. *In* T. A. Thorpe (ed.). In Vitro Embryogenesis in Plants. Kluwer Acad. Publ., Dordrecht, Netherlands.

Rajasekaran, K., M. B. Hein, and G. C. Davis, M. G. Carnes and J. K. Vasil. 1987. Endogenous growth regulators in leaves and tissue cultures of *Pennisetum purpureum* Schum. J. Plant Physiol., 130(1): 13-25.

Rajora, O. P., J. W. Barrett, B. P. Dancik and C. Strobeck. 1992. Maternal transmission of mitochondrial DNA in interspecific hybrids of *Populus*. Current Genetics, 22: 141-145.

Raju, M. V. S. 1961. Morphology and anatomy of Saururaceae. I. Floral anatomy and embryology. Ann. Missouri Bot. Garden, 48(2): 107-124.

Ram, M. 1956. Floral morphology and embryology of *Trapa bispinosa*, with a discussion on the systematic position of the genus. Phytomorph., 6(3-4): 312-323.

Randal, J. A. 1982. The transmitting tract in *Gladiolus*. I. The stigma and pollen-stigma interaction. Amer. J. Bot., 69: 389-401.

Randolph, L. F. 1936. Developmental morphology of the caryopsis in maize. J. Agric. Res., 53(12): 881-916.

Rao, C. V. 1959. Contributions to the embryology of Palmae. II. Ceroxylineae. J. Indian Bot. Soc., 38(1): 46-75.

Rao, P. N. and B. V. N. Reddy. 1979. Tricotyledony in *Indigofera linifolia* Retz. Geobios, 6(1): 29-30.

Rao, P. N. and D. C. Narayana. 1980. Occurrence and identification of semigamy in Coi8x aquatica. J. Heredity, 71: 117-120.

Rathore, K. S. and K. R. Robinson. 1989. Ionic currents around developing embryos of higher plants in culture. Biol. Bull. 176(2), suppl.: 46-48.

Rathore, K. S., T. K. Hodges, and K. R. Robinson. 1988. A refined technique to apply electrical currents to callus cultures. Plant Physiology, 88(3): 515-517.

Rau, M. A. 1950. Endosperm in *Cassia tora* Linn. Nature, 165(4187): 157.

Rau, M. A. 1951. The endosperm in some species of *Cassia* L. Svensk Bot. Tidskrift, 45(3): 516-522.

Rau, M. A. 1953. Some observations on the endosperm in Papillionaceae. Phytomorph., 3(2): 209-222.

Rau, M. A. 1970. Saxifragaceae. Bull. Indian Natl. Sci. Acad., 41: 90-92.

Rau, M. A. and V. K. Sharma. 1967. Early embryogeny in *Saxifraga* Linn. Phytomorph., 17(1-4): 252-255.

Rauh, W. 1937. Die Bildung von Hypokotyl und Wurzelsprossen und ihre Bedeutung für die Wuchsformen der Pflanzen. Nova Acta Leopoldina, 4(24): 395-553.

Ray, S., S.S. Park and A. Ray. 1997. Pollen tube guidance by the female gametophyte. Development, 124: 2489-2498.

Razumova, M. V. 1978. Endogenous seed dormancy and possibility of its overcoming by treating with gibberellin and kinetin. Ph.D. thesis, Leningrad, Russia.

***Razumova, M. V. 1987.** Biology of seed germination in species of the genus *Sorbus*. Bot. Journal (St. Petersburg), 72(1): 77-83.

Razumova, M. V. and M. G. Nikolaeva. 1981. The effect of gibberellins and cytokinins on the germination of seeds with different types of dormancy, pp. 56-75. *In* M. G. Nikolaeva (ed.). The Role of Temperature and Phytohormones in Breaking of Seed Dormancy. Nauka, Leningrad, Russia.

Reed, C. F. 1954-1955. The comparative morphology of the Olacaceae, Opiliaceae and Octoknemaceae. Mem. Sociedade Broteriana, 10: 29-80.

Reeder, J. R. 1957. The embryo in grass systematics. Amer. J. Bot., 44(9): 756-768.

Reeder, J. R. 1961. The grass embryo in systematics. Recent Advances Bot., 1(2): 91-96.

Reger, B. J., R. Pressey and R. Chaubal. 1992. Chemotropic responses of pearl millet pollen tubes. Sexual Plant Reprod., 5: 47-56.

Richard, M. L. C. 1808. Demonstrations botaniques ou analyse du fruit. Paris, France.

Richter-Landmann, W. 1959. Der Befruchtungsvorgang bei *Impatiens grandulifera* Royle unter Berucksichtigung der plasmatischen Organelle von Spermzelle, Eizelle und Zygote. Planta, 53: 162-177.

Rijven, A .H. G. C. 1952. In vitro studies on the embryo of *Capsella bursa-pastoris.* Acta Bot. Neerlandica, 1(2): 157-200.

Robbins, W. J. and A. Harvey. 1969. Auxin and growth of excised roots of *Bryophyllum calycinum.* Proc. Natl. Acad. Sci. USA, 64(2): 495-497.

Robertson, D. S. 1955. The genetics of vivipary in maize. Genetics, 40: 745-760.

Robertson, D. S. 1984. A study of heterofertilization in diverse lines of maize. J. Heredity, 75: 457-462.

Robinson, K. R. 1977. Reduced external calcium or sodium stimulates calcium influx in *Pelvetia* eggs. Planta, 136(2): 153-158.

Robinson, K. R. and L. F. Jaffe. 1975. Polarized fucoid eggs drive a calcium current through themselves. Science, 187: 70-72.

Robinson, K. R. and C. McCaig. 1980. Electrical fields, calcium gradients and cell growth. Ann. New York Acad. Sci., 132-138.

Rocén, T. 1927. Zur Embryologie der Centrospermen. Ph.D. thesis, Appelbergs Boktryckeri Aktiebolag, Uppsala, Sweden.

***Rodionova G. B. 1966.** Relative to embryogenesis of *Lunaria annua* L. Bot. Journal (St. Petersburg), 51(10): 1506-1511.

Rodionova G. B. 1983a. Family Brassicaceae, pp. 154-164. *In* M. S. Yakovlev (ed.). Comparative Embryology of Flowering Plants. Phytolaccaceae—Thymelaeaceae. Nauka, Leningrad, Russia.

Rodionova G. B. 1983b. Family Capparaceae, pp. 150-154. *In* M. S. Yakovlev (ed.). Comparative Embryology of Flowering Plants. Phytolaccaceae—Thymelaeaceae. Nauka, Leningrad, Russia.

Rodkiewicz, B. 1973. Embryologia roślin kwiatowych [Embryology of Flowering Plants]. Państwowe Wydawnictwo Naukowe. Warsawa, Poland.

Rodkiewicz, B. 1984. Embryologia roślin nagozalążkaowych [Embryology of Gymnosperms]. Państwowe Wydawnictwo Naukowe. Warsawa, Poland.

Rodkiewicz, B., B. Fyk and E. Szczuka. 1994. Chlorophyll and cutin in early embryogenesis in *Capsella, Arabidopsis* and *Stellaria* investigated by fluorescence microscopy. Sexual Plant Reprod., 7: 287-289.

Rogers, S. O. and R. S. Quatrano. 1983. Morphological staging of wheat caryopsis development. Amer. J. Bot., 70: 308-311.

Roman, H. 1948. Directed fertilization in maize. Proc. Natl. Acad. Sci. USA, 34: 36-42.

Rondet, P. 1965. Mise en place des méristems ches les Angiospermes au cours de l'embryogénése. Bull. Soc. Française, Physiol. Végétale, 11(2): 175-183.

Rosén, W. 1932. Zur Embryologie der Campanulaceen und Lobeliaceen. Acta Horti Gothoburgensis. Meddelanden från Göteborgs Botaniska Trädgärd, 7: 31-42.

Rosén, W. 1949. Endosperm development in Campanulaceae and closely related families. Bot. Notiser, 2: 137-147.

Rosén, W. G. 1964. Chemotropism and fine structure of pollen tubes, pp. 159-166. *In* H. F. Linskens (ed.). Pollen Physiology and Fertilization. North-Holland Publ. Co., Amsterdam, Netherlands.

Rosenberg, O. 1904. Über die Individualität der Chromosomen im Pflanzenreich. Flora, pp. 3-8.

Rosenblum J. M. and D. V. Basile. 1984. Hormonal regulation of morphogenesis in *Streptocarpus* and its relevance to evolutionary history of the Gesneriaceae. Amer. J. Bot., 71: 52-64.

Ross, H. A. and T. W. Hegarty. 1979. Sensitivity of seed germination and seedling radicle growth to moisture stress in some vegetable crop species. Ann. Bot., (London), 43(2): 241-243.

Roth, I. 1955. Zur morphologischen Deutung des Grasembryos und verwandter Embryotypen. Flora, 142(4): 564-600.

Roth, I. 1957. Histogenese und Entwicklungsgeschichte des *Triticum*-Embryos. Flora, 144(2): 163-212.

Roth, I. 1961. Histogenesis and morphological interpretation of the grass embryo, pp. 96-99. *In* Recent Advances in Botany. Lectures and Symposia IX International Botanical Congress, Montreal, August 1959, vol. I. Univ. Toronto Press, Montreal, Canada.

Roth, J. 1977. Über die anisokotylen Keimlinge von Gyranthera caribensis, Bombacaceae. Struktur der Samenschale des Makro- und Mikrokotyledons und der Primarblätter. Bot. Jahrb. Systematik, Pflanzengeschichte, Pflanzengeographie, 97(4): 515-575.

Rothe, H. 1951. Morphologisch-entwicklungsgeschichtliche und genetische Analyse einer sich variabel manifestierenden Mutation von *Antirrhinum majus* L. Zeitschrift Inductive Abstammungs, Vererbungslehre, 84: 74-132.

Rougier, M., N. Inoud and C. Said, S. Russell and C. Dumas. 1991. Male gametophyte development and formation of the male germ unit in *Populus deltoides* following compatible pollination. Protoplasma, 162(2-3): 140—150.

Roux, W. 1912. Terminologie der Entwicklungsmechanik der Tiere und Pflanzen. Verlag von Wilhelm Engelmann, Leipzig.

Ruijter, N. and A. M. Emons. 1993. Immunodetection of spectrum antigenes in plant cells. Cell Biol. Int. Reports, 17(2): 169-182.

Rusche, M. L. and H. L. Mogensen. 1988. The male germ unit of *Zea mays*: quantitative ultrastructure and three-dimensional analysis, pp. 221-226. *In* M. Cresti, P. Gori and E. Pacini (eds). Sexual Reproduction in Higher Plants. Springer-Verlag, Berlin.

Russell, S. D. 1980. Participation of male cytoplasm during gamete fusion in angiosperm *Plumbago zeylanica*. Science, 210(4466): 200-201.

Russell, S. D. 1982. Fertilization in *Plumbago zeylanica*: entry and discharge of the pollen tube in the embryo sac. Canadian J. Bot., 60(11): 2219-2230.

Russell, S. D. 1983. Fertilization in *Plumbago zeylanica*: gametic fusion and fate of the male cytoplasm. Amer. J. Bot., 70(3): 416-434.

Russell, S. D. 1984. Ultrastructure of the sperm of *Plumbago zeylanica*: 2. Quantitative cytology and three-dimensional reconstruction. Planta, 162(4): 385-391.

Russell, S. D. 1985a. Preferential fertilization in *Plumbago*: ultrastructural evidence for gamete-level recognition in an angiosperm. Proc. Natl. Acad. Sci. USA, 82(18): 6129-6132.

Russell, S. D. 1985b. Sperm specificity in *Plumbago zeylanica*, pp. 145-146. *In* M. T. M. Willemse and J. L. Van Went (eds). "Sexual Reproduction in Seed Plants, Ferns and Mosses". Proc. 8th Int. Symp. Sexual Reproduction, August 20-24, 1984. PUDOC, Wageningen, Netherlands.

Russell, S. D. 1987. Quantitative cytology of the egg and central cell of *Plumbago zeylanica* and its impact on cytoplasmic inheritance patterns. Theor. Appl. Genet., 74: 693-699.

Russell, S. D. 1991. Isolation and characterization of sperm cells in flowering plants. Ann. Rev. Plant Physiol., Molec. Biol., 42: 189-204.

Russell, S. D. 1992. Double fertilization. Int. Rev. Cytol., 140: 357-388.

Russell, S. D. 1993. The egg cell: development and role in fertilization and early embryo development. Plant Cell, 5: 1349-1359.

Russell, S. D. 1994. Fertilization in higher plants, pp. 140-152. *In* A. G. Stephenson and T. H. Kao (eds). Pollen–Pistil Interactions and Pollen Tube Growth. Current Topics in Plant Physiology. Amer. Soc. Plant Physiologist Series, vol. 12. Beltsville, MD. USA.

Russell, S. D. 1996. Attraction and transport of male gametes for fertilization. Sexual Plant Reprod., 9: 337-342.

Russell, S. D. and D. D. Cass. 1981a. Ultrastructure of fertilization in *Plumbago zeylanica*. Acta Soc. Bot. Poloniae, 50(1-2): 185-190.

Russell, S. D. and D. D. Cass. 1981b. Ultrastructure of the sperm of *Plumbago zeylanica*. I. Cytology and association with vegetative nucleus. Protoplasma, 107(1-2): 85-107.

Russell, S. D. and D. D. Cass. 1988. Fertilization in *Plumbagella micrantha*. Amer. J. Bot., 75(6): 778-781.

Russell, S. D., M. Cresti and C. Dumas. 1990a. Recent progress on sperm characterization in flowering plants. Physiol. Plantarum (Copenhagen), 80: 669-676.

Russell, S. D., M. Rougier and C. Dumas. 1990b. Organization of the early post-fertilization megagametophyte of *Populus deltoides*. Ultrastructure and implications for male cytoplasmic transmission. Protoplasma, 155(1-3): 153-165.

Rutishauser, A. 1969. Embryologie und Fortpflanzungsbiologie der Angiospermen. Springer-Verlag, Berlin.

Ryczkowski, M. 1980. Physico-biochemical and physiological gradients in the ovule during embryogenesis. Bull. Soc. Bot. France. Actualities Bot., 127(3/4); 51-58.

Ryczkowski, M. and E. Szewczyk. 1972. Changes of the chlorophylls content in the developing embryo of *Haemanthus katharinae* Bak. Bull. l'Acad. Polonaise Sci., Série Sci. Biol., 20(1): 49-53.

Ryczkowski, M. and W. Reczynski. 1988. Concentration of elements in the central vacuole sap during proembryogenesis *Clivia* sp. Ann. Sci., 23: 176-177.

***Rytova, N. G. and N. N. Tsvelev. 1982.** On the homology of embryos and lateral buds in cereals (Poaceae) and on the nature of monocotyledony. Bull. Moscow Soc. Naturalists, Biol. Dept., 87(1): 78-85.

Sabet, Y. S. 1931. Development of the embryo sac in *Calotropis procera* with special reference to endosperm formation. Ann. Bot. (London), 45(179): 503-518.

Sachar, R. C. and H. Y. Mohan Ram. 1958. The embryology of *Eschscholzia californica*. Phytomorph., 8(1-2): 114-124.

Sachs, J. 1874. Lehrbuch der Botanik. Verlag von Wilhelm Engelmann, Leipzig.

Sachs, J. 1882. Text-book of Botany Morphological and Physiological. S. H. Vines (ed.). Clarendon Press, Oxford, UK.

Saha, B. 1956. Studies on the development of the embryo of *Oryza sativa* L. and the homologies of its parts. Proc. Natl. Inst. Sci. India, B., 22(2): 86-101.

Sakurai, N., K. Shibata and S. Kamisaka. 1974. Stimulation of cucumber hypocotyl elongation by dihydroconiferyl alcohol. Interaction between dihydroconiferyl alcohol and auxin or gibberellin. Plant Cell Physiol., 15(4): 709-716.

***Salamatova, T. S. 1994.** The effect of red light on gibberellin content in maize seedlings during deetiolation. Vestnik Sankt-Peterburg. Gosudarstvennogo Univ., Series 3, 4(24): 97-102.

Salamatova, T. S. and O. A. Zauralov. 1991. Physiology of secretion of substances by plants. Publ. Leningrad State Univ., Leningrad, Russia.

Salisbury, E. 1942. The Reproductive Capacity of Plants. G. Bell and Sons, Ltd., London, UK.

Salmenkallio, M. and T. Sopanen. 1989. Amino acid peptide uptake in the scutella of germinating grains of barley, wheat, rice and maize. Plant Physiol., 89: 1285-1291.

Samuelsson, G. 1913. Studien über die Entwicklungsgeschichte einiger Bicornes-Typen. Ein Beitrag zur Kenntnis der systematischen Stellung der Diapensiaceen und Empetraceen. Svensk Bot. Tidskrift, 7(2): 97-188.

Sargant, E. 1900. Recent work on the results of fertilization in angiosperms. Ann. Bot. (London), 14: 689-712.

Sargant, E. 1902. The origin of the seed-leaf in monocotyledons. New Phytologist, 1: 25-48.

Sargant, E. 1903. A theory of the origin of monocotyledons founded on the structure of their seedlings. Ann. Bot. (London), 17: 1-92.

Sargant, E. 1908. The reconstruction of a race of primitive angiosperms. Ann. Bot. (London), 22: 121-186.

Sastri, R. L. N. 1958. Floral morphology and embryology of some Dilleniaceae. Bot. Notiser, 111(3): 495-511.

Satarova, T. N. 1990. Family Asparagaceae, pp. 114-124. *In* T. B. Batygina and M. S. Yakovlev (eds). Comparative Embryology of Flowering Plants. Butomaceae—Lemnaceae. Nauka, Leningrad, Russia.

Satina, S. and J. Rietsema. 1959. Seed development, pp. 181-195. *In* A. G. Avery, S. Satina, J. Rietsema and A. F. Blakeslee (eds). The Genus *Datura*. Ronald, New York, NY.

Sato, N. and M. Furuya. 1985. Synthesis of translatable mRNA for phytochrome during imbibition in embryonic axes of *Pisum sativum* L. Plant, Cell Physiol., 26(8): 1511-1517.

Satoh, S., A. Sturm, T. Fujii and M. J. Chispeels. 1992. cDNA cloning of an extracellular dermal glycoprotein of carrot and its expression in response to wounding. Planta, 188: 432-440.

***Savina, G. I. 1964.** Fertilization in *Cypripedium calceolus* L. Bot. Journal (St. Petersburg), 49(9): 1317-1322.

Savina, G. I. 1965. Fertilization in orchids (Orchidaceae). Ph.D. thesis, Leningrad, Russia.

***Savina, G. I. 1972.** Fertilization process in *Ophrys insectifera* L. Bot. Journal (St. Petersburg), 57(3): 382-392.

Savina, G. I. 1978. Certain peculiarities in the embryology of orchids. Proc. Indian Natl. Sci. Acad., 44(2): 141-145.

***Savina, G. I. 1979.** Peculiarities of the embryology in Orchidaceae family, pp. 19-28. *In* M. S. Yakovlev (ed.). Actual Questions on Embryology of Angiosperms. Nauka, Leningrad, Russia.

Savina, G. I. and G. Ya. Zhukova. 1983. Ultrastructural aspects of the embryo sac development in *Epipactis atrorubens* (Hoffm.) Schult., pp. 207-209. *In* O. Erdelská (ed.). Proc. 7th Int. Symp. "Fertilization and Embryogenesis in Ovulated Plants" High Tatra (Račkova dolina), June 14-17, 1982, Veda, Bratislava, Czechoslovakia.

Savina, G. I. and V. A. Poddubnaya-Arnoldi. 1990. Family Orchidaceae, pp. 172-179. *In* T. B. Batygina and M. S. Yakovlev (eds). Comparative Embryology of Flowering Plants. Butomaceae—Lemnaceae. Nauka, Leningrad, Russia.

Sax, K. 1918. Behaviour of the chromosomes in fertilization. Genetics, 3: 309-327.

Scanlon, M. J., M. James, P. S. Stinard and D. S. Robertson. 1991. Genetic placement and mapping of mutator-induced detective kernel mutants. Maize Genet. Coop. Newsletter, 65: 11.

Schacht, H. 1850. Entwicklungsgeschichte des Pflanzen-Embryon. Ph.D. thesis, Amsterdam.

Schaffner, J. H. 1896. The embryo sac of *Alisma plantago*. Bot. Gazette (Chicago), 21: 123-132.

Schaffner, J. H. 1897. The life history of *Sagittaria variabilis*. Bot. Gazette (Chicago), 23: 252-273.

Schaffner, J. H. 1904. Some morphological peculiarities of the Nymphaeaceae and Helobiae. Ohio Naturalist, 4: 83-92.

Schaffner, M. 1906. The embryology of the shepherd's purse. Ohio Naturalist, 7(1): 1-8.

Schel, J. H. N., H. Kieft and A. A. M. Van Lammeren. 1984. Interactions between embryo and endosperm during early developmental stages of maize caryopses (*Zea mays*). Canadian J. Bot., 62: 2842-2853.

Scheres, B., H. Wolkenfelt, V. Willemsen, M. Terlouw, E. Lawson, C. Dean and P. Weisbeek. 1994. Embryonic origin of the *Arabidopsis* primary root and root meristem initials. Development, 120: 2475-2487.

Scheres, B., L. Di Laurenzio, V. Willemsen, M.-T. Hauser, K. Janmaat, P. Weisbeek and P. N. Benfey. 1995. Mutations affecting the radial organisation of the *Arabidopsis* root display specific defects throughout the embryonic axis. Development, 121: 53-62.

Schiavone, F. M. and R. H. Racusen. 1989. The development of a new root at the apical end of surgically-transected embryo: reestablishment of the original root-shoot pattern. Amer. J. Bot., 76(6) suppl.: 69.

Schiavone, F. M. and R. H. Racusen. 1990. Microsurgery reveals regional capabilities for pattern reestablishment in somatic carrot embryos. Developm. Biol., 141: 211-219.

Schleiden, M. J. 1837. Einige Blicke auf die Entwicklungsgeschichte des Vegetablischen Organismus bei den Phanerogamen. Archiv Naturgeschichte (Wiegmann), 3(1): 289-320.

Schleiden, M. J. 1839a. Beiträge zur Anatomie der Cacteen, pp. 335-380. Mémoires des Savants Étrangers, vol. 4. Saint-Pétersbourg, Russia.

Schleiden, M. J. 1839b. Über die Bildung des Eichens und Entstehung des Embryos bei den Phanerogamen. Nova Acta Physico-medica Academiae Caesareae Leopoldino — Carolinae nature curiosorum. Exhibentia Ephemerides Sive Observationes Historias et Experimenta, 19: 27-58.

Schmalhausen, I. I. 1939. Pathways and Patterns of Evolutionary Process. Publ. USSR Acad. Sci., Moscow-Leningrad, Russia.

***Schmalhausen, I. I. 1958.** The control and regulation in evolution. Bull. Moscow Soc. Naturalists, Biol. Dept., 63(5): 93-121.

Schmalhausen, I. I. 1968. Factors of Evolution. The Theory of Stabilized Selection. Nauka, Moscow, Russia.

Schmid, B. 1902. Beiträge zur Embryo-Entwicklung einiger Dicotylen. Bot. Zeitschrift (Berlin), 60: 207-230.

Schmidt, E. D. L., A. J. De Jong and S. C. De Vries. 1994. Signal molecules involved in plant embryogenesis. Plant Molec. Biol., 26: 1305-1313.

Schmit, A. C. and A. M. Lambert. 1990. Microinjected fluorescent phalloidin in vivo reveals the F-actin dynamics and assembly in higher plant mitotic cells. Plant Cell, 2(2): 129-138.

Schnall, J. A., T. J. Cooke and D. E. Cress. 1988. Genetic analysis of somatic embryogenesis in carrot cell culture: initial characterization of six classes of temperature-sensitive variants. Developm. Genet., 9: 49-67.

Schnarf, K. 1917. Beiträge zur Kenntnis der Samenentwicklung der Labiaten. Denkschriften der Kaiserlichen Akademie der Wissenschaften, Wien. Mathematisch-Naturwissenschaftliche Klasse, 94: 211-275.

Schnarf, K. 1926. Kleine Beiträge zur Entwicklungsgeschichte der Angiospermen. VI. Über die Samenentwicklung einiger Gramineen. Österreichische Bot. Zeitschrift, 75(4-6): 105-113.

Schnarf, K. 1929. Embryologie der Angiospermen. Gebrüder Bornträger, Berlin.

Schnarf, K. 1931. Vergleichende Embryologie der Angiospermen. Gebrüder Bornträger, Berlin.

Schneider, E. L. 1978. Morphological studies of the Nymphaeaceae. IX. The seed of *Barclaya longifolia* Wall. Bot. Gazette, 139(2): 223-230.

Schneider, T., R. Dinkins, K. Robinson, J. Shellhammer and D. W. Meinke. 1989. An embryo-lethal mutant of *Arabidopsis thaliana* is a biotin auxotroph. Developm. Biol., 131: 161-167.

Schnepf, E. and W. Nagl. 1970. Über einige Strukturbesonderheiten der Suspensorzellen von *Phaseolus vulgaris*. Protoplasma, 69(1); 133-143.

Schötz, F. 1954. Über Plastidenkonkurrenz bei *Oenothera*. Planta, 43: 182-240.

Schröder, M.-B. 1984. Ultrastructural studies on plastids of generative and vegetative cells in the family Liliaceae. 1. *Lilium martagon* L. Biol. Zentralblatt, 103(5): 547-555.

Schröder, M.-B. 1985a. Ultrastructural studies on plastids of generative and vegetative cells in the family Liliaceae. 2. *Fritillaria imperialis* and *F. meleagris*. Biol. Zentralblatt, 104(1): 21-27.

Schröder, M.-B. 1985b. Ultrastructural studies on plastids of generative and vegetative cells in Liliaceae. 3. Plastid distribution during the pollen development in *Gasteria verrucosa* (Mill.) Duval. Protoplasma, 124(1-2): 123-129.

Schröder, M.-B. 1986a. Ultrastructural studies on plastids of generative and vegetative cells in Liliaceae. 4. Plastid degeneration during generative cell maturation in *Convallaria majalis* L. Biol. Zentralblatt, 105(4): 427-433.

Schröder, M.-B. 1986b. Ultrastructural studies on plastids of generative and vegetative cells in Liliaceae. 5. The behaviour of plastids during pollen development in *Chlorophytum comosum* (Thunb.) Jacques. Theor. Appl. Genet., 72(6): 840-844.

Schröder, M.-B. 1988. The cytological basis of the plastid distribution during the formation and development of the male gametophyte in Liliaceae, p. 144. *In* Abstracts 10th Int. Symp. "Sexual Reproduction in Higher Plants", May 30-June 4 1988. Univ. Siena, Siena, Italy.

Schulz, P. and W. A. Jensen. 1969. *Capsella* embryogenesis: the suspensor and the basal cell. Protoplasma, 67(2-3): 139-163.

Schulz, P. and W. A. Jensen. 1973. *Capsella* embryogenesis: the central cell. J. Cell Sci., 12(3): 741-763.

Schulz, P. and W. A. Jensen. 1974. *Capsella* embryogenesis: the development of the free nuclear endosperm. Protoplasma, 80(1-3): 183-205.

Schulz, P. and W. A. Jensen. 1977. Cotton embryogenesis: the early development of the free nuclear endosperm. Amer. J. Bot., 64(4); 384-394.

Schulz, P. and W. A. Jensen. 1978. Ultrastructural studies of early endosperm development in fertilized and unfertilized cotton ovules. Bull. Soc. Bot. France, 125(1/2): 267-271.

Schulz, R., W. A. Jensen. 1968a. *Capsella* embryogenesis: the early embryo. J. Ultrastructure Res., 22(5-6): 376-392.

Schulz, R. and W. A. Jensen. 1968b. *Capsella* embryogenesis: the egg, zygote and young embryo. Amer. J. Bot., 55(7): 807-819.

Schürhoff, P. N. 1919. Zur Phylogenie des Angiospermen-Embryosackes. Berichte der Deutschen Bot. Gesellschaft, 37: 160-168.

Schuster, W. 1956. Untersuchungen über die Wirkung der 2,4-D und der α-NAA auf die Blüte und den Samen der Sonnenblume sowie die Nachwirkungen in den folgenden Generationen. Der Züchter, 26(3): 78-83.

Schwartz, B. W., E. C. Yeung and D. W. Meinke. 1994. Disruption of morphogenesis and transformation of the suspensor in abnormal suspensor mutants of *Arabidopsis*. Development, 120: 3235-3245.

Schwartz, B. W., D. M. Vernon and D. W. Meinke. 1995. Development of the suspensor: differentiation, communication, and programmed cell death during plant embryogenesis. *In* B. A. Larkins and I. K. Vasil (eds). Cellular and Molecular Biology of Plant Seed Development. Kluwer Acad. Publ., Dordrecht, Netherlands.

Schwarz-Sommer, Z., P. Huijser, and W. Nacken, H. Saedler and H. Sommer. 1990. Genetic control of flower development by homeotic genes in *Antirrhinum majus*. Science, 250: 931-936.

Scott, F. M. 1944. Cytology and microchemistry of nuclei in developing seed of *Echinocystis macrocarpa*. Bot. Gazette (Chicago), 105(3): 329-338.

Scott, T. K. and M. B. Wilkins. 1969. Auxin transport in roots. IV. Effect of light on IAA movement and geotropic responsiveness in *Zea* roots. Planta, 87(2): 249-258.

Sedley, M. 1976. Control by the embryo sac over pollen tube growth in the style of the avocado (*Persea americana* Mill.). New Phytologist, 77: 149-152.

Sehgal, C. B. 1965. The embryology of *Cuminum cyminum* L. and *Trachyspermum ammi* (L.) Spragne (=*Carum copticum* Clarke). Proc. Natl. Inst. Sci. India, B., 31: 175-201.

***Selyaninova-Korchagina. M. V. 1968.** On the code of primitivism and specialization in phylogenetic system. Zh. obshchei Biol., 29(5): 574-586.

Serdjuk, L. S. and V. G. Konarev. 1970. Fluorescent analysis of the structural state of DNA in nucleus, pp. 235-238. *In* I. B. Zbarsky (ed.). Cellular Nucleus and Its Ultrastructure. Nauka, Moscow, Russia.

***Serdjuk, L. S., O. A. Khvedynich and V. P. Bannikova. 1978**. Investigation on functional state of nucleic acids of fusing gamete nuclei in tobacco. Reports USSR Acad. Sci., 241(3): 730-732.

Serebryakov, I. G. 1952. Morphology of Vegetative Organs in Higher Plants. Sovetskaya nauka, Moscow, Russia.

Sethumadhavan, P. and K. Gopakumar. 1973. Occurrence of tricotyly in brinjal (*Solanum melongena* Linn.) and sweet potato (*Ipomoea batatas* Lam.). Agric. Res. J. Kerala, 11(2): 181.

Severtsov, A. N. 1939. Morphological Patterns of Evolution. Publ. USSR Acad. Sci., Moscow-Leningrad, Russia.

Shah, C. K. 1963. The life history of *Juncus bufonius* Linn. J. Indian Bot. Soc., 42(2): 238-251.

Shamrov, I. I. 1981. Embryology of family Ceratophyllaceae in connection with its systematics. Postdoctoral thesis, Leningrad, Russia.

***Shamrov, I. I. 1990.** The ovule of *Gentiana cruciata* (Gentianaceae): structural-functional aspects. Bot. Journal (St. Petersburg), 75(10): 1363-1379.

Shamrov, I. I. 1991. The ovule of *Swertia iberica* (Gentianaceae): structural and functional aspects. Phytomorph., 41(3-4): 213-229.

Shamrov, I. I. 1995. Ovule of flowering plants. Principles of organization and typification. Postdoctoral thesis, St. Petersburg, Russia.

Shamrov, I. I. 1996. Genealogical and dynamic approaches to embryo study, p. 26. *In* Abstracts Plant Embryogenesis Workshop, 12-14 Sept. 1996. Centre for Applied Plant Molecular Biology, Univ. Hamburg, Hamburg, Germany.

Shamrov, I. I. 1997a. Nucellus typification and ovule classification. Bull. Polish Acad. Sci., Biol. Sci., 45(2-4): 1-10.

***Shamrov, I. I. 1997b.** Ovule and seed development in *Ceratophyllum demersum* (Ceratophyllaceae). Bot. Zhurnal, 82: 1-13.

***Shamrov, I. I. 1998a.** Formation of hypostase, podium and postament in the ovule of *Nuphar lutea* (Nymphaeaceae) and *Ribes aureum* (Grossulariaceae). Bot. Zhurnal, 83: 3-14.

Shamrov, I. I. 1998b. Ovule classification in flowering plants—new approaches and concepts. Bot. Jahrb. Systematik, Pflanzengeschichte, Pflanzengeographie, 120(3): 377-407.

Shamrov, I. I. 2000. Translocation pathways for metabolites in developing ovules of *Gentiana cruciata* L., *Gymnadenia conopsea* (L.), R. BGr., *Gagea stipitata* Merclin et Luzula, *Pedemontana* Boiss. et Reut. Acta Biol. Cracovensia, Series Bot., 42(1): 61-77.

***Shamrov, I. I. and T. B. Batygina. 1984.** The development of the embryo and endosperm in representatives of the family Ceratophyllaceae. Bot. Journal (St. Petersburg), 69(10): 1328-1335.

Shamrov, I. I. and T. B. Batygina. 1988. The family Ceratophyllaceae, pp. 153-156. *In* A. L. Takhtajan (ed.). Comparative Anatomy of Seeds. Magnoliidae—Ranunculidae. Nauka, Leningrad, Russia.

***Shamrov, I. I. and Z. I. Nikiticheva. 1992.** The ovule and seed morphogenesis in *Gymnadenia conopsea* (Orchidaceae): the structural and histochemical investigation. Bot. Journal (St. Petersburg), 77(4): 45-60.

***Shamrov, I. I. and G. M. Anisimova. 1993a.** The development of ovule and seed in *Juncus filiformis* (Juncaceae). Bot. J. (St. Petersburg), 78(11): 18-33.

***Shamrov, I. I. and G. M. Anisimova. 1993b.** Transformation patterns of ovule into seed in *Luzula pedemontana* (Juncaceae). Bot. Journal (St. Petersburg), 78(12): 24-44.

***Shamrov, I. I. and N. A. Zhinkina. 1994.** Ovule development in *Azorina vidallii* (Campanulaceae). Bot. Journal (St. Petersburg), 79(6): 19-34.

***Shamrov, I. I., T. I. Djachuk, T. B. Batygina and P. A. Djachuk. 1988.** Embryoidogenous type of asexual reproduction and classification of anomalies in the anther culture on the example of wheat, pp. 210-211. *In* Proc. Int. Conf. "Biology of Cultivated Cells and Biotechnology", Novosibirsk, Russia.

Shamrov, I. I., G. K. Alimova, B. R. Koudarov, T. L. Dyachuk, I. D. Nikiforova and T. B. Batygina. 1992. Some morphogenetic aspects of secondary embryoidogeny (somatic embryogeny) in tissue culture, pp. 387-388. *In* Book of Poster Abstracts, XIII Eucarpia Congress "Reproductive Biology and Plant Breeding", July 6-11, 1992. Angers, France. Imprimiere MALGOGNE, Angers, France.

Sharma, H. C. and B. S. Gill. 1982. Effect of embryo age and culture media on plant growth and vernalization response in winter wheat. Euphytica, 31(3): 629-634.

Sharp, W. R., D. A. Evans and M. R. Sondahl. 1982. Application of somatic embryogenesis to crop environment, p. 839. *In* A. Fujuwara (ed.). Plant Tissue Culture. Proc. 5th Int. Cong. Plant Tissue and Cell Culture. Japanese Assoc. Plant Tissue Culture, Tokyo, Japan.

Sharp, W. R., M. R. Sondahl, L. S. Cadlas and S. B. Maratta. 1980. The physiology of in vitro asexual embryogenesis. Hortic. Rev. (Cairo), 2: 268-310.

Shattuck, C. H. 1905. A morphological study in *Ulmus americana*. Bot. Gazette (Chicago), 20: 209-223.

Shepherd T. D. and W. S. Alverson. 1981. A new *Gatostemma* (Bombacaceae) from Columbia. Brittonia, 33(4).

Szcziparev, S. M. and G. V. Chuprova. 1980. Secretion of acids by germinating scutellums of maize caryopsises and its biological role, pp. 149-154. *In* Photosynthesis, Respiration and Organic Acids. Publ. Voronezh State Univ., Voronezh, Russia.

***Szcziparev, S. M. and V. V. Ilchukov. 1983.** Effect of gibberellin on protein secretion by maize scutellums. Vestnik Leningrad. Gosudarstvennogo Univ., Biol., 21: 66-69.

***Szcziparev, S. M. and F. V. Myachin. 1990.** Studies on ferrecyanid reductase activity of maize caryopsis scutellums. Vestnik Leningrad. Gosudarstvennogo Univ., Biol., 4: 89-93.

***Szcziparev, S. M., G. V. Chuprova and V. V. Polevoy. 1976.** Secretion of acids by isolated maize scutellums. Vestnik Leningrad. Gosudarstvennogo Univ., Biol., 21: 130-133.

Szcziparev, S. M., V. V. Polevoy and I. V. Podsoblyaeva. 1990. Uptake of sugars by maize scutellums, pp. 16-20. *In* Membrane Transport and Bioelectrical Activity of Plants. Publ. Gorky State Univ., Gorky, Russia.

***Szcziparev, S. M., E. A. Alyeva and A. N. Shinkarev. 1993.** Effect of ATPase inhibitors on acidification of medium by maize scutellums. Vestnik Sankt-Peterburg. Univ., 10: 86-88.

Shellhammer, J. and D. Meinke. 1990. Arrested embryos from the bio 1 auxotroph of *Arabidopsis thaliana* contain reduced levels of biotin. Plant Physiol., 93: 1162-1167.

Sheridan, W. F. 1988. Maize developmental genetics: genes of morphogenesis. Ann. Rev. Genet., 22: 353-385.

Sheridan, W. F. and M. G. Neuffer. 1980. Defective kernel mutants of maize. II. Morphological and embryo culture studies. Genetics, 95: 945-960.

Sheridan, W. F. and M. G. Neuffer. 1981. Maize mutants altered in embryo development, pp. 1137-1156. *In* S. Subtelney and U. Abbot (eds). Levels of Genetic Control in Development. Alan R. Liss, New York, NY.

Sheridan, W. F. and M. G. Neuffer. 1982. Maize developmental mutants. J. Heredity, 73: 318-329.

Sheridan, W. F. and J. K. Clark. 1993. Mutational analysis of morphogenesis of the maize embryo. Plant Journal, 3(2): 347-358.

Sherwood, R. T. 1995. Nuclear DNA amount during sporogenesis and gametogenesis in sexual and aposporous buffelgrass. Sexual Plant Reprod., 8: 85-90.

Shevell, D. E., W.-M. Leu, C. S. Gilimor, and G. Xia, K. A. Feldmann and N.-H. Chua. 1994. EMB30 is essential for normal cell division, cell expansion, and cell adhesion in *Arabidopsis* and encodes a protein that has similarity to Sec7. Cell, 77: 1051-1062.

***Shevtsova, G. G., T. B. Batygina and A. N. Lavrentyeva. 1986.** Some aspects of orchid reproduction system on the example of *Cymbidium hybridum* (Orchidaceae). Bot. Journal (St. Petersburg), 71(11): 1457-1467.

Shi, L., T. Zhu, H. L. Mogensen and S. E. Smith. 1991. Paternal plastid inheritance in alfalfa—plastid nucleoid number within generative cells correlates poorly with plastid number and male plastid transmission strength. Current Genetics, 19: 399-402.

Shibaoka, H. 1991. Microtubules and the regulation of cell morphogenesis by plant hormones, pp. 159-182. *In* C. W. Lloyd (ed.). The Cytoskeletal Basis of Plant Growth and Form. Academic Press, London, UK.

Shibaoka, H. 1993. Regulation by gibberellins of the orientation of cortical microtubules in plant cell. Australian J. Plant Physiol., 20: 461-470.

Shivanna, K. R. and D. C. Sastri. 1981. Stigma-surface esterase activity and stigma receptivity in some taxa characterized by wet stigmas. Ann. Bot. (London), 47(1): 53-64.

Shivanna, K. R., Y. Heslop-Harrison and S. S. Heslop-Harrison. 1978. Inhibition of the pollen-tube in the self-incompatibility response of grasses. Incompatibility Newsletter, 10: 5-7.

Shivanna, K. R., H. Xu, P. Taylor and R. B. Knox. 1988. Isolation of sperms from the pollen tubes of flowering plants during fertilization. Plant Physiol., 87: 647-650.

***Shmaraev, I. G. 1983.** Ultrastructural investigation of the early stages of endosperm development in *Triticum aestivum* L. Trudy po prikladnoy botanike, genetike i selektsii, 74: 15-21.

Shmidt, G. A. 1968. Embryogenesis Types and Their Adaptive Significance. Nauka, Moscow, Russia.

Shreve, F. 1905. The development of *Sarracenia purpurea*. Johns Hopkins Univ. Circ., 178: 31-34.

Shugaeva, E. V. 1985. Cytoembryology of some medicinal species of genus *Valeriana* L. Ph.D. thesis, Leningrad, Russia.

Shumilina, Z. K. 1949. The Preparation to the Sowing of Woody and Shrub Sorts. Goslesbumizdat, Moscow-Leningrad, Russia.

Simola, L. K. 1974. The ultrastructure of dry and germinating seeds of *Pinus sylvestris* L. Acta Bot. Fennica, 103:

Simoncioli, C. 1974. Ultrastructural characteristics of *Diplotaxis erucoides* (L.) DC. suspensor. Giornale Bot. Italiano, 108(34): 175-189.

Simpson, M. J. A. 1977. Fruit characters and seed germination of *Euphrasia disperma* Hook f. New Zealand J. Bot., 15(1): 181-183.

Singh, A. P. and H. L. Mogensen. 1975. Fine structure of the zygote and early embryo in *Quercus gambelii*. Amer. J. Bot., 62(2): 105-115.

Singh, A. P. and H. L. Mogensen. 1976. Fine structure of early endosperm in *Quercus gambelii*. Cytologia, 41: 345-361.

Singh, D. and Y. S. Gupta. 1956. Cytoplasmic nodules in the endosperm of *Coriandrum sativum* L. Sci. Cult., 22: 342-343.

Singh, R. P. and Y. S. Chauhan. 1977. Experimental production of pleiocotyly in *Chenopodium album* L., pp. 517-522. Recent Trends and Contacts between Cytogenetics, Embryology and Morphology.

Sinha, N. R., R. E. Williams and S. Hake. 1993. Overexpression of the maize homeobox gene, KNOTTED-*1*, causes a switch from determinate to indeterminate cell fates. Genes Dev., 7: 787-795.

Sinnott, E. W. 1960. Plant Morphogenesis. McGraw-Hill Book Company Inc., New York, NY.

Sitton, D., A. Richmond and Y. Vaadia. 1967. On the synthesis of gibberellins in roots. Phytochemistry, 6(8): 1101-1105.

Smart, M. G. and T. P. O'Brien. 1983. The development of wheat embryo in relation to the neighbouring tissue. Protoplasma, 114: 1-13.

Smirnov, A. G. 1982. Female Gametophyte of Angiosperms and Its Evolution. Izd. Kazanskogo Univ., Kazan, Russia.

***Smirnov, A. G. and L. Sh. Grakhantseva. 1986.** Double fertilization in members of the genus *Peperomia* L., 169. *In* Abstracts IX All-Union Conf. Plant Embryology "Gametogenesis, Fertilization and Embryogenesis of Seed Plants, Ferns and Mosses", Shtiintsa, Kishinev, Moldova.

***Smirnov, P. A. 1953.** Morphological investigations of cereals. Bull. Moscow Soc. Naturalists, Biol. Dept., 58(6): 71-75.

***Smirnova, E. S. 1965.** Types of flowering plant seed structure in phylogenetic aspect. Zh. obshchey Biol., 26(3): 310-325.

Smith, B. E. 1934. A taxonomic and morphological study of the genus *Cuscuta*, dodders in North Carolina. J. Elisha Mitchell Sci. Soc., 50: 283-302.

Smith, S. E. 1989. Biparental inheritance of organelles and its implications in crop improvement. Plant Breeding Rev., 6: 361-393.

Smith, S. E., E. T. Bingham and R. W. Fulton. 1986. Transmission of chlorophyll deficiencies in *Medicago sativa*—evidence for biparental inheritance of plastids. J. Heredity, 77: 35-39.

Snell, W. J. 1990. Adhesion and signaling during fertilization in multicellular and unicellular organisms. Current Opinion Cell Biol., 2: 821-832.

***Snigirevskaya, N. S. 1964.** Materials on the morphology and systematics of genus *Nelumbo* Adans. Trans. Komarov Bot. Inst. USSR Acad. Sci. 13: 104-172.

***Sokolov, I. D. 1968.** Cytology of onion endosperm, pp. 209-211. *In* Materials All-Union Symposium on Plant Embryology. Naukova dumka, Kiev, Ukraine.

***Sokolov, I. D. 1975.** The pathways of the formation of nuclei of different shapes in angiosperm endosperm. Manuscript presented by Donetsk Univ. in VINITI on 12 May 1975, N 1297-75. Donetsk, Ukraine.

Sokolov, I. D. 1978a. Cytology of the endosperm nuclear type at the cellular stage of development, pp. 91-92. *In* Abstracts VI Cong. All-Union Bot. Soc., Leningrad, Russia.

***Sokolov, I. D. 1978b.** Spontaneous disturbances of mitosis and their consequences in the endosperm of certain monocots, pp. 52-54. *In* Abstracts VII All-Union Symp. Plant Embryology, vol. 2. Naukova dumka, Kiev, Ukraine.

Sokolov, I. D. 1980. The development of endosperm of nuclear type. Ph.D. thesis, Leningrad, Russia.

***Sokolov, I. D., I. D. Romanov and N. Kh. Aminov. 1980.** Cytology of Flowering Plant Endosperm. "Visha shkola", Kiev-Donetsk, Ukraine.

*Sokolovskaya, T. B. 1965. On the question about epiblast nature. Bot. Journal (St. Petersburg), 50(12): 1686-1693.

*Sokolovskaya, T. B. 1967. On the coleorhiza nature. Bot. Journal (St. Petersburg), 52(12): 1773-1784.

Sokolovskaya, T. B. 1968. The nature of embryo organs in cereals. Ph.D. thesis, Leningrad, Russia.

Sokolovskaya, T. B. 1981. Family Ranunculaceae, pp. 130-138. *In* M. S. Yakovlev (ed.). Comparative Embryology of Flowering Plants. Winteraceae—Juglandaceae. Nauka, Leningrad, Russia.

Sokolovskaya, T. B. 1983. Family Bombacaceae, pp. 256-259. *In* M. S. Yakovlev (ed.). Comparative Embryology of Flowering Plants. Phytolaccaceae—Thymelaeaceae. Nauka, Leningrad, Russia.

Sokolovskaya, T. B. and B. E. Shpilevoy. 1990. Family Iridaceae, pp. 129-134. *In* T. B. Batygina and M. S. Yakovlev (eds). Comparative Embryology of Flowering Plants. Butomaceae—Lemnaceae. Nauka, Leningrad, Russia.

Solms-Laubach, H. 1878a. Monographia Pandanacearum. Hermanno camite de Solnms-Laubach. Bri., Wiegandt, Hempelu. Parey.

Solms-Laubach, H. 1878b. Über monocotyle Embryonen mit Scheitelburtigen Vegetationspunkt. Bot. Zeitung (Berlin), 36: 65-74, 81-93.

*Solntseva, M. P. 1961. Some data on the fertilization process in *Fragaria ananassa* Duch. Bot. Journal (St. Petersburg), 46(3): 371-377.

*Solntseva, M. P. 1968. On the behaviour of sperms in embryo sac of *Rudbeckia laciniata* L., 211-214. *In* Materials All-Union Symposium Plant Embryology. Naukova dumka, Kiev, Ukraine.

*Solntseva, M. P. 1971. On the behaviour of sexual cell nuclei and the formation of mosaic embryo under semigamy in *Rudbeckia laciniata* L. Reports USSR Acad. Sci., 197(1): 234-237.

Solntseva, M. P. 1973a. On the behaviour of sexual nuclei and the formation of mosaic embryo under hemigamy in *Rudbeckia laciniata* L., pp. 161-169. *In* D. F. Petrov (ed.). Problems of Apomixis in Plants and Animals. Nauka, Novosibirsk, Russia.

*Solntseva, M. P. 1973b. Semigamy and fertilization in *Rudbeckia laciniata* L. Bot. Journal (St. Petersburg), 58(9): 1261-1277.

*Solntseva, M. P. 1973c. Types of embryo sac development and embryo morphology in *Rudbeckia laciniata* L., pp. 47-56. *In* A. I. Kovarsky et al. (eds). Embryology of Angiosperms. Shtiintsa, Kishinev, Moldova.

Solntseva, M. P. 1974. Disturbances in the process of fertilization in angiosperms under hemigamy: hemigamy and its manifestations in plants, pp. 311-324. *In* H. F. Linskens (ed.) Fertilization in Higher Plants. North-Holland Publ. Co., Amsterdam, Netherlands.

Solntseva, M. P. 1976. Hemigamy and its manifestation in plants, pp. 73-81. *In* D. F. Petrov (ed.). Apomixis and Its Use in Selection. Nauka, Novosibirsk, Russia.

Solntseva, M. P. 1979. Hemigamy and possibilities of its use as one of apomixis forms, pp. 38-46. *In* M. S. Yakovlev (ed.). Actual Questions of Embryology of Angiosperms. Nauka, Leningrad, Russia.

Solntseva, M. P. 1982. Evolutionary significance of hemigamy, pp. 134-136. *In* Materials VI Moscow Conference Plant Phylogeny, Moscow, Russia.

*Solntseva, M. P. and M. S. Yakovlev. 1964. Prerequisites of monocotyledony in early embryogenesis of feather-grasses. Bot. Journal (St. Petersburg), 49(5): 625-683.

***Solntseva, M. P. and L. I. Vorsobina. 1972.** Semigamy in *Zephyranthes macrosiphon* Herb. Reports USSR Acad Sci., 206(4): 1006-1009.

***Solntseva, M. P. and N. V. Dgebuadze. 1974.** On the embryology of pecan *Carya illinoinensis* (Wangenheim) K.Koeh (Juglandaceae). Bot. Journal (St. Petersburg), 59(12): 1794-1804.

***Solntseva, M. P. and O.P. Kamelina. 1976.** Double fertilization and embryo in *Cephalaria gigantea* (Ledeb.) Bobr. (Dipsacaceae). Bot. Journal (St. Petersburg), 61(2): 254-263.

***Solntseva, M. P. and L. I. Vorsobina. 1979.** Endosperm development in the genus *Zephyranthes* (family Amaryllidaceae Jaume St. Hilaire). Reports USSR Acad. Sci., 245(2): 509-512.

Sood, R. K., K. Prabha and S. C. Gupta. 1982. Is the "rejection reaction" inducing ability in sporophytic self-incompatibility system restricted only to pollen and tapetum? Theor. Appl. Genet., 63: 27-32.

Souéges, R. 1910. Recherches sur l'embryogénie des Renonculacées—Clematidees. 1. L'embryon. Bull. Soc. Bot. France, 57: 242-252, 266-275.

Souèges, R. 1913. Recherches sur l'embryogénie des Renonculacées—Renonculées (genre *Ranunculus*). Bull. Soc. Bot. France, 60: 506, 542, 615.

Souèges, R. 1914a. Nouvelle recherche sur le développement de l'embryon chez les Cruciferes. Ann. Sci. Naturelles.Bot., 19(9): 311-339.

Souèges, R. 1914b. Nouvelles observations sur l'embryogénie des Cruciferes. C.R. Hebdomadaires Séances l'Acad. Sci., 158: 1356-1359.

Souèges, R. 1916. Les principales divisions de l'oeuf et l'origine de l'hypophyse chez le *Capsella bursa-pastoris*. C.R. Hebdomadaires Séances l'Acad. Sci., 163: 158-160.

Souèges, R. 1919. Les premières divisions de l'oeuf et les differenciations du suspenseur ches le *Capsella bursa-pastoris* Moench. Ann. Sci. Naturelles.Bot., ser. 10, 1: 1-28.

Souèges, R. 1920a. Développement de l'embryon chez le *Chenopodium bonus-henricus* L. Bull. Soc. Bot. France, 67: 233-297.

Souèges, R. 1920b. Embryogénie des Composées. I. Les premiers stades du développement de l'embryon chez le *Senecio vulgaris* L. II. Les derniers stades du développement de l'embryon chez le *Senecio vulgaris* L. C.R. Hebdomadaires Séances l'Acad. Sci., 171: 254-256; 356-357.

Souèges, R. 1920c. Embryogénie des Oenotheracees: Développement de l'embryon chez l'*Oenothera biennis*. C.R. Hebdomadaires Séances l'Acad. Sci., 170: 946-948.

Souèges, R. 1920d. Embryogénie des Solanacées. Développement de l'embryon chez le *Nicotiana*. C.R. Hebdomadaires Séances l'Acad. Sci., 170: 1125-1127.

Souèges R. 1921. Embryogénie des Boragacées. Les premier terms du développement de l'embryon chez le *Myosotis hispida*. C.R. Hebdomadaires Séances l'Acad. Sci., 173: 726-728.

Souèges, R. 1922a. Embryogénie des Caryophyllacées. Les premiers stades du développement de l'embryon chez le *Sagina procumbens*. Les derniers stades du développement chez le *Sagina procumbens*. C.R. Hebdomadaires Séances l'Acad. Sci., 175: 709-711, 894-896.

Souèges, R. 1922b. Recherches sur l'embryogenie des Solanacées. Bull. Soc. Bot. France, 69: 163-178, 236-241, 352-365, 555-585.

Souèges, R. 1923a. Développement de l'embryon chez le *Geum urbanum* L. Bull. Soc. Bot. France, 70: 645-660.

Souèges, R. 1923b. Embryogénie des Plantaginacées. Développement de l'embryon chez le *Plantago lanceolata* L. C.R. Hebdomadaires Séances l'Acad. Sci., 177(20): 964-967.

Souèges, R. 1924a. Développement de l'embryon chez le *Myosotis hispida* Schlecht. Bull. Soc. Bot. France, 70: 385-401.

Souèges, R. 1924b. Développement de l'embryon chez le *Sagina procumbens* L. Bull. Soc. Bot. France, 71: 590-614.

Souèges, R. 1924c. Embryogénie des Graminées. Développement de l'embryon chez le *Poa annua* L. C.R. Hebdomadaires Séances l'Acad. Sci., 178: 1307-1310.

Souèges, R. 1924d. Embryogénie des Linaceae. Développement de l'embryon chez le *Linum catharticum* L. C.R. Hebdomadaires Séances l'Acad. Sci., 178: 1307-1310.

Souèges, R. 1927. Développement de l'embryon chez le *Sedum acre* L. Bull. Soc. Bot. France, 74: 234-251.

Souèges, R. 1931a. L'embryon chez le *Sagittaria sagittifolia*. Le cone végétativ de la tige et l'extrémite radialaire chez le monocotyledones. Ann. Sci. Naturelles. Bot., 13(1-2): 353-402.

Souèges, R. 1931b. Recherches sur l'embryogenie des Liliacées. *Anthericum ramosum* et *Allium ursinum*. Bull. Soc. Bot. France, 78(9-10): 662-681.

Souèges, E. C. R. 1932. Recherches sur l'embryogénie des Liliacees, *Muscari comosum*. Bull. Soc. Bot. France, 79(1): 11-23.

Souèges, R. 1934. L'hypophyse et l'epiphyse: les problems d'histogenese qui leur sont lies. Bull. Soc. Bot. France, 81: 737-748, 769-778.

Souèges, R. 1935a. Embryogénie des Oenotheracées. Les principaux termes du développement de l'embryon chez le *Ludwigia palustris* Elliott. C.R. Hebdomadaires Séances l'Acad. Sci., 200: 1626-1628.

Souèges, R. 1935b. Exposes l'embryologie et de morphologie végétales. IV. La segmentation. Premier fascicule: I. Les fondaments. II. Les phenomenes internes. Actualités Scient. Industrielles, 266: 1-88.

Souèges, R. 1936a. Embryogénie des Droseracées. Développement de l'embryon chez le *Drosera rotundifolia* L. C.R. Hebdomadaires Séances l'Acad. Sci., 202: 1457-1459.

Souèges, R. 1936b. Exposes l'embryologie et de morphologie végétales. La segmentation. Deuxieme fascicule: III. Les phenomenes externes. IV. Les blastomeres. Actualités Scient. Industrielles, 318: 1-80.

Souèges, R. 1937a. Embryogénie des Dipsacacées. Développement de l'embryon chez le *Scabiosa succisa* L. C.R. Hebdomadaires Séances l'Acad. Sci., 204: 292-294.

Souèges, R. 1937b. Exposes d'embryologie et de morphologie végétales VII. Les lois du développement. Actualités Scient. Industrielles, 521: 1-94.

Souèges, R. 1937c. Développement de l'embryon chez le *Chelidonium majus* L. Ann. Sci. Naturelles. Bot., 19: 445.

Souèges, R. 1939a. Embryogénie des Oxalidacées. Développement de l'embryon chez l'*Oxalis corniculata* L. C.R. Hebdomadaires Séances l'Acad. Sci., 209(19): 698-700.

Souèges, R. 1939b. Embryogénie des Polémoniacées. Développement de l'embryon chez le *Polemonium coeruleum* L. C.R. Hebdomadaires Séances l'Acad. Sci. (Paris), 208(17): 1338-1340.

Souèges, R. 1939c. Exposes d'embryologie et de morphologie végétales. X. Embryogénie et classification. Deuxieme fascicule: Essai d'un systeme embryogénique (Partie générale). Actualités Scient. Industrielles, pp. 1-85.

Souèges, R. 1940. Embryogénie des Potamogetonacées. Développement de l'embryon chez *Potamogeton natans* L. C.R. Hebdomadaires Séances l'Acad. Sci., 211: 232-233.

Souèges, R. 1941. Embryogénie des Fumariacées. La differenciation des rigions fondamentales du corps chez le *Fumaria officinalis* L. C.R. Hebdomadaires Séances l'Acad. Sci., 213(1-26): 699-701.

Souéges, R. 1948. Exposes d'embryologie et de morphologie végétales. Embryogénie et classification. 3. Essai d'un systéme embryogénique (Partie speciale: Premiere periode du systeme). Actualités Scient. Industrielles, 1060: 1-101.

Souèges, R. 1949. Embryogénie des Papaveracées. Développement de l'embryon chez l'*Eschscholzia californica* Cham. C.R. Hebdomadaires Séances l'Acad. Sci., 229(9): 485-487.

Souèges, R. 1952. L'albumen et l'embryon chez le *Callitriche vernalis* (*C. verna*). C.R. Hebdomadaires Séances l'Acad. Sci., 235: 453-456.

Souèges, R. 1954. Embryogénie et classification. Huitieme Congrés Int. Bot., Sect.8. Paris: 23.

Souèges, R. 1957. Embryogénie des Dipsacacées. Développement de l'embryon chez le *Scabiosa columbaria* L. C.R. Hebdomadaires Séances l'Acad. Sci., 245(5): 465-468.

Souèges, R. 1958a. Embryogénie des Commelinacées. Développement de l'embryon chez le *Commelina communis* L. C.R. Hebdomadaires Séances l'Acad. Sci., 246(14): 2082-2086.

Souèges, R. 1958b. Embryogénie des Commelinacées. Développement de l'embryon chez le *Rhoeo discolor* Hance. C.R. Hebdomadaires Séances l'Acad. Sci., 246(17): 2436-2440.

Souèges, R. 1959. Embryogénie des Hydrocharitacées. Développement de l'embryon chez l'*Hydrocharis morsus-ranae* L. C.R. Hebdomadaires Séances l'Acad. Sci., 248 (I): 45-49.

Souèges, R. 1961. Embryologie végétale: Reponse aux observations critiques de D.A. Johansen relatives a mon essai de classification embryogénique. C.R. Hebdomadaires Séances l'Acad. Sci., 252(22): 3359-3364.

Souèges, R. 1963a. Embryogénie des Dipsacacées. Développement de l'embryon chez le *Cephalaria tatarica*. C.R. Hebdomadaires Séances l'Acad. Sci., 256(1): 45-48.

Souèges, R. 1963b. Embryogénie des Dipsacacées. Développement de l'embryon chez le *Knautia arvensis*. C.R. Hebdomadaires Séances l'Acad. Sci., 256(6): 1190-1194.

Souèges, R., J. L. Guignard and J. C. Méstre. 1967. Développement de l'embryon chez le *Laurus nobilis* L. (Lauracées). Phytomorph., 17(1-4): 255-266.

Spemann, H. 1938. Embryonic Development and Induction. Yale Univ. Press, New York, NY.

Sperber, D. and K. Dransfeld. 1981. Oriented growth of pollen tubes in strong magnetic fields. Naturwissenschaftliche Wochenschrift, 68: 40-41.

Sponsel, V. M. and J. MacMillan. 1980. Metabolism of 13c1-gibberellin A29 to 13c1-gibberellin catabolite in maturing seeds of *Pisum sativum* cv. Progress N9. Planta, 150(1): 46-52.

Sporne, K. R. 1954. A note on nuclear endosperm as a primitive character among dicotyledons. Phytomorph., 4(3-4): 275-278.

Sporne, K. R. 1980. A re-investigation of character correlations among dicotyledons. New Phytologist, 85: 419-445.

Stebbins, G. L. 1974. Flowering Plants: Evolution above the Species Level. Harvard Univ. Press, Cambridge, MA,USA.

Steeves, T. A. and I. M. Sussex. 1989. Patterns in Plant Development. Cambridge Univ. Press, Cambridge, UK.

Steffen, K. 1951. Zur Kenntnis des Befruchtungsvorganges bei *Impatiens glandulifera* Lindl. Cytologische Studien am Embryosack der Balsamineen. Planta, 39(3): 175-244.

Steffen, K. 1953. Zytologische Untersuchungen an Pollenkorn undschlauch. 1. Phasenkontrastoptische Lebensuntersuchungen an Pollenschläuchen von *Galanthus nivalis.* Flora, 140(1): 140-174.

Sterckx, R. 1900. Recherches anatomiques sur l'embryon et les plantules dans la famille Renonculacées. Archives l'Inst. Bot. l'Univ. Liège, 2: 1-120.

Sterling, C. 1955. Embryogeny in the lima bean. Bull. Torrey Bot. Club, 82: 325-338.

Steward, C. R. 1971. Some characteristics of the uptake of glutamine by corn scutellum. Plant Physiol., 47(2): 157-161.

Stockard, Ch. R. 1921. Developmental rate and structural expression: an experimental study of twins, double monsters and single deformities, and interaction among embryonic organs during their origin and development. Amer. J. Bot., 28(2): 115-226.

Stoddart, J. L., H. Thomas and A. Robertson. 1973. Protein synthesis patterns in barley embryos during germination. Planta, 112(3): 309-321.

Stolt, K. A. H. 1921. Zur Embryologie der Gentianaceen und Menyanthaceen. Kongliga Svenska Vetenskapsakademiens Handlingar, N.S., 61(14): 1-56.

Stone, D. E. 1970. Evolution of cotyledonary and nodal vasculature in the Juglandaceae. Amer. J. Bot., 57(10): 1219-1225.

***Stozharova, I. A. and V. A. Poddubnaya-Arnoldi. 1977.** Role of synergids in fertilization, embryo- and endospermogenesis in *Allium nutans* L. Bull. Main Bot. Garden USSR Acad. Sci., 105: 70-78.

Strasburger, E. 1877. Über Befruchtung und Zellteilung. Jenaische Zeitschrift für Naturwissenschaft herausgegeben von der Medicinisch-naturwissenschaftlichen. Gesellschaft zu Jena, 11: 435-536.

Strasburger, E. 1884. Neue Untersuchungen über den Befruchtungsvorgang bei den Phanerogamen als Grundlage für eine Theorie der Zeugung. G. Fischer, Jena.

Strasburger, E. 1900. Einige Bemerkungen zur Frage nach der "doppelten Befruchtung" bei den Angiospermen. Bot. Zeitung (Berlin), 58(19-20): 293-316.

Strasburger, E. 1902. Ein Beitrag zur Kenntnis von *Ceratophyllum submersum* und phylogenetische Erörterungen. Jahrb. Wissenschaftliche Bot., 37(3): 477-526.

Straub, J. 1948. Zur Genetik der Trikotylie. Zeitschrift induktive Abstammungs und Vererbungslehre, 88(1-2): 331-338.

Straus, D. and F. M. Ausubel. 1990. Genomic subtraction for cloning DNA corresponding to deletion mutations. Proc. Natl. Acad. Sci. USA, 87: 1889-1893.

Stuart, D. A., D. J. Durnam and R. L. Jones. 1977. Cell elongation and cell division in elongating lettuce hypocotyle sections. Planta, 135(3): 249-255.

Subramanyam, K. 1951. Embryology of *Oxyspora paniculata* DC. Phytomorph., 1(3-4): 205-212.

Suessenguth, K. 1921. Beiträge zur Frage des systematischen Anschlusses der Monocotylen. Beihefte Bot. Centralblatt, 38, II (1): 1-79.

Sung, Z. R., A. Belachew, B. Shunong and R. Bertrand-Garcia. 1992. EMF, an *Arabidopsis* gene required for vegetative shoot development. Science, 258: 1645-1647.

Svensson, H. G. 1925. Zur Embryologie der Hydrophyllaceen, Borraginaceen und Heliotropiaceen. Uppsala Universitetsrsskrift, 2: 1-176.

***Svetlov, P. G. 1923.** Early developmental stages in *Rhynchelmis limosella* Hoffmeister (preliminary report). Reports boil. Inst. Biol. Stat. Perm State Univ., 2(4): 141-152.

Svetlov, P. G. 1960. The theory of critical periods and its significance for understanding of principles of environmental effect on ontogenesis, pp. 263-285. *In* Problems of Cytology and General Physiology. Moscow-Leningrad, Russia.

Svetlov, P. G. 1978. Physiology (Mechanics) of Development. Nauka, Leningrad, Russia.

Swamy, B. G. L. 1942. Female gametophyte and embryogeny in *Cymbidium bicolor* Lindl. Proc. Indian Acad. Sci., 15: 194-201.

Swamy, B. G. L. 1943. Gametogenesis and embryogeny of *Eulophea epidendrae* Fischer. Proc. Natl. Inst. Sci. India, 9: 59-65.

Swamy, B. G. L. 1946. Inverted polarity of the embryo sac of angiosperms and its relation to the archegonium theory. Ann. Bot. (London), N. S., 10: 171-183.

Swamy, B. G. L. 1947. On the life history of *Vanilla planifolia*. Bot. Gazette (Chicago), 108: 449-456.

Swamy, B. G. L. 1948a. Agamospermy in *Spiranthes cernua*. Lloydia, 11: 149-162.

Swamy, B. G. L. 1948b. A contribution to the life history of *Casuarina*. Proc. Amer. Acad. Arts Sci., 77(1): 1-32.

Swamy, B. G. L. 1948c. The embryology of *Epidendrum prismatocarpum*. Bull. Torrey Bot. Club, 75(3); 245-249.

Swamy, B. G. L. 1948d. On the post-fertilisation development of *Trillium undulatum*. Cellule, 52(1): 7-14.

Swamy, B. G. L. 1949a. Embryological studies in Orchidaceae. I. Gametophytes. II. Embryogeny. Amer. Midland Naturalist, 41(1): 184-201, 202-232.

Swamy, B. G. L. 1949b. Further contributions to the morphology of the Degeneriaceae. J. Arnold Arboretum, 30: 10-38.

Swamy, B. G. L. 1962. Contribution to the embryology of the Najadaceae. J. Indian Bot. Soc., 41 (2): 247-267.

Swamy, B. G. L. 1963. The origin of cotyledon and epicotyl in *Ottelia alismoides*. Beiträge Biol. Pflanzen, 39(1): 1-16.

Swamy, B. G. L. 1966. The origin and the organization of the embryonic shoot apex in *Eichhornia crassipes*. Bull. Torrey Bot. Club., 93 (I): 20-34.

Swamy, B. G. L. and K. Periasamy. 1955. Contributions to the embryology of *Acrotrema arrnottianum*. Phytomorph., 5(2-3): 301-314.

Swamy, B. G. L. and P. M. Ganapathy. 1957a. A new type of endosperm haustorium in *Nothapodytes foetida*. Phytomorph., 7(3-4): 331-336.

Swamy, B. G .L. and P. M. Ganapathy. 1957b. On endosperm in dicotyledons. Bot. Gazette, 119(1): 47-50.

Swamy, B. G. L. and N. A. Parameswaran. 1960. A contribution to the embryology of *Begonia crenata*. J. Indian Bot. Soc., 39(4): 140-148.

Swamy, B. G. L. and D. Padmanabhan. 1961. Notulae embryologicae. 1. The functions of endosperm in *Avicennia officinalis*. Current Sci., 30(11): 424-425.

Swamy, B. G. L. and K. K. Lakshmanan. 1962a. Contribution to the embryology of the Najadaceae. J. Indian Bot. Soc., 41 (2): 247-267.

Swamy, B. G. L. and K. K. Lakshmanan. 1962b. The origin of epicotylary meristem and cotyledon in *Halophila ovata* Gaudich. Ann. Botany (London), 26(102): 243-249.

Swamy, B. G. L. and D. Padmanabhan. 1962. A reconnaissance of angiosperm embryogenesis. J. Indian Bot. Soc., 41: 422-437.

Swamy, B. G. L. and N. A. Parameswaran. 1962. On the origin of cotyledon and epicotyl in *Potamogeton indicus*. Österreichische Bot. Zeitschrift, 109(3): 344-349.

Swamy, B. G. L. and N. A. Parameswaran. 1963. The helobial endosperm. Biol. Rev., 38(1): 1-50.

Swamy, B. G. L. and K. V. Krishnamurthy. 1973. The helobial endosperm: a decennial review. Phytomorph., 23(1,2): 74-79.

Swamy, B. G. L. and K. V. Krishnamurthy. 1975. Certain conceptual aspects of meristems. I. On hypophysis and quiescent centre. Phytomorph., 25(1): 60-65.

Swamy, B. G. L. and K. V. Krishnamurthy. 1977. Certain conceptual aspects of meristems. II. Epiphysis and shoot apex. Phytomorph., 27(1): 1-8.

Swamy, B. G. L. and K. V. Krishnamurthy. 1981. On embryos and embryoids. Proc. Indian Acad. Sci., 96: 401-414.

Swaraupanahdan, K. 1986. Late embryogenesis and morphology of mature embryos in three species of Dipterocarpaceae. Canadian J. Bot., 64(11).

Swift, H. 1953. Quantitative aspects of nuclear nucleoproteins. Int. Rev. Cytol., 2: 1-76.

Swift, I. G. and T. P. O'Brien. 1970. Vascularization of the scutellum of wheat. Australian J. Bot., 18(1): 45-53.

Swift, I. G. and T. P. O'Brien. 1971. Vascular different action in the wheat embryo. Australian J. Bot., 19(1): 63-71.

Swift, I. G. and T. P. O'Brien. 1972a. The fine structure of the wheat scutellum before germination. Australian J. Biol. Sci., 25(1): 9-22.

Swift, I. G. and T. P. O'Brien. 1972b. The fine structure of the wheat scutellum during germination. Australian J. Biol. Sci., 25(5): 469-486.

Symons, S. J., R. E. Angold, M. Black and J. M. Chapman. 1983. Changes in the growth capacity of the developing wheat embryo. J. Exper. Bot., 34(148): 1541-1550.

***Symonyan, E. G. 1955.** Fertilization in rye. Reports Armenian SSR Acad. Sci., 8(11): 61-68.

Szabo, S. P. and D. H. O'Day. 1983. The fusion of sexual nuclei. Biol. Rev., 58(3): 323-342.

Szczuka, E., B. Fyk and B. Rodkiewicz. 1995. Cutin and chlorophyll fluorescence in developing embryos of some angiosperms. Platinum Jubilee Volume Indian Botanical Society. 74A: 273-282.

Szewczyk, E. and M. Ryczkowski. 1977. Changes in the chlorophyll content in the developing embryo (dicotyledonous plants). Bull. l'Acad. Polonaise Sci., Série Sci. Biol., 25(10): 679-683.

Takhtajan, A. L. 1948. Morphological Evolution of Angiosperms. Moscow Soc. Naturalists, Moscow, Russia.

Takhtajan, A. 1959. Die Evolution der Angiospermen. Fischer, Jena.

Takhtajan, A. L. 1964. Foundations of Evolutionary Morphology of Angiosperms. Nauka, Moscow-Leningrad, Russia.

Takhtajan, A. L. 1966. The System and Phylogeny of Flowering Plants. Nauka, Moscow-Leningrad, Russia.

Takhtajan, A. L. 1970. The Origin and Dispersal of Flowering Plants. Nauka, Leningrad, Russia.

Takhtajan, A. L. 1980. Outline of the classification of flowering plants (Magnoliophyta). Bot. Rev., 46(3): 225-359.

Takhtajan, A. L. 1987. The System of Magnoliophyta. Nauka, Leningrad, Russia.

Takhtajan, A. L. 1991. Evolutionary Trends in Flowering Plants. Columbia Univ. Press, New York, NY.

Takhtajan, A.L. 1978. Gametophyte, gametangies and fertilization, p. 32. *In* A.L. Takhtajan (ed.). Life of Plants. Introduction. Prosveshchenie, Moscow, Russia, vol. 1.

Tamura, M., J. Mizumoto and H. Kubota. 1977. Observations in seedlings of the Ranunculaceae. Japanese J. Bot., 52(10): 233-304.

Tan, K. 1980. Studies in the Thymelaeaceae. 1. Germination, seedling, fruit and seeds. Notes Royal Bot. Garden Edinburgh, 38(1): 149-164.

Tang, Pei-Hua, Sun De-Lan, Zhao Jing and Su Xiu-Zheu. 1989. Light and electron microscopic observation on the developing pattern of cotyledons of *Nelumbo nucifera*. Acta Bot. Sinica, 31(10); 750-756.

Tang, X., P. K. Hepler and S. P. Scordilis. 1989. Immunocytochemical identification of myosin, heavy chain polypeptide in *Nicotiana* pollen tubes. J. Cell Sci., 92: 569-574.

Tanimoto, E. 1992. Gibberellin requirement for the normal growth of roots, pp. 229-240. *In* N. Takahashi et al. (eds.). Gibberellins. Springer, New York, NY.

***Tarasenko, L. V. and V. P. Bannikova. 1986.** The ultrastructure of the three-celled pollen grain of *Triticale*. Ukrainian Bot. Journal, 43(1): 71-73.

***Tarasenko, L. V., V. P. Bannikova and K. M. Sytnik. 1988.** Interconnection of cells in the pollen grains of *Secale* and *Mattiola*. Reports Ukrainian SSR Acad. Sci., Series B, 3: 81-84.

Tateoka, T. 1964. Notes on some grasses. XVI. Embryo structure of the genus *Oryza* in relation to the systematics. Amer. J. Bot., 51(5): 539-543.

***Tatintseva, S. S. 1975.** The formation and development of sperms in some angiosperms. I. Poaceae (Gramineae). II. Asteraceae (Compositae). Bot. Journal (St. Petersburg), 60(12): 1707-1720.

***Tatintseva, S. S. 1988.** The formation and development of sperms in some angiosperms. III. Boraginaceae. IV. Najadaceae. Reports of the Turkmenian SSR Acad. Sci., Series Biol. Sci., 1: 3-9.

Taylor, A. and D. J. Cosgrove. 1989. Gibberellic acid stimulation of cucumber hypocotyl elongation. Effect on growth, turgor, osmotic pressure and cell wall properties. Plant Physiol., 90: 1335-1340.

Taylor, A. R. A. 1957a. Studies of the development of *Zostera marina* L. 1. The embryo and seed. Canadian J. Bot., 35(4): 477-499.

Taylor, A. R. A. 1957b. Studies of the development of *Zostera marina*. L. 2. Germination and seedling development. Canadian J. Bot., 35(5): 681-695.

Taylor, D. W. and L. J. Hickey. 1992. Phylogenetic evidence for the herbaceous origin of angiosperms. Plant System. Evol., 180: 137-156.

Taylor, F., J. Kenrick, Y. Li, V. Kaul, B. E. S. Gunning and R. B. Knox. 1989. The male germ unit in *Rhododendron*: quantitative cytology, three-dimensional reconstruction, isolation and detection using fluorescent probes. Sexual Plant Reprod., 2: 254-264.

Taylor, I. H., P. H. Woods and W. L. Huges. 1957. The organization and duplication of chromosomes as revealed by autographic studies using tritium labelled thymidine. Proc. Natl. Acad. Sci., USA, 43: 122-134.

Taylor, M. G. and J. K. Vasil. 1995. The ultrastructure of zygotic embryo development in pearl millet (*Pennisetum glaucum*, Poaceae). Amer. J. Bot., 85(2): 205-219.

***Teryokhin, E. S. 1962.** Influence of some ecological factors on the development of embryonal structures in Pyroleae–Monotropoideae. Bot. Journal (St. Petersburg), 47(4): 571-577.

***Teryokhin, E. S. 1965.** On the terms "saprophyte", "hemisaprophyte" and "hemiparasite" (in connection with the character of biotic relations in some angiosperms). Bot. Journal (St. Petersburg), 50(1): 60-69.

***Teryokhin, E. S. 1968a.** Metamorphosis in the ontogenesis of angiosperm parasitic plants. Reports USSR Acad. Sci., 178(4): 657-659.

Teryokhin, E. S. 1968b. On the problem of embryo reduction in angiosperm parasitic plants, pp. 37-48. *In* L. V. Kudryashev (ed.). Morphology of Higher Plants. Nauka, Moscow, Russia.

***Teryokhin, E. S. 1972.** On the possibility of using the term "behaviour" in investigation of plant evolution. Bot. Journal (St. Petersburg), 57(1): 75-89.

***Teryokhin, E. S. 1973.** Embryology of Orobanchaceae. II. On embryology and ecology of *Phacellanthus tubiflorus* S. et Z. Bot. Journal (St. Petersburg), 58(12): 1775-1888.

***Teryokhin, E. S. 1975.** On pathways, trends and evolutionary significance of the reduction of biological structures. Bot. Journal (St. Petersburg), 60(10): 1401-1412.

Teryokhin, E. S. 1977. Parasitic Flowering Plants. The Evolution of Ontogenesis and the Mode of Life. Nauka, Leningradskoe otdelenie, Leningrad, Russia.

Teryokhin, E. S. 1984. Peculiarities of embryo structure and evolution in some monocotyledonous plants, pp. 153-154. *In* Abstracts of the I All-Union Conf. Plant Anatomy. Komarov Bot. Inst., Leningrad, Russia.

Teryokhin E. S. 1985a. Family Corsiaceae, pp. 140-141. *In* A. L. Takhtajan (ed.). Comparative Anatomy of Seeds. Monocotyledons. Nauka, Leningrad, Russia, vol. 1.

Teryokhin E. S. 1985b. Family Droseraceae, pp. 48-52. *In* M. S. Yakovlev (ed.). Comparative Embryology of Flowering Plants. Brunelliaceae—Tremandraceae. Nauka, Leningrad, Russia.

Teryokhin E. S. 1985c. Family Hydrocharitaceae, pp. 38-43. *In* A. L. Takhtajan (ed.). Comparative Anatomy of Seeds. Monocotyledons. Nauka, Leningrad, Russia, vol. 1.

Teryokhin E. S. 1985d. Family Ruppiaceae, pp. 53-55. *In* A. L. Takhtajan (ed.). Comparative Anatomy of Seeds. Monocotyledons. Nauka, Leningrad, Russia, vol. 1.

Teryokhin E. S. 1985e. Family Zannichelliaceae, pp. 55-57. *In* A. L. Takhtajan (ed.). Comparative Anatomy of Seeds. Monocotyledons. Nauka, Leningrad, Russia, vol. 1.

Teryokhin E. S. 1985f. Family Zosteraceae, pp. 59-62. *In* A. L. Takhtajan (ed.). Comparative Anatomy of Seeds. Monocotyledons. Nauka, Leningrad, Russia, vol. 1.

Teryokhin E. S. 1987a. Family Balanophoraceae, pp. 89-93. *In* T. B. Batygina and M. S. Yakovlev (eds). Comparative Embryology of Flowering Plants. Davidiaceae—Asteraceae. Nauka, Leningrad, Russia.

Teryokhin E. S. 1987b. Family Orobanchaceae, pp. 268-272. *In* T. B. Batygina and M. S. Yakovlev (eds). Comparative Embryology of Flowering Plants. Davidiaceae—Asteraceae. Nauka, Leningrad, Russia.

Teryokhin, E. S. 1987c. Family Viscaceae, pp. 82-86. *In* T. B. Batygina and M. S. Yakovlev (eds). Comparative Embryology of Flowering Plants. Davidiaceae—Asteraceae. Nauka, Leningrad, Russia.

Teryokhin E. S. 1988a. Order Balanoforales, pp. 120-132. *In* A. L. Takhtajan (ed.). Comparative Anatomy of Seeds. Dicotyledons. Magnoliidae—Ranunculidae. Nauka, Leningrad, Russia, vol. 2.

Teryokhin E. S. 1988b. Order Hydnorales, pp. 112-113. *In* A. L. Takhtajan (ed.). Comparative Anatomy of Seeds. Dicotyledons. Magnoliidae—Ranunculidae. Nauka, Leningrad, Russia, vol. 2.

Teryokhin E. S. 1988c. Order Rafflesiales, pp. 114-119. *In* A. L. Takhtajan (ed.). Comparative Anatomy of Seeds. Dicotyledons. Magnoliidae—Ranunculidae. Nauka, Leningrad, Russia, vol. 2.

Teryokhin, E. S. 1990a. Family Burmaniaceae, pp. 165-169. *In* T. B. Batygina and M. S. Yakovlev (eds). Comparative Embryology of Flowering Plants. Butomaceae—Lemnaceae. Nauka, Leningrad, Russia.

Teryokhin E. S. 1990b. Family Zosteraceae, pp. 52-54. *In* T. B. Batygina and M. S. Yakovlev (eds). Comparative Embryology of Flowering Plants. Butomaceae—Lemnaceae. Nauka, Leningrad, Russia.

***Teryokhin, E. S. 1991.** The problems of of evolution of seed plant ontogenesis. Trans. Komarov Bot. Inst. USSR Acad. Sci. Morphology and Anatomy of Plants, 2: 1-69.

Teryokhin, E. S. 1994. Hierarchic classification of embryo sacs, pp. 239-248. *In* T. B. Batygina (ed.). Embryology of Flowering Plants. Terminology and Concepts. 1. Generative Organs of Flower. Mir i Semya, St. Petersburg, Russia.

Teryokhin, E. S. 1996. Seed and Seed Reproduction. Mir i Semya, St. Petersburg, Russia.

***Teryokhin, E. S. and M. S. Yakovlev. 1967.** Embryology of Balanophoraceae (on the question about homology of *Balanophora* "flower"). Bot. Journal (St. Petersburg), 52(6): 745-758.

***Teryokhin, E. S. and Z. I. Nikiticheva. 1968.** Postseminal development of parasitic Angiospermae. 1. Metamorphosis. Bot. Journal (St. Petersburg), 53(1): 39-57.

***Teryokhin, E. S. and O. P. Kamelina. 1969.** Endosperm of Orchidaceae (on the question about reduction). Bot. Journal (St. Petersburg), 54(5): 657-669.

***Teryokhin, E. S. and O. P. Kamelina. 1972.** Endosperm of *Habenaria linearifolia* Maxim. and *Neottia nidus-avis* (L.) Rich. (on the question about the evolution of endosperm in orchids). Bot. Journal (St. Petersburg), 57(11): 1373-1383.

***Teryokhin, E. S. and Z. I. Nikiticheva. 1976.** The development of generative shoot and pollination in *Cynomorium songaricum*. Bull. Moscow Soc. Naturalists, Biol. Dept., 81(3): 74-89.

Teryokhin, E. S. and Z. I. Nikiticheva. 1979. On principles of the creation of ecological classification of angiosperm embryos, pp. 120-130. *In* M. S. Yakovlev (ed.). Actual Questions of Embryology of Angiosperms. Nauka, Leningrad, Russia.

Teryokhin, E. S. and Z. I. Nikiticheva. 1981. Family Orobanchaceae. Ontogenesis and Phylogenesis. Nauka, Leningrad, Russia.

***Teryokhin, E. S., Z. I. Nikiticheva and M. S. Yakovlev. 1975.** Development of seed, embryo and endosperm in *Cynomorium songaricum* (Cynomoriaceae). Bot. Journal (St. Petersburg), 60(11): 1651-1660.

***Teryokhin, E. S., T. B. Batygina and I. I. Shamrov. 1993.** The classification of microsporangium wall types in angiosperms. Terminology and conception. Bot. Journal (St. Petersburg), 78(6): 16-24.

Thathachar, T. 1952. Morphological studies in the Euphorbiaceae: 1. *Acalypha lanceolata* Willd. Phytomorph., 2: 197-201.

Theunis, C. H., E. S. Pierson and M. Cresti. 1991. Isolation of male and female gametes in higher plants. Sexual Plant Reprod., 4: 145-154.

Thiselton-Dyer, W. T. 1902. Morphological notes. VIII. On polycotyledony. Ann. Bot. (London), 16(63): 553-558.

Thoma, S., U. Hecht, A. Kippers, J. Botella, S. C. De Vries and C. Somerville. 1994. Tissue-specific expression of a gene encoding a cell wall-localized lipid transferprotein from *Arabidopsis*. Plant Physiol., 105: 35-45.

Thomas, E. N. 1907. Some aspects of "double fertilisation" in plants. Sci. Progress, 1: 420-426.

Thomas, E. N. 1914. Seedling anatomy of Ranales, Rhoedales and Rosales. Ann. Bot. (London), 28: 695-733.

Thomas, T. L. 1993. Gene expression during plant embryogenesis and germination: an overview. Plant Cell, 5: 1401-1410.

Thorpe, T. A. and C. Stasolla. 2001. Somatic embryogenesis, pp. 279-336. *In* S. S. Bhojwani and W. Y. Soh (eds). Current Trends in the Embryology of Angiosperms. Kluwer Acad. Publ., Dordrecht, Netherlands.

Tiagi, B. 1951a. A contribution to the morphology and embryology of *Cuscuta hyalina* Roth. and *C. planiflora* Tenore. Phytomorph., 1(1-2): 9-21.

Tiagi, B. 1951b. Studies in the family Orobanchaceae. III. A contribution to the embryology of *Orobanche cernua* Loefl. and *O. aegyptiaca* Pers. Phytomorph., 1(3-4): 158-169.

Tiagi, B. 1952a. Studies in the family Orobanchaceae. II. A contribution to the embryology of *Aeginetia indica* L. Bull. Torrey Bot. Club, 79(1): 63-78.

Tiagi, B. 1952b. Studies in the family Orobanchaceae. I. A contribution to the embryology of *Cistanche tubulosa*. Lloydia., 15(3): 129-148.

Tiagi, B. and H. Sankla. 1963. Studies in the Orobanchaceae. V. A contribution to the embryology of *Orobanche lucorum*. Bot. Magazine (Tokyo), 76: 81-88.

Tiagi, Y. D. 1970. Cactaceae. Bull. Indian Natl. Sci. Acad., 41: 29-35.

Tiezzi, A., A. Moscatelli, F. Ciampolini, C. Milanesi, M. Murgia and M. Cresti. 1988. The cytoskeletal apparatus of the generative cell in several angiosperm species, pp. 215-220. *In* M. Cresti, P. Gori and E. Pacini (eds). Sexual Reproduction in Higher Plants. Springer, Berlin.

Tillich, H.-J. 1977. Vergleichend-morphologische Untersuchungen zur Identität der Gramineen-Primarwurzel. Flora, 166: 415-421.

Tilney-Bassett, R. A. E. 1986. Plant Chimeras. Arnold, London, UK.

Tilton, V. R. and H. T. Horner, Jr. 1980. Stigma, style and obturator of *Ornithogalum caudatum* (Liliaceae) and their function in reproductive process. Amer. J. Bot., 67: 1113-1131.

Tilton, V. R., L. W. Wilcox and R. G. Palmer. 1984. Postfertilization wandlabirinthe formation and function in the central cell of soybean, *Glycine max* (L.) Merr. (Leguminosae). Bot. Gazette, 145(3): 334-339.

Timofeev-Resovsky, N. V., N. N. Vorontsov and A. V. Yablokov. 1977. Brief Story on the Theory of Evolution. Nauka, Moscow, Russia.

Tirlapur, U. K., J. L. Van Went and M. Cresti. 1993. Visualization of membrane calcium and calmodulin in embryo sacs in situ and isolated from *Petunia hybrida* L. and *Nicotiana tabacum* L. Ann. Bot. (London), 71(2): 161-167.

Tirlapur, U., E. Kranz and M. Cresti. 1995. Characterization of isolated egg cells, in vitro fusion products and zygotes of *Zea mays* L. using the technique of image analysis and confocal laser scanning microscopy. Zygote, 3: 57-64.

Titova, G. E. 1988. The embryology of some water plants (*Nelumbo nucifera* Gaertn., *Cabomba caroliniana* A. Gray, *Trapa astrachanica* (Fler.) Winter). Ph.D. thesis, Komarov Botanical Institute, Leningrad, Russia.

Titova, G. E. 1990. The Nelumbonales and Nymphaeales embryogenesis and the origin of the monocotyledony, p. 173. *In* Abstracts XI Int. Symp. "Embryology and Seed Reproduction". Leningrad, USSR, July 3-7, 1990. Nauka, Leningrad, USSR.

Titova, G. E. 1992a. The embryo development of *Trapa astrachanica* (Fler.) Winter (Trapaceae), pp. 69-70. *In* Abstracts XII Int. Cong. Sexual Plant Reproduction. "Plant Reproductive Biology. Pollen, Ovules and Seeds". July 19-23, 1992, Columbus, OH, USA. Ohio State Univ., Columbus, OH, USA.

Titova, G. E. 1992b. The Nelumbonales and Nymphaeales embryogenesis and the origin of the monocotyledony, pp. 560-561. *In* Proc. XI Int. Symp. "Embryology and Seed Reproduction". July 3-7, 1990, Leningrad. Nauka, St. Petersburg, USSR.

Titova, G. E. 1998. Morphogenesis of embryos with different types and degrees of cotyledon fusion. Biologia, 1: 38-44.

***Titova, G. E. 2000.** On the nature of pseudomonocotyly in flowering plants. Bot. Journal (St. Petersburg), 85(7): 76-91.

Titova, G. E. 2002. Algorithms of embryo morphogenesis in *Agapanthus praecox* Willd. (Alliaceae) at mono-, dicotyly and transitional forms, p. 163. *In* Abstracts XVII Int. Cong. Sexual Plant Reproduction "Sexual Plant Reproduction in Nature and the Laboratory", 9-13 July, 2002, Lublin, Poland. Maria Curie Sklodovska Univ. Press, Lublin, Poland.

Titova, G. E. 2003. Algorithms of embryo morphogenesis in *Agapanthus praecox* Willd. (Alliaceae) at mono-, dicotyly and transitional forms. Acta Biol. Cracovensia, Series Bot. 45(1): 161-165.

Titova, G. E. and T. B. Batygina. 1996. Is the embryo of Nymphaealean plants (Nymphaeales s.l.) dicotyledonous? Phytomorph., 46(2): 171-190.

Titova, G. E., I. I. Shamrov and T. B. Batygina. 1993. Comparative embryology of Nymphaeales s.l. in relation to phylogeny and taxonomy, p. 235. *In* Abstracts XV Int. Bot. Cong., Yokohama, Japan.

Titova G. E., I. I. Shamrov and A. A. Zakharova. 1997. Ovule and seed development in *Trapa natans* L. in connection with the specific embryo sac structure, absence of endosperm and pseudomonocotyly. Bull. Polish Acad. Sci., Biol. Sci., 45 (2-4): 81-92.

Tobe, H. and P. H. Raven. 1984. An embryological contribution to systematics of the Chrysobalanaceae. I. Tribe Chrysobalaneae. Bot. Magazine, (Tokyo), 97(1047): 397-411.

Tohda, H. 1968. Development of the embryo of *Bletilla striata*. Sci. Reports Tôhoku Imper. Univ. Ser. 4 (Biol.), 34(2-4): 125-131.

Tokin, B. P. 1987. General Embryology. Vysshaya shkola, Moscow, Russia.

Toko, K., T. Fujiyoshi, C. Tanaka, S. Iiyama, T. Yoshida, K. Hayashi and K. Yamafuji. 1989. Growth and electric current loops in plants. Biophys. Chem., 33(2): 161-176.

Toks, E. 1976. Endosperm and embryo development in *Cynoglossum officinale* L. Acta Biol. Cracoviensia, Series Bot., 19: 47-59.

Torosian C.D. 1971. Ultrastructural study of endosperm haustorial cells of *Lobelia dunnii* Green (Campanulaceae, Lobelioideae). Amer. J. Bot., 58(5, pt. 2 (Abstr.)): 456-457.

Torres Ruiz, R. A. and G. Jürgens. 1994. Mutations in the FASS gene uncouple pattern formation and organogenesis in *Arabidopsis thaliana*. Development, 120: 2967-2978.

Toyomasu, T., H. Yamane, I. Yamaguchi, N. Murofushi, N. Takahashi and Y. Inoue. 1992. Control by light of hypocotyl elongation and levels of endogenous gibberellins in seedlings of *Lactuca sativa* L. Plant Cell Physiol., 33(6): 695-701.

Trela, Z. 1963a. Embryological studies in *Anemone nemorosa* L. Acta Biol. Cracovensia, Series Bot., 6(1): 1-14.

Trela, Z. 1963b. Cytological studies in the differentiation of the endosperm in *Anemone nemorosa* L. Acta Biologica Cracovensia, Series Bot., 6(2): 177-183.

Trela-Sawicka, Z. 1978. Embryology studies of *Trapa natans* L. Acta Biol. Cracovensia, Series Bot., 21-22(2): 101-103.

Treub, M. 1879. Notes sur l'embryogénie de quelques Orchidées. Verhandelingen der Koninklijke Nederlandsche Academie van Wetensschappen; Afdeeling Natuurkunde Tweede Sectie, Amsterdam, 19: 1-50.

Treub, M. 1891. Sur les Casuarinées et leur place dans le système naturel. Ann. Jardin Bot. Buitenzorg, 10: 145-231.

Treviranus, L. Ch. 1815. Von der Entwicklung des Embryo und seiner Umhüllungen in Pflanzen-Ey. Realschulbuchhandlung, Berlin.

Trifonova, V. I. 1988. Family Ranunculaceae, pp. 176-181. *In* A. L. Takhtajan (ed.). Comparative Anatomy of Seeds. Dicotyledons. Magnoliidae—Ranunculidae. Nauka, Leningrad, Russia, vol. 2.

Tronchet, A. 1925a. La polycotylie et la schizocotylie dans la *Dimorphotheca pluvalis* Moench. C.R. Hebdomadaires Séances l'Acad. Sci., 180: 1862.

Tronchet, A. 1925b. L'Acceleration vasculaire dans la schizocotylie. C.R. Hebdomadaires Séances l'Acad. Sci., 181: 73.

Tronchet, A. 1928. Sur la reduction du nombre des convergents chez les phanerogames, ses rapports avec la polycotylie et le développement vasculaire. Rev. Gén. Bot., 40: 1.

Tschirch, A. 1891. Physiologische studien über die Samen, insbesondere die Saugorgane derselben. Ann. Jardin Bot. Buitenzorg, 9: 143-183.

Tsien, R. W. and R. Y. Tsien. 1990. Calcium channels, stores and oscillations. Ann. Rev. Cell Biol., 6: 715-760.

Tucker, L. R. 1938. Relation between number of carpels and number of cotyledons of the apple, pp. 9-11. Proc. Amer. Soc. Hortic. Sci. 1937, 1938.

Tupy, J. 1961. Changes in glucose and fructose level in *Nicotiana alata* styles ovaries accompanying compatible and incompatible pollen tube growth. Biol. Plantarum, 3: 1-14.

Turcotte, E. L. and C. V. Feaster. 1967. Semigamy in Pima cotton. Heredity, 58(2): 55-57.

Turcotte, E. L. and C. V. Feaster. 1973. The origin of 2n and n sectors of chimeral Pima cotton plants. Crop Sci., 13(1): 111-112.

Turcotte, E. L. and C. V. Feaster. 1974. Semigametic production of cotton haploids, pp. 53-64. *In* Haploids in Higher Plants. Advances and Potential. Publ. Univ. Guelph, Guelph, Canada.

***Turssunov, D. Yu. 1986.** Biology of flowering and embryology of saponiferous Middle Asia species of family Caryophyllaceae. Ph.D. thesis, Tashkent, Uzbek Republic.

***Turssunov, D. Yu. and G. Ju. Savinova. 1979.** Endospermo- and embryogenesis in *Acanthocarpus adenophorum*, pp. 101-104. *In* Biological and Structural Peculiarities of Economic Plants. FAN, Tashkent, Uzbek Republic.

Tykarska, T. 1979. Rape embryogenesis. II. Development of embryo proper. Acta Societatis Bot. Poloniae, 48: 391-421.

***Tzvelev, N. N. 1975.** On the nature of embryo parts in cereals (Poaceae) in connection with monocotyledony origin. Bull. Moscow Soc. Naturalists, Biol. Dept., 80(3): 68-75.

Tzvelev, N. N. 1982. Family of cereals (*Poaceae*), pp. 314-378. *In* A. L. Takhtajan (ed.). The Life of Plants. Prosveshchenie, Moscow, Russia, vol. 6.

***Vaintraub, I. A., A. D. Shutov and E. S. Ropot. 1990.** Regulation of reserve protein lysis in germinating seeds, pp. 142-149. *In* Seed Physiology: Formation, Germination, Applied Aspects. Donish, Dushanbe, Tajikistan.

***Valishina, V. P. and N. V. Zinger. 1952.** Dependence of aconite seed germination on embryo sizes. Bull. Main Bot. Garden USSR Acad. Sci., 13: 45-47.

Vallade, J. 1966. Organisation et functionnement de l'apex radiculaire *l'Elaeis guineensis* Jaca. au cours de la germination. C.R. Hebdomadaires Séances l'Acad. Sci., ser. D. 263(25): 1961-1964.

Vallade, J. 1972. Structure et fonctionnement du méristème lors de la formation de la jeune racine primaire chez un *Petunia hybrida* Hort. C.R. Hebdomadaires Séances l'Acad. Sci., ser. D. 274(7): 1027-1030.

Vallade, J. 1976. Contributions à la connaissance des problèmes fondamentaux de l'embryogénèse chez les Angiosperms. Recherches sur le *Petunia hybrida* Hort. et quelques autres espèces. Thèse, Dijon.

Vallade, J. 1980. Données cytologiques sur la proembryogénese du Pétunia: intérêt pour une interprétation morphogénétique du développement embryonnaire. Bull. Soc. Bot. France, 127(3-4): 19-37.

Vallade, J. 1989. Embryogenesis and research of fundamental morphogenetic processes, pp. 171-187. *In* J. Pare and M. Bugnicourt (eds). Some Aspects and Actual Orientation in Plant Embryology. Univ. Picardie, Amiens, France.

Vallade, J. 1992. Different kinds of growth processes during embryogenesis, pp. 584-585. *In* Proc. XI Int. Symp. "Embryology and Seed Reproduction", July 3-7, 1990, Leningrad. Nauka, St. Petersburg, USSR.

Van Aelst, A. C., C. H. Theunis and J. L. Van Went. 1990. Freeze-fracture studies on isolated sperm cells of *Spinacia oleracea* L. Protoplasma, 153(3): 204-207.

Van Der Pijl, L. 1934. Die Mycorrhiza von *Burmannia* und *Epirrizanthes* und die Fortpflanzung ihrer Endophyten. Recueil des Travaux Bot. Neerlandais, 31: 761-779.

Van Der Pijl, L. 1955. Sarcosta, aril, pulpa and the evolution of the Angiosperm fruit. Proc., Koniklijke Nederlandse Akad. van Wetenschappen, Series C, 58(2-3): 154-161.

Van Lammeren, A. A. M. 1986a. A comparative ultrastructural study of the megagametophytes in two strains of *Zea mays* L. before and after fertilization. Agric. Univ., Wageningen, Netherlands.

Van Lammeren, A. A. M. 1986b. Developmental morphology and cytology of the young maize embryo (*Zea mays* L.). Acta Bot. Neerlandica, 35: 169-188.

Van Lammeren, A. A. M. 1988. Structure and function of the microtubular cytoskeleton during endosperm development in wheat: an immunofluorescence study. Protoplasma, 146: 18-27.

Van Lammeren, A. A. M. and J. H. N. Schel. 1983. Embryogenesis in *Zea mays* L. Development of structure and polarity during proembryo formation, pp. 283-285. *In* O. Erdelská (ed.). Proc. VII Int. Cytoembryological Symp. "Fertilization and Embryogenesis in Ovulated Plants". High Tatra (Račkova dolina), 14-17 June, 1982. Veda, Bratislava, Czechoslovakia.

Van Lijsebettens, M., R. Vanderhaeghen and M. Van Montagu. 1991. Insertional mutagenesis in *Arabidopsis thaliana*: isolation of a T-DNA-linked mutation that alters leaf morphology. Theor. Appl. Genet., 81: 277-284.

Van Tieghem, P. E. L. 1897. Morphologie de l'embryon et de la plantule chez les Graminées et les Cyperacées. Ann. Sci. Naturelles (Paris), 8(3):259-309.

Van Tieghem, Ph. 1900. Sur Ie prothalle femelle des stigmatées. J. Bot., appliq. l'Agric. Pharma. Méd., Arts, 14: 101.

Van Went, J. L. 1970a. The ultrastructure of the egg and central cell of *Petunia*. Acta Bot. Neerlandica, 19(3): 313-322.

Van Went, J. L. 1970b. The ultrastructure of the fertilized embryo sac of *Petunia*. Acta Bot. Neerlandica, 19(4): 468-480.

Van Went, J. L. 1970c. The ultrastructure of the synergids of *Petunia*. Acta Bot. Neerlandica, 19(2): 121-132.

Van Went, J. L. 1984. Unequal distrubution of plastids during generative cell formation in *Impatiens*. Theor. Appl. Genet., 68(4): 305-309.

Vasil, I. K. and A.C. Hildebrandt. 1966. Variations of morphogenetic behaviour in plant tissue cultures. II. *Petroselinum hortense*. Amer. J. Bot., 25: 7-11.

***Vasilevskaya, V. K. 1961.** On the relation between root structure and cotyledon venation. Vestnik Leningrad. Gosudarstvennogo Univ., Biol. Series, 21(4): 14-22.

Vasilyeva, V. E. 1988. Morpho-physiological correlations in the development of the reproductive structures of *Nelumbo nucifera* Gaertn., pp. 168-170. Ann. Scient. l'Univ. Reims Champagne-Ardenne et de l'A.R.E.R.S.

Vasilyeva, V. E. 1993a. Endogenic rhythmicity of the processes in developing seed, p. 272. *In* Abstracts of III Cong. All-Russian Botanical Society Plant Physiologists, vol. 3. St. Petersburg Univ. Publ., St. Petersburg, Russia.

Vasilyeva, V. E. 1993b. Some morphophysiological aspects of plant embryogenesis, p. 36. *In* Abstracts II Cong. All-Russian Botanical Society Plant Physiologists, 1990, Minsk, Russia. Inst. Bot. Byelorussia Acad. Sci, Minsk, USSR.

***Vasilyeva, V. E. and T. B. Batygina. 1981.** *In vitro* cultivation of lotus embryos and ovules isolated at various stages of development. Physiol. rastenyi, 28(2): 319-327.

***Vasilyeva, V. E., T. E. Freiberg and T. B. Batygina. 1978.** The development of peony embryos in *in vitro* culture, 9-11. *In* V. P. Bannikova (ed.). Abstracts VII All-Union Symp. Plant Embryology, Kiev, Ukraine, Naukova dumka, Kiev, Ukraine.

Vasilyeva, V.E., T. B. Batygina and G. E. Titova. 1987. Morpho-physiological correlations in the development of the reproductive structures of *Nelumbo nucifera* Gaertn. Phytomorph., 37(4): 349-358.

***Vasilyeva, V. E., G. E. Kolesova, K. A. Chochya and T. B. Batygina. 1984.** Morpho-physiological correlations in the development of lotus ovule, 11. *In* Materials

All-Union Symp. "Morpho-functional Aspects of Development of Female Generative Structures in Seed Plants", Telavi, Georgia.

Vasilyeva, V. E., G. E. Titova, I. P. Ermakov, N. M. Morozova and K. A. Chochya. 1988. Time and space concordance of the growth, differentiation and metabolism in the developing seed of *Nelumbo nucifera* Gaertn., p. 487. *In* M. Cresti, P. Gori and E. Pacini (eds). Proc. 10th Int. Symp. "Sexual Reproduction in Higher Plants", May 30-June 4, 1988, University of Siena, Siena, Italy. Springer-Verlag, Berlin.

***Vassileva, O. A. 1954.** Nucleic acid changes in the course of fertilization in pea. Reports USSR Acad. Sci., 95(1): 163-166.

***Vassilyev, A. E. 1970.** Ultrastructure of stigmatoid cells of *Lilium*. Physiol. rastenyi, 17(6): 1240-1248.

Vaughn, K. C., L. R. DeBonte, K. G. Wilson and G. W. Schaffer. 1980. Organelle alteration as a mechanism for maternal inheritance. Science, 208: 196-198.

Vazart, B. 1955. Contribution à l'étude caryologique des elements reproducteurs et de la fécondation chez lex végétaux angiospermes. Rev. Cytol. Biol. Végétales, 16(3-4):209-406.

Vazart, B. 1958. Differenciation des cellules sexualles et fécondation chez les Phanerogames. Protoplasmatol., 7(3a): 1-158.

Vazart, B. and J. Vazart. 1966. Deposition et infrastructure des cellules du sac embryonaire du lin (*Linum usitatissimum* L.). Rev. Cytol. Biol. Végétales, 24(3-4): 251-266.

Vegis, A. 1964. Dormancy in higher plants. Ann. Rev. Plant Physiol., 15: 185-224.

Velenovsky, J. 1907. Vergleichende Morphologie der Pflanzen, Bot. 2. Verlagsbuchhandlung von Fr. Řivnáč, Prague.

***Vereshchagina, V. A. 1976.** Some embryological peculiarities of gynodioecious species of *Echium* (Boraginaceae). Bot. Journal (St. Petersburg), 61(11): 1582-1589.

Vereshchagina, V. A. 1983a. Family Violaceae, pp. 111-115. *In* M. S. Yakovlev (ed.). Comparative Embryology of Flowering Plants. Phytolaccaceae—Thymelaeaceae. Nauka, Leningrad, Russia.

***Vereshchagina, V. A. 1983b.** Development of endosperm in *Pulmonaria* (Boraginaceae). Bot. Journal (St. Petersburg), 68(8): 1107-1112.

Vernon, D. M. and D. W. Meinke. 1994. Embryogenic transformation of the suspensor in twin, a polyembryonic mutant of *Arabidopsis*. Developm. Biol., 165: 566-573.

Vesque, J. 1879. Nouvelles recherches sur le développement du sac embryonnaire des Phanerogames Angiospermes. Ann. Sci. Naturelles, Bot., 6(8): 261-390.

Veyret, Y. 1958. Embryogénie des Orchidacées. Développement de l'embryon chez l'*Angraecum distichum* Lindl. Bull. Soc. Bot. France, 105(5-6): 227-230.

Veyret, Y. 1965. Embryogénie comparée et blastogénie chez les Orchidaceae—Monandrae. ORSTOM, Paris, 106 pp.

Veyret, Y. 1974. Development of the embryo and the young seedling stages of orchids, pp. 223-265. *In* C. L. Withner (ed.). The Orchids Scientific Studies. Acad. Press, New York, NY.

Vijayaraghavan, M. R. and K. Prabhakar. 1984. The endosperm, pp. 319-376. *In* B. M. Johri (ed.). Embryology of Angiosperms. Springer-Verlag, Berlin.

***Vishnyakova, M. A. 1986.** Investigation of progamic phase of fertilization in *Medicago* in relation to self-incompatibility. Trans. Appl. Bot., Genetics, Selection (Leningrad), 99: 17-22.

*Vishnyakova, M. A. 1987. Growth of pollen tubes in pistils of apple tree (*Malus domestica* Borkh.) in relation with self-incompatibility. Bull. All-Union Inst. Plant Breeding, 170: 34-38.

*Vishnyakova, M. A. 1989. Structural-functional bases of self-incompatibility in flowering plants. Bot. Journal (St. Petersburg), 74(2): 137–152.

*Vishnyakova, M. A. 1991. The role of stigma and style in interactions of pollen with pistil in *Trifolium pratense* (Fabaceae). Bot. Journal (St. Petersburg), 76(5): 677-686.

*Vishnyakova, M. A. 1994. Structural bases of activity of self-incompatibility genes. Genetica, 30(10): 107-117.

Vishnyakova, M. A. and M. T. M. Willemse. 1994. Pollen-pistil interaction in wheat. Acta Botanica Neerlandica, 43(1): 51-64.

Vogelmann, T. C. 1993. Plant tissue optics. Ann. Rev. Plant Physiol. Molec. Biol., 44: 231-252.

*Volotovsky, I. D. 1992. Phytochrom—A Regulatory Photoreceptor of Plants. Nauka i Tekhnika, Minsk, Belorussia.

Von Engelen, F. A., P. Sterk, H. Booij, J. H. G. Cordewener, W. Rook, Ab. Van Kammen and S. C. De Vries. 1991. Heterogeneity and cell typespecific localization of a cell wall glycoprotein from carrot suspension cells. Plant Physiol., 96: 705-711.

Von Guttenberg, H. 1957. Embryologische und histogenetische Untersuchungen an Monocotyleledonen. Bot. Studien, 7: 1-161.

Von Guttenberg, H. 1960. Gründzuge der Histogenese von höherer Pflanzen. I. Die Angiospermen. Handbuch der Pflanzenanatomie, Gebrüder Bornträger, Berlin. 8(3): 1-315.

Von Guttenberg, H. 1968. Der primäre Bau der Angiospermenwurzel. Handbuch der Pflanzenanatomie, Gebrüder Bornträger, Berlin, 9(5).

Von Guttenberg, H. and A. Semlow. 1957. Die Entwicklung des Embryos und des Keimpflanze von Cyperaceen. Bot. Studien, 7.

Von Guttenberg, H., H.-R. Heydel and H. Pankow. 1954. Embryologische Studien an Monokotyledonen. I. Die Entstehung der Primarwurzel bei *Poa annua*. Flora, 141: 298-311.

Von Guttenberg, H., J. Burmeister, and H. J. Brosell. 1955. Studien über der Entwicklung die Wurzelvegetationspunktes der Dikotyledonen. Planta, 46(2): 179-222.

Von Müller, F. 1882. Plurality of cotyledons in the genus *Persoonia*. New Zealand J. Sci., 115.

Von Rübsamen, T. 1988. Morphologische, embryologische und systematische Untersuchungen an Burmanniaceae und Corsiaceae (Mit Ausblick auf die Orchidaceae). J. Cramer (ed.). Diss. Botanicae, Berlin.

*Voronin, N. S. 1956. On the evolution of roots in plants. 1. Apical root meristems. Bull. Moscow Soc. Naturalists, Biol. Dept., 61(5): 47-58.

*Voronin, N. S. 1964. Evolution of primary structures in plant roots. Uchenye zapiski Kaluzh. Gosudarstvennogo Pedagogich. Inst., 13: 3-179.

Vorosbina, L. I. and M. P. Solntseva. 1979. The development of helobial endosperm in Amaryllidaceae family in the genera *Zephyranthes* and *Sternbergia*, pp. 75-81. *In* M. S. Yakovlev (ed.). Actual Problems of Angiosperm Embryology. Nauka, Moscow-Leningrad, Russia.

Vorsobina, L. I. and M. P. Solntseva. 1990. Family Amaryllidaceae, pp. 87-96. *In* T. B. Batygina and M. S. Yakovlev (eds). Comparative Embryology of Flowering Plants. Butomaceae—Lemnaceae. Nauka, Leningrad, Russia.

Vuilleumier, B. S. 1967. The origin and evolutionary development of heterostyly in angiosperms. Evolution, 21: 210-226.

Vyshenskaya, T. D. 1985. Family Centrolepidaceae, pp. 187-189. *In* A. L. Takhtajan (ed.). Comparative Anatomy of Seeds. Monocotyledons. Nauka, Leningrad, Russia, vol. 1.

Vyshenskaya, T. D. 1991. Embryo development trends in the family *Myrtaceae* in connection with its systematics, pp. 28-30. *In* Materials VIII Moscow Conf. Plant Phylogeny "General Biological Aspects of Plant Phylogeny", Nauka, Moscow, Russia.

Vyshenskaya, T. D. 1992. Family Bombacaceae, pp. 305-325. *In* A. L. Takhtajan, (ed.) Comparative Anatomy of Seeds. Nauka, Leningrad, Russia, vol. 4.

Wade Roush, 1995. Embryos travel forking path as they tell left from right. Science, 269 (5230): 1514-1515.

Wagner, V. T. and H. L. Mogensen. 1987. The male germ unit in the pollen and pollen tubes of *Petunia hybrida*: ultrastructural quantitative and three-dimensional features. Protoplasma, 143: 93-100.

Wagner, V. T., M. Cresti, P. Salvatica and A. Tiezzi. 1990. Changes in volume, surface area, and frequency of nuclear pores on the vegetative nucleus of tobacco pollen in fresh, hydrated, and activated conditions. Planta, 181: 304-309.

Walia, K. and R. N. Kapil. 1965. Embryology of *Frankenia* Linn. with some comments on the systematic position of the Frankeniaceae. Bot. Notiser, 118(4): 412-429.

Walker, R. I. 1947. Megasporogenesis and embryo development in *Tropaeolum majus* L. Bull. Torrey Bot. Club, 74: 240-249.

Walthall, E. D. and T. Brady. 1986. The effect of suspensor and gibberellic acid on *Phaseolus vulgaris* embryo protein synthesis. Cell Differentiation, 18: 37-44.

Wang, M., S. Heimovaara-Dijkstra and B. Van Duijn. 1995. Modulation of germination of embryos isolated from dormant and non-dormant barley grains by manipulation of endogenous abscisic acid. Planta, 195(4): 586-592.

Wardlaw, C. W. 1955. Embryogenesis in Plants. Methuen, London.

Wardlaw, C. W. 1963. Plant embryos as reaction systems. *In* P. Maheshwari (ed.). Recent Advances in the Embryology of Angiosperms. Univ. Delhi, Delhi, India.

Wardlaw, C. W. 1965. Physiology of embryonic development in Cormophytes, pp. 844-965. *In* E. Ruhland (ed.). Encyclopedia of Plant Physiology, Vol. XV(1). Springer-Verlag, Berlin.

Wareing, P.F. and J. D. J. Phillips. 1978. The Control of Growth and Differentiation in Plants. Pergamon Press, Oxford, New York, NY.

Warren Wilson J. and P. M. Warren Wilson. 1993. Mechanisms of auxin regulation of structural and physiological polarity in plants, tissues, cells and embryos. Australian J. Plant Physiol., 20(4-5): 555-571.

Warren Wilson J., P. M. Warren Wilson and E. S. Walker. 1991. Patterns of trachel differentiation in lettuce pith explants: positional control and temperature effects. Ann. Bot. (London), 68: 109-128.

Watanabe K. 1934. Geschlechtliche Fortpflanzung von *Mitrastemon* Yamamotoi. Proc. Imper. Acad. Japan, 10 (4): 421-423.

Weatherwax, P. 1919. Gametogenesis and fecundation in *Zea mays* as the basis of xenia and heredity in the endosperm. Bull. Torrey Bot. Club, 46: 73-90.

Webb, D. P., J. Van Staden and P. F. Wareing. 1973a. Seed dormancy in *Acer*: changes in endogenous cytokinins, gibberellins and germination inhibitors during the breaking of dormancy in *Acer saccharum* Marsh. J. Exper. Bot., 24(78); 105-116.

Webb, D. P., J. Van Staden and P. F. Wareing. 1973b. Seed dormancy in *Acer*: changes in endogenous germination inhibitors, cytokinins and gibberellins during the breaking of dormancy in *Acer pseudoplatanus* L. J. Exper. Bot., 24(81): 741-750.

Webb, M. C. and B. E. S. Gunning. 1991. The microtubular cytoskeleton during development of the zygote, proembryo and free-nuclear endosperm in *Arabidopsis thaliana* (L.) Heynh. Planta, 184: 187-195.

Weber, M. 1988a. Formation of sperm cells in *Galium mollugo* (Rubiaceae), *Trichodiademe setuliferum* (Aizoaceae), and *Avena sativa* (Poaceae). Plant System. Evol., 161: 53-64.

Weber, M. 1988b. Unusual sperm cells in Apiaceae. Plant System. Evol., 159: 273-276.

Weigel, D. and E. M. Meyerowitz. 1994. The ABCs of floral homeotic genes. Cell, 78: 203-209.

Weiland, U. and A. J. Müller. 1972. In-vitro-Kultur der Wurzeln von letalen fusca-Mutanten von *Arabidopsis thaliana*. Kulturpflanze, 20: 151-164.

Weisenseel, M. N., R. Nuccitelli and L. F. Jaffe. 1975. Large electrical currents traverse growing pollen tubes. J. Cell Biol., 66: 556-567.

Weisenseel, M. N., A. Dorn and L. F. Jaffe. 1979. Natural currents traverse growing roots and root hairs of barley (*Hordeum vulgare* L.). Plant Physiol., 64(3): 512-518.

Weiβe, A. 1930. Blattstellungsstudien an Samlingen abnorm Keimender Dikotylen. Beitrage Biol. Pflanzen, 18(1): 17-80.

Went, F. W. 1944. Morphological observations on the tomato plant. Bull. Torrey Bot. Club, 71: 77-92.

West, M. A. L. and J. J. Harada 1993. Embryogenesis in higher plants: an overview. Plant Cell, 5: 1361-1369.

Westermaier, M. 1876. Die ersten Zellteilungen im Embryo von *Capsella bursa-pastoris*. Flora, 59(31-33): 483-520.

Whatley, J. M. and D. N. Price. 1983. Do lemon cotyledons green in the dark? New Phytologist, 94(1): 19-27.

Wheeler, H. and T. Humphreys. 1979. Localization of a phosphatase (ATPase) on the plasmalemma of the maize scutellum. Phytochem., 18: 549-554.

White, J. M. 1992. Membrane fusion. Science, 258: 917-924.

White, O. E. 1948. Fasciation. Bot. Rev., 14: 319-358.

Wiegand, K. M. 1900. Notes on the embryology of *Potamogeton*. Bot. Gazette, 30(1): 25-47.

Wiens, D., D. L. Nickrent, C. I. Davern, C. L. Calvin and N. J. Vivrette. 1989. Developmental failure and loss of reproductive capacity in the rare palaeoendemic shrub *Dedeckera eurekensis*. Nature, 338: 65-67.

Wiens, D., C. L. Calvin, C. A. Wilson, C. I. Davern, D. Frank and S. R. Seavey. 1987. Reproductive success, spontaneous embryo abortion, and genetic load in flowering plants. Oecologia, 71: 501-509.

Williams, E. G. and G. Maheshwaran. 1986. Somatic embryogenesis: factors influencing coordinated behaviour of cells as an embryogenic group. Ann. Bot. (London), 57(4): 443-462.

Williams, E. G., R. B. Knox and V. Kaul. 1984. Post-pollination callose development in ovules of *Rhododendron* and *Ledum* (Ericaceae): zygote special wall. J. Cell Sci., 69: 127-135.

Wilms, H. J. 1981a. Pollen tube growth through nucellus into embryo sac of *Spinacia*, an ultrastructural investigation. Acta Bot. Neerlandica, 50(1): 191-194.

Wilms, H. J. 1981b. Pollen tube penetration and fertilization in spinach. Acta Bot. Neerlandica, 30(1-2): 101-122.

Wilms, H. J. 1981c. Ultrastructure of the developing embryo sac of spinach. Acta Bot. Neerlandica, 30(1-2): 75-99.

Wilms, H. J. 1981d. Ultrastructure of the egg apparatus of *Spinacia*. Acta Soc. Bot. Poloniae, 50(1-2): 165-168.

Wilms, H. J. 1985. Behaviour of spinach sperm cell in pollen tubes prior to fertilization, pp. 143-144. *In* M. T. M. Willemse and J. L. Van Went (eds). Sexual Reprod. in Seed Plants, Ferns and Mosses. Proc. 8th Int. Symp. Sexual Reproduction, August 20-24, 1984. PUDOC, Wageningen, Netherlands.

Wilms, H. J. 1986. Dimorphic sperm cells in the pollen grain of *Spinacia*, pp. 193-198. *In* M. Cresti and R. Dallai (eds). Biology of Reproduction and Cell Motility in Plants and Animals. Univ. Siena, Siena, Italy.

Wilms, H. J. and A. C. Van Aelst. 1983. Ultrastructure of spinach sperm cells in mature pollen, pp. 105-112. *In* O. Erdelská (ed.). Proc. 7th Int. Symp. 'Fertilization and Embryogenesis in Ovulated Plants" High Tatra (Račkova dolina), June 14-17, 1982. Veda, Bratislava, Czechoslovakia.

Wilson, K. A. 1986. Role of proteolytic enzymes in the mobilization of protein reserves in the germinating dicot seed, pp. 19-47. *In* Plant Proteolytic Enzymes. CRC Press, Boca Raton, FL, USA.

Winkler, A. 1874. Über die Keimblätter der deutschen Dicotylen. Verhandlungen Bot. Vereins Provinz Brandenburg, angrenzenden Länder, 16: 6-21, 54-56.

Winkler, A. 1894. Anomale Keimungen. Verhandlungen Bot. Vereins Provinz Brandenburg, angrenzenden Länder, 36: 127.

Winkler, H. 1931. Die Monokotylen sind monokotyl. Beitrage Biol. Pflanzen, 19: 29.

Winter, A. N. 1987. Embryology of the genus *Hydrostemma* (*Barclaya*) in connection with its systematics. Ph.D. thesis, Leningrad, Russia.

Winter, A. N. and T. B. Batygina. 1988. Family Barclayaceae, pp. 147-152. *In* A. L. Takhtajan, (ed.) Comparative Anatomy of Seeds. Dicotyledons. Magnoliidae—Ranunculidae. Nauka, Leningrad, Russia, vol. 2.

Wodzicki, T. J. and S. Zajaczkowski. 1989. Auxin waves in cambium and morphogenetic information in plants, pp. 45-64. *In* J. Krekule and F. Sedlova (eds). Proc. 14th Biochem. Cong. "Signals in Plant Development", Prague, July 10-15, 1988. SPB Acad. Publ., The Hague, Netherlands.

Wojciechowicz, M. K. and K. Strzelecka. 1997. Antipodal embryos formation in *Allium odorum*, p. 93. *In* The Scientific Programme and Abstracts VIII Conf. Plant Embryologists, 16-18 September, 1997, Gdańsk, Poland, Gdańsk Univ., Gdańsk, Poland.

Wolf, M. J., U. Khoo and H. L. Seckinger. 1969. Distribution and subcellular structure of endosperm protein in varieties of ordinary and high-lysine maize. Cereal Chem., 46: 253-263.

Wolfsberg, M. F. 1964. Cell population kinetics in the epithelium of the forestomach of the mouse. Exper. Cell Res., 35: 119-131.

Woodard, J. W. 1956. DNA in gametogenesis and embryogeny in *Tradescantia*. J. Biophys. Biochem. Cytol., 2: 765-776.

Wunderlich, R. 1959. Zur Frage der Phylogenie der Endospermtypen bei den Angiospermen. Österreichische Bot. Zeitschrift, 106(3-4): 203-293.

Wunderlich, R. 1967. Ein Vorschlag eu einer natürlichen Gliederung der Labiaten auf Grund der Pollenkorner, der Samenentwicklung und des reifen Samens. Österreichische Bot. Zeitschrift, 114(4-5): 383-483.

Wylie, R. B. 1904. The morphology of *Elodea canadensis*. Bot. Gazette, 37: 1-22.

Xi, X.-J. and B.-X. Ye. 1994. Embryo development research and accumulation of nutrients in *Coix lacryna-jobi*. Acta Bot. Sinica, 36(8): 573-580.

XuHan, X. 1995. Seed development in *Phaseolus vulgaris* L., *Populus nigra* L., and *Ranunculus sceleratus* L. with special reference to the microtubular cytoskeleton. Ph.D. thesis, Wageningen, Netherlands.

XuHan, X. and A. A. M. Van Lammeren. 1993. Microtubular configurations during the cellularization of coenocytic endosperm in *Ranunculus sceleratus* L. Sexual Plant Reprod., 6: 127-132.

XuHan, X. and A. A. M. Van Lammeren. 1994. Microtubular configurations during endosperm development in *Phaseolus vulgaris*. Canadian J. Bot., 72: 1489-1495.

XuHan, X., A. Souvre, C. Granier and M. Petitprez. 1995. An improved immunolabelling method for microtubular cytoskeleton in poplar (*Populus nigra* L.) free nuclear endosperm. Biotech. Histochem.

Yadegari, R., G. R. De Paiva, T. Laux, A. M. Koltunow, N. Apuya, J. L. Zimmerman, R. L. Fischer, J. J. Harada and R. B. Goldberg. 1994. Cell differentiation and morphogenesis are uncoupled in *Arabidopsis* rasberry embryos. Plant Cell, 6: 1713-1729.

***Yakobson, L. Ya. 1976.** Embryological investigation of rhododendrons. Reports Latvian Acad. Sci., 352(11): 84-96.

***Yakovlev, M. S. 1946a.** Monocotyledony in the light of embryology data. Sovetskaya bot., 14(5): 351-363.

Yakovlev, M. S. 1946b. The experience of construction of cereal system on the basis of embryo morphology and endosperm structure. Ph.D. thesis, Leningrad, Russia.

***Yakovlev, M. S. 1950.** The structure of endosperm and embryo of cereals as a systematic feature. Trans. Komarov Bot. Inst. USSR Acad. Sci., Series VII, 1: 121-218.

***Yakovlev, M. S. 1951.** On the uniformity of embryogenesis in angiosperms and gymnosperms. Trans. Komarov Bot. Inst. USSR Acad. Sci., Series VII, 2: 231-242.

***Yakovlev, M. S. 1958.** Principles of distinguishing of main embryonal types and their significance for the phylogeny of angiosperms. Problems Bot., 3: 168-195.

Yakovlev, M. S. 1969. Embryogenesis and some problems of phylogenesis. Rev. Cytol. Biol. Végétales, 32: 325-330.

***Yak vlev, M. S. 1971.** Fundamental tasks in the research of embryogenesis, pp. 152-169. *In* V. P. Zosimovich (ed.). Problems of Embryology. Naukova dumka, Kiev, Ukraine.

Yakovlev, M. S. 1972. The embryo, pp. 370-371. *In* Large Soviet Encyclopaedia. Sovetskaya entsiklopedia, Moscow, Russia.

Yakovlev, M. S. 1981. The glossary of main terms, pp. 7-25. *In* M. S. Yakovlev (ed.). Comparative Embryology of Flowering Plants, Winteraceae—Juglandaceae. Nauka, Leningrad, Russia.

Yakovlev, M. S. 1983a. Family Dilleniaceae, pp. 65-68. *In* M. S. Yakovlev (ed.). Comparative Embryology of Flowering Plants. Phytolaccaceae—Thymelaeaceae. Nauka, Leningrad, Russia.

Yakovlev, M. S. 1983b. Family Paeoniaceae, pp. 70-77. *In* M. S. Yakovlev (ed.). Comparative Embryology of Flowering Plants. Phytolaccaceae—Thymelaeaceae. Nauka, Leningrad, Russia.

Yakovlev, M. S. and M. D. Yoffe. 1957a. A new type of embryogenesis among angiosperms, pp. 213-214. *In* O. T. Pomalenkaya (ed.). Abstracts of 2nd Conference of USSR Embryologists. Moscow, Russia, Moscow State Univ. Publ., Moscow USSR.

Yakovlev, M. S. and M. D. Yoffe. 1957b. On some peculiar features in the embryology of *Paeonia*. Phytomorph., 7(1): 74-82.

***Yakovlev, M. S. and M. D. Yoffe. 1957c.** Peculiarities of embryogenesis in the genus *Paeonia* L. Bot. Journal (St. Petersburg), 42(10): 1491-1502.

***Yakovlev, M. S. and M. D. Yoffe. 1961.** Further investigation of a new type of embryogenesis in angiosperms. Bot. Journal (St. Petersburg), 46(10): 1402-1421.

Yakovlev, M. S. and M. P. Solntseva. 1965. Some questions about flower morphology and embryology of feather-grasses, pp. 61-73. *In* M. S. Yakovlev (ed.). Flower Morphology and Reproductive Process in Flower Plants. Nauka, Moscow-Leningrad, Russia.

Yakovlev, M. S. and M. D. Yoffe. 1965. Embryology of some representatives of genus *Paeonia* L., pp. 140-176. *In* Flower Morphology and Reproductive Process in Angiosperms. Nauka, Moscow-Leningrad, Russia.

***Yakovlev, M. S. and O. P. Kamelina. 1973.** The development of embryo in *Melocanna bambusoides* Trin. Bot. Journal (St. Petersburg), 58(2): 248-263.

Yakovlev, M. S. and G. Ya. Zhukova. 1973. Angiosperms with green and colourless embryo (Chloro- and Leucoembryophytes). Nauka, Leningrad, Russia.

Yakovlev, M. S. and G. Ya. Zhukova. 1975. On the presence of chlorophyll in embryos of angiosperm seeds, pp. 223-229. *In* H. Y. Mohan Ram, J. J. Shah and C. K. Shah (eds). Form, Structure and Function in Plants. Sarita Prakashan, Nauchandi, Meerut, UA, India.

***Yakovlev, M. S. and G. K. Alimova. 1976.** Embryogenesis in *Arabidopsis thaliana* (L.) Heynh. (Cruciferae). Bot. Journal (St. Petersburg), 61(1): 12-24.

Yakovlev, M. S. and I. A. Chaban. 1979. Ultrastructural aspects of embryogenesis in *Reseda lutea* L., pp. 96-104. *In* M. S. Yakovlev (ed.). Actual Problems of Angiosperm Embryology. Nauka, Moscow-Leningrad, Russia.

Yakovlev, M. S. and G. Ya. Zhukova. 1980. Chlorophyll in embryos of angiosperm seeds, a review. Bot. Notiser, 133: 323-336.

Yamashita, T. 1983. Embryo- und Wurzelentwicklung bei *Commelina communis* mit besonderer Berucksichtigung der Coleorhiza Wurzelokologie und ihre Nutzanwendung, pp. 21-32. *In* Proc. Int. Symp. "Root Ecology and Its Practical Application", Gumpenstein, 1982, Bundesansalt Gumpenstein, A-8952 Irdning.

Yamazaki, T. 1957. Taxonomical and phylogenetic studies of Scrophullariaceae—Veronicae, with special reference to *Veronica* and *Veronicastrum* in Eastern Asia. J. Fac. Sci., Univ. Tokyo, Section 3, 7: 91-162.

Yamazaki, T. 1963. Embryology of *Mitrosacme alsinoides* var. *indica*. Science Reports Tokyo Imperial University. Ser. 4 (Biology), 29: 201-205.

Yamazaki, T. 1974. A system of Gamopetalae based on the embryology. J. Fac. Sci., Univ. Tokyo, Sect. 3, 11: 263-281.

Yamazaki, T. 1975. Embryogeny of *Elaeagnus umbellata* Thunb. Japanese J. Bot., 50(9): 281-284.

Yamazaki, T. 1982. Recognized types in early development of the embryo and the phylogenetic significance in the dicotyledons. Acta Phytotax. Geobot., 33: 400-409.

Yan Hua and Hong-Yuan Yang. 1991. Ultrastructure of the developing embryo sac of sunflower (*Helianthus annuus*) before and after fertilisation. Canadian J. Bot., 69: 191-202.

Yanofsky, M. F., H. Ma, J. L. Bowman, G. N. Drews, K. A. Feldmann and E. M. Meyerowitz. 1990. The protein encoded by the *Arabidopsis* homeotic gene agamous resembles transcription factors. Nature, 346: 35-39.

Yeung, E. C. 1980. Embryogeny of *Phaseolus*: the role of the suspensor. Zeitschrift Pflanzenphysiol., 96(1): 17-28.

Yeung, E. C. 1995. Structural and developmental patterns in somatic embryogenesis, pp. 205-247. *In* T. A. Thorpe (ed.). In vitro Embryogenesis in Plants. Kluwer Acad. Publ., Dordrecht, Netherlands.

Yeung, E. C. and M. E. Clutter. 1978. Embryology of *Phaseolus coccineus*: growth and microanatomy. Protoplasma, 94(1-2): 19-40.

Yeung, E. C. and M. E. Clutter. 1979. Embryogeny of *Phaseolus coccineus*: the ultrastructure and development of suspensor. Canadian J. Bot., 57(2): 120-136.

Yeung, E.C. and I. M. Sussex. 1979. Embryogeny of *Phaseolus coccineus*: the suspensor and the growth of the embryo proper in vitro. Zeitschrift Pflanzenphysiol., 91(5): 423-433.

Yeung, E. C. and M. J. Cavey. 1988. Cellular endosperm formation in *Phaseolus vulgaris*. I. Light and scanning electron microscopy. Canadian J. Bot., 66(6): 1209-1216.

Yeung, E. C. and S. K. Law. 1992. Embryology of *Calypso bulbosa*. II. Embryo development. Canadian J. Bot., 70(3): 461-468.

Yeung, E. C. and D. W. Meinke. 1993. Embryogenesis in angiosperms: development of the suspensor. Plant Cell, 5: 1371-1381.

Yeung, E. C., S. Y. Zee and X. Z. Ye. 1996. Embryology of *Cymbidium sinense*: embryo development. Ann. Bot. (London), 78(1): 105-110.

***Yoffe, M. D. 1952.** On the occurrence of chlorophyll in endosperm of Cruciferae. Reports USSR Acad. Sci., 82(3): 473-476.

***Yoffe, M. D. 1957.** The development of embryo and endosperm in wheat, horse bean and radish. Trans. Komarov Bot. Inst. USSR Acad. Sci., Series VII, 4: 211-269.

***Yoffe, M. D. 1962.** On the embryology of *Trochodendron arabioides* Sieb. et Zucc. Trans. Komarov Bot. Inst. USSR Acad. Sci., Series VII, 5: 250-259.

***Yoffe, M. D. 1968.** Fertilization process in *Anemone nemorosa* L. and *A. ranunculoides* L. Bot. Journal (St. Petersburg), 53(7): 891-905.

Yoffe, M. D. 1969. Fertilization and formation of the primary endosperm nucleus in *Anemone nemorosa* L. Rev. Cytol. Biol. Végétales, 32: 319-323.

***Yoffe, M. D. 1971a.** Polyploidy in the endosperm of flowering plants, pp 170-196. *In* V. P. Zosimovich (ed.). Problems of Embryology. Naukova dumka, Kiev, Ukraine.

***Yoffe, M. D. 1971b.** Embryo and endosperm development in *Anemone nemorosa* L. Bot. Journal (St. Petersburg), 56(11): 1582-1592.

***Yoffe, M. D. 1976.** Peculiarities of double fertilization in the genus *Melampyrum* L. (Scrophulariaceae) (in connection with the formation of diploid endosperm). Bot. Journal (St. Petersburg), 61(11): 1515-1530.

***Yoffe, M. D. and G. Ya. Zhukova. 1965.** Culture of angiosperm isolated embryos on artificial medium. Bot. Journal (St. Petersburg), 50(8): 1157-1182.

***Yoffe, M. D. and G. Ya. Zhukova. 1974.** Modification of zygote wall during its development in the genus *Dillenia* (Dilleniaceae). Bot. Journal (St. Petersburg), 59(10): 1409-1416.

Yoshida, O. 1957. Embryologische Studien über die Ordnung Piperales. 1. Embryologie von *Chloranthus japonicus*. J. College Arts Sciences, Chiba Univ. (Nat. Sci.), 2(2): 172-178.

Yoshida, O. 1959. Embryologische Studien über die Ordnung Piperales. 2. Embryologie von *Chloranthus serratus*. J. College Arts Sciences, Chiba Univ. (Nat. Sci.), 2(2): 295-303.

You, R. and W. A. Jensen. 1985. Ultrastructural observation of the mature megagametophyte and fertilisation in wheat (*Triticum aestivum*). Canadian J. Bot., 63(2): 163-178.

Yu, H. S. and S. D. Russell. 1992. Male cytoplasmic diminution and male germ unit in young and mature pollen of *Cymbidium goeringii*: a three-dimensional and quantitative study. Sexual Plant Reprod., 5: 169-181.

Yu, H. S. and S. D. Russell. 1993. Three-dimensional ultrastructure of generative cell mitosis in the pollen tube of *Nicotiana tabacum*. European J. Cell Biol., 61: 338-348.

Yu, H. S. and S. D. Russell. 1994a. Male reproductive cell development in *Nicotiana tabacum*: male germ unit associations and quantitative cytology during sperm maturation. Sexual Plant Reprod., 7: 324-332.

Yu, H. S. and S. D. Russell. 1994b. Occurrence of mitochondria in the nuclei of *Nicotiana tabacum* sperm cells. Plant Cell, 6: 1477-1484.

Yu, H. S. and S. D. Russell. 1994c. Quantitative differences in sperm cells and organelles of tobacco (*Nicotiana tabacum* L.) grown under differing environmental conditions. Theor. Appl. Genet., 89: 814-817.

Yu, H. S., S. Y. Hu and S. D. Russell. 1992. Sperm cells in pollen tubes of *Nicotiana tabacum* L.: three-dimensional reconstruction, cytoplasmic diminution and quantitative cytology. Protoplasma, 168: 172-183.

Yu, H. S., B. Q. Huang and S. D. Russell. 1994. Transmission of male cytoplasm during fertilization in *Nicotiana tabacum*. Sexual Plant Reprod., 7: 312-323.

Zaman, B. 1950. The embryology of *Juncus prismatocarpus* Br. and *J. effusus* Linn. Proc. Indian Acad. Sci., 31(4): 223-234.

***Zamotailov, S. S. 1958.** On the peculiarities of sexual cell nucleus fusion in groundnut (*Arachis hypogaea* L.). Reports USSR Acad. Sci., 123(5): 941-943.

Zamotailov, S. S. 1960. On the fertilization in groundnut, pp. 59-60. *In* Abstracts III All-Union Conference Embryologists, Izd. Moskov. Univ., Moscow, Russia.

***Zamotailov, S. S. 1968.** Fertilization in groundnut and influence of temperature on this process. Tr. Kuban. selsko-chozya' stvennogo Inst., 17(45): 134-142.

Zamski, E. 1975. Light and electron microscope observation on scutellum epithelial cells during germination of *Zea mays* seeds. Israel J. Bot., 22: 211-230.

Zavala, M. E. and D. L. Branden. 1983. Localization of a phytohormone using immunocytochemistry. J. Cell Biol., 97(4): 1235-1239.

***Zazhurilo, K. K. 1940.** On the anatomy of seed coats in Magnoliaceae (*Liriodendron tulipifera* L.). Bull. Soc. Naturalists, Voronezh Univ., 4(1): 33-40.

Zee S. Y. and Aziz-Un-Nisa. 1991. Mitosis and microtubule organization changes in isolated generative cells of *Allemanda neriifolia*. Sexual Plant Reprod., 4: 132-137.

Zenkteler, M. 1990. In vitro fertilization and wide hybridization in higher plants. Crit. Rev. Plant Sci., 9: 267-279.

Zhang, G., D. J. Gifford and D. D. Cass. 1993. RNA and protein synthesis in sperm cells isolated from *Zea mays* L. pollen. Sexual Plant Reprod., 6: 239-243.

Zhang, H. Q., J. Bohdanowicz, E. S. Pierson, and Y. Li, A. Tiezzi and M. Cresti. 1995. Microtubular organization during asymmetrical division of the generative cell in *Gagea lutea*. J. Plant Cell, 108: 269-276.

Zhang, J. and M. A. Maun. 1989a. The role of embryo and endosperm in the determination of seedling size in two dune species. (Abstracts 1989 joint Meeting Canadian Bot. Assoc. and Bot. Soc. America with other BSA-Affiliated Soc., Univ. Toronto, Toronto, 6-10 August, 1989). Amer. J. Bot., suppl., 76(6): 132.

Zhang, J. and M. A. Maun. 1989b. Effect of partial removal of endosperm on seedling sizes of *Panicum virgatum* and *Agropyron psammophilum*. Oikos, 56(2): 250-255.

Zhang, Z., G. W. Struot, H. Zhu and S. D. Russell. 1996. Cell surface biology of male gametes of *Plumbago zeylanica*, p. 42. *In* Abstracts 14th Int. Cong. Sexual Plant Reproduction "Plant Reproduction' 96", February 18-23, 1996, Lorne, Victoria, Cumberland Resort, Lorne, Victoria, Australia.

***Zhgenti, L. P. 1986.** Ultrastructure of zygote and central cell in *Saxifraga flagellaris* Willd, pp. 91-92. *In* Abstracts IX All-Union Conf. Plant Embryology "Gametogenesis, Fertilization and Embryogenesis of Seed Plants, Ferns and Mosses". Shtiintsa, Kishinev, Moldova.

Zhu, T., H. L. Mogensen and S. E. Smith. 1990. Generative cell composition and its relation to male plastid inheritance patterns in *Medicago sativa*. Protoplasma, 158: 66-72.

Zhu, T., H. L. Mogensen and S. E. Smith. 1991. Quantitative cytology of the alfalfa generative cell and its relation to male plastid inheritance patterns in three genotypes. Theor. Appl. Genet., 81: 21-26.

Zhu, T., H. L. Mogensen and S. E. Smith. 1992. Heritable paternal cytoplasmic organelles in alfalfa sperm cells. Ultrastructural reconstruction and quantitative cytology. European J. Cell Biol., 59: 211-218.

Zhu, T., H. L. Mogensen and S. E. Smith. 1993. Quantitative, three-dimensional analysis of alfalfa egg cells in two genotypes. Implications for biparental plastid inheritance. Planta, 190: 143-150.

***Zhukova, G. Ya. 1963.** On qualitative analysis on pigment complex of green embryo plastids in *Vicia faba* L. Bot. Journal, 48(9): 1380-1392.

***Zhukova, G. Ya. 1965.** On qualitative analysis on pigment complex of green embryo plastids in *Vicia faba* L. Qualitative analysis of xantophylles. Bot. Journal, 50(11): 1601-1602.

***Zhukova, G. Ya. 1967.** On the qualitative composition of pigment complex of green embryo plastids in some angiosperms. Bot. Journal (St. Petersburg), 52(8): 1085-1095.

Zhukova, G. Ya. 1968. The occurrence of chlorophyll in embryo of angiosperms. Ph.D. thesis, Leningrad, Russia.

***Zhukova, G. Ya. 1971.** Electron microscopy data on the active state and morphological peculiarities of embryo chloroplasts in some Angiospermae, pp. 56-57. *In* Materials V All-Union Conference Plant Embryology. Shtiintsa, Kishinev, Moldova.

***Zhukova, G. Ya. 1972.** On the shape of embryo chloroplasts in some species of angiosperms and its significance. Bot. Journal (St. Petersburg), 57(3): 290-298.

***Zhukova, G. Ya. 1975a.** Plastid apparatus of embryo in chloro- and leucoembryophytes and its functional significance. Reports USSR Acad. Sci., 224(6): 1425-1427.

***Zhukova, G. Ya. 1975b.** Problem of origin and evolution of plastids in the view of data of plant embryology. Bot. Journal, 60(5): 713-738.

Zhukova, G. Ya. 1978. Ultrastructural aspects of formation of some storage substances in the embryo of angiosperms. Proc. Indian Natl. Sci. Acad., 44: 133-140.

Zhukova, G. Ya. 1979. Peculiarities of embryo plastid apparatus in chloroembryophytes and leucoembryophytes, pp. 104-119. *In* M. S. Yakovlev (ed.). Actual Problems of Embryology of Angiosperms. Nauka, Leningrad, Russia.

Zhukova, G. Ya. 1981. Embryogenesis of angiosperms: ultrastructural transformations of plastids. Acta Soc. Bot. Poloniae, 50(1-2): 303-306.

***Zhukova, G. Ya. 1986.** Ultrastructure of embryo sac in *Valeriana officinalis* L. after fertilization, pp. 91-92. *In* Abstracts IX All-Union Conf. Plant Embryology "Gametogenesis, Fertilization and Embryogenesis of Seed Plants, Ferns and Mosses". Shtiintsa, Kishinev, Moldova.

Zhukova, G. Ya. 1987. Family Polemoniaceae, pp. 206-209. *In* T. B. Batygina and M. S. Yakovlev (eds). Comparative Embryology of Flowering Plants. Davidiaceae—Asteraceae. Nauka, Leningrad, Russia.

Zhukova, G. Ya. 1988. Plastids of embryo in flowering plants. Ph.D. thesis, Leningrad, Russia.

Zhukova, G. Ya. 1990. On the correlation of endospermogenesis type with one of alternative features of embryogenesis, i.e., occurrence or absence of chlorophyll in flowering plant embryo, p. 196. *In* Abstracts XI Int. Symp. "Embryology and Seed Reproduction". July 3-7, 1990, Leningrad. Nauka, St. Petersburg, USSR.

Zhukova, G. Ya. 1992a. On the correlation of endospermogenesis type with one of alternative features of embryogenesis, i.e., occurrence or absence of chlorophyll in angiosperm embryo, pp. 634-635. *In* Proc. XI Int. Symp. "Embryology and Seed Reproduction", July 3-7, 1990, Leningrad. Nauka, St. Petersburg, USSR.

Zhukova, G. Ya. 1992b. Plastids of embryo in flowering plants. Abstracts Ph.D. thesis, Leningrad, Russia.

***Zhukova, G. Ya. 1993.** Plastids of gametes and embryo in flowering plants. Bot. Journal (St. Petersburg), 78(4): 18-35.

Zhukova, G. Ya. 1994. Egg cell, pp. 188-192. *In* T. B. Batygina (ed.). Embryology of Flowering Plants. Terminology and Concepts. 1. Generative Organs of Flower. Mir i Semya, St. Petersburg, Russia.

Zhukova, G. Ya. 1998. Occurrence of chlorophyll in the embryo. *In* Fabaceae of North Eurasia: Information system on disc, St. Petersburg, Russia. Departed in NTZ "Informregistr" 29.10.98, N 0229804660.

***Zhukova, G. Ya. 2000.** Occurrence of chlorophyll in the embryo of flowering plants and perspectives of its application in taxonomy. Bot. Journal (St. Petersburg), 85(6): 67-75.

***Zhukova, G. Ya. and M. S. Yakovlev. 1966.** Electron microscopy investigation of embryo chloroplasts in *Nelumbo nucifera* Gaertn. Bot. Journal (St. Petersburg), 51(11): 1542-1546.

***Zhukova, G. Ya. and G. I. Savina. 1978.** Electron microscopy investigation of embryo in *Epipactis atrorubens* (Hoffm.) Schult. (Orchidaceae) at the early stages of its development. Bot. Journal (St. Petersburg), 63(9): 1241-1246.

***Zhukova, G. Ya. and E. V. Dragunova. 1980.** Ultrastructure of chloroplasts in embryo cotyledons of *Acer platanoides* (Aceraceae). Bot. Journal (St. Petersburg), 65(10): 1404-1411.

Zimmer, E. A., R. K. Hamby, M. L. Arnold, D. A. LeBlanc and E. C. Theriot. 1989. Ribosomal RNA phylogenies and flowering plant evolution, pp. 205-214. *In* B. Fernholm, K. Bremer and H. Jornvall (eds). The Hierarchy of Life. Elsevier Science, Amsterdam, Netherlands.

Zinger, N. V. 1958. The seed, its development and physiological characteristics. Publ. USSR Acad. Sci., Moscow, Russia.

***Zinger, N. V. and T. P. Petrovskaya-Baranova. 1961.** The wall of pollen grain—a living physiologically active structure. Reports USSR Acad. Sci., 138: 466-469.

***Zubkus, L. P. 1950.** On the viability of haricot bean immature embryos on nutritive media. Agrobiologia, 1: 137-141.

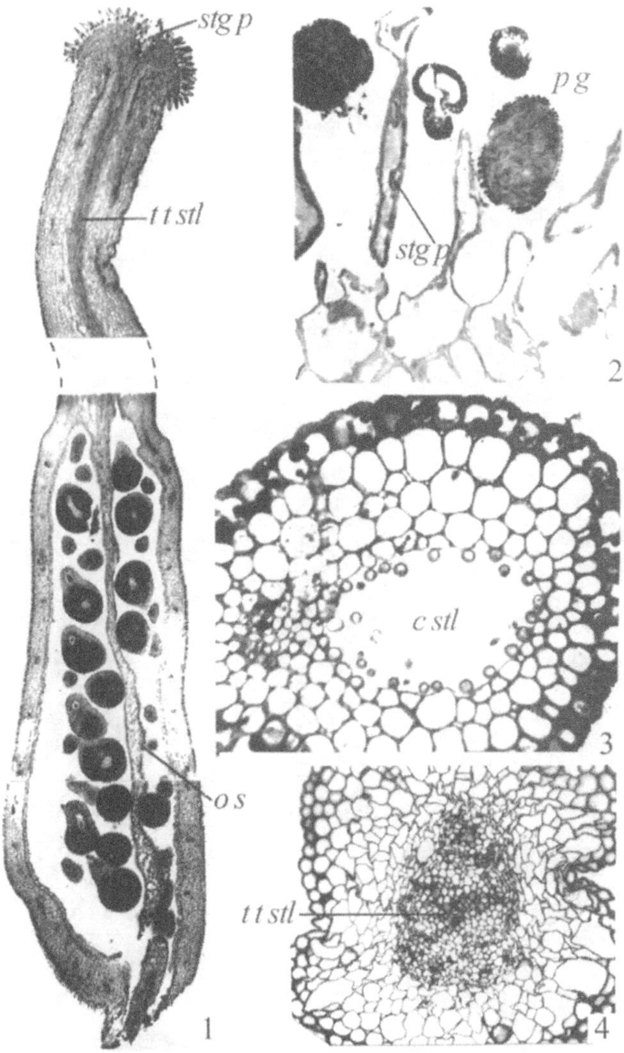

Plate I: Transmitting tract in pistil.
1, 2 -structure of pistil in *Brassica oleracea*; 3, 4—open in *Fifolium pratense* (3) and closed in *Yonicera turczaninowii* (4) types of style, transverse sections; 5-10—transmitting tract in *Medicago sativa*: 5—longitudinal section of pollen tube in secret of wet stigma, 6—transverse section of transmitting tissue in apical portion of style; 7—electron micrograph of the same part of style, 8—cross-section of the stylar channel, 9—longitudinal section through obturator and micropyle of ovule; 10—electron micrograph of obturator cells.
c c—channel cells, *c stl*—channel of style, *m*—micropyle, *ob*—obturator, *o s*—ovary septa, *p g*—pollen grain, *p t*—pollen tube, *stg p*—stigma papilla, *stg s*—stigma secret, *t t stl*—transmitting tissue of style.

1, 2, 4, 6, 7, 10—after Vishnyakova, orig.; 3—after Vishnyakova, 1991; 5, 8, 9—after Vishnyakova, 1986.

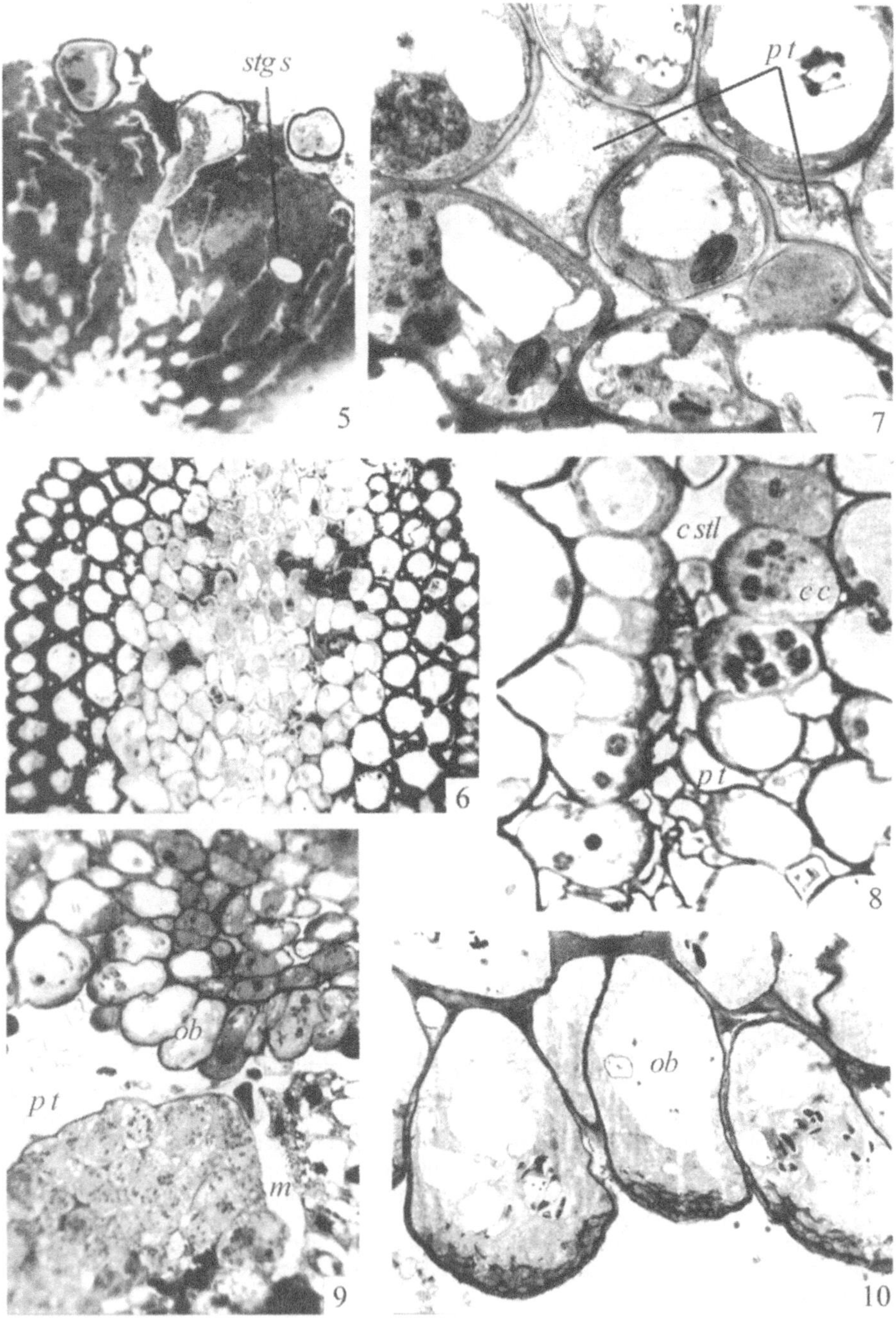

Plate I (*Contd.*)

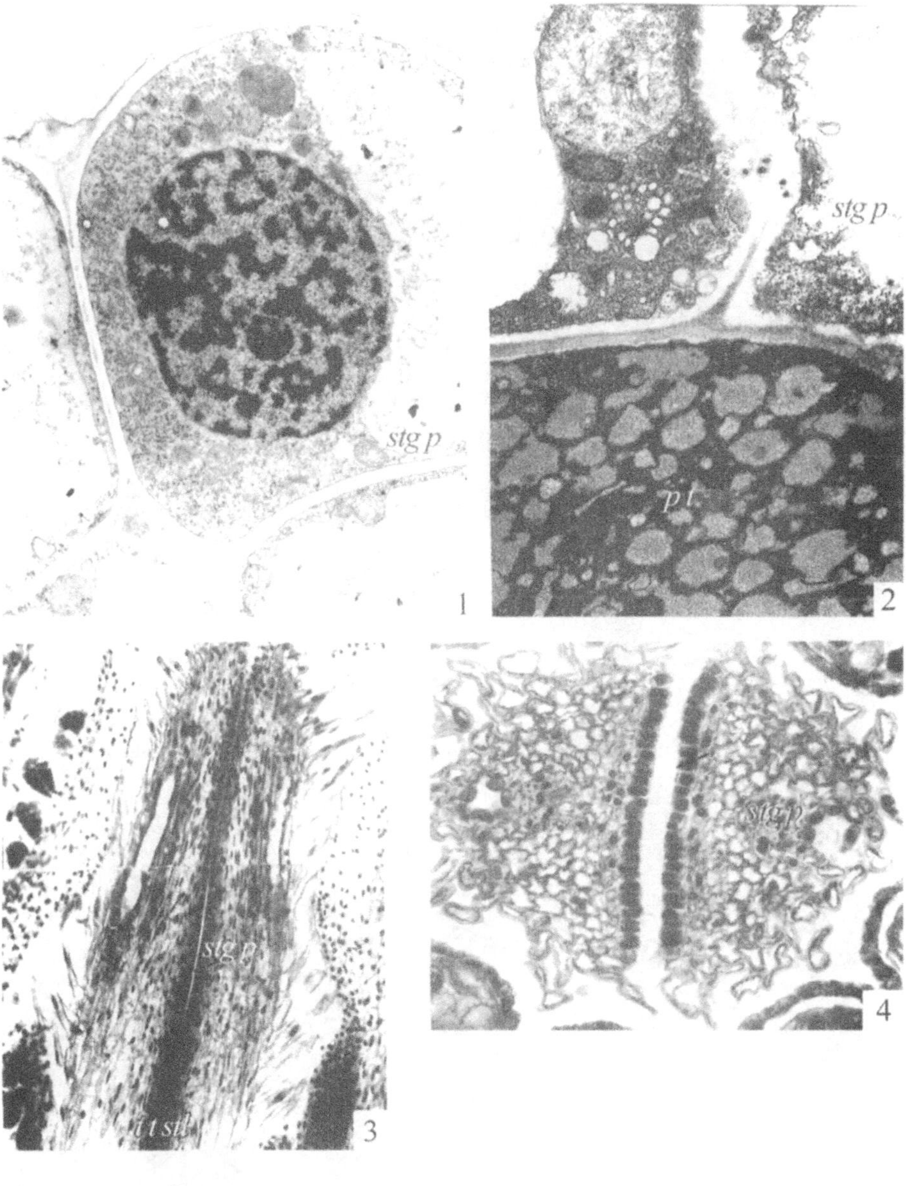

Plate II: Stigma.

1, 2—electron micrographs of transverse sections of stigma receptive hairs (papillae) in *Triticum aestivum* before (1) and after (2) pollination; 3, 4—longitudinal (3) and transverse (4) sections of stigma lobes in *Helianthus annuus* bud.

p t—pollen tube, *stg p*—stigma papilla, *t t stl*—transmitting tissue of style.

1, 2—after Vishnyakova and Willemse, 1994; 3, 4—after Vishnyakova, orig.

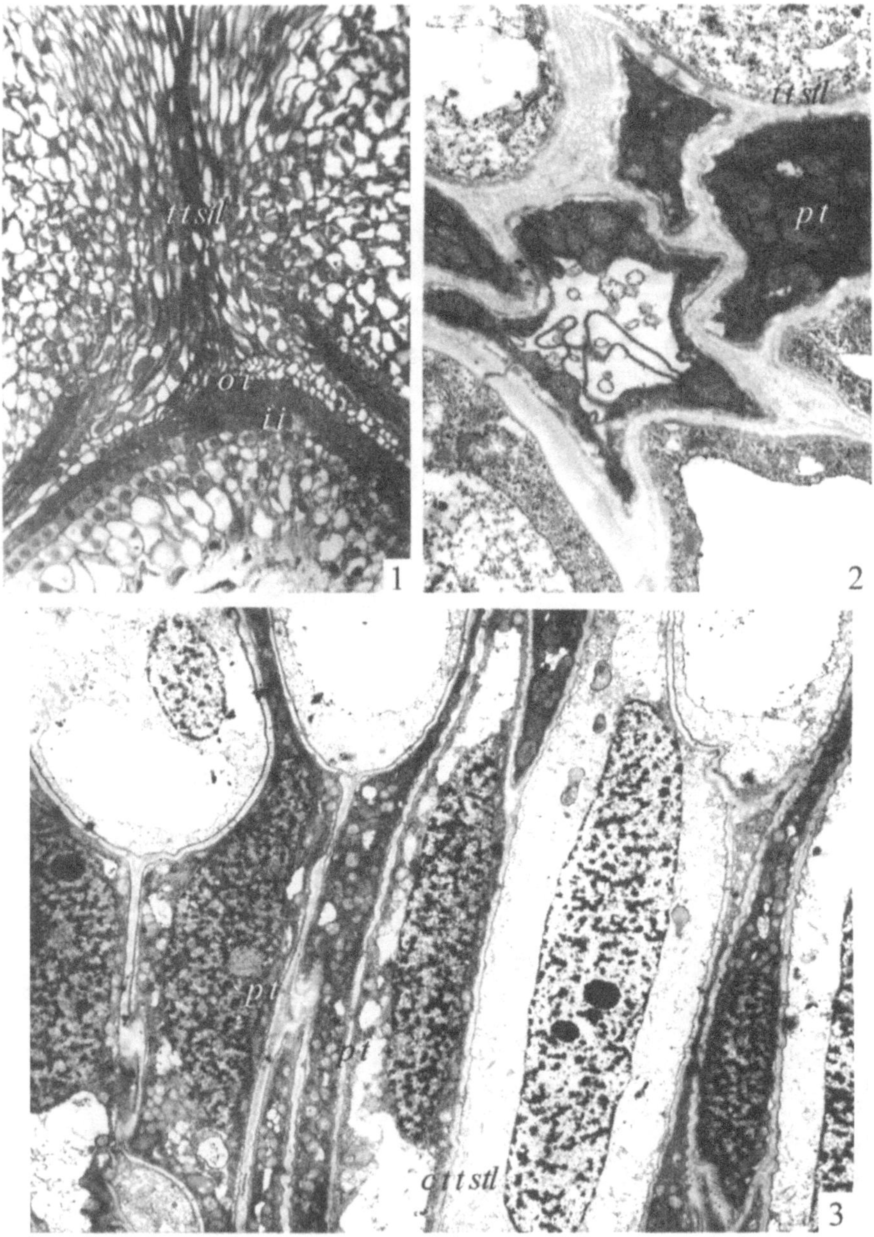

Plate III: Style.

1-3—style and apical portion of ovary with fragment of ovule (1), transmitting tissue of style with pollen tubes (2, 3) in *Triticum aestivum*.

c t t stl—cell of style transmitting tissue, *c stl*—channel of style, *i i*—inner integument, *o i*—outer integument, *p t*—pollen tube, *t t stl*—transmitting tissue of style.

1-3—after Vishnyakova and Willemse, 1994.

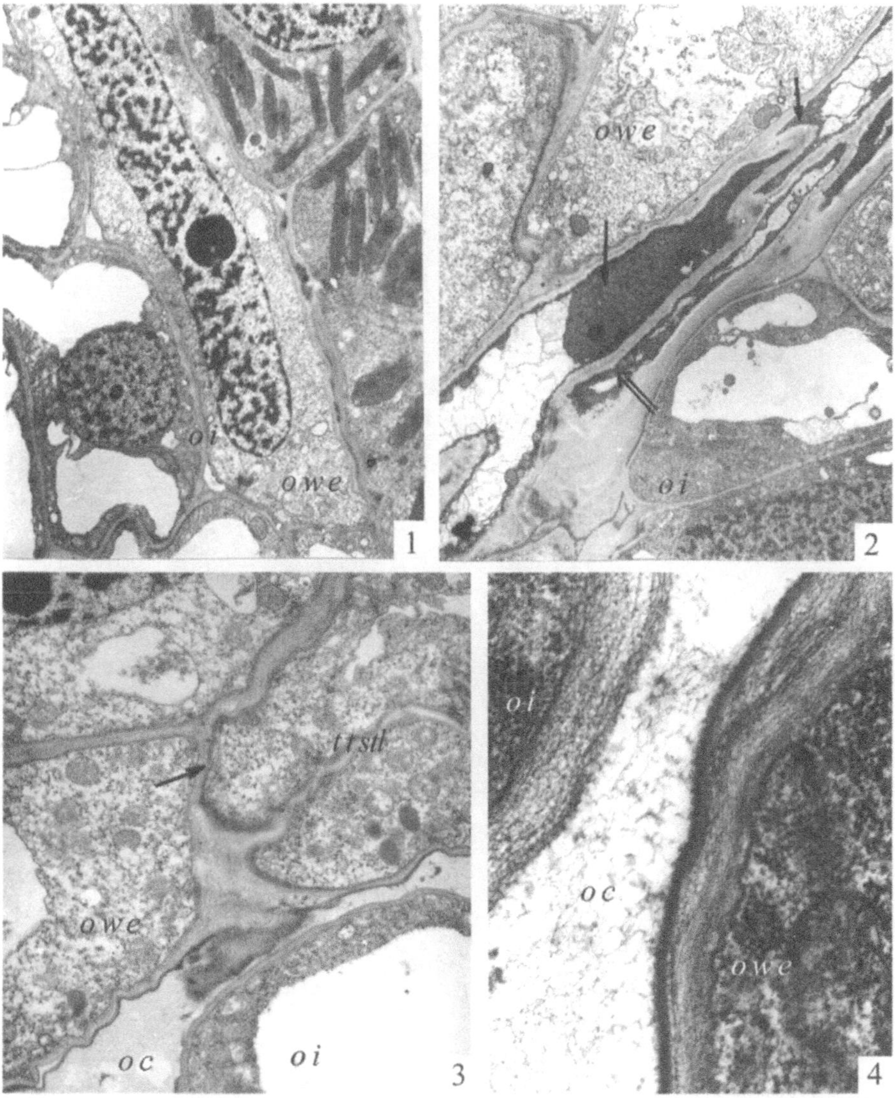

Plate IV: Ovary.

1-4—electron micrographs of ovary cross-sections in *Triticum aestivum*: 1—epidermis of ovary wall and ovule outer integument in unpollinated pistil; 2—growing (single arrows) and one stopping (double arrow) pollen tubes between ovary wall and ovule outer integument; 3—boundary of cells of style transmitting tissue with inner epidermis of ovary wall, plasmodesmata are shown by arrow, 4 –ovary cavity with fibrillar substance.

o c—ovary cavity, *o i*—outer integument, *o w e*—ovary wall epidermis, *t t stl*—transmitting tissue of style.

1-3—after Vishnyakova and Willemse, 1994; 4—after Vishnyakova, orig.

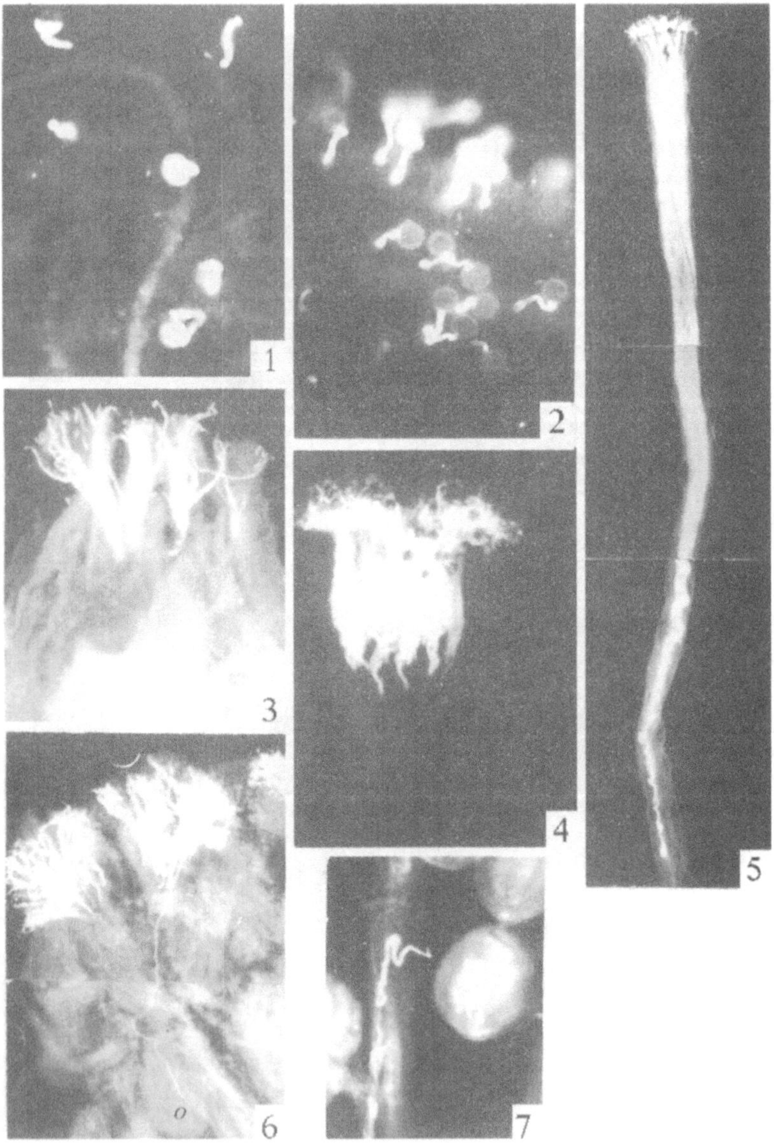

Plate V: Self-incompatibility.

1-7—gametophytic (1, 3-7) and sporophytic (2) self-incompatibility, manifesting in inhibition of pollen tube growth (luminescent microscopy): on stigma of *Hordeum bulbosum* (1, × 150) and *Helianthus annuus* (2, × 180), in style of *Viburnum opulus* (3, × 100), in the apical portion of *Cerasus maximowiczii* style (4, × 320), in basal portion of *C. concinna* style (5, × 100), in style of *Sambucus nigra* where growth of only one pollen tube is observed, as a result of which fertilization of single ovule occurs (6, × 70), in the ovary in front of micropyle of ovule in *Medicago sativa* (7, × 280).

o—ovule.

1-6—after Vishnyakova, orig.; 7—after Kazachkovskaya *et al.*, 1991.

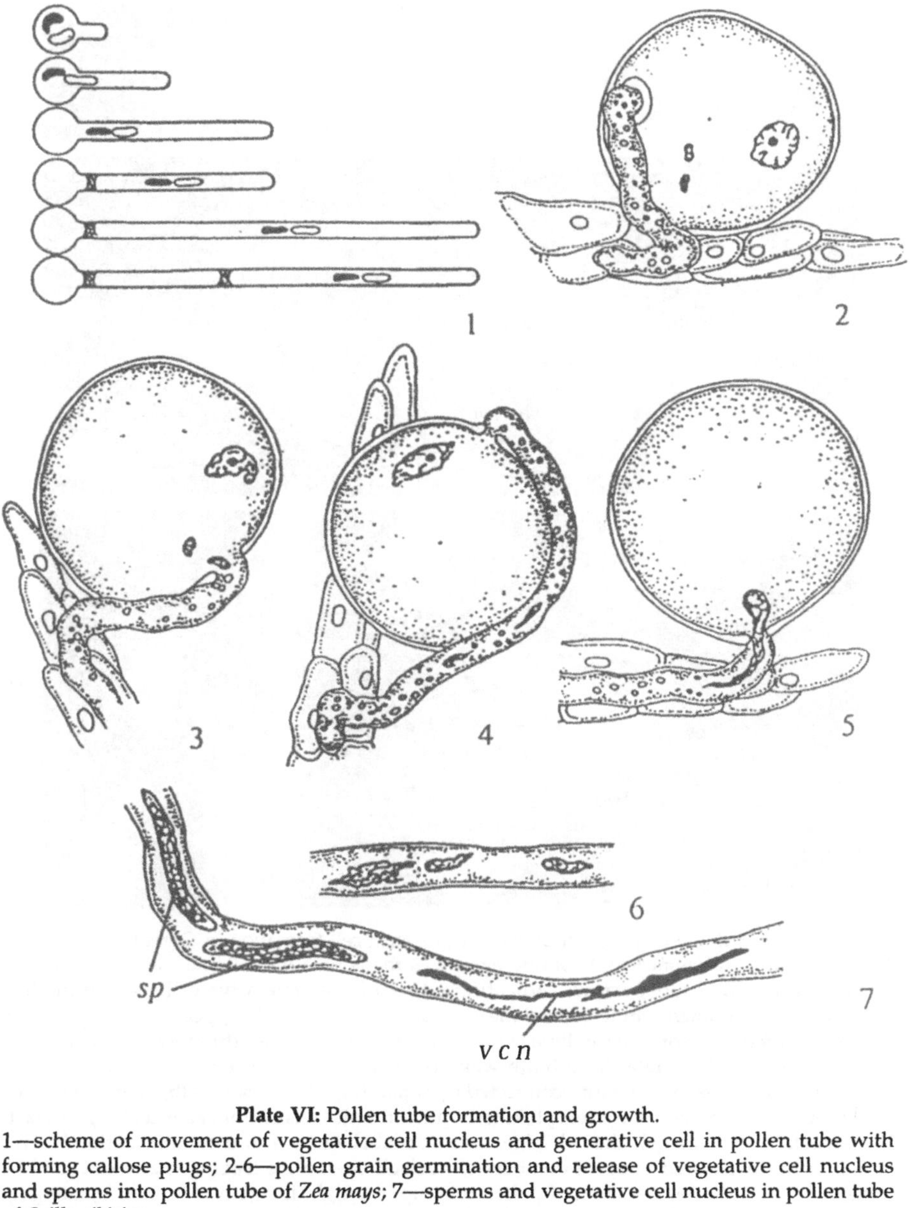

Plate VI: Pollen tube formation and growth.
1—scheme of movement of vegetative cell nucleus and generative cell in pollen tube with forming callose plugs; 2-6—pollen grain germination and release of vegetative cell nucleus and sperms into pollen tube of *Zea mays*; 7—sperms and vegetative cell nucleus in pollen tube of *Scilla sibirica*.
sp—sperms, *v c n*—vegetative cell nucleus.

1—after Korobova, 1977; 2-6—after Korobova, 1962; 7—after Korobova, 1979.

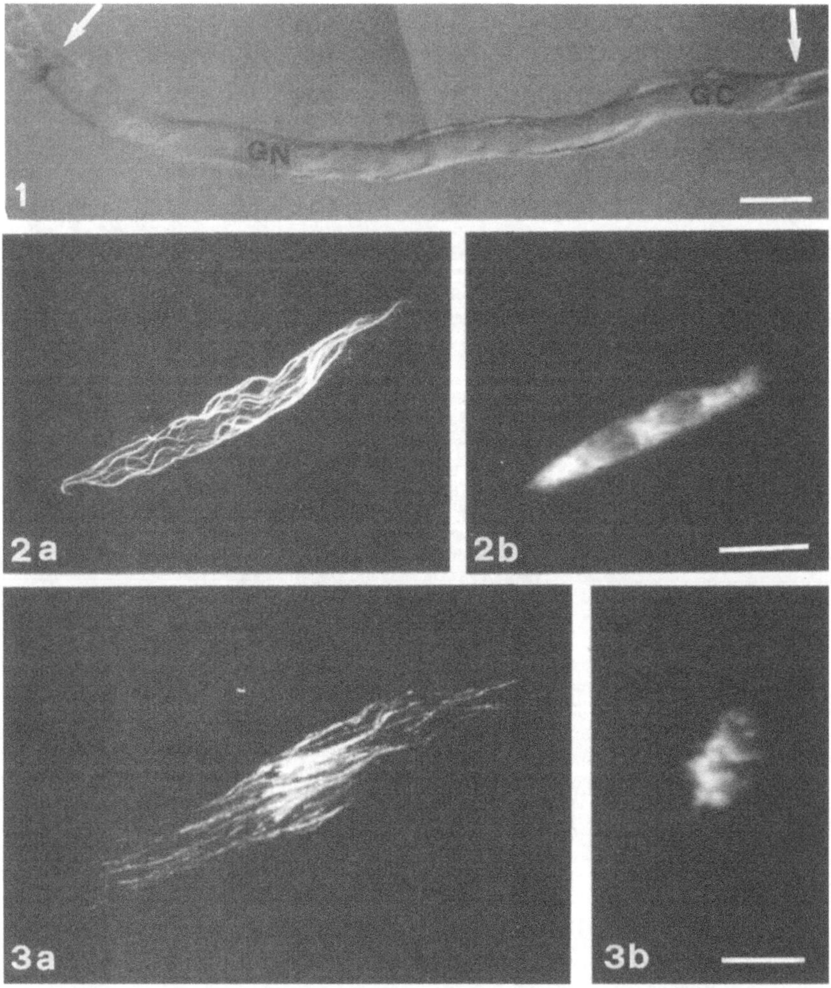

Plate VII: Generative cell division and sperm cell formation in *Gagea lutea*.
1—differential interference contrast micrograph of the generative cell and the generative cell nucleus in a pollen tube, arrows indicating edges of the generative cell; 2—spindle-like organization of the microtubules (a) and the generative cell nucleus (b) appeared at late prophase in a pollen tube; 3—short microtubule spindle (a) and chromosomes (b) showed at metaphase in a pollen tube, the metaphase plate was not labelled by anti-α-tubulin antibody; 4—microtubule bundles (a) were interconnected by separating chromosomes (b) at anaphase in a pollen tube; 5—in pollen tube after cytokinesis: microtubules were arranged in a phragmoplast-like structure (left image), the cell plate was not labelled by anti-α-tubulin antibody, two daughter nuclei showed unequal numbers of nucleoli, less in the leading sperm cell (middle image), arrow indicates direction of pollen tube tip, the right image was formed by merging the other two images showing the cell plate is close to the leading sperm cell.

Bar: 1, 2—20 μm, 3-5—10 μm.

G C—generative cell, *G N*—generative cell nucleus.

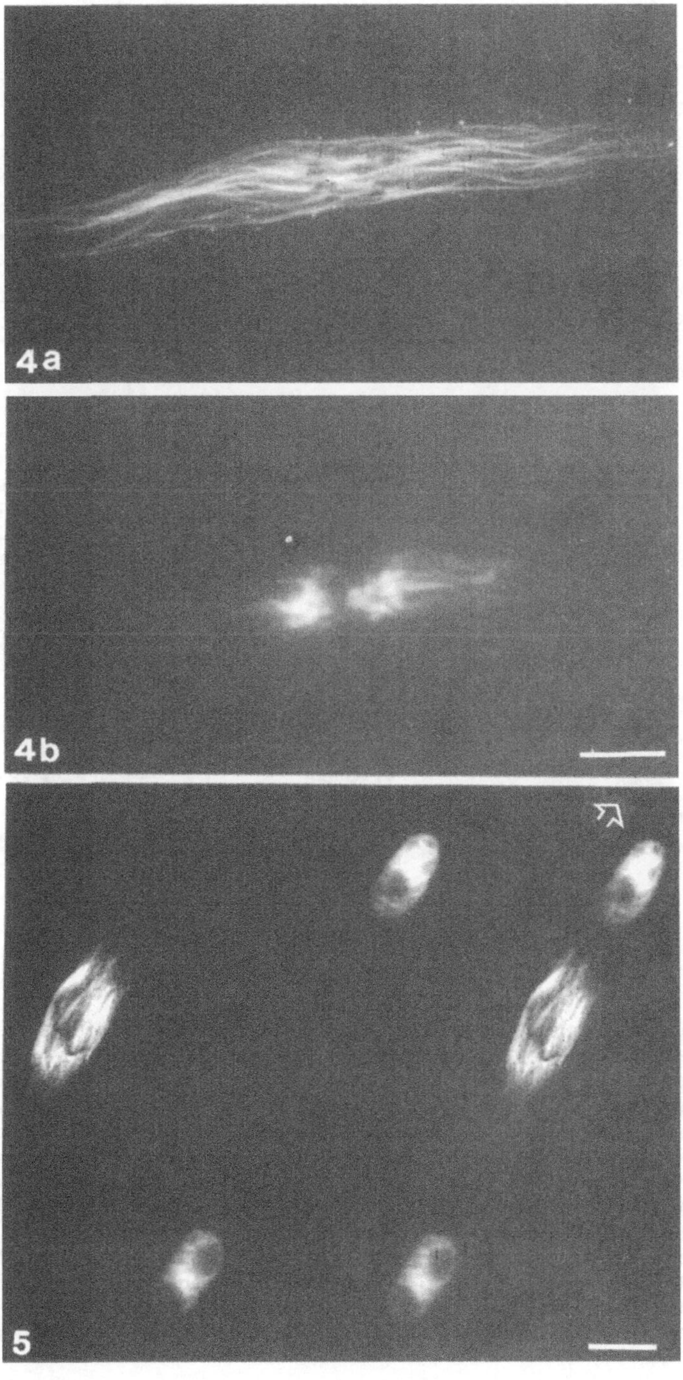

Plate VII (*Contd.*)

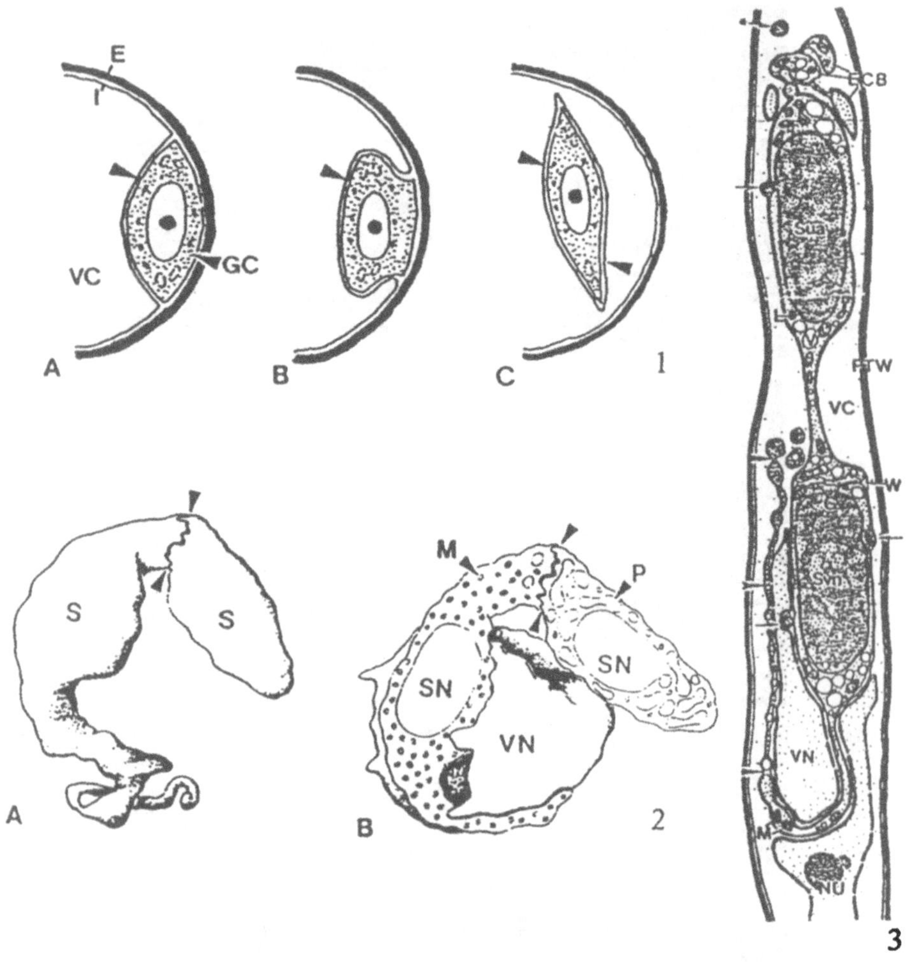

Plate VIII: Male germ unit.

1—formation and immersion of generative cell in pollen grain and origin of internal pollen plasma membrane: A—lens-shaped generative cell adjacent to vegetative cell in contact with inner pollen wall (intine), vegetative cell plasma membrane indicated by unlabelled arrow-head, B—generative cell separates from intine, vegetative cell plasma membrane (arrow-head) remains associated with cell, C—generative cell enters vegetative cell and becomes fully immersed in vegetative cytoplasm, plasma membrane of the vegetative cell (arrow-head) moves inward with it, forming inner pollen plasma membrane; 2—reconstruction of male germ unit of *Plumbago zeylanica* inside a pollen grain at anthesis: A—two sperm cells connected by common crosswall (unlabelled arrow-head), sperm cell to left associated with vegetative nucleus, B—sectional view of two sperm cells showing their cytoplasmic heterospermy, sperm cell associated with vegetative nucleus has majority of mitochondria, whereas other sperm cell contains majority of plastids; 3—reconstruction of male germ unit of *Nicotiana tabacum*; 4—cytokinesis and sperm cell separation in *Galium*: A—centripetal cell plate extension followed by

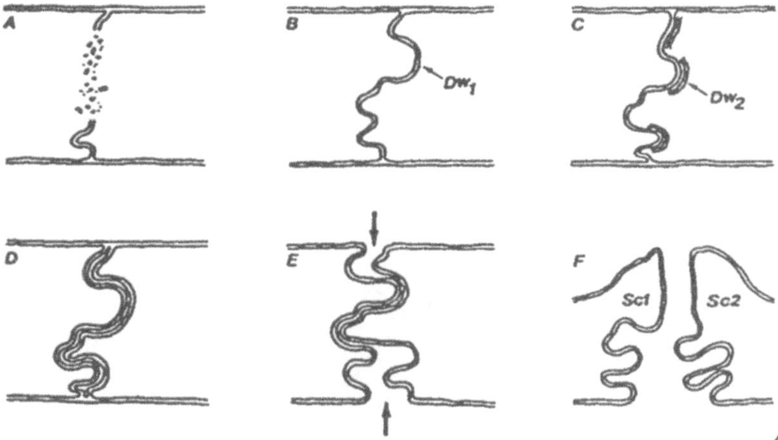

4

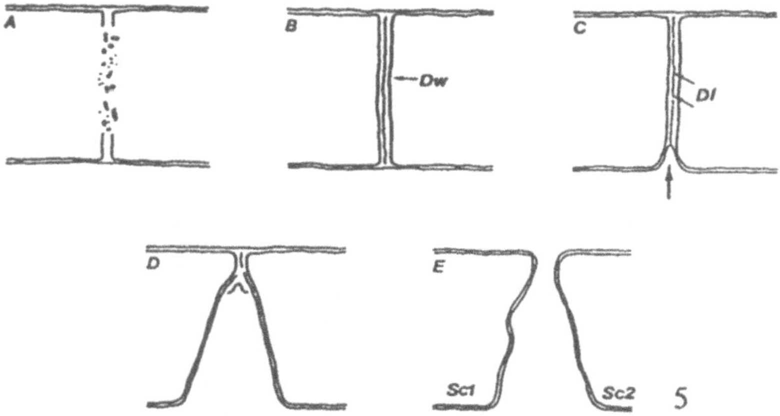

5

Plate VIII (*Contd.*)

(B) completion of first division wall, C—second division wall then forms and is completed (D), followed by (E) cell wall separation (arrows) from the edge to the centre and ultimately (F) causes separation of two sperm cells; 5—cytokinesis and sperm cell separation in *Trichodiadema* and *Avena*: A—centripetal cell plate extension followed by (B) completion of division wall, C—dense layer present at centre of wall is the site of cell wall separation (arrows), which (D) proceeds from one edge of cell to the other, ultimately (E) causes separation of the two sperm cells.

Dl—dense layer, *Dw*—division wall, Dw_1, Dw_2—first and second division walls, *E*—exine, *ECB*—enucleated cytoplasmic bodies, *G*—Golgi bodies, *GC*—generative cell, *I*—intine, *L*—lipid body, *M*—mitochondrion, *NU*—nucleolus, *P*—plastid, *PTW*—pollen tube wall, *S*—sperm cells, *Sc1*, *Sc2*—two sperm cells, *SN*—sperm nucleus, *Sua*—sperm cell unassociated with vegetative nucleus, *Svn*—sperm cell, associated with vegetative nucleus, *V*—vacuole, *VC*—vegetative cytoplasm, *VN*—vegetative nucleus, *W*—cell wall.

1—after Russell, 1994; 2—after Russell, 1984; 3—after Yu *et al.*, 1992; 4,5—after Weber, 1988a.

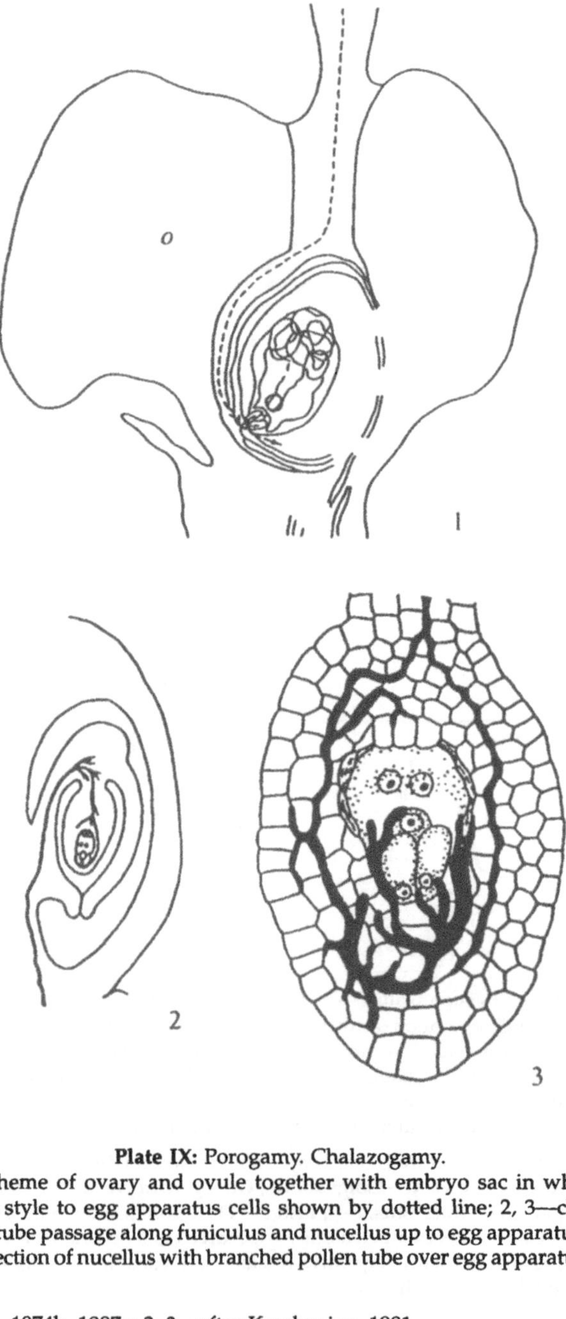

Plate IX: Porogamy. Chalazogamy.

1—porogamy: scheme of ovary and ovule together with embryo sac in wheat, pathway of pollen tube from style to egg apparatus cells shown by dotted line; 2, 3—chalazogamy: 2—scheme of pollen tube passage along funiculus and nucellus up to egg apparatus (*Betula ermani*), 3—longitudinal section of nucellus with branched pollen tube over egg apparatus (*Betula costata*). *o*—ovary.

1—after Batygina, 1974b, 1987a; 2, 3—after Korchagina, 1981.

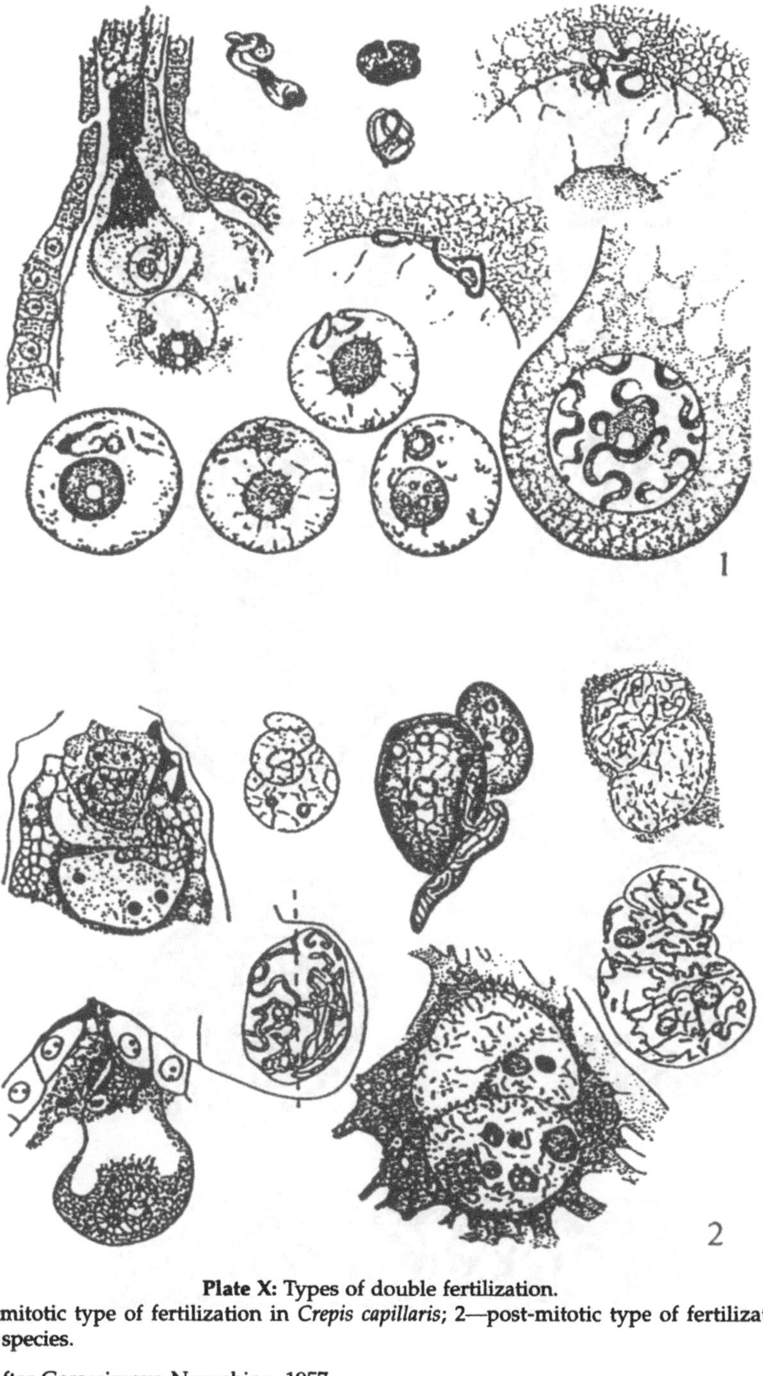

Plate X: Types of double fertilization.

1—premitotic type of fertilization in *Crepis capillaris*; 2—post-mitotic type of fertilization in *Lilium* species.

1, 2—after Gerassimova-Navashina, 1957.

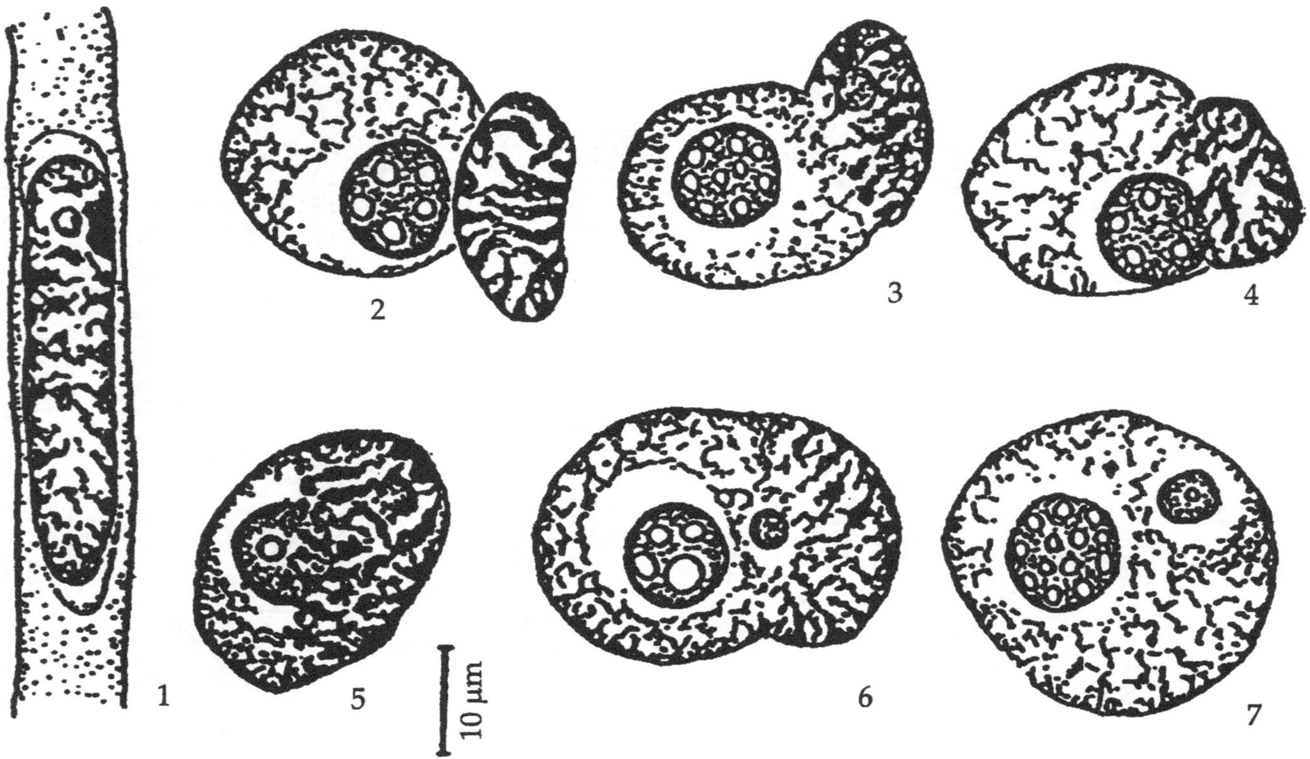

Plate XI: Syngamy in *Listera ovata*.

1—sperm cell in pollen tube; 2-7—fusion of sexual nuclei.

1-7—after Savina, 1978.

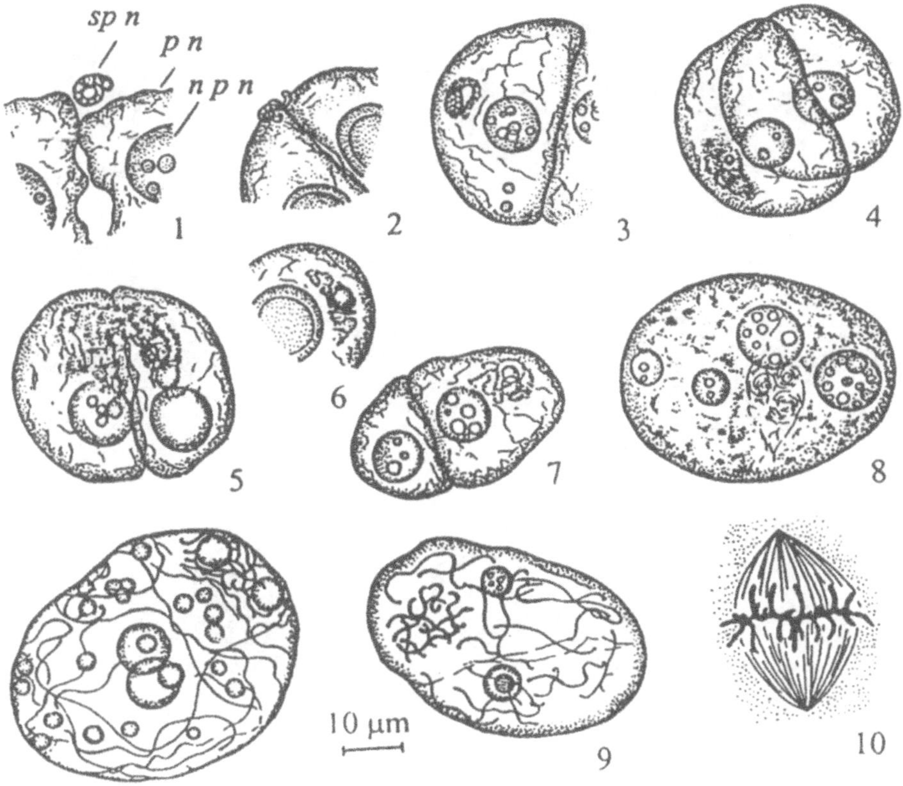

Plate XII. Triple fusion.

1-18—subsequent phases of triple fusion in cereals, beginning from the approach of sperm nucleus to nuclei of central cell up to their complete fusion and first division of primary endosperm nucleus: 1-2—sperm nucleus at border of two polar nuclei, 3,4,6,7—sperm nucleus unifies with one of polar nuclei, 5—sperm nucleus unifies with both polar nuclei at once, 8-10—nucleus of primary endosperm cell in interphase (8), prophase (9) and metaphase (10), 11-15—unification of sperm nucleus with one of polar nuclei; 16-18—division of primary endosperm cell nuclei in prophase, (1-5, 8, 9—*Triticum aestivum*, 6, 7, 10—*T. monococcum*, 11-18—*Zea mays*); 19-22—double fertilization and first phases of endosperm development in *Melampyrum nemorosum*: 19—nucleus of one of two sperms situated at egg cell nucleus, nucleus of second sperm at micropylar polar nucleus, 20—sperm nucleus immersion into micropylar polar nucleus, 21, 22—second nucleolus has distinguished in zygotic nucleus, unifying primary endosperm nucleus and chalazal polar nucleus passing prophase; 23-25—embryo sac of *Carya illinensis*: 23—polar nuclei in contact, 24—sperms have gone out of synergids, vegetative nucleus of pollen tube has remained in synergid, polar nuclei have fused, secondary nucleus of central cell has been formed, 25—gamete fusion; 26-32—subsequent phases of fusion of sperm nucleus with secondary nucleus of central cell in *Anemone nemorosa*, 32—first two endosperm nuclei;

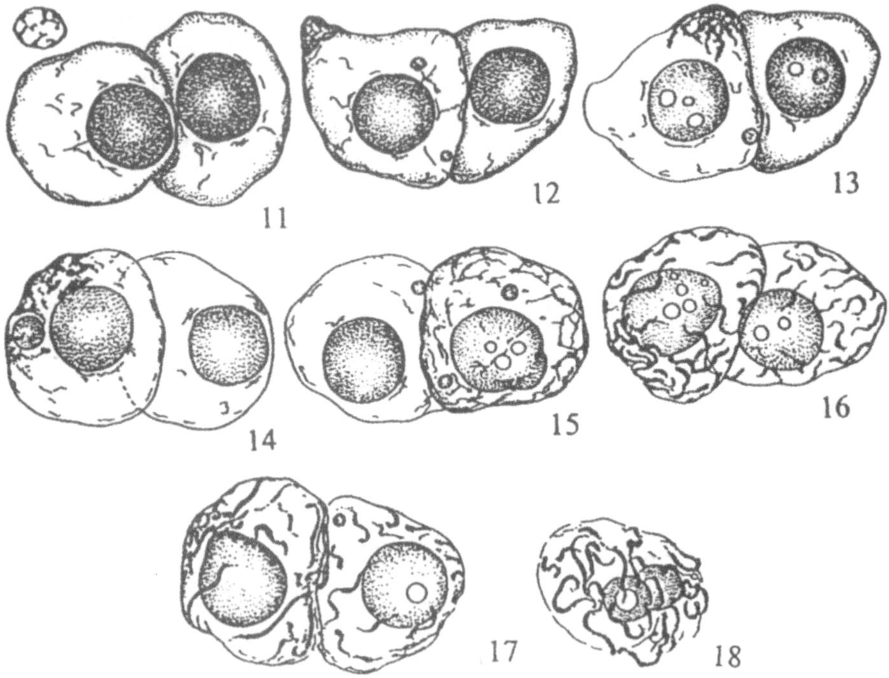

Plate XII (*Contd.*)

33-41—double fertilization in *Neottia nidus avis* (33-38) and *Gymnadenia conopsea* (39-41); 42-50—single fertilization in *Calanthe veitchii*: 42—pollen tube in embryo sac, one of synergids destroyed, 43, 44—sperms near egg cell and polar-antipodal nucleus group, 45—sperm nuclei near egg cell and one synergid nuclei, 46, 47—fusion of egg cell and sperm nuclei, nuclei of polar-antipodal group and sperm nucleus do not fuse with each other, moreover, in case 46 both synergids appeared not to be destroyed, because pollen tube passed near them laterally, 48, 49—zygotes, 50—nucleus of substantially increased zygote dividing, nuclei of polar-antipodal group remain unfused; 51—fertilization process in *Dactylorhiza romana* with disturbances—zygote dormant, primary endosperm nucleus formed, antipodes degenerating; 52-55—fertilization process demonstrating lack of triple fusion in *Trapa astrachanica* as a result of delay of one of sperms in synergid or in fissure between synergid and central cell; 56, 57—double fertilization in *Arachis hypogea*: 56—sperm nuclei have come to egg cell nucleus and polar nuclei, 57—syngamy and triple fusion.

ch p n—chalazal polar nucleus, *e c n*—egg cell nucleus, *e*—embryo, *m p n*—micropylar polar nucleus, *n p n*—nucleolus of polar nucleus, *n v c*—nucleus of vegetative cell, *p n*—polar nucleus; *p en n*—primary endosperm nucleus, *sp*—sperm, *sp n*—sperm nucleus; *s c c n*—secondary central cell nucleus, *sy n*—synergid nucleus, *v n*—vegetative nucleus.

1-10—after Batygina, 1987a; 11-18 —after Korobova, 1982; 19-22—after Yoffe, 1976; 23-25—after Solntseva and Dgebuadze, 1974; 26-32—after Yoffe, 1968; 33-41—after Savina and Poddubnaya-Arnoldi, 1990; 42-50—after Poddubnaya-Arnoldi, 1964a; 51—after Lagutova, orig; 52-54—after Titova, 1988; 55, 56—after Zamotailov, 1958.

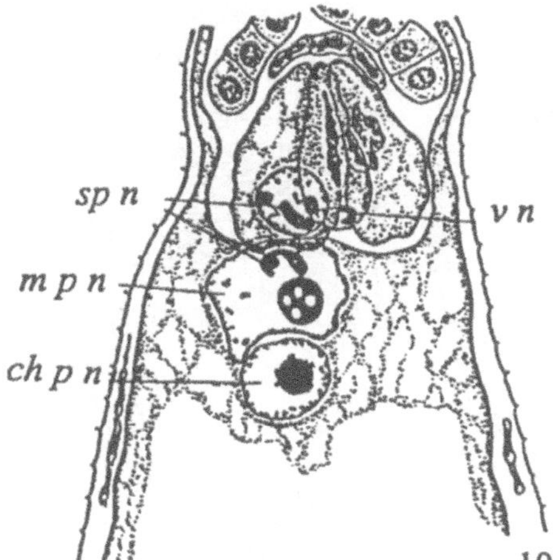

19

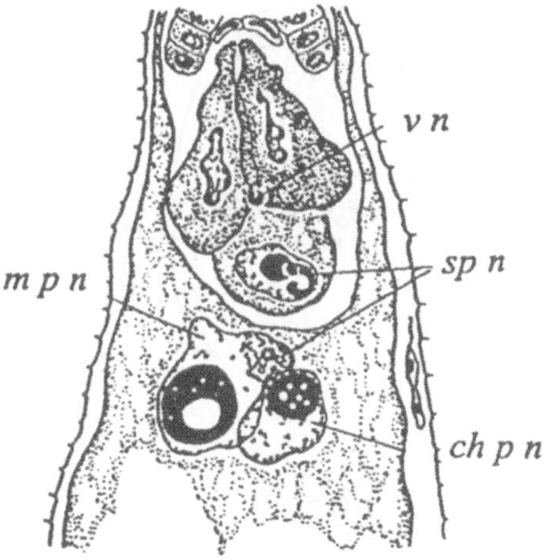

20

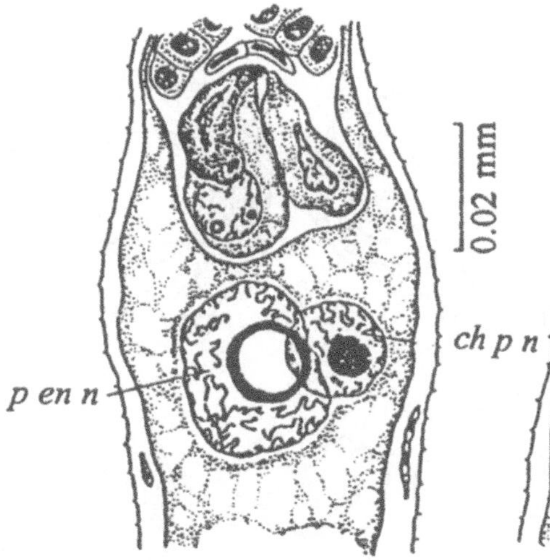

21

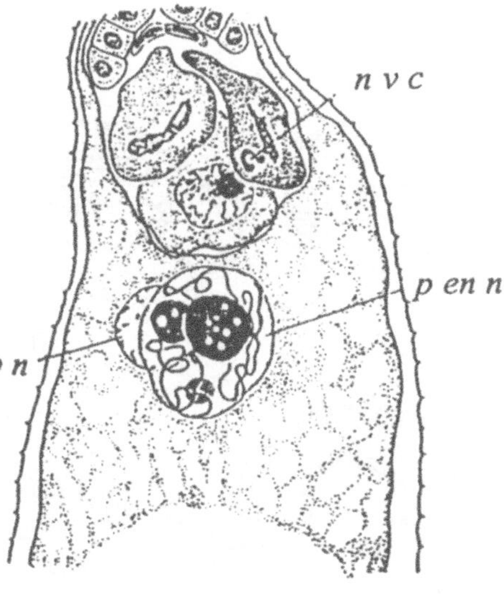

22

Plate XII (*Contd.*)

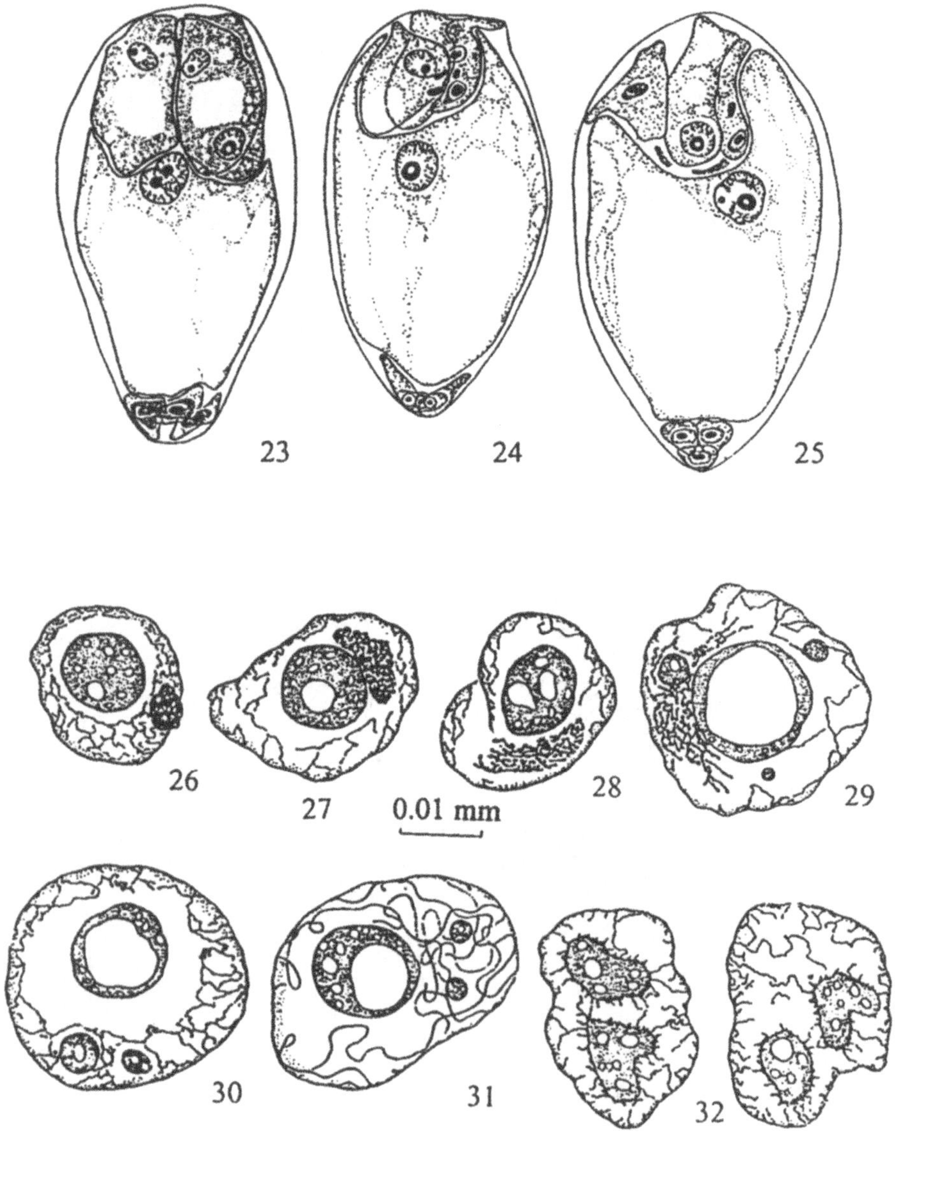

Plate XII (*Contd.*)

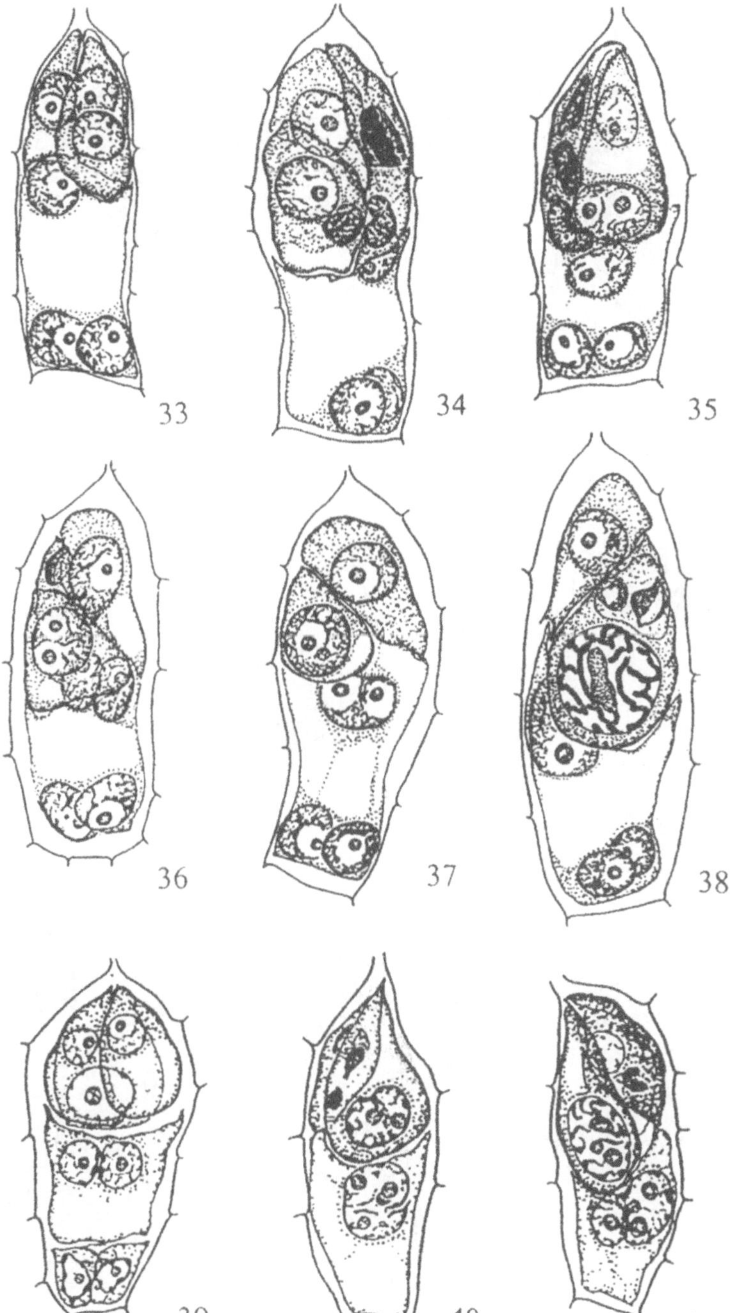

Plate XII (*Contd.*)

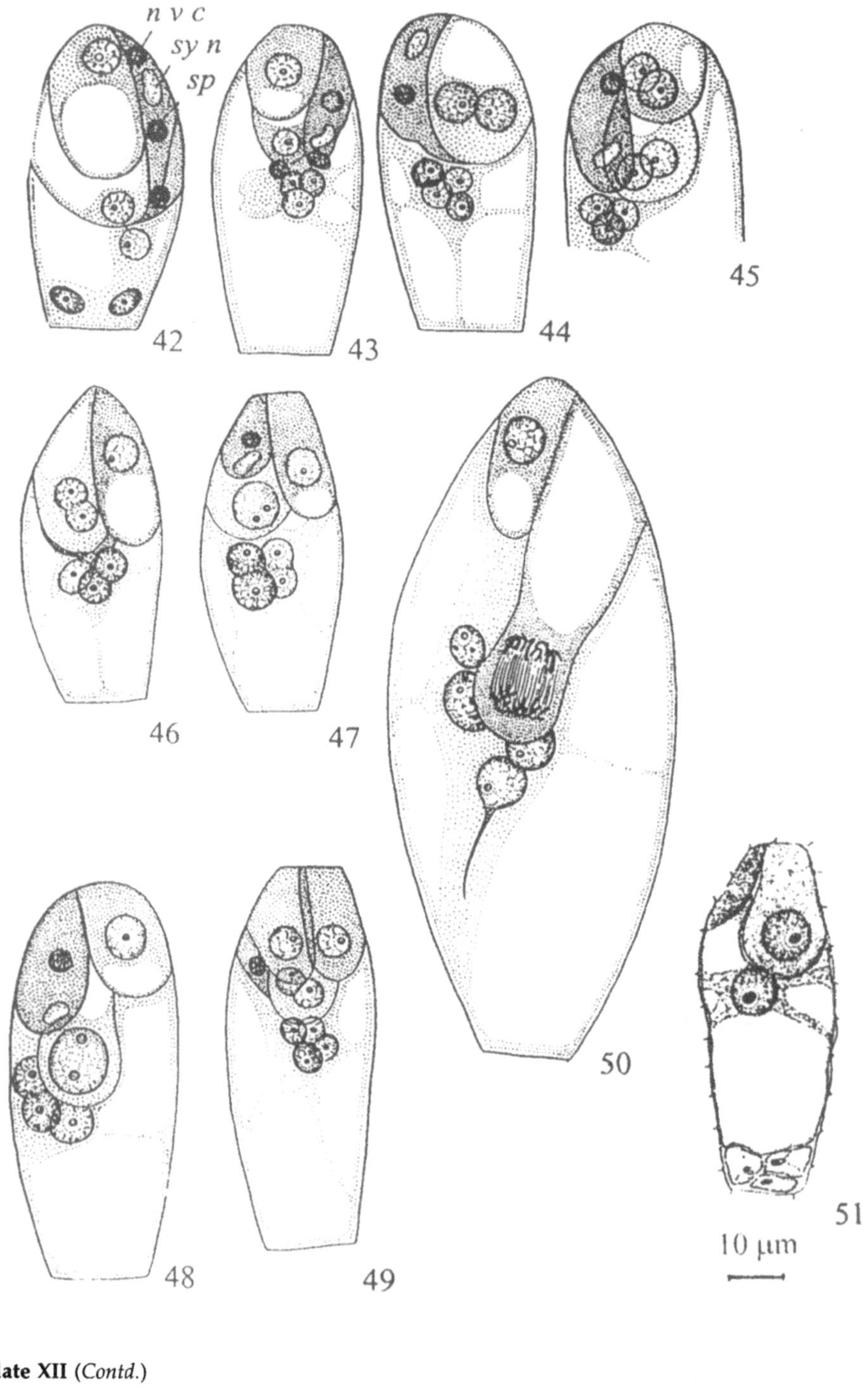

Plate XII (*Contd.*)

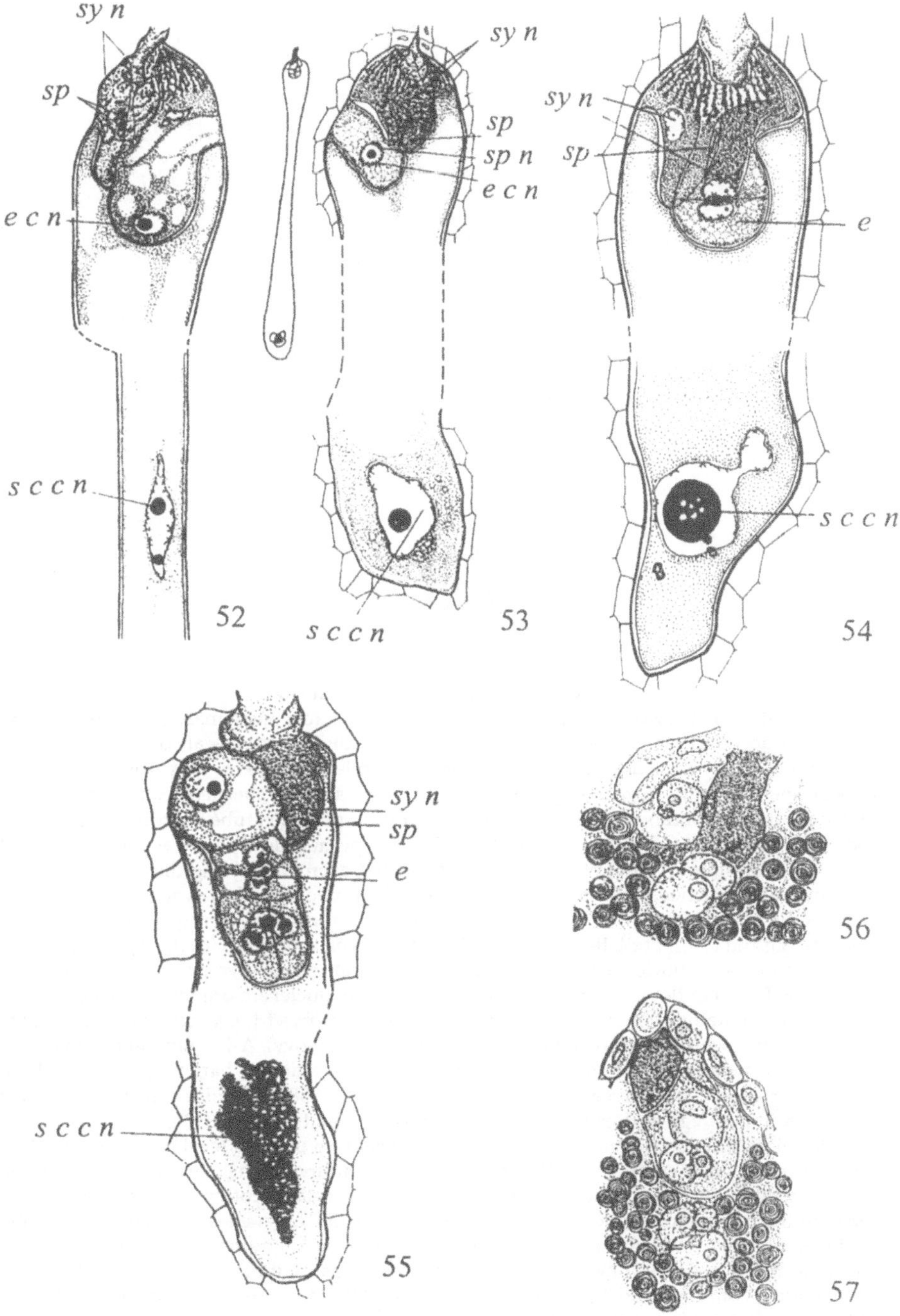

Plate XII (*Contd.*)

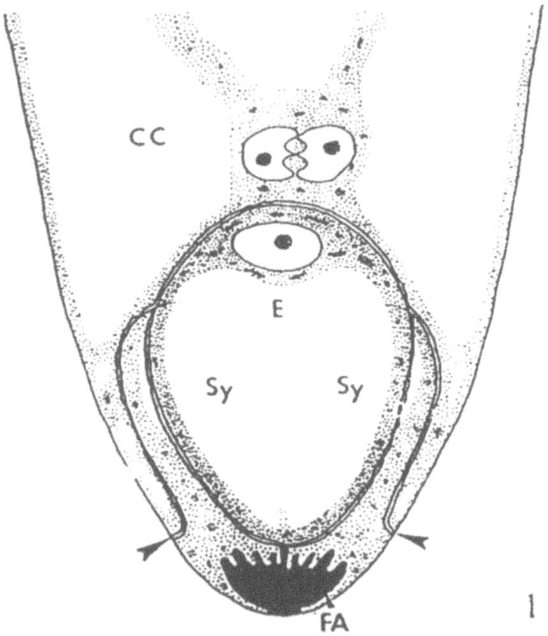

Plate XIII: Fertilization mechanisms.

1—representation of egg cell, central cell and two synergids, forming female germ unit—arrow-heads indicate attachments of synergids to embryo sac wall forming so-called synergid 'hooks' enclosing apical pockets of central cell cytoplasm; 2—release of sperm cells from pollen tube: A—one sperm associated with vegetative nucleus, whereas other sperm cell unassociated, B—pollen plasma membranes break down quickly after tip of pollen tube ruptures, releasing its contents and breaking the MGU association, C—two sperm cells separate as they are ejected from tube, forming cytoplasmic bodies; 3—postulated events involving discharge of pollen tube and its interactions within embryo sac: A—pollen tube arrival in synergid releases two sperm cells and vegetative nucleus which attach to actin coronas (arrow-heads) and move towards egg and central cell, B—migration of sperm cells on actin trails to site of gamete fusion, C—fusion of sperm nuclei with nuclei of egg and central cell; 4—five proposed mechanisms for gamete fusion in flowering plants, in all figures sperm nuclei are originally surrounded by a cytoplasm containing organelles (top, centre); darkened object the vegetative nucleus (egg cell located below in each drawing and central cell located above): A-D—uniparental maternal transmission patterns: A—unidirectional cytoplasmic restriction in barley, sperm cytoplasm excluded from egg, but not central cell (Mogensen, 1982, 1988), B—bidirectional cytoplasmic restriction in both egg and central cell in cotton (Jensen and Fisher, 1968b), C—bidirectional cytoplasmic restriction in spinach, but two sperm cells fuse prior to fertilization followed by dissolution of male cytoplasm (Wilms, 1981b), D—unidirectional cytoplasmic restriction accompanied by extreme vacuolization in unfused sperm cells of cottonwood; this appears to separate the cell from much of its surrounding cytoplasmic organelles (Russell *et al.*, 1990b), E—biparental transmission—essentially all of cytoplasm surrounding sperm cells enters egg and central cell in cases of cytoplasmic permissiveness, as examplified by *Petunia, Plumbago* and *Nicotiana* (van Went, 1970c; Russell, 1983; Yu *et al.*, 1994); 5—preferential double fertilization in *Plumbago* (a plant in which synergids are absent), cytoplasmic organelles of both sperm cells transmitted in this model: A—single membrane bound sperm cells deposited between egg and

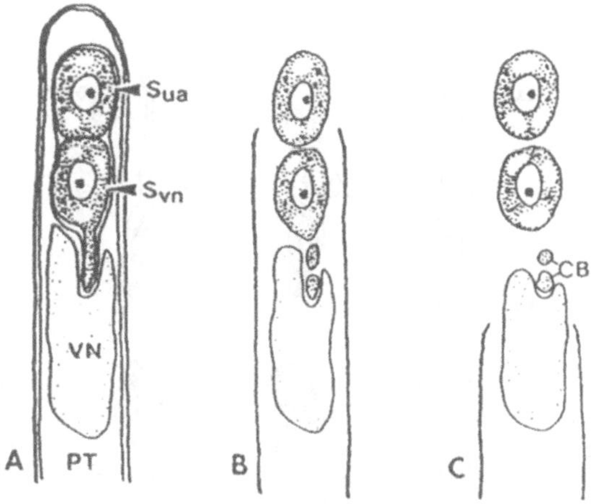

2

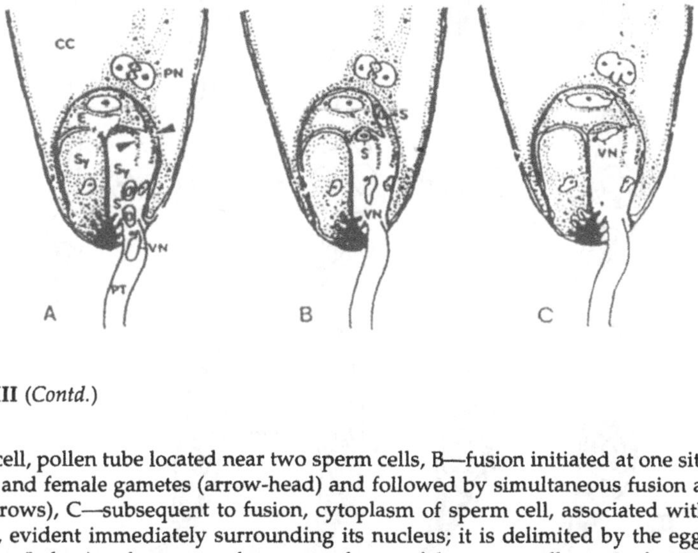

3

Plate XIII (*Contd.*)

central cell, pollen tube located near two sperm cells, B—fusion initiated at one site on surface of male and female gametes (arrow-head) and followed by simultaneous fusion at additional sites (arrows), C—subsequent to fusion, cytoplasm of sperm cell, associated with vegetative nucleus, evident immediately surrounding its nucleus; it is delimited by the egg-central cell boundary (below) and remnant plasma membrane of the sperm cell near its fusion point in the central cell (arrows); fusion membranes undergo progressive vesiculation, and disintegrate rapidly enough that they become difficult to detect several minutes after fusion (arrow-head); mitochondria of sperm cell surround sperm nucleus in central cell and sperm plastids transmitted into egg; presence of sperm plastids in egg cell indicates that sperm cell unassociated with vegetative nucleus preferentially fused with this cell.

CB—cytoplasmic body, *CC*—central cell, *E*—egg cell, *FA*—filiform apparatus, *Ms*—mitochondria of sperm cell, *Ps*—plastids of sperm cell, *PN*—polar nucleus, *PT*—pollen tube, *S*—sperm cell, *Sua*—sperm cell unassociated with vegetative nucleus, *Svn*—sperm cell associated with vegetative nucleus, *Sy*—synergid, *VN*—vegetative nucleus.

1—after Russell, 1993; 2, 3—after Russell, 1994; 4—after Russell *et al.*, 1990a; 5—after Russell, 1983.

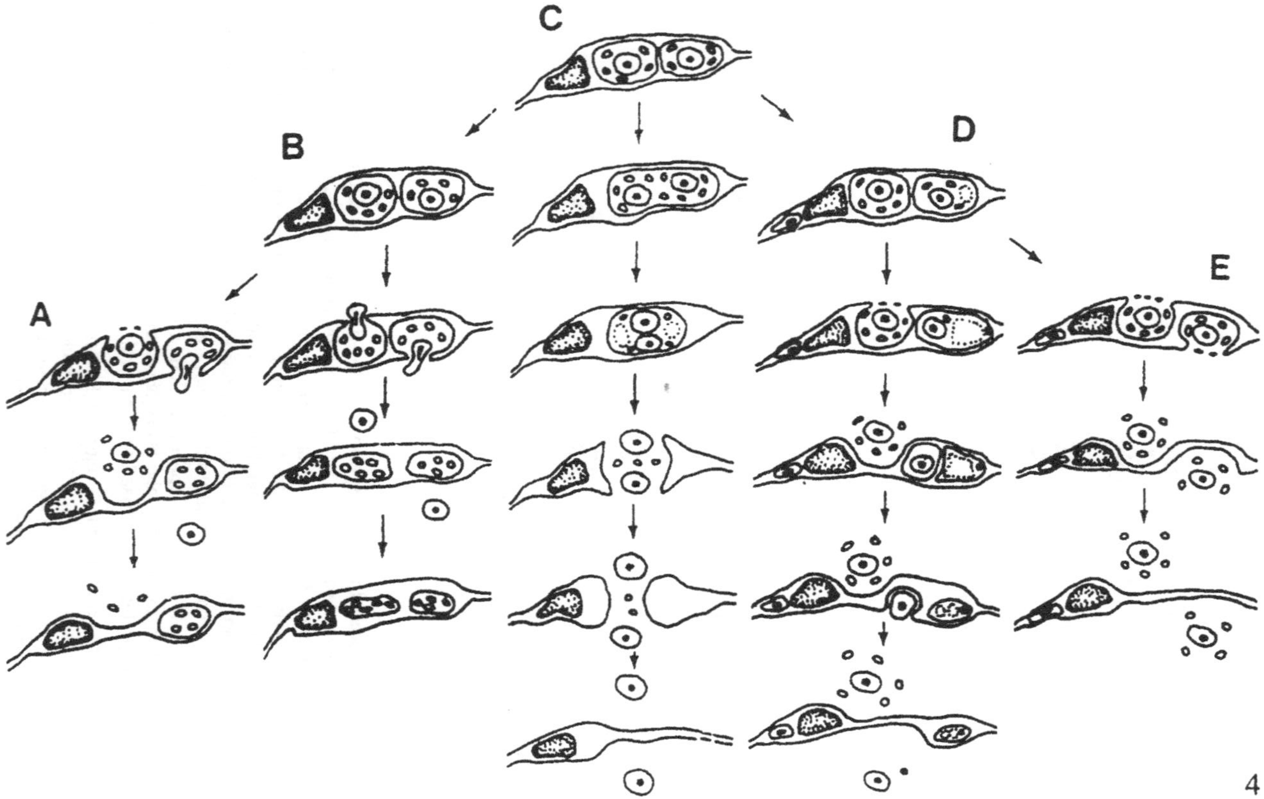

Plate XIII (*Contd.*)

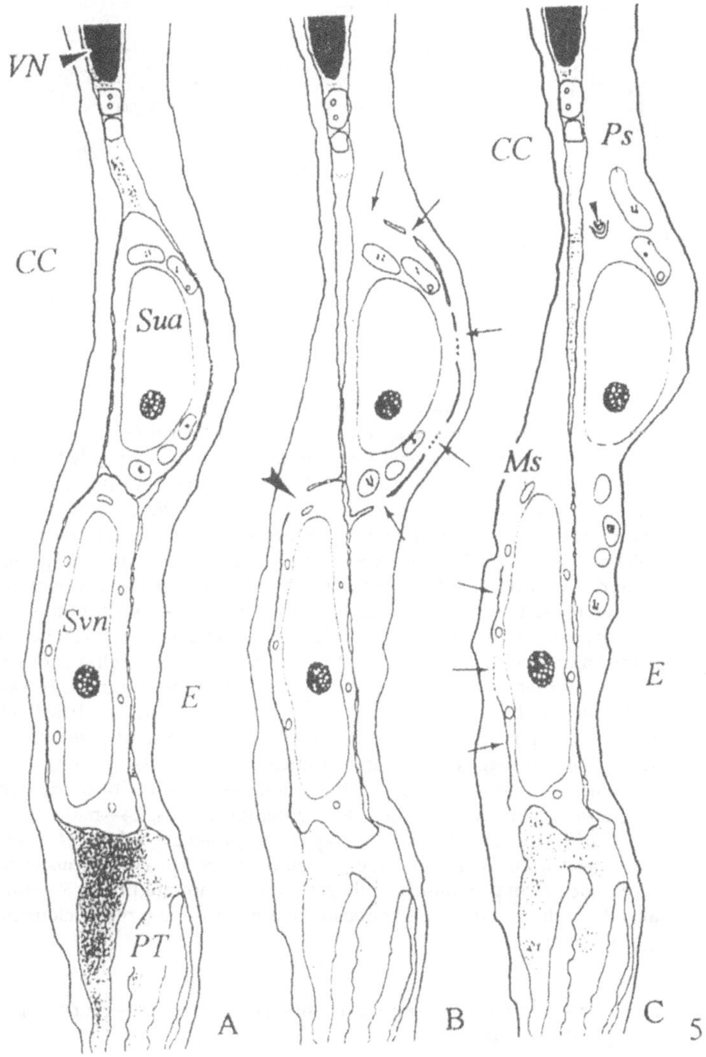

Plate XIII (*Contd.*)

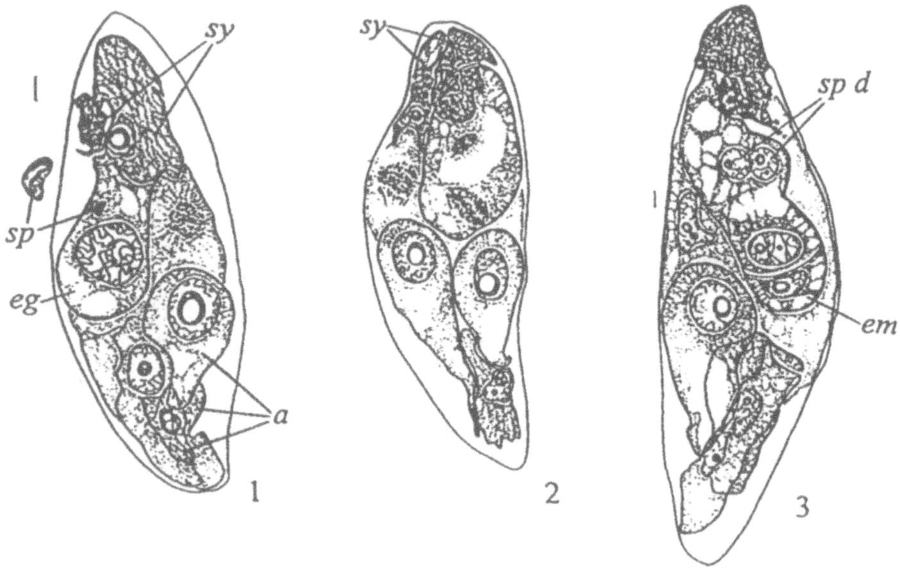

Plate XIV: Hemigamy.

1-17—*Rudbeckia laciniata*: 1—nuclei of egg cell (female) and sperm (male) in prophase, 2—synchronous figures of division in egg cell and sperm nuclei, 3—two-celled proembryo, in basal cell of which two small male nuclei observed, 4—synchronous division of female and male nuclei in basal cell, resulted in formation of 'hybrid' (triploid) nuclei, 5—creation of nuclei of different ploidy in proembryo cells (scheme), 6, 7—proembryo with nuclei of different origin and ploidy (7a—large triploid nucleus conspicuous), 8—nuclei (derivates of sperms) located in lateral part of embryo, 9—division of male nuclei with creation of restitutive nuclei of paternal origin, 10—restitutive nucleus formed during zygote division, 11—synchronous divisions of male and female nuclei in basal cell of proembryo, 12—triploid nucleus in apical part of embryo, 13—male nuclei and micronuclei in basal cells of proembryo, 14—egg cell nucleus and sperm nucleus in prophase, 15—metaphase of division of egg cell and sperm nuclei, 16—proembryo with nuclei of different ploidy: haploid (small, derivatives of sperm)—in basal cell, triploid (large)—in cell of central part, 17—more advanced in development proembryo; 18-28—*Zephyranthes macrosiphon*: 18—egg cell with sperm nucleus close to its nucleus, 19—anaphase of sperm nucleus division in basal cell of proembryo, 20-22—egg cell with sperm in basal part, 23—telophase of sperm nucleus division in one of proembryo cells, 24—male nuclei in basal part of proembryo, 25-27—male nuclei in various parts of proembryo, 28—restitutive nucleus formed from sperm derivatives in apical cell of proembryo; 29-30—*Habranthus andersonii* var. *texanus*: 29—egg cell nucleus in prophase, small sperm nucleus in interphase, 30—multicellular embryo, in basal cell of which 4 small male nuclei situated close to large nucleus of maternal origin; (14-19 light micrographs).

Bar 10 μm.

a—antipodal cells, *em*—embryo, *eg*—egg cell, *sy*—synergids, *sp*—sperm, *sp d*—sperm derivatives, *t n*—triploid nucleus.

1-3, 8, 11, 13—after Solntseva, 1974; 4, 6, 7—after Solntseva, 1971; 5, 12, 16—after Solntseva, 1973a,b,c; 9, 10, 18, 19, 23, 24—after Solntseva, 1976; 14, 15—after Solntseva, 1979; 17, 30—after Solntseva, orig.; 20-22, 25-28—after Solntseva and Vorsobina, 1972; 29—modified from Vorsobina and Solntseva, 1990.

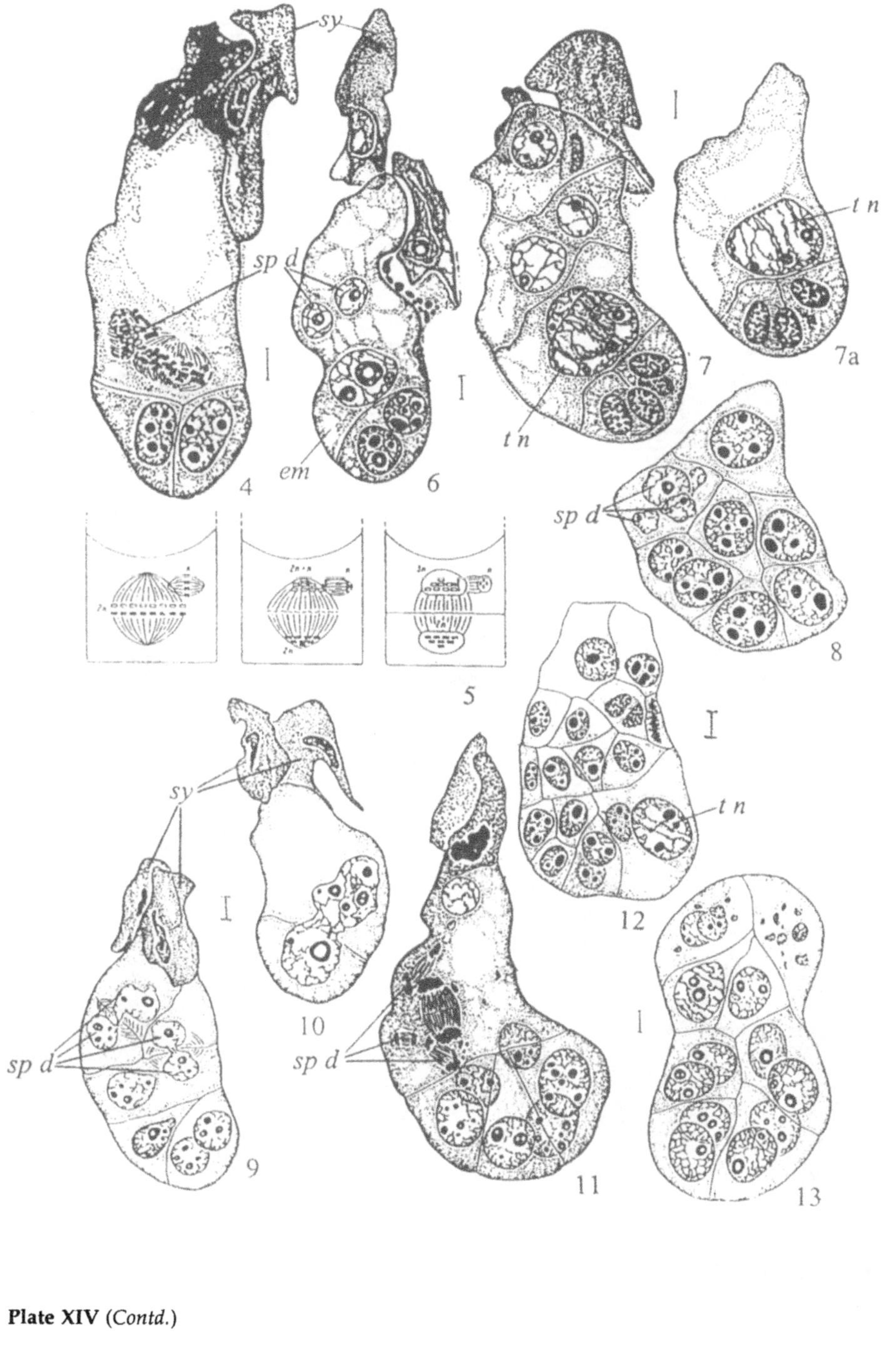

Plate XIV (*Contd.*)

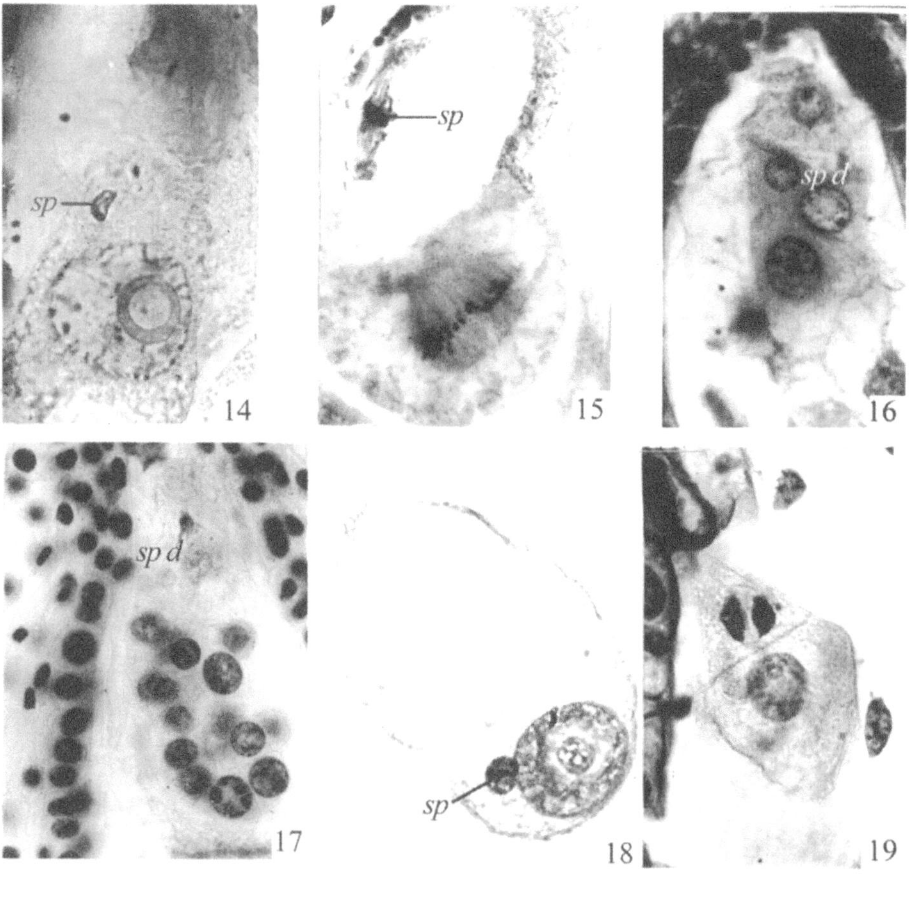

Plate XIV (*Contd.*)

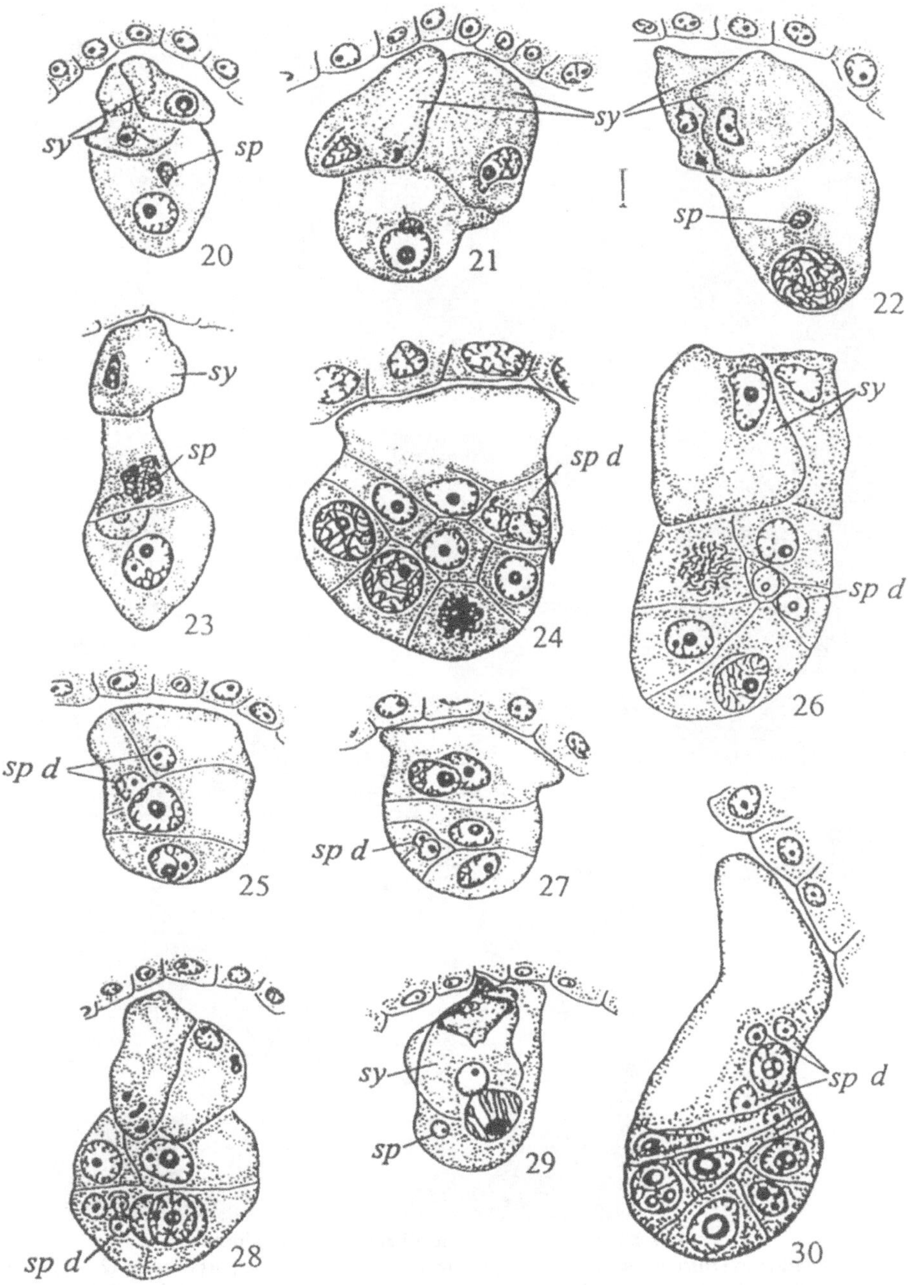

Plate XIV (*Contd.*)

Plate XV: Pollination and fertilization in Gramineae.

1-9 pollination in *Friticum aestivum:* 1-3—opening of flowers in middle part of spike (2—separate flower with pistil and stamens); 4,5—throwing out of stamens and opening of anthers; 6,7—pollination, 8—stigmatic part, pollen with germinating pollen grains can be seen; 9—lodicule; 10-11—fertilization process in *Zea mays*: 10—micropylar part of embryo sac in 19 h 30 min after pollination, sperm nuclei near female nuclei, 11—same in 23 h, fusion of sperm nucleus with one of polar nuclei

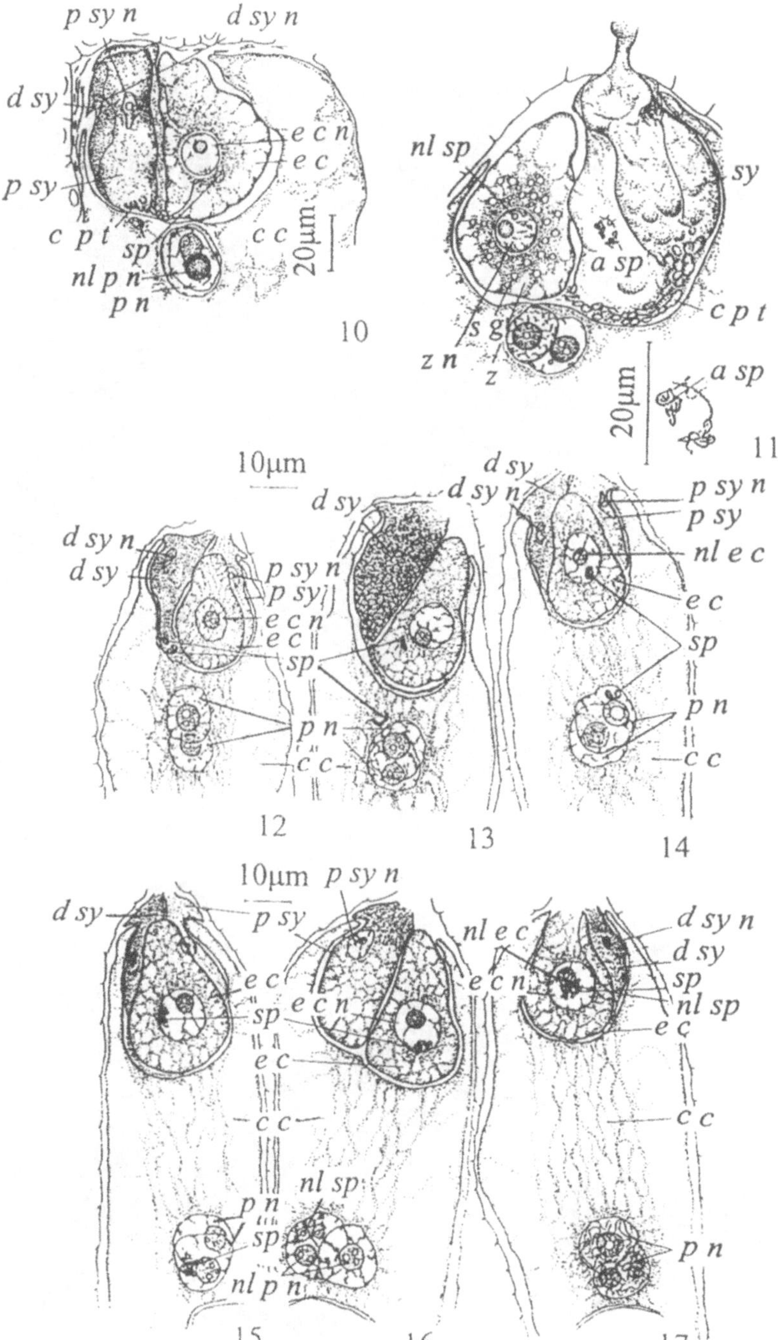

Plate XV (*Contd.*)

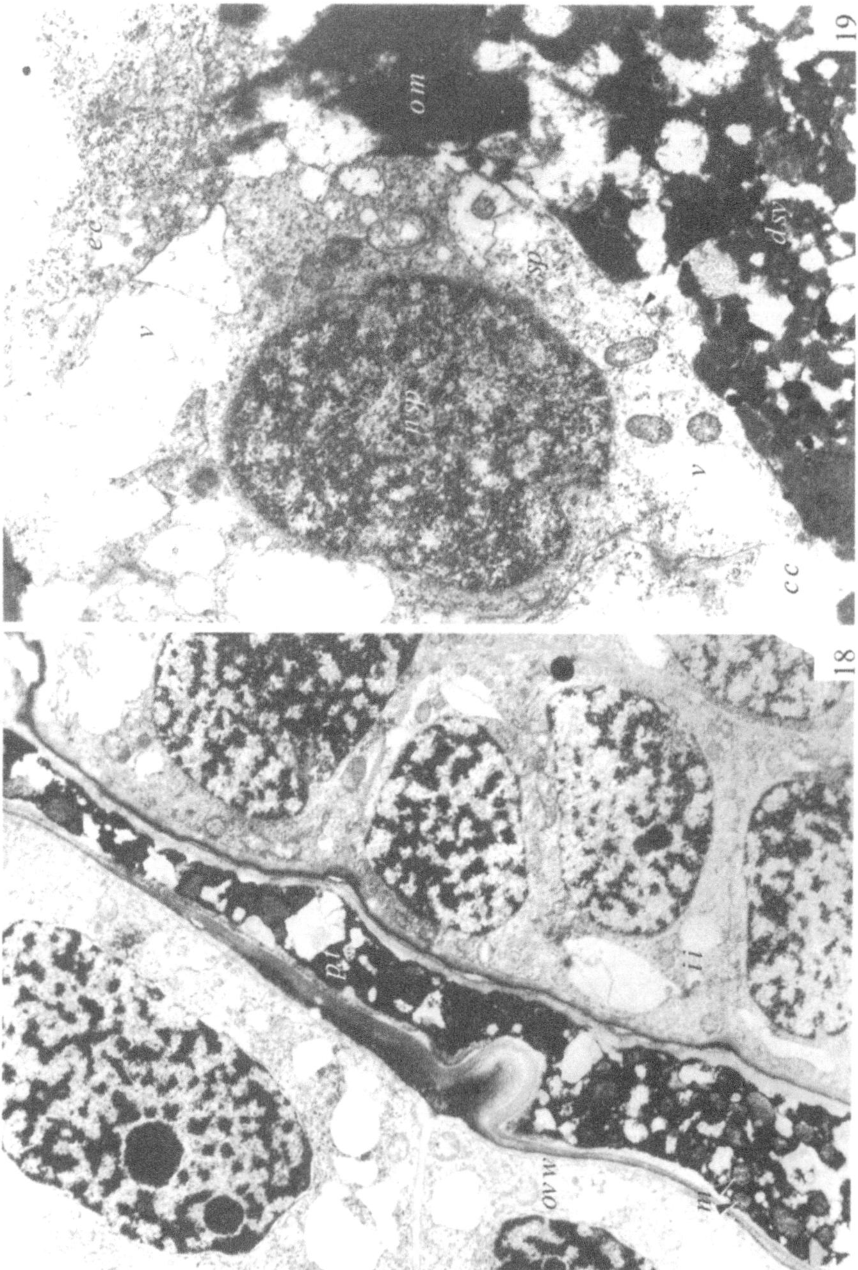

Plate XV (*Contd.*)

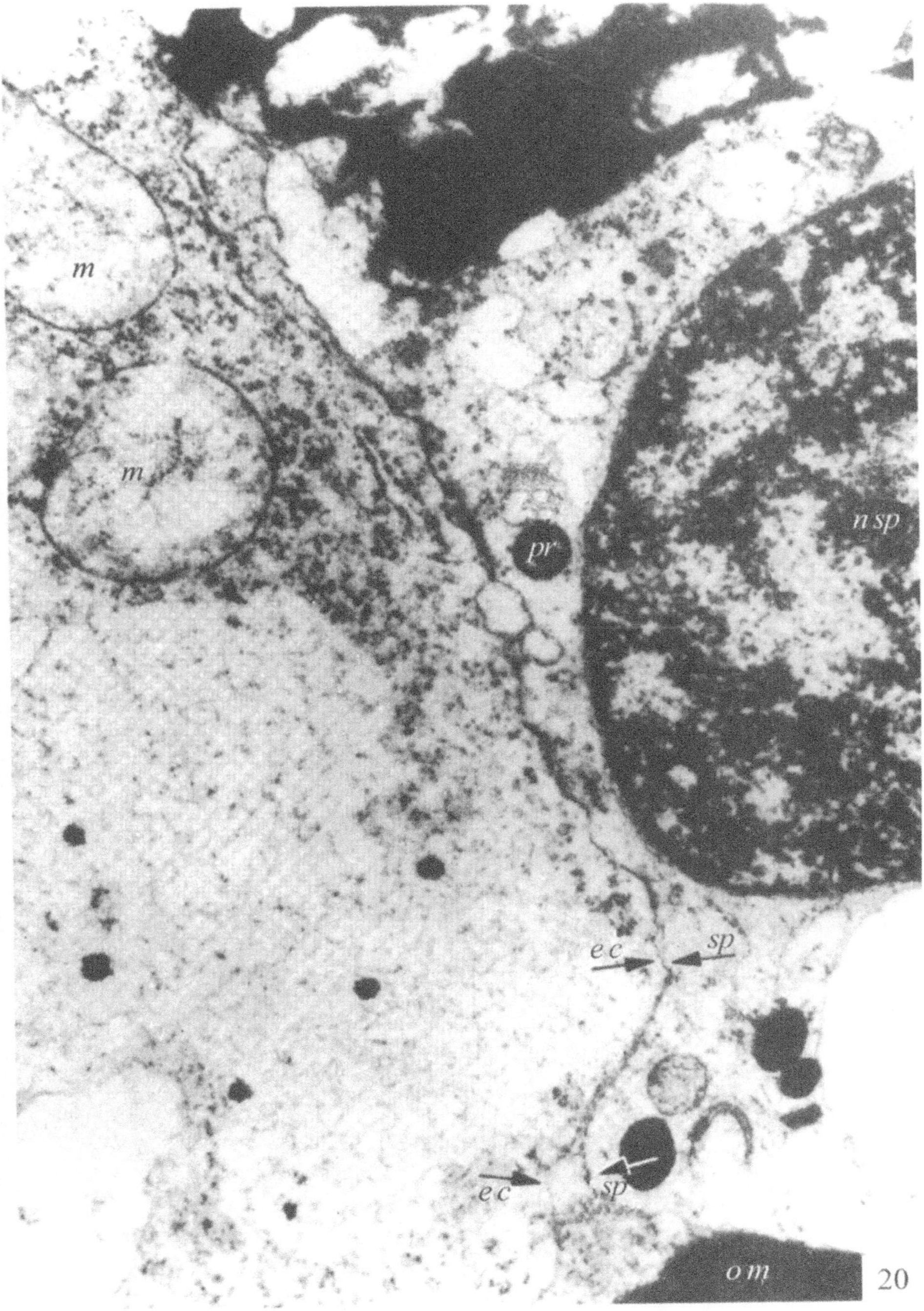

Plate XV (*Contd.*)

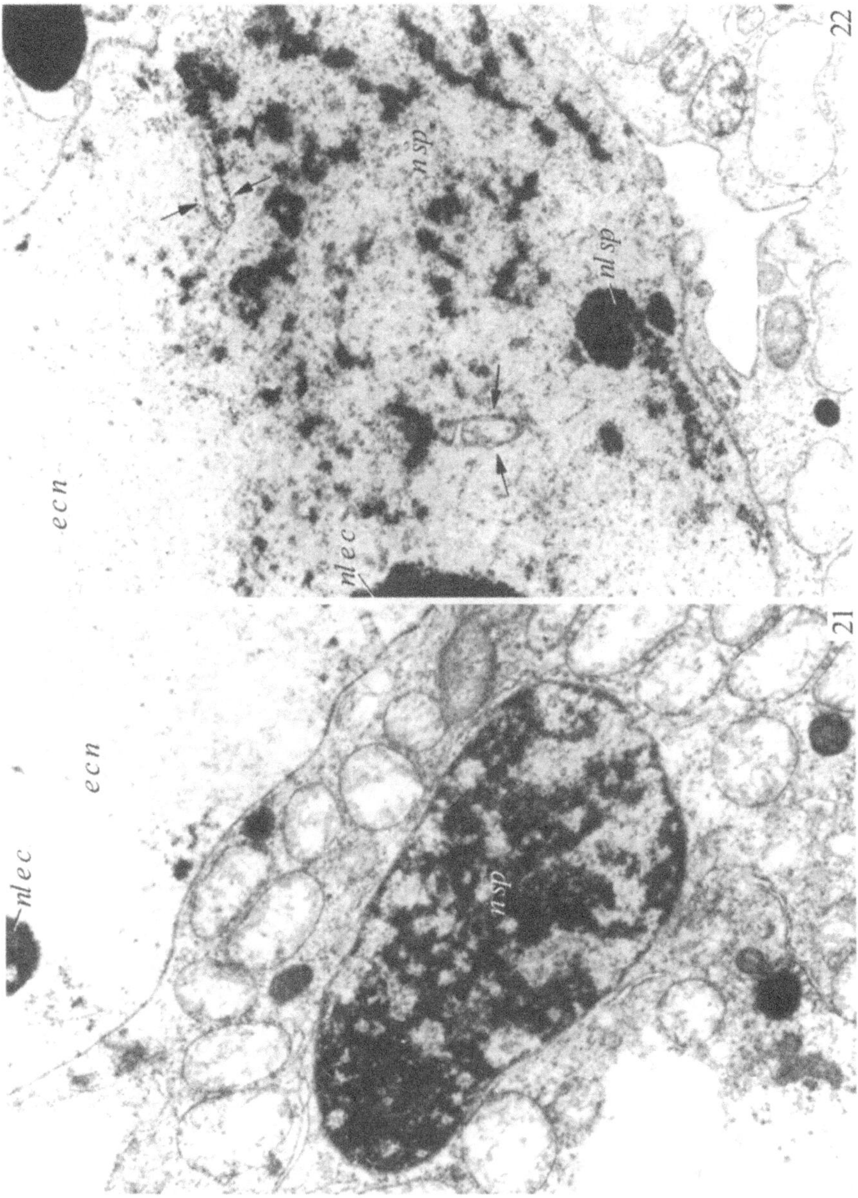

Plate XV (*Contd.*)

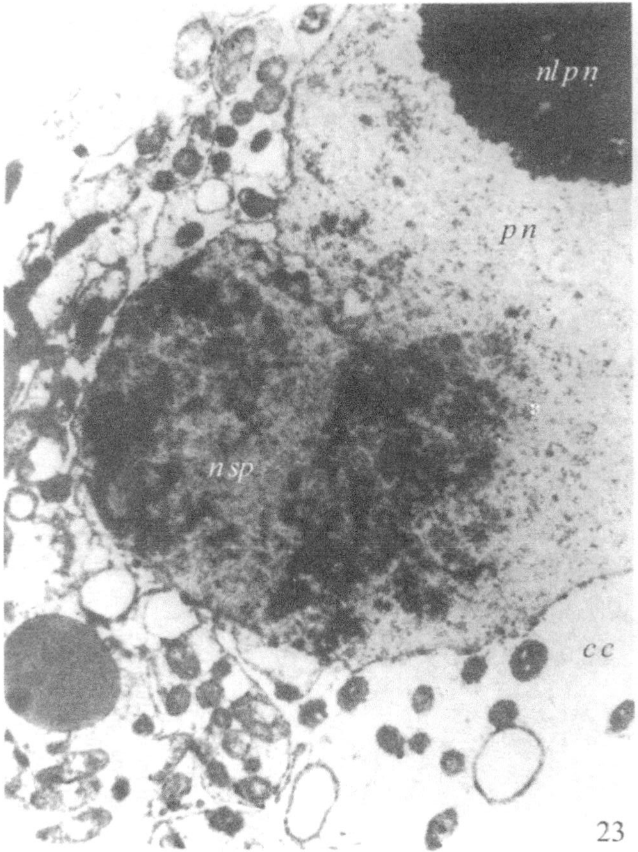

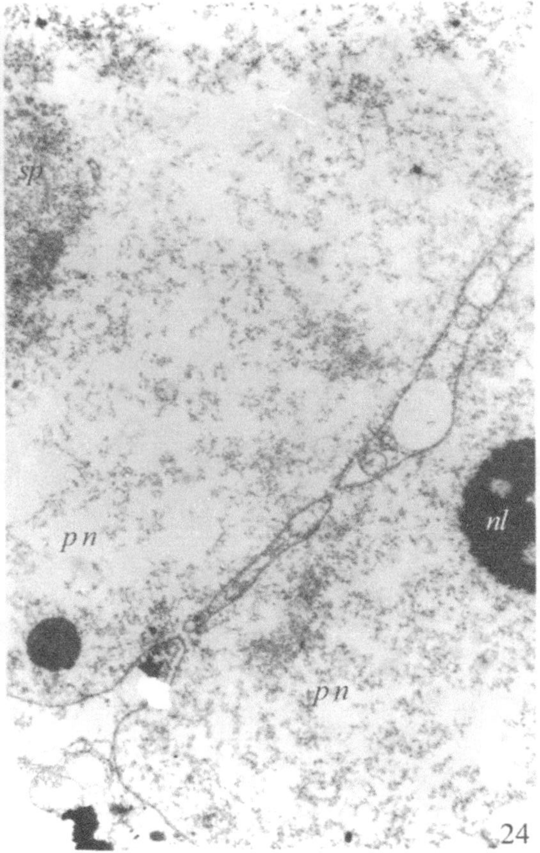

Plate XV (*Contd.*)

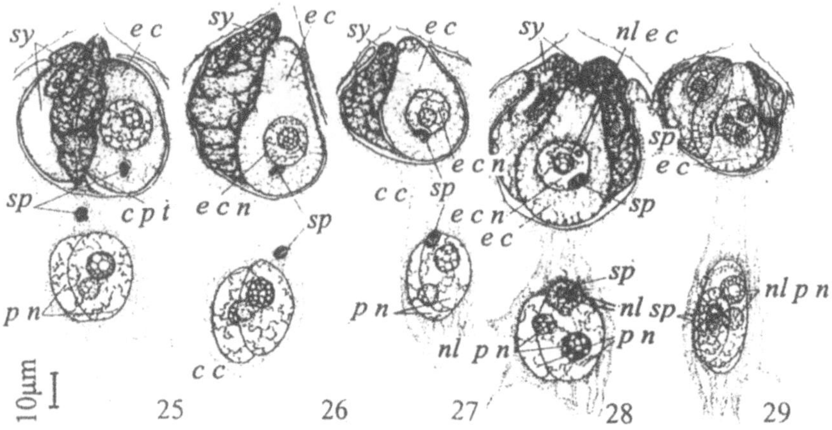

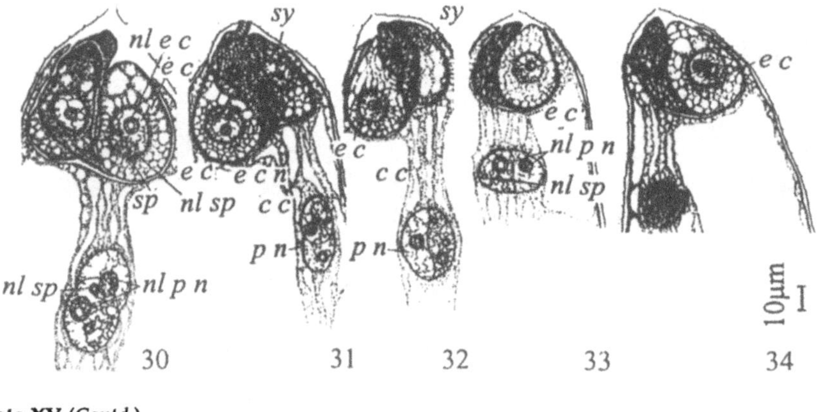

Plate XV (*Contd.*)

has been completed, male and female nucleoli have fused (characteristic for *Zea mays* fertilized nuclei chromatinic net is visible, process of nuclei fusion in zygote not yet finished, two nucleoli observed; two pollen tubes have entered this embryo sac: the first one, whose sperms have carried out fertilization, has entered via right synergid, second one—additional—has entered embryo sac via left synergid, its sperms shown with great magnification, have degenerated); 12-17—double fertilization in *Hordeum vulgare*: micropylar part of embryo sac in 20 min (12), 30 min (13), 1 h (14), 2 h (15), 4 h (16), 6 h (17) after pollination; 18-24—separate stages of double fertilization in *Hordeum vulgare* (TEM): 18—pollen tube in micropylar region between ovary wall and inner integument in 2-2 h 45 min after pollination, degenerated cytoplasm of pollen tube visible, 19—sperm in intercellular space between egg cell and central cell in 1 h 45 min after pollination, plasma membrane observed (arrow), 20—same slide contact of plasma membranes of egg cell and sperm shown (arrows), 21—sperm nucleus in cytoplasm of egg cell near its nucleus, 22—late stages of fusion of egg cell and sperm nuclei in 4-4 h 45 min after pollination, *de novo* formed sperm nucleolus and remnants of nuclear membranes

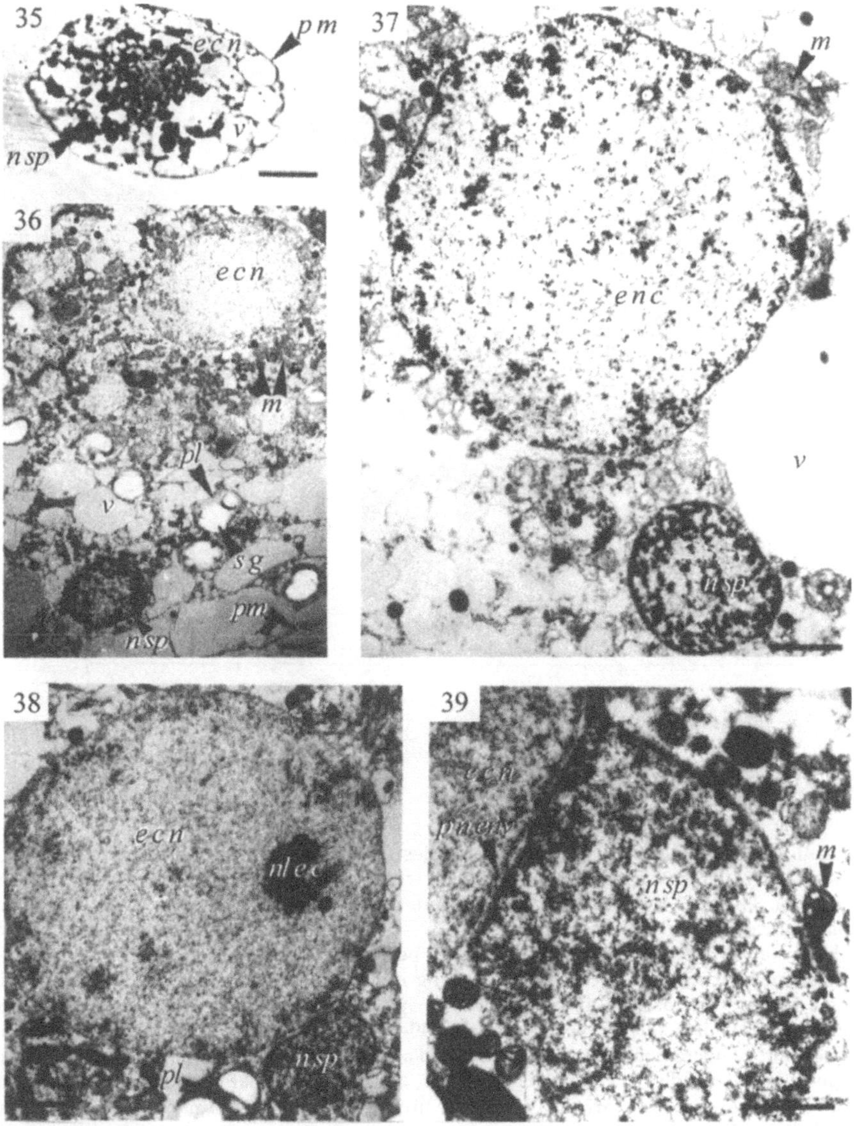

Plate XV (*Contd.*)

of egg cell and sperm observed (arrow), 23—sperm nucleus fusing with polar nucleus in 2-2 h 45 min after pollination, 24—polar nuclei contact, remnants of cytoplasm from central cell which were captured during fusion visible; 25-34—fertilization process in *Triticale* AD 537: micropylar part of embryo sac in 60 min after pollination (25), 70 min (26), 80 min (27), 1 h 50 min (28), 3 h (29), 4 h (30), 5 h (31), 6 h (32), 8 h (33) (disturbances), 9 h (34) after pollination; 35-39—fusion process of isolated gametes in *Zea mays* in case of induction with electric impulses (36-39—TEM): 35—fusion product in 8 min after electroinduction (phase-contrast microscopy), arrow marks sperm nucleus, cytoplasm contains numerous peripheral vacuoles, and organelles around

40 41

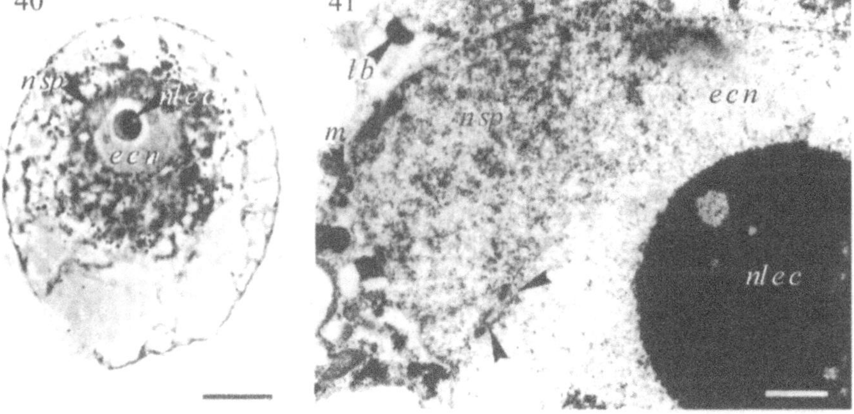

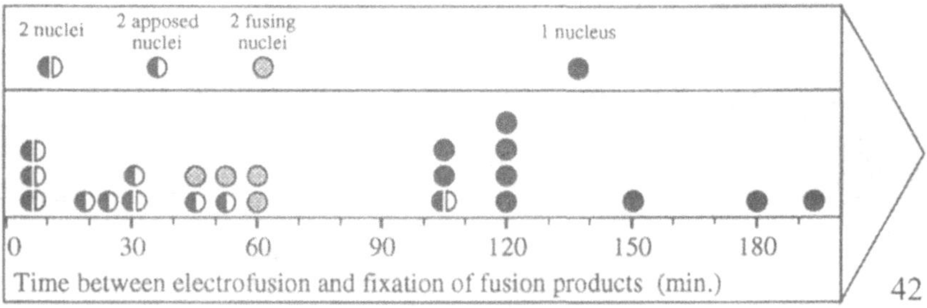

42

Plate XV (*Contd.*)

egg cell nucleus, 36—same, sperm nucleus lies in egg cell cytoplasm, nucleus chromatin dense, 37—in 10 min 30 s after electroinduction with electrondense sperm nucleus and nucleus of egg cell, 38—in 21 min after electroinduction, contact of sperm and egg cell nuclei, association of nuclei at serial sections not observed, 39—in 30 min after induction, sperm and egg cell nuclei in contact with each other, nuclear envelopes with nuclear pores are situated quite close, but not unified, small mitochondria lie along envelope of sperm nucleus, mitochondrial cristae highly contrasting; 40-42—fusion of sperm nucleus with egg cell nucleus after electroinduction of isolated *Zea mays* gametes: 40—product of fusion in 58 min after induction (phase-contrast microscopy), nuclei of sperm and egg cell (arrows) fuse, 41—same (TEM), fusion of sperm nucleus and egg cell nucleus observed, density of chromatin in nuclei different, remnants of nuclear envelope (arrows) near male hromatin visible; 42—time sequence of events after fusion of isolated gametes.

a sp—additional sperm, *c c*—central cell, *c p t*—contents of pollen tube, *d sy*—degenerated synergid, *d sy n*—degenerated synergid nucleus, *e c*—egg cell, *e c n*—egg cell nucleus, *i i*—inner integument, *l b*—lipid body, *m*—mitochondrion, *n sp*—nucleus of sperm, *nl e c*—nucleolus of egg cell, *nl p n*—nucleolus of polar nucleus, *nl sp*—nucleolus of sperm, *o m*—osmiophil material, *ov w*—ovary wall, *pl*—plastid, *p m*—plasma membrane, *p n*—polar nucleus, *p n env*—pores of nuclear envelope, *p sy*—persistent synergid, *p sy n*—persistent synergid nucleus, *p t*—pollen tube, *s g*—starch grain, *sp*—sperm, *sy*—synergid, *v*—vacuole, z—zygote, *z n*—zygote nucleus.

1-9—modified from Batygina, 1974b; 10-11—after Korobova, 1982; 12-17—after Khvedynich, 1969; 18-24—modified from Mogensen, 1982; 25-34—after Khvedynich, 1983 (see Batygina 1987a); 35-39 after Faure *et al.*, 1993; 40-42—after Kranz *et al.*, 1991b.

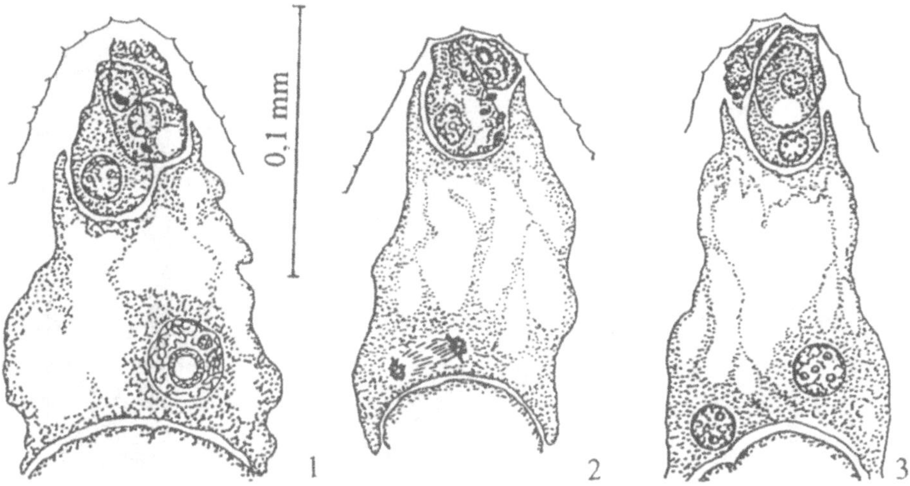

Plate XVI: Nuclear type of endosperm development.

1-8—endosperm development in *Anemone nemorosa*: 1-3—division of primary endosperm nucleus and formation of 2-nucleate endosperm, 4, 5—4- and 8-nucleate endosperm, 6—formation of multinucleate coenocytic endosperm, zygote does not start with division, persistent multinucleate antipodal cells visible, 7—laying down of cell walls between endosperm nuclei, beginning from micropylar pole, embryo is 6-celled, 8—two-layered endosperm in process of development on cross-section of carpel, fragmoplasts visible, 8a—fragment of developing two-layered endosperm at larger magnification; 9-17—schematic representation of successive stages of formation of central vacuole in endosperm coenocyte in species of *Triticum* (9-14) and its gradual disappearance in the course of cellularization (15-17); 18-22—endospermogenesis of *Diplacrum caricinum*, proceeding without formation of large central vacuole at free-nucleate stage of development: fig. 20 shows beginning of cellularization in coenocytic endosperm (initiation of cellularization process by means of 'free growth' of anticlinal cell walls—noted by G.Zh.); 23, 24—endosperm development in *Melo sativus* (2 ×) with haustorium remaining free-nucleate also in period of cellularization in endosperm, embryo at heart-shape stage of development (24); 25-27—schematic representation of endosperm (and its fragments—25) with chalazal coenocytic haustoria of various forms and extention in certain species at cellular stage of its development (25—*Cassia miangustifolius*, 26, 27—*Echinocystis lobata*); 28-30—stages of endosperm development in *Elaeagnus argentea*: 28—fragment of developing seed at mace-like stage of embryo development, in period of cellularization beginning in central part of endospermal coenocyte, anucleate cytoplasmic outgrowths ("vesicles") are observed, 29, 30—chalazal endospermal haustorium at free-nucleate and cellular stages of endospermogenesis; 31-37—endosperm development in *Alcea rosea*: 31-34—longitudinal sections of micropylar portion of developing seed at stages of zygote, globular and heart-shape embryo, demonstrating location of daughter nuclei of endosperm primary nucleus (31) and the transition of endosperm from free-nucleate (32) to cellular state (33), 34—case of embryo development without endosperm up to stage of 3-tiered proembryo (undivided primary nucleus of endosperm situated close to proembryo), 35-37—protoplast fragments of free-nucleate endosperm (in micropylar—35, 37, and middle—36, regions of endosperm layer, adjacent to cell wall), at early globular stage of embryo development, outgrowths of cell wall observed at early stages of their development leucoplasts with scant starch, mitochondria in close contact with plasma membrane

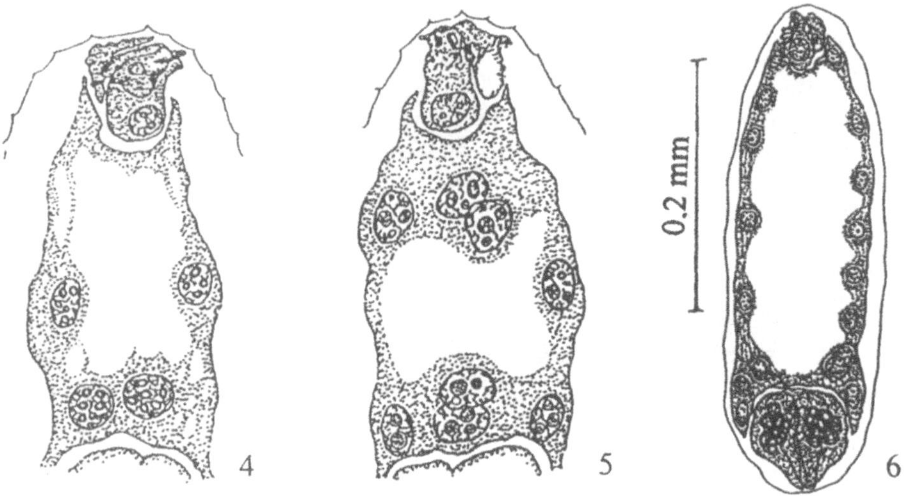

Plate XVI (*Contd.*)

invaginations—(arrows); 38-44—protoplast fragments of free-nucleate endosperm in *Epipactis atrorubens*: 38-40—near subapical cell of early proembryo, part of persistent synergid also visible (38), near basal part of proembryo, leucoplasts as well as fragment of endosperm nucleus (39) are shown (40—part of one of subsequent sections of the same endosperm portion, evagination of outer membrane of nuclear envelope can be seen, noted with double arrow); 41, 42—near persistent antipodes (developing seed at stage of multicelled proembryo), leucoplasts both in mutual association (41) and in association with mitochondria (41, 42), ER poorly developed, fragment of one of endosperm nuclei (41), rather small starch grain at one of leucoplast sections visible (42); 43—fragments of near-wall layer (at stage of multicelled proembryo), tonoplast of central vacuole thicker than plasma membrane, solitary thylakoid can be seen lying in parallel to plastid envelope and vacuole tonoplast (44—fragment of one of subsequent sections of very same endosperm part, together with solitary peripheral thylakoid other plastid membrane structures and plastoglobules are shown); 45-46—chloroplasts of endospermal coenocyte in *Iberis umbellata* (at stage of multicelled proembryo); 47, 48—fragments of near-wall layer of protoplast in the submicropylar zone of free-nucleate endosperm in *Paeonia peregrina* (embryo at stage of the four-nucleate coenocyte); rough ER well developed, ergastoplasma not observed, Golgi vesicles near plasma membrane and in contact with it (arrow) (32, 33—LM, 35-48—TEM).

Bar—1 µm (35-44, 47), 0.5 µm (45-46); *a c*—antipodal cell, *ch*—chloroplast, *c o*—cytoplasmic outgrowth ("vesicle"), *c v en*—central vacuole of coenocytic endosperm, *c w o*—cell wall outgrowth, *d*—dictyosome, *e*—embryo, *en*—endosperm, *en n*—endosperm nucleus, *e r*—endoplasmic reticulum, *e s w*—cell wall of embryo sac, *h*—haustorium of endosperm, *i i*—inner integument, *l b*—lipid body, *lp*—leucoplast, *m*—mitochondrion, *nc*—nucellus, *pe*—proembryo, *p en n*—primary endosperm nucleus, *pg*—plastoglobule, *p m*—plasma membrane, *p sy*—persistent synergid, *s g*—starch grain, *t c*—tubular complex, *thl*—thylakoid, *v*—vacuole, *z*—zygote.

1-8—after Yoffe, 1971b; 9-17 after Batygina, 1974b; 18-22—after Nijalingappa and Devaki, 1978; 23, 24—after Dzevaltovsky, 1983; 25—after Rau, 1951; 26, 27—after Chopra and Seth, 1967—cited by: Maheshwari, modified from Kapil, 1956; 28-30—after Kamelina and Proskurina, 1987; 31-37, 39-48—after Zhukova, orig; 38—modified from Zhukova, 1994.

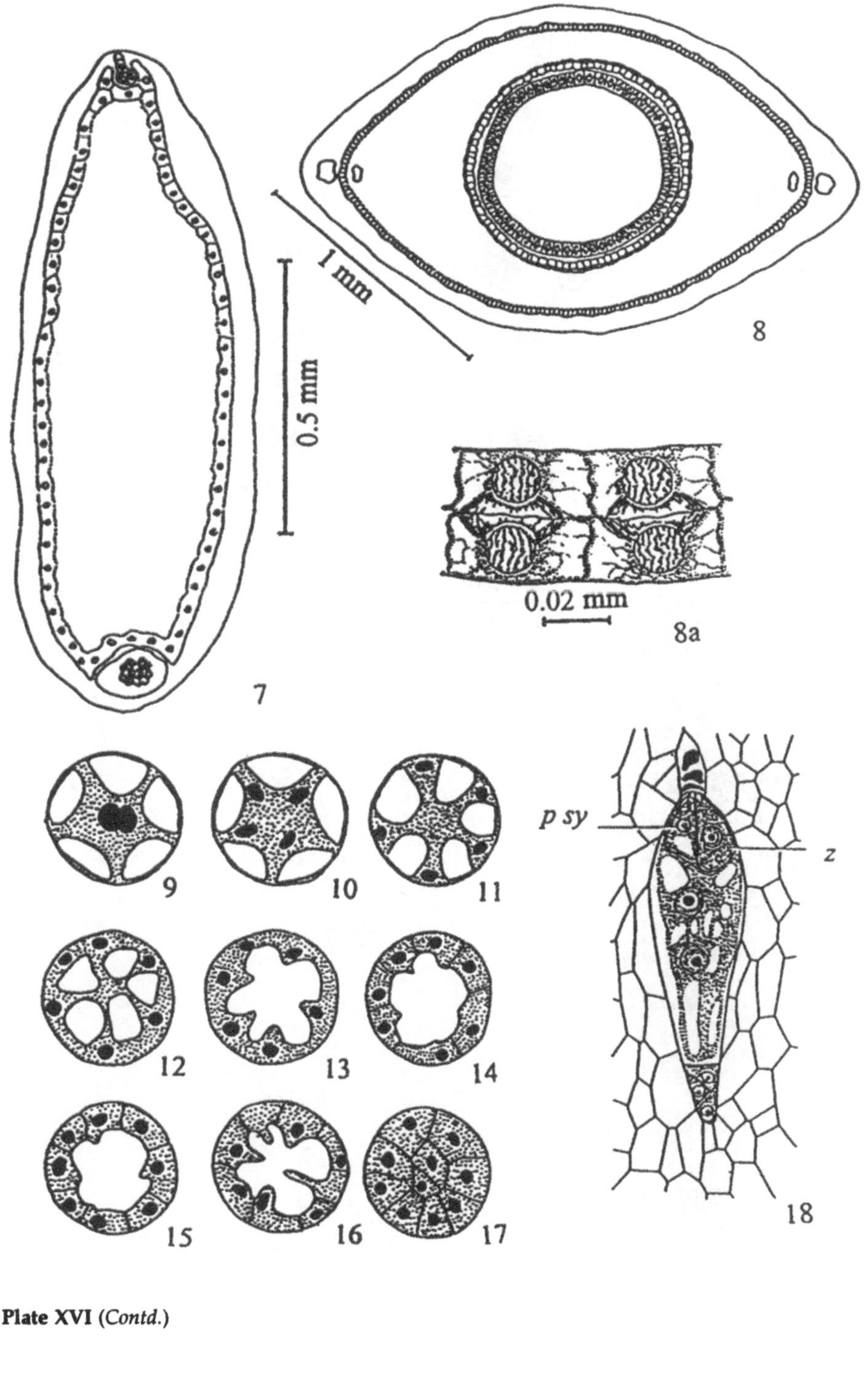

Plate XVI (*Contd.*)

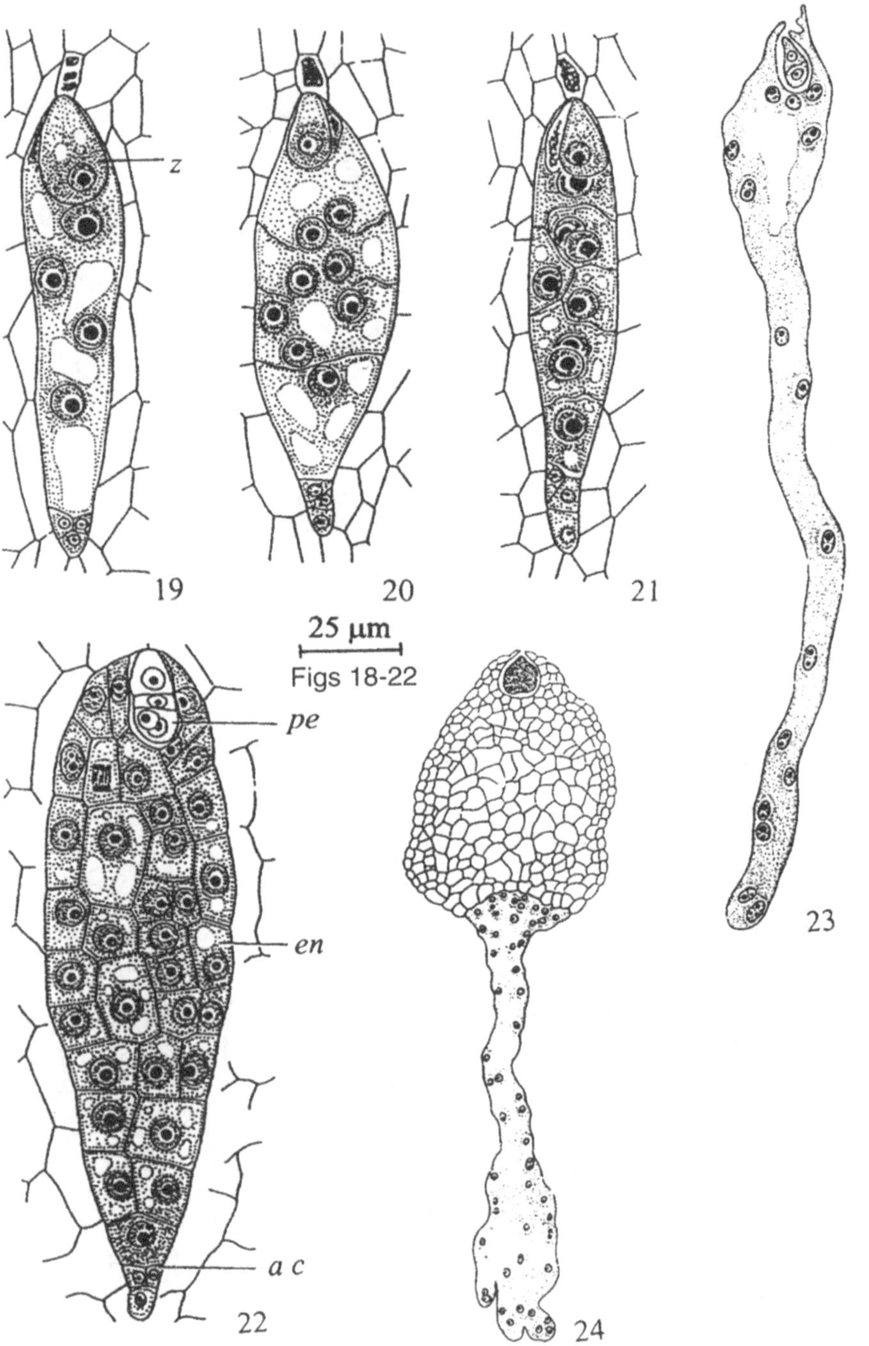

Plate XVI (*Contd.*)

25

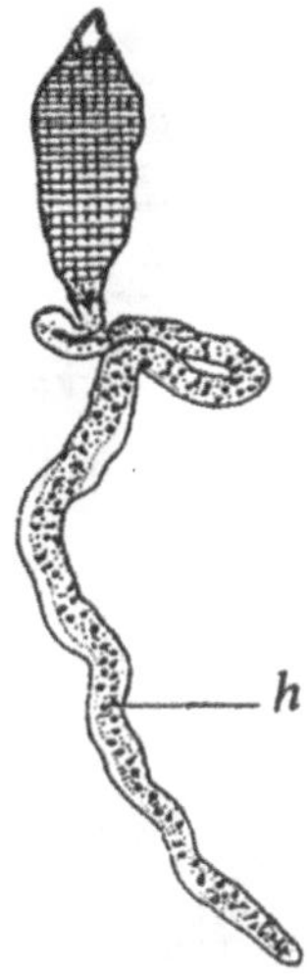

26

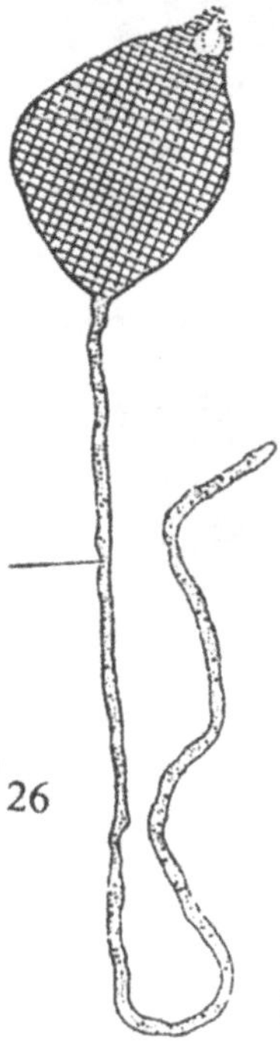

27

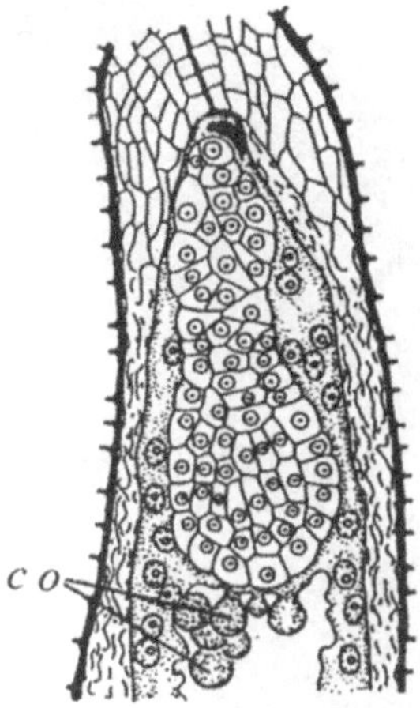

28

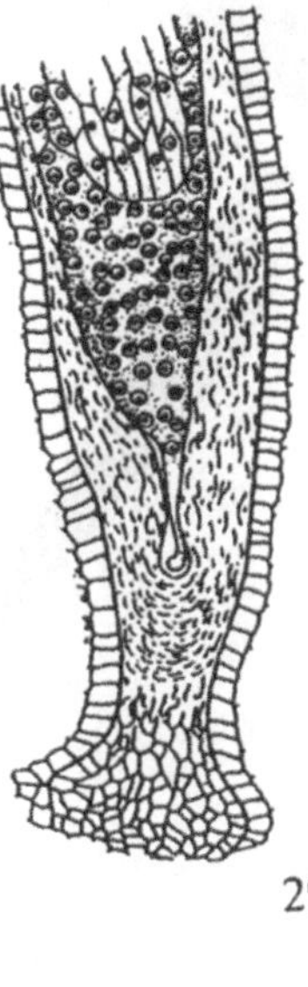

29

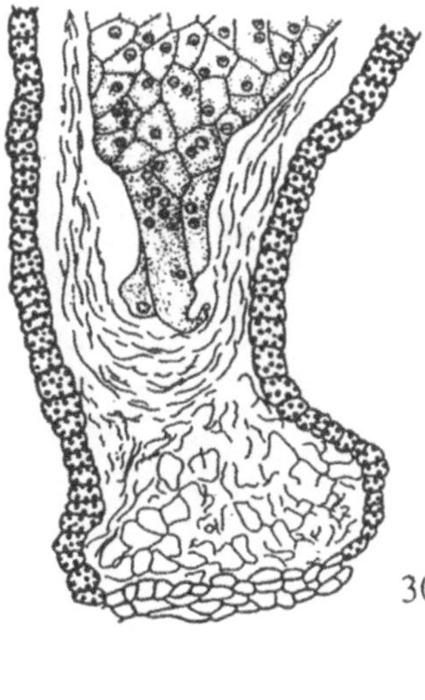

30

Plate XVI (*Contd.*)

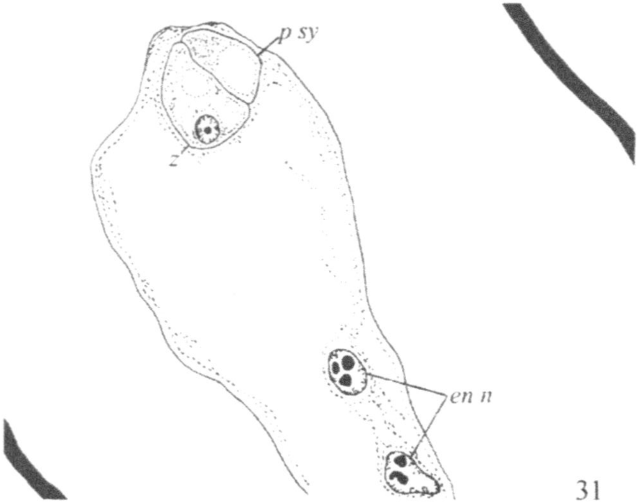

31

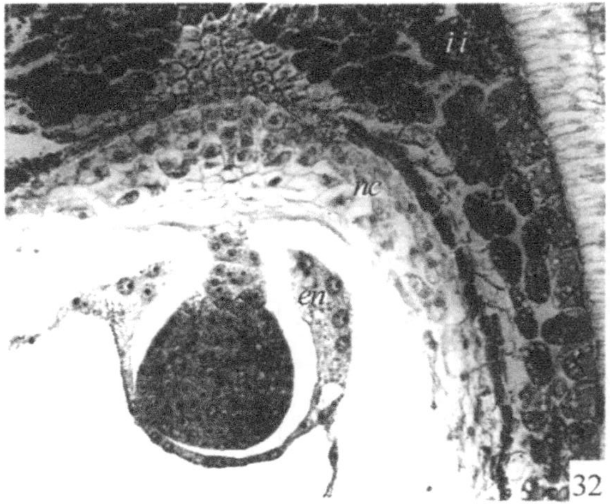

32

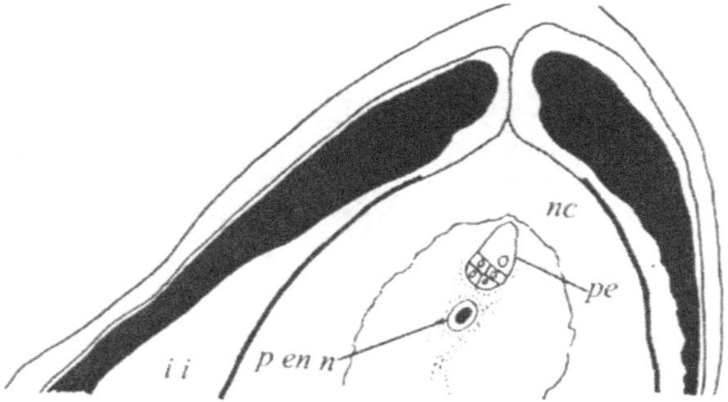

34

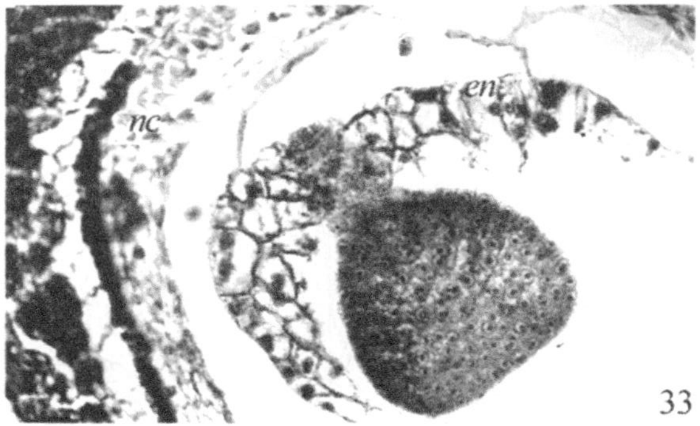

33

Plate XVI (*Contd.*)

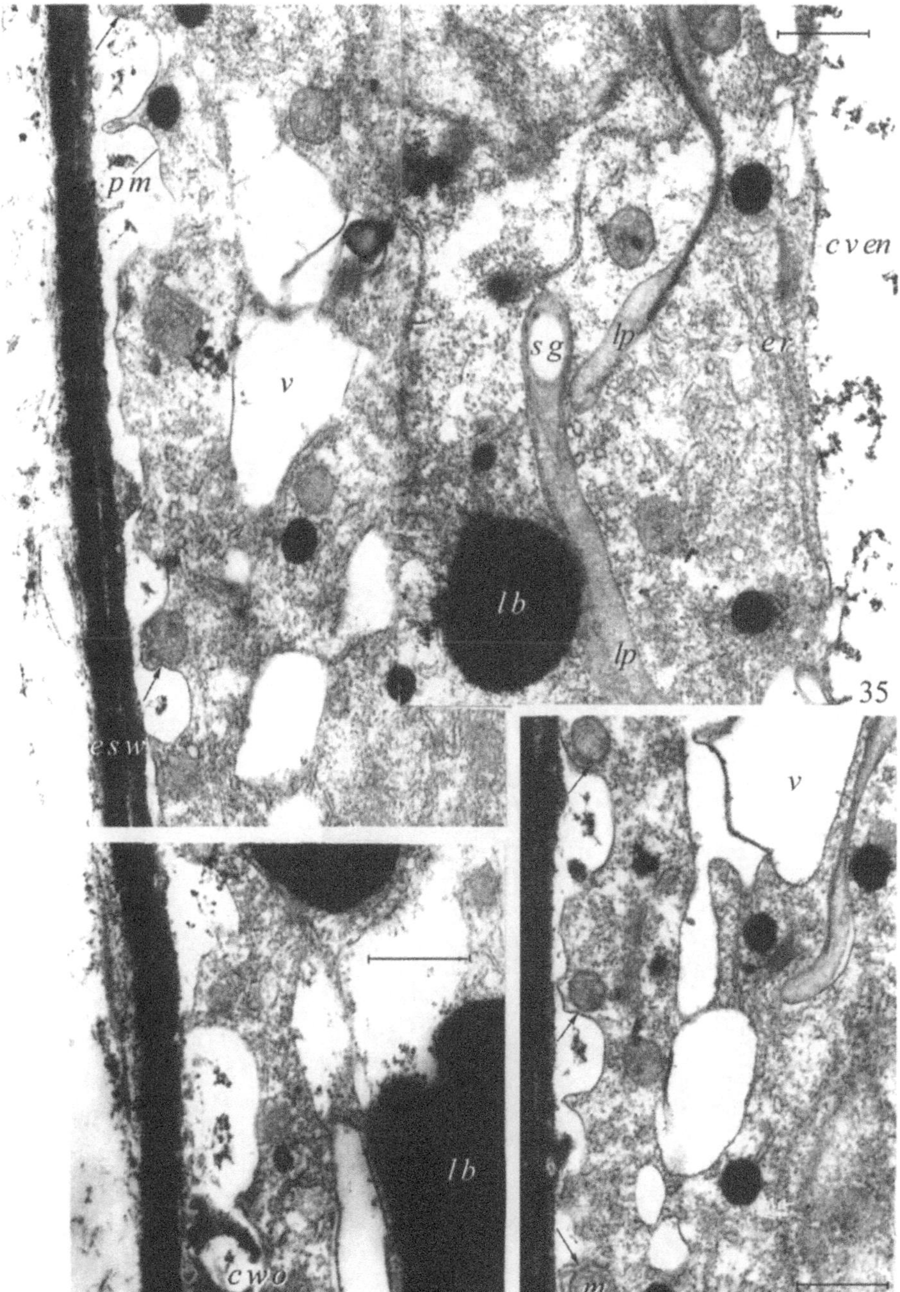

Plate XVI (*Contd.*)

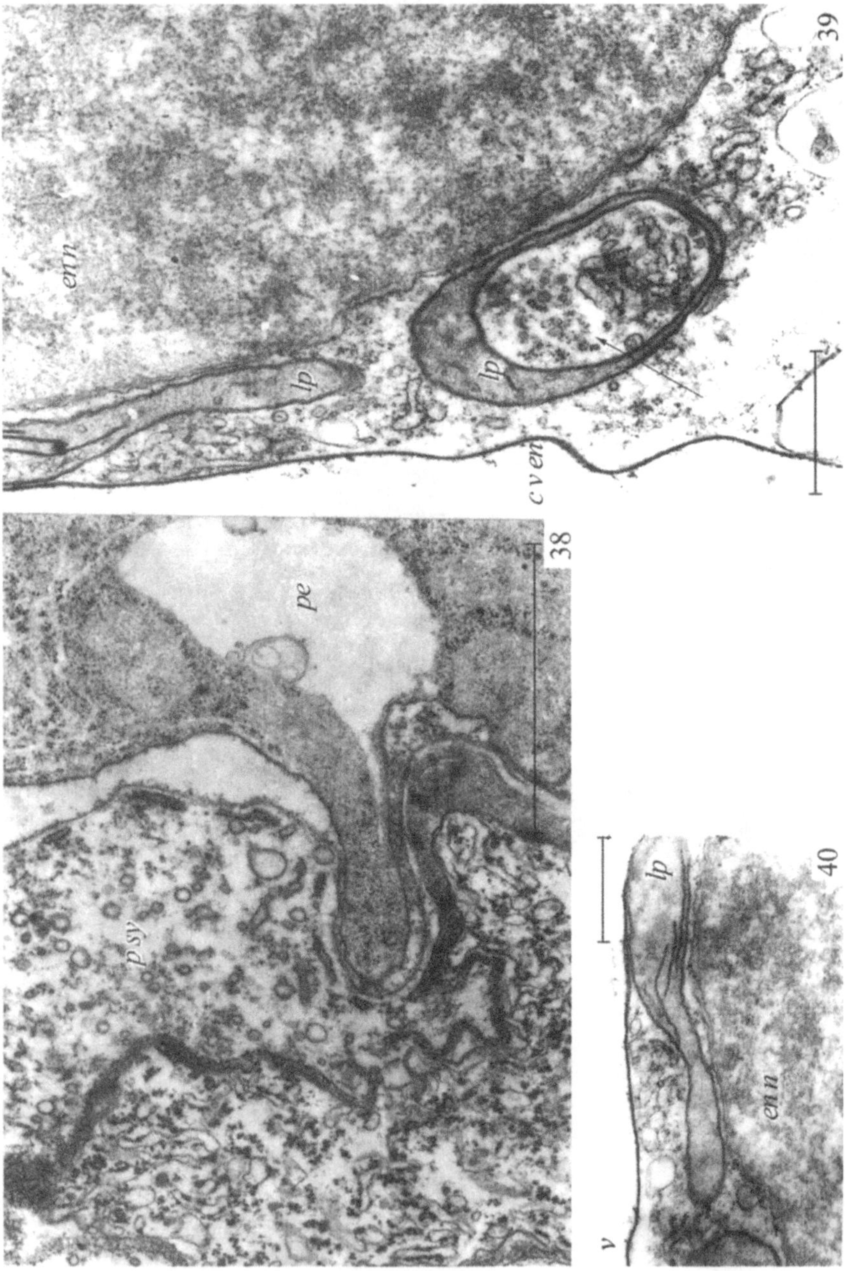

Plate XVI (*Contd.*)

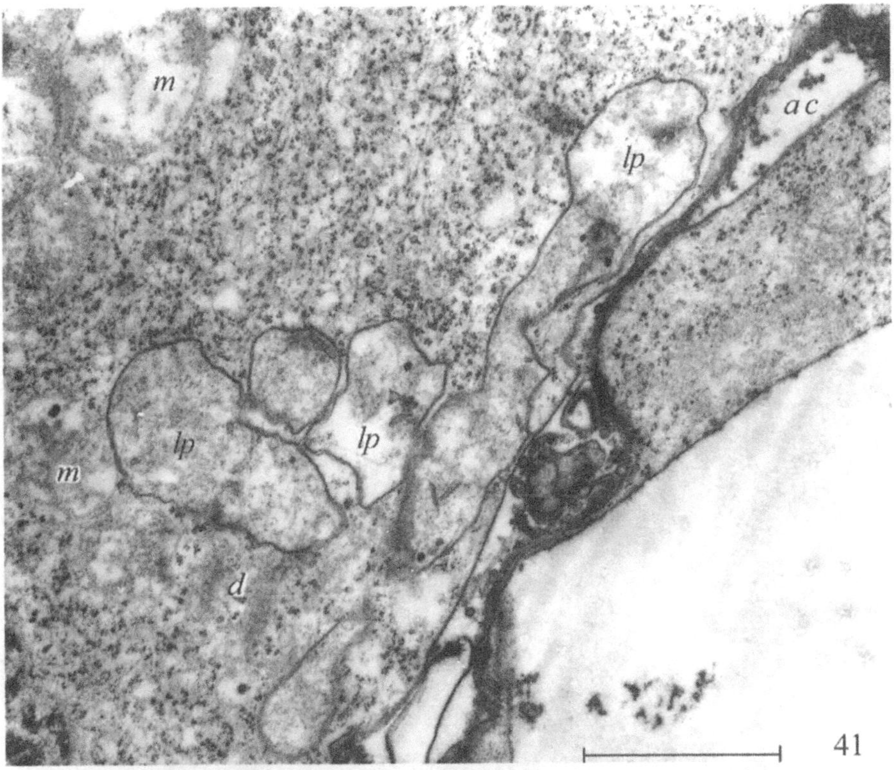

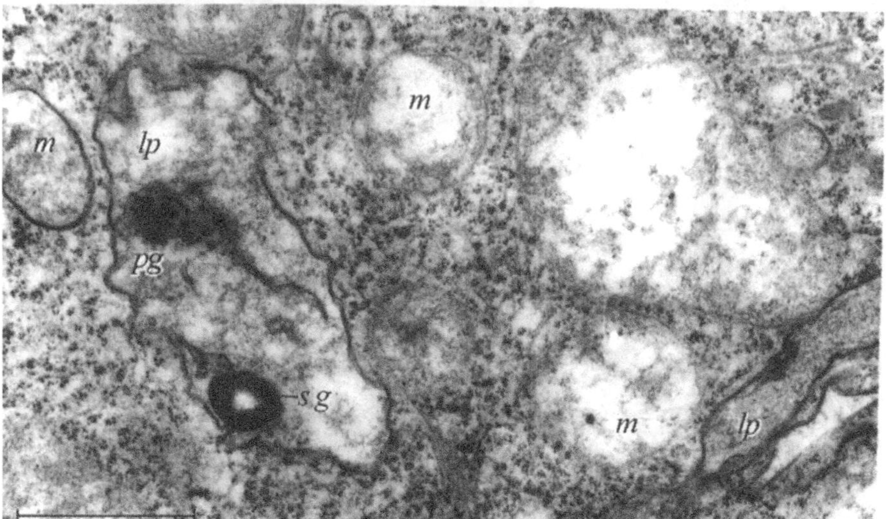

Plate XVI (*Contd.*)

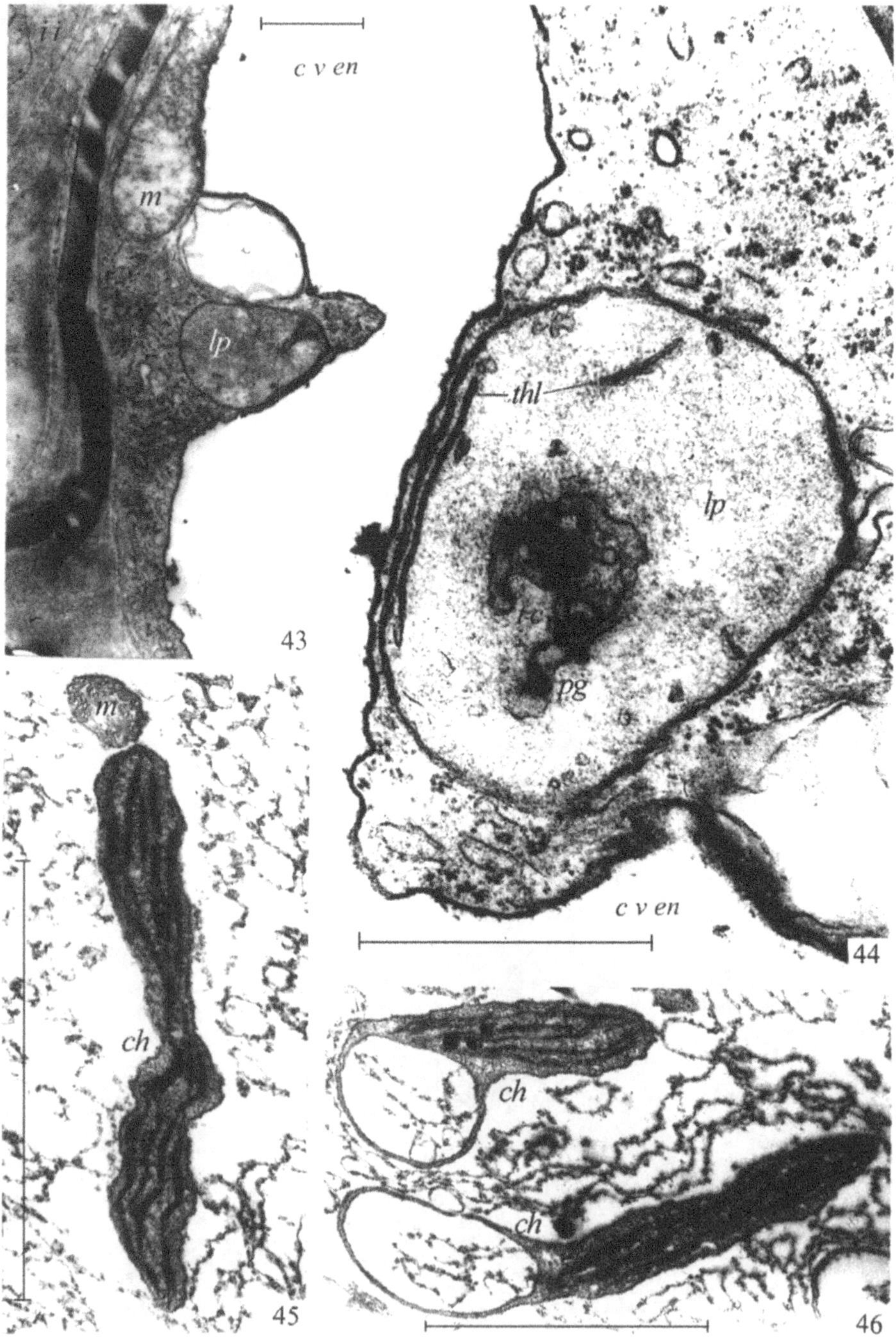

Plate XVI (*Contd.*)

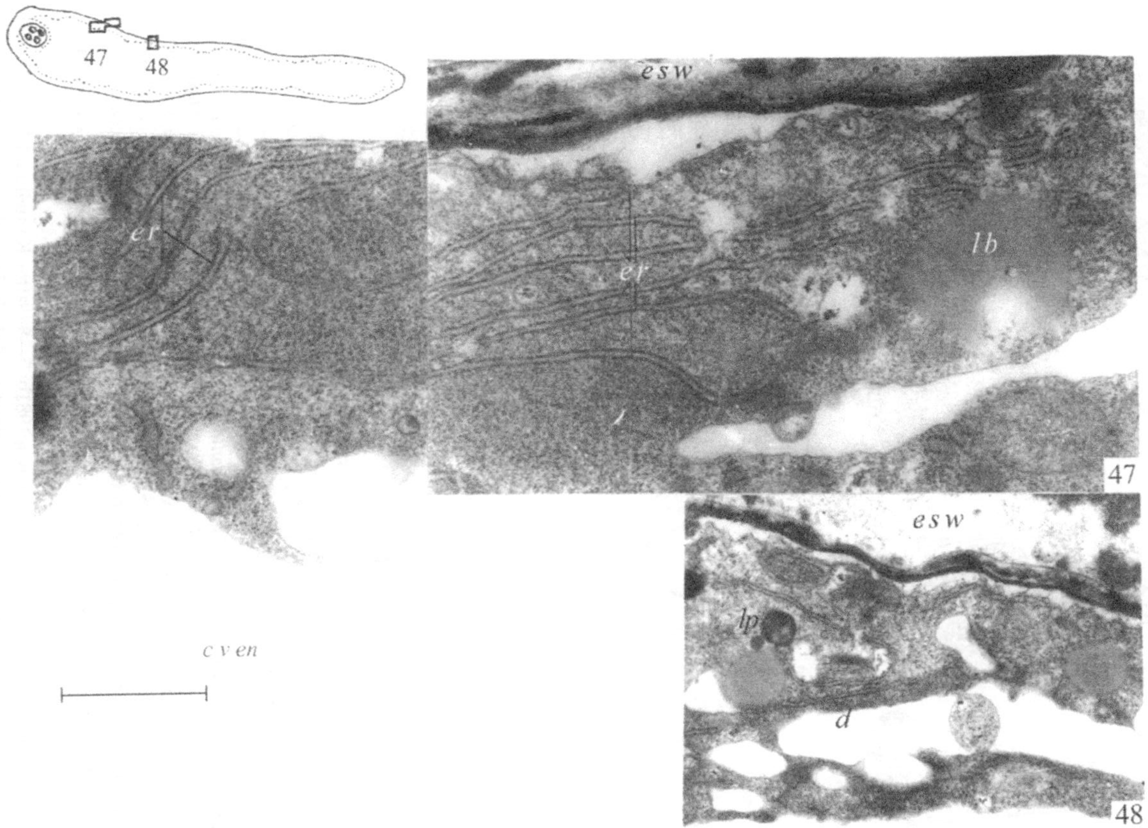

Plate XVI (*Contd.*)

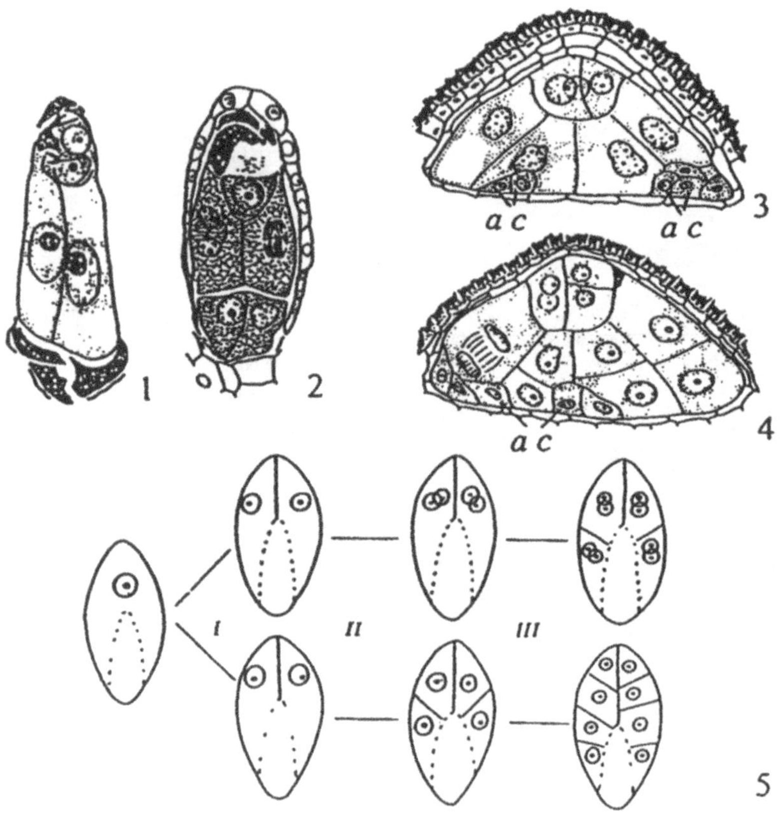

Plate XVII: Cellular type of endosperm development.
1-31—different variations of endosperm development: 1-4—Adoxa-variation in *Circaeaster agrestis* (1, 2) and *Peperomia blanda* (3, 4); 5-8—Lappula-variation in Dipsacaceae members: schematic representation of endosperm development at first stages (5), in *Succisa pratensis* (6, 7), in *Cephalaria brevipalea* (8); 9-11—Prunella-variation in *Orobanche cumana*; 12-16—Annona-variation in *Bragantia wallichii* (12, 13), *Menyanthes trifoliata* (14), *Villarsia reniformis* (15, 16); 17, 18—Phyteuma-variation in *Trochodendron aralioides*; 19, 20—Ericaceae-variation in *Rhododendron mucronulatum*; 21-25—Ceratophyllum-variation in *Ceratophyllum pentacanthum* (21, 22), C. *demersum* (23) and C. *submersum* (24, 25); 26-30—Azorina-variation in *Azorina vidalii*; 31—variations in endosperm development in families Campanulaceae and Lobeliaceae.

a c—antipodal cell, *en is*—endosperm isthmus, *ch h*—chalazal haustorium, *i en c b*—initials of endosperm central body, *m h*—micropylar haustorium.

1, 2—after Junell, 1931; 3, 4—after Nikiticheva, 1981b; 5—8—after Kamelina, 1987; 9-11—after Anisimova, 1983; 12, 13—after Nair and Narayanan, 1961; 14-16—after Stolt, 1921; 17, 18—after Yoffe, 1962; 19, 20—after Yakobson, 1976; 21-25—after Shamrov and Batygina, 1984; 26-30—after Shamrov and Zhinkina, 1994; 31—after Korobova and Zhinkina, 1987.

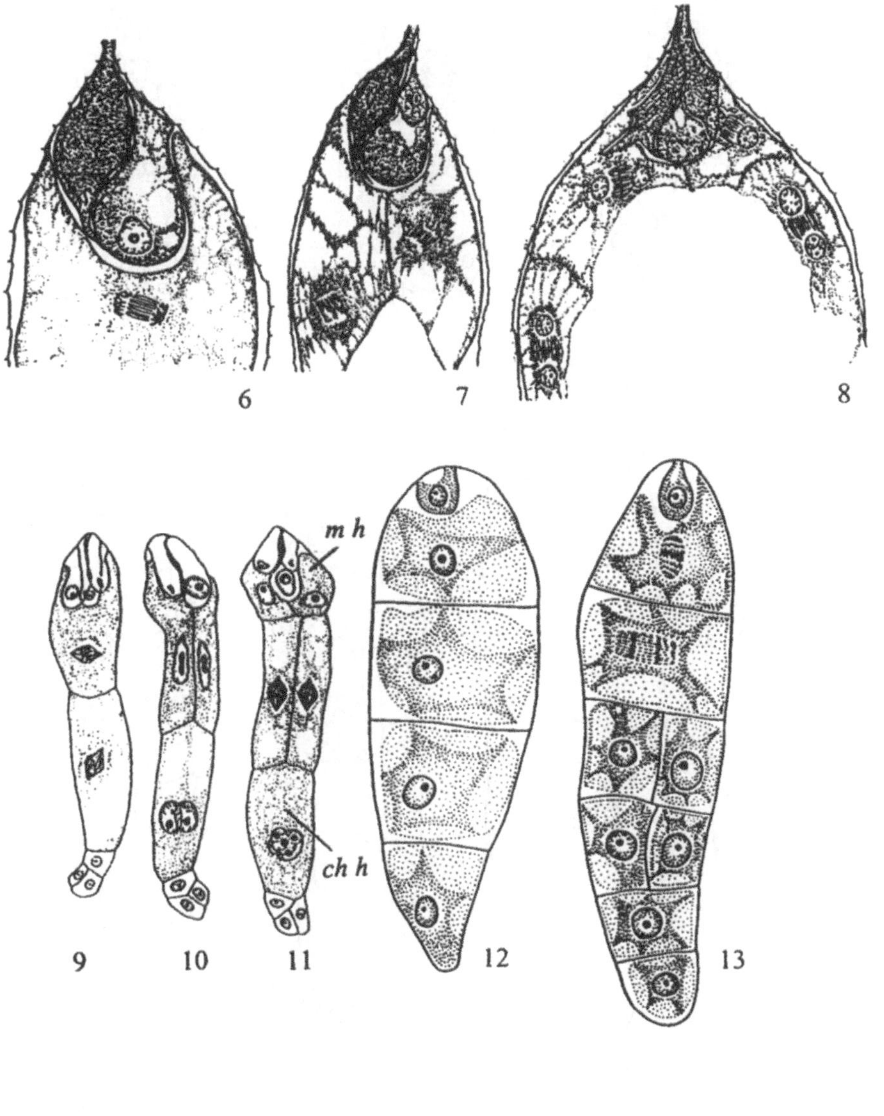

Plate XVII (*Contd.*)

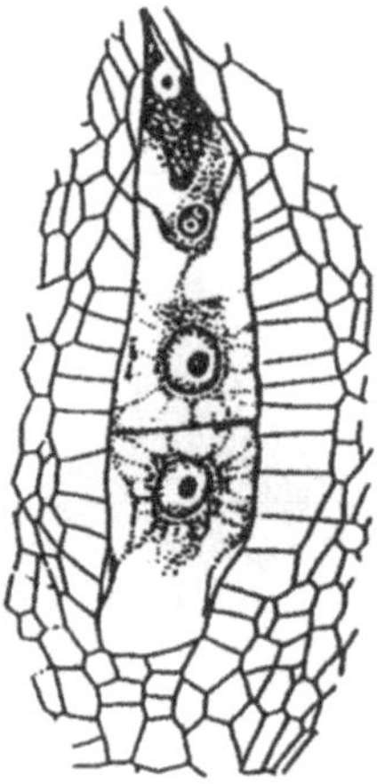
14

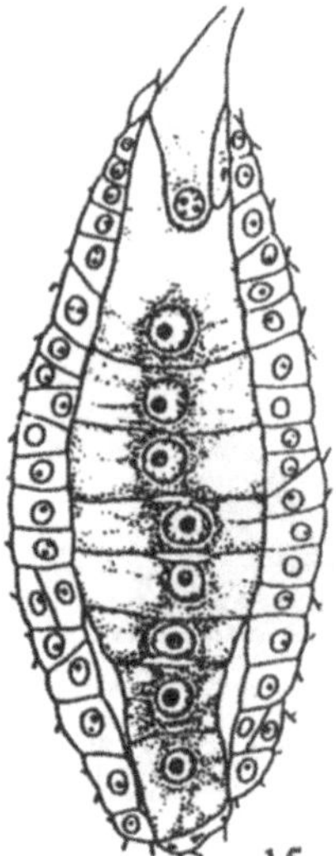
15

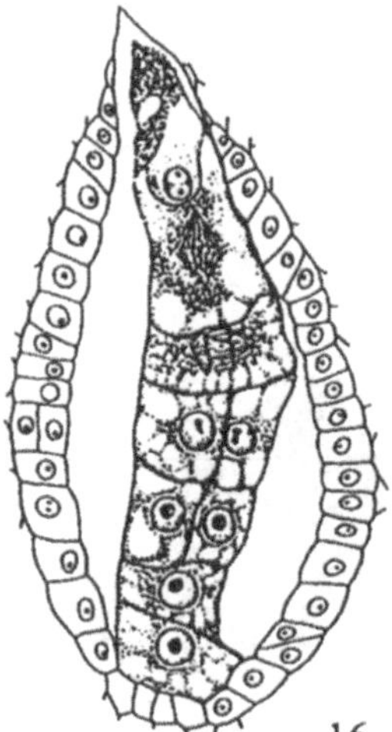
16

17

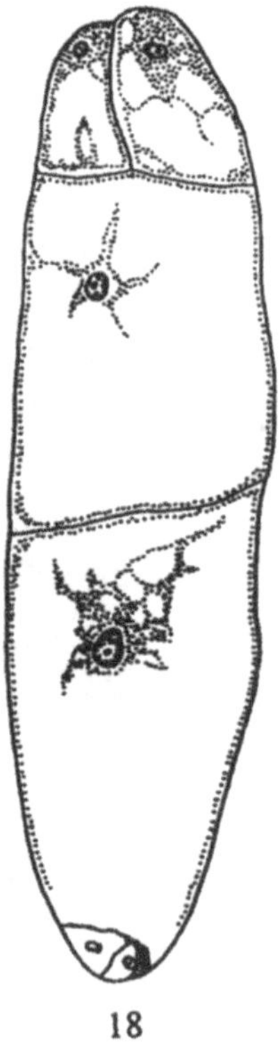
18

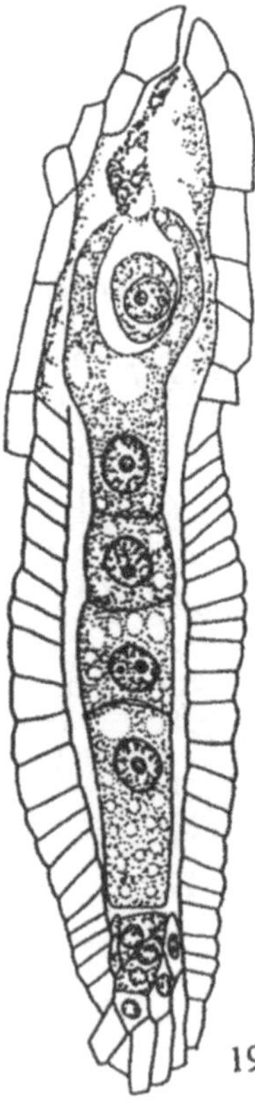
19

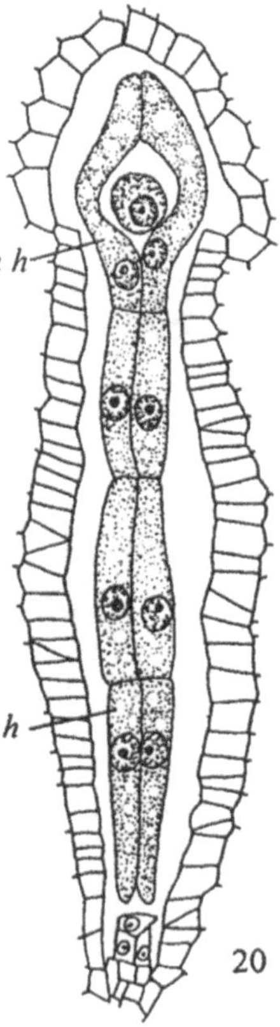

20

Plate XVII (*Contd.*)

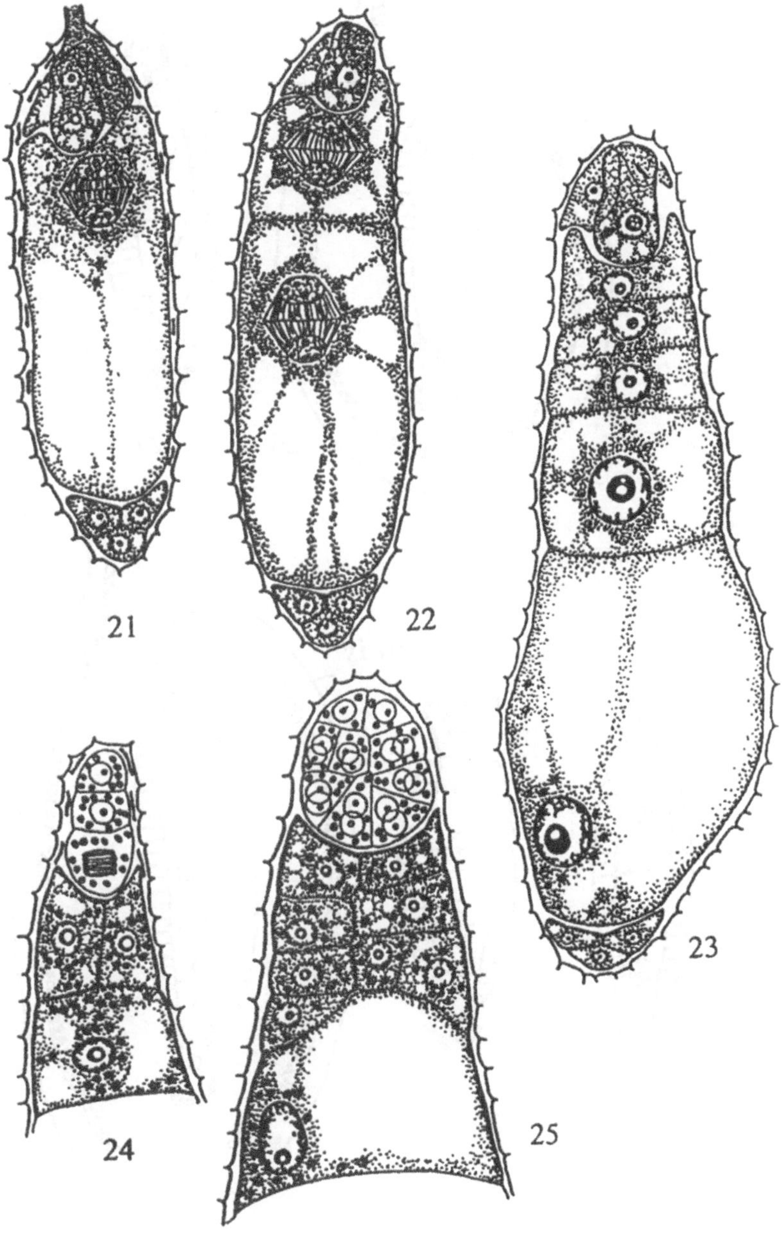

Plate XVII (*Contd.*)

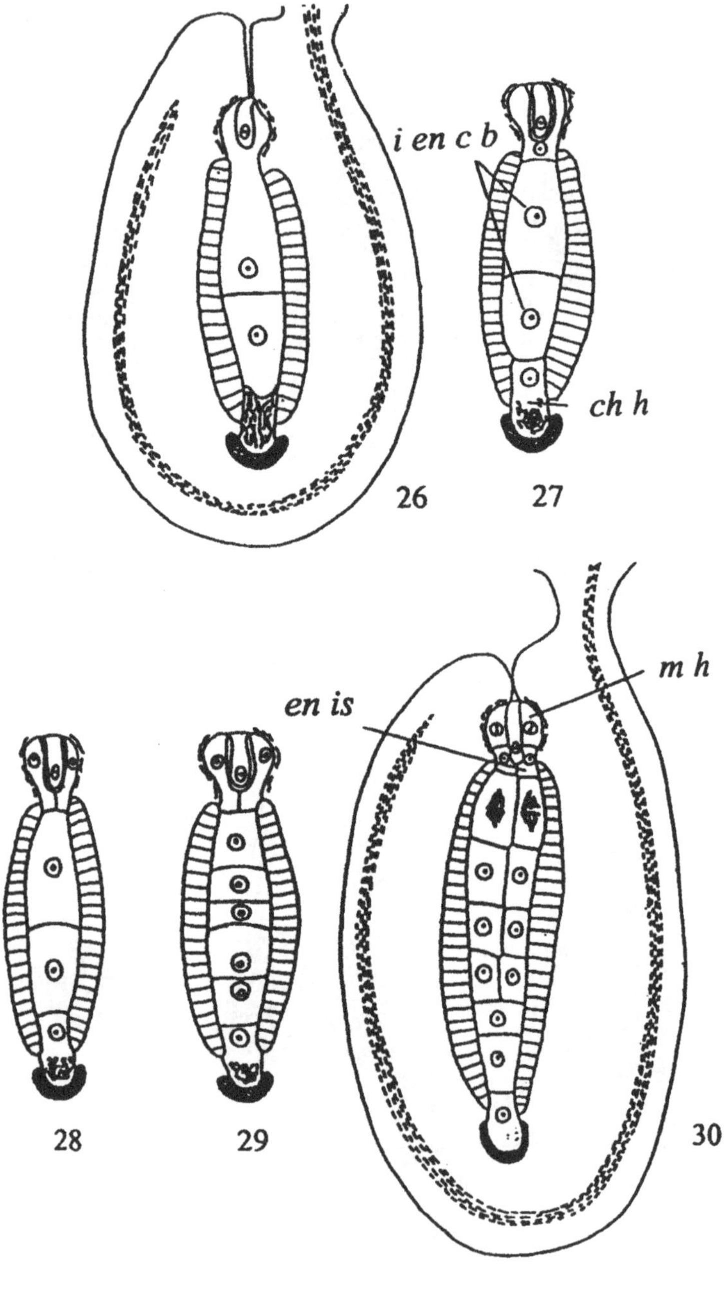

Plate XVII (*Contd.*)

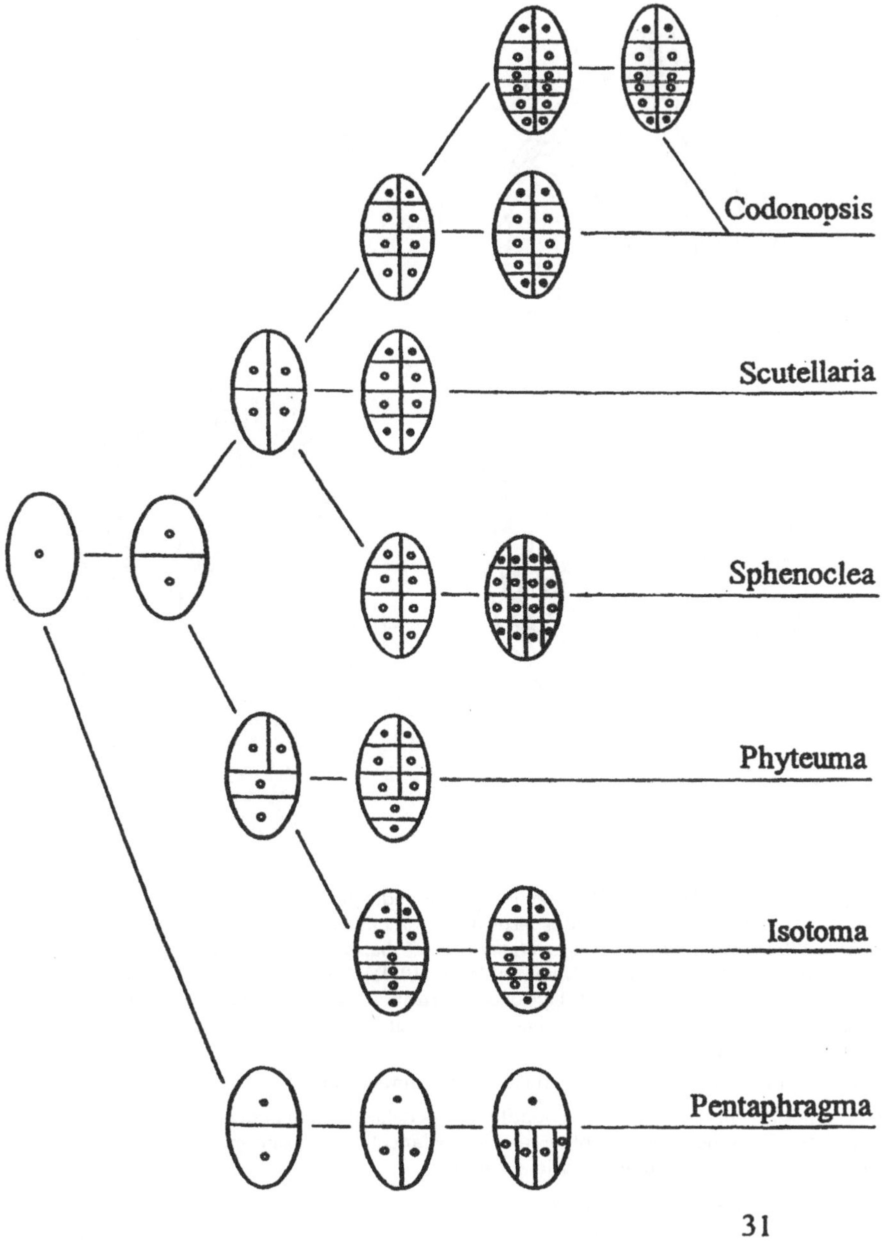

Plate XVII (*Contd.*)

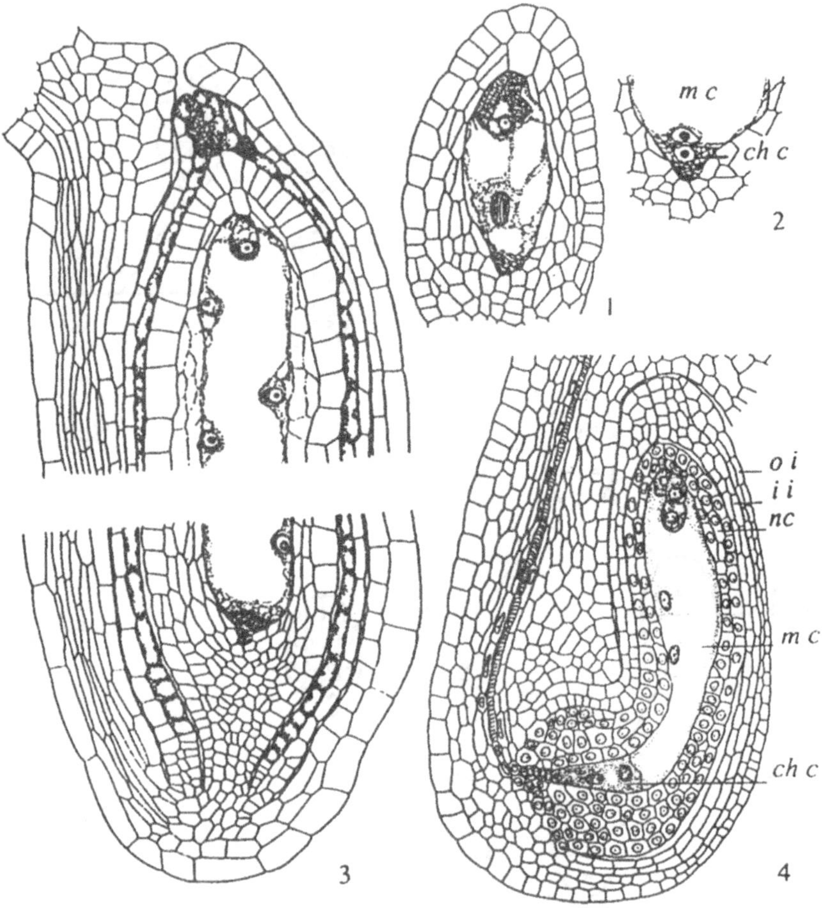

Plate XVIII: Helobial type of endosperm development.

1-11—main steps of helobial endosperm development in monocotyledons —*Juncus filiformis* (1-3), *Sagittaria sagittifolia* (4), *Polygonatum roseum* (5, 5a, 5b, 8), *Tofieldia japonica* (6, 7), *Asphodelus tenuifolius* (9) and dicotyledons *Nymphaea stellata* (10, 11): 1—7, 10, 11—early development, 8, 9—chalazal portion of endosperm at stage of cellularization; 12-19—protein dynamics in developing seed of *Luzula pedemontana* (dark shading—high amount of protein; dots—average amount; white—low amount).

Bar—0.03 mm (12, 17a, 18a), 0.1 mm (13-16), 0.3 mm (17-19).

a—antipodal cell, *c b*—conductive bundle, *ch c*—chalazal cell of endosperm, *em*—embryo, *hy*—hypostase, *i i*—inner integument, *m c*—micropylar cell of endosperm, *nc*—nucellus, *o i*—outer integument, *pd*—podium, *p t*—parietal tissue.

1-3—after Anisimova, 1990; 4—after Nikiticheva, 1990; 5, 8—after Satarova, 1990; 6, 7—after Ono, 1929; 9—after Eunus, 1952; 10, 11—after Khanna, 1967; 12-19—after Shamrov and Anisimova, 1993b.

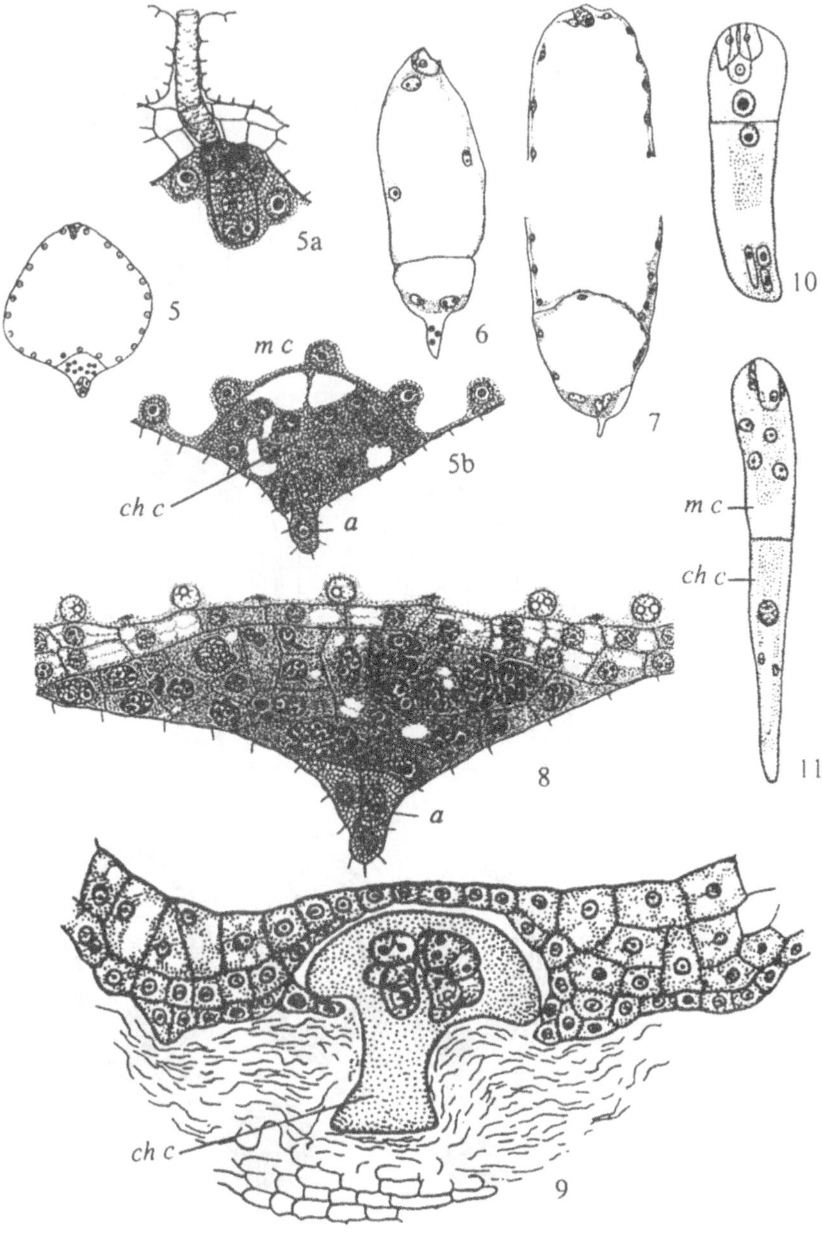

Plate XVIII (*Contd.*)

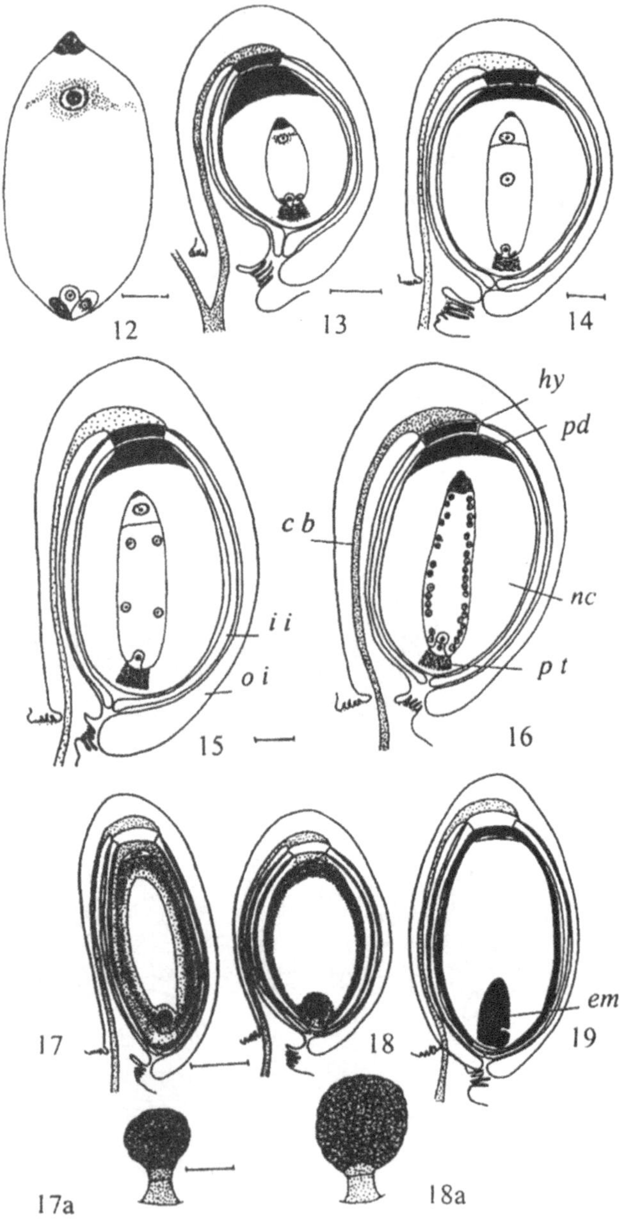

Plate XVIII (*Contd.*)

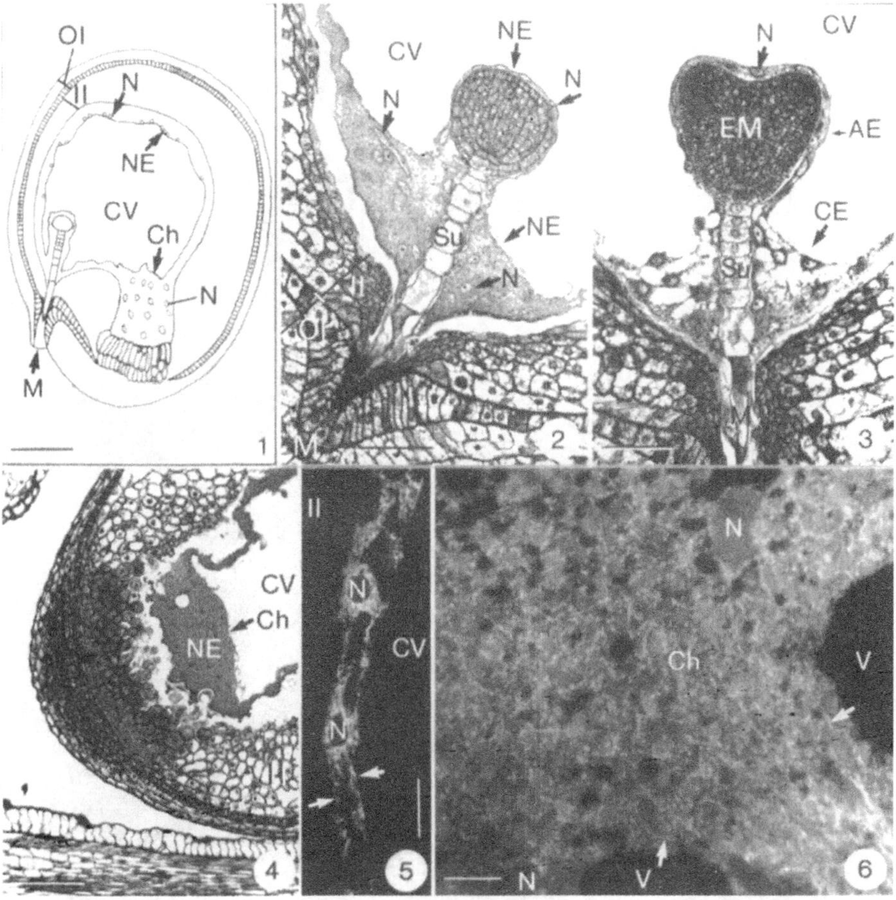

Plate XIX: Light microscopical study of endosperm formation in *Brassica napus* L. 1—topographic overview of median section of rapeseed at 6 days after anthesis showing globular embryo and nuclear endosperm; 2-4—light micrographs of median sections of Technovit-embedded rapeseed: 2—(micropylar part) at 6 days after anthesis, 3—at 9 days after anthesis, note vacuolation and onset of cellularization of endosperm surrounding suspensor, embryo proper is still surrounded by layer of nuclear endosperm, 4—at 6 days after anthesis showing nuclear endosperm at chalazal region of embryo sac; 5, 6—fluorescent micrographs of microtubular cytoskeleton: 5—in sectioned alveolar endosperm of rapeseed at 12 days after anthesis, note fluorescent bundles of microtubules surrounding the endosperm nuclei (arrows), 6—in nuclear endosperm of sectioned rapeseed (chalazal region) at 6 days after anthesis, note short arrays of fluorescent Mts distributed throughout the cytoplasm; 7—topographic overview of a median section of rapeseed at 14 days after anthesis showing torpedo-shaped embryo and cellular endosperm; 8—light micrograph of median section of Technovit-embedded rapeseed at 14 days after anthesis showing cellular endosperm and base of embryo; 9—detail of cellular endosperm at 14 days after anthesis, note presence of arrays of

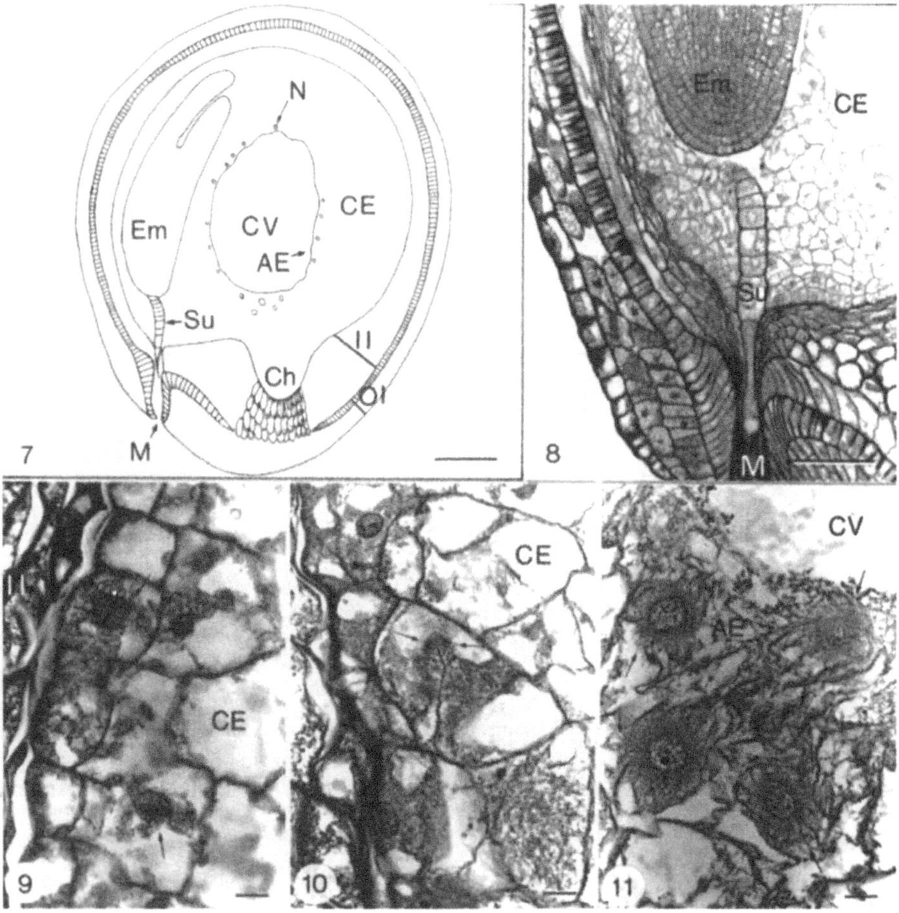

Plate XIX (*Contd.*)

Mts in mitotic spindles (arrows); 10—detail of cellular endosperm at 14 days after anthesis, outer cell layers of endosperm exhibit cell divisions, note presence of arrays of Mts at outer margins of phragmoplast which forms cell plate (arrows); 11—light micrograph of median section of Technovit-embedded rapeseed at 14 days after anthesis showing area of alveolar endosperm bordered by cellular endosperm and central vacuole, arrays of Mts radiate from nuclear surfaces and sometimes point to place where freely growing wall-ends are being formed.

Bar—200 µm (1, 7), 100 µm (2-4, 8), 20 µm (5), 10 µm (6,9,10,11). AE—alveolar endosperm, Ch—chalazal region of endosperm, CE—cellular endosperm, CV—central vacuole of endosperm, En—endosperm, Em—embryo proper, II—inner integument, M—micropyle, Mt—microtubule, OI—outer integument, N—nucleus, NE—nuclear endosperm, Su—suspensor, V—vacuole.

1-11—after van Lammeren, Kieft, Ma and van Veenendaal, orig.

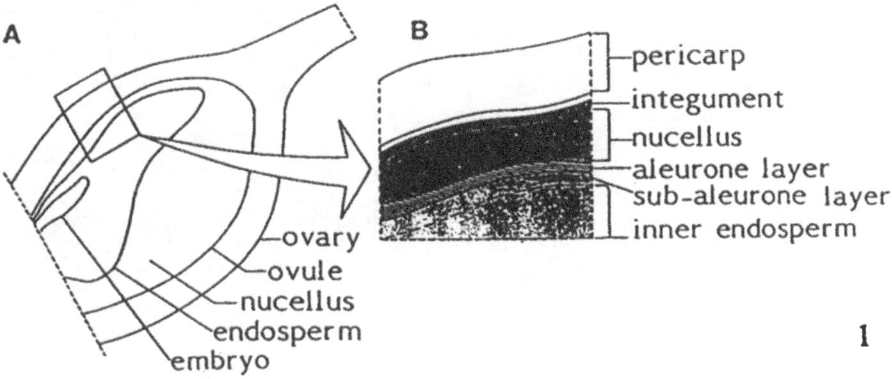

1

Plate XX: Cell differentiation in pericarp and endosperm of developing maize kernels (*Zea mays* L.).

1—schematic representation of median section of young caryopsis with sampling area boxed (A) and anatomy of sampling block (B); 2—low magnification light micrographs of 2 µm Epon sections from sampling area of developing caryopsis, note enlargement, differentiation and subsequent degeneration of endocarp, mesocarp and exocarp cells, integument and subepidermal nucellus cells degenerate (a-b), endosperm accumulates proteins and starch: a—3 DAP, b—8 DAP, c—17 DAP, d—20 DAP; 3—electron micrographs of developing maize caryopsis, conspicuous changes in aleurone and subaleurone layers include differentiation of epidermal-like tissue (a-b) and accumulation of starch (b-c) and storage products in two types of spherosomes, aleurone bodies and endosperm protein bodies: a—5 DAP, b—12 DAP, c—22 DAP; 4—high magnification electron micrographs of aleurone cells of endosperm, note accumulation of spherosomes and development of aleurone bodies from aleurone vacuoles: a—8 DAP, b—12 DAP, c-d—19 DAP, e—22 DAP; 5—high magnification electron micrographs of cytoplasm of subaleurone and inner endosperm cells, note changes in ER configuration, development and accumulation of endosperm protein bodies from ER and appearance of amyloplasts: a—5 DAP, b—8 DAP, c—12 DAP, d—17 DAP, e—19 DAP, f—22 DAP; 6—electron micrographs of cytoplasm and cortical microtubules (arrows) in ultrathin Epon sections of endosperm. Glutaraldehyde-OsO_4 fixation and Pb-citrate and Ur-acetate post staining: a—inner endosperm at 8 DAP, b—aleurone cells at 8 DAP, c—aleurone cells at 17 DAP.

Bar—100 µm (2), 10 µm (3), 1 µm (4-6). AB—aleurone body, AL—aleurone layer, AV—aleurone vacuole. CW—cell wall, D—dictyosome, DAP—days after pollination, EC—endocarp, EM—embryo, END—endosperm, EXC—exocarp, EPB—endosperm protein body, I—integument, IE—inner endosperm, IS—intercellular space, M—mitochondrion, MC—mesocarp, N—nucleus, NUC—nucellus, PC—pericarp, R—ribosome, RER—rough endoplasmic reticulum, S—starch, SAL—subaleurone layer, S1—spherosome type 1 presents in aleurone and subaleurone layer, S2—spherosome type 2 presents in the aleurone layer from 8 DAP onwards.

1-6 — after van Lammeren, Kieft and Schel, orig.

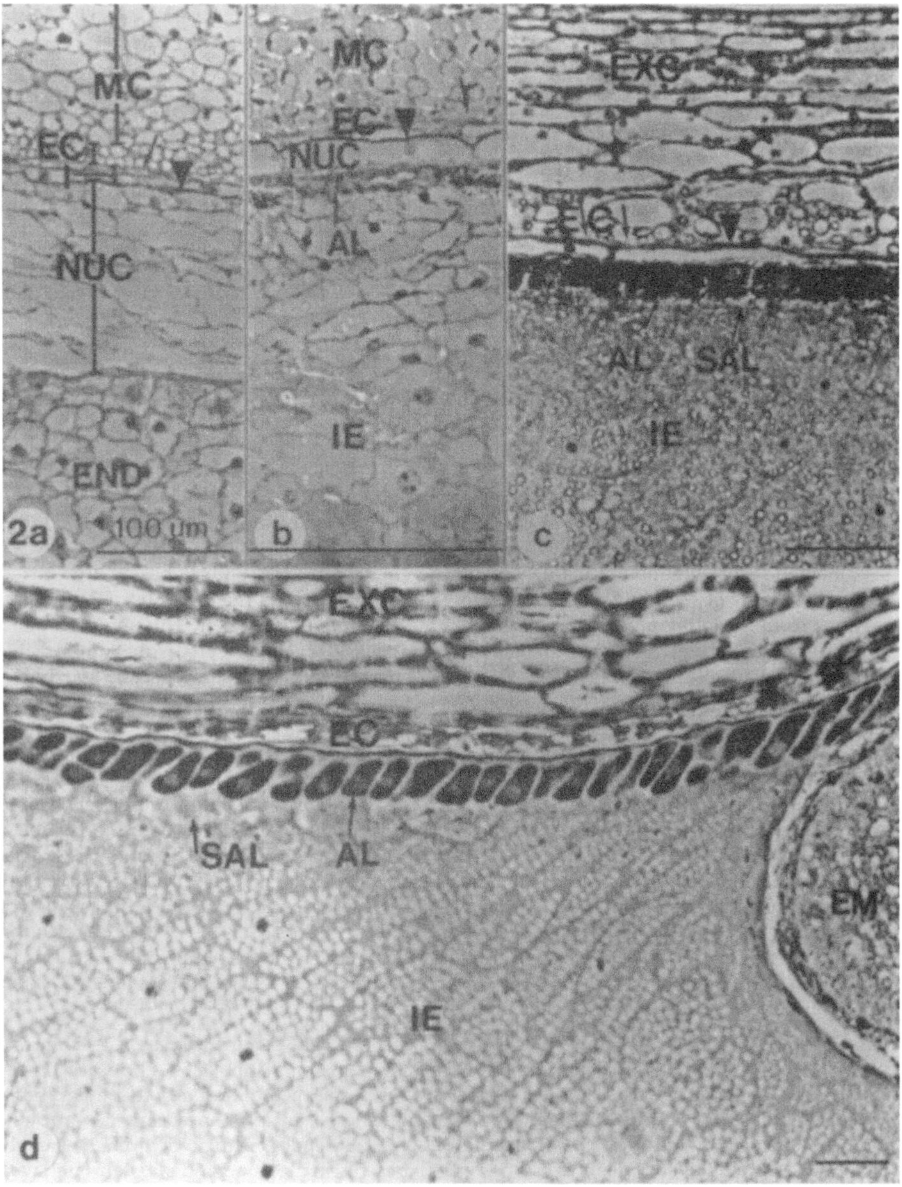

Plate XX (*Contd.*)

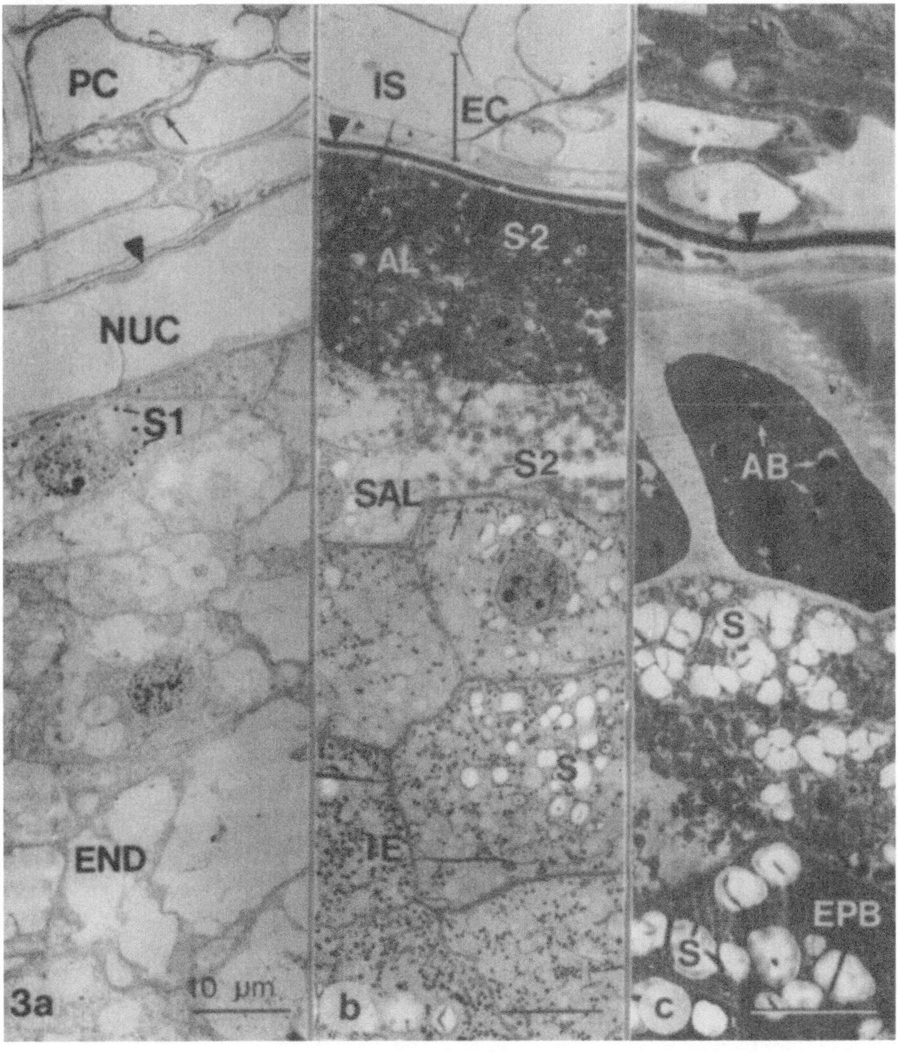

Plate XX (*Contd.*)

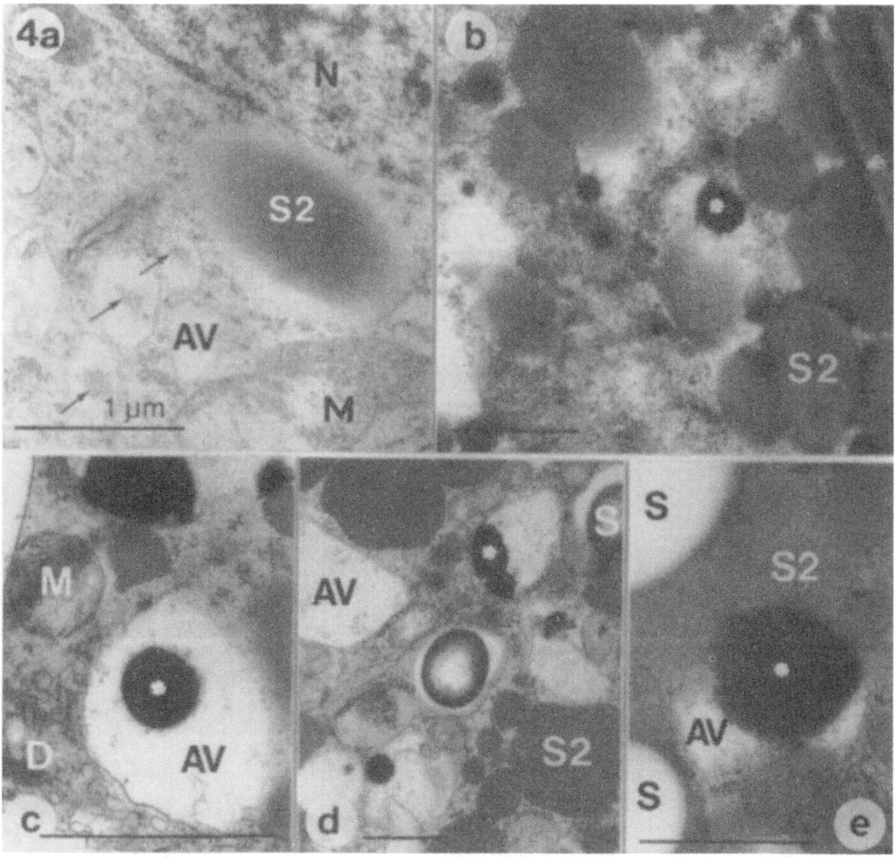

Plate XX (*Contd.*)

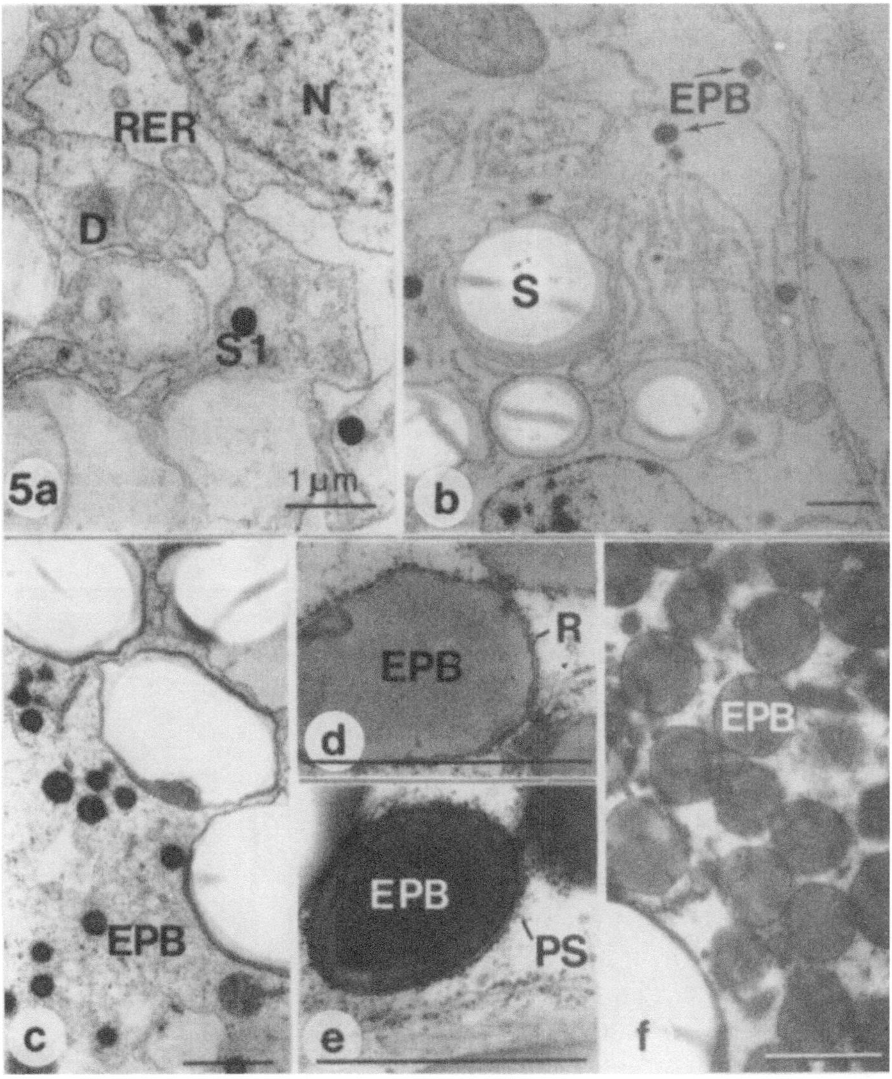

Plate XX (*Contd.*)

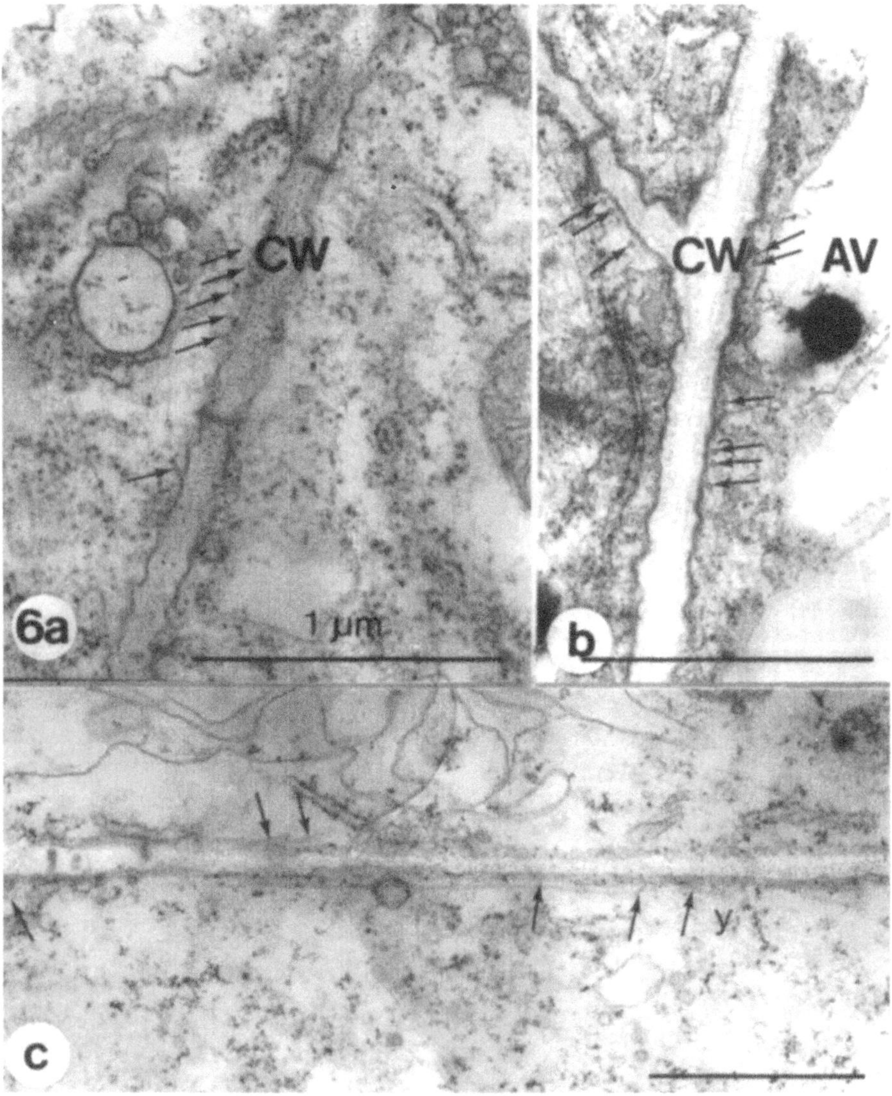

Plate XX (*Contd.*)

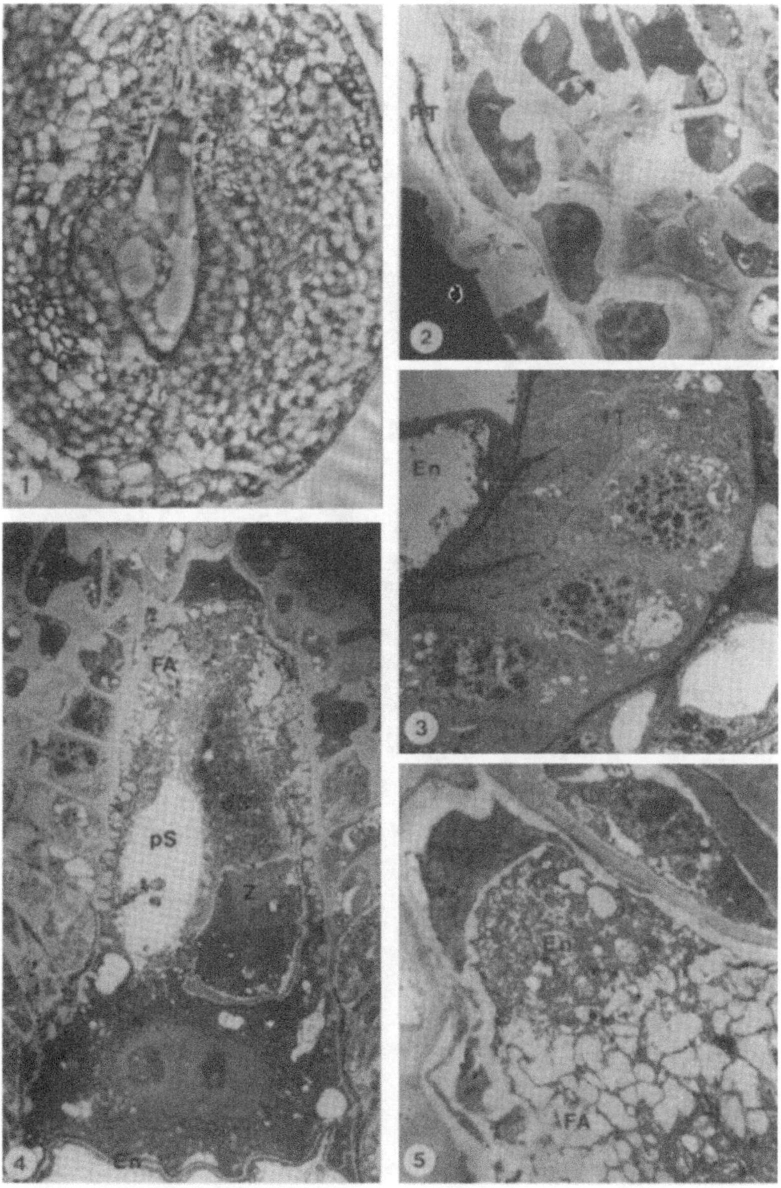

Plate XXI: Micropylar haustorium of *Impatiens walleriana* Hook. ex Arn.
1—longitudinal section through ovule, showing unfertilized embryo sac, integumentary tapetum and thick-walled integumentary cells near micropyle, ×300; 2—thick-walled integumentary tissue, ×2200; 3—integumentary tapetum shortly after fertilization, ×2200; 4—zygote, two-cellular endosperm, persistent and degenerated synergid, and pollen tube; micropylar endosperm cell grows through filiform apparatus into micropyle, ×1600;

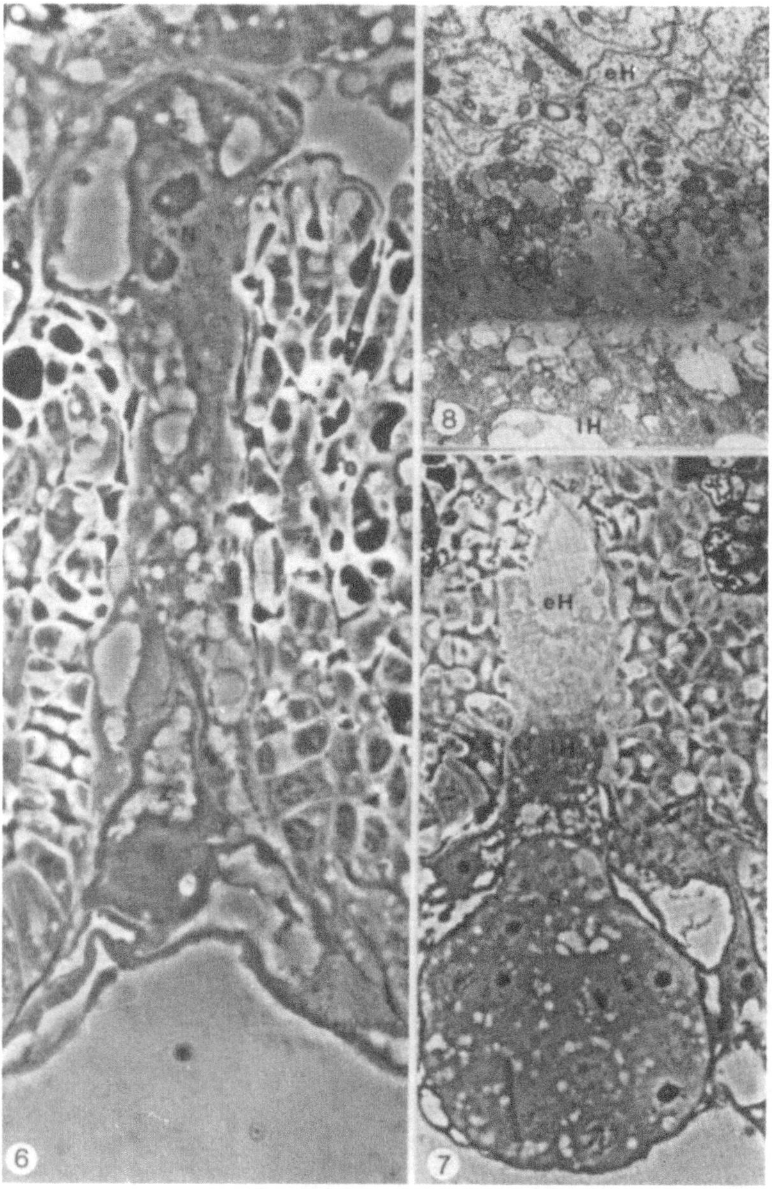

Plate XXI (*Contd.*)

5—endosperm cell growing into micropyle, ×7500; 6—haustorium protruding from micropyle, containing large nucleus, zygote has not yet divided, ×1500; 7—globular embryo with short suspensor, haustorium with internal and external parts, separated by a wall, ×1000; 8—wall between internal and external haustorium parts, ×6000; 9—haustorium growing into funiculus,

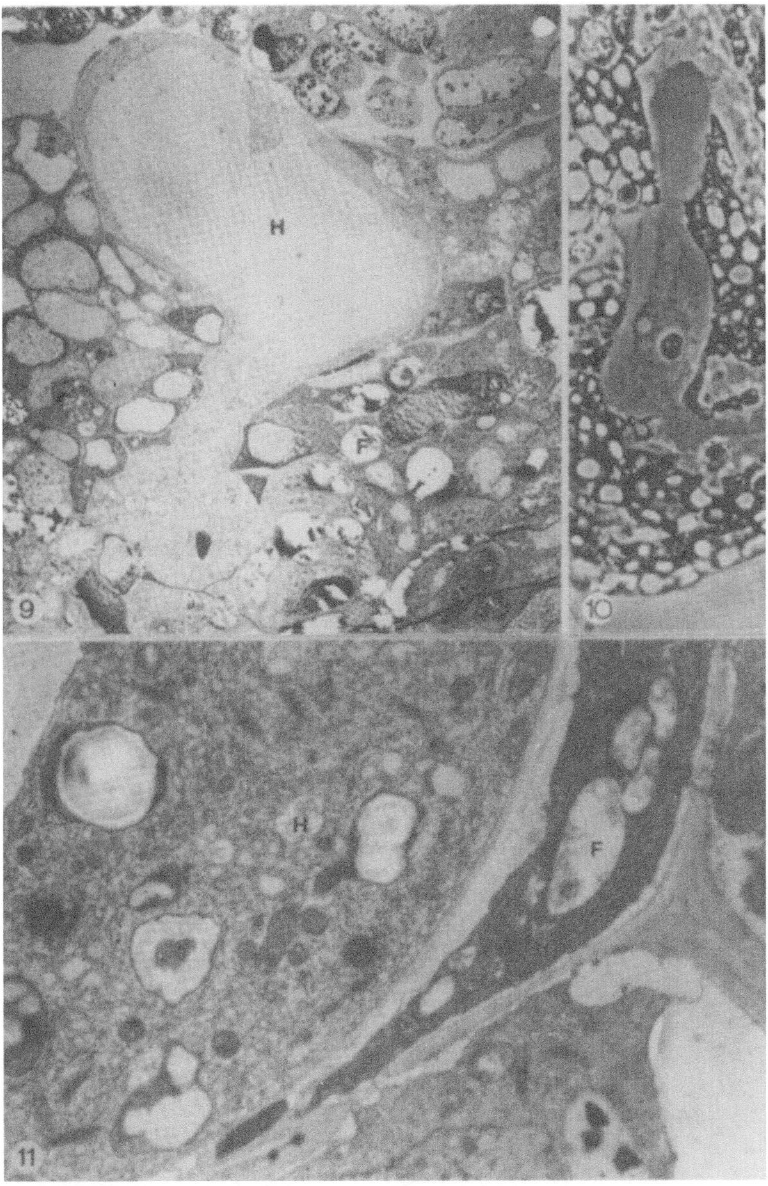

Plate XXI (*Contd.*)
×2200; 10—survey of haustorium, showing its large nucleus, ×400; 11—portion of haustorium and adjacent degenerating funiculus cell, ×13000.

dS—degenerated synergid, *eH*— external part of haustorium, *En*—endosperm, *F*—funiculus, *FA*—filiform apparatus, *H*—haustorium, *iH*—internal part of haustorium, *IT*—integumentary tapetum, *N*—nucleus, *pS*—persistent synergid, *PT*—pollen tube, *S*—suspensor, *Z*—zygote.

1–11—after van Went, orig.

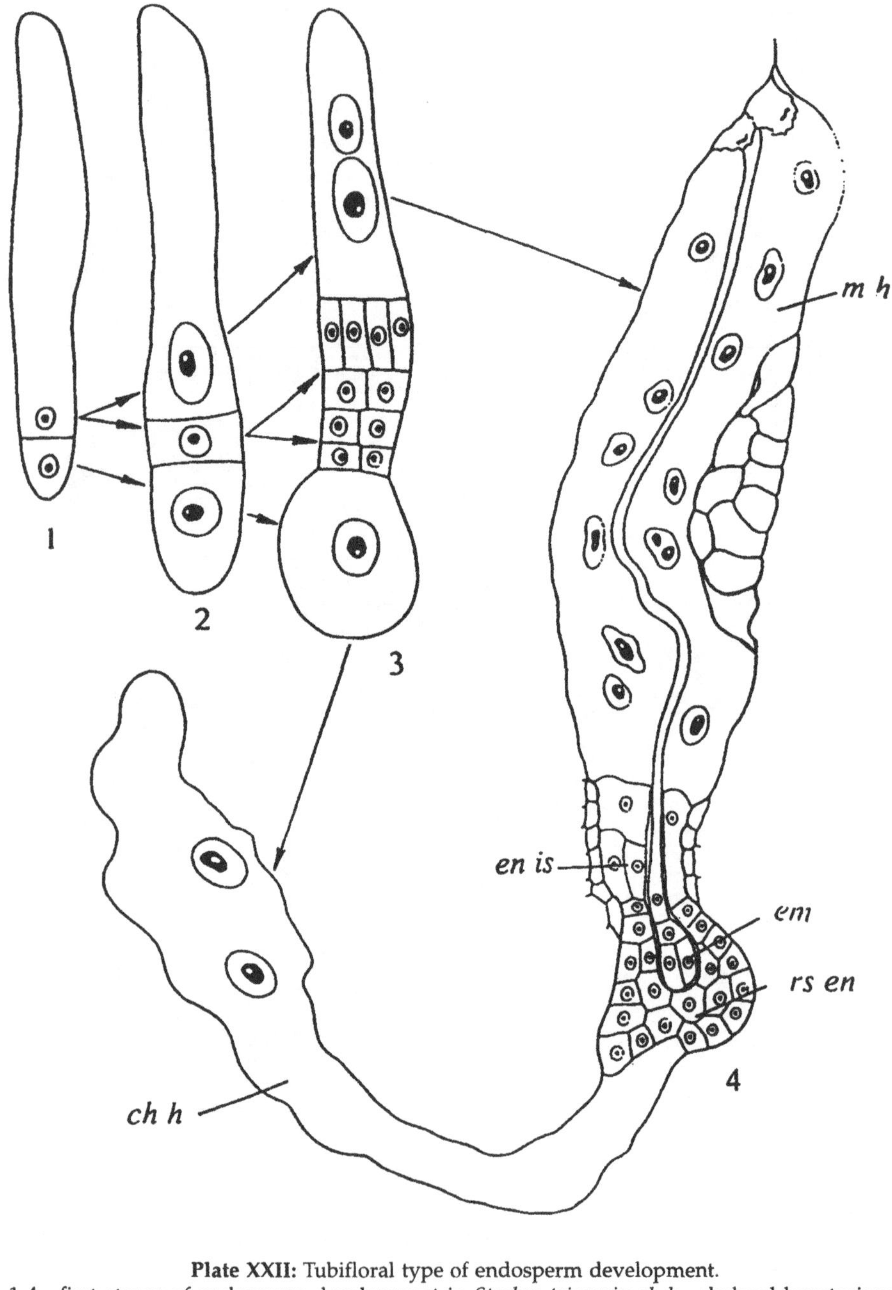

Plate XXII: Tubifloral type of endosperm development.

1-4—first stages of endosperm development in *Stachys trinervis*; *ch h*—chalazal haustorium, *em*—embryo, *en is*—endosperm isthmus, *m h*—micropylar haustorium, *rs en*—reserved endosperm.

1-4—after Kamelina, orig.

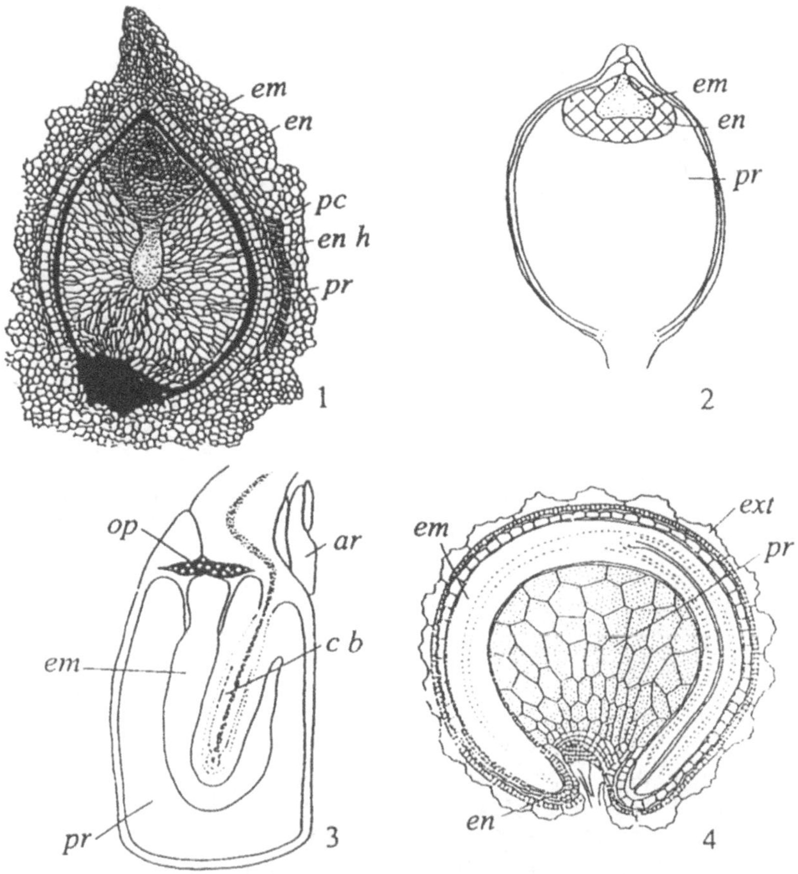

Plate XXIII: Perisperm.

1-6—longitudinal median sections (and section schematic representations) of seeds with perisperm, applied to different groups (explanation in text): first group (*Saururus cernuus*—1, *Houttuynia cordata*—2, *Calathea picturata*—3), second group (*Stellaria media*—4), third group (*Heliconia psittacorum*—5), fourth group (*Climacoptera crassa* —6); 7-10—*Nuphar lutea* (first group of seeds): 7—mature ovule, 8—mature seed (area of perisperm shown by small dots, central portion of podium with tannins and lignified cell walls shown by large dots, lateral portion of podium with thin cell walls—by net, narrow cells in lateral portion of nucellus—by slanting line), 9, 10—section fragments of lateral portions of podium and perisperm, transforming from certain part of nucellus at stage of globular embryo (9), and fully formed in mature seed (10, detail of fig. 8); these fragments show occurrence of starch in their cells (transitory—small grains and reserved—large ones).

ar—arilloid, *c b*—conductive bundle, *em*—embryo, *en*—endosperm, *end*—endocarp, *en h*—endosperm haustorium, *exc*—exocarp, *ext*—exotesta, *hy*—hypostase, *i i*—inner integument, *m*—micropyle, *nc c*—nucellar cup, *n c nc*—narrow cells of nucellus, *o i*—outer integument, *op*—operculum, *pc*—pericarp, *pd*—podium, *pr*—perisperm, *pst*—postament.

1—after Jöhnson, 1900a; 2—after Nikiticheva, 1981c; 3—after Grootjen, 1983; 4—after Gibbs, 1907; 5—after Humphrey, 1896; 6—after Butnik, 1991; 7-10—after Shamrov, 1998a.

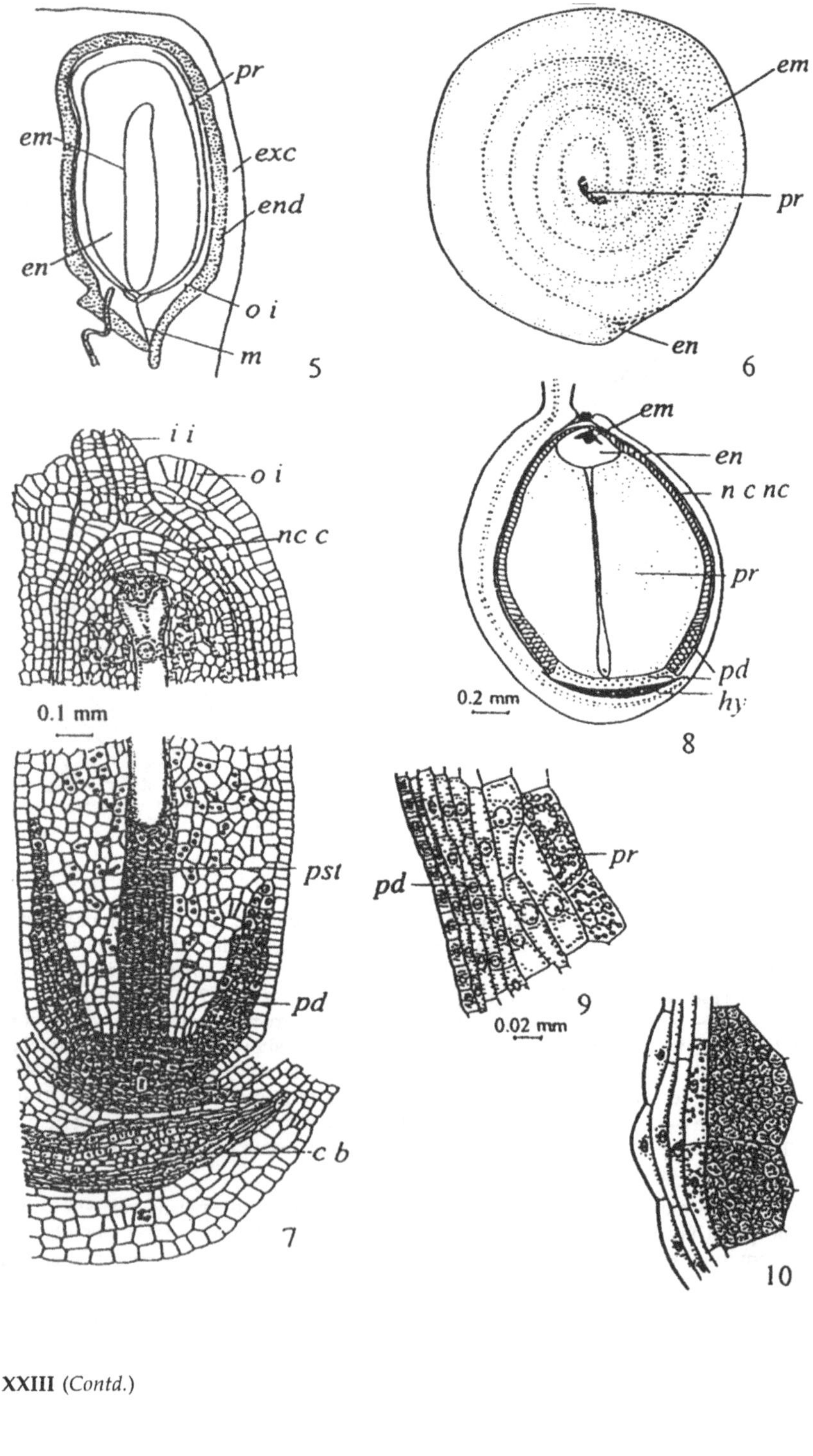

Plate XXIII (*Contd.*)

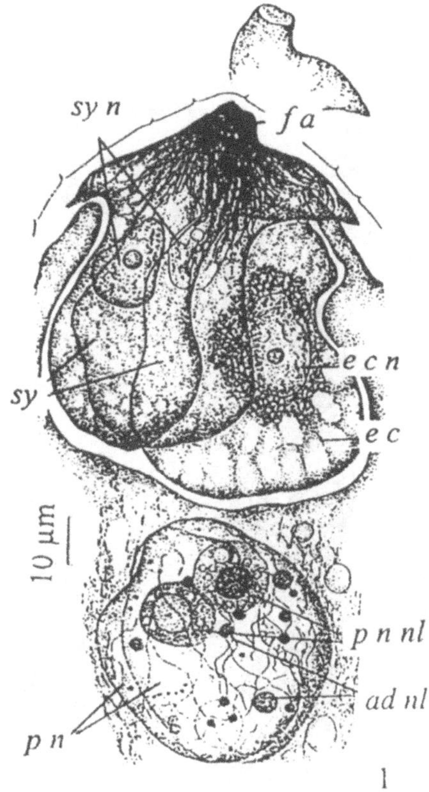

Plate XXIV: Zygote.
1—egg apparatus and part of central cell with polar nuclei (one lies on the other) in the mature embryo sac of *Triticum aestivum*, variety Diamant, above to the right—detail of upper part of synergid ('funnel' and tooth-like outgrowth); 2-5—structural peculiarities of cell walls of zygote (*Dillenia suffruticosa*—2) and embryo (*Acrotrema arnottianum*—3-5): 2—cell wall of zygote substantially thickened, with large outgrowths, 3—two-cellular proembryo surrounded with swollen cover of zygote, outgrowths penetrate into endosperm, 4—end of proembryonic phase, outgrowths of cover mainly in proembryo basal part, 5—embryo with rudiments of cotyledons, remnants of cover around suspensor visible; 6—synthetic scheme representing zygote and proembryo development in *Petunia hybrida*: A—egg cell (after van Went, 1970b), B—young zygote in presynthetic period, C—zygote in post-synthetic period, D—two-cellular proembryo, E—four-cellular proembryo, arrows note the direction of growth (unidirected in zygote, radial at the *ca* level and later on in tiers *l* and *l′* of proembryo).

ad nl—additional nucleolus, *e c*—egg cell, *e c n*—egg cell nucleus, *f a*—filiform apparatus, *p n*—polar nuclei, *pn nl*—polar nuclei nucleoluses, *sy*—synergid, *sy n*—synergid nuclei; rest of legend explained in text (see Proembryo).

1—after Batygina, 1974b; 2—after Yoffe and Zhukova, 1974; 3-5—after Swamy and Periasamy, 1955; 6—after Vallade, 1980.

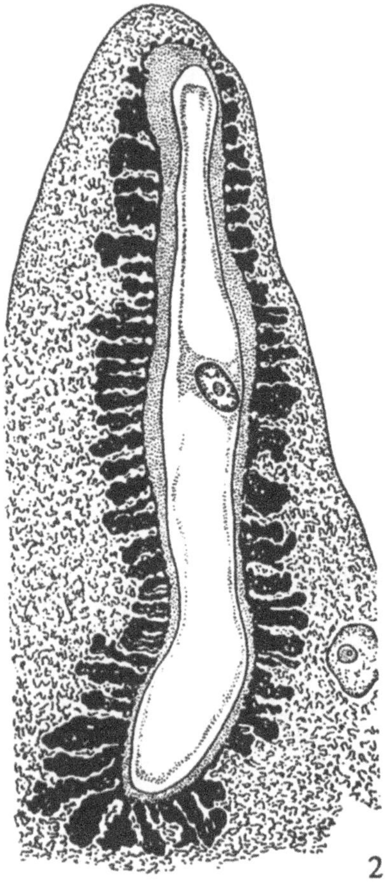

2

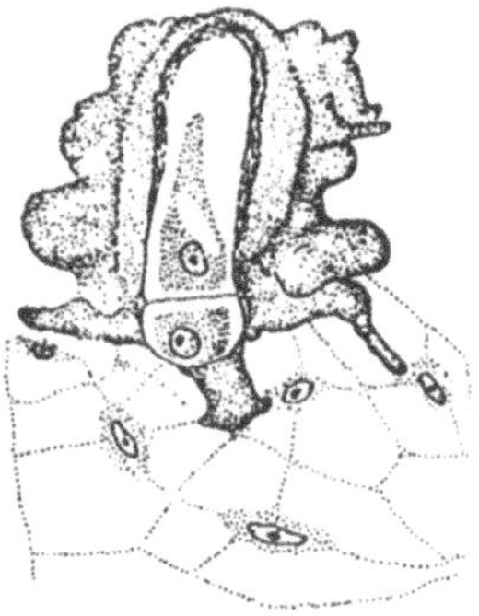

3

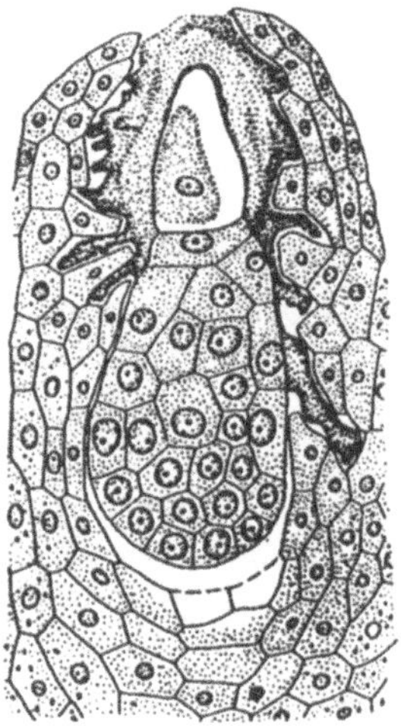

4

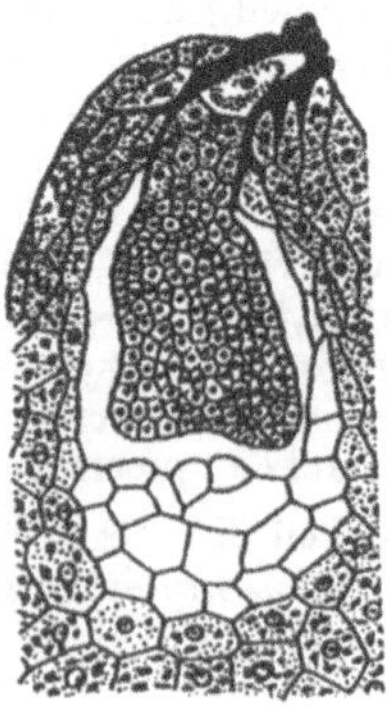

5

Plate XXIV (*Contd.*)

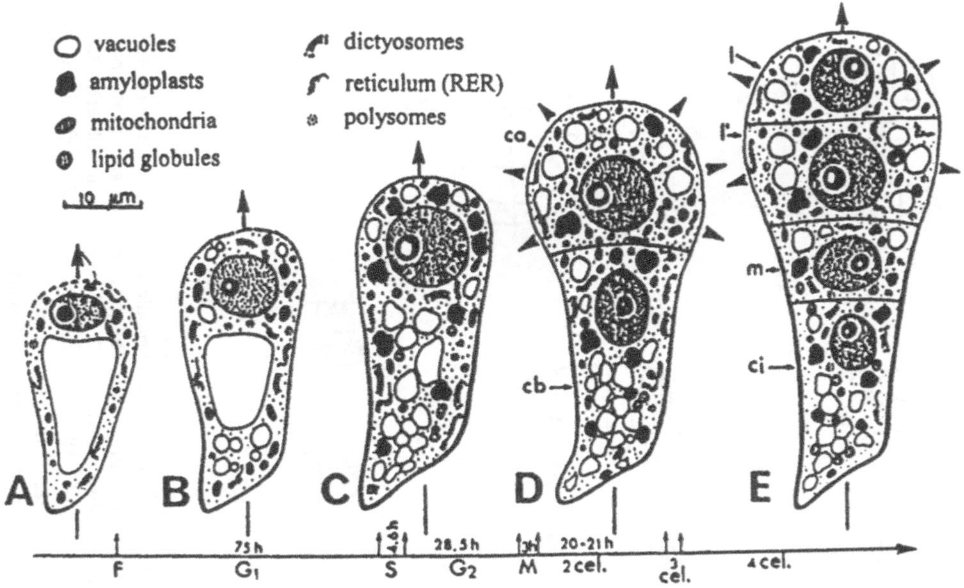

6

Plate XXIV (*Contd.*)

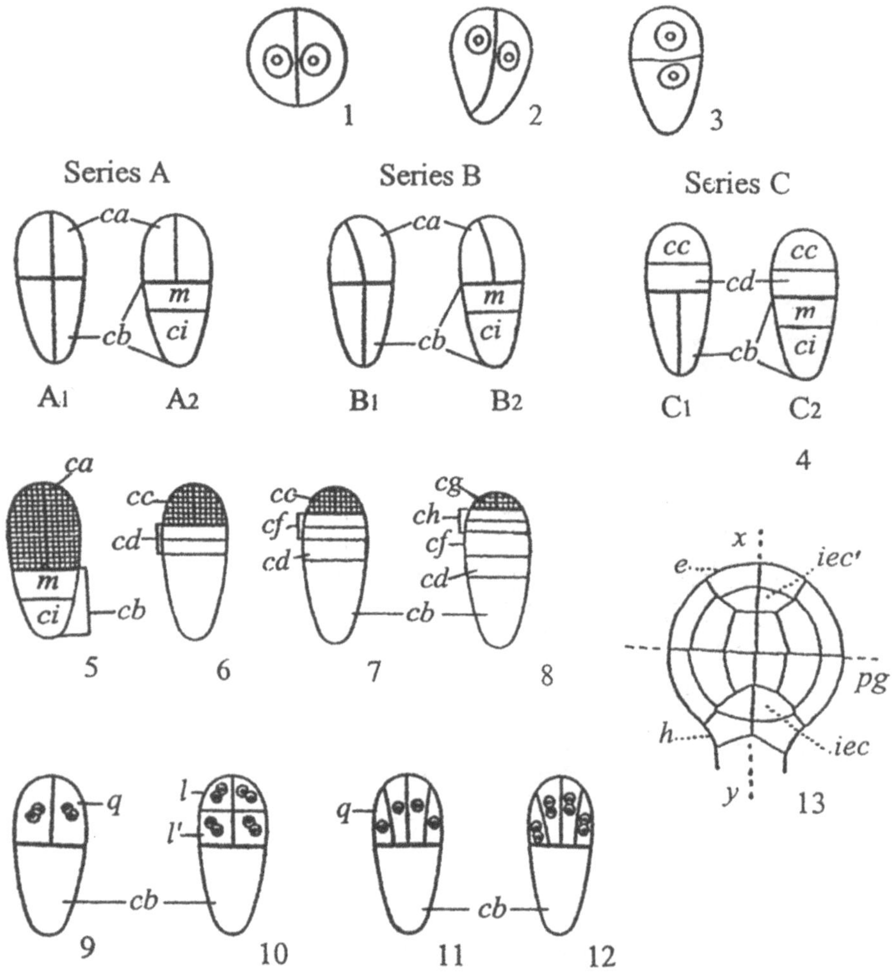

Plate XXV: Proembryo.

1-3—types of two-cellular proembryo; 4—types of proembryonic tetrads, as based on orientation of division plane in apical cell (longitudinal in A series, oblique in B series and transverse in C series) and basal cell (longitudinal in subseries A1, B1, C1 and transverse in subseries A2, B2, C2); 5-8—types of proembryonic tetrads of 1-4th orders in C series of Souèges classification of embryogenesis types; 9-12—modes of octant formation via organization of two tiers *l* and *l'* (9, 10) and of one tier *q* (11,12); 13—scheme of disposition of main histogenes in proembryo at final stages of its formation (*x-y*—longitudinal and *pg*—equatorial axes of symmetry in proembryo); 14-61—geometrical types of proembryo: 14-24—ortoradial (*Myosurus minimus*), 25-35—linear (*Solanum sisimbrifolium*), 36-45—spiral (*Adonis autumnalis*), 46-61—irregular (*Datura stramonium*).

Rest of legend explained in text (see Proembryo).

1-3—after Souèges, 1937b; 4-12—after Natesh and Rau, 1984; 13—after Crété, 1963; 14-61—after Souèges, 1936b.

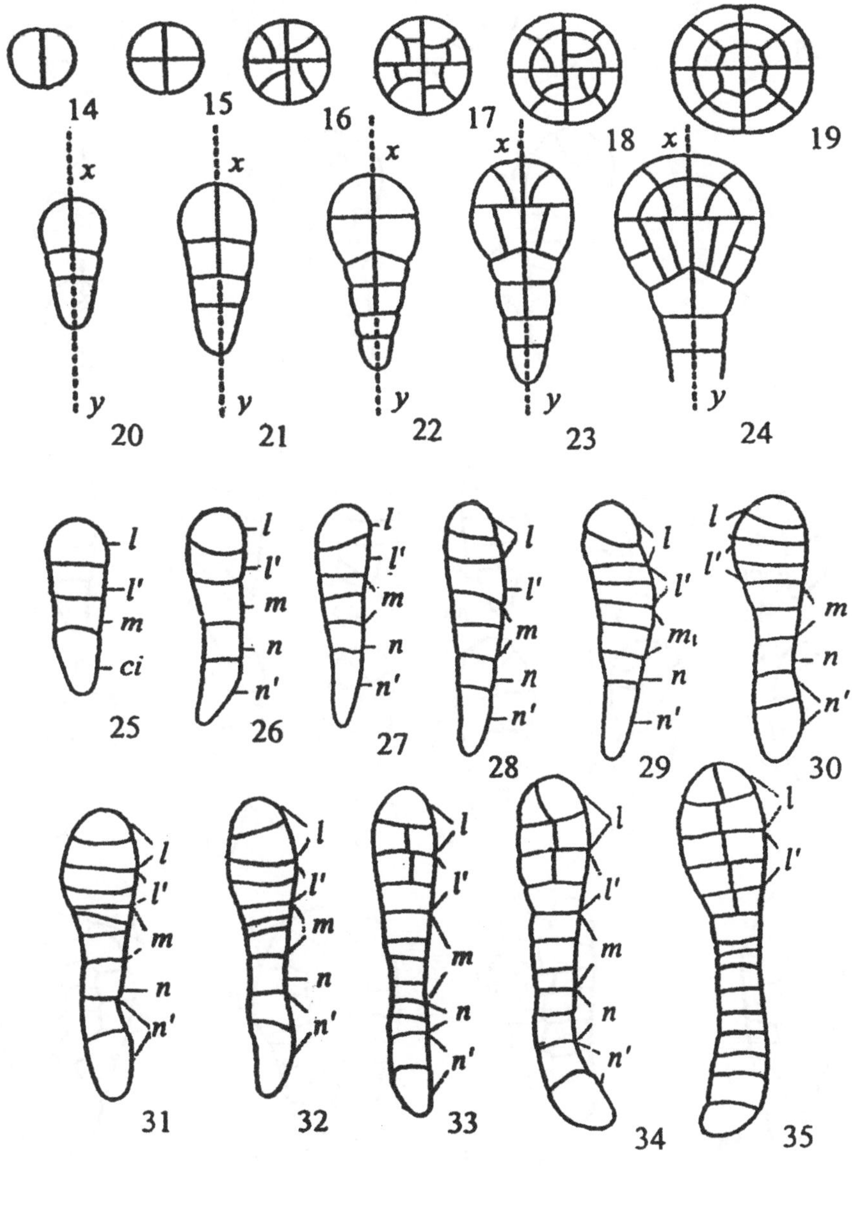

Plate XXV (*Contd.*)

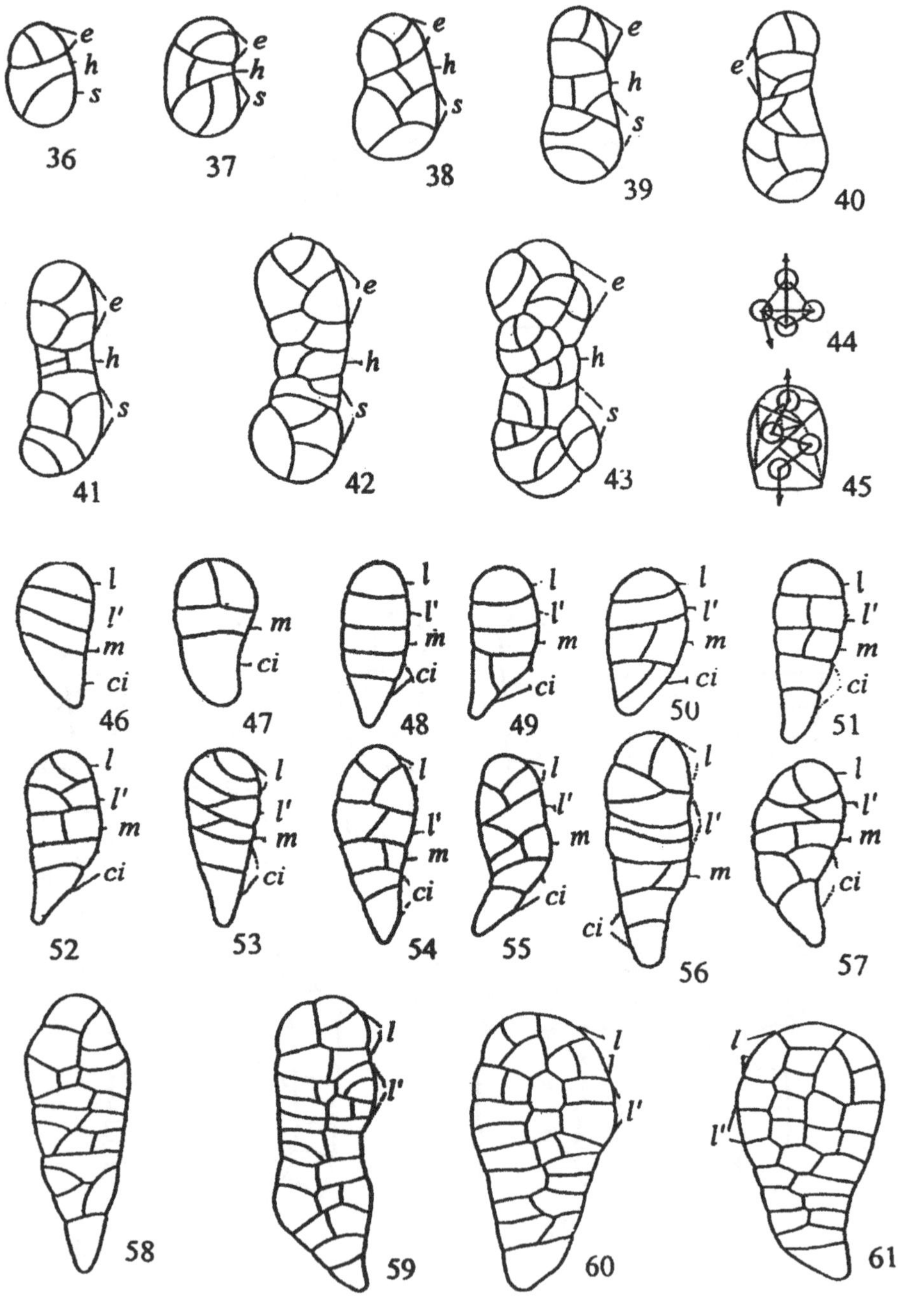

Plate XXV (*Contd.*)

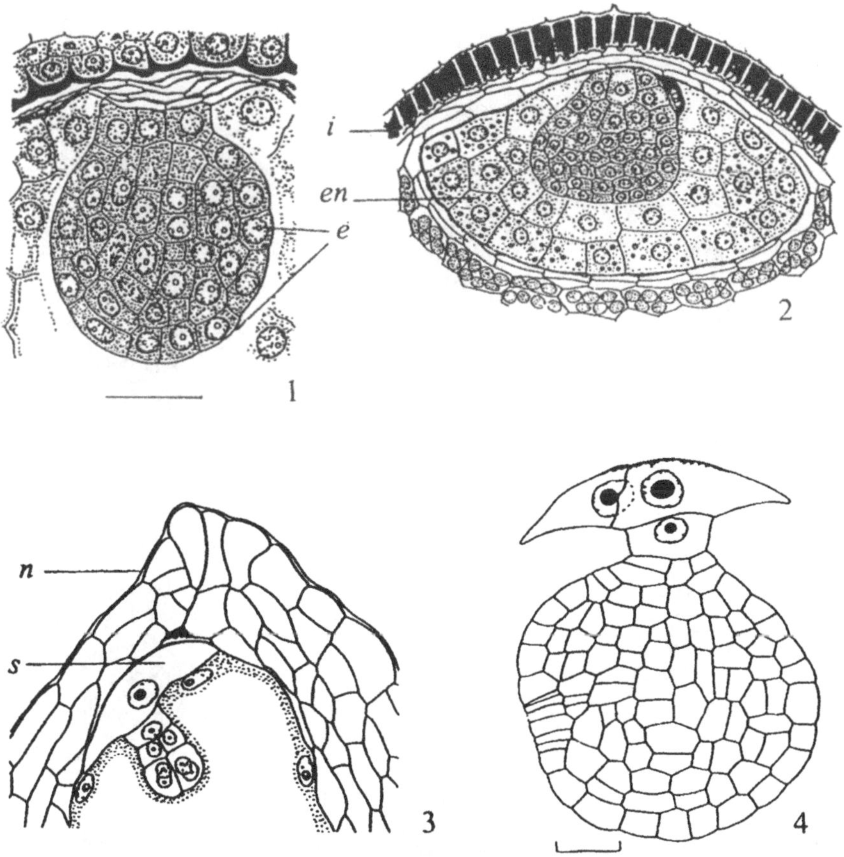

Plate XXVI: Suspensor.

1,2—proembryo with weakly pronounced suspensor in *Cynomorium songaricum* (1) and *Peperomia eburnea* (2); 3,4—suspensor with overgrown haustorial cells along border of nucellus, with labyrinth of basal cell wall and with endopolyploid nuclei in *Scheuchzeria palustris*; 5—filamentous suspensor dipping embryo proper into central part of endosperm in *Pinguicula vulgaris*; 6—filamentous suspensor outgoing from ovule and instilling itself in placenta in *Gymnadenia conopsea*; 7—suspensor differentiated to several zones: hypertrophic multinuclear cells in basal part, large uninucleate cells in middle part and small cells in 'isthmus' zone of suspensor, which are adjacent to embryo proper in *Cuscuta monogyna*; 8—cluster-like multicellular suspensor of *Convolvulus scammonia*; 9—complicated suspensor in *Sedum hispanicum* with small cells of 'isthmus' and main hypertrophic cell having endopolyploid nucleus, labyrinth of cell wall and forming narrow haustorial outgrowths in integument; 10—suspensor of *Cuscuta japonica*, formed by vesicle-like multinuclear cells.

e—embryo, *en*—endosperm, *i*—integument, *n*—nucellus, *p*—placenta, *s*—suspensor. Bars: 1,5,9—0.05 mm, 4—0.02 mm, 6—0.03 mm.

1—after Teryokhin *et al.*, 1975; 2—after Nikiticheva, 1981b; 3,10—after Nikiticheva and Proskurina, 1992; 4,9—after Nikiticheva, 1985a; 5—after Nikiticheva, 1987c; 6—after Shamrov and Nikiticheva, 1992; 7—after Nikiticheva and Teryokhin, 1987; 8—after Nikiticheva and Proskurina, 1989.

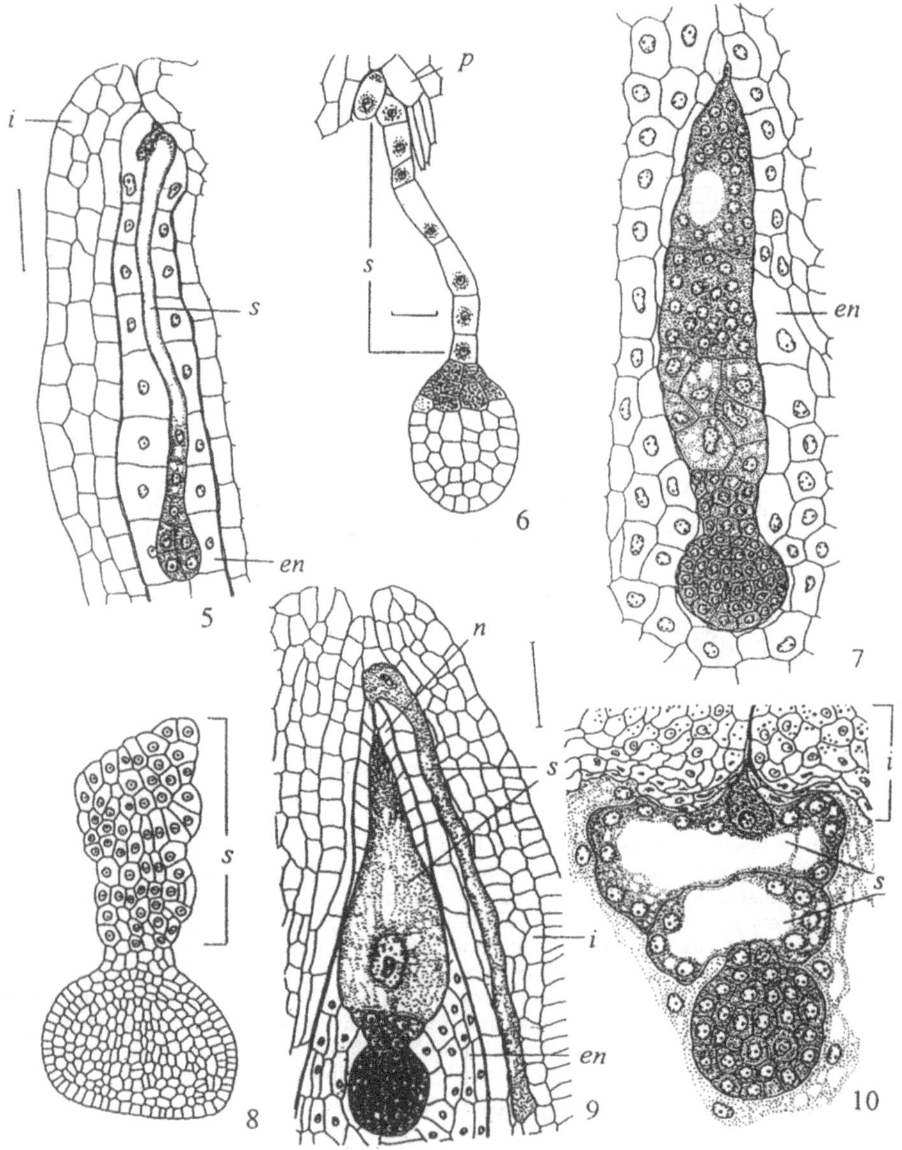

Plate XXVI (*Contd.*)

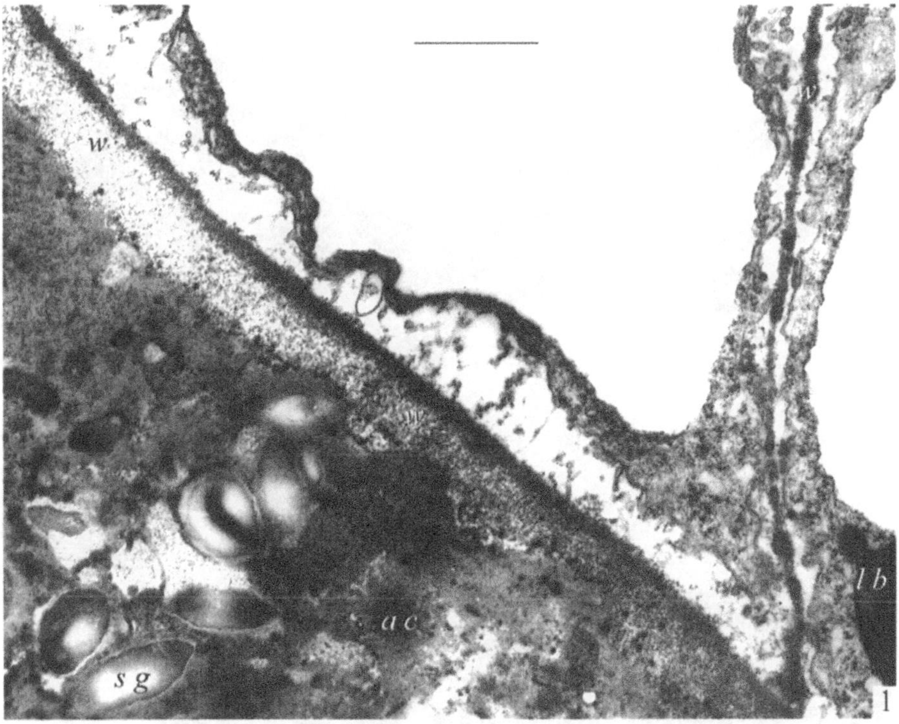

Plate XXVII: Ultrastructural characteristics of suspensor.
1-8—two-celled proembryo (4), three-celled proembryo (1-3), early and late globular embryo (5-8) of *Alcea rosea:* 4—basal cell much larger than apical one, highly vacuolized, nuclei of these cells differ in size and form, 1,3—fragments of apical and divided basal cell of proembryo, difference in starch grain size in amyloplasts of these cells observed, 2—fragment of proembryo basal part, storage substances in divided basal cell—starch and lipids, 5 and 6—two successive sections of early globular embryo, 7,8—two successive sections of late globular embryo, heteromorphism of suspensor cells shown, endosperm still free-nucleate; 9-14—two-celled (9-11) and four-celled (12-14) proembryo of *Epipactis atrorubens:* 9,12—general view of proembryo, 10—fragments of both proembryo cells, coenocytic endosperm and cells of inner integument, 11—fragments of proembryo basal cell and coenocytic endosperm, 13—fragments of basal and median proembryo cells, 14—fragments of basal and median cells of proembryo and coenocytic endosperm, at zone of contact of proembryo basal cell with endosperm lomasoma (arrow) observed; (4-9, 12—LM, 1-3, 10, 11, 13, 14—TEM).

Bars: 1 μm, 9,12—×500, 10—×9400, 11—×10,700, 13— ×8800, 14—× 9000; *ap*—amyloplast, *a c*—apical proembryo cell, *b c*—basal proembryo cell, *en*—endosperm, *e r*—endoplasmic reticulum, *l b*—lipid body, *m*—mitochondrion, *m c*—median cell of four-celled proembryo, *n*—nucleus, *nc*—nucellus, *n e*—nuclear envelope, *o w*—outgrowth of cell wall, *p m*—plasma membrane, *s g*—starch grain, *w*—cell wall.

1-8—Zhukova, orig.; 9-14—after Zhukova and Savina, 1978.

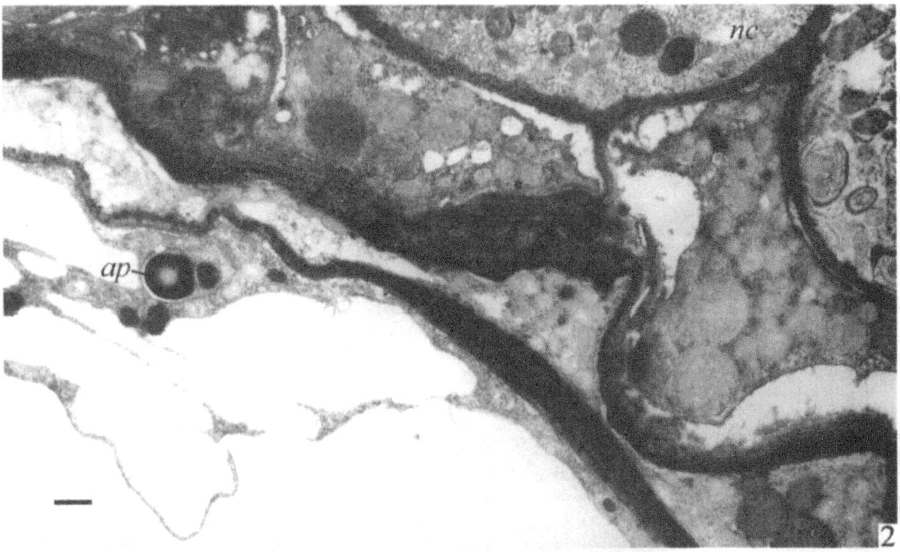

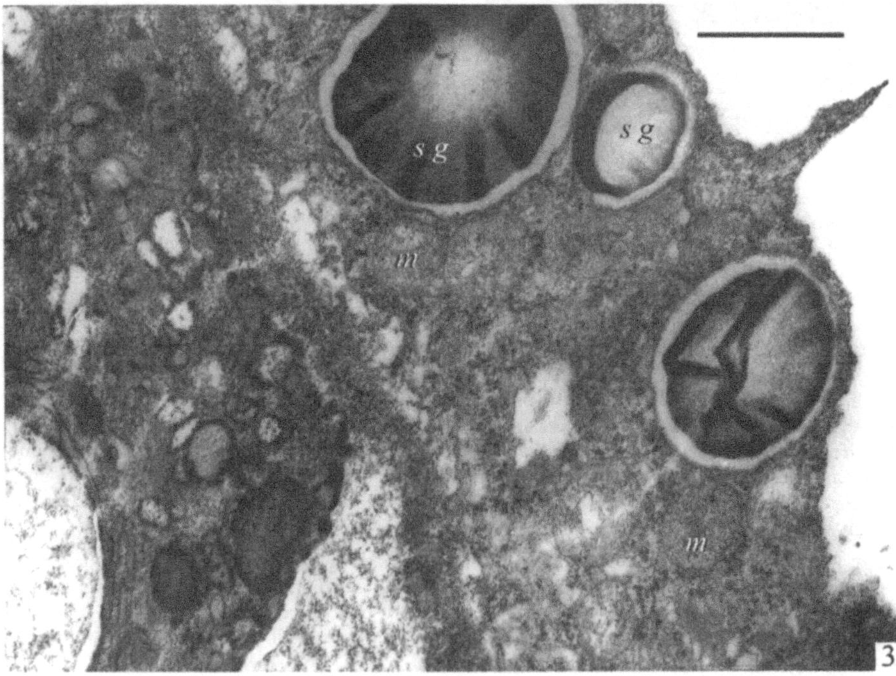

Plate XXVII (*Contd.*)

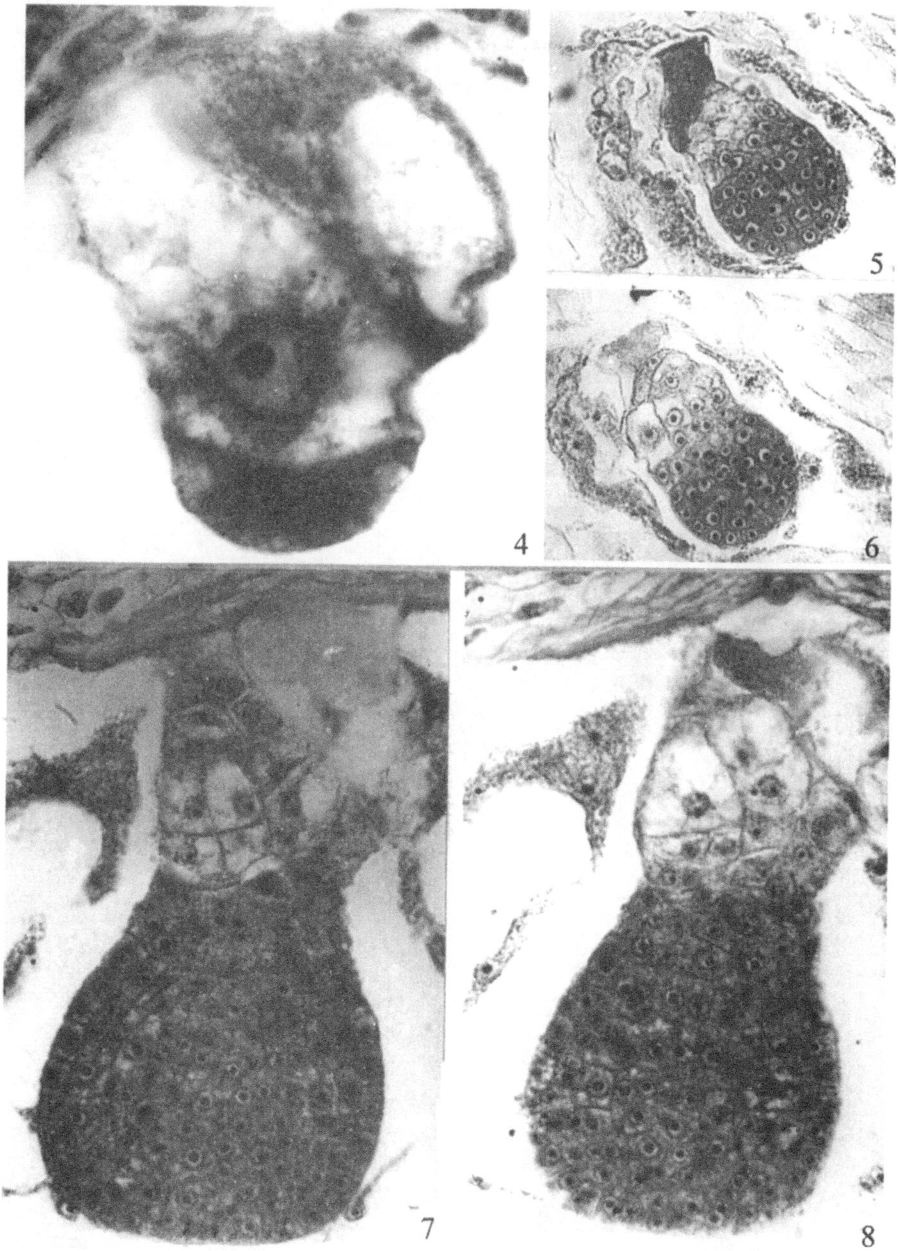

Plate XXVII (*Contd.*)

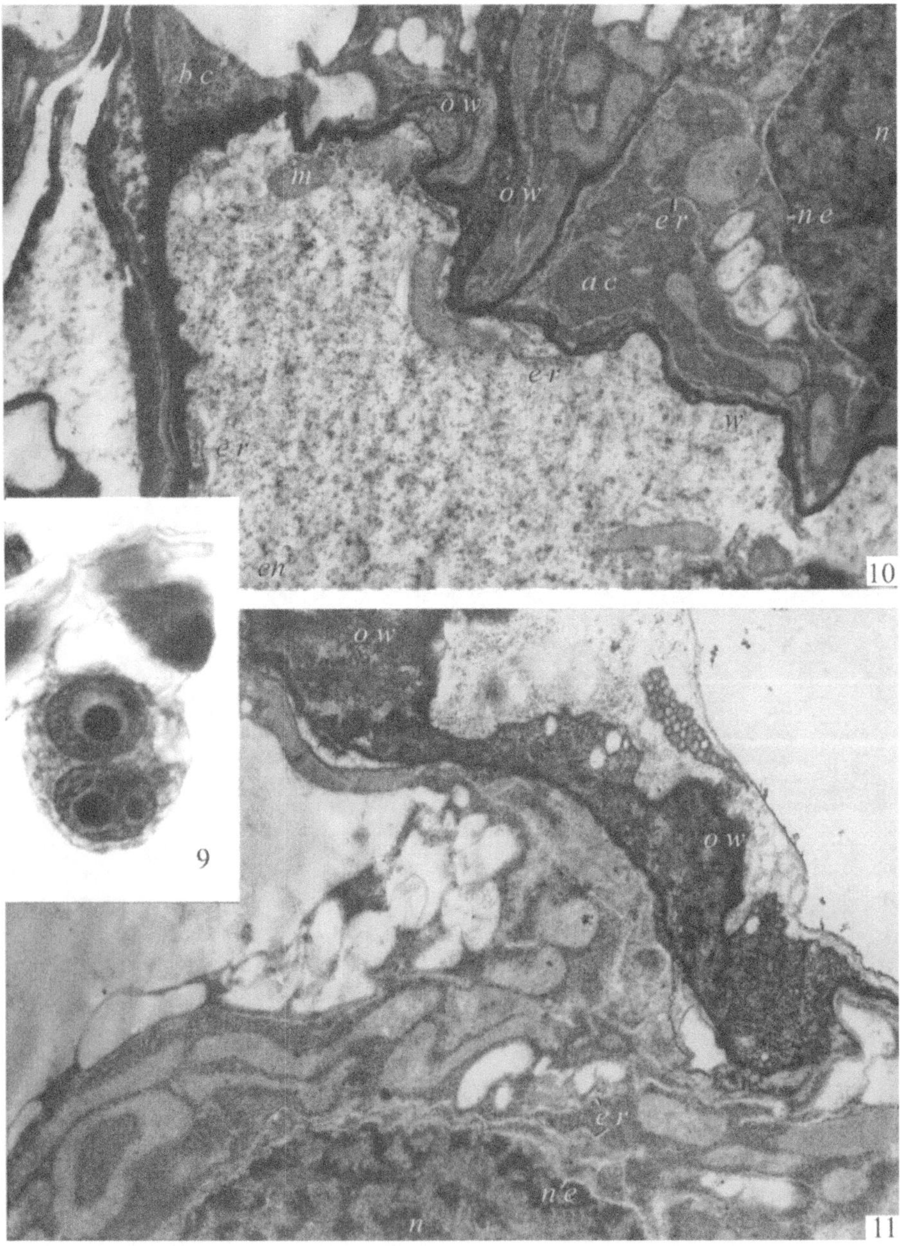

Plate XXVII (*Contd.*)

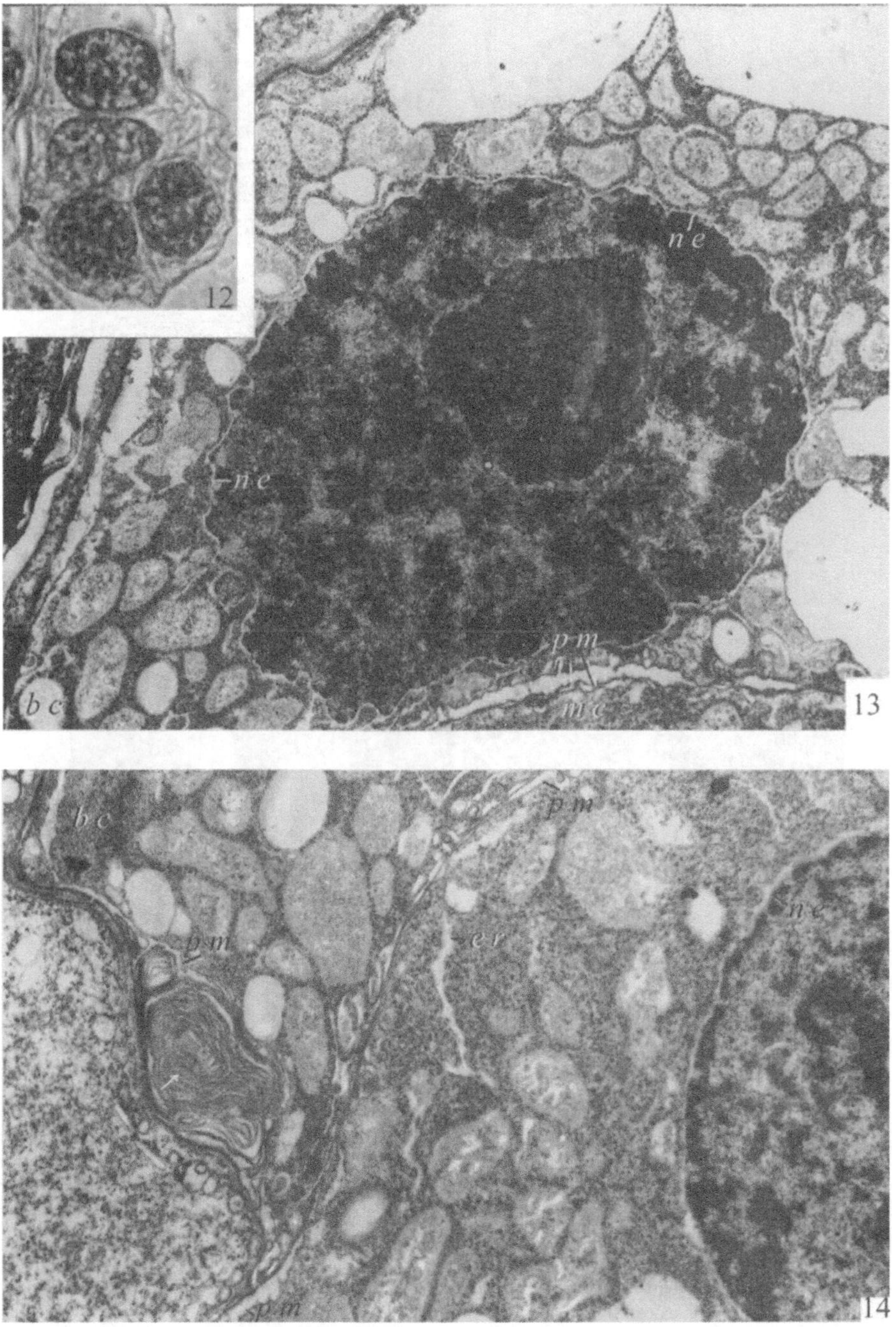

Plate XXVII (*Contd.*)

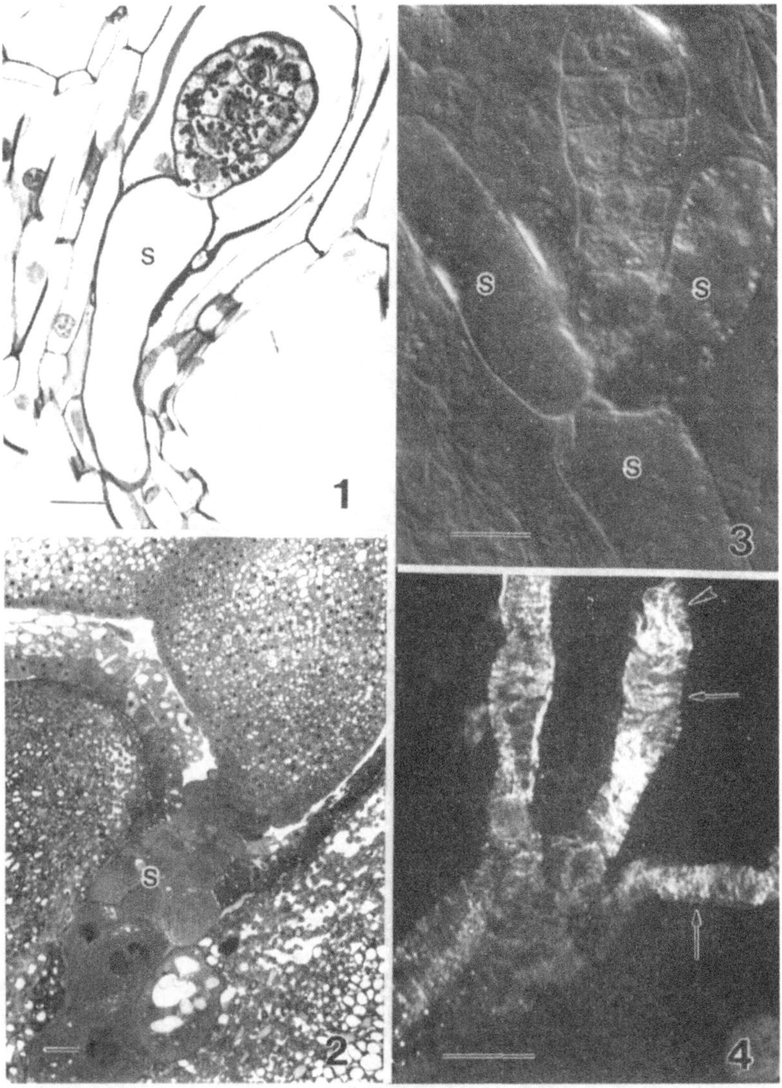

Plate XXVIII: Suspensor functions.

1—longitudinal section through *Phaius tankervilliae* embryo with single large, vacuolated suspensor cell; 2—longitudinal section through *Phaseolus vulgaris* embryo at cotyledon stage, multicellular suspensor large, cells have dense cytoplasm; 3, 4 –*Cymbidium sinense* embryo with large elongating suspensor cells: 3—embryo dissected from seed coat and viewed with interference contrast microscope, 4—microtubule configuration within suspensor cells, microtubules (arrow) tend to align at right angles to long axis of cell, in tip region (arrow-head) microtubules appear to be more abundant and arranged in the same direction as cell axis, embryo was fixed, dissected from seed, and stained with FITC-conjugated secondary antibodies against tubulin and viewed with a MRC-600 BioRad confocal laser scanning microscope (for

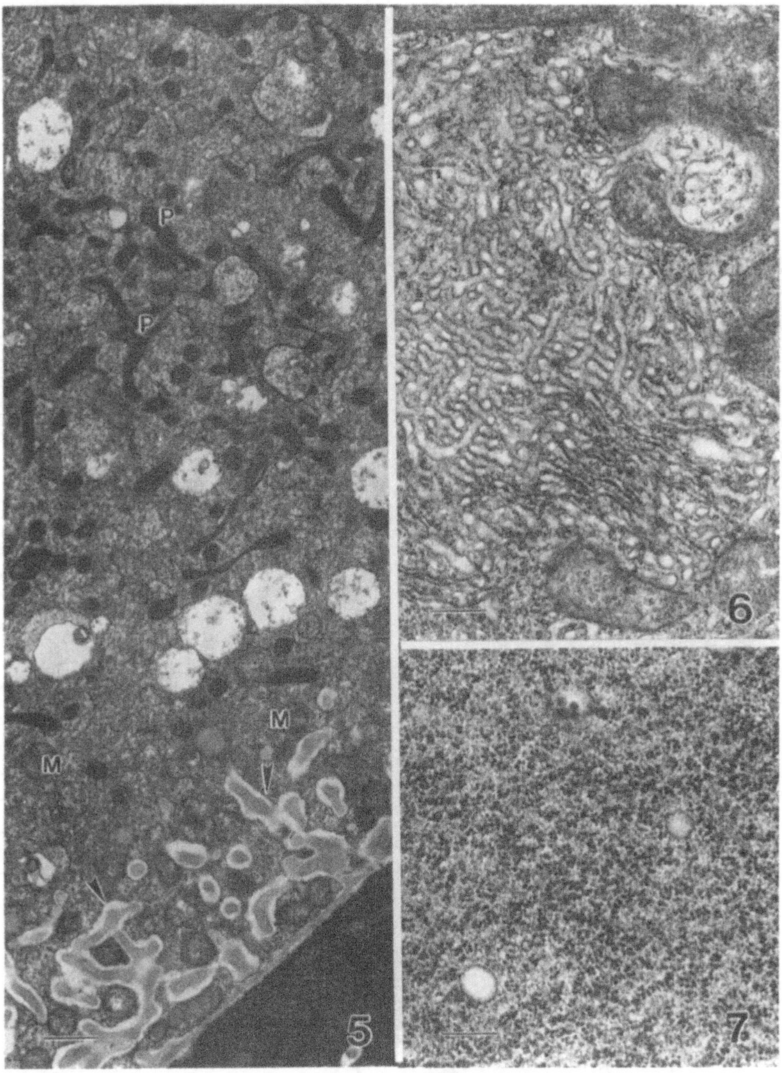

Plate XXVIII (*Contd.*)

details, see Zee and Aziz–Un–Nisa, 1991); 5—7—*Phaseolus coccineus*: 5—electron micrograph showing transfer cell morphology of suspensor cell at globular stage of embryo development, wall ingrowths (arrow-heads), mitochondria located near wall ingrowths, electron dense plastids more centrally located, 6—smooth endoplasmic reticulum very abundant within cytoplasm of suspensor cells, 7—fields of polysomes in suspensor cells as early as proembryo stage. M—mitochondria, P—plastids, S—suspensor.

Bars: 1-3—20 µm, 4—25 µm, 5—1 µm, 6—0.4 µm, 7—0.2 µm.

1—7—after Yeung, Zee and Ye, orig.

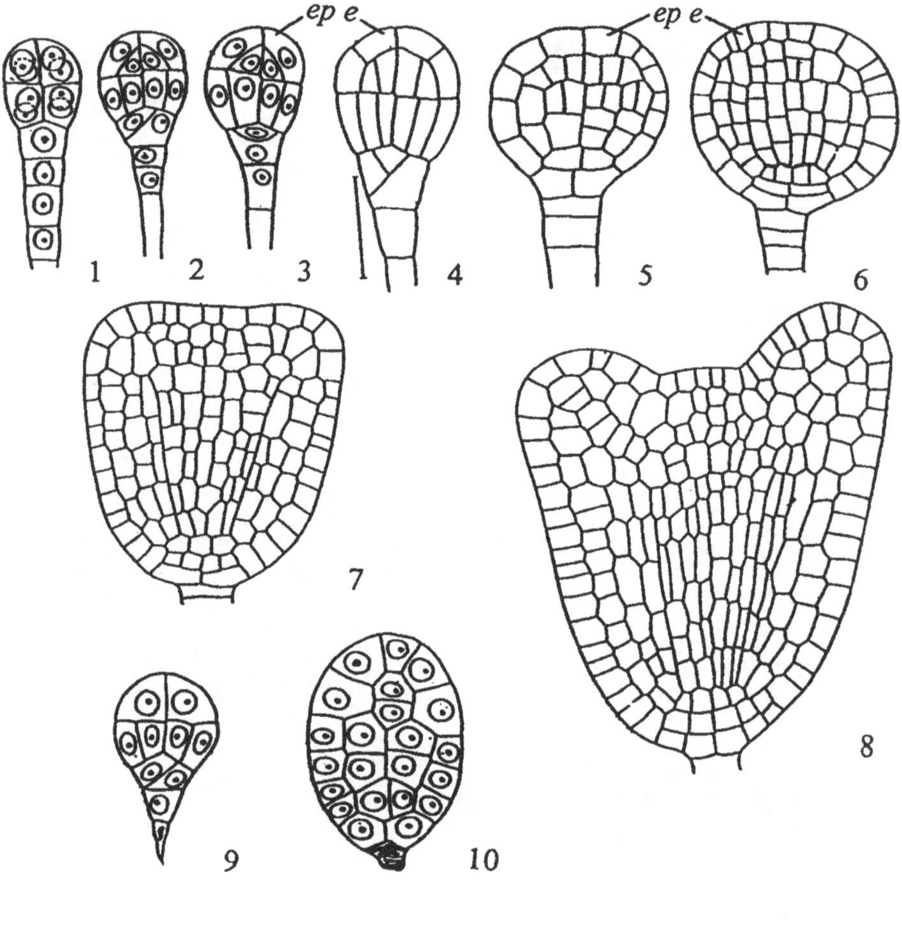

Plate XXIX: Epidermis of embryo.

1-8—formation of epidermis of embryo in *Bartsia alpina:* 1—proembryo at stage of quadrants, 2-8—demarcation of epidermis initials at octant stage (2,3) and their anticlinal divisions at subsequent stages of embryo development (4-8); 9, 10—irregularities in process of embryo epidermis differentiation in *Orobanche pallidiflora*: epidermis initials are not formed owing to absence of periclinal divisions in tier *l* (9), outer cells in zone of epidermis differentiation disturbance (apical part of embryo) larger than cells of basal part (10); *ep e*—epidermis of embryo.

Bar: 0.03 mm

1-10—after Nikiticheva, 1979a.

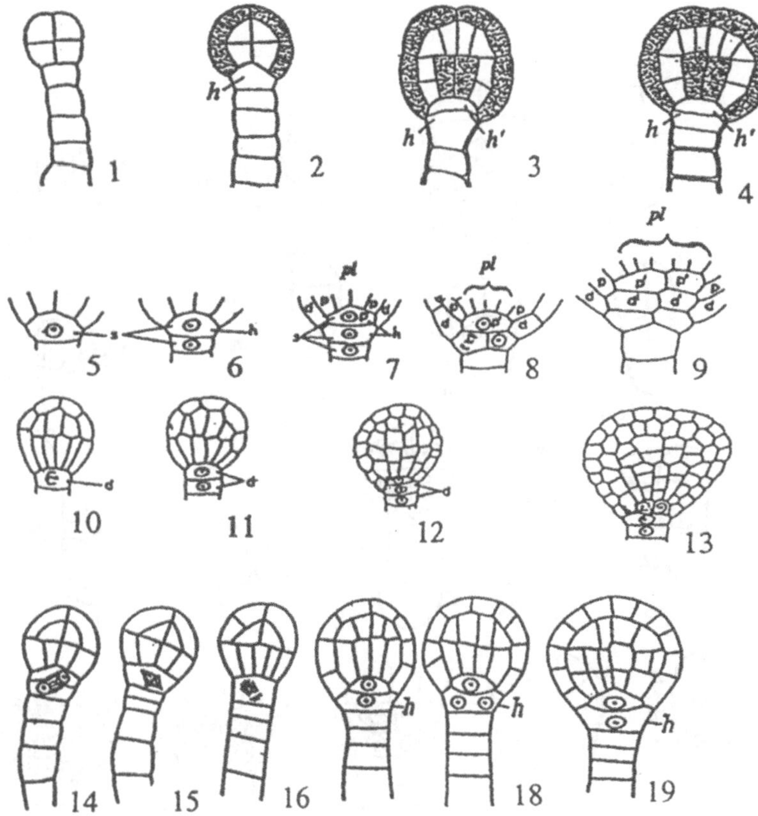

Plate XXX: Hypophysis.

1-19—formation of hypophysis during embryogenesis in *Capsella bursa-pastoris* according to notions of Hanstein (1-4), Coulter and Chamberlain (5-9), Schaffner (10-13) and Souèges (14-19); 20-46—different modes of hypophysis formation according to Souèges (explanation in text): 20-23 (*A.a*)—*Oenothera biennis*, 24-27 (*A.b*)—*Myosurus minimus*, 28-32 (*B*)—*Nicotiana acuminata*, 33-37 (*C*)—*Senecio vulgaris*, 38-42 (*D.a*)—*Erodium cicutarium*, 43-46 (*D.b*)—*Trifolium minus*; 47—scheme of cell wall position in hypophysis initial cell at different orientation of latter after first (I), second (II), and third (III) divisions; 48-60—character of initial hypophysial cell division in *Lathraea squamaria*: 48-53—early embryogenesis stages, 54-56—detachment of dermatogen and beginning of hypophysis differentiation, 57-60—differentiation of periblem and plerome initials in globular embryo; 61-66—origin of hypophysis/dormant centre in main root of *Sphenoclea zeylanica* embryo (61-65) and seedling (66); 67-70—origin of hypophysis/dormant centre (xxx) in lateral root of *Commelina bengalensis*.

d, d', de—dermatogen and *d*—one of tiers derived from basal cell in detail 10-12, 30, 31, 51, *e*—epiphysis, *h, h'*—hypophysis, *ico*—initials of root cap, *iec*—initials of root cortex, *p, p'*—periblem, *pl*—plerome, *s*—suspensor. Rest of legend explained in text (see Proembryo).

1-4—after Hanstein, 1870; 5-9—after Coulter and Chamberlain, 1903, cited by Swamy and Krishnamurthy, 1975; 10-13—after Schaffner, 1906, cited by Swamy and Krishnamurthy, 1975; 14-19—after Souèges, 1919; 20-46—after Souèges, 1934; 47-60—after Nikiticheva, 1971; 61-70—after Swamy and Krishnamurthy, 1975.

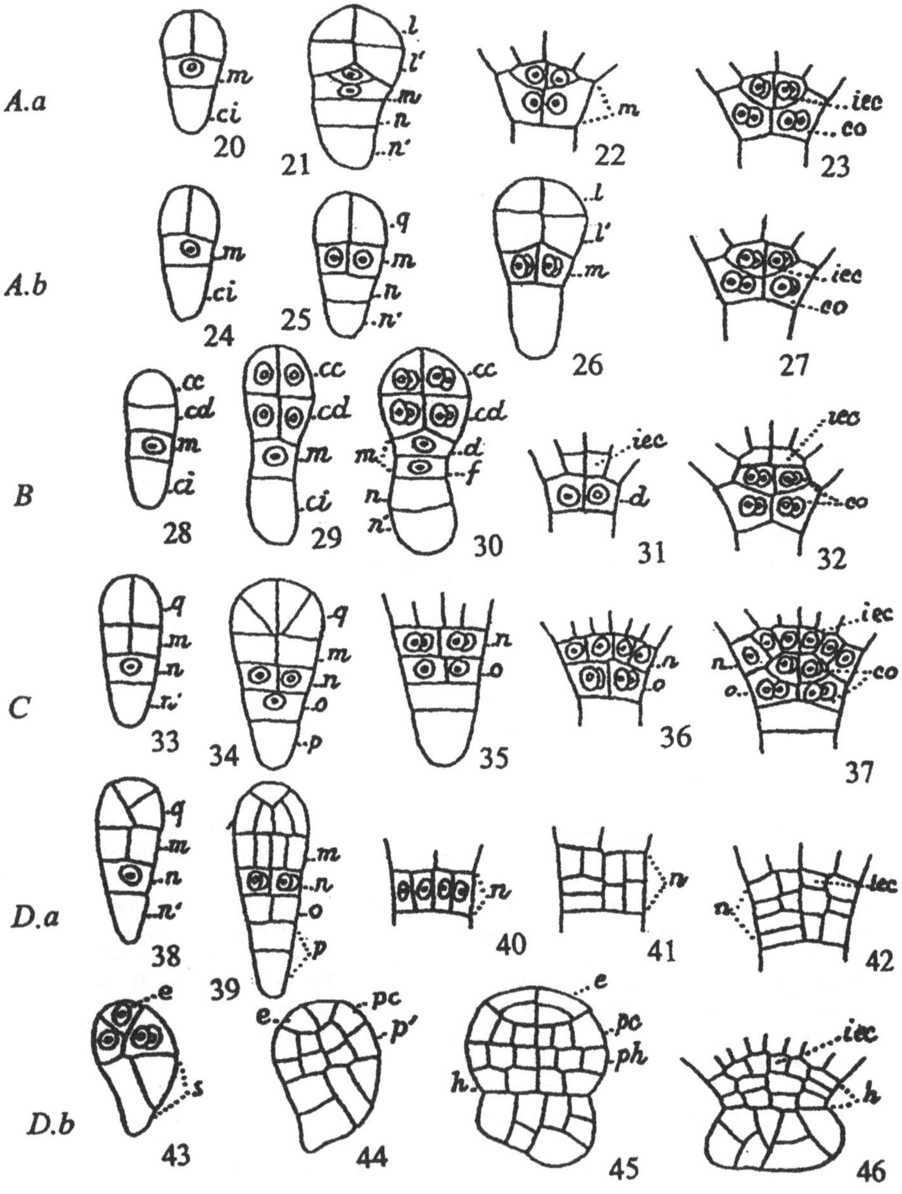

Plate XXX (*Contd.*)

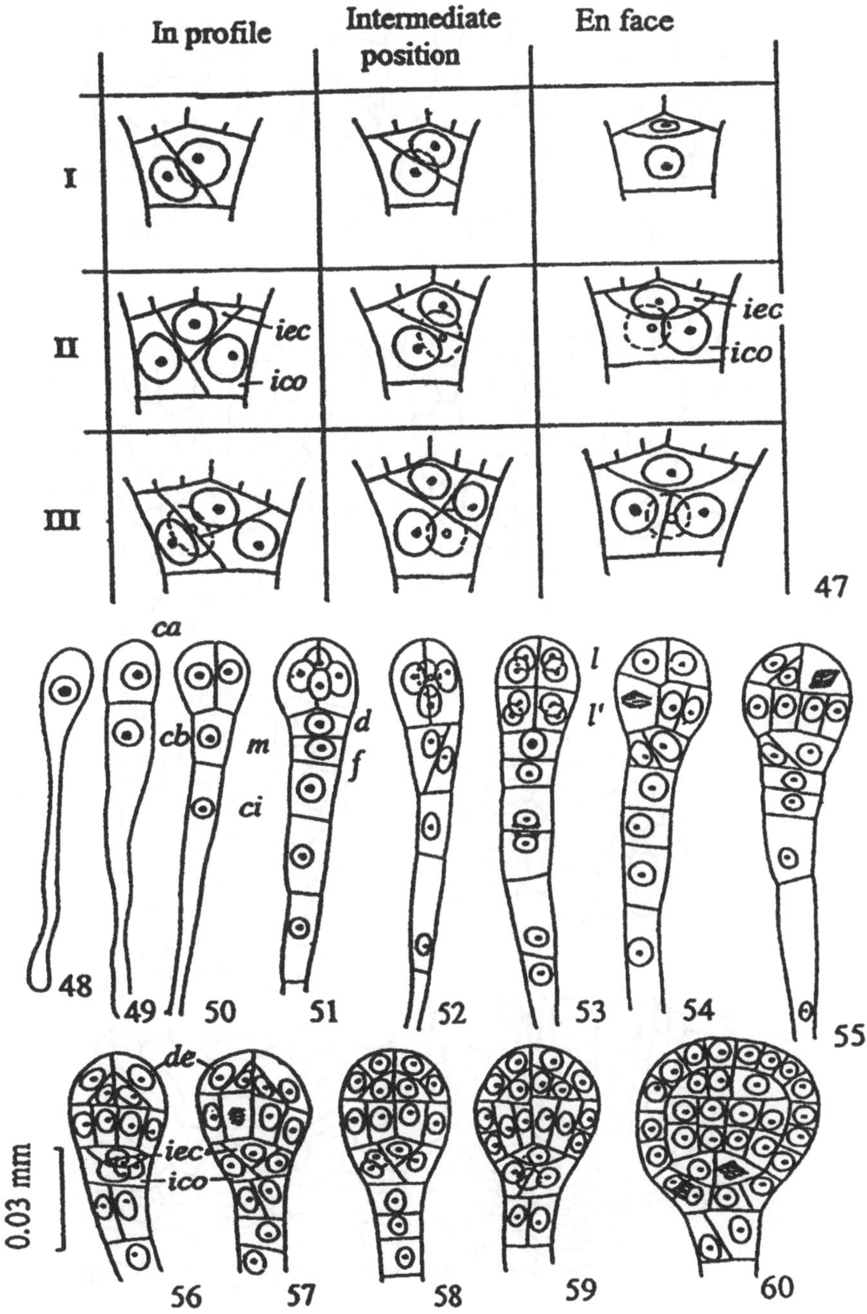

Plate XXX (*Contd.*)

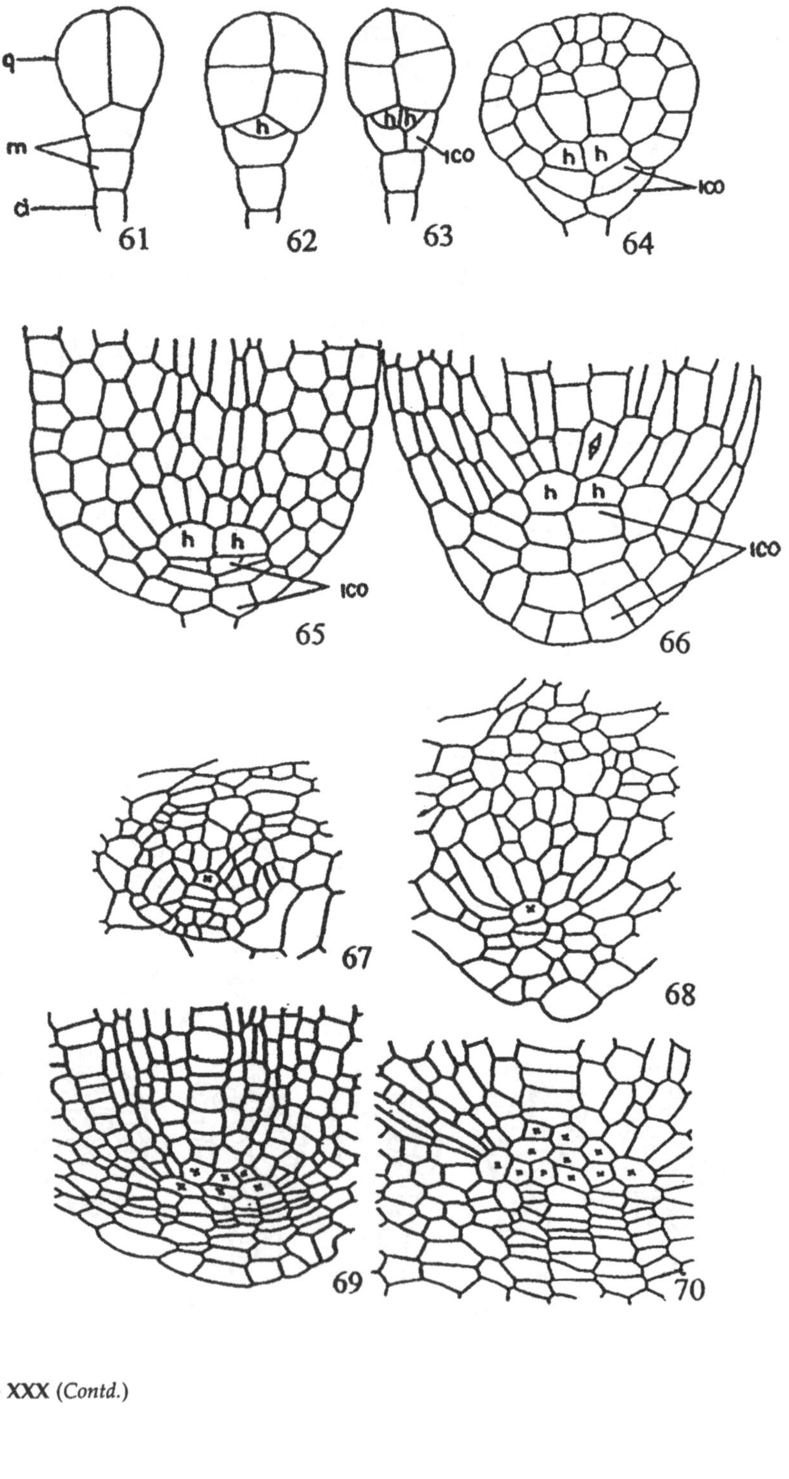

Plate XXX (*Contd.*)

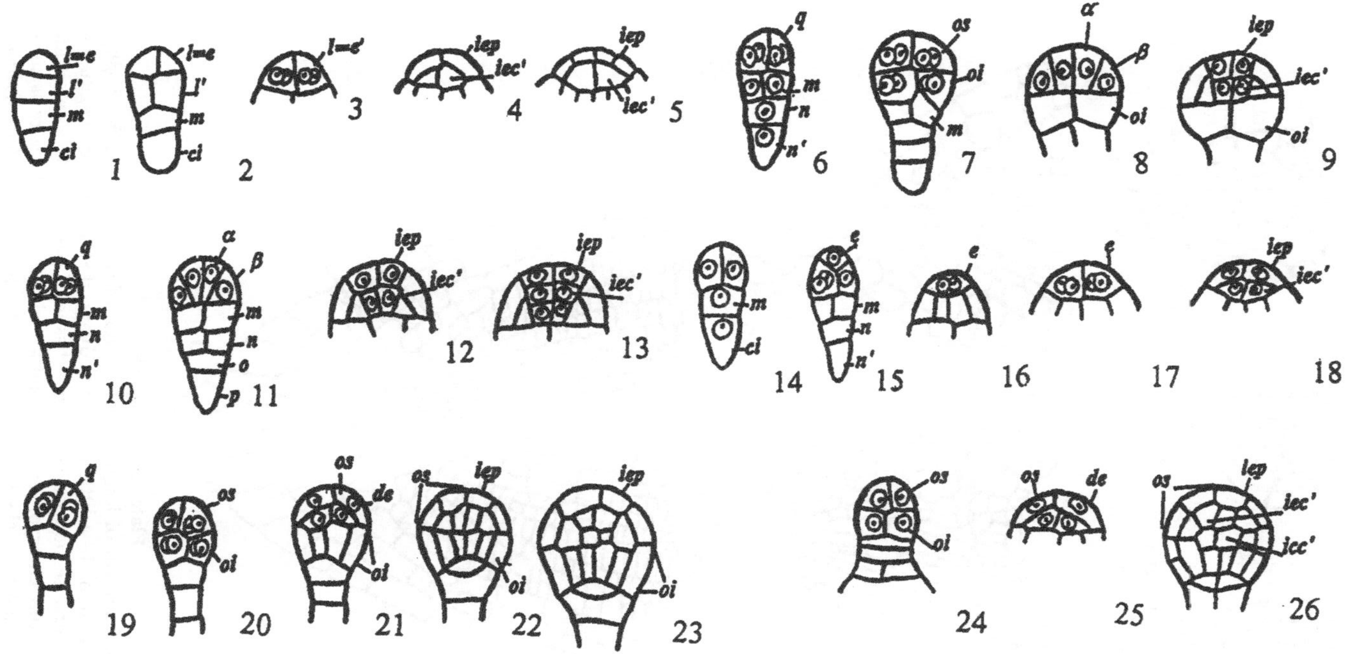

Plate XXXI: Epiphysis.

1-26—different modes of formation of epiphyseal tissue in *Papaver rhoeas* (1-5), *Myosurus minimus* (6-9), *Senecio vulgaris* (10-13), *Erodium cicutarium* (14-18), *Lepidium sativum* (19-23), *Sedum acre* (24-26).

l and *l′*—two uppermost cells of tetrad, laying one under the other, *m*—middle cell, *ci*—lower cell of tetrad, *e*—epiphysis, *n*—upper derivative of the cell *ci* or the tier *ci*, *o* and *p*—derivatives of the cell *n′*, which is the lower derivative of *ci*, *q*—quadrants, *iep*—initials of the stem epiderm, *iec′*—initials of stem cortex, α and β—daughter cells of quadrants or upper octants, *icc′*—initials of stem central cylinder, *os*—upper octants, *oi*—lower octants, *de*—dermatogen.

1-26—after Souèges, 1934.

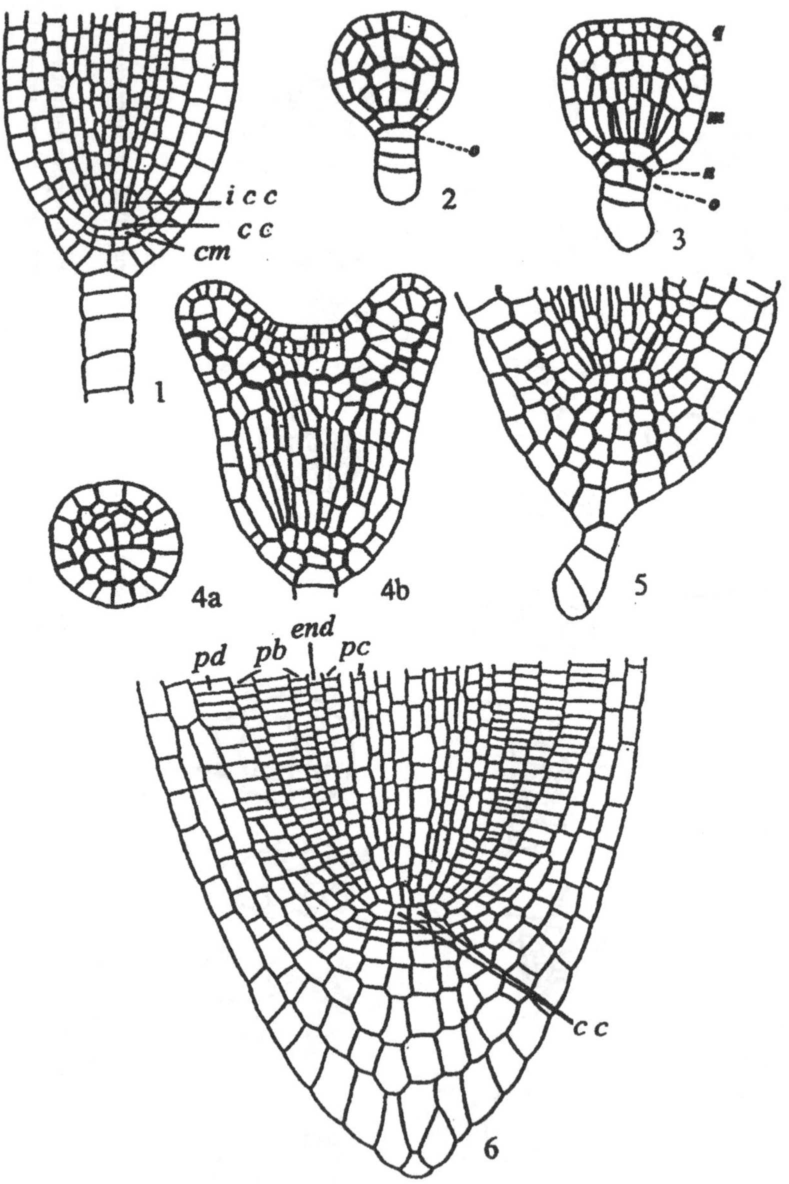

Plate XXXII: Radicle

1—*Capsella bursa-pastoris:* fragment of longitudinal section of the embryo; 2-5—*Biophytum dendroides:* 2—globular embryo, 3—a little bit more advanced stage, 4a, 4b –beginning of organogenesis (4a—transverse section via connective tissue of periblem, 4b—longitudinal section), 5—root pole of mature embryo (median longitudinal section); 6—*Brassica napus:* root apex of mature embryo, connective cells underlaid in bottom by root cap; 7-10—*Helianthus*

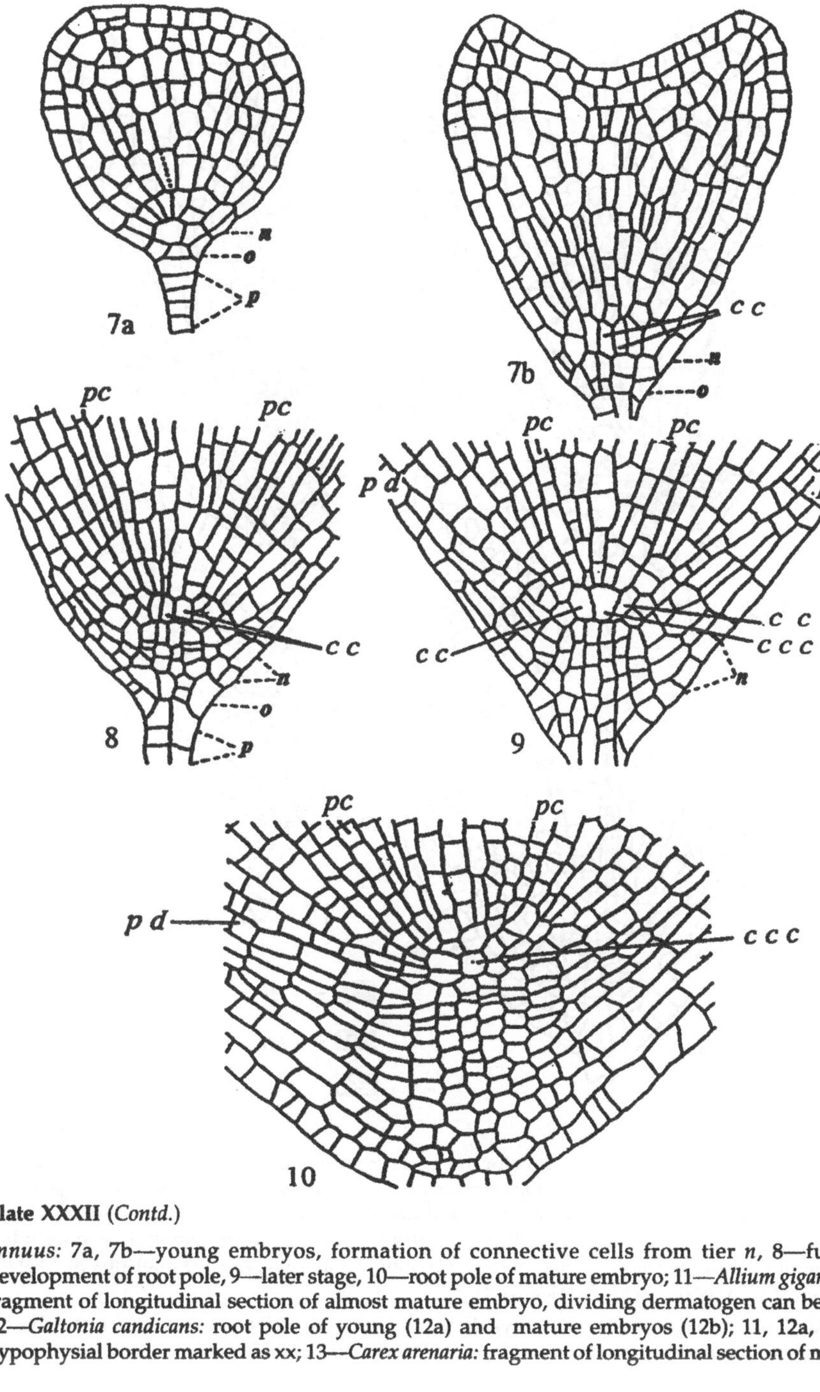

Plate XXXII (*Contd.*)

annuus: 7a, 7b—young embryos, formation of connective cells from tier *n*, 8—further development of root pole, 9—later stage, 10—root pole of mature embryo; 11—*Allium giganteum:* fragment of longitudinal section of almost mature embryo, dividing dermatogen can be seen; 12—*Galtonia candicans:* root pole of young (12a) and mature embryos (12b); 11, 12a, 12b—hypophysial border marked as xx; 13—*Carex arenaria:* fragment of longitudinal section of mature

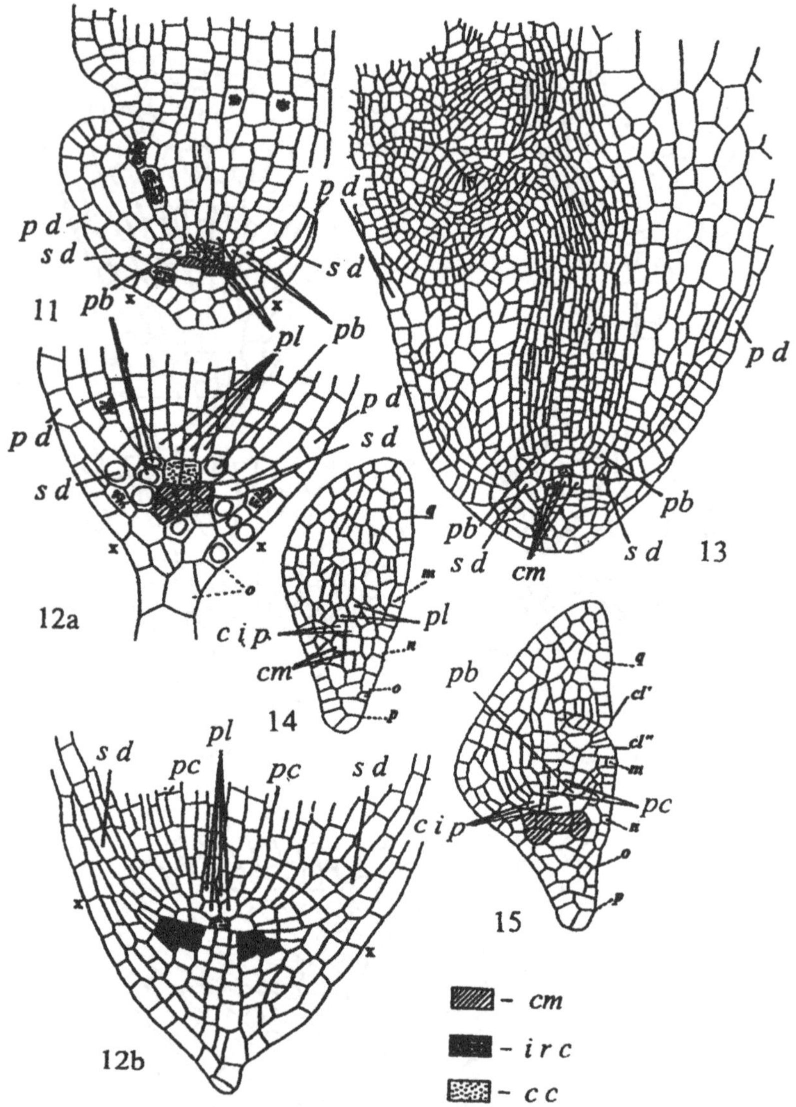

Plate XXXII (*Contd.*)

embryo; 14,15—*Poa annua:* 14—developing embryo with periblem, plerome and columella initials, 15—a little bit more developed embryo with root rudiment; 16-20—development of root cap in *Commelina communis*: 16-18—longitudinal sections of embryo at different stages of development, 19,20—root tips of seedlings at different stages of germination; 21—longitudinal section of root tip in *Washingtonia robusta;* 22—longitudinal sections via root tips in *Magnolia grandiflora* (22a) and *Paeonia hybrida* (22b).

cl' and *cl"*—part of coleoptyle, *cot*—cotyledon, *c c*—connective cells, *c c c*—central connective cell, *c c e r*—central cylinder of embryo root, *c i p*—central initials of periblem, *cm*—columella, *co*—collar,

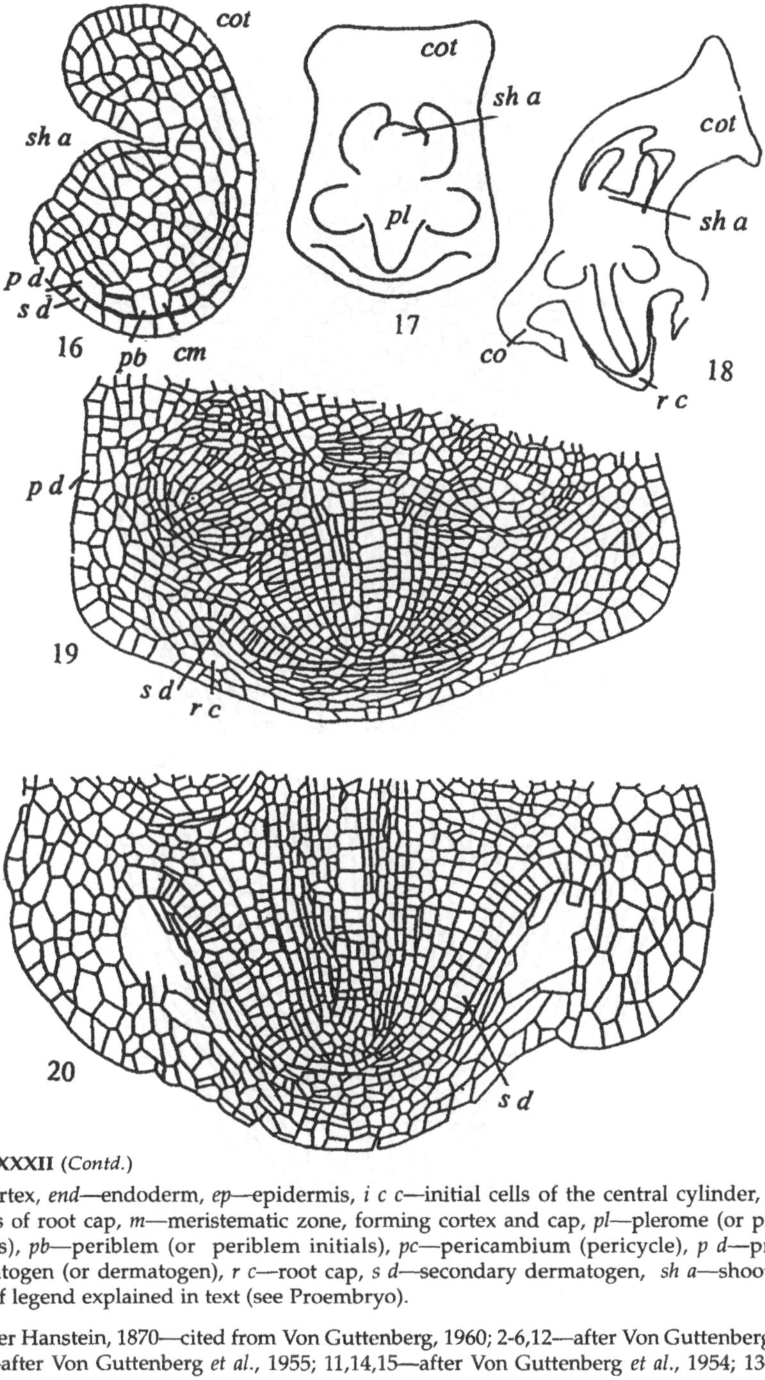

Plate XXXII (*Contd.*)

ct—cortex, *end*—endoderm, *ep*—epidermis, *i c c*—initial cells of the central cylinder, *i r c*—initials of root cap, *m*—meristematic zone, forming cortex and cap, *pl*—plerome (or plerome initials), *pb*—periblem (or periblem initials), *pc*—pericambium (pericycle), *p d*—primary dermatogen (or dermatogen), *r c*—root cap, *s d*—secondary dermatogen, *sh a*—shoot apex. Rest of legend explained in text (see Proembryo).

1—after Hanstein, 1870—cited from Von Guttenberg, 1960; 2-6,12—after Von Guttenberg, 1960; 7-10—after Von Guttenberg *et al.*, 1955; 11,14,15—after Von Guttenberg *et al.*, 1954; 13—after

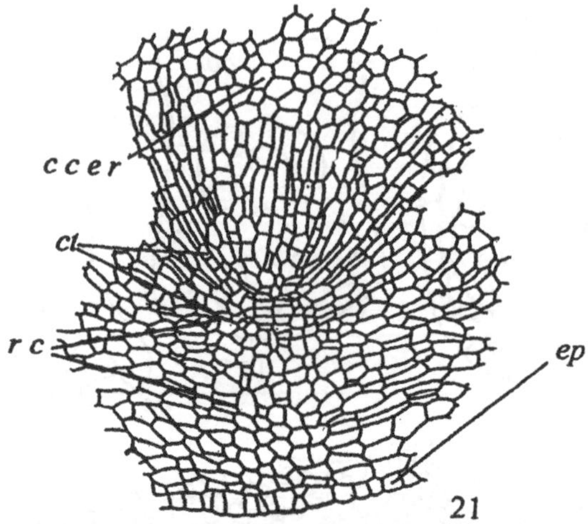

21

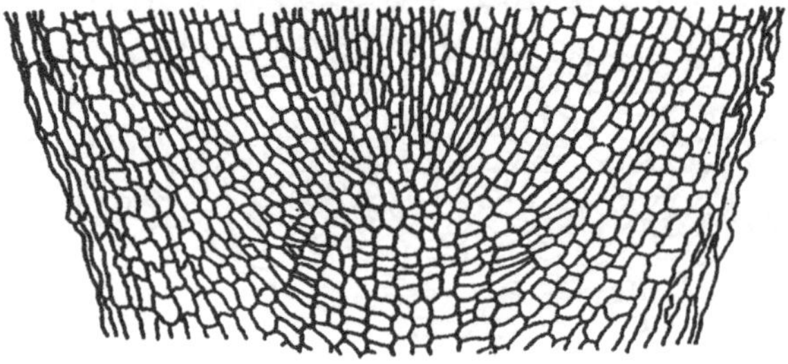

22a

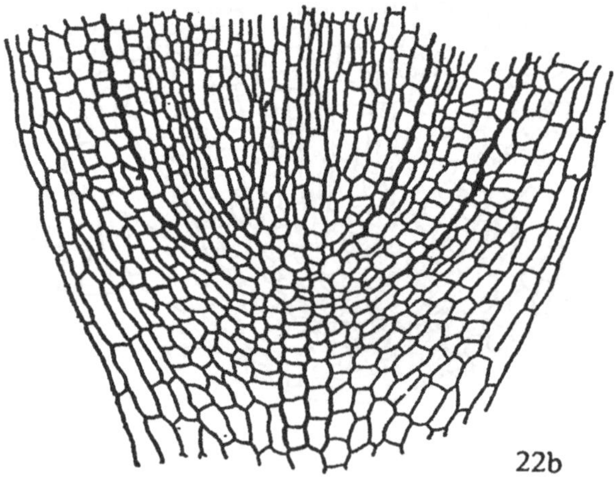

22b

Plate XXXII (*Contd.*)

Von Guttenberg and Semlow, 1957; 16-20—after Yamashita, 1983; 21—after Gatin, 1906; 22a, b—after Voronin, 1964.

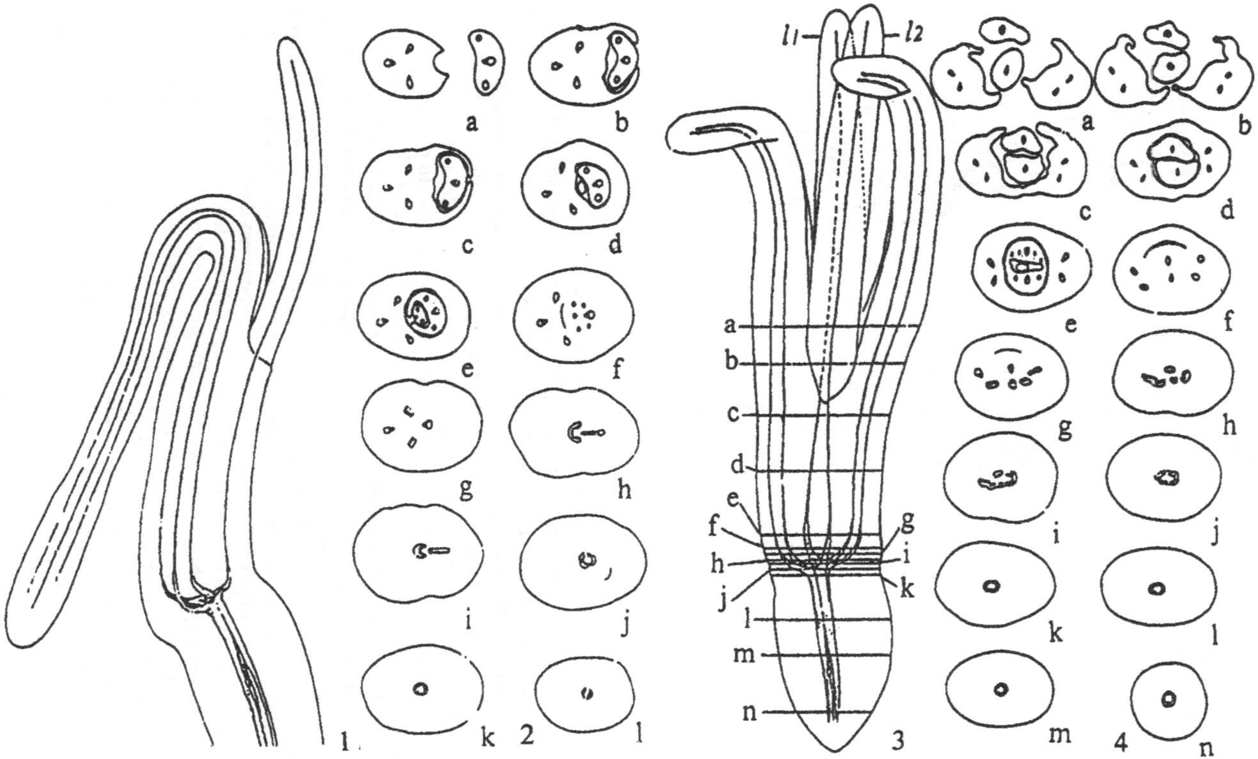

Plate XXXIII: Pseudodicotyly.

1-4—structure of germinating monocotyledonous (1,2) and dicotyledonous (3,4) embryos of *Agapanthus umbellatus:* (1,3—longitudinal sections, 2,4—cross-sections in region of cotyledonary sheath (a-e), cotyledonary node (f-i) and hypocotyl (j-n); l_1, l_2—first leaves of plumule.

1-4—after Coulter and Land, 1914.

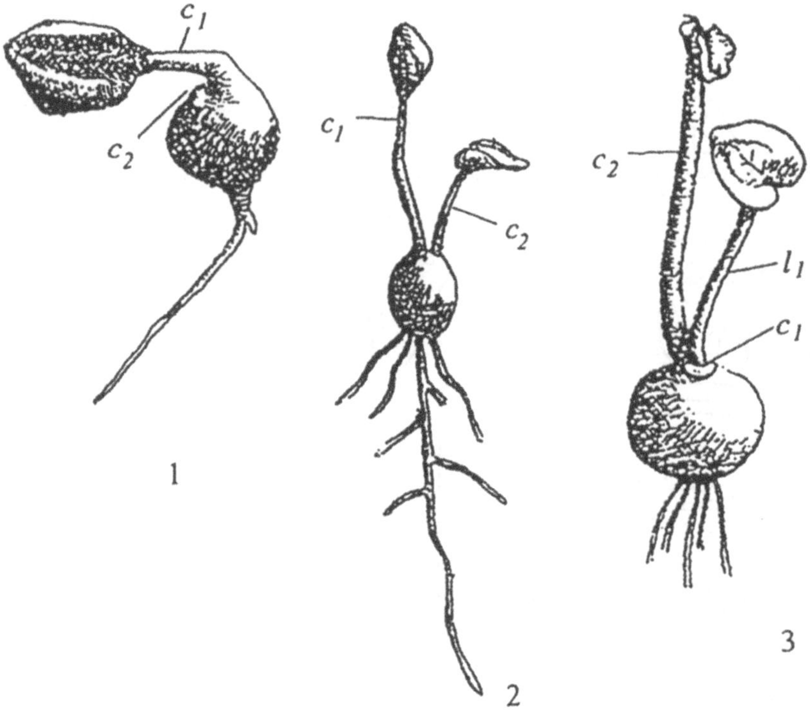

Plate XXXIV: Pseudomonocotyly.

1-3—structure of seedling, heterocotylar after Hill, in *Cyclamen persicum* (1,3) and *C. neapolitanum* (2),—with one, normal (large) cotyledon (c_1) and second, rudimentary (c_2), retention of bigger cotyledon in seed (2) and also its artificial removal (3) stimulate development of rudimentary one; 4-17—development of embryo, 'monocotyledonous dicotyledons' after Hagemann, in *C. persicum*: laying down of a single cotyledon with no traces of second one; 18-23—structure of abnormal embryo of *C. persicum* arising after treatment with phenylboric acid (19-22) compared with normal pseudomonocotyledonous (18,21) and also with typical dicotyledonous (20,23) embryos, formation of two unequal cotyledons coalesced by common cotyledonary base; 24-53—development of anizocotylar embryo in *Trapa astrachanica*: 24-25—structure of seedling with two sharply unequal cotyledons—large, preserving in seed, and small scale-like, going out from seed, 26-33—proembryo formation, 34-53—organogenesis: heterochrony in laying down of cotyledons, unequal size of their primordia (clasping angle—dotted line—240º and 120º -48), decrease of divergence angle between them up to 172º approximately (thin solid line -48) and different depth of sinuses between them (thick line -48, 49), testifying to partial unimarginal fusion of unequal cotyledons (anizocotyly on background of unimarginal syncotyly); 4-16, 26-42, 51-53—longitudinal sections, 17-23—cross-sections (14-16,18-20,38b-42—schematized), 43-50—SEM, 51-53—LM.

Bars: 26-42—0.01 mm, 43, 47b—10 µm; 44-47a, 48-50—100 µm. *a b*—axillary bud, *ap b*—apical bud, *a r*—adventive root, *c*, c_1, c_2—cotyledons, *d z*—dormant zone, *h*—hypocotyl, l_1—first leaf, *pc*—procambium, *sh a*—shoot apex, *sus*—suspensor, *v b*—vascular bundle. Rest of legend explained in text (see Proembryo).

1-3—after Hill, 1920; 4-17—after Hagemann, 1959; 18-23—after Haccius and Lakshmanan, 1967; 24-53—after Titova, 2000.

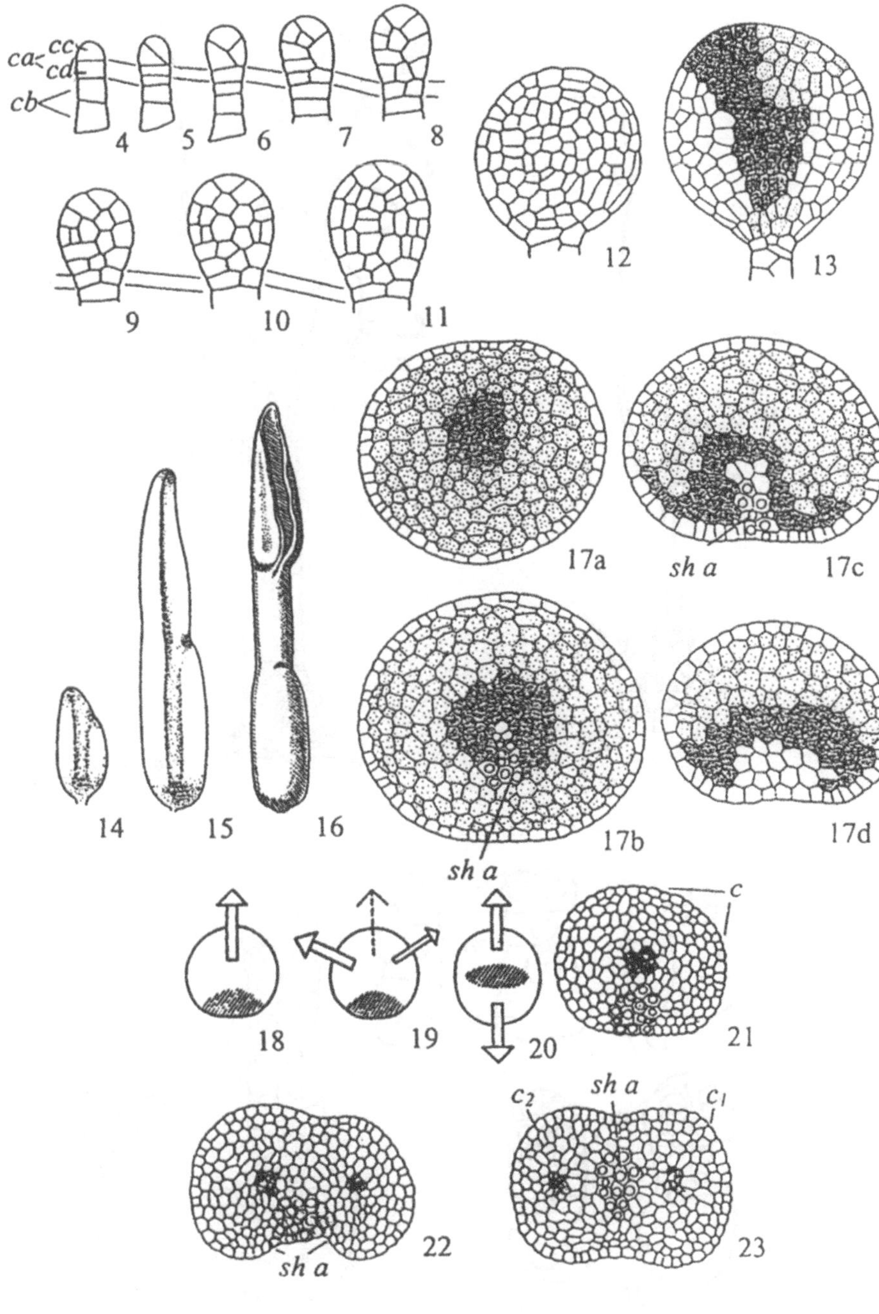

Plate XXXIV (*Contd.*)

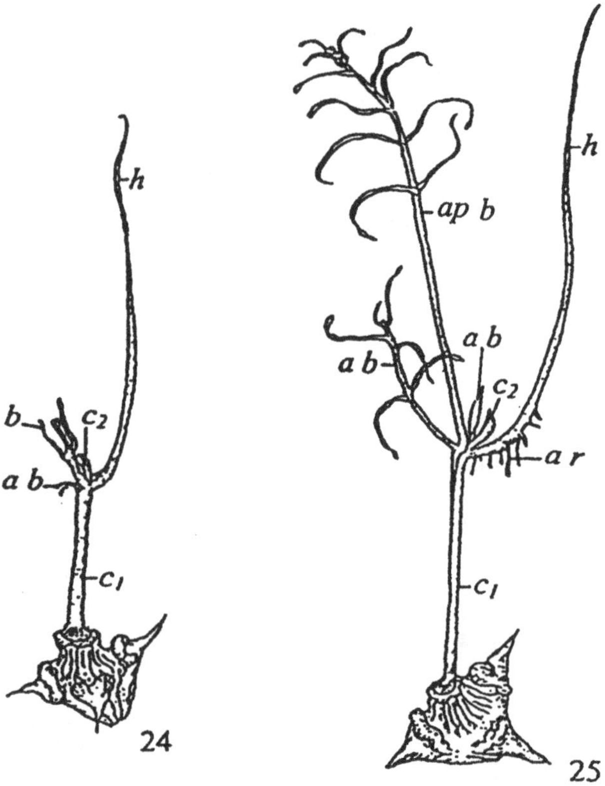

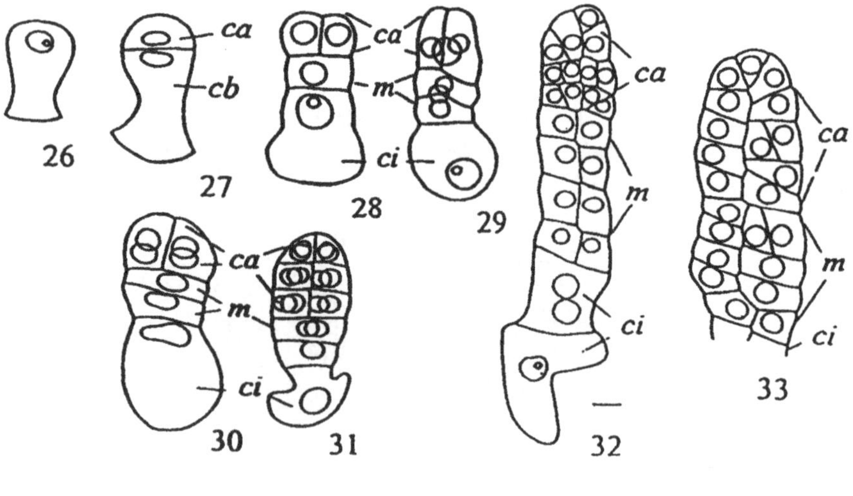

Plate XXXIV (*Contd.*)

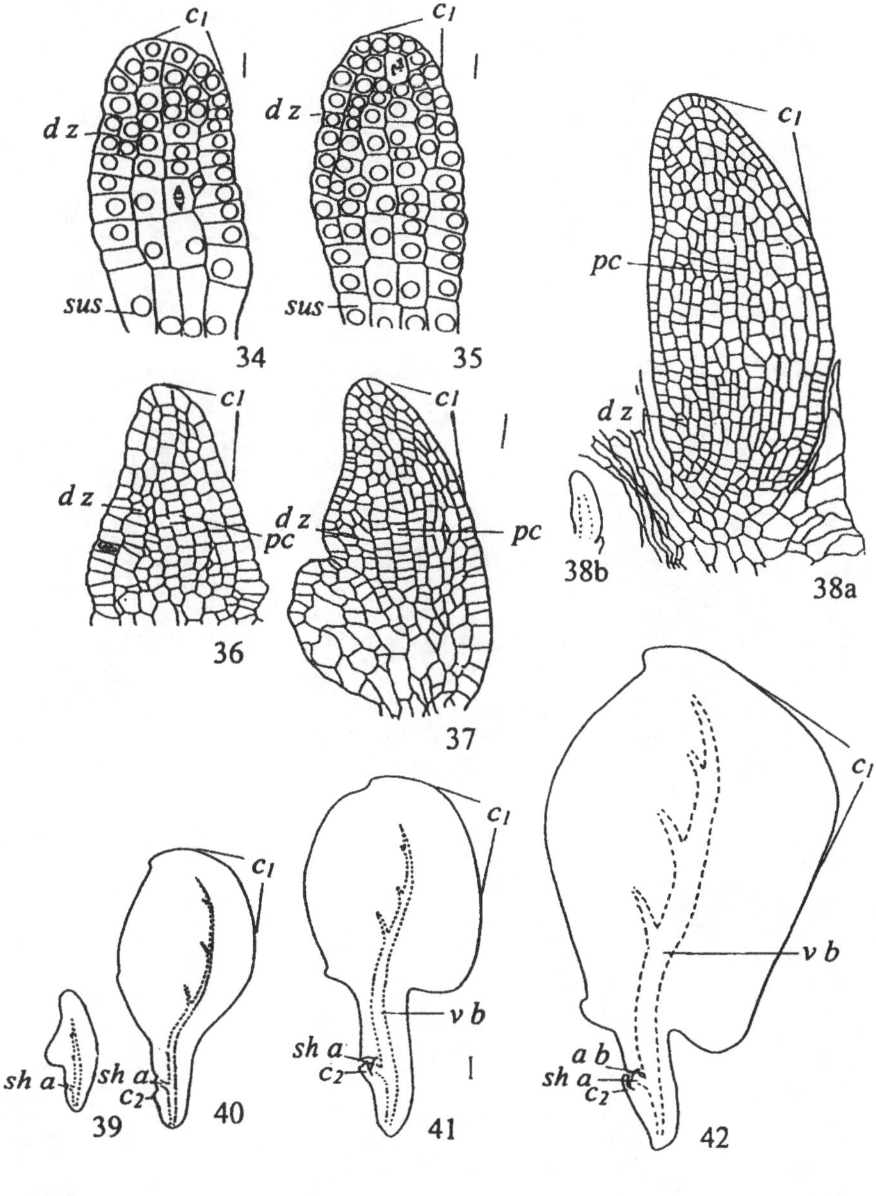

Plate XXXIV (*Contd.*)

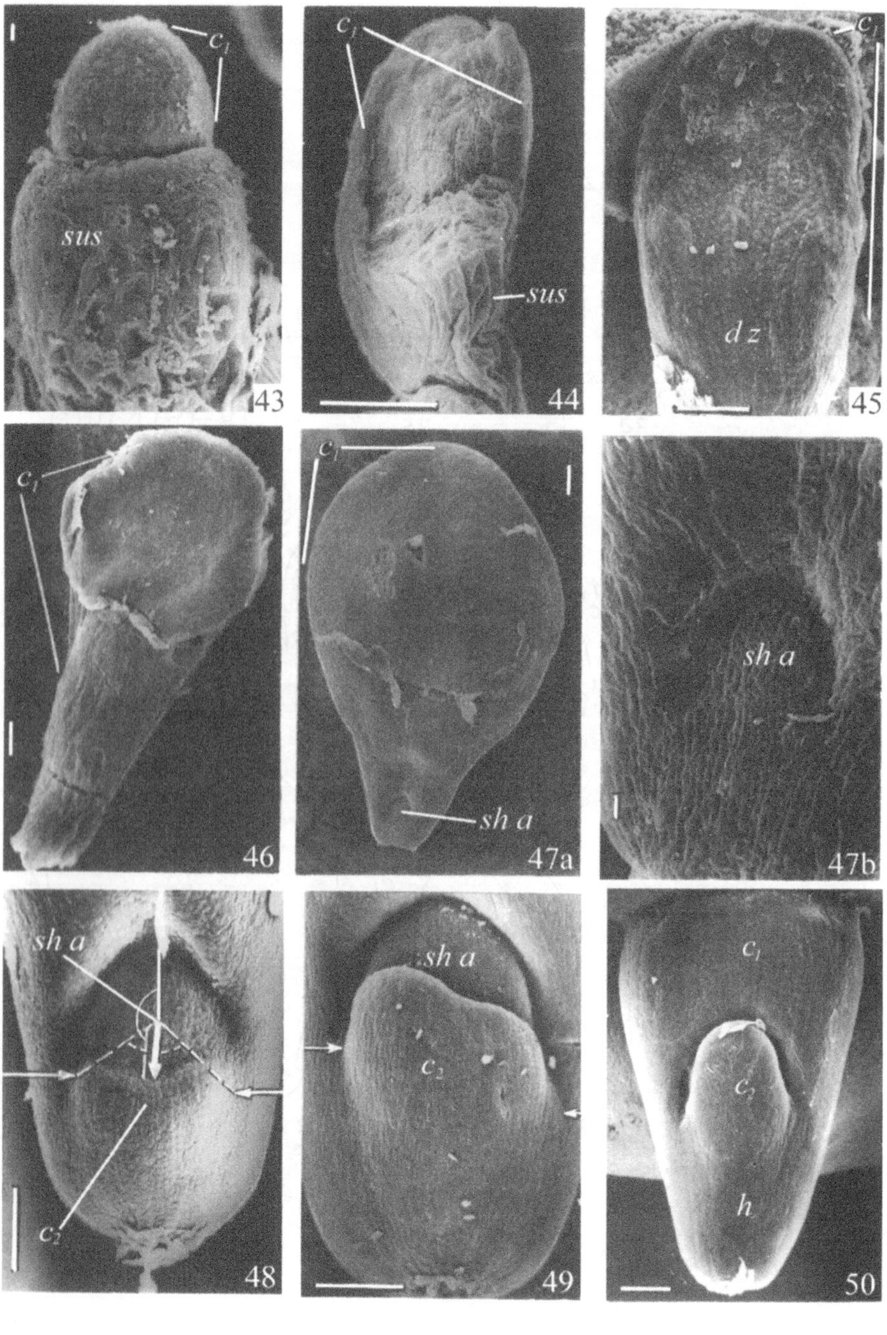

Plate XXXIV (*Contd.*)

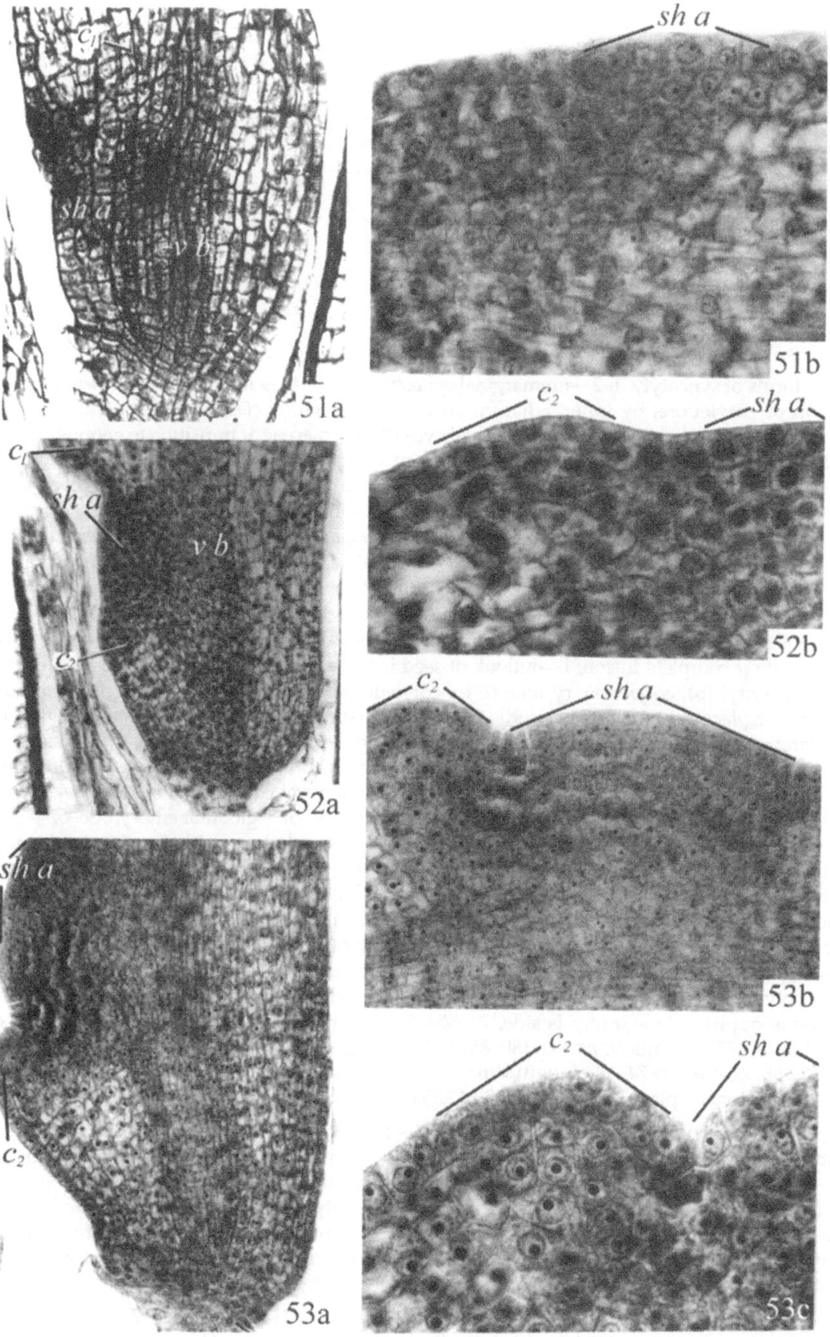

Plate XXXIV (*Contd.*)

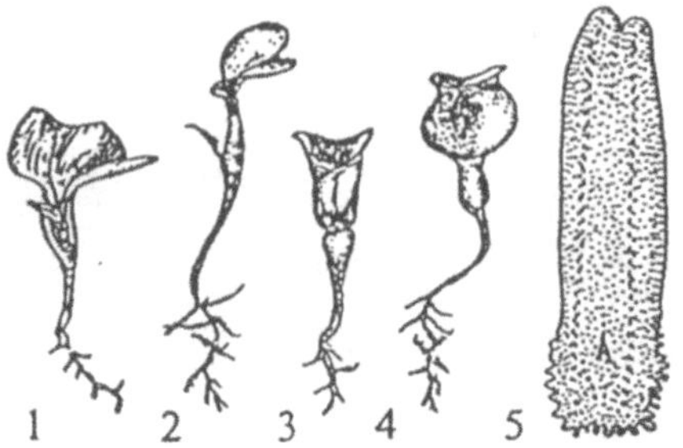

Plate XXXV: Syncotyly

1–5—forms of syncotyly: 1–2—unimarginal syncotyly and 3–4—gamocotyly (*Oenothera glauca*), 5—fusion of cotyledons by inner, adjacent surfaces (*Helicanthes elastica*); 6—scheme illustrating evolutionary trends of seedlings (A) and embryos (B) in Apiaceae at transition to geophylous mode of life (single arrow marks position of growing point in respect of soil surface, little cross the boundary of geotropically negative and positive parts of cotyledonary tube); 7–27—laying down of cotyledons by the crescentic ridge at natural partial unimarginal syncotyly (7–11—*Sanguinaria canadensis*, 12–14—*Actaea alba*, 15—*Cabomba piauhiensis*, 16–20—*Nymphaea advena*, 21–27—*Magnolia grandiflora*); 28–43—seedling (28–30, 36–38) and embryo (31–35, 39–43) structure at natural gamocotyly: post-genial fusion of cotyledons by adaxial sides—partial (*Podophyllum emodi*, 28-35) or complete (*Eranthis hiemalis*, 36-43); 44–47—anatomical structure of teratological syncotylous seedlings of *E. hiemalis*, influenced by 2,4D (schematized): 44—normal seedling, 45—partial fusion of cotyledonary blades, 46–47—their complete fusion; I—outlook of seedling, II a–f—cross-sections at the level of radicle (a), hypocotyl (b), cotyledonary tube (c–e), cotyledonary blades (f), III—changes of conductive system: displacement of median bundle (M_1) in dominating cotyledon and of plane of xylem poles of diarch stele in radicle to symphysis side, gradual reduction of median bundle (M_2) in suppressed cotyledon and one (*l*) of two lateral bundles (*l, l'*) in both cotyledons in region of their fusion; 48–76—natural partial unimarginal syncotyly coupled with amphisyncotyly in *Nelumbo nucifera*: 48–55—formation of proembryo with dorsoventral structure, 56–57—globular embryo, 58–67—change of symmetry of apical embryo part, laying down of cotyledons with formation of common cotyledonary zone and its subsequent 'division' into free, unfused cotyledonary parts and common cotyledonary base (66,67), 68—formation of monocotyledonous embryo, 69–70—torpedo-shaped and 71–72—mature embryo (noncoincidence of its morphological axis (A–A') with geometrical (B–B') one, displacing of shoot apex into oblique ('lateral') position (D–D') relative to common cotyledonary base (C–C'), higher level of entrance of bundles of one cotyledon (black colour) in hypocotyls, 73–76—successive initiation of first (73,74) and second (75, 76) leaf primordia in plumule, first one opposite to symphysis side; 77–85—'lateral' position of shoot apex in embryos of *C. piauhiensis* (77–79) and *N. advena* (80–85) with partial unimarginal syncotyly; 7,31,32,34,35,39-41,43,58b,69,70b,71b,74,76—longitudinal sections in cotyledonary plane, 48-58a,70a,71b—in intercotyledonary plane, 8-11,13,15-27,33,42,55a-d,59-61a-c,65,67,72a–k,77–85—cross-sections.

Bars: 48–62—0.01 mm, 63,64—10 µm, 66,73,75—100 µm, 68—1000 µm. c, c_1, c_2—cotyledons, *c b*—cotyledonary blade, *c c b*—common cotyledonary base, *cr*—coleorhiza, *c t*—cotyledonary tube, *d*—dermatogen, *ed*—exodermis, *f c*—fissure-like canal of cotyledonary tube, *i c*—initials of unfused parts of cotyledons, *i c b*—initials of common cotyledonary base, *p*—plumule, *r*—radicle, *r c*—root cap, *rd*—rhizodermis, *sh a*—shoot apex, *t*—tubercle, *u p c*—upper unfused parts of cotyledons, , *v*—vegetative growing point, *v m*—ventral meristem. Rest of legend explained in text (see Proembryo).

1–4—after De Vries, 1903; 5—after Johri *et al.*, 1957; 6—after Haccius, 1952b; 7–14—after Mottier, 1905; 15–20—after Cook, 1906, 21–27—after Earle, 1938; 28–43—after Haccius, 1953b; 44–47—after Haccius and Trompeter, 1960; 48–67, 69–75—after Titova and Batygina, 1996; 68,76—Titova and Vasylieva, orig.; 77–85— after Cook, 1906.

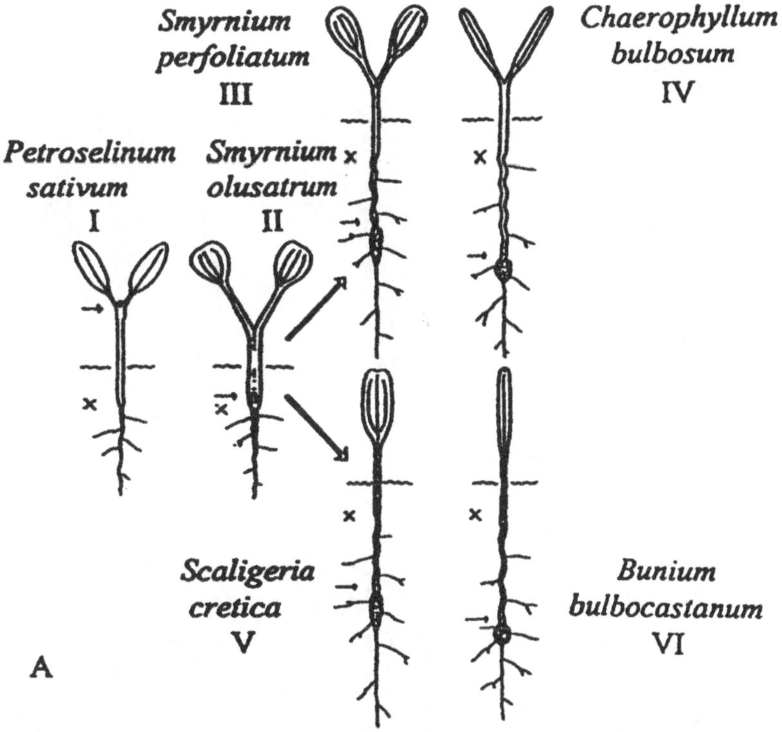

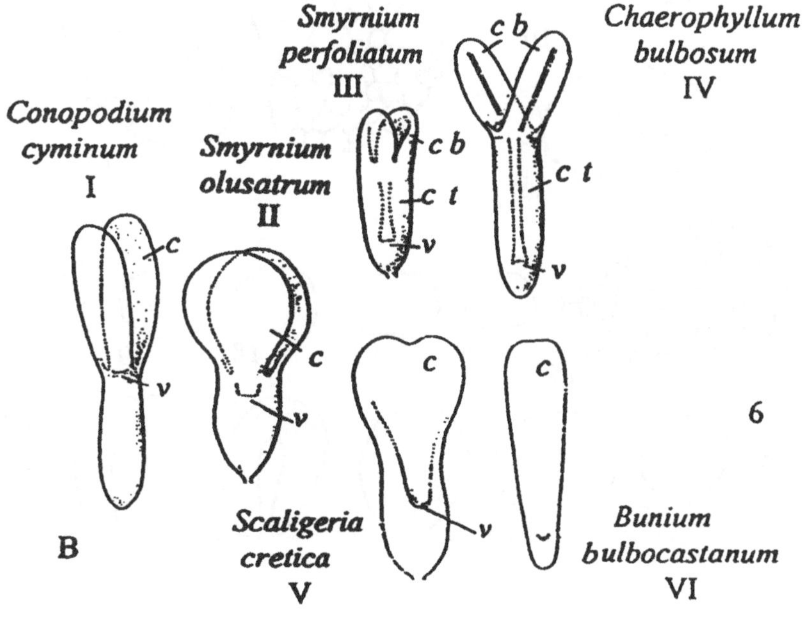

Plate XXXV (*Contd.*)

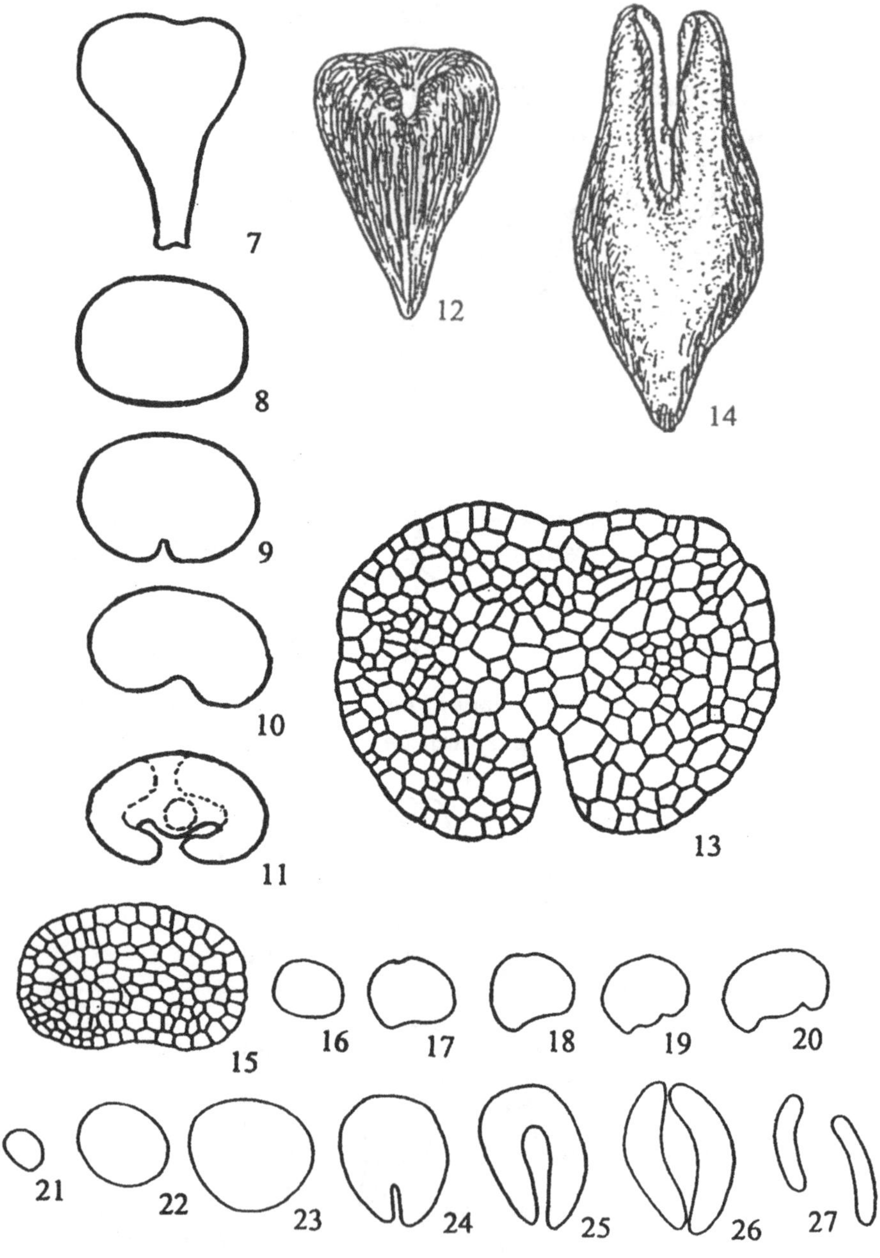

Plate XXXV (*Contd.*)

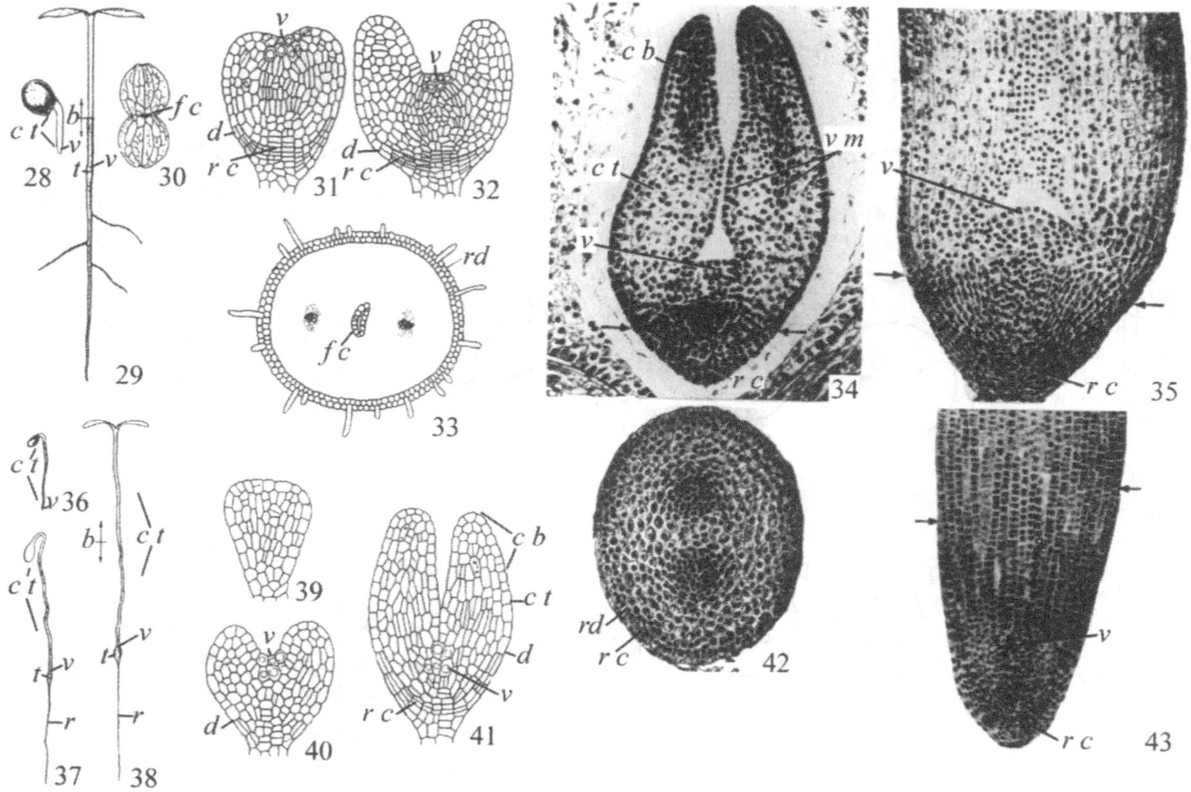

Plate XXXV (Contd.)

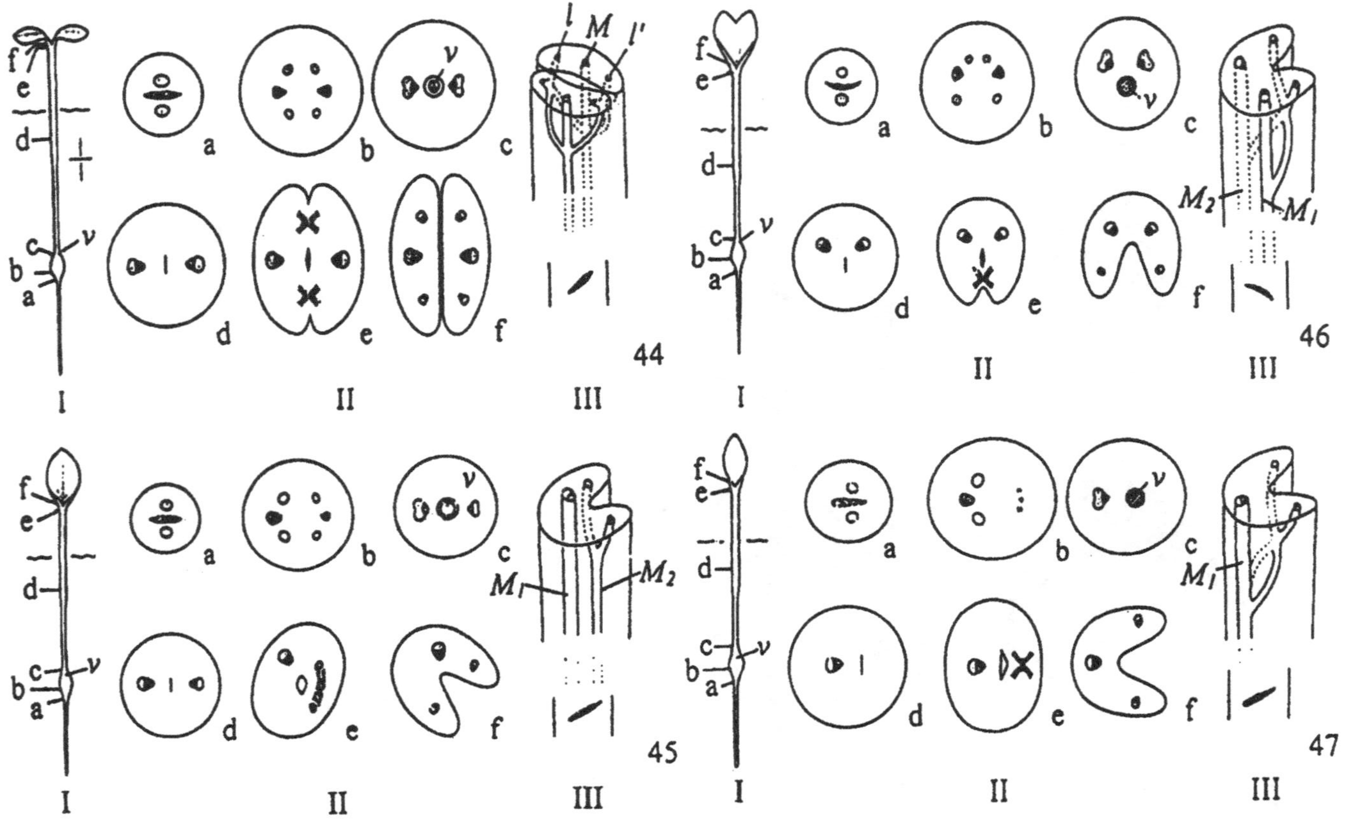

Plate XXXV (*Contd.*)

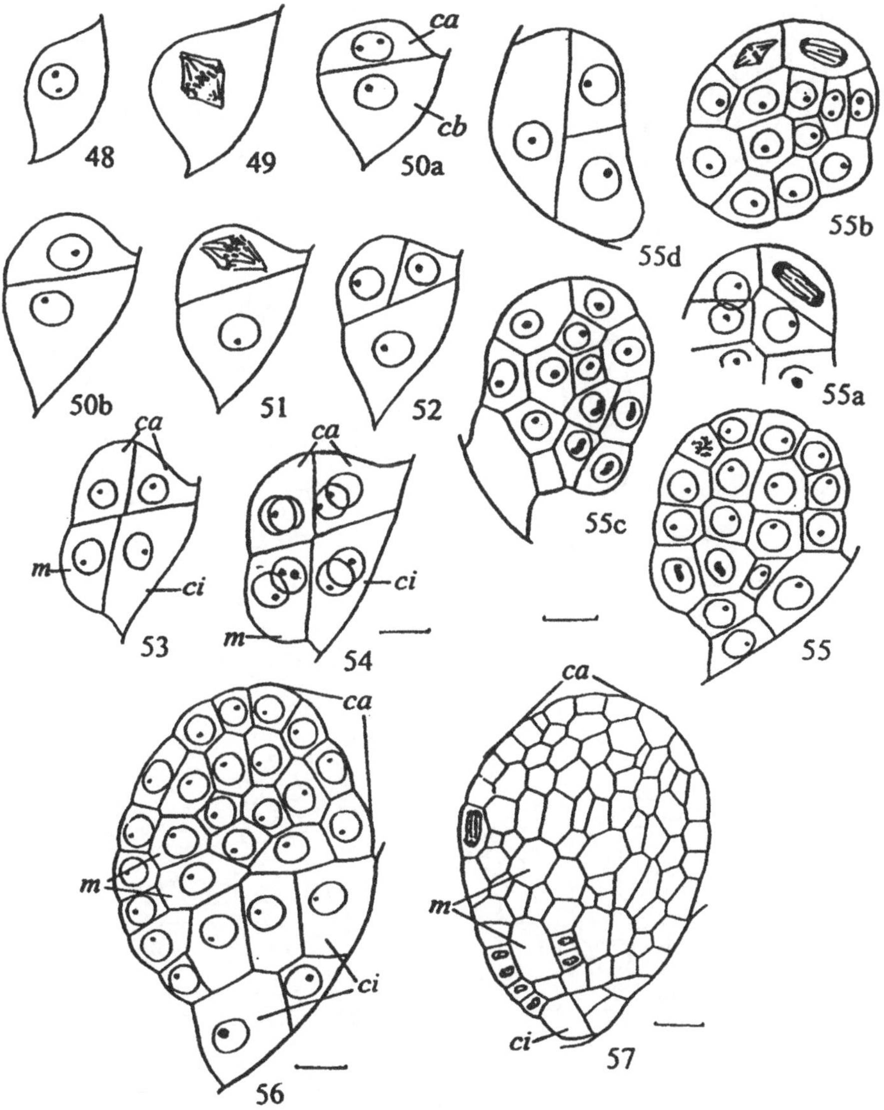

Plate XXXV (*Contd.*)

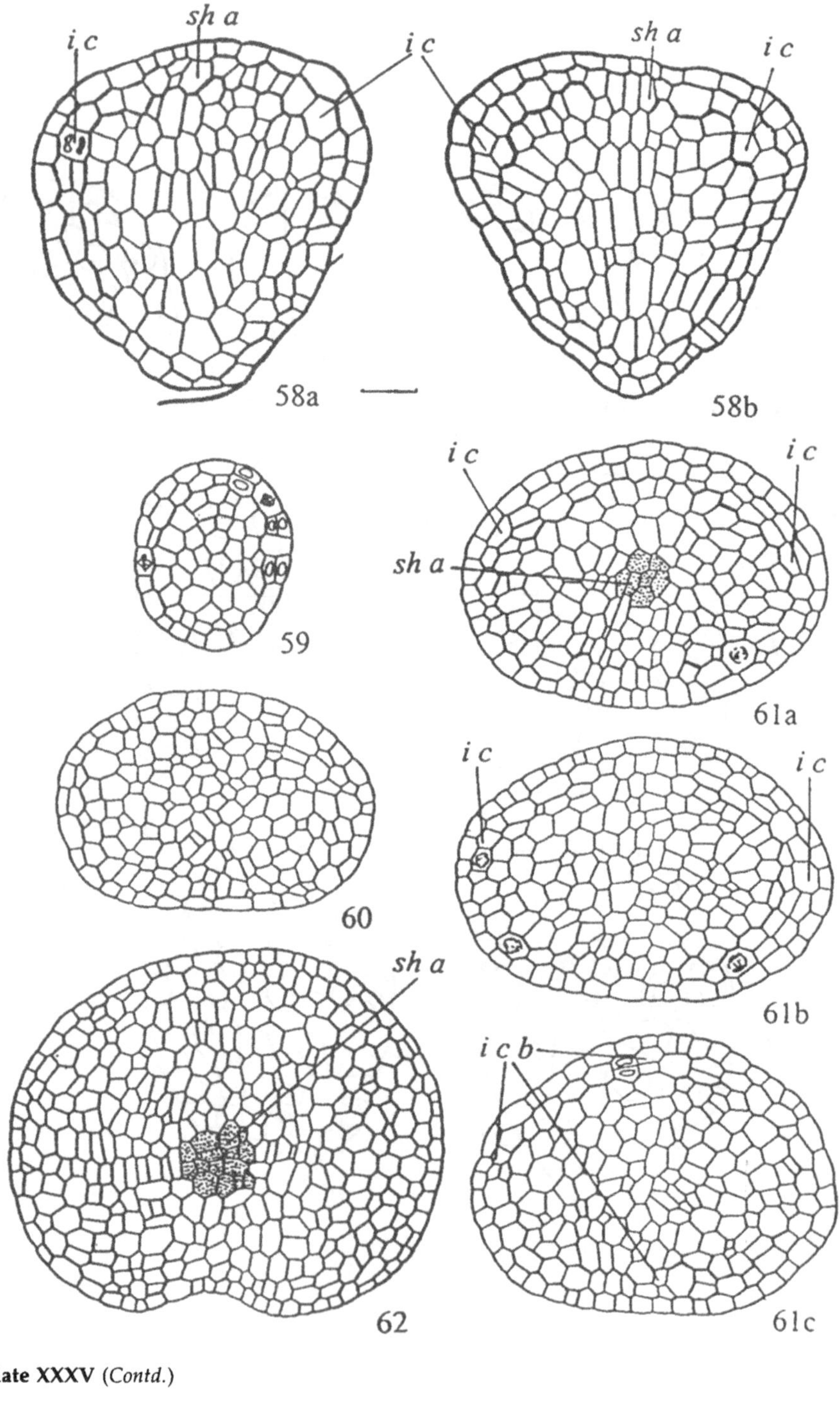

Plate XXXV (*Contd.*)

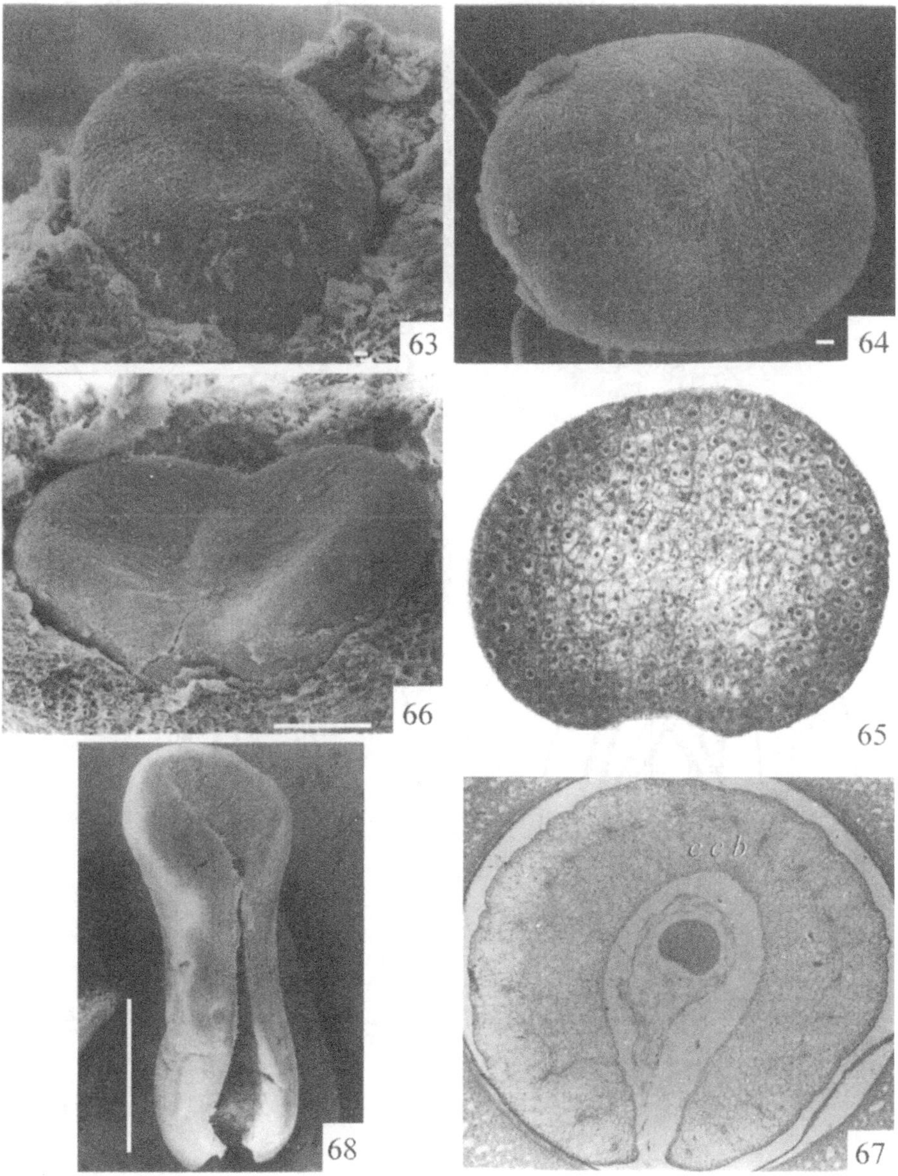

Plate XXXV (*Contd.*)

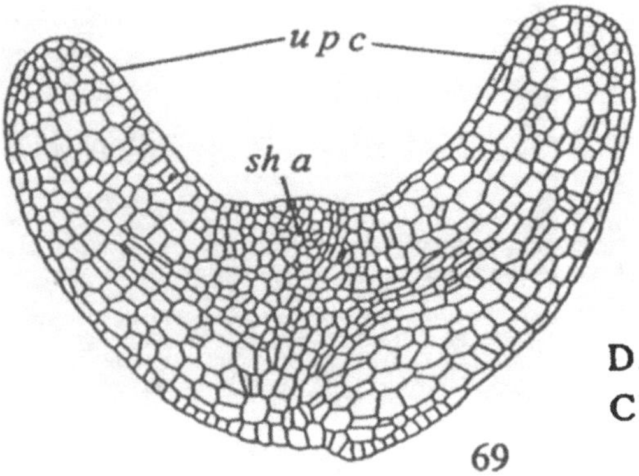

69

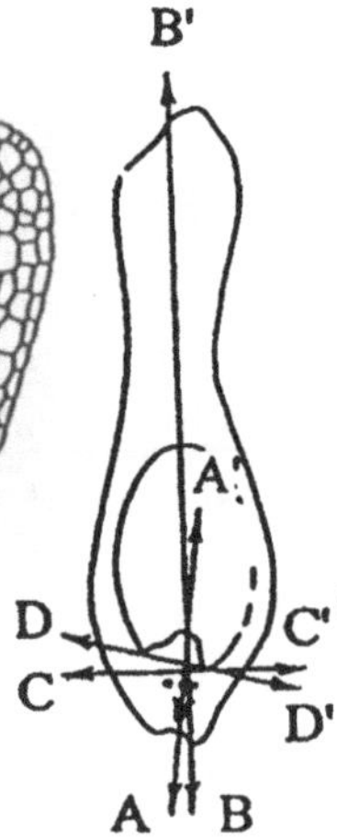

70a

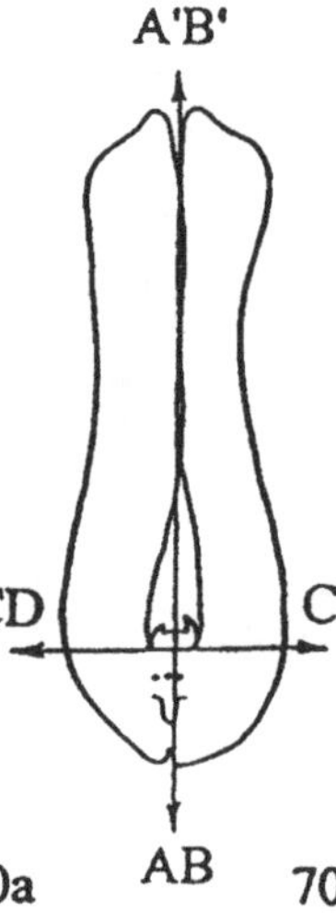

70b

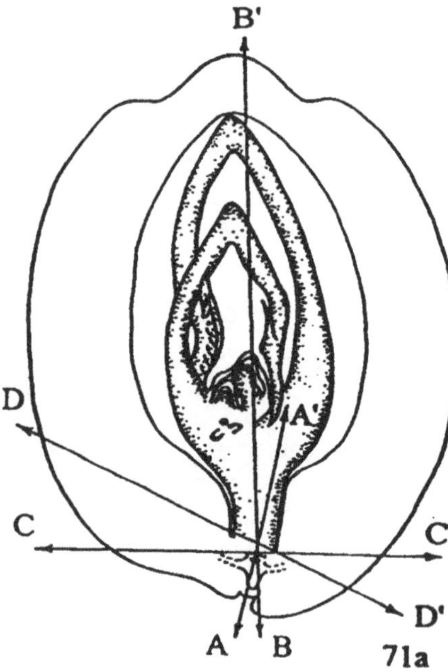

71a

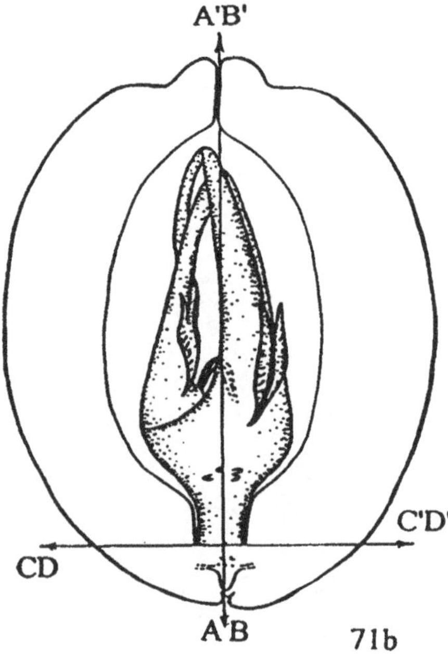

71b

Plate XXXV (*Contd.*)

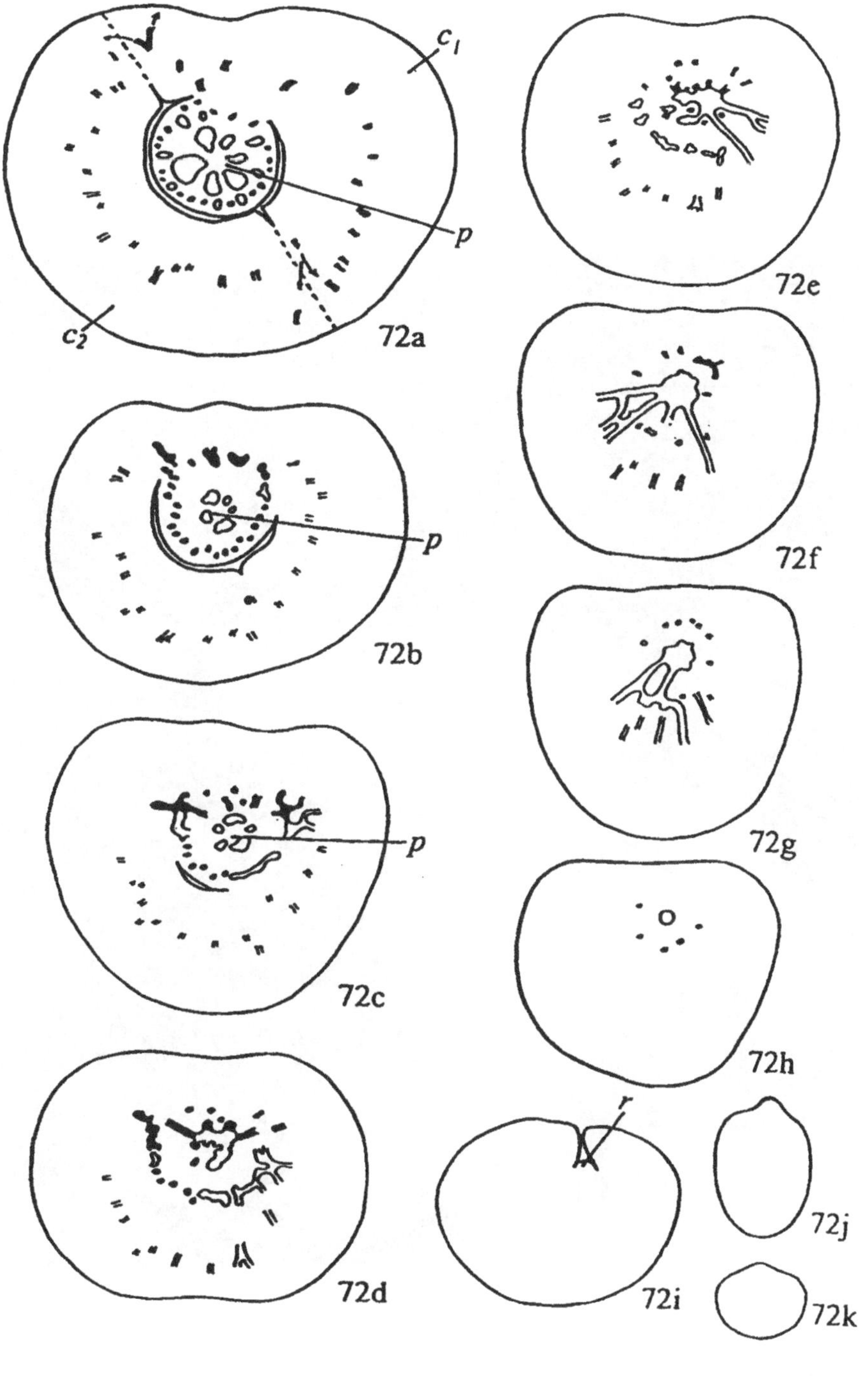

Plate XXXV (*Contd.*)

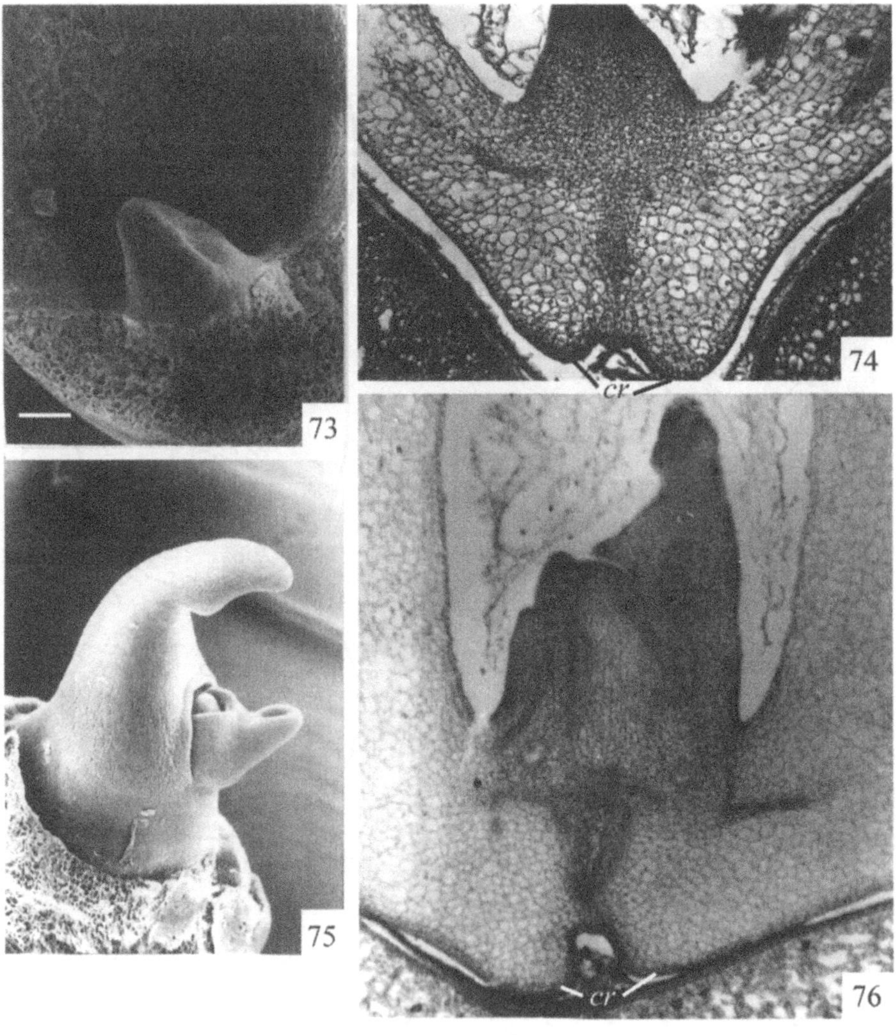

Plate XXXV (*Contd.*)

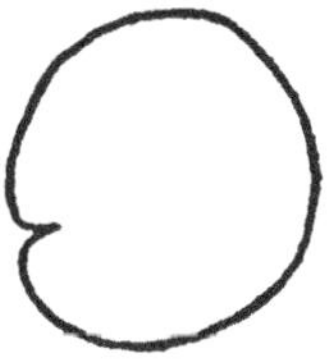

77

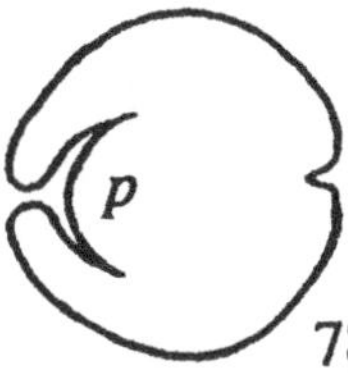

78

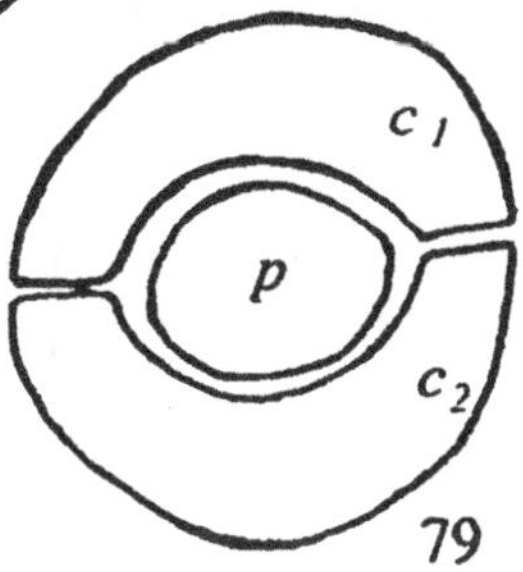

79

80

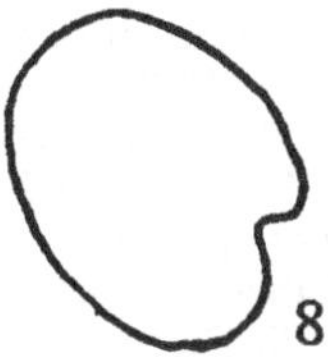

81

82

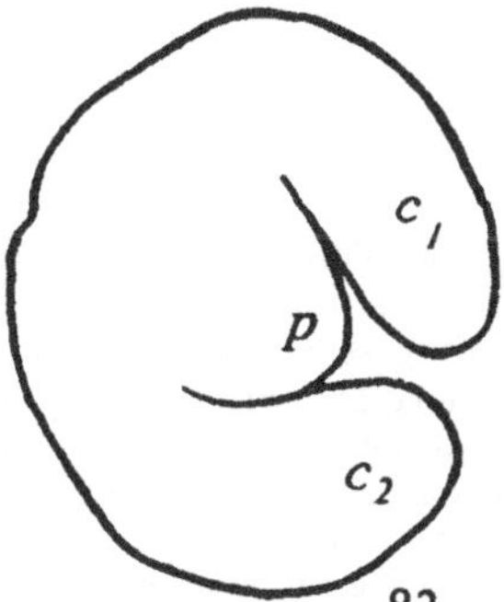

83

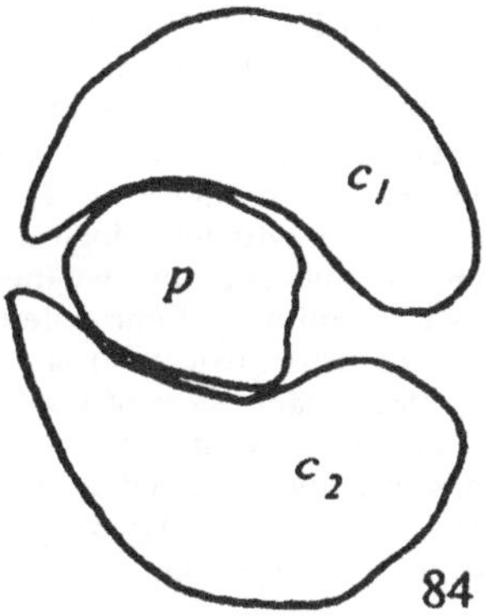

84

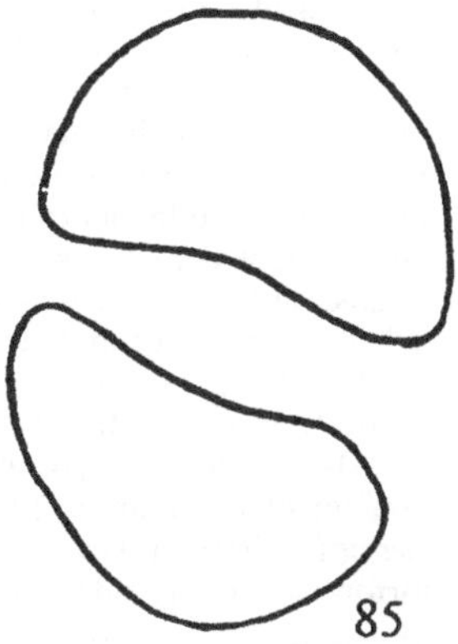

85

Plate XXXV (*Contd.*)

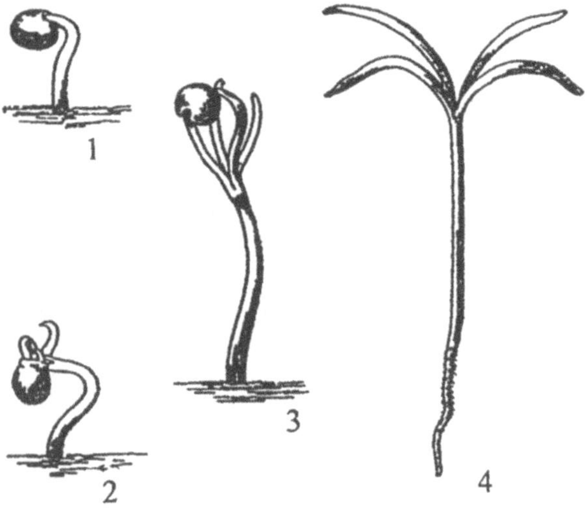

Plate XXXVI: Polycotyly.

1-35—germination and embryo development in plant species with natural schizocotyly: 1-4—going out of narrow cotyledonary lobes of seedling in turn from seed in *Schizopetalon walkeri*, 5-20—seedling structure (5), formation of proembryo (6-13), laying down of two initial cotyledonary primordia (14-18) and their subsequent bifurcation (19-20) in embryo of *Eschscholzia californica*, 21-23—formation of deeply twice-bifurcated cotyledons with equal (21,22) and unequal (23) lobes in embryo of *Pterocarya caucasica* (21), *Engelhardia roxburghianum* (22) and *Platycarya strobilacea* (23), 24-33—formation of initial cotyledonary primordia with apical meristem at their tops in embryo of *Juglans mandshurica* (24) and their primary (25-28) and secondary (29-33) division into 2 and 4 lobes respectively in *J. regia* (25-33), 34-35—filling with cotyledonary lobes of seed (34) and fruit (35) cavities in *P. caucasica;* 36-47—structure of teratological schizocotyledonous seedlings of *Oenothera hirtella* (36-42) and *Papaver rhoeas* (43-47): range of transitional forms from dicotyledonous to tricotyledonous seedling through division of one of initial cotyledonary primordia into two (36-45; 36-42—second cotyledon is removed), tetracotyledonous (46) and pentacotyledonous (47) seedlings, derived as a result of division of both initial cotyledonary primordia into two and into three lobes; 48-56—changes in conductive system structure of seedlings in *Cannabis sativa* with teratological schizocotyly of different degree: 48-51—partial division of one from two cotyledons into two (hemitricotyly), increasing xylem ray number in stele of embryo root by splitting of one pole of originally diarch stele, 52-54—complete splitting of one from two cotyledons (tricotyly), absence of traces of original dicotyly in conductive system structure, three main traces of leaf primordia of plumule seen (54), 55, 56—tricotyly, arising of cotyledonous traces *de novo;* 57-59—diagrams showing character of vascular supplying cotyledons in truly tricotyledonous seedlings of *Degeneria vitiensis* (57, 58) compared with dicotyledonous seedlings of *Magnolia grandiflora* (59). *a f*—apical furrow, *a m c*—apical meristem of cotyledon, *c*—cotyledons.

1-5, 21, 34-35—after Lubbock, 1892; 6-20—after Sachar and Mohan Ram, 1958; 22-23—after Stone, 1970; 24—after Langdon, 1934; 25-33—after Nast, 1941; 36-47—after De Vries, 1903; 48-56—after Compton, 1913; 57-59—after Swamy, 1949b.

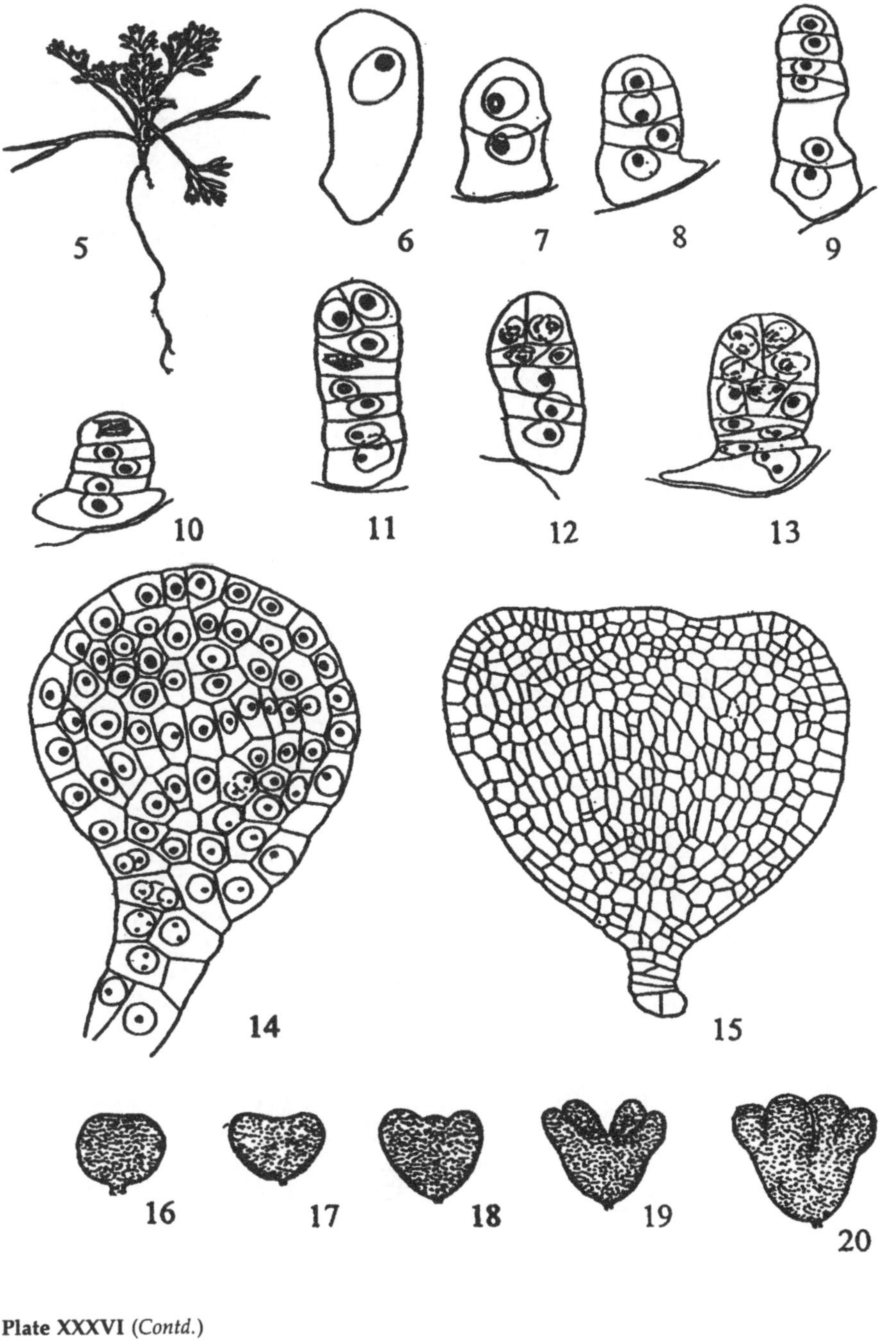

Plate XXXVI (*Contd.*)

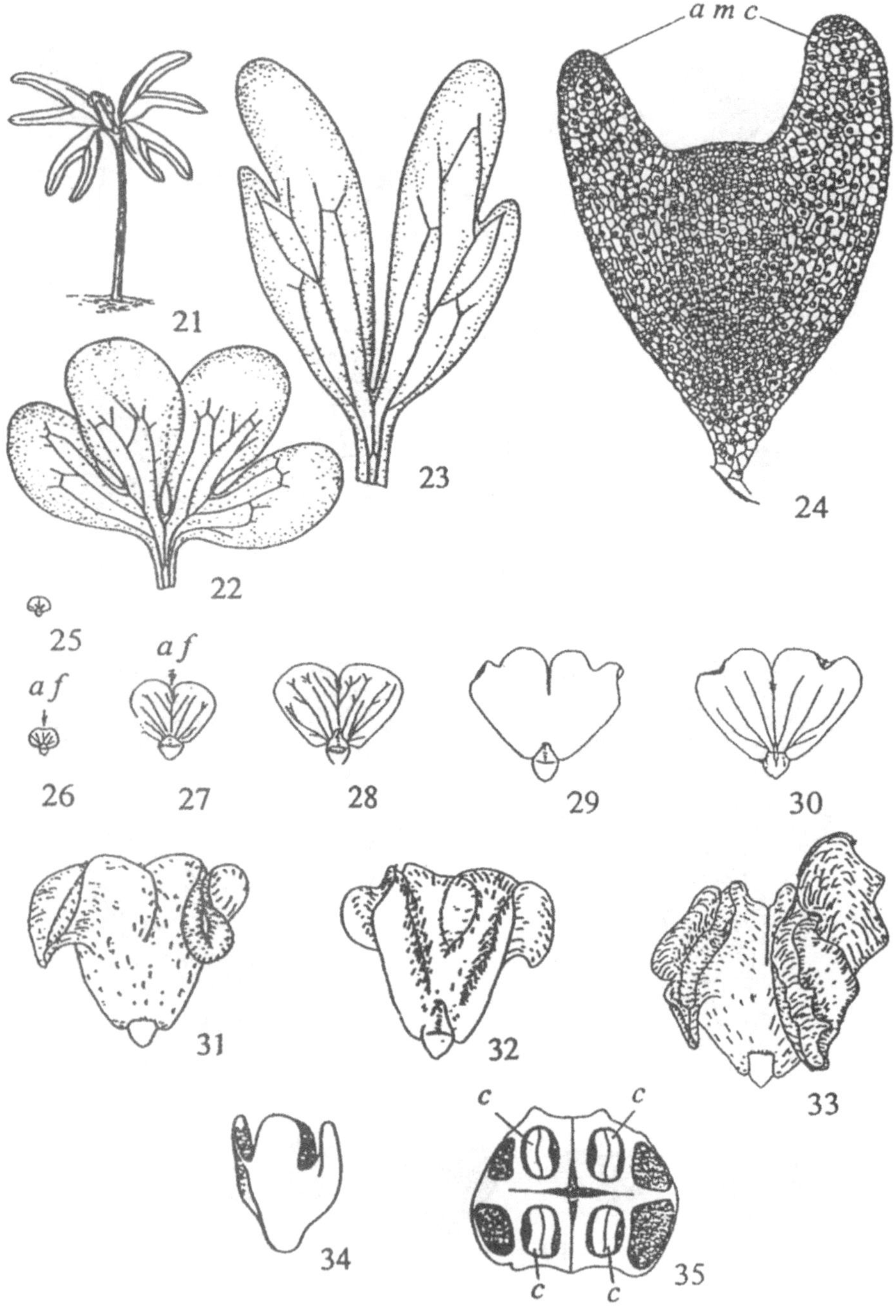

Plate XXXVI (*Contd.*)

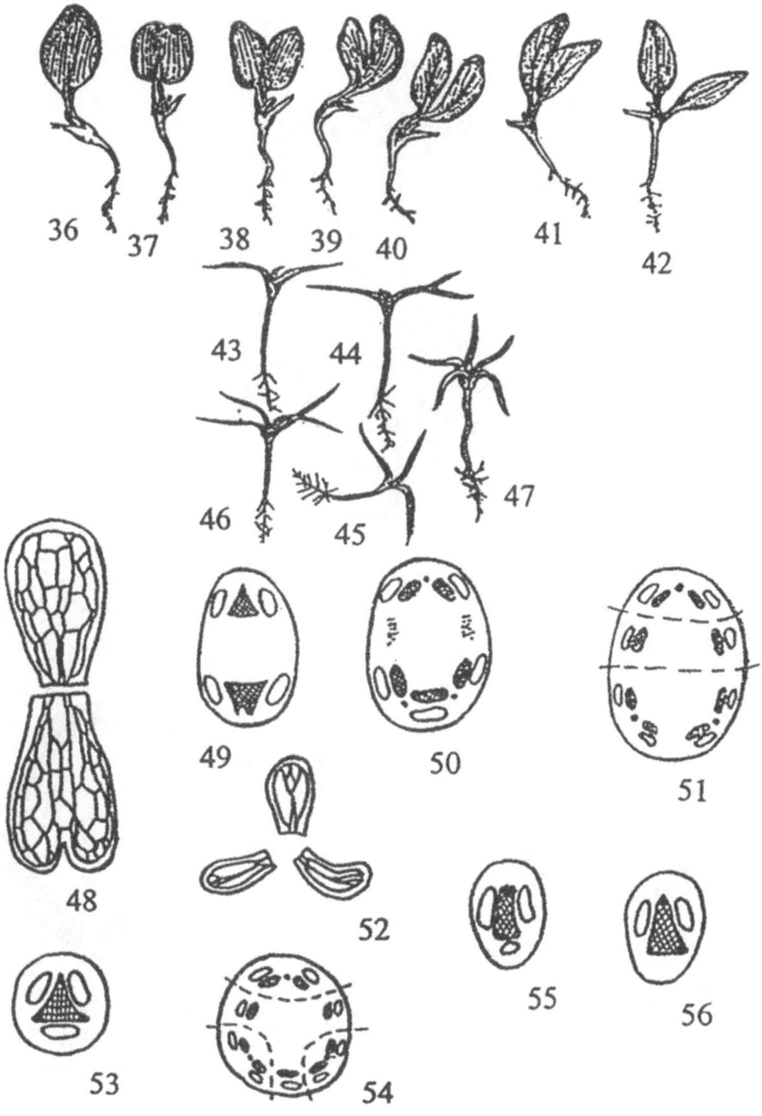

Plate XXXVI (*Contd.*)

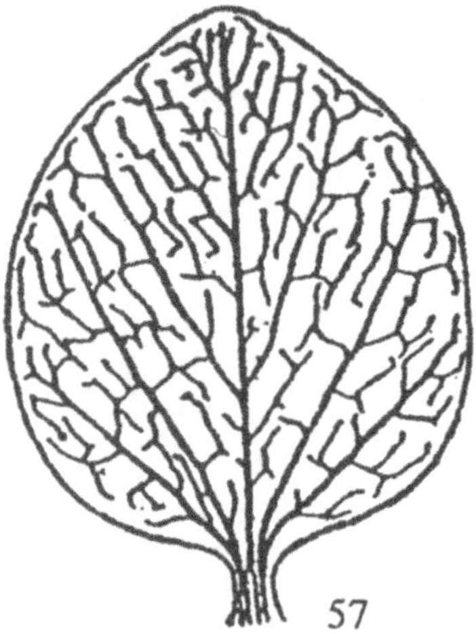

57

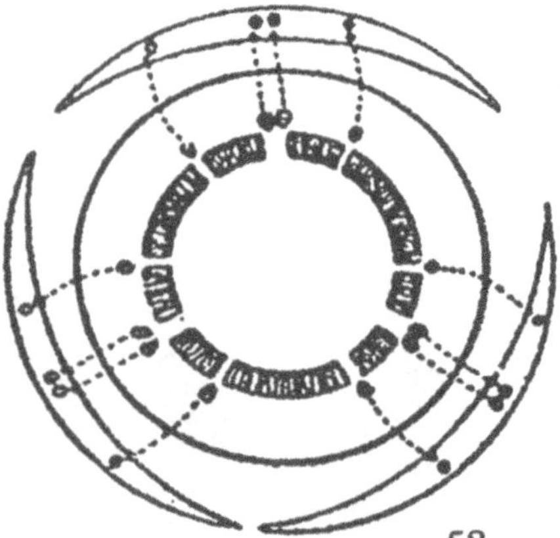

58

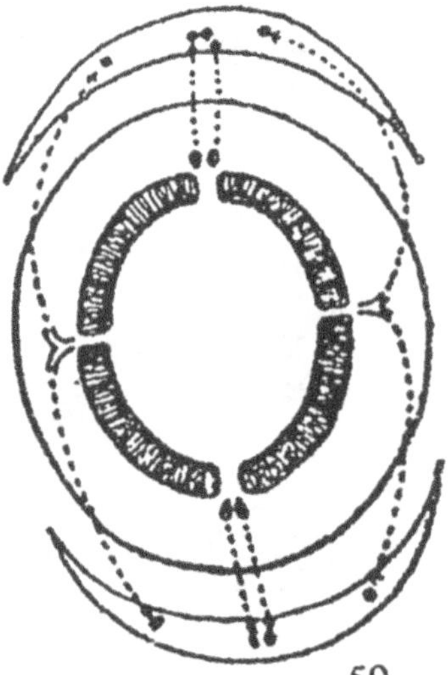

59

Plate XXXVI (*Contd.*)

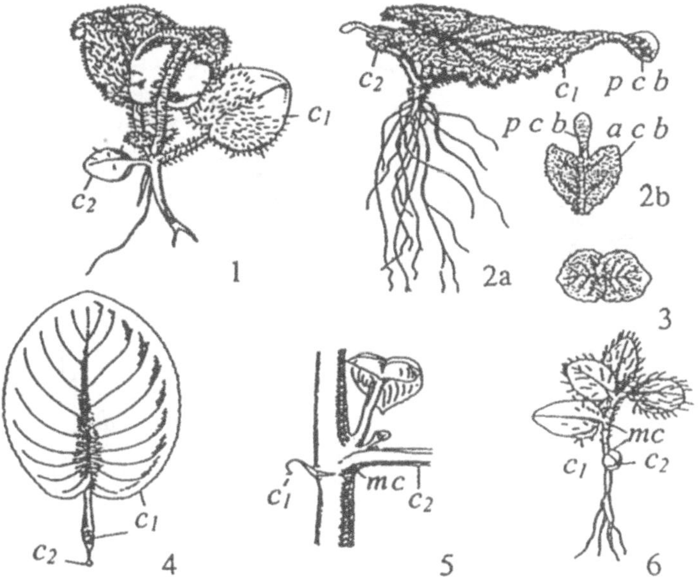

Plate XXXVII: Heterocotyly.

1-20—seedling structure (1-13) and embryogenesis (14-20) in the Gesneriaceae species with different fate of cotyledons at germination: 1—*Oreocharis primuloides*, rosette species with two unequal cotyledons—the large remaining during development, with pubescent petiole *(c_1)* and the smaller abortive one, without hairs on the petiole (c_2), plumule with several leaf pairs; 2,3—*Streptocarpus* sp., unifoliate species, remaining cotyledon (c_1) forms adventive blade, primary blade seen at tip, plumule development not observed (2b,3—formation of adventive blade in smaller, usually abortive cotyledon (c_2); 4,5—seedling *Chirita lavandulacea* with two unequal cotyledons and removed plumule (4) and its part on later stage of development with axillary and extra-axillary buds in greater cotyledon (c_1) and short mesocotyl (5); 6—*Didymocarpus platypus*, formation of mesocotyl between remaining (c_1) and abortive (c_2) cotyledons; 7-13—*S. fanniniae*, diagrams representing development of cotyledonary phyllomorph (7-9), topographical relationship of meristems in longitudinal section (10) and general morphology of phyllomorph (11), structure of seedling with embryonic meristem displaced to base of large cotyledon (c_1) (12) and young phyllomorph on level of two new phyllomorph primordia (13); 14-20—*S. rexii*, laying down of two identical cotyledons during embryogenesis; 21-25—structure of seedlings in *Peperomia* species: 21,22—normal dicotyledonous one (*P. pellucida*), 23-25—heterocotyledonous one (*P. peruviana*), 22,24,25—cross-sections of seed up to (22,24) and after (25) cotyledon(s) going out from seed coat.

a—aerenchyma, *a c b*—adventive cotyledonary blade, *a r*—adventive root, b_1, b_2—detached meristems, *b as c*—blade of assimilating cotyledon, *b ab c*—blade of absorbing cotyledon, *b m*—basal meristem, c_1,c_2—cotyledons, *c s*—cavity in seed, remaining after going out of the assimilating cotyledon, *en*—endosperm, *g m*—groove meristem, *hc*—hypocotyl, *l*—lamina, *mc*—mesocotyl, *p c*—procambial strand, *p c b*—primary cotyledonary blade, *pd*—petiolode, *ph*—additional phyllomorphs from adaxial surface of petiolode, *pl*—plumule, *p m*—petiolode meristem, *p p*—new phyllomorph primordia, *p r*—primary root, *ps*—perisperm, *p t*—palisade tissue, *s c*—seed coat, vs—vascular tissue.

1-6—after Hill, A., 1938; 7-13—after Jong and Burtt, 1975; 14-20—after Alimova and Yakovlev, 1982; 21-25—after Hill A., 1906.

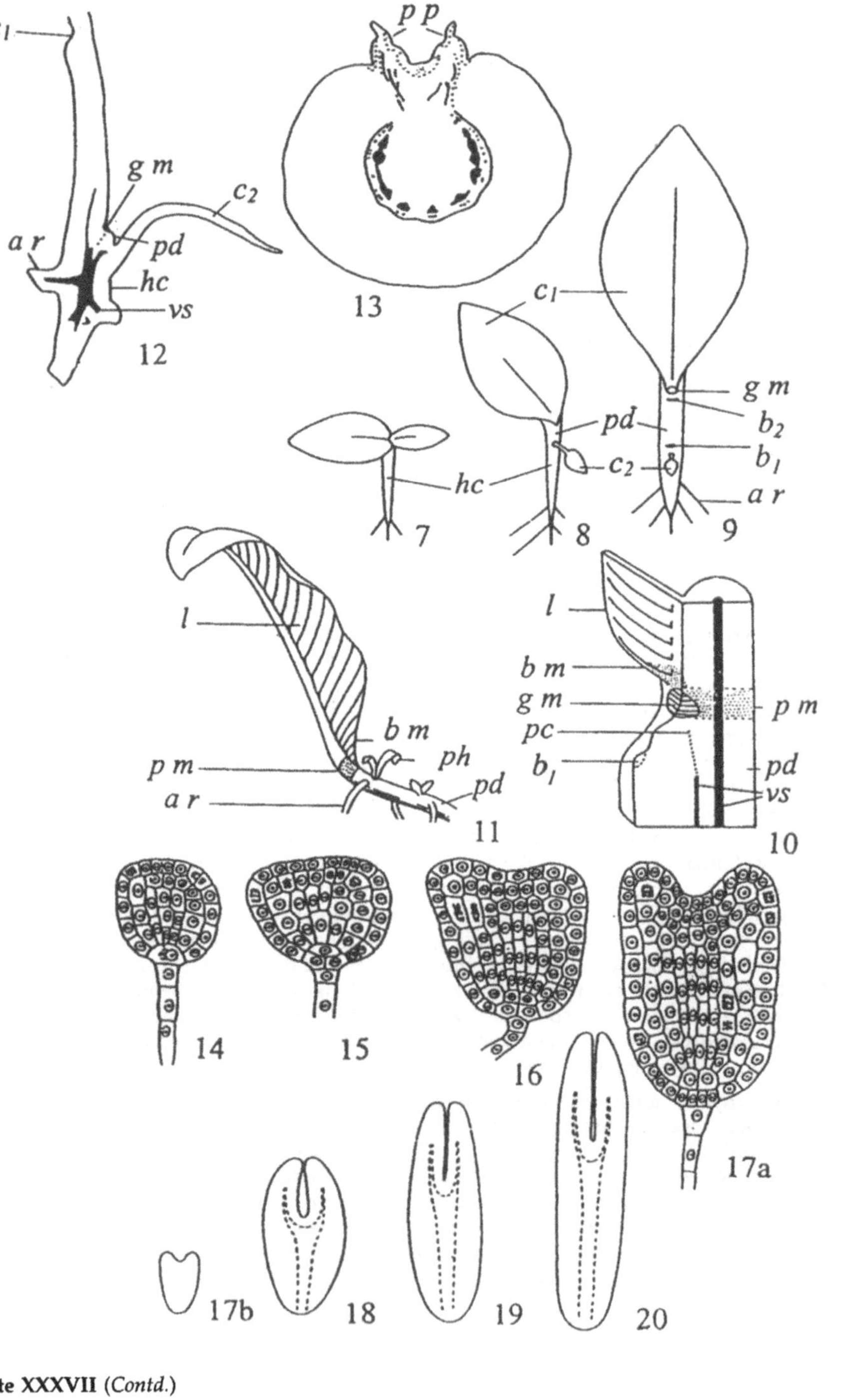

Plate XXXVII (*Contd.*)

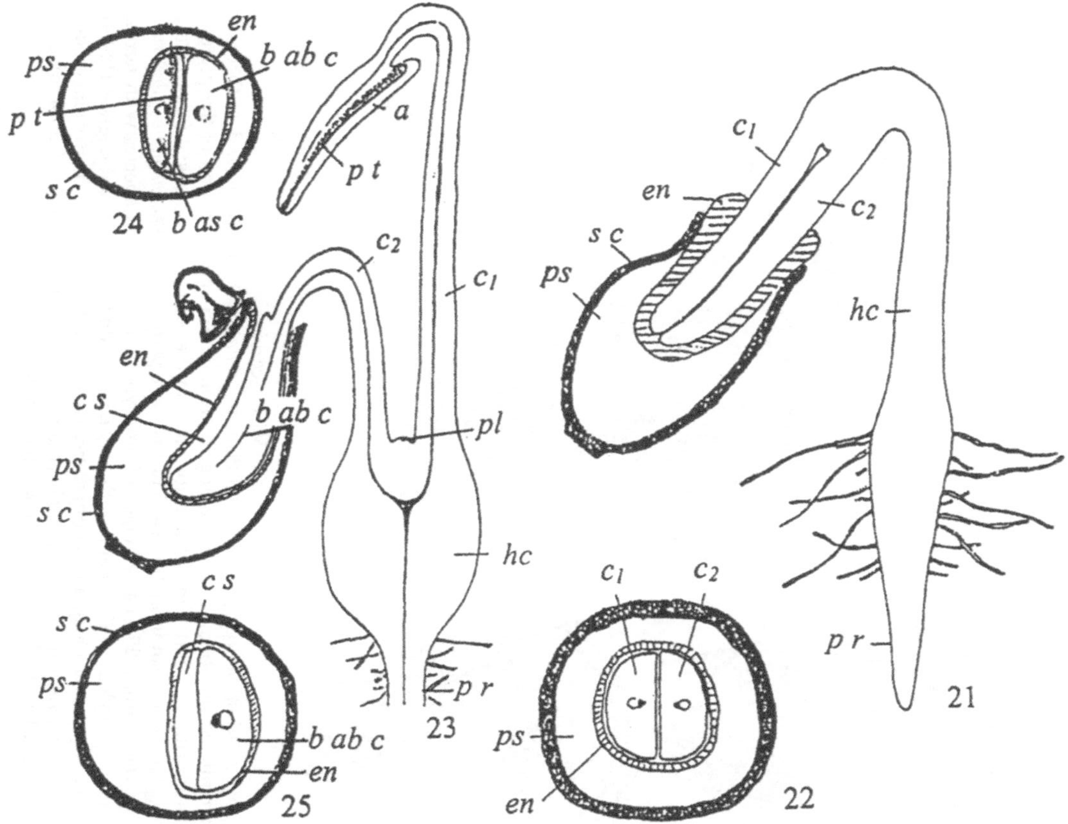

Plate XXXVII (*Contd.*)

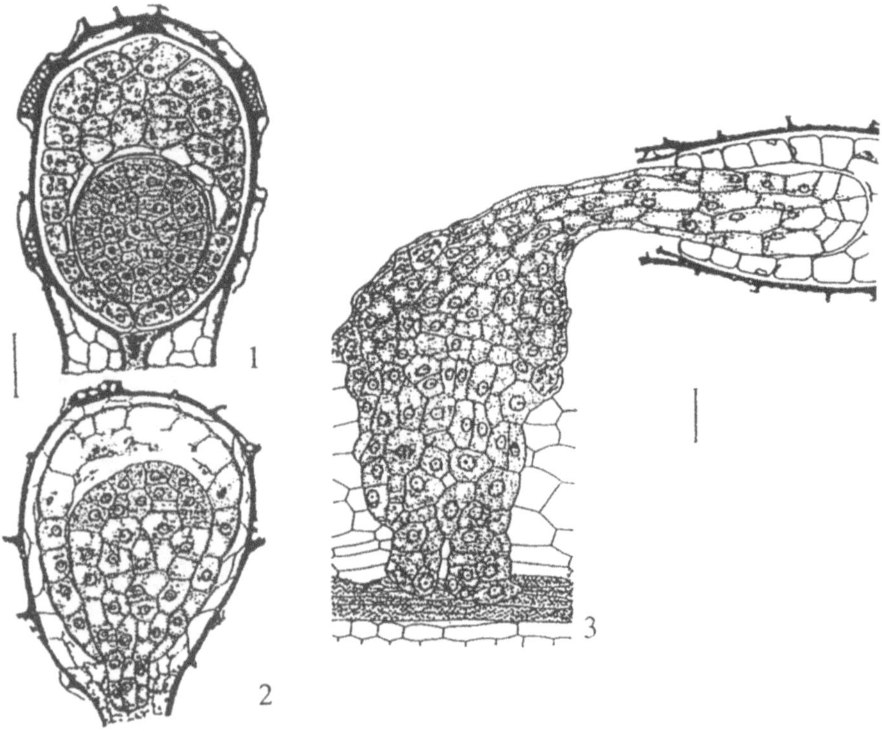

Plate XXXVIII: Acotyly.

1-3—post-seminal development of *Orobanche crenata*: 1—mature seed with embryo undifferentiated into organs, 2 –beginning of germination and formation of the protosome with haustorium, 3—intruding of haustorium into the host-plant tissues; 4-6—structure of embryo and endosperm in mature seed of *Cistanche phelypaea* (4), *Phacellanthus tubiflorus* (5), *Aeginetia indica* (6); 7-29—early stages of embryo development in species with normally differentiated cotyledons (7-13) and with cotyledons reduced to different degrees (14-29): 7-13—*Lathraea squamaria* (Onagrad - type of embryogenesis and normal formation of embryoderm), 14-20—*Ph. tubiflorus* (disturbance of embryoderm initiation in apical part of embryo), 21-29—*A. indica* (disturbance of succession in *ca* cell divisions).

a p—apical pole of embryo, *b p*—basal pole of embryo, *en*—endosperm, *s c*— seed coat. Bars: 1-3—0.05 mm.

1-6—after Teryokhin, 1977; 7-20—modified from Teryokhin and Nikiticheva, 1981; 21-29—after Tiagi, 1952a.

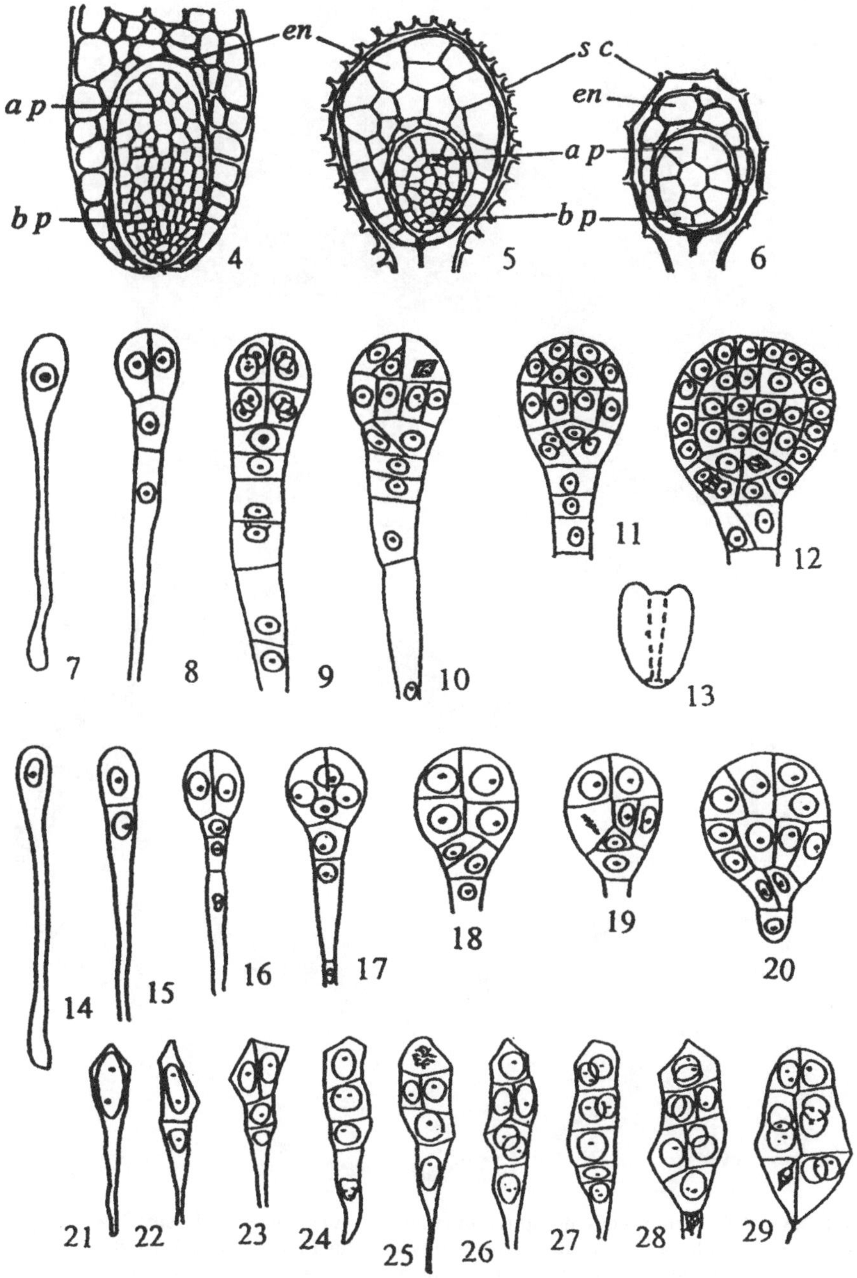

Plate XXXVIII (*Contd.*)

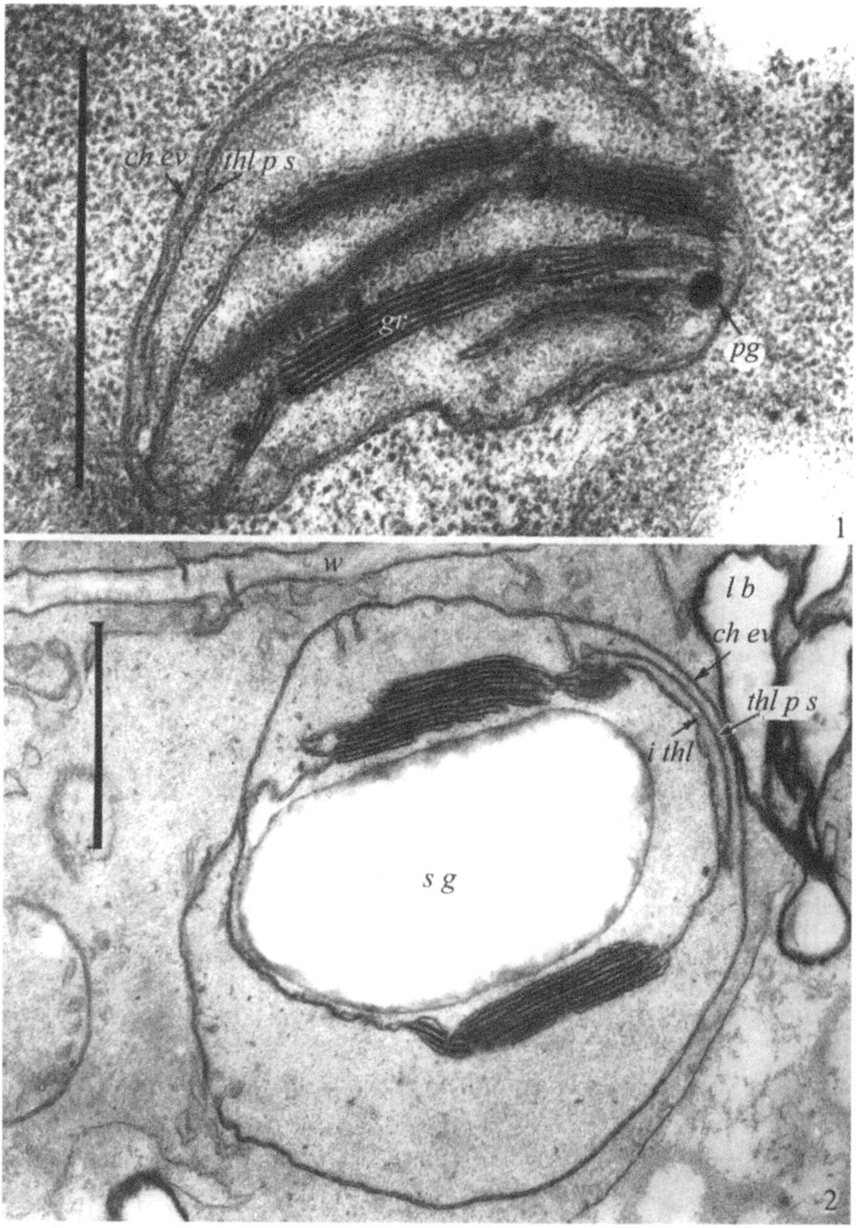

Plate XXXIX: Occurrence of chlorophyll in embryo as a character to classify flowering plants. 1-21—embryo chloroplasts in chloroembryophytes of the first (main) (1-12) and second (13-21) types: 1-4—*Alcea rosea*: 1—chloroplast in ground meristem of embryo with cotyledon primordia, 2-4—chloroplasts and fragment of chloroplast of cotyledon mesophyll in embryo of immature seed; 5-12—chloroplasts and fragments of chloroplasts of cotyledon mesophyll of embryo in

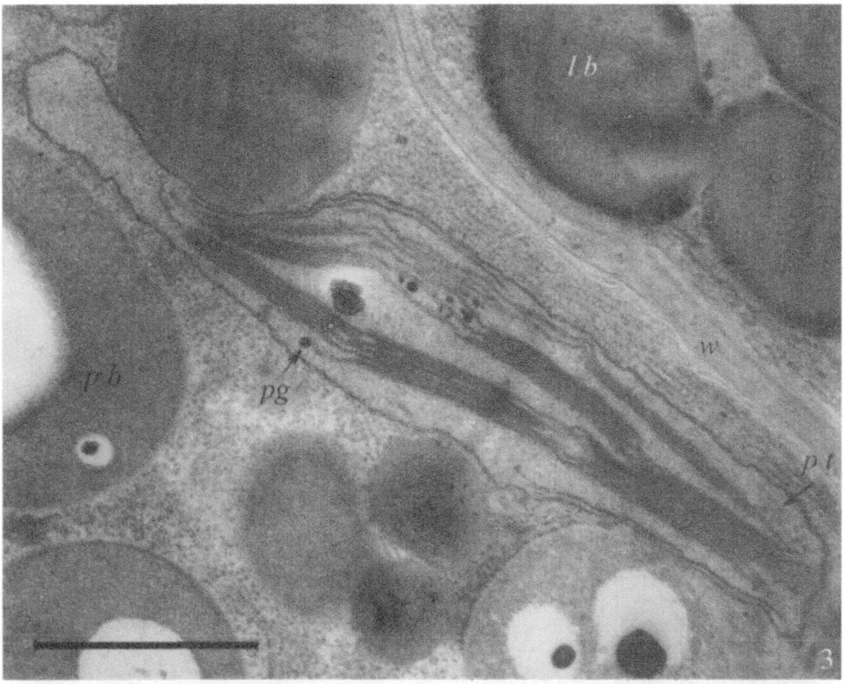

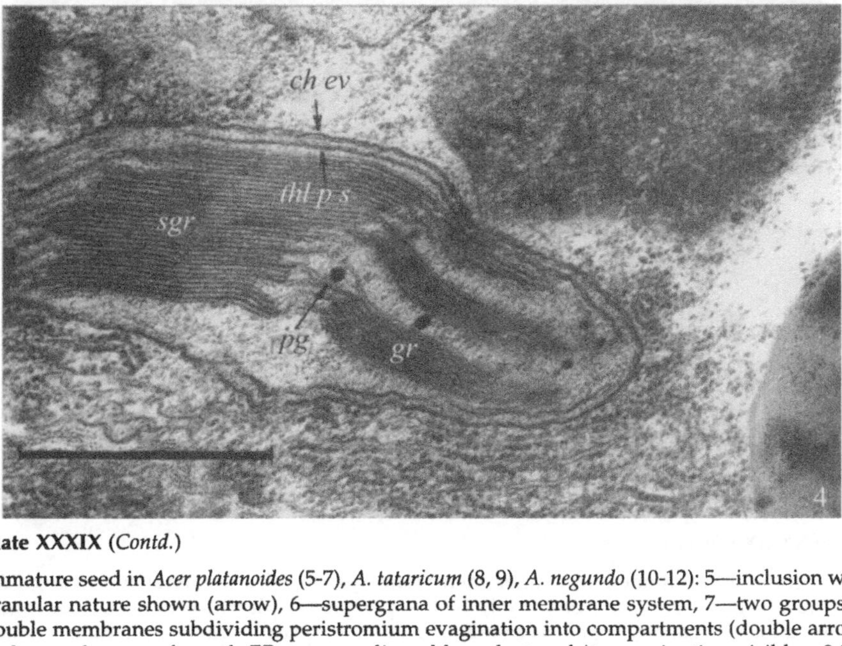

Plate XXXIX (*Contd.*)

immature seed in *Acer platanoides* (5-7), *A. tataricum* (8, 9), *A. negundo* (10-12): 5—inclusion with granular nature shown (arrow), 6—supergrana of inner membrane system, 7—two groups of double membranes subdividing peristromium evagination into compartments (double arrow) and membranes of rough ER, surrounding chloroplast and its evagination visible, 8,9—

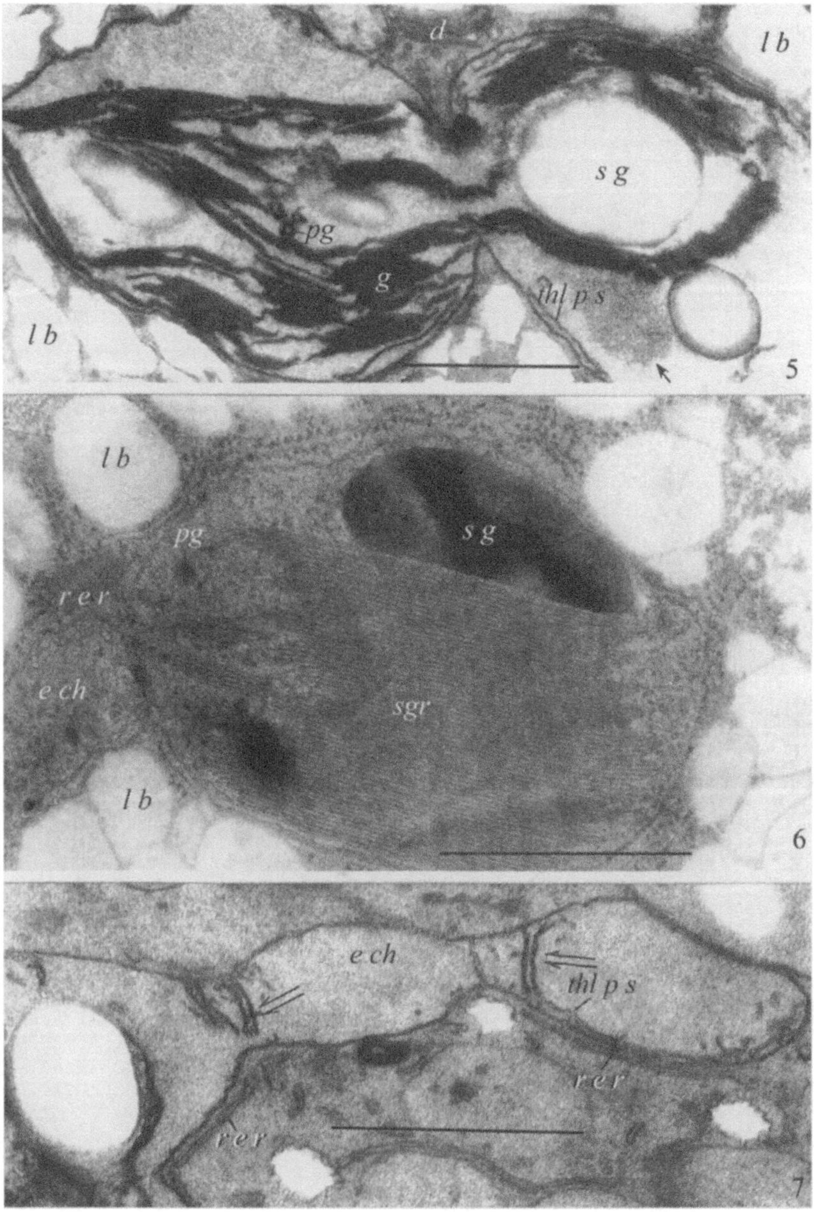

Plate XXXIX (*Contd.*)

supergrana as components of inner membrane system, also membranes of rough ER, surrounding chloroplasts, observed, 10-12—peripheral plastid reticulum, concentric membranous formations, supergrana absent; 13,14 –chloroplasts of embryo plumule of mature seed in *Nelumbo nucifera*, membranes of ER surrounding chloroplast visible (13), site of

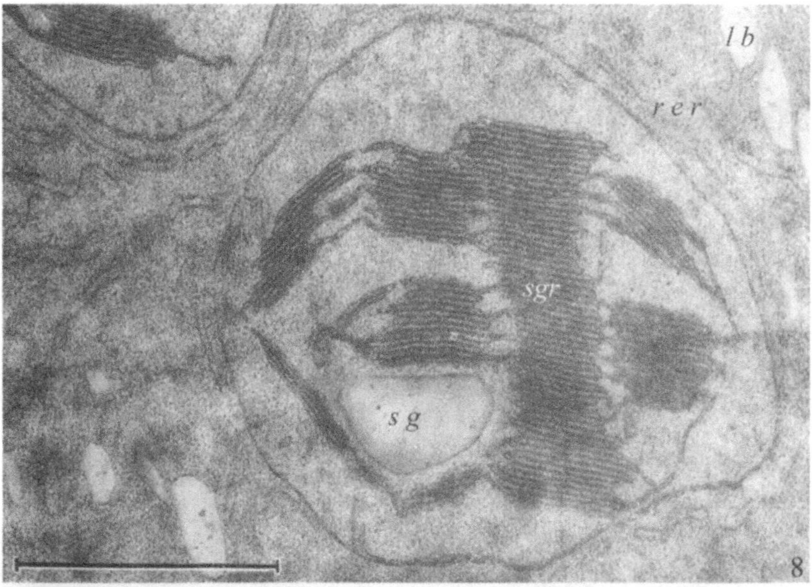

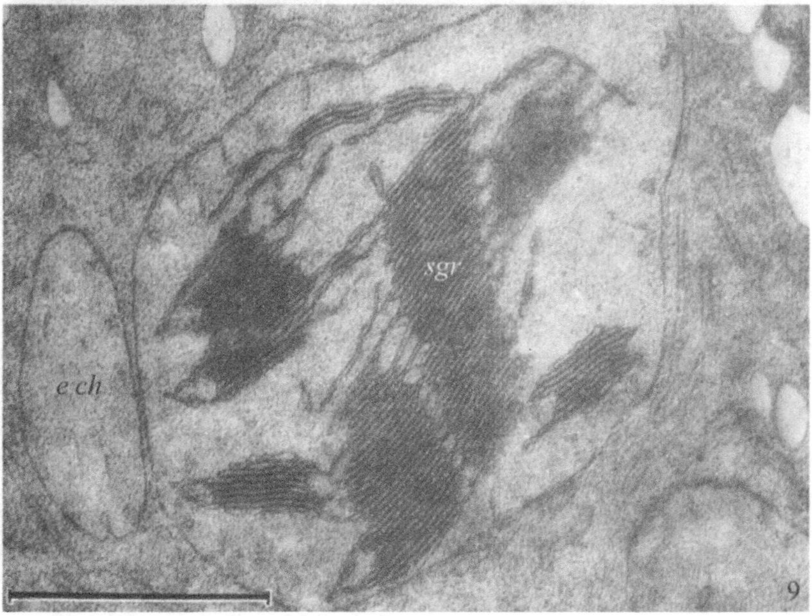

Plate XXXIX (*Contd.*)

invagination of inner membrane of chloroplast envelope shown (arrow); 15—chloroplast of adult above-water leaf of *N. nucifera*; 16,17—plastids with prolamellar bodies (plumule of embryo of immature seed in *N. nucifera* begins to turn green 20 days after flower opening); 18,19— chloroplast fragments from embryo of immature seed in *N. nucifera* (27 days after flower

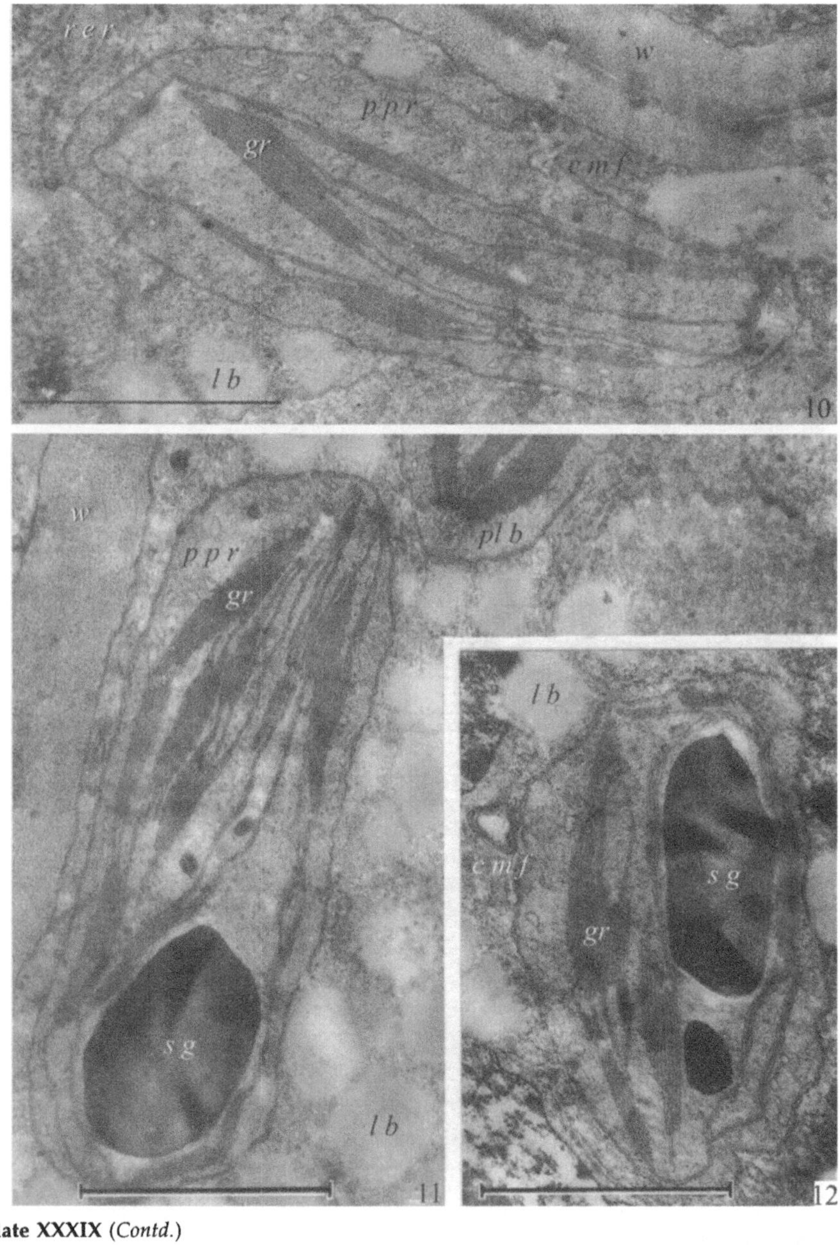

Plate XXXIX (*Contd.*)

opening), thylakoid of peripheral part of inner membrane system situated parallel to chloroplast envelope (arrow), place of divergence of peripheral thylakoid and chloroplast envelope shown (double arrow); 20,21—chloroplast fragment and chloroplast of the embryo plumule from mature seed of *N. nucifera*, moving aside of peripheral thylakoid from plastid envelope and its

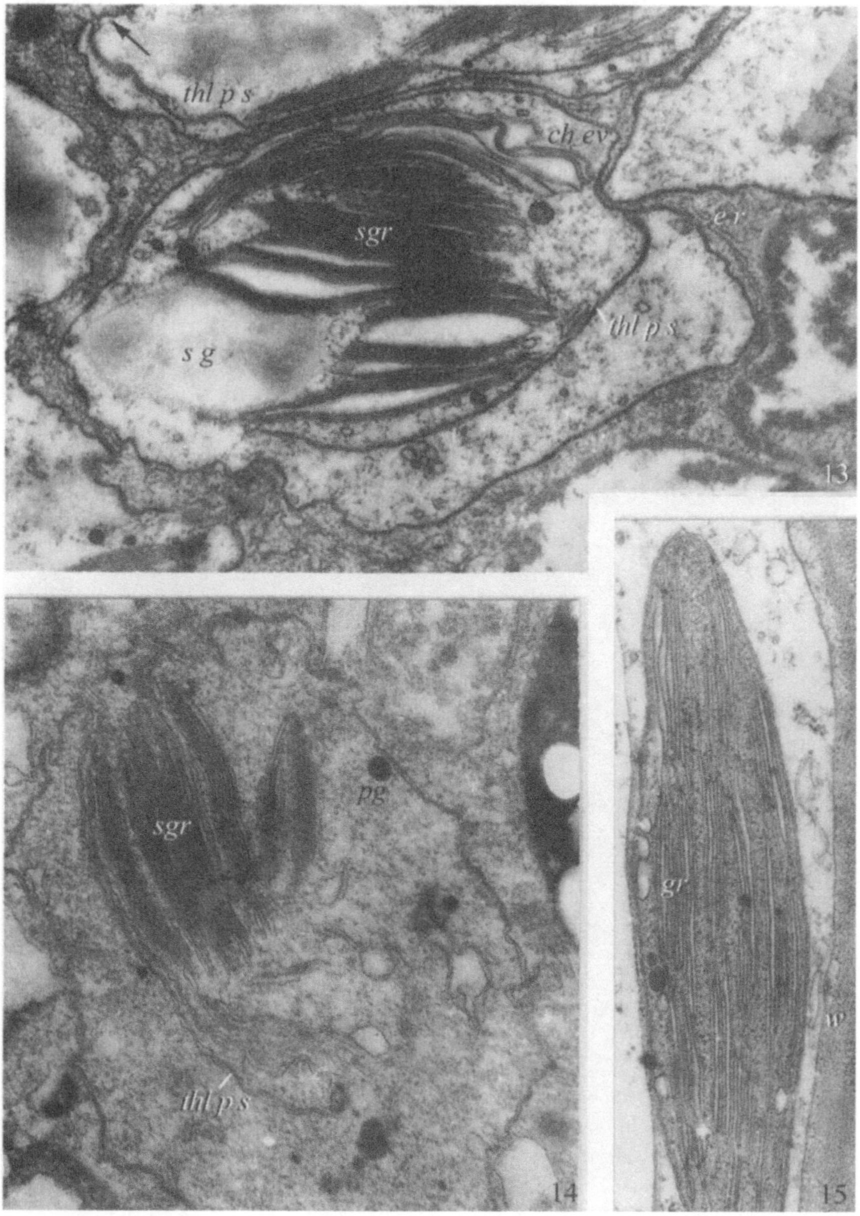

Plate XXXIX (*Contd.*)

disordered position in plastid stroma (arrows) observed, place of connection of this thylakoid with envelope shown (double arrow); 1-21—TEM.

Bar—1 μm; 13—×36,000, 14—×34,800, 15—×38,900, 18—×45,000, 19—×36,300, 20—×39,600, 21—×39,000; *ch ev*—chloroplast envelope, *c m f*—concentric membranous formation, *ch st*—

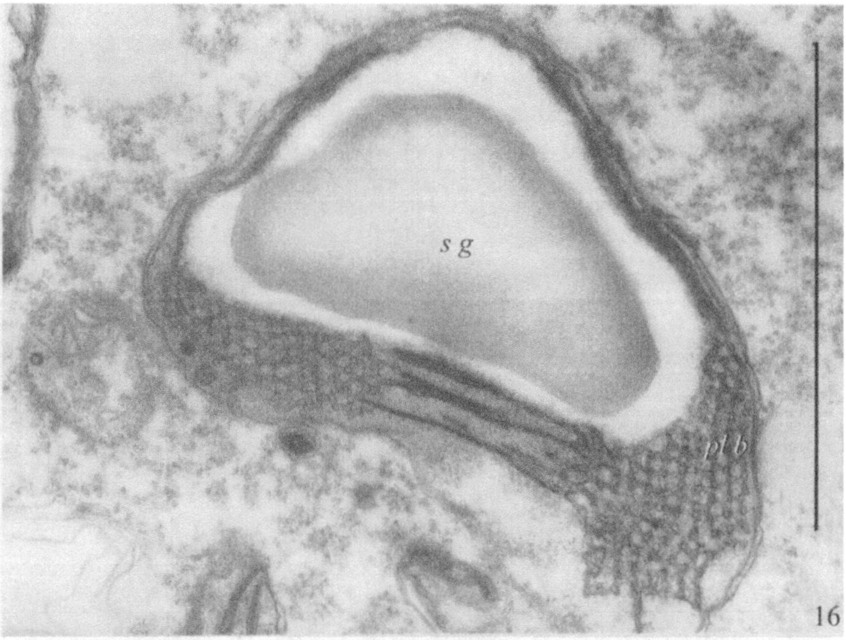

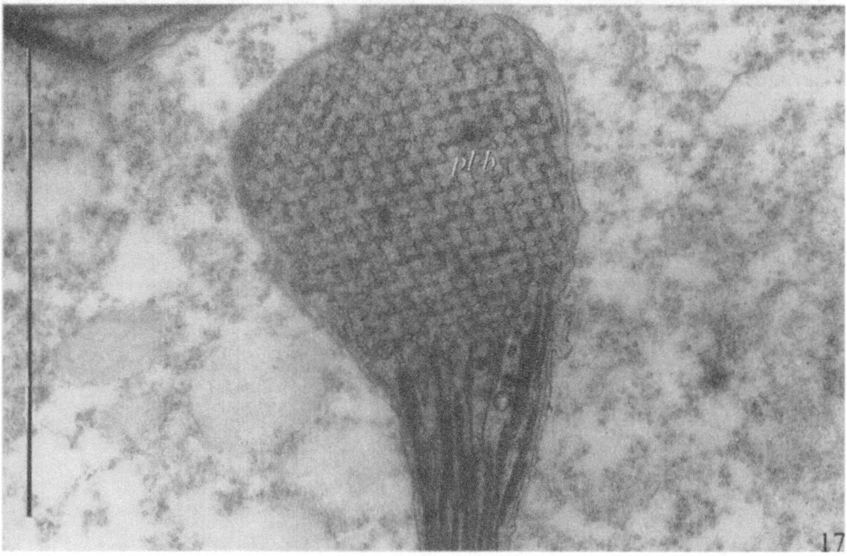

Plate XXXIX (*Contd.*)

chloroplast stroma, *d*—dictyosome, *e ch*—evagination of chloroplast peristromium, *e r*—endoplasmic reticulum, g, *gr*—grana, *i thl*—intergrana thylakoid, *l b*—lipid body, *p b*—protein body, *pg*—plastoglobule, *pl b*—prolamellar body, *p p r*—peripheral plastid reticulum, *p t* —plastid tubules, *r e r*—rough endoplasmic reticulum, *sgr*—supergrana, *s g*—starch grain, *thl p s*—thylakoid of peripheral part of inner membrane system, *w*—cell wall.

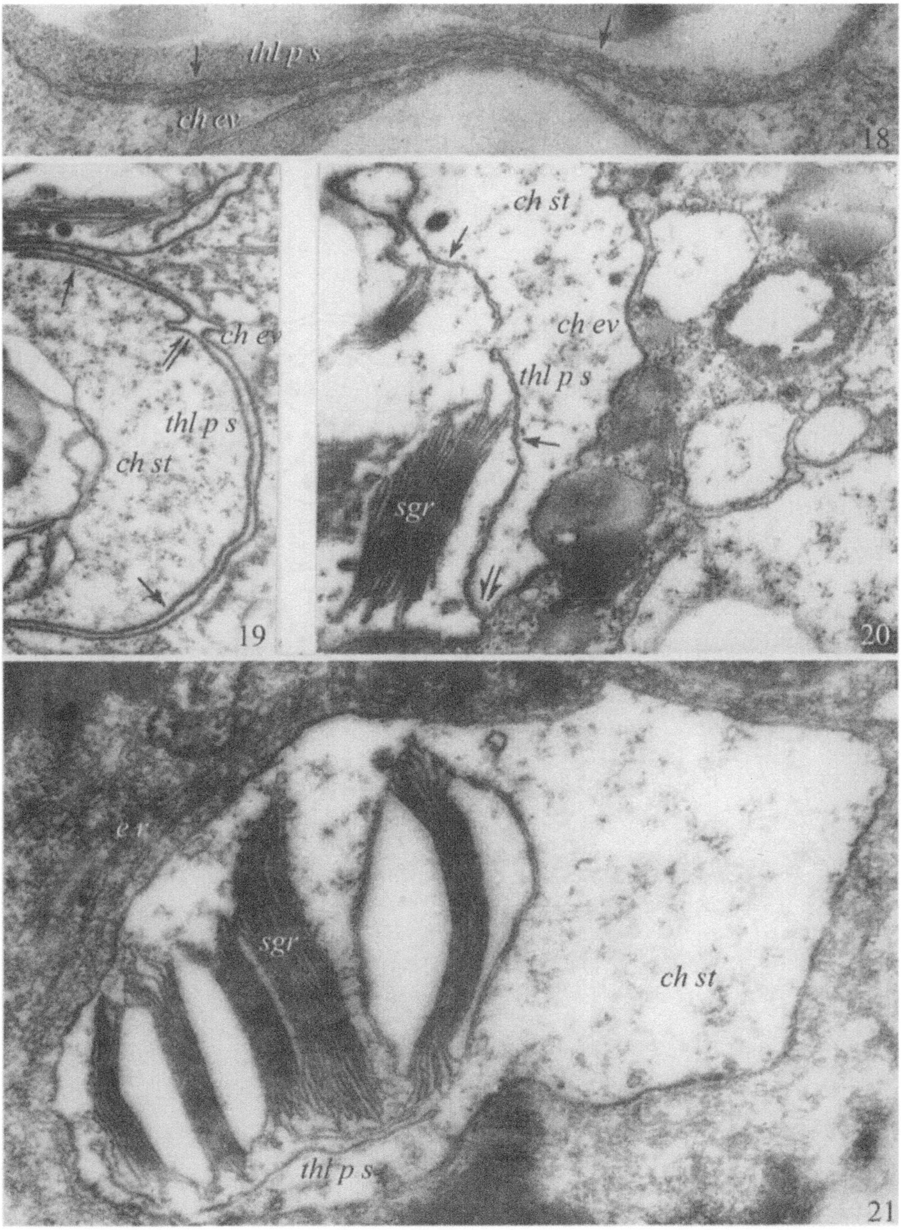

Plate XXXIX (*Contd.*)

1-4—after Kamalova *et al.*, 1983; 5-7—modified from Zhukova and Dragunova, 1980; 8-12—Zhukova, orig. electron-micrographs see: Alimova, 1985a; 13-15, 18-20—after Zhukova, 1972; 16—after Yakovlev and Zhukova, 1973; 17—after Zhukova, 1979; 21—modified from Zhukova and Yakovlev, 1966.

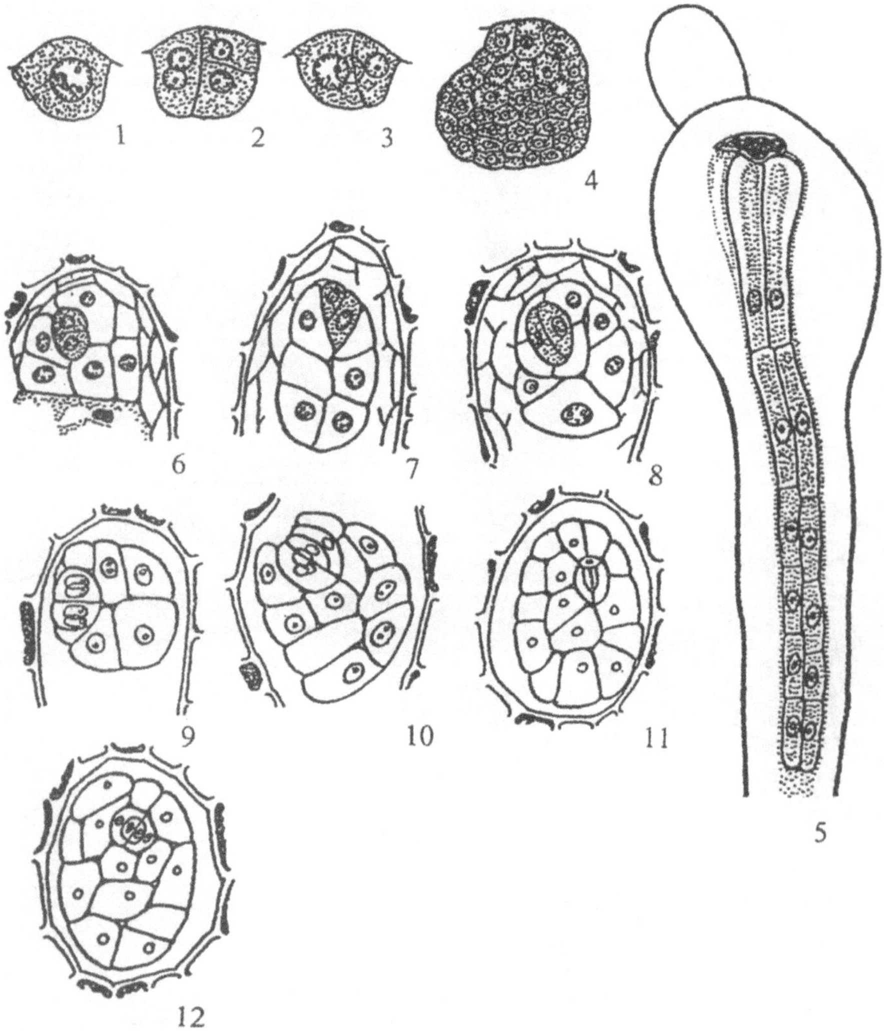

Plate XL: Piperad-type of embryogenesis.

1-4—typical variation (*Peperomia blanda*); 5—Dendrophtoe-variation (*Nuytsia floribunda*); 6-12—Balanophora-variation (*Balanophora laxiflora*, 6-10 and *B. poliandra*, 11-12); 13-30—Scabiosa-variarion (*Cephalaria gigantea*).

a r—apical meristem of root, *e—embryo*, *en*—endosperm, *p b*—procambial bundle, *r c*—root cap, *sh a*—shoot apex; *a, b, c, c', c'', d, d', d'', e, e', e'', o, p,*—different tiers of proembryo.

1-4—after Nikiticheva, 1981b; 5—after Narayana, 1958; 6-12—modified from Teryokhin and Yakovlev, 1967; 13-30—after Kamelina, 1980.

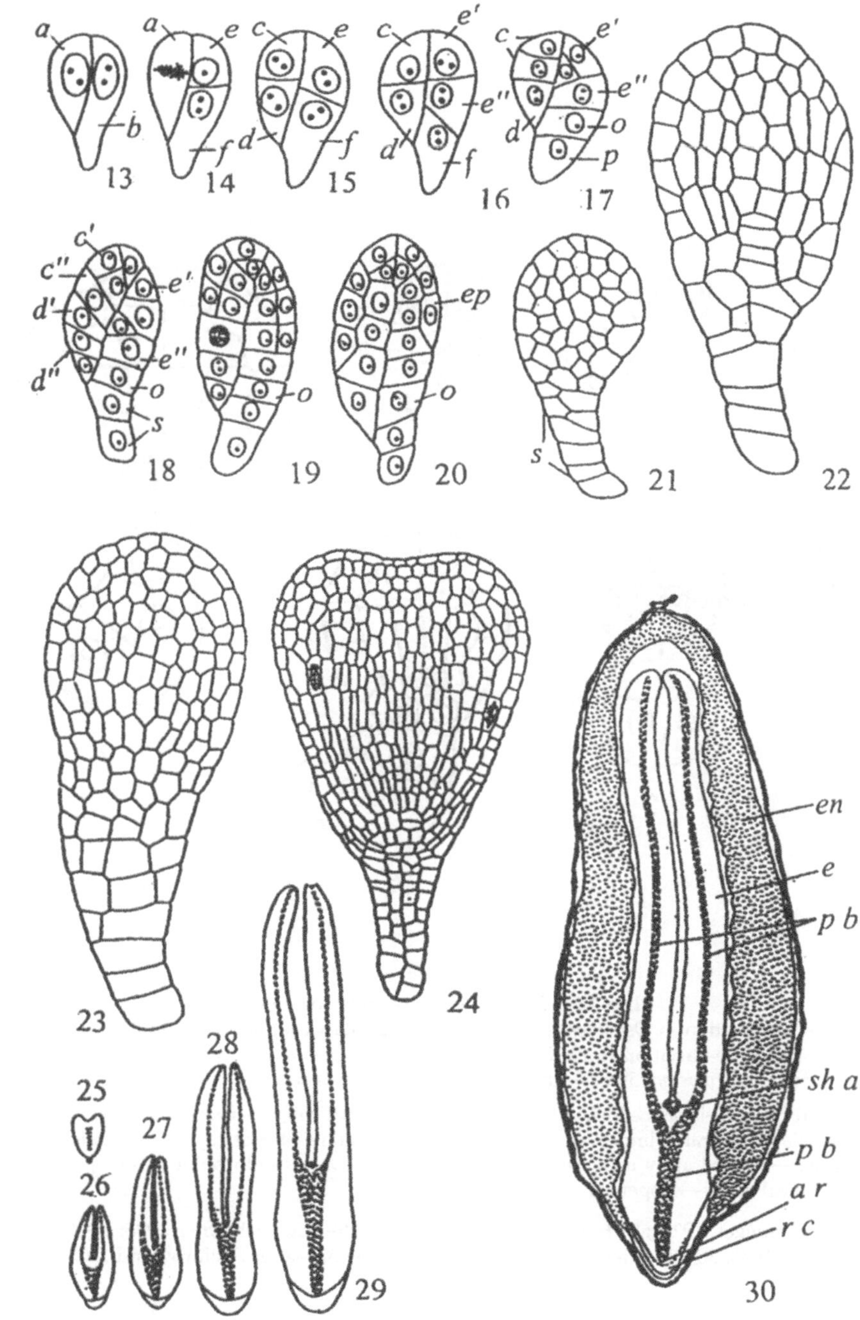

Plate XL (*Contd.*)

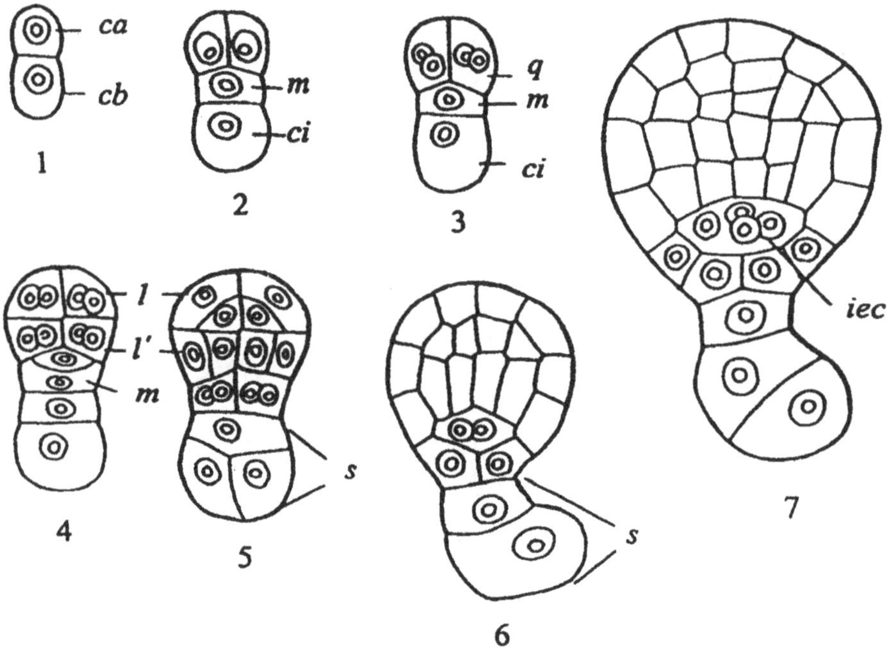

Plate XLI: Onagrad-type of embryogenesis.
1-7—typical variation in *Oenothera biennis*; 8-19—Capsella-variation in *Capsella bursa-pastoris*; 20-29—Orobanche-variation in *Orobanche cumana* subsp. *cernua*; 30-38—Philydrum-variation in *Philydrum lanuginosum*; 39-52—Juncus-variation in *Luzula pedemontana*.

ca—apical cell; *cb*—basal cell; *ci, d, f, m, n, n'*—tiers of basal cell; *co*—root cap; *h*—hypophysis; *icc*—initials of central cylinder of stem and root; *ico*—initials of root cap; *iec*—initials of root cortex; *pco*—cotyledonary area; *phy*—hypocotylar area; *pvt*—growth point of stem; *q, l, l',*—tiers of apical cell; *s*—suspensor.

1-7—after Souèges, 1920c; 8-16—after Souèges, 1919; 17-19—after Schaffner, 1906; 20-29—after Anisimova, 1983; 30-38—after Kapil and Walia, 1965; 39-52—after Shamrov and Anisimova, 1993b.

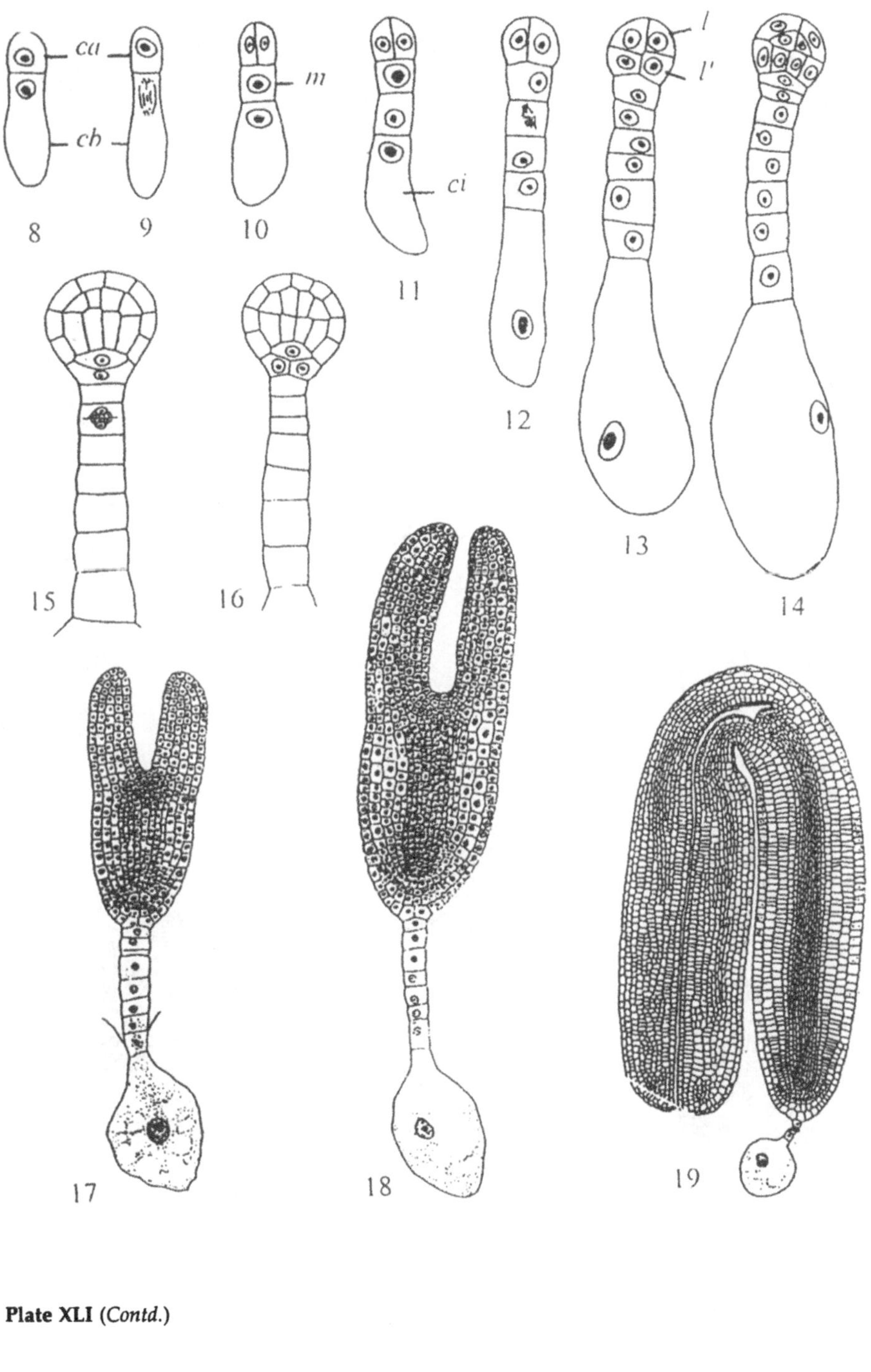

Plate XLI (*Contd.*)

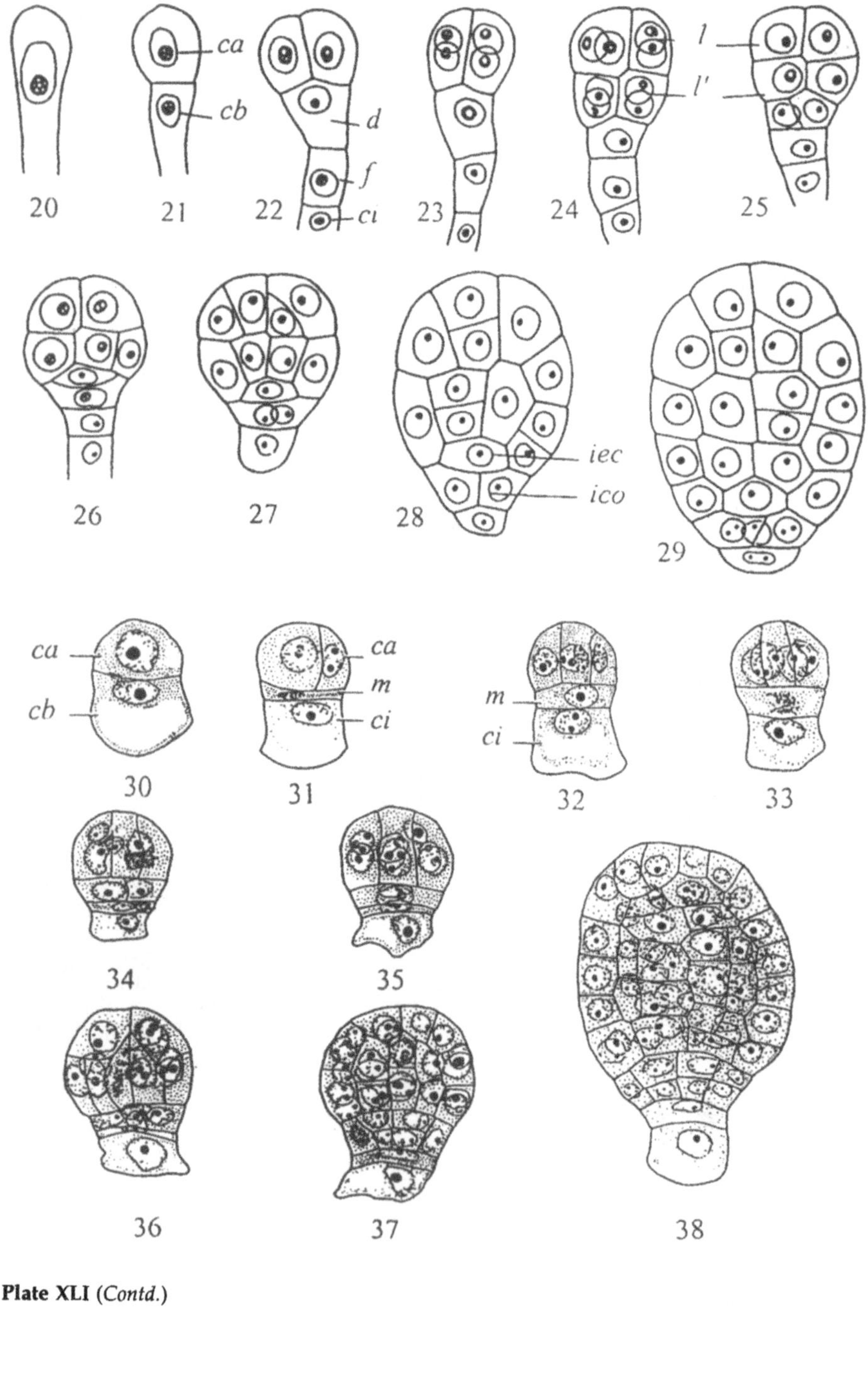

Plate XLI (*Contd.*)

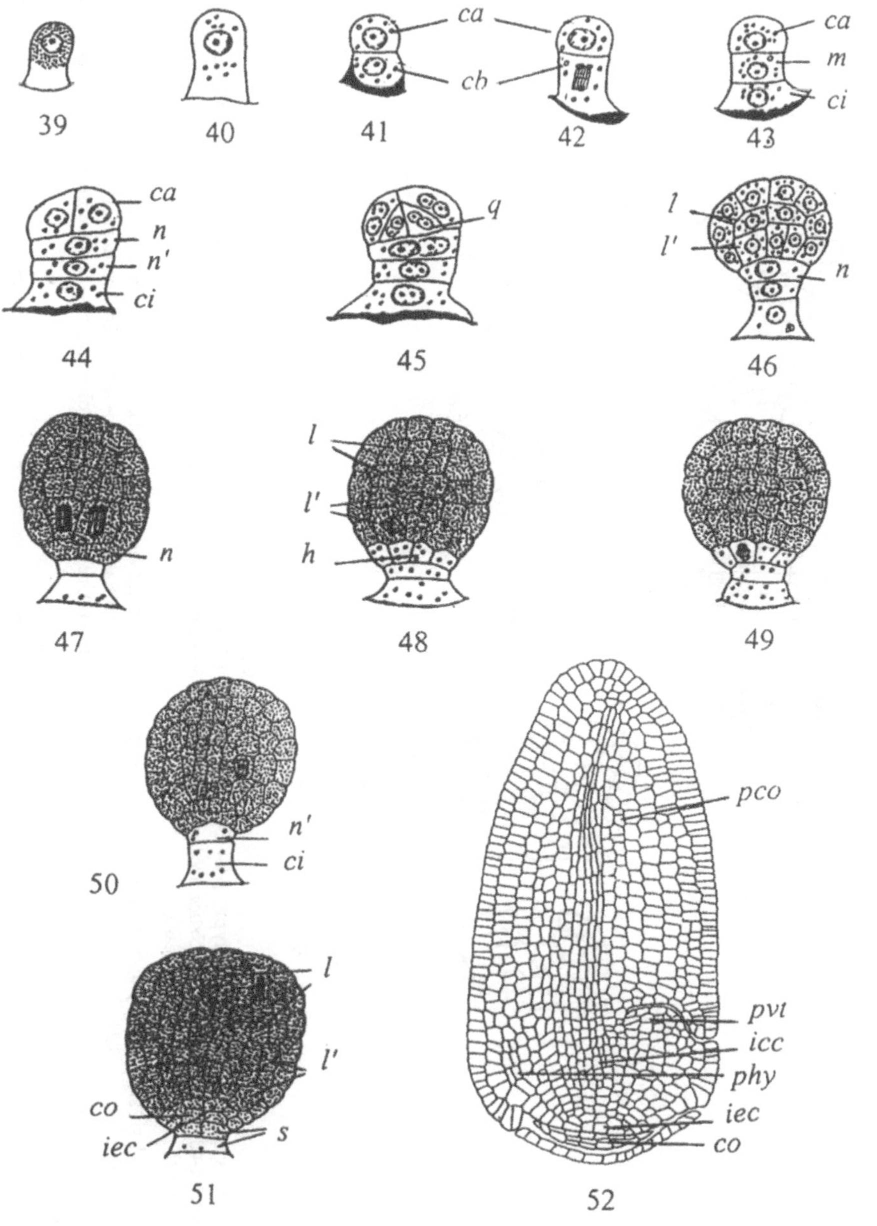

Plate XLI (*Contd.*)

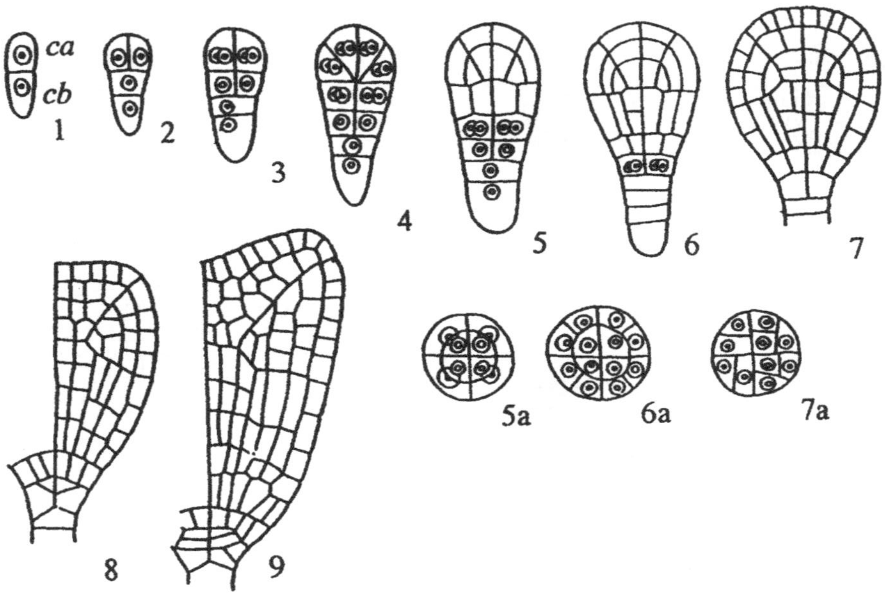

Plate XLII: Asterad-type of embryogenesis.

1-76—different variations of Asterad-type: 1-9—Senecio–variation (*Senecio vulgaris*, Asteraceae), 10-23—Oxalis–variation (*Oxalis corniculata*, Oxalidaceae), 24-40—Urtica–variation (*Urtica urens*, Urticaceae), 41-56—Polygonum-variation (*Ceratophyllum demersum*, Ceratophyllaceae), 57-76—Geum-variation (*Geum urbanum*, Rosaceae).

α, β, *a*, *a'*, *b'*, *b''*—derivatives of tier *q*; *ci*, *m*, *n*, *n*[1], *o*, *p*, *r*, *t*—derivatives of apical and basal cells; *cot*—cotyledon; *de*—dermatogen; *e*—epiphysial cell; *h*—hypophysial cell; *ico*—initials of root cap; *iec*—initials of root cortex; *iep*—initials of plerome; *mc*—méristème cortical externe du cotylédon (periblem of cotyledon); *mv*—méristème vas-culaire du cotylédon (plerome of cotyledon); *pe*—periblem; *pl*—plerome; *pr*—procambium. Rest of legend explained in text (see Proembryo).

1-9—after Souèges, 1920b; 10-23—after Souèges, 1939a; 24-40—after Lebègue and Poix, 1956; 41-56—after Shamrov and Batygina, 1984; 57-76—after Souèges, 1923a.

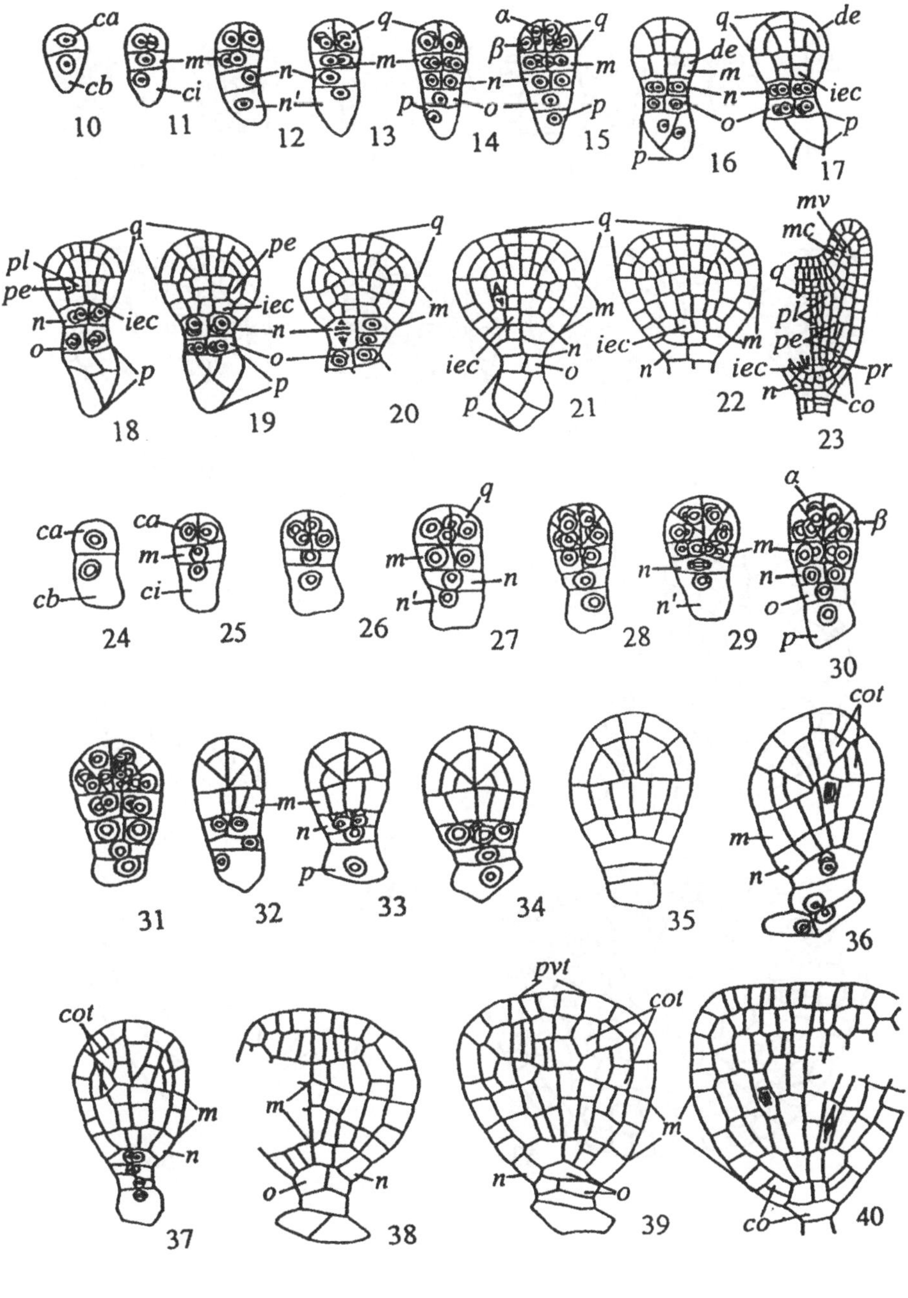

Plate XLII (*Contd.*)

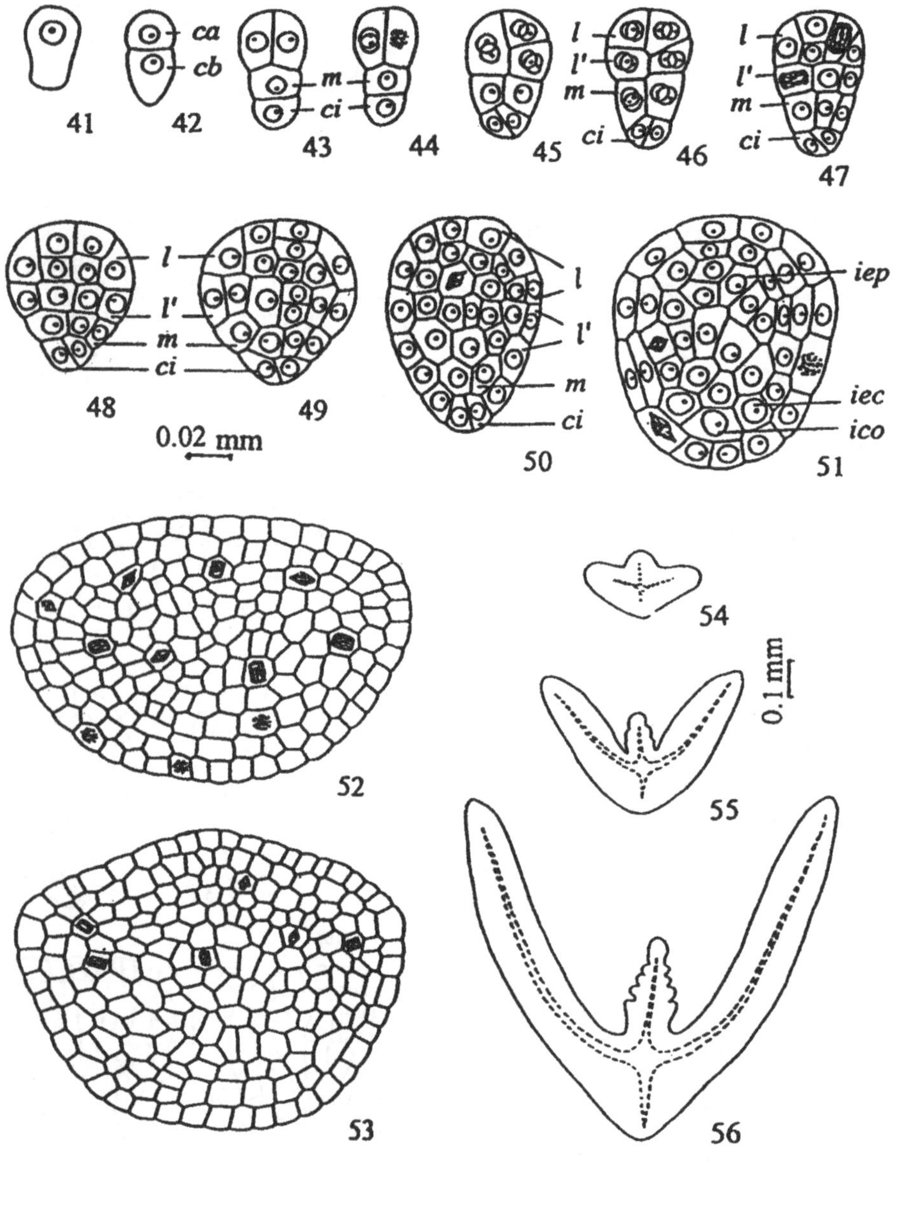

Plate XLII (*Contd.*)

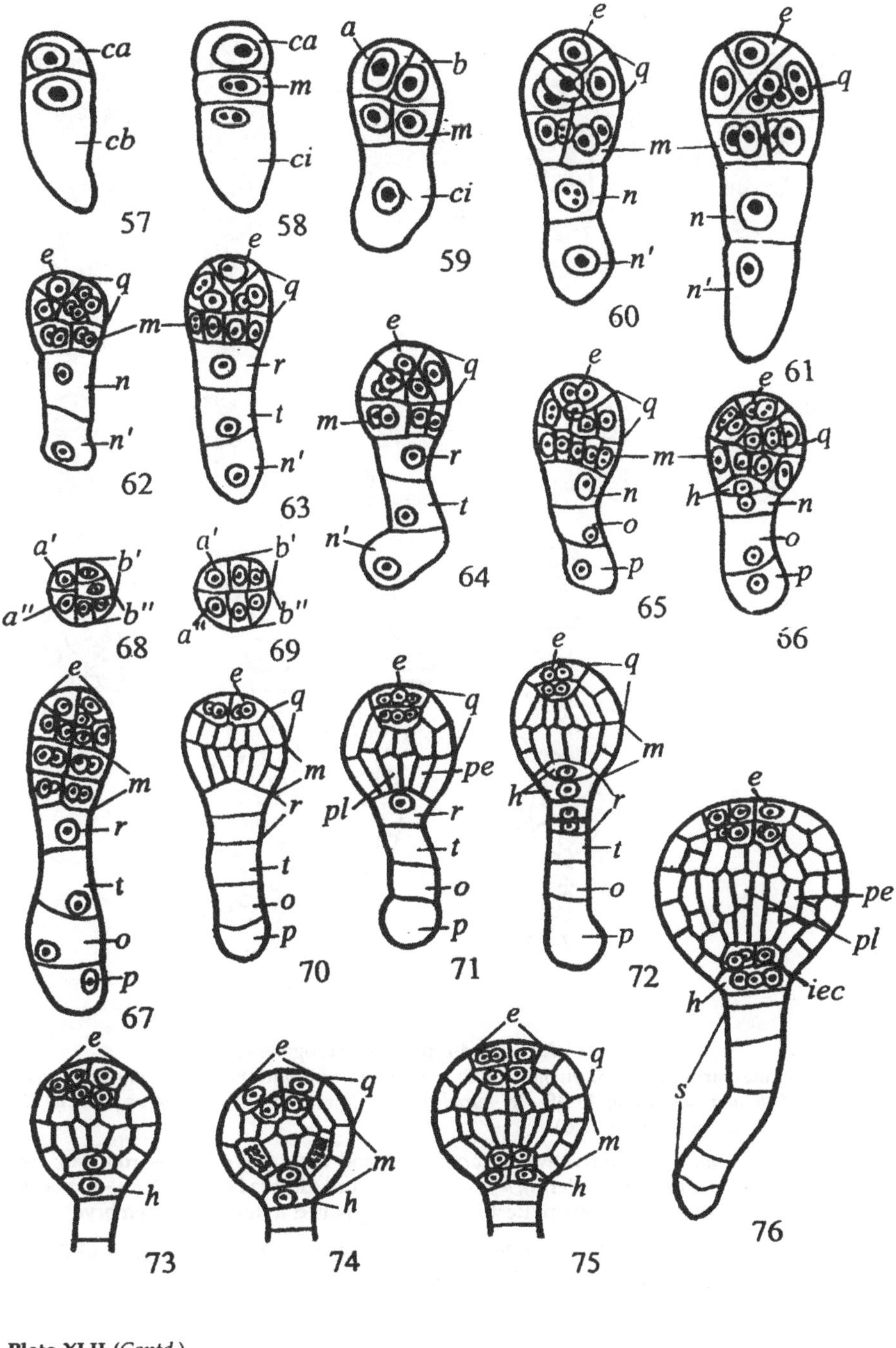

Plate XLII (*Contd.*)

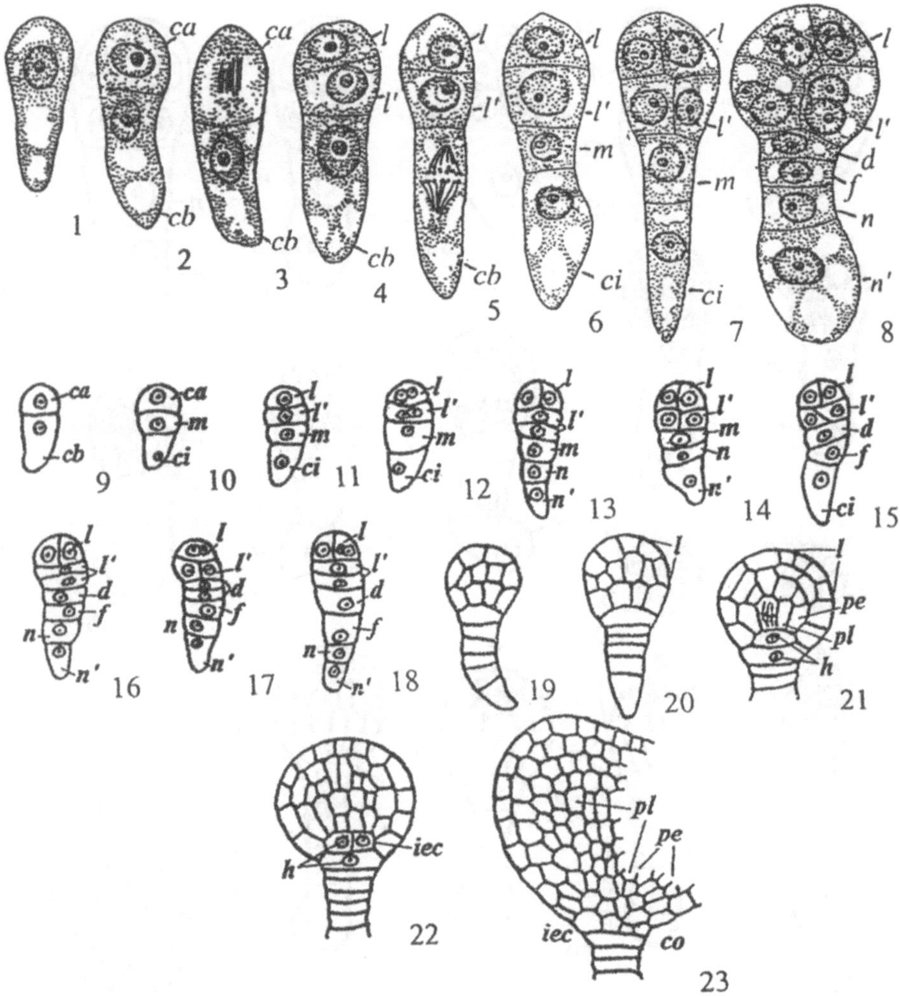

Plate XLIII: Solanad-type of embryogenesis.
1–8—Nicotiana-variation in *Nicotiana tabacum*; 9–23—Linum-variation in *Linum catharticum*; 24–39—in *Solanum tuberosum*; 40–45—in *Tetradiclis tenella* and 46–50—in *Ruscus ponticus* (24–50—variations are not determined).

ci, d, f, h, h', l, l', m, n, n'—derivatives of apical (*ca*) and basal (*cb*) cells; *co*—root cap; *cot*—cotyledon; *epi*—epidermis (protoderm); *h*—hypophysial cell (details 21 and 22); *hy*—hypocotyl; *pe*—periblem; *pl*—plerome; *st*—stem. Rest of legend explained in text (see Proembryo).

1–8—after Souèges, 1922b; 9–23—after Souèges, 1924d; 24–39—after Dnyansagar and Cooper, 1960; 40–45—after Kamelina, 1994; 46–50—after Satarova, 1990.

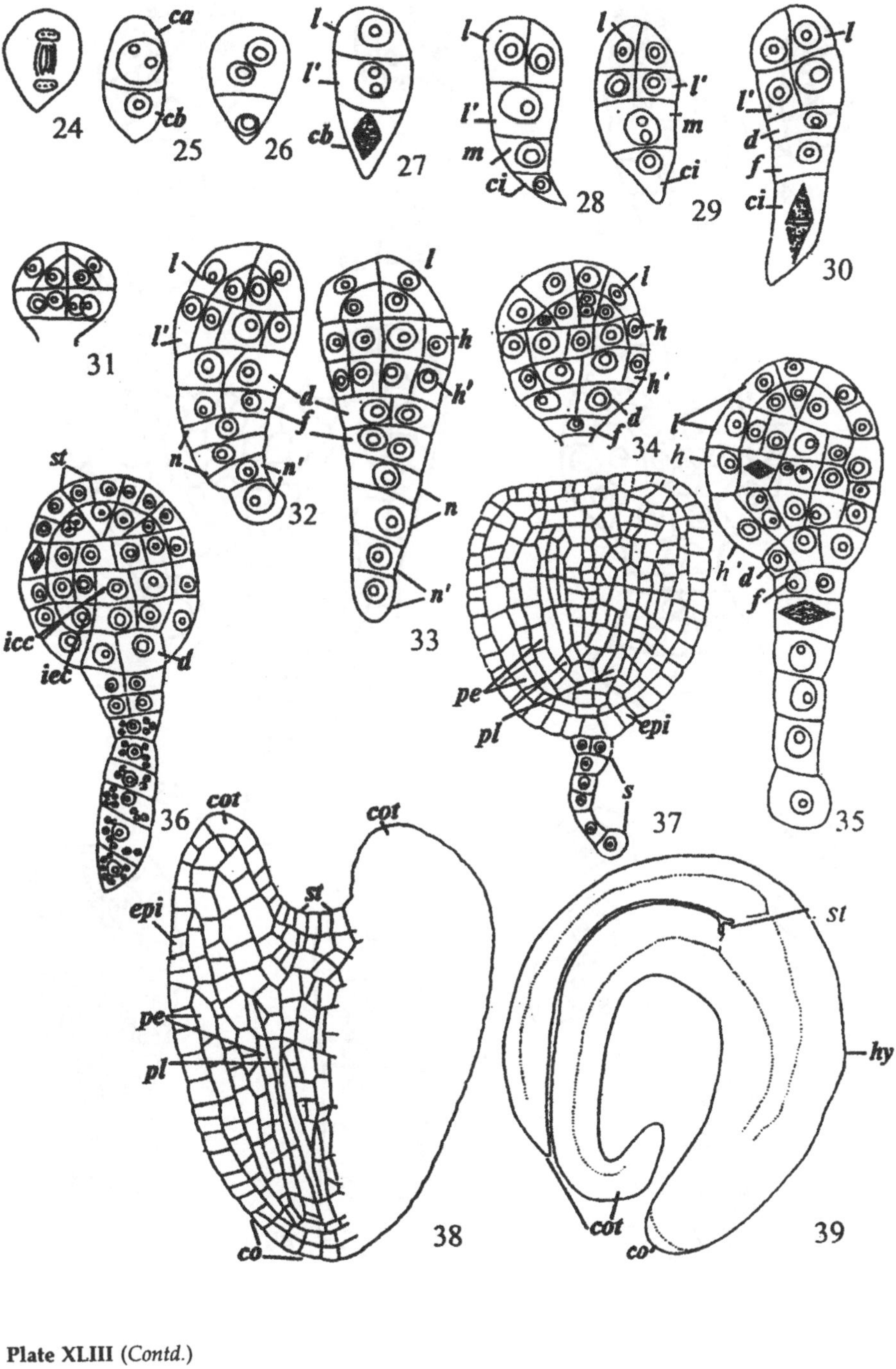

Plate XLIII (*Contd.*)

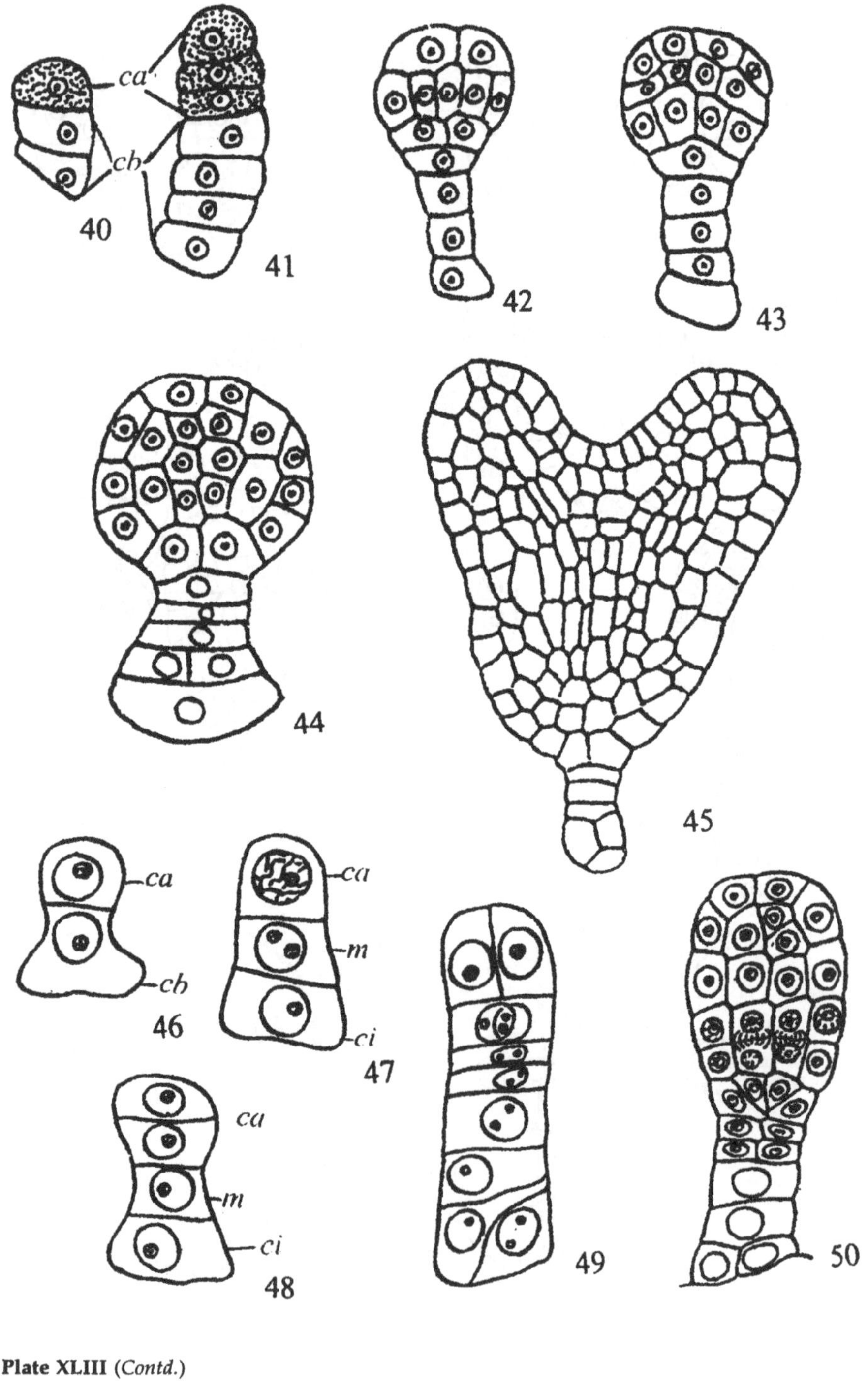

Plate XLIII (*Contd.*)

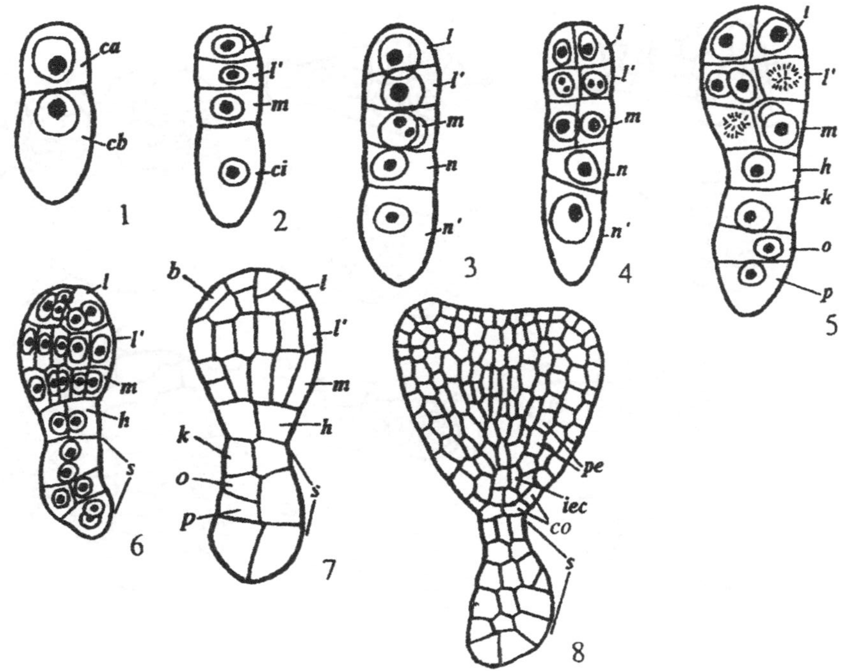

Plate XLIV: Chenopodiad-type of embryogenesis.

1-8—typical variation (*Chenopodium bonus-henricus*); 9-15—Polemonium-variation (*Polemonium caeruleum*); 16-22—Myosotis-variation (*Myosotis hispida*).

a, b—tiers of apical cell (*ca*); *e*—epiphysis; *h*—hypophysial cell; *k, l, l′, m, n, n′, o, p*—derivatives of apical (*ca*) and basal (*cb*) cells; *pe*—periblem; *pl*—plerome. Rest of legend explained in text (see Proembryo).

1–8—after Souèges, 1920a; 9–15—after Souèges, 1939b; 16–22—after Souèges, 1921.

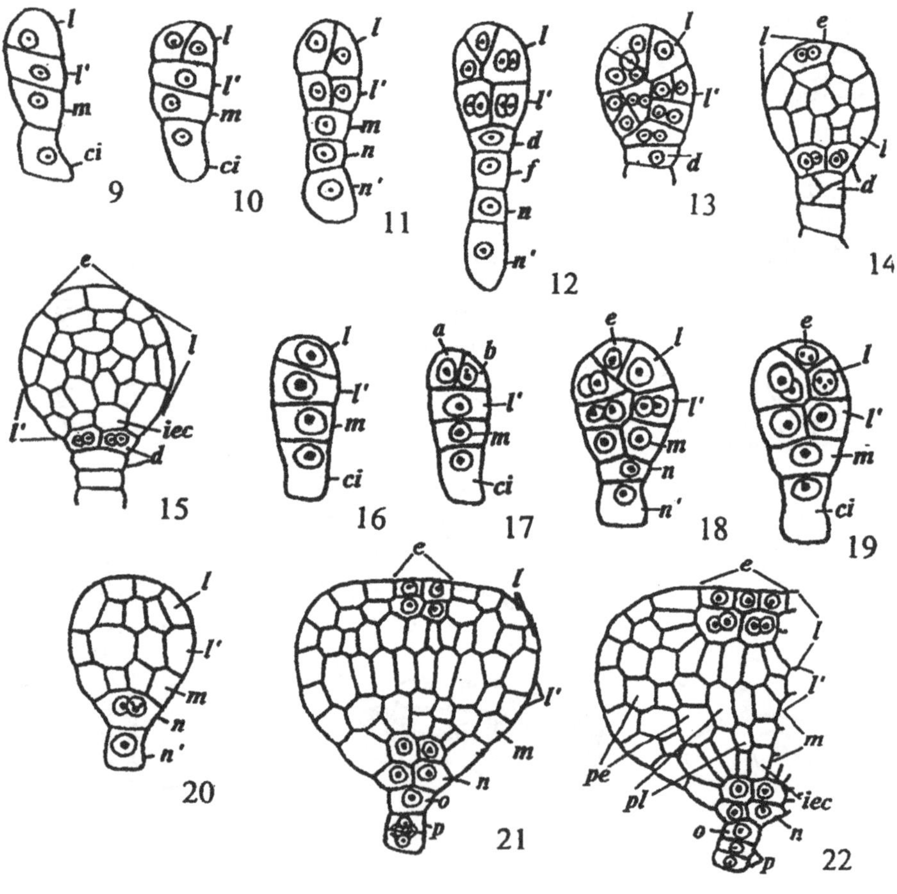

Plate XLIV (*Contd.*)

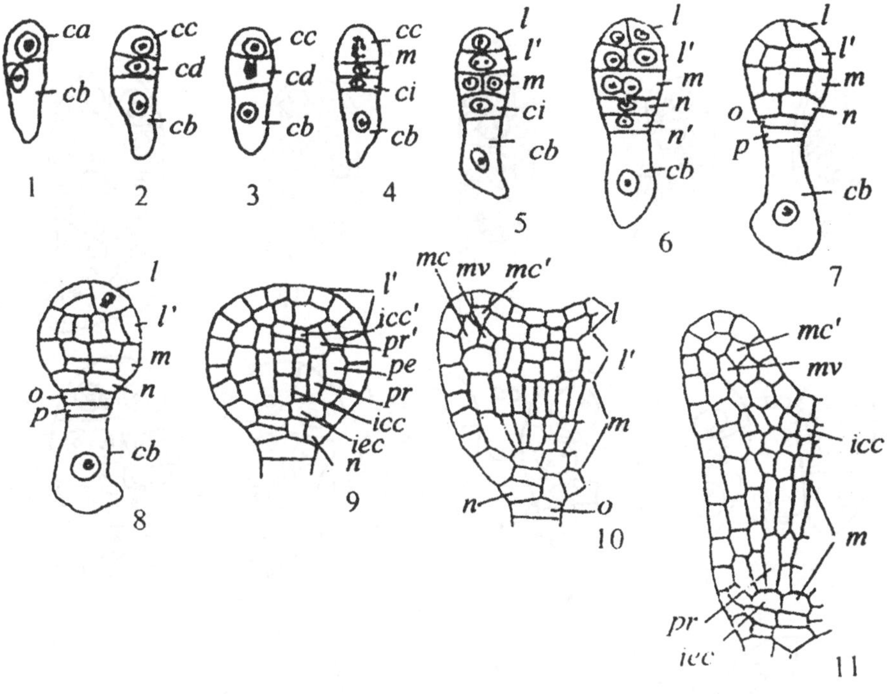

Plate XLV: Caryophyllad-type of embryogenesis.

1–11—Sagina-variation in *Sagina procumbens*; 12–16—Phytolacca-variation in *Phytolacca dioica*; 17–25—Drosera-variation in *Drosera rotundifolia*; 26–41—Fumaria-variation in *Fumaria officinalis*; 42–49—Sagittaria-variation in *Sagittaria quayanensis*; 50–58—development of embryo in *Sedum acre*, variation not determined, thread-like outgrowths of suspensor basal cell shown.

ca—apical cell of bicellular proembryo; *cb*—basal cell of bicellular proembryo; *cc, cd, ce, cf, l, l'*—upper derivatives of cell *ca; pco*—cotyledonary part; *ch*—haustorium; *ci, f, m, n, o, p*—lower derivatives of cell *ca; e*—epiphysis; *end*—endosperm; *h, h'*—hypophysis; *icc, icc'*—initials of stem stele; *iec*—intials of root cortex; *mc, mc'*—internal and external periblem of cotyledons; mv—plerome of cotyledon; *pe*—periblem; *ph*—hypocotyls; *pr, pr'*—procambium.

1–11—after Souèges, 1924b; 12–16—after Kajale, 1954; 17–25—after Souèges, 1936a; 26–41—after Souèges, 1941; 42–49—after Johri, 1935; 50–58—after Souèges, 1927.

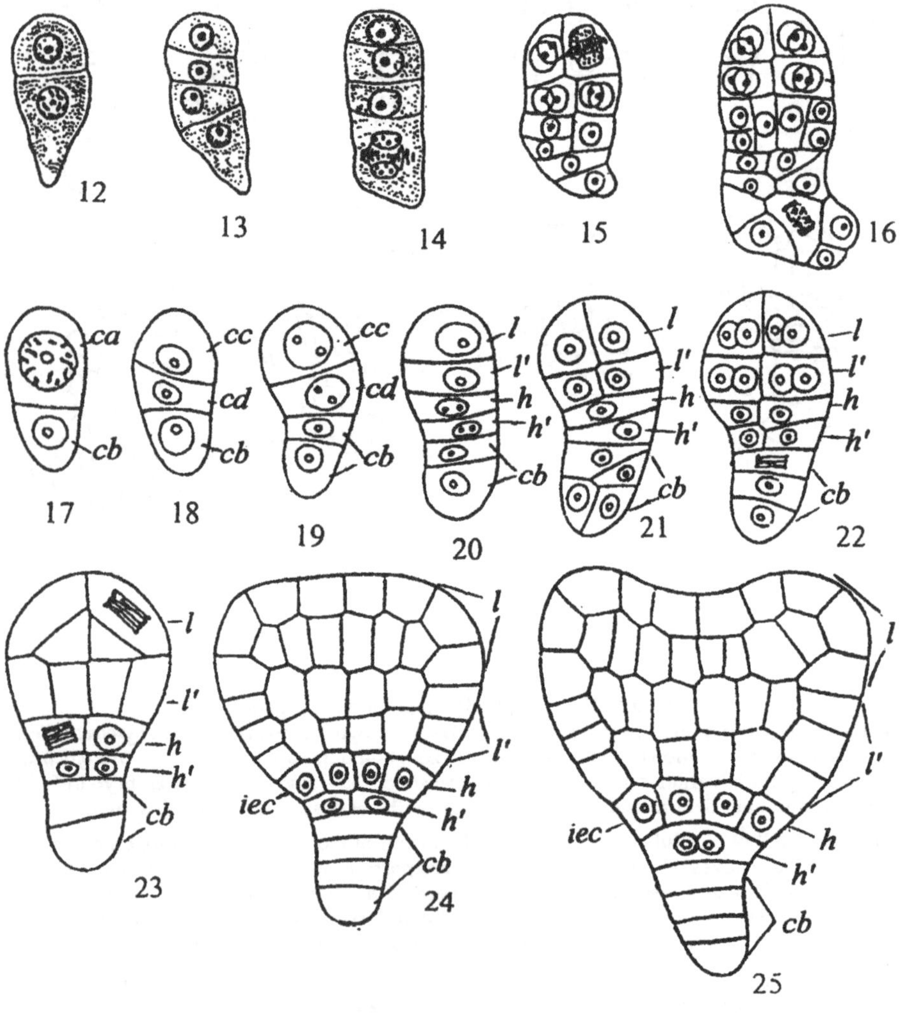

Plate XLV (*Contd.*)

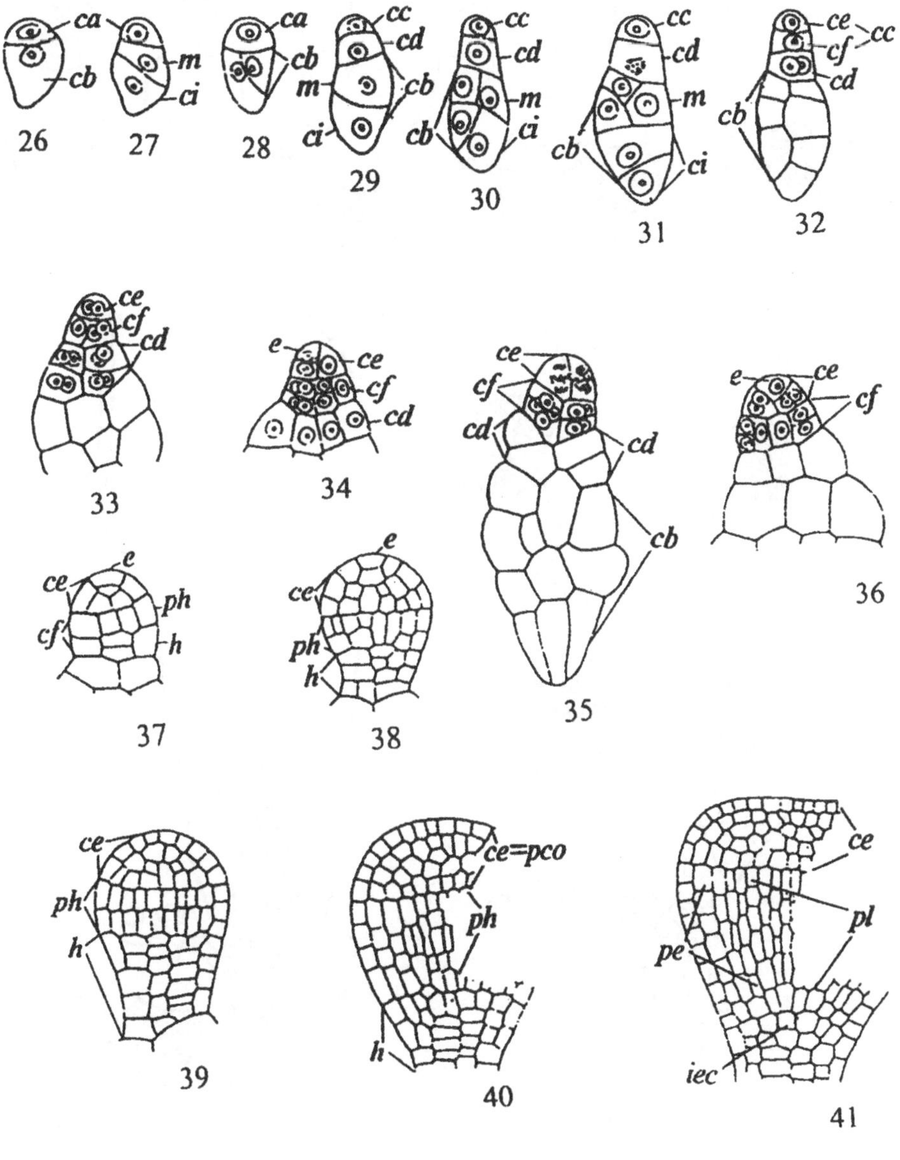

Plate XLV (*Contd.*)

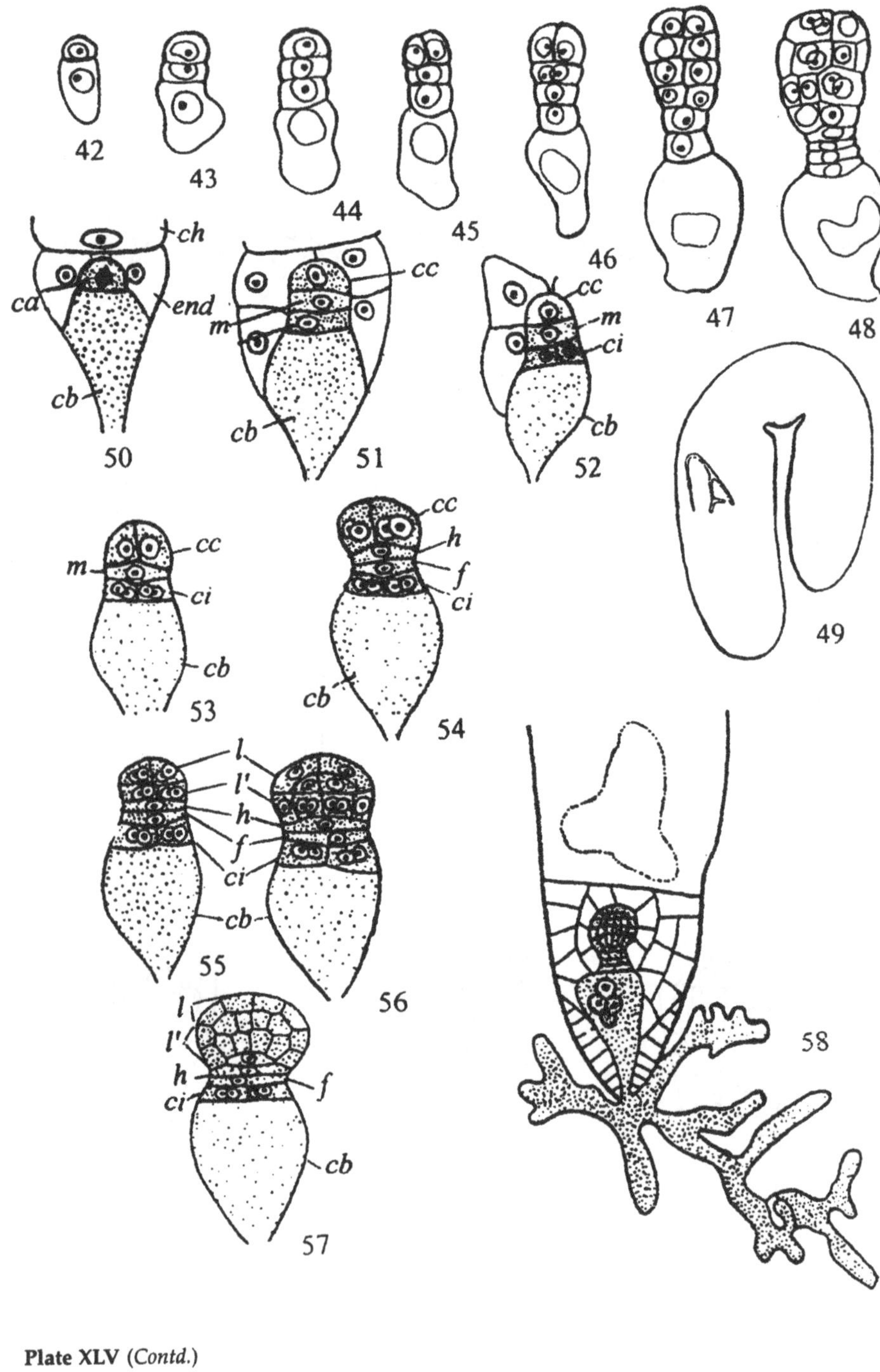

Plate XLV (*Contd.*)

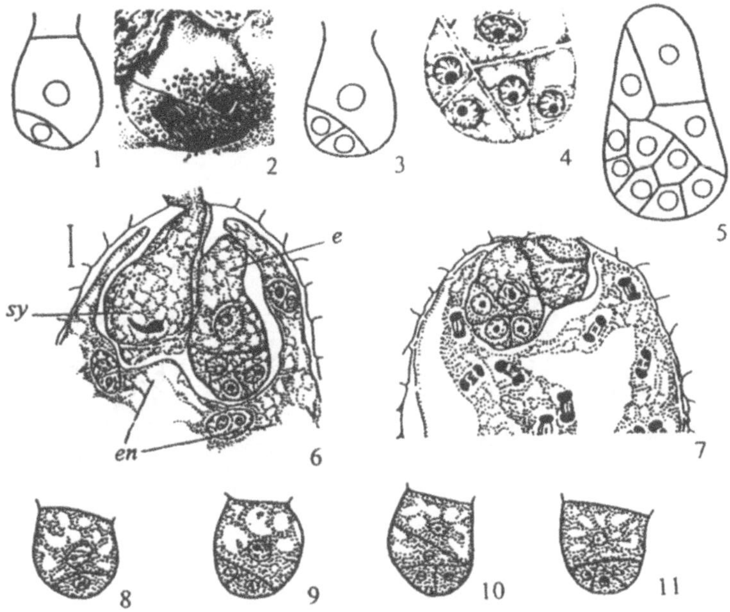

Plate XLVI: Graminad-type of embryogenesis.

1-31—early stages of embryo development in Poaceae, illustrating position of first cellular partitions: 1-11—*Zea*, 12-15—*Sorghum*, 16-25—*Poa*, 26—*Triticum*, 27,28—*Stipa*, 29—*Hordeum*, 30,31—*Avena* (16-20—dorsoventral profile, 21-25—different profiles, 23,24—'straight' embryos and 'seeming' transverse partition); 32-60—development of embryo (schematized) in *Triticum aestivum* (32-50) and *T. monococcum* (51-60) (32-50, 51-60—dorsoventral profile, dots mark the zone of mitotic activity, 33a-37a, 54a-57a—stereoscopic image, 51 –scheme of development of *Triticum* embryo), 61—structure of early proembryo of wheat in various projections: A—various sections of kernel (a—dorsoventral, b—lateral, c—transverse), B—upper part of embryo sac in dorsoventral (1-4) and lateral (1a-4a) projections before fertilization (1,1a), at moment of fertilization (2,2a), at stage of two- (3,3a), three- (4,4a) and four-celled embryo, C—successive stages of early proembryo development in dorsoventral (1-13) and lateral (1a-13a) projections, D—structure of four-celled embryo in lateral, dorsoventral and transversal projections (A, B 1-4a, C, D—schematized); 62-79—development of embryo in *Commelina forskalai*: 62—zygote, 63, 64—two-, 65, 66—four-, 67—six-, 68, 69—eight-(quadrants formed from upper layer), 70—ten-, 71—thirteen-, and 72—seventeen-celled proembryo, 73, 74—formation of hollow at ventral side of cotyledon, 76-78—late stages of embryo development, 79—mature embryo with well-developed conductive system (75-79—schematized).

ad b—additional bud, *ad e r*—adventive embryo root, *c*—cotyledon, *ca*—apical cell, *cb*—basal cell, *cl*—coleorhiza, *clp*—coleoptile, *c v*—cotyledonary vagina, *e*—embryo, *en*—endosperm, *ep*—epiblast, *e r*—embryo root, *g p*—growing point, l_1, l_2, l_3—the first leaves of plumule, *m, n, n', o, p*—tiers of basal cell; *q*—upper tiers of apical cell; *r c*—root cap, *sc*—scutellum, *sy*—synergid.

1, 3, 5—after Yakovlev, 1946b; 2, 4—after Chebotaru, 1961; 6,7—after Korobova, 1962; 8-11—after Randolph, 1936; 12-15—after Khazova, 1970; 16-25—after Batygina, 1974b; 26, 29—after Modilevsky *et al*., 1958; 27, 28—after Yakovlev and Solntseva, 1965; 30, 31—after Kihara and Nishiyama, 1932; 32-61—after Batygina, 1968a, 1987a; 62-79—after Maheshwari and Baldev, 1958.

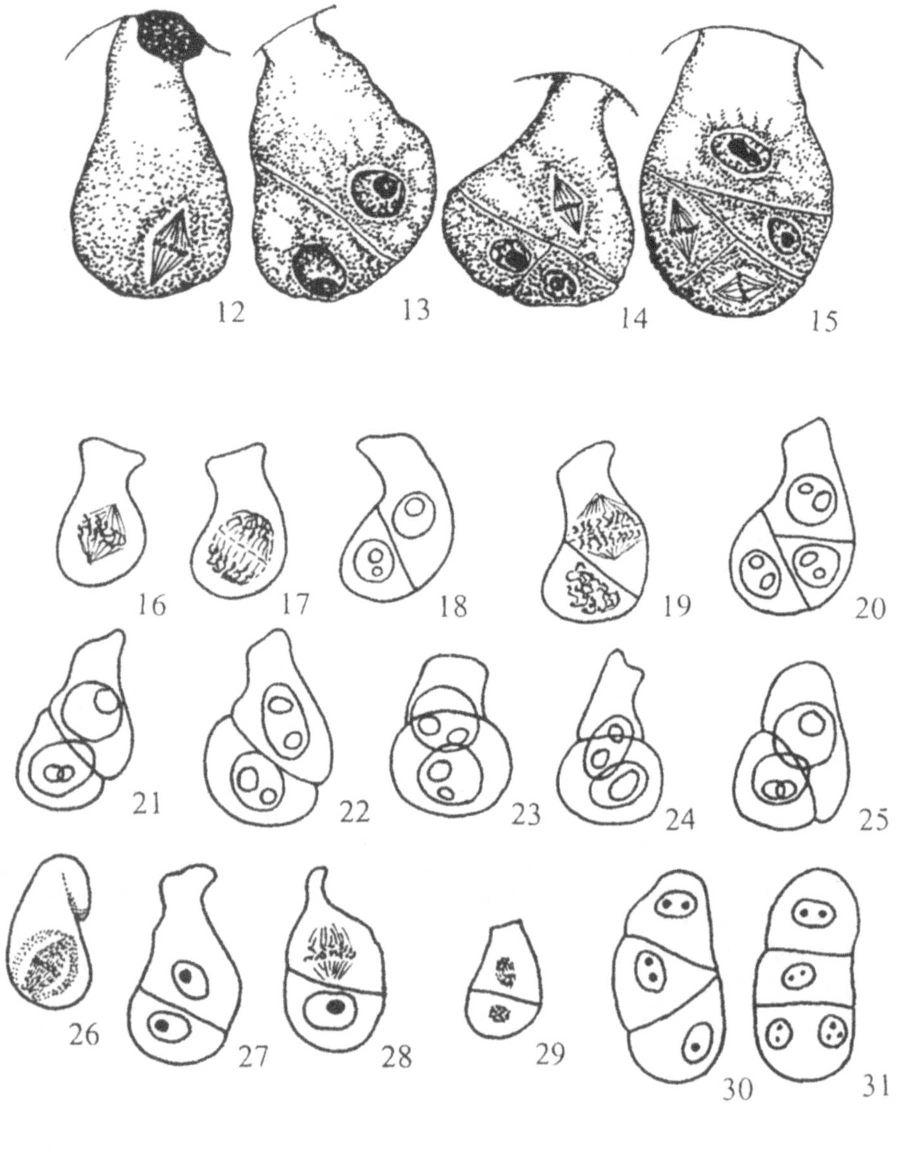

Plate XLVI (*Contd.*)

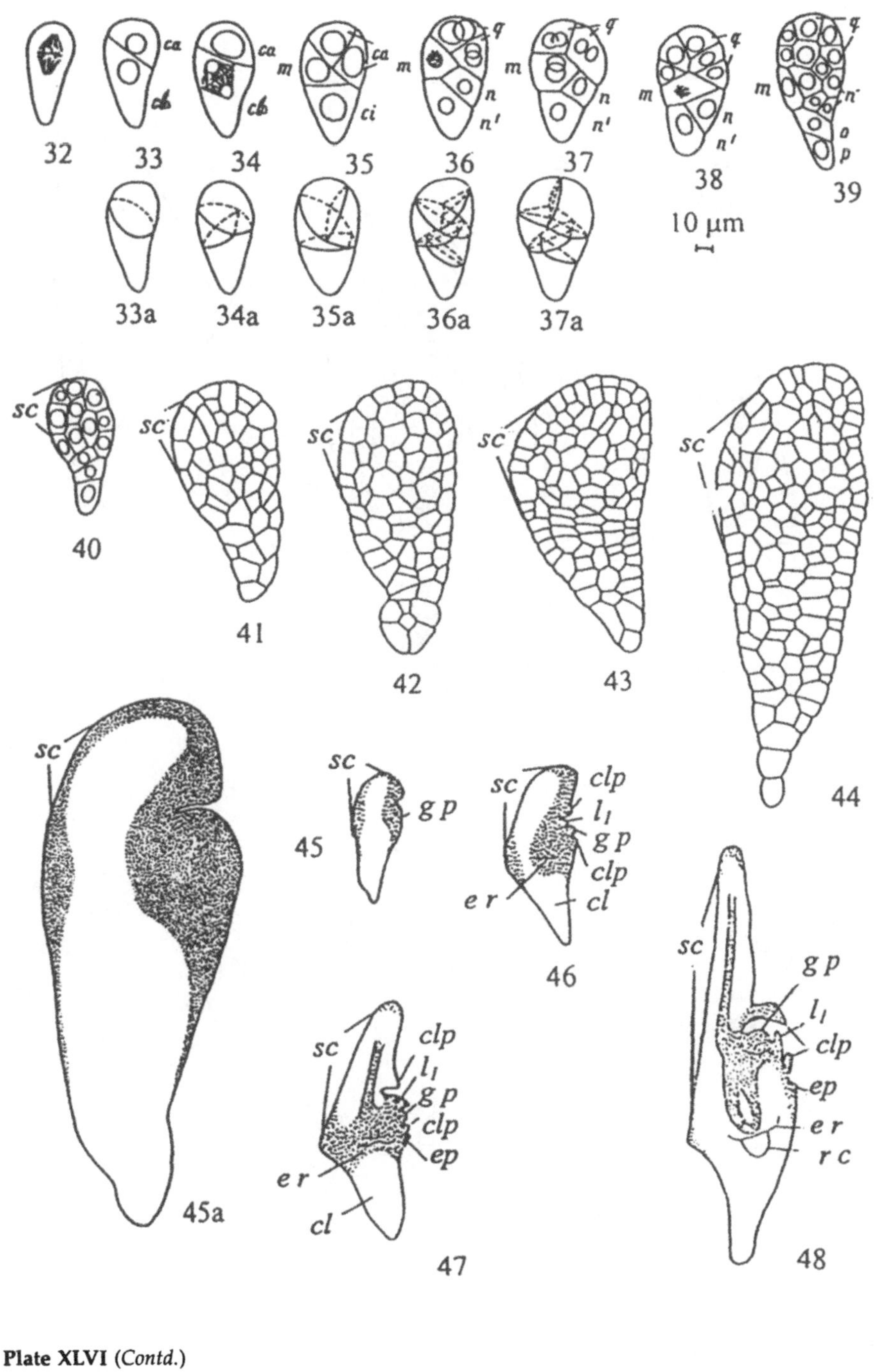

Plate XLVI (*Contd.*)

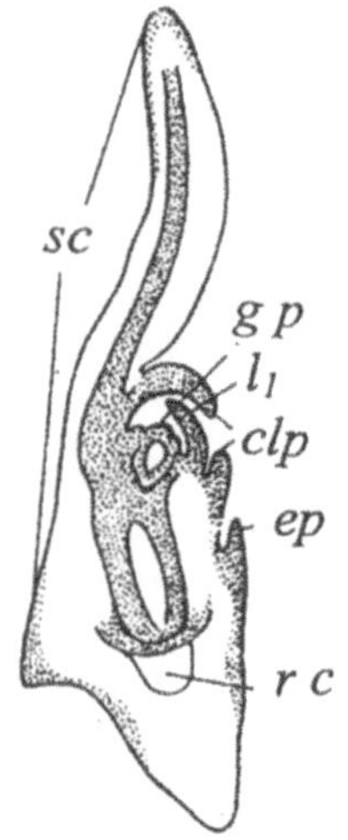

49

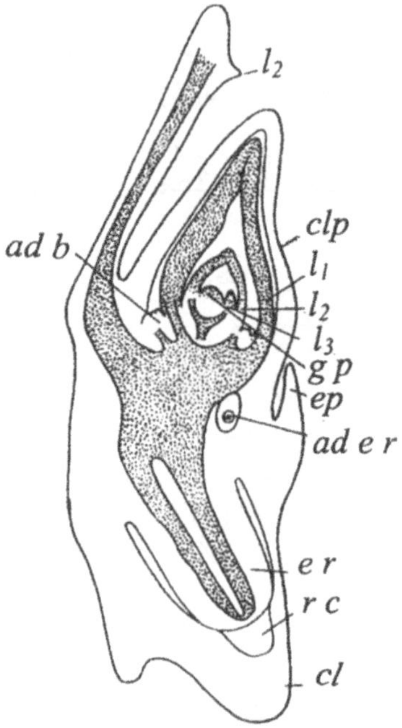

50

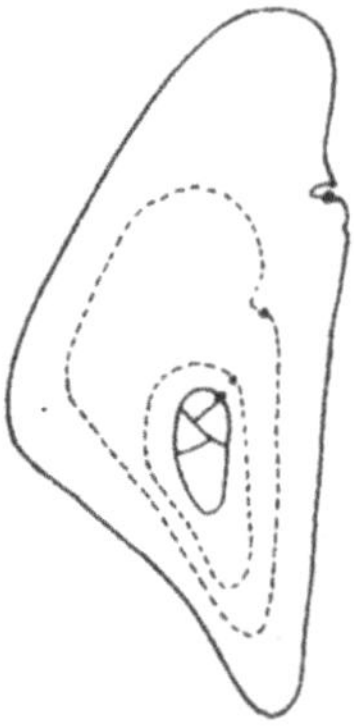

51

52

53

54

55

56

57

58

59

60

54a

55a

56a

57a

Plate XLVI (*Contd.*)

61 A

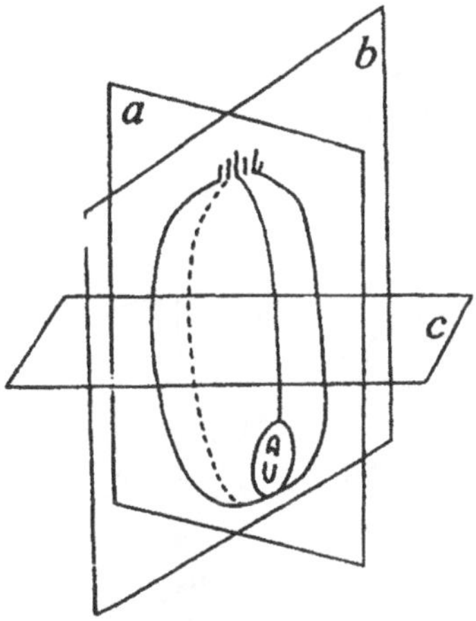

61 B

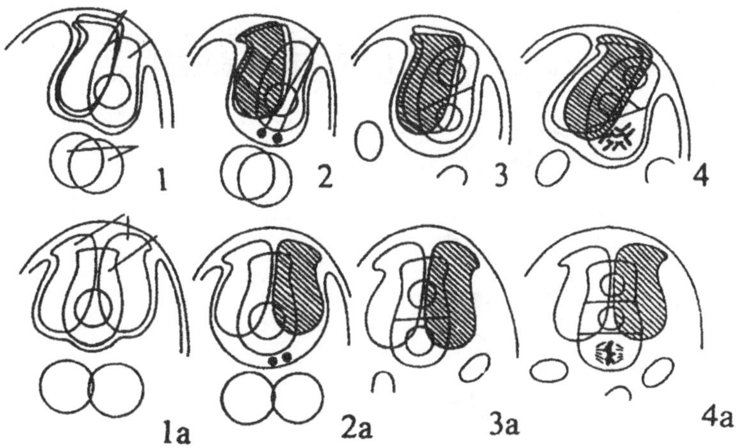

Plate XLVI (*Contd.*)

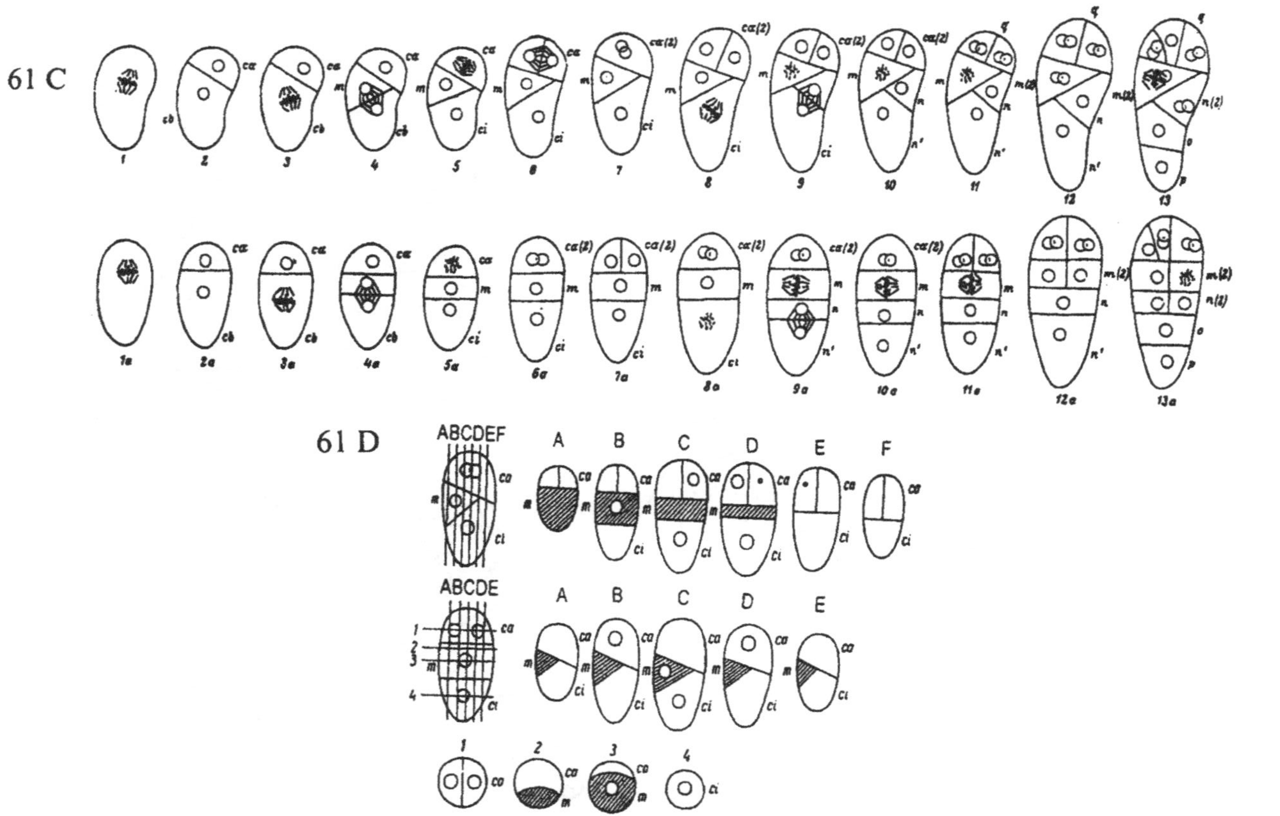

Plate XLVI (*Contd.*)

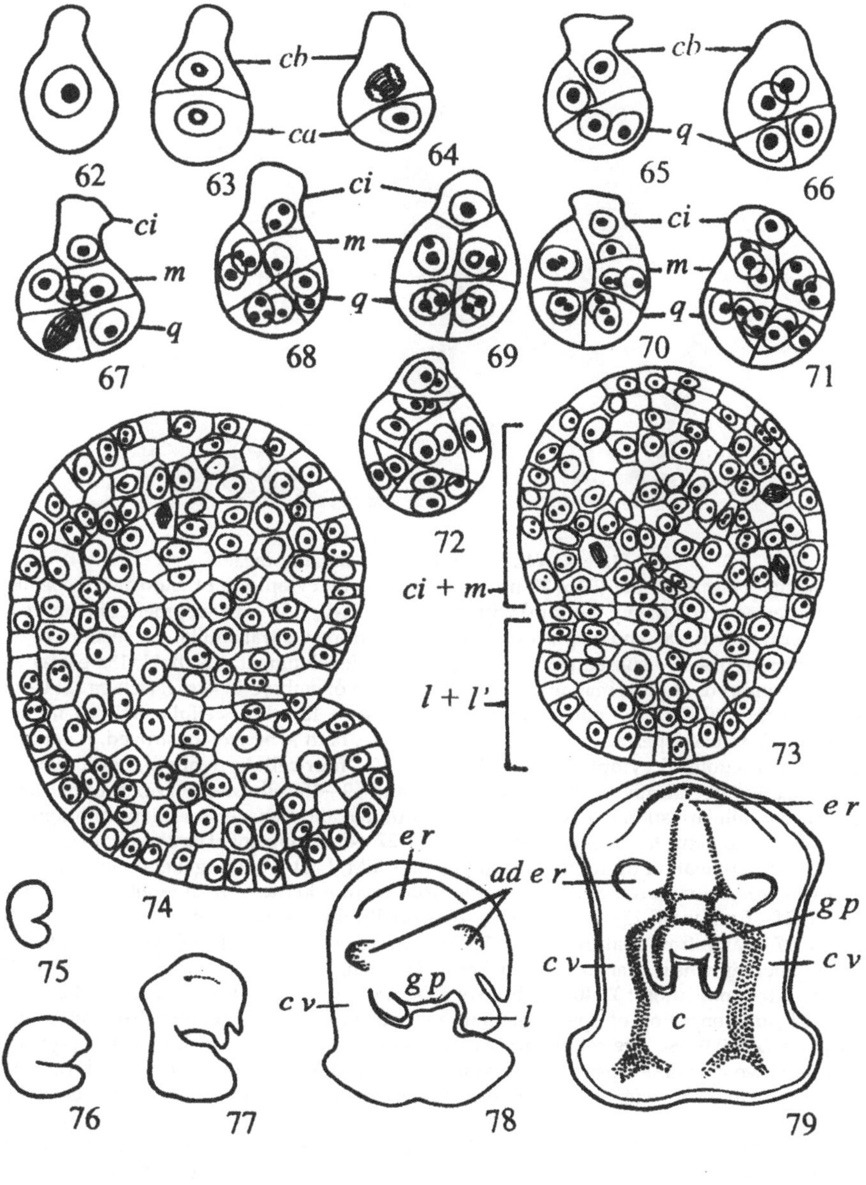

Plate XLVI (*Contd.*)

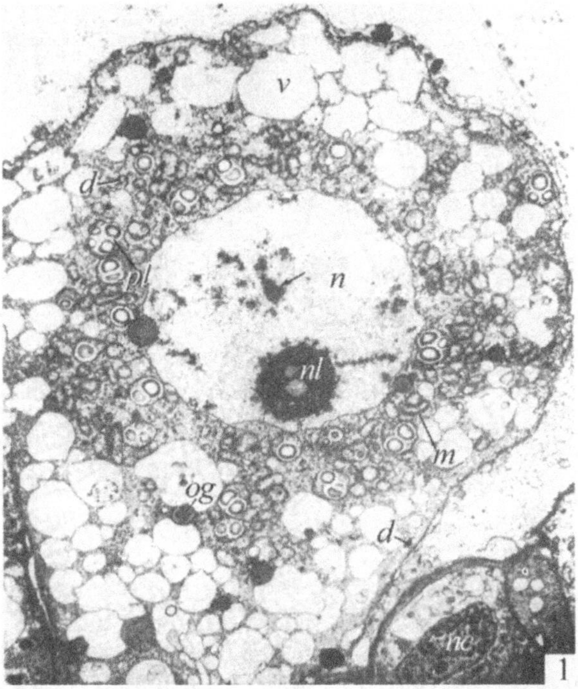

Plate XLVII: Embryogenesis in Gramineae.

1—*Hordeum vulgare* zygote (52 × 70 µm), surrounded by cytoplasm of endosperm coenocyte and bordering immediately (bottom) with nucellus (TEM); 2-36—development of embryo and endosperm in *Triticum*, and its position in seed: 2,3—dorsoventral and lateral sections of four-cellular embryo, 4-13—subsequent stages of embryo development, 10a, b, 11, 13a-e—separate details of embryos at great magnification, 12—position of embryo in kernel (schem.), 14—two-cellular embryo (full face), endosperm nuclei can be seen, 15—three-cellular embryo, initiation of cell walls in endosperm, 16—four-cellular embryo, both synergids destroyed, polarization of greatly vacuolized cytoplasm in endosperm, 17—four-cellular embryo, continuation of cellularization in endosperm, 18—eight-cellular and 19—15-cellular embryo, in upper part of endosperm cellularization has passed, 20—longitudinal and transverse sections of outgrowths of endospermal tissue towards central vacuole, 21-27—development of embryo from first stages of organ differentiation up to its full completion (SEM, LM): 21, 22—general view of embryo removed from immature kernel, 23—same from mature kernel, 24, 25—embryo position in kernel (24—front view, 25—lateral view), 26—longitudinal section of embryo from immature kernel, 27—same from mature kernel, 28-36—successive phases of embryo and endosperm development in wheat Diamant variety kernel (at 5-6 h, 1, 2, 3-4, 7-8, 9-10, 11-12, 13-15 and 25-30 days after pollination); 37-40—embryo of *Hordeum vulgare* (TEM): 37—three-cellular embryo (84 × 44 µm) consisting of apical, median and suspensor cells, vacuoles distributed at cell periphery (× 1600), 38—five-cellular embryo, contents of pollen tube and degenerating synergids seen near two-cellular suspensor, arrows mark longitudinal wall in suspensor cell (× 11,600), 39—13-cellular embryo (45 × 84), in figure, 8 cells can be seen, endosperm surrounds embryo, one suspensor cell in contact with nucellus (arrow) (× 2200), 40—20-cellular embryo, figure shows 11 cells, vacuole distribution in embryo cells varies, greater vacuoles in suspensor area, endosperm surrounds embryo, no direct contact of embryo and nucellus (×2500); 41-69—embryo

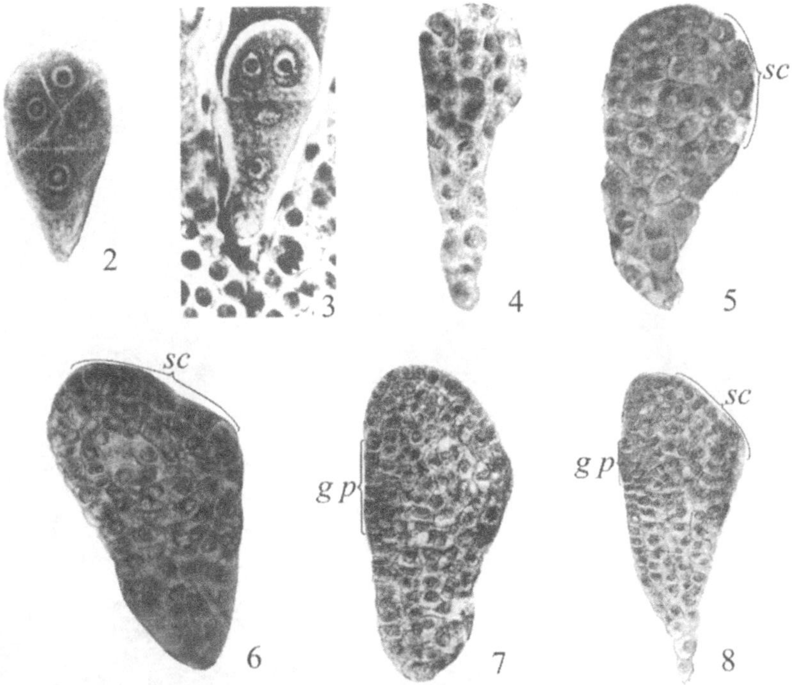

Plate XLVII (*Contd.*)

of some cereals and its transformations during kernel germination: 41-50—*Avena sativa*, 51-59—*Zea mays*, 60-69—*Triticum vulgare* (41-45, 51-55, 60-64—different stages of kernel germination, 46, 56, 65—lateral and 47, 57, 66—dorsoventral sections of embryo, 46a, 56a, 65a—kernel and embryonic position in it (front view), 47a, 57a, 66a—kernel and embryonic position in it (view of one side), 48, 58—section of embryo at 24-36 h after beginning of kernel germination, 49—seedling at 5 days after beginning of kernel germination, 50, 59, 69—fragments of seedlings illustrating intercalary growth and position of nodes, shoot apex, scutellum and other structures, 67—seedling 7 days after beginning of kernel germination, beginning of axial elongation directly above scutellar node noticeable, 68—seedling 1 month after beginning of kernel germination, internodes elongate substantially); 70—scheme of evolutionary establishment of monocotyly: I—development of cotyledons in dicotyledonous plant *Capsella bursa-pastoris*, II—predominating development of only one cotyledon over another in pseudomonocotyledonous plant *Pinguicula vulgaris*, III-V—cotyledon expanding and shoot apex displacing into lateral position in monocotyledonous plants (III—*Ottellia alismoides*, IV—*Eliusina coracana*, V—*Triticum aestivum*).

Bar—10 µm (14 -19); *ad r*—adventive root, *c*—cotyledon, *cl*—coleorhiza, *clp*—coleoptile, *c p t*—contents of pollen tube, *c s*—conductive system, *d*—dictyosome, *e*—embryo, *en*—endosperm, *en n*—endosperm nucleus, *ep*—epiblast, *f ph*—fibres of phragmoplast, *g p*—growing point, l^1, l^2, l^3—first little leaves of plumule, *lg*—ligula, *m*—mitochondrion, *m a*—meristematic area, *n*—nucleus, *nc*—nucellus, *nl*—nucleolus, *o*—ovule, *o g*—osmiophilic globule, *pl*—plastid, *p r*—primary root (embryonic root), *r c*—root cap, *sc*—scutellum, *sy*—synergid, *sus*—suspensor, *s g*—starch grain, *sc n*—scutellar node, *v*—vacuole.

1, 37-40—modified from Norstog, 1972; 2-20, 26, 27—after Batygina, 1974b; 21-25—Rytova, orig.; 28-36, 70—after Batygina, 1987a; 41-69—modified from Avery, 1930; McCall, 1934.

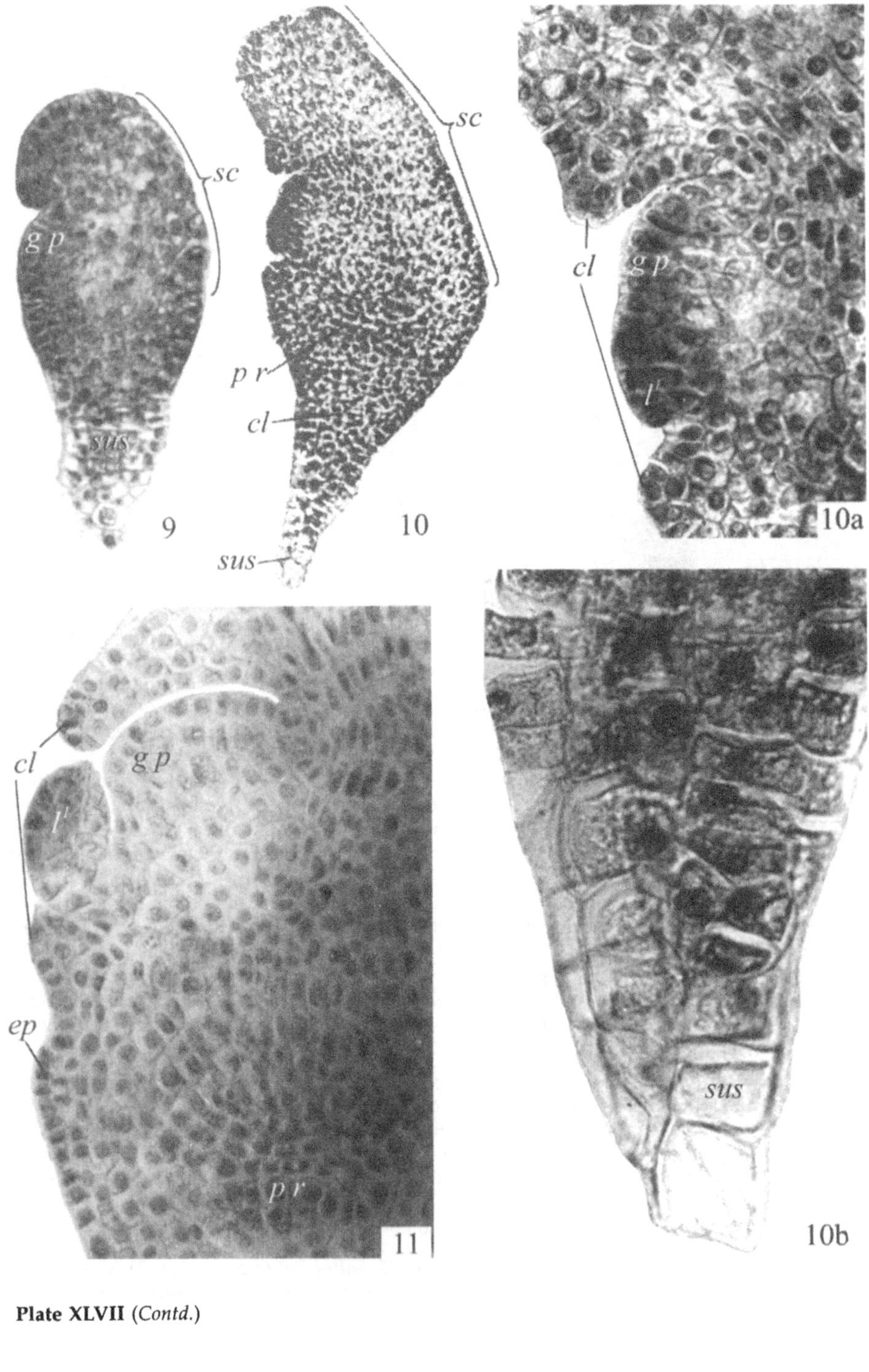

Plate XLVII (*Contd.*)

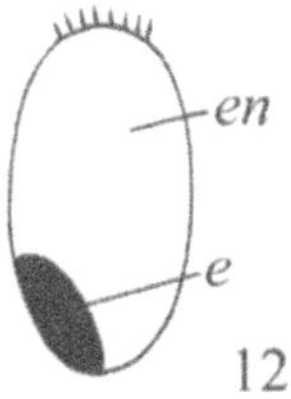

12

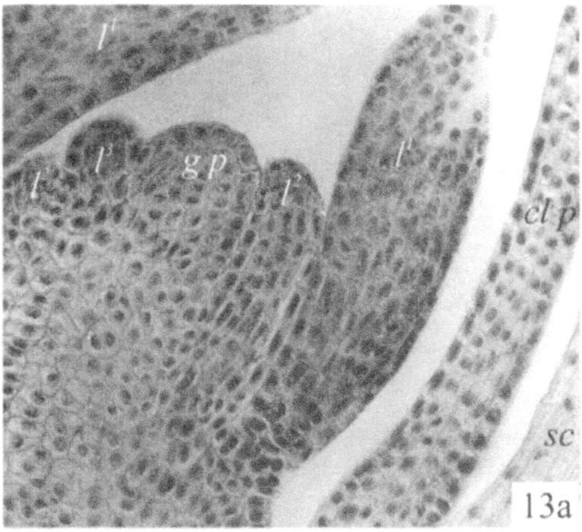

13a

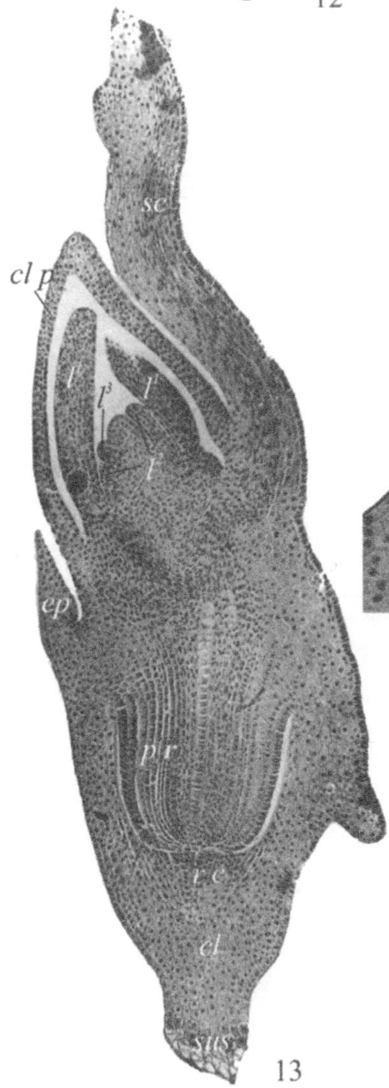

13

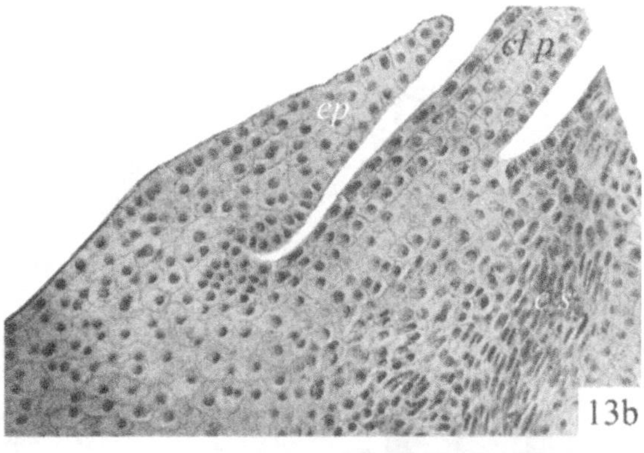

13b

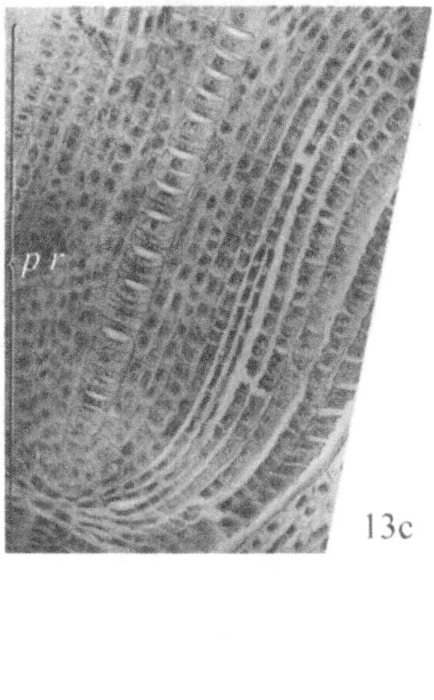

13c

Plate XLVII (*Contd.*)

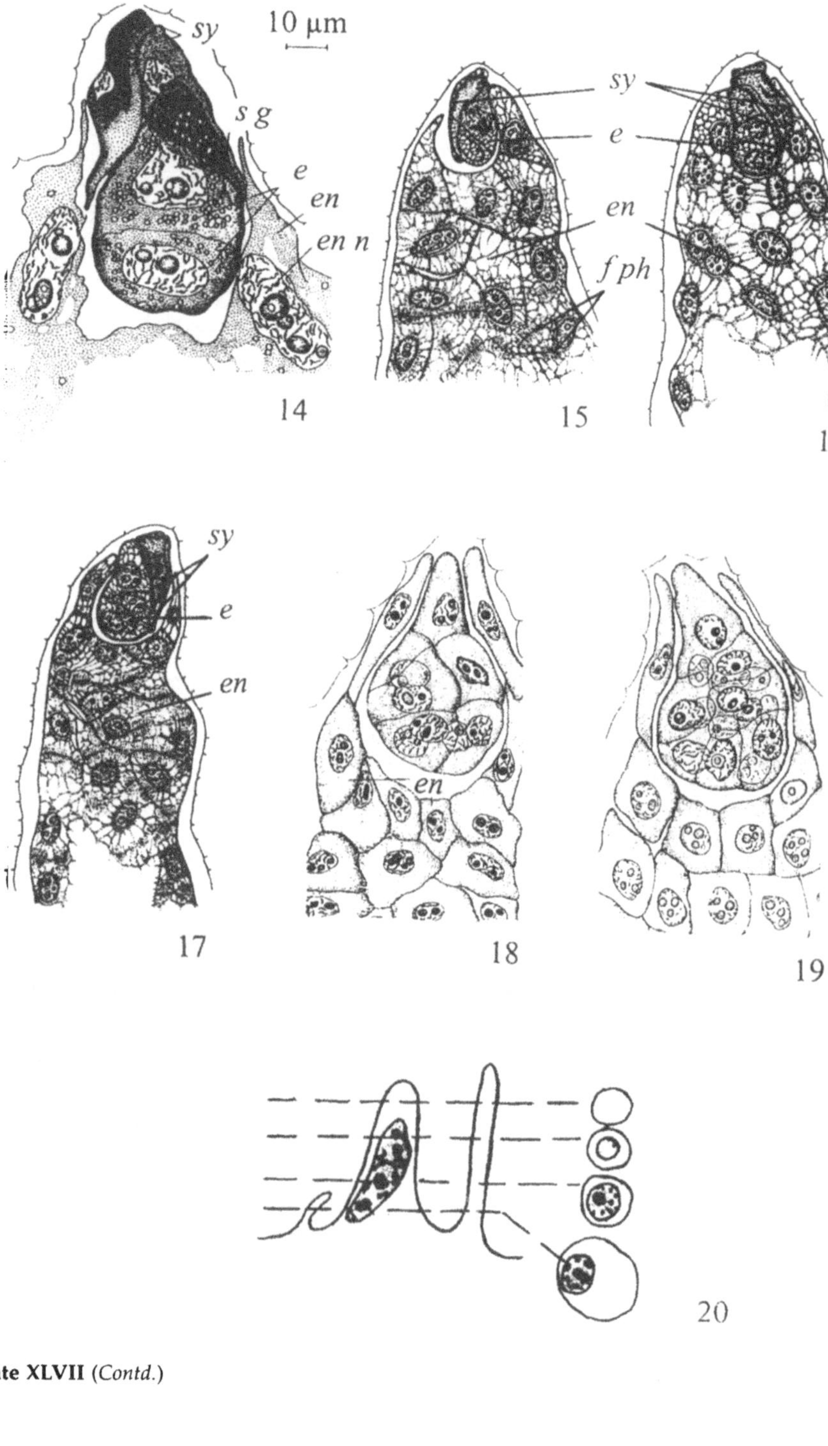

Plate XLVII (*Contd.*)

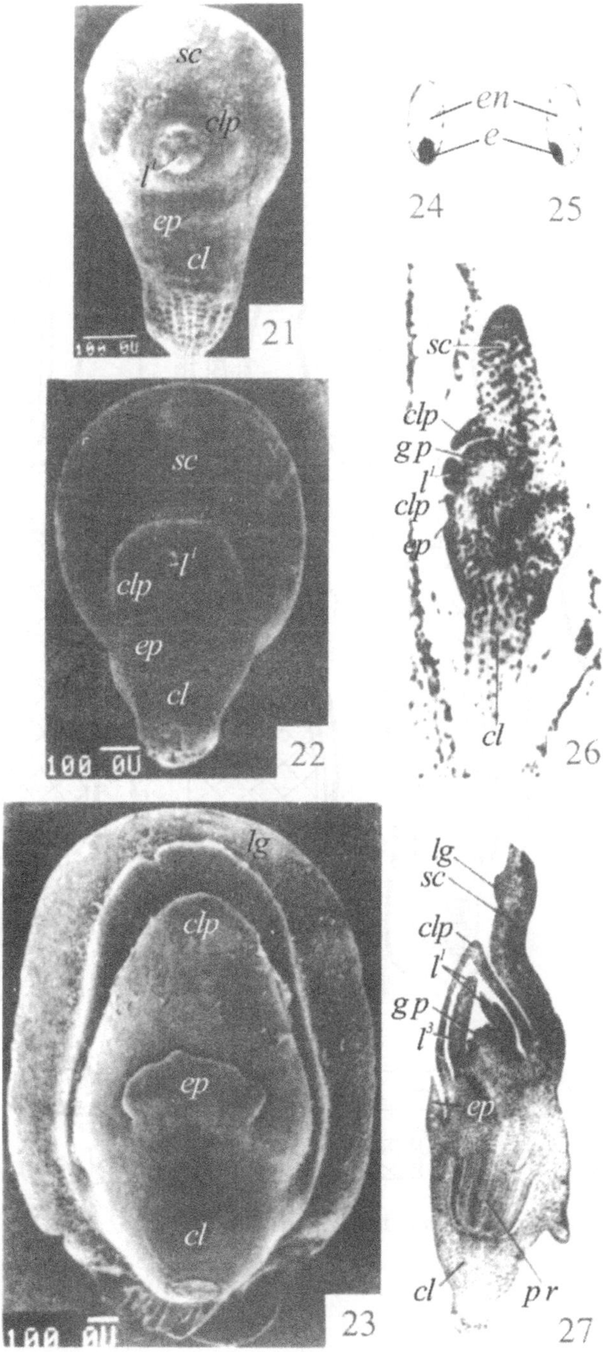

Plate XLVII (*Contd.*)

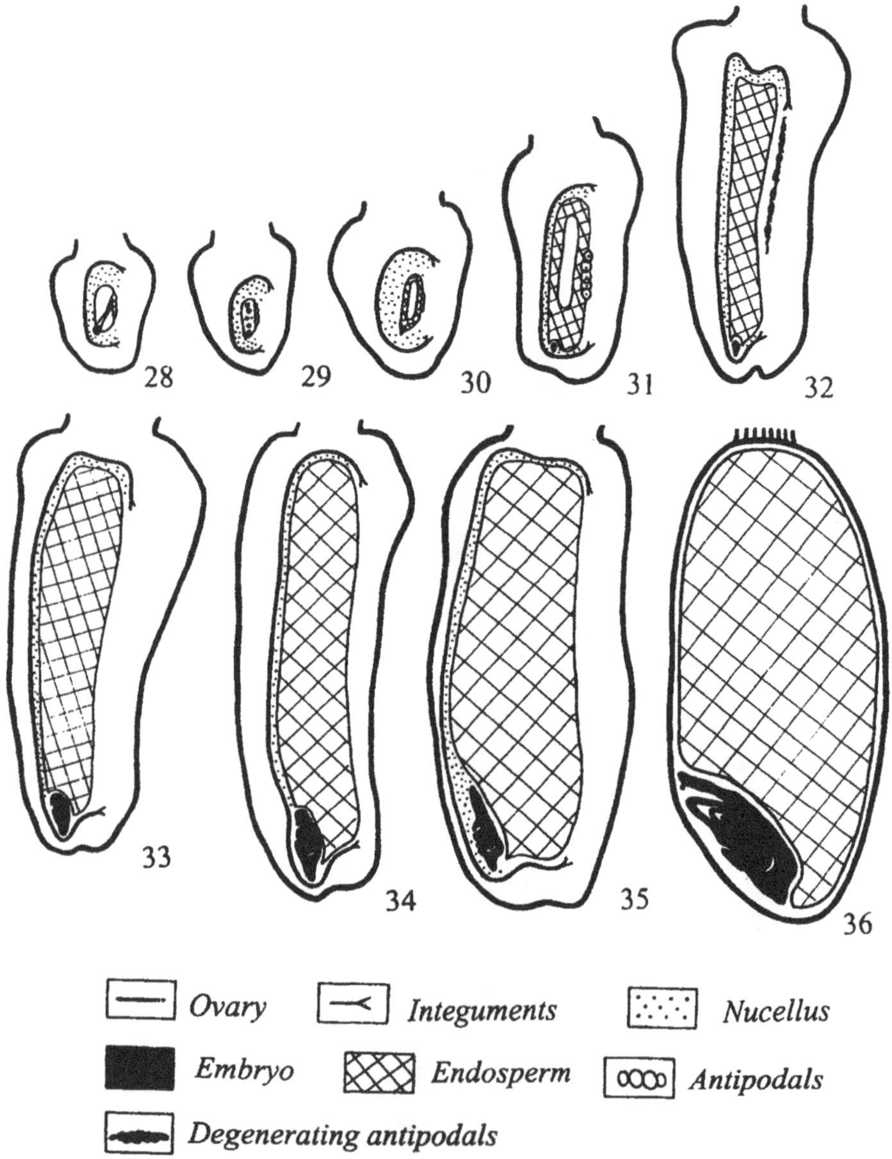

Plate XLVII (*Contd.*)

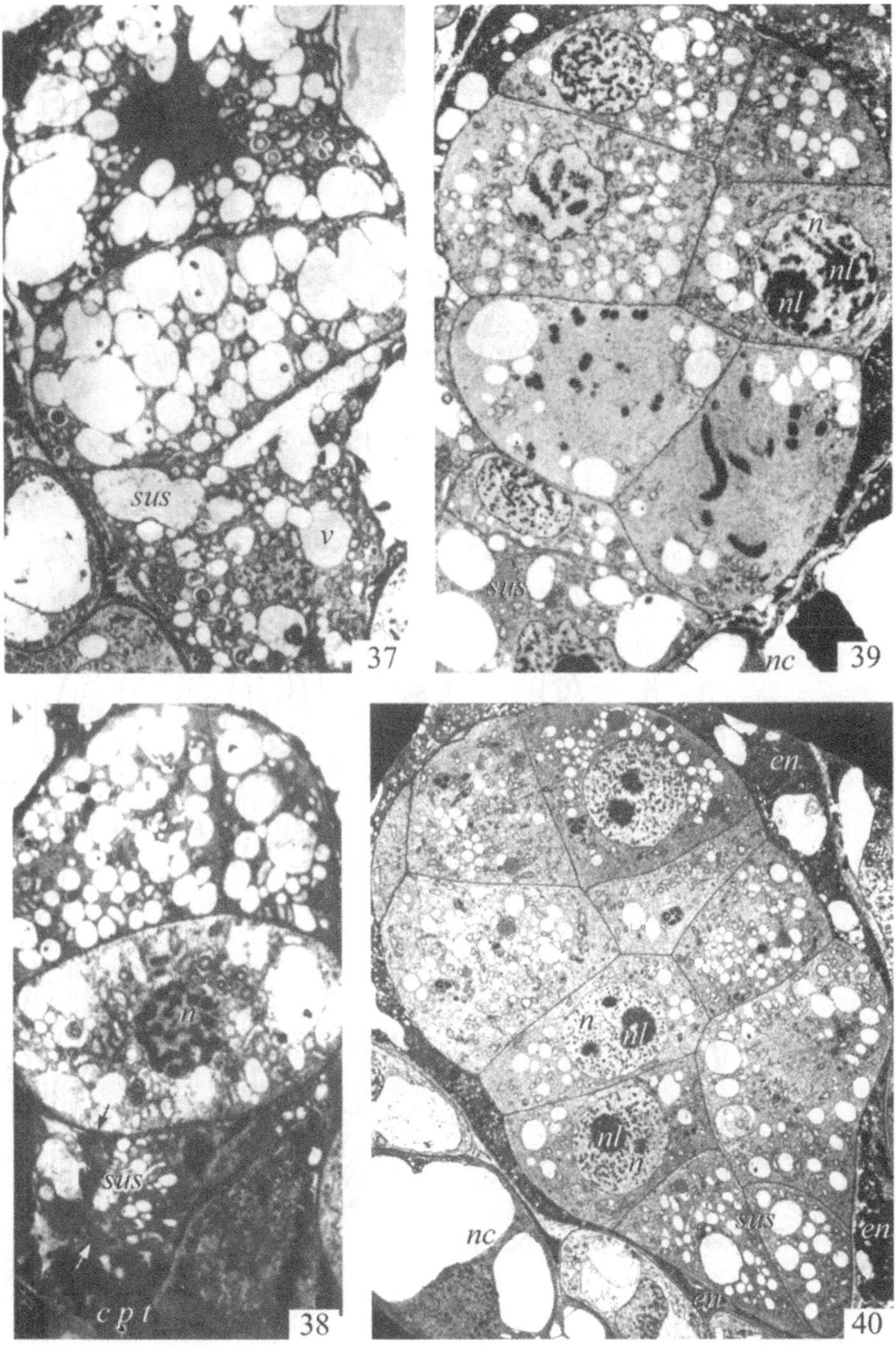

Plate XLVII (*Contd.*)

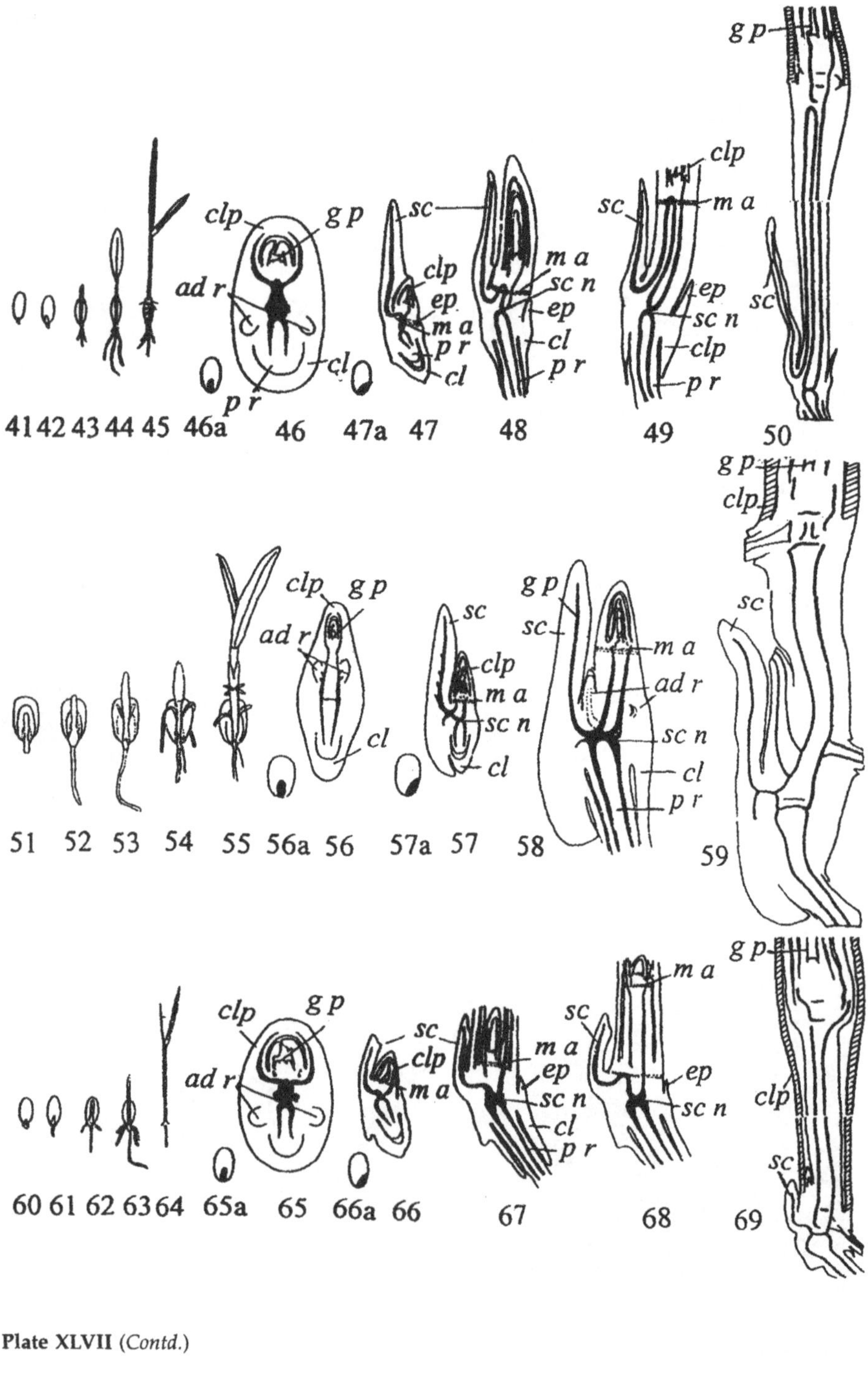

Plate XLVII (*Contd.*)

I *Capsella bursa-pastoris*
(Souèges, 1914-19)

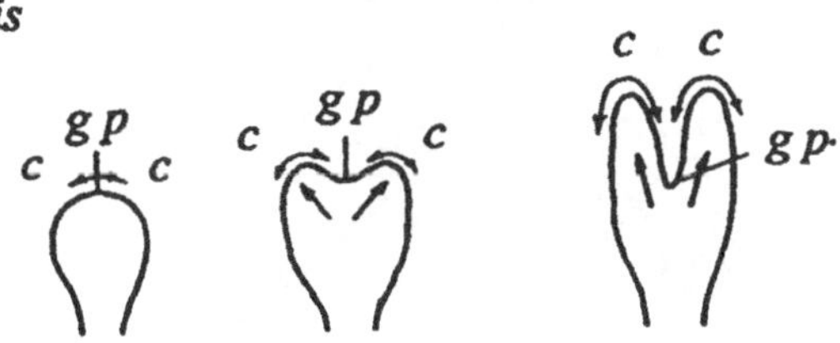

II *Pinguicula vulgaris*
(Haccius a. Hartl-Baude, 1957)

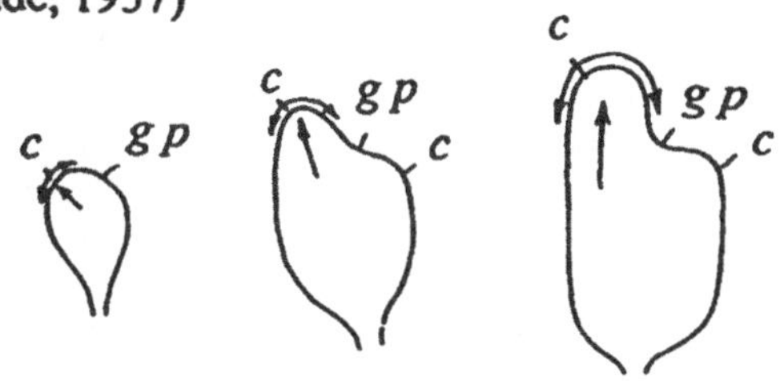

III *Ottelia alismoides*
(Haccius, 1952)

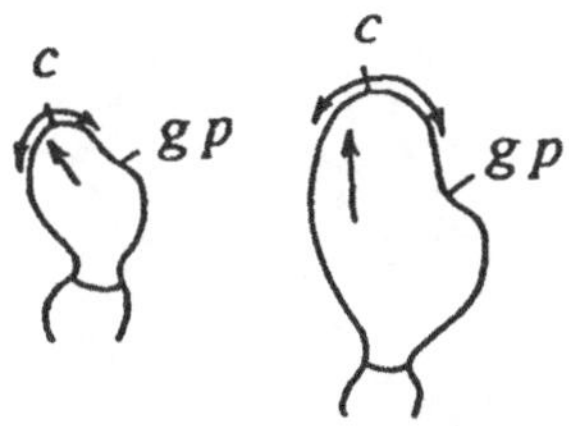

IV *Eleusine coracana*
(Guignard, 1961)

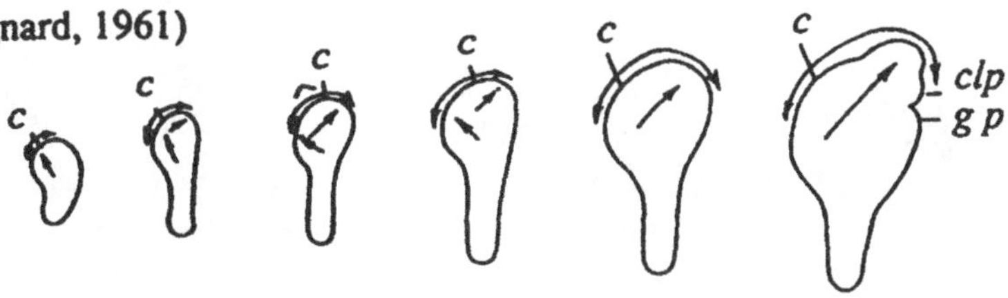

V *Triticum aestivum*
(Batygina, 1968)

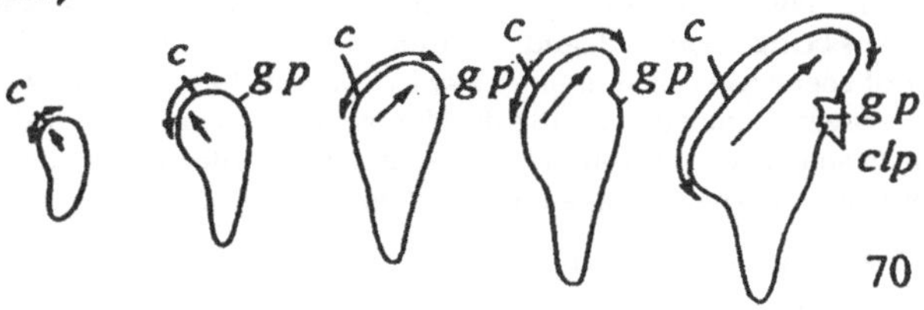

70

Plate XLVII (*Contd.*)

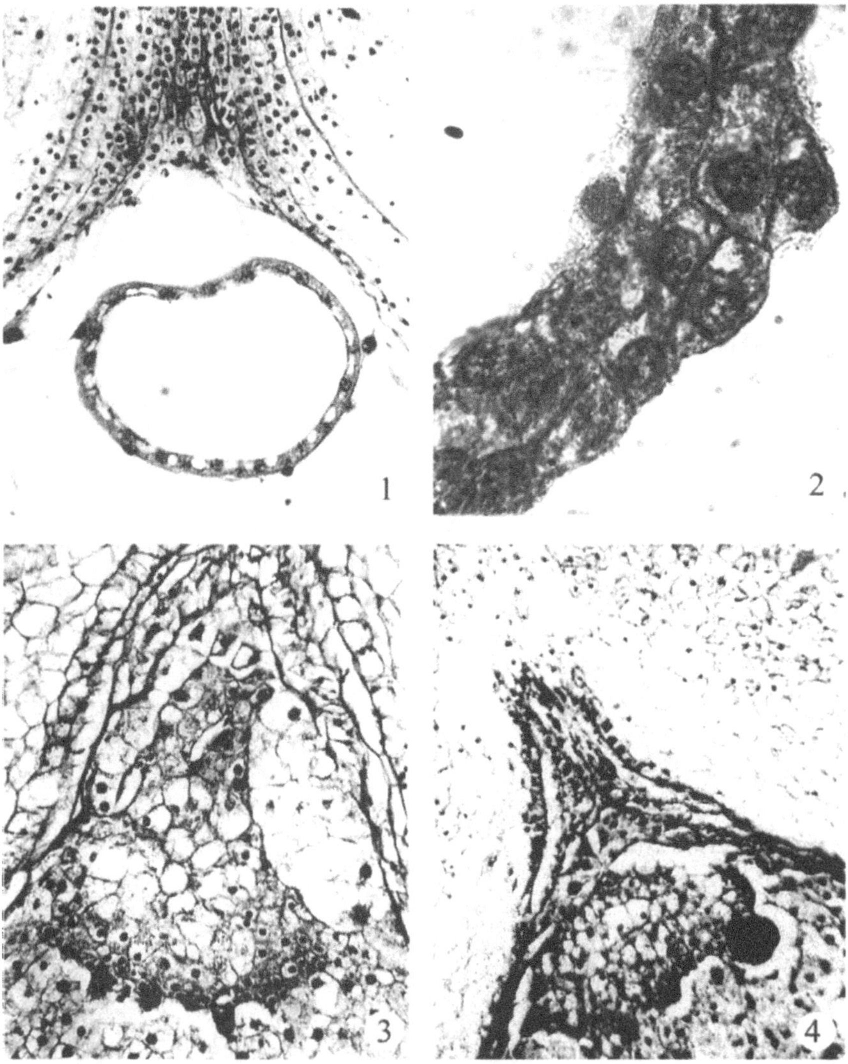

Plate XLVIII: Paeoniad-type of embryogenesis.

1-4.—stages of zygotic and somatic embryo development in *Paeonia*: 1—proembryo coenocyte in *Paeonia*, note peripheral layer of cytoplasm, containing nuclei and large central vacuole; 2—cell walls being organized, and initiation of somatic embryos (embryoids) occurs in epidermal layer of coenocyte-celled embryo's apical part; 3, 4—numerous somatic embryos have arisen; 5-10—stages of single embryoid's development on a coenocyte-celled proembryo, other embryoids have degenerated, note cellular endosperm, inner integument almost destroyed: 5, 6—globular embryoid, 7, 8—heart-shape embryoid; 9, 10—torpedo-shape embryoid.

1-5, 7, 9—LM, magnification × 400 (1, 4, 7), × 1000 (2), × 600 (3, 5), × 200 (9), 6, 8, 10—SEM.

After Batygina and Brukhin, orig.

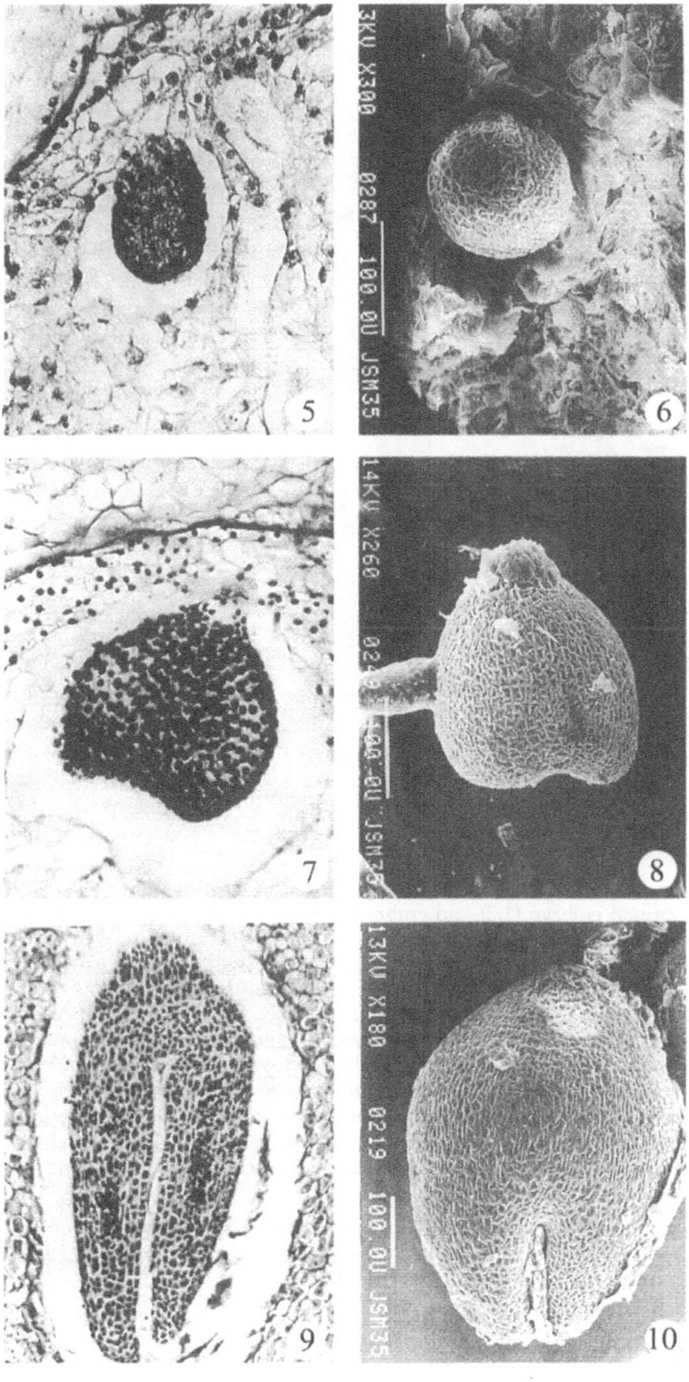

Plate XLVIII (*Contd.*)

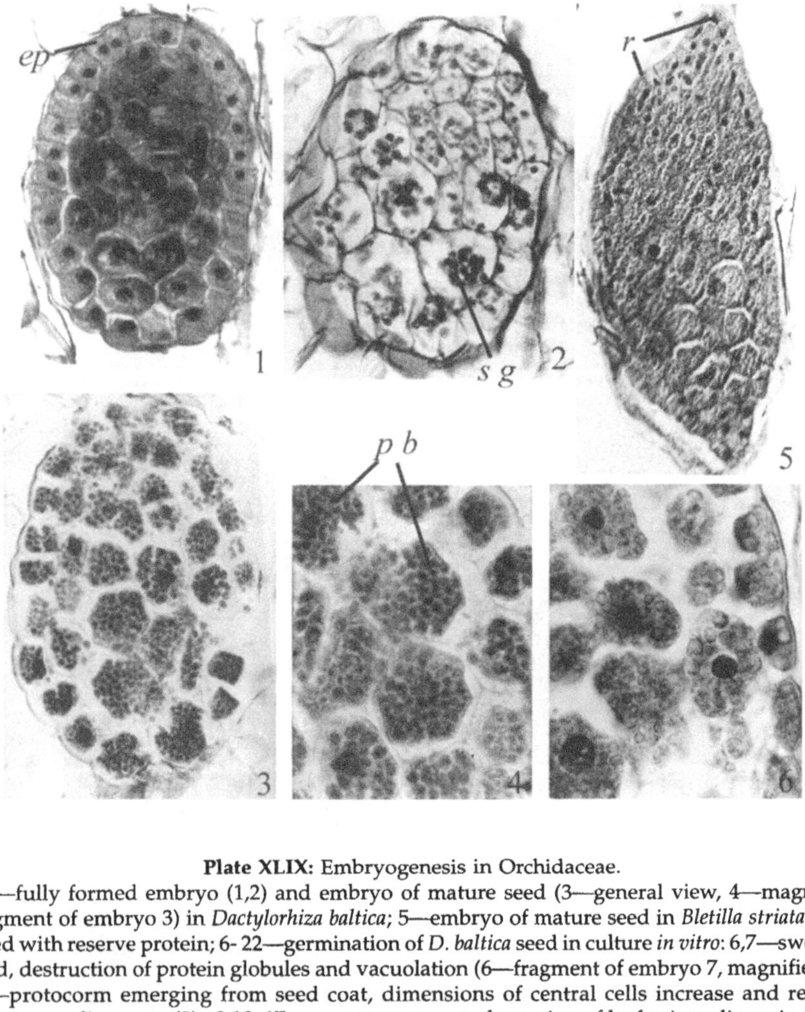

Plate XLIX: Embryogenesis in Orchidaceae.

1-4—fully formed embryo (1,2) and embryo of mature seed (3—general view, 4—magnified fragment of embryo 3) in *Dactylorhiza baltica*; 5—embryo of mature seed in *Bletilla striata*, cells filled with reserve protein; 6- 22—germination of *D. baltica* seed in culture *in vitro*: 6,7—swelling seed, destruction of protein globules and vacuolation (6—fragment of embryo 7, magnified), 8, 16—protocorm emerging from seed coat, dimensions of central cells increase and reserve substances disappear (8), 9,10, 17— protocorms up to formation of leaf primordia, apical pole consists of small cells, large parenchymal cells filled with starch, 11-15, 18-22—seedlings with leaves; 23-25—developmental stages of protocorms in *Phalaenopsis hybridum*; 26-28—embryo (26) and protocorms (27,28) in *Bletilla striata*; 29-33—embryo (29) and protocorms (30-33) in *Thunia alba*; 1-15—LM, 16-18, 23-33—SEM; 1,8,11,13-15—triple staining with fuchsin-sulphuric acid, hematoxylin and alcian blue; 2,9,10,12—PAS reaction, 3-7—procion bright blue RS by Ivanov and Litinskaya (1967).

a r—adventive root, *ep*—epidermis of embryo, *d r s*—degenerated region of seedling, *l1, l2, l3*—1st-3rd leaves, *p b*—protein bodies, *r*—ridge of leaf like primordium, *s g*—starch grains, *s c r*—small-celled region, *v b*—vascular bundle.

After Andronova, 1988.

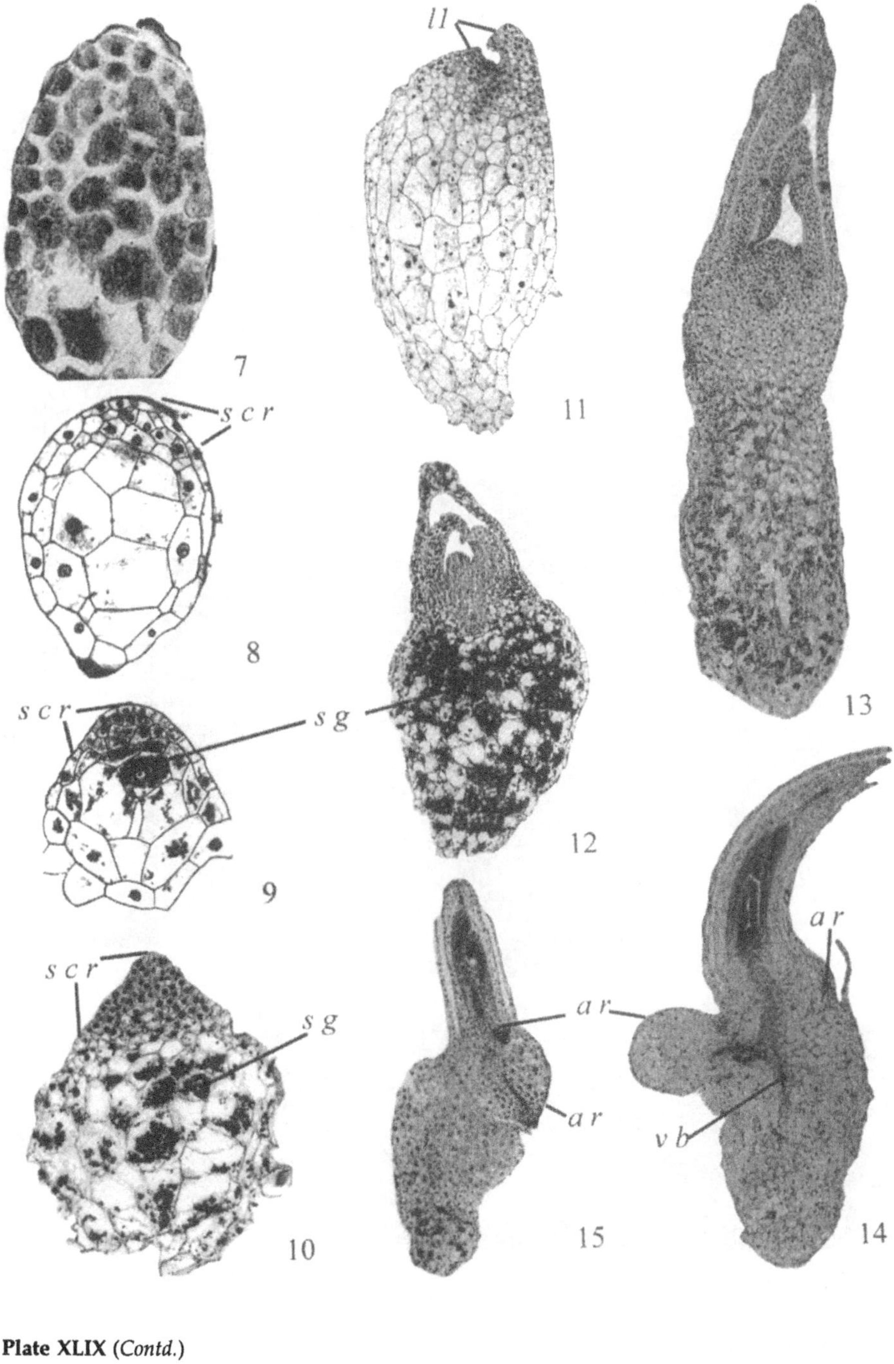

Plate XLIX (*Contd.*)

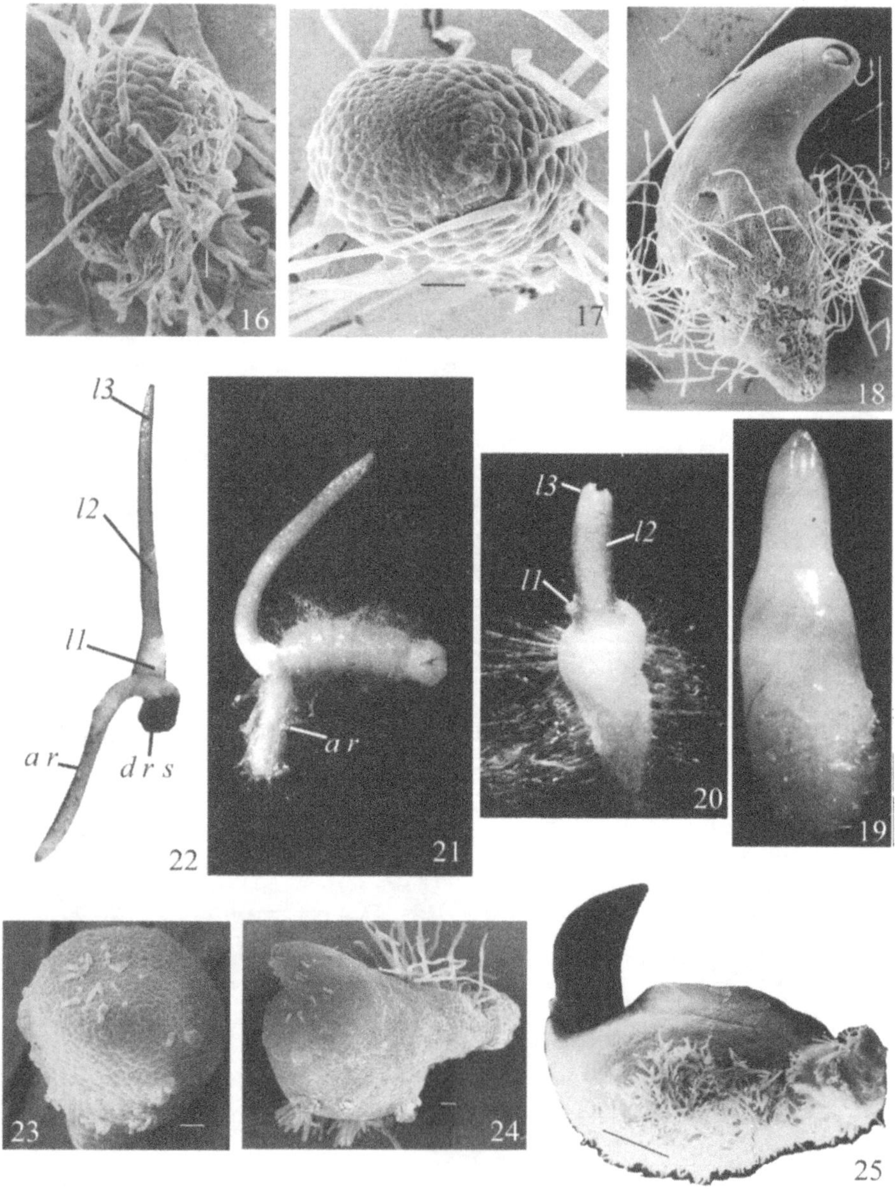

Plate XLIX (*Contd.*)

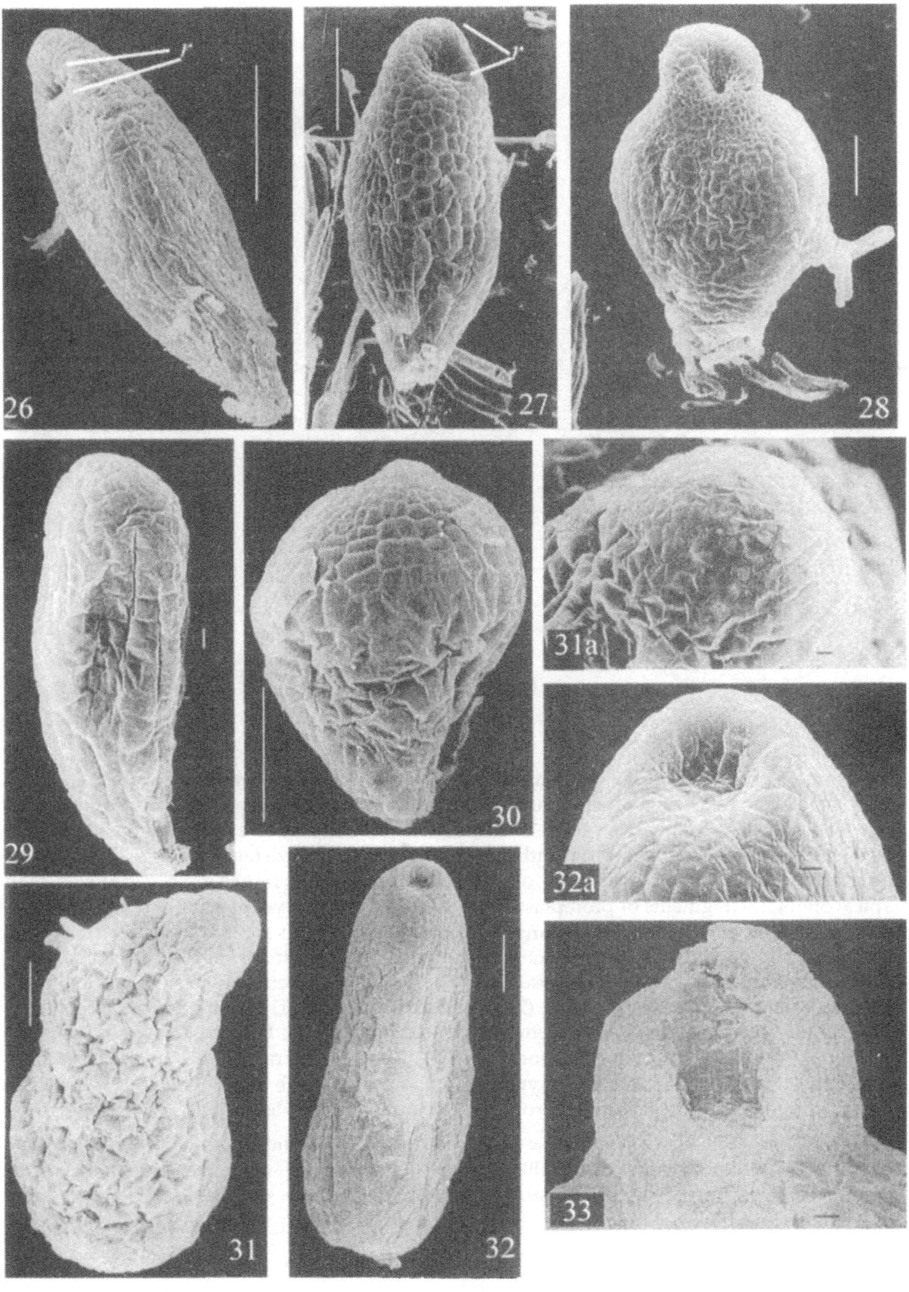

Plate XLIX (*Contd.*)

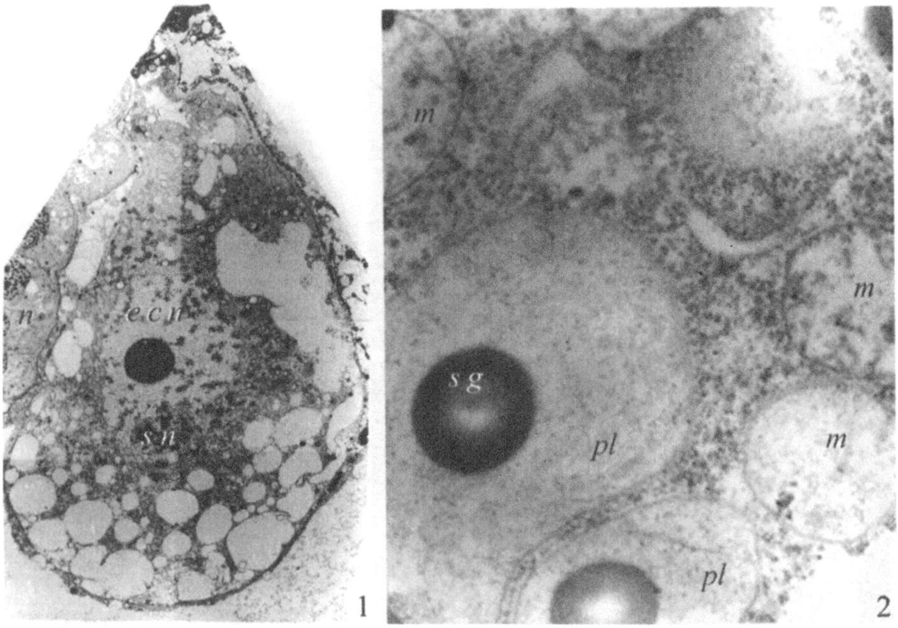

Plate L: Ultrastructural aspects of embryogenesis.

1-6—zygote and two-celled proembryo in *Triticum aestivum*, alloplasmatic line *aS*: 1—micropylar part of fertilized ovule with zygote, 17 hours after pollination, 2—zygote fragment, plastids with starch grains, mitochondria with developed system of cristae, majority of ribosomes grouped in polysomes, 3—basal cell of two-celled proembryo, rather more vacuolized than apical one, 4,5—fragments of protoplast in apical cell of proembryo, plastids and mitochondria compared with zygotic ones not changed, ribosomes mainly as monosomes, outer cell wall thickened and devoid of plasmodesmata, sedimentation of electron-dense substance between plasma membrane and cell wall observed (5), 6—cell wall between cells thin with much plasmodesmata; 7-12—early globular (7-10) and torpedo - like (11,12) embryos in *Reseda lutea*: 7—general view of embryo with demarcated embryoderm (× 1200), 8—hypophysial zone (× 7500), 9—cells of embryo apical zone (× 7500), 10—cellular fragments of globular embryo (× 8500), 11—cellular fragments of hypocotyl (chloroplasts with starch grains, × 9000), 12—cellular fragments of cotyledon (chloroplasts without starch, × 7500).

chl—chloroplast, *d*—dictyosome, *e c n*—egg cell nucleus, *en*—endosperm, *e r*—endoplasmic reticulum, *l b*—lipid body, *l c*—lens-like cell, *lp*—leucoplast, *m*—mitochondrion, *n*—nucleus, *nc*—nucellus, *nl*—nucleolus, *pl*—plastid, *pd*—plasmodesma, *sus*—suspensor, *s g*—starch grain, *s n*—sperm nucleus, *tp*—tonoplast, *v*—vacuole, *w*—cell wall.

1-6—after Naumova and Matzk, orig.; 7-12—after Yakovlev and Chaban, 1979.

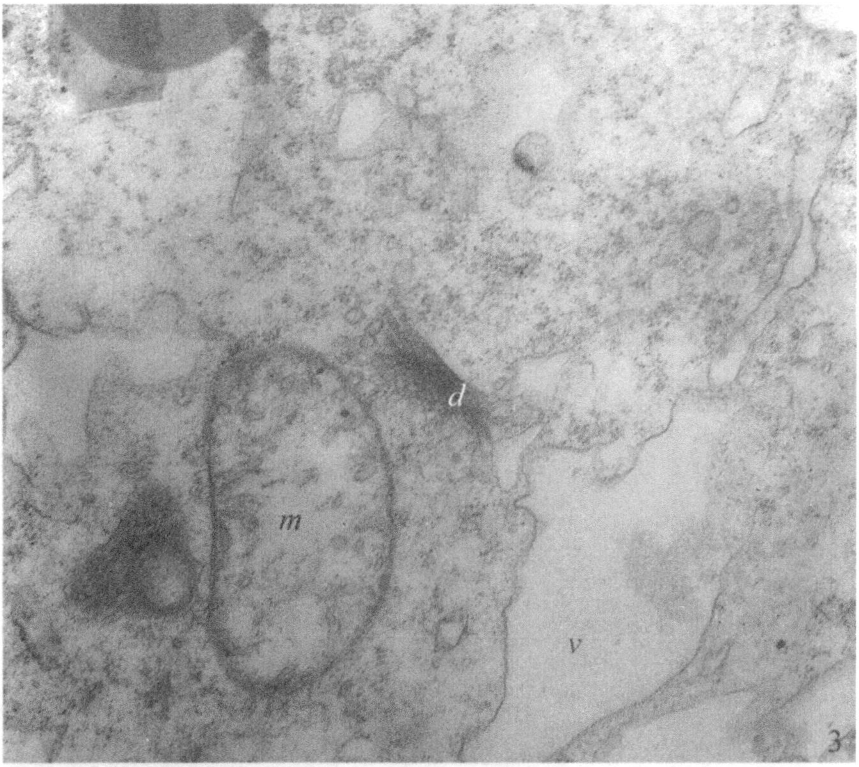

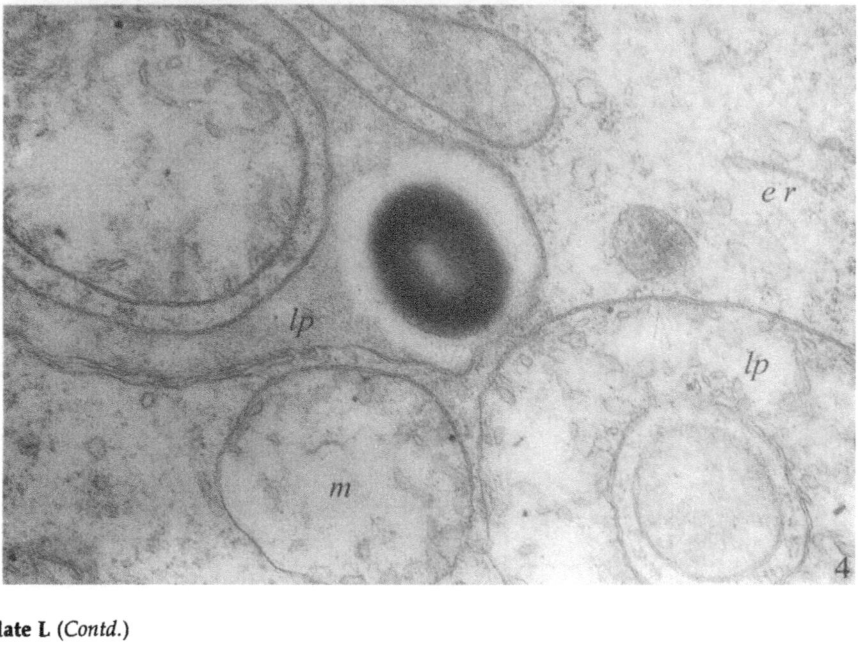

Plate L (*Contd.*)

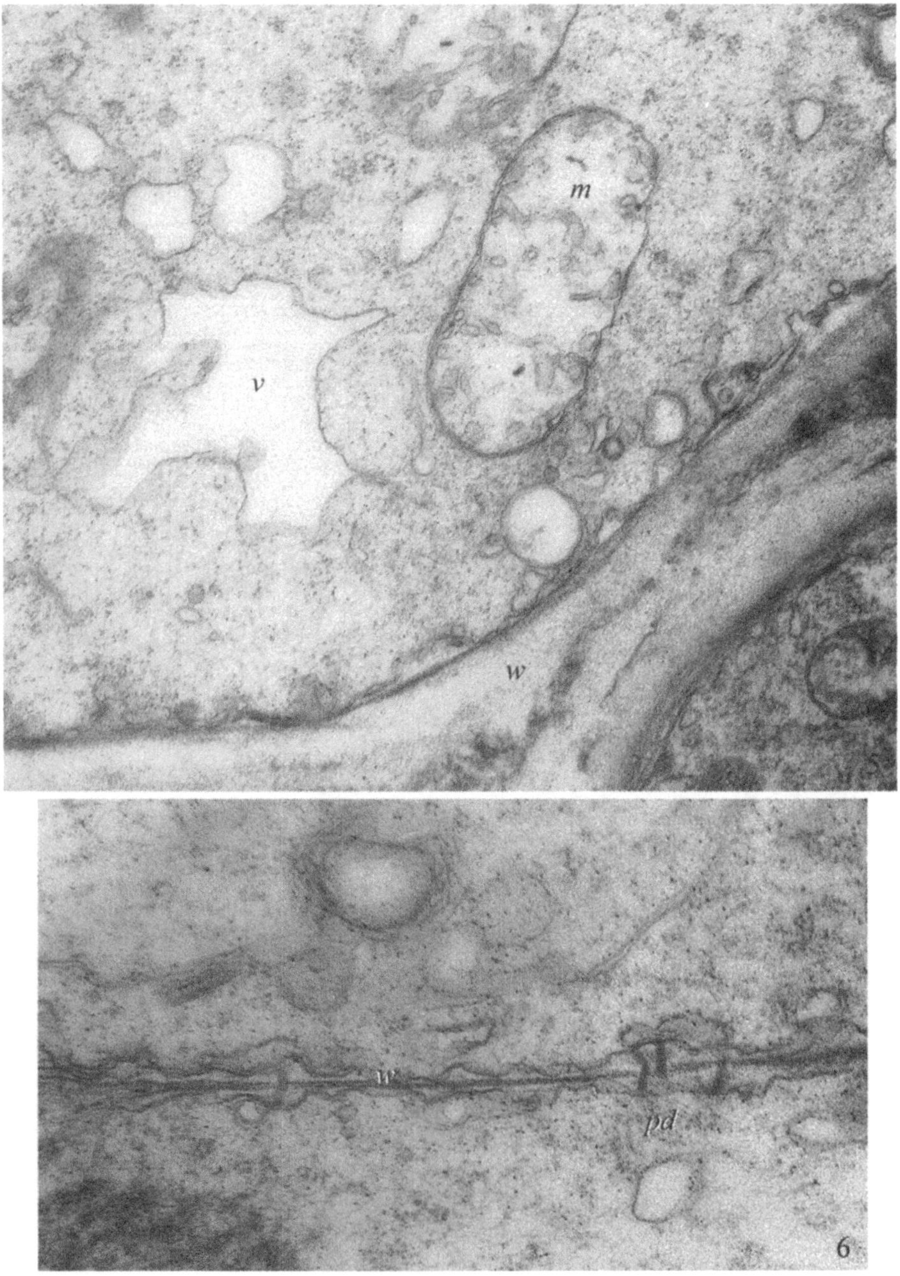

Plate L (*Contd.*)

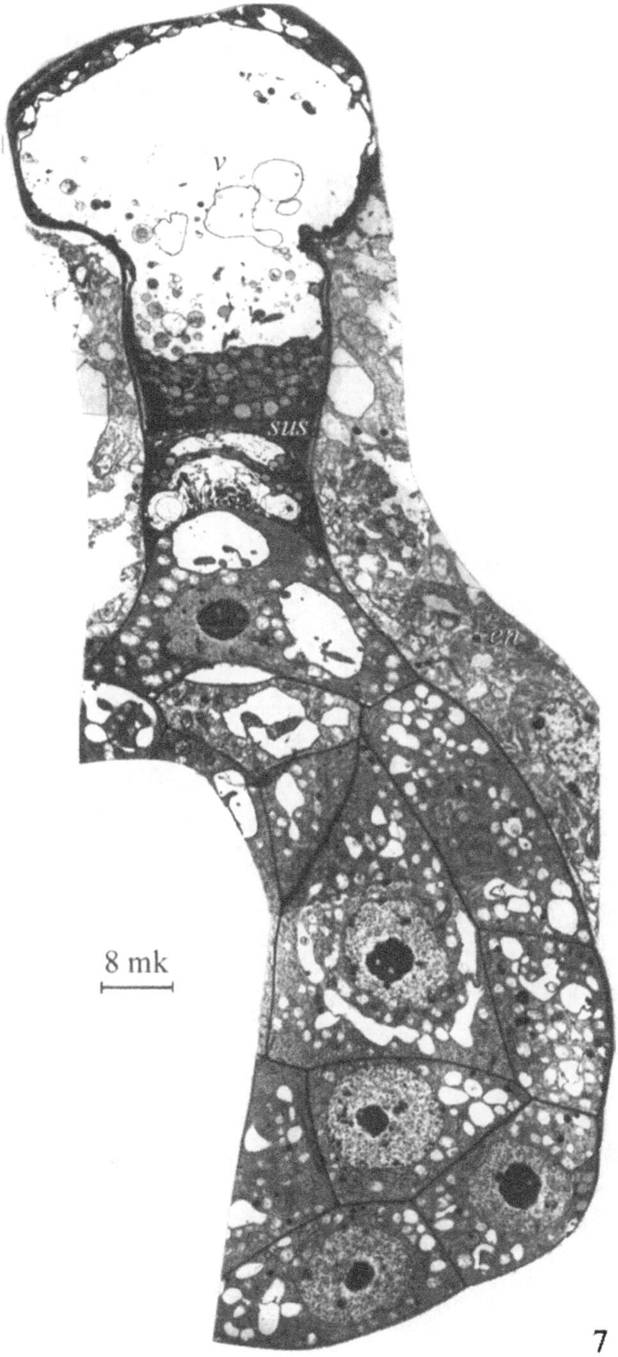

Plate L (*Contd.*)

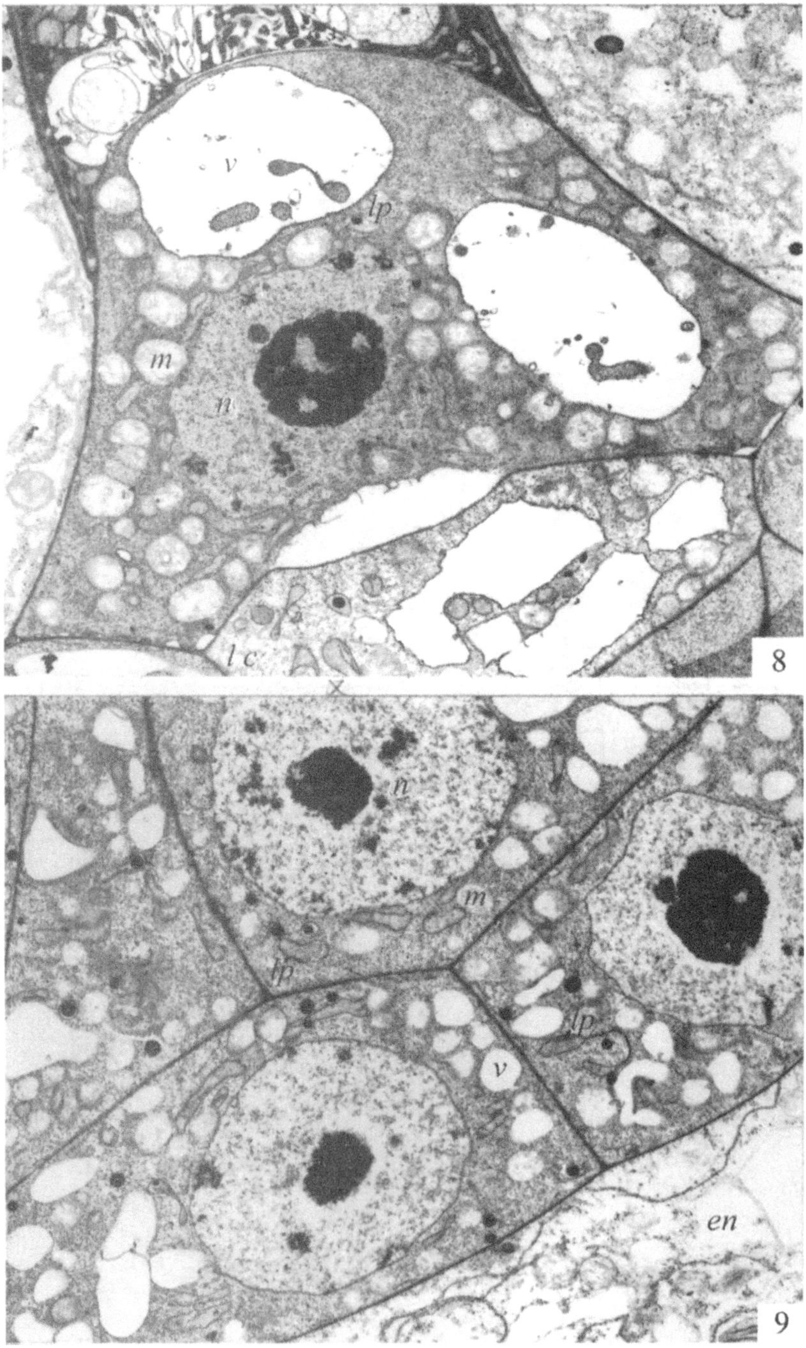

Plate L (*Contd.*)

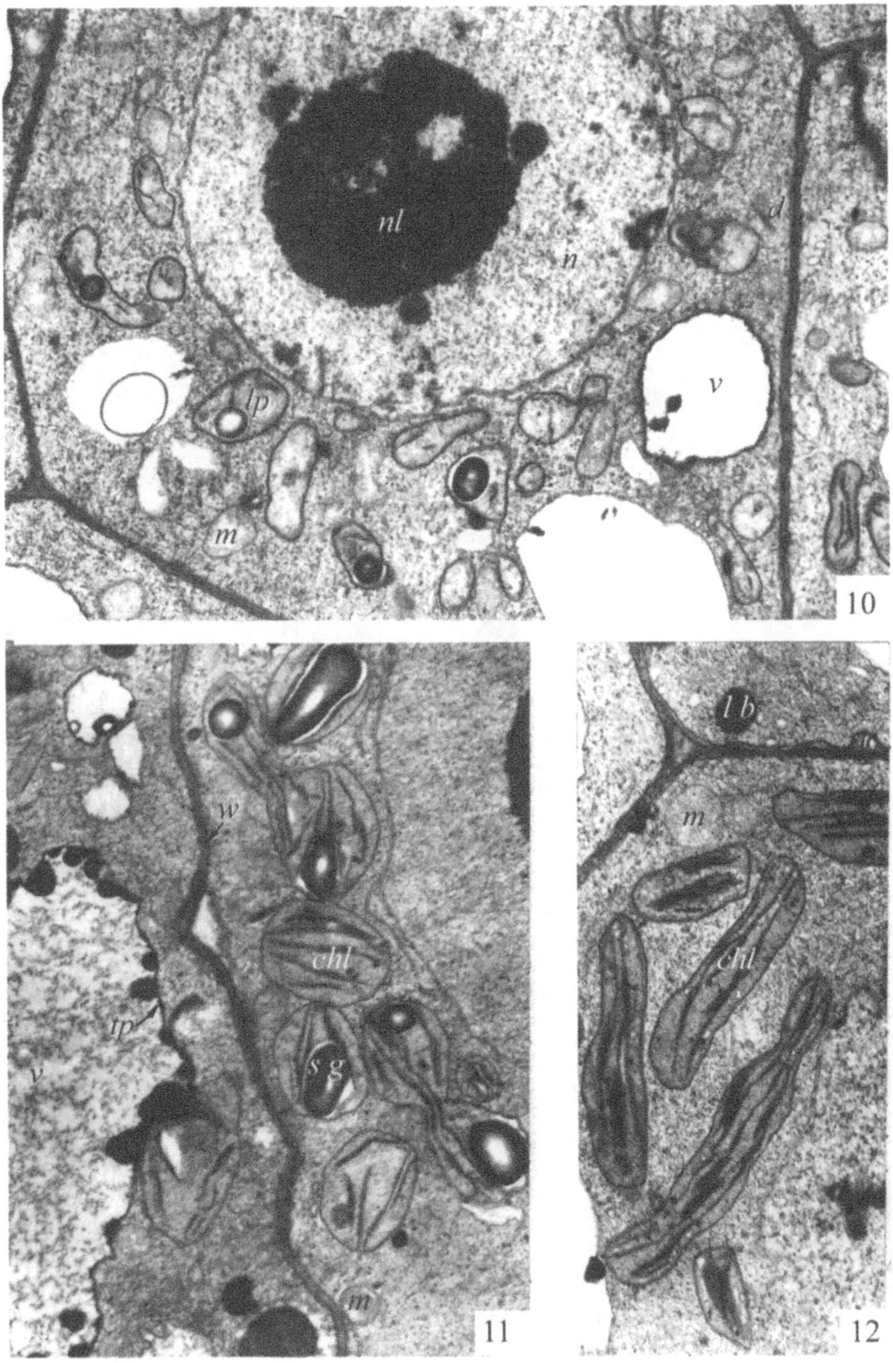

Plate L (*Contd.*)

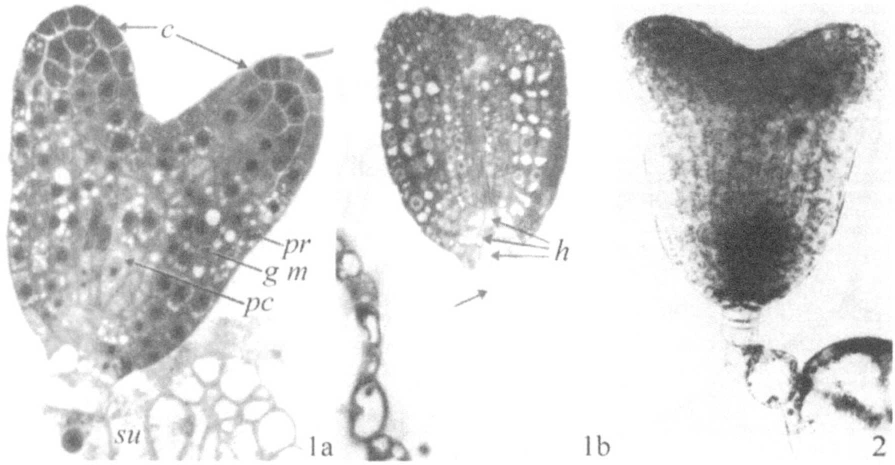

Plate LI: Histochemical and immunohistochemical aspects of embryogenesis.
1a, b—localization of nuclein acids (a) and proteins (b) in heart-shaped embryo of *Capsella bursa-pastoris* (stained with alcian black); 2—distribution of cytochromoxidase in heart-shaped embryo of *Stellaria media*; 3–6—distribution of spectrin C in cells of somatic embryo of *Daucus carota* in suspension culture: 3—isolated long cells of suspension culture, strong connection of anti–spectrin C. 1/100 marker with plasmatic membrane observed, nuclei reveal slight fluorescence; 4—proembryonal mass after marking by anti–spectrin C. 1/160; 5—somatic embryo at globular stage after preparation of cryosections and marking by anti–spectrin C. 1/125, slight luminescence on plasmatic membrane and powerful luminescence of nuclei; 6—somatic embryo at the torpedo-shape stage (in transverse section), enclosed in polyethylenglycol and marked by anti-spectrin C. 1/180, epidermal cells demonstrate powerful luminescence in region of plasmatic membrane, cytoplasm and nuclei; intensity of autofluorescence in vascular tissue differs from that of FITC-conjugant; 7—distribution of F-actin at moment of thelophase during microinjection with rodamin–phalloidin in endosperm cell of *Haemanthus katherinae*, 5 minutes after injection short actin fascicles, forming ring near cell equator, where cellular plate develops, revealed (shown by arrow); 8a, b—lateral part of alveolar endosperm in *Brassica napus*: fluorescence of cortical microtubules in peripheral areas, radial microtubules run from surface of nucleus into cytoplasm, phragmoplast microtubules participate in cytokinesis (a); after treatment with 10 μM taxol solution for 24 hours, tubuline microtubules became more prominent, longer and thicker running as bright crown arrays from endosperm nuclei (b).

c—cotyledon, *g m*—ground meristem, *h*—hypophysis, *pc*—procambium, *pr*—protoderm, *su*—suspensor, *v b*—vascular bundle.

1a, b—after Schulz and Jensen, 1968a; 2a,b—after Pritchard and Bergster, 1969; 5, 6—after Ruijter and Emons, 1993; 7—after Schmit and Lambert, 1990; 8a, b—after Brukhin and Van Lammeren, orig.

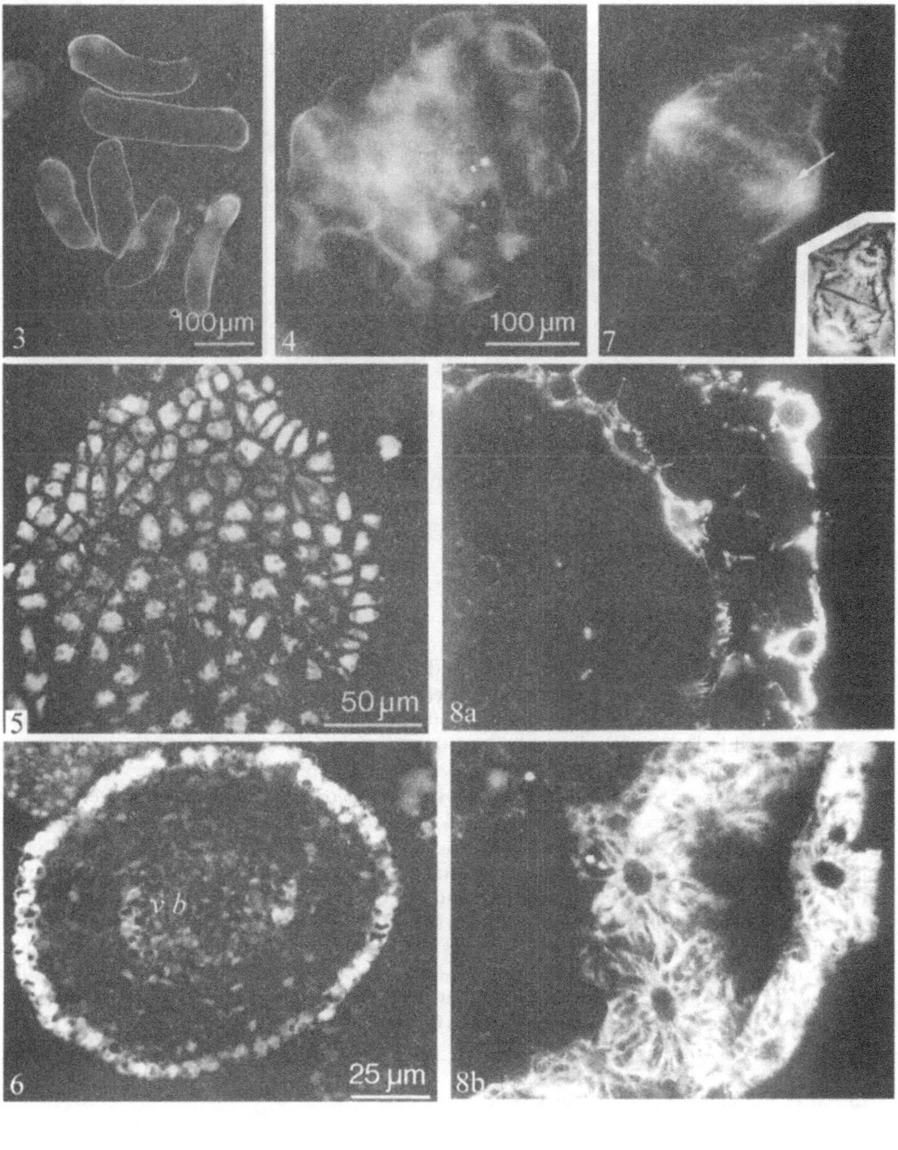

Plate LI (*Contd.*)

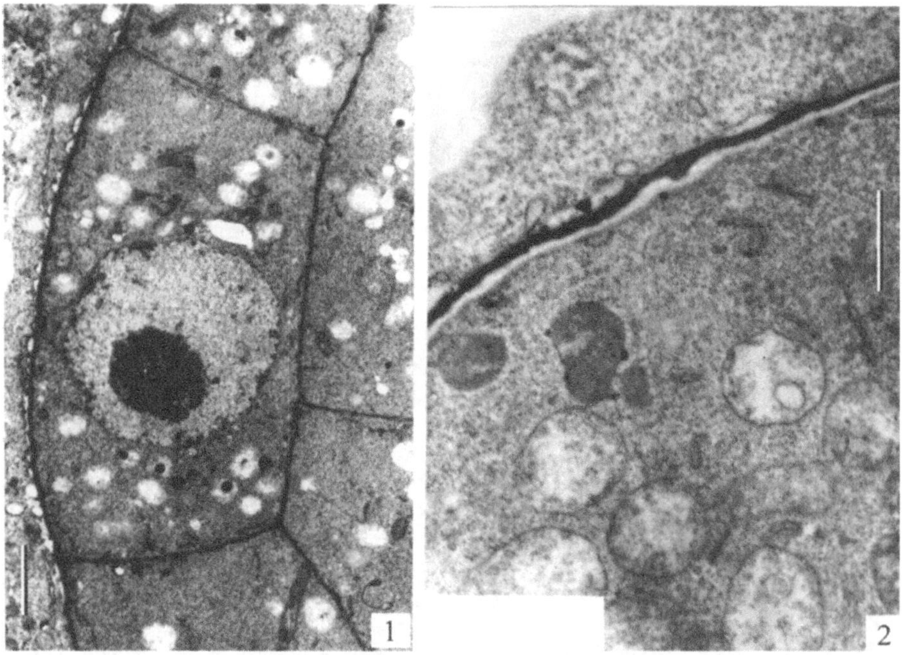

Plate LII: Cuticle of developing embryo.

1,2—*Stellaria media:* 1—protodermal cell wall of early globular embryo, thin homogeneous outer cell walls, 2—late globular embryo, outer wall with distinct cuticular layer; 3-5—*Alisma plantago-aquatica*: 3—multicellular, globular embryo with protoderm, 4—cutin fluorescence on apex of same embryo as in fig 3, 5—cutin fluorescence around ovoid embryo proper; 6,7—*Stellaria media:* 6—about ten-celled proembryo, Nomarski microscope, 7—weak cutin fluorescence on apical surface of same proembryo as in fig. 6; 8—*Capsella bursa-pastoris,* cutin fluorescing surface of globular embryo proper and non-fluorescing cell walls of suspensor; 9-11—*Linum usitatissimum:* 9—cutin fluorescence in heart-shaped embryo, 10—basal region of heart embryo proper, fluorescing cutin on embryo surface and upper part of suspensor cell, 11—basal region of squashed heart embryo, cutin fluorescence in side wall of suspensor cell and a non-fluorescing cell wall between suspensor and embryo proper; 12—*Sinapis alba,* late torpedo stage, very distinct fluorescence of embryo surface; 13,14—*Capsella bursa-pastoris:* 13—strongly fluorescent radicular region at late torpedo stage embryo, 14—upper part of curved hypocotyl; 15—*Iberis umbellata,* part of cotyledon of U-shaped embryo—cutin fluorescence.

Bars: 1—2 μm, 2—1 μm, 3-7, 9—100 μm, 8, 10, 11—50 μm, 12—500 μm, 13—150 μm, 14, 15—200 μm.

1-15—after Rodkiewicz and Szczuka, orig.

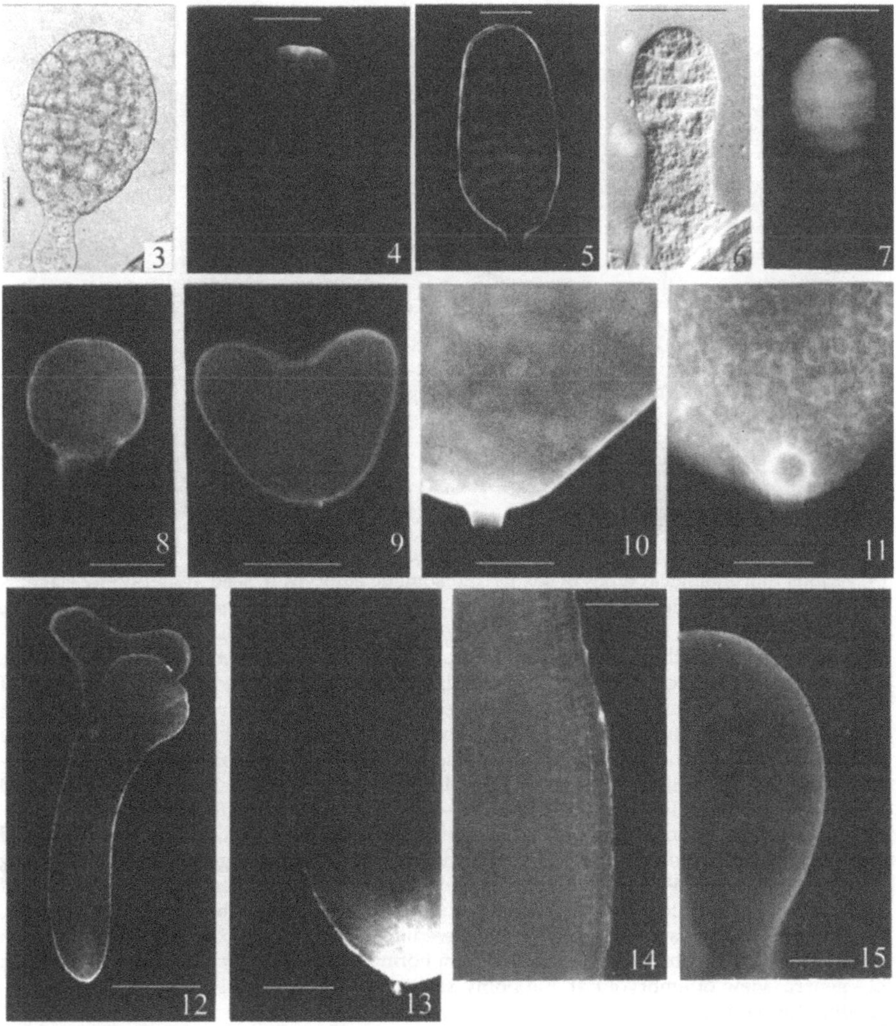

Plate LII (*Contd.*)

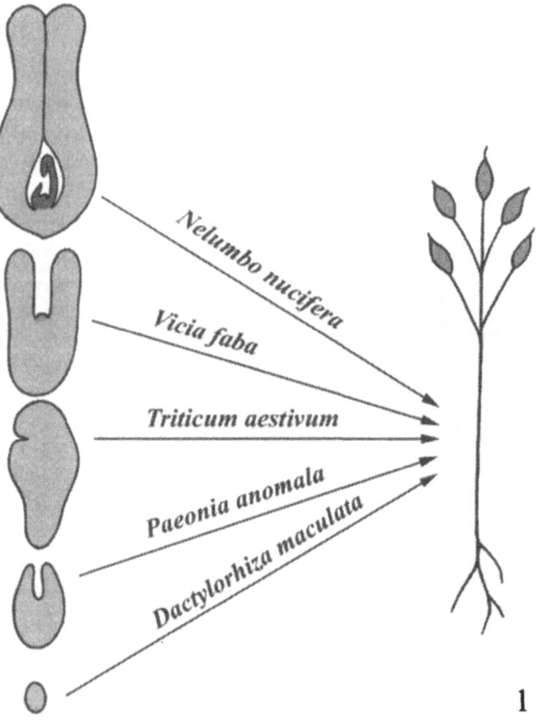

Plate LIII: Embryo autonomy.

1—various morphological types of sexual and somatic embryos in seed at autonomy stage; 2-14—morphophysiological correlations during embryogenesis in *Nelumbo nucifera*: 2—relative rates of growth and total soluble carbohydrate accumulation in embryo *in vivo*, 3—morphogenetic events throughout embryogenesis, 4—dynamics of structure development inside seed (schematic), vertical dot line marks fertilization (6-8 h after pollination), 5—dynamics of total nucleic acid concentration in embryo body, plumule and endosperm (conventional values), 6-9—endosperm structure at globular (6), early heart-shaped (7), and autonomy (8—increase in nucleus number and thickening of endosperm cell walls in chalazal part of seed, 9—endosperm cells magnified) stages of embryo development, 10,11—embryo isolated from seed at autonomy stage (10—general view, 11—plumule, one cotyledon removed), 12,13—'germination' of autonomous embryo and plant development in modified non-hormonal medium by Murashige and Skoog, 14—scheme presenting embryo 'germination' depending on its isolation stage; 15-21—*in vitro* cultivation of *Dactylorhiza maculata* embryo in non-hormonal medium by Knudson: 15—embryo in seed at autonomy stage, 16,17—'budding' of protocorms, 18—formation of shoots on protocorm, 19-21—seedling development; 22-25—*in vitro* cultivation of *Paeonia anomala* embryoids in modified non-hormonal medium by Murashige and Skoog: 22—general view of embryoid at autonomy stage, 23—longitudinal section of seed, 24,25—seedling formation.

a z—apical zone, *b z*—basal zone, *e*— embryo, *em*—embryoid, *end*—endosperm, *cot*—cotyledon, *c z en*— central zone of endosperm, *p z en*—peripheral zone of endosperm; 2,5—days after

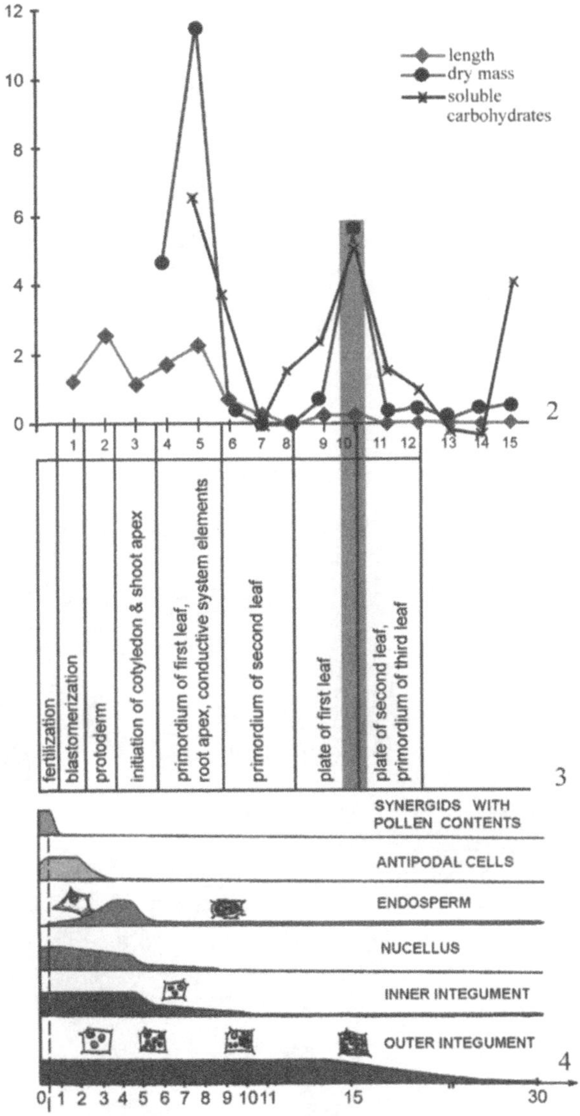

Plate LIII (*Contd.*)

pollination plotted as abscissa, 2—ordinate: relative rate of processes $(l_2—l_1)/l_1 \cdot t$, l_1 and l_2 – preceding and following values, *t*—time period (in days); 2,3,14—solid column denotes autonomy stage.

1,4,14—after Batygina and Vasilyeva, 1981; 2,3,10, 12,13—after Vasilyeva and Titova, orig.; 5—after Morozova, 1993; 6,11—after Vasilyeva *et al.*, 1987; 7-9—after Titova, orig., 15-17, 19—after Batygina and Vasilyeva, 1983c; 18,20,21—after Vasilyeva and Batygina, orig.; 22, 24, 25—after Vasilyeva and Freiberg, orig.; 23—after Batygina and Butenko, 1981.

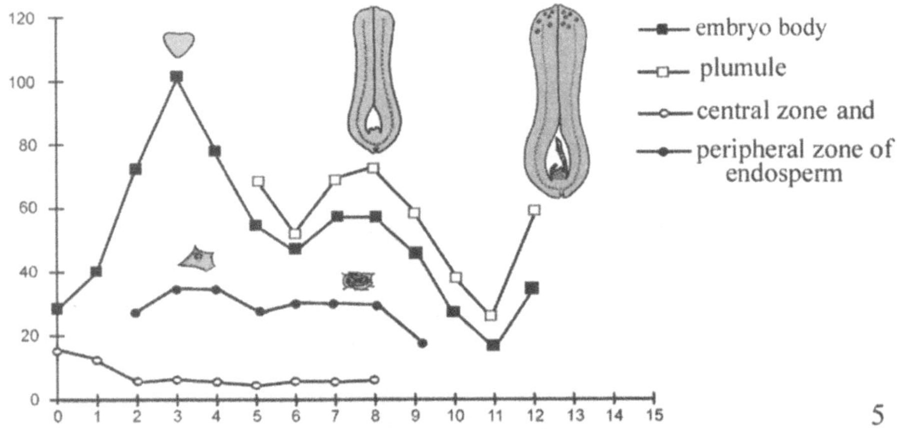

5

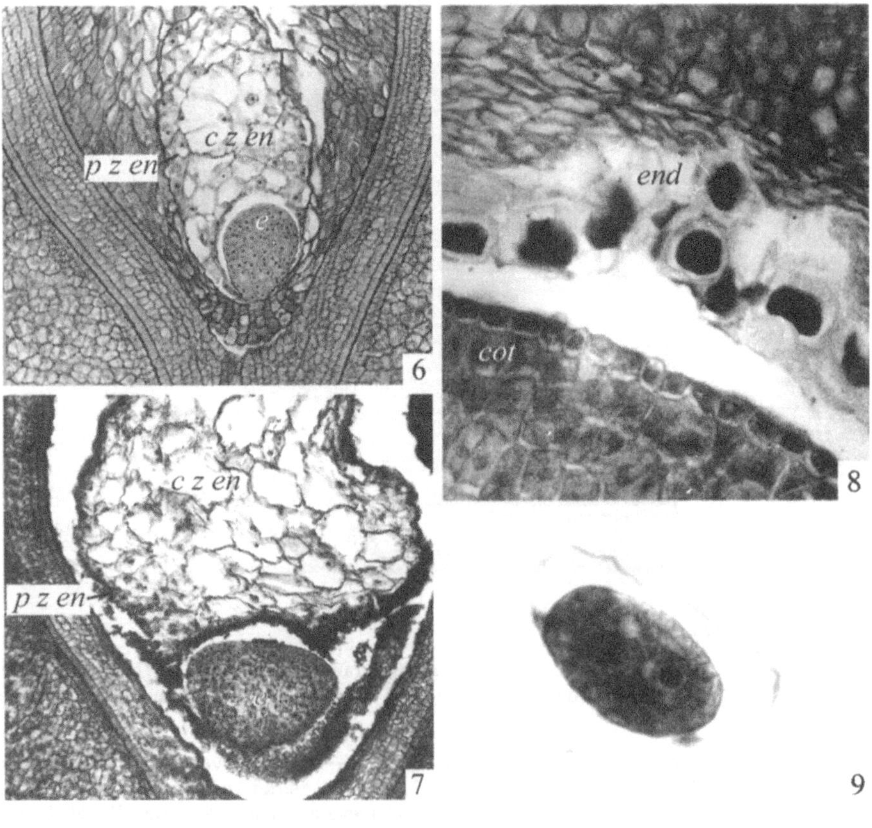

6 7 8 9

Plate LIII (*Contd.*)

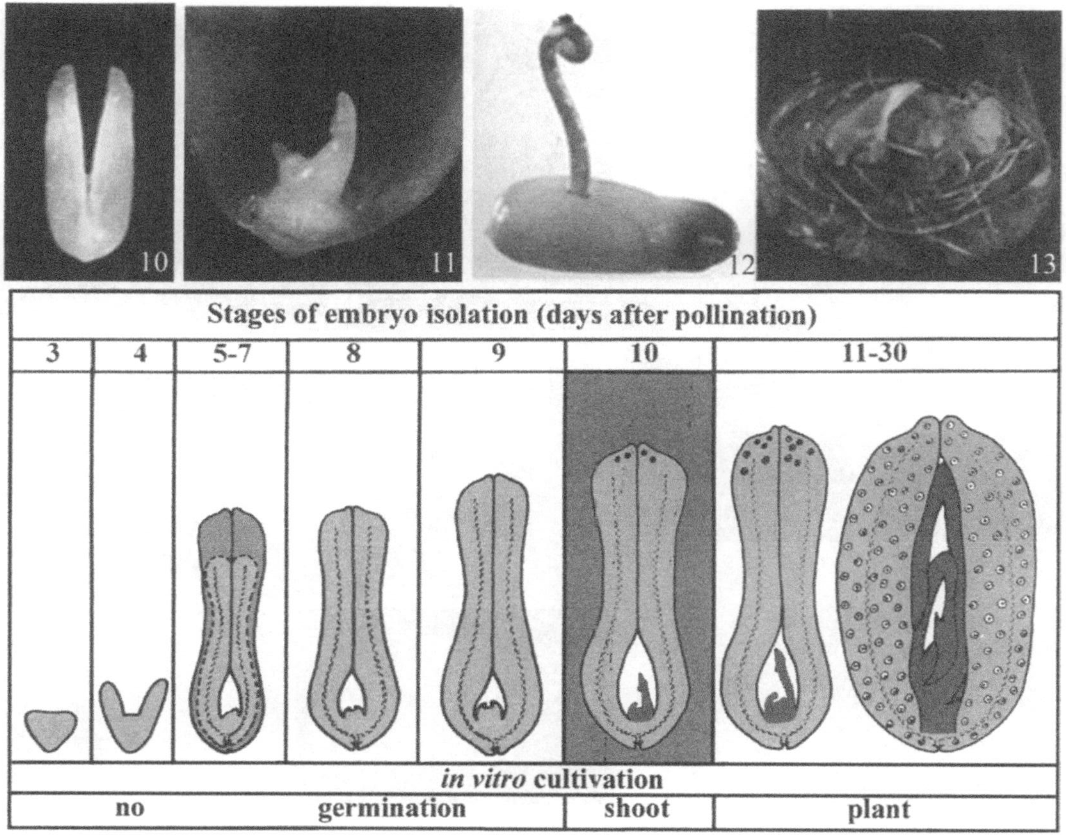

14

Plate LIII (*Contd.*)

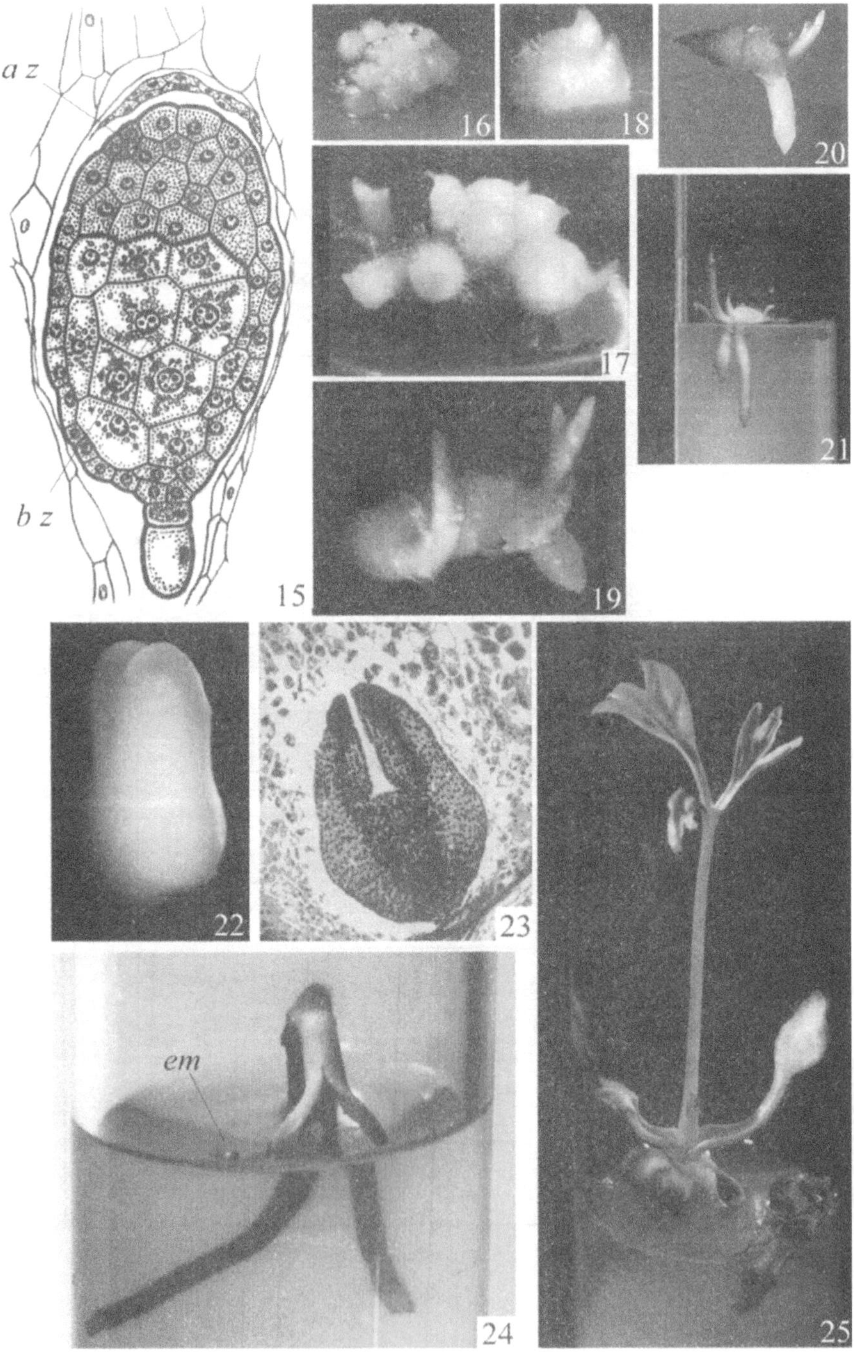

Plate LIII (*Contd.*)

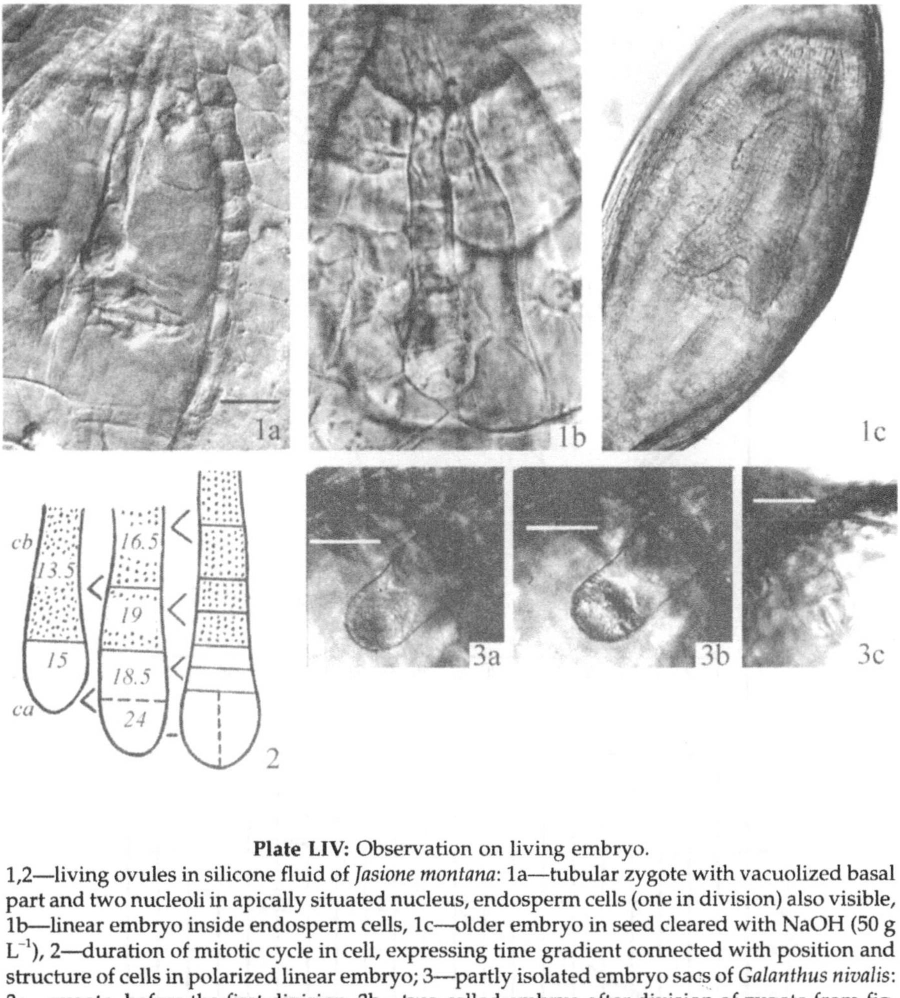

Plate LIV: Observation on living embryo.

1,2—living ovules in silicone fluid of *Jasione montana*: 1a—tubular zygote with vacuolized basal part and two nucleoli in apically situated nucleus, endosperm cells (one in division) also visible, 1b—linear embryo inside endosperm cells, 1c—older embryo in seed cleared with NaOH (50 g L^{-1}), 2—duration of mitotic cycle in cell, expressing time gradient connected with position and structure of cells in polarized linear embryo; 3—partly isolated embryo sacs of *Galanthus nivalis*: 3a—zygote before the first division, 3b—two-celled embryo after division of zygote from fig. 3a, 3c—four-celled embryo with longitudinal wall between two apical cells; 4,5—*Linum usitatissimum*: 4—growth of embryo *in vivo*, 5—growth of embryo excised as heart-shape (8 days after pollination) and cultured *in vitro* (medium after Monnier, 1976, with 120 g L^{-1} sucrose), after 72 hours of cultivation embryo becomes torpedo-shape.

Bars: 1 a-c—25 µm, 3 a-c—100 µm. *ca*—apical cell, *cb*—basal cell.

1 a-c, 2—after Erdelská, orig.; 3—after Erdelská, 1982; 4,5—after Erdelská, 1991.

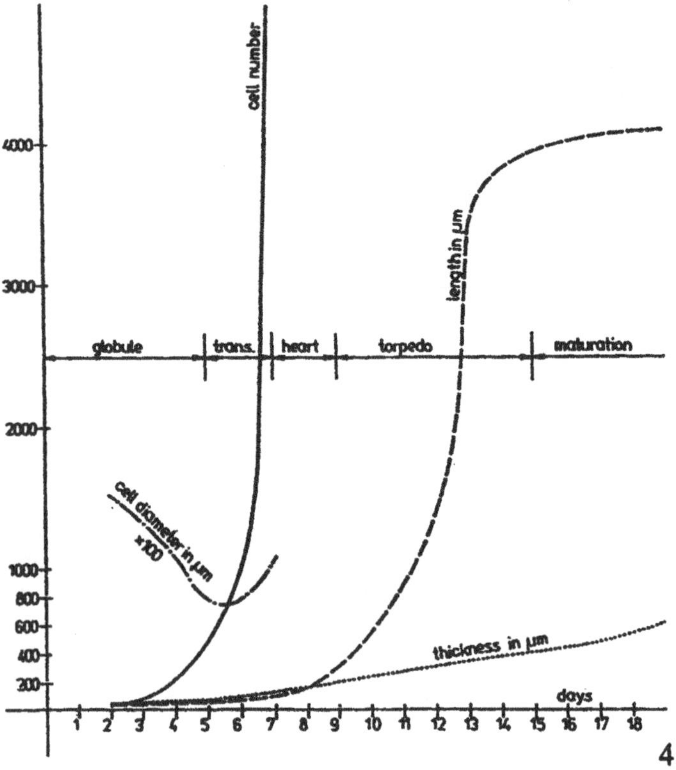

4

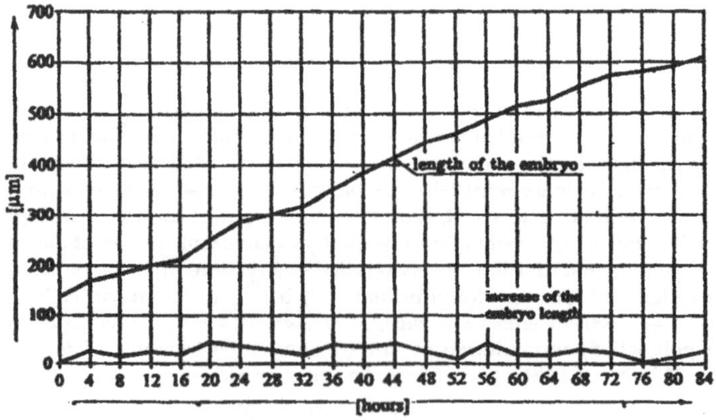

5

Plate LIV (*Contd.*)

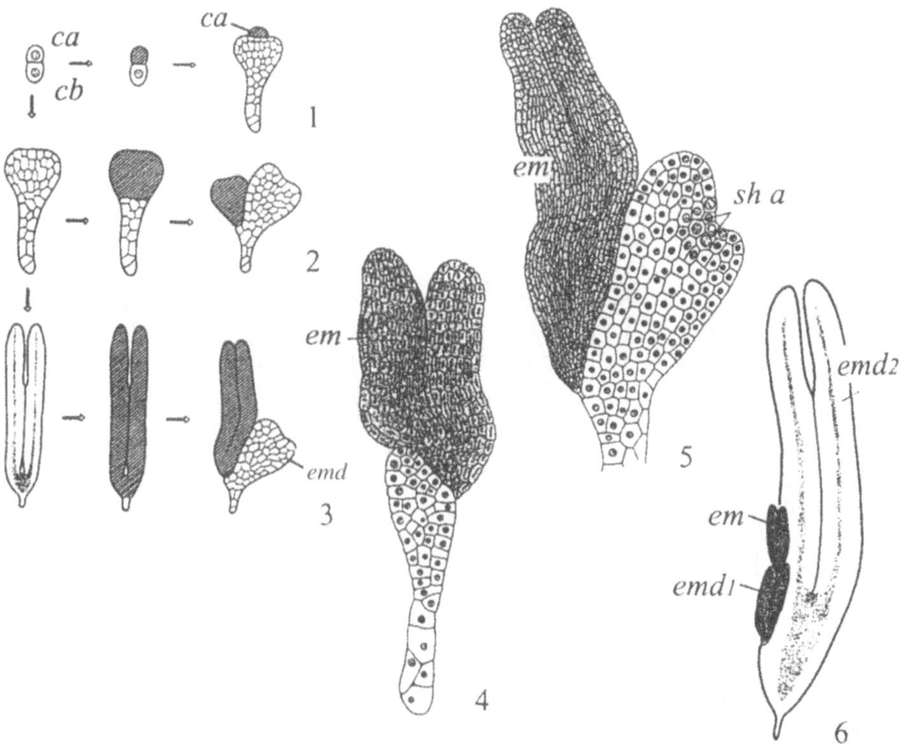

Plate LV: Embryoid.

1-6—formation of somatic embryo (embryoid) from suspensor cells of sexual embryo at different stages of its development in experimental conditions in *Eranthis hiemalis*; 7-13—development of sexual embryo in *Caltha palustruis* in seed according to Onagrad-type; 14-33—embryoid development in *Aconitum heterophyllum* in culture *in vitro* according to Onagrad-type, abnormalities observed: premature differentiation of xylem in the region of central cylinder (22, 23, 30, 32), multinuclear initial cells (30,31), formation of monocotyledonous embryo (33); 34-39—early stages of sexual (34) and somatic (35-39) embryo development in *Triticum aestivum*: callus with meristematic zone, in which initiation of embryoids takes place (36), orientated in different ways towards callus surface embryoids (37-39); 40-52—development of vegetative propagule on leaf of *Bryophyllum calycinum* (41-46) and *B. daigremontianum* (47-52) *in vivo*: 41—margin of young leaf, 42—margin of adult leaf, 43—vegetative propagule at the 'heart-shaped' stage of development, 44-46—vegetative propagule at torpedo-shaped stage (45—view from above, 46—view from below), 47—margin of young leaf, 48—vegetative propagule at 'globular' stage of development, 49-51—vegetative propagule at 'torpedo-shaped' stage, 52—zone remaining after propagule separation; (34, 45, 46, 50-52—SEM, 35-37, 41-44, 47-49—LM, 38—schematic).

ad r—adventive root, *c*—cotyledon, *ca*—apical cell, *cb*—basal cell, *h*—hollow, *em*—embryo, *emd, emd1, emd2* —embryoid, *m z*—meristematic zone, *sh a*—shoot apex, *v b*—vascular bundle.

1-6—after Haccius, 1965; 7-13—after Bhandari and Asnani, 1968; 14-33—after Titova, Ahuja and Batygina, orig.; 34-39—after Batygina, Shamrov, Titova and Nikiphorova, orig.; 40—after Batygina *et al.*, 1996; 41-52—after Bragina, orig.

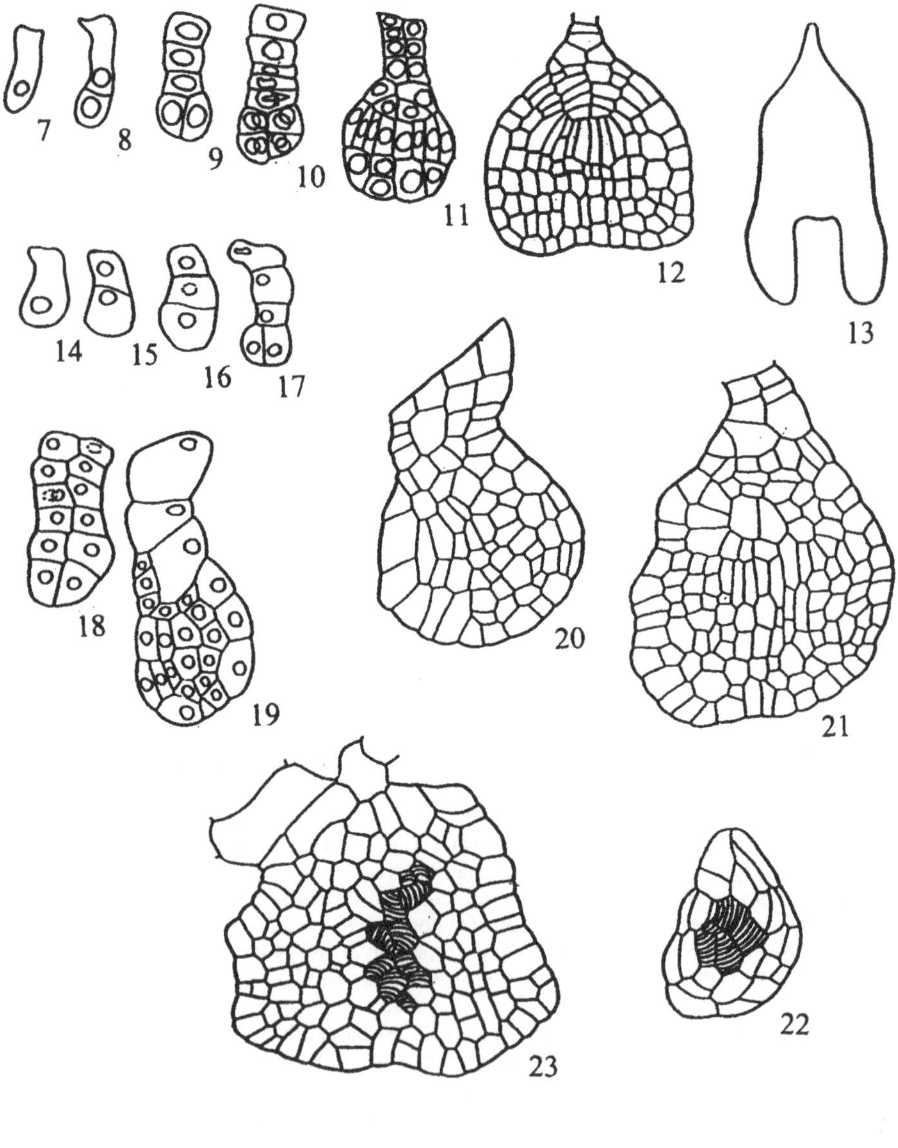

Plate LV (*Contd.*)

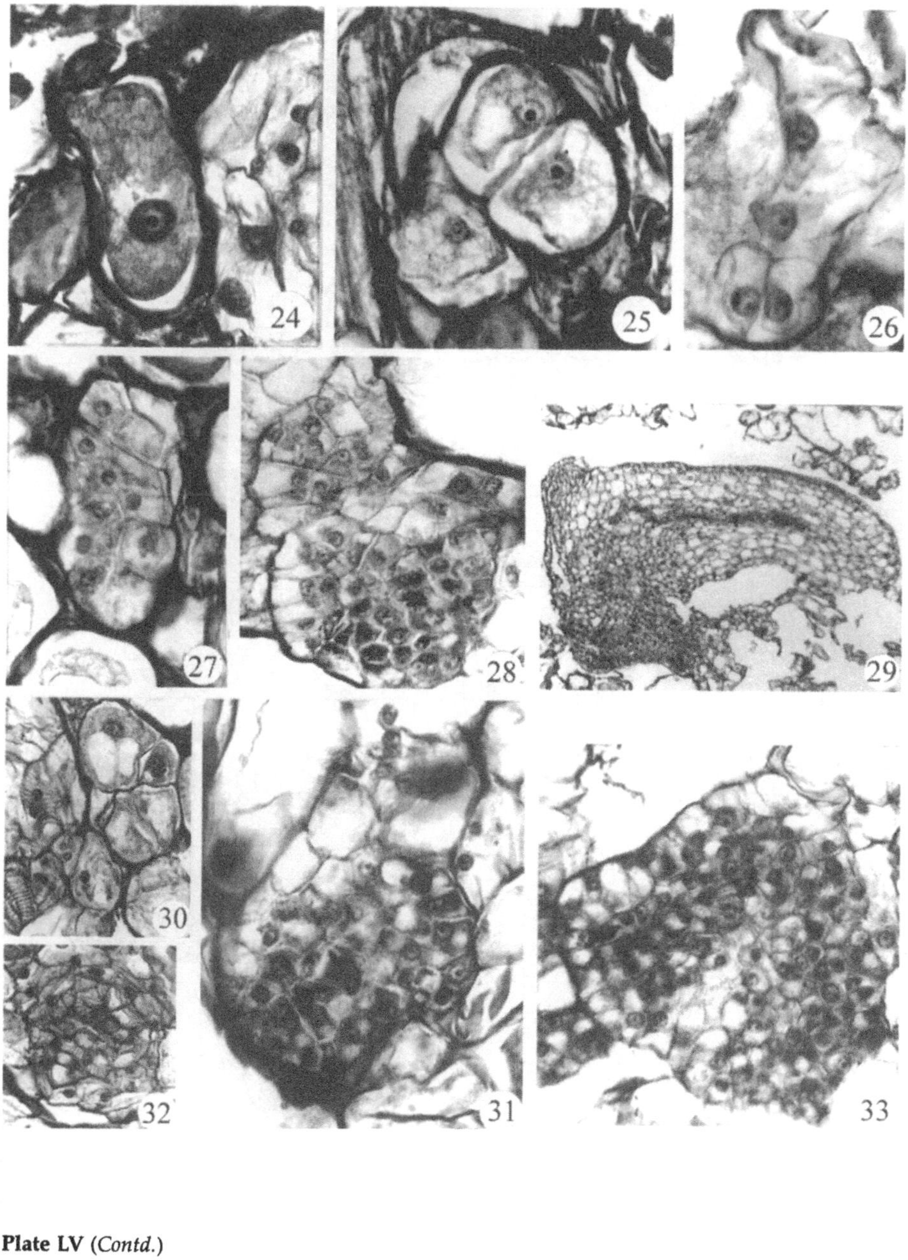

Plate LV (*Contd.*)

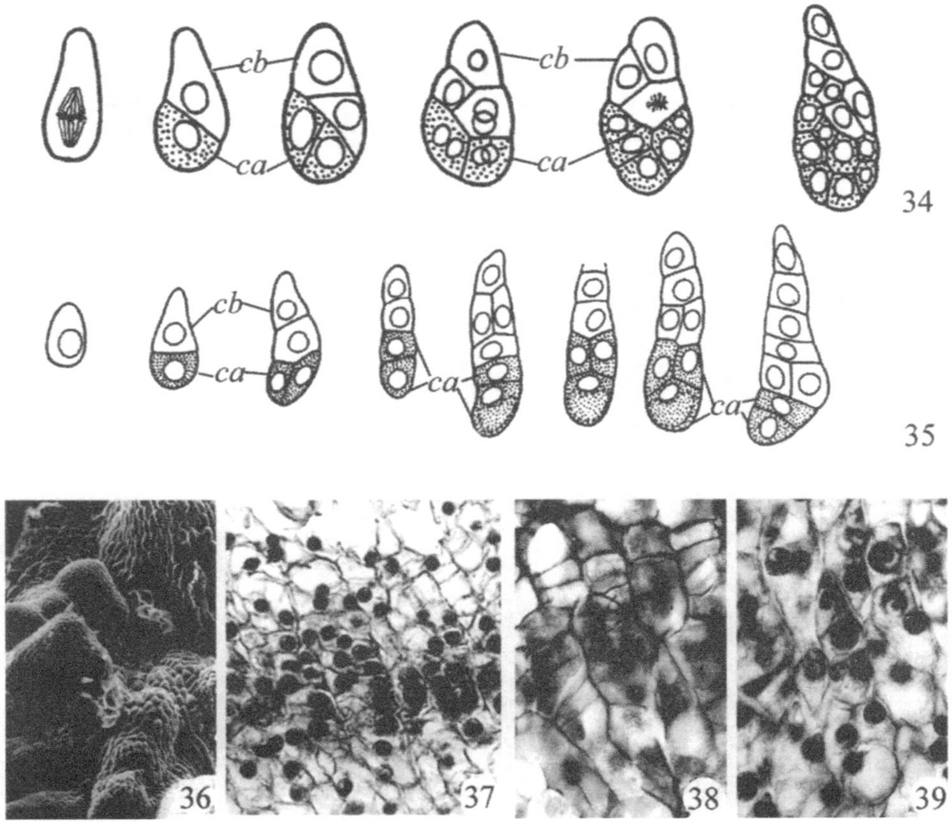

Plate LV (*Contd.*)

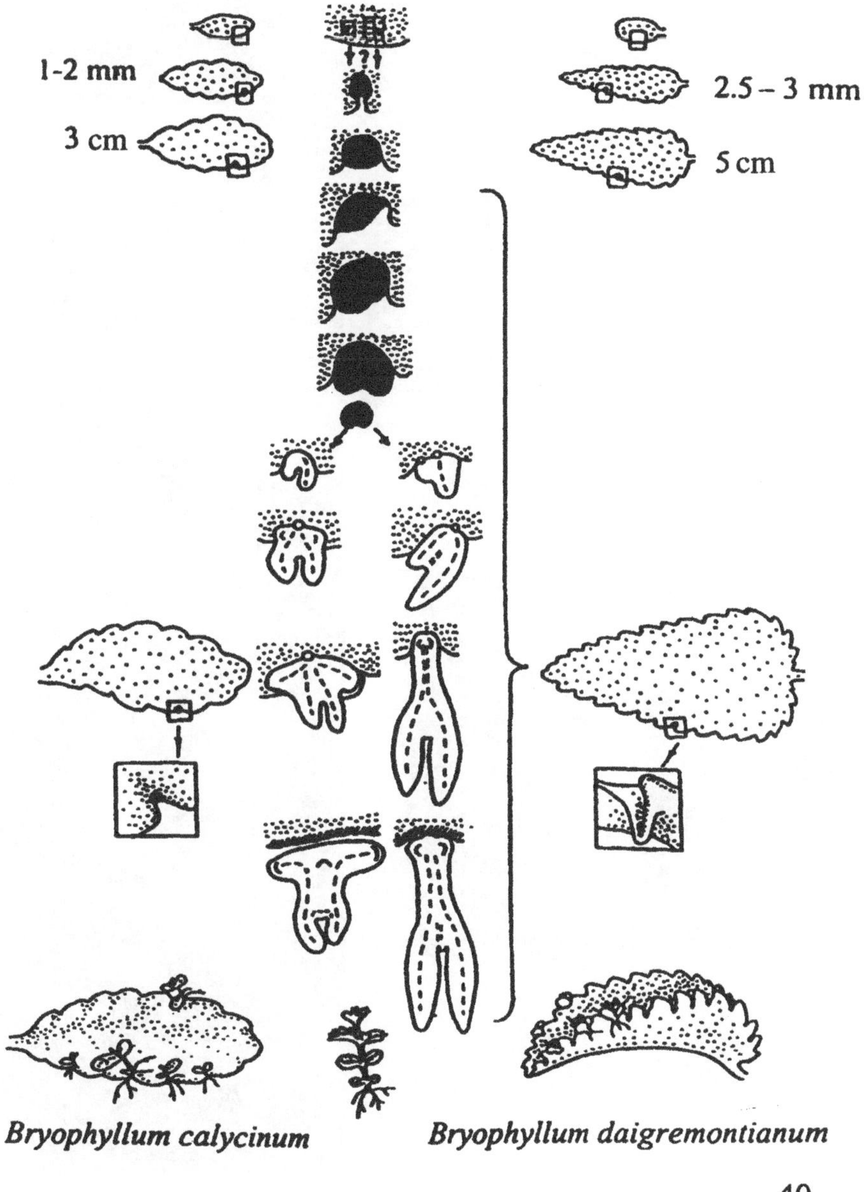

40

Plate LV (*Contd.*)

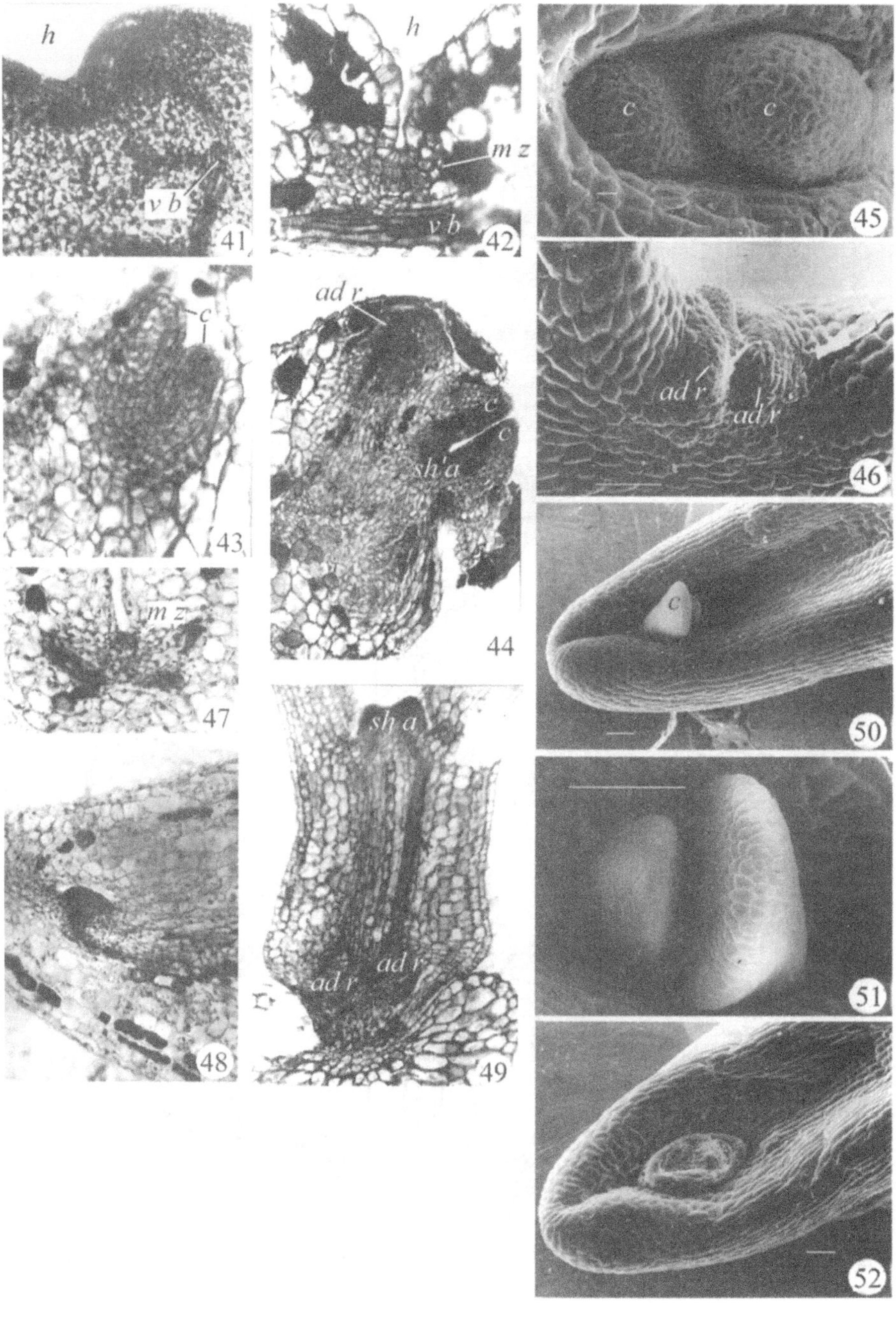

Plate LV (*Contd.*)

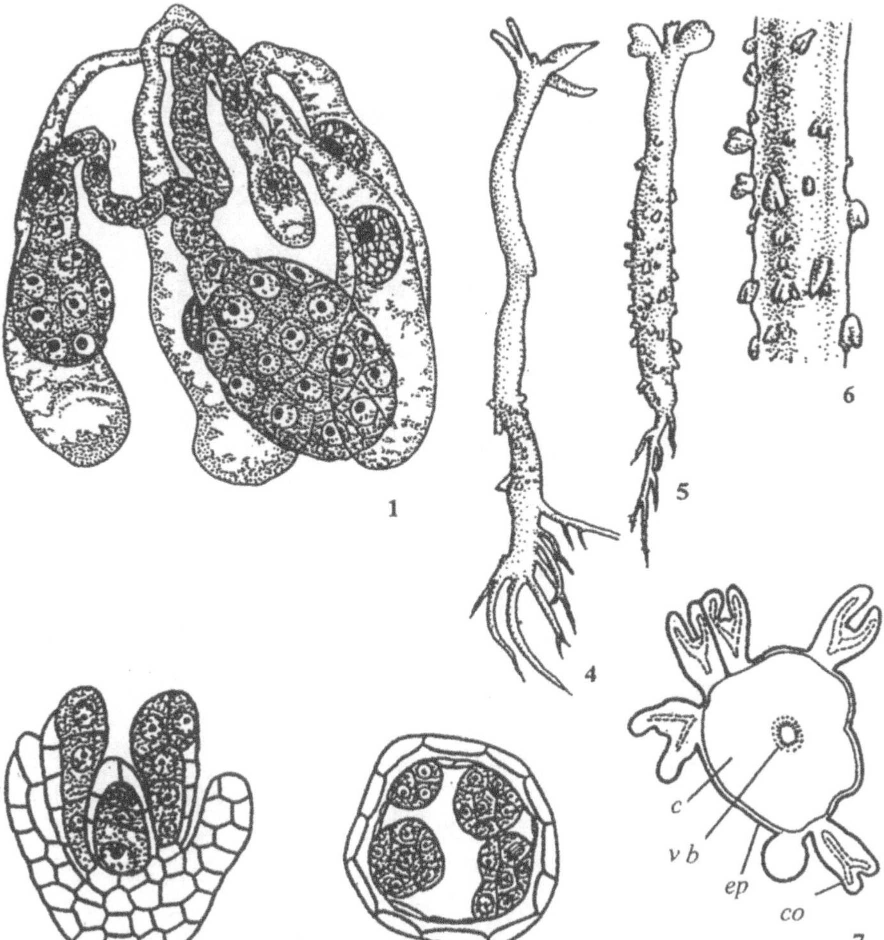

Plate LVI: Embryoidogeny.

1—polyembryony in *Eulophea epidendraea* (sketched from total slides of ovule): two embryos, probably originating as a result of cleavage of one embryo, large vacuolated cells are cells of suspensor; 2,3—integumentary somatic embryos (embryoids) in micropylar region of bitegmic ovule in *Spiranthes cernua* at stage of tetrad (2—longitudinal and 3—transverse sections); 4-7—development of embryos from hypocotyl epidermal cells in *Ranunculus sceleratus*: from lower (4) and upper (5,6) parts of hypocotyl, 6—part of hypocotyl with embryos, magnified, 7—transverse section of stem with developing embryos.

c—cortex, *co*—cotyledon, *ep*—epidermis, *v b*—vascular bundle.

1—after Swamy, 1943; 2,3—after Swamy, 1948a; 4-7—after Konar and Nataraja, 1965.

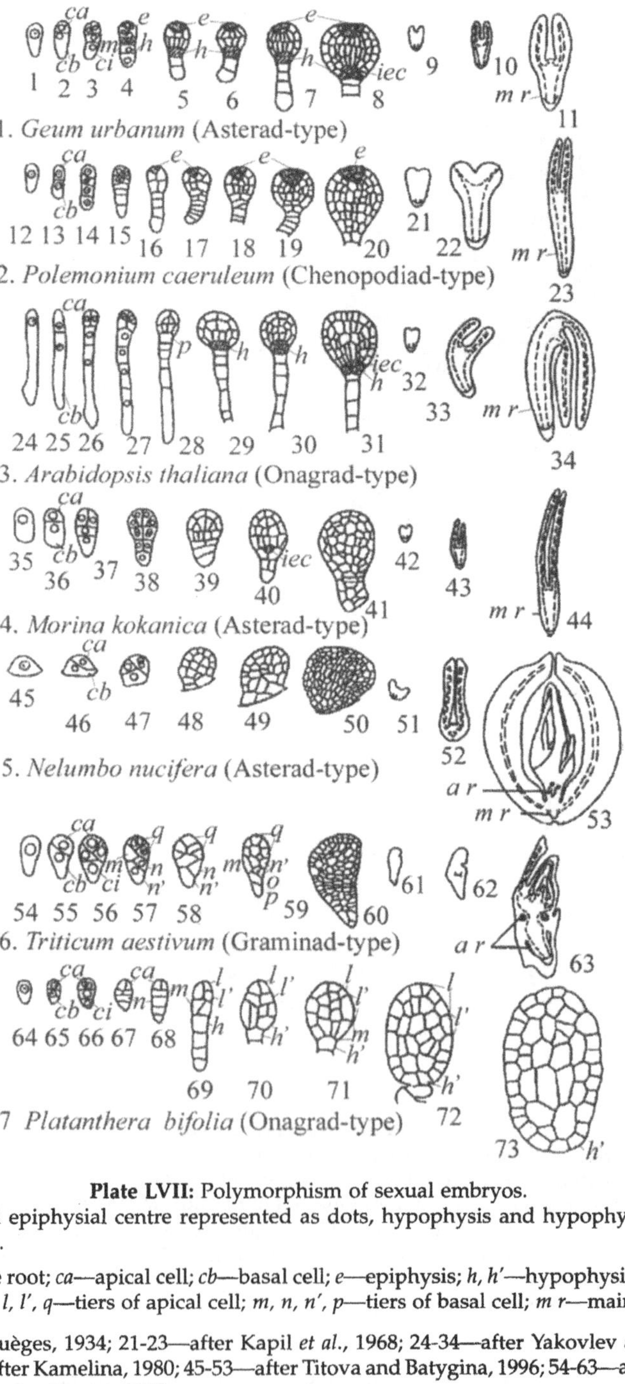

Plate LVII: Polymorphism of sexual embryos.

Epiphysis and epiphysial centre represented as dots, hypophysis and hypophysial centre as shaded points.

a r—adventive root; *ca*—apical cell; *cb*—basal cell; *e*—epiphysis; *h, h'*—hypophysis; *iec*—initials of root cortex; *l, l', q*—tiers of apical cell; *m, n, n', p*—tiers of basal cell; *m r*—main root.

1-20—after Souèges, 1934; 21-23—after Kapil *et al.*, 1968; 24-34—after Yakovlev and Alimova, 1976; 35-44—after Kamelina, 1980; 45-53—after Titova and Batygina, 1996; 54-63—after Batygina, 1987a; 64-73—after Veyret, 1965.

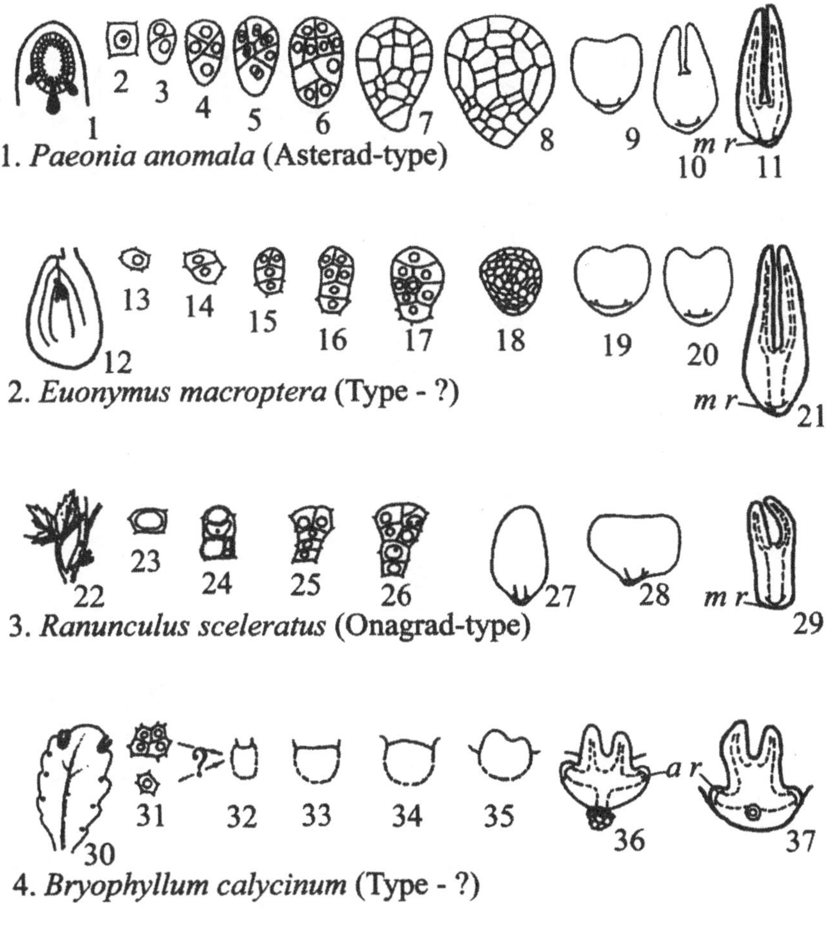

Plate LVIII: Polymorphism of somatic embryos in natural conditions.

1-37—developmental stages of different kinds of embryoids: embryonic cleavage, monozygotic (1-11), integumental (12-21), cauligenic (22-29), foliar (30-37).

a r—adventive root, *m r*—main root.

1-11—after Brukhin and Batygina, 1994; 12-21—after Naumova, 1987; 22-29—modified from Konar *et al.*, 1972; 30-37—after Batygina *et al.*, 1996.

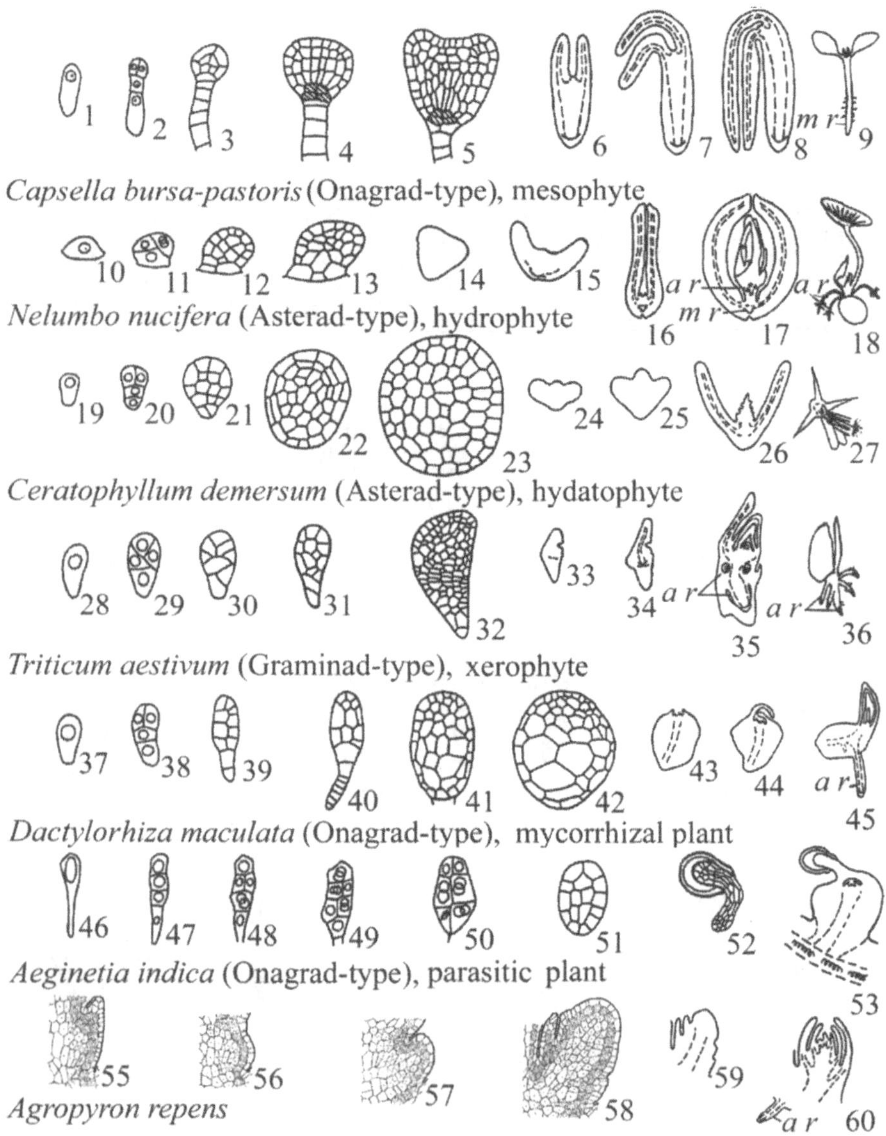

Plate LIX: Parallelism in early stages of morphogenesis during seed (1-53) and vegetatative (55-60) propagation.

Hypophysis centre shown by shading; *a r*—adventive root, *m r*—main root.

1–8—after Maheshwari, 1950; 9—after Batygina and Zakharova, orig.; 10–18—after Titova and Batygina, 1996; 19–27—after Shamrov and Batygina, 1984, 1988; 28–36—after Batygina, 1987a; 37–45—after Batygina and Vasilyeva, 1983a; 46–54—after Teryokhin and Nikiticheva, 1981; 55–60—after Esau, 1965.

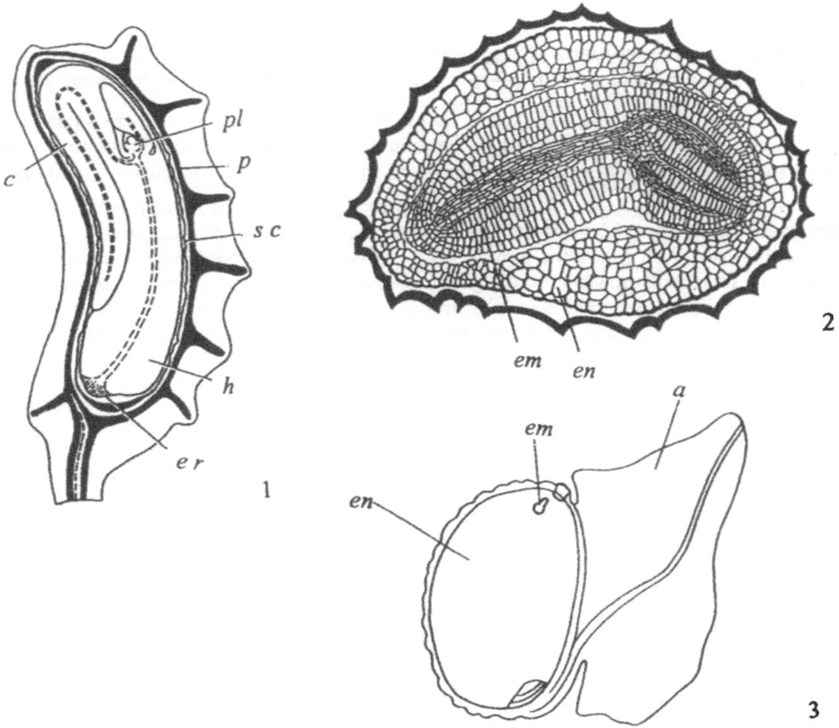

Plate LX: Seed

1-3—various types of seeds: 1—with filmy seed coat and without endosperm in *Zannichellia palustris* (longitudinal section of fruit), 2—with endosperm and embryo differentiated into organs in *Nicotiana tabacum*, 3—with aril in *Trillium ovatum* (2,3—longitudinal section).

a—aril, *c*—cotyledon, *em*—embryo, *en*—endosperm, *e r*—embryonic root, *h*—hypocotyl, *p*—pericarp, *pl*—plumule, *s c*—seed coat.

1—after Teryokhin, 1985e; 2—after Alexandrov, 1966; 3—after Berg, 1958.

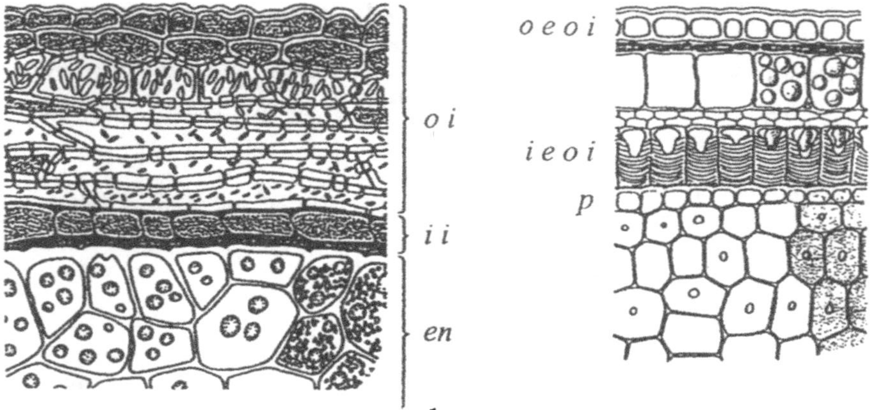

1 2

Plate LXI: Seed coat.

1-6—seed coat with various disposition of mechanical and other specialized layers: 1—exo-mesotestal in *Stemona tuberosa*, 2—exo-endotestal with various specialization of separate layers in *Elettaria cardamomum*, 3—testal in *Allium cardiostemon*, 4—endotegminal in *Centrolepis drummondii*, 5—unstructured in *Smilax excelsa*, 6—with crystal layer in *Dioscorea elephantipes;* 7—seed coat with succulent and mucous exotesta in *Aporosa frutescens;* 8—10—chalazal and micropylar structures in monocotyledon seeds: 8—wing-like outgrowth in *Dioscorea camestica*, 9—operculum in *Stanfieldiella imperforata*, 10—micropylar collar, operculum and arilloide in *Stromanthe lutea;* 11-12—structural diversity of dicotyledon seeds: 11—developing seed of *Gossypium hirsutum*, 12—fragment of exotegminal seed coat in *Mallotus* sp.

ar—aril, *c*, *c1*, *c2*, *c3* —cuticular layers, *e*—embryo, *en*—endosperm, *i e i i*—inner epidermis of inner integument, *i e o i*— inner epidermis of outer integument, *i i*—inner integument, *n*—nucellus, *n c*—nucellar cap, *o*—operculum, *o e o i*—outer epidermis of outer integument, *o i*—outer integument, *p*—perisperm, *st*—sarcotesta, *s c*—seed coat, *w*—wing, *m c*—micropylar collar.

1—after Fedotova, 1985; 2—after Loesener, 1930; 3, 5—after Komar, 1985a,b; 4—after Vyshenskaya, 1985; 6, 8—Savchenko, orig., cited from Petrova and Safina, 1985; 7, 12—after Corner, 1976; 9—after Grootjen and Bouman, 1981; 10—after Mauritzon, 1936; 11—after Baranov, 1937.

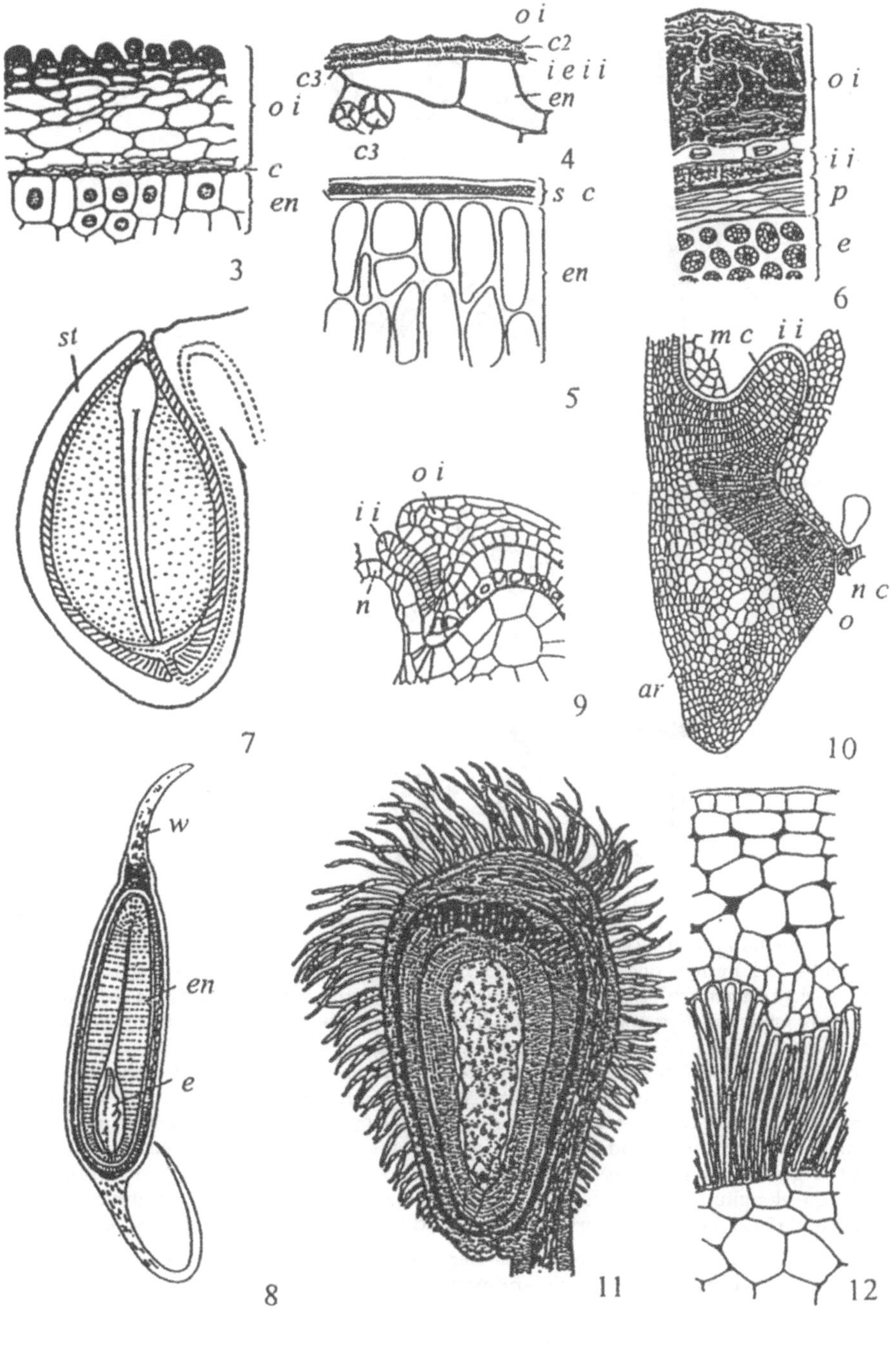

Plate LXI (*Contd.*)

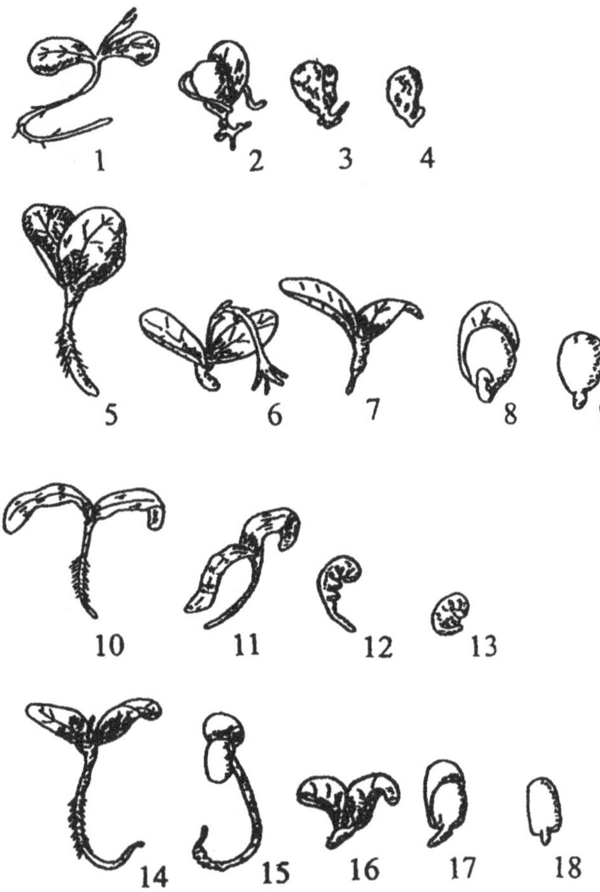

Plate LXII: Seed dormancy.

1-18—external appearance of seedlings and embryos from seeds after cold stratification (1,5,10,14) and dormant ones (2-4, 6-9, 11-13, 15-18): 1-4—*Acer tataricum*, 5-9—*Euonymus europaea*, 10-13—*Acer platanoides*, 14-18—*Sorbus aucuparia*; 19-22—structure of mature seeds with different degree of embryo differentiation (longitudinal sections, schematized): 19—*Panax ginseng*, 20—*Aralia continentalis*, 21—*Fraxinus excelsior*, 22—*Euonymus europaea*; 23-37—embryo post-development in seed of *Aconitum soongaricum* at various temperatures: 23—schematized, 24-33—embryo structure after 40 (24-26) and 90 (27-33) days from beginning of germination at temperature 9-10ºC, 31-33—fragments of cotyledons with complexes of tabular cells, protoxylem elements in central vein and cells of mesophyll palisade; 34-37—embryo structure after 100 days of germination at temperature 18-20ºC, cell degeneration in hypocotyl (37) observed (30,31,33—longitudinal sections, 24-29, 32,34-37—transverse sections of the embryo at level of hypocotyl (24,27,34,37), shoot apex (25,28,35) and cotyledonous plates (26,29,32,36).

c c—central cylinder, *c t c*—complex of tabular cells, *en*—endosperm, *m e r*—main embryo root, *m p*—mesophyll palisade layer, *p c*—primary cortex, *px*—protoxylem elements, *sh a*—shoot apex, *v b*—vascular bundle.

1—after Nikolaeva *et al.*, 1985; 23-37—after Pozdova, Butuzova and Titova, orig.

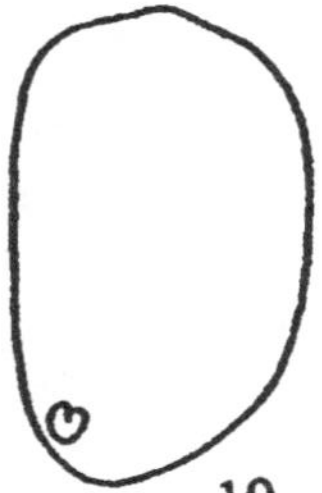

19

20

21

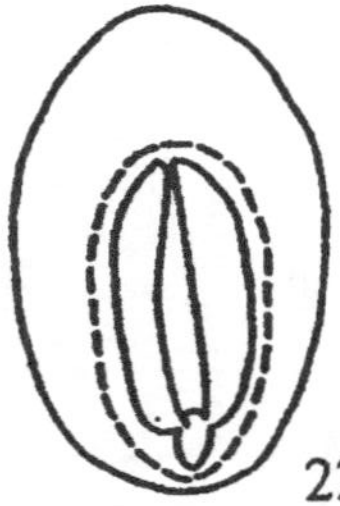

22

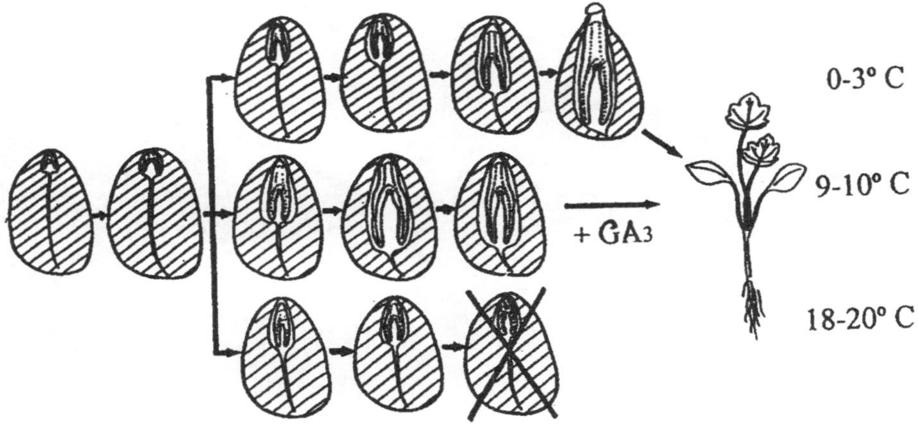

23

Plate LXII (*Contd.*)

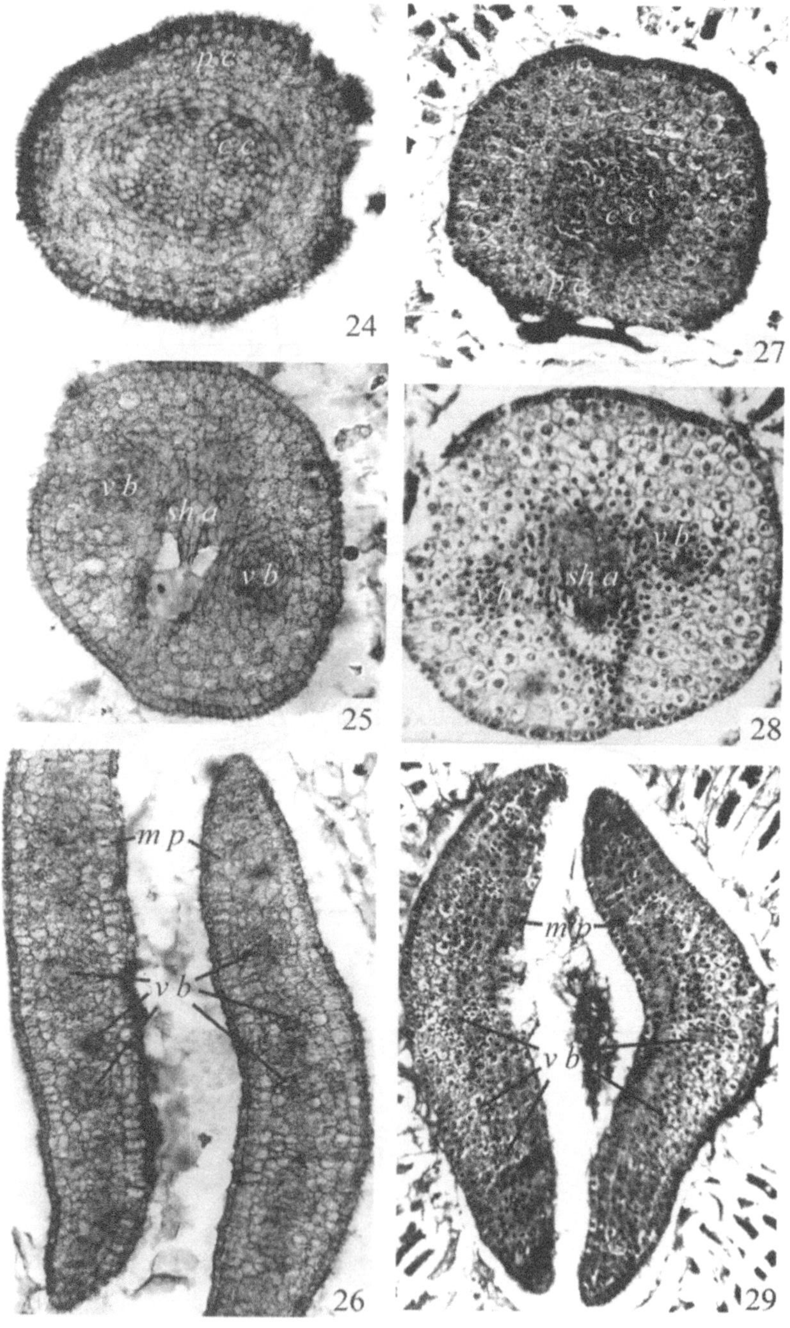

Plate LXII (*Contd.*)

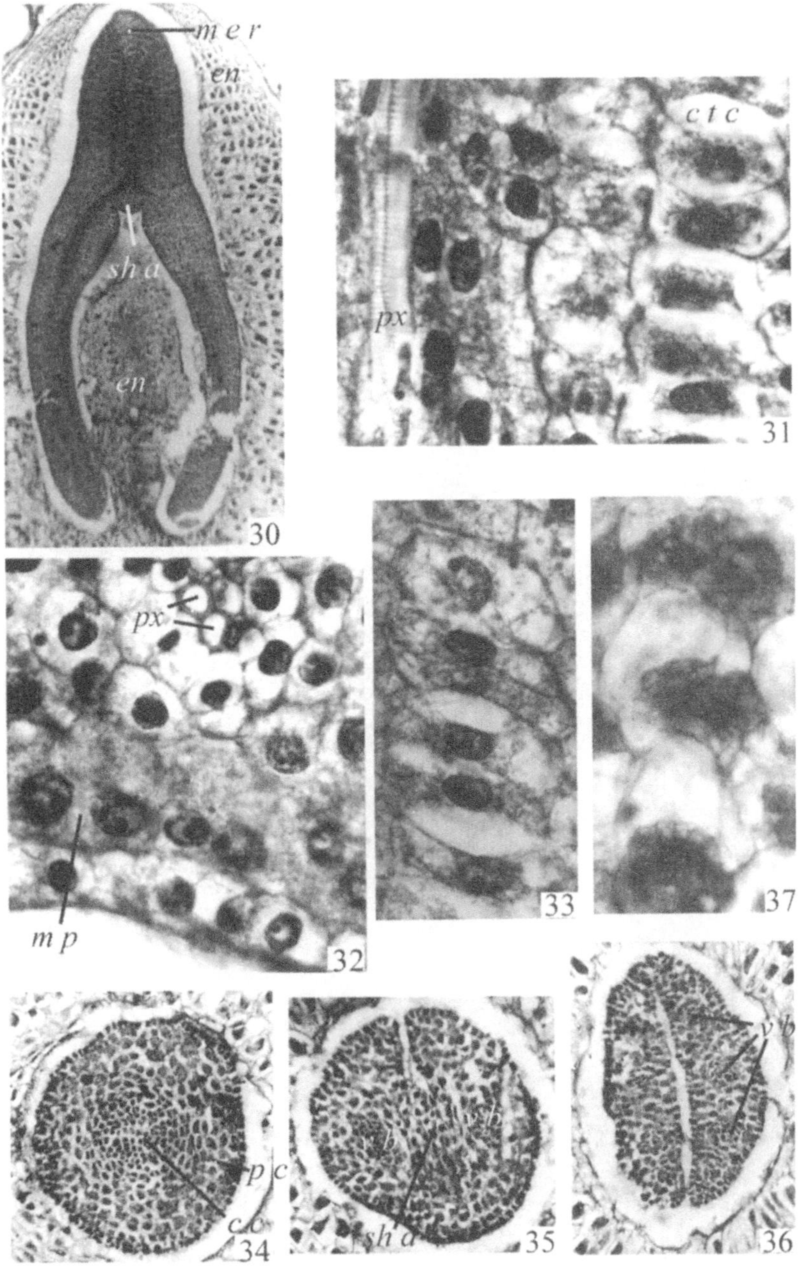

Plate LXII (*Contd.*)

Index

AUTHORS OF VOLUME 2

ANDRONOVA E. V.

ANISIMOVA G. M.

BATYGINA T. B.

BRAGINA E. A.

BRUKHIN V. B.

CRESTI M.

DANILOVA M. F.

ERDELSKA O.

GERASSIMOVA-
NAVASHINA E. N.

KAMELINA O.P.

KIEFT H.

KOROBOVA S. N.

MEDVEDEV S. S.

NAUMOVA T. N.

NIKITICHEVA Z. I.

OBRUCHEVA N. V.

POZDOVA L. M.

RAZUMOVA M. V.

RODKIEWICZ B.

RUSSEL S. D.

SCHEL J. H. N.

SHAMROV I. I.

SOLNTSEVA M. P.

SZCZIPAREV S. M.

SZCZUKA E.

TERYOKHIN E. S.

TITOVA G. E.

VAN LAMMEREN A. A. M.

VASILYEVA V. E.

VISHNYAKOVA M. A.

YE X. L.

YEUNG E. C.

ZEE S. Y.

ZHINKINA N. A.